# 2nd International Conference on Computing and Applied Informatics (ICCAI 2017)



Journal of Physics: Conference Series Volume 978

Medan, Indonesia
28 – 30 November 2017

ISBN: 978-1-5108-5988-3
ISSN: 1742-6588

**Printed from e-media with permission by:**

Curran Associates, Inc.
57 Morehouse Lane
Red Hook, NY 12571

CURRAN ASSOCIATES INC. proceedings.com

**Some format issues inherent in the e-media version may also appear in this print version.**



Printed by Curran Associates, Inc. (2018)



Institute of Physics
Dirac House, Temple Back
Bristol BS1 6BE UK

Phone: 44 1 17 929 7481
Fax: 44 1 17 920 0979

techtracking@iop.org

**Additional copies of this publication are available from:**

Curran Associates, Inc.
57 Morehouse Lane
Red Hook, NY 12571 USA
Phone: 845-758-0400
Fax: 845-758-2633
Email: curran@proceedings.com
Web: www.proceedings.com

# TABLE OF CONTENTS

**IMPROVED LINE SEGMENTATION FRAMEWORK FOR SUNDANESE OLD MANUSCRIPTS** ........................ 1
*E Paulus, M Suryani, S Hadi*

**COMPARATIVE ANALYSIS OF GAMEPLAY AND PLAYERS EMOTION IN THE MOST POPULAR GAMES FROM PLAY STORE** ........................ 7
*Riwinoto, N Mahfud, L Lumombo*

**VI-DA: VITILIGO DIAGNOSTIC ASSISTANCE MOBILE APPLICATION** ........................ 13
*G A Nugraha, A Nurhudatiana, R Bahana*

**PERFORMANCE TEST FOR PROTOTYPE GAME FOR CHILDREN WITH ADHD** ........................ 19
*R Bahana, F L Gaol, T Wiguna, S W H L Hendric, B Soewito, E Nugroho, B P Dirgantoro, E Abdurachman*

**DESIGNING INDONESIAN TEACHER ENGAGEMENT INDEX (ITEI) APPLICATIONS BASED ON ANDROID** ........................ 26
*S R Manalu, Sasmoko, S D Permai, S A Widhoyoko, Y Indrianti*

**IMPLEMENTATION OF UTAUT MODEL TO UNDERSTAND THE USE OF VIRTUAL CLASSROOM PRINCIPLE IN HIGHER EDUCATION** ........................ 32
*B R Aditya, A Permadi*

**USABILITY TESTING OF INDONESIA TOURISM PROMOTION WEBSITE** ........................ 38
*U Yunus, Anindito, E Tanuar, Maryani*

**AUTOMATED COLOR CLASSIFICATION OF URINE DIPSTICK IMAGE IN URINE EXAMINATION** ........................ 46
*R F Rahmat, Royananda, M A Muchtar, R Taqiuddin, S Adnan, R Anugrahwaty, R Budiarto*

**THE DEVELOPMENT OF INDONESIAN TRADITIONAL BEKEL GAME IN ANDROID PLATFORM** ........................ 54
*R F Rahmat, O R Fahrani, S Purnamawati, M F Pasha*

**USABILITY EVALUATION OF USER INTERFACE OF THESIS TITLE REVIEW SYSTEM** ........................ 63
*Y Tri, A Erna, U Gellysa*

**DIAGNOSTICS VEHICLE'S CONDITION USING OBD-II AND RASPBERRY PI TECHNOLOGY: STUDY LITERATURE** ........................ 69
*J V Moniaga, S R Manalu, D A Hadipurnawan, F Sahidi*

**INDONESIA KNOWLEDGE DISSEMINATION: A SNAPSHOT** ........................ 75
*M K M Nasution*

**ASSESSING ERP SAP IMPLEMENTATION IN THE SMALL AND MEDIUM ENTERPRISES (SMES) IN INDONESIA** ........................ 82
*Yohannes, W Gunawan, R B Ikhsan, Aries*

**AUTOMATED HYDROPONICS NUTRITION PLANTS SYSTEMS USING ARDUINO UNO MICROCONTROLLER BASED ON ANDROID** ........................ 89
*P Sihombing, N A Karina, J T Tarigan, M I Syarif*

**REAL-TIME DETECTION WITH ADABOOST-SVM COMBINATION IN VARIOUS FACE ORIENTATION** ........................ 95
*R P Fhonna, M K M Nasution, Tulus*

**DELAY AND COST PERFORMANCE ANALYSIS OF THE DIFFIE-HELLMAN KEY EXCHANGE PROTOCOL IN OPPORTUNISTIC MOBILE NETWORKS** ........................ 102
*B Soelistijanto, V Muliadi*

**APPLICATION MODELING IPV6 (INTERNET PROTOCOL VERSION 6) ON E-ID CARD FOR IDENTIFICATION NUMBER FOR EFFECTIVENESS AND EFFICIENCY OF REGISTRATION PROCESS IDENTIFICATION OF POPULATION** ........................ 108
*A M H Pardede, Y Maulita, R Buaton*

**SUPPLY CHAIN MANAGEMENT USING FP-GROWTH ALGORITHM FOR MEDICINE DISTRIBUTION** ........................ 115
*A Wahana, D S Maylawati, M Irfan, H Effendy*

**THE IMPLEMENTATION OF TWO STAGES CLUSTERING (K-MEANS CLUSTERING AND ADAPTIVE NEURO FUZZY INFERENCE SYSTEM) FOR PREDICTION OF MEDICINE NEED BASED ON MEDICAL DATA** ........................ 121
*A M Husein, M Harahap, S Aisyah, W Purba, A Muhazir*

**IDENTIFICATION OF HAND MOTION USING BACKGROUND SUBTRACTION METHOD AND EXTRACTION OF IMAGE BINARY WITH BACKPROPAGATION NEURAL NETWORK ON SKELETON MODEL** ........................ 129
*Fauziah, E P Wibowo, S Madenda, Hustinawati*

**IMPLEMENTATION OF SUPPLY CHAIN BUSINESS APPLICATION THROUGH BUSINESS MODEL CANVAS AND WATERFALL FRAMEWORK COLLABORATIONS FOR FISH FARMERS SMES IN ULEKAN MARKET BANDUNG** ... 143
*Y Priyadi, A Prasetio*

**FACIAL EXPRESSION SYSTEM ON VIDEO USING WIDROW HOFF** ... 150
*M Jannah, M Zarlis, H Mawengkang*

**W.A.W (WE ARE WATCHING) SMART APP: ACCOMMODATING SOCIAL PERCEPTION TOWARDS PUBLIC OFFICERS' PERFORMANCE** ... 156
*S A Widhoyoko, Sasmoko, L A Nasir, S R Manalu, Y Indrianti*

**IMPLEMENTATION AND EVALUATION OF LMS MOBILE APPLICATION: SCELE MOBILE BASED ON USER-CENTERED DESIGN** ... 166
*R D Banimahendra, H B Santoso*

**METACOGNITIVE COMPONENTS IN SMART LEARNING ENVIRONMENT** ... 173
*M Sumadyo, H B Santoso, D I Sensuse*

**IMPROVING DATA QUALITY IN THE LINKED OPEN DATA: A SURVEY** ... 180
*A Hadhiatma*

**MAPPING OF MEDICINE DATA WITH K-MEANS AND APRIORI COMBINATIONS BASED ON PATIENT DIAGNOSIS** ... 187
*N P Dharshinni, H Mawengkang, M K M Nasution*

**E-LEARNING PROCESS MATURITY LEVEL: A CONCEPTUAL FRAMEWORK** ... 194
*A Rahmah, H B Santoso, Z A Hasibuan*

**DEVELOP APPLICATIONS BASED ON ANDROID: TEACHER ENGAGEMENT CONTROL OF HEALTH (TECH)** ... 201
*Sasmoko, S R Manalu, S A Widhoyoko, Y Indrianti, Suparto*

**AN ANALYSIS OF ABSORBING IMAGE ON THE INDONESIAN TEXT BY USING COLOR MATCHING** ... 207
*G A Hutagalung, Tulus, Iryanto, Y F A Lubis, M Khairani, Suriati*

**MICROCONTROLLER BASED AUTOMATIC TEMPERATURE CONTROL FOR OYSTER MUSHROOM PLANTS** ... 212
*P Sihombing, T P Astuti, Herriyance, D Sitompul*

**FEATURE WEIGHTING USING PARTICLE SWARM OPTIMIZATION FOR LEARNING VECTOR QUANTIZATION CLASSIFIER** ... 219
*A Dongoran, S Rahmadani, M Zarlis, Zakarias*

**SOCIAL NETWORK EXTRACTION BASED ON WEB: 3. THE INTEGRATED SUPERFICIAL METHOD** ... 225
*M K M Nasution, O S Sitompul, S A Noah*

**DEVELOPMENT OF EXCAVATOR TRAINING SIMULATOR USING LEAP MOTION CONTROLLER** ... 232
*F Fahmi, F Nainggolan, U Andayani, B Siregar*

**DEVELOPING CLOUD-BASED BUSINESS PROCESS MANAGEMENT (BPM): A SURVEY** ... 238
*Mercia, W Gunawan, A N Fajar, H Alianto, Inayatulloh*

**ANALYSIS OF FLOOD VULNERABILITY IN URBAN AREA; A CASE STUDY IN DELI WATERSHED** ... 245
*I Indrawan, R I Siregar*

**STUDY TO THE CURRENT PROTECTION OF PERSONAL DATA IN THE EDUCATIONAL SECTOR IN INDONESIA** ... 251
*E Rosmaini, T F Kusumasari, M Lubis, A R Lubis*

**A REAL TIME MOBILE-BASED FACE RECOGNITION WITH FISHERFACE METHODS** ... 258
*D Arisandi, M F Syahputra, I L Putri, S Purnamawati, R F Rahmat, P P Sari*

**ANALYSIS OF FACTORS THAT INHIBITING IMPLEMENTATION OF INFORMATION SECURITY MANAGEMENT SYSTEM (ISMS) BASED ON ISO 27001** ... 265
*R Tatiara, A N Fajar, B Siregar, W Gunawan*

**AUGMENTED REALITY SOCIAL STORY FOR AUTISM SPECTRUM DISORDER** ... 271
*M F Syahputra, D Arisandi, A F Lumbanbatu, L F Kemit, E B Nababan, O Sheta*

**HYBRID CRYPTOSYSTEM RSA – CRT OPTIMIZATION AND VMPC** ... 277
*R Rahmadani, H Mawengkang, Sutarman*

**INSIGHTS TO DEVELOP PRIVACY POLICY FOR ORGANIZATION IN INDONESIA** ... 283
*E Rosmaini, T F Kusumasari, M Lubis, A R Lubis*

**IDENTIFICATION TIBIA AND FIBULA BONE FRACTURE LOCATION USING SCANLINE ALGORITHM** ... 290
*M A Muchtar, S E Simanjuntak, R F Rahmat, H Mawengkang, M Zarlis, O S Sitompul, I D Winanto, U Andayani, M F Syahputra, I Siregar, T H Nasution*

**BIOINFORMATICS APPROACH OF THREE PARTIAL POLYPRENOL REDUCTASE GENES IN KANDELIA OBOVATA** ..... 297
*M Basyuni, R Wati, H Sagami, H Oku, S Baba*
**AN OPTIMAL GENERIC MODEL FOR MULTI-PARAMETERS AND BIG DATA OPTIMIZING: A LABORATORY EXPERIMENTAL STUDY** ..... 303
*D N Utama, N Ani, M M Iqbal*
**READINESS FACTORS FOR INFORMATION SYSTEM STRATEGIC PLANNING AMONG UNIVERSITIES IN DEVELOPING COUNTRIES: A SYSTEMATIC REVIEW** ..... 310
*M Irfan, S J Putra, C N Alam, A Subiyakto, A Wahana*
**IMPROVING THE ACCURACY OF K-NEAREST NEIGHBOR USING LOCAL MEAN BASED AND DISTANCE WEIGHT** ..... 317
*K U Syaliman, E B Nababan, O S Sitompul*
**MOBILE/ANDROID APPLICATION FOR QRS DETECTION USING ZERO CROSS METHOD** ..... 323
*M I Rizqyawan, A I Simbolon, M A Suhendra, M F Amri, D E Kusumandari*
**MAIN FACTORS IN E-LEARNING FOR THE EQUIVALENCY EDUCATION PROGRAM (E-LEEP)** ..... 329
*M B Yel, Sfenrianto*
**BIOINFORMATICS ANALYSIS OF THE PREDICTED POLYPRENOL REDUCTASE GENES IN HIGHER PLANTS** ..... 336
*M Basyuni, R Wati*
**THE RECOGNITION OF FEMALE VOICE BASED ON VOICE REGISTERS IN SINGING TECHNIQUES IN REAL-TIME USING HANKEL TRANSFORM METHOD AND MACDONALD FUNCTION** ..... 345
*R Meiyanti, A Subandi, N Fuqara, M A Budiman, A P U Siahaan*
**MOBILE TECHNOLOGY FOR EXPANSION OF SERVICE RANGE MEDAN PUBLIC LIBRARY** ..... 351
*A R Siregar, H Dewiyana*
**MULTI-DOCUMENTS SUMMARIZATION BASED ON CLUSTERING OF LEARNING OBJECT USING HIERARCHICAL CLUSTERING** ..... 356
*M Mustamiin, I Budi, H B Santoso*
**SKIPPING STRATEGY (SS) FOR INITIAL POPULATION OF JOB-SHOP SCHEDULING PROBLEM** ..... 364
*M Abdolrazzagh-Nezhad, E B Nababan, H M Sarim*
**HIERARCHICAL RHETORICAL SENTENCE CATEGORIZATION FOR SCIENTIFIC PAPERS** ..... 371
*G H Rachman, M L Khodra, D H Widyantoro*
**APPLICATION OF THE FUZZY TOPSIS MULTI-ATTRIBUTE DECISION MAKING METHOD TO DETERMINE SCHOLARSHIP RECIPIENTS** ..... 380
*I Irvanizam*
**DECISION SUPPORT MODEL FOR MOSQUE RENOVATION AND REHABILITATION (CASE STUDY: TEN MOSQUES IN JAKARTA BARAT, INDONESIA)** ..... 387
*D N Utama, Y S Triana, M M Iqbal, M Iksal, I Fikri, T Dharmawan*
**PCA BASED FEATURE REDUCTION TO IMPROVE THE ACCURACY OF DECISION TREE C4.5 CLASSIFICATION** ..... 394
*M Z F Nasution, O S Sitompul, M Ramli*
**SPECIFIC ACOUSTIC MODELS FOR SPONTANEOUS AND DICTATED STYLE IN INDONESIAN SPEECH RECOGNITION** ..... 400
*C B Vista, C H Satriawan, D P Lestari, D H Widyantoro*
**ATTENDANCE FINGERPRINT IDENTIFICATION SYSTEM USING ARDUINO AND SINGLE BOARD COMPUTER** ..... 407
*M A Muchtar, Seniman, D Arisandi, S Hasanah*
**THE GAP VALUES IN THE PROFILE MATCHING METHOD BY FUZZY LOGIC** ..... 414
*S A Sitepu, S Efendi, Z Situmorang*
**SIMULATION EVALUATION OF CAPACITOR BANK IMPACT ON INCREASING SUPPLY CURRENT FOR ALUMUNIUM PRODUCTION** ..... 420
*S Hasan, K Badra, R Dinzi, Suherman*
**NO RESEARCH WITHOUT PUBLICATION: EARLY MINING** ..... 424
*M K M Nasution*
**IMAGE PROCESSING ANALYSIS OF GEOSPATIAL UAV ORTHOPHOTOS FOR PALM OIL PLANTATION MONITORING** ..... 429
*F Fahmi, D Trianda, U Andayani, B Siregar*
**SIMULATING 3D DEFORMATION USING CONNECTED POLYGONS** ..... 436
*J T Tarigan, I Jaya, S M Hardi, E M Zamzami*

**CONTROL OF MOTION STABILITY OF THE LINE TRACER ROBOT USING FUZZY LOGIC AND KALMAN FILTER** ..... 441
*M S Novelan, Tulus, E M Zamzami*
**DEVELOPING AND PILOT TESTING M-HEALTH CARE APPLICATION FOR PREGNANT AND TODDLERS BASED ON USER EXPERIENCE** ..... 447
*I D Lestantri, Putrima, A Sabiq, E Suherlan*
**HYBRID CRYPTOSYSTEM FOR IMAGE FILE USING ELGAMAL AND DOUBLE PLAYFAIR CIPHER ALGORITHM** ..... 458
*S M Hardi, J T Tarigan, N Safrina*
**OPENLOBBY: AN OPEN GAME SERVER FOR LOBBY AND MATCHMAKING** ..... 464
*E M Zamzami, J T Tarigan, I Jaya, S M Hardi*
**DESIGN AND ARCHITECTURE OF RETAILAPP: AN APPLICATION TO SUPPORT CONVENTIONAL RETAILERS** ..... 469
*I Jaya, J T Tarigan, S M Hardi, M K M Nasution*
**INVESTIGATION OF THE RIPENESS OF OIL PALM FRESH FRUIT BUNCHES USING BIO-SPECKLE IMAGING** ..... 474
*R Salambue, A Adnan, M Shiddiq*
**MOLYBDENUM-99 PRODUCTION CALCULATION ANALYSIS OF SAMOP REACTOR BASED ON THORIUM NITRATE FUEL** ..... 481
*Syarip, E Togatorop, Yassar*
**APPLYING TRANSPOSE MATRIX ON ADVANCED ENCRYPTION STANDARD (AES) FOR DATABASE CONTENT** ..... 487
*E B P Manurung, O S Sitompul, Suherman*
**SIMULATION COMPARISON OF PROPORTIONAL INTEGRAL DERIVATIVE AND FUZZY LOGIC IN CONTROLLING AC-DC BUCK BOOST CONVERTER** ..... 492
*A Faisal, S Hasan, Suherman*
**COMPUTER SIMULATION AND IMPLEMENTATION OF DEFECTED GROUND STRUCTURE ON A MICROSTRIP ANTENNA** ..... 497
*H Adrian, A H Rambe, Suherman*
**IMPROVED HYBRIDIZATION OF FUZZY ANALYTIC HIERARCHY PROCESS (FAHP) ALGORITHM WITH FUZZY MULTIPLE ATTRIBUTE DECISION MAKING - SIMPLE ADDITIVE WEIGHTING (FMADM-SAW)** ..... 501
*B E Zaiwani, M Zarlis, S Efendi*
**PREDICTED CYCLOARTENOL SYNTHASE PROTEIN FROM KANDELIA OBOVATA AND RHIZOPHORA STYLOSA USING ONLINE SOFTWARE OF PHYRE2 AND SWISS-MODEL** ..... 508
*M Basyuni, N Sulistiyono, R Wati, Sumardi, H Oku, S Baba, H Sagami*
**5W1H INFORMATION EXTRACTION WITH CNN-BIDIRECTIONAL LSTM** ..... 514
*A Nurdin, N U Maulidevi*
**IMPLEMENTATION OF CHAOTIC GAUSSIAN PARTICLE SWARM OPTIMIZATION FOR OPTIMIZE LEARNING-TO-RANK SOFTWARE DEFECT PREDICTION MODEL CONSTRUCTION** ..... 524
*M A Buchari, S Mardiyanto, B Hendradjaya*
**MODELING OF SEDIMENTATION PROCESS IN WATER** ..... 531
*Tulus, I. Z. Sefnides, Sawaluddin, Suriati, M Dwiastuti*
**ONLINE THESIS GUIDANCE MANAGEMENT INFORMATION SYSTEM** ..... 536
*T H Nasution, F Pratama, K Tanjung, I Siregar, A Amalia*
**PLAGIARISM DETECTION FOR INDONESIAN LANGUAGE USING WINNOWING WITH PARALLEL PROCESSING** ..... 542
*Y Arifin, S M Isa, L A Wulandhari, E Abdurachman*
**FEATURE DETECTION OF CURVE TRAFFIC SIGN IMAGE ON THE BANDUNG - JAKARTA HIGHWAY** ..... 548
*M Naseer, I Supriadi, S H Supangkat*
**NEARBY SEARCH INDEKOS BASED ANDROID USING A STAR (A*) ALGORITHM** ..... 554
*B Siregar, Eb Nababan, Ja Rumahorbo, U Andayani, F Fahmi*
**ANALYSIS BACKPROPAGATION METHODS WITH NEURAL NETWORK FOR PREDICTION OF CHILDREN'S ABILITY IN PSYCHOMOTORIC** ..... 560
*F Izhari, H W Dhany, M Zarlis, Sutarman*
**AN IMPLEMENTATION OF RC4+ ALGORITHM AND ZIG-ZAG ALGORITHM IN A SUPER ENCRYPTION SCHEME FOR TEXT SECURITY** ..... 566
*M A Budiman, Amalia, N I Chayanie*

**COMPARISON OF NAIVE BAYES AND DECISION TREE ON FEATURE SELECTION USING GENETIC ALGORITHM FOR CLASSIFICATION PROBLEM** ........ 572
*S Rahmadani, A Dongoran, M Zarlis, Zakarias*

**IMPLEMENTATION OF SUPER-ENCRYPTION WITH TRITHEMIUS ALGORITHM AND DOUBLE TRANSPOSITION CIPHER IN SECURING PDF FILES ON ANDROID PLATFORM** ........ 578
*M A Budiman, D Rachmawati, Jessica*

**ANALYSIS SPECTRUM OF ECG SIGNAL AND QRS DETECTION DURING RUNNING ON TREADMILL** ........ 584
*M. Agung Suhendra, M. Ilham R., Artha I. Simbolon, M. Faizal A., A. Munandar*

**AN IMPLEMENTATION OF SUPER-ENCRYPTION USING RC4A AND MDTM CIPHER ALGORITHMS FOR SECURING PDF FILES ON ANDROID** ........ 590
*M A Budiman, D Rachmawati, M R Parlindungan*

**APPLICATION OF OPEN GARDEN SENSOR ON HYDROPONIC MAINTENANCE MANAGEMENT** ........ 594
*S Nasution, B Siregar, M Kurniawan, H Pranoto, U Andayani, F Fahmi*

**CLASSIFICATION OF STROKE DISEASE USING CONVOLUTIONAL NEURAL NETWORK** ........ 601
*J T Marbun, Seniman, U Andayani*

**AUDITING ALBAHA UNIVERSITY NETWORK SECURITY USING IN-HOUSE DEVELOPED PENETRATION TOOL** ........ 607
*M E Alzahrani*

**A LOW-COMPLEXITY SUBGROUP FORMATION WITH QOS-AWARE FOR ENHANCING MULTICAST SERVICES IN LTE NETWORKS** ........ 612
*M Algharem, M H Omar, R F Rahmat, R Budiarto*

**PROTEIN MODELLING OF TRITERPENE SYNTHASE GENES FROM MANGROVE PLANTS USING PHYRE2 AND SWISS-MODEL** ........ 624
*M Basyuni, R Wati, N Sulistiyono, R Hayati, Sumardi, H Oku, S Baba, H Sagami*

**FLUID FLOW AND HEAT TRANSFER CHARACTERISTICS OF AN ENCLOSURE WITH FIN AS A TOP COVER OF A SOLAR COLLECTOR** ........ 632
*H Ambarita, A D Ronowikarto, R E T Siregar, E Y Setyawan*

**EFFECTS OF THE INCLINATION ANGLE ON THE PERFORMANCE OF FLAT PLATE SOLAR COLLECTOR** ........ 638
*H Ambarita, R E T Siregar, A D Ronowikarto, E Y Setyawan*

**THE OPTIMUM INTERMEDIATE PRESSURE OF TWO-STAGES VAPOR COMPRESSION REFRIGERATION CYCLE FOR AIR-CONDITIONING UNIT** ........ 646
*H Ambarita, H V Sihombing*

**NUMERICAL STUDY ON THE EFFECTS OF ABSORPTIVITY ON PERFORMANCE OF FLAT PLATE SOLAR COLLECTOR OF A WATER HEATER** ........ 652
*D R S. Tambunan, Y P Sibagariang, H Ambarita, F H Napitupulu, H Kawai*

**DESIGNING A HOLISTIC END-TO-END INTELLIGENT NETWORK ANALYSIS AND SECURITY PLATFORM** ........ 660
*M Alzahrani*

**VIRTUAL REALITY INTERACTIVE MEDIA FOR UNIVERSITAS SUMATERA UTARA – A CAMPUS INTRODUCTION AND SIMULATION** ........ 669
*R F Rahmat, Anthonius, M A Muchtar, A Hizriadi, M F Syahputra*

**AN INTERACTIVE AUGMENTED REALITY IMPLEMENTATION OF HIJAIYAH ALPHABET FOR CHILDREN EDUCATION** ........ 678
*R F Rahmat, F Akbar, M F Syahputra, M A Budiman, A Hizriadi*

**RANGKU ALU – A TRADITIONAL EAST NUSA TENGGARA GAME IN ANDROID PLATFORM** ........ 685
*R F Rahmat, R Ramadhan, D Arisandi, M F Syahputra, O Sheta*

**MEASURING THE POWER CONSUMPTION OF SOCIAL MEDIA APPLICATIONS ON A MOBILE DEVICE** ........ 697
*A I M Dunia, Suherman, A H Rambe, R Fauzi*

**DEVELOPMENT OF BUILDING SECURITY INTEGRATION SYSTEM USING SENSORS, MICROCONTROLLER AND GPS (GLOBAL POSITIONING SYSTEM) BASED ANDROID SMARTPHONE** ........ 701
*P Sihombing, Y M Siregar, J T Tarigan, I Jaya, A Turnip*

**HYPERTENSIVE RETINOPATHY IDENTIFICATION THROUGH RETINAL FUNDUS IMAGE USING BACKPROPAGATION NEURAL NETWORK** ........ 708
*M F Syahputra, C Amalia, R F Rahmat, D Abdullah, D Napitupulu, M I Setiawan, W Albra, Nurdin, U Andayani*

**IDENTIFICATION MALE FERTILITY THROUGH ABNORMALITIES SPERM BASED MORPHOLOGY (TERATOSPERMIA) USING INVARIANT MOMENT METHOD** ............ 715
*M F Syahputra, R Chairani, Seniman, R F Rahmat, D Abdullah, D Napitupulu, M I Setiawan, W Albra, C I Erliana, U Andayani*

**IMPLEMENTATION OF AUGMENTED REALITY TO MODELS SULTAN DELI** ............ 722
*M F Syahputra, N P Lumbantobing, B Siregar, R F Rahmat, U Andayani*

**IMPLEMENTATION OF AUGMENTED REALITY TO TRAIN FOCUS ON CHILDREN WITH S PECIAL NEEDS** ............ 732
*M F Syahputra, P P Sari, D Arisandi, D Abdullah, D Napitupulu, M I Setiawan, W Albra, Asnawi, U Andayani*

**REAL-TIME MONITORING SYSTEM FOR ELDERLY PEOPLE IN DETECTING FALLING MOVEMENT USING ACCELEROMETER AND GYROSCOPE** ............ 739
*B Siregar, U Andayani, R P Bahri, Seniman, F Fahmi*

**CLEAN WATER BILLING MONITORING SYSTEM USING FLOW LIQUID METER SENSOR AND SMS GATEWAY** ............ 747
*F Fahmi, A Hizriadi, F Khairani, U Andayani, B Siregar*

**KOREAN LETTER HANDWRITTEN RECOGNITION USING DEEP CONVOLUTIONAL NEURAL NETWORK ON ANDROID PLATFORM** ............ 754
*S Purnamawati, D Rachmawati, G Lumanauw, R F Rahmat, R Taqyuddin*

**ADVERTISEMENT SCHEDULING ON COMMERCIAL RADIO STATION USING GENETICS ALGORITHM** ............ 763
*S Purnamawati, E B Nababan, B Tsani, R Taqyuddin, R F Rahmat*

**FILE TEXT SECURITY USING HYBRID CRYPTOSYSTEM WITH PLAYFAIR CIPHER ALGORITHM AND KNAPSACK NACCACHE-STERN ALGORITHM** ............ 771
*Amalia, M A Budiman, R Sitepu*

**PDF FILE ENCRYPTION ON MOBILE PHONE USING SUPER-ENCRYPTION OF VARIABLY MODIFIED PERMUTATION COMPOSITION (VMPC) AND TWO SQUARE CIPHER ALGORITHM** ............ 778
*D Rachmawati, M A Budiman, F Atika*

**A COMPARATIVE STUDY OF MESSAGE DIGEST 5(MD5) AND SHA256 ALGORITHM** ............ 782
*D Rachmawati, J T Tarigan, A B C Ginting*

**DETERMINATION OF QUALITY TELEVISION PROGRAMMES BASED ON SENTIMENT ANALYSIS ON TWITTER** ............ 788
*A Amalia, W Oktinas, I Aulia, R F Rahmat*

**DATA PRE-PROCESSING IN RECORD LINKAGE TO FIND THE SAME COMPANIES FROM DIFFERENT DATABASES** ............ 794
*D Gunawan, M S Lubis, D Arisandi, B Azzahry*

**BUILDING AUTOMATIC CUSTOMER COMPLAINTS FILTERING APPLICATION BASED ON TWITTER IN BAHASA INDONESIA** ............ 801
*D Gunawan, R P Siregar, R F Rahmat, A Amalia*

**THE IMPLEMENTATION OF COSINE SIMILARITY TO CALCULATE TEXT RELEVANCE BETWEEN TWO DOCUMENTS** ............ 808
*D Gunawan, C A Sembiring, M A Budiman*

**SSL/TLS VULNERABILITY DETECTION USING BLACK BOX APPROACH** ............ 814
*D Gunawan, E H Sitorus, R F Rahmat, A Hizriadi*

**STANDALONE MOBILE APPLICATION FOR SHIPPING SERVICES BASED ON GEOGRAPHIC INFORMATION SYSTEM AND A-STAR ALGORITHM** ............ 818
*D Gunawan, I Marzuki, A Candra*

**PRIME NUMBERS COMPARISON USING SIEVE OF ERATOSTHENES AND SIEVE OF SUNDARAM ALGORITHM** ............ 824
*D Abdullah, R Rahim, D Apdilah, S Efendi, T Tulus, S Suwilo*

**Author Index**

IOP Conf. Series: Journal of Physics: Conf. Series **978** (2018) 011001 doi:10.1088/1742-6596/978/1/011001

# Introduction from the Chair

The 2nd International Conference on Computing and Applied Informatics organized by Faculty of Computer Science and Information Technology, Universitas Sumatera Utara (USU) Medan, Indonesia is published by IOPP in the UK and indexed by Scopus. This international forum provides an opportunity for the community of computer science, information technology and engineering to share and discuss their current area of research.

Area of research in this conference includes: (1) Artificial Intelligence, Machine Learning, Big Data, Data Mining, (2) Smart City, E-government, E-commerce, E- health (3) Bio Informatics, Socio Informatics, GIS, Information Security, (4) Human Computer Interaction, Augmented Reality, Image Vision, Natural Language Processing, Natural Language Generation, and (5) Software Engineering, Computer Graphics and Networking.

There are 248 papers submitted to this conference and have reviewed by reviewers according to their respective field of interest. The acceptance rate of the paper is 123 (49.5%). Accepted papers come from Indonesia, Malaysia, Japan, Germany, Saudi Arabia, Iran and Egypt.

We would like to express our sincere thanks to our keynote speakers: Prof Dr. Graham Kendall and Prof. Dr. Masafumi Nishimura. Moreover, our sincere gratitude Prof. Dr. Runtung Sitepu, Rector of Universitas Sumatera Utara, Mahyudin Nasution, M.IT, Phd, Vice Rector for Research, Community Service and Cooperation, Universitas Sumatera Utara, Prof. Dr. Opim Salim Sitompul, MSc, Dean of Faculty of Computer Science and Information Technology Universitas Sumatera Utara, all presenters, all participants, all reviewers, all the committee members, faculty staff and students for their effort and supports in this conference.

Lastly, on behalf of the organizing committee, I would like to welcome you to our second international conference on computing and applied informatics, ICCAI 2017. Our hope is that you have an enjoyable and inspiring as well as a successful moment during our conference. Thank you for your participation on ICCAI 2017.

**Dr. Erna Budhiarti Nababan, M.IT**
Chair, ICCAI 2017

IOP Conf. Series: Journal of Physics: Conf. Series **978** (2018) 011002 doi:10.1088/1742-6596/978/1/011002

# Committee

**Honorary Chair**

Prof. Dr. Runtung Sitepu,S.H., M.Hum. (Rector University of Sumatera Utara, Indonesia)

**International Board**

Prof. Hiromi Homma (Toyohashi University of Technology, Japan)
Prof. Madya Dr. Rahmita Wirza O. K. Rahmat (Universiti Putra Malaysia, Malaysia)
Prof. Dr. Salwani Abdullah (The National University of Malaysia, Malaysia)
Prof. Dr. Masashi Daimaruya (Muroran Institute of Technology, Japan)
Prof. Dr. Rahmat Budiarto (Albaha University, Saudi Arabia)
Muhammad Fermi Pasha, B.Comp.Sc., M.Sc., Ph.D. (Monash University, Malaysia Campus)
Dr. Osama Sheta (Zagazig University, Egypt)
Dr. Mu'awya Al Dalaien, (Abu Dhabi and Khalifa City Womens Colleges, UAE)

**Steering Committees**

Prof. Dr. Hideki Kawai (Muroran Institute of Technology, Japan)
Prof. Zainal A.Hasibuan, Ir., MLS, Ph.D. (University of Indonesia, Indonesia)
Prof. Dr. Opim Salim Sitompul (Universitas Sumatera Utara, Indonesia)
Prof. Dr. Herman Mawengkang (Universitas Sumatera Utara, Indonesia)
Prof. Dr. Saib Suwilo (Universitas Sumatera Utara, Indonesia)
Prof. Dr. Muhammad Zarlis, M.Sc. (Universitas Sumatera Utara, Indonesia)
Prof. Dr. rer.nat. Dedi Rosadi, M.Sc. (University of Gadjah Mada, Indonesia)
Dr. Ahmed H. Alahmadi (Albaha University/Taibah University, KSA)
Dr. Mohammed Y. Alzahrani (Albaha University, KSA)
Dr. Adil F. Alharthi (Albaha University, KSA)
Drs. Mahyuddin K. M. Nasution, M.IT, Ph.D. (Universitas Sumatera Utara, Indonesia)
Dr. Ir. Rinaldi Munir, MT (Bandung Institute of Technology, Indonesia)
Dr. Fitri Arnia, M.Eng.Sc (Syiah Kuala University, Indonesia)
Dr. Eng. Himsar Ambarita (Universitas Sumatera Utara, Indonesia)
Benny B. Nasution, Ph.D. (Politeknik Negeri Medan, Indonesia)
Tami Alwajeeh (Albaha University, KSA)
Ahmed Alghamdi (Albaha University, KSA)

IOP Conf. Series: Journal of Physics: Conf. Series **978** (2018) 011002 doi:10.1088/1742-6596/978/1/011002

**Chair**

Dr. Erna Budhiarti Nababan, M.IT

**Technical Chair**

Romi Fadilah Rahmat, B.Comp.Sc., M.Sc.

**Co. Chair**

Dian Rachmawati, S.Si., M.Kom
Jos Timanta Tarigan, S.Kom., M.Sc.
Mohammad Andri Budiman, S.T., M.Comp.Sc., M.E.M.
Dani Gunawan, S.T., M.T.

**Members**

Prof. Dr. Muhammad Zarlis
Dr. Ir. Fahmi, M.Sc., IPM
Emerson Sinulingga, Ph.D
Dr. Elviawati Zamzami, S.T., M.T., M.M.
Dr. Sawaluddin, M.IT
Dr. Maya Silvi Lydia, M.Sc.
Dr. Poltak Sihombing, M.Kom.
Dr. Syahril Efendi, M.IT
Mohammad Fadly Syah Putra, M.Sc.
Muhammad Anggia Muchtar, ST, M.MIT
Amalia, S.T., M.T.
Indra Aulia, S.Ti., M.Kom.
Ivan Jaya, S.Si., M.Kom.
Sarah Purnamawati, S.T., M.Sc.
Dedy Arisandi, S.T., M.Kom.
Ulfi Andayani, S.Kom., M.Kom.
Melvani Hardi, S.Kom., M.Kom.
Tigor Hamonangan Nasution, S.T., M.T.
Taufiq Bin Nur, S.T., M.Eng.
Baihaqi Siregar, S.Si., M.T.
Amer Sharif, S.Si., M.Kom.

IOP Conf. Series: Journal of Physics: Conf. Series **978** (2018) 011002 doi:10.1088/1742-6596/978/1/011002

# Venue Map

**Hotel GranDhika Setiabudi Medan**

Jalan Dr. Mansyur no. 169

Medan, Sumatera Utara

Indonesia

Phone : +62-61-42081234

Fax : +62-61-42081234

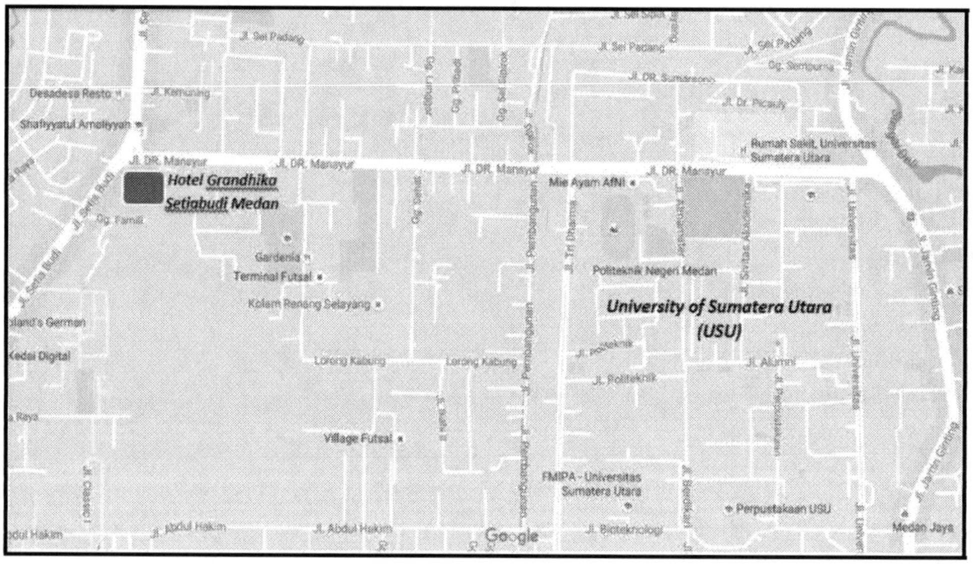


*ICCAI 2016 Venue*

IOP Conf. Series: Journal of Physics: Conf. Series **978** (2018) 011002 doi:10.1088/1742-6596/978/1/011002

# Floor Plan

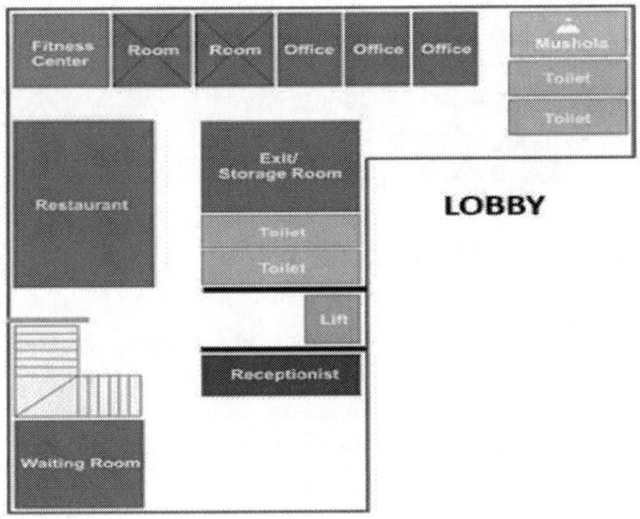


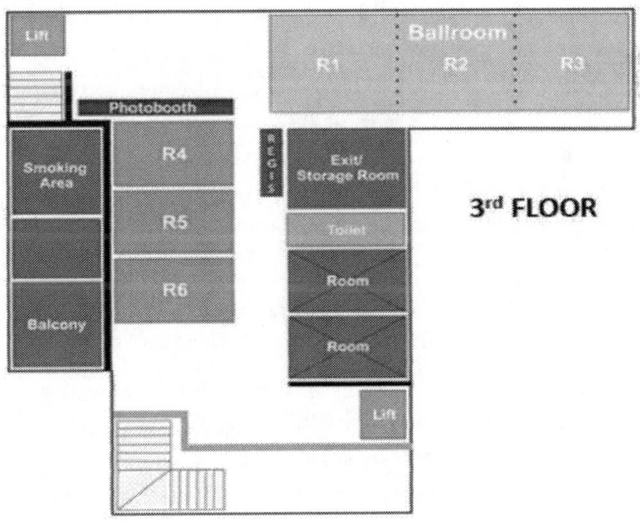

IOP Conf. Series: Journal of Physics: Conf. Series **978** (2018) 011002 doi:10.1088/1742-6596/978/1/011002

# Keynote Speaker

## About Prof. Dr. Graham Kendall

The University of Nottingham Malaysia Campus (UNMC), Malaysia

Professor Graham Kendall is the Provost and CEO of The University of Nottingham Malaysia Campus (UNMC). He is also a Pro-Vice Chancellor of the University of Nottingham. He has been in Malaysia since August 2011 and was previously the Vice-Provost for Research and Knowledge Exchange. He is a Director of MyResearch Sdn Bhd, Crops for the Future Sdn Bhd. and Nottingham Green Technologies Sdn Bhd. He was awarded a BSc (Hons) First Class in Computation from the University of Manchester Institute of Science and Technology (UMIST), UK in 1997 and received his PhD from The University of Nottingham (School of Computer Science) in 2000. Professor Kendall is a Fellow of the British Computer Society (FBCS) and a Fellow of the Operational Research Society (FORS). He has published over 90 refereed journal papers (the vast majority in ISI ranked journals) and over 230 peer reviewed papers in total. He has edited 11 books and authored almost 20 book chapters. He is currently an Associate Editor of ten journals and the Editor-in-Chief of the IEEE Transactions of Computational Intelligence and AI in Games. As a Professor of Computer Science at the University of Nottingham he is a member of the Automated Scheduling, Planning and Optimisation Group (ASAP) in the School of Computer Science. His research interests include Operations Research, Scheduling, Logistics, Vehicle Routing, Meta- and Hyper-heuristics, Evolutionary Computation and Games.

## Keynote Abstract

### Is evolutionary computation evolving fast enough?

Evolutionary Computation (EC) has been part of the research agenda for at least 60 years but if you ask the average EC researcher to name three examples of EC being used in real world applications they might struggle. Other technologies have seen much wider adoption. 3D printing is changing the way that manufacturing is done, moving some of that functionality into the home. Immersive reality is on the verge of changing society, in ways that are not totally clear yet. Ubiquitous computing is becoming more prevalent, enabling users to access computing resources in ways that were unimaginable even just a few years ago. It might be argued that EC has not had the same penetration as other technologies. In this talk, we will look back at what EC promised, see if it has delivered on that promise and compare its progress with other technologies. Finally, we will suggest some challenges that might further advance EC and enable its wider adoption.

## About Prof. Dr. Masafumi Nishimura

Shizuoka University - Hamamatsu Campus, Japan

Masafumi Nishimura received his B.E. and M.E. degrees in Biophysical Engineering from Osaka University in 1981 and 1983, and his Dr. of Engineering degree from Toyohashi University of Technology in 1998. In 1983 he joined the IBM Japan Science Institute and worked in speech recognition area. Since 2003, he had been a Senior Technical Staff Member of IBM and a Manager of the Speech Technology Group at the IBM Research - Tokyo, Japan. In 2014, he moved to the Shizuoka University as a Professor of Informatics. His research interests include speech signal processing, robust speech recognition, statistical language modeling, expressive speech synthesis, emotion detection and speech analytics. Dr. Nishimura became Editor of the Institute of Electronics, Information and Communication Engineers Transactions in 2010. He received the SIG Research Award from the Information Processing Society of Japan in 1998, and the Prize for Outstanding Technological Development in Acoustics from the Acoustical Society of Japan in 1999. He is a Senior Member of the IEEE, the IEICE and the IPSJ, and a Member of the ASJ and the JSAI.

## Keynote Abstract

### Use of Sound Information for Supporting the Elderly and the Disabled

In many developed countries including Japan, population is very rapidly aging. Considering the lack of welfare and long-term care human resources, maximum utilization of information-communication technologies (ICT) such as IoT is an urgent need. Microphone is one of the most useful and conventional sensors of IoT because it is cheap and easy to use, and sound carries various kinds of information than we imagine. In recent years, speech to text technologies become to be widely used but text information is just a part of sound information. Speech also includes non-verbal information such as emotion, and information of environmental sounds is indispensable for our daily life though we are not aware of its importance. In addition, it is well known that biological sounds are very useful for medical check-up besides auscultation. In this talk we will introduce some sound information-based applications and technologies for supporting the elderly and the hearing-disabled in their daily life.

IOP Conf. Series: Journal of Physics: Conf. Series **978** (2018) 011002 doi:10.1088/1742-6596/978/1/011002

**PROGRAM SCHEDULE**

Internasional Conference on Computing and Applied Informatics (ICCAI) 2017

GranDhika Hotel Medan, Indonesia

## WORKSHOP

**Tuesday, 28th November 2017**

| Time | Session Workshop |
|---|---|
| 12.00 – 13.30 am | **Lunch** |
| 13.30 – 13.35 am | **Opening and Welcome Speech** |
| 13.35 – 13.40 am | • Singing Indonesia National Anthem: Indonesia Raya |
| 13.40 – 13.45 am | • Opening Speech : Chairman Of ICCAI 2017 |
| 13.45 – 13.50 am | • Prayer |
| 13.50 – 15.20 am | **Workshop (Speaker I)**<br>• Prof.Dr. Salwani Abdullah<br>The National University of Malaysia<br>• Moderator:<br>Amer Sharif, S.Si., M.Kom.<br>Dian Rachmawati, S.Si., M.Kom. |
| 15.20 – 15.25 am | • Souvenirs Handover |
| 15.25 – 16.00 am | **Coffee Break** |
| 16.00 – 17.30 am | **Workshop (Speaker II)**<br>• Dr. Benny Benyamin Nasution, Dipl.–Ing. M.Eng<br>Politeknik Negeri Medan<br>• Moderator:<br>Mohammad Andri Budiman, S.T., M.Comp.Sc., M.E.M.<br>Sarah Purnamawati, ST., MSc. |
| 17.30 – 17.35 am | • Souvenirs Handover |
| 17.35 - 18.00 am | **Closing** |

IOP Conf. Series: Journal of Physics: Conf. Series **978** (2018) 011002 doi:10.1088/1742-6596/978/1/011002

**PROGRAM SCHEDULE**

Internasional Conference on Computing and Applied Informatics (ICCAI) 2017

GranDhika Hotel Medan, Indonesia

## INTERNATIONAL CONFERENCE

**Wednesday, 29th November 2017**

| Time | Session |
|---|---|
| 07.00 – 09.00 am | **Registration** |
| 07.45 – 09.00 am | **Parallel Session I** |
| 09.00– 09.05 am | **Opening and Welcome Speech** |
| 09.05 – 09.10 am | • Singing Indonesia National Anthem: Indonesia Raya |
| 09.10 - 09.20 am | • Traditional Dance Performance |
| 09.20 – 09.25 am | • Opening Speech : Chairman Of ICCAI 2017 |
| 09.25 – 09.30 am | • Opening Speech : Rector of USU |
| 09.30 – 09.40 am | • Dies Natalis Fasilkom-TI 2017 |
| 09.40 – 09.50 am | • Photo Session |
| 09.50 – 10.00 am | • Prayer |
| 10.00 – 10.30 am | Coffee Break |
| 10.30 – 11.10 am | **( Keynote Speech I )**<br>• **Prof. Graham Kendal**<br>G of Nottingham, Malaysia<br>• Moderator :Prof. Dr. Drs. Opim Salim Sitompul, M.Sc |
| 11.10 – 11.15 am | • Souvenirs Handover to Keynote Speech ICCAI 2017 |
| 11.15 – 11.55 am | **(Keynote Speech II)**<br>• **Prof. Masafumi Nishimura**<br>• *Shizuoka university, Japan*<br>• Moderator : Dr. Ir. Fahmi, ST., M.Sc., IPM |
| 11.55 – 12.00 am | • Souvenirs Handover to Keynote Speech ICCAI 2017 |
| 12.00 - 13.00 pm | **Lunch** |
| 13.00 - 15.30 pm | **Parallel Session II** |
| 15.30 - 16.00 pm | **Coffee Break** |
| 16.00 - 18.00 pm | **Parallel Session III** |
| | Gala Dinner |
| 19.00 - 21.00 pm | • Welcome Speech from ICCAI 2017 Chief<br>• Conference awards for best paper and best presentation<br>• Acoustic Performance<br>• Conference dinner<br>• Closing ceremony |

# The 2[nd] International Conference on Computing and Applied Informatics (ICCAI) 2017

SOCIETY
SCIENCE

IOP Conf. Series: Journal of Physics: Conf. Series **978** (2018) 011003 doi:10.1088/1742-6596/978/1/011003

IOP Conf. Series: Journal of Physics: Conf. Series 978 (2018) 011003 doi:10.1088/1742-6596/978/1/011003

UNIVERSITAS

IOP Conf. Series: Journal of Physics: Conf. Series **978** (2018) 011003 doi:10.1088/1742-6596/978/1/011003

ICCAI
The 2nd International Conference on
Computing and Applied Informatics
(ICCAI) 2017

ICCAI
The 2nd International Conference on
Computing and Applied Informatics
(ICCAI) 2017

IOP Conf. Series: Journal of Physics: Conf. Series **978** (2018) 011003 doi:10.1088/1742-6596/978/1/011003

Introduction

developme
Construct2
create thei

IOP Conf. Series: Journal of Physics: Conf. Series **978** (2018) 011003 doi:10.1088/1742-6596/978/1/011003

IOP Conf. Series: Journal of Physics: Conf. Series **978** (2018) 011004 doi:10.1088/1742-6596/978/1/011004

# Peer review statement

All papers published in this volume of *Journal of Physics: Conference Series* have been peer reviewed through processes administered by the proceedings Editors. Reviews were conducted by expert referees to the professional and scientific standards expected of a proceedings journal published by IOP Publishing.

IOP Conf. Series: Journal of Physics: Conf. Series **978** (2018) 012001 doi:10.1088/1742-6596/978/1/012001

# Improved Line Segmentation Framework for Sundanese Old Manuscripts

**E Paulus, M Suryani, S Hadi**

Department of Computer Science, University of Padjadjaran, Jalan Raya Bandung Sumedang KM21 Jatinangor 45363, INDONESIA

E-mail: erick.paulus@unpad.ac.id, mira.suryani@unpad.ac.id, setiawanahdi@unpad.ac.id

**Abstract.** Line segmentation can be a useful process for further text segmentation. There are some certain line segmentation framework that use binarization method as an initial step. But binarization process is still facing a major challenge, especially on old document palm-leaf manuscripts. As the quality of the image has varying degrees of noises in the non-text region. Seam Carving method, one of line segmentation methods that uses binarization-free approach, can be an alternative solution. However, this method can separate the incorrect text line on small element text located at the bottom or at the top of a main character contour. Therefore, an improvement on line segmentation framework is proposed by using hybrid binarization and its implemented on the smallest energy function to separate out the text-lines. The proposed framework have been evaluated on 44 Sundanese old manuscript images that consist of true color and binary images. The evaluation matrix shows that this framework can improve Niblack binarization process up to 50%. In addition, our framework does not only generate the number of text-lines to come near to the number of target lines, but it also can separate the text-lines well on small element text. Overall, the expected result can in the end be produced from the proposed line segmentation framework.

## 1. Introduction

Preservation programs on old manuscripts have become important issues for the government, library, and society. Digitization process, which is one of preservation types, have produced many manuscript images. OCR system is needed to extract the text contained in the manuscript images. To recognize the text pattern, binarization and segmentation processes are required as fundamental steps. Some line segmentation methods can produce text line using binarization-free approach. It means that we can separate text line directly from true color image. However, whether or not line segmentation using binarization-free approach can deliver the best result compared to line segmentation using binarization approach is still uncertain. Therefore, we evaluate the performance of Seam Carving method. Then, we construct binarization and line segmentation framework to obtain better solution in separating text line. In manuscript image analysis, binarization process has an important role to transform true color or grey scale image into binary image[1]. There are some challenges to binarize the old document image such as low contrast, stain, smear, shadow, and irregular illumination[1-2]. Diversiform binarization algorithm has been formulated to convert true color image into binary images using global, local, or hybrid thresholding [3-9]. Five steps of binarization process are proposed by Ntogas

[6] for Byzantine old manuscripts. For old manuscripts, many studies show that binarization method using hybrid thresholding is better [11]. There is binarization algorithm that uses whale optimization approach and implement fuzzy c-means as objective function[8], contour model[3] and stroke connectivity[9]. Local thresholding such as Niblack, Sauvola[5] and Howe[15] can also give contribution in binarization. The binarization performance of Sauvola and Howe show better result than Niblack. But Niblack method can keep the solid contour of the text, though a lot of subtle noise still remain on binary image. Hence, we improve hybrid binarization framework using the advantage of Niblack method, edge detection, color map, and filtering method. After binarization process, we can continue to line segmentation process[12]. This step eases segmentation process for the character, syllable and word. Projection profile method[13], as the simplest line segmentation, can identify the line of a text by using histogram profile. The profile presents valley points at line boundaries and the location of these minimum points mark the line boundaries. This method runs well if the input image is binary image and there is no waved text. Adaptive projection profile [14] is used to deal with waved text, by dividing the image into some columns. Seam Carving method implements the minimum energy function to separate the lines of a text. The advantage of Seam Carving method works well on the true color image and waved text. But, this method does not work well on some small element text located at the bottom or at the top of the main character. Figure 1 shows the incorrect line segmentation. In this paper, binarization and line segmentation frameworks are exhibited in detail. Systematically, the paper is divided into sections. The next section of this paper (the second section) describes the collection of Sundanese old manuscripts and the challenges for binarization and line segmentation. Following this, our proposed binarization and line segmentation frameworks are shown on the third section. Some experiments and results of our proposed framework are explained in the fourth section. The last section describes the conclusions and some potential research for future study.

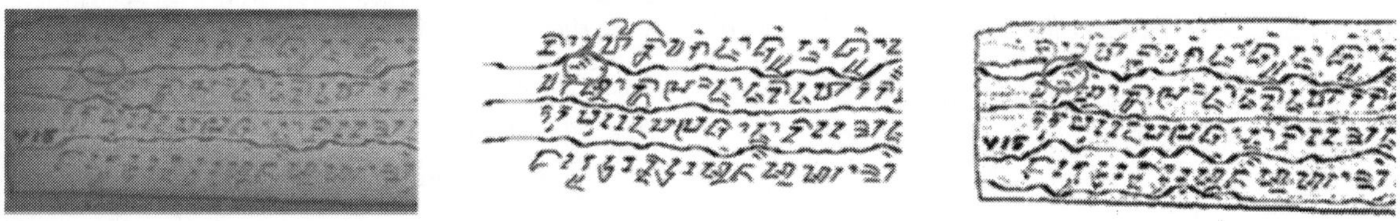

**Figure 1.** The examples of incorrect line segmented both on color and binary image.

## 2. The Condition and Challenges of Sundanese Old Manuscripts

Several mediums are used in old manuscripts such as paper, palm leaf, metal and stone. Generally, Sundanese old manuscriptss in museums are of many mediums. But, there are some other places that keep both the original and duplication of the manuscripts in a private place. Kabuyutan Ciburuy, one of Indonesia's private heritage place, stores the collection of Sundanese old manuscripts. These collections, known as Sundanese Lontar uses palm leaf as its writing medium. The length and width of Sundanese Lontar is about 25 to 45 cm and 10 to 15 cm respectively. Each page consists of four to six lines, and it is estimated there are 15 to 20 words in each line. The Sundanese Lontar has a hole in the center and it is rope-bound. It consists of a variety of writings such as Ramayana epoch, farm and medical formulae, and social life. This made the Sundanese Lontar extremely valuable for Sundanese people. The quality of Sundanese old manuscript images are poor, as they contain various noises such as smears, non-uniform illumination, shadow, and many random noises. All of the characteristics are the challenges for every step in image processing including binarizarion and text line segmentation.

## 3. The Proposed Line Segmentation Framework

Our proposed framework, is of two part steps, namely: pre-processing and line segmentation process. We also proposed novel filtering to increase the performance of Niblack binarization. This framework is designed for the designated Sundanese old manuscripts.

### 3.1. Pre-processing

The goal of pre-processing step is to minimize the noise that is distributed on the original image and transform it to binary image. The first step, We convert RGB color map into the HSV color map and Grayscale. The second step is to remove Non-Lontar Area. In digitizing process, the photographer used fabric (dark or light color) as background so we needed to remove the non-lontar area. We used hue or saturation channel to transform it into white color.

The third step, we evaluated several thresholding methods such as Otsu[6], Niblack[4][10], Sauvola[5], and Howe[15] to our framework. Then, the result showed that Niblack is suitable for our framework. Even though Niblack still leaves noise everywhere. But, the contour of the text is still solid. There were no missing texts. This advantage of Niblack was chosen and implemented in our framework. The binarization process can be formulated on equation (1).

$$T = m + k * s \tag{1}$$

where s stands for standard deviation, and m for average of color intensity I in the sub window. Also, k is a constanta a value in [-1,0) and (0,1]. Based on the experiment, k= + 0.2 is appropriate for white object detection and k = -0.2 is appropriate for black object detection.

The last step, Filtering was used to remove noise that is distributed in the digital image area. We proposed a novel filtering to improve Niblack method, as presented in equation (2) and (3). The goal of this filtering is to remove the tiny group of noise. It is represented by the sum of black area which is less than a half of white area.

$$E(i,j) = \sum_{x=-r}^{r} \sum_{y=-r}^{r} f(i+x, j+y) \tag{2}$$

$$bw_{filter}(i,j) = \begin{cases} 1\,(white\,color) & , if\, E(i,j) \geq \lfloor (2*r+1)^2/2 \rfloor \\ 0\,(black\,color) & , otherwise \end{cases} \tag{3}$$

where r is a half-length (N) of sub window (N x N); E is the cumulative sum of white region.

### 3.2. Line Segmentation

The basic method for line segmentation is projection profile[13]. Projection profile can be processed vertically or horizontally. Vertical projection profiles can segment word or syllable while horizontal projection profile can segment text line. Nowadays, there are several line segmentation algorithm adopted from Seam Carving method [16-18]. Some studies explain that this algorithm is free from binarization. Hence, it can be an alternative solution to our old manuscript segmentation. The following is a segmentation algorithm.

The first process, the image was divided vertically into several pieces. Then, we implemented Sobel Operator as edge detection. After that, we calculate the smoothed horizontal projection profile for each peace. Finally, we found and connected the nearest local maximum of each piece.

The second process is to compute the Separator of Text-line. This process implements modified Seam Carving procedure. We calculated Energy Function as initial step. Then, we calculated the cumulative of minimum energy M. After that, we calculated the optimal path from cumulative energy M. The high-energy regions represent the foreground (text component) and the low-energy regions represent the background (non-text component). The sample image of separating text-line is shown in Figure 1. Sometimes there are some text components which were not included in their lines.

IOP Conf. Series: Journal of Physics: Conf. Series **978** (2018) 012001 doi:10.1088/1742-6596/978/1/012001

## 4. Experiments and Result

We used datasets which consist of 22 true color images of Sundanese old manuscripts and 22 ground truth binary images. True color images consisted of 12 true color images from Kropak 18 and 10 true color images from Kropak 22. In general, every image contains four lines. Then, we used PixLabelers[19] tools to construct the ground truth (GT) images. After that, we segmented it to become text-line GT images using Alethea Lite[20] tools and built an application program of image cropping based on XML file using Scilab[21]. This experiment implemented the evaluation matrix that was used in ICDAR contest in 2013.

### *4.1. Experiment I - Pre-processing Phase*

The first sub framework is proposed to increase the performance of binarization for Sundanese manuscript. We investigated some common binarization methods such as Otsu, Niblack, Sauvola, and Howe. Figure 2 presents the result of pre-processing phase. We found that Niblack binarization method resulted in many noises. Apart from that, the contour text is still visually solid. Therefore, we added some filtering method to improve the pre-processing phase. Our proposed framework succeeded in getting the highest score on binarization evaluation, as presented in table 1–2.

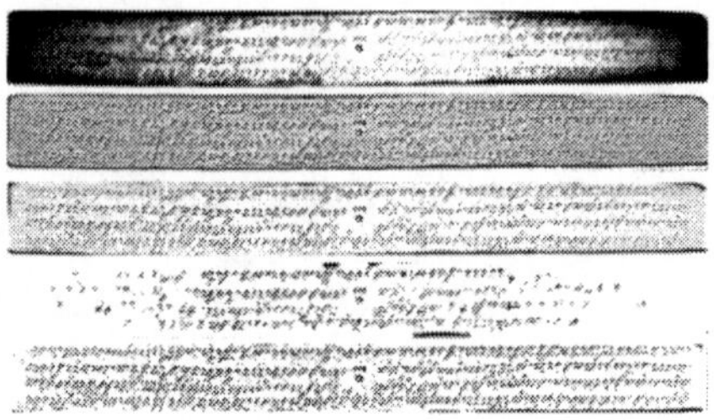

**Figure 2.** Example of binarized images (up to bottom) using the methods of Otsu[11], Niblack[4], Sauvola[5], Howe[15], and proposed scheme.

**Table 1.** The binarization evaluation matrix based on ICDAR competition[11] for 3309.tif

| Methods | Precision | Recall | F-m (%) | pF-m (%) | NRM | PSNR | DRD | MPM (x1000) |
|---|---|---|---|---|---|---|---|---|
| Otsu | 0.71 | 0.19 | 30.37 | 31.76 | 0.43 | 4.92 | 25.38 | 109.97 |
| niblack | 0.24 | 0.71 | 35.74 | 36.43 | 0.27 | 5.98 | 43.76 | 151.99 |
| sauvola | 0.36 | 0.46 | 40.74 | 43.20 | 0.31 | 8.80 | 20.60 | 64.08 |
| Howe | 0.30 | 0.56 | 39.38 | 40.49 | 0.29 | 7.71 | 27.58 | 62.34 |
| OurProposed | 0.46 | 0.71 | 56.03 | 57.53 | 0.19 | 9.61 | 16.33 | 18.01 |

Table 2 describes the binarization performance for Kropak 18 and Kropak 22 using some common binarization methods. For Koropak 22, Niblack, Sauvola, Howe and our proposed method result in the average F-measure above 50%. But, the highest value of F-measure was obtained by our proposed method, which amounted to 58,51, while Sauvola acquired the highest PSNR of about 11.79, followed by our proposed method of about 10.86. In this case, the Sauvola binarized image is closer to ground truth image. The F-measure average of Kropak 18 is still below 50% because of the poor physical condition of the manuscripts: smudges, black spot, and low contrast. Furthermore, the MPM average of Kropak 18 decreased from 0.146 point to 0.018 point. It shows that the resulted image is closer to ground truth image, but the result is not

**Table 2.** The binarization evaluation matrix based on ICDAR Competition

| | Koropak 18 | | | | Koropak 22 | | | |
|---|---|---|---|---|---|---|---|---|
| Methods | F-m (%) | pF-m (%) | PSNR | MPM (x1000) | F-m (%) | pF-m (%) | PSNR | MPM (x1000) |
| Otsu | 21.44 | 23.78 | 4.23 | 158.68 | 26.03 | 26.21 | 4.11 | 304.41 |
| Niblack | 29.12 | 29.56 | 5.78 | 146.18 | 50.54 | 51.32 | 8.72 | 68.60 |
| Sauvola | 30.82 | 32.30 | 8.88 | 64.46 | 51.65 | 55.62 | 11.79 | 11.29 |
| Howe | 30.58 | 31.32 | 8.02 | 63.97 | 51.82 | 52.53 | 9.90 | 44.13 |
| Our proposed | 46.59 | 47.76 | 9.74 | 18.12 | 58.51 | 59.83 | 10.86 | 17.40 |

good enough. On the next sub section, the impact of these binarized images will be further discussed.

### *4.2. Experiment II -Segmentation Process*
This sub-section discusses text line segmentation using true color image and binary image as input image. Then we manually evaluate the sum of the lines for each manuscript, as shown in Table 3. For this experiment, we use a number of input images like ground truth images($I_{GT}$), Niblack's binarized images($I_{BN}$), Sauvola's binarized image($I_{BS}$), our proposed's binarized image($I_{BP}$), and original true color image($I_{TC}$). Each lontar leaf on Kropak 22 consists of four lines. First, we do line segmentation of the ground truth images. The result is shown that the number of lines from ground truth images is as same as the number of the target line. Niblack's binarized image produces five lines on six lontars, while the original true color image gained five lines on four lontars. It explains that when the quality of binarized image is poor, a better solution is to use the original image. For our proposed binarized image, the evaluation showed that the number of lines is the same as the target line number. Segmentation evaluation for Kropak 18 shows that there is no segmentation method that can perfectly reach the number of the target lines, as shown in Table 3. For document of 3306.tif, line segmentation that used a binarized image from Niblack and Sauvola generated the number of lines which are different from the target line number. Line segmentation using true color image as input produces fewer errors on the number of rows, while line segmentation of our proposed binarized image can produce nearly the same number of rows as the target.

## 5. Experiments and Result
We have built line segmentation framework that consists of three parts. Although the binarization process cannot entirely solve all the challenges of old Sundanese manuscripts, our proposed framework succeeded in improving the Niblack binarization method. We built a novel filtering to eliminate the groups of little noise so the binarized image was cleaner than the previous one. Hence, line segmentation for our proposed binarized image can closely achieve the target lines number.

## Acknowledgments
We thank to Universitas Padjadjaran, for funding this research under the Internal Fundamental Research grant. We also thank to the Sundanese Culture Study and Kabuyutan Ciburuy for supplying the Sundanese old manuscripts. The authors is also deeply thankful to all suggestions given by Prof. Jean-Christophe Burie on old manuscript image analysis research.

**Table 3.** The comparison of the number of detected line for Kropak 18 and Kropak 22

| Images | Target Line | ($I_{GT}$) | ($I_{BS}$) | ($I_{BN}$) | ($I_{BP}$) | ($I_{TC}$) |
|---|---|---|---|---|---|---|
| CB-3-22-90-1 | 4 | 4 | 4 | 4 | 4 | 4 |
| CB-3-22-90-4 | 4 | 4 | 4 | 5 | 4 | 5 |
| CB-3-22-90-5 | 4 | 4 | 4 | 5 | 4 | 4 |
| CB-3-22-90-23 | 4 | 4 | 5 | 4 | 4 | 5 |
| CB-3-22-90-30 | 4 | 4 | 4 | 5 | 4 | 4 |
| CB-3-22-90-33 | 4 | 4 | 4 | 5 | 4 | 4 |
| CB-3-22-90-39 | 4 | 4 | 4 | 5 | 4 | 4 |
| CB-3-22-90-40 | 4 | 4 | 4 | 5 | 4 | 5 |
| CB-3-22-90-61 | 4 | 4 | 4 | 4 | 4 | 5 |
| CB-3-22-90-62 | 4 | 4 | 4 | 4 | 4 | 4 |
| 3306 | 4 | 4 | 6 | 1 | 4 | 5 |
| 3308 | 4 | 4 | 6 | 6 | 5 | 5 |
| 3309 | 4 | 4 | 6 | 6 | 4 | 4 |
| 3310 | 4 | 4 | 5 | 6 | 4 | 3 |
| 3311 | 4 | 4 | 5 | 5 | 4 | 4 |
| 3312 | 4 | 4 | 4 | 5 | 4 | 4 |
| 3313 | 4 | 4 | 5 | 5 | 4 | 4 |
| 3316 | 4 | 4 | 4 | 6 | 4 | 3 |
| 3322 | 4 | 4 | 5 | 4 | 4 | 4 |
| 3323 | 4 | 4 | 5 | 5 | 4 | 4 |
| 3349 | 4 | 4 | 5 | 1 | 4 | 5 |
| 3372 | 2 | 2 | 4 | 1 | 4 | 4 |

## References

[1] Gatos B, Pratikakis I, and Perantonis S J 2006 *Pattern Recognit.* **39** 317-327
[2] Kesiman M W A et al. 2017 *Int. Jour. of Electronic Imaging* **26** 1-15
[3] Hadjadj Z, Cheriet M, Meziane A and Cherfa Y 2017 *Signal, Image and Video Process* 1-8.
[4] Khurshid K, Siddiqi I, Faure C and Vincent N 2009 *SPIE Proc.* **7247** 1-9
[5] Sauvola J and PietikÃďinen M 2000 *Pattern Recognit.* **33** 225-236
[6] Ntogas N and Veintzas D 2008 *Proc. 12th WSEAS Int. Conf. Commun.* 41-51
[7] Chamchong R, Fung C C, Wong K W 2014 *Cultural Comp* 55-64
[8] Hassanien A E, Elfattah M A, Aboulenin S and Schaefer G 2016 *IEEE Int. Conf. on Systems, Man, and Cybernetics* 3842-3846
[9] Chen Y and Wang L 2017 *Neurocomputing* **237** 272-280
[10] Pratikakis I, Gatos B and Ntirogiannis K 2013 *Proc. Int. Conf. Doc. Anal. Recognit.* 1471-1476
[11] Su B, Lu S and Tan C L 2013 *IEEE Trans. Image Process.* **22** 1408-1417
[12] Louloudis G, Gatos B, Pratikakis I and Halatsis C 2008 *Pattern Recognit.* **41** 3758-3772
[13] Mamatha H R and Srikantamurthy K 2012 *Int. J. Appl. Inf. Syst.* **4** 13-19
[14] Bar-Yosef I and Hagbi N 2009 *Proc. Int. Conf. Doc. Anal. Recognit.* 1161-1165
[15] Howe N R 2013 *Int. J. Documant Anal. Recognit.* **16** 247-258
[16] Arvanitopoulos N and SÃijsstrunk S 2014 *Int. Conf. Front. Handwrit. Recognit.* 726-731
[17] Garz A, Fischer A, Sablatnig R and Bunke H 2012 *Proc. - 10th IAPR Int. Work. Doc. Anal. Syst.* 95-99
[18] Boiangiu C A, Ioanitescu R and Tanase M C 2014 *Int. J. Comput. Commun. Control* **9** 8-15
[19] Saund E, Lin J and Sarkar P 2009 *Proc. Int. Conf. Doc. Anal. Recognit.* 646-650
[20] Clausner C, Pletschacher S and Antonacopoulos A 2011 *Proc. Int. Conf. Doc. Anal. Recognit.* 48-52
[21] Liao W, Dong N and Fan T 2009 *2009 IEEE Int. Works. on Open-source Soft. for Scient. Comput.* 88-90

# Comparative analysis of gameplay and players emotion in the most popular games from play store

**Riwinoto, N Mahfud and L Lumombo**

Batam State Polytechnic
Informatics Department, Multimedia and Networking Engineering Study Program
Jalan Ahmad Yani, Batam Centre, Batam 29461, Indonesia

Email: riwi@polibatam.ac.id, liony@polibatam.ac.id

**Abstract**. The development of the android mobile operating system and the presence of Play Store services poses challenges for developers to produce exciting mobile games. Although publishing games in Play Store is not difficult, in fact developers have to face tough competition to make homemade games can become popular. This study analyzes the gameplay of three popular paid-free games in Play Store that can survive for a period of one year from the top 10 positions in October 2015- November 2016. Analysis performed on 8 elements of the game based on the definition of Fullerton and emotional expression analysis of respondents who appeared while playing the three choosen games. The analysis shows that scarce resources are the main attraction of all three games because they create conflicts, giving constraints and challenges to players. The multiplayer game has a pattern of results and the result makes the opponent become more negative than the player to win. While the single player game is analyzed has a pattern to make the player get positive to win. There are 3 basic emotions that most often appear that is joy, disgust and surprise. Multiplayer games tend to emphasize the emotions of joy players, while single player games tend to bring disgust emotions.

## 1. Introduction

The rise of the Android system is affecting the development of mobile games. Android was first publicly recognized in 2005, after Google acquired a Start-up company called Android Inc. According to [1], Google's speculation step shows Google's interest to enter in mobile devices field because of the advantage of mobile game [2]. Android became one of the operating systems that are very supportive of gaming applications on mobile devices. Even with Play Store, Android system users can easily find games based on the best categories, including Top Free, Top Paid, Top Grossing, Top New Paid, Top New Free and Trending.

Android system also stimulate the increasing number of game developers, including in Indonesia. Game Association of Indonesia [3] noted the number of local game developers continue to rise significantly. In its prediction in 2016 the number of local game developers will rise 50 percent compared to 2015. Based on the survey Newzoo [4] Indonesia is referred to as a promising market for the online game industry and mobile. With a population of 255.7 million, Indonesia has an online population of 66 million and gamers of 42.8 million. The amount of market potential is open to both local and foreign game developers. As well as Swedish game developers who have captured the opportunity with the release of mobile games called Duel Otak. The game successfully attracted tens of millions of players and became a popular game in Play Store.

According to [5], popular games on mobile devices have gameplay features with player touches like, tapping, shifting or drawing. In addition popular games at least have the features: easy rules, social interaction, or get rid of the enemy. The game designer's understanding of market tastes is the main thing needed to be able to design a game that can occupy a top free position in Play Store.

Therefore, in this study, authors wanted to know the gameplay that affects an incoming game in Play Store's top ranking list and creates an initial reference for game design to be marketed through Play Store. To find a good gameplay, authors performs an analysis based on 8 game elements defined by [6]. In addition, authors also want to know how the emotional characteristics that occur in game players when playing these popular games. According to [7], emotion expression can be used to improve the gameplay ability in responding to players' emotions.

## 2. Theories

### *2.1. Gameplay*

Ref [8] says gameplay is the most important pillar of game design activity. A good number of good mechanics make it simple but not rich if they do not involve gameplay, able to give players the challenges, skills and rewards they seek, without unnecessary trouble, so as to maintain and enhance their motivation. [9] describes Gameplay consists of two things: challenge and action. Challenge that must be passed by the player to arrive at the destination while playing the game. Action that players can take to take and overcome the challenges provided. Ref [10] describes gameplay experience have immersion which compound sensory, challenge-based and imaginative immersion. Ref [11] says that not all games with good graphics and advanced features are fun but game that deliver think and feel the value o game will be popular game. Meanwhile, according to Csikszentmihalyi (1990) cited by [12], the success of digital games has been linked to the ability of game designers to design games that meet the player's enjoyment experience while playing. This experience is important to be placed in the mental state of players called "flow" as people who care less about their condition when they are in a state of pleasure.

### *2.2. Game Elements*

According to [13], there are three basic elements that are used as a framework by game developers to form a game, which is mechanics, dynamics and aesthetics abbreviated as MDA. Ref [14] was the first book that describe that the proper way to understand games is from an aesthetic perspective. Ref [15] give another perspective about game design that consist of mechanics,aestethic, story and technology. The fun that players can get when playing games are narrative, challenges, fellowship, discovery, expression, sensation, fantasy and submission. Meanwhile, according to [6] there are 8 formal elements in the design of a game, i.e: player, goal, procedure, rule, resources, conflict, limitation, results and consequences.

### *2.3. Emotion*

Based on [7], there are 7 types of emotions that are recognized based on the combination of facial movement. The types of emotions are Anger, Disgust, Fear, Joy, Sadness, Surprise and Contempt. The application used to recognize emotions is AFFDEX SDK that available in Play Store and http://www.affectiva.com/sdk

## 3. Comparative Analysis of Gameplay and Players Emotion

### *3.1. Three Popular Games Election*

From observations during the period of October 1, 2015 to September 1, 2016 using the site www.appannie.com, there are 43 games that appear in the top 10 free positions that are recorded every 1st of each month. Table 1 shows the rank of the game with its occurrence frequency. Game that is in position after 6 has frequency less than 6 frequency of occurrence. Of all the games there are 3 most prominent games with the frequency of occurrence most often is Clash of Clans, Pou and Subways surfers. This game has a different genre, but is a popular game with the number of downloads of more than 100 million people.

**Table 1** Frequency of Popular Games Occcurrences

| No | Name of Games | Frequency |
|---|---|---|
| 1 | Clash of Clans | 12 |
| 2 | Pou | 12 |
| 3 | Subway Surfers | 11 |
| 4 | Tebak Gambar | 9 |
| 5 | Piano Tiles 2 (Don't Tap) | 9 |
| 6 | My Talking Tom | 6 |

### *3.2. Game Elements Analysis*

#### *3.2.1. Player*
Of the three games analyzed, only Clash Of Clans can be played multiplayer online to attack/be attacked. While Pou and Subway Surfers is a single player game. The similarity of the three games is that players can share their achievements in the game directly to social media.

#### *3.2.2. Goals*
The three games analyzed have different game objectives, this is influenced by different genres of each game. Clash Of Clans with strategy genre, Pou with casual genre and Subway Surfers with arcade genre. But the goal of the three games has the same equation that is unlimited and sustainable. When the player has reached the peak, developers usually always update for the next mission / level. The entire game also has 8 elements of fun from the aesthetic concept based on the MDA game design model.

#### *3.2.3.Procedure*
Each game has different ways of playing procedures to move characters or games. Clash Of Clans game has 5 basic movements and has no combination movements. While Pou has 2 basic movements and 2 movement combinations, so have a total movement of 4 movements. While Subway Surfers has 2 basic movements, and 4 movement combinations, so the total movement amounted to 6 movements. All three games have an easy way to play procedures, where players simply click, tap or drag.

#### *3.2.4. Rules*
From the rule side, Clash Of Clans and Pou have similarity to the level system when played. While Subway Surfers do not use it. Also the rules in Clash Of Clans and Pou also make the player will never be in a game over state, in contrast to Subway Surfers that allows players to experience game over when failing to overcome obstacles in a game challenge. From the results of analysis, Subway Surfers has relations between the rules and the most movement compared to the other two games, i.e as many as 13 movements. While Clash Of Clans only has 12 movement relation and Pou with relation 8 movement.

#### *3.2.5. Resources*
In all three games every major resource is instrumental in generating support resources. While some categories of support resources may also have a role to maintain the availability of key resources. In Clash of Clans and Pou games that have a level system, it requires players to be at a certain level to have the item locked. While in Subway Surfers, players simply have the coins and keys to get the items.

#### *3.2.6. Conflict*
In-game conflict becomes a challenge that makes players more interested in playing the game. Of the three games analyzed, Pou has a conflict that tends to be flat. But Pou has a mini games feature in which

there is a conflict each to help players not get bored quickly when playing. As for the game Clash Of Clans and Subway Surfers have an increasingly difficult conflict along with the longer playing players.

*3.2.7. Limitation*

The most complicated limits of these three games are in the Clash Of Clans game. For players are limited by time when upgrading. So players can only wait if they do not have support resources to speed up the upgrade process. While in Pou, players can still play when Pou is not prime and looking for Coins through mini games. Similarly with Subway Surfers, although players do not find other characters, players can still play with the characters that have been provided since the beginning. The similarity of these three games is that each game provides an app in purchase feature that can make it easier for players to skip game limits.

*3.2.8. Results and Consequences*

In Clash of Clans, the ability to build and defend the kingdom will make players rank. While in Pou, the player's ability to take care of the pou will make the player has an excellent status indicator and an increased level. While in Subway surfers, the more players can run, the higher the high score will be obtained. From the results of comparative analysis, Pou and Subway Surfers have the same positive pattern > on result and result. Both of these single player games have increasing results and consequence parameters, such as Subway Surfers with high score and Pou game with level. Whereas in Clash Of Clans has a negative opponent pattern > on result and consequences. Because while fighting, the winner in the Clash Of Clans is determined by the player's success in making the opposite negative > compared to the player's negative.

Game elements can have an effect on game player emotions. Positive emotions happen if the game allows players to present their achievements to social media, providing challenging goals, easy play procedures, player actions for existing rules, payment facilities. Negative emotions happens if games provide conflicting resource management. While the pattern of positive results or negative enemies gives effect of increasing the level of emotion that exists.

*3.3. Players Emotion*

Testing is done to determine the emotions of players while playing the three popular games are analyzed. Testing is done through surveys of the correspondent ever and have experience playing the three games. Surveys conducted in the form of observation of player expression through the application of emotion recognition, Affdexme [16].

The number of respondents obtained is 7 people, as explained by [17] that the number of samples used in qualitative research is range from 4-10 informants by seeing if the data has been saturated, if sample less than 10 has reached saturation point then the researcher can stop sample search.

In the Clash Of Clans, the emotions of players that most often appear is disgust as much as 34 times. While under it is surprise with 28 times and joy as much as 14 times. While 3 other emotions, namely sadness, anger and fear only appear less than 10 times. Although disgust is the most common emotion when summed from 7 respondents, the comparative test between the average emotional emergence and the standard deviation shows the joy emotions having a better homogeneity level of opinion.

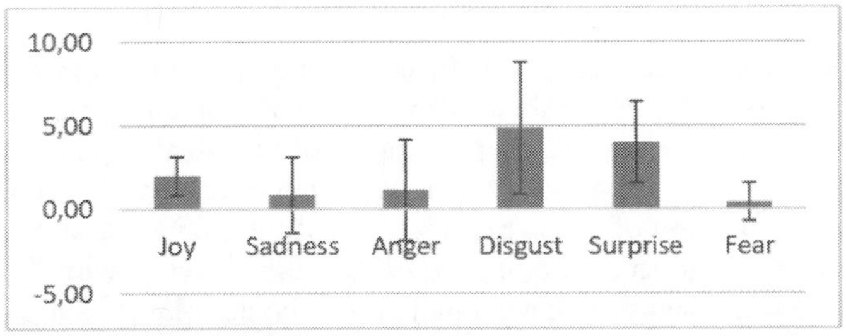


**Figure 1.** Comparative of occurrences number average with standard deviation of respondent's emotional expressions on Game COC (Clash of Clans)

In the game Pou emotions players most often appear is surprise as much as 36 times. While the thin difference below it is disgust with 35 occurrences and joy 28 times. While the emotions of sadness and anger both appear 13 times. The least emotion appears is the fear that as much as 5 times. When summed from 7 respondents, the comparative test between the average emotional emergence and the standard deviation shows the disgust emotion also has a better homogeneity level of opinion.

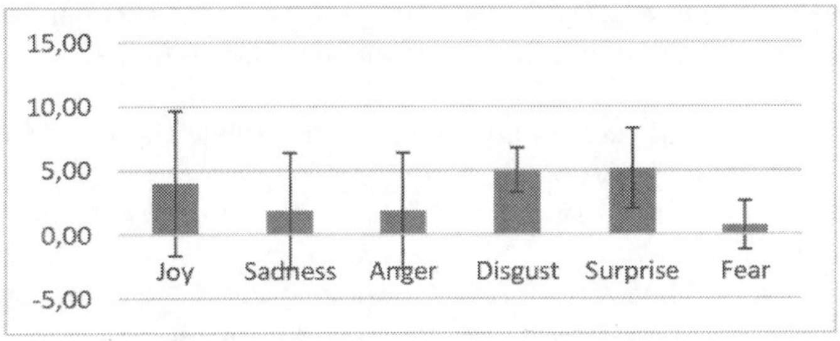


**Figure 2.** Comparative of occurrences number average with standard deviation of respondent's emotional expressions on Game Pou

In the Subway Surfers, the emotions of players that most often appear is disgust as much as 48 times. While under it is a surprise with 46 times and joy as much as 39 times. While the emotions of sadness and anger both appear 15 times. The least emotion arises is fear that is as much as 10 times. When summed from 7 respondents, the comparative test between the average emotional emergence and the standard deviation shows the disgust emotion also has a better homogeneity level of opinion.

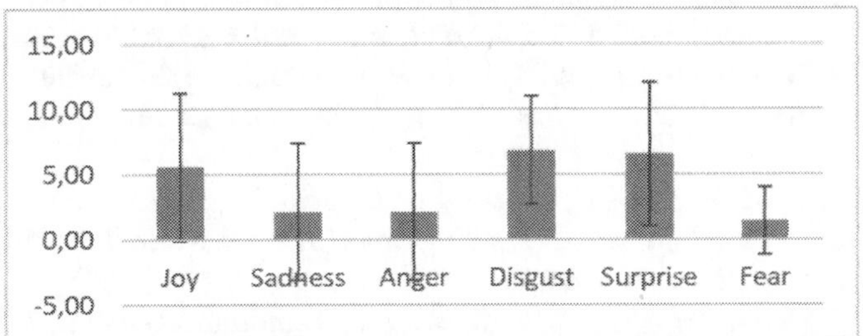


**Figure 3.** Comparative of occurrences number average with standard deviation of respondent's emotional expressions on Game Subway Surfers

## 4. Conclusion

This study has produced early indications of the form of gameplay and emotion that a game needs to become popular in the Play Store. From gameplay analysis, every popular game on the Play Store is not limited to a single player mode. Game with single player or multiplayer can both be a game that is popular within Android users. In the design of gameplay, resources that are scarce and become the needs of the players become the attraction of all three games. The multiplayer game has a pattern of results and the result makes the opponent become more negative to the player to win. While the single player game on popular games that are analyzed have a pattern of making players get positive to win. Based on the results of emotional observation, there are 3 basic emotions that most often arise are joy, disgust and surprise.

## References

[1] Zechner M and Green R 2012 *Beginning Android Games* (New York: Apress)
[2] Jeong E J and Kim J 2009 *Definitions, Key Characteristics, and Generations of Mobile Games* (Pennsylvania: IGI Global)
[3] Jamaludin F 2016 *Tahun 2016, Industri Game Tanah Air Makin Cerah*. Merdeka. 5 Jan. 2016. Web. 10 Oct. 2016. <https://www.merdeka.com/teknologi/tahun-2016-industri-game-tanah-air-makin-cerah.html>
[4] Cosseboom L 2015 *Indonesia Is The Fastest Growing Market For Games In Southeast Asia.* 12 Jan. 2015. Web 10 Oct. 2016. <https://www.techinasia.com/indonesia-games-southeast-asia-newzoo-study>
[5] Kim HM 2013 Mobile Media Technology And Popular Mobile Games In Contemporary Society *Int. J. of Mobile Marketing* pp 42-54.
[6] Fullerton T 2008 *Game Design Workshop: A Playcentric Approach to Creating Innovative Games* (Burlington : Morgan Kaufmann)
[7] McDuff D, Mahmoud A, Mavadati M, Amr M, Turcot J and Kaliouby R 2016 A Cross-Platform Real-Time Multi-Face Expression Recognition Toolkit *Proc. of the 2016 CHI Conference Extended Abstracts on Human Factors in Computing Systems* pp 3723-3726.
[8] Fabricatore C 2007 *Gameplay And Game Mechanics: A Key To Quality In Videogames* OECD-CERI Expert Meeting on Videogames and Education
[9] Adams E 2014 *Fundamentals of Game Design: Third Edition* (USA: New Riders)
[10] Ermi L and Mäyrä F 2005 Fundamental Components Of The Gameplay Experience: Analysing Immersion *Digital Games Research Association's (DiGRA) Second Int. Conf.*
[11] Lazzaro N 2004 *Why We Play Games: Four Keys To More Emotion In Player Experiences* (San Jose: Game Developers Conf.)
[12] Yannakakis G, Karpouzis K, Paiva A and Hudlicka E 2011 Emotion in Games *Proc. of the 4th Int. Conf. on Affective Computing and Intelligent Interaction* **2** pp 497-497.
[13] Hunicke R, LeBlanc M and Zubek R 2004 *MDA: A Formal Approach to Game Design and Game Research* (San Jose: Game Design and Tuning Workshop at the Game Developers Conference).
[14] Salen K and Zimmerman E 2003 *Rules of Play: Game Design Fundamentals* (USA: The MIT Press)
[15] Schell J 2008 *The Art of Game Design a Book of Lenses: 1st Edition* (Florida: CRC Press)
[16] Groce J 2011 *New Technology Measures Emotional Responses to Web Videos* Mashable.com. Mashable, 31 Jan. 2011. Web. 12 Oct. 2016.
[17] Moleong L J 2007 *Metodologi Penelitian Kualitatif* (Bandung: Remaja Rosdakarya)

# Vi-da: vitiligo diagnostic assistance mobile application

**G A Nugraha, A Nurhudatiana and R Bahana**

Computer Science Program/Faculty of Computing and Media, Bina Nusantara University, Jakarta, Indonesia

E-mail: arfika.nurhudatiana@binus.ac.id, rbahana@binus.edu

**Abstract**. Vitiligo is a skin disorder in which white patches of depigmentation appear on different parts of the body. Usually, patients come to hospitals or clinics to have their vitiligo conditions assessed. This can be very tiring to the patients, as vitiligo treatments usually take a relatively long period of time, which can range from months to years. To address this challenge, we present in this paper a prototype of an Android-based mobile application called *Vi-DA*, which stands for *Vitiligo Diagnostic Assistance*. Vi-DA consists of three subsystems, which are user sign-up subsystem, camera and image analysis subsystem, and progress report subsystem. The mobile application was developed in Java programming language and uses MySQL as the database system. Vi-DA adopts a vitiligo segmentation algorithm to segment input image into normal skin area, vitiligo skin area, and non-skin area. Results showed that Vi-DA gave comparable results to the previous system implemented in Matlab. User acceptance testing results also showed that all respondents agreed on the usefulness of the system and agreed to use Vi-DA again in the future. Vi-DA benefits both dermatologists and patients as not only a computer-aided diagnosis (CAD) tool but also as a smart application that can be used for self-assessment at home.

## 1. Introduction

Amongst skin diseases, vitiligo is a rare skin disorder in which white patches of depigmentation appear on the skin. Vitiligo is an abnormality affecting melanocytes, which are the pigment-producing cells in the skin, where the cells are destroyed, resulting in depigmentation on different parts of the body. This pigment loss appears as brighter mottles in the skin and sometimes might also take place on the mucous membranes, retina, and hair follicles [1]. Though vitiligo patients generally do not feel ill, the disease can have profound psychological impact as the appearance of white patches can be very distressing. This emotional distress is more common in patients with darker skin tone as the white patches tend to be more noticeable. Currently there are several treatment options for vitiligo. Common treatments include topical corticosteroid, oral medication, UVB light therapy, and surgery. Due to the relatively long duration of the treatments, which may range from months to years, patients might feel discouraged. Some patients have even quitted treatments midway from not seeing any visible progress. In addition, having to physically come to the hospitals or clinics frequently to get their vitiligo conditions assessed can be very tiring.

To address this challenging situation, a prototype of a vitiligo diagnostic assistance system is proposed in this research. The system is called Vi-DA, which stands for *Vitiligo Diagnostic Assistance*. The mobile application was developed on Android in Java programming language and uses a MySQL database. Vi-DA automatically segments vitiligo-affected skin from the normal skin and calculates the ratio between vitiligo-affected skin area and the whole skin area for specific parts of the body, for example face, hand, and chest. As Vi-DA provides quantitative assessment, it can help dermatologists and patients discover minor progress which might have been left unnoticed if the skin examination is performed manually. Vi-DA benefits both dermatologists and patients in two ways. Firstly, the system can be used as a computer-aided diagnosis (CAD) tool. Secondly, the mobile application can also be used by vitiligo patients to perform self-assessment at home. In addition, patients can come to the clinic less frequently as patients can capture photographs of their skin and perform analysis by using the mobile application. The data are stored in a database and can be sent to the attending dermatologists for consultations.

Today, there are many mobile applications dedicated to dermatology. These applications can generally be categorized into two types— social networking platforms [2] and computer-aided diagnostic systems [3]. The first type is usually educative in nature, and several apps provide features for consultations with dermatologists. *First Derm* app [4] is a mobile application with which users can take pictures of their skin problems, input descriptions of the problems, and press 'send'. Within 24 hours, there will be a dermatologist responding with a medical opinion and suggestion for the first step treatment options and whether the problem requires an immediate visit to a doctor in-person. *Vitiligo Treatment + Symptoms* app [5], *Vitiligo Handbook* app [6], and *Skin Disease and Treatment* app [7] provide information about vitiligo and treatment options. Some of these apps aim to build a community and support groups for people with vitiligo. For example, *Vitiligomatch* app [8] provides vitiligo sufferers with a social networking platform to interact with each other.

This research focuses on the second type, which is as a computer-aided diagnosis (CAD) system. To the best of our knowledge, there is still no Android application for vitiligo image analysis. However, there are existing mobile applications embedded with image analysis algorithms which are targeted to other skin diseases. *Molexplore* [9], *Skin Vision* [10], *Miiskin - Melanoma Skin Cancer* [11], and *UM Skin Check* [12] analyze skin lesions for skin cancer detection. If the lesion is categorized as suspicious or at a risky level, the user is recommended to follow up with a dermatologist. The *Molexplore Melanoma Skin Cancer* app [9] allows users to perform a comprehensive monitoring of moles on different parts of the body. The app also includes an approximation of the UV index at user's location. The *Skin Vision* app [10] automatically assesses the risk level of a mole through its camera algorithm. It also provides expert review which refers to assessments given by dermatologists. The *Miiskin* app [11] focuses on mole tracking. The *UM Skin Check* app [12] provides several functions in the app, including guidance for self-exam, full-body survey, lesion tracker and sun safety information. In general, all these apps have image capture and analysis capability. Some of the apps focus on quantitative analysis, while some others focus on telemedicine practice by providing consultations with dermatologist through the apps.

## 2. System Design & Implementation

The overall system architecture is illustrated in Fig. 1. The mobile app was developed using Android Studio 2.1 in Java programming language with a MySQL database [13,14]. The system was developed using the waterfall approach. The main functions of Vi-DA include: 1) user account creation and login, 2) user profile management, 3) image capture and retrieval, 4) image analysis (which includes skin segmentation and vitiligo area percentage calculation), and 5) data visualization.

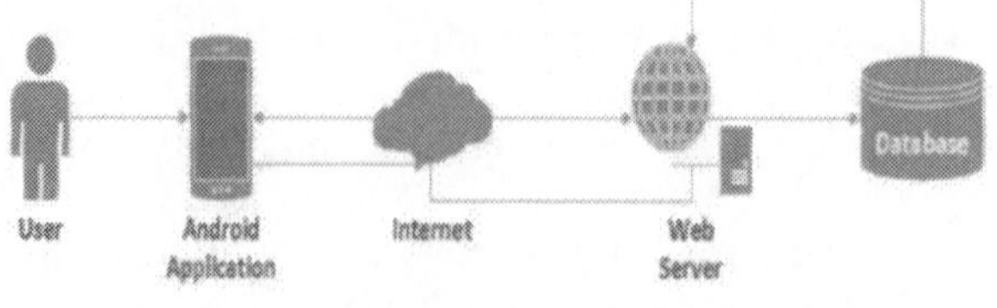


**Figure 1**. *Vi-DA* app System Architecture

This application needs the user to register to access the content. Username and password are required to login to the system, and for a new user, who does not have an account yet, he/she will be directed to the user registration screen which requires user to enter username, password and other basic information. Successful authentication will bring the user to the home screen which shows a list of photographs which have been analyzed and the corresponding analysis dates and results. User can access his/her profile and change his/her particulars and login information.

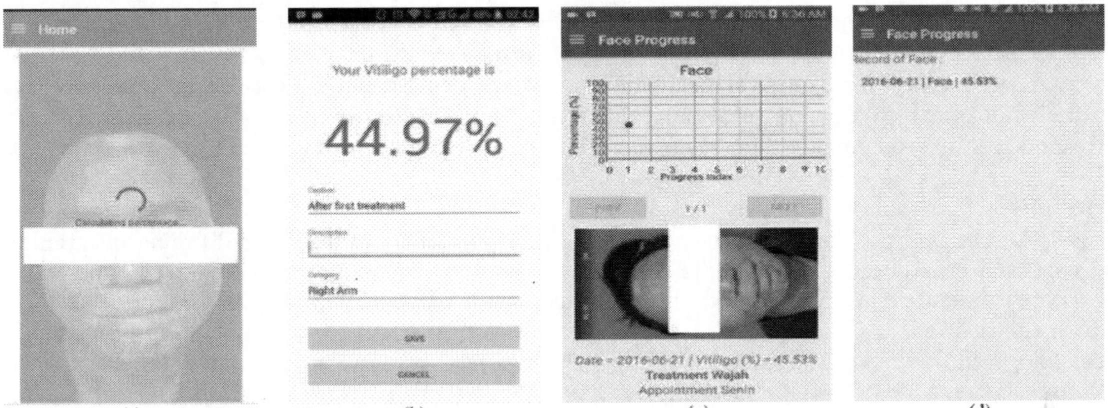


(a) (b) (c) (d)

**Figure 2.** User Interface design for Vi-DA app: (a) image analysis progress, (b) image analysis result, (c) progress visualization, (d) record history.

There are two sources for image analysis: from the device's photo gallery through the image retrieval function in Android or from device's camera by using the photo capture function in Android. The photograph is further processed by performing skin segmentation followed by vitiligo segmentation using an algorithm developed by Nurhudatiana [15]. The ratio between vitiligo segment and whole skin segment is then calculated (Fig. 2(a)). The segmentation algorithm uses the Fuzzy C-Means (FCM) clustering algorithm to segment input image into three clusters (i.e., background/non-skin, normal skin, and vitiligo skin), followed by vitiligo percentage calculation by taking the ratio between the total number of vitiligo skin pixels and the total number of skin pixels [15]. The algorithm was originally written in Matlab and was rewritten in Java for the Android app. The algorithm is described in detail as follow:

**Step 1:** Read the input RGB image as Bitmap file.

**Step 2:** Convert the RGB image into YCbCr image using the formula:

$$Y = 0.299*R + 0.587*G + 0.114*B$$
$$Cb = -0.1687*R - 0.3313*G + 0.5*B + 128 \quad (1)$$
$$Cr = 0.5*R - 0.4187*G - 0.0813*B + 128$$

**Step 3:** Extract the Cr channel from the YCbCr image.

**Step 4:** Apply FCM algorithm onto the Cr channel (grayscale image) obtained from Step 3. The algorithm segments the Cr image into two clusters - background area (darker region) and skin area (brighter region). The skin cluster is represented by higher intensity values, whose indexes are then stored into skin index array *S*.

**Step 5:** Extract the Blue channel from the original RGB image given in Step 1.

**Step 6:** Apply FCM algorithm for the second time onto the Blue channel (grayscale image) obtained from Step 5 by taking only the pixels included in *S*. In this round, the pixels are further separated into two clusters - normal skin area (darker region) and vitiligo area (brighter region).

**Step 7:** Calculate the depigmentation percentage:

$$depigmentation\ \% \approx \frac{vitiligo}{vitiligo + normal\ skin}$$

The analysis result is displayed as depigmentation percentage (see Fig. 2(b)). User is allowed to type additional information related to the analyzed image in text form including caption (e.g., after first treatment), description, and the body part category from which the image was captured (e.g., right arm). When the save button is clicked, the system automatically saves the data and timestamp. Fig. 2(c) shows the progress tracking visualization screen, which shows the depigmentation percentage values from time to time in the form of a scatter plot diagram. Each data point is linked to the photograph and the corresponding analysis. Finally, Fig. 2(d) shows record history screen categorized by body part.

## 3. Results

Because Vi-DA app was developed using an algorithm previously written in a different programming language and environment, two results are reported here: 1) the performance of the image analysis algorithm embedded in the Android app compared with the performance of the previous Matlab implementation, and 2) the result of user acceptance testing (UAT) involving vitiligo patients in a dermatology clinic in North Jakarta, Indonesia. A questionnaire-based survey based on the information systems success model was distributed to the respondents to collect UAT results [16].

### *3.1. Algorithm result*

The comparison result in Fig. 3 shows that in general, the percentage difference between the Matlab implementation and the Vi-DA app is small, except that for the arm body part, noise was observed at the arm boundary for the Matlab implementation.

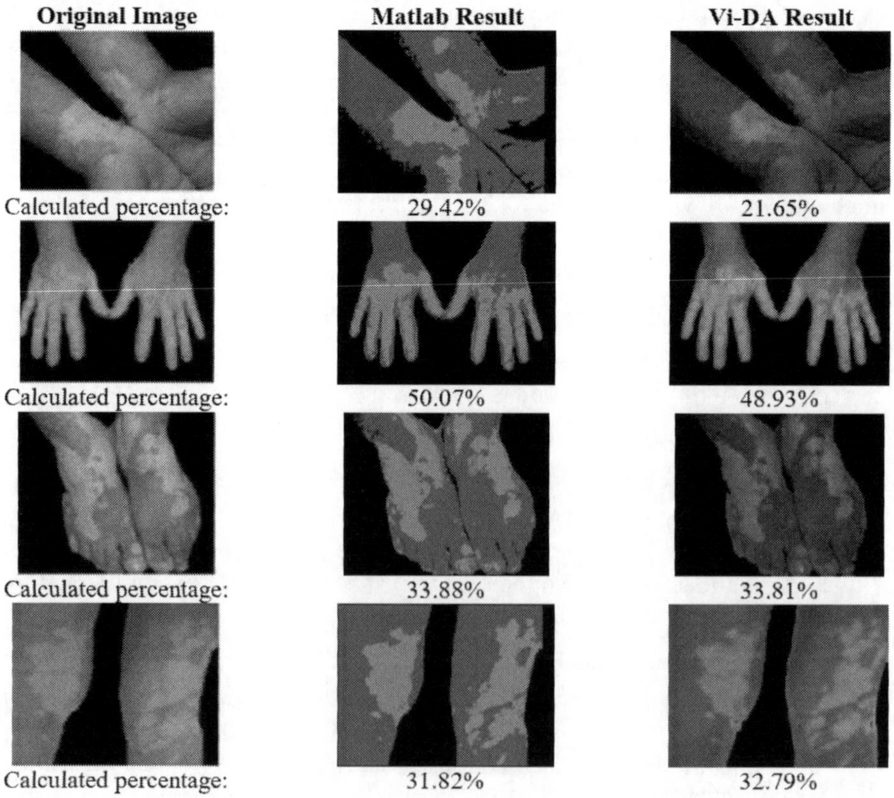


**Figure 3.** Comparison of the segmentation output given by previous Matlab implementation [15] and Vi-DA app.

It can be observed that both implementations give consistent performance. In addition, the weakness of the previous Matlab implementation could be handled well by the Vi-DA app. The Matlab implementation produced rough edges at the arm boundary and the middle part of the palm was wrongly segmented as background. The Vi-DA app fixed the problem and gave smoother segmentation output for all four different body parts.

### *3.2. User Acceptance Testing (UAT) Result*

To test system usability, user acceptance testing (UAT) was conducted. The UAT involved three patients in the clinic who were diagnosed with vitiligo and were undergoing treatments. These respondents were asked to use Vi-DA app for a day and surveyed regarding their uses of the app through questionnaires. The questionnaires consisted of ten questions and used 5-scale Likert scale to indicate each respondent's level of agreement with the statements provided in the questionnaire. The responds collected were "1" for strongly disagree, "2" for disagree, "3" for neutral, "4" for agree, and "5" for strongly agree. As the app was still in prototype phase, the respondents were also asked to provide input on how to improve the system. The responses are provided in Fig. 4.

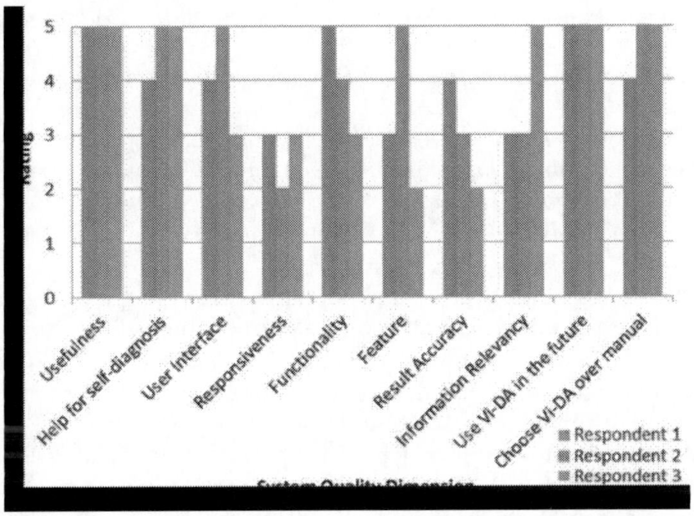


**Figure 4.** Responses obtained from user acceptance testing (UAT)

All respondents strongly agreed that the Vi-DA app is useful for them. They also agreed that the app was helpful as self-diagnosis assistance. However, these respondents generally disagreed on the responsiveness of the app. This is because it could take up to 60 seconds to process a single image. This was considered too slow for them and requires improvement. When these respondents were asked whether they would like to use Vi-DA in the future, all of them agreed to use this app and they would choose to use Vi-DA over conventional visit to the clinic.

## 4. Conclusion & Future Work

As reported in the UAT survey results, some improvements need to be made to the system. The first priority is to reduce the processing speed. In average, a photograph requires a minute to process, and this can be a problem since user generally expects to wait for 5-10 seconds for a live (real-time) processing. One way to reduce the time consumption for image analysis is by reducing the number of iterations in the clustering process during the image segmentation. However, further experiment is required to determine the proper number of iterations to achieve optimum segmentation output. Another way is to further reduce the image size to be processed, and further experiment is required to determine the optimum image size which will not compromise the segmentation output quality.

There are also several variability factors during the image acquisition stage which could affect the calculation of the results. Those factors include illumination, distance, and viewpoint differences. These inconsistencies could result in high variability in the depigmentation percentage. To overcome the problem, a preprocessing module will be added to the system before the image segmentation module. This preprocessing step will help to normalize the lighting/color in the image before analysis. In addition, a functionality to guide the image capture process in terms of lighting (e.g., white light and not yellow light), viewpoint (e.g., straight pose), and distance (e.g., to meet certain value for face-to-frame ratio) will be added so that the variability can be further minimized.

**Acknowledgment**
The authors would like to thank Dr. Bing Wibisono, Sp.KK, the nurses, and the patients at the Laser Skin Center, Gading Pluit Hospital, Jakarta, for their tremendous support and contributions for this research.

**References**
[1] National Institute of Arthritis and Musculoskeletal and Skin Diseases 2017 *What is Vitiligo?* [Online] http://www.niams.nih.gov/Health_Info/Vitiligo/vitiligo_ff.asp
[2] Kochmann M and Locatis C 2016 Telemedicine in the Apple App Store: an Exploratory Study of Teledermatology Apps *Proc. of the Int. Conf. on Collaboration Tech. and Systems*.
[3] Steward D 1996 *What is Computer-aded Diagnosis?* (Cambridge: Massachusetts Institute of Technology).
[4] Google Play 2017 *First Derm: Online Dermatology* [Online] https://play.google.com/store/apps/details?id=com.idoc24.firstderm. [Accessed: 5-Jul-2017].
[5] Google Play 2015 *Vitiligo Treatment + Symptoms* [Online] https://play.google.com/store/apps/details?id=vitiligo.treatment.symptoms. [Accessed: 5-Jul-2017].
[6] Google Play 2015 *Vitiligo Handbook* [Online] https://play.google.com/store/apps/details?id=com.wVitiligoHandbook. [Accessed: 5-Jul-2017].
[7] Google Play 2016 *Skin Disease and Treatment* [Online] https://play.google.com/store/apps/details?id=com.jdsk.discj. [Accessed: 5-Jul-2017].
[8] Google Play 2015 Vitiligomatch [Online] https://play.google.com/store/apps/details?id=com.vitiligomatch.com.app. [Accessed: 5-Jul-2017].
[9] Google Play 2017 *Molexplore Melanoma Skin Cancer* [Online] https://play.google.com/store/apps/details?id=com.borealos.medical.molexplore. [Accessed: 5-Jul-2017].
[10] Google Play 2017 *SkinVision - Skin Cancer Detection App* [Online]. https://play.google.com/store/apps/details?id=com.rubytribe.skinvision.ac. [Accessed: 5-Jul-2017].
[11] Google Play 2017 *Miiskin - Melanoma Skin Cancer* [Online] https://play.google.com/ store/apps/details?id=com.rubytribe.skinvision.ac. [Accessed: 5-Jul-2017].
[12] Google Play 2016 *UM Skin Check* [Online] https://play.google.com/store/apps/details?id=edu.umich.MySkinCheck. [Accessed: 5-Jul-2014]
[13] Android Developers 2015 *Android Studio Overview* [Online] https://developer.android.com/studio/intro/index.html . [Accessed: 5-Jul-2017].
[14] Gosling J, Joy B, Steele G, Bracha G and Buckley A 2015 *The Java® Language Specification - Java SE 8th Edition* (California: Oracle America, Inc.).
[15] Nurhudatiana A 2015 A Computer-aided Diagnosis System for Vitiligo Assessment: a Segmentation Algorithm *Comm. in Comp. and Info. Sc.* **516** pp 323-331.
[16] Petter S, DeLone W and McLean E 2008 Measuring Information Systems Success: Models, Dimensions, Measures, and Interrelationships *European J. of Info. Systems* **17** pp 236-263.

IOP Conf. Series: Journal of Physics: Conf. Series **978** (2018) 012004 doi:10.1088/1742-6596/978/1/012004

# Performance test for prototype game for children with adhd

**R Bahana[1,2], F L Gaol[1], T Wiguna[3], S W H L Hendric[1], B Soewito[4], E Nugroho[5], B P Dirgantoro[6] and E Abdurachman[1]**

[1]Computer Science Department, BINUS Graduate Program - Doctor of Computer Science, Bina Nusantara University, Jakarta, Indonesia 11480
[2]Computer Science Department, Faculty of Computing and Media, Bina Nusantara University, Jakarta, Indonesia 11480
[3]Faculty of Medicine, University of Indonesia, Jakarta, Indonesia
[4]Computer Science Department, BINUS Graduate Program - Master of Computer Science, Bina Nusantara University, Jakarta, Indonesia 11480
[5]Kummara Game Design Studio, Bandung, Indonesia
[6]Graphic Design and New Media Program, Visual Communication Design Department, BINUS Northumbria School of Design, Bina Nusantara University, Jakarta, Indonesia 11480

E-mail: rbahana@binus.edu

**Abstract.** The gaming industry has become one of the fastest growing industries in the world. In recent years there have been several studies that have used games, especially computer-based ones to improve the executive function of children with Attention Deficit Hyperactivity Disorder (ADHD) especially in working memory improvements. There are still several obstacles in using computer-based games that have been circulating today in Indonesia. The aim of this research is to develop and create a prototype game that can be used to improve working memory of children with ADHD using Bahasa Indonesia. The prototype game application is a race game with matching colors and memory work tasks; developed using Construct2. Players will play as a driver, delivering certain fruits to specific homes with different colors. The game also included visuospatial tasks such as remembering the objects. For performance test, CPU utilization, average FPS and image memory usage has been tested in several browsers. In conclusion, the performance tests results show that the first type of the prototype game is good enough to be played in desktop or laptop.

## 1. Introduction

The game industry has turned out to be one of the fastest growing industries in the world. According to Newzoo, the provider of market intelligence that covers e-sports, the global games, and mobile markets [1], shows their Global Games Market Report for 2017. It states that around 2 billion gamers over the globe are created almost $109 billion in diversion incomes in year 2017. That is $7.8 billion or 7.8% more than 2016. Portable is the most lucrative fragment, with cell phone and tablet gaming growing 19% year on year to $46.1 billion, asserting 42% of the market. [2]. Newzoo predicts that the worldwide market will grow around 6.2% toward 2020 to reach $128.5 billion [2]. In addition, with the development of open source applications for game development, such as Unity, Game Maker and Construct2, everyone could have the opportunity to create their own games more easily [3].

The rapid development of games does not all have a positive impact. Many popular media issue warnings against the dangers of addiction to this game, as well as games can lead to violence and aggression, especially in children and adolescents. Most of the psychological research on gaming effects is more focused on negative impacts such as potential dangers associated with aggression, addiction, and depression [4,5,6]. But in the field of research, many researchers are seeing the diverse potential benefits of cognitive-related, motivational, social, and emotional development of children's games. In the field of medicine began to elevate this positive effect [7], this can be seen with the publication of Games for Health Journal. Researchers and practitioners in medicine are beginning to realize the power of video games to encourage and, ultimately, to improve the patients' health [8].

In recent years there have been several studies that have used games, especially computer-based ones to improve the executive function (EF) of children with Attention Deficit Hyperactivity Disorder (ADHD) especially in working memory improvements, but have not been able to deliver consistent results. Research conducted by Nouchi [9] in healthy young adult individuals using computer games applications such as 'Tetris' and 'Brain age' for 5 times a week; within 4 weeks of training it improves the executive function of working memory. Meanwhile, in a meta-analysis study on the effectiveness of training with computer-based games in children with ADHD with a duration of 4-14 weeks or 3-36 sessions with an average duration of 30 minutes [10]; the results concluded that the training had effect on working memory of children with ADHD but did not affect the ADHD clinical symptoms so it was suggested that training using computer-based games should also be accompanied by drug administration. Computer-based game training to improve executive function including working memory for children is the right choice because childhood is an explorative and playing period is one of the activities favoured by every child especially with computer-based games. By giving stimulation in the form of the computer-based game repeatedly then there are two things can be achieved at once that is to give them a fun time as well as stimulate their executive function. For that, it is necessary to create a game application to give impact to brain development of children with ADHD.

There are still obstacles in using computer-based games that have been circulating today in Indonesia, including: incompatibility with the culture and background of Indonesia, the use of English or other foreign languages as the language of instruction, and the use of the application is also required a fee that is not cheap. The aim of this research is to test the performance of a prototype game that has been developed and created using Bahasa Indonesia. Then it can be used to help to improve working memory of children with ADHD. In this research paper, the prototype game that has been tested only contains one type (with visual direction) and the performance tests have been done to several browsers in desktop/personal computer (PC) and laptop.

## 2. Theoretical Framework

### *2.1. Attention Deficit Hyperactivity Disorder (ADHD)*

ADHD is one of the most common health disorders found among children. Around the world ranges from 3-15% [11,12]. In Indonesia until now there is still no definite data. As an illustration, the proportion of primary school children with ADHD of 26%, and the proportion of children with ADHD who visited the Child and Adolescent Psychiatric at Cipto Mangunkusumo Hospital in 2015 also around that number [13]. This disorder has a major impact on children and adolescents, especially the impact on their daily functions. Children experience learning difficulties, interactions with disturbed social environment and often fail to continue education to a higher level. Therefore, effective handling of these conditions is absolutely necessary for the quality of life of children with ADHD to be maintained as much as possible. A study using the Behavior Rating Inventory for Executive Function tool indicates that the executive function impairment in children with ADHD is largely a work-related memory disorder (56.8%) [13]. Current research suggests that intelligence and memory capacity in working memory is an important factor in determining children's learning abilities. Working memory is defined as the ability to process and remember information for a short time. Every person including a child has an unequal working memory capacity and this capacity is not influenced by social or economic status [14]. For a child with ADHD, working memory greatly determines the ability to process all the information obtained from the surrounding environment so that the existence of interference in working memory is certainly complicate the learning ability of children.

### *2.2. Performance Test*

Some companies already have started to utilize software engineering techniques in the game development process. Most of the companies separated their game application developer to several teams that manage different parts of the application, so every single process could be tested to make sure every step meet the expectation. Many kind of testing techniques can be used right now [15]. One of non functional testing techniques which have a significant impact on the final product quality is performance test. Performance tests will detect how fast the system of the product or some parts of it can work under the specific workload. Stress testing, configuration testing, load testing, and spike testing are some of example of performance testing [16]. In this research paper, Central Processing Unit (CPU) utilization, average (in one minute) frame per seconds (FPS) and image memory usage will be tested in several browsers.

### *2.3. Construct2*

Construct2 is an HTML5 game engine made by Scirra. Construct2 has a distinguished feature compared to other HTML game engines: developers do not have to code to create games. Some coding logics still exists, but in a form of click-and-drag. Sprites and entities can be drag-and-dropped to the stage, called layout, while the game logic is put on event sheet. The basic component of display is called a sprite object, and behaviours can be added to supported objects, sprite included [17]. Construct2 also supports AJAX request, collision detection, multiple animations, path finding and AI (from behaviour), and some WebGL effects. Using the behaviour-concept for objects, Construct2 enables a wide open extensibility support for custom objects and behaviour [18].

## 3. Related Work

### *3.1. Re-Mission video game*

Perhaps a well-known success story of games that have a big impact on health-related condition is the Re-Mission video game [19], a video game designed for pediatric cancer patients [20]. Players are asked to control the nanobot by shooting cancer cells, overcoming bacterial infections, and knowing signs of nausea and constipation (common in cancer patients). A study conducted globally in 34 health centers, comparing children playing Re-Mission with children playing other computer games [19]. The result is that children who play Re-Mission more obediently follow their cancer treatment. This game has been distributed to more than two hundred thousand patients and continues to be referred to as a successful treatment approach using game assistance.

### *3.2. Braingame Brian*

The Braingame Brian is a computer-based training with three EFs: visuospatial, inhibition, and cognitive flexibility. One may choose to train one, two, or all three EFs. Braingame Brian consists of 25 training sessions of about 40–50 minutes. Every session contain two sets of three training tasks in a fixed order. The first training task is always a training task; the second and third task, an inhibition training task and a cognitive flexibility task, are presented in changing order. In the current double-blind, placebo-controlled study by Prins et al using this game, they researched, 5-week at home, multiple executive function training intervention in children with ADHD [21]. The result of their study is promising although Braingame Brian is better not to be considered as a stand-alone treatment module. It should be

IOP Conf. Series: Journal of Physics: Conf. Series **978** (2018) 012004 doi:10.1088/1742-6596/978/1/012004

combined with existing, empirically supported treatments for ADHD such as parent and teacher training programs, and/or medication, to have full impact [21].

### *3.3. Performance test by Scirra*

Scirra has been done three performance tests in the past for Construct2 game application, once in 2012, 2014 and 2016 [22]. The test is using an automated version of Space Blaster that included in Construct2 demo program. The performance test that they did is by taking an average frame rate over a period of about one minute. The reason is because usually browsers spend the first few seconds optimising Javascript. The higher FPS (average frame rate) is better. Below 20 FPS means it is questionable if the game will be playable; 30 FPS is perfectly playable; 60 FPS is the target. Based on their tests, all of the devices that they test with Space Blaster game performed well (38 FPS – 61 FPS).

## 4. Development of the Prototype

The prototype game application is a race game with matching colors and memory work tasks. Players will play as a driver, delivering certain fruits to specific homes with different colors. In level one or first location, the player should deliver two types of fruits (total is three kind of fruits) to specific homes (total is three kind of houses) as many as possible in two minutes. The direction is in visual images. This prototype game was developed in several months, started from collected user requirements from Doctors at Cipto Mangunkusumo hospital and based on several papers that discussed about this kind of game for therapy. Some inputs from the doctors are very interesting such as, it is better the game does not show the score because maybe it will create an anxiety for the children with ADHD. The score itself with others data (deliver correctly, deliver wrongly, total correct etc) will be recorded in several variables (and arrays) and all data will be transfer to a database.

The game also included visuospatial tasks such as remembering the objects. Visuospatial is a loss of the sense of where about in the relation of environment [23]. The player is given an object that must be remembered, then associate the object in accordance with the given task. This game is trying to stimulate players on the side of accuracy and speed in completing the mission. Only one type has been tested for this prototype, three more types will be added after beta testing. This game is intended for children aged between 7 to 12 years. Therefore, the design of this game follows the rating used under the Indonesian Game Rating System (IGRS) for 7 years and above. According to the Regulation of the Minister of Communication and Information Technology of the Republic of Indonesia Number 11 of 2016 on the classification of electronic interactive games, the rating for 7 years and above must follow Article 6 of the rule [24].

## 5. Testing and Analysis

For performance test, CPU utilization, average (in one minute) FPS and image memory usage has been tested in several browsers. In this testing, the devices for testing are desktop (Windows XP SP3, Dual-Core E5400 @ 2.70 GHz; 2GB RAM) and laptop (Windows 8.1 64-bit Edition, Core i3-2357M @ 1.30 GHz; 2GB RAM). Browsers tested include Google Chrome (ver. 60.0.3112.90), Firefox (ver. 54.0.1) and Opera (ver. 46.0.2597.32). The testing is only for those two devices because for actual testing to the children, it will only use desktop and/or laptop. Total size of the file is around 450 KB for browser or around 1,4 MB for execution file. In this test, the CPU utilisation measurement only indicates time spent in the main Javascript thread because Javascript does not provide direct access to the CPU usage. As an alternative, Construct2 estimates it using timers. ImageMemoryUsage is an estimate of the total memory used by the currently loaded WebGL textures in the game. Usually textures exist in GPU memory, which will not appear in Task Manager's measurements. Games may also need a lot of other memory for everything else other than images. Basically, ImageMemoryUsage is an indication of GPU texture memory usage, which is just one aspect of the overall memory usage.

Figure 1 shows the game page of the game application. For the purpose of the performance tests, several statements were added to show the result (Figure 2). In actual game application, the results of performance tests are not appeared. First line in Figure 2 indicates FPS at that time, second line indicates CPU Utilisation, test score in third line indicates the average of FPS and the fourth line indicates image memory usage of the game. The last line is the status that FPS already reaches one minute. Table 1 shows the result of three kinds of the performance tests with two different devices and several browsers. Based on the result, all of the devices and the browsers performed well for average FPS; near 60 FPS. CPU utilisation is low and also the ImageMemoryUsage is low for both devices.

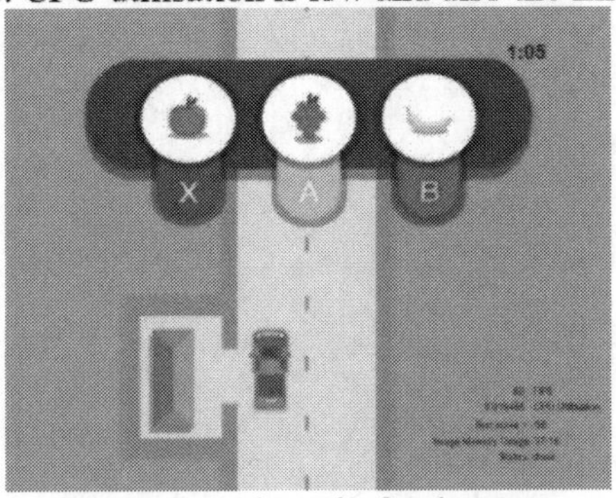


**Figure 1.** Using Firefox browser

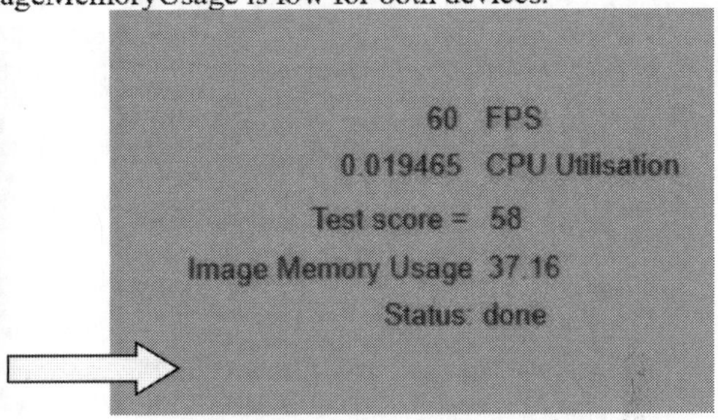


**Figure 2.** Several tests conducted

**Table 1.** Comparison between browsers

| Browser | FPS 1 minute avg | | CPU Utilisation (%) | | ImageMemoryUsage (MB) | |
|---|---|---|---|---|---|---|
| | PC | Laptop | PC | Laptop | PC | Laptop |
| **Chrome** | 57 | 58 | 3.9 | 2.5 | 37.34 | 37.23 |
| **Firefox** | 58 | 58 | 2.1 | 1.9 | 37.22 | 37.16 |
| **Opera** | 57 | 59 | 1.7 | 1.8 | 37.11 | 37.15 |

## 6. Conclusion

The prototype game for children with ADHD has been tested. In conclusion, the performance tests results show the first type of the prototype game is good enough to be played in desktop or laptop. In the next three types, audio, voice over and obstacles will be added, so the performance tests results should be different from the first type.

**Acknowledgment**
The authors would like to thank Christopher Lychealdo, Steven Muliamin and Gwenda Wibowo for their tremendous support and contributions for this research.

## References

[1] Newzoo [Internet]. [cited 2017 August 1]. Available from: https://newzoo.com/about
[2] McDonald E 2017 *The Global Games Market Will Reach $108.9 Billion in 2017 With Mobile Taking 42%.* [Internet]. [Cited 2017 May 27]. Available from: https://newzoo.com/insights/articles/the-global-games-market-will-reach-108-9-billion-in-2017-with-mobile-taking-42/
[3] Takahashi D 2015 *US Games Industry Forecast to Grow 30 Percent to $19.6 B by 2019* [Internet]. [cited 2015 October]. Available from: https://venturebeat.com/2015/06/02/u-s-games-industry-forecast-to-grow-30-to-19-6b-by-2019/
[4] Anderson CA, Shibuya A, Ihori N, Swing EL, Bushman BJ, Sakamoto A, Rothstein HARI and Saleem M 2010 Violent Video Game Effects on Aggression, Empathy, and Prosocial Behavior

in Eastern and Western Countries: A Meta-Analytic Review *Psychological Bulletin* **136** pp 151-173.

[5] Ferguson CJ 2013 Violent Video Games and the Supreme Court: Lessons for the Scientific Community in the Wake of Brown v. Entertainment Merchants Association. *American Psychologist* **68** pp 57-74.

[6] Lemola S, Brand S, Vogler N, Perkinson-Gloor N, Allemand M and Grob A 2011 Habitual Computer Game Playing at Night is Related to Depressive Symptoms. *Personality And Individual Differences* **51** pp 117-22.

[7] Ritterfeld U, Cody M, Vorderer P 2009 *Serious games: Mechanisms and effects* (Routledge).

[8] Kato PM 2010 Video Games in Health Care: Closing the Gap *Review of General Psychology* **14** p 113.

[9] Nouchi R, Taki Y, Takeuchi H, Hashizume H, Nozawa T, Kambara T, Sekiguchi H, Miyauchi C M, Kotozaki Y, Nouchi H and Kawashima R 2013 Brain Training Game Boost Executive Functions, Working Memory and Processing Speed in The Young Adults: A Randomized Controlled Trail *PloS One* **8**.

[10] Cortese S, Ferrin M, Brandeis D, Buitelaar J, Daley D, Dittmann RW, Holtmann M, Santosh P, Stevenson J, Stringaris A and Zuddas A 2015 Cognitive Training for Attention-Deficit/Hyperactivity Disorder: Meta-Analysis of Clinical and Neuropsychological Outcomes from Randomized Controlled Trials *J. of the American Academy of Child & Adolescent Psychiatry* **54** pp 164-74.

[11] Polanczyk G, de Lima MS, Horta BL, Biederman J and Rohde LA 2007 The Worldwide Prevalence of ADHD: A Systematic Review and Metaregression Analysis. *American J. of Psychiatry* **164**.

[12] Willcutt EG 2012 The Prevalence of DSM-IV Attention-Deficit/Hyperactivity Disorder: A Meta-Analytic Review. *Neurotherapeutics* **9**.

[13] Suryani E 2011 Gambaran Fungsi Eksekutif Pada Anak Sekolah Dasar Dengan Gangguan Pemusatan Perhatian/Hiperaktivitas (GPPH) di Wilayah DKI Jakarta [thesis] (Indonesia: University of Indonesia).

[14] Colom R, Abad FJ, Quiroga MÁ, Shih PC and Flores-Mendoza C 2008 Working Memory and Intelligence are Highly Related Constructs, But Why? *Intelligence* **36** pp 584-606.

[15] Redavid C and Adil F 2011 An Overview of Game Testing Techniques (Västerås: Mälardalen University).

[16] Avritzer A, Kondek J, Liu D and Weyuker EJ 2002 Software Performance Testing Based on Workload Characterization *Proc. of the 3rd International Workshop on Software and Performance ACM* pp 17-24.

[17] Scirra [Internet]. [cited 2016 June 10]. Available from: https://www.scirra.com/ manual/1/ construct-2

[18] Gunawan K, Bahana R. AethelmE, HTML5 Game Engine with Multiple Canvas Elements. Journal of Game, Game Art, and Gamification. 2016 Dec;1(2):57-68.

[19] Kato PM, Cole SW, Bradlyn AS, Pollock BH. A video game improves behavioral outcomes in adolescents and young adults with cancer: a randomized trial. Pediatrics. 2008 Aug 1;122(2):e305-17.

[20] Re-Mission [Internet]. [cited 2017 March 16]. Available from: http://www.re-mission.net/

[21] Prins PJ, Brink ET, Dovis S, Ponsioen A, Geurts HM, De Vries M and Van Der Oord S 2013 Braingame Brian: Toward an Executive Function Training Program With Game Elements for Children With ADHD and Cognitive Control Problems *GAMES FOR HEALTH: Research, Development, and Clinical Applications* **2**.

[22] Ashley 2016 The Great HTML5 Gaming Performance Test: 2016 edition [Internet]. [cited 2017 July 10]. Available from: https://www.scirra.com/blog/175/the-great-html5-gaming-performance-test-2016-edition

[23] Cogan DG 1979 Visuospatial Fysgnosia *American J. of Ophthalmology* **88**.

[24] Peraturan Menteri Komunikasi dan Informatika Nomor 11 Tahun 2016 tanggal 20 Juli 2016 [internet]. [cited 2017 July 15]. Available from: https://jdih.kominfo.go.id/produk_hukum/view/id/540/t/peraturan+menteri+komunikasi+dan+informatika++nomor+11+tahun+2016+tanggal+20+juli+2016

# Designing indonesian teacher engagement index (itei) applications based on android

**S R Manalu[1,5], Sasmoko[2,5], S D Permai[3,5] , S A Widhoyoko[4,5] and Y Indrianti[5]**

[1]Mobile Application & Technology Department, School of Computer Science, Bina Nusantara University, Jakarta, Indonesia
[2]Faculty of Humaniora, Bina Nusantara University, Jakarta, Indonesia
[3]Mathematics and Statistics Department, School of Computer Science, Bina Nusantara University, Jakarta, Indonesia
[4]Podomoro University, Jakarta, Indonesia
[5]Research Interest Group in Educational Technology, Bina Nusantara University, Jakarta, Indonesia

Email : smanalu@binus.edu, sasmoko@binus.edu, spermai@binus.edu, samuel.anidyo@podomorouniversity.ac.id, yasintaindrianti@gmail.com

**Abstract.** Teachers who have a good level of engagement will be able to produce students who engage and excel. Level of national teachers' engagement needs to be a reference to the level of educational success and equity of national education. The spread of geographically inaccessible Indonesian teachers is a barrier to these measurements. ITEI Android application developed by analysing the geographical problem, so that each teacher can participate wherever they are. The ITEI app is designed by implementing Android on the client side and load balancer on the server side. Android ITEI will feature a number of questions questionnaire to teachers. Meanwhile, the load balancer will distribute the answers to each server for processing. Load Balancer ensures fast data processing and minimize server failure. The results of the processing on the server will be sent back to Android in the form of profiling themselves ITEI teachers. While the data obtained and stored in the server can be used to measure the level of national teachers' engagement. The result of this research is the design of ITEI application ready to be implemented in order to support the data collection process of teacher national engagement level.

## 1. Introduction

Teacher development efforts in Indonesia has strived so well by the Indonesian government through the Ministry of Education and Culture of the Republic of Indonesia. The basis for developing such coaching efforts is preceded by performance measurement and teacher competency testing each year. But if we look at the implementation of the measurement, there is a suspicion that the measurement of performance only contain questions that measure a norm that should be done by teachers, but do not measure the performance of each teacher what should be done (process) through established standards [1]. The same is true in terms of taking an approach in measuring teacher competence. In examining the competencies that are routinely performed by the government, it is suspected that testing of teacher competence is limited to test his mastery of learning theory, how to become a teacher, how to run the profession as an educator, how to teach, and others.

Teacher competence testing is not measured up to measure the capacity aspect of the taught science field and its qualifications as a teacher and educator [2]. The meaning, the two facts of the measurement approach (teacher performance and competence) are not designed to have a profile that impacts the actual size of a professional teacher, a self-developed teacher that can impact learning. Indeed the results of the measurement of the performance and competence of teachers throughout Indonesia that has been implemented is the basis for the Ministry of Education and Culture of the Republic of Indonesia to provide educator profession allowance to the teachers. But the facts on the ground show that the two measurements seem to be an additional burden on teachers, resulting in not increasing capacity, but decreasing the capacity of teachers. Measuring the profile of Indonesian teachers through performance and competence instruments is a good concept and step, but if the concept is top down, it will be a heavy burden for all actors both government and teachers themselves. For that ITEI a new solution that is the concept of measurement that is not just measuring the performance and competence by way of self-assessment through the Android based Apps.

The Indonesian Teacher Engagement Index (ITEI) is an instrument that provides a new alternative to a measurement approach of teachers as adults are to build self-measurement through self-evaluation as a form of humanizing an adult human. Basically, an adult who works as a teacher-educator, in himself, wherever the teacher is, what his or her existence is, is the form of the totality of the profession as an educator. The condition of the adult teacher is, in fact, able to build his own positive psychological condition (healthy), able to give the meaning of education that always have success and positive, have high performance, able to develop basic competence as the main requirement of a teacher, able to show Character of nationality based on Pancasila values and shows itself as a teacher who is not only a leader in the school and local duties but himself as a teacher throughout Indonesia (nationalism leadership engagement). [4] - [10], [2] The ITEI-based App approach is what will enable teachers to introspect their professions, which ultimately improve what teachers have to do for the improvement of their profession, performance, competence, As well as its leadership that reflects Indonesian teachers (not teachers who are limited only in the duties of devotion in their respective schools).

Because the nature of ITEI is to evaluate oneself as one form of introspection of the teacher's self-achievement, the necessary tools are privately owned by the teacher inherent in his gadget. This form of self-introspection not only shows the score or its value but ITEI Apps is developed up to the suggestion of intervention through artificial intelligence that serves to help teachers in developing themselves according to priority. Thus, with ITEI apps this is not just performance and competence but includes building the character of teachers as educators in total (engagement).

ITEI Apps is a representation of the rapid growth of information technology that influences the way humans collect, process and share their information as teachers/educators to the wider community. This growth is supported by the implementation of the Internet in almost all teacher activities throughout Indonesia. This includes supporting mobility with available online bookings such as Uber, and Grab or as in research [11] which uses information technology to facilitate the search for boarding. If it refers to the statistics that can be derived from [12] Internet user activity is dominated by the workers/entrepreneurs as much as 82.2 million (62%) of the total 132.7 million Internet users in Indonesia by 2016. Internet is accessed bigger in smartphone than others. Total internet smartphone is 89.9 million (67.8%). Referring to data shown on [13] since 2011 smartphone with Android operating system is the most widely sold operating system in the market [14].

Information technology is also used in the field of education, which in this study focuses on the use of information technology for the profiling of Indonesian teachers to see the teacher's engagement level as it is based on research [15], [16] the engage teacher proved able to produce students who are engaged and have achievement Which is good [15], [17], [18]. This study was designed using information technology in collecting data from each teacher from all over Indonesia, doing data processing in accordance with ITEI benchmarks, and distributing and displaying the results to the teacher. Processed data is also used by related parties as a measure of teacher engagement based on school, city, to the provincial level. However, this study will only focus on the design of Android ITEI applications to be built.

The process of measuring the engagement of Indonesian teachers is still done manually and cannot reach all Indonesian teachers. If you want to make teacher engagement measurements, it is usually only done if there are national high-level teachers or special teacher meetings. This process takes a long time, this is because the first is still done by filling out the questionnaire on the paper which for processing and processing the data takes a long time, the two geographical Indonesia wide so that required cost and proper scheduling when will do data collecting, third Data validation is still manual by checking data one by one. That's why an application is needed that can accommodate the teacher's engagement measurement needs by solving the problem.

The Ministry of Education and Culture of the Republic of Indonesia noted that there are 2,926,565 teachers spread across 216,666 schools across Indonesia [19]. It takes a long time to collect data and profiling throughout the teacher. With the Android-based ITEI application will facilitate in collecting data from all teachers and with high server capacity will perform the process well and correctly.

The ITEI application was developed based on six main dimensions of ITEI, namely the positive psychology dimension, the positive education dimension, the dimension of good performance, the teacher competence dimension, the dimension of national character, the dimension of nationalism leadership engagement. Each ITEI dimension has an indicator and each indicator has an item. This item will be presented in the form of an assessment quiz to the teachers, where teachers will provide answers following the Likert scale (1- 4-) for each question.

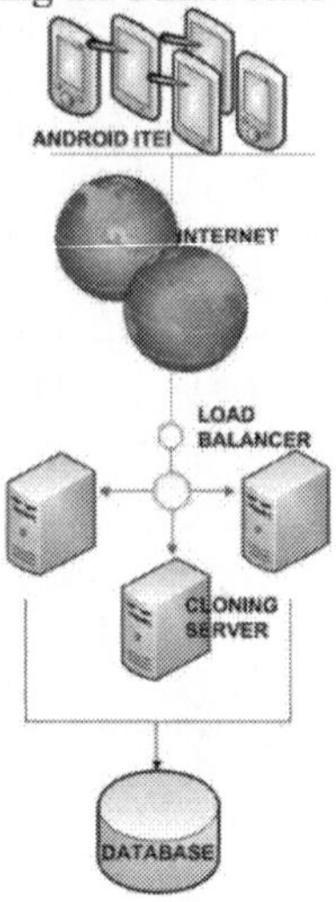


(a)

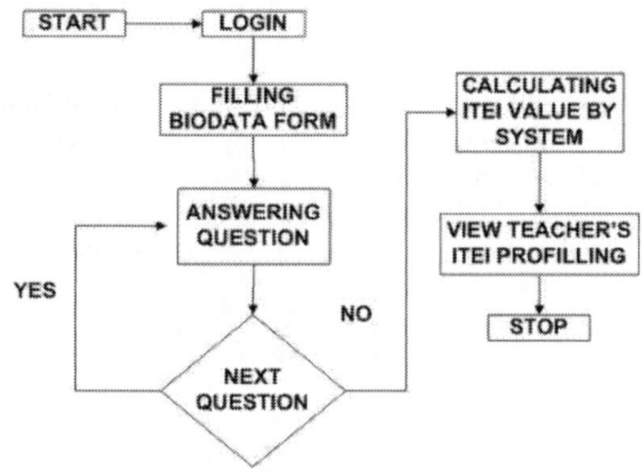


(b)

**Figure 1.** (a) ITEI Architecture; (b) Android ITEI Application Work Flow

Teachers can access the questionnaire material through Android smartphones, the process on ITEI described as in Figure 4 before filling out the questionnaire needs to be authenticated first. This authentication is used as the identity of each teacher. Teachers are required to log in using google account which by default is used on Android smartphone, Figure 2a. After login, the teachers are then asked to fill in self-data in the form of school origin, school location, educational status and employment status, Figure 2b. The application will not solicit or record Master Master's data as well as Teacher Name, this is done to ensure the confidentiality of teacher's data and comfort when filling out the questionnaire. After the data has been submitted successfully, it will display questionnaire questions consisting of fifty-four total questions, Figure 2c. The answer to the questionnaire will be processed following weighted index weighting rules. Weights based on the weighted index are selected because each dimension has a different level of importance [20].

Using Internet data and teacher questionnaire answers will be sent to the server as shown in Figure 1a. On the server side is implemented load balancer to ensure the traffic input from all Android ITEI smartphones can be processed correctly and data processing time is fast. A load balancer is a tool that acts as a reverse proxy, to efficiently distribute all the incoming network traffic to a number of backend servers where each backend server is a cloning of a server application. Cloning this server usually called an instant server [21] - [24]. Access and manipulation of the same database in an effort to ensure the data accessed on Android ITEI is on the backend server and is real-time data Figure 1a.

Teacher self-data, answers questionnaire, and the results of data processing will be stored on the database. The Database used is a relational database that is MySQL. Selection of this database because Android ITEI applications require data storage in a structured table. Each data has a special identity

IOP Conf. Series: Journal of Physics: Conf. Series **978** (2018) 012005 doi:10.1088/1742-6596/978/1/012005

where the value should not be the same as other data, this identity is called the primary key. Similar data will be grouped in the same table. When one table requires another table for the two tables it will communicate using a foreign key in the destination table [25].

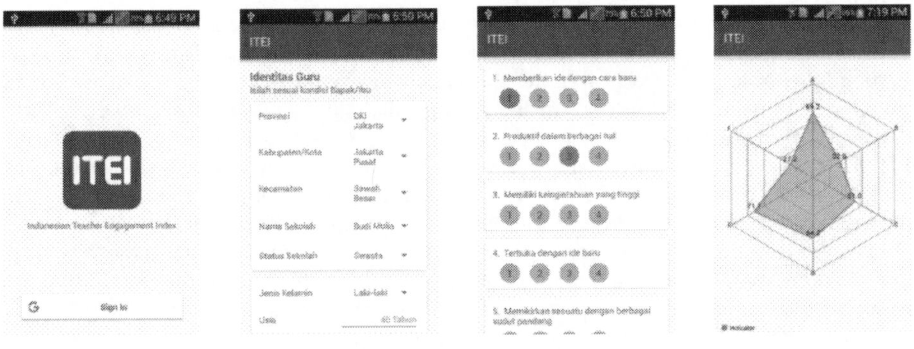


**(a)** **(b)** **(c)** **(d)**

**Figure 2.** Android ITEI Application: (a) Login Page; (b) Entry Personal Data; (c) Filling Questionnaire; (d) Personal Profiling

The results of data processing performed on the server then sent back to the Android smartphone each teacher. The app on Android will generate a self-profiling graph of self-profiling, Figure 2d. Each point represents the six dimensions of ITEI, and the results obtained will be displayed in the average value of each rib. In addition to display profiling, ITEI calculations are also displayed conclusions that can be drawn from the results of these calculations.

## 2. Research Methods

The current research focuses only on designing ITEI application architecture including server design, database, and Android ITEI application display. This research follows the method of waterfall development. The waterfall is a sequential application development model [26], [27]. The stages consist of requirements definition, system and software design, implementation and unit testing, integration and system testing, and operation and maintenance [26]. The waterfall motive ensures the need for software to be developed is well defined, clear and each sequence is sequentially done. This method is quite easy to implement and well documented [27] - [30]. The designs were based on study literature from various sources using the same techniques as this study. From the five stages in a waterfall, research on this paper only until there is a second stage, system and software design.

Requirements definition is done in two stages, the first stage by first validating questionnaire question as many as 165 questions with answer choice 1 - 4. Of the 165 questions validated to the experts in the field of education and outstanding teachers, then generated twenty-eight questions the final questionnaire, which are grouped into indicators and dimensions. At this stage also determined the weighted index of each dimension and indicator, which will determine how data processing and teacher profiling conclusion. The second stage is the collection of needs related to the application. Based on data recapitulation on [19] required client server applications that can manage all data and display profiling results without the risk of server down. At this stage, it is also concluded that the teacher can only perform a one-time profiling, using a certain identity, not including Master's Number or Master's Name. By using this identity the teacher can re-login and see the profiling results in the future.

System and software design is the design stage of ITEI system architecture and software as a whole. The architecture of the ITEI system follows Figure 2a, where after the teacher completes the questionnaire the data is not directly sent to the server, but must first through the load balancer. The load balancer will divide equally all incoming traffic to each server and each server will access the same database. The next design is database design. This database has eighteen tables that relate to each other using a certain key. ITEI software in the form of Android ITEI application interface design in accordance with the collection needs to be done in the previous stage, designed as in Figure 3. With the application steps as in Figure 2 that has been described in the previous chapter.

## 3. Discussion

The ITEI Android app was developed to facilitate the measurement of Indonesian teachers' engagement. Questionnaires will be grouped into six dimension groups. Teachers are asked to answer each question by choosing between 1-4. After the teacher answers twenty-eight questions, eat all data and answers will be sent to the server. On the server will be done processing answers by grouping answers based on dimensions.

The ITEI server application will calculate the average value generated in each dimension. Then the average value will be multiplied by the weight of each dimension in Table 1. The final value obtained in each dimension will be stored in the database. The ITEI server application sends two analytical values to the ITEI Android app; First the end result of each dimension will be used as the value to be displayed on the self-profiling chart as shown in Figure 3d, the two results of the analysis of the narrative obtained by mapping the final value of the dimension where if the calculation value is approaching four then the teacher's engagement level On the dimension the more positive.

**Table 1.** Weighted Dimensions of ITEI

| Dimension | A | B | C | D | E | F |
|---|---|---|---|---|---|---|
| Weight | 0.108 | 0.343 | 0.199 | 0.17 | 0.087 | 0.092 |

Implementing load balancers on the server side ensures the server process will not be hampered despite large data processing since load balancer will share the incoming traffic to each server equally. And when it was discovered that there was a full load of data processing on each server, the load balancer would automatically create a new server instantiation to handle the processing of new data. Thus the entire input data can be processed properly.

## 4. Conclusions

The ITEI Android app will facilitate the process of collecting Indonesian teachers' engagement data without geographical limitations because the Android ITEI app can be accessed anywhere and data processing can be done as long as the smartphone has an Internet connection [31]. Data processing on the server will be easier and faster with the load balancer. With the help of this tool will minimize the occurrence of server processing failure. The results of data collected from all Indonesian teachers can be used by stakeholders as a consideration for the development of Indonesian teachers. However, for the next development, it should be noted also to reach teachers who do not use Android smartphones. In addition, stress testing is required to prove load balancer resistance in processing multiple data simultaneously, to ensure that the server is eligible for national use.

## Acknowledgment.

This work is partially supported by Ministry of Research, Technology and Higher Education of the Republic of Indonesia and The Ministry of Education and Culture as a partner of research. The authors also gratefully acknowledge the helpful comments and suggestions of the reviewers, which have improved the presentation.

## References

[1] D J P M P dan T K Kementerian Pendidikan Nasional 2011 Pedoman Pelaksanaan Penilaian Kinerja Guru.
[2] Pemerintah RI 2005 Undang-undang Republik Indonesia Nomor 14 Tahun 2005 tentang Guru dan Dosen, *Prod. Huk.* p 54.
[3] Darling-Hammond L 2000 How teacher education matters, *J. Teach. Educ.* **51** pp 166–173.
[4] Seligman M E P, Steen T A, Park N and Peterson C 2005 Positive psychology progress:

Empirical validation of interventions., *Am. Psychol.* **60** pp 410–21.
[5] Seligman M E P and Csikszentmihalyi M 2000 Positive psychology: An introduction, *Am. Polit. Sci. Rev.* **55** pp 5–14.
[6] Seligman M E P, Ernst R M, Gillham J, Reivich K and Linkins M 2009 Positive education: positive psychology and classroom interventions *Oxford Rev. Educ.* **35** pp 293–311.
[7] Norrish J 2013 An applied framework for Positive Education, *Int. J. Wellbeing* **3** pp 147–16.
[8] Koopmans L, Bernaards C M, Hildebrandt V H, Schaufeli W B, de Vet Henrica C W and van der Beek A J 2011 Conceptual Frameworks of Individual Work Performance, *J. Occup. Environ. Med.* **53** pp 856–866.
[9] Koopmans L, Bernaards C M, Hildebrandt V H, Van Buuren S, Van der Beek A J and De Vet H C W 2013 Development of an individual work performance questionnaire, *Int. J. Product. Perform. Manag.* **62** pp 6–28.
[10] Yusoff R B, Ali A and Khan A 2014 Assessing reliability and validity of job performance scale among university teachers *J. Basic Appl. Sci. Res.* **4** pp 35–41.
[11] Manalu S R, Wibisurya A, Chandra N, Oedijanto A P and Service A W 2016 Development and Evaluation of Mobile Application for Room Rental Information With Chat and Push Notification pp 7–11.
[12] Data Statistik Pengguna Internet Indonesia Tahun 2016 Pembicara Internet Marketing SEO | Konsultan.
[13] Total Jumlah Pengguna Smartphone Seluruh Dunia - ET Studios Jogja. .
[14] Xu Q, Erman J, Gerber A, Mao Z M, Pang J and Venkataraman S 2011 Identifying Diverse Usage Behaviors of Smartphone Apps.
[15] Cardwell M E 2011 Patterns of Relationships Between Teacher Engagement and Student Engagement, *Fish. Digit. Publ.*
[16] Basikin 2007 Vigor, Dedication and Absorption: Work engagement among secondary school English teachers in Indonesia, *AARE Int. Conf. Perth, Aust. 27-28 Novemb. 2007* pp 25–29.
[17] Burns A 2010 Teacher engagement in research: Published resources for teacher researchers Books from regional locations Gregory Hadley (ed.). Action research in action, Singapore: SEAMEO Regional Language Centre (2003). Pp. iv + 54. ISBN 9971-74-081-8. Gertrude Tinker , *Lang. Teach.* **43** pp 527–536.
[18] Timms C and Brough P 2013 I like being a teacher *J. Educ. Adm.* **51** pp 768–789.
[19] K. Pendidikan dan Kebudayaan, Rekapitulasi Data Pokok Pendidikan Nasional. .
[20] McKinnon R I 2014 Money and Capital in Economic Development *The Canadian J. of Econ.* **7** pp 333–335.
[21] Wang R, Butnariu D and Rexford J 2011 OpenFlow-Based Server Load Balancing Gone Wild Into the Wild : Core Ideas pp 1–6.
[22] Sarood O and Kale L V 2011 A cool load balancer for parallel applications *Sc11.* pp 1–11.
[23] Amazone W S AWS | Elastic Load Balancing - Cloud Network Load Balancer. .
[24] Handigol N and Seetharaman S 2009 Plug-n-Serve: Load-balancing web traffic using OpenFlow, *Acm Sigcomm.*
[25] Whitten J L and Bentley L D 2007 System Analysis and Design Methods 7th- Whitten & Bentley.pdf.
[26] Verma J, Bansal S and Pandey H 2014 Develop Framework for Selecting Best Software Development Methodology *Int. J. Sci. Eng. Res.***5** pp 1067–1070.
[27] Balaji S 2012 Waterfall vs. V-Model vs. Agile: A comparative study on SDLC *Int. J. Inf. Technol. Bus. Manag.* **2** pp 26–30.
[28] Pressman R S 2009 *Software Engineering: A Practioner's Approach.*
[29] Sommerville I 2010 *Software Engineering.*
[30] Hong J C, Hwang M Y, Liu M C, Ho H Y and Chen Y L 2014 Using a 'prediction-observation-explanation' inquiry model to enhance student interest and intention to continue science learning predicted by their Internet cognitive failure, *Comput. Educ.* **72** pp 1–11.
[31] Dayer L, Heldenbrand S, Anderson P, Gubbins P O and Martin B C 2013 Smartphone medication adherence apps : Potential benefits to patients and providers.

IOP Conf. Series: Journal of Physics: Conf. Series **978** (2018) 012006 doi:10.1088/1742-6596/978/1/012006

# Implementation of utaut model to understand the use of virtual classroom principle in higher education

**B R Aditya**[1] **and A Permadi**[2]

[1] School of Applied Science, Telkom University, Bandung, Indonesia
[2] Center of E-learning Development, Padjadjaran University, Sumedang, Indonesia

E-mail: [1]bayu@tass.telkomuniversity.ac.id, [2]aditya@unpad.ac.id

**Abstract**. This paper describes implementation of Unified Theory of Acceptance and User of Technology (UTAUT) model to assess the use of virtual classroom in support of teaching and learning in higher education. The purpose of this research is how virtual classroom that has fulfilled the basic principle can be accepted and used by students positively. This research methodology uses the quantitative and descriptive approach with a questionnaire as a tool for measuring the height of virtual classroom principle acception. This research uses a sample of 105 students in D3 Informatics Management at Telkom University. The result of this research is that the use of classroom virtual principle are positive and relevant to the students in higher education.

## 1. Introduction

Internet technology today has all the advantages, the education sector with e-learning is an advantage to facilitate and enhance learning through computer and information technology. The use of e-learning as a tool can vary, from the simplest and isolated activities or the online exam. A comprehensive approach by educational institutions in general the use of e-learning software to support the overall learning activities as much as possible. In the current e-learning trend, learning can take place inside / outside the classroom, students can be online / offline [1][2]. This learning model creates ideas and assets that are not visible [1]. Knowledge can be readily accessed and shared to form students' intellectual acceleration through a virtual learning environment [1].

Virtual classroom is a synchronous teaching and learning process in internet-based. The teacher and the students interact with each other over the computers [3][4]. Virtual classroom is a powerful media for learning [5]. Challenges in virtual classroom utilization in higher education still exist ease of use [6]. Beside that, a lot of learning process that does not pay attention to the basic principles that must be done when implementing a virtual classroom [7][8]. Implementation of virtual classroom must be fulfill the basic principle of virtual classroom technology [5]. Integrating basic principles into a virtual classroom is important when preparing effective virtual classroom [5].

The purpose of this study is how virtual classroom that has fulfilled the basic principle of virtual calssroom can be accepted and used by students positively. This research result will be used sustainably to support effective virtual classroom at higher education level. For the method of measuring factors that affect student acceptance of virtual classroom, authors use the acceptance model of Unified Theory Technology of Acceptance and User of Technology (UTAUT).

## 2. Virtual Classroom Principle

Virtual classroom as instructor-led synchronous computer learning environments attended by participants online at the same time but in different locations [5]. Virtual classroom has more multimedia contents, functions of queries, transmissions and data integrations when compared with traditional classroom [4] offers synchronous tools for the activities of teaching, discussing, reading and testing [9] [10]. According to Clark and Kwinn [5], the basic principles for virtual classroom success can be divided into seven principles, that are:

- Pre-engineering lesson.
  Preliminary planning and preparation of all key elements including slide content, teaching video scenarios, interactions (storyboards), participant handouts, e-learning media use manuals for students and facilitators. In this principle, authors create a storyboard for multimedia teaching video, teaching slide and technical guidebook for student.
- Include diverse media: diverse delivery media complement one another.
  Learning goals are best served by a blended approach. In this principle, synchronous learning that used is online chat in Edmodo Learning Management System (LMS), Whatsapp Messenger, and Skype. Teaching materials consist of slide and teaching video.
- Good virtual events are explicitly relevant job.
  This principle means the ritual event in the teaching topic must be in accordance with the semester learning plan / RPS (The institutions call it RPS) or syllabus that can be combined with the use of asynchronous learning events. In this principle, authors add technical guidance in the learning media column in the standard syllabus / RPS. Because the needs and learning media have evolved, the instructional designer has positioned itself as an important player in the design and delivery of learning experiences.
- Learning is interactive.
  Open interaction is essential to learn new knowledge and skilll, the virtual classroom environment offers many facilities for student interactivity. In this principle, authors use free interactive media for quiz activity in Kahoot game based learning (https://kahoot.it/#/).
- Social presence promotes learning.
  Discussion-based learning activities, group work, group assignments. If this feature is well managed it will increase students' opportunities for interaction that can improve learning. In this principle, authors use the group tools on Edmodo LMS (https://edmodo.com).
- Appropriate visual and verbal modalities.
  The appropriate visual and verbal use makes learning effectively. Appropriate visualization described (audio) can effectively improve learning by more than 80% compared to relying on textual explanations only. People learn better from words and pictures than from words alone, People learn better when the narration in multimedia lessons is spoken in a friendly human voice rather than a machine voice. In this principle, authors create multimedia content in some learning topics.
- Cognitive load must be managed in all instructor-led events.
  Every interactivity in the event (tools, features, teaching content, e-learning manuals and help-desk support) should be manageable so that the learning process becomes easy to be understood and used by students. In this principle, authors make all things related to the interactivity of the learning process into a content that is very easily understood by students to create a guide book and to socialize to students in the classroom at the first meeting.

## 3. Research Methodology

This research is a kind of quantitative descriptive research. Quantitative descriptive method is a method that only provides an overview or description of the variables of a phenomenon studied.

### *3.1. Learning Process Scenario*
In this research, three classes with different course were choosen to case study with flex model blended learning. Each class had 16 meetings in one semester with the following scenario details were 3 meetings (with 2 topics for knowledge learning concept and facts, and 1 topic for learning evaluation that was interactive quiz based on kahoot.it) were conducted fully online distance learning.

### *3.2. Research sample*
Measuring of learning experiences through virtual classroom principles was made to 105 students of D3 Informatics Management at Telkom University. The number of students who participated in this research can be seen in Table 1 and Figure 1.

**Table 1.** Percentage of participant by courses.

| Courses | N | % |
|---|---|---|
| ICT Literacy | 35 | 34 |
| Professional Ethics | 34 | 32 |
| Information System Audit | 36 | 34 |

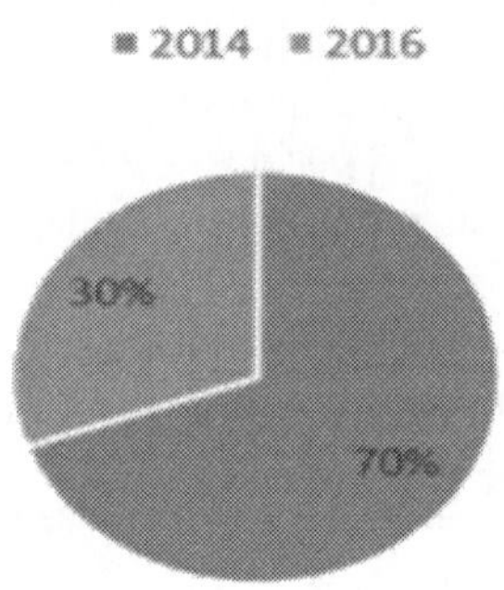


**Figure 1.** Percentage of participants by class.

### *3.3. Instrument and procedures*
The questionnaire in this research was designed to measure the acceptance level of virtual classroom using 4 dimensions according to UTAUT model that are:

- Performance Expectancy
- Effort Expectancy
- Social Influence
- Fasilitating Condition

The design of the questionnaire was done by several stages:

- Making preliminary questionnaire: the preliminary questionnaire was designed by the research team with the aim of describing the structure and content of the questionnaire. The structure of the questionnaire referred to the four dimensions of the previously mentioned UTAUT. The content of each structure was designed in the likert-scale form. Variables of each dimension can be shown in Table 2.

**Table 2.** Variable of each UTAUT dimensions.

| Dimensions | Variables |
|---|---|
| Performance Expectancy | Improved achievement<br>Motivation to learn<br>Compatibility in learning<br>Improved way of learning<br>Learning outcomes |
| Effort Expectancy | The effort required<br>Complexity<br>Ease of learning |
| Social Influence | The necessity of using ICT media<br>Habit or culture<br>Social status |
| Fasilitating Condition | Self-efficacy<br>Condition of resources<br>Experience |

- Testing validity: validity test was conducted to ensure that the questionnaire (preliminary questionnaire) is accurate and appropriate to measure the level of student acceptance of virtual classroom technology. Validity test was conducted on 15 respondents and processed by using spreadsheet software. Based on the validation test, the results showed that the whole question items were valid (corrected item-total correlation> 0.4973).
- Testing reliability : reliability test was performed to determine the level of accuracy (reliability) of the questionnaire made (preliminary questionnaire). Reliability test used split-half technique. Based on reliability test, the results showed value of reliability coefficient of 0.8868 (very high reliability).
- Finalizing the questionnaire: based on the results of validity and reliability test of the preliminary questionnaire, it could be concluded that the questionnaire was really ready to be used as a data collecting tool in this research.

## 4. Research Result

The results obtained from the assessment of the implementation of virtual classroom in 105 students based on four dimensions of UTAUT can be seen in Figure 2.

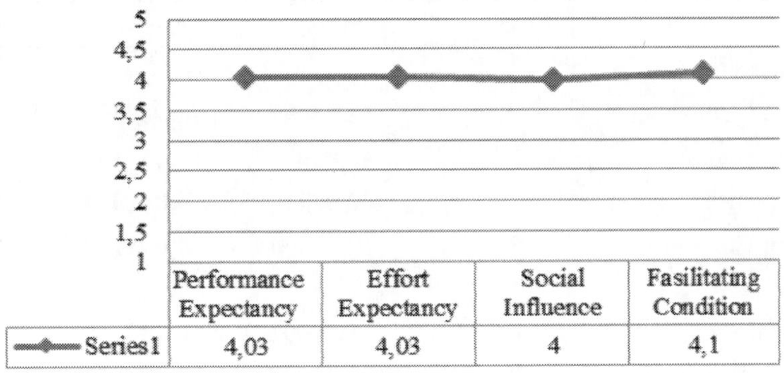


**Figure 2.** The average value for each dimension.

IOP Conf. Series: Journal of Physics: Conf. Series **978** (2018) 012006 doi:10.1088/1742-6596/978/1/012006

Interpretation of measurement results using Likert scale as shown in Table 3.

**Table 3.** Interpretation of Likert Scale.

| Scale | Information |
|---|---|
| 5 | Strongly Agree |
| 4 | Agree |
| 3 | Neutral |
| 2 | Disagree |
| 1 | Strongly Disagree |

For an average value based on the courses and dimensions of UTAUT can be seen in Figure 3.

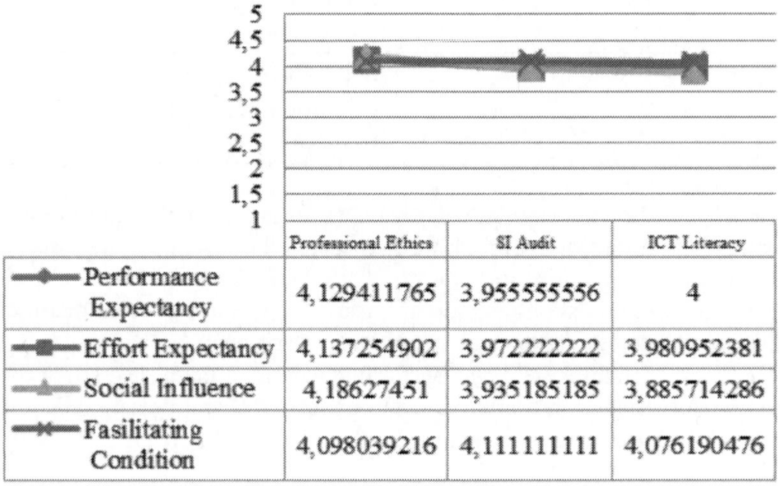


**Figure 3.** The average value based on courses and dimensions of UTAUT.

Two Way ANOVA test result to test the difference of mean value between courses and dimensions of UTAUT can be seen in Table 4 and Table 5.

**Table 4.** Descriptive analysis of UTAUT and course variables

| Summary | Count | Sum | Average | Variance |
|---|---|---|---|---|
| UTAUT Dimensions | | | | |
| Performance Expectancy | 3 | 12,08497 | 4,028322 | 0,008158 |
| Effort Expectancy | 3 | 12,09043 | 4,030143 | 0,008624 |
| Social Influence | 3 | 12,00717 | 4,002391 | 0,025972 |
| Fasilitating Condition | 3 | 12,28534 | 4,095114 | 0,000311 |
| Course | | | | |
| Professional Ethics | 4 | 16,55098 | 4,137745 | 0,001334 |
| Information System Audit | 4 | 15,97407 | 3,993519 | 0,006375 |
| ICT Literacy | 4 | 15,94286 | 3,985714 | 0,006138 |

**Table 5**. Anova analysis between UTAUT and course variables

| Source of Variation | SS | df | MS | F | P-value | F crit |
|---|---|---|---|---|---|---|
| Course | 0,014044 | 3 | 0,004681 | 1,021564 | 0,446742 | 4,757063 |
| UTAUT Dimensions | 0,058634 | 2 | 0,029317 | 6,397517 | 0,032533 | 5,143253 |
| Error | 0,027495 | 6 | 0,004583 | | | |
| Total | 0,100174 | 11 | | | | |

## 5. Discussion and Conclusion

This research has assessed the use of virtual classroom in a sample at D3 Management of Informatics at Telkom University. The assessment were made by using questionnaires that have been evaluated through validity and reliability tests using four dimensions of UTAUT.

Our findings indicate that the average value for each dimensions of UTAUT in a pilot sample is above four, this indicate that the virtual classroom has been positively accepted by class of 2014 and class of 2016. In addition, based on each course, we get the average value for each dimensions of UTAUT is around 4, this means that virtual classroom can be accepted in three different courses that are sampled in this study.

Additionally, through detailed analysis with Two-Way ANOVA test using spreadsheet application, some interesting conclusions were drawn on the outputs, obtained p-value for course dimensions of 0.4467 and p-value for UTAUT dimensions of 0.0325. Since the desired level of confidence in this study is 95%, the significance level is 100-95 = 5% or 0.05. So for the course variables, because p-value> 0.05, it means that there is no significant difference among the average count of the three courses.

In sum, we conclude that the use of classroom virtual are positive and relevant to the students in higher education with the high acceptance value of each dimension of UTAUT namely performance expectancy, effort expectancy, social influence and fasilitating condition. We have also oberved that success in the use of classroom virtual depends on how the basic principle in virtual classroom is fulfilled.

## References

[1] Harasim L 2000 Shift Happens: Online Education as a New Paradigm in Learning *The Internet and Higher Education* **3** pp 41-61.

[2] Jiang Z and Guo X 2012 A miniautre e-learning portal of full teaching and Learning Capacity *IEEE Computer Society Press*.

[3] Turoff M 1995 Designing a Virtual Classroom *Int. J. of Educ. Telecommunications* **1** pp 245-262.

[4] Chen RC and Liu SC 2009 Applying concept mapping on the influent learning in virtual classroom *Int. Conf. on Hybrid Intelligent System. IEEE Computer Society Press*.

[5] Clark R C and Mayer R E 2011 *E-Learning and the Science of Instruction: Proven Guidelines for Consumers and Designers of Multimedia Learning 3rd Edition*. Pfeiffer. Willey Inprint.

[6] Owston R, York D, & Murtha S 2012 Student Perceptions And Achievement In A University Blended Learning Strategic Initiative *The Internet and Higher Education* **18** pp 38–46.

[7] Su T A 2007 Distant vcr - a virtual classroom for distance learning *IEEE* pp 1-2.

[8] Debnath B C, Rahman M M, Bashir G M M and Hossain M J 2014 Learning and Evaluation of Blended Learning Approach for ICT Undergraduate Students *Int. Conf. on Elect. Eng. and Info. & Comm. Technology (ICEEICT) IEEE Comp. Society Press*.

[9] Hiltz S R and Barry W 1994 Asynchronous Learning Networks as a Virtual Classroom *Communication of the ACM* **40** pp 44-49.

[10] Marco F A, Penichet V M R, & Lázaro J A G 2013 Drawer: an Innovative Teaching Method for Blended Learning *Proc. of the 2013 Federated Conf. on Comp. Sc. and Info. Systems IEEE Computer Society Press*.

IOP Conf. Series: Journal of Physics: Conf. Series **978** (2018) 012007 doi:10.1088/1742-6596/978/1/012007

# Usability testing of indonesia tourism promotion website

**U Yunus[1], Anindito[2], E Tanuar[3] and Maryani[2]**

[1]Communications Department, Bina Nusantara University, Indonesia
[2]School of Information Systems, Bina Nusantara University, Indonesia
[3]School of Computer Science, Bina Nusantara University, Indonesia

uyunus@binus.edu

**Abstract**. Indonesian tourism has been viewed as one of the source of foreign exchange, therefore visit Indonesia campaign through Wonderful Indonesia should be given attention. This research aims to test the usability of web http://www.indonesia.travel/en, at the end can give input to the Tourism Ministry of the Republic of Indonesia. The concept used based on usabilitygeek.com which mentions the three main categories for usability testing, i. e. explorative, assessment and comparative with the users are three classes of Communication Department students of Bina Nusantara University with a total of 120 students, however finally only 22 students were elected because of the criteria are the students should be like travelling and access to the tourism website more than 12 times. The methodology used is qualitative descriptive and experiments, with data collection techniques through observation and interviews after participants' access to the web: Due to their one-on-one nature, interviews enable the observer to ask direct questions to the user and the participants posted their opinion in online discussion forum. The results mentioned that the elements web of Wonderful Indonesia has a high learning ability and memorability but less efficiency and satisfaction, as well as several times errors when used. In conclusion, to maximize about the visit of tourism to Indonesia, this web is only able to learn and impress the visitor, but do not give satisfaction to users.

## 1. Introduction

Currently, Indonesia's tourism sector contributes to approximately 4% of the total economy. By 2019, the Government of Indonesia wants to increase this figure to double to 8% of Gross Domestic Product Revenue, meaning the number of visitors needs to be doubled to approximately 20 million. In order to achieve this target, the Government will focus on improving Indonesia's infrastructure, including information and communication technology infrastructure), access, health & hygiene and also enhancing online promotional campaigns (overseas). The government has also revised its free visa access policy since 2015 to attract more foreign tourists [1]. The development of tourism is quite encouraging seen from year to year about tourist visits such as depicted in Table 1. The amount at the end of 2016 reaches one million per month. Expected in 2017 and then its numbers continue to increase.

**Table 1.** The Number of Tourists to Indonesia each Month From 2013 till 2016 [1].

| Month | 2013 | 2014 | 2015 | 2016 |
|---|---|---|---|---|
| **January** | 614,328 | 753,079 | 723,039 | 814,303 |
| **February** | 678,415 | 702,666 | 786,653 | 888,309 |
| | 725,316 | 765,607 | 789,596 | 915,019 |

| | | | | |
|---|---|---|---|---|
| **April** | 646,117 | 726,332 | 749,882 | 901,095 |
| **May** | 700,708 | 752,363 | 793,499 | 915,206 |
| **June** | 789,594 | 851,475 | 815,148 | 857,651 |
| **July** | 717,784 | 777,210 | 814,233 | 1,032,741 |
| **August** | 771,009 | 826,821 | 850,542 | 1,031,986 |
| **September** | 770,878 | 791,296 | 869,179 | 1,006,653 |
| **October** | 719,900 | 808,767 | 825,818 | 1,040,651 |
| **November** | 807,422 | 764,461 | 777,976 | |
| **December** | 766,966 | 915,334 | 913,828 | |

Viewed from the origin of the country, the most tourists who come to Indonesia in March 2017 compared to March 2016, i.e. from China are 140.98 thousand visitors, which is increasing from previously only 94.22 thousand visitors. While the number of tourists from Singapore are 128.38 thousand or decreasing from 130.80 thousand visitors. Tourists from Malaysia, Australia, Japan, each with 117.26 thousand, 88.20 thousand, and 43.85. The number of tourists from abroad increases, the occupancy rate of star-rated hotels in Indonesia in March 2017 averaged 54.70 percent or an increase of 1.82 points compared to March 2016 which was recorded at 52.88 percent [1]. Review the record of number, we can optimist, Indonesian tourists will continue to increase. Furthermore, it is necessary to find out the dissemination of information gained by each country to tell about Indonesian tourism.

Based on the observation result in this research, the promotional tools made by the Ministry of Tourism of Indonesia through Wonderful Indonesia program is through TV Ads in various pay channels in Asia and overseas; branding through digital media tunnel LED in several subway screen door located in Seoul Metro, outdoor media advertisements such as billboards, advertising on subway screen door and One Vision Way, also the web becomes an important information tools for potential tourists as mentioned by researcher from Lithuania which states that in the recent years, a considerable number of scientists have shown interest in the significance of innovations in tourism sectors, have analyzed the impact of information technologies to their services and competitiveness. Unfortunately, such studies in Eastern European countries, including Lithuania are not available of such a service. In this competitive era, promotional strategies are an important part of the organization, students majoring in communication learn this, in some of their courses, and they are energetic youngsters who love to enjoy the life, tend to move dynamically from one place to another, this characteristic somewhat different from students with other majors.

We see that the existences of website for tourism is so important that another researcher calls it destination marketing as an investment described as follows: destination marketing organizations invest considerable amounts of money in the development of websites as part of their overall promotion efforts [2]. Based on this background, the research focused on questioning how is usability testing website of promotion of Indonesia Tourism by communication 'students? The focus of this study is broken down into three questions. How does Indonesia's web culture work as a medium for learning about Indonesian destinations and cultures? How can the web capable to be memorable for the users? What is the level of errors experienced by users when accessing the web of Wonderful Indonesia? The results of this study are expected to be input for the Ministry of Tourism of Indonesia or other general public regarding the usability of the website for its organization.

## 2. Literature Review

Usability testing refers to evaluating a product or service by testing it with representative users. Typically, during a test, the participants will try to complete typical tasks while observers watch, listen and takes notes. The goal is to identify any usability problems, collect qualitative and quantitative data and determine the participants's satisfaction with the product. In this study more qualitatively tested with participants who are accustomed access to web tourism.

Web site evaluation has been proposed in many ways and various contexts over the past decade. An unified framework of commonly used web site success factors (1) information quality; (2) ease of use;

(3) responsiveness; (4) security / privacy; (5) visual appearance; (6) trust; (7) interactivity; (8) personalization; And (9) fulfillment. Additional factors to further inform Werthner and Klein's dimensions of web communication [3]. In this study the elements studied are the learning media for students and how interesting for students to learn the destination of Indonesia through the web under study. If associated with the above factors are about information quality and visual appearance, whether can create memorable for the user and how high the level of errors experienced by the user. Finally, it is mean ease of use of one of the factor above.

Other researchers have learned that a good appearance will cause proximity to the user as stated that the results indicate that perceived fashion ability has a positive impact on the user's stickiness to the website, and perceived fashion ability also performs a mediating role between website security and stickiness. This study contributes to concept by explaining that website stickiness is not entirely driven by rationality but also perceived fashion ability. In this study, we find out the reasons why the users choose a particular destination [4].

The way to evaluate tourism websites has also been done by researchers from China who said that the research on the evaluation of tourism websites can measure the websites' development in the same industry, and it is helpful to improve the quality of the websites. In view of the complexity of the tourism website evaluation. The evaluation index system of tourism websites, including subjective indicators and objective indicators [4]. Tourism website comprehensive expectation is calculated and the website ranking according to comprehensive expectation is given from high to low. In this research more subjective indicator because every student was asked to give his/her opinion personally and then validated to web builder and marketing communication experts, data is discussed and categorized by researcher.

## 3. Methodology

This research used descriptive qualitative method completed with experiment three times by the same participants. Data collected through interviewing student who love traveling and looking for tourist destination information through website of tourism. There are 22 students selected because according to the criteria of was 120 students who will be studied determined purposively i.e. three classes of communication science majors of Bina Nusantara University. They are welcome and briefed to access the website http://www.indonesia.travel/en in three times. These 22 students have the category of "frequent" access to the tourism website (more than 12 times a year to access the tourist partial web). While they use the website, they are observed, how the feeling expressions when they access the web and they are asked to give their opinion about the web being researched every time they access. They are given 20 minutes each time access to the Web was studying. The place of research is in the classroom using each smartphone. The classroom has very stable internet access so it is worthy to be a computer lab. As in previous studies in the way mentioned by Norman & Panizzi in 2006, the study was conducted on a study of different variants of TA methods and the fact that TA testing methods were typically applied in usability laboratory settings [5]. Data are collected, they posted the opinion also in the online discussion forum; their opinions are examined by the website builder for triangulation of source as qualitative paradigm used. Their opinions to be checked for validity. Another observation is that researchers try to access the web under investigation to check the validity of the answers of the participants. They also asked the origin of the culture so it can be analyzed why they love the destination of a particular area to be a priority such as tourist destinations on the web http://www.indonesia.travel/en.

The data obtained is processed and discussed between research teams with web experts and marketing communications expert. Data are categorized and analyzed for further conclusions. The study lasted for 6 months. The study began with secondary data collection in the form of Indonesian tourism data, selection of participants and triangulation of sources, implementation of website usage for three times, interviews and discussions between research teams.

## 4. Results and Discussions

Website of Wonderful Indonesia consists of "Start Your Journey Here" menu which consists of three pictures with nuances of Medan, Bali and Jakarta. When clicked will appear information about each destination of the three places. Equipped with reviews from TripAdvisor, Garuda flights, hotel selection, maps and any kind of tours in the city. When we scroll the page, it will appear "Experience 5 Wonders" menu, "Destination Highlight", "What's News", "Discover Indonesia", "What's Happening", the last is about "This Month Events" with menu: "Experience" "Travel Guidance" and "Facts". The describe menu is like the following Figure 1.

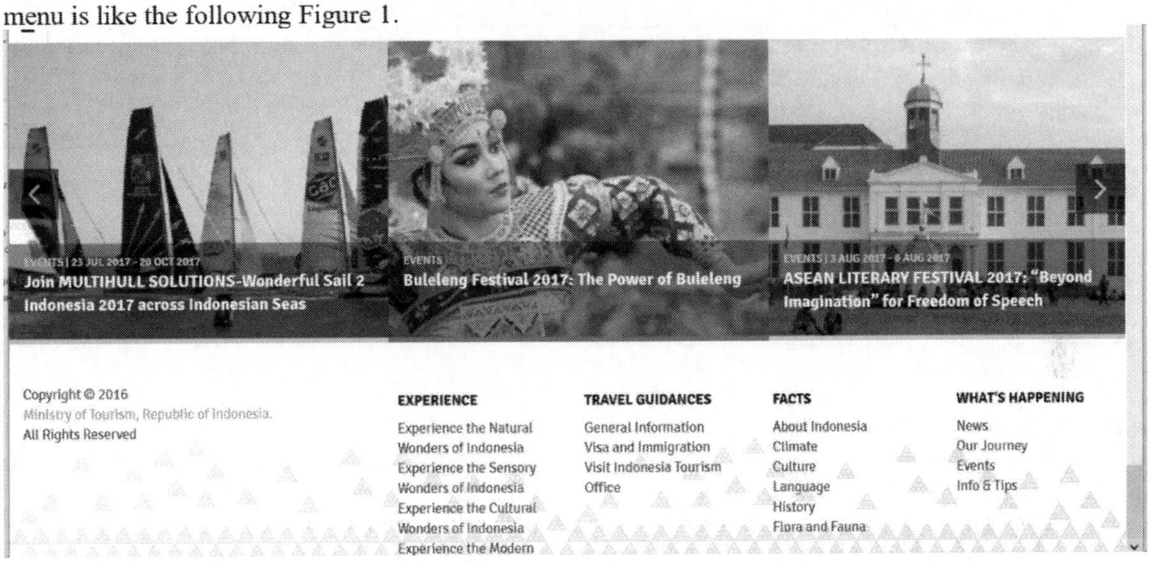


**Figure 1.** Menu of Web: This Month Events (http://www.indonesia.travel/en, 2017)

Based on the observations and interviews completed with the literature, the discussion will cover three things that the purpose of this research, such as to know how website of Wonderful Indonesia as a medium for learning about Indonesia's destinations and culture; how the web can be memorable to users; and how the error rate experienced by the users when access to web of Wonderful Indonesia. Website as a learning media for the participants felt that they had an idea of each of the tourist destinations listed on the web, although it took them a long time to learn one by one from the menu provided by the web. The participants said that: the titles written are very interesting to visit, there are even video shows that tell the destination. But the information submitted is too much so the website becomes too complicated as more stories are written about the tours. Typically, the testing method has tested the participants on their experiences, thoughts, actions, and feelings whilst interacting with the system [6]. The advantages of the web are the story of the culture of each region so that becoming a science enhancer not only about the location but also the local culture of the area. On the next layer there is Tripadvisor's application which helps the user to decide his choice. Disadvantages of this web is the top menu of "Start Your Journey" is too quickly shifted so it makes the user must race in speed when clicking the button, although after successfully clicking one menu will appear information about the three destinations with complete. One of the participants mentioned "I were position if I do not know Indonesia well, I am somewhat disturbed by the shift of the menu". Then, the other layers menu is difficult to open so it takes a long time to wait for further information. Then still found the hotel reviews in Indonesian so it is not suitable for foreigners. This data have been asked repeatedly to participants and found an agreement about the advantages and disadvantages. The examination of this data is important because is given the limited memory of the participants as mentioned by Alhadreti and Mayhew who said according to Ericsson and Simon (1993), in tests utilizing the traditional CTA method, verbalizations can only be valid if they are represented directly accessible information contained in the participant's short-term memory. Such verbalizations do not alter the sequence of information comprehended by participants, and so do not affect the tasks that participants performed during TA sessions [6]. Based on participant interviews and validated by web builder expert and marketing communications expert, conclusions are made about the website http://www.indonesia.travel/en in Table 2 below:

**Table 2.** Lessons Learned from Use of the Web Top of Form (Source: Processed from Research, 2017)

| No | Advantages of Web | Disadvantages of Web |
|---|---|---|
| 1 | The existence of Culture learning that existed in the Tourism Destination | Tourism Destination The "Start Your Journey" menu moves too fast |
| 2 | Complete information including maps, how to get and nearby hotels | There are still hotels reviews of Indonesian language, it is not suitable for foreign tourists |
| 3. | The existence of Tripadvisor application, it can help the decision based on the experience of others | Other menus are layered so many, it is difficult to access the information |

### *4.1. The Elements of Memorable Creator*

In this study, the session began with welcoming each participant and asked them to read and sign a consent form. Participants were informed that the purpose of the study was to evaluate tourism website. Following this, the concept of thinking aloud was introduced using Ericsson and Simon's instructions (1993) [6]After they get together and try to access the web. Upon completion of the one-day time access, they are identified with data to determine the ethnicity and destination of the city choice and the reason for selecting without view the web again. The answers are as follows.

Table 3. Respondents Data and Choice of City According to Memory (Source: Processed from Research, 2017)

| Characteristics | Gender | | Access to tourism web/year | | Ethnicity | | | | Destination Choice (each student is asked to choose 2 destinations that are remembered in his memory) | | | | | |
|---|---|---|---|---|---|---|---|---|---|---|---|---|---|---|
| | F | M | Less or 12 times | More 12 times | Jakarta | Javanese | Sumatera | Borneo | Bali | Papua | Jogjakarta | Labuan Bajo | Bandung | Bangka |
| | 12 | 10 | 7 | 15 | 14 | 2 | 5 | 1 | 14 | 10 | 8 | 6 | 4 | 2 |

Participants choose the cities interested because of information provided by the web and they have heard about the city before and there is a desire to come to the place of choice. But by looking at the images displayed on the web as well as information about the place makes they are more interested. When asked what is remembered from the web about these places, here are the results at Table 4.

**Table 4.** The City and Students Memory (Source: Processed from Research, 2017)

| No | Preferred Destination | Memorable Things |
|---|---|---|
| 1 | Bali | The beach and the place of the Gods |
| 2 | Papua | Favorites for Overseas Tourists |
| 3 | Jogjakarta | Really Indonesia; Culture and Food |
| 4 | Labuan Bajo | In contrast to other areas |
| 5 | Bandung | The air is cool and friendly to the tourists |
| 6 | Bangka | Coastal tour |

It appears that no participants have chosen the city of their parents. For the destination of tourism there is a variety of things in cultural choice, not always according to their original culture. But there is a tendency that they choose the same culture for cultural recognition. For example, the Javanese tend to make the first choice of Jogjakarta because they love culture and are interested in the information available on the Web. For the second option they choose places that are popular and promoted more by the government. On the Web there are other destinations such as Lampung, Malang etc. however from the participants no one remembers it. Only those that have been in their previous memory and the names of these cities are already popular among the student population. Other people's experiences or experiences will be more supportive to remember as mentioned in the previous research. These findings are supported by Brakus et al., (2009), Zarantenello and Schmitt, (2000); Ha and Perks, (2005), brand experience creates and develops trust-based relationship platform between brand and customer. Brand experiences arise in a variety of settings when consumers search for, shop for, and consume brands [7]. When interviewed further to the participants and examined the validity of the answers to marketing communications experts it is true that what is listed on the web will be more memorable if appropriate experience or hear stories of the experience of people close to them. It should therefore be able to display tourist profiles that match the targets of the Indonesian government, such as China, Singapore, Malaysia, Australia and Japan as identified from the data of origin of major tourist countries coming to Indonesia.

Previous research adds RADAR as a memorable creator as mentioned. Researcher wrote, I believe that this is significant because, while the students may not recall instantly each of the RADAR criteria, the term itself is memorable, as is its associated concept, that is, the need to be constantly alert to possible information dangers. Radar stands for Relevance, Authority, Date, Appearance, and Reason for writing [8]. So for the web memorable then a web needs to update and hold the information and write down the reasons why Indonesia should be visited.

### *4.2. Web Error Level*

Discussion of the web error rate under study, there was a difference in the three experiments. In the first experiment only 2 people said there was no problem with web access, but 20 other students said that the web is difficult to open. Internet network does not experience any constraints. It cannot be explained why in the first experiment the web was difficult to access by the majority of participants. But in the second experiment, all participant said that the web can be opened quickly as well as the third experiment. According to the web expert who was interviewed that in the first experiment too much access to the web and it happened in early May. The period before the fasting or Ramadan in Indonesia where some people make use of holiday information for before fasting and for Eid. The second experiment takes place a week later and the third is the day after. In the second experiment, 16 of the participants still mentioned the difficulty of opening some of the selected icons on the web, 6 others said quite quickly open for access to information. In the third experiment, more participants said the destination icon is easy to open as many as 17 people while 5 others still say the icon is difficult to open. With different participants so that the constraints not on the person or his smartphone but more to aksebilitas his web. Here is an overview presented by Table 5.

**Table 5.** Web Error Level (Source: Processed from Research, 2017)

| Experimental | Access to Web | | Access to Icon | |
|---|---|---|---|---|
| | Easy | Difficult | Easy | Difficult |
| First | 2 | 20 | 0 | 22 |
| Second | 18 | 4 | 6 | 16 |
| Third | 21 | 1 | 17 | 5 |

Thus, the error level is high and this requires the web manager to pay attention. The cause of this error can be caused by the following three things we can consider the following failure sources:

a) Host, network, or browser failures: These failures are similar to regular systems, networks, or software failures, which can be analyzed by existing techniques.
b) Source or content failures caused by web pages and other resources
c) User errors may also cause the problem: they are controlled by web contents providers [9].

In this case after being examined by the research team and experts, the error was caused by the point b, i.e. source or content failures caused by web pages. To know more can be into a general procedure consisting of the following five stages:

*(1) In the pre-classi fi cation stage, preliminary analysis of the given site and its log data is carried out to de fi ne error measurements, to establish an analysis baseline and to customize the classi fi cation scheme. (2) One-way analysis examines one attribute at a time to identify problematic areas for corrections or for further analyses. Detailed error trends and reliability analyzes can be performed by stamped error data with detailed usage measurements (Kallepalli and Tian, 2001; Tian et al., 2004). (3) Two-way analysis of the interaction between two error attributes to address the problems identified above. (4) Multi-way analysis focuses on the remaining problems by the combination of three or more attributes. (5) The above analysis results are consolidated to provide feedback and recommendations. [9]*

The above conditions cannot be allowed to remember that the demands of today who want to be fast and smooth in the activities should be the organization's consideration is now coupled with the characteristics of the millennial children who by described the results of research PricewaterhouseCoopers which states that the particular characteristics of millennial - such as their ambition and desire to keep learning and move quickly upwards through an organization, as well as their willingness to move on quickly if their expectations are not being met. This depiction is to tell how they are in different workspaces with Generation X and Y who tend to try adaptation and still remain calm for something that does not run fast [10].

Responding to the present circumstances suggested by a researcher that advances in internet technologies have led to the popularity of technology-based self-services, with the design of such services becoming increasingly important [9]. This research has been conducted on proven government web using technology-based services in the public sector as the setting, researcher identified the key service attributes driving the adoption and use of transactional e-government services, and citizens' preference structures across these attributes. They research by four key attributes, i.e., usability, computer resource requirements, technical support provision and security provision. They do a web-based survey and a conjoint experiment between 2465 citizens. 4 months later, the results show that the four key attributes influenced citizens' intentions, subsequent use and satisfaction. In the conjoint experiment, citizens reported their preferences for key service attributes for two transactional e-government services [9].

## 5. Conclusions

Based on the above discussion, then in this study can be concluded that Web of Wonderful Indonesia as a medium for learning about Indonesian destinations and culture has been successful on the subject studied so that content on the web contains cultural information and tourist destinations can be maintained but there are still shortcomings in some respects such as the Indonesian language in hotel reviews. Then the Web of Wonderful Indonesia has a memorable ability for the users because of the similarity of cultural background and experience stories of close friends, thus in this web if the goal of increasing the most tourists from China can display the tourist pictures of the happy and comfortable Chinese Tourist in Indonesia. Similarly, other tourist profiles that are targeted by Indonesian tourism market, such as from Singapore, Australia and Japan, bringing the closeness and confidence in potential tourists who will come to Indonesia. The study also mentions that there is still a high level of error in participant experiments when they access to the web destination or icon. This needs to be a correction

IOP Conf. Series: Journal of Physics: Conf. Series **978** (2018) 012007 doi:10.1088/1742-6596/978/1/012007

for the Ministry of Tourism of Indonesia in its web management considering that promotions made through other media are only branding and do not contain detailed information like the one on the web.

**References**

[1] Indonesia Investments, Industri Pariwisata Indonesia, 16 December 2016. [Online]. Available: https://www.indonesia-investments.com/id/bisnis/industri-sektor/pariwisata/item6051?.

[2] Fominiene V B 2016 The Importance of Tourism Websites to Tourism Services And Assurance Of Their Competitiveness, *Montenegrin J. of Economics* **12** pp 41-51.

[3] Park Y A and Gretzel U 2007 Success Factors for Destination Marketing Web Sites: A Qualitative Meta-Analysis, *J. of Travel Research* **46** pp 46-63.

[4] Chen Y and Lin Z 2015 Fashionability vis-à-vis rationality: investigating factors driving users' e-tourism website stickiness *Curent Issues in Tourism*.

[5] Lian T, Yu C and Wang Z 2016 The evaluation study on tourism websites: from the perspective of triangular intuitionistic fuzzy multiple attribute group decision making, *J. of Applied Statistics* pp 1-13.

[6] Alhadreti O and Mayhew P 2017 To Intervene or Not to Intervene: An Investigation of Three Think-Aloud Protocols in Usability Testing *J. of Usability Studies* **12**.

[7] Sahin A, Zehir C and Kitapçı H 2011 The Effects of Brand Experiences, Trust and Satisfaction on Building Brand Loyalty; An Empirical Research On Global Brands *Procedia - Social and Behavioral Sciences* **24** pp 1288-1301.

[8] Mandalios J 2013 RADAR: An approach for helping students evaluate Internet sources *J. of Information Science* **39**.

[9] Ma L and Tian J 2007 Web error classification and analysis for reliability improvement *J. of Systems and Software* **80** pp 795-804.

[10] PWC 2011 *Millenials at work: Reshaping the workplace*.

[11] Venkatesh V, Chan F K Y and Thong J Y L 2012 Designing e-government services: Key service attributes and citizens' preference structures *J. of Operations Management* **30** pp 116-133.

# Automated color classification of urine dipstick image in urine examination

**R F Rahmat[1], Royananda[1], M A Muchtar[1], R Taqiuddin[1], S Adnan[2], R Anugrahwaty[2] and R Budiarto[3]**

[1]Department of Information Technology, Faculty of Computer Science and Information Technology, Universitas Sumatera Utara, Medan, Indonesia
[2]Faculty of Engineering, Politeknik Negeri Medan, Medan, Indonesia
[3]Department of Information System, College of Computer Science and Information Technology, Albaha University, Saudi Arabia.

Email: romi.fadillah@usu.ac.id

**Abstract**. Urine examination using urine dipstick has long been used to determine the health status of a person. The economical and convenient use of urine dipstick is one of the reasons urine dipstick is still used to check people health status. The real-life implementation of urine dipstick is done manually, in general, that is by comparing it with the reference color visually. This resulted perception differences in the color reading of the examination results. In this research, authors used a scanner to obtain the urine dipstick color image. The use of scanner can be one of the solutions in reading the result of urine dipstick because the light produced is consistent. A method is required to overcome the problems of urine dipstick color matching and the test reference color that have been conducted manually. The method proposed by authors is Euclidean Distance, Otsu along with RGB color feature extraction method to match the colors on the urine dipstick with the standard reference color of urine examination. The result shows that the proposed approach was able to classify the colors on a urine dipstick with an accuracy of 95.45%. The accuracy of color classification on urine dipstick against the standard reference color is influenced by the level of scanner resolution used, the higher the scanner resolution level, the higher the accuracy.

## 1. Introduction

Health is one of the essential essences of human life. Performing an early examination of the symptoms of an illness that appears is recommended to receive proper medical treatment from the medical staff, especially. The result of initial examination of the symptoms of an illness considerably affects the subsequent doctor action against a patient. Therefore, precise and accurate of an early examination result is needed in order to obtain the right result of early diagnosis or screening test.

The symptoms of an illness can be detected through various examinations performed on the body or from the results of human excretion such as examination through urine called urinalysis.

Urinalysis is an integral part of the early evaluation of kidney disease and urinary tract infections [1]. Urine examination using urine dipstick is performed to determine some chemical substances contained in the urine like proteins, glucose, ketones, hemoglobin, bilirubin, urobilinogen, acetone, nitrites, and leucocytes as well as pH testing and its density [2].

Urine examination using dipstick urine can be accomplished by matching the color change on the reagent paper that has been dipped into the urine sample with the standard color chart of urine examination. This is done to determine the level of elements content contained in the urine [3]. Color matching between reagent papers and standard color chart of urine examination is still done manually using the eyes, causing differences in color perception that may affect urine examination results using a urine dipstick.

In this research, Euclidean Distance is proposed to resolve differences in color perception that occurs in urine examination using a dipstick urine, i.e., at the time of performing color matching between reagent paper with a standard color chart of urine examination.

## 2. Problem identification

Content test of chemical elements in the urine examination using a urine dipstick has been used by doctors and health practitioners as a screening test (early diagnosis) to determine the potential of the disease suffered by a person before further testing. The result of urine examination using urine dipstick is still read manually using naked eyes by comparing reagent paper with a standard color chart of urine examination. This resulted perception differences in the color reading of the examination results. Therefore, an approach to classifying the color on the urine dipstick image is required so that the results of urine examination become more accurate.

## 3. Previous research

Several other types of research are talking about digital image processing on urine dipstick. For example, research conducted by Wijaya (2013) which discussed the color blend of Sa*b* elements on urine dipstick analysis of smartphone camera images using Backpropagation neural network. The color blend of Sa*b* elements is a combination of the color space of HSV and La* b*. Backpropagation neural network was used to obtain the most optimal value for each unit of the Sa * b * value. From the test results conducted on the color blend of Sa * b * elements on urine dipstick analysis of the smartphone camera images using Backpropagation neural network has proven to produce a better level of color matching accuracy compared to the color space of La*b* optimized and HSV [4].

Next research was conducted by Shanty (2014) using color interpretation method of urine dipstick test results in quantitative and qualitative methods. Stepwise linear interpolation with color distance trigonometric approach has proposed a method to determine the proportional position of test data. The proposed method can be used to assess the substance of test data quantitatively. Qualitative assessment of the test data is performed by initializing threshold with a round to nearest and round to down from a proportional position of specified test data. The research used the images of the smartphone camera [5]. Another study was conducted by Selvarasu (2010) using Euclidean Distance to calculate the proximity of RGB colors on the thermal image of the skin temperature or thermographic [6].

## 4. Methodology

The proposed method to conduct the color classification of urine dipstick images in this research consists of several steps. The steps are as follows: image acquisition of urine dipstick using scanner; image cropping; RGB feature extraction for color images; Otsu threshold for textured images; and Euclidean Distance calculation. After all the steps are performed, will obtain the color classification result of urine dipstick image. The general architecture of this system can be seen in Fig 1.

### *4.1. Image Acquisition*

Image acquisition is the initial step to obtain digital images. Image acquisition is performed to achieve the desired image so that can be prepared in the next step. In this step, digital image acquisition will be performed on urine dipstick that has been dipped into a urine sample using a scanner with a resolution of 4800 x 1200 dpi and stored in a joint photographic group (.jpg) format.

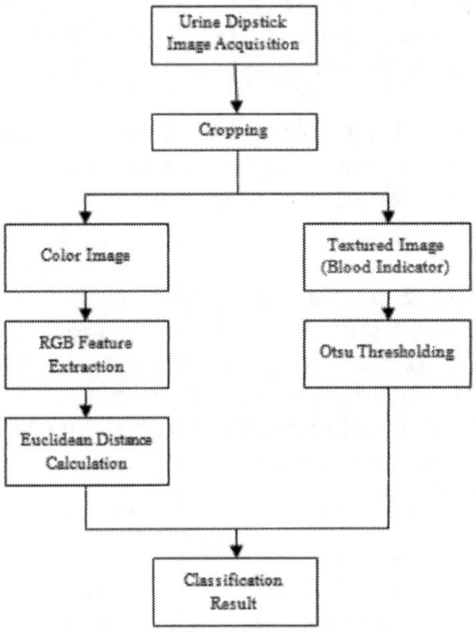


**Figure 1.** General Architecture.

*4.2. Dataset*

*4.2.1. Urine dipstick image.* Dataset used in this research is urine dipstick images that have been dipped into 120 urine sample from the General Hospital Haji Medan. The scanning result of urine dipstick images used in this research can be seen in Fig 2 and Fig 3.

**Figure 2.** Urine dipstick before dipped into the urine sample.

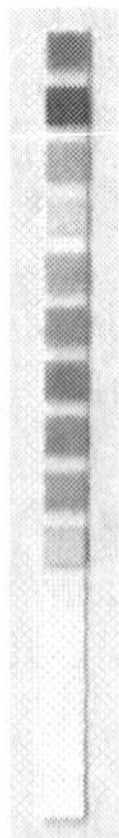

**Figure 3.** Urine dipstick after dipped into the urine sample.

*4.2.2. Color chart of urine dipstick.* Color chart of urine dipstick is a chart that contains information in the form of urine examination indicator, rine examination time and the value of urine examination indicator that is interpreted in colors to be used as a reference for the urine examination using urine dipstick [7]. Color chart of urine dipstick can be seen in Fig 4.

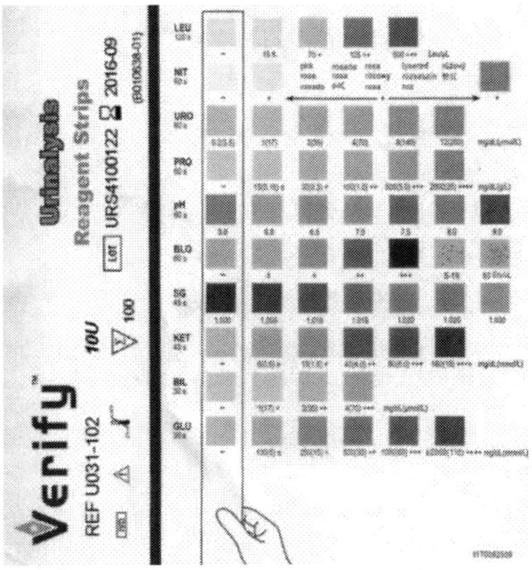


**Figure 4.** Color chart of urine dipstick.

*4.3. Image pre-processing*

Image pre-processing is a process performed to prepare the images before to be processed in further image processing. In this research, the image pre-processing performed is image cutting or cropping. Image pre-processing is a process performed to prepare the images before to be processed in further image processing. In this research, the image pre-processing performed is image cutting or cropping.

*4.3.1. Cropping.* Urine dipstick image required for the next step is the part attached to the celluloid pad. For that the bottom of the urine dipstick is discarded since it is not needed for the following process. Cropping result of urine dipstick image can be seen in Fig 5.

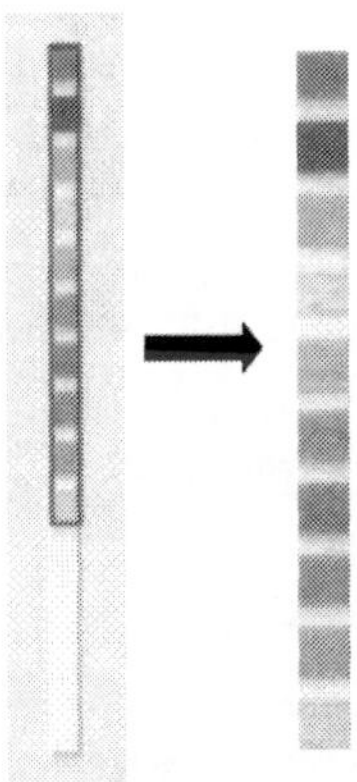

**Figure 5.** Cropping result of urine dipstick image.

*4.4. RGB color feature extraction*

RGB color feature extraction step is performed to obtain the characteristic of the color of inputted urine dipstick images so that the result of the feature extraction can be compared with the reference data. Steps performed in RGB color feature extraction for this research are as follows:

1) Determine each of the three values of red, green and blue that most appear in a single urban dipstick indicator.

IOP Conf. Series: Journal of Physics: Conf. Series **978** (2018) 012008 doi:10.1088/1742-6596/978/1/012008

2) After the three most common RGB values are specified, then the mean value of each RGB value will be determined [9]. To determine the mean of the RGB value is used equation 1 to equation 3 as follows:

$$R = \frac{R_1+R_2+R_3}{3} \tag{1}$$

$$G = \frac{G_1+G_2+G_3}{3} \tag{2}$$

$$B = \frac{B_1+B_2+B_3}{3} \tag{3}$$

3) The result of the calculation using equation 1 to equation 3 obtained a new RGB value which is the mean of the RGB value of the inputted image. This RGB value will be set as the RGB value to be equal to the reference image.

### 4.5. *Otsu Threshold*

Usually, universal thresholding will convert a grayscale image into a black and white image [10]. However, we used Otsu thresholding in this case to distinguished blood spots. Otsu method in this research was used for blood indicator with values of 5-10 and 50 Ery / µL. Because in the blood indicator, there are dots that indicate the presence of blood spots on the urine sample examined. Otsu method used in this research will separate the foreground and background of the blood indicator image. Where the dots in the image of blood indicator are defined as the foreground and are white, while the background of the image is black. Otsu method is applied to the blood indicator image with values of 5-10, and 50 Ery / µL can be seen in Figure 6 and Figure 7.

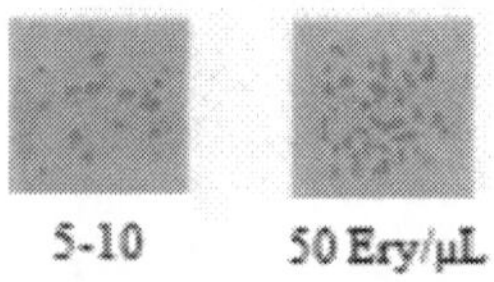


**Figure 6.** indicator with values of 5-10 and 50 Ery / µL.

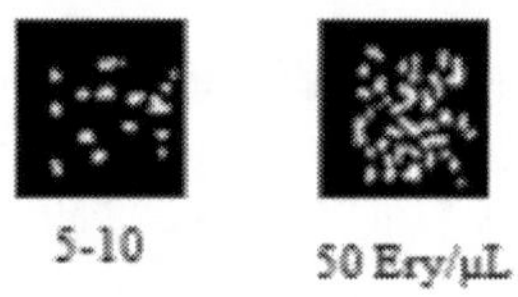


**Figure 7.** The result of Otsu method on blood indicator image with values of 5-10.

Calculation of foreground and background areas can be calculated using equation 4 below.

$$L = \frac{Foreground\ area}{background\ area}\ x\ 100\% \tag{4}$$

### 4.6. *Classification using Euclidean Distance*

After feature extraction on urine dipstick urine image performed, the next step is classification. Classification using Euclidean Distance is performed by matching the color distance using the calculation of red, green and blue (RGB) color-coded differences in the corresponding pixel. Then the distance difference is calculated using Euclidean distance equation as in the following equation 5 [8].

$$d_{xy} = \sqrt{(x_1-y_1)^2+(x_2-y_2)^2+(x_3-y_3)^2+\cdots+(x_n-y_n)^2} \tag{5}$$

After calculating Euclidean Distance using equation 5 is performed, then will obtain the smallest RGB value difference. The RGB value with the smallest difference is the value closest to the RGB value of the reference color.

## 5. Result and analysis

The testing is performed to determine the ability of the system in color classification of urine dipstick image. System testing was done with 120 urine samples. Urine samples used in this research were obtained from RSU Haji Medan.

### *5.1. Urine dipstick image testing*

In the system testing process, the thing that affects the classification result of urine dipstick image is the resolution of the scanner tool used when scanning the image. Testing was performed by using two different scanner resolutions which are 4800 x 1200 dpi and 600 x 600 dpi.

The result of the accuracy of each urine examination indicator using 4800x1200 dpi scanner can be seen in Table 1.

**Table 1**. Image testing accuracy using 4800 X 1200 dpi scanner

| No. | Indicator | Testing result | | Accuracy rate (%) |
|---|---|---|---|---|
| | | Succeed | Failed | |
| 1 | LEU | 119 | 1 | 99 |
| 2 | NIT | 120 | 0 | 100 |
| 3 | URO | 118 | 2 | 98 |
| 4 | PRO | 101 | 19 | 84 |
| 5 | pH | 119 | 1 | 99 |
| 6 | BLO | 119 | 1 | 99 |
| 7 | SG | 120 | 0 | 100 |
| 8 | KET | 117 | 3 | 97.5 |
| 9 | BIL | 113 | 7 | 94 |
| 10 | GLU | 119 | 1 | 99 |

From Table 1 can be calculated the average accuracy rate in classifying the data resulted from urine examination using the proposed method using equation 6 as follows:

$$\bar{A} = \frac{99+100+98+84+99+99+100+97.5+94+99}{10} = 95.45\% \quad (6)$$

### *5.2. Image testing with lower scanner resolution*

The lower scanner resolution used to test the image in this research is 600 x 600 dpi. The test was performed using ten urine samples taken at a lower resolution. Based on the image testing, the accuracy rate of each urine examination indicator can be seen in Table 2.

**Table 2**. Image testing accuracy using 600 X 600 dpi scanner

| No. | Indicator | Testing result | | Accuracy rate (%) |
|---|---|---|---|---|
| | | Succeed | Failed | |
| 1 | LEU | 9 | 1 | 90 |
| 2 | NIT | 9 | 1 | 90 |
| 3 | URO | 9 | 1 | 90 |
| 4 | PRO | 6 | 4 | 60 |
| 5 | pH | 10 | 0 | 100 |
| 6 | BLO | 9 | 1 | 90 |
| 7 | SG | 8 | 2 | 80 |

| | | | | |
|---|---|---|---|---|
| 8 | KET | 8 | 2 | 80 |
| 9 | BIL | 6 | 4 | 60 |
| 10 | GLU | 9 | 1 | 90 |

### *5.3. Textured image (blood indicator) testing*

We used urine samples with artificial blood spots for image testing for blood indicator. This resulted due to the unavailability of urine samples that has blood spots to be tested. The test using samples with artificial blood spotting is performed only to determine the ability of the proposed Otsu method to classify the textured image that existed in blood indicator. Testing result of textured images for blood indicator can be seen in Table 3 below.

**Table 3**. Testing result of textured image (blood indicator) testing

| No | Image name | Blood image | Otsu image | Result (Ery/μL) | Notes |
|---|---|---|---|---|---|
| 1 | DUT1 |  |  | 5-10 | Succeed |
| 2 | DUT2 | 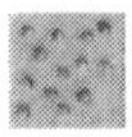 |  | 5-10 | Succeed |
| 3 | DUT3 |  |  | 50 | Succeed |

The test result performed in Table 3 shows that the Otsu method proposed in this research was able to classify the textured image on the blood indicator into 5-10 and 50 Ery/μL values.

## 6. Conclusion and future research

Based on the system testing of color classification on urine dipstick images, obtained several conclusions which are the proposed method in this research is able to classify the color of urine dipstick images to determine the concentration level of chemical elements in urine samples tested. Resolution on the scanner used when taking urine dipstick images can affect the result of image classification. The accuracy rate using the images taken by a scanner with 4800 x 1200 is higher at 95.45%, compared to images taken by a scanner with lower resolution that is at 83%.

Further research can perform the process of image cropping on urine dipstick image automatically so that the urine dipstick image can be directly processed to the next step. In order to improve the analysis result of the color similarity on the digital image can be applied methods of measuring distance between other colors.

## References

[1] Singh K A and Loscalzo J 2014 *The Brigham Intensive Review of Internal Medicine Second Edition* (UK: Oxford University Press)
[2] Strasinger S K and Marjorie S D L 2008 *Urinalysis and Body Fluids Fifth Edition* (Florida: F A Davis Company)
[3] Schumann G B and Friedman S K 2003 *Wet Urinalysis. Amer Society of Clinical* (Hong Kong)
[4] Wijaya T A 2013 *The Color Blend of Sa*b* Element on Urine Dipstick Analysis of Smartphone Camera Images Using Backpropagation Neural Network* (Surabaya: Institut Teknologi Sepuluh November)

[5] Shanty R N T 2014 *Color Interpretation Model of Urine Dipstick Test Results Qualitatively and Quantitatively for Mobile Devices* (Surabaya: Institut Teknologi Sepuluh November)

[6] Selvarasu N, Nachiappan A, and Nandhitha N M 2010 Euclidean Distance Based Color Image Segmentation of Abnormality Detection from Pseudo Color Thermographs vol 4 *Int. J. of Computer Theory and Engineering 2* **4** p 514

[7] Estridge B H, Reynolds A and Walters N J 2000 *Basic Medical Laboratory Techniques Fourth Edition* (US: Delmar)

[8] Pradana I H 2015 *Classification of Fingerprint Images Based on Six Types of Pattern Using Euclidean Distance Method* (Semarang: Universitas Dian Nuswantoro)

[9] Rahmat R F, Chairunnisa T, Gunawan D, and Sitompul OS 2016 Skin Color Segmentation Using Multi-Color Space Threshold *3rd Int. Conf. on Computer and Information Sciences, ICCOINS 2016 – Proceedings* pp 391–396.

[10] Rahmat R F, Pulungan A F, Faza S and Budiarto R 2017 Image classification of ribbed smoked sheet using Learning Vector Quantization *J. of Physics: Conference Series* **801** p 12050

IOP Conf. Series: Journal of Physics: Conf. Series **978** (2018) 012009 doi:10.1088/1742-6596/978/1/012009

# The development of indonesian traditional bekel game in android platform

**R F Rahmat[1], O R Fahrani[1], S Purnamawati[1] and M F Pasha[2]**

[1]Department of Information Technology, Universitas Sumatera Utara, Medan, Indonesia
[2]School of Information Technology, Monash University Malaysia Campus, Malaysia

Email: romi.fadillah@usu.ac.id

**Abstract**. Bekel is one of traditional Indonesian game that is rarely played nowadays. Bekel is a game to test dexterity level using a bekel ball and 6 to10 seeds. The game is played by throwing the ball up in the air, spreading the seeds randomly on the floor then picking the seeds up until the ground is clear. This game application is an adaptation of Bekel game focusing on the movements of the ball and the randomization of the seed positions. This game application has three levels of difficulty based on the basic rules of the actual Bekel game. The focus of the study is the free fall method of the ball and the random function of the seeds in the Android environment. The result show the Bekel application has sensitivity level of 71% for the ball movements and the probability rate of the random event occurrence is at 23%.

## 1. Introduction

Bekel is one of the traditional Indonesian games to test the level of dexterity. The name of Bekel is originated from the Dutch language, *bikkelen*, means fighting spirit. In some area, Bekel calls in different names such as bekelan, beklen, and bekles. The game is usually played by children using a Bekel ball and 6-10 seeds.

The way to play Bekel is, the ball is thrown up, while the ball is still up in the air, the player has to pick the seeds up then catch the ball again. Firstly, the seed is taken one at a time until the floor is clear. Then it will continue by taking two seeds at a time, then continue with three seeds and so on until the seeds are grasped at once in a single try. Afterwards, the player has to flip the seeds to the other side; then the seeds are taken one at the time, two at a time and continued until the seeds are taken at once. After finishing this step, continue with flipping over the seeds to the first side then pick the seeds up as before [1].

Traditional games have a simple concept to teach the children. However, nowadays, classical games are rarely played because of an online game that can be played individually by anyone and anywhere using a device such as a mobile or PC. The study certainly has an impact on the traditional games that began to lose its existence as a medium of entertainment in the past.

A previous research of the game bekel in 2014 was conducted by Alamsyah et al. they built and Android-based Bekel game with five steps and system implementation using Adobe Flash. The results obtained including the accelerometer for the ball movement that can move up and down also touch response to seed extraction that works [2].

Other research related to the traditional game has also been conducted by Juardi et al. on 2013. The research is in the development of Karapan Sapi game. Karapan Sapi game is based on an actual sport from Madura, Indonesia. The resulted study is an educational game of "Karapan Sapi Race" with an interesting, entertaining and has "easy to play" gameplay [3].

In 2013, Farizi et al. also researched Angklung, one traditional musical culture from Java. He builds a game simulator on an Android-based smartphone with Game Design and Technical Design phases. These phases then being implemented and developed using AndEngine in Java programming language. According to the unit and integration testing result, it can be concluded that the module unit of the game application has fulfilled the functional requirements of the development. The effect of performance testing shows an optimal performance when the game runs on a smartphone with a minimum processor of 600 MHz [4].

Numerous interactive game applications have been developed for Android. Android is a suite of software designed for mobile devices. [5], our previous interactive application also built in Android Platform, such as Augmented Reality for potential asset [12]. In this research, we develop a game adapted from Bekel game with some adjustments in order to be able to run on Android-based mobile devices as a means to restore the existence of one of the traditional Indonesian games via mobile devices. The complexity of the ball movements and the seed randomization is varied in the implementation of Bekel game application. Therefore, a method is required for the implementation of ball movements and the randomization of the seed positions so that the bekel game can be built.

## 2. Methodology

### 2.1. General architecture

The development of Bekel adaptation game consists of 4 main principles. Those functions are User Interface for the interface on the device screen, Object Data Model for ball movement and seeds randomization in the game, Game Logic for the game flow, and Database for the storage of the game score. The general architecture of the game can be seen in Figure 1.

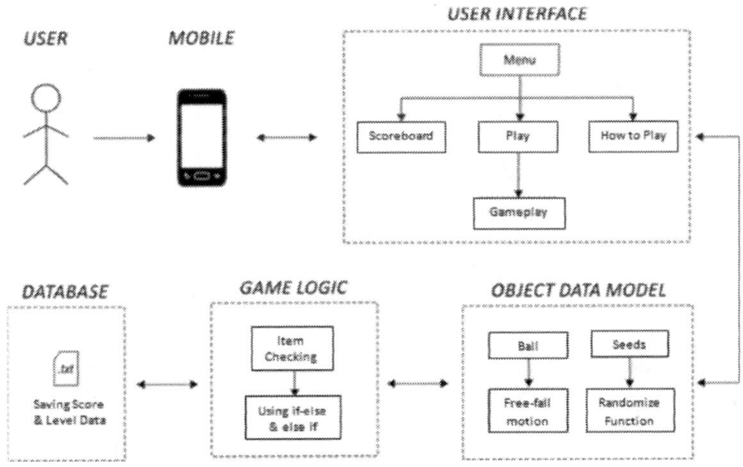


**Figure 1.** General Architecture.

### 2.2. Object data model analysis of ball and seeds

Free fall motion can be implemented as movement method on ball object. Initially, the ball will be given mileage to reach a certain height. When the ball reaches a predetermined height limit, then the free fall method plays a role in adjusting the ball path until the ball reaches the initial position.

The random function will be applied as the seed position randomization so that the seeds appear in a random position on each game. The location of the seeds is deliberately made random to avoid the player memorize the position of seeds scattered on the game board.

*2.2.1. Logic description of ball object motion using free fall motion.* Free fall motion is the motion of falling objects in the vertical direction of a particular height without the initial velocity (v0 = 0), so the movement of the object is only affected by the earth gravity [6].

Free fall motion (FFM) is included in accelerated uniformly accelerated motion (UCM). The acceleration experienced by this object is due to the gravitational pull of the earth. Since the condition of free fall motion is its initial velocity is zero, then UCM equation applies as follows:

$$v = v_0 + at \tag{1}$$

$$s = v_0 t + ½ at^2 \tag{2}$$

$$v^2 = 2as \tag{3}$$

Since the initial velocity of free fall motion is $(v_0) = 0$ so that in formula $v_0$ is omitted, $s$ in the above equation replaced by $h$ which states the height and acceleration experienced by the object is the acceleration of earth gravity, then applies the formula [7]:

$$v = gt \tag{4}$$

$$h = ½ gt^2 \tag{5}$$

$$v^2 = 2gh \tag{6}$$

Where:
$v$ = Object final speed(m/s).
$t$ = Time(s).
$h$ = Object height(m).
$g$ = Gravitational acceleration (m/s$^2$).

From the equation of falling time, it can be seen that the free-falling object time is only affected by two factors, namely height ($h$) and earth gravitational acceleration ($g$). Other quantities do not jeopardize its fall time, which means that despite the different weights, two objects that fall from the same distance and in the same place will fall at the same time. But if the objects do not fall together then it is due to the air friction [8]. Pseudocode of the free fall motion used in this application can be seen in Figure 2.

*2.2.2. The description of a random function.*
The random function is used to randomize the seed position during the game. Random function is formed from a random number. A random number is a number taken from a set of numbers, where each element of the set of numbers has an equal chance of being drawn [9]. The process of picking random numbers itself is a random event, meaning the events where processes and outcomes cannot be predicted.

Java provides two approaches in generating random numbers through the java.lang.Math and java.util.Random classes. Java.util.Random class can be used to get a better result of the random number. This class provides a more practical approach and supports other types except for double, including the Gaussian distribution [10]. Random Class contained in the java.util package is a pseudo-random number generator using an algorithm derived from The Art of Computer Programming by Donald Knuth [11].

In Bekel game, the seed positions are randomized as a variation of the game. Seed randomization used a random function to ensure every appeared seed is not monotonous with the same position on each level. A random function that is used also come with the provisions of the randomized seed that will appear in the coordinates which are restricted in length and width. It is to prevent the appearance of randomized seeds beyond the 450 x 350 pixel board's resolution. Pseudocode of random function for seed position can be seen at in Figure 3.

```
begin
Set x as object.
Declare ballPos in initial-position.
Declare highball as height.
Declare gravityBall as gravitation.
if(x = clicked){
    if(x = drag) {
        if(ballReturn == false) {
            if(highBall <= 0)
                ballReturn = true;
            else {
                ballPos -= highBall;
                highBall--;
                }
        }
        else if(ballReturn == true) {
            if(ballPos >= 0) {
                ballReturn = false;
                ballClicked = false;
                ballPos = 0;
                highBall = 25;
                gravityBall = 1;
            }
            else {
                ballPos += gravityBall;
                gravityBall++;
            }
        }
        showBall(g, world);
    }
    if(x = touch) {
        if(ballReturn == false) {
            if(highBall <= 15)
                ballReturn = true;
            else {
                ballPos -= highBall;
                highBall--;
            }
        }
        else if(ballReturn == true) {
            if(ballPos >= 0) {
                ballReturn = false;
                ballClicked = false;
                ballPos = 0;
                highBall = 25;
                gravityBall = 1;
            }
            else {
                ballPos += gravityBall;
                gravityBall++;
            }
        }
        showBall(g, world);
    }
}
end
```

**Figure 2.** Pseudo-code of the free fall motion.

```
begin
Determine the randomization limit coordinate.
Declare batasX as coordinate-limitX.
Declare batasY as coordinate-limitY.
while(biji.israndom()) do
        batasX = biji_random(maxX-minX)+ minX
        batasY = biji_random(maxY-minY)+ minY
end
```

**Figure 3.** Pseudocode of random function for seed position.

### *2.3. Game logic*

In game logic, item-checking function is utilized to determine whether the players play the game right or not. Item checking is a function used as a trigger between interconnected items, in this case, the ball movements, the number of seeds on the board, the number of seeds taken and the number of lives available. This function is performed if the ball moves.

Item Checking function is utilizing if-else and else-if structure to execute a statement if one of the expressions gives true value and also includes another alternative if the evaluated expression gives a false value. In another word, item checking is a game rule that determines the player to win or lose (game over).

The process of item checking is also affected by the difficulty level of existing games. The difficulty level of the game is designed based on the number of seeds used. The more seeds to collect in the game, the more difficult the players are to win. In Bekel game application, there are three types of the difficulty

IOP Conf. Series: Journal of Physics: Conf. Series **978** (2018) 012009 doi:10.1088/1742-6596/978/1/012009

level of the game. Explanation of the number of available seeds and the level of taking on each level of the game can be seen in Table 1.

**Table 1.** Difficulty level

| Difficulty levels | Number of seeds | Number of taking |
|---|---|---|
| Easy | 6 | 6 |
| Normal | 8 | 8 |
| Hard | 10 | 10 |

*2.4. Database*

The database is containing the score's value and level data. The data is stored and will be displayed on the scoreboard interface menu. For storage, scores and levels are only saved to the database if the player's total score is higher than the previous scores. Score and level storage are done when the game ends. As for the reading, when initialization, the application will read the stored file of players score and level. If the file is found, then the contents of the file will be saved to the storage variable to be displayed. If the score file is not found, then the storage variable will be filled with the default value. The design of scores table can be seen in Figure 4.

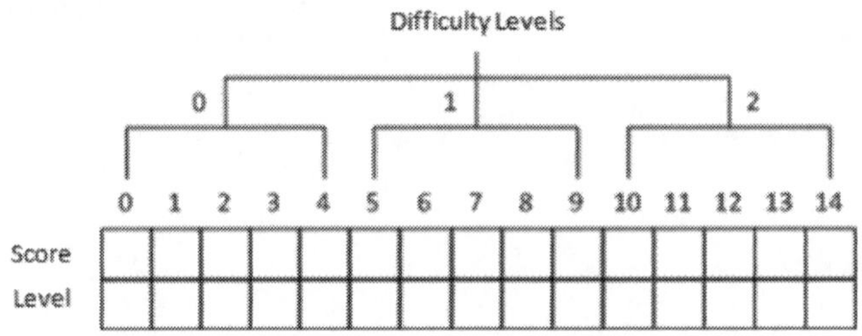


**Figure 4.** Game scoreboard storage block.

## 3. Result and analysis

When the player starts the application, the audio will automatically turn on, and the music will play. Players can choose to mute the sound by pressing the button on the top right. To start the game, the 'play' button has to be clicked. Afterwards, will appear select grain to be used and difficulty levels displays. Main menu and item selection menu displays can be seen in Figure 5 and Figure 6 respectively,

**Figure 5.** Main menu display.

**Figure 6.** Item selection menu display.

When the player selects the difficulty level on the Item selection menu displays, then the system will show the Game Arena display. The Game Arena display can be seen in Figure 7.

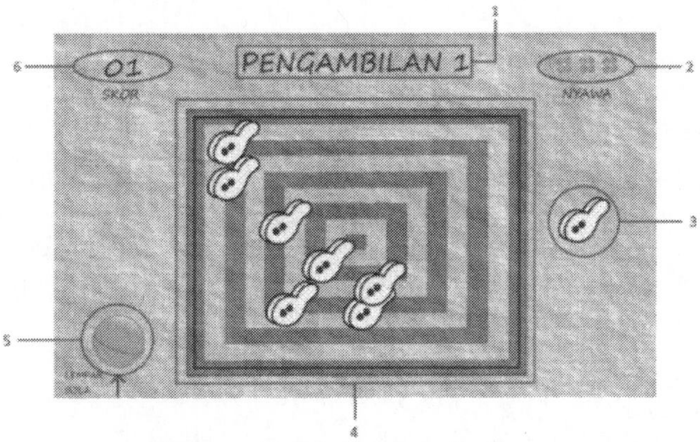


**Figure 7.** Game arena display.

The numbering description in Figure 7 is (1) the round of taking (2) the player's life (3) the taken seed (4) the game board (5) drag the ball (6) the player's score. Some testing has been conducted to find out whether the game application can run well or not, such as methods testing, test respondents, and test cases.

### *3.1. Testing*

*3.1.1. Ball movements.* The ball movements testing is conducted to assess how much the percentage of the ball object movements succeed when it is moved. The parameter of the object movements and its sensitivity level of the ball moved to reach its maximum height limit by the dragging action. If the ball is moved only in short range then considered as failed, otherwise it is succeeded. The test is performed with the ball being driven 100 times by the tester.

The test results of sensitivity level on the ball object are at 71% where 71 times the ball movements successfully done and 29 times the movements of the ball were failed. The existence of the ball movements that failed to be performed is classified as bugs which is caused by the weak system of the coordinate range settings on the ball movements that has been designed. The x,y coordinate range setting plays a significant role in the implementation of the ball movement method. The improvement in this area is a must for the further research.

#### *3.1.2. Seeds randomization function*

In the random process where the length and width have been limited, sometimes random combinations can also produce the same coordinate point so that if the seeds appear on the board, there is a circumstance wherein the naked eye is only 1 visible seed but when the seed is taken it turns out to be 2 or more. This is called an 'unpredictable' random event.

In "unpredictable" random circumstances events require a test to calculate how likely the occurrence of random events in a game application is created. It aims to find out how significant the impact of random event conditions on the performance of seed randomization function.

The possibilities of ‘unpredictable’ random event was tested with 30 times repetition of seed randomization function on the board game. The results obtained are 23 times the normal seed condition and seven times the state of random event occurs. The result shows that the probability rate of the random event occurrences is 23%. Although the rate of the probability of random event occurrence is considered as a small category still the random event is one of the shortcomings of this game application which in the future must be improved.

IOP Conf. Series: Journal of Physics: Conf. Series **978** (2018) 012009 doi:10.1088/1742-6596/978/1/012009

*3.2. User interface testing.*
The testing of Bekel game application user interface was conducted on 20 respondents by filling out the questionnaire. The trial of respondents undertaken to determine whether the game is made feasible to play or not. The result of this application user interface assessment can be seen in Table 2.

**Table 2**. The result of application user interface evaluation.

| | | Evaluation scores | | | | |
|---|---|---|---|---|---|---|
| No. | Evaluation targets | Excellent | Good | Fair | Bad | Very bad |
| 1 | Application design in general | 4 | 15 | 1 | - | - |
| 2 | The displays of ball, seeds, and gameboard | 2 | 18 | - | - | - |
| 3 | Item colors | 3 | 14 | 3 | - | - |
| 4 | Game animation display | 2 | 15 | 3 | - | - |
| 5 | Menu button design | 1 | 16 | 3 | - | - |
| 6 | Audio | 1 | 16 | 3 | - | - |

Then obtained the overall value of the user interface aspect as follows:

$$S_{total} = \frac{4.15+4.1+4+3.7+3.9+3.9}{6} = 3.93 \quad (7)$$

From the overall value equation, obtained value of 3.93 for the user interface aspect. It can be concluded that the user assessment result of the Bekel game interface aspect is Fair.

*3.3. System testing*
The result of this game application difficulty level can be seen in Table 3.

**Table 3**. Result of system testing

| | | Game difficulty levels | | | | | |
|---|---|---|---|---|---|---|---|
| | | Easy | | Normal | | Hard | |
| No | Respondents | S | L | S | L | S | L |
| 1 | R1 | x | x | 48 | 6 | 13 | 1 |
| 2 | R2 | 12 | 3 | x | x | x | x |
| 3 | R3 | 23 | 4 | 32 | 4 | 6 | 1 |
| 4 | R4 | x | x | x | x | 16 | 2 |
| 5 | R5 | 33 | 6 | 25 | 3 | 57 | 5 |
| 6 | R6 | 23 | 3 | x | x | x | x |
| 7 | R7 | 29 | 5 | 36 | 5 | 64 | 7 |
| 8 | R8 | 19 | 3 | x | x | x | x |
| 9 | R9 | 25 | 3 | x | x | x | x |
| 10 | R10 | 14 | 3 | 11 | 1 | 13 | 1 |
| 11 | R11 | 24 | 5 | x | x | x | x |
| 12 | R12 | 30 | 6 | 37 | 5 | 36 | 4 |
| 13 | R13 | 17 | 3 | x | x | 9 | 1 |
| 14 | R14 | x | x | x | x | 23 | 3 |
| 15 | R15 | 18 | 4 | 8 | 1 | 24 | 2 |

| | | | | | | | |
|---|---|---|---|---|---|---|---|
| 16 | R16 | x | x | 16 | 3 | x | x |
| 17 | R17 | 36 | 6 | 43 | 6 | x | x |

In table 3, S is the total player score while P is the level of seeds taking achieved by the players. Based on the assessment result, it can be concluded that the average succeeds rate of the game difficulty level of 20 respondents are as follows:

- Easy level average achievement rate is 4.1 out of 5 in 6 levels.
- Normal level average succeeds rate is 3.8 out of 5 in 8 levels.
- Hard level average succeeds rate is 2.8 out of 5 in 10 levels.

From the data above, it can be concluded that the success rate of Bekel game application has a has a medium-high level of difficulties.

## 4. Conclusion and future research

In accordance with Bekel game application system analysis and testing can be drawn some findings in this study as follows:

- The ball-catching movement after the seeds are taken has not been able to be implemented in this Bekel game applications.
- The testing conducted to the ball movement is obtained the sensitivity level of 71% of a total 100 times the ball being moved. While the test result on the percentage of the possibility of random events in the random function is at 23% and considered as a low category and often occurs in random functions because every element of the set of random numbers has an equal opportunity to be drawn so that the results cannot be predicted.
- The testing results of game difficulties of 20 respondents are as follows. Easy level average succeeds rate 4.1 out of 5 in 6 levels; normal level average succeeds rate is 3.8 out of 5 in 8 levels, and hard level average succeed rate is 2.8 out of 5 in 10 levels.
- In general, Bekel traditional game has many variations on the levels. The authors only adapt the basic level of Bekel game called the Mi round into the application.

Based on the conclusions drawn from this research, the future study is as follows:

- In future research, the ball-catching movement after the seeds taken can be implemented into the application by expanding the touch area of moving object.
- The ball movements can be developed using swipe gesture by implementing speed function so that the ball speed can be adjusted by players desire.
- The random function used for the seeds can be upgraded by giving a distance limit to every item to be randomized so that the seeds will not be overlapped at the same coordinates.

## References

[1] Husna A M 2009 *100 Indonesian Traditional Games 1st Edition* (Yogyakarta: ANDI)
[2] Alamsyah R Y R and Gunandi A H 2014 *Android Based Bekel Game Application Development* (Bandung: STMIK LPKIA)
[3] Juardi A, Tommy and Salim W 2013 *Racing Game Application "Karapan Sapi Racing" On Android-Based Smartphone Using LIBGDX Engine* (Jakarta: Bina Nusantara University)
[4] Farizi HR, Aryadita H & Adam JEM 2013 Simulator Mobile Game Development of Angklung on Android-Based System *Repository of PTIIK UB Student Journal* **1** 5
[5] Zechner M 2011 *Beginning Android Games* (New York: Apress)
[6] Coletta and Vincent P 1995 *College Physics* (USA: McGraw-Hill)
[7] Sutrisno 1986 *Basic of Physics* (Bandung: Mekanika Bandung)

IOP Conf. Series: Journal of Physics: Conf. Series **978** (2018) 012009 doi:10.1088/1742-6596/978/1/012009

[8] Mansfield, Michael and O'Sullivan C 1998 *Understanding Physics Wiley*
[9] *LavaRnd, Terms & Condition: Pseudo-Random Number Generator* Retrieved from http://www.lavarnd.org/faq/prng.html
[10] Pardede and Jasman 2010 Implementation of Random Numbers Java to Creation Missing Value of the Table SENAPATI (Bali: National Seminar of Informatics Technique of Education)
[11] Naughton P 1996 The Java Handbook (USA: McGraw-Hill, Inc)
[12] Muchtar M A, Syahputra M F, Syahputra N, Ashrafia S and Rahmat R F 2017 Augmented Reality for Searching Potential Assets in Medan using GPS based Tracking *J. of Physics: Conference Series* **801** p 12010

# Usability evaluation of user interface of thesis title review system

**Y Tri, A Erna and U Gellysa**

Dumai High School Of Technology (Sekolah Tinggi Teknologi Dumai), Utama Karya Street, Bukit Batrem, Dumai

Email: triyuliati00@gmail.com

**Abstract.**
Presentation of programs with user interface that can be accessed online through the website of course greatly provide user benefits. User can easily access the program they need. There are usability values that serve as a benchmark for the success of a user accessible program, ie efficiency, effectiveness, and convenience. These usability values also determine the development of the program for the better use. Therefore, on the review title thesis program that will be implemented in STT Dumai was measured usability evaluation. It aims to see which sides are not yet perfect and need to be improved to improve the performance and utilization of the program. Usability evaluation was measured by using smartPLS software. Database used was the result of respondent questionnaires that include questions about the experience when they used program. The result of a review of thesis title program implemented in STT Dumai has an efficiency value of 22.615, the effectiveness of 20.612, and satisfaction of 33.177.

## 1. Introduction

Thesis is obligatory by a student to finish their education in college. The filling process the title of the thesiswhich was exists in STT Dumai is not relevant demandingto the repair and the speed of service process, because it has some weaknesses. Manual process requires a large amount of resources and the process will not going faster. Problems often faced by students in completing the thesis title review such as poor time management, ineffective utilization of time, low variation of the title of the thesis, the reading interest of students is very minimal to expand the insight and indiscipline is the cause of the delay of the thesis of the students. Thus, this research motivated to make thesis title review system. It also helps head of department and academicsection avoid the redundancy of the final assignment of the students.

Review thesis title system that now exist is in local intranet STT Dumai help students in the process of thesis proposal. The valuable thing in this process is to ease and smooth the process of thesis proposalin STT Dumai, in order to develop the software or application to have a better and can be useful optimallyneeds to be evaluated.Some previous research have made the proposal thesis sytem. Slamet, et al. [1]. Have made a proposal thesis system which didn't have ability to detect the similar title with previous thesis that had been submitted. Thus, this research purposed to develop the proposal thesis system with this ability. This research aim is analyse usability that affects attitude revenue using (acceptability) the system to application. The benefits of the application of parameters measured according to the satisfaction of users against the application .A tendency the attitude of rejection or acceptance against application through the activities of the survey is expected to provide input for the improvement of this application at a later date .The presentation of the application of a user interface on the overhead of a website is very conducive to provide facilities (usability) to the users [2]. This has resulted in current STT Dumai with efficient aplication, but less efficient interface design. Performance of a software can be measured by it usability, and one way to define quality of a software is by its "ease of use". Software that is not easy to use by its user can be considered of lower quality.

The word "usability" has become a catch-phrase for products that work better for their users, but it is difficult to pin down just what people mean by it [3]. Usability means :

- A result, software that is usable
- A process, also called user-centered design, for creating usable software
- A set of techniques, such as contextual observation and usability testing, used to achieve that result or
- A philosophy of designing to meet user needs

The definition of usability is the extent to which a product can be used by specified users to achieve specified goals with effectiveness, efficiency, and satisfaction in a specified context of use [4]. This definition can be expanded, and made more comprehensive, by including three characteristics which must be met for the users of a product :

- Effectiveness defined as how well users attain their goals by using system and completeness it can be obtained in complete task. user assistance quality of interface can have a big influence on effectiveness. Optional of presentation influences the effectiveness of an interface, which is clearly understandable to the user. Users are able to do an activity without problems by using a more informative interface. An user's language and appropriate to the task used in good interface terminology.
- Efficiency defined as resources issued to achieve accuracy and comprehensiveness of the purpose. View point of user is main point to define the interaction is not a single and granular. The Information of knowledge will be very effficient if it share little answer but it will describe users question.
- Satisfaction defined freedom from discomfort and positiveness on the use of products or size subjective how users feel about the use of system. That Impact of efficiency is design of navigation elements such as keyboard shortcuts, menus, links and other buttons. User just need less time and effort to use navigation and action choices if performance of well designed[5].

Usability is the extent towhich a computer system enables users, in a given context of use, to achieve specified goals effectively and efficiently while promoting feelings of satisfaction. Usability Evaluation (UE) consists of methodologies for measuring the usability aspects of a system's user interface (UI) and identifying specificproblems [6]. A wide range of usability evaluation techniques have been proposed, and a subset of these is currently in common use [7]. Some evaluation techniques, such as formal user testing, can only be applied after the interface design or prototype has been implemented. Others, such as heuristic evaluation, can be applied in the early stages of design. Each technique has its own requirements, and generally different techniques uncover different usability problems [8]. This paper reviews the aplication with of the review thesis system in STT Dumai. This article also presents a discussion which can provide future directions in STT Dumai development with usability context.

## 2. Research Methodology

There are several steps in doing this research as shown in Figure 1. This is a general explanation of the steps in the research methodology.

- Study literature term weighting method.
  At this stage a literature study is conducted by collecting existing literature on studies that have relevance to the detection of word equality. This is done because the information system submission thesis title is designed to be able to detect the similarity of the title of the proposed thesis with a title that has been there before. So the newly submitted title will not be similar to the existing title, thus the research conducted by the student will be more developed.
- Examining the existing conditions in STT Dumai by observing activities based on the specific conditions that exist in the process of review thesis title manually.

- Creating a thesis title system review based on the manual thesis title review process that is running.
- Conducting a thesis title system review with target students, head of study program, and BAAK.
- Evaluation of the usability of the results of applying the thesis title system review by testing the usability side. Usability evaluation measured, ie effectiveness, efficiency, and satisfaction [2].
- Analysis of usability evaluation results by smartPLS [9] is based on effectiveness, efficiency, and satisfaction values to measure the success of system implementation.

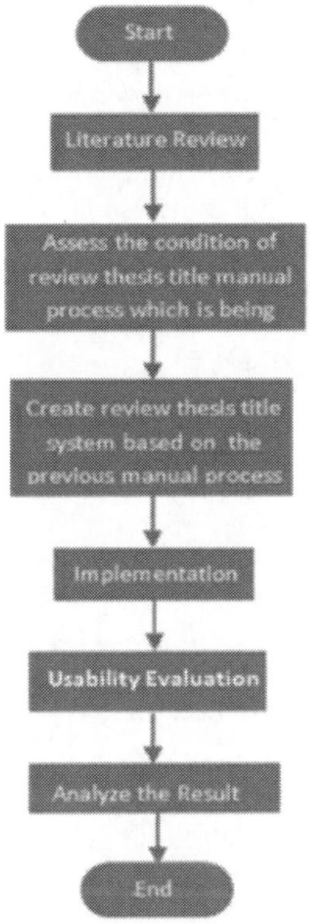


**Figure 1.**Research Methodology.

In evaluating usability result of applying thesis title system review required research sample. The sample of research in question is the respondent which representsystem users. The respondents were 58 students, head of study program, and 1 person form academic section. In one semester, there are at least 20-30 student who want to submit thesis proposal. The respondents are final year students who have submitted the title of thesis through the manual process. It is intended that the respondents selected as research samples are better able to measure the effectiveness, efficiency, and satisfaction when using the review thesis title system because previously been through the process of submitting the title thesis manually.

Respondents will be given a questionnaire to measure evaluation of usability. Questions given about experience when using the system. The assessment used is taken in terms of effectiveness, efficiency, and user comfort. Therefore, questionnaire questions given to the respondent about the user interface, the ease of using the service, the ease of following the process, and the ease of get information.

## 3. Result and Discussion

The first step is to prepare the questionnaire data in *.csv* file format to be used as an active data file. Table 1 showsthe coefficients used in active data file. The coefficients are based on the question items in the questionnaire. Where construction used for usability evaluation is effectiveness, efficiency, and satisfaction (user satisfaction). The result is shown by Table 1.

**Table 1.** Initial List of Construct and Item for Questionnaire Instrument

| Construct | Code | Item | Loading Factor |
|---|---|---|---|
| Efficiency | EN1 | Upon opening the system did not wait too long to reponse back from the system | -0.835 |
| | EN 2 | Easily get the required information from the system | 0,002 |
| | EN 3 | The use of the system works more efficiently and quickly | 0,493 |
| | EN 4 | Information needed more efficiency in finding data system | 0,986 |
| | EN 5 | menu content Available is not difficult in its use | -0,155 |
| Effectiveness | EF1 | The system used makes it easier to access data sought from anywhere | -0.205 |
| | EF 2 | Guidelines in system usage are not difficult for users | 0.875 |
| | EF 3 | Rapid System Response check for data search | -0.822 |
| | EF 4 | The time required is not long for the results of the inputted data | -0.814 |
| | EF 5 | This information is effective in helping user to get things done tasks and scenarios | 0.816 |
| Satisfaction | ST1 | I feel comfortable using this site | 0.939 |
| | ST 2 | The system interface user meets the expected usage | 0,259 |
| | ST 3 | The output of the system as expected | 0,866 |
| | ST 4 | user satisfied for the use of existing data system | 0,874 |

The second step is to create a model file. Where there are four latent variables used are, Effectiveness (EN), Efficiency (EF), Satisfaction (ST), and Usability. The three variables are EN <EF, and ST will connect (connect) to Usability. The file model can be seen in Figure 2.

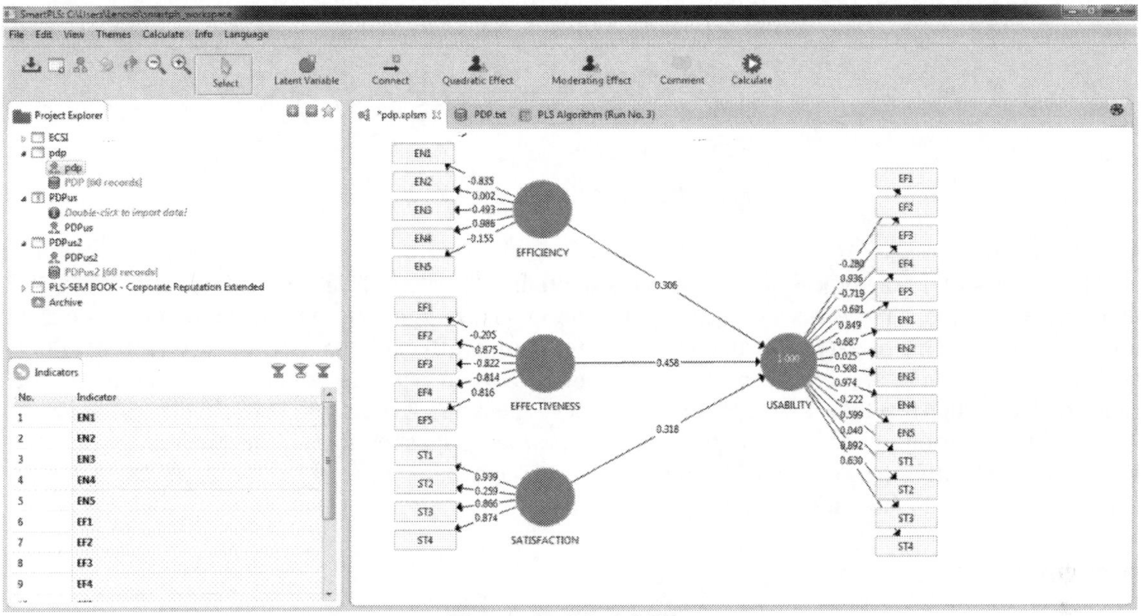


**Figure 2**. Model File

The third step is calculated PLS algorithm. Then the fourth step, outer loading /loading factor results are reviewed. Coefficient with below-standard loading factor is less than the rule of tumbs 0.70 [10] will be removed or deleted from the active data file then evaluated cronbach's Alpha and Path Coefficient Result. The results can be seen in Figure 3.

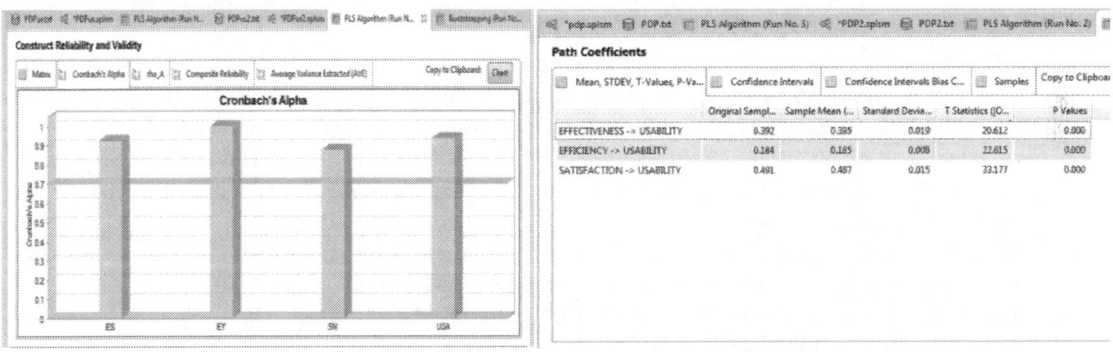


**Figure 3.** Cronbach's Alpha and Path Coefficient Results

The Cronbach Alpha score was evaluated, where the standard of measurement model relativity was good, ie more than the rule of tumbs 0.60 [11]. Thus it can be concluded that the measurement model of all variables have good reliability. The rule of tumbs from the support of a research hypothesis is if the statistical T value is more than 1.64 (two-tiled) or 1.96 (one-tiled) and the probability value (p-value) is less than 0.05 or 5%. From the above coefficient path can see that the T statistic value for effectiveness, efficiency, and satisfaction is more than 1.64 (two-tiled) or 1.96 (one-tiled) and probability value (p-value) less than 0.05 or 5%. This shows that the three constructs affect the usability value.The rule of tumbs from the support of a research hypothesis is if the statistical T value is more than 1.64 (two-tiled) or 1.96 (one-tiled) and the probability value (p-value) is less than 0.05 or 5%. From the above coefficient path we can see that the T statistic value for effectiveness, efficiency, and satisfaction is more than 1.64 (two-tiled) or 1.96 (one-tiled) and probability value (p-value) less than 0.05 or 5%. This shows that the three constructs affect the usability value. The order of construction of the largest statistical T value to the small can be seen in Table 2.

IOP Conf. Series: Journal of Physics: Conf. Series **978** (2018) 012010 doi:10.1088/1742-6596/978/1/012010

**Table 2.** Construction and T statistic value

| Construction | T statistic |
|---|---|
| Effectiveness (EF) | 20.612 |
| Efficiency (EN) | 22.615 |
| Satisfaction (ST) | 33.177 |

## 4. Conclusion

The importance of the process is to ease and smooth in filing their thesis title student at STT Dumai, therefore evaluation is necessary in order to develop products or application to better system and can be useful optimally. This research was done to analyse usability affecting the user revenue (acceptability) system of application. Using 60 respondent, using regression analysis double by 3 parameter uses smartPLS with the results are 20.612 the efficiency, the effectiveness of 22.615and satisfaction of 33.177. The rejection or accept of respondents reaction to application through the survey expected to inform the applicationimprovement.The presentation of application a user interface on a website is very conducive to provide facilities (usability) to the users.

## References

[1] Billah S A and Hariadi S 2010 Rancang Bangun Aplikasi Sistem Sidang Skripsi Menggunakan Metode Object Oriented *Jurnal Ilmu computer dan Teknologi Informasi Universitas Islam Negeri (UIN) Maulana Malik Ibrahim Malang* **4**.

[2] Albers M and Kim L 2002 Information Design For The Small-Screen Interface: An Overview of Web Design Issues For Personal Digital Assistants *Technical Communication* pp 45-60.

[3] International Organization for Standardization 1998 *ISO 9241-11 1998 Ergonomic Requirements for Office Work with Visual Display Terminals (VDTs)-Part 11 Guidance on Usability* (International Organization for Standardization)

[4] Nayebi F, Desharnais JM and Abran A 2012 *The state of the art of mobile application usability evaluation (*Canada: In Electrical & Comp. Engineering (CCECE), 2012 25th IEEE) pp 1-4.

[5] Nielsen J 1990 *Heuristic Evaluation of User Interface* (Denmark:CHI'90 Proc.) pp 25-62.

[6] Jeffries R, Miller J R, Wharton C and Uyeda K 1991User Interface Evaluation in the Real World: A Comparison of Four Techniques *Proc. of the SIGCHI Conf. on Human factors in computing systems, ACM* pp 119-124.

[7] Ivory M and Hearst M 2001 The State of the Art in Automating Usability Evaluation of User Interfaces (*Berkeley:ACM Computing Surveys*) **33** pp 470–516.

[8] Benbunan R 2001 *Using Protocol Analysis to Evaluate The Usability Of A Commercial Web Site* (Information & management) *39*(2) pp 151-163

[9] Halawi L and McCarthy R 2008 Measuring Students Perceptions Of Blackboard Using The Technology Acceptance Model: A PLS Approach (*America:Issues in Information Systems*) **9** p 95.

[10] Chin WW 1998 *The partial least squares approach to structural equation modelling Modern Methods For Business Research* (New Jersey: Lawrence Erlbaum Associates Publiser) pp 295-336.

[11] Horrocks S, Anderson E and Salisbury C 2002 *Systematic Review of Whether Nurse Practitioners Working in Primary Care Can Provide Equivalent Care to Doctors* BMJ 324(7341) pp 819-823.

# Diagnostics vehicle's condition using obd-ii and raspberry pi technology: study literature

**J V Moniaga, S R Manalu, D A Hadipurnawan and F Sahidi**

Computer Science Department, School of Computer Science, Bina Nusantara University, Jakarta, Indonesia 11480

Email: jurike@binus.edu, smanalu@binus.edu dionisius.hadipurnawan@binus.ac.id firda.sahidi@binus.ac.id

**Abstract.** Transportation accident rate are still being a major challenge in many countries. There are many factors that could be cause transportation accident, especially in vehicle's internal system problem. To overcome this problem, OBD-II technology has been created to diagnostics vehicle's condition. OBD-II scanner plugged to OBD-II port or usually called Data Link Connector (DLC), and after that it sends the diagnostics to Raspberry Pi. Compared from another microcontrollers, Arduino, Raspberry Pi are chosen because it sustains the application to receive real-time diagnostics, process the diagnostics and send command to automobiles at the same time, rather than Arduino that must wait for another process finished to run another process. Outcome from this application is to enable automobile's user to diagnostics their own vehicles. If there is found something unusual or a problem, the application can told the problem to user, so they could know what to fix before they use their vehicle safely.

## 1. Introduction

On-Board Diagnostic is a computer-based system developed by Automobile manufacturers for diagnosing car. It focuses on diagnosing the performance of car's engine to check any errors in car's engine components [1]. OBD-II connect with DLC in the vehicle. Almost all the problem with the vehicle can be detect by OBD-II, such as exhaust manifold, ABS brake function, airbag (iSRS & SRS), inle performance, and so on. However, OBD was only for scanning the vehicle's diagnostics. To reading, processing, and showing the data, it needs Microcontroller. Microcontroller is a microprocessor system that installed in the chip. Microcontroller is different with microprocessor that used by PC because usually general microcontroller already filled with supporting minimal component system, such as memory and I/O interface. In another side, microprocessor usually only filled by CPU [2]. Nowadays a lot of tools that can control device digitally. However, we have choosen two of the popular models and compared their feature. Those two models are, Arduino Uno and Raspberry Pi. Arduino is an open source microcontroller board that able to connect to computer and others devices via universal serial bus (USB). Arduino can be powered from any power supply such as computer, battery, and more. [3]. There a lot of various of Arduino to choose for every needs, to conform with our research we chose to use Arduino Uno.

**Figure 1.** From left to right: An Arduino Uno [4], a Raspberry Pi [4], iSaddle Bluetooth OBD-II Scanner[5]

The Raspberry Pi is an inexpensive credit-card sized computer, that has 700Mhz processor, 512MB of SDRAM, on board Ethernet, HDMI and RCA interfaces, a dual USB connector and an SD card slot. In addition is maintains an array of input/output pins that make the Raspberry Pi extraordinarily conductive to tinkering and device innovation [6].

**Table 1.** Comparison of Arduino Uno and Raspbery Pi [4]

| Indicator | Arduino Uno | Raspberry Pi |
|---|---|---|
| Model Tested | R3 | Model B |
| Processor | ATMega 328 | ARM11 |
| Clock Speed | 16Mhz | 700Mhz |
| RAM | 2KB | 256MB |
| Ethernet | N/A | 10/100 |
| USB Master | N/A | 2 USB 2.0 |

Raspberry Pi has been chosen rightly as the single board computer for this application because it has the highest performance compared to Arduino (Arduino still use ATMega 328 for its processor and Raspberry Pi already use ARM11 which has higher specification) and Arduino cannot run multiple process at the same time since it only has 2KB size of RAM. In addition, Raspberry Pi has input port for Ethernet and USB, which make it easier to make it connected to internet connection and connect to another device.

## 2. OBD-II Scanners

Recently, On Board Diagnosis-II (OBD-II) scanner have been introduced to use a vehicle network (OBD-II) for the diagnosis and checking of a vehicle, which has previously used many devices such as black boxes, vehicle diagnostic equipment and vehicle supply management [7]. OBD is a standard tool that allows driver to check car's engine status by plug the OBD to the porst that placed under the driving wheel. Mostly OBD can get the status of emission control, driving speed, voltage control, engine coolant temperature, idle time, and more [8]. OBD-II is mostly used with extension tools such as gas analysers and vehicle telematics [9]. The OBD scanner is a costly device and is mainly owned by workshops [10]. We are able to access this car's ECU with mobile device by connected the OBD-II to the port and connected to the mobile device via OBD-II connector. Then OBD-II will export the data to the mobile device after CAN-Bus convert the data. [11]. There are many scanners that could be connected to OBD-II, for the example, one of the most basics scanner, U480 CAN OBD-II Diagnostic Scanner. This scanner has a basic LCD display to give information about vehicle on board diagnostics that connected to it, but need wire to connect to the vehicle. The developed scan tool is compliant with various SAE (CAN) standard communication protocols to connect with the ECU installed in the vehicle [12]. On board diagnostics is a tool that have a function to diagnosing and reporting the car's engine condition. It is the interface between cars Controller Area Network (or CAN bus) and external devices, e.g., a OBD scan tool that connects the OBD port and a laptop. All components of car's vehicle are able to communicate with each other with CAN-Bus. For example, a position value is sent to the Electronic Control Unit (ECU) after human driver press the gas pedal, and the ECU adjusts air/fuel injections according to the position value. Therefore, we can read this position message transmitted over CAN bus from the OBD port [13]. Now scanners mostly are wireless for example there are 3 scanners that most used for now, first is ERUSUN Bluetooth Mini OBD-II Scanner, that serving a wireless connection between user vehicle's computer and user device's via Bluetooth, then there is iSaddle Bluetooth / Wi-Fi Wireless

OBD-II Scanner, is a scanner that serving a connection both Wi-Fi or Bluetooth and can connect to tablet, smartphone and computer. ELM-327, is one such adapter where the data read from the OBD2 port are transmitted via Bluetooth upon pairing [14]. Last, there is iSaddle OBD-II Wi-Fi / Bluetooth Scanner that connect also with both Bluetooth or Wi-Fi [15].

## 3. iSaddle Bluetooth OBD-II Scanner on Android

It is necessary to know the OBD2 protocols and convert the OBD2 signals to the serial data standard for the PC [16]. iSaddle Bluetooth OBD-II Scanner is more suitable to this application than iSaddle WI-FI OBD-II Scanner because this application require internet connection from the devices. When the device connect to this application via Wi-Fi, it would make impossible for the device to connect to the internet via Wi-Fi. In this case, user must choose between connect into this application but using mobile data, or not using this application at all. Furthermore, Bluetooth wireless technology is an open specification that enables low-power and short range wireless connection [17]. Since this application will be used while user in the vehicle that does not has any electricity to charge the smartphone, low-power wireless technology would be a great solutions. By using the OBD2 Bluetooth interface, the information on real instantaneous fuel consumption is also acquired from the vehicle engine [18]. OBD2 speed is taken as benchmark for true vehicle speed. Bluetooth based OBD2 device is installed in the vehicle, which collects vehicle speed at 1sec interval and transmits it a mobile phone [19]. OBD-II does not provide a user friendly data and meaningful, although there are a lot of research that claim there is a possibility to analyse driver behaviour using real-time data from OBD-II's diagnosing data. [20]. Nowadays those researches focusing on to use multi-source information fusion to get better interpret of relationship of driver's behaviour and car's engine performance.[21] Unfortunately the analysis part is done off-line after the data gather, so OBD-II can't give a real-time driver's support [22].

iSaddle Bluetooth OBD-II Scanner has a small size, compact structure, and supports all OBD-II protocols. Besides, stable wireless connection, no wires, and no batteries are some extra feature that iSaddle OBD-II has [23]. To work, iSaddle Bluetooth OBD-II Scanner require the vehicle that has 16-pin Data Link Connector (DLC) under the vehicle's dash [24]. Also, that must be an information on vehicle's emission control label that the vehicle is OBD-II compliant. This scanner then get plugged to DLC and it connects to any computer, tablet, or smartphone [25]. There are many diagnostic software that associated with iSaddle Bluetooth OBD-II scanner. With these software, it would be able to read diagnostic trouble codes, clear trouble codes, turn off the check engine light, set the maximum number of responses to obtain, removes spaces from Engine Control Unit (ECU) responses (to make data transmission more speed), store unique identifier in the Electronically Erasable Programmable Read Only Memory (EEPROM), set the CAN mask and filter with one command, easily switch between variable and fixed CAN message lengths, send CAN remote frames (RTRs), and many more [23].

The car's engine trouble indications or commonly called by Diagnostic Trouble Codes (DTC) is save in the system. Those codes are not the same in all vehicles, and not all country that produce vehicle use the manufacturer codes. However, now everyone is able to connect OBD-II to the port and diagnosing the car's engine trouble then identify the trouble data [26]. Nowadays freeware and commercial software packages are available, that allow to monitor OBD-II vehicle data by using just a smartphone and off-the-shelf scan tools. There a lot of software that able to monitor car's engine by using smartphone and some external tools. However all on-board diagnosing systems only display the low-level data. OBD-II doesn't provide a user friendly data and meaningful, although there are a lot of research that claim there is a possibility to analyse driver behaviour using real-time data from OBD-II's diagnosing data. Nowadays those researches focusing on to use multi-source information fusion to get better interpret of relationship of driver's behaviour and car's engine performance. Unfortunately the analysis part is done off-line after the data gather, so OBD-II can't give a real-time driver's support.

## 4. Process in OBD-II

To set up the OBD scanner tool is not a uniform process since there are many varieties of types and brands of vehicle and the OBD scanner itself. But generally the process can be described like this, first user must download the application for the mobile and it depends on the OBD scanner tool itself, because different tool may require different application. This step should be done well before user's smartphone connect to the OBD scanner tool. Next step user must connect the OBD scanner to the DLC (Data Link Connector), usually it was located under the steering wheel. DLC consist of 16 pins. This step also must be done well because it will connect directly to the OBD scanner tool. As a concern when user is installing the OBD scanner to the DLC, this is the following list of 16 Pin DLC and its function, These 16 pins include: (1) Manufacturer reserved pin, (2) The J1850, (3) Manufacturer reserved pin , (4) The chassis ground connector,(5) The signal ground , (6) The CAN high, J-2284, (7) The k line, ISO9141-2/DIS 14230-4connector port, (8) Manufacturer reserved pin, (9) Manufacturer reserved pin, (10) J1850 Bus, (11) Manufacturer reserved port, (12) Manufacturer reserved port,(13) Manufacturer reserved port,(14) CAN, J2284 PORT,(15) The L line, ISO 9141 and ISO/ DIS 14230-4, (16) The battery power connection. Since there are many varieties of types and brand of vehicle also the OBD scanner itself and this process is a general process, there are some

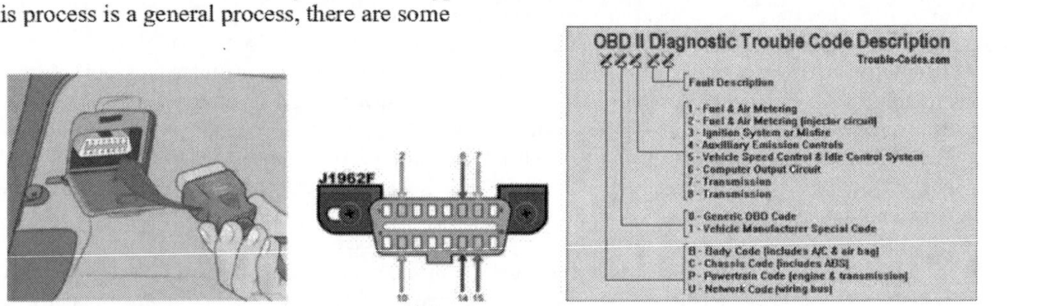


(a) (b) (c)

**Figure 2.** (a) Data Link Connector and OBD Wire, (b) 16-Pins of OBD-II [28], (c) OBD-II Diagnostics Trouble Code Description [29]

The other concern is DTC (Diagnostic Trouble Code), DTC is a set of code consisting of letters and numbers, where each code represents type of fault on user's vehicle. DTC originally made by SAE (Society Automotive Engineers) for use by all vehicle manufacturers who have to comply with OBD 2 emissions regulations in the U.S and adopted by European and Asian market. There are 4 basic categories of DTC, Powertrain (P) category covers functions that include engine, transmission and associated drivetrain accessories, Body (B) category refers to functions that are, generally, inside of the passenger compartment. These functions provide the driver with assistance, comfort, convenience, and safety. Chassis (C) category represent functions that are, generally, outside of the passenger compartment. These functions typically include mechanical systems such as brakes, steering and suspension. Network and Vehicle Integration (U) category is for functions that are shared among computers and systems on the vehicle. DTC also divided to 2 major groups, Generic OBD Codes are code that have "0" as second digit to represent they are common to all makes and models of vehicles also for basic emissions fault diagnosis, Vehicle Manufacturer Special Codes are code that have 1 as second digit to represent unique vehicle to a particular vehicle make or model [30].

## 5. Process in Raspberry Pi

The Raspberry PI is not a fixed function device, unlike many mobile devices. Although a mobile device often has a camera, it is usually impossible to add extra hardware to most mobile devices. Using the Raspberry Pi, it is possible to play with rudimentary hardware. There are multitudes of the resources devoted to creating innovative systems using the Raspberry Pi as a foundation, from supercomputers to time-lapse cameras. One of the advantages of the Raspberry Pi is that they limit the resources available to the programmer, who must then develop the cleaner and better programming habits as the programming projects gain complexity. The restriction on resources will likely force the design of more efficient programs [6]. To work with the iSaddle Bluetooth OBD-II scanner, the Raspberry Pi needs a Bluetooth scanner to be pairing with. There are several commercial Bluetooth scanner, but in this application will use the Raspberry Pi Nano Bluetooth Dongle, which is the perfect low cost solution and easy to install. The first step before using The Bluetooth Dongle is to install the latest Raspbian Operating System (OS) in the Raspberry Pi which where the data will be shown and processed. Raspbian OS is a free operating system based on Debian, optimized for the Raspberry Pi hardware [31]. The next step is download pyOBD. pyOBD is an open source OBD-II compliant scan tool software that written entirely in Phyton. It is made with OBD-II diagnostic interfaces. It will allow the user to see their vehicle's ECU data, display DTC, measured values, status tests, and more [32].

## 6. Application Process

The recommend system is able of diagnosis the vehicles, collecting, storing, and also retrieve the command from mobile application. Vehicle diagnosis data are collected using an iSaddle OBD2 Bluetooth scanner and are sent to Raspberry Pi via Bluetooth. Raspberry Pi needs Nano Bluetooth Dongle as extra tools that enabled it to retrieve and send data using Bluetooth. After the diagnostics data retrieved to Raspberry Pi, pyOBD will display the interface of the diagnostic data based from OBD-II diagnostics trouble codes and process it. To process the diagnostics data, this application is using Node.JS as server-side language. Compared to PHP that only handles small requests well, but struggles with large requests Node.JS has much better performance in high concurrency situation, no matter in benchmark test or scenario tests [33]. These data are encapsulated to JSON Objects, it will be store to MongoDB (NoSQL document oriented database management system [34]). MongoDB provides flexibility during the development process and easy to deploy and copy databases from one server another server compared than its competitor, Oracle Databases [34], which makes MongoDB is more suitable in this application. By connecting to internet connection, JSON Objects that stored in MongoDB can be parsed to user's smartphone through this application.

Users are also enabled to send their own command to their vehicle's. User must input their command in this application to do so. Next, the application will generated the input into JSON Objects and stored to MongoDB. The Node.JS API will process the JSON Objects into ELM-USB command list. These ELM-USB command lists are send to iSaddle OBD-II Bluetooth Scanner, and it will triggered the vehicle to do the command from the user.

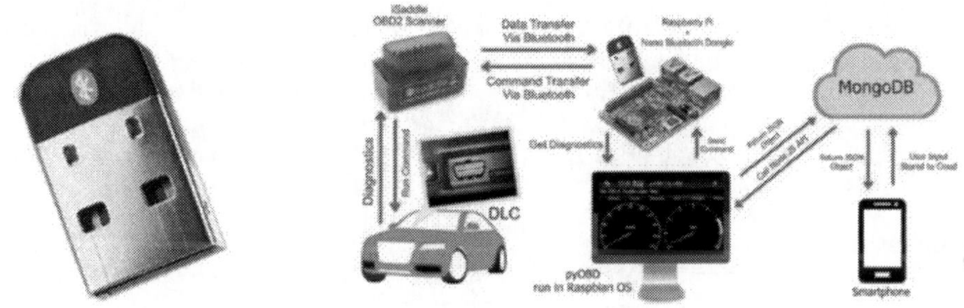


(a) (b)

**Figure 3.** (a) Raspberry Pi Nano Bluetooth Dongle [31]; (b) Application Process Flow

## 7. Conclusion

Vehicle's diagnostic get much attention from industry and researchers in recent years. The variety and heterogeneity of vehicle diagnostics implementation has been the major reason which makes it interested. This paper presents a technique to analyse diagnostics from vehicle that connected to OBD-II and process the diagnostics data using Raspberry Pi. People can also send a command to their own vehicle by using this application. Although the process occurs only the delivery of vehicle diagnostic data to user's smartphone, Raspberry Pi is more suitable viewing the ability of Raspberry Pi that can be multitasking. In the future research that related to OBD-II scanner, its recommended to use OBD-II scanner with Bluetooth rather than Wi-Fi, because it's more energy saving, easier to use and more stable.

## References

[1] Graham N J 2008 *Wireless Automotive Data Link Connector* US7584030 B1.
[2] Nurfalah M 2014 *Makalah Arduino dan Raspberry Pi Mikrokontroler* p 9.
[3] Monk S 2012 *Programming Arduino Getting Started with Sketches* (McGraw-Hill Education)

IOP Conf. Series: Journal of Physics: Conf. Series **978** (2018) 012011 doi:10.1088/1742-6596/978/1/012011

[4] Brandon 2013 Arduino Uno vs Raspberry Pi vs BeagleBone Black *MCM Electronics*. [Online]. Available: http://blog.mcmelectronics.com/post/Arduino-Uno-Raspberry-Pi-and-BeagleBone-Black#.Ua753EDLpWI.
[5] Amazon 2017 iSaddle Super Mini Bluetooth OBD2 OBDII Scan Tool Check Engine Light & CAN-BUS Auto Diagnostic Tool for Windows & Android Torque (Blue Color, Super Mini). [Online]. Available: https://www.amazon.com/iSaddle-Bluetooth-CAN-BUS-Diagnostic-Windows/dp/B00OMT6UEK. [Accessed: 07-Apr-2017].
[6] Wirth M and McCuaig J 2014 Making Programs With The Raspberry Pi, *Proc. West. Can. Conf. Comput. Educ. - WCCCE '14* pp 1-5.
[7] Baek S H and Jang J W 2015 Implementation of Integrated OBD-II Connector With External Network *Inf. Syst.* **50** pp 69–75.
[8] Wengert J 2005 On-board diagnostics in engine management systems for motorcycles, *SAE Tech. Pap.* pp 1–6.
[9] Hilpert H, Thoroe L and Schumann M 2011 Real-Time Data Collection for Product Carbon Footprints in Transportation Processes Based on OBD2 and Smartphones *Proc. of the Annual Hawaii Int. Conf. on System Sciences*.
[10] Hasan N N, Arif A, Pervez U, Hassam M and Ul Husnain S S 2011 Micro-Controller Based On-Board Diagnostic (OBD) System for Non-OBD Vehicles *Proc. - 2011 UKSim 13th Int. Conf. on Modelling and Simulation* pp 540–544.
[11] Türker G F and Kutlu A 2016 Survey of Smartphone Applications Based on OBD-II for Intelligent Transportation Systems *J. Eng. Res. Appl. www.ijera.com* **6** p 2248.
[12] Villatico F and Zuccari F 2008 Efficiency comparison between FC and ice in real urban driving cycles *Int. J. Hydrogen Energy* **33** pp 3235–3242.
[13] Kang L, Qi B, Janecek D and Banerjee S 2015 EcoDrive: A Mobile Sensing and Control System for Fuel Efficient Driving *Proc. of the 21st Annual Int. Conf. on Mobile Computing and Networking* pp 358-371.
[14] Amarasinghe M, Kottegoda S, Arachchi A L, Muramudalige S, Bandara H L M D and Azeez A 2015 Cloud-Based Driver Monitoring and Vehicle Diagnostic With OBD2 Telematics *IEEE Int. Conf. on Electro Information Technology*.
[15] Ntziachristos L, Fragkiadoulakis P, Samaras Z, Janka K and Tikkanen J 2011 Exhaust Particle Sensor for OBD Application *SAE Tech. Pap.*
[16] Hermann A and Desel J 2008 Driving Situation Analysis in Automotive Environment *Proc. of the 2008 IEEE Int. Conf. on Vehicular Electronics and Safety ICVES* pp 216–221.
[17] Yu SN and Cheng JC 2005 A Wireless Physiological Signal Monitoring System with Integrated Bluetooth and WiFi Technologies *Conf. Proc. IEEE Eng. Med. Biol. Soc.* **3**.
[18] Astarita V, Guido G, Mongelli D and Giofrè V P 2015 A Co-operative Methodology to Estimate Car Fuel Consumption by Using Smartphone Sensors *Transport* **30** pp 307–311.
[19] Chowdhury A, Chakravarty T and Balamuralidhar P 2014 Estimating True Speed of Moving Vehicle Using Smartphone-Based GPS Measurement *IEEE Int. Conf. Syst. Man, Cybern.* pp 3348–3353.
[20] Choi S, Kim J, Kwak D, Angkititrakul P and Hansen J H L 2007 Analysis and Classification of Driver Behavior Using In-Vehicle Can-Bus Information *Bienn. Work. DSP In-Vehicle Mob. Syst.* pp 17–19.
[21] Quintero G C M, Lopez J A O and Rua J M P 2010 Intelligent Erratic Driving Diagnosis Based on Artificial Neural Networks *IEEE Andescon* pp 1–6.
[22] Ruta M, Scioscia F, Gramegna F, Loseto G and Di Sciascio E 2012 *Knowledge-based Real-Time Car Monitoring and Driving Assistance* (Padova, Italia: Edizioni Libreria Progetto).
[23] Amazon *iSaddle Bluetooth OBDII Scan Tool User Manual*.

IOP Conf. Series: Journal of Physics: Conf. Series **978** (2018) 012012 doi:10.1088/1742-6596/978/1/012012

# Indonesia knowledge dissemination: a snapshot

**M K M Nasution**

Fasilkom-TI, Universitas Sumatera Utara, Padang Bulan 20155 USU, Medan, Indonesia.

Email: mahyuddin@usu.ac.id

**Abstract.** The educational progress of a country or educational institution is measured through the implementation of knowledge dissemination. Evidence of knowledge dissemination has carried out be in form of the type of published document, which is based on the databases of the index of scientific publications: Scopus. This paper expresses a simple form of knowledge dissemination based on document type. Although the growth of knowledge dissemination does not have the same pattern based on the appearance of document types, the general implementation is almost the same. However, maximum effort needs to be done by PTN-bh to support Indonesia knowledge dissemination.

## 1. Introduction

In the concept of knowledge management, the knowledge must always be available to all those who need it with the principle that the knowledge be an appropriate for the right person at the appropriate place and time [1], in order to increase public welfare [2]. Most people today get information about something for their knowledge using search engines whereby they access information resources, such as through the Web [3]. However, the Web contains not only truly reliable information, but contains other elements that undermine the welfare of society such as slander and false news. Therefore, knowledge dissemination is an approach in managing knowledge so that the right information grows faster than incorrect information [4].

Dissemination of knowledge, on the one hand, is to transfer knowledge within and across settings, so that all people that have the knowledge also to have competitiveness in their lives [5]. However, on the other hand, knowledge dissemination aims to reduce the duplication of knowledge development where the concepts, ideas, science, and technologies be overlap with each other. In addition to broadening the insight and improving the quality of ideas [6]. Today, information technology has become part of knowledge management [7], and as a medium for delivering and disseminating knowledge [8]. Various classical approaches in knowledge dissemination have been adapted for ease of administration: from registration, submit, commenting, acceptance, improvement, implementation, publishing, and recognition [9]. Therefore, the existence of knowledge development today can be seen from the existence or absence of scientific papers in trusted database. This paper aims to recognize the dissemination of knowledge based on reputable indexer databases for that scientific publications.

## 2. Basic Concept and Motivation

In higher education and/or research institutions, knowledge dissemination is evidence of the existence or absence of knowledge development carried out at the institution [10]. Knowledge development is done from the results of research and implementation of education, namely the output in the form of

IOP Conf. Series: Journal of Physics: Conf. Series **978** (2018) 012012 doi:10.1088/1742-6596/978/1/012012

scientific paper and patent [5]. In dissemination of knowledge, the implementation of education resulted in each participant conceptually obtaining learning, enlightenment, acquisition of new perspectives or attitudes, while from an instrumental vocational point each participant obtain modification, new practice, or expertise [6]. Therefore, knowledge dissemination has impact not only academic environment but also social through third pillar of university, that is community service with downstream task. Through downstream will be increased social awareness of every member of the community on the impact of emerging technologies. Through downstream will be enhanced the ability to make a choice based on available alternative information. In general, knowledge dissemination causes the exchange of information, capital, matter and perspective [8].

The dissemination of knowledge has different meanings for every institution, any field of knowledge and each branch of knowledge, and consequently is implemented differently by everyone [4]. Yet classically, knowledge dissemination is carried out by

a. the resource person writes the knowledge, sends it to the reader, and the reader interprets and understands it, or
b. the speaker talks with the audience, and conducts the discussion, and the audience interprets and understands it.

Although there is a possibility of discussion in the classical dissemination of knowledge, but in general knowledge dissemination takes place in one direction. Therefore, its interpretation and execution always depends on the ability of the absorber of the listener or reader [11].

Conversely, with the presence of information technology, like the Web, has caused the dissemination of knowledge into two directions directly, by which analysis can be done by audiences or readers about the truth of knowledge conveyed or transmitted [12,13]. However, the implementation of knowledge dissemination recognized in the information age involves interrelated components such as authors; committee; publisher; media: magazine, journal, proceeding, or book; reviewers; editors; and indexer. Newspapers or magazines either online or not within certain limits are recognized as scientific publications, whereas blogs in certain circumstances can be only considered as a means of knowledge dissemination. Types of recognized scientific publication media include [14]:

- *Journal* is one of the media of scientific papers recognized, is also sometimes known as periodical scientific or scientific magazine, is a form of publication that serves to register scholarly activities, certify the results of activities that meet minimum scientific requirements, disseminate widely to the public, and archive all findings of researches activities, and the scientists and the discourse that it contains. In general journal/magazine/scientific periodical is marked by ISSN (International Standard Serial Number).
- *Book* is one of the scientific works which is the result of original research or thought, in the form of reference books, monographs, or other. Generally a book has an ISBN (International Standard Book Number).
- *Proceeding* in book form are collections of scientific papers presented at conferences, seminars, or scientific meetings. Proceeding possess ISBN or ISSN.

All documents of scientific work are deemed reputable when indexed by reputable indexers and have an impact factor. Reputable indexers recognized by the Indonesia Ministry of Higher Education and Research are the Web of Science (https://clarivate.com/products/web-of-science/), Scopus (www.scopus.com), or Microsoft Academic Search (http://academic.research.microsoft.com/), while the impact factor comes from the ISI Web of Science (Thomson Reuters) or from the Scimago Journal Rank (SJR) [15]. In addition, there are some well-recognized indexers like DOAJ, CABI, Copernicus and others, but the recognition unlike the three indexing databases above. However, in the dissemination of knowledge there are various documents recognized by that criterion.

## 3. The Adaptive Approach

Identification of the existence of knowledge dissemination in the information age is done through information available online [16]. The recognized dissemination is evidenced by the publication of scientific paper and indexed by reputable indexer databases. The knowledge dissemination data of an institution or the country is derived from reputable indexer databases, whereas behaviour is derived by timeline (year by year) following the number of documents and types of documents published.

The types of documents that are evidence of the implementation of knowledge dissemination are as follows (http://www.bmj.com/about-bmj/resources-authors/article-types, http://retractionwatch.com/2013/02/25/is-an-article-in-press-published-a-word-about-elseviers-withdrawal-policy/ ) [17]:

- *Article* is a classic document that changes a factual essay be complete with a certain length (usually between five and twenty pages) to convey ideas and facts that can convince, educate and increase knowledge (contains full descriptions of the latest research finding original).
- *Letter* also called communications is a document containing a brief description (as information) of the latest findings that are considered urgent.
- *Note* is a document containing a brief description that is considered a bit urgent or important than a letter.
- *Review* is a document that contains the accumulation of the results (research) of different articles on a topic that is poured into a coherent narrative about state of the art of a field.
- *Editorial* is the document as the main article (leading or leader) written by senior editorial staff.
- *Report* is a document as an information article created to convey information or tell a specific event in a form that can be presented widely and researched.
- *Book* is a document consisting of a set of sheets of paper, parchments, or similar material bound in such a way as to comprise parts of the thinking for the stated subject matter.
- *Book Chapter* is a document that contains the story of something in particular, complete and finished as an intermediary for the reader.
- *Conference Paper* is a document as an article written to be accepted for oral presentation or as a poster on a scientific activity.
- *Conference Review* is a document as unpublished articles in proceedings to review issues related to the conferences.
- *Erratum* is a document containing corrections to a publication text.
- *Short Survey* is a document that explains the sampling process briefly to conduct a survey.
- *Data Paper* is a document as articles that describe dataset.
- *Abstract Report* is a document written briefly and clearly.
- *Opinion* is a document that is written as a view or statement that is not conclusive.
- *Extended Abstract* is a concise document derived from the abstract.
- *Article in Press* is an article that has been accepted for publication but may still contain errors or duplications from other publications so that article may be withdrawn from publishing, so the article does not have complete volume/issue/page information.
- *Business Article* is a document issued by mutual agreement between the author, translator, and publisher.
- *Retracted* is a document containing public statements made to withdraw previous statements.

So the line-shaped timeline, for example, contains year information on the right or top, while on the left or bottom with the document type information, or in the order of colours that make up the letter type identifier of the document type. The comparison between document types is normalized for every type document for between institutions by using similarity formulation [18].

## 4. A Selective Approach as Discussion

Indonesia knowledge dissemination intends to see the behaviour of knowledge dissemination from Indonesia, whole implementation activities carried out inside or abroad. This behaviour is based on the type of document published, i.e. through reputable indexing databases such as Scopus [15]: The collected data on the number of scientific publication documents per year and per document type of institution in the top rankings in Scopus. From the Scopus database, we selected the top 20 document-producing institutions from Indonesia, and selected several institutions from other countries. This is a selective approach done to obtain knowledge dissemination behaviour based on comparison. Data retrieved until 2016.

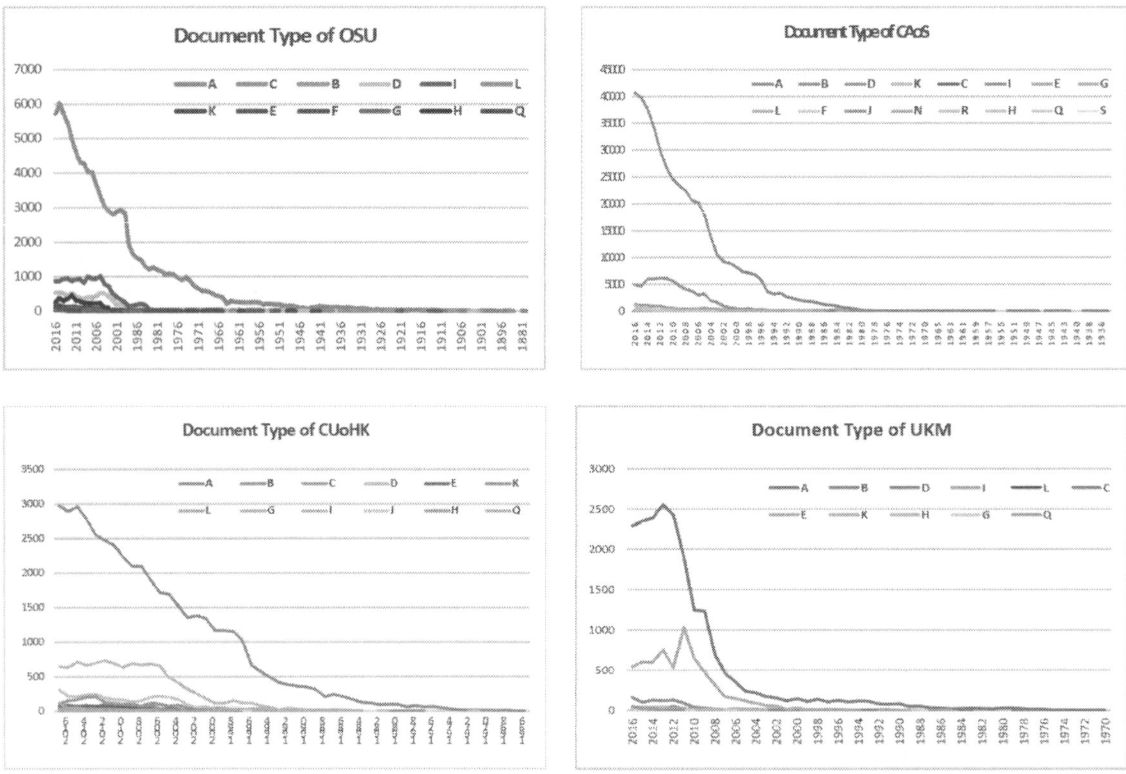


**Figure 1.** Pattern of Knowledge dissemination for OSU, CAoS, CUoHK, and UKM

The number of documents per year based on the type of document has affiliation of Ohio State University, Chinese Academy of Sciences, Chinese University of Hong Kong, and Universiti Kebangsaan Malaysia. All of them are used as the comparison of knowledge dissemination: Ohio State University (OSU) is a college that has produced the highest number of the Scopus-indexed documents in United State America (until August 2017), OSU has produced the article as the first Scopus-indexed document in 1880, OSU was established in 1870; Chinese Academy of Sciences (CAoS) is one of institutions which produces the largest number of Scopus-indexed scientific document (until August 2017) in China, CAoS was established in 1949 but has been discourse in Scopus-indexed first scientific work of 1935 whereby CAoS as an affiliation; Chinese University of Hong Kong (CUoHK) (until August 2017) is ranked 19$^{th}$ in the producer of Scopus-indexed documents in China, CUoHK was founded in 1963 and produced the first Scopus-indexed paper in 1965; and then Universiti Kebangsaan Malaysia (UKM) is one of the leading universities in Malaysia, established in 1970 and produced the first paper in Scopus indexing in the same year. The pattern of knowledge dissemination development for the four different institutions abroad can be seen in Fig. 1. In general, knowledge dissemination involves publishing journals, where articles are the medium of communication to readers, and "Article" contributes most of the scientific documents of institutions. The second largest contributor to the number of scientific documents comes from conference

IOP Conf. Series: Journal of Physics: Conf. Series **978** (2018) 012012 doi:10.1088/1742-6596/978/1/012012

(scientific meetings): In OSU, 1903 had known Scopus-indexed "Conference Paper", but CAoS was new having to Scopus-indexed "Conference Paper" in 1981. In general, conferences become the second largest contributor of Scopus-indexed documents in each institution begin 1980s. In addition, knowledge dissemination is also carried out in a variety of ways as evidenced by different types of documents: The greatest contributors to scientific documents other than articles and conference papers are "Review" and "Book Chapter", but in the acceleration of the knowledge dissemination there will be an incomplete article like "Article in Press" the number continuous to increase in CAoS as a result of potential presence about duplication. This means that knowledge dissemination needs to be done more quickly to take the first opportunity so that the same idea be thought by others can be muted. For that reason, always the conference as the first tool used to claim an idea.

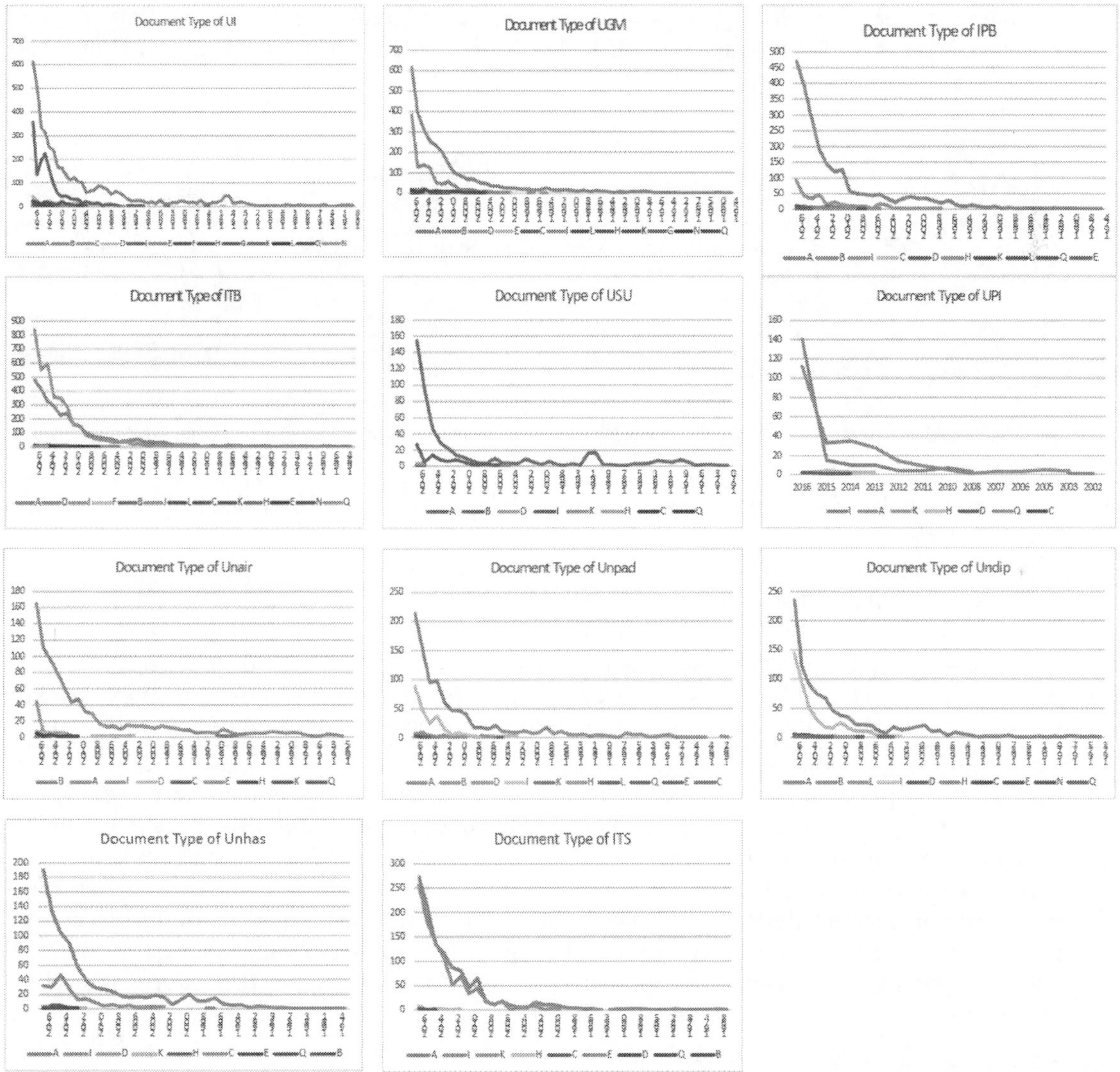


**Figure 2.** Documents type of 11 PTN-bh as representation of Indonesia knowledge dissemination.

The Indonesian knowledge dissemination snapshot is represented by all state-owned universities (Perguruan Tinggi Negeri badan-hukum (PTN-bh)): 11 institutions, i.e. Universitas Indonesia (UI), Universitas Gadjah Mada (UGM), Institute Pertanian Bogor (IPB), Institut Teknologi Bandung (ITB), Universitas Sumatera Utara (USU), Universitas Pendidikan Indonesia (UPI), Universitas Airlangga (Unair), Universitas Padjadjaran (Unpad), Universitas Diponegoro (Undip), Universitas Hasanuddin

(Unhas), dan Institut Teknologi Sepuluh Nopember (ITS). The pattern of knowledge dissemination of each PTN-bh can be seen in Fig. 2. With the exception of ITB and UPI, the knowledge dissemination begins with the publishing of articles in journals. Article is a major contributor of Scopus-based scientific documents, and the conference papers are the second largest contributor. While in ITB and UPI, the conference papers contribute the most compared to articles in journals. The similarity of the knowledge dissemination pattern between 11 PTN-bh as illustrated in Fig. 3a, whereas the similarity of knowledge dissemination is based on the type of document for 11 PTN-bh as shown in Fig. 3b [18, 19]: The growth pattern of Indonesia knowledge dissemination is generally not the same between 11 PTN-bh, it is shown from the emergence of document types on each PTN-bh, there are only two highest similarities between USU and Unpad (0.78) or between USU and Undip (0.56). Other than that it is below 0.50. However, with the exception of ITB, USU and UPI, the growth of the Scopus-indexed document type between PTN-bh is very close to one another, for example there are several proxies worth between some of PTN-bh: Unpad and IPB, Unair and Unhas, Unair and ITS, and Unhas and ITS.

| | 1 | 2 | 3 | 4 | 5 | 6 | 7 | 8 | 9 | 10 | 11 |
|---|---|---|---|---|---|---|---|---|---|---|---|
| 1 UI | | 0.32 | 0.17 | 0.08 | 0.19 | 0.00 | 0.18 | 0.17 | 0.17 | 0.09 | 0.18 |
| 2 UGM | 0.32 | | 0.18 | 0.24 | 0.30 | 0.00 | 0.10 | 0.36 | 0.18 | 0.19 | 0.19 |
| 3 IPB | 0.17 | 0.18 | | 0.17 | 0.33 | 0.12 | 0.21 | 0.30 | 0.40 | 0.11 | 0.11 |
| 4 ITB | 0.08 | 0.24 | 0.17 | | 0.08 | 0.00 | 0.08 | 0.19 | 0.10 | 0.09 | 0.09 |
| 5 USU | 0.19 | 0.30 | 0.33 | 0.08 | | 0.13 | 0.00 | 0.78 | 0.56 | 0.35 | 0.24 |
| 6 UPI | 0.00 | 0.00 | 0.12 | 0.00 | 0.13 | | 0.13 | 0.00 | 0.24 | 0.00 | 0.29 |
| 7 Unair | 0.18 | 0.10 | 0.21 | 0.08 | 0.00 | 0.13 | | 0.00 | 0.00 | 0.00 | 0.22 |
| 8 Unpad | 0.17 | 0.36 | 0.30 | 0.19 | 0.78 | 0.00 | 0.00 | | 0.40 | 0.32 | 0.21 |
| 9 Undip | 0.17 | 0.18 | 0.40 | 0.10 | 0.56 | 0.24 | 0.00 | 0.40 | | 0.11 | 0.11 |
| 10 Unhas | 0.09 | 0.19 | 0.11 | 0.09 | 0.35 | 0.00 | 0.00 | 0.32 | 0.11 | | 0.44 |
| 11 ITS | 0.18 | 0.19 | 0.11 | 0.09 | 0.24 | 0.29 | 0.22 | 0.21 | 0.11 | 0.44 | |

(a)

| | 1 | 2 | 3 | 4 | 5 | 6 | 7 | 8 | 9 | 10 | 11 |
|---|---|---|---|---|---|---|---|---|---|---|---|
| 1 UI | | 0.96 | 0.96 | 0.92 | 0.76 | 0.50 | 0.82 | 0.87 | 0.87 | 0.82 | 0.82 |
| 2 UGM | 0.96 | | 0.95 | 0.83 | 0.84 | 0.56 | 0.90 | 0.95 | 0.95 | 0.90 | 0.90 |
| 3 IPB | 0.96 | 0.95 | | 0.87 | 0.89 | 0.59 | 0.95 | 1.00 | 0.90 | 0.95 | 0.95 |
| 4 ITB | 0.92 | 0.83 | 0.87 | | 0.76 | 0.50 | 0.82 | 0.87 | 0.87 | 0.82 | 0.82 |
| 5 USU | 0.76 | 0.84 | 0.89 | 0.76 | | 0.67 | 0.94 | 0.89 | 0.89 | 0.94 | 0.94 |
| 6 UPI | 0.50 | 0.56 | 0.59 | 0.50 | 0.67 | | 0.63 | 0.59 | 0.47 | 0.63 | 0.63 |
| 7 Unair | 0.82 | 0.90 | 0.95 | 0.82 | 0.94 | 0.63 | | 0.95 | 0.84 | 1.00 | 1.00 |
| 8 Unpad | 0.87 | 0.95 | 1.00 | 0.87 | 0.89 | 0.59 | 0.95 | | 0.90 | 0.95 | 0.95 |
| 9 Undip | 0.87 | 0.95 | 0.90 | 0.87 | 0.89 | 0.47 | 0.84 | 0.90 | | 0.84 | 0.84 |
| 10 Unhas | 0.82 | 0.90 | 0.95 | 0.82 | 0.94 | 0.63 | 1.00 | 0.95 | 0.84 | | 1.00 |
| 11 ITS | 0.82 | 0.90 | 0.95 | 0.82 | 0.94 | 0.63 | 1.00 | 0.95 | 0.84 | 1.00 | |

(b)

**Figure 3.** Similarity between 11 PTN-bh in pattern and document type of knowledge dissemination

**Table 1.** Similarity between 11 PTN-bh and 4 overseas institutions.

| | UI | UGM | IPB | ITB | USU | UPI | Unair | Unpad | Undip | Unhas | ITS |
|---|---|---|---|---|---|---|---|---|---|---|---|
| **OSU** | 0.96 | 0.92 | 0.91 | 0.88 | 0.80 | 0.74 | 0.86 | 0.91 | 0.82 | 0.86 | 0.86 |
| **CAoS** | 0.90 | 0.86 | 0.77 | 0.90 | 0.67 | 0.61 | 0.72 | 0.77 | 0.77 | 0.72 | 0.72 |
| **CUoHK** | 0.88 | 0.92 | 0.91 | 0.88 | 0.80 | 0.74 | 0.86 | 0.91 | 0.82 | 0.86 | 0.86 |
| **UKM** | 0.92 | 0.96 | 0.95 | 0.83 | 0.84 | 0.78 | 0.90 | 0.95 | 0.86 | 0.90 | 0.90 |

In general, all institutions have the potential to produce scientific works that are not published perfectly (Article in Press). There is no significant difference in knowledge dissemination between PTN-bh and other institutions abroad. However, some of PTN-bh has not used the maximum potential to implement knowledge dissemination, as indicated by not all type of document have become a knowledge dissemination medium, see Table 1. In addition, the dissemination of knowledge in principle should be made with the actual effort to the type of "Article in Press" document through the knowledge management, especially from the point of the author to increase the references reading related to the study being written.

## 5. Conclusion

Although there are differences in patterns between 11 PTN-bh in knowledge dissemination. Commonly the knowledge dissemination used media is almost identical. The type of documents that are major contributors to knowledge dissemination are articles, and then conference papers (except ITB and UPI), but other type of documents also significantly give meaning like "Review" and "Book Chapter". The maximum effort of knowledge dissemination by involving all potentials needs to be done, but by avoiding duplication and minimum errors.

IOP Conf. Series: Journal of Physics: Conf. Series **978** (2018) 012012 doi:10.1088/1742-6596/978/1/012012

**References**

[1] Kingston J 2012 Choosing a Knowledge Dissemination Approach, *Knowledge and Process Management* **9**(3)

[2] Standfort J R 2000 Moving Beyond Discretion and Outcomes: Examining Public Management from the Front Lines of the Welfare System *J. of Public Administration Research and Theory* **10**(4).

[3] Nasution M K M 2017 Modelling and Simulation of Search Engine *J. of Physics: Conf. Series* **801**(1)

[4] Nasution M K M, Hardi M and Sitepu R 2016 Using Social Networks to Assess Forensic of Negative Issues *IEEE Proc. of Int. Conf. on Cyber and IT Service Management* (CITSM).

[5] Matusik S F and Hill C W L 1998 The Utilization of Contingent Work, Knowledge Creation, and Competitive Advantage *Acad Manage Rev* **23** (4).

[6] Eraut M 1994 *Developing Professional Knowledge and Competence* (UK, London: RoutledgeFalmer, Taylor & Francis Group).

[7] McDermott R 1999 Why Information Technology Inspired But Cannot Deliver Knowledge Management *California Management Review* **41**(4).

[8] Tanriverdi H 2005 Information Technology Relatedness, Knowledge Management Capability, and Performance of Multibusiness Firms *Information Tech. and Knowledge Manag.* **29**(2).

[9] Marwick A D 2001 Knowledge Management Technology *IBM System Journal* **40**(4).

[10] Gold A H, Malhotra A and Segars A H 2001 Knowledge Management: An Organizational Capabilities Perspective *J. of Manag. Information Systems* **18**(1).

[11] Nasution M K M Theory of sharing knowledge: An introduction *Bulletin of Mathematics*.

[12] Nasution M K M, Elfida M and Mahfudz S 2010 Diskoveri Pengetahuan: Suatu Kritik *Prosiding Seminar Nasional Ilmu Komputer* (SNIKOM 2010).

[13] Nasution M K M 2013 Teknologi Pengetahuan *Dies Fasilkom-TI.*

[14] Tichenor P J, Donohue G A and Olien C N 1970 Mass Media Flow and Differential Growth in Knowledge *Public Opinion Quarterly* **34**(2).

[15] Guz A N and Rushchitsky J J 2009 Scopus: A System for the Evaluation of Scientific Journals *Int. Applied Mechanics* **45**.

[16] Alavi M and Leidner D E 2001 Review: Knowledge Management and Knowledge Management Systems: Conceptual Foundations and Research Issues *MIS Quarterly* **25**(1).

[17] Flowerdew J and Dudley-Evans T 2002 Genre Analysis of Editorial Letters to International Journal Constributors *Applied Linguistics* **23**(4).

[18] Nasution M K M, Sitompul O S, Nasution S and Ambarita H 2017 New Similarity *IOP Conf. Series: Materials Science and Engineering* **180**(1).

[19] Nasution M K M 2011 Kolmogorov Complexity: Clustering and Similarity *Bulletin of Mathematics* **3**(1).

IOP Conf. Series: Journal of Physics: Conf. Series **978** (2018) 012013 doi:10.1088/1742-6596/978/1/012013

# Assessing ERP SAP implementation in the small and medium enterprises (SMEs) in Indonesia

**Yohannes[1*], W Gunawan[1], R B Ikhsan[2] and Aries[2]**

[1]Information Systems Management Department, BINUS Graduate Program – Master of Information System Management Bina Nusantara University, Jakarta, Indonesia
[2] Management Department, Business Management Program, BINUS Online Learning, Bina Nusantara University, Indonesia

[*]Email: yohanes@intidatautama.com

**Abstract.** The consistent growth of Indonesian SMEs has received attention of global IT vendors. One of major ERP player in Indonesian market such as SAP, has introduced ERP solutions for SMEs through SAP Business One. Due to its fast growth, unfortunately, there are still very few articles can be found to understand the ERP implementation in Indonesian SMEs. The article addressed the common factors that affect the successful of the ERP implementation at SMEs. They are: identifying vendor quality, opening balance data migration, and A-SAP methodology. The article applied case study method to examine the fast growth companies owned by same owners, such as: GRK, MMM, and KM25; with the objectives to identify the practical pattern of ERP implementation in SMEs that will be useful for SAP project manager and consultants.

## 1. Introduction

Enterprise resource planning (ERP) system is known as one of the most popular business solutions, enabling real-time benefits and unlimited communication. Michael [1] explained that some advantages of companies that have implemented ERP can improve access to provide accuracy and timely information, increased workflow, increased efficiency, reduce the use of paper to print, monitoring and automation of information via e-mail, ease of user interface, simplicity of process and ease of use of business practice. Bradley [2] revealed the phenomenon of successful implementation of ERP is influenced by the completion of critical success factors. Companies that implement ERP attempt to reduce redundancy and inconsistency data through procurement and maintenance of enterprise information databases [3].

The article examines the market leader of ERP software such as SAP ERP in the global market. SAP solutions have been widely applied in major leading multinational enterprises and SMEs. The popular SAP solutions for SMEs is SAP Business One. SAP Business One inherits the complexity of SAP R/3 with customization to the needs of SMEs [4]. SAP Business One involves a complex process design and requires a detailed work plan and well-planned timetable to ensure that all critical steps in the ASAP method are properly implemented. Since its introduction in 2012, SAP Business One has become major ERP provider for a low cost solution to SMEs in global market [5]. Major leading SMEs in Indonesia has shown increasing trends to adopt SAP Business One as their preference ERP solutions [6].

With consistent growth of Indonesian SMEs has made the needs of SAP Business One becomes inevitable [7]. Aisyah [8] examined the common problems faced by ERP implementation in SMEs such as: people-related factors, and followed with data, process, and management/organisation respectfully. Although those problems are important, Unfortunately, there are still very few articles can be found to address these factors. For this reason, the article examined those issues in fast growth companies under the same owners to identify the characteristics of ERP implementation in SMEs. The study showed although with the same management style, however, with different business environments need different way to address the ERP needs. They are: GRK, a distributor company of four-wheeled vehicles; MMM, a medium-size construction company; and KM25, an automotive spare-part distributor. There are several factors to examine that affecting successful implementation of SAP Business One in those companies are summarised as: quality of implementation (vendor capability), initial data balance migration, and implementation of ASAP methodology.

## 2. Theoretical Foundation

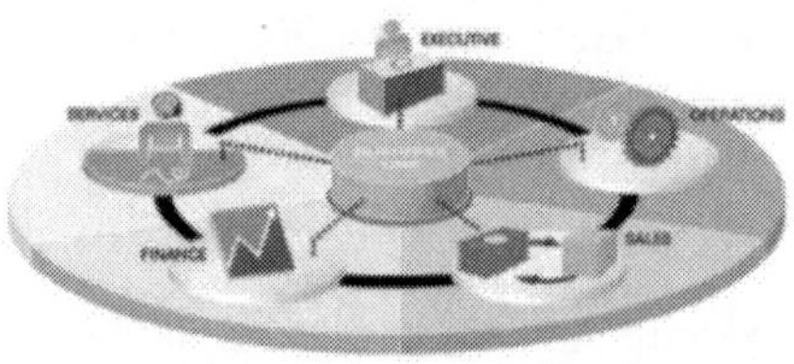

**Figure 1**. Integration SAP SAP Business One

SAP Business One is one of SAP important solution designated to address business needs of SMEs (9). SAP Business One is designed in a single and affordable solution that covers finance, sales, customer relationship management, supply chain management, inventory, purchasing, small-scale manufacturing, project tracking, planning, and reporting [10].

### *2.1. SAP Business One Implementation*

ERP Implementation is the execution of basic policy decisions, usually in the form of laws, but can also take the form of important executive orders or decisions or judicial decisions [11]. Implementation is actions undertaken by individuals / officials or government or private groups directed at achieving the objectives outlined in policy decisions [12]. SAP Business One implementation can be interpreted as a group of people who are working together according certain steps with careful plan. SAP Business One implementation comprised of two main tasks: business functional and technical tasks.

### *2.2. Initial Balance Migration Data*

Data migration is defined as a process of transferring business data from other application systems to SAP Business One system. Data migration referred to migration of data that stored in legacy system to new SAP Business One environment. Data migration stage of a project can consume up to 30% of the total project resources, therefore it is important to plan the data migration on each project [13]. Zaitar et.al. addressed that data migration as an important factor that has risks in the implementation of ERP [14]. The beginning account balance and previous month account balance status are the common problems faced by SAP consultants when applying SAP Business One for SMEs.

## 3. Methodology

### *3.1. SAP Accelerated Methodology*

SAP has established methodology known as accelerated SAP method, to assist business consultant to applying ERP solutions (see Figure 2), comprised of: (1) project preparation; (2) establishing business blueprint (project preparation and refine goals); (3) project realization (implement system, test and configure); (4) final preparation (system testing, end-user training and fine tuning); and (5) go-live and support (move from pre-live to a live-operation, including end-user support).

**Figure 2.** ASAP Methodology

### *3.2. Research method*

The article applies the multiple-case studies for SAP-ERP implementation in the SMEs, that enables to examine the differences within and between cases [15] [16]. The case study method involves using interviews and focus group meeting with the selected respondents, with the aims to replicate the findings with other Indonesian SMEs. The article applies the following propositions: (1) Proposition 1. The quality of the vendor (capability) in managing the project affects the implementation of SAP Business One; (2) Proposition 2. The initial balance migration data affects the implementation of SAP Business One; (3) Proposition 3. Accelerated SAP methodology has maximum effects on the implementation of SAP Business One.

### *3.3. Data gathering method*

The questionnaires were designed with the Likert scale 1-5, and distributed to 75 respondents in: GRK, MMM and KM25. The questionnaires are delivered to each project manager and key users that involved in SAP Business One implementation. The important key users to address such as: supply chain management (SCM)-purchasing division, SCM-warehouse division (logistic), SCM-sales division, finance and accounting division. The interviews and focus group also conducted to the selected respondents in those SMEs to explore their thoughts and data findings. The distribution of respondents is shown in Table 1 below.

**Table 1.** Distribution the respondents

| No | Responden | GRK | MMM | KM25 | Total | Percentage |
|---|---|---|---|---|---|---|
| 1 | **Sex** | | | | | |
| | • Male | 1 | 3 | 2 | 6 | 40% |
| | • Female | 4 | 2 | 3 | 9 | 60% |
| 2 | **Age** | | | | | |
| | • < 25 year | | | | | |
| | • 26 – 35 year | 3 | 4 | 4 | 11 | 73% |
| | • 36 – 45 year | 2 | 1 | 1 | 4 | 27% |
| | • > 45year | | | | | |
| 3 | **Education** | | | | | |
| | • High School | 1 | 2 | 2 | 5 | 33% |
| | • Vocational | 1 | | | 1 | 7% |
| | • Bachelor | 3 | 3 | 3 | 9 | 60% |
| | • Post Graduate | | | | | |
| 4 | **Working Experience** | | | | | |
| | • < 3 years | 1 | 1 | 4 | 6 | 40% |
| | • 3 – 5 years | 2 | 2 | | 4 | 27% |
| | • 6 – 10 years | 2 | 2 | 1 | 5 | 33% |
| | • > 10 years | | | | | |
| 5 | **Title** | | | | | |
| | • ERP Project Manager | 1 | 1 | 1 | 3 | 20% |
| | • Department Head | 4 | 4 | 4 | 12 | 80% |

## 4. Case Study

### *4.1. Overview Enterprise*

The article examined the three companies belong to the same owners to illustrate the difference characteristics of ERP implementation in SMEs. Although they have same owners, it does not necessary have same characteristics of ERP implementation pattern in each company. The article examined the unique characteristics of Business One implementation in those companies. Those companies are: (1) GRK is a distributor of spare-part four-wheeled vehicles and two wheels. Before using SAP Business One, GRK uses Accurate applications to run its business. Due to advance development of the company size, increasing customer needs and complexity business processes have influenced the management to switch legacy Accurate application to SAP Business One. Legacy applications are considered no longer to address the needs of the company; (2) MMM is a construction company and has implemented SAP Business One in 2013. Like GRK, MMM also utilised the Accurate application to support financial reporting, and decided to switch to SAP Business One; KM25 is a distributor company of two-wheeled vehicles. Before using SAP SAP Business One, KM25 applied manual recording by using Microsoft Excel. With fast development of the company and the complexity of business processes in KM25, the management decided to replace entire manual recording system with SAP Business One. KM25 completed SAP Business One implementation in 2014.

### *4.2. The Result Findings*

Based on interviews and observations, the result findings of the three companies are summarized as follows:

#### *4.2.1. SAP Business One Project Execution.*

Major respondents agree that project manager and business function consultant shared the similar roles to support the SAP Business One implementation, with following percentages: (1) GRK, with 71% respondents agree; (2) MMM, with 60% respondents agree; (3) KM25, with 60% respondents agree. The respondents addressed the characteristics of the project execution (see Table 2) below:

**Table 2.** Characteristics of SAP Business One Project Execution.

| GRK | MMM | KM25 |
|---|---|---|
| Equal role of PM and FC is needed to execute the project. | FC to be more dominant than PM. FC needs to manage complexities of business process, although with better preparation. | PM to be more dominant than FC. Having better experiences enables PM to execute the project faster. |

Note: PM: Project Manager, and FC: Functional Consultant.

#### *4.2.2. Initial Data Migration Balance*

**Table 3.** Comparative Results of Initial Data Migration Balance.

| | Source and Target System | |
|---|---|---|
| | **Similarities** | **Differences** |
| **GRK** | Data source: ending balance of legacy system to be initial balance in the new SAP Business One system. | Legacy system from SAP Business One. |
| **MMM** | | Legacy system from Accurate software. |
| **KM25** | | No system, manual recording (MS.Excel). |

| | **Extract Data** | |
|---|---|---|
| | **Similarities** | **Differences** |
| **GRK** | Perform data revision into DTW templates. | Extract data is done from the legacy system of SAP Business One, so the template is in line with DTW. |
| **MMM** | | Extract data is done from the legacy system Accurate, so it needs to be mapped to DTW template. |
| **KM25** | | Extract data is done manually, data entry into DTW templates manually, so it takes longer effort. |
| | **Data Cleansing** | |
| | **Similarities** | **Differences** |
| **GRK** | Data cleansing before uploading into SAP Business One system. | Data cleaning is done after extracting from legacy system (directly inside DTW template). |
| **MMM** | | Data cleaning is done before mapping into DTW templates. |
| **KM25** | | Data cleaning is done simultaneously when going to mapping into DTW templates. |
| | **Data Upload** | |
| | **Similarities** | **Differences** |
| **GRK**<br>**MMM**<br>**KM25** | The upload process uses the Data Work Workbench (DTW) application already provided by SAP. | There is no difference in the point of upload data, because both pass testing and real upload using DTW applications. |

*4.2.3. Accelerated SAP Model*

**Table 4.** Results Comparison of ASAP Method Implementation

| | **Project Preparation** | |
|---|---|---|
| | **Similarities** | **Differences** |
| **GRK** | Project preparation phase had been properly according to procedure | Server installation for SAP database and remote application were done in the preparation phase. |
| **MMM** | | The server of MMM and KM25 were combined into one with GRK server. |
| **KM25** | | The server of MMM and KM25 were combined into one with GRK server. |
| | **Business Blueprint Establishment** | |
| | **Similarities** | **Differences** |
| **GRK** | Blueprint establishment was done through accelerated methodology, where important tasks such as identifying user requirement and supporting documents were produced. | This phase identifies requirement report and type of documents to be used. |
| **MMM** | | Requirement report and documents were taken from GRK. |
| **KM25** | | Similar requirement report and document taken from GRK, and PM assured the blueprint from best practices. |
| | **Project Realization** | |
| | **Similarities** | **Differences** |
| **GRK** | Database setting and configuration, create report and document, upload master data and other configurations. | Internal testing and monitoring implementation process. |
| **MMM** | | Idem |

| | | |
|---|---|---|
| **KM25** | | Focus only at database setting and configuration. |
| | **Final Preparation** | |
| | **Similarities** | **Differences** |
| **GRK** | Critical activities such user-acceptance test (UAT), training, upload opening balance, decision for go-live and restore database to production server. | PM ensured the data is valid. |
| **MMM** | | After opening balance was uploaded. There was no supporting document stated data is valid. PM convinced all data are valid. |
| **KM25** | | There was no documentation is created. PM convinced all data is working well. |
| | **Go Live and Support** | |
| | **Similarities** | **Differences** |
| **GRK** | | Supporting documentation is made to confirm Go-Live status. |
| **MMM** | Final phase and system is ready to run. Users are ready to operate SAP Business One. | MMM and KM25 did not create supporting documents for Go-Live status, since it had been ignored in previous phase; and users are ready to operate SAP Business One. |
| **KM25** | | |

## 5. Lessons Learned

The success of the SAP Business One project in the SMEs sector depends on clear objectives, good implementation quality, structured migration data processes, and maximization of implementation methods used. Sumner [17] stated the critical success factors in system management projects depends on the selection of vendors, needs to be properly identified. In appropriate vendor selection commonly contributes high failures in the ERP implementation. Paul Dorsey [13] examined that migration data as an essential factor that affecting the success of project implementation. So the initial balance-migration data should be planned as early as possible before the project begins. With lack of preparation, well-planned and documentation may lead to project failure. SAP accelerated methodology requires each critical phase should be proceeded properly such as: preparing blueprint, processing steps, user acceptance test and training. Jose and Joan [18] stated that failing to address these critical phases may lead to creating implementation failures.

## 6. Summary

Based on the results of questionnaires, interviews and observations are expected to be the quality factor of the implementation, initial balance data migration and implementation of A-SAP implementation method to be more focused to support the successful implementation of SAP Business One in the SMEs sector. In addition, other non- technical factors such as project management, user engagement, communication, and top management support can influence the success of SAP Business One implementation in the SMEs sector.

## References

[1] Frimpon MF 2011 A Re-Structuring of the Enterprise Resource Planning Implementation Process *Int. J. Bus Soc. Sci.* **2**(24).
[2] Bradley J, Lee CC 2007 ERP Training and User Satisfaction: A Case Study. *Int. J. Enterp. Inf. Syst* **3**(4) pp 33–50.
[3] Poston R and Grabski S 2001 Financial Impacts of Enterprise Resource Planning Implementations *Int. J. Account Inf. Syst.* **2**(4) pp 271–94.
[4] Niefert W 2009 *SAP Business ONE Implementation: Bring the Power of SAP Enterprise Resource Planning to Your Small-to-midsize Business* (Packt Publishing Ltd).

[5] SAP 2017 *History of SAP Business One*; Available from: http://sap-business-one.us/history- sap-business-one-chapter-7-building-global-team/
[6] Handayani PW, Hidayanto AN and Budi I 2013 Business Process Requirements for Indonesian SmallMedium Enterprises (SMEs) in Implementing EnterpriseResource Planning (ERP) *Int. J. Innov Manag Technol* **4**(1) p 93.
[7] Tambunan THT 2011 Development of Small and Medium Enterprises in a Developing Country: The Indonesian Case. *J. Enterprising Communities People Places Glob Econ* **5**(1) pp 68–82.
[8] Aisyah MN 2011 Using Enterprise Resource Planning (ERP) for Enhancing Business Processes in Small and Medium Enterprises (SMEs). *J. Pendidik Akunt Indonesia* **9**(2).
[9] Overview S 2011 *SAP Business One: Solution Overview* pp 1–28.
[10] Nemy J 2008 SAP Business Planning and Consolidation *DM Rev.* **18**(2) p 42.
[11] Bingi P, Sharma MK, Godla JK 1999 Critical Issues Affecting an ERP Implementation. *IS Manag.* **16**(3) pp 7–14.
[12] Van Meter DS and Van Horn CE 1975 The Policy Implementation Process: A Conceptual Framework. *Adm Soc.* **6**(4) pp 445–88.
[13] Koletzke P and Dorsey P 1999 *ORACLE Designer Handbook.* (McGraw-Hill Osborne Media)
[14] Zaitar Y and Ouzarf M. ERP 2012 Projects: Key Success Factors and Risk of Failure a Proposed Model of Governance of Enterprise Resource Planning. *Int. J. Comput Appl* **46**(8) pp 34–9.
[15] Baxter P and Jack S 2008 Qualitative Case Study Methodology: Study Design and Implementation for Novice Researchers Qualitative Case Study Methodology : Study Design and Implementation **13**(4) pp 544–59.
[16] Yin RK 2013 *Case Study Research: Design and Methods* (Sage publications).
[17] Sumner M 2007 *Enterprise Resource Planning* (Pearson Education).
[18] Esteves J and Pastor J 1999 An ERP Lifecycle-Based Research Agenda *1st Int. Workshop in Enterprise Management & Resource Planning.*

IOP Conf. Series: Journal of Physics: Conf. Series **978** (2018) 012014 doi:10.1088/1742-6596/978/1/012014

# Automated hydroponics nutrition plants systems using arduino uno microcontroller based on android

**P Sihombing, N A Karina, J T Tarigan and M I Syarif**

Faculty of Computer Science and Information Technology, Universitas Sumatera Utara, Jl. Dr. Mansyur No.9, Padang Bulan, Medan, Sumatera Utara 20155

Email: poltakhombing@yahoo.com

**Abstract**. Technological developments today make the combination of science is very common, including in Computer Science and Agriculture to make both of science need each other. This paper aims to develop a control tool for the flow of nutrients of hydroponic plants automatically using Arduino microcontroller and controlled by smartphone. We use an Arduino Uno microcontroller to automatically control the flow of nutrient solution with *logic if else*. The microcontroller can also send data of fluid level (solution) and temperature around the plant to smartphone android of the owner of the hydroponics plant. The height of the nutrient solution (water) is detected by the Ultrasonic sensor HC-SR04 and the temperature is detected by the temperature sensor LM35. Data from the sensor will forward into Arduino Uno and displayed in liquid crystal display (LCD) then via wireless fidelity (WIFI) ESP8266 module will transmit the height of the nutrient solution and the temperature around of the plants to Android smartphone.

## 1. Introduction

Plant care is a routine and important activity to keep plants healthy and well groomed. Plant care includes many aspects i.e. watering, rejuvenation, fertilizer, and others. So many types of plants with different forms of treatment are different and all treatments are usually done manually. Although the types of plants are very diverse, water remains the main source of life for all plants to help the photosynthesis process, especially in plants Hydroponics that lives rely on nutrients from the water.

Hydroponics plants have various types of planting media such as Rockwell, sponge, coconut and other coconut powder. In the treatment of plants, Hydroponics is very important to consider the timing of when the water should be added and replaced nutrients, and it would be very inconvenient if the plant owner has Hydroponics plants with a lot. To watering a lot of hydroponics plants, need to do an automatic system that can automatically to hydrate the hydroponics if necessary.

With the development of computer science especially on microcontroller, then automatic watering system for plant hydroponics is very possible to do. Base on the description above, is needed a revolution in agriculture industry. In related with this description, will be built the project of automated tools based on the Arduino Uno Microcontroller and the ATmega328 board as a data sheet. The Arduino Uno as the brain of the tool will monitor the hydroponics plants, assisted by proximity sensor (Ultrasonic HC - SR04) to detect the height of water. The temperature around the plants will be detected by sensors (LM - 35). The height of water and the temperature room will be displayed on the LCD microcontroller. All of the data will be sent to the smartphone Android by using the WIFI (Wireless Fidelity) network.

## 2. Related Work

In some literature, Hydroponics is one of the best alternatives for plants on narrow land. There have been several papers published in several journals in hydroponics systems in recent years; they suggest how hydroponic plant systems work. To supply hydroponic plant nutrients is very appropriate if done by using microprocessors for nutrient control [2]. According to Kumar and Cho, that waste nutrients from hydroponics plants can be reused [3]. In terms of processor-based hydroponic growth spaces, it is more easily controlled via a virtual instrument system using Lab VIEW [4].

Another paper is the development of an automatic microcontroller system for Deep Water Culture (DWC), this paper provides a basic idea of hydroponic water culture. This paper also discusses the methodology used to measure pH values of the sensors and also maintains water levels in hydroponic reservoirs [5]. Other studies are related to the hydroponic nutrient solution control system and factor analysis that influences it. This paper also addresses the difficulties of control that occurred during automation [6].

The other work related to this project is work with control and monitoring of plants in real time has been performance by Dan Wang, Jinling Zhao, Linsheng Huang, Deheng Xu in 2015 [7]. This research focuses on monitoring the data of aquaponic plants which will be sent to smartphones trough the internet of things. The next related is works about automatic addition of nutrients have been made by Rajeev Lochan Mishra and Preet Jain in 2015 [8] which uses electrodes as a benchmark of the amount of nutrients, then if the nutrients under the pump conditions will turn on to add the nutrient. Related works about aotomatic plant watering has been perfomance by Devika et al [9], and the watering object using a plant grown on soil where the humidity sensor will detect the soil moisture level, if the soil is not moist / dry then the system will automatically do the watering.

## 3. Hydroponics Nutrition Plants Systems and prototype

This automatic hydroponics system plant is placed in a special chamber or vessel and the nutrients are sent directly to the hydroponic roots at any given time. Arduino Uno microcontroller will control the flow of water (nutrients) on the vessel automatically, and the microcontroller can be controlled from Android smartphone. To run this project, the program module has been embedded in this system. Arduino Uno microcontroller is used in real-time to set Alarms on nutrient pumps. When the alarm is on Enabled, Relay will be activated and the nutrient pump will drain the Nutrition solution on the hydroponics plant. When the alarm is deactivated, the relay will be extinguished and the nutrient pump will stop supplying.

To support this project, has designed a virtuino application on Android smartphone that serves to see the water level and temperature around the Hydroponics plants. Before starting the design of a virtuino application, it first provides the data storage using the features of thingspeak.com.

In Figure 1, it is shown that a simple of the hydroponic flow system starts from the detection of a proximity sensor and a temperature sensor. The sensor will detect the water level in the hydroponic tube and the temperature sensor to detect the room temperature. The sensor detection results will be connected with a Relay connected to the microcontroller port. When the relay port pin is lower than the specified height, the water flow will be run on the water pump to irrigate the hydroponics plant.

When the relay of port pin is high, it means the water level is above of the specified height, and then the water pump will stop running water. Thus the water circulation will flow regularly. To support this system, we also use the LCD screen to display the time and date at any time of the process. The nutrient pumps are used not only to increase water but also to add nutrients to the hydroponic tube. The water pumps are used for water recirculation and relays are used to regulate both pumps (pumps for water entry and exit of water).

### *3.1 General Architecture and program module*

The general architecture of automated hydroponics nutrition plants systems is shown in figure 1 below.

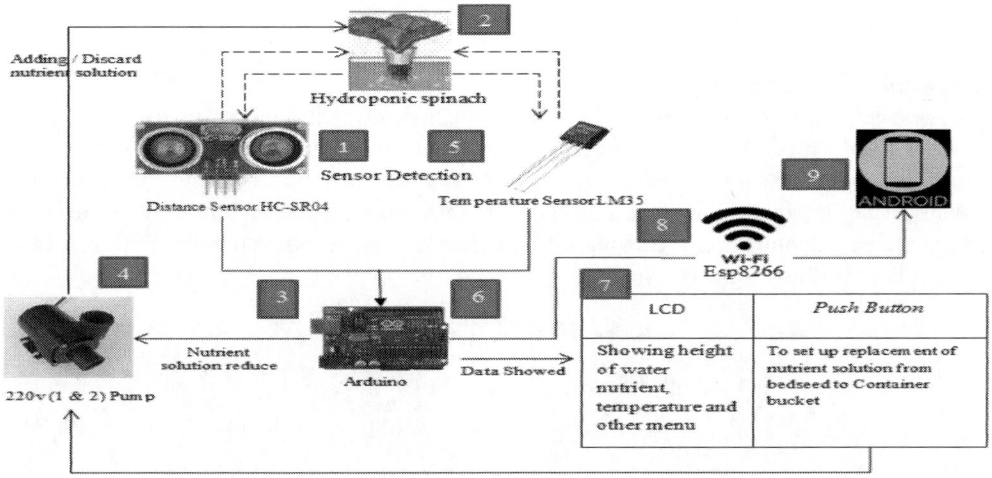


**Figure 1.** The General Architecture of Automated Hydroponics Nutrition Plants Systems

The system mechanism is follows; Ultrasonic sensor (HC - SR04) detects the height value of nutrient solution in hydroponic plants by the parameter of the high of water (in Cm) unit and the temperature. The sensor (LM-35) will detects the temperature in units of °C range. The ultrasonic sensors will measure the distance of water based on ultrasonic wave. The distance between transmit time and receive time becomes a representation of the distance of the water.

In Figure 3 shown above is a representation of the hydroponic water flow system. The system starts working by using of sensors connected to the electrode. The output of this electrode will be forwarded to the microcontroller as an entry point to be processed by the Arduino Uno Microcontroller system's brain. The microcontroller receives this voltage and compares it with the preceding value and makes a decision based on that input. Based on this input the decision is made by the microcontroller whether to drain the water at the pump or not. All of the commands on Arduino Uno microcontroller can be controlled from a smartphone based Android.

### *3.2 The Program module*

The program module to support this project has been embedded in Arduino Uno microcontroller is presented in the Figure 2 below.

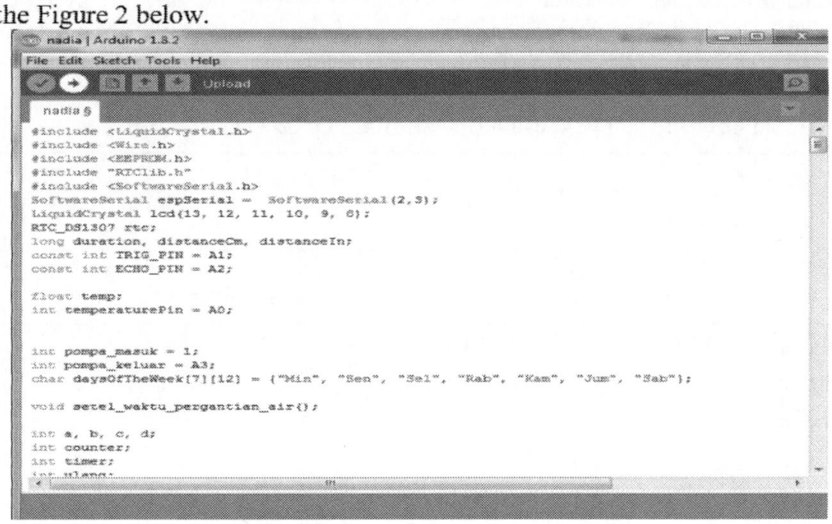


**Figure 2.** The program module

## 4. The Experimental set up and Result

The design and implementation of the hydroponics nutrition system automatically using arduino uno microcontroller performed is shown in Figure 3 below. Generally, this hydroponic plant requires five things: nutrition food, water, light, air, and support. We can provide everything in this hydroponics system. We see that this hydroponic plant grows well with proper water and nutrient usage because it is controlled by the microcontroller. We observed that the rate of hydroponic plant growth was faster when compared to plants with soil-grown systems.

**Figure 3.** The Experimental set up

In the table 1 below shows the WiFi module will send the water level information and the temperature values of the plants area. This value will be compared to the value in LCD microcontroller (in thingspeak.com module) and on Android smartphone application. This value is directly obtained from the sensor and sent to the Arduino. The data resulted shown in the Table 1 below. Table 1 also shows a variation of water level. After the water level 5 cm in the nutrient tube then the pump stops and water does not flow anymore. The average temperature for 5 number of test is obtained 28,426 °C.

**Table 1**. The Data resulted in Arduino Uno Microcontroller.

| Number of tests | Time (WIB) | Benchmark (LCD) | | Data in our Arduino Microcontroller (*thingspeak.com*) | | | Data in Android App | | |
|---|---|---|---|---|---|---|---|---|---|
| | | Temperature | Height (Cm) | Temperature | Height | Delay/ Second | Temperature | Height (Cm) | Delay/ Second |
| 1 | **06.00** | 27,30°C | 0 | 27,30°C | 0 Cm | 3 | 27,30°C | 0 | 5 |
| 2 | **12.00** | 34,41°C | 5 | 34,41°C | 5 Cm | 2 | 34,41°C | 5 | 6 |
| 3 | **18.00** | 29,20°C | 4 | 29,20°C | 4 Cm | 3 | 29,20°C | 4 | 4 |
| 4 | **00.00** | 25,10°C | 5 | 25,10°C | 5 Cm | 1 | 25,10°C | 5 | 2 |
| 5 | **06.00** | 26,12°C | 5 | 26,12°C | 5 Cm | 3 | 26,12°C | 5 | 4 |

To detect the height of water used ultrasonic sensors. Ultrasonic sensor is the main component used in determining the flow of hydroponic nutrition. Ultrasonic sensors work by detecting physical quantities into electrical quantities. The result of sensor detection is shown in Table 2 below.

**Table 2.** The result of sensor detection

| Number of tests | Time (WIB) | Height (Cm) | Information |
|---|---|---|---|
| 1 | 06.00 | 0 | The sensor can detect that the water spool is empty so the pump is on and fills the water up to 6 cm |
| 2 | 12.00 | 5 | At the first 6 a clock the sensor has detected that the water has been reduced in the nutrient tube. |
| 3 | 18.00 | 4 | At the second 6 a clock the sensor has detected that the water has been reduced in the nutrient tube. |
| 4 | 00.00 | 5 | At the third 6 a clock the sensor has detected that the water has been reduced in the nutrient tube. |
| 5 | 06.00 | 5 | At the first 6 a clock after one circle the sensor has detected that the water has been reduced in the nutrient tube |

The relationship of water height in the nutrient tube with time is recorded continuously by the ultrasonic sensor on a number of measurements and the results are shown in figure 4.

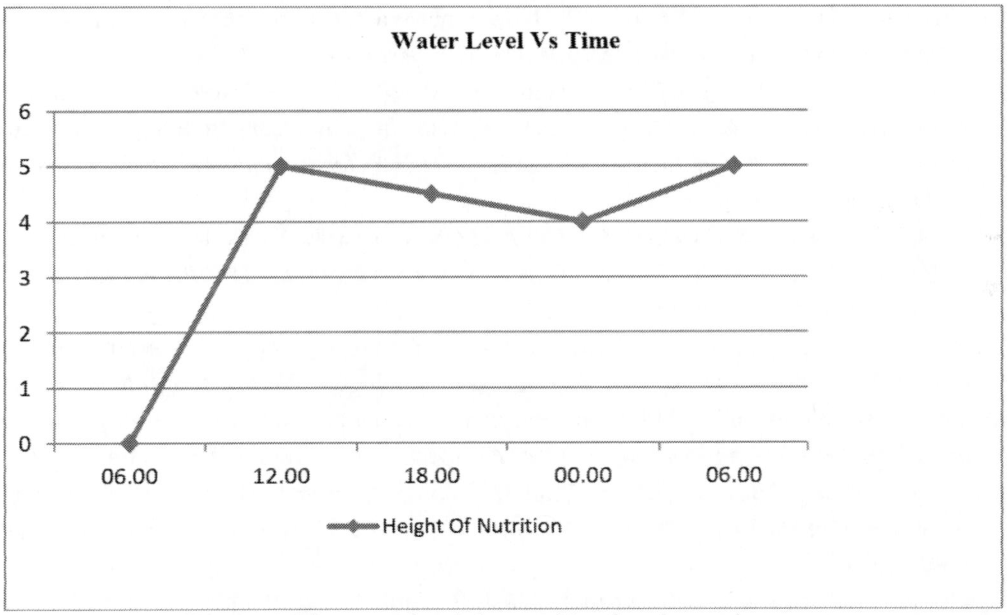


**Figure 4.** The water level detected by the ultrasonic sensor

In the hydroponic nutrient tube has been determined that the maximum water level is 6 cm. The sensor will detect if there is a decrease the water level in hydroponic nutrition tube. If the water level has decreased, then the sensor will report the resulted and automatically the water pump will be turned on to increase the water level on the hydroponic nutrient tube.

## 5. Conclusions

An automation system for hydroponic plant irrigation using Arduino Uno Microcontroller based on Android is presented in this paper. Nutritional water flow system by utilizing sensor distance has been successfully done. The water level in the hydroponic tube can be adjusted according to the need of hydroponic, and this is made into a basic guideline for watering or not. The sensor results are sent to

IOP Conf. Series: Journal of Physics: Conf. Series **978** (2018) 012014 doi:10.1088/1742-6596/978/1/012014

Arduino Uno microcontroller and communication with an Android smartphone. Likewise, the temperature setting has been successfully done well. The room temperature can be determined according to his needs. In our experiments, it was shown that the mixture of water and nutrients was automatically transferred to plant roots in hydroponic tubes. The water level is maintained up to 6 cm in hydroponic tubes because the system will drain water periodically if there is a water shortage.
For future research, it is expected that future developers can to detect acidity levels of pH solution, and viscosity, oxygen, and other aspects. Our automation system is still on a small scale, so it is expected to be developed to make it on a wider scale. In this project, the system only uses the program on Android as a real-time information panel. So it is expected that in advanced research make the tool and system able to make information panel with other operating systems that can be used as a standard system.

**References**

[1] Wang D, Zhao J, Huang L and Xu D 2015 Design of A Smart Monitoring and Control System for Aquaponics *Proc. of the 5th Int. Conf. on Information Engineering for Mechanics and Materials (ICIMM 2015)* pp 937-942.

[2] Velazquez L A, Hernandez M A, Leon M, Dominguez R B and Gutierrez J M 2013 First Advances on the Development of a Hydroponic System for Cherry Tomato Culture *Proc. of IEEE 10th Int. Conf. on Elect. Eng. Comp. Sc. and Automatic Control (CCE)*.

[3] Kumar RR and Cho JY 2014 Reuse of Hydroponic Waste Solution *Environ Sci Pollut Res Int* http://link.springer.com/article/10.1007/s11356-01430243.

[4] Asumadu J A, Smith B, Dogan N S, Loretan P A and Aglan H 1996 Microprocessor-Based Instrument for Hydroponic Growth Chambers Used in Ecological Life Support Systems Instrumentation and Measurement Technology *IEEE Instrumentation and Measurement Technology Conf.* **1** pp 325-329.

[5] Saaid M F, Yahya N A M, Noor M Z H and Ali M S A M 2013 A Development of an Automatic Microcontroller System for Deep Water Culture (DWC) *IEEE 9th Int. Colloquium on Signal Processing and its Applications*.

[6] Chenzhong Y, Yinchun H and Weihong Z 2004 Research of Hydroponics Nutrient Solution Control Technology *Proc. 5' World Congress on Intelligent Control and Automation* **1**.

[7] Shafahi M and Woolston D 2014 Aquaponics: A Sustainable Food Production System *Proc. of the ASME 2014 International Mechanical Engineering Congress and Exposition*.

[8] Mishra R L and Jain P 2015 Design and Implementation of Automatic Hydroponics System Using ARM Processor *Int. J. of Advanced Research in Electrical, Electronics and Instrumentation Engineering* **4**.

[9] Devika S V, Khamuruddeen S, Khamurunnisa S, Thota J and Shaik K 2014 Arduino Based Automatic Plant Watering System *Int. J. of Advanced Research in Comp. Sc. and Software Engineering* **4** pp 449-456.

# Real-time detection with AdaBoost-svm combination in various face orientation

**R P Fhonna[1], M K M Nasution[1*] and Tulus[2]**

[1]Department of Information Technology, Faculty of Computer Science and Information Technology, Universitas Sumatera Utara, Medan 20155, Indonesia
[2]Department of Mathematics, Faculty of Mathematics and Natural Sciences, Universitas Sumatera Utara, Medan 20155, Indonesia

*Email : mahyuddin@usu.ac.id

**Abstract**. Most of the research has used algorithm AdaBoost-SVM for face detection. However, to our knowledge so far there is no research has been facing detection on real-time data with various orientations using the combination of AdaBoost and Support Vector Machine (SVM). Characteristics of complex and diverse face variations and real-time data in various orientations, and with a very complex application will slow down the performance of the face detection system this becomes a challenge in this research. Face orientation performed on the detection system, that is $90^0$, $45^0$, $0^0$, $-45^0$, and $-90^0$. This combination method is expected to be an effective and efficient solution in various face orientations. The results showed that the highest average detection rate is on the face detection oriented $0^0$ and the lowest detection rate is in the face orientation $90^0$.

**Keywords:** face detection, AdaBoost-SVM, real-time, face orientations.

## 1. Introduction

Detection and recognition of objects are widely used in our daily lives for the purposes of inspection, registration, and manipulation [1]. Face detection, in particular, is probably one of the most interesting object detection that allows the machine to interact with human beings in face recognition. The face is a distinguishing feature that is often used as a key to get to know one's identity [2].The Face detection system is complicated due to the level in various of the face are complex, so real-time system [9], the application of the approach so complex will slow system performance, and instead the application of an approach that is not so complex will affect the level of accuracy in the detection of the face. To date more than 150 approaches that have been applied to building a system of detection of the face.

There are several approaches to the problem of recognition of facial expression classification [3] and gender [4] which uses Support Vector Machine (SVM), which the SVM showed better results in terms of the level of truth but requires more intensive computing, while on a research [5]. AdaBoost is much faster with a little more accuracy level decreased. Boosting is a common method to improve the accuracy of the algorithm learning and designed to deal with the problem of classification. The most popular Boosting algorithms based on Freund and Schapire [4] is AdaBoost.

In research conducted by [1] have tried to combine the methods of AdaBoost-SVM to detect faces and this method proved to be able to have a higher detection rate and a lower error rate, and reducing the time calculation in training. But the facial data detected with image data, not in real-time and face detection system conducted in this study only detect the orientation of 0°.

Based on the above research, this time, researchers tried to use a combination of methods AdaBoost and Support Vector Machine (SVM) in real-time data and tries to detect the face at the various orientations, that is the orientation of the faces of the $90^0$, $45^0$, $0^0$, $-45^0$, and $-90^0$. The combination of this method is expected to be the most efficient and effective method for face detection in real-time in a variety of orientations of the face.

## 2. Face Detection

There are a number of approaches that have been done by some researchers in building a system of detection of the face. In general, the methods applied on the face detection system can be classified into [6].

- *Knowledge-based Method*, this method uses the basic rules that are commonly used by humans to determine what form a face. On this approach, the face detection method developed based on rules (rule) obtained from the knowledge of the researchers about the human face.
- *Featured Invariant Approach,* algorithm in this method aims to find the structural features of the face that still exist, although there are variations in pose, viewpoints, and light conditions. On this approach, the researchers tried to find the features that do not change (invariant) on the face.
- *Template Matching*, this method will be kept in several standard patterns to describe the face as a whole nor its parts. At the time of detection will be calculated the correlation between the input image with the image patterns stored earlier.
- *Appearance-Based Method,* in this method, the model face learned through the process of training by using a set of training data that contains instances of the face. Then the results of this training are used to detect faces. In general, this method use techniques of statistical analysis and machine learning to discover the characteristics of the relevant characteristics of the face. The challenge facing the problem of face detection caused the following factors [7] the position of the face. The position of the faces in the image can vary because of the positions may be upright, tilting, turning, or viewed from the side.

1. The components of the face can exist or not exist, for example, a mustache, beard, and glasses.
2. Facial expression. The facial appearance is strongly affected by a facial expression, such as smiling, laughing, sad, talk and so on.
3. Obstructed by other objects. The image of a face can be deterred in part by an object or another face, for example on the image contains a group of people.
4. Image retrieval Conditions. The image obtained is highly influenced by factors such as light intensity, a direction of the light source, and the characteristics of the sensor and the quality of the camera.

## 3. AdaBoost

AdaBoost short for Adaptive Boosting is a machine learning algorithm that was created from a collection of a weak classifier which formed into being strong classifier [4]. AdaBoost by Freund and Schapire in 1995 [8]. Called adaptive because this algorithm can confirm with the data and the method of classifying. Named after boosting algorithms because it can reduce errors of a weak classifier and increase the accuracy of any given learning algorithm [12]. This is because basically, AdaBoost appears to be able to reduce errors in the process of learning[5]. AdaBoost algorithm for feature selection in large quantities, with only select certain features.

Boosting is a meta-algorithm in machine learning to perform the supervised learning. AdaBoost function to look for features that have a high level of distinction [10]. This is done by evaluating every feature against the training data by using the value of the feature. The features that have the biggest limit between objects and non-object is considered the best features [11].

## 4. Support Vector Machine (SVM)

Support Vector Machine (SVM) is a technique of performing a prediction which could only distinguish two classes, techniques using a formula which would later form a hyper plane that separates a data into two classes. In General, SVM linear in nature, due to a nonlinear SVM is the result of the modification to include the functions of the kernel. This technique was very tough, but the disadvantage SVM classification in the process still doesn't have rapid performance [3].

Support Vector Machine (SVM) is a learning system using space hypothesis in the form of linear functions in a space feature (feature space) high-dimensional, trained with the learning algorithm based on optimization theory by implementing learning bias derived from statistical learning theory [1].

## 5. Research purposes

The purpose of this research is to measure the performance of the combination method of AdaBoost-SVM using real-time data in a variety of orientations of the face.

### 5.1 AdaBoost-SVM

For AdaBoost-SVM later begins with the selection of a single trait, after each selection process with each error, the smallest error is selected to be used as ingredients for classification SVM training data. The process of AdaBoost-SVM can be described in described below:

- Input sample
- Initiation of weights for each sample

$$\mathbf{D_i = \frac{1}{n} \ (i = 1, 2, \ldots n)} \tag{1}$$

- For t = 1 ... ... ... T:
  For each h(xi), calculate the weighted error

$$\boldsymbol{\varepsilon_t^{(i)} = \sum_{j=1}^{n} D_i \left| h\left(x_j - y_{(j)}\right) \right| \pi r^2} \tag{2}$$

- Then do the calculation of the weighted error in weak classifier AdaBoost

$$\mathbf{a_t = \frac{1}{2} \ln\left(\frac{1-\varepsilon_t}{\varepsilon_t}\right)} \tag{3}$$

- After calculating the weighted error above, then done updated on weights for the next iteration process error

$$\mathbf{D_{t+1}(i) = D_t(i)^{\frac{-a_t y_i h_t(x_i)}{Z_t}}} \tag{4}$$

- Sort the data from lowest to highest error.
- Take the number of data selection as much as.
- At this point, enter the SVM classification which should determine the initial value of initiation, i.e. alpha, maximum iterations, C, epsilon, lambda, and gamma.
- Calculate the value of a linear function of the Kernel

$$\mathbf{K(x, y) = x, y} \tag{5}$$

- Calculate The Hessian Matrix

$$\mathbf{D_{ij} = y_i y_j \left(K\left(x_i, x_j + \lambda^2\right)\right)} \tag{6}$$

IOP Conf. Series: Journal of Physics: Conf. Series **978** (2018) 012015 doi:10.1088/1742-6596/978/1/012015

- Do as much Iteration n times

$$\mathbf{E_n} = \sum_{j=1}^{l} \mathbf{a_j D_{ij}} \tag{7}$$

- Calculate the maximum value of an alpha,

$$\boldsymbol{\delta} \mathbf{a_j} = \mathbf{min}\{\mathbf{max}[\boldsymbol{\gamma}(\mathbf{1} - \mathbf{E_i}), -\mathbf{a_j}], \mathbf{c} - \mathbf{a_j}\} \tag{8}$$

If the maximum value $|\delta a_j|$ more than epsilon, then iteration continues. However, if the maximum value $|\delta a_j|$ less than epsilon, then the iteration will stop. After getting the process of AdaBoost-SVM above obtained the data model will be tested with data testing.

## 6. Research Methodology

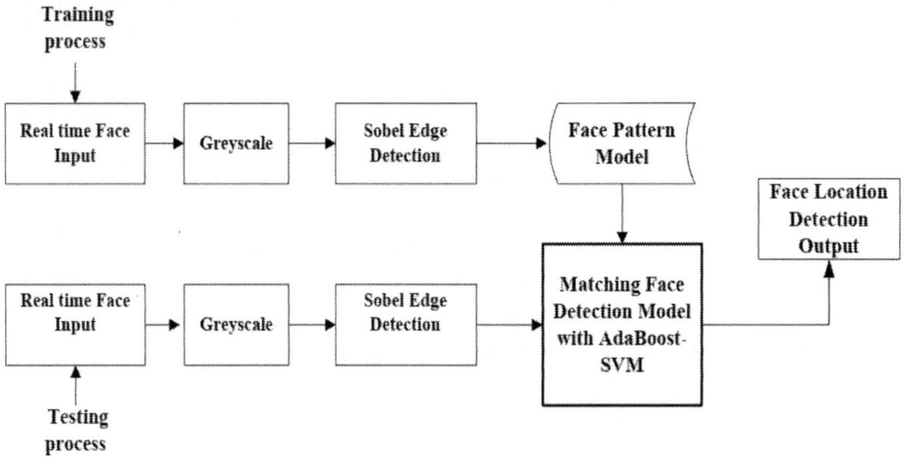


**Figure 1.** The Scheme Research

Figure 1 describes the research scheme, where there are two processes, namely the process of training and testing process, training process on the input image done pre-processing with Greyscale and Sobel operator using edge detection and continued localized patterns as a model that will be detected at the stage of testing data, and at the stage of testing, the input image done pre-processing with using edge detection of Greyscale and Sobel operator and forwarded to the stages using pattern matching AdaBoost-SVM method, If the pattern is similarly or close to the pattern of training then the pattern will be in blocking as an output location detection is a face.

## 7. Results and Discussion

Face detection system scheme that was built using a combination of AdaBoost-SVM in this study are illustrated in the following figure:

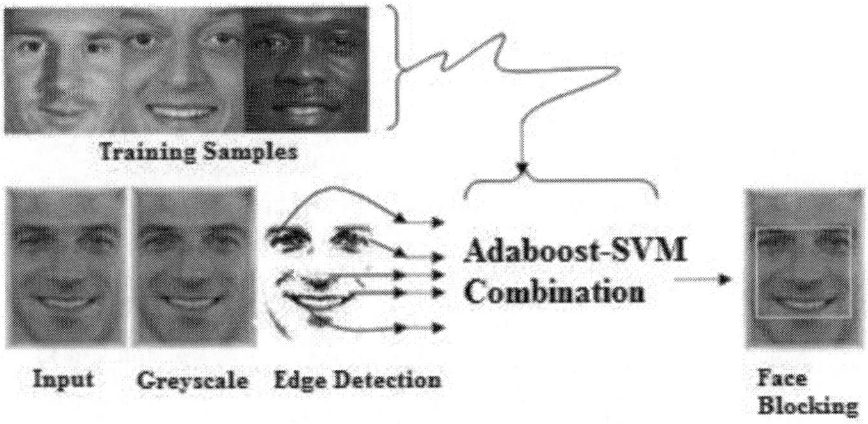


**Figure 2.** The Overall Scheme of the System Face Detection

Face detection system performance based on the parameters of the true and false detection. Table 1 illustrates some of the results of measurement performance AdaBoost-combination method of SVM on face detection in real-time in a variety of orientations.

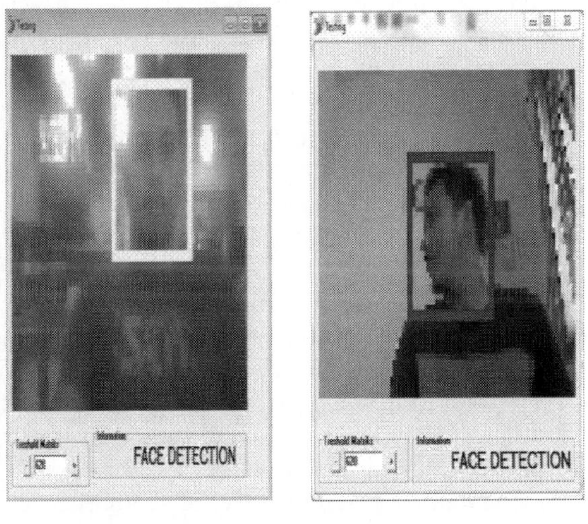


**Figure 3a.** Training Processing

**Figure 3b.** Testing Processing ($0^0$) and ($90^0$)

**Table 1.** The results of the combination method AdaBoost-SVM show on real-time face detection in orientations

| The number of training | The number of testing | True detection *real-time* face detection with *Adaboost-SVM* combination in orientation (%) | | | | | The average detection rate | |
|---|---|---|---|---|---|---|---|---|
| | | $90^0$ | $45^0$ | $0^0$ | $-45^0$ | $-90^0$ | *true detection* | *false detection* |
| 10 | 20 | 5 | 25 | 40 | 20 | 2 | 50% | 50% |
| 20 | 40 | 9 | 32 | 55 | 40 | 7 | 62% | 38% |
| 40 | 60 | 15 | 45 | 75 | 44 | 16 | 78% | 22% |
| 60 | 70 | 23 | 53 | 87 | 67 | 22 | 89% | 13% |
| 80 | 100 | 25 | 64 | 92 | 72 | 28 | 95% | 5% |

Table 1 describes the results of research in which the detection rate of the highest average in the detection of facial orientated $0^0$ is the position of the front view and the detection rate is the lowest in the orientation of the faces of $90^0$. True the detection rate reached 95% on 100 test data.

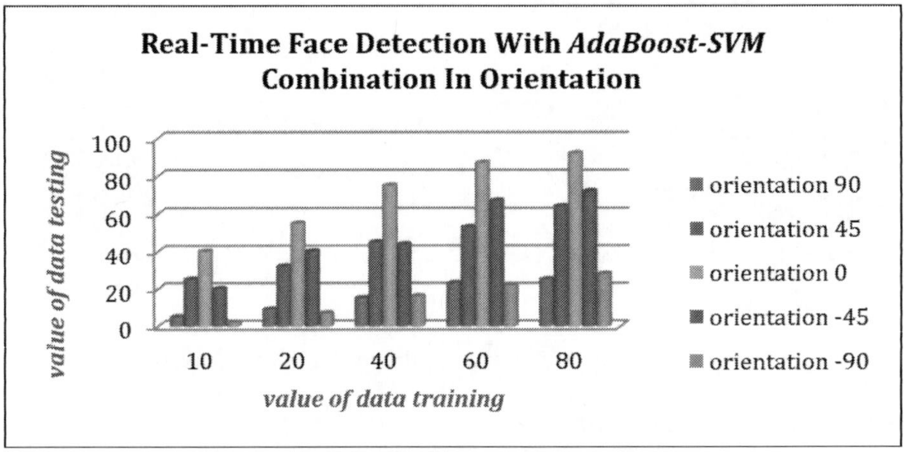


**Figure 4.** Graph of *AdaBoost-SVM* performance

Based on Figure 4 graphs illustrates the measurement of performance the work done after 20 sample faces are trained to face pattern characteristics using a combination of methods AdaBoost-SVM. The number of samples testing as many as 100 patterns done gradually in real-time on the orientation $90^0$, $45^0$, $0^0$, $-45^0$, and $-90^0$ , in the early stages of testing done detection system that has trained 10 facial samples taken in real-time in a variety of combinations using the method of face orientation AdaBoost-SVM and obtained results as has been described previously on the table and look at the graph of AdaBoost-SVM performance. The highest detection of the orientation of $0^0$ is the orientation of the face position looks ahead to reach 95% the level of accuracy of detection.

## 8. Conclusion and Future Work

On the research of detection system face trying to build in real-time using a combination of methods AdaBoost-SVM in various orientations to the face. The results of the evaluation of the system show that the increased detection rate is highly associated with an increased amount of training. Based on the results of detection rate face detection in real-time using a combination of methods AdaBoost-SVM in various orientation accuracy rate reaches 95%. There is still room for improvement and some recommendations by adding specific feature extraction and recognition of facial characteristics as complex detail in order to facilitate the detection of the face either in real time or not, it can also combines the methods of AdaBoost with other classification methods.

## References

[1] Mingxing J, Junqiang D, Tao C, Ning Y, Yi J and Zhen Z 2013 An Improved Detection Algorithm Of Face With Combining AdaBoost And SVM *Chinese Control and Decision Conf. (CCDC)* pp 2459-2463.

[2] Gonzalez R C and Woods R E 1992 *Digital Image Processing* (USA: Addison - Wesley Publishing Company).

[3] Tsai HH and Chang YC 2017 *Facial Expression Recognition Using A Combination Of Multiple Facial Futures And Support Vector Machine* (Springer-Verlag Berlin Heidelberg).

[4] Ruan C, Ruan Q and Li X 2010 Real AdaBoost Feature Selection For Face Recognition *Int. Conf. on Signal Processing (ICSP)* pp 24-28.

[5] Zhao XC, Yuan JZ, Liu HZ and Zhou JS 2017 Improved AdaBoost Algorithm for Robust Real-Time Multi-Face Detection *J. of Software* **12** (1) pp 53-61.

[6] Huiyu Z, Wu J and Zhang J 2010 *Digital Image Processing Part I,* bookboon,com The Ebook Company.

[7] Gonzalez R C and Woods R E 2008 *Digital Image Processing Third Edition* (USA: Addison - Wesley Publishing Company).

[8] Ma S and Bai L 2016 A Face Detection Algorithm Based on AdaBoost and New Haar-Like Feature *Int. Conf. on Software Engineer and Service Science (ICSESS)* pp 651-654.

[9] Viola P and Jones M J 2004 Robust Real-time Face Detection *Int. J. of Computer Vision* **57** pp 137–154.

[10] Niyomugabo C, Choi HT and Kim T Y 2016 A Modified AdaBoost Algorithm to Reduce False Positive in Face Detection *Mathematical Problems in Engineering*.

[11] Tofighi A and Monadjemi S A 2011 Face Detection And Recognition Using Skin Color And AdaBoost Algorithm Combined With Gabor Features And SVM Classifier *Int. Conf. On Multimedia And Signal Processing* **1** pp 141-145.

[12] Viola P and Jones M J 2001 Rapid Object Detection Using A Boosted Cascade Of Simple Features *Proc. of IEEE Comp. Society Conf. on Comp. Vision and Pattern Recognition* pp 511–518.

IOP Conf. Series: Journal of Physics: Conf. Series **978** (2018) 012016 doi:10.1088/1742-6596/978/1/012016

# Delay and cost performance analysis of the diffie-hellman key exchange protocol in opportunistic mobile networks

**B Soelistijanto and V Muliadi**

Department of Informatics, Sanata Dharma University, Yogyakarta, Indonesia

Email: b.soelistijanto@usd.ac.id, vmuliadi@max-metal.us

**Abstract.** Diffie-Hellman (DH) provides an efficient key exchange system by reducing the number of cryptographic keys distributed in the network. In this method, a node broadcasts a single public key to all nodes in the network, and in turn each peer uses this key to establish a shared secret key which then can be utilized to encrypt and decrypt traffic between the peer and the given node. In this paper, we evaluate the key transfer delay and cost performance of DH in opportunistic mobile networks, a specific scenario of MANETs where complete end-to-end paths rarely exist between sources and destinations; consequently, the end-to-end delays in these networks are much greater than typical MANETs. Simulation results, driven by a random node movement model and real human mobility traces, showed that DH outperforms a typical key distribution scheme based on the RSA algorithm in terms of key transfer delay, measured by average key convergence time; however, DH performs as well as the benchmark in terms of key transfer cost, evaluated by total key (copies) forwards.

## 1. Introduction

The subject of key distribution is one of the important issues in key management [1]. An obvious method of distribution of keys is simply by hand. This method was frequently used in the days of couriers. However, it is used only infrequently nowadays, since most key distribution is performed automatically. Automatic distribution is not only more convenient, but often even essential; for instance, in telephone or computer networks, which require two parties to transmit their security keys along the same communication line. In these cases often two types of keys are employed: keys which are used for the actual security of the data (so-called *session keys*), and keys which are used for the security of these session keys during transmission (so-called *transportation keys*).

The Diffie-Hellman (DH) protocol [2] is a method for exchanging keys in the network. In this algorithm, two parties unknown to one another can set up a private however arbitrary key for their symmetric key cryptosystem. Along these lines, there is no requirement for Alice and Bob to meet ahead of time, or utilize a safe dispatch, or utilize some other secret means, to choose a key. DH was the first practical method for setting up a "shared secret" over an unsecured communication channel. The security of this algorithm is based on the hardness of a certain computational problem. This paper discusses the delay and cost performance analysis of the DH key exchange protocol in opportunistic mobile networks (OMNs), a class of delay-tolerant networks (DTNs) [3] where nodes come into contact with each other at unpredictable intervals and the duration of each contact is also unpredictable. The area of key management in DTNs is relatively new and many research challenges remain to date [4]. Traditional key management is not suitable for DTNs due to the environment limitations and technical constraints, e.g. long round trip delays and frequent link disconnections. The author in [5] identified some requirements for key management in DTNs. The works in [6,7,8] proposed new solutions to address the security issues of key management in DTNs.

In this paper, however, we discuss the key transfer delay and cost performance analysis of DH in OMNs. To best of our knowledge, the performance evaluation of DH in OMNs in terms of key transfer delay and cost has not been discussed before. In delay-tolerant networks, such as OMNs, where end-to-end delay is very large, key transfer delay is an important aspect that directly affects the performance of DH. Indeed, a low key transfer latency is required to achieve high performance of DH. In this analysis, we consider average key convergence time to identify the key transfer delay performance of DH in OMNs. Besides transfer delay, delivery cost is an important parameter in mobile communication networks, since mobile devices typically have limited resources, e.g. storage and power. Furthermore, we use total key (copies) forwards as a metric to quantify the key transfer cost of DH in OMNs.

The rest of the paper is structured as follows. Section 2 gives an overview of the DH key exchange algorithm. Section 3 describes our experimental setup for evaluating the key transfer delay and cost performance of DH in OMNs. A performance analysis of the protocol compared to a typical key distribution scheme based on the RSA algorithm is reported in Section 4. Finally, Section 5 concludes the paper.

## 2. The Diffie-Hellman Key Exchange Protocol

The Diffie-Hellman (DH) protocol offers an efficient method for exchanging keys in the network. Consider a network of $n$ nodes (parties) who communicate bilaterally in an enciphered form, controlled by a key. In the network, the total number of keys is $n(n-1)$ in which two nodes are involved during each communication. Each node must therefore have $(n-1)$ keys available to be able to communicate with the other $(n-1)$ nodes. This kind of communication will require a large number of keys and managing these keys correctly can be very complicated.

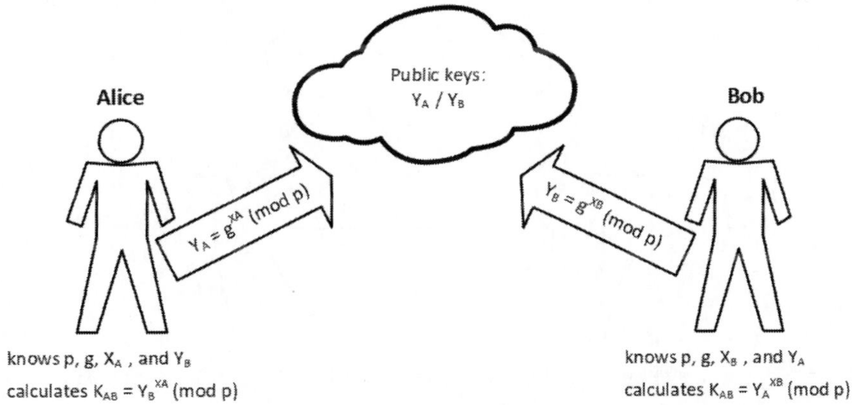


**Figure 1.** The Diffie-Hellman key exchange protocol

Diffie and Hellman (1976) have devised a method with which the number of keys can be reduced drastically. Their system is based on the exponential and logarithmic functions. Given that all nodes in the network select the same prime number $p$ and a base $g$. Each node $N_i$ may now choose a secret key $x_i$ and calculates $y_i$ according to $y_i = g^{x_i} \pmod p$; the value of $y_i$ is then made public. Suppose a node $N_i$ wishes to communicate with a node $N_j$. This will require a mutual session key $K_{ij}$, with which the message can be enciphered and deciphered. Node $N_i$ must first calculate $K_{ij} = {y_j}^{x_i} \pmod p$. $N_i$ can perform this calculation since $p$ is published, $x_i$ is $N_i$'s own key, and $y_j$ was disclosed to $N_i$ by $N_j$. On the other hand, node $N_j$ can calculate the session key $K_{ji} = {y_i}^{x_j} \pmod p$. This will result in exactly the same session key for both nodes ($K_{ij} = K_{ji}$), due to symmetry of the power function. In figure 1, we illustrate the DH key exchange protocol between two parties, Alice and Bob.

## 3. Simulation Setup for Evaluating the Diffie-Hellman Protocol in OMNs

In this section, we discuss the choice of simulation scenarios, evaluation metrics, and benchmark protocols to evaluate the delay and cost performance of DH in OMNS. We implement DH using the ONE simulator [9], an event-driven simulator for mobile opportunistic networks. In our simulations, the number of nodes and the length of simulation time vary depending on the node mobility scenarios. The node buffer size and the message size are set to 10 MB and 4 kB, respectively. Moreover, we assume that a message can only carry a single key. The simulations were run 10 times for both DH and its protocol benchmark with different random number seeds.

For the simulation's node mobility scenario, we use a random movement model and real human mobility traces. For the latter case, we exploit the Infocomm [10], Reality [11] and Sassy [12] human contact datasets. In Infocomm, 41 iMote Bluetooth-enabled devices were distributed to attendees at the IEEE Infocomm conference in Miami in 2005. These devices recorded the human contacts occurred during the 3-day seminar. In Reality, on the other hand, 100 smart phones were deployed among the students and staffs of MIT over period of 9 months. These phones were running software that logged contacts with other phones, capturing academic activities in the campus over an academic year. The Sassy trace, in contrast, was collected using a mobile sensor network with TMote invent devices carried by 25 participants from the University of St. Andrews for period of 74 days.

For performance analysis, we use two evaluation metrics as follows:

1. **Average key convergence time**: the mean of the times required before all the nodes in the network can reach knowledge regarding the latest updated of the key of a particular node.
2. **Total key (copies) forwards**: the total number of key (copies) forwarded during node contacts throughout the simulation time.

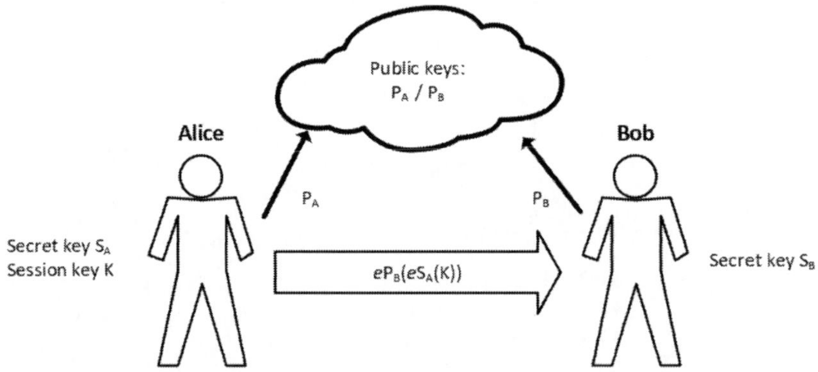


**Figure 2.** A key distribution scheme based on the RSA algorithm

In order to benchmark the performance of DH, we consider a typical key distribution scheme based on the RSA algorithm (hereafter called RSA) as follows. In figure 2, consider two parties, Alice and Bob, whose public keys $P_A$ and $P_B$, respectively, are made available in the network, and secret keys are denoted by $S_A$ and $S_B$. Suppose that Alice wishes to transfer a (session) key $K$ to Bob. Alice may do this by first enciphering the key $K$ with her own secret key and subsequently enciphering this result with Bob's public key. The total result can be written as $(K^{S_A})^{P_B}$. Then, Bob can find the original key $K$ by decrypting the received message with his own secret key followed by a second decipherment with Alice's public key.

In our simulations, a new public key is generated at a randomly selected node at a rate of one key per hour for both DH and RSA. Moreover, in RSA, we assume that after a node broadcast its new public key to all nodes in the network, the node promptly sends an individual encrypted session key $K$

to each the network node. Therefore, the average key convergence time in RSA is calculated as the mean of the times when all the network nodes are able to find the original key *K* of a node by decrypting the received messages minus the time when *K* is created in the original node. Furthermore, the peer node is able to disclose the encrypted session key *K* whenever it has knowledge of the latest updated public key of the original node. This is however not the case in DH since *K* is never transported over the network but is calculated autonomously on both communicating end nodes, providing that each end node has information of the public key of the other end node (as shown in figure 1). As a result, the key convergence time in DH can be calculated as the average of times when all nodes in the network receive a particular node' public key minus the time when the key is created in the original node.

Finally, to broadcast nodes' public keys in the network (in DH and RSA) and to deliver session keys *K* to the particular destination (in RSA) we use Epidemic routing [13]. The algorithm is flooding-based in nature, as nodes continuously replicate and transmit messages to newly discovered contacts that do not already possess a copy of the message. Epidemic routing results in a high per message delivery probability and the lowest delay. Despite its benefits, this oblivious forwarding strategy consumes much the constraint resources of mobile nodes, such as storage and power.

## 4. Simulation Results

In this section, we present the simulation results that compare the key transfer delay and cost performance of DH with that of RSA in OMNs. In the first experiment, we consider a random node movement model in the simulation. In the second experiment, we use real human contact data traces as the simulation's node mobility scenario.

### *4.1. Random Node Movement Scenario*

We now discuss the key transfer delay and cost performance comparison results between DH and RSA in a random node movement scenario. We used a synthetic random walk model available in the ONE simulator's library. In this setting, node contacts are spread evenly between the nodes over the network. From the simulation results, in figure 3 we depict the key transfer delay and cost performance comparison results of DH and RSA in the random node scenario. We defined a fixed simulation area at 2500m x 2500m but varied the number of nodes in the network from 50 up to 150 nodes. The length of the simulation time was 2500ks for both protocols.

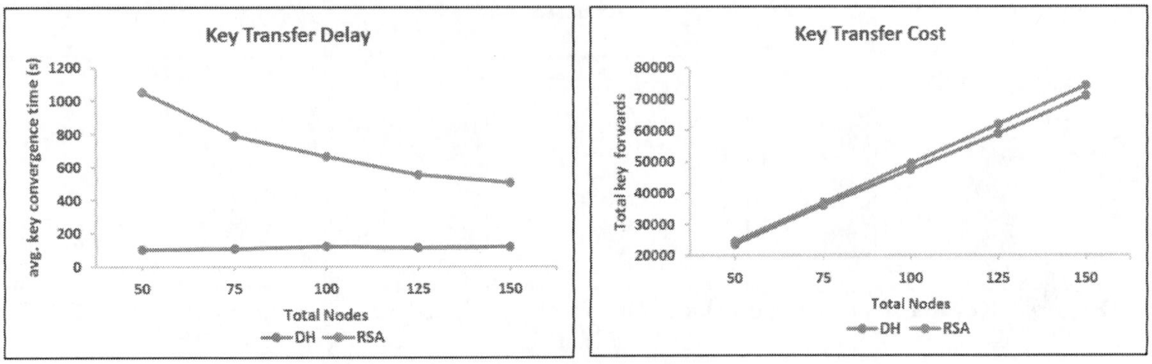


**Figure 3.** Key transfer delay and cost performance comparison of DH and RSA in the random node mobility scenario

Figure 3 shows that DH outperforms RSA in terms of key transfer delay, measured by average key convergence time. This is because the key convergence time in DH relies merely on the speed of public keys spreading in the network. In RSA, on the other hand, the key convergence time depends not only on the broadcast rate of public keys to all the network nodes, but also on the delivery time of session keys to their destinations (unicast transmissions). Moreover, the increase of total nodes in the network significantly reduces the key transfer delay in RSA, but it gives little impact in DH. In terms of key transfer cost performance, on the other hand, both protocols perform almost similarly. Indeed, DH slightly reduces the total key (copies) forwards of RSA. In RSA, the majority of key (copies) distributed in the network are nodes' public keys and the delivery of session keys consequently less contributes on the total key (copies) forwarded during the simulation time. In line with this, the absence of the unicast transmissions of session keys in DH slightly decreases the key transfer cost below that of RSA. Finally, the key transfer cost grows linearly with the increasing of total nodes in the network for both DH and RSA.

### *4.2. Real Human Mobility Scenario*

In the second experiment, we consider a real human mobility scenario to evaluate the key transfer delay and cost performance of DH compared to that of RSA. We used Reality, Sassy and Infocomm as the simulation's node mobility scenario. We again used the simulation settings of the previous experiment, except for the number of nodes and the length of simulation time which vary depending on the datasets.

From the simulation results, in figure 4 we depict the key transfer delay performance comparison of DH and RSA, measured by avg. key convergence time, for all datasets. It is clear that DH outperforms RSA in this performance metric (i.e. a lower avg. key convergence time of DH compared to that of RSA) in all there datasets. Moreover, the key transfer delay performance's difference between DH and RSA is more obvious in Reality and Sassy. In these datasets, node contacts are sparser (than Infocomm's) and, as a result the delivery time of session keys to the destinations is larger, leading to the significant increase of the key transfer delay of RSA beyond that of DH in both Reality and Sassy.

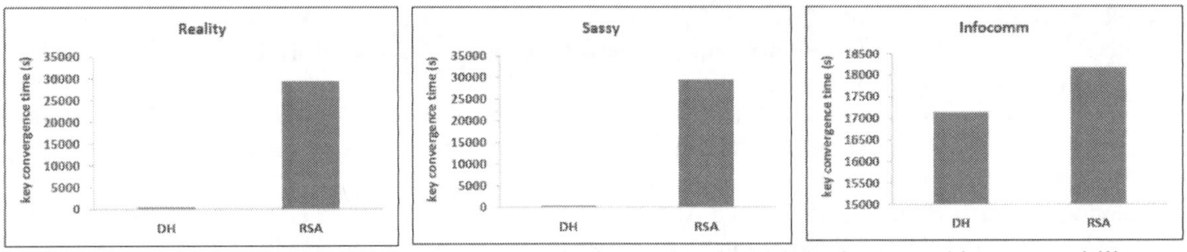


**Figure 4.** Key transfer delay performance comparison of DH and RSA in the real human mobility scenario

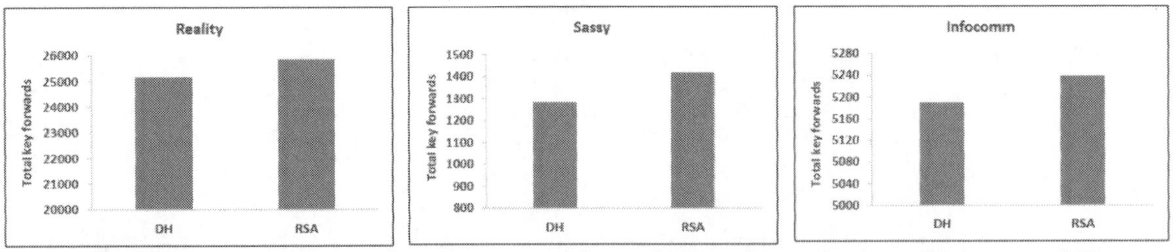


**Figure 5.** Key transfer cost performance comparison of DH and RSA in the real human mobility scenario

Finally, in figure 5 we show the key transfer cost performance comparison results, measured by total key (copies) forwards, for DH and RSA in all the datasets. As in the random node mobility scenario, both protocols perform nearly equally in this evaluation metric and DH slightly reduces the total key (copies) forwards of RSA in all three datasets. The explanation of this is similar to that given in the key transfer cost performance evaluation in the previous scenario as follows: in RSA, the unicast transmissions of session keys give a little impact on the total key (copies) forwarded during the simulation time since the majority of the key (copies) distributed in the network are nodes' public

IOP Conf. Series: Journal of Physics: Conf. Series **978** (2018) 012016 doi:10.1088/1742-6596/978/1/012016

keys. According this, the absence of the unicast transmissions of session keys in DH results in a small decrease of the key transfer cost below that of RSA in all the datasets.

## 5. Conclusion

We have discussed the delay and cost performance evaluation of the DH key exchange protocol in OMNs. We have demonstrated that DH outperforms a typical key distribution scheme based on RSA in terms of key transfer delay, measured by average key convergence time, in both the random node mobility and real human mobility scenarios. However, in the key transfer cost performance DH performs as well as RSA in both scenarios. Indeed, DH slightly reduces the total key (copies) forwards below that of RSA.

## References

[1] Van der Lubbe J C A 1998 *Basic Methods of Cryptography* (London: Cambridge University Press) chapter 8 pp 194-199

[2] Diffie W and Hellman M E 1976 New Directions in Cryptography *IEEE Trans. on Information Theory* **IT-22** (New Jersey: IEEE) pp 644-650

[3] Fall K 2003 A Delay-Tolerant Network Architecture for Challenged Internets *Proc. ACM SIGCOMM* (Karlsruhe: ACM) pp 27-34

[4] Clarke N L, Katos V, Menesidou S-A, Ghita B and Furnell S 2012 A Novel Security Architecture for a Space-Data DTN *Proc. Int. Conf. on Wired/Wireless Internet Communications* (Santorini: Springer) pp 342-349

[5] Menesidou S A, Katos V and Kambourakis G 2017 Opportunistic Key Management in Delay-Tolerant Networks *Int. J. Information and Computer Security* **9** (Geneva:Inderscience) pp 212-228

[6] Menesidou S A and Katos V 2012 Authenticated Key Exchange (AKE) in Delay-Tolerant Networks *Proc. Int. Conf. on Information Security* (Creta: Springer) pp 49-60

[7] Andrade D and Albini L C P 2016 Fully Distributed Public Key Management through Digital Signature Chains for Delay and Disrupt Tolerant Networks *Proc. Int. Conf. on Mobile Ad Hoc and Sensor Systems* (Brasilia: IEEE) pp 316-324

[8] Lv X, Mu Y and Li H 2014 Non-Interactive Key Establishment for Bundle Security Protocol of Space DTNs *IEEE Trans. on Information Forensics and Security* **9** (New Jersey: IEEE) pp 5-13

[9] Keranen A, Ott J, and Karkkainen T 2009 The ONE Simulator for DTN Protocol Evaluation *Proc. Int. Conf. on Simulation Tools and Techniques* (Rome: IEEE) pp 1-10

[10] Scott J, Gass R, Crowcroft J, Hui P, Diot C and Chaintreau A 2009 CRAWDAD Dataset Cambridge/Haggle/Infocom (online: http://crawdad.cs. dartmouth. edu/cambridge/haggle/ imote/infocom)

[11] Eagle N and Pentland A 2006 Reality Mining: Sensing Complex Social Systems *J. Personal and Ubiquitous Computing* **10** (London: Springer-Verlag) pp 255-268

[12] Bigwood G, Henderson T, Rehunathan D, Bateman M, and Bhatti S 2011 CRAWDAD Dataset st_andrew/sassy v. 2011-06-03 (online: http://crawdad.org/st_andrews/sassy/20110603/)

[13] Vahdat A and Becker D 2000 Epidemic Routing for Patially Connected Ad Hoc Networks *Tech. Report CS-200006 Duke Univ.* (Durham: Duke University)

IOP Conf. Series: Journal of Physics: Conf. Series **978** (2018) 012017 doi:10.1088/1742-6596/978/1/012017

# Application modeling ipv6 (internet protocol version 6) on e-id card for identification number for effectiveness and efficiency of registration process identification of population

**A M H Pardede[1,2], Y Maulita[1] and R Buaton[1,2]**

[1]STMIK Kaputama, Jl. Veteran No. 4A-9A,Binjai- Sumatera Utara, Indonesia
[2]Department of Computer Science, Faculty of Computer Science and Information Technology, Universitas Sumatera Utara, Indonesia

Email: akimmhp@live.com

**Abstract**. When someone wants to be registered in an institution such as Birth Certificate, School, Higher Education, e-ID card, Tax, BPJS, Bank, Driving License, Passport and others then have to register and do registration one by one and have registration number or account respectively agency. It may be said that everyone is bothered with the registration process, from the moment of birth must be registered to be registered as a resident, to enter the school must also registration, it is considered ineffective and efficient because one must continue to register one by one and there is repetition of ownership registration number which vary each agency. Seeing these problems need to find a solution or attempt how to keep the affairs of registration is not repetitive and quite once and the number applies to all agencies. The presence of the latest technology that IPv6 brings opportunities for the efficiency and effectiveness of the registration system. The method used in this research is the exploration and modeling of system development with NDLC (Network Development Life Cycle) to produce a model to build IPv6 implementation on e-ID card. The results of the study will show that the public has one registration number.

## 1. Introduction

The development of the age and technology today brings great impact and changes to human life, including the increasing needs, lifestyle and activities of everyone. Humans are always faced with the registration process, it can be seen from when a person is born must register to obtain the parent number and then registered as a resident on the family card, entering the world of education faced with registration of the parent number of students for elementary, junior, senior high school and student number for university. Likewise for banking, taxes, BPJS, social media such as email, FB, Twitter and others all have to be registered to get an identity number in other words each person to currently have dozens of registration number or identification number (ID) as identifier or identity . This is not efficient because for a person's identification number requires tens or even hundreds of ID number or registration number.

Along with the development of technology that is the presence of IPv6 (Internet Protocol Version 6) which replaces the predecessor version of IPv4 brings the impact of major changes to the development of internet protocol addressing at this time. With IPv4 Internet addressing protocols can only address 4,294,967,296 hosts worldwide or 232 with a total length of 32 bits, but with the birth of IPv6 addressing increasingly adds address space with a total length of 128 bits or 2128 or 340,282,366,920,938,463,374,607,431,768. 211,456 hosts worldwide.

Seeing this opportunity, as a researcher interested in designing a model, how to address this IPv6 address can be applied as a e-ID card number, as well as the number can be used for registration or master number at school, Student Identification Number at University, bank account number, BPJS (Social Security Health and Labor), driving license, passport and even mobile phone number. So one day want to know data or someone information simply by typing IP number, then automatically all information can be presented.

## 2. State Of The Art

Related to this research, first Hutagalung M Kifli, 2012 has done research with the title "Designing E-Voting Devices Based E-ID card". The background of the number of people who are not registered in the voter list even though the relevant has e-ID card or has been registered but did not get an invitation to vote on the day of the incident. The effect of this is the occurrence of electoral fraud and social upheaval in the community. Another constraint is the distribution of election devices such as paper and ink that often experience interference. The purpose of his research is to design an e-voting device that can solve the above problems. The designed e-voting is based on Electronic Identity Card (E-ID). This means the ID card is used as a basis for someone to be able to vote. ID Card is used as voter registration card to make it easier to verify voter list[1].

According to Ade Frihadi, 2015 has conducted a study entitled "Indonesian Government Network Using Ineternet VPN And IPv6 Technology To Support National E-Government" on the back of increasingly high technological developments, and the need for large data exchange both private sector, government and stackholder then technology is needed that allows more efficiency and security to perform storage and exchange of data[2]. Research Objectives Design of interconnection design of intra network in government institutions using IPv6-based internet VPN, Then perform ineterkoneksi security analysis of network of government institutions and the results obtained are analyzed by comparing the interconnection cost of intra-government network that is designed or built itself with inetrkoneksi service owned provider and paid like: VPN-IP, MPLS, Leased Line, etc. The results of this study Using VPN Virtual private network communications that pass through the public network (internet) will be more secure so as to feel the transaction data exchange owned by the government of Indonesia, the cost for interconnection between government agencies geographically distances far apart will be cheaper and implementation if using service providers such VPN-IP, MPLS, Leased line, etc., the quality of data exchange, picture and voice using VPN Vtun for network interconnection among government institutions is good with good internet condition and quality as well as interconnection of Indonesian government network with interconnection system using IPv6 will be able to reach all government agencies in Indonesia.

From this problem, it is necessary to design an e-voting device that can solve the above problems. The designed e-voting is based on Electronic Identity Card (E-ID). This means the ID card is used as a basis for someone to be able to vote. The rationale for the use of ID cards as voting cards is that the government is now preparing the National Identity Card (E-KTP) equipment and infrastructure nationally. The conclusion of this research is that e-Voting Application will save the cost of the election (Presidential) and Regional Head Elections (Pilkada).

## 3. The advantages and disadvantages of e-ID card

Based on the statement of the Minister of Internal Affairs Gamawan Fauzi on the official website of e-ID card, Electronic Identity Card (e-ID card) applied in Indonesia has advantages compared to e-ID card applied in People's Republic of China and India. e-ID card in Indonesia is more comprehensive. In the People's Republic of China, its electronic identity card (e-IC) is not equipped with biometrics or fingerprint records. There, e-IC is only equipped with chips that contain limited personal data. While in India, the system used for the management of population data is UID (Unique Identification Data) system, while in Indonesia the name is NIK (Population Identity Number). UIDs are published through

IOP Conf. Series: Journal of Physics: Conf. Series **978** (2018) 012017 doi:10.1088/1742-6596/978/1/012017

enrollment at 68 service points, while e-ID card programs in Indonesia are implemented in more than 6,214 sub-districts. Thus, the e-ID card applied in Indonesia is a composite of India's UID and People's Republic of China's e-ID, as e-ID cards are equipped with biometrics and chips.

e-ID card also has advantages compared to the national ID card / national ID card, the advantages include:

- Identity of a single identity
- Can not be falsified
- Can not be duplicated
- Can be used as a card in the Election or Pilkada (E-voting)

The Disadvantages of e-ID card can not be used as a single identity number on the student ID number, Student Identification Number at the University, bank account number, BPJS (Social Security Health and Workers), driver's license, passport and even mobile phone number.

## 4. IPv6 Protocol

In one of the discussions proposed by the IETF, which was stated in RFC 1550, several series of proposals were obtained in December 1992. From these proposals came three journals published by the IEEE Network, namely S. E. Deering, and D. Katz and P. S. Ford. Once done discussion & revision, then the combined version of Deering & Francis proposals later called SIIP (Simple Internet Protocol Plus), selected and named IPv6. IPv6 meet the requirements are quite good, including[3][4]:

- IPv6 has a longer IP address that is 16 bytes.
- IPv6 has a simpler header, which is 7 fields (versus 13 on IPv4).
- IPv6 has good support for options. Previous versions of the required fields, now changed to optimal. Option is also represented differently, making it easy for the router to pass an option that is not intended for it.
- IPv6 gives great advances in security

IPv6 has some fictures that can anticipate future application developments and overcome the shortcomings of its predecessor, the IPv4. IPv6 is designed as an improvement over IPv4. The header in IPv6 consists of two types, the first is the field required by each packet called the base header, while the second is the field that is not always required in the packet called the extension header, and the header is defined separately from the base header. The base header is always present on every packet, whereas additional headers only when required are inserted between the base headers and the data. Additional headers, currently defined in addition to security functions and others [5].

## 5. IPv6 Addressing

An IPv6 address is a 128-bit identifier for interfaces and a set of interfaces. There are three types of IPv6 addresses [5] :

- Unicast addresses: A Unicast address acts as an identifier for a single interface. An IPv6 packet sent to a Unicast address is delivered to the interface identified by that address.
- Multicast addresses: A Multicast address acts as an identifier for a group/set of interfaces that may belong to the different nodes. An IPv6 packet delivered to a Multicast address is delivered to the multiple interfaces.
- Anycast addresses: Anycast addresses act as identifiers for a set of interfaces that may belong to the different nodes. An IPv6 packet destined for an Anycast address is delivered to one of the interfaces identified by the address.

## 6. Research purposes

The purpose of this research are:

- Describe IPv6 addressing so that it can provide registration number up to 3,4x1038
- Creating an Analysis Model to determine one registration number able to represent and apply to all agencies without having to re-register with the same data.

- Determine and apply IPv6 IP address can be used as e-ID number so that double data does not occur.
- Identify someone with just one registration number and avoid someone double ID number with one registration number.

## 7. Benefits of research

The benefits of this research are :

- Helping the public to have only one registration number for all affairs of registration both for governmental affairs and personal purposes.
- Helping the government to more easily identify a person with just one registration number.
- Helping the government to avoid multiple e-ID numbers on the community

With this IPv6, bringing the opportunity for the development of e-KTP technology today that is in addition to use for internet protocol addressing can also be applied to technology e-ID card (electronic identity card) or id number of all humans around the world, because IPv6 addressing is ensured enough for all people in the world, the long-term plan is that once the child is born, they are given an IP number with IPv6, then the number becomes the registration number for all administrative or registration affairs starting from ID card, Student ID Number, Employee Identity Number, bank account , BPJS, Driver License, Passport and others.

## 8. Research Framework

To obtain results in accordance with the problems and objectives to be achieved, the following determine the research framework as a stage in the implementation of research from the beginning to the achievement of the end result. The study is divided into 4 stage :

I. The first stage : Analyze problems, collect and study literature
At this stage a specific problem is formulated, how to connect IPv6 to a registration number, including :

- What issues arise related to e-ID card registration
- Describe addressing or numbering with IPv6
- Formulate the possibility of e-ID format
- Divide IPv6 into blocks according to numbering needs

II. The second stage :
Based on the identified problem, and formulation of e-KTP format then build the implementation model to be simulated in the field :

- Registration on the education system
- Registration in banking
- Registration on BPJS health field
- Registration in the field of driving license
- Registration in the field of immigration or passport
- Registration in the field of communication

III. The third stage :
At this stage is to build a numbering simulation with IPv6, the system is built online

IV. The fourth stage :
At this stage, publish the results of research that is Draft Journal, Proceedings, Seminar of teaching materials.

## 9. IPv6 Concept For e-ID Numbering

In accordance with the concept of IPv6 consists of 8 blocks, shown in table 1 ie :

**Table 1.** Concept of IPv6 consists of 8 blocks

| VIII | | | VII | | | | | VI | | | | V | | | | IV | | | | III | | | | II | | | | I | | | |
|---|---|---|---|---|---|---|---|---|---|---|---|---|---|---|---|---|---|---|---|---|---|---|---|---|---|---|---|---|---|---|---|
| Province ID | | Regency / City ID | | | Districts ID | | | | Village ID | | | | | Sequence Number | | | | | | | | | | | | | | | | | |
| | | | | | | | | | | | | | | | | | | | | | | | | | | | | | | | |
| 32 | 31 | 30 | 29 | 28 | 27 | 26 | 25 | 24 | 23 | 22 | 21 | 20 | 19 | 18 | 17 | 16 | 15 | 14 | 13 | 12 | 11 | 10 | 9 | 8 | 7 | 6 | 5 | 4 | 3 | 2 | 1 |

Province ID = 2 BITS = 00-FF = can accommodate the number of provinces = 255 Provinces
Regency / City ID = 3 BITS = 000-FFF = 4,095 Regencys / Citys
Districts ID = 4 BITS = 0000-FFFF = 65.535 Districts
Village ID = 5 BITS = 00000-FFFFF = 1.048.575 Villages
Sequence Number = 18 BITS = 4.72237E + 21 Numbers
With the above numbering combinations then IPv6 can generate monitors = 4,095 X 65,535 X 1,048,575 X 4,72237E + 21= 3.4X1038
With capacity to reach trillion trillion trillion numbering.

For example a resident named Randy registered with the address in the village Tangkahan, Medan Labuhan Sub-district, Medan City, SUMUT.
For North Sumatera province serial number 1, Hexa conversion becomes 01.
City 3 digit, Medan city serial number 1, Hexa becomes 001.
District Id 4 digits, Medan Labuhan District serial number 1, Hexa become 0001
Village Id 5 digits, Tangkahan Village Id serial number 1, Hexa become 00001
The 18 digit serial number, on behalf of Randy registered serial number 1, then the e-ID number is : **01001000100001000000000000000001**, , shown in table 2 ie :

**Table 2.** Examples of population registration

| SUMUT | | MEDAN | | | Medan Lambuhan | | | | Tangkahan | | | | | Sequence Number | | | | | | | | | | | | | | | | | |
|---|---|---|---|---|---|---|---|---|---|---|---|---|---|---|---|---|---|---|---|---|---|---|---|---|---|---|---|---|---|---|---|
| 0 | 1 | 0 | 0 | 1 | 0 | 0 | 0 | 1 | 0 | 0 | 0 | 0 | 1 | 0 | 0 | 0 | 0 | 0 | 0 | 0 | 0 | 0 | 0 | 0 | 0 | 0 | 0 | 0 | 0 | 0 | 1 |

## 10. Implementation Model of IPv6 e-ID card In Institutions

In Figure 10.1. Model of IPv6 e-ID card In Institution described starting from the time of registration of the identity card, then the numbering process in accordance with IPv6 format that has been determined, and for subsequent data has been entered stored into the database, and so on to produce the output of the ID card number that has been formatted IPV6. And subsequently every resident wishing to register for education, banking and other institutions will be processed in the agency system, by simply adding the data required by each of the institutions and without creating a new identification number. And then all the data can be connected into one main data storage.

The following is presented flowchart of Model of IPv6 e-ID card In Institutions.

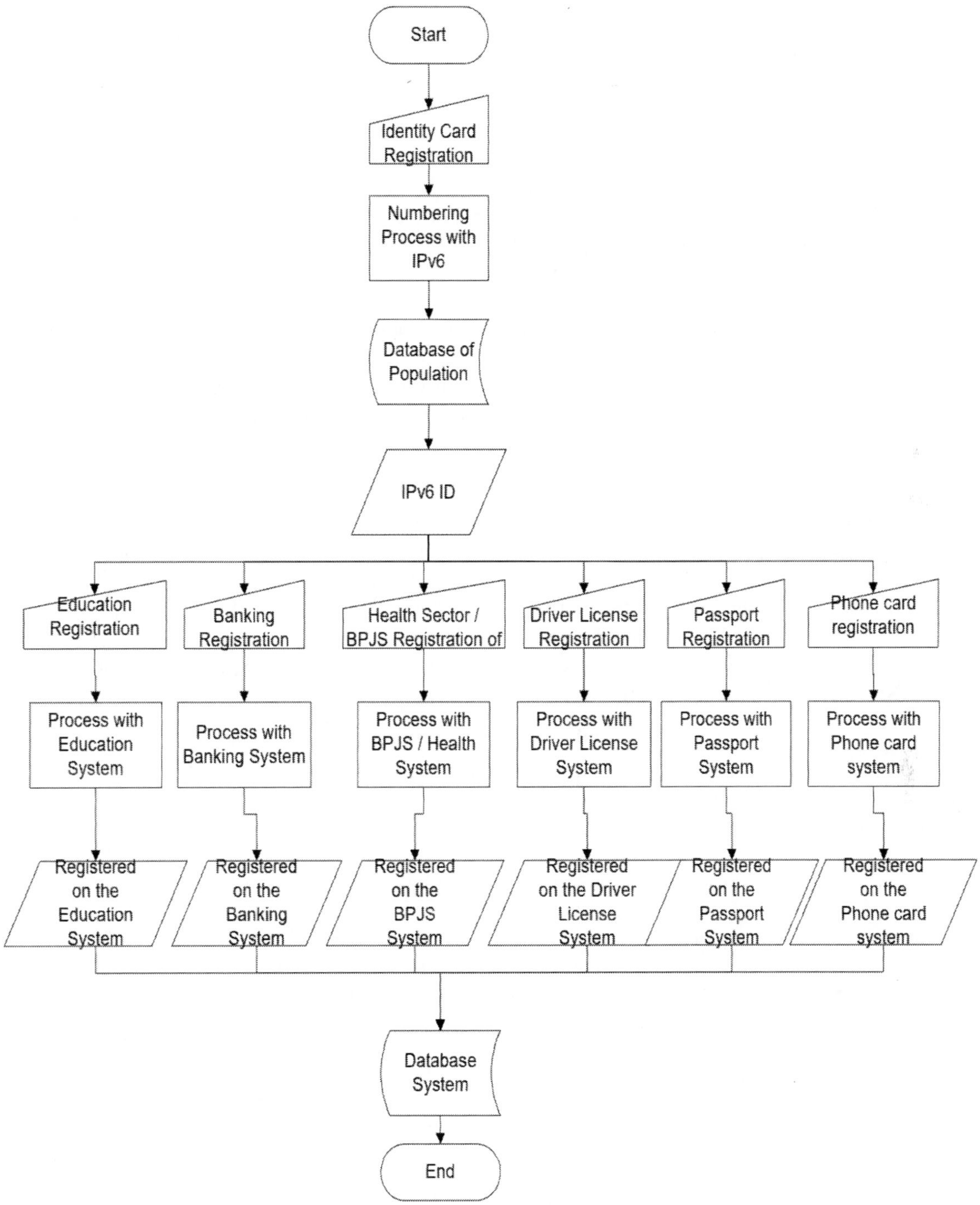


**Figure 1.** Model of IPv6 e-ID card In Institution

## 11. Future Research Plans

As for future research plan are:

- Make a system simulation online
- Complete the model and simulation e-ID card to be applied in banking and taxation

- Complete the model and simulation e-ID card to be applied in the field of communication or the use of e-ID card as the IP address number

## 12. Conclusion

After describing the model results and simulating the implementation of e-ID card, can be obtained some conclusions as follows :

- Internet Protocol version 6 numbering can be applied for e-ID numbering.
- With IPv6 numbering, the identity number or e-ID card can generate numbers with a total length of 128 bits or 2128 or 340,231,366,920,938,463,374,607,431,768,211,456 numbers.
- The residents are very helpful in terms of registration services because each resident has only one number and applies to all agencies.
- Regulatory agency management system is very helpful because there is no repetition of data or redundancy of data.
- Easy to know and get the population data when needed by the relevant agencies.
- No repetition of data, data errors, because the data centered in one system for use in every agency.

## 13. Recommendations

For the perfection of this research, there are several suggestions ie :

- Model and simulation is complemented by implementation of other institutions such as junior high school, high school and all institutions related to the needs of the population.
- Discussion of data security to ensure the data of the population is not misused by certain parties.
- Discussion or draft regulation is required if the system is used.
- Each model and simulation designed, should be discussed in more detail to find the point or thing that is not possible to apply.

## 14. Acknowledgments

We would like to extend our thanks to Direktorat Riset dan Pengabdian Kepada Masyakat (DRPM) Ditjen Penguatan Riset dan Pengembangan Kementerian Riset, Teknologi, dan Pendidikan Tinggi for the funding support of the Fundamental Fund research grant in 2017.

## References

[1] Hutagalung M K 2012 Perancangan Perangkat E-Voting Berbasis E-KTP *SAINTIKOM* **11**.
[2] Frihadi A 2015 *Jaringan Pemerintah Indonesia Menggunakan VPN Internet dan Teknologi IPv6 Untuk Mendukung E-Goverment Nasional* (Jakarta: Tesis Universitas Mercu Buana).
[3] Deering S E 1993 SIP: Simple Internet Protocol *IEEE Network* **7** pp 16-28.
[4] Katz D and Ford P S 1993 TUBA: Replacing IP with CLNP *IEEE Network* **7** pp 38-47.
[5] Li Q, Jinmei T and Shima K 2007 *IPv6 Advanced Protocols Implementation 3th. Ed.* (San Fransisco: Morgan Kaufman).

# Supply chain management using fp-growth algorithm for medicine distribution

**A Wahana, D S Maylawati, M Irfan and H Effendy**

Departement of Informatics, UIN Sunan Gunung Djati Bandung, Indonesia

Email: wahana.agung@gmail.com

**Abstract.** Distribution of drugs evenly in accordance with the needs of Public Health Center (Puskesmas) become one of the responsibilities by the Health Office in Indonesia. This study aims to provide recommendations for distribution of drugs from the Department of Health according to the needs of each Puskesmas. Because often the distribution of drugs is not in accordance with the needs of medicines stock of each Puskesmas. This causes the possibility of drug stock void in Puskesmas in need, while there is excess stock of drugs in Puskesmas that do not need. Supply Chain Management (SCM) is applied as a controlling drug stock at Puskesmas in order to avoid drug vacuum or excess drug that eventually unused. In addition, the Frequent Pattern Growth (FP-Growth) algorithm that generates frequent item sets is used to provide drug distribution recommendations by looking at the highest frequency of drug occurrences and frequent drug frequencies. Based on testing black box system conducted in Health Office Purwakarta regency of Indonesia which oversees 20 Puskesmas with 100 data of drug distribution transactions, it can be concluded that system functionality is running well and SCM successfully implemented to arrange distribution process of medicine well. Furthermore FP-Growth algorithm was able to provide recommendations for distribution of drugs with a high success rate. This is evidenced by the test results with various combinations of input parameters, FP-Growth is able to produce the right frequent item sets.

## 1. Introduction

The Health Office as a government apparatus is in charge of organizing, fostering and supervising the implementation of equitable, quality and affordable health efforts, mobilizing community participation in health development, and providing health facilities and resources accessible to all levels of society. Distribution of drugs evenly to the health center became one of the responsibilities of the Health Office. The process of managing and controlling incoming and outgoing drugs at the Puskesmas needs to be done automatically so that errors in managing the drug data. In addition, the Health Office can control the stock of drugs directly at each Puskesmas so that the distribution of drugs performed according to the needs of each Puskesmas.

Distribution and control of drugs needs to be managed properly, one of them by applying SCM. SCM differs from logistics management where logistics management is only part of SCM [1] [2]. More than simply controlling incoming and outgoing goods, SCM connects all business processes from the provider of goods to services and information to consumers and related stakeholders [1]. Along with its development, SCM has 8 processes within its framework, including Customer Relationship Management, Supplier Relationship Management, Customer Service Management, Demand Management, Order Fulfillment, Manufacturing Flow Management, Product Development and Commercialization, and Return Management [2].

In this research, SCM is used to manage all drug procurement activities among drug providers in this case the Health Office and Puskesmas as recipients of the drug. In addition to SCM, drug distribution control is assisted by track record of drug transactions at Puskesmas so that the distribution of drugs is evenly in accordance with the needs of each Puskesmas. This can streamline and streamline the distribution and control of drug stocks. The history of drug transactions is presented in the form of frequent item sets, where drugs as frequently occurring items can be recommended for the provision of medicinal stock from the Health Office to the Puskesmas. It is said to often arise when it meets the minimum value of support or the minimum value of occurrence of all transactions.

This research uses the FP-Growth algorithm as an algorithm for frequent pattern mining that generates frequent item set without the need to generate feature candidates [3] [4] [5] [6] [7]. FP-Growth is an Apriori algorithm development, where there is a difference in database scanning [5] [8]. FP-Growth only does one scanning database, while Apriori needs to do scanning database repeatedly. This is because the Apriori algorithm requires the generation of feature candidates to get frequent item sets [9] [3]. However, the FP-Growth algorithm does not generate feature candidates because FP-Growth uses the concept of tree development in the search for frequent item sets [5] [8]. This is what causes the FP-Growth algorithm faster than Apriori algorithm.

In the upcoming section, we describe the concept of Supply Chain Management for Medicine Distribution in Section 2, Frequent Pattern Growth Algorithm for Extracting Frequent Medicine Sets in Section 3. Experiment and Result in Section 4, then Evaluation and Discussion in Section 5. Finally, conclusion of this research stated in Section 6.

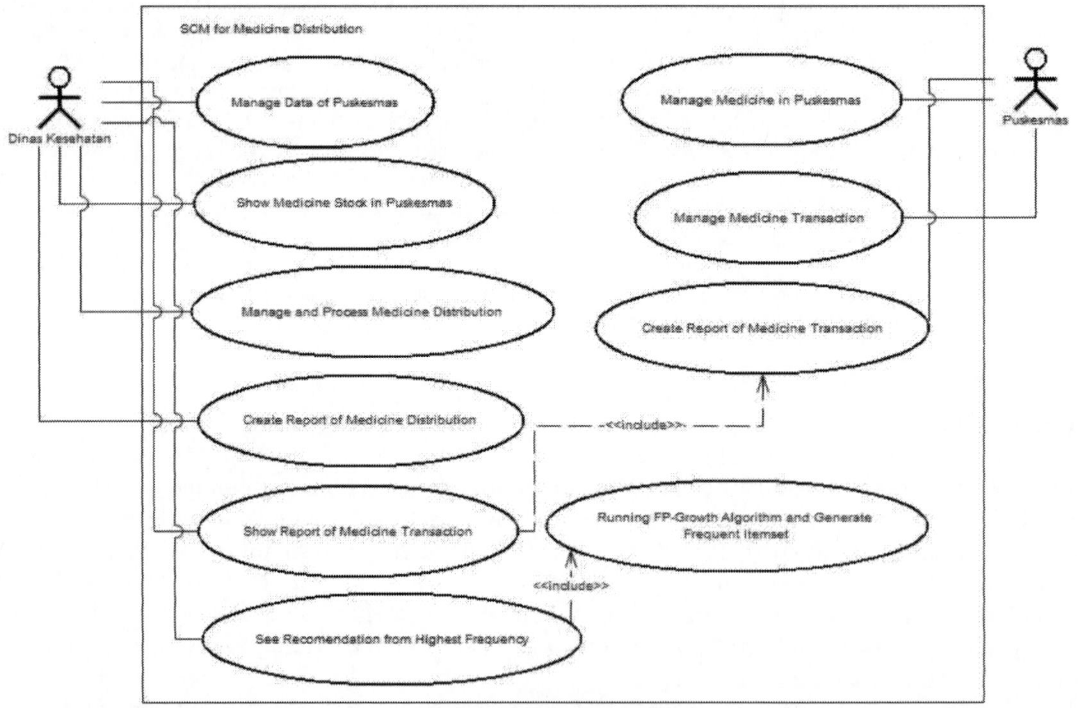


**Figure 1.** Use Case Diagram for SCM of Drug Distribution using FP-Growth Algorithm

## 2. Supply Chain Management for Medicine Distribution

SCM is wider than its logistics oriented product flow only [10]. SCM connects and coordinates every business process between both the supplier and the consumer or the organization itself. In this study, the Health Office is the supplier that distributes the drug, while the Puskesmas is the consumer receiving the drug. To meet the needs of drugs in Puskesmas as well as vice versa, both Health Office and Health Center must manage the flow of drug transactions and drug stock needs well. Based on the SCM process framework, SCM implementation in this study is modeled with the use case diagrams in Figure 1.

## 3. Frequent Pattern Growth Algorithm for Extracting Frequent Medicine Sets

Frequent pattern mining is a process or technique that generates frequent itemsets that do not pay attention to the order of occurrence of items on a transaction, in contrast to the sequential pattern that takes into account the order of occurrence of items on transactions [9] [3] [7]. FP-Growth is one of the algorithms that generate frequent itemset by compressing the database into a tree structure called FP-Tree [11] [12] [5]. There are several major steps in generating frequent itemsets without generating feature candidates, among others [6] [11] [13] [14] [15] [16]: Perform a database scan and find support from each item; Eliminates all items that are not frequent; Sort frequent items based on their support. Starting from the highest to lowest support; Any transactions that are read and saved in the tree; The first root is labeled with null, then the items of each transaction are stored in a node that contains item-name, count, and node-link; FP-Growth extracts frequent itemsets from FP-Tree; Read from leaves to root (buttom-up algorithm); Devide and conquer.

**Table 1.** An Example of Medicine Transactions

| ID | Item sets DB |
|---|---|
| 1 | Klorfeniramin Maleat (CTM) Tablet 4 mg, Parasetamol Tablet 500 mg |
| 2 | Piridoksin HCl. Tablet, Klorfeniramin Maleat (CTM) Tablet 4 mg, Parasetamol Tablet 500 mg |
| 3 | Kotrimoksazol Syr, Klorfeniramin Maleat (CTM) Tablet 4 mg |
| 4 | Kotrimoksazol Syr, Parasetamol Tablet 500 mg |

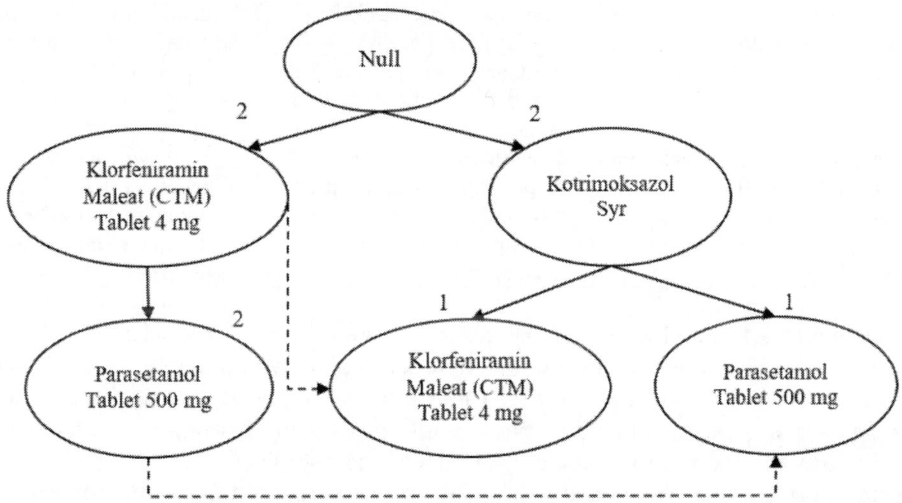


**Figure 2.** FP-Tree formed from Example of Drug Transaction on Table 1

Figure 2 illustrates the FP-Tree and FP-Growth algorithms for generating frequent itemsets of drug transactions in Table 1. Figure 2 is a FP-Tree with a minimum value of 50% support (from 4 transactions appearing at least 2 times). Where, the frequency of occurrence of Klorfeniramin Maleat (CTM) Tablet 4 mg is 3, Paracetamol Tablet 500 mg is 3, and Srimrimazsazole Syr is 2, whereas Piridoksin HCl Tablet will be eliminated because it does not meet the minimum support with frekuendi occurrence once. After each item is sorted by the largest frequency, then the FP-Tree is established in Figure 2. Furthermore, from FP-Tree generated frequent itemset with conditional pattern base and conditional FP-Tree [5] [11] which is described in Table 2.

The emergence of frequent itemset drugs indicates that the drug is the most widely used and needed, of course every puskesmas has different needs. In addition, often concurrent rug ems indicate that these drugs are often needed imultaneously. or example, for a sample transaction on Table 1 and generating a frequent pattern <Paracetamol Tablet 500 mg, Klorfeniramine Maleat (CTM) Tablet 4 mg> indicates that people tend to use Paracetamol always together with CTM. So when it will distribute Paracetamol, it is necessary to check also the stock CTM, vice versa. This is because based on the history of Paracetamol transactions often appear simultaneously with CTM.

**Table 2.** Extraction Processes of Frequent Itemset or Frequent Pattern from FP-Tree on Figure 2

| Item | Freq. | Conditional Pattern Base | Conditional FP-Tree | Frequent Pattern |
|---|---|---|---|---|
| Kotrimoksazol Syr | 2 | {} | <> | {< Kotrimoksazol Syr >}: 2 |
| Parasetamol Tablet 500 mg | 3 | { Klorfeniramin Maleat (CTM) Tablet 4 mg } : 2; { Kotrimoksazol Syr } : 1 | < Klorfeniramin Maleat (CTM) Tablet 4 mg > : 2 | {< Parasetamol Tablet 500 mg >}: 3; {< Parasetamol Tablet 500 mg , Klorfeniramin Maleat (CTM) Tablet 4 mg >} : 2 |
| Klorfeniramin Maleat (CTM) Tablet 4 mg | 3 | {}; { Kotrimoksazol Syr } : 1 | <> | {< Klorfeniramin Maleat (CTM) Tablet 4 mg >}: 3 |

## 4. The experiment and result

**Table 3.** Scenarios and Black Box Test Result

| Scenario | Result | Expected Output | Real Output |
|---|---|---|---|
| **Clear the form then submit** | Success | The system gives an error message. Column can not be empty. | The system gives an error message. Column can not be empty. |
| **Displays the highest frequency of drug use** | Success | The system successfully displays the highest drug frequency data. | The system successfully displays the highest drug frequency data. |
| **Displays the highest frequency of drug use with minimum support** | Success | The system successfully displays the highest drug frequency data with support ≥ minimum support that has been specified. | The system successfully displays the highest drug frequency data with support ≥ minimum support that has been specified. |
| **Displays the highest frequency of drug use with minimum length** | Success | The system succeeded in displaying the highest drug frequency data with a length of ≥ minimum length that has been specified. | The system succeeded in displaying the highest drug frequency data with a length of ≥ minimum length that has been specified. |

The test is done in black box and white box. Black box testing is done to test whether the SCM system is built in accordance with the specification of functional system requirement specification that is modeled on the use case diagrams in Figure 1. Table 3 shows the black box testing scenario and the result of testing each system functionality. While white box testing is done to see the accuracy and accuracy of FP-Growth algorithm in generating frequent item set from drug transactions. White box testing performed a test scenario to compare the results of the number of frequent item sets obtained by entering the minimum support varies with the same number of datasets. Testing is done by using 100 drug transaction data from Health Office Purwakarta regency of Indonesia which oversees 20 Puskesmas with minimum support and minimum length or minimum length of varied items. Table 4 to Table 7 show the result of frequent itemset produced by FP-Growth algorithm with variation of input parameter.

**Table 4.** Test Result on the Amount of requent Itemset with Minimum Support ≥ 20% and Lenght = 1, 2, and 3

| No | *Frequent Itemsets* | *Support* | *Length* |
|---|---|---|---|
| 1 | Parasetamol Tablet 500 mg | 69% | 1 |
| 2 | Klorfeniramin Maleat (CTM) Tablet 4 mg | 51% | 1 |
| 3 | Klorfeniramin Maleat (CTM) Tablet 4 mg, Parasetamol Tablet 500 mg | 41% | 1 |
| 4 | Piridoksin HCl. Tablet | 37% | 1 |
| | Piridoksin HCl. Tablet, Parasetamol Tablet 500 mg | 36% | 1 |
| | Gliseril Guaikolat Tablet 100 mg | 27% | 1 |
| 7 | Gliseril Guaikolat Tablet 100 mg, Parasetamol Tablet 500 mg | 25% | 1 |
| 8 | Piridoksin HCl. Tablet, Klorfeniramin Maleat (CTM) Tablet 4 mg | 25% | 1 |
| 9 | Piridoksin HCl. Tablet, Klorfeniramin Maleat (CTM) Tablet 4 mg, Parasetamol Tablet 500 mg | 25% | 1 |
| 10 | Gliseril Guaikolat Tablet 100 mg, Klorfeniramin Maleat (CTM) Tablet 4 mg | 23% | 1 |

| No | *Frequent Itemsets* | *Support* | *Length* |
|---|---|---|---|
| 11 | Kotrimoksazol Syr | 23% | 1 |
| 12 | Kotrimoksazol Syr, Parasetamol Tablet 500 mg | 21% | 1 |
| 13 | Kotrimoksazol Syr, Klorfeniramin Maleat (CTM) Tablet 4 mg | 20% | 1 |
| 14 | Antasida tab | 20% | 1 |
| 15 | Klorfeniramin Maleat (CTM) Tablet 4 mg, Parasetamol Tablet 500 mg | 41% | 2 |
| 16 | Piridoksin HCl. Tablet, Parasetamol Tablet 500 mg | 36% | 2 |
| 17 | Gliseril Guaikolat Tablet 100 mg, Parasetamol Tablet 500 mg | 25% | 2 |
| 18 | Piridoksin HCl. Tablet, Klorfeniramin Maleat (CTM) Tablet 4 mg | 25% | 2 |
| 19 | Piridoksin HCl. Tablet, Klorfeniramin Maleat (CTM) Tablet 4 mg, Parasetamol Tablet 500 mg | 25% | 2 |
| 20 | Gliseril Guaikolat Tablet 100 mg, Klorfeniramin Maleat (CTM) Tablet 4 mg | 23% | 2 |
| 21 | Kotrimoksazol Syr | 23% | 2 |
| 22 | Kotrimoksazol Syr, Parasetamol Tablet 500 mg | 21% | 2 |
| 23 | Kotrimoksazol Syr, Klorfeniramin Maleat (CTM) Tablet 4 mg | 20% | 2 |
| 24 | Piridoksin HCl. Tablet, Klorfeniramin Maleat (CTM) Tablet 4 mg, Parasetamol Tablet 500 mg | 25% | 3 |

**Table 5.** Test Result on the Amount of requent Itemset with Minimum Support ≥ 30% and Lenght = 1 and 2

| No | *Frequent Itemsets* | *Support* | *Length* |
|---|---|---|---|
| 1 | Parasetamol Tablet 500 mg | 69% | 1 |
| 2 | Klorfeniramin Maleat (CTM) Tablet 4 mg | 51% | 1 |
| 3 | Klorfeniramin Maleat (CTM) Tablet 4 mg, Parasetamol Tablet 500 mg | 41% | 1 |
| 4 | Piridoksin HCl. Tablet | 37% | 1 |
| 5 | Piridoksin HCl. Tablet, Parasetamol Tablet 500 mg | 36% | 1 |
| 6 | Klorfeniramin Maleat (CTM) Tablet 4 mg, Parasetamol Tablet 500 mg | 41% | 2 |
| 7 | Piridoksin HCl. Tablet, Parasetamol Tablet 500 mg | 36% | 2 |

**Table 6.** Test Result on the Amount of requent Itemset with Minimum Support ≥ 40% and Lenght = 1 and 2

| No | *Frequent Itemsets* | *Support* | *Length* |
|---|---|---|---|
| 1 | Parasetamol Tablet 500 mg | 69% | 1 |
| 2 | Klorfeniramin Maleat (CTM) Tablet 4 mg | 51% | 1 |
| 3 | Klorfeniramin Maleat (CTM) Tablet 4 mg, Parasetamol Tablet 500 mg | 41% | 1 |
| 4 | Klorfeniramin Maleat (CTM) Tablet 4 mg, Parasetamol Tablet 500 mg | 41% | 2 |

**Table 7.** Test Result on the Amount of requent Itemset with Minimum Support ≥ 50% and Minimum Lenght ≥ 1

| No | *Frequent Itemsets* | *Support* | *Length* |
|---|---|---|---|
| 1 | Parasetamol Tablet 500 mg | 69% | 1 |
| 2 | Klorfeniramin Maleat (CTM) Tablet 4 mg | 51% | 1 |

## 5. Evaluation and Discussion

Based on the SCM framework [2] [10], each process is largely represented by each use case in this study. Customer Relationship Management, Supplier Relationship Management and Customer Service Management even though it is technically more done outside the system, but through the system automatically also woke up the communication between the Department of Health as a supplier and health center as customers were represented by the use case "Show Report of medicine Transaction" And "Create Report of Medicine Transaction". Especially the Demand Management, Order Fulfillment and Manufacturing Flow Management processes that are directly met by all use cases. Especially for Demand Management and Order Fulfillment facilitated by appropriate drug experts on the needs of each health center with implementing the FP-Growth algorithm that is modeled in use case "See Recommendation from the Highest Frequency" and "Running FP-Growth Algorithm and Generate frequent item set". For Product Development and commercialization process, and Return Management is not directly represented by the use case is there, but the use case "Manage Medicine Transaction" there are controls on drugs that enter and exit, including the expired drugs and conditions can be restored if there is a mismatch.

Furthermore the FP-Growth algorithm has been implemented well in accordance with the literature [5] [6] [11]. Evident from the test results with several scenarios variation input parameters, including minimum support and minimum length, frequent item set of drugs produced in accordance with existing transaction data. In addition, the test results show the consistency of frequent item set produced. Seen in Table 4 to Table 7 although the minimum support and minimum length values entered are different (with the same transaction data), as long as the item is said to be frequent it will still produce the same frequent pattern. It also proves that function and structure of FP-Growth algorithm has been applied correctly.

## 6. Conclusion

The conclusion that can be drawn from this research is SCM can be applied to manage the whole process of distribution and management of drugs between the Health Office and Puskesmas. Much of this research fills the SCM framework by modeling and implementing it into system functionality. In addition, based on the test results indicate that the FP-Growth Algorithm can be used for the recommendation of drug needs planning in the Department of Health so that the drug distribution can be done effectively and efficiently according to the needs of each Puskesmas. For future research can be developed the process of SCM framework that has not been implemented, among others Product Development and Commercialization. Furthermore, in testing the algorithm needs more data because it can affect the accuracy of the test. In addition, it can also use a frequent pattern algorithm other than FP-Growth so that the formation process of frequent item set can be optimized in terms of time and effort required.

## References

[1] Lambert D M and Cooper M C 2000 Issues in Supply Chain Management *Industrial Marketing Management* **29** pp 65-83.

[2] Lambert D M and Enz M G 2017 Issues in Supply Chain Management: Progress and potential *Industrial Marketing Management* **62** pp 1-16.

[3] Han J and Kamber M 2006 *Data Mining: Concepts and Techniques* (USA: Morgan Kaufmann Publisher).

[4] Han J, Cheng H, Xin D and Yan X 2007 Frequent Pattern Mining: Current Status and Future *Data Mining Knowledge Discovery* **15** pp 55-86.

[5] Jiawei H, Pei J and Yin Y 2000 Mining Frequent Patterns without Candidate Generation *Proc. of the 2000 ACM SIGMOD Int. Conf. on Management of data.*

[6] Maylawati D S 2015 Development of Pre-processing Library with Set of Frequent Word Item Set Representation for Text Mining Application Case Study: Indonesian's Slang Language *Thesis Report Bandung Institute of Technology*.

[7] Maylawati and Saptawati G A P 2016 Set of Frequent Word Item sets as Feature Representation for Text with Indonesian Slang *1st Int. Conf. on Computing and Applied Informatics (ICCAI).*

[8] Lin KC, Liao IE and C ZS 2011 an Improved Frequent Pattern Growth Method For Mining Association Rules *Expert System with Applications* **38** pp 5154-5161.

[9] Agrawal R and Srikant R 1995 *Mining Sequential Pattern* IBM Research Center.

[10] Christoper M 2011 *Logistics and Supply Chain Management Fourth Edition* (Edinburg: Prentice Hall).

[11] Pawar D 2013 *The Comparative Study of Apriori and FP-Growth Algorithm for Acossiation Rule Mining* [Online] source: https://wwwslidesharenet/deepti92pawar/the-comparative-study-of-apriori-and-fpgrowth-algorithm.

[12] Miraldi R N, Rahmat A and Susanto B 2014 Implementasi Algoritma FP-Growth untuk Sistem Rekomendasi Buku di Perpustakaan UKD *Informatika*, **10**.

[13] Ardani N R and Fitrian N 2016 Sistem Rekomendasi Pemesanan Sparepart dengan Algoritma FP-Growth (Studi Kasus PT Rosalia Surakarta) (Yogyakarta: *STMIK AMIKOM Yogyakarta*).

[14] Ririanti 2014 Implementasi Algoritma PF-Growth pada Aplikasi Prediksi Persediaan Sepeda Motor (Studi Kasus PT Pilar Deli Labumas*) Pelita Informatika Budi Darma* **VI** .

[15] Arifin R N 2014 Implementasi Algoritma Frequent Pattern Growth (FP-Growth) Menentukan Asosiasi Antar Produk (Studi Kasus Nadiamart) (Yogyakarta: *STMIK AMIKOM Yogyakarta*).

[16] Sari E W, Rianto A and Andarawarih S D 2014 Decision Support on Supply Chain Management System Using Apriori Data Mining Algorithm *College of Informatics and Computer Engineering Indonesia (STIKI).*

IOP Conf. Series: Journal of Physics: Conf. Series **978** (2018) 012019 doi:10.1088/1742-6596/978/1/012019

# The implementation of two stages clustering (k-means clustering and adaptive neuro fuzzy inference system) for prediction of medicine need based on medical data

**A M Husein[1], M Harahap[1], S Aisyah[2], W Purba[2] and A Muhazir[3]**

[1] Computer Science Departement, Faculty of Technology and Computer Science, Universitas Prima Indonesia, Medan, Sumatera Utara 20111, Indonesia
[2] Information System Departement, Faculty of Technology and Computer Science, Universitas Prima Indonesia, Medan, Sumatera Utara 20111, Indonesia,
[3] Computer Science Departement, Faculty of Industry Technology, Institut Teknologi Medan, Medan, Sumatera Utara 20217, Indonesia

Email :
amirmahmud@unprimdn.ac.id, mawaddah@unprimdn.ac.id, sitiaisyah@unprimdn.ac.id, windania@unprimdn.ac.id, muhazir@itm.ac.id

**Abstract.** Medication planning aim to get types, amount of medicine according to needs, and avoid the emptiness medicine based on patterns of disease. In making the medicine planning is still rely on ability and leadership experience, this is due to take a long time, skill, difficult to obtain a definite disease data, need a good record keeping and reporting, and the dependence of the budget resulted in planning is not going well, and lead to frequent lack and excess of medicines. In this research, we propose Adaptive Neuro Fuzzy Inference System (ANFIS) method to predict medication needs in 2016 and 2017 based on medical data in 2015 and 2016 from two source of hospital. The framework of analysis using two approaches. The first phase is implementing ANFIS to a data source, while the second approach we keep using ANFIS, but after the process of clustering from K-Means algorithm, both approaches are calculated values of Root Mean Square Error (RMSE) for training and testing. From the testing result, the proposed method with better prediction rates based on the evaluation analysis of quantitative and qualitative compared with existing systems, however the implementation of K-Means Algorithm against ANFIS have an effect on the timing of the training process and provide a classification accuracy significantly better without clustering.

## 1. Introduction

Medicines is one of the items needed by the sick community[31]. The sale of medicines in the hospital 50%-60% affects the budget and expenditure of the hospital, so it is needed to predict the optimal medicine needs and because of an error in planning will affect the medical and economical treatment [31]. Medicine planning aims to get the type of the number of medicines in accordance with the requirements and avoid the void based on the pattern of disease and the frequency of the number of patients visit [31]. In general, morbidity method is used for planning of medicines because the expected needs approach the truth and support the improvement of the use of medicines appropriate pattern standard treatment [31]. But it needs a long time and skilled labor, it is difficult to obtain definite disease data and need good recording and reporting [31].

Easy consumption method used does not need data of disease and lack of or excess medicines very small, but it is difficult to get medicine data and the number of patients contacts, lack, excess and loss of medicines are not reliable [31]. Composite method is a method of improvement of the weakness of the method morbidity and consumption, uncertainty number of visits and the frequency of illnesses make this method is less accurate used for the prediction of medicine needs.

Fuzzy Inferences System (FIS) can resolve the problem of uncertainty factors [19], but membership function dependence on the FIS affect the result of the decision-making [32]. Divination with a combination of data mining such as Genetic Algorithm with BP Neural, SOP with Neural Network, SVR with RBFNN outlined on [30] and ARIMA with Neural Network [7]. Neural Network (NN) is a method used by many researchers for divination or prediction [35], NN is better and accurate for divination gold prices compared with ARIMA [13], the ability of NN generalize, identify non-linier relationship and implementation in various applications are very useful [9]. But NN has the weakness of the need for large training data and requires a long computing time to identify the pattern on the data in high dimension[27].

Adaptive Neuro Fuzzy Inference System (ANFIS) is one of the algorithm that combines fuzzy system with artificial nerve network [1]. ANFIS method can be used for divination or prediction [25][1][3][4][6][18], ANFIS method is more accurate in predicting compared ARIMA method and NN [33][24]. But the characteristics of the data used for training will affect the structure of the ANFIS then it affects the accuracy prediction [14].

Clustering is one of the techniques for the classification of objects or patterns in one association so each cluster possess cluster similarity of one another [20]. There are several algorithms in clustering such as K-means, Fuzzy C means (FCM) and Possibilistic c-means (PCM) [26]. FCM is one of the cluster algorithm optimal enough in the classification, but it needs a long computing time [16]. PCM is very strong against noise [17], but it's too sensitive to the cluster initialization [22]. The K Means Clustering is the algorithm that efficient in grouping a large data set and fast in the computing [16], but limited to numeric data [10].

From the results of several research, combination of K-means Clustering and ANFIS Algorithm is proposed for the prediction of medicine needs. We know with good prediction results will greatly affect the decision-making to improve the medical and medicine need budget more optimal. Data analysis framework needs prediction of medicine based on ANFIS method. The Data is derived from the hospital database. The first stage we use the year 2015-2016 medical data sources to predict the need of medicines with standard ANFIS structure and then calculated the value of the Root Mean Square Error (RMSE) for training and testing. For the second approach, we still use ANFIS algorithm, but after the process of grouping with the algorithm K-Means. The second is the end result of this approach will we compare to know the impact of the classification of accuracy ANFIS prediction. This paper is organized as follows: Part 2 related research. In section 3 the methods used. Section 4 explains the model used, the results and discussion are presented in section 5 and section 6 conclusion.

## 2. Related Research

Research of Zubir Haider Khan [23], used Artificial Neural Network (ANN) for stock price prediction using two models, one model is for training by using backpropagation and second model is multilayer Feedforward network for testing. Fauzi Yudhi Septiawan [15], applied the Genetic Algorithm to predict the movement of the currency Forex which circulates in Indonesia, USDJPY USDCHF, GBPUSD and EURUSD. Imam Baihaqi Siregar et all [8], performed the comparison method of single ICE, double exponential smoothing holt, triple exponential smoothing based on Exponential Smoothing for palm oil production forecasts and analyzes the accuracy of the model and characteristics of production data forecasts, result of divination single ICE lower based on the value of the RMSE. David Ijegwa Acheme [2], developed a share price prediction using Fuzzy Logic method for decision-making sale and purchase of shares based on the market trends with four indicators are moving average convergence/divergence (MACD), relative strength index (RSI), stochastic oscillator (SO) and on-balance volume (OBV).

Mustain Billah et all [8], did research prediction of share price closing with ANFIS method and ANN on the case of a Securities Exchange Dhaka (BED). Paul Sunjay [28] using ANFIS approach for optimal supply forecasts. Research of Abbasi [1] for stock price prediction on Iran Khodro Coporation on long-term period by using the triangle membership and four independent variables namely dividends per share (FIXED), price to earning ratio (P/E), closing price and the stock price are used as the independent variables short term. Research of Hien Nguyen Nhu [25], applied the combination of Firefly and ANFIS algorithm. Network model ANFIS trained using the Firefly algorithm and assigned to predict the share price in Vietnam Stock Market with the purpose of comparing the performance among ANFIS with Hybrid Algorithm, Back Propagation and Particle Swarm Optimization (PSO). Research of Osman assured Hegazy had [18] proposed the combination of Genetic Algorithm (GA) with Adaptive Neuro Fuzzy Inference System (ANFIS) for stock price prediction for optimizations. The combination of the previous stock prices as input model for the best stock trend prediction on the next day to see the results of the mean square error (MSE). GA optimize fuzzy paremeter to overcome constraints statistics method. Research of Georgia Makridou[24] used ANFIS method for predicting the price of gold and compares with the method 'Buy and Hold' (B&H) strategy. The result of the test accuracy of ANFIS is superior performance (B&H). Ehsan Lotfi [21] proposed ANFIS method can be applied to the oil price forecasts earth by considering input parameters on ANFIS network.

## 3. The method used

### *3.1 Adaptive Neuro-Fuzzy Inference System (ANFIS)*

ANFIS is Neuro-Fuzzy approach which was introduced by the outskirts (Outskirts et all 1997,1993) which explores fuzzy system Takagai and Sugeno model (1985) and can be used to control application, prediction or divination. ANFIS architecture is similar with neural network, radial function and a little certain restrictions. The process of learning neural network is used to optimize the parameter values of fuzzyfication. In Neuro-Fuzzy consists of five layers of process. The first layer is the parameter of the membership function fuzzy that his nature is nonlinier association of Exodus system. The learning process on the parameters us error Back-Propagation (EBP) to renew the value of parameters. On the fourth layer, the parameter is a linier parameter against the exodus of the system that makes up the basis of fuzzy rules. On this layer is using Least-Squares Estimotor method (LSE).

### *3.2 K-Means Clustering*

The K means clustering is one of the most popular clustering algorithm used because it has simple algortihm, easy to implement and efficient in complexity [5]. Algortihm K-Means determines the cluster of the object based on the attribut/features from the object into the cluster (K). These groups differentiated by its center. This algorithm is very sensitive to the placement of the initial values cluster center. The determination of the partition number of cluster (k) is very important in this algorithm, but there is no terms applicable to determine the amount of cluster(K) formed [34]. The steps in the algorithm K-means are as follows:

1. Prepare the training data.
2. Set the value of the K cluster.
3. Set the initial values centroids.
4. Count the distance between the data and using the formula (Euclidean centroid Distance).
5. Data partition based on the minimum value.
6. And then perform iteration during the data partition is still moving (no longer moving objects to other partitions), when it continues then to points 3.
7. If a group of current data is similar with the previous one, then stop iteration.
8. The data has been partitioned according to the value of the final centroid.

### *3.3 Dataset*

Data need for training and testing ANFIS originates from the previous year data obtained from actual data 2015 and 2016 on one of the Hospital in the City of Medan. Input Data 2015 will be used for the prediction of 2016 and compare the results of ANFIS prediction with the actual data 2016 and so on until the comparison 2017 prediction data with the actual one 2017.

*3.4 Accuracy Calculation*
ANFIS method prediction is based on the training and testing using the Root Mean Square Error (RMSE). RMSE is an alternative method for evaluation techniques of divination. If the value of the RMSE is getting smaller, and the estimation of the model or variable is more valid.

$$RMSE = \sqrt{\frac{\sum_{i=1}^{n}(y_i - \hat{y}_i)^2}{N}} \quad (1)$$

## 4. The proposed Model

The proposed model for the prediction of medicine needs at the hospital is based on the results of these observations from two hospitals. In the figure 1 illustrates the framework of research. The first step is to make observations on the private and government hospital, the results of the interview with the head of IFRS, pharmacy and director become as input material, there are three main categories which are used as indicator parameter namely patients, diagnosis and inventory data.

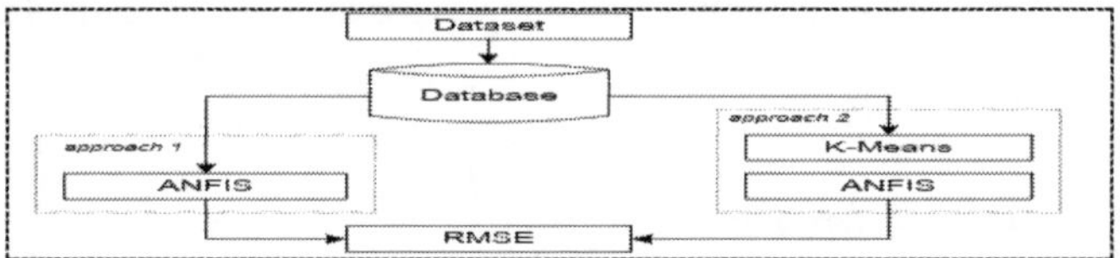


**Figure 1.** Research framework.

The purpose of this research is to see the influence of the classification of the accuracy of the prediction of ANFIS method sugeno model order one, therefore it uses two different approach. The first stage approach ANFIS method directly applied to perform the training and testing data based medical data. For the second approach, data are grouped with the algorithm k-means as input parameters. The second result of this approach will be made as the material inputs in planning medicine IFRS needs.

## 5. The results and Discussion
Prediction of medicine needs in IFRS uses the data result of observation at the hospital in 2015 and 2016. The determination of the medical data variable consists of gender, age and ICD patients.

*5.1 Prediction with ANFIS without grouping*
The training and testing on the first approach directly applies ANFIS method on the data sources with five indicator inputs namely gender, age patients, diagnosis code, long hospitalized, stock, and one of the exodus. There are 34 rules used on the model of the ANFIS Sugeno Order one. In the first stage is to seek the learning rate for optimal learning. This testing uses the amount of data trained as much as 50% of the total data.

RMSE produced by the least is 0.004. The value of the smallest RMSE will be used for the test RMSE and accuracy with a combination of data trained and test data. For the test RMSE and accuracy on the training used the amount of data as much as 85% of the total amount of data. To test the data used 70% of the total amount of data. Each combined data is done by the repetition of ten times to take the value of the RMSE and the accuracy of the test results. The results of the training and testing can be seen in figure 2 and 3.

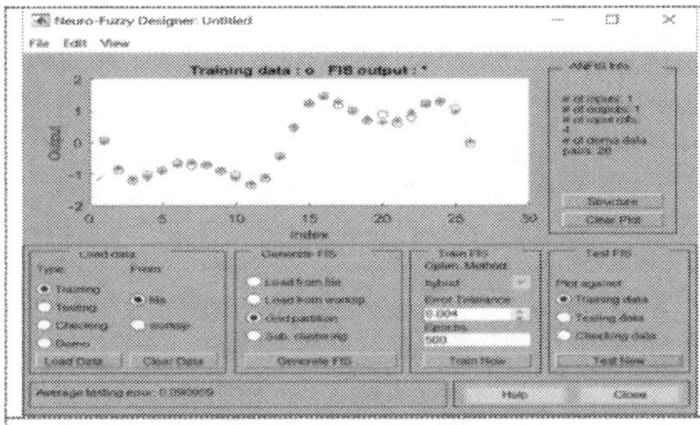


**Figure 2.** The Training .

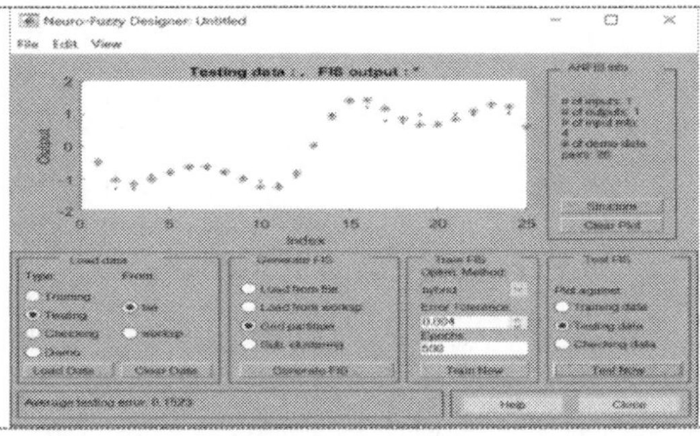


**Figure 3.** Testing.

In the figure 3 is the result of the training data 80% of the total data, with the value of the error 0.004, epoch 500, the results of the training will be made as a test for the prediction of medicine needs with error 0.004, epoch 500 test results seen in fig. 4 for predicting the needs of 2016 and 2017.
In the fig.4 the test result 70% data from the total amount of data for the prediction of medicine needs on 2016 using data 2015, while 2017 uses data 2016. Overall results of the prediction of medicine needs 2016 and 2017, accurate ANFIS method based on the comparison of the predictions with the actual data, but there is still value RMSE data above numbers 0, the value of the average RMSE 0,0669 for 2016 prediction and 0,0594 for prediction of 2017. Graph prediction is shown in figure 4.

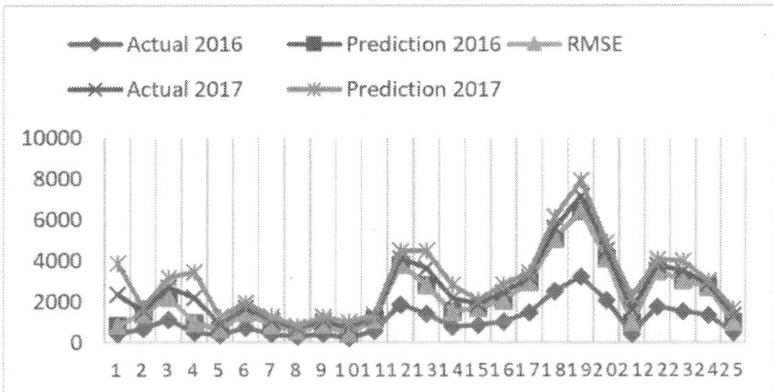


**Figure 4.** Graph prediction year 2016 and 2017

*5.2 Prediction with K-Means ANFIS (K-ANFIS)*

The second approach to prediction of medicine needs is done with the grouping of data. The Data is grouped with the algorithm k-means and then applied to the ANFIS. In table 5 is the result of a cluster of medical data with three groups, namely gender, ICD and age. The process of classification is used in two stages, the first phase of the medical data grouping is 2015 and the second grouping is 2016. $Cluster_3$ is the most dominant group in the determination of the medicine used, it is very reasonable for an indication of the highest determination of the medicine is the result of disease diagnosis from the doctor. $Cluster_1$ is the lowest domain because there are no rules of the determination of the medicines for gender. While in $cluster_2$ is the classification based on age with domain number 2 is the most dominant, this is due to the fact that there are different types of medicines for each of the age of the patient.

The process of computing time on the classification of this data is still too high, this can be seen the result of time which is used for the classification of a domain takes 0.09 seconds while the lowest 0.02 seconds with average 0.05 seconds for one-time iteration from 60 percent of the total data. The number of the cluster of the international Classification of Diseases (ICD) produced 139 clusters isn't defined and 23 isn't detected for data 2015, while 2016 produced 145 clusters and isn't known 12 clusters. On the table 1 explains the distance and the ratio between the cluster.

IOP Conf. Series: Journal of Physics: Conf. Series **978** (2018) 012019 doi:10.1088/1742-6596/978/1/012019

**Table 1.** Medical cluster and their number of instances of.

| | # Of instances of | Ratio % |
|---|---|---|
| Clusterl | 31,66666667 | 24.1 |
| Cluster2 | 46,47058824 | 46,5 |
| Cluster3 | 55,29411765 | 77,8 |
| Clusterl | 31,66666667 | 24.1 |

Based on the results of the medical data classification (gender, age, ICD) will be made as ANFIS input parameters for training and testing the predictions of medicine need. The test result can be seen in figure 6.

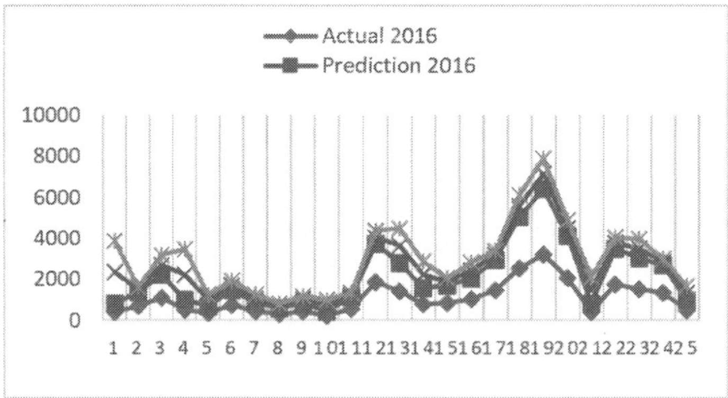


**Figure 5.** Prediction K-ANFIS year 2016 and 2017

In the figure 6 shows the prediction results with applying k-means algorithm for data grouping, significant prediction is more accurate and more quickly in the training process and testing. In the figure 7 and 8 the results of prediction of the comparison ANFIS with K-ANFIS for prediction of medicine needs 2016 and 2017.

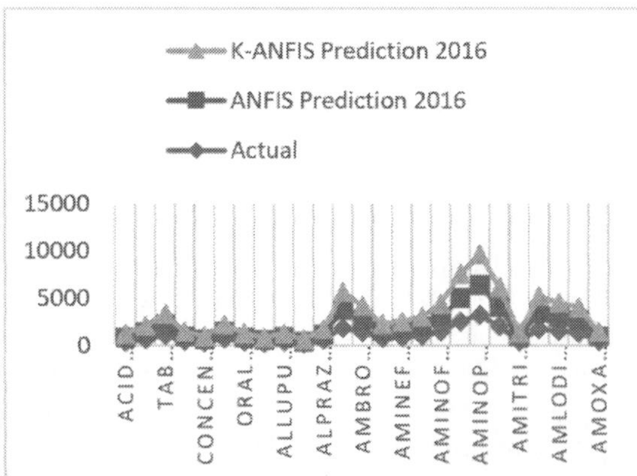


**Figure 6.** ANFIS and K-ANFIS Prediction 2016

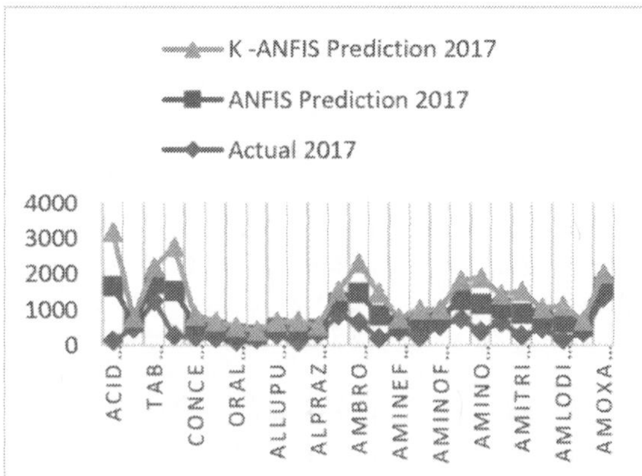


**Figure 7.** ANFIS and K-ANFIS Prediction 2017

## 6. Conclusion

Based on the results of the test with the use of medical data of patients 2015 and 2016, for prediction of medicine needs 2016 and 2017, overall ANFIS method is accurate in the prediction, this is seen on the value of the average RMSE, both of the approach is under 0 and comparison of the actual data with the results of prediction is accurate enough. The second approach is by applying the algorithm k-means of ANFIS which is having an effect on the time of the training process and it provides the accuracy of classification which is significantly better than the first approach without grouping, but the computing time used the algorithm k-means is still too high with an average of 0.05 seconds for each iteration so that it needs further research for computing time optimization by applying the other classification algorithm.

**Acknowledgments**
This paper is funded by Research Institutes and Community Service(LPPM) of Universtas Prima Indoesia, and supported by one of the private hospital in Medan city and the goverment hospital at one of district Sumatera Utara for decision making in medicine planning.

**References**

[1] Abbasi E and Abouec A 2008 Stock Price Forecast by Using Neuro-Fuzzy Inference System *Int. J. of Social, Behavioral, Educational, Economic, Business and Industrial Engineering* **2**.
[2] Ijegwa A D, Rebecca V O, Olusegun F and Isaac O O 2014 A Predictive Stock Market Technical Analysis Using Fuzzy Logic *Computer and Information Science* **7**.
[3] Agrawal S, Jindal M and Pillai G N 2010 Momentum Analysis based Stock Market Prediction using Adaptive Neuro-Fuzzy Inference System (ANFIS) *Proc. of the Int. Multi Conf. of Engineers and Computer Scientists IMECS* **I**.
[4] Agrawal A T and Ashtankar P S 2013 An Expert System for Home Health Monitoring: The ANFIS Approach *Int. J. of Scientific and Research Publications* **3**.
[5] Aggarwal N and Aggarwal K 2012 Comparative Analysis of kmeans and Enhanced K-means clustering algorithm for data mining *Int. J. of Scientific & Eninnering Research* **3**(3).
[6] Ayat S and Khosravanian A 2014 Identification and Classification of Coronary Artery Disease Patients using Neuro-Fuzzy Inference Systems *J. of Mathematics and Computer Science* **13** pp 136-141.
[7] Babu C N and Reddy B E 2014 A Moving-Average Filter Based Hybrid ARIMA – ANN Model for Forecasting Time Series Data *Applied Soft Computing J.* **23** pp 27–38.
[8] Siregar B, Butar-Butar I A, Rahmat RF, Andayani U and Fahmi F 2016 Comparison of Exponential Smoothing Methods in Forecasting Palm Oil Real Production *Int. Conf. on Computing and Applied Informatics* **801**.
[9] Bennett C, Stewart R and Lu J 2014 Forecasting Low Voltage Distribution Network Demand Profiles Using a Pattern Recognition Based Expert System *Energy* **67**.
[10] Bai L, Liang J and Dang C 2011 An Initialization Method to Simultaneously Find Initial Cluster Centers and the Number of Clusters for Clustering Categorical Data *Knowledge-Based Systems* **246** pp 785–795.
[11] Billah M and Waheed S 2015 Predicting Closing Stock Price using Artificial Neural Network and Adaptive Neuro Fuzzy Inference System ANFIS: The Case of the Dhaka Stock Exchange *Int. J. of Computer Applications* **129**.
[12] Budiono S, Suryawati S and Sulanto SD 1999 *Manajemen Obat Rumah Sakit Magister Manajemen Rumah Sakit* (Yogyakarta: Fakultas Kedokteran UGM).
[13] Chen H H, Chen M and Chiu C C 2013 The Integration of Artificial Neural Networks and Text Mining to Forecast Gold Futures Prices *Communications in Statistics-Simulation and Computation*.
[14] Damayanti A, Cahyadi R 2016 Algoritma Genetika Untuk Optimasi Struktur Anfis *Seminar Riset Teknologi Informasi SRITI.*
[15] Septiawan F Y, Dewa C K and Afiahayati 2017 Prediction Of Currency Exchange Rate In Forex Trading System Using Genetic Algorithm *Int. Interdisciplinary Conf. on Sc. Technology Engineering Management Pharmacy and Humanities Held on 22nd – 23rd.*

[16] Ghosh S, Dubey SK 2013 Comparative Analysis of K-Means and Fuzzy C Means Algorithms *IJACSA Int. J. of Advanced Computer Science and Applications* **4**.
[17] Hamasuna Y, Endo Y and Miyamoto S 2009 On Tolerant Fuzzy C-Means Clustering and Tolerant Possibilistic Clustering *Soft Computing* **145** pp 487–494.
[18] Hegazy O, Soliman O S and Toony A A 2014 Hybrid Of Neuro-Fuzzy Inference System and Quantum Genetic Algorithm for Prediction in Stock Market *Issues in Business Management and Economics* **2** pp 094-102.
[19] Husein A M, Mawengkang H, Budiarthi E 2010 *Teknik Fuzzy Inferences System Pada Manajemen Persediaan,* (Medan: Magister Teknik Informatika Universitas Sumatera Utara)
[20] Izakian H and Abraham A 2011 Fuzzy C-Means And Fuzzy Swarm for Fuzzy Clustering Problem *Expert Systems With Applications* **383** pp 1835–1838
[21] Lotfi E and Karimi M R 2014 OPEC Oil Price Prediction Using ANFIS *J. of Mathematics and Comp. Sc.* **10** pp 286-296.
[22] Ji Z, Sun Q and Xia D 2011 A Modified Possibilistic Fuzzy C -Means Clustering Algorithm for Bias Field Estimation and Segmentation of Brain MR Image *Comp. Medical Imaging and Graphics* **35** pp 383–397.
[23] Khan Z H, Alin T S, Hussain M A 2011 Price Prediction of Share Market using Artificial Neural Network ANN *Int. J. of Computer Applications* **22**.
[24] Makridou G, Atsalakis GS and Zopounidis C 2013 Gold Price Forecasting With a Neuro-Fuzzy-Based Inference System *Int J Financial Engineering and Risk Management* **1**.
[25] Nhu H N, Nitsuwat S and Sodanil M 2013 Prediction of Stock Price Using an Adaptive Neuro Fuzzy Inference System Trained by Firefly Algorithm *Int. Computer Science and Engineering Conf. ICSEC.*
[26] De Oliveira, J V and Pedrycz W 2007 Advances in Fuzzy Clustering and its Applications *J. Valente de Oliveira and W Pedrycz, Eds Chichester* (UK: John Wiley and Sons, Ltd)
[27] Pan S, Iplikci S, Warwick K and Aziz T Z 2012 Parkinson's Disease tremor classification – A comparison between Support Vector Machines and neural networks *Expert Systems with Applications* **39** pp 10764–10771.
[28] Paul, S K, Azeem A and Ghosh A K 2015 Application Of Adaptive Neuro-Fuzzy Inference System And Artificial Neural Network In Inventory Level Forecasting *Int. J. Business Information Systems* **18**.
[29] Quick JP, Rankin JR, Laing RO, O'Cornor RW 2012 *Managing Medicine Supply, The Selection, Procurement, Distribution and Use of Pharmaceutical, Third Edition* (USA: Kumarin Press, Conecticus).
[30] Suryani I, Wahono R S 2015 Penerapan Exponential Smoothing untuk Transformasi Data dalam Meningkatkan Akurasi Neural Network pada Prediksi Harga Emas *J. of Intelligent Systems* **1**.
[31] Satibi 2014 *Manajemen Obat di Rumah Sakit* (Yogyakarta: Fakultas Farmasi Universitas Gadjah Mada)
[32] Siringoringo R and Situmorang Z 2014 Optimasi Fungsi Keanggotaan Fuzzy Berbasis Algoritma Modified Particle Swarm Optimization *Jurnal Ilmiah Komputer dan Informatika KOMPUTA* **3**.
[33] Yazdani-Chamzini A, Yakhchali S H, Volungevičienė D and Zavadskas E K 2012 Forecasting Gold Price Changes by Using Adaptive Network Fuzzy Inference System *J. of Business Economics and Management* **135** pp 994-1010.
[34] Zhang C and Fang Z 2013 An Improved K-means Clustering Algorithm Traditional K-mean Algorithm *J. of Information and Computational Science* **1** pp 193–199.
[35] Zhang G P 2004 *Neural Networks in Business Forecasting* (USA: IGI Global).

IOP Conf. Series: Journal of Physics: Conf. Series **978** (2018) 012020 doi:10.1088/1742-6596/978/1/012020

# Identification of hand motion using background subtraction method and extraction of image binary with backpropagation neural network on skeleton model

**Fauziah[1, 2*], E P Wibowo[3], S Madenda[4] and Hustinawati[5]**

[1]Information Technology Doctoral Programme in Gunadarma University
[2]Lecturer in Faculty of Informatic and Communication Technology, Nasional University, Indonesia
[3]Microelectronic & Images Processing Studies, Gunadarma University
[4]Professor in Gunadarma University, Indonesia
[5] Head Laboratory of Application and Programming, Gunadarma University

Email: fauziah@civitas.unas.ac.id, mstiziah@gmail.com

**Abstract.** Capturing and recording motion in human is mostly done with the aim for sports, health, animation films, criminality, and robotic applications. In this study combined background subtraction and back propagation neural network. This purpose to produce, find similarity movement. The acquisition process using 8 MP resolution camera MP4 format, duration 48 seconds, 30frame/rate. video extracted produced 1444 pieces and results hand motion identification process. Phase of image processing performed is segmentation process, feature extraction, identification. Segmentation using bakground subtraction, extracted feature basically used to distinguish between one object to another object. Feature extraction performed by using motion based morfology analysis based on 7 invariant moment producing four different classes motion: no object, hand down, hand-to-side and hands-up. Identification process used to recognize of hand movement using seven inputs. Testing and training with a variety of parameters tested, it appears that architecture provides the highest accuracy in one hundred hidden neural network. The architecture is used propagate the input value of the system implementation process into the user interface. The result of the identification of the type of the human movement has been clone to produce the highest acuracy of 98.5447%. The training process is done to get the best results.

## 1. Introduction

Computer vision a field that aims to make a useful decision about a real physical object and state based on an image. Computer Vision is a combination of image processing and pattern regocnition the output of the computer vision has been processed into the result of image understanding. Development of this field is done adapting the ability of human vision in taking information. In the dicipline, computer vision deals with supporting theories such as artificial intellegent can extract information from the displayed image. Image data can be obtained with some from video, with some camera or multidimensional data from scanner. Area contained in computer vision include: pattern recognition,and classification, image segmentation, image restoration, and other area related to computer vision. Digital Processing is widely used to study matter relating to image quality improvement contrast enhacement, image transformation, colour, image restoration, selection of image characteristic (feature image) is optimal for the purpose analysis, making the process of with drawing information or object. Description or the introduction of objects contained in the image, compression, or reduction and for data processing time and input from image processing result [13]. The feature extraction process in image processing is used to perform the recognition process the objects in the image, or the video used by using feature extraction can be known to the existing classes and search for significant feature areas on an image or video object on the intrinsic elemen used [14]. Feature Extraction is general used to perform the identification process of the charateristic can form the best representation of an object to distinguish one object from another object. The feature extraction of forms is used to match an image or object with another object exists with the region used, the process of feature extraction involves the computation of a number of feature characteristic, values of an object shape independent of size or orientation, form feature extraction can be calculated for each object can be identified on each image or image stored, two types of feature are used global features (feature extraction that include aspect ratio, cilcularity and moment invariant) local feature extraction is a sequential boundry segment). Background method is one of the methods used to perform the separation process of one object with another object, doing background image reduction process. Background subtraction known as foreground detection is on of the tecniques in digital image processing and computer vision to perform the process of detecting or detecting an image to detect or retrieve the foreground of the approriate background based on human objects, text and another object movements [3]. In general background subtraction method is used to perform the process of moving object detection of an image or object on the video based on the difference between the reference background with the frame or reduce from the video, identification process is a determination of the identify of people, and objects [15][10].The definition of identification in general is the provision of sign on the class of goods or something with the aim of providing one component with another. Artificial Neural Network is one system that process information that has information idential with biological nerves. Artificial Neural Networks are used as pattern regocnition, signal processing and forecasting [6].

## 2. Theory

### *2.1. Digital Image Processing.*

Digital image processing is a dicipline that studies things to image quality improvement(contrast enhacement, transformation, colour( image restoration ), image transformation, selection of image feature ( featuring image ) is optimal for the purpose of analysis, making the process of with drawing information or object description or the introduction of objects contained on the image, perform compression or reduction and for the purposes of data storage, data transmission, and data processing time.

### *2.2 Computer Vision*

Computer Vision is a combination of image processing and pattern regocnition, the output of the computer vision process has been processed into the result of image understanding. Area contained in the computer vision include recognition, motion, image restoration, and other areas related to Computer Vision the most common function performed by include: image acquisition, pre processing, feature extraction, detection, segmentation, high level processing, decision making [7].

IOP Conf. Series: Journal of Physics: Conf. Series **978** (2018) 012020 doi:10.1088/1742-6596/978/1/012020

*2.3. Stage Of image Processing*

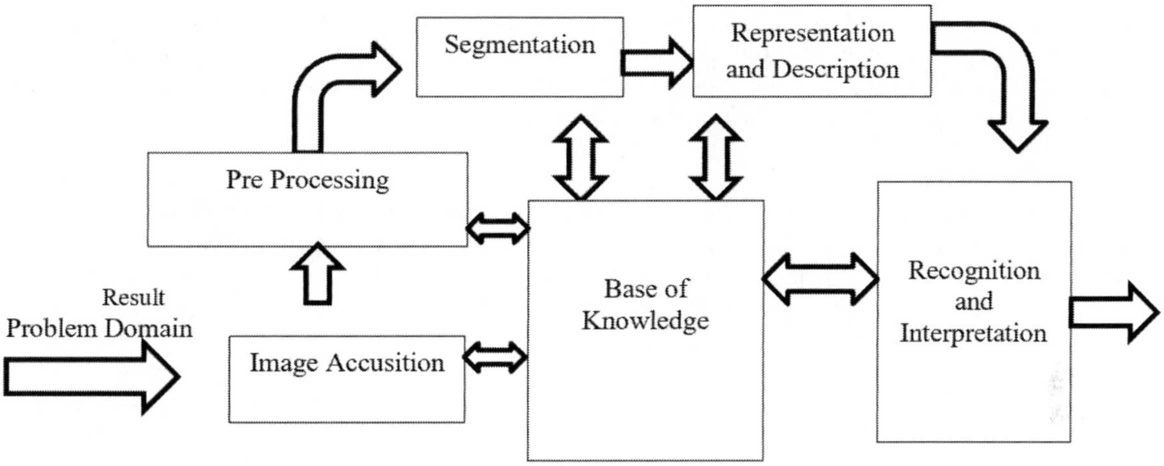


**Figure 1.** Stage of Image Processing [3]

Figure 1. are the stage of image processing, the initial stage is to acquire the image with the aim of obtaining a digital image to determine the data needed and choose the method of recording digital image at this stage steps taken to take the object to be taken pictures, and also to the preparation of tools, actors to imaging, the result of image acqusition is strongly influenced by the ability of the sensor to perform the process of digitizing the signal access. Preprocessing is related to image quality improvement, noise remove process, image transformation, determined the image to be observed. Segmentation is a stage that aim to partition the image into sub section that have important information such a sparating objects and backgrounds. Representation and description by representating a region with a list of coordinate points in a close curve and performing image description by performing feature selection and feature extraction. Recognition and Interpretation provide labeling objects whose information for the interpretation to giving meaning to groups of recognition objects. The last stage useful for based knowlegde to guide the operation of each process module and the performa the process of controlling the interaction to template matching or pattern recognition.

*2.4. Background Subtraction*

Background subtraction method is on of the methods used to perform the sparation process of one object with another object by doing background image reduction process. Background subtraction known as foreground detection is one of the techniques in digital image processing and computer vision to perform the process of detec or retrieve the foreground of the appropriate background based on human object, text and other object movements [3]. In general background subtraction method is used to perform the process of moving object detection of an image or object on the video based on the difference between the reference background with the frame produced from the video.

*2.5. Artificial Neural Network*

Artificial Neural Network is one system process information that has information identical with biological nerves. Artificial Neural Networks are used as pattern recognition, signal processing and forecasting [6].

*2.6. Feature Extraction*

The feature extraction process in image processing is used to perform the recognition process and the classification processing of objects in the image, or video used by using feature extraction can be known to exsiting classes and search for significant feature areas on an image or video object and rely on the intrinsic elements used to perform the identification process of the characteristics can form the best representation of and object to distinguish one object from another object, the feature extraction of form is used to match an image or object exists within in the region used, the process of feature extraction involves the computation of a number of feature characteristic values of an object shape independent of size or orientation. Feature extraction can be calculated for each object that can be identified on each image and on image stored.

*2.7. Skeletonization*

There are saveral ways used to form skeleton. Skeleton is a unique form of an object, which the order of an object. One way to get skeleton is the trough thinning. Thinning is a morfological operation is use to minimizing the geometric size of an object with the end result of a skeleton. Skeleton using an image to express the topology and characteristic model a grass field [16].

*2.8. Previous Research*

**Table 1. Previous Research**

| NO | Reserchers | Method | Result | Feature Work |
|---|---|---|---|---|
| 1. | Syed Sohaib Ali, Zafar M.F, Moeen Tayyab, 2009 | "Neural Network" | Detection people in video with five part and good performa for classification and the correct classification rate 85%, the results show better speed of and accuracy | From the system can be tracking, detec and recognize from side view and identification for human action |
| 2. | Huimin Qian, Yaobin Mao, 2010 | " SVM Decission Tree " | Identification of movement in walking, running, stand to sit, stand to squat, fall . | Extraction is still on the foot, not in the hands, head and the others body, and for identification of human action (hand, head and the others) |
| 3. | Alessio Ferone, Lucia Maddalena, 2014 | "Neural Background Subtraction" | Moving object detection with Zoom Cameras and solve the problem with background subtraction and good result | From the result limited comparasion of average precision, still moving object, in the system can be crate for human action and identification |
| 4. | Graciela Ramirez-Alonso and Mario I. Chacon - Murguia, 2014 | "Background Subtraction Model" | The Algorithm good parameters from the result combine background subtraction and Self Organized Map Neural Network the result good threshold | Implementation in the system for dynamic background subtraction, in the system good for identification of human action |

| | | | | |
|---|---|---|---|---|
| 5. | Nasrullah Ratu BSL, Deby Faisol Akbar, 2014 | "Background Subtraction and Kalman Filter " | Generate process of tracking and identification of objects by combining the method background subtraction and Kalman Filter Method and managed to track the object | Not seen the accuracy of the filter results obtained only limited to the process of tracking object only and the identification value is also not seen, for the future in the system good for identification human action |
| 6. | Lyra Vega Ugi, Bambang Hidayat, Suryo Adhi Wibowo, 2015. | "Variabel Modul Graph" | The system is made with the purpose of performing the process of identification and detection through a person walking gesture without having to pay attention to the individual by using the classification method and the result obtained with an average accuracy of 86.67% | The process of taking video is strongly influenced by the intensity of light so that affect the process of identification of objects used, can be created for identification of human action(hand, hand and the others) |
| 7. | Zhang Di Ban Xiajuan and Wu Di, Hao Wang, 2016 | "Spherical Coordinate and Angular Velocity (SCAV) Feature" | From the result good classification for action can be represented as spatial temporal, and the high accuracy 92.45% | From the system can be tracking, detect and recognize from side view and identification for human action(hand, head and the others) |
| 8 | Archita Tah, Sudipta Roy, Prasenjit Das and Anirban Mitra, 2017 | "Background Subtraction and Kalman Filter" | In the result good method and for background subtraction increase the computer vison, the result good similarity 93.1% | From the result limited feature extraction and multimodal feature, Extraction can be create for identification of human action |

## 3. Proposed Method

Stages performed in this study include:

1. Preparing from the human as actor who will give examples of movement hands: hands down, hands to the side and hands up.
2. Conducting Process Preparation H/W and Tools consisting of:
   Hardware processor Core i5, with 16 GB memory and 500 GB harddisk capacity,
   Nvdia Geforce 820 monitor 14.0 ", with windows OS 8.64 bit.
3. The other device used is Camera support smartphone 8MP resolution, Video acquisition using the camera that has been prepared.
4. Conducting video frame extraction process to image.
5. Perform background subtraction process on object component to obtain a binary image
6. Conducting the process of skeleton and do the process of extraction of binary image characteristics using invariant moment.

7. Perform the process of identification of hand movements using artificial back propagation. The steps of the research undertaken and determine the renewal or contribution of proposed research

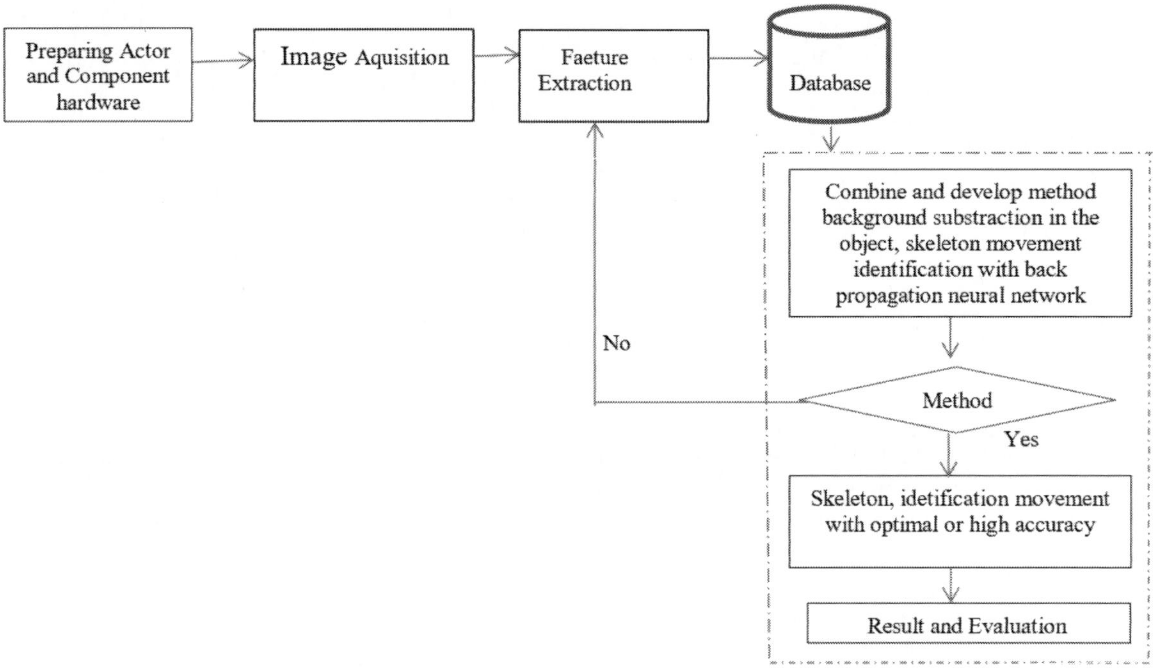


**Figure 2.** Research Methodology

*3.1. Data Acquisition*

In this research the data acquisition process is done using camera 8 Mega Pixel resolution smartphone. Data is a video that contains the movement of humans with various movement including hands down, hands to side and hands up. The video specification of the acquisition results are shown in table 2 below

**Table 2.** Video specifications of acquisition

| No | Specifications | Description |
|---|---|---|
| 1 | Name | Video |
| 2 | Format | Mp4 |
| 3 | Duration | 48 seconds |
| 4 | Frame Rate | 30 times/sec |

*3.2. Video Frame Extraction*

The next acquisition video is extracted into multiple frames. In this study, the total frame of extracted video totaled 1444 pieces. In each frame extraction results are then performed image processing to detect the type hand movement. The following is a coding for extracting video frames:

```
vid = ('Video 1.mp4');

Vid width = vid Width;
Vid height = vid. E Height
Move ('c data zero (vid height, vid Width,3,'uint8'),...
   'colormap',[]);

k = 1
while has Frame(vid)
   move (k).c data = read Frame(vid);
```

**Table 3.** Some Extraction represents the result of frame extraction

| Frame 1 | Frame 100 | Frame 200 |
|---|---|---|
|  |  |  |
| Frame 300 | Frame 400 | Frame 500 |
|  | 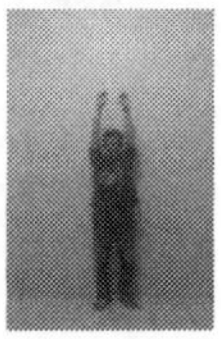 |  |

### 3.3. *Image Processing*

Image processing is performed on each frame of extraction result in order to identify the type of human movement. Stages of image processing implemented include: image segmentation, feature extraction, and identification

#### 3.3.1. *Image segmentation*

In this research, image segmentation is done using background subtraction method to separate between foreground / human and background / object other than human.
The steps of image segmentation are as follows:

1. Define the frame containing the background (background frame) and the frame you want detect (current frame).
2. Convert the image color space that was originally in the RGB color space (red, green, blue to grayscale against both frames. The equation used to convert RGB image to grayscale image is Grayscale = 0.2989 * R + 0.5870 * G + 0.1140 * B. [1] (Rafael C Gonzalez, Richard E Woods, 2002).
3. Perform a subtraction operation (subtraction) between the grayscale current frame images with the image grayscale background frame.
4. Convert grayscale image of subtracted result into binary image through thresholding operation.
   The binary image that the thresholding result is defined as: $g(x,y) = \begin{cases} 1 \text{ if } f(x,y) \geq T \\ 0 \text{ if } f(x,y) < T \end{cases}$ [2]
   Where g (x, y) is the binary image of the threshold result, f (x, y) is the grayscale image, and T is the threshold value [11].
5. Perform morphological operations on binary imagery in the form of filling holes and opening area for eliminating noise and perfecting segmentation results

IOP Conf. Series: Journal of Physics: Conf. Series **978** (2018) 012020 doi:10.1088/1742-6596/978/1/012020

6. Skeletonize the binary image of the segmentation result.
Coding to do image segmentation is:

```
% Read the background frame image
Background =in read ('Frame 1.jpg');
Background = in rotate(Background,90);
Background = in resize(Background,.1);

% Read the current frame image
Foreground = in read ('Frame 600.jpg');
Foreground = in rotate (Foreground,90i
Foreground = in resize (Foreground,

% Convert image to grayscale
Background_ gray = rgb2gray(Background);
Current Frame _gray = rgb2gray(Foreground);

% Grayscale image reduction
Subtraction = (double (Background _gray)-double (Current Frame_ ))
Min S = min (Subtraction (:));
Max _S = max (Subtraction (:));
Subtraction = ((Subtraction-Min _S)/(Max _S-Min _S))*255;
Subtraction = uint8(Subtraction);
```

The display of image segmentation process shown in Table 4

**Table 4.** Display Image Segmentation Process

| No | Image Processing | *Current Frame* | *Background Frame* | **3. Process Subtraction** (current Frame with Background Frame) | |
|---|---|---|---|---|---|
| 1 | **Read Image** | 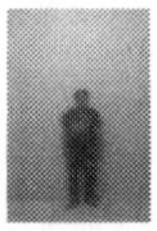 |  |  | |
| 2 | **Convert RGB to GrayScale** |  |  |  | **4.Convert Binary Image with Tresholding)** |
| 5 | **Motphology aperation** *filling holes* and *area opening* for good segmentation process |  | |  | **6.Skeleton** |

*3.3.2. Feature Extraction*

After the object in the image separated with background then performed the feature extraction step. Extracted traits are used to distinguish the type of movement between one object to another object. In this study feature extraction is done through morphological analysis (form) based on seven invariant moment values. The values are extracted from each image to distinguish the 4 output classes: class no object, hand-down class, hand-to-side class, and hands-up class. The calculation of the invariant moment value is based on the set of moments of a function f (x, y) of two variables defined as follows:

$$m_{pq} = \int_{-\infty}^{\infty}\int_{-\infty}^{\infty} x^p y^q f(x,y)dxdy \quad pq = 1,2,3,\ldots \qquad [3]$$

The two-dimensional moment with the order (p + q) of a digital image of size M x N is defined as:

$$m_{pq} = \sum_{x=0}^{M-1}\sum_{y=0}^{N-1} x^p y^q f(x,y) \qquad [4]$$

where p = 0.1,2, ... and q = 0.1,2, ... are integers. Furthermore, the central moments of the order (p + q) are calculated on the basis of the equation

$$\mu_{pq} = \sum_{x=0}^{M-1}\sum_{y=0}^{N-1} (x - \bar{x})^p (y - \bar{y})^q f(x,y) \qquad [5]$$

for p and q is 0.1,2, ..., where:

$$\bar{x} = \frac{m_{10}}{m_{00}} \text{ dan } \bar{y} = \frac{m_{01}}{m_{00}} \qquad [6]$$

Then normalized central moments, denoted by hpq, are defined as follows.

$$\eta = \frac{\mu_{pq}}{\mu_{00}^{\gamma}} \quad \gamma = \frac{p+q}{2} + 1 \qquad [7]$$

For p + q = 2, 3, 4, etc

So the seven invariant moments can be derived from the second and third moments through the following equation

$$M_1 = \eta_{20} + \eta_{02}$$

$$M_2 = (\eta_{20} + \eta_{02})^2 + 4\eta_{11}^2$$

$$M_3 = (\eta_{30} - 3\eta_{12})^2 + (3\eta_{21} - \eta_{03})^2$$

$$M_4 = (\eta_{30} + \eta_{12})^2 + (\eta_{21} + \eta_{03})^2$$

$$M_5 = (\eta_{30} - 3\eta_{12})(\eta_{30} + \eta_{12})[(\eta_{30} + \eta_{12})^2 - 3(\eta_{21} + \eta_{03})^2] + (3\eta_{21} - \eta_{03})(\eta_{21} + \eta_{03})[3(\eta_{30} + \eta_{12})^2 - (\eta_{21} + \eta_{03})^2]$$

$$M_6 = (\eta_{20} - \eta_{02})[(\eta_{30} + \eta_{12})^2 - 3(\eta_{21} + \eta_{03})^2] + (4\eta_{11}(\eta_{30} + \eta_{12})(\eta_{21} + \eta_{03})$$

$$M_7 = (3\eta_{21} - \eta_{03})(\eta_{30} + \eta_{12})[(\eta_{30} + \eta_{12})^2 - 3(\eta_{21} + \eta_{03})^2] + (3\eta_{12} - \eta_{30})(\eta_{21} + \eta_{03})[3(\eta_{30} + \eta_{12})^2 - (\eta_{21} + \eta_{03})^2] \qquad [8]$$

M1 - M7: Invariant Moment 1-7 (Hu, M. K, 1961)

The seven values of the moment do not change with respect to translation, scale change, reflection, and rotation. The extracted moment invariant is then used as input in the identification algorithm. Image processing is performed on each frame of extraction purpose to identify the type of human movement. Stages image processing implemented among others are: image segmentation, feature

IOP Conf. Series: Journal of Physics: Conf. Series **978** (2018) 012020 doi:10.1088/1742-6596/978/1/012020

extraction, and identification. Image segmentation are done using background subtraction method to separate between foreground human with background (object other than human).

**Table 5.** Extraction Results

| No | Real Class | Image Skeltonization | Moment Invariant | | | | | | |
|---|---|---|---|---|---|---|---|---|---|
| | | | M1 | M2 | M3 | M4 | M5 | M6 | M7 |
| 1 | No Object | 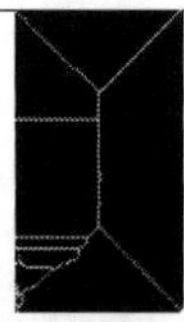 | 7.9632 | 26.1607 | 58.718 | 52.7462 | 3307.9726 | 260.9713 | 1381.9107 |
| 2 | Hand-down class |  | 6.7129 | 41.4148 | 21.5833 | 21.6338 | 317.9023 | 138.2061 | -5.2105 |
| 3 | Hand-to-side class |  | 3.7781 | 0.8647 | 61.9913 | 10.2550 | 157.1118 | 9.3815 | -7.2060 |
| 4 | Hands up class |  | 9.1374 | 78.5791 | 86.8490 | 77.9693 | 5872.0859 | 690.9475 | 50.6525 |

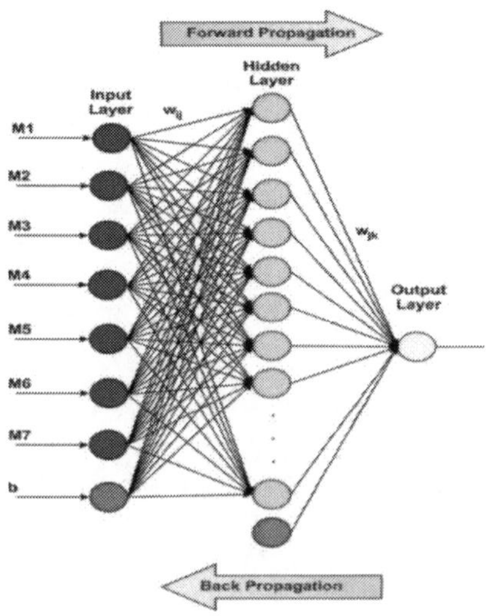

**Figure 3.** Artificial propagation neural network architecture

### *3.3.3. Identification process*

The artificial neural network used consists of three layers that include one input layer, one hidden layer, and one output layer. The identification process is performed to identify the type of human movement based on the input value in the form of seven invariant moment values previously extracted. The identification process is done using a back propagation neural network algorithm with an architecture like that shown in Figure 3 below: Figure 3 Identification process using Artificial Neural Network Back Propagation The training process uses the function activation of bipolar sigmoid on layer hidden and training functions Levenberg-Marquardt. In this process the seven invariant moment values are propagated progress through the initial weights had previously been initialized. Propagation the input value is made the neuron contained on the layer hidden. After arriving at the layer hidden, the total value by received by each neuron is processed use the activation function. So obtained the value of neurons in the layer hidden

The value of the neuron then again flipped towards layer output to obtain values output. The output value is compared with the target value. If the resulting error is smaller than target error that has been previously set, then the propagation process will be Stop. But if on the contrary, is done back propagation process by updating weight value. The back propagation neural network algorithm is used as follows:

Algorithm 1. Training

Step 1: Initialize, initialize all the weights on the hidden layers and output layers, and define the activation functions used at each layer and the rate of learning. Initialize weights use random numbers within range in a small range.

Step 2: Activation, activate network by applying input and expected ouput

a. Calculating the output obtained from neurons in hidden layers.
b. Calculating the output obtained from neurons in the output layers

Step 3: The weights are updated when the error is receded in Artificial Neural Network the returned error corresponds to the output signal.

a. Calculates the gradient of errors for neurons in the output layer, calculating weight correction. Updating weights on output layer neurons:
b. Counting gradient errors for neurons in hidden layers: calculating weight correction, updating weights on hidden layer neurons

Step 4: Iterate, raise one for the iteration p, goback to step 2 and repeat the process until the error criteria is reached. The artificial neural network training process is done by varying the number of neuron in the hiden layers, the result obtained are shown in Table 6

**Table 6.** Result Tarining with Back Propagation Nueral Network

| Neural totals | | | epoch | accuracy (%) |
|---|---|---|---|---|
| **Input Layer** | **Hidden Layer** | **Output Layer** | | |
| 7 | 10 | 1 | 718 | 97.5745 |
| 7 | 20 | 1 | 1000 | 97.8517 |
| 7 | 30 | 1 | 963 | 98.4061 |
| 7 | 40 | 1 | 730 | 98.3368 |
| 7 | 50 | 1 | 623 | 98.3368 |
| 7 | 60 | 1 | 661 | 98.4061 |
| 7 | 70 | 1 | 374 | 98.2675 |
| 7 | 80 | 1 | 302 | 98.4754 |
| 7 | 90 | 1 | 323 | 98.2675 |
| 7 | 100 | 1 | 258 | 98.5447 |
| 7 | 110 | 1 | 274 | 98.2675 |
| 7 | 120 | 1 | 255 | 98.2675 |

| | | | | |
|---|---|---|---|---|
| 7 | 130 | 1 | 236 | 98.1289 |
| 7 | 140 | 1 | 254 | 98.1289 |
| 7 | 150 | 1 | 201 | 98.2675 |

In Table 6 it appears that the architecture that provides the highest accuracy is the architecture with the number of neurons 100 on the hidden layer. The architecture is used to propagate the input value to the output value of the system implementation process into the user interface.

1. Main Program

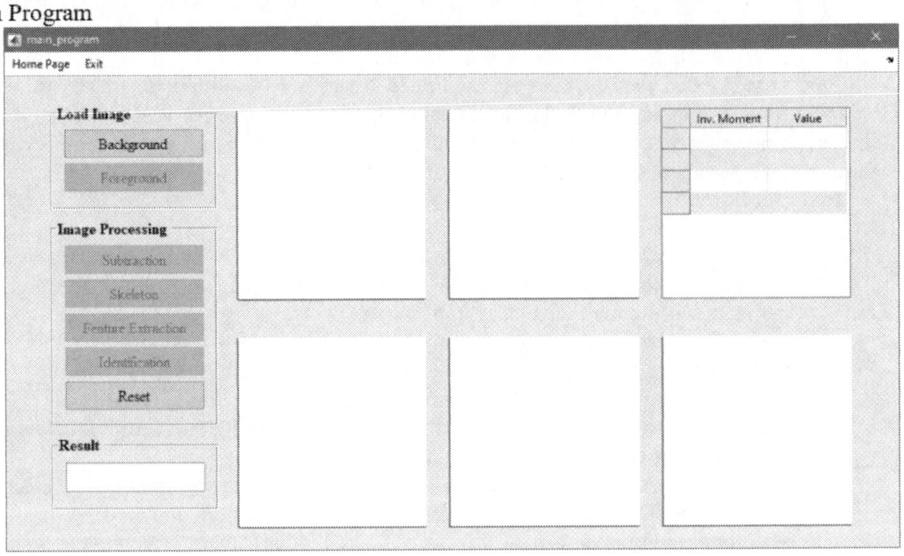


**Figure 4.** User Interface Main Program

2. Identification of images using back propagation neural network algorithm (down hand class and side-to-side hand movements).

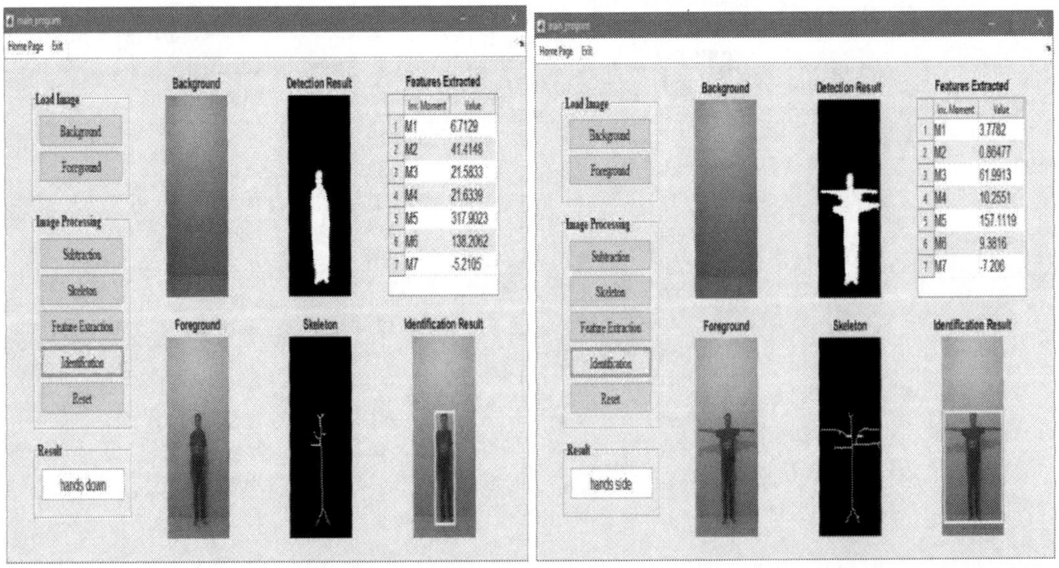

IOP Conf. Series: Journal of Physics: Conf. Series **978** (2018) 012020 doi:10.1088/1742-6596/978/1/012020

**Figure 5.** User interface down hand class **Figure 6.** User interface site to side hand movement

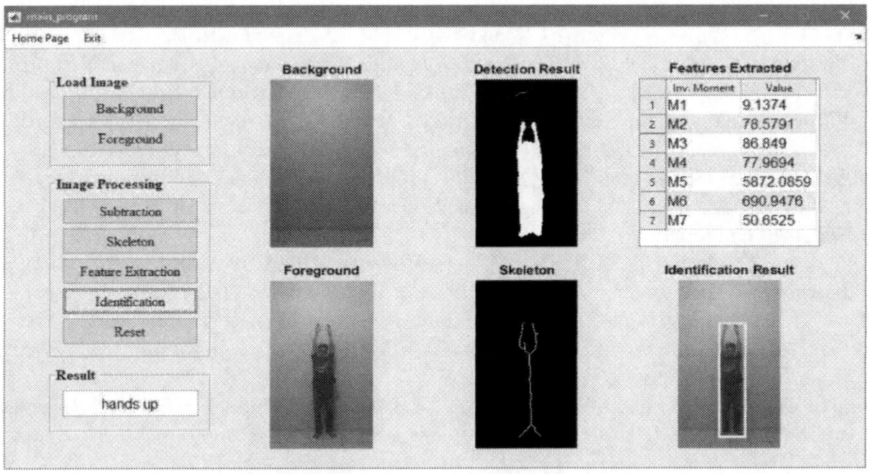


**Figure 7.** user interface hands up movement

## 4. Conclusions

1. Background subtraction method in this research is used to separate between foreground (human) with background (object other than human), and succeed are done by producing good skeleton.
2. Video frame extracted amounted to 1444 pieces. Each frame the extraction results, then performed image processing to detect the type of hand movement, with the process identification using back propagation neural network using 7 pieces invariant moment value input to produce an accuracy of 98.5557% for the process of identification of hand movements based on 4 pieces of class, one class no object, hand-down class, hand-to-side class and hands-up class.
3. Further development can be tested with various movements, and types of videos with various formats, so it can be compared the results and can be recommended the result with high identification and accuracy value and better, and for future work can be done with other action for identification.

## References

[1] Ferone A and Maddalena L 2014 Neural Background Subtraction for Pan-Tilt-Zoom Cameras *IEEE Transactions on Systems, Man, and Cybernetics: Systems* **44** pp 571-579.

[2] Tah A, Roy S, Das P and Mitra A 2017 Moving Object Detection and Segmentation using Background Subtraction by Kalman Filter *Indian J. of Science and Tech.* **10**

[3] Ramirez-Alonso G and Chacon-Murguia M I 2014 Segmentation of Dynamic Objects in Video Sequences Fusing the Strengths of a Background Subtraction Model, Optical Flow and Matting Algorithms *Conf.: 2014 IEEE Southwest Symp. on Image Analysis and Interpretation (SSIAI)* pp 33-36.

[4] Hu, M. K 1961 Pattern Recognition by Method of Moments *IRE Transactions of Information Theory* **49** pp 179-187.

[5] Qian H, Mau Y, Wang H and Wang Z 2010 On Video-Based Human Action Classification By SVM Decission Tree *Intelligent Control and Automation (WCICA), 2010 8th World Congress on Jinan, China, 7-9 July 2010.*

[6] Fausett L 1994 *Fundamentals of Neural Networks, Architectures, Algorithms, and Applications* (USA: Prentice-Hall).

[7] Shapiro L G, Stockman G C 2001 Computer Vision ( University of Washington, USA: Prentice Hall).

[8] Ugi L V, Hidayat B, Wibowo S A 2015 Individu Identification System Running Video-Based Processing Using the Variable Module Graph Method *e-Proceeding of Engineering* **2** p 331.

[9] Talib M R, Shafique A, Hanif M K and Sarwar M U 2017 Gait Identification using Neural Network *(IJACSA) Int. J. of Advanced Computer Science and Applications* **8.**

[10] BSL N R and Akbar D F 2014 *Object Tracking Based Background Subtraction and Kalman Filter* [Thesis] yudistira.lecture.ub.ac.id/files/2014/01/paper-pcd-object-tracking.pdf

[11] Gonzalez R C, Woods R E 2002 *Digital Image Processing* (New Jersey: Prentice Hall).

[13] Szeliski R 2011 *Computer Vision Algorithms and Applications* (London: Springer).

[14] Aviv R 2011 Algorithms for Testing Connectivity – Implementation on Binary Images (Israel: Tel-Aviv University).

[15] Bouwmans T, Porikli F, Höferlin B and Vacavant 2014 *A Background Modeling and Foreground Detection for Video Surveillance* (Chapman and Hall/CRC).

[16] Chen Y, Drechsler K and Zhao W and Laura C O 2011 A Thinning-based Liver Vessel Skeleton Method *Int. Conf. on Internet Computing and Information Services,IEEE, Computer Society* pp 152-155.

[17] Zang D, Ban X, Wu D and Wang H 2016 3D-LSR: A New Model For Skeletal Representation and Segmentation *IEEE Int.Conf. on Internet of things (I things) and IEEE Green Comp. and Comm. (Green Com) and IEEE Cyber, Physical and Social Computing (CPS Com) and IEEE Smart Data(Smart Data) IEEE Computer Society* pp 80-86.

[18] Zhu Z 2008 *Computer Vision Research Progress* (UK: Nova Science Publishers).

# Implementation of supply chain business application through business model canvas and waterfall framework collaborations for fish farmers SMEs in ulekan market bandung

**Y Priyadi and A Prasetio**

Departement of Telecommunication and Informatics Business Management, Telkom University.

Email: whyphi@telkomuniversity.ac.id, adhipras@gmail.com

**Abstract.** This research resulted in the development of e-SCM application, in small-scale group of fish farmers based on Open Source technology in Ulekan Market Bandung, by collaborating the implementation of e-SCM and Data Management. Then proceed with the application of supply chain business through collaboration Business Model Canvas and Waterfall Framework. For the design of business process reengineering in this activity, it produces a context diagram called e-SCM SME Fish consisting of five entities directly involved with the system, namely: fish shop supervisor, fish shop retailer, employees, fish farmers, and customers. Referring to the Context Diagram, decomposition process of Level 0 e-SCM SMEs Fish. The decomposition results in Data Flow Diagram Level 1 for four sub processes, namely: business partners, transactions, retailer stock, and documentation. Result of nine blocks on Business Model Canvas on e-SCM activity, its category consist of Priority 1, Priority 2, Direct, Indirect, Purchase / e-SCM, Transactional, Community, Asset Sale, Physical Asset, Human, Production, Strategic Alliance -competitors, Coopetition, Buyer supplier relationship, Fixed Cost, Variable Cost. For integration of data management on Localhost Server media on e-SCM using http: // whyphi: 8080 address, as prototype which will soon be adopted by farmer fish farmer.

## 1. Introduction

There are 42 million data on the implementers of Small and Medium Enterprises (hereinafter referred to as SMEs), spread over 23 government agencies. Referring to the Law of the Republic of Indonesia Number 20 Year 2008, concerning small and medium enterprises in business development activities, to be able to carry out improvement is done through several ways, namely: improving network marketing, distribution, technical skills, quality control, cooperation and transfer of technology.

The basic reference of Supply Chain Management (hereinafter referred to as SCM) is the cycle of activities in the business processes of an organization, in coordinating the flow management for raw materials, finance, and information that adds value to the achievement of the organization's objectives. To achieve its goals, the information technology is very important role in improving the performance of SCM. There are four driving factors for the performance, namely: information, transportation, inventory, and facilities. [3].

An activity that can be collaborated to handle SME business processes is Open Source technology and Electronic Supply Chain Management (hereinafter referred to as e-SCM). In doing business process adjustment and implementation of such activities, a business model design strategy is required. Business Model Canvas (hereinafter referred to as BMC), is a strategic mapping tool for business development that can describe, design, and focus on strategic aspects, thus becoming a unified business strategy. Referring to the BMC device then the results of a business activity, can be analyzed and designed as a solution to business processes that have a problem.

In the tabulation data of Spatial Planning (RTRW) Bandung 2011-2013, Ulekan Market including the location that became the development of Sub Service Center City. This market stagnated in increasing revenue, due to difficulties in adopting supply chain technology.

## 2. Discussion

All these activities, is a continuation of research that has been done in the previous year. This section is an implementation, which aims to improve the current system.

### *2.1. Stages Waterfall Framework*

Referring to the specific objectives of the study, the following describes the stages of developing BMC collaboration with Open Source e-SCM that adopt the waterfall framework as its development cycle. Stages performed for this second year of research, including the flow that can be observed in Figure 1.

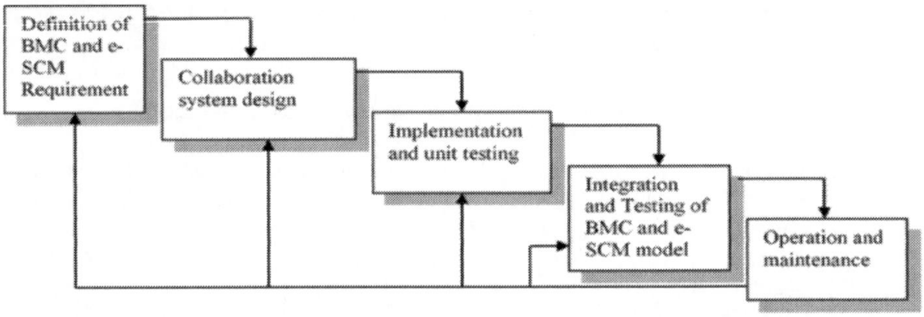


**Figure 1.** Adoption of Waterfall Stages for Second Year Research

Through this step approach, a series of process activities are described and presented in separate processes, such as BMC and e-SCM requirements specifications, design and implementation for design collaboration, and piloting. After each step is defined, the step is signed off and development resumes in the next step. This model has been obtained from the reengineering process. [18].

### *2.2. Business Process Reengineering for CD/DFD and E-R Diagram*

The next activity is to create Business Process Reengineering (BPR) based on the existing system that has been described previously, by adopting the scheme Open Source e-SCM. Referring to Figure 2, there is a change in the flow of data and the entities involved. This happens from Context Diagram engineering with e-SC M scheme. In this figure, there are five entities involved directly with the system, namely: supervisor toko ikan (supervisor fish shop), retailer toko ikan (fish shop retailers), employees, petani ikan (fish farmers), and customers.

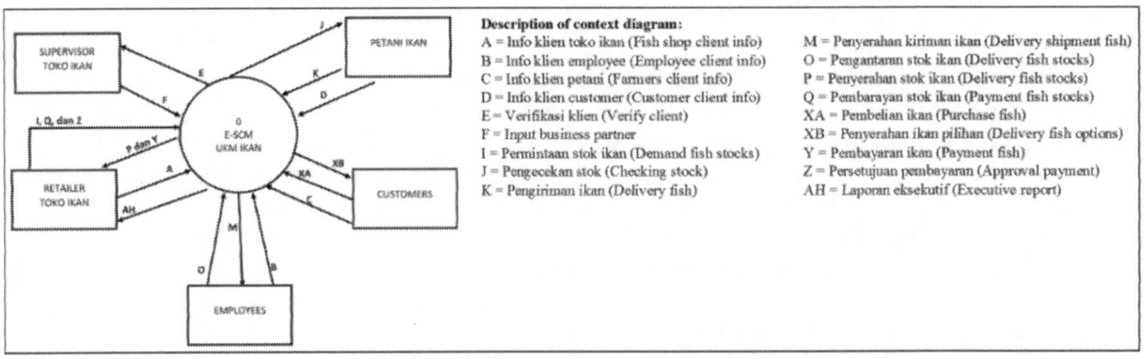


**Figure 2.** CD of E-SCM UKM Ikan [15]

Based on Figure 3, there is a breakdown DFD results for level 1. [9]. In the diagram, there are four sub-processes, namely: business partner, transaksi (transaction), stoknya retailer (the stock of retailers), and dokumentasi (documentation).

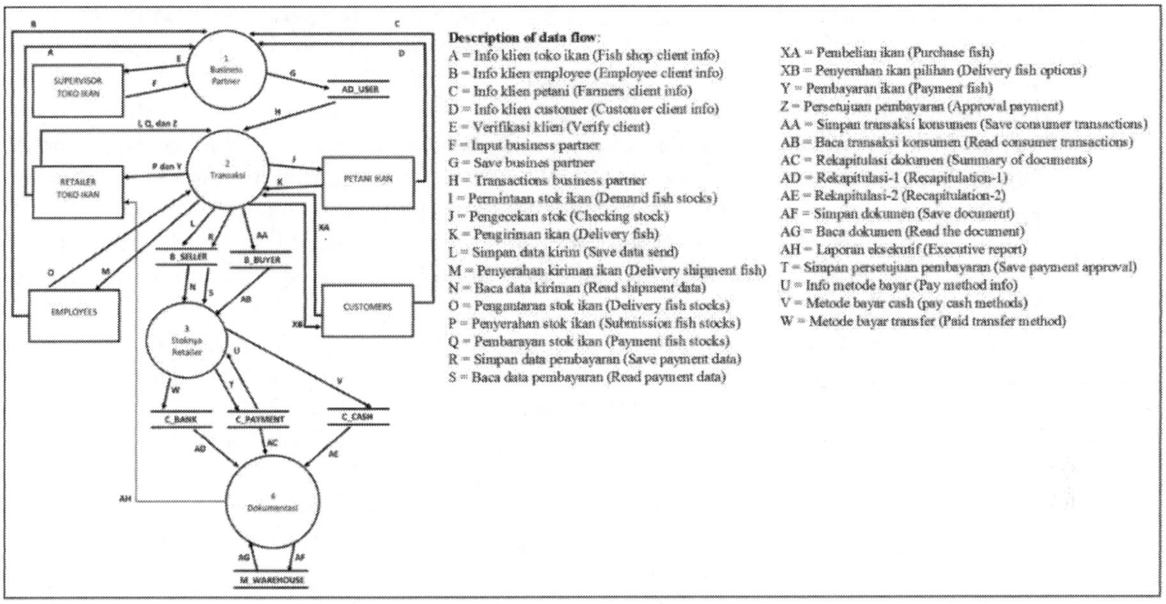


**Figure 3.** DFD of E-SCM UKM Ikan [15]

Designing models for data management, data storage is engineered to the process of e-SCM. This design, built based on the existence of the data store on the DFD for all levels. [9].[14].

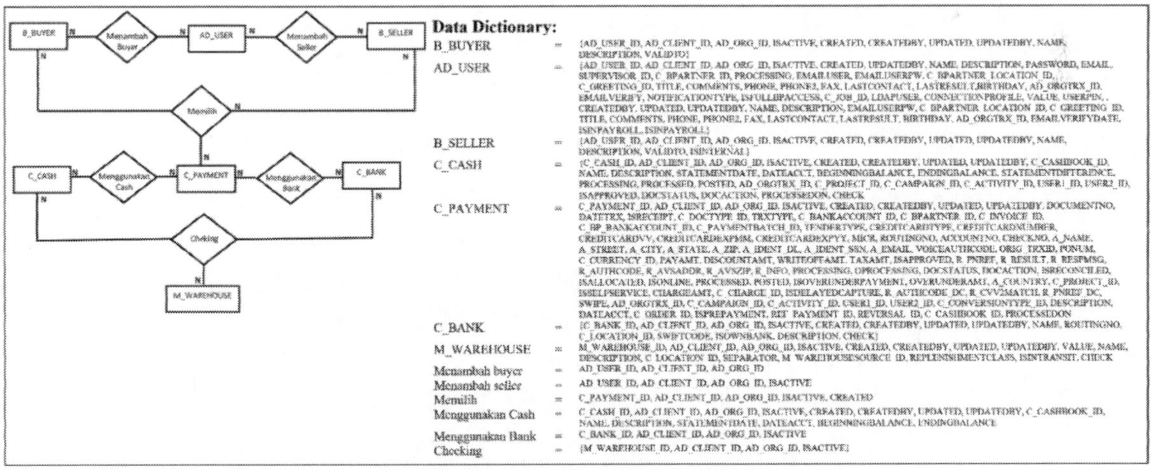


**Figure 4.** Data Dictionary in the E-R Diagram [15]

When referring to the design of DFD modified, then all data store contained in the DFD will be the entity on the E-R Diagram. The entity comprised of B_BUYER, AD_USER, B_SELLER, C_CASH, C_PAYMENT, C_BANK, and M_WAREHOUSE. Use of the data dictionary at E-R Diagram, aims to show the relationships between entities without being disturbed by the presence of key attributes and descriptive attributes. This can be observed in Figure 4, which presents the use of the data dictionary to E-R Diagram.

### *2.3. Business Model Canvas e-SCM*

Referring to survey results from fish farmers in Ulekan Market, the mapping of the business model in the environment is culminated with nine blocks from BMC, namely: Customers Segment, Value Proposition, Customer Relationship, Channel, Revenue Stream, Key Resource, Key Activities, Key Partnership, and Cost Structure. [13]. For details of the explanation of the nine blocks, are as follows:

- Customer Segment. (Priority 1: An ornamental fish hobbyist, Priority 2: Reseller)
- Value Propositions. (e-SCM SME Fish Farmers is an application used by smallholders of ornamental fish farmers. This activity can be implemented, because the application has been through the adjustment of system requirements. Its development aims to improve the supply chain business processes that have been adapted to the conditions in the market Ulekan)
- Channels. (Direct: Breeder Place, Indirect: Reseller, Purchase / e-SCM: http://whyphi:8080)
- Customer Relationships. (Transactional: purchase broken / restricted for customer segments. Community: social media group)
- Revenue streams. (Asset Sale: Sale of fish, medicine / vitamins, and aquascape supplies)
- Key Resources. (Physical Asset: Fish breeder, breeder room, aquarium, water, pump. Human: e-SCM operational staff, distribution personnel and asset sale maintainers)
- Key Activities. (Production: Fish supply chain activities from fish farmers to Customer segments)
- Key Partnerships. (Strategic Alliance between non-competitors: Provider of fish feed, Provider of medicines and vitamins, Aquascape equipment dealer. Coopetition: Reseller, Buyer supplier relationship: Ornamental fish hobbyist)
- Cost Structure. (Fixed Cost: purchase of feed, electricity payment, purchase of fish farming equipment, worker's wage. Variable Cost: Delivery fee via courier or COD)

Summary mapping of nine blocks and all its categories, can be observed in Figure 5. In the business model seen some linkages and slices of all categories in the Ulekan Market.

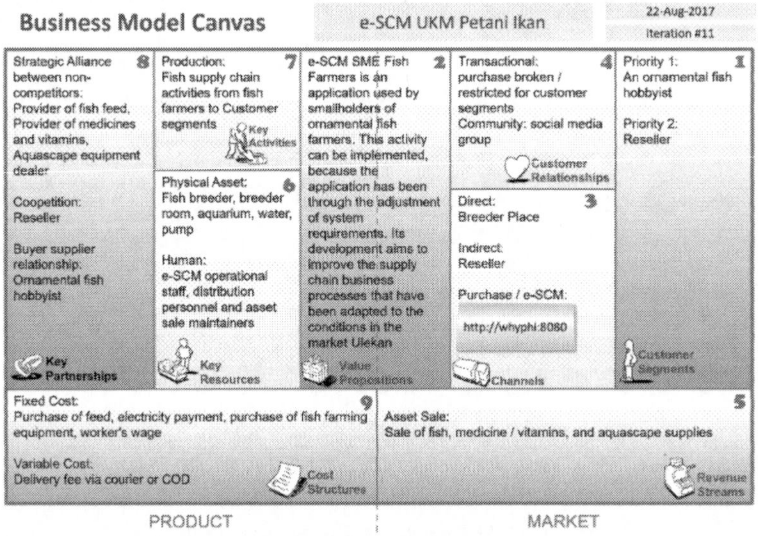


**Figure 5.** BMC for e-SCM

### *2.4. Results of Prototype Engineering Applications e-SCM*

The following are some of the results of open source e-SCM application engineering, adapted to business processes in the Fish Market of Ulekan, among others, as follows:

- Localhost address. For testing this application engineering, used server that is local with address http: // whyphi: 8080

- Database used in this application using PostgreSQL which is open source, so it is suitable with e-SCM application. Referring to the database, there are several core tables that are modified according to system requirements.

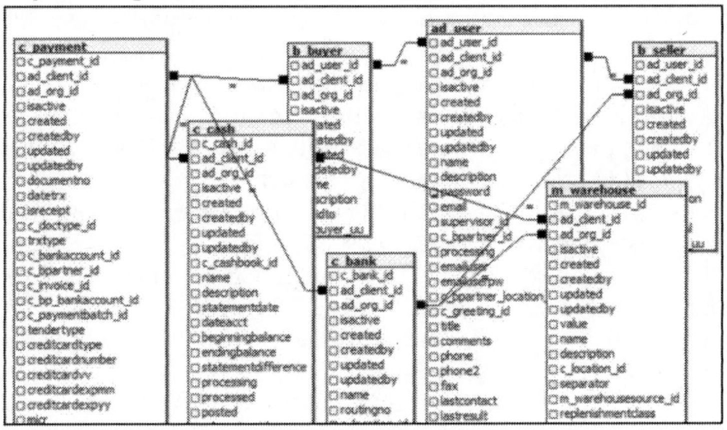


**Figure 6.** Relation on Physical Table [15]

- Login form for e-SCM, using an input view of the application tailored to the needs of the fish supply system.

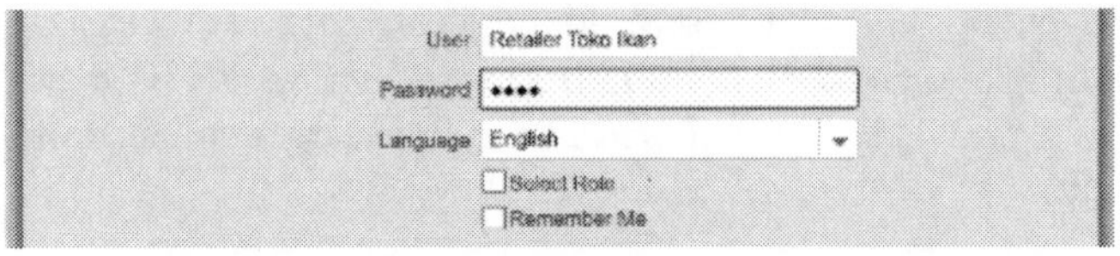


**Figure 7.** Prototype for e-SCM Login Form

- Business partner transactions for Fish Farmers, tailored to the needs of users.

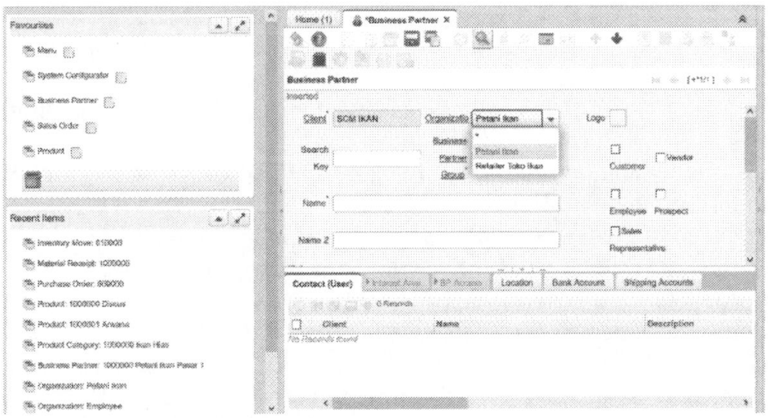

**Figure 8.** Business partner transactions

## 3. Conclusions

This research, in general, resulted in the development of e-SCM application in small-scale fish farmers based on Open Source technology in Ulekan Market Bandung, by collaborating the implementation of e-SCM and Data Management. Then proceed with application of supply chain business through collaboration Business Model Canvas and Waterfall Method. In addition, specifically this research has achieved the following results:

- In the existing system model design, for Context Diagram Level 0 there are four entities directly involved, ie fish farmers, fish shop, courier, and buyer. Referring to the Context Diagram of the Fish Supply System, the decomposition process of Level 0. The decomposition results in Data Flow Diagram Level 1 for 3 sub processes, namely: stock transactions, courier transactions, and consumer transactions.
- In existing system modeling for data management, there is the use of a single data store named "simpan". The role of data store on the existing system, just as a medium to record transactions and buy only fish, often not even documented for transactions. Based on these facts, then the use of ERD concept for existing system can not be used.
- To design business process reengineering in this activity, generate a context diagram called e-SCM SME Fish consisting of five entities directly involved with the system, namely: fish shop supervisor, fish shop retailer, employees, fish farmers, and customers. Referring to the Context Diagram, decomposition process of Level 0 e-SCM SMEs Fish. The decomposition results in Data Flow Diagram Level 1 for four sub processes, namely: business partners, transactions, retailer stock, and documentation. In addition, for data storage activities in Data Flow Diagram Level 1, the seven data store results modified from the e-SCM process, as follows: consists of B_BUYER, AD_USER, B_SELLER, C_CASH, C_PAYMENT, C_BANK, and M_WAREHOUSE.
- The results of nine blocks on Business Model Canvas on e-SCM activities, the categories consist of Priority 1, Priority 2, Direct, Indirect, Purchase / e-SCM, Transactional, Community, Asset Sale, Physical Asset, Human, Production, Strategic Alliance between non-competitors, Coopetition, Buyer supplier relationship, Fixed Cost, Variable Cost
- For integration of data management on Localhost Server media on e-SCM using http: // whyphi: 8080 address, as a prototype that will soon be adopted by small scale fish farmers in the second year of this study.

**Acknowledgments**
This work was supported by Ristek Dikti and Telkom University.

**References**

[1] Ascef R, Ferrer G & Mullins S 2014 Supply Chain Management and its Route to Normal Science: a Kuhnian Analysis *J. of Operations and Supply Chain Manag.* **7** pp 59-74.
[2] Blanchard D, 2010, Supply chain management best practices 2nd ed,, Hoboken NJ: John Wiley & Sons,
[3] Chopra S, and Meindl P 2001 *Supply chain management: Strategy planning and operations* (New Jersey - Prentice-Hall).
[4] CSCMP 2004 *Council of supply chain management professionals* [Online] http://cscmp,org/about-us/supply-chain-management-definitions
[5] Durkin M, McGowan P and McKeown N 2013 Exploring social media adoption in small to medium-sized enterprises in Ireland *J. of Small Bus. and Enterprise Develop.* **20** pp 716-734.
[6] Friedman A L, Miles S and Adams C 2000 Small and Medium-Sized Enterprises and the Environment: Evaluation of a Specific Initiative Aimed at All Small and Medium-Sized Enterprises *J. of Small Bus. and Enterprise Develop.* **7** pp 325-342.
[7] Giménez C and Lourenço H R 2008 e-SCM: Internet's Impact on Supply Chain Processes *The Int. J. of Logistics Management* **19** pp 309-343.
[8] Halley A and Guilhon A 1997 Logistics Behaviour of Small Enterprises: Performance Strategy and Definition *Int. J. of Physical Distribution & Logistics Management* **27** pp 475-495.
[9] Hapsari K and Yudi P 2017 Perancangan Model Data Flow Diagram Untuk Mengukur Kualitas Website Menggunakan Webqual 4.0 *JSINBIS Jurnal Sistem Informasi Bisnis* **7** pp 66-72.
[10] Laseter T & Oliver K 2003 When will Supply Chain Management Grow Up? Strategy+Business [Online] http://www,strategy-business,com/article/ 03304?gko= 54182.

[11] Lummus R R and Vokurka R J 1999 Defining Supply Chain Management: a Historical Perspective and Practical Guidelines *Industrial Manag. & Data Systems* **99** pp 11-17.
[12] McLeod R Jr 1995 *Management Information System, 6th ed* (NJ: Prentice – Hall).
[13] Osterwalder A 2010 *Business Model Generation* (Jakarta: Elex Media Komputindo).
[14] Priyadi Y 2014 Kolaborasi SQL & ERD Dalam Implementasi Database (Yogyakarta: Andi Publisher).
[15] Priyadi Y and Hatammimi J 2016 Development of e-SCM Open Source as Integration of Data Management For Fish Farmers SMEs in Pasar Ulekan Bandung *Int. Conf. on Science and Technology ICST* **1**.
[16] Ross D F 2003 *Introduction to E-supply Chain Management: Engaging Technology to Build Market-winning Business Partnerships* (Florida: CRC Press).
[17] Rencana Kerja Pemerintah RKP Bidang Pemberdayaan Koperasi Dan Ukm Tahun 2014
[18] Turban E, McLean E and Wetherbe J 2007 Information Technology for Management (John Willey & Sons Inc).
[19] Undang-Undang Republik Indonesia Nomor 20 Tahun 2008 Tentang Usaha Mikro Kecil Dan Menengah.

# Facial expression system on video using widrow hoff

**M Jannah[1], M Zarlis[2] and H Mawengkang[3]**

[1]Department of Computer Science, Universitas Sumatera Utara, Medan, 20155.
[2]Department of Computer Science, Universitas Sumatera Utara, Medan, 20155.
[3]Department of Mathematics, Universitas Sumatera Utara, Medan, 20155.

Email: miftahuljannah0077@students.usu.ac.id, m.zarlis@usu.ac.id, hmawengkang@yahoo.com

**Abstract**. Facial expressions recognition is one of interesting research. This research contains human feeling to computer application Such as the interaction between human and computer, data compression, facial animation and facial detection from the video. The purpose of this research is to create facial expression system that captures image from the video camera. The system in this research uses Widrow-Hoff learning method in training and testing image with Adaptive Linear Neuron (ADALINE) approach. The system performance is evaluated by two parameters, detection rate and false positive rate. The system accuracy depends on good technique and face position that trained and tested.

## 1. Introduction

Facial expression includes rich information regarding individual's behaviour. In addition, face is an indicator of human reactions and feelings.[6]. The Automatic facial expression has been applied to many practical applications Fields such as human-computer interactions. In recent years, the study of facial expression recognition has become the more important topic to many researchers from different research fields [3].In the development, image processing as a replacement of human eye while computer vision as a brain that used to process the information. In this case, the computer can detect an image object that moves anywhere (real time) and the computer can decide an expression from an image (happy and sad). In previous research, many researchers find facial expression without real-time image while in this research, the researcher focus on the video image.

## 2. Theoretical Basis

### *2.1. Face Detection*

The first step of face detection is preprocessing. The step of feature extraction is detecting feature points then rotating to line up then locating and cropping the facial image using a rectangle form. In the process of face area detecting, a single image involves some methods knowledge based that is known by the facial invariant process, template matching and appearance based [2]. Many factors that must be considered in detecting of face such as hairstyles, glasses, this is some variables that vary significantly, so this is not a trivial task in getting face pattern [3]. This is a step of image process Firstly, convert into a binary scale and scan the image for forehead and then find out for a maximum width of continuous white pixels of image until reach eyebrows and last cut the face in such a way to perform face detection image [4].

### *2.2. Expression Detection*

Facial expression recognition is related o the problem of classifying facial images into human expressionclasses. Expression recognition contains a kind of subjects such as perceptual recognition, affective computing and machine learning[7]. It is very difficult to define the meaning of emotions

IOP Conf. Series: Journal of Physics: Conf. Series **978** (2018) 012022 doi:10.1088/1742-6596/978/1/012022

even that almost everyone knows intuitively about the intent of emotion. The emotion as a stimulus that evaluates experiences based on the potential for gain or pleasure[1].

*2.3. Video Image*
In the object movement detection, there is a matching process called by Frame Difference which reduces one frame to other frame and labeled the larger frame. This process will capture the border of the moving object. First of all the video image will display the image captured by the webcam then converted to 16 grayscale images with the aim to facilitate image processing.When a new image is captured by the webcam, the program will calculate the difference between two images by calculating the average value of all the gray value.

*2.4. Adaptive Linear Neuron*
Adaline (Adaptive Linear Neuron) is simple linear model two-layer neural network. One layer refers to input and another layer refers to output layer which has a single output neuron.The function of all linear model neurons transfer (y = kx+n) and network exercise the result with Least Mean Squares (LMS) algorithm for learning. The scopes of Adaline network are recognizing patterns, data filtering, or to approximate linear function.
Calculate the net input to the output unit "y"

$$y_{in} = b + \sum_{i=1}^{n} = x_i w_i \tag{1}$$

Where :
$y_{in}$ : The number of input
b : bias
wi : weight

Update the weights $w_i$ and bias $b_i$ for i=1 to n

$$w_i \text{ (new)} = w_i\text{(old)} + \alpha(t - y_{in})\, x_i \tag{2}$$

$$b_i \text{ (new)} = b\text{(old)} + \alpha(t - y_{in}) \tag{3}$$

Here't' represents the target value. Apply the activation function over the net input calculate

$$Y = \begin{cases} 1 \text{ if } y_{in} > 0 \\ -1 \text{ if } y_{in} < 0 \end{cases} \tag{4}$$

In Adaline neural network, the learning process held by the adjustment of the weights of the neurons as per the weighted summation of the net inputs. Each neuron in the Adaline neural network accepts more than one input but generates one single output [5]. Adaline network architecture is shown in the following figure.

**Figure 1.** The Adaline architecture

*2.5. Widrow Hoff*
The least mean square error (LMS or Widrow-Hoff) algorithm is an example of supervised training, in which the learning rule is provided with a set of examples of desired network behavior:

$$\{p1, t1\} , \{p2, t2\} , \ldots, \{pQ, tQ\} \tag{5}$$

As each input is applied to the network, the network output is compared to the target. the difference between the target output and the network output can be calculated to be an error in this method. it is used to double the average number of errors in the network, the target will be compared with the network output.

## 3. The Scheme Of System

The scheme of System is shown in Figure 2.

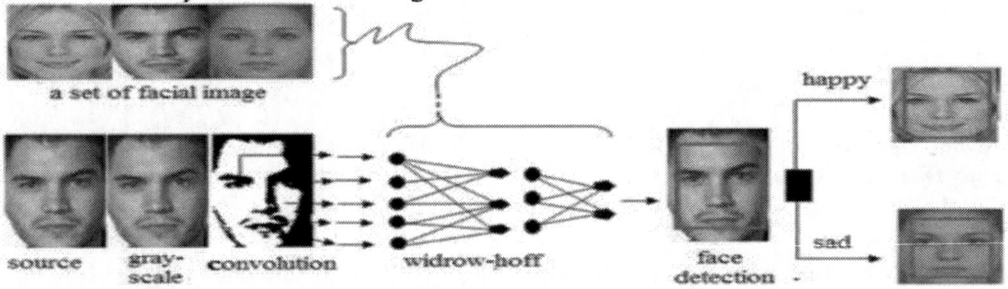


**Figure 2.** The scheme of facial expression system in video

The step is performed after the system receives video input is gray-scale, convolution, and face pattern recognition test through ADALINE network. In the pre-processing step, the video will be represented in the form of a single channel and ends with edge detection through the convolution process. In the main process, computing using the ADALINE network, the facial pattern vector will be trained to obtain a weight matrix, which then the weight matrix is used as the testing matrix to get the expression present in the face area by inputing the facial data in real-time

### *3.1 Gray-Scale*

The gray-scale Flowchart processes are developed based on Figure 3:

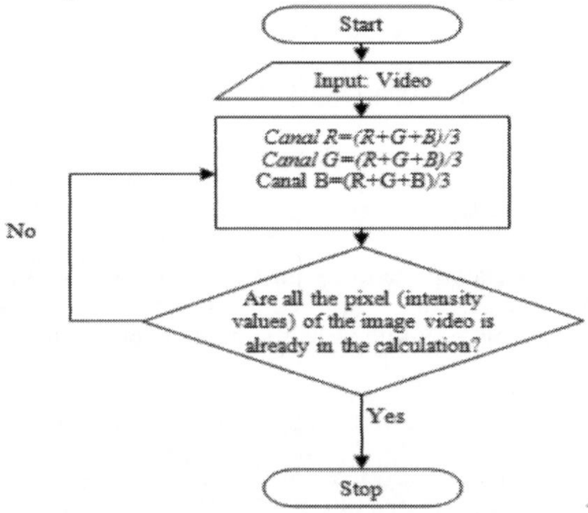


**Figure 3.** Flowchart of the *grey-scale* process

The grayscale step is required to align the intensity value of a 24-bit video. Consecutive values in channel R, G and B are summed and divided by the number of channels so that it can be calculated by the number of each channel divided by the number of channels. The process of the gray-scale video really helps the acceleration in the next concept.

### *3.2. Convolution*

The convolution Flowchart processes are developed based on Figure 4:

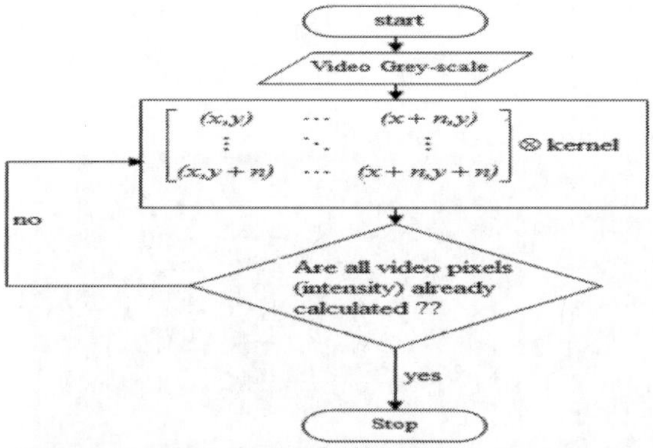


**Figure 4.** Flowchart of convolution process

The convolution process is involved in transforming the conditioned intensity values from the previous stage, into the intensity values that represent the edge of the object (face). Each intensity value will be partitioned into a 3x3 matrix and consonat with a Sobel kernel.

### *3.3. Widrow Hoff*

Flowchart of Widrow Hoff proces in the system is shown in Figure 5:

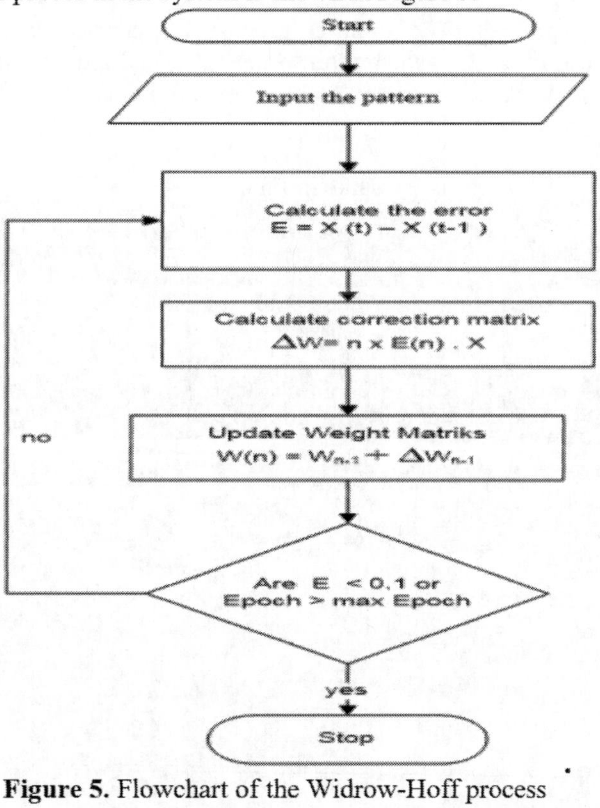


**Figure 5.** Flowchart of the Widrow-Hoff process

In this stage, the network receives a pattern structure input with a sequence of the edges of objects that may contain the face or not the face, and then the network performs an error calculation. After the error calculation is performed, the network performs a correction matrix calculation and ends with a weighted matrix renewal. These steps continue to be repeated until the minimum error or have met the maximum iteration limit.

## 4. *Result and Discussion*

### *4.1 View of the training form*

The training form in the system is look like in figure 6 :

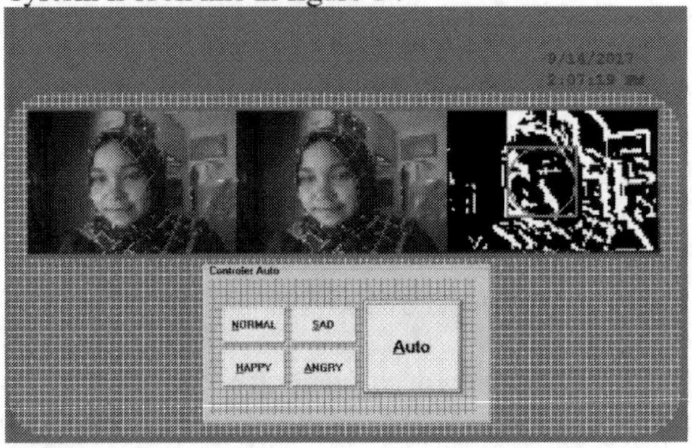


**Figure 6.** View of the example the Training form

In this form, the facial image was captured by webcam and face area located by red shape to lock the pixels of the face area and will save the matrix weigh of the pixel to storage and later use as a reference to the testing process. The training process is carried out in accordance with each expression.Firstly, We can click the auto button to get the facial matrix then click every expression button ((happy, sad, angry and normal) to get the facial expression matrix.

### *4.2 View of testing form*

Testing or result form in the system is look like in figure 7:

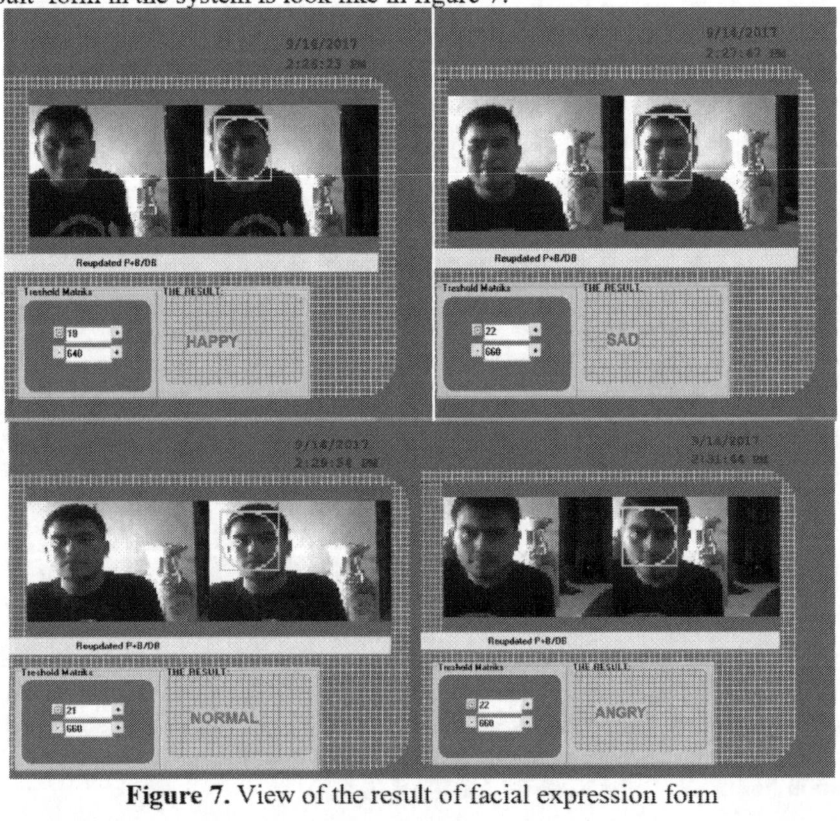


**Figure 7.** View of the result of facial expression form

In the testing process, the system will read the value of the input video image that will be compared with the value that has been stored during the training process and where the matrix value is close.

## 5. Conclusions

From the results of research and discussion undertaken conclusions can be drawn as follows:

1. The light that even affects the detection, the intensity is not too bright and not too dark so that the test in the system can work well.
2. for accurate and precise results, the conditions or environments of the image capture must be observed. The conditions in question include the light, place, distance, and position of the object image.
3. Detection of facial expression is very dependent on the existing threshold value.
4. The accuracy of the system depends on the number of samples, the more accurate the sample of faces being trained and tested, the more accurate the expression facial recognition results and the impact on the high percentage.
5. The threshold matrix setting depends on the face when tested, if the face tested shows a higher matrix threshold then approaches the false positive rate, if not too low then approaching the detection rate, the threshold matrix is adjusted from the tested face.
6. Accuracy results in detecting video objects have a range of 85% in detecting happy expression, 80 % in detecting sad expression, 75 % in detecting normal expression and 85 % in detecting angry expression.

## References

[1] Singh D 2012 Human Emotion Recognition System *Int. J. Image , Graphics and Signal Processing* **8** pp 50-56.

[2] Hemalatha G and Sumathi C P 2014 A Study of Techniques for Facial Detection and Expression Classification *Int. J. of Comp. Sc. & Engineering Survey (IJCSES)* **5** pp 27-37

[3] Su MC, Yang CK, Lin SC, Huang DY, Hsieh YZ and Wang PC 2013 An SOM-based Automatic Facial Expression Recognition System *Int. J. on Soft Comp., Artificial Intell. and App. (IJSCAI)* **2** pp 45-57.

[4] Chavan P M, Jadhav M C, Mashruwala J B, Nehete A K and Panjari P A 2013 Real Time Emotion Recognition through Facial Expressions for Desktop Devices *Int. J. of Emerging Science and Engineering (IJESE)* **1** pp 104-108.

[5] Saranya R and Kuppusamy K 2016 Performance analysis of ADALINE and MADALINE Network in Classification of Thyroid Disease *Int. J. of Advanced Research Trends in Engineering and Technology (IJARTET)* **3** pp 117-119.

[6] Shbib R and Zhou S 2015 Facial Expression Analysis using Active Shape Model *Int. J. of Signal Processing, Image Processing and Pattern Recognition* **8** pp 9-22.

[7] Ghosh S and Bandyopadhay S K 2015 Method for Human Emotion Recognition System *British J. of Math. & Comp. Sc.* **11** pp 1-27.

# W.a.w (we are watching) smart app: accommodating social perception towards public officers' performance

**S A Widhoyoko[1,5], Sasmoko[2,5], L A Nasir[3], S R Manalu[4,5] and Y Indrianti[5]**
[1]Podomoro University, Jakarta, Indonesia
[2]Faculty of Humaniora, Bina Nusantara University, Jakarta, Indonesia
[3]University of Indonesia, Jakarta, Indonesia
[4]Mobile Application & Technology Department, School of Computer Science, Bina Nusantara University, Jakarta, Indonesia
[5]Research Interest Group in Educational Technology, Bina Nusantara University, Jakarta, Indonesia

Email: samuel.anindyo@podomorouniversity.ac.id, sasmoko@binus.edu, laode.arrahman@ui.ac.id, smanalu@binus.edu, yasintaindrianti@gmail.com

**Abstract.** This research is a continuation of previous research that is corruption early prevention is expanded by using expert system to analyze data and produce information to build decision. The research method used is neuroresearch method through three stages of research, namely exploratory stage, explanatory stage and confirmatory stages. The exploratory research finds W.aW's Principles and W.a.W's Units of Assessment as the basis for the preparation of the application. Stages of explanatory research in the form of W.a.W's design of IT and confirmatory research stages are the design of expert system W.a.W. Expert System uses this formulation to generate dynamic standard value for each category and current social perception.

## 1. Introduction

The impact of good governance value in reducing the public officers' misconduct has been a trending topic over the last ten years. However, not many discussions include the role of society as one of the objective components although, some scholars set aside this element as the secondary consideration in terms of impartial respondents resulting bias information. Moreover, some developed countries with certain governmental ideology tend to separate society from political environment to neutralize any possibility of conflict. On the other hand, Organisation for Economic Co-operation and Development (OECD) has been developing a framework emphasizing government's transparency to public in which societal perception towards government is built amongst the society.

Accordingly, there are three elements that trigger good societal perception: (i) law enforcement, (ii) infrastructure development, and (iii) fair treatment. In its political application, these aspects shall be distorted by corrupted nature infecting both personally and systemically. In this context, there is a belief to transform social perception into a more supervision function as it is the most independent opinion towards government's performance. Therefore, a tool in form of a smart application corresponding modern world requirements must be built to facilitate societal perception. Furthermore, the tool used must be able to present an impartial information because political partiality does exist.

W.a.W Smart App. is the extension of CEP's theoretical construct for presidential Decision Support System (DSS). This research aims to build an Information and Technology (IT) construct of W.a.W Smart App. including the assessment system with accountable data and resources. This IT model is expected to accommodate public's perception towards the government performance (personal and organizational) through an application without sacrificing the reliability of information. This research is expanded from previous research that using DSS to build decisions, but for this research uses Expert System (ES) to analyze the data and generate information to build decision. Furthermore, this application is designed to support the CEP that is constructed for presidential Expert System (ES) purposes [1].

## **2.** Research Method

This research is developed using Neuroresearch. Neuroresearch is a research methodology approach that combine quantitative and qualitative techniques. This research conducts using three steps namely exploratory, explanatory, and confirmatory. This research methodology already use in previous research, that focus on Corruption Early Prevention blue print by implementing decision support system [2]. This research is also conducted using exploratory research. That is doing literature review regarding to W.a.W to produce construct theoretical W.a.W's technology principles to be implemented. This research use Expert System to analyze and summarize the data.

## **3. Result and Analysis**

### *3.1. W.aW's Principles*

As an information system, W.a.W is grounded on two frameworks that was developed from basic internal control designed by COSO (Committee of Sponsoring Organizations of the Treadway Commission). In terms of framework development, W.a.W is not built on COBIT (Control Objective for Information and related Technology) due to its purposes that emphasises on the validation of data source. In this aspect, COBIT is mainly designed to bridge the gap between control requirements, technical issues and business risks which enables clear policy development and good practice for IT control throughout organizations. COBIT also emphasizes regulatory compliance, helps organizations to increase the value attained from IT, simplifies implementation of the enterprises' IT governance and control framework [2].

Due to its needs to trace both users and targeted personnel, W.a.W is equipped with a Occupational Fraud Detection by Social Network Analysis (OFD-SNA). According to Lookman and Nurcan (2015), SNA is an emerging framework that enables fraud examiner to inter-connect interaction patterns between person-to-person. This method is successful for a problem-solving application, mainly in several economic crime issues i.e. business strategy, fraud detection, and counter terrorism. In other words, SNA provides interaction patterns between people in the forms of data collection and graph visualization [3]. Furthermore, as the SNA's coverage enabling fraud examiners to detect the insider fraud through networking organization, computer algorithms pattern, and auditable raw data [4], it is assumed that SNA will also be able to uncovered the users' data based on the tips [5].

Another framework is also implemented for the purpose of data base security named CIA (Confidentiality-Integrity-Availability) triad [6]. Confidentiality is the first requirement for data security which protects the data from thieving or reaching wrong target through restricted access [2]. Second, the data security must also built in a procedure that support consistency and accuracy [7]. According to Houck et all., (2006) Other requirements for data integrity are backup files for restoration system and permission of access control. The last component that is the most important thing for fraud detection is availability, as it is a fundamental aspect that protect both integrity and confidentiality [8]. In this context, availability comprises all aspects including the originality of the software adopted which has been the requirement for admissibility of forensic reporting [9].

As an easy-use application, W.a.W offers a platform offers both respondents and analysts a set of investigative tool that lead respondents to provide their accounts which are arranged unbiasedly. Through investigative interviewing interface, respondents are allowed to choose and raise their interested cases [10]. On the other hand, W.a.W attempts to treat respondents as independent whistle-blower to be interviewed as witness interviewee [9]. According to Wells (2013), there are two hints in gathering comprehensive materials from interview: (i) informal approach is prioritized and (ii) the use of introductory and informational questions are necessary to be conducted in sequence. However, well-performed closing stage of interview must be conducted to reconfirm facts, obtain additional clues, and retain goodwill [11].

Therefore, According to Ord, Shaw and Green (2011) W.a.W's questions' configuration is grounded on opened and closed questions. Closed question is a type of inquiry that is able to lead respondents' focuses into a certain field of information with minimum distraction [12], [13] and detect respondents' inconsistencies which eliminates biased informants [13], [14]. Based on these traits, the closed question is suitable for both introductory and closing stages which are also equipped with multi-part-patterned question between both stages to validate the information [15]. Meanwhile the opened question is more flexible type which allows users to information related to person-to-person linkages, alleged motivation, and other related societies' suspicions [13]. At the end, the system is set to evaluate all information through automated summarizing system and syncing all summarized information to specific target in terms of producing decisive output.

Last but not least, filtering system is pivotal for the information objectivity. The aspect of objectivity has been an issue for all surveying tools in political corruption fields. A research done in China finds that there are gaps and discrepancies between corruption perception between national, regional, and local areas even though, the questions lead to the same purposes [16]. Another finding related to the deficiency of perception is the critiques addressed to Corruption Perception Index (CPI) published by Transparency International (TI). According to Andersson and Heywood (2009), the main issue of CPI is in the unclarified definition of corruption as a factor to deteriorated corporate governance which led to diffused data articulation [17]. However, the use of perception has been considered as an empirical approach in the field of research. The use of Delphi method in both research applications and techniques [18], [19] and investigative auditing especially in anti-fraud risk assessment program [20] have one main requirement which is identifiable respondents with sufficient capacity.

Corruption perception is a very subjective type of information which needs a rigorous-selective filtering system that reduces bewilderment in decision making process in terms of matching perception and reality [21]. This can easily be done through social-partiality and personal-biodata analysis [4]. Both of these techniques aim to both (i) reveal respondents' inclination in either striking-down or elevating a public officer [22], [23]and (ii) form an output contained with subjective information from respondents that possess sufficient capacity in providing such information [17], [24], [25]. In processing the big data collection, Stanfield (2009) provides simplest clues related to big-volume data discovery in the field of computer forensics. In reacting upon wide range of data discovery, forensic analysts usually undertake two alternatives: (i) gathering similar output from varied perspectives (in this context: all society elements, groups, and organization) and/or deleting similar output from similar perspectives [26].

### *3.2. W.a.W's Units of Assessment*

As it is revealed, that the nature of society perception is limited only as a set of supporting information for direct information provided through industry and retail society which means that, numerical analysis towards the social perception must get through the screening system. There are commonly three aspects to assess the public officers' performance: (i) growth of public infrastructure, (ii) convenience of public services, and (iii) financial transparency of officers [27]. All these facts must be viewed in a framework that represents three areas in which society is always concern at i.e. politics, economics, and regulations. According to Adams (1981), reflecting from the politics circumstance United States of America (USA), which is a fair representation of the majority of politic practices [24], "Iron Triangle" can be used to view a modern politics that embraces politicians, regulators, and certain industrial parties [28].

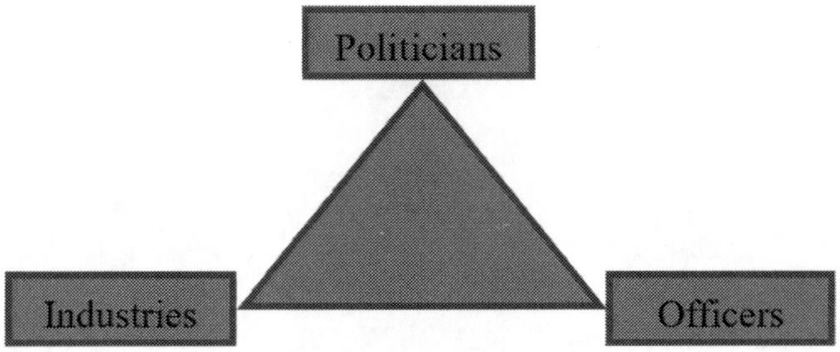


**Figure 1. Iron Triangle: Political Economy of Regulation**

Based on three indicators, it is revealed that the dimensions and indicators for the assessment system are completely structured. The table below depicts the relationship between the both groups.

**Table 1. Relationship Between Dimension and Indicators of W.a.W Assessment System**

| W a W | | Dimensions | | |
|---|---|---|---|---|
| | | D.1 | D.2 | D.3 |
| | | Infrastructure Growth | Public Services Quality | Financial Transparency |
| Indicators | | I of D.1 | I of D.2 | I of D.3 |
| I.1 | Politics | I.1 related with D.1 | I.1 related with D.2 | I.1 related with D.3 |
| I.2 | Economics | I.2 related with D.1 | I.2 related with D.2 | I.2 related with D.3 |
| I.3 | Regulation | I.3 related with D.1 | I.3 related with D.2 | I.3 related with D.3 |

*3.3 W.a.W's Design of IT*

Expert System is a computer program that act as an expertise to analyse and assist problems, the program has a specific expertness [29] such as to diagnose mental health disorder [30], [31] or to discover a knowledge [32]. Expert system is one of artificial intelligence branch that aim to understand the expertise knowledge and build the computer program from it. This knowledge is transforming into rule-based and used while generate inference or reasoning. Previous research [27] focusing on how to generate decision from respondent tends, but this research focusing on how to transform expertise knowledge in Table 1 to inference of Expert System (ES), while all the data input are dynamic from smart phone as shown in Figure 2a .

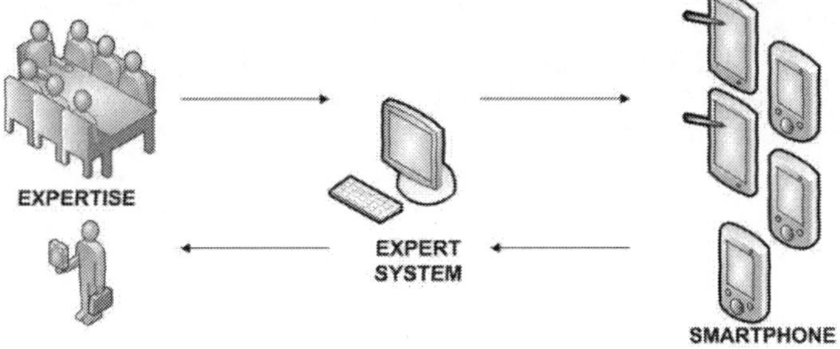


**(a)**

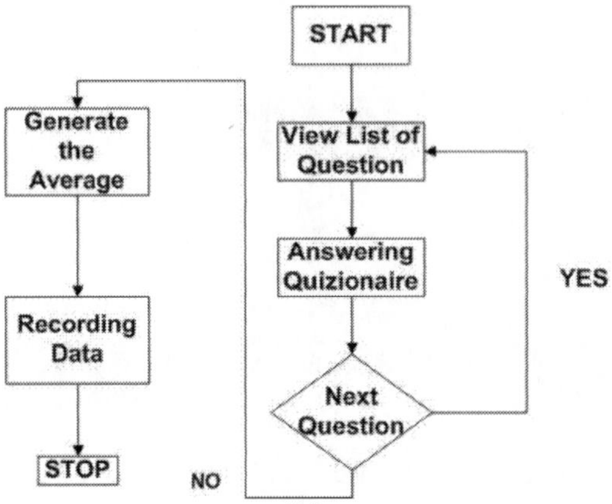


**(b)**

**Figure 2. (a)** Architecture W.a.W; **(b)** Forward Chaining W.a.W

W.a.W Expert System are using forward chaining, it starts from a set of condition (in this case, condition is rule that has been transform into questions) and generate conclusions according to answer for each conditions. Each of rules will ask and all the answers computes and at the end average value of answer and all the answers are recorded, as shown in Figure 2b. They are recorded at relational database.

*3.3.1. Generate Rule-Based*

Rule-based are specific factor as a measurement of social perception to government, this rules are divided into nine, which is explain below in detail:

*Rule 1: IF "Government related with Infrastructure Growth and Politics"*
*Rule 2: IF "Government related with Infrastructure Growth and Economics"*
*Rule 3: IF "Government related with Infrastructure Growth and Regulation"*
*Rule 4: IF "Government related with Public Services Quality and Politics"*
*Rule 5: IF "Government related with Public Services Quality and Economics"*
*Rule 6: IF "Government related with Public Services Quality and Regulation"*
*Rule 4: IF "Government related with Financial Transparency and Politics"*
*Rule 5: IF "Government related with Financial Transparency and Economics"*
*Rule 6: IF "Government related with Financial Transparency and Regulation"*

Rules are displayed in several questions in expert system. Questions are accessed using smart phone by respondent. Respondent must be answered between 1 to 4 for each question, 1 is very negative and 4 is very positive.

*3.3.2. Generate Reasoning (Inference)*

Reasoning is the main expertise knowledge in term of how to generate the conclusions. This research uses probabilistic as basic reasoning. As mention before each of respondent answered are recorded at database, the conclusion generates dynamically according to current respondent answer.

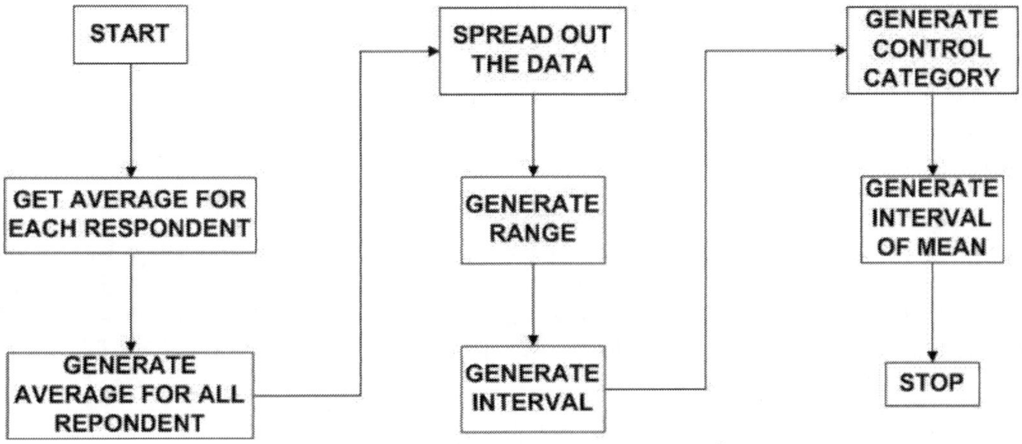


**Figure 3.** W.a.W Expert System Reasoning

W.a.W Expert System's Reasoning starts with getting the average data for each respondent, this average calculates with $X_i = \sum_1^n A_{1+\dots+}A_n$ (1). 'A' is answered for each question. Next step is generating average all the respondent using formula $\bar{X} = \sum_{i=0}^{n} X_1 + \cdots + X_n$ (2). While 'X' is average value from each respondent in (1). To get the amount of respondent diversity this research uses formula of standard deviation $s = \sqrt{\frac{\Sigma(X_i-\bar{X})^2}{n}}$ (3). Value of '$X_i$' comes from (1) and '$\bar{X}$' comes from (2). Table 2 describe sample data from respondents. There are 10 respondents that answer each rule, from rule number 1 until 10.

**Table 2.** Sample Data from Respondents

| | Rule | | | | | | | | | $X_i$ | $X_i - \bar{X}$ |
|---|---|---|---|---|---|---|---|---|---|---|---|
| Sample | 1 | 2 | 3 | 4 | 5 | 6 | 7 | 8 | 9 | | |
| 1 | | 3 | 4 | 4 | 4 | 3 | 4 | 4 | 4 | 3.6 | 1 |
| 2 | 3 | 3 | 3 | 3 | 3 | 3 | 4 | 3 | 3 | 3.1 | 0.25 |
| 3 | 4 | 4 | 3 | 3 | 3 | 4 | 3 | 3 | 3 | 3.3 | 0.49 |
| 4 | 2 | 2 | 3 | 3 | 2 | 3 | 2 | 2 | 2 | 2.3 | 0.09 |
| 5 | 2 | 3 | 2 | 3 | 3 | 3 | 2 | 2 | 2 | 2.4 | 0.04 |
| 6 | 3 | 3 | 3 | 3 | 3 | 3 | 3 | 3 | 3 | 3 | 0.16 |
| 7 | 1 | 1 | 1 | 1 | 1 | 1 | 1 | 1 | 2 | 1.1 | 2.25 |
| 8 | 3 | 2 | 3 | 2 | 2 | 3 | 3 | 2 | 2 | 2.4 | 0.04 |
| 9 | 2 | 2 | 2 | 2 | 2 | 2 | 2 | 2 | 2 | 2 | 0.36 |
| 10 | 3 | 3 | 3 | 2 | 2 | 3 | 3 | 3 | 3 | 2.8 | 0.04 |

Inference generates average value of answer for each respondent using (1), and label as yellow in Table 2. The average for all the respondent data is formulated by (2), and by using data in Table 2 will simulate as below:

$$\bar{X} = 3.6 + 3.1 + 3.3 + 2.3 + 2.4 + 3 + 1.1 + 2.4 + 2 + 2.8$$
$$\bar{X} = 2.6$$

Inference needs to generate amount of respondent diversity using (3), and after calculating the deviation between $X_i - \bar{X}$ as label as green at Table 2, the calculation for standard deviation (3) as below:

$$s = \sqrt{\frac{0.45 + 0.25 + 0.49 + 0.09 + 0.04 + 0.16 + 2.25 + 0.04 + 0.36 + 0.04}{10}}$$

$s = 0.69$

Next step Inference will generate control value, but Inference needs to generate each factor of control value namely 'range' and 'interval'. This control value helps to make a standard value for each category in (5). Inference generates 'range' value by subtracting the maximum and minimum value of $X_i$, to get the 'range' value is following the formula $Range = X_{i\,max} - X_{i\,min}$ (4). The calculation by using data above as below:

$Range = 3.6 - 1.1$
$Range = 2.5$

After calculating 'range', 'interval' generates by using this formula $interval = \frac{Range}{\Sigma\, category}$ 5). By using (5) according to data before, Inference generates the 'interval' as below:

$Interval = \frac{2.5}{4}$
$Interval = 0.51$
$Interval = 1$

Category at (5) is number of choices for questions. Inference generates control value by using $Interval * Category \geq Range + 1$ (6). If it is true and $(Interval * Category) - Range + 1 = 2$ than first category (in this case is 1, that indicate very negative) is $X_{i\,min} - 1$ and last category (in this case is 4, that indicate very positive) is $X_{i\,max} + 1$. If it is true and $(Interval * Category) - Range + 1 = 3$ than first category (in this case is 1, that indicate very negative) is $X_{i\,min} - 1$ and last category (in this case is 4, that indicate very positive) is $X_{i\,max} + 2$. If it is true and $(Interval * Category) = Range + 1$ than first category (in this case is 1, that indicate very negative) is $X_{i\,min}$ and last category (in this case is 4, that indicate very positive) is $X_{i\,max}$. From the data above, standard value for each category as below:

$Interval * category \geq Range + 1$
$1 * 4 \geq 2.5 + 1$

$1\ (very\ negative)$ : 1.1 – 1.9
$2\ (negative)$ : 1.9,5 – 2.75
$3\ (positive)$ : 2.8 – 3.55
$4\ (very\ positive)$ : 3.6 – 4.45

Final Inference calculation is generating the current social perception to government according to data in Table 2. This perception value is following this formula $\bar{X} - t_p\frac{s}{\sqrt{n}} < \mu < \bar{X} + t_p\frac{s}{\sqrt{n}}$ (7). While $t_p$ is the value from 't Table' where $df = n - 1$. By using data we can get range for social perception as below:

$$2.6 - \left(2.262 * \left(\frac{0.69}{\sqrt{10}}\right)\right) < \mu < 2.6 + \left(2.262 * \left(\frac{0.69}{\sqrt{10}}\right)\right)$$

$$2.6 - \left(2.262 * \left(\frac{0.69}{3.16}\right)\right) < \mu < 2.6 + \left(2.262 * \left(\frac{0.69}{3.16}\right)\right)$$

$$2.6 - (2.262 * 0.22) < \mu < 2.6 + (2.262 * 0.22)$$

$$2.6 - 0.5 < \mu < 2.6 + 0.5$$

$$2.1 < \mu < 3.1$$

To generate the conclusion, Expert System Inference is mapping value $\mu$ at (7) into category standard value and decided this value is in which category, for this case the social perception is in third category (positive).

Expert System uses this formulation to generate dynamic standard value for each category and current social perception. This value will change dynamically according to the newest input from customer. By using this Expert System social perception to government is valid.

## 4. Conclusions and Recommendation

Expert System is a full power computer program to mimic knowledge skills to summarize conclusions. This knowledge then turns into useful conclusions in decision making. The rules for social perception produce nine rules, each rule has four categories of answers 1 (very negative), 2 (negative), 3 (positive), and 4 (very positive). An arrangement is approximated by probabilistic to ensure each valid value matches the current data, this inference can result in a tendency for a condition of social perception dynamically.

Through expert systems, communities respond as their perceptions of government performance over time. Automatically the system built in the W.a.W apps will generate any incoming information and provide conclusions and types of social interventions that should be implemented by the government. Any addition of a community response will alter the fluctuating and dynamic conclusions that give the consequences of the current state. Up to date conclusions will help the government in responding rapidly to the phenomena that occur in the community. With this pattern, W.a.W application becomes a smart system in implementing new policies to prevent unhealthy practices affecting government performance in Indonesia.

## References

[1] Sasmoko, Widhoyoko S A, Ariyanto S, Indrianti Y, Noerlina, Muqsith A M and Alamsyah M 2017 Corruption Early Prevention: Decision Support System for President of the Republic of Indonesia *J Phys Conf Ser* **801** pp 1–8.

[2] Feruza S and Kim T H 2007 IT Security Review: Privacy, Protection, Access Control, Assurance and System Security *Int. J. Multimed Ubiquitous Eng.* **2** pp 17–32.

[3] Scott J and Carrington P J 2011 Crime and Social Network Analysis *Sage Handb. Soc. Netw. Anal.* pp 236–255.

[4] Lookman S and Nurcan S 2015 A Framework for Occupational Fraud Detection by Social Network Analysis *CEUR Workshop Proc* **1367** pp 221–228.

[5] Association of Certified Faud Examiners 2016 *Report to Nations.*

[6] Qadir S and Quadri S M K 2016 Information Availability : An Insight into the Most Important Attribute of Information Security *J. Inf. Secur.* **7** pp 185–194.

[7] Sridhar S and Manimaran G 2010 Data Integrity Attacks and Their Impacts on SCADA Control System *IEEE PES General Meeting.*

[8] Houck M M, Kranacher MJ, Morris B, Riley R A Jr, Robertson J and Wells J T 2006 Forensic Accounting as an Investigative Tool *CPA J.* **76** pp 68–70.
[9] Pedneault S, Silverstone H, Rudewicz F and Sheetz M 2012 *Forensic Accounting and Fraud Investigation for Non-Experts* (USA: Wiley)
[10] Alexander R 2004 The Role of Whistleblower in the Fight Against Economic Crime, *J. Financ. Crime* **12** pp 131–138.
[11] Wells J T 2003 There is an Art on Conducting Interview *J. Account.*
[12] Schuman H and Presser S 1979 The Open and Closed Question *Am Sociol Rev* **44** pp 692–712.
[13] Ord B, Shaw G and Green T 2011 *Investigative Interviewing Explained 3rd ed.* (Australia: LexisNexis).
[14] Dohrenwend B 1965 Some Effects of Open and Closed Questions on Respondents' Answers *Human Organization* **24** pp 175–184.
[15] Conway J M and Peneno G M 1999 Comparing Structured Interview Question Types: Construct Validity and Applicant Reactions *J. Bus. Psychol.* **13** pp 485–506.
[16] Song X and Cheng W 2012 Perception of Corruption in 36 Major Chinese Cities: Based on Survey of 1,642 Experts *Soc. Indic. Res.* **109** pp 211–221.
[17] Andersson S and Heywood P M 2009 The Politics of Perception: Use and Abuse of Transparency International's Approach to Measuring Corruption *Polit. Stud.* **57** pp 746–767.
[18] Linstone H A and Turoff M 2002 *The Delphi Method Techniques and Applications* (London: Addison-Wesley Publishing Company).
[19] Pawlowski S D and Okoli C 2004 The Delphi Method as a Research Tool : An Example , Design Considerations and Applications *Inf. Manag.* **42** pp 15–29.
[20] Verwey I G F, Universiteit N B and Asare S K 2016 The Effect of Forensic Expertise and Time Pressure on Fraud Risk Assessment and Responsiveness *AAA Forensic Account Res. Conf.* pp 1–30.
[21] Olken B A 2009 Corruption Perceptions vs Corruption Reality *J. Public Econ.* **93** pp 950–964.
[22] Robertson-Snape F 1999 Corruption, Collusion and Nepotism in Indonesia *Third World Q* **20** pp 589–602.
[23] Mietzner M 2007 Party Financing in Post-Soeharto Indonesia: Between State Subsidies and Political Corruption *Contemp Southeast Asia* **29** pp 238–263.
[24] Freie J F 2012 Postmodern Politics in America *Society* **49** pp 323–327.
[25] Ernstson H, Sörlin S and Elmqvist T 2008 Social Movements and Ecosystem Services - The Role of Social Network Structure in Protecting and Managing Urban Green Areas in Stockholm *Ecol Soc* **13** pp 38–48.
[26] Stanfield A 2009 *Computer Forensics, Electronic Discovery, and Electronic Evidence 1st Ed.* (LexisNexis).
[27] Sasmoko, Widhoyoko S A, Ariyanto S, Indrianti Y, Noerlina, Muqsith A M and Alamsyah M 2017 Corruption Early Prevention: Decision Support System for President of the Republic of Indonesia *J. of Physics: Conf. Series* **801**.
[28] Adams G 1981 *The Politics of Defense Contracting* (New York: Council on Economic Priorities)
[29] R S Design, Fundamentals of Expert System, pp 1–26, 1987
[30] Singh P K and Sarkar R 2015 A Simple and Effective Expert System for Schizophrenia Detection *Int. J. Intell. Syst. Technol. Appl.* **14** p 27.

IOP Conf. Series: Journal of Physics: Conf. Series **978** (2018) 012023 doi:10.1088/1742-6596/978/1/012023

[31] Manalu S R, Abbas B S and Gaol F L 2017 An Expert System to Assist with Early Detection of Schizophrenia *Intell. Inf. Database Syst.* **10191** pp 802–812.
[32] Abouelhoda M and Ghanem M 2010 Scientific Data Mining and Knowledge Discovery *Knowl Creat Diffus Util* **19** pp 207–247.

IOP Conf. Series: Journal of Physics: Conf. Series **978** (2018) 012024 doi:10.1088/1742-6596/978/1/012024

# Implementation and evaluation of LMS mobile application: scele mobile based on user-centered design

**R D Banimahendra and H B Santoso**

Faculty of Computer Science, Universitas Indonesia, Depok, Indonesia

Email: riandy.dimas@ui.ac.id, harrybs@cs.ui.ac.id

**Abstract**. The development of mobile technology is now increasing rapidly, demanding all activities including learning should be done on mobile devices. It shows that the implementation of mobile application as a learning medium needs to be done. This study describes the process of developing and evaluating the Moodle-based mobile Learning Management System (LMS) application called Student Centered e-Learning Environment (SCeLE). This study discusses the process of defining features, implementing features into the application, and evaluating the application. We define the features using user research and literature study, then we implement the application with user-centered design basis, at the last phase we evaluated the application using usability testing and system usability score (SUS). The purpose of this study is to determine the extent to which this application can help the users doing their tasks and provide recommendation for the next research and development.

## 1. Introduction

Currently smartphone devices and the accessibility of the internet has become the basic needs of the world community including Indonesia. The Internet sparked new innovations in various fields, one of the emerging innovations in education is e-Learning. An e-Learning process' has four basic components of infrastructure technology which are e-Learning content, e-Learning users, and the most vital is the e-Learning software platform or so-called Learning Management System (LMS) [1].

The Faculty of Computer Science Universitas Indonesia e-Learning team has developed an LMS known as Student Centered e-Learning Environment (SCeLE) based on Moodle. Until now SCeLE Mobile application development has never been done, but one of the strategies to increase the use of LMS is by developing a mobile application [2]. At the time of developing the application, the risk of failure also needs to be considered. According to Weinschenk, three of the 12 common causes of failure of software development projects (applications) are closely related to the user experience [3]: the needs are not well defined; Lack of communication between developers and users; And stakeholder politics. Weinschenk also added that the user experience problem can be overcome with user-centered design, which focusing on gaining a deep understanding of the pre-users throughout design and development life-cycle [4].

Therefore, it is necessary to implement and evaluate the SCeLE Mobile design with user-centered design into mobile application. In this study, we will develop the mobile applications and then evaluating for known what needs to be improved. The main purpose of this research is to continue the implementation of SCeLe Mobile from previous research, evaluation as improvement material for further research, to see suitability of application to user requirement, and get feedback for future research and development.

## 2. Relevant Literature Review

### *2.1. SCeLE Based on User-Centered Design*
User Centered Design (UCD) is a framework of designing and developing products that focus on understanding the potential users [4]. This framework also guarantees the product (software or website) will easy to be used [5]. The International Usability Standard ISO 13407, which is the basis for UCD, stated six things that must be pointed out in UCD [5]. It emphasizes the user involvement in design, development, and evaluation.

The design prototype of SCeLE based on user-centered design was introduced by Ibadurrahman [6]. He developed the prototype by improving a previously made prototype which designed based on Moodle Mobile, a mobile app created using HTML 5. By using UCD he constructs the prototype using Material Design principal, which is commonly used on Android devices. Using commonly used design principal will make the user easier learning the application.

### *2.2. Android Platform-Based Application*
Platform-based apps are an application that is specifically created and used on a specific platform, for example only on the Android operating system [7]. This type of application is developed using a specific programming language (e.g. Java), but is more dependent on the use of application frameworks, supporting libraries and runtimes in the form of thousands of lines of code typically written in C and C ++ [8]. In the Android operating system, applications are usually developed using the Java programming language with the help of application frameworks that provide access to the API framework so that it can be used by core applications [9].

### *2.3. Mobile Usability*
Usability is a qualitative assessment of the user interface in a product such as websites and mobile applications in addition to its purpose that allows users to achieve specific goals [10]. Usability can also be described as "the extent to which a product can be used by a particular user to achieve its goals with effectiveness, efficiency and satisfaction in a particular usage context" [11]. There are three (3) usability aspects on any type of software: 1) more efficient to use: user takes less time to work on a task; 2) easier to learn: the software can be learned by simply observing and trying them; and 3) more user satisfaction: the software used meets the expectations of its users.

Small screen size, limited interaction model, unreliable network connection, limited processing and storage space, and dynamic usage environment and context make mobile usability an important factor in mobile application development [10]. In addition, users prefer mobile apps that are easy to learn, take less time when completing a task, and are easier to use [12]. Therefore, evaluating the mobile application especially in terms of usability needs to be done to know the quality of the application from the user side.

### *2.4. Usability Evaluation*
Evaluation is an activity related to data collection related usability of a product, involving a group of users in a specific environment and context [13]. While the usability evaluation is a series of procedures performed to collect data related to the interaction between users with software products, to find out how big the contribution of a software in helping users achieve goals [14]. So, in the end the results of the usability evaluation can be used to predict the success of a product when it is launched [15]. There are 12 steps that should be done to conduct usability testing [16]: 1) define the usability evaluation's goal; 2) specify user interface aspects that will be evaluated; 3) identify the target users; 4) determine the usability metric that will be used; 5) decide the evaluation method; 6) select the tasks that will be executed; 7) plan the experiment; 8) gather the usability evaluation data; 10) give improvement recommendation by criticizing the user interface; 11) reiterate the step (when needed); and 12) summarize the result.

## 3. Method

Methods conducted during the study are as follows. First, we do the review of relevant literatures and make online surveys about the use of SCeLE in general. The survey also contains user-owned smartphone information that is used as the target of the application development platform. The results of the literature study and online survey are used to design the information architecture including what features that will be implemented into the application. The next step is to implement the application on the platform in accordance with the data analysis results. This implementation uses a prototyping methodology.

The prototype is made in the form of an Android platform-based application. Application development is carried out based on prototypes that have been made from previous research with changes tailored to the results of this research. The prototype development process started by modifying the application that has been made. The original application is an open source application called MDroid (https://github.com/praveendath92/MDroid) developed by Praveen Kumar Pendyala. The principle of the eight golden rules is also used in this development. We use the principle because it has often been adapted into various types of systems to identify usability problems [17].

After the prototype of application is made, we conduct the usability evaluation in the form of usability testing and SUS to 14 participants. Usability testing is used to obtain qualitative and quantitative data used as input application development. While SUS is used to obtain quantitative data from tested applications that are used as a reference value of usability of SCeLE Mobile. Participants consisting of 10-15 peoples were asked to run nine (9) tasks and then we ask them to assess the application through SUS. We also request feedbacks from participants regarding the experience of using the application. Table 1 shows the template task scenario used during usability testing.

**Table 1.** Template scenario for usability testing

| | Degree of Success | | | Difficulty Level (*range*) | Notes |
|---|---|---|---|---|---|
| **See All Courses** | S | PS | F | 1 – 4<br>(easy to difficult) | |
| **Mark the course to Starred Courses** | S | PS | F | 1 – 4<br>(easy to difficult) | |
| **Read the forums on a course** | S | PS | F | 1 – 4<br>(easy to difficult) | |
| **View the assignment** | S | PS | F | 1 – 4<br>(easy to difficult) | |
| **Read the material** | S | PS | F | 1 – 4<br>(easy to difficult) | |
| **View the list of participants in the same course** | S | PS | F | 1 – 4<br>(easy to difficult) | |
| **View the list of events in SCeLE** | S | PS | F | 1 – 4<br>(easy to difficult) | |
| **Read the Vacancy Forum** | S | PS | F | 1 – 4<br>(easy to difficult) | |
| **See notifications** | S | PS | F | 1 – 4<br>(easy to difficult) | |

After completing the task, participants were asked to rate the difficulty level on a scale of 1 (easy) to 4 (difficult) when working on the requested task. The degree of success is also written with the 'S' code if successful, 'PS' if successful but requires guidance, and 'F' if it does not complete the task. The degree of success is an indicator of success seen from the researcher, while the difficulty level is an indicator of success seen from the participant side. Both indicators are used so that the analysis results are seen from two directions, so the results can be more accurate. At the end of the usability evaluation we ask the participant to fill in the SUS questionnaire which has been translated into Bahasa Indonesia [15]. At last, we performed a data analysis of the findings from the usability testing and SUS score to be concluded for the next research guidelines.

## 4. Results

The results of online survey analysis from 107 respondents and literature studies shows that the general activities of the user are accessing the tasks and materials of a course. Previously, we have been divided the user activities into five classification which is material, course, assignment, forum, and calendar. Then from these activities, the main problem which faced by the user is a complex display when they look for information [6]. Figure 1 shows the detail of the general activity classification performed by the user. The dominant category of activity that was carried out by 107 respondents is Material (57%) and Assignment (55%). Then the next common category is Course (42%) where 45 of the total respondents do the activity. Subsequently a recessive category was the Forum of 37 out of 107 (35%) of respondents and Calendar 24 out of 107 (24%) respondents.

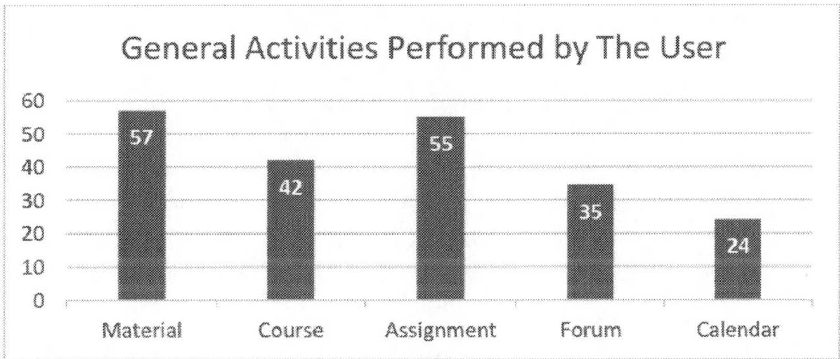


**Figure 1.** General activities that has been classified into categories in percent

Further data on the smartphone information owned by users (Figure 2) shows that 85% of total participant is using Android, make it become the most widely used platform.

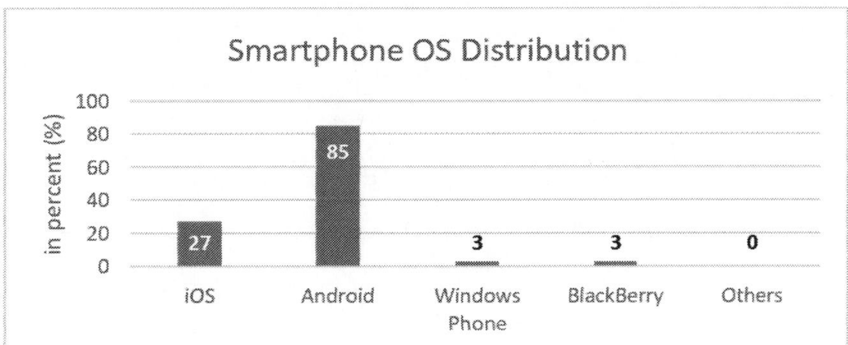


**Figure 2.** Smartphone OS distribution

Implementation is then performed on the Android platform and focuses on the navigation features related to the course including tasks, materials, and information about the course. The example of implemented application can be seen on Figure 3.

**Figure 3.** The example of SCeLE Mobile view

Due to the limited API provided by Moodle, not all features that have been planned before can be implemented. One of the features is uploading assignment on the course. An alternative solution to this limitation is giving users access through a browser that leads to the SCeLE website. The differences between planned features (from design prototype [6]) and implemented features can be seen on Table 2.

**Table 2.** The differences of features between the prototype and the application

| Feature | Design Prototype of SCeLE Mobile | Application Prototype of SCeLE Mobile |
|---|---|---|
| **Dashboard** | ✓ | × |
| **Course** | ✓ | ✓ |
| **Star Course** | × | ✓ |
| **Messages** | ✓ | × |
| **Calendar** | ✓ | ✓ |
| **Forum** | ✓ | ✓ |
| **Assignment** | ✓ | ✓ |
| **Notification** | × | ✓ |

## 5.Conclusion and Discussion

Overall, the results of usability testing of UCD-based SCeLE Mobile applications among 14 participants are good enough and suitable for use by users. This can be indicated from the dominant positive values for each task scenario in terms of the degree of success, as well as the difficulty level of task work according to the participants.

The result of SUS score processing according to SUS calculation rule shows that SCeLE Mobile application get 71.25 point. Table 3 below explained the results from each participant. SUS scores with these values can be categorized as having a good usability as they are over 68, as defined by Sauro [15]. Then, the SUS score can also be an indicator that the SCeLE Mobile application has a 'good' property of grade 'C', if judged by the Bangor categorization method [18].

**Table 3.** SUS results from each participant

| Participant | SUS Score |
|---|---|
| **1** | 72.5 |
| **2** | 77.5 |
| **3** | 62.5 |
| **4** | 75.0 |
| **5** | 72.5 |
| **6** | 57.5 |
| **7** | 67.5 |
| **8** | 62.5 |
| **9** | 80.0 |
| **10** | 72.5 |
| **11** | 70.0 |
| **12** | 75.0 |
| **13** | 90.0 |
| **14** | 62.5 |
| **Total Score** | **71.25** |

Although the results of the usability evaluation are good enough, there needs to be improvements that must be done. These improvements include hiding unnecessary progress status at login, changing the content writing on the confirmation box when marking the course into the starred course, changing the icon and mentioning the starred course become more representative, making the thread reply feature, changing the content icon inside the course, rearrange the layout of the content in the form of a list or thread, provide a download indicator in progress or finish, add photos to the participant list of the course, click date mechanism and range of events that appear on the calendar, add search features in the forum, and notification feature improvement.

The enthusiasm of the participants of this study when trying to use SCeLE Mobile application is quite high. Many of them are looking forward to the development of this application. In addition, the applications developed in this study are still in prototype form, so there is still an opportunity for further development. Therefore, the next SCeLE Mobile application development should continue to be implemented.

**Acknowledgments**

This publication is within the research program funded by Universitas Indonesia – International Indexed Publications for Students Final Project Grant with number: 403/UN2.R3.1/HKP.05.00/2017

IOP Conf. Series: Journal of Physics: Conf. Series **978** (2018) 012024 doi:10.1088/1742-6596/978/1/012024

**References**

[1] Harrati N, Bouchrika I and Mahfouf Z 2016 e-Learning : on the Uptake of Modern Technologies for Online Education *Proc. of the 6th Int. Conf. on Communication and Management*.

[2] Mtebe J S and Kondoro A W 2016 Using Mobile Moodle to Enhance Moodle LMS Accessibility and Usage at the University of Dar Es Salaam *IST-Africa Week Conf. Durban.*

[3] Weinschenk S 2011 *ROI of User Experience* YouTube.

[4] Usabilitygov [Online] Available: https://wwwusabilitygov/what-and-why/user-centered-design.html

[5] Userfocus [Online] Available: http://wwwuserfocuscouk/consultancy/ucd.html [Accessed May 2017]

[6] Ibadurrahman H 2016 *Pengembangan Alternatif Desain Interaksi SCeLE Mobile dalam Perspektif Mahasiswa Berdasarkan User-Centered Design* (Depok: Universitas Indonesia).

[7] Rajapakse D 2008 Techniques for De-fragmenting Mobile Applications: A Taxonomy *Proc. of 20th Int. Conf. on Software Engineering and Knowledge Engineering Conf. (SEKE'08).*

[8] Tarkoma S and Lagerspetz E 2011 Arching over the Mobile Computing Chasm: Platforms and Runtimes *Computer IEEE* **44** pp 22-28.

[9] Okediran O O, Arulogun O T, Adesina G R and Akinwale O C 2014 Mobile Operating Systems and Application Development Platforms: A survey *Int. J. Advanced Networking and Applications (IJANA)* **6** pp 2195-2201.

[10] Az-zahra H M, Pinandito A and Tolle H 2015 Usability Evaluation of Mobile Application in Culinary Recommendation System *IEEE Asia Pacific Conf. on Wireless and Mobile (APWiMob).*

[11] ISO 9241-11:1998 [Online] Available: https://www.iso.org/standard/16883.html.

[12] Nayebi F, Desharnais J-M and Abran A 2012 The State of the Art of Mobile Application Usability Evaluation *25th IEEE Canadian Conf. on Electrical and Comp. Eng. (CCECE).*

[13] Yvonne R, Sharp H, Preece J, Benyon D, Holland S and Carey T 1994 *Human-Computer Interaction* (Essex: Addison-Wesley Longman Ltd).

[14] Fernandez A, Insfran E and Abraho S 2011 Usability Evaluation Methods for the Web: A Systematic Mapping Study *Information and Software Technology* **53** pp 789-817.

[15] Sharfina Z and Santoso H B 2016 An Indonesian Adaptation of the System Usability Scale (SUS) *Int. Conf. on Advanced Comp. Sc. and Information Systems (ICACSIS).*

[16] Nielsen J 1993 *Usability Engineering* (Boston, MA: Academic Press).

[17] Sharp H, Rogers Y and Preece J 2007 *Interaction Design: Beyond Human-computer Interaction 4th ed.* (Chichester: Wiley).

[18] Bangor A, Kortum P and Miller J 2009 Determining What Individual SUS Scores Mean: Adding an Adjective Rating Scale *J. of Usability Studies* **4** pp 114-123.

# Metacognitive components in smart learning environment

**M Sumadyo[1,2], H B Santoso[2] and D I Sensuse[2]**

[1]Faculty of Engineering, Universitas Islam 45, Bekasi, Jawa Barat, Indonesia
[2]Faculty of Computer Science, Universitas Indonesia, Depok, Jawa Barat, Indonesia

**Abstract**. Metacognitive ability in digital-based learning process helps students in achieving learning goals. So that digital-based learning environment should make the metacognitive component as a facility that must be equipped. Smart Learning Environment is the concept of a learning environment that certainly has more advanced components than just a digital learning environment. This study examines the metacognitive component of the smart learning environment to support the learning process. A review of the metacognitive literature was conducted to examine the components involved in metacognitive learning strategies. Review is also conducted on the results of study smart learning environment, ranging from design to context in building smart learning. Metacognitive learning strategies certainly require the support of adaptable, responsive and personalize learning environments in accordance with the principles of smart learning. The current study proposed the role of metacognitive component in smart learning environment, which is useful as the basis of research in building environment in smart learning.

## 1. Introduction

The new paradigm of today's intelligent education focuses on contextual, personal and unlimited learning issues to improve student intelligence and facilitate problem-solving skills in intelligent environments [1]. Seamless learning can be done collectively and individually, not limited by space and time. Individual learning takes into account the personal characteristics of students, learning styles and students' cognitive abilities. Contextual learning takes into account the background, condition and location of learning. So intelligent learning is a convergence of development in epistemology, psychology, and technology. Implementation of the convergence is a smart learning environment [2].

The international association for the Smart Learning Environment argues that the learning environment is considered smart if it utilizes adaptive technology or is designed to include innovative features and capabilities that enhance understanding and performance [3]. The innovative and adaptive capabilities of a technology in learning are technological skills that involve new needs in learning. To meet those needs intelligent learning has two types of technology that must exist, namely smart devices and intelligent technologies [4].

The intelligence of technology to respond to students 'personal abilities is the intelligence to understand learning styles and assess students' cognitive skills. In addition there are aspects involved are the metacognitive ability of students [5]. In the field of educational technology and information science it is stated that metacognitive strategies are important in overcoming problem-solving barriers [6]. Even in modeling of students in digital environments requires the expansion of discrete knowledge statistics of students' holistic assessment of student profiles in the cognitive, metacognitive and affective domains [5]. With metacognitive ability, students are not only aware of strategies but also knowing when and how to use problem-solving strategies. Without using metacognitive strategies, one does not know how to analyze the answer, fails to allocate enough time to study, or does not want to know how much he has learned about a subject, then, every learning task seems as though it was encountered for the first time [7]. Various metacognitive components will be studied in depth along with its function. This component can be part of the development of the next smart learning environment design.

Section 2 in this paper is a review of the relationship between the smart learning environment and the students' metacognitive abilities. Section 3 contains metacognitive roles and functions in the smart learning environment to enhance learning service, Section 4 highlights several technologies to facilitate the service, and Section 5 illustrates the conclusions and work to be done next.

## 2. Related Work

This section is a review of metacognitive research in e-learning and the smart learning environment and the linkage between the two from a general perspective.

### *2.1. An Introduction to Metacognitive*

Brown (1977) in [8] revealed that metacognitive skills are skills in solving problems of self-awareness in efficient problem solving which includes planning and scheduling problem-solving strategies, monitoring and evaluation of problem solving. Even disclosed earlier by Flavel (1976) in [8] that the inability of students to solve problems is more due to metacognitive deficiency than cognitive.

Metacognitive has two groups of components namely metacognitive knowledge and metacognitive regulation. Metacognitive knowledge refers to one's own cognitive processes consisting of knowledge of personal variables, knowledge of task variables and knowledge of strategy variables, while metacognitive regulation refers to one's procedural knowledge to organize cognitive processes with four components of planning, monitoring, evaluating and revise [9]. Metacognitive ability means the awareness of one's self towards his knowledge so as to understand the tasks to be performed and the strategies to be taken.

A person who has metacognitive skills means having the ability to know how to learn. Skills include the ability to question his/herself against self-study resources, the ability to look back at self-checking resources [10], the ability to monitor self-learning progress (self- monitoring), the ability to explain self-learning materials in the domain of self-explanation and the ability to learn from analogical reasoning [11]. The ability to know how students learn independently, procedurally can be done with components in the metacognitive regulatory group.

### *2.2. Improved Metacognitive Ability*

There are several techniques that can help students in improving metacognitive skills. Scaffolding technique is a technique for the learning process with the gradual aid to students who then gradually reduce the aid to increase the knowledge and competence of students. Scaffolding theory has become a major topic since it was discovered in Vygotsky's research, which became a key topic in modern learning research [12]. This technique is directives, questions and commands that involve students to improve their cognitive abilities. Because each student has different metacognitive abilities the facilitator will provide assistance according to his / her abilities.

In addition to scaffolding techniques, there is a help-seeking technique, with this technique, the facilitator will monitor the help options that will indicate students' cognitive abilities so as to stimulate their metacognitive ability to plan for problem solving. The research on help-seeking was developed by Roll in [13] as part of the tutor system to assess the students' quiet moment-learning behavior. Assessment of feedback outcomes consisting of general instruction and self-assessment combined into help-seeking error, i.e. assessing the student's mistake in choosing each help-seeking step. The help-seeking behavior measurement technique is performed by Vaessen in [14], i.e. calculating frequency and orderliness in selecting clues to solve the problem.

Some examples of such techniques are directed to capture the stages of students' metacognitive abilities composed on certain smart devices and of course with improved accuracy later. Data from smart devices are processed by intelligent technology to provide valid conclusions about the stages of metacognitive ability and then respond according to metacognitive strategies developed.

### *2.3. Smart Learning*

Electronic-based learning technology or e-learning at first just an information technology-based learning, radio and broadcasting but continues to be integrated with the development of technology until the emergence of intelligent technology [15]. At the time of the development of mobile technology, learning can be done with notebooks or mobile devices, but using wireless internet, making the form of learning becomes unlimited by place and time, which then becomes m-learning [16]. There are four components to be studied in the development of mobile-based learning communities, namely human factors, learning subjects, mobile technology support systems, and human support systems [17].

The next emerging technology is ubiquitous computing, this technology integrates learning into ubiquitous learning or u-learning. Learning can be done anywhere anytime with any device without having to use personal computer, just use internet connection [1], [18]. The latest developments are learning with intelligent technology, the technology emphasizes student-centered learning, which tends to be informal [15]. Learning that facilitates sensitivity, responsiveness and adaptive to student conditions. Smart learning is not just serving the interaction of students but also captures all interactions to be managed in the analysis tool then gives conclusions about the condition of student development to provide appropriate response to the condition of students.

### *2.4. Smart Learning Environment (SLE)*

Hwang defines SLE as a learning environment technology that makes adaptation and other provision support at the right place and time according to student needs. The framework has been compiled by him illustrating that the learning environment contains five module components, such as, student state detection module, student performance evaluation module, adaptive learning task module, adaptive learning content module and personal learning support module as well as an inference engine. The personal learning support module is supported by data on learning ppifolio, student duty sheet and student profile [16].

Like Hwang, Gros also agrees that SLE is not just a learning system that can be used in any place and at all times, but also actively provides learning guides, guides and adaptive support tools according to the right place, time and form [4].

SLE not only contains the components of intelligence tools, learning resources, the student community, and the teaching community, but also the way of learning and teaching. These two components make the learning environment "intelligent" according to the different needs of learning. [19]. Huang and colleagues have proposed further research to improve the quality of SLE with a more comprehensive coverage area covering education, computer science, psychology and engineering. Of course with the support of the latest technologies such as intelligent systems, internet of things and cloud computing. SLE focuses on user experience as a developmental principle that prioritizes personalized learning and a shift in the role of teacher from instructor to facilitator.

From the two points above, the problem raised in this paper is that metacognitive is an important factor in the learning process that emphasizes the focus on the personalization of learning, [9], [11], [12], and if the smart learning environment is Adaptive technology serves students 'personalization needs [16], [4], then how SLE becomes adaptive to the improvement of students' metacognitive abilities, what components support them.

## 3. A Proposed Smart Learning Environment Components

Metacognitive is a learning factor that emphasizes the personalization of learning. As mentioned above, students' ability to solve problems is strongly influenced by metacognitive ability. Techniques used in the process of developing metacognitive abilities, such as scaffolding and help seeking require students' responses to assess their development. Another section of the above review revealed that smart learning not only serves interaction, but also has to capture student interaction. The problem of the development of metacognitive ability and the need of smart learning is the component of learning environment which always catch the interactive and adaptive response to the development of students' metacognitive ability.

The SLE component proposed in this paper is an environment that focuses on adaptive services on improving students' metacognitive abilities. Any student who has the same cognitive ability may have far different metacognitive abilities that lead to different cognitive development. Awareness activities of self-knowledge and ability for self-learning can be done with metacognitive skills improvement techniques. Components in SLE-metacognitive are arranged in the form of modules connecting students, inference engines, environments and supporting databases.

The first SLE component is the student's cognitive ability detection module is a component to capture prior knowledge possessed by students. The result of this component becomes a learning portfolio and a benchmark of student development. The form can be done with pre-test or feedback to the questions posed.

The second component is the metacognitive technique module, this module contains steps that provide direction for improvement of metacognitive ability. If the technique used is scaffolding, this module contains gradual help with feedback to evaluate its metacognitive enhancement. If the technique used is help-seeking, this module contains sequences and hint branches that can be tracked for their accuracy so that they can be evaluated for their metacognitive capabilities.

The third component is the learning content management module. This module provides instructional materials that are sequential and staged according to the level of content understanding. Each content has relevance to the stages in the metacognitive technique module.

The fourth component is the adaptive assignment module. This module receives prior knowledge information, and metacognitive level information. Then with the help of the inference engine analyzing the prior knowledge level of the students, and making the decision to provide appropriate metacognitive technique steps then re-evaluating the feedback, the feedback evaluation results are re-used for subsequent assignment according to the student's ability. This is done in an iterative way until the students are able to solve the problem.

The fifth component is the inference engine as part of intelligent technology is an intelligent device that contains various algorithms to define the student's cognitive level status, determines the metacognitive status of the students so that students can be helped to decide the choice in planning their knowledge enhancement.

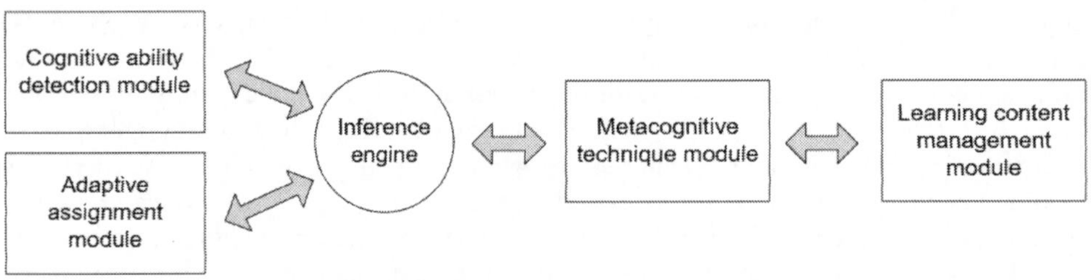


**Figure 1:** Relationships between components

## 4. Technologies for Smart Learning Environment

Smart Learning Environment for supporting students' metacognitive skill is greatly helped by the rapid development of technology. Completeness of features of mobile devices such as tablets, smartphones and various wearable computers supported by wireless communication networks and sensing technologies such as Computer Vision, Voice Capturing, RFID, and GPS [18], [20], can meet SLE-metacognitive needs.

For example, a group of students from different places and at the same time, can study the same subjects even with different devices. Suppose they will study math subjects on integrals, the cognitive ability detection module will detect their prior knowledge of integrals. A variety of user interfaces can help capture the cognitive abilities of every student. Through the monitor screen on a tablet or smartphone can provide questions and receive student answers. Sound counseling technology can conduct interviews. Computer vision technology can capture the faces of students when given a particular problem. The GPS device can record the student's environmental conditions. All data about students along with initial cognitive abilities become student profile notes.

The metacognitive technique module, referring to the inference given by the inference engine will provide knowledge assistance. The extent to which assistance will be provided, the inference engine provides a decision based on the results of the cognitive ability detection analysis. Certain student groups may be helped with understanding the basic formula to help resolve integrals. Other groups of students may receive the help of differential understanding or even start from basic numerology. All the subject matter along with the level of help is managed by the learning content management module. To capture the development of capabilities after the provision of assistance, the adaptive assignment module will provide tasks according to the level of assistance provided.

The SLE-metacognitive component will be greatly helped by Augmented Reality (AR) technology. AR integrates problems into the digital world just like the real world [21]. When students face their own problems, students will encourage themselves to learn independently, the above four modules will help to self-questioning and self-checking and monitor his ability with cell-monitoring. The integration of real-world and virtual presence in AR will deepen the metacognitive role in the learning environment.

## 5. Implications

As mentioned above, metacognitive is an important part of the learning process, individually or in groups from various places and times. Crucial elements needed to support metacognitive modules are smart devices such as Internet of Things and other wearable technologies that support and intelligent technology such as learning analytics or big data that focuses on learning data [4].

Therefore, the procurement of metacognitive components in learning becomes a logical implication in the development of the smart learning environment, it is an investment in social capital and communication infrastructure that plays a major role in improving the quality of human resources [22]. Smart infrastructure with intelligent components built by the government into the building base in building smart city [3].

Thus the smart learning environment with metacognitive components can be implemented in the context of smart city governance. Governance that provides a holistic focus for decision making that defines and delivers all educational components and even all other components of the smart city including mobility, environment, economy and society [22]

## 6. Conclusion

Information and communication technology is growing rapidly accompanying the development of the concept of intelligent learning. The presence of smart devices and intelligent technology add to the discourse of the development of smart learning environment, forming new ideas for improving the quality of learning.

The role of the field of study of educational psychology and computer science is very strong to form smart learning environment, but still have to involve again other fields such as education technology, this field will play a lot on instructional instructional, curriculum and some other learning technology.

According to bloom, learning is not just for cognitive development, but also student behavior, social ethics and psychomotor ability. This will be the field of future work in the smart learning environment involving computer science experts, educational psychology and teaching technology.

**Acknowledgement**
This article's publication is partially supported by Universitas Indonesia's Scientific Modeling, Application, Research and Training for City-centered Innovation and Technology (SMART CITY) Project.

**References**

[1] Zhu ZT, Yu MH and Riezebos P 2016 A Research Framework of Smart Education *Smart Learn. Environ.* **3** p 4.
[2] Spector J M 2014 Conceptualizing the Emerging Field of Smart Learning Environments *Smart Learn. Environ.* **1** pp 1–10.
[3] Nikolov R, Shoikova E, Krumova M, Kovatcheva E, Dimitrov V and Shikalanov A 2016 Learning in a Smart City Environment *J. Commun. Comput.* **13** pp 338–350.
[4] Gros B 2016 The design of smart educational environments *Smart Learn. Environ.* **3** pp 1–11.
[5] Jang E E, Lajoie S P, Wagner M, Xu Z and Poitras E 2017 Person-Oriented Approaches to Profiling Learners in Technology-Rich Learning Environments for Ecological Learner Modeling *J. of Educational Comp. Research* **55**.
[6] Blummer B and Kenton J M 2014 Metacognitive Skills and Online Search Behavior *Improv Student Inf. Search a Metacognitive Approach* pp 75–88.
[7] Abdolhossini A 2012 The Effects of Cognitive and Meta-Cognitive Methods of Teaching in Mathematics *Procedia - Soc Behav Sci.* **46** pp 5894–5899.
[8] Blummer B and Kenton J M 2014 Problem Solving and Metacognition *Improv Student Inf. Search* pp 33–43.
[9] An YJ and Cao L 2014 Examining the Effects of Metacognitive Scaffolding on Students' Design Problem Solving and Metacognitive Skills in an Online Environment *J. Online Learn. Teach.* **10** pp 552–568.
[10] Bayat S and Meamar A 2016 Predicting Algebra Achievement: Cognitive and Meta-cognitive Aspects *Procedia - Soc Behav Sci* **217** pp 169–176.
[11] Conati C 2002 Intelligent Tutoring Systems: New Challenges and Directions *Proc. of the Twenty-First Int. Joint Conf. on Artificial Intelligence (IJCAI-09)* pp 2–7.
[12] Wesiak G, Steiner C M, Moore A, Dagger D, Power G, Berthold M, Albert D and Conlan O 2014 Iterative Augmentation of a Medical Training Simulator: Effects of Affective Metacognitive Scaffolding *Comput Educ* **76** pp 13–29.
[13] Roll I, Aleven V, Mclaren B M and Koedinger K R 2011 Improving Students ' Help-Seeking Skills Using Metacognitive Feedback in an Intelligent Tutoring System *Learn Instr* **21** pp 267–280.
[14] Vaessen B E, Prins F J and Jeuring J 2014 Computers & Education University Students' Achievement Goals and Help-seeking Strategies in an Intelligent Tutoring System *Comput. Educ.* **72** pp 196–208.
[15] Lee M and Son Y 2013 A Study of Learning System for Smart Learning Using BYOD *SERSC Sci. Eng. Res. Support Soc.* **19** pp 106–111.
[16] Hwang GJ 2015 Definition, Framework and Research issues of Smart Learning Environments - A Context-aware Ubiquitous Learning Perspective *Smart Learn Environ* **1** pp 1–14.
[17] Lee J E 2015 A Formative Intervention on Mobile Learning Community *Procedia - Soc. Behav Sci.* **191** pp 68–76.
[18] Hwang G H, Chen B, Chu H C and Cheng Z S 2012 Development of a Web 20-based Ubiquitous Learning Platform for Schoolyard Plant Identification *Proc. 2012 17th IEEE Int. Conf. on Wireless, Mobile and Ubiquitous Technology in Education, WMUTE* pp 259–263.
[19] Huang R, Yang J and Zheng L 2013 The Components and Functions of Smart Learning Environments for Easy, Engaged and Effective Learning *Int. J. Educ. Media Technol* **8** pp 4–14.
[20] Bekara C 2014 Security Issues and Challenges for the IoT-based Smart Grid *Proc. Comput. Sci.*

**34** pp 532–537.

[21] Dunleavy M, Dede C and Mitchell R 2009 Affordances and Limitations of Immersive Participatory Augmented Reality Simulations for Teaching and Learning *J. Sci. Educ. Technol.* **18** pp 7–22.

[22] Hammad R and Ludlow D 2016 Towards A Smart Learning Environment for Smart City Governance *Proc. of the 9th Int. Conf. on Utility and Cloud Computing* pp 185–190.

IOP Conf. Series: Journal of Physics: Conf. Series **978** (2018) 012026 doi:10.1088/1742-6596/978/1/012026

# Improving data quality in the linked open data: a survey

**A Hadhiatma**[1]

[1]Department of Informatics, Sanata Dharma University, Yogyakarta, Indonesia

Email: agunghad@usd.ac.id.

**Abstract**. The Linked Open Data (LOD) is "web of data", a different paradigm from "web of document" commonly used today. However, the huge LOD still suffers from data quality problems such as completeness, consistency, and accuracy. Data quality problems relate to designing effective methods both to manage and to retrieve information at various data quality levels. Based on review from papers and journals, addressing data quality requires some standards functioning to (1) identification of data quality problems, (2) assessment of data quality for a given context, and (3) correction of data quality problems. However, mostly the methods and strategies dealing with the LOD data quality were not as an integrative approach. Hence, based on those standards and an integrative approach, there are opportunities to improve the LOD data quality in the term of incompleteness, inaccuracy and inconsistency, considering to its schema and ontology, namely ontology refinement. Moreover, the term of the ontology refinement means that it copes not only to improve data quality but also to enrich the LOD. Therefore, it needs (1) a standard for data quality assessment and evaluation which is more appropriate to the LOD; (2) a framework of methods based on statistical relational learning that can improve the correction of data quality problems as well as enrich the LOD.

## 1. Introduction

The Linked Open Data (LOD) project is the implementation of the semantic web idea [1], shifting representation from openly "web of document" to "web of data". The LOD can be derived from either structured [2], semi-structured [3] or unstructured data format [4]. The LOD, which is developed, linked and integrated from diverse data sources, has been evolving to be open, large distributed knowledge representation. As a consequence, it results several major issues, e.g. data quality, data interoperability, data management and scalability. One of these challenging issues in the LOD is about data quality. In general, data quality is defined as the fitness for use [5]. Several researches, e.g. [6, 7, 8], indicate that the LOD still inherits with data quality problems especially in terms of completeness, consistency, and accuracy.

Data quality in the semantic web can be classified into dataset quality [9], link quality [10] and schema quality [11].

- Dataset quality issues show that firstly, dataset or instances may be incorrect (noisy data), and incomplete (missing data) and secondly, regarding to their relations, datasets may be incorrect, incomplete, and inconsistent
- Link quality refers to the measured level that links between instances are not missing and weather every link is useful, and appropriate or not. Interlinking in the LOD means that dataset triples in Resource Description Framework (RDF), consisting a various of Uniform Resource Identifiers (URIs), form links from an entity as a subject to an entity as an object.

- Schema quality consists of completeness and consistency. Completeness of schema is the level that entities, attributes, classes, links are not missing regarding to their schema. Consistency of schema is the degree that entities, attributes, classes and links are coherent and free of logical conflict or contradiction with respect to their schema. In the LOD, a schema is represented by its ontology.

The challenges of data quality in the LOD are how to efficiently manage and to effectively analyse data and knowledge in big volume, since managing big data becomes more complex (that data is expanding massively, is mostly unstructured format as distributed graph, is controlled autonomously and has various levels of quality).

Dealing data quality means to address of some issues e.g. assessing information in various data quality levels, detecting errors, finding missing data, and improving data quality. Hence, the issues of data quality correspond to establishing methods and models to manage and to retrieve information. The poorer the quality of data, the worse the information can be retrieved. In the other side, according to [12],the growing volume of distributed semantic web leads to the more complex distributed graph and consequently becomes the greater potential information (hidden information or implicit knowledge) to be extracted but the harder to be derived.

## 2. Improving data quality in semantic web (the LOD): strategies and methods.

As the use of semantic web has been growing fast, it is necessary and challenging to deal with data quality on the semantic web. Consistency, accuracy and completeness of the semantic web have been recognized as an important problem in the recent decades. Searching and retrieving information in various data quality levels in semantic web may produce inaccurate or irrelevant results.

There have been many efforts to deal with consistency, accuracy and completeness in the Semantic Web. The author in [13] developed a model to address imprecise or uncertain data in order to improve recall and to achieve a better performance matrix in information retrieval. Several authors used logical reasoning model to predict missing data. Based on ontological background knowledge and a dataset of statements, logical reasoning can find inconsistency and infer new statements as predictions of missing data. Semantic web has framework including both syntax and model-theoretic semantics and inherits Description Logics, a knowledge representation supporting for logical reasoning. Logical reasoning plays an important role on semantic web. But logical reasoning methods have two main drawbacks. Firstly, for computation, it is hard to follow the growing of the web size [14]. Secondly, it is difficult to infer information on noisy, probabilistic and uncomplete data, identified in [15]. To deal with uncertainty in semantic web, many researchers have proposed OWL (Ontology Web Language, a family of knowledge representation languages) combined with special mathematical approaches, such as fuzzy extensions [16]. With extended OWL, the new approach of reasoning methods and algorithms have been specifically developed by[17] to derive relevant and unambiguous information from uncertain and unorganized data, but they still have some limitations.

Meanwhile, in many applications of data mining researches, successful approaches to predict missing data can be achieved by inductive learning from the huge data. Then, for the web search engines and applications, there have been increasing researches combining two research areas: web and data mining, well known as web mining. The author [18] successfully used inductive learning methods for data exploration on semantic web, namely semantic web mining. Subsequently, several approaches followed to use inductive learning methods for predicting data and finding missing data in the area of semantic web; there are listed as follow: Paulheim [4] proposed specifically association rule algorithm to detect similar co-occurrence patterns between types on DBpedia dataset. The authors in [19] employed k-nearest neighbours for predicting types, based on feature input of class and related resources. Sleeman and Finin [20] exploited a labelled training set to predict the type of instances. For inductive learning methods, it can be stated that several past studies tried to detect patterns and find missing information using the concepts including machine learning, statistical methods and exploitation of external knowledge.

Traditional data mining explores hidden knowledge by utilizing vector data input with independent variables. All data in the form of a vector input look like a single table, which feds to a machine learning for deriving hidden information or finding missing data. However, if the such machine learning try to search patterns that it need relational tables or graph as inputs (for instance web and linked data), the accuracy of learning would reduce and even be unreliable. These causes may be that in the traditional data mining, the machine learning does not consider about information of linkages and relationship, and it is hard to utilize rich information of a graph structure or schema. Unlike traditional data mining algorithms which find for patterns in a single table, relational data mining algorithms search for patterns among multiple relational tables. Considering the graph structure of multiple relational tables, relational learning algorithms will extract more rich information than learning on a single table. The most extensively applicable method and research for relational learning is Inductive Logic Programming (ILP) [21]. ILP is a relational learning approach which constitutes intersection of inductive learning and logic programming, employing concepts and methods from both machine learning and logic programming. However, ILP works well with deterministic structured data, but not with probabilistic structured data.

To handle probabilistic data, the ILP and the statistical machine learning concept are combined, involving the concept of machine learning, probability theory, and logic, which trigger to the emerging research area, namely Statistical Relational Learning (SRL). SRL methods and techniques evolve to represent, reason, and learn on data or knowledge with complex relational and probabilistic structure [22]. Indeed, SRL functionally works on interlinked, sparse, noisy, incomplete, and conflicting data. Researches in [23,24] resumed that SRL can operate in semantic web for extracting hidden information and finding missing data. Several SRL representations, frameworks and models have been proposed, for examples, Probabilistic Relational Models, Relational Dependency Network, Bayesian Logic Programming, and Markov Logic Networks. Among SRL models, we focus on Markov Logic Networks. Markov logic networks (MLNs) are a statistical relational model that generalizes weight first-order logic and Markov Networks [46]. MLN leaning consists of 3 main components: structural learning, weight learning, and Maximum a Posteriori (MAP) inference. MLN structure learning for huge data is hard and challenging. Structure learning should repeatedly learn the structure of graph or schema and weight learning process computes the weight of the structure with expensive statistical inference (NP-Hard MAP inference). Both structure and weight learning work to evaluate its relational structure. Several researches have tried to improve MLN computation performances of weight and structure learning. The author [25] conducted efforts to reduce computation load of the pseudo log likelihood and its gradient in the Weighted Pseudo Log Likelihood (WPLL) evaluation function. The authors [26] proposed Markov Logic Structure Learner (MSL) model for learning data structure by optimizing a likelihood-type measure.

Meanwhile, the LOD is incorporated with its schemas and ontologies, aiming both to function machine data interpretation and to enhance data accessibility. Ontology learning can supports for semi or automatic support ontology construction [11] and can partially replace the hard effort of manual ontology construction [27]. Those researches show that recent ontology learning framework have attempted to establish ontology by learning from various types of data sets for examples: database, text, html, and xml, but they seldom learning and exploring from the LOD.

Research by [28] proposed Statistical Relational Learning (SRL) methods to inductively learn ontologies from a repository containing noisy data. Similarly, SRL can learn Ontology Web Language (OWL) axioms from the Linked Data which has probabilistic unstructured graph representation. For an example, Zhu [29] suggested ontology learning approach that Description Logic and Bayesian Network are incorporated to deal with incomplete data in semantic web for constructing schema axioms from data axioms. Exploring the LOD by relational learning with especially focusing to its ontology is challenging and necessary because of some reasons. According to Buhmann [30], in the LOD, there are still less datasets which are structured and referred to their ontologies. The LOD research community still put less attention to the ontological layer and the correspondences between datasets and their ontology. Moreover, Ontology expressiveness can still be enhanced for more useful and powerful applications.

Only few researches used some guidelines as data quality assessment methodologies for improving data quality in the LOD, although several data quality assessment methodologies have been proposed. Research in [31] presented a systematic review of data quality assessment methodologies especially to the LOD. The review has resumed several assessment approaches, extracted a number of data quality dimensions and investigated relevant methods and tools. This research also showed that the flexibility of the tools with regard to enhance the level of automation still needs improvement. Several researchers also proposed data quality assessment and evaluation particularly in Semantic Web [32][33] and methods for detecting error presented by [34,35,36]. Furber and Hepp[32] presented Semantic Web Information Quality Assessment (SWIQA), a methodology that adopts rule templates to represents quality requirements. The rule templates are automatically used to detect error as well as to measure quality scores for five dimensions.

## 3. Discussion

From the literature review above, we have identified some potential issues as a background to define objectives of our next research, as below:

Schema and ontology quality is a degree to which entities, attributes, classes, links are complete, accurate, and consistent or coherent in the integration context. On the other hand, according to Zhu [35 29], the LOD is still lack of expressive ontologies, and instance datasets based on their ontologies. The LOD itself consists of mainly instance data, but still less representing data to ontological layer. Based on these facts, we intend to investigate some main research questions. Firstly, can ontology expressiveness be enhanced from data underlying on it? Secondly, does ontology expressiveness become a significance factor to improve data quality? Finally, how efficiently detect and curate incomplete, inaccurate, inconsistent data and links relating to other data, their schema and ontology in a context of integrated data? Hence, it may be significant that dealing with data quality should be corresponded to schema and ontology of that LOD.

For detecting and curating data quality, Zaveri [31] presented a comprehensive systematic review of data quality assessment methodologies and tools (methods) applied to the LOD. He showed that enhancing flexibility of the data quality tools still requires much effort and open discussion. The flexibility regards subsequent works to increase the level of automation and minimize user involvement. In addition, some tools are hard to configure and require a considerable amount of setting. Although the tools are simple to use, but they still work with limited scope or need much human interpretation. Also, in general, those other previous papers presents that the LOD quality is an essential but efforts to standardize data quality tracking and assurance are still less and how the data quality dimensions and metrics should be defined is still open discussion.

We suggest that firstly, to effectively and efficiently improve data quality needs the assessment methodologies as a guide and secondly, our next research should focus on dealing data quality in the LOD, considering relational data, schema and ontology in integrative context. Therefore, the main purposes of the future research are to achieve:

- A selected or modified standard for data quality assessment and evaluation which is more appropriate to facilitate the identification of the data quality problems and the assessment of data quality for an integration context
- A framework of methods based on statistical relational learning that can improve the correction of data quality problems in the LOD through ontology learning. The measured performances for handling data quality in huge LOD by statistical relational learning are learning time, level of learning automation (flexibility), ontology expressiveness, scalability, and number of data and links which can be repaired.

Using existing data quality assessment methodologies, we would identify to which data quality problems in the LOD (in the terms of completeness, consistency, and accuracy) correspond to their

IOP Conf. Series: Journal of Physics: Conf. Series **978** (2018) 012026 doi:10.1088/1742-6596/978/1/012026

own relational data, schema and ontology in integration context. Also, we would study which data quality assessment methodology is appropriate to asses these problems or whether the methodology requires to be updated.

For improving data quality, we are interested to focus on the concept of ontology enrichment and refinement. Ontology enrichment techniques are a process to generate and add data and link extensions from and to the linked data which is bounded with its ontology. The extensions can include new entities, concepts and relations. Meanwhile, ontology refinement relates to improving and correcting an existing ontology. Ontology refinement process consist of several tasks such as updating, adding, deleting axioms and links in existing ontology, aiming to make semantic web to be more detailed and relevant. Ontology enrichment and refinement can be established through ontology learning.

Then, we would propose several strategies to address data quality problems which have been identified by a suitable data quality assessment. Firstly, improving data quality in the LOD, considering in its schema and ontology, can be approached by ontology learning namely ontology refinement. Secondly, enhancing ontology expressiveness in the LOD may be addressed by ontology learning namely ontology enrichment. Finally, by referring to [23, 24], it is believed that relational learning such as Markov Logic Network (MLN) may be adjusted as a tool to handle ontology refinement and enrichment. The existing RDF data and ontology in the LOD can be enriched and refined into a new one, which aims to improve data quality as well as to enhance ontology expressiveness.

Based on the strategies, we would propose a new framework of methods to improve data quality as well as learning performances. In the framework, the detected data quality problems are then repaired by ontology learning, adopting some techniques/methods of relational machine learning. Subsequently, these learning performances in some variables are measured and compared. Then, based on evaluation of some chosen techniques/methods in relational machine learning, we would propose two main steps to develop a new framework, as presented below:

- Studying and developing suitable techniques of the relational learning to address data quality problems in integrative context: Relational learning consists of structure learning and weight learning. To deal with relational learning in the LOD is a challenging task because structure learning should search in a huge search space and a weight learning process requires computationally expensive statistical inference[37][26].
- Establishing certain domain knowledge and rules as a background knowledge represented by ontology: a background knowledge is incorporated into a relational machine learning aims to enhance a level of automation, control learning process, improve effective learning, and constrain learning spaces. In an addition, background knowledge represented as ontology would support relational machine learning in logical reasoning. Background knowledge could be captured from LOD itself or others.

The result of Framework would be evaluated by a selected assessment and evaluation methodology. Weight learning and structural learning performance can be measured using average conditional log-likelihood of test atoms (CLL) and average area under the precision-recall curve (AUC).

## 4. Conclusion

Most previous researches have conducted research to improve data quality not as considering an integrative view but as a partial approach. Considering integrative view, it is opportunity to improve data quality in the LOD in the term of incompleteness, inaccuracy and inconsistency referring to its schema and ontology through ontology learning, approached with relational leaning. Relational learning utilizes the combination of machine learning, probability theory, and logic. In addition, relational learning in big data should consider huge searching space for its structural learning and computationally expensive statistical inference for its weight learning.

Investigating the characteristic of data quality in the LOD is by adopting appropriate assessment methodology, purposed to develop mechanism and knowledge to detect and evaluate the quality dimension/criteria. The mechanism and the knowledge are used formally to curate and enrich dataset and links. Alternative methods to curate and enrich dataset and links could be established by modified relational learning. Existing RDF data and underlying ontology could be enriched and refined into a new one which aims to improve data quality as well as to enhance ontology expressiveness. The performances of a method in relational learning for improving the LOD quality are measured by a number of data and links which can be improved as well as enriched, learning time, a level of learning automation, ontology expressiveness, and scalability.

We highlight that the quality of the LOD includes a number of novel aspect such as coherence and consistency with regard to the ontology and the schema in the context of integrated data. However, there are still research opportunities to formulize how the data quality dimensions to be determined and how data quality tracking, assessment, and assurance to be implemented.

**References**

[1] Augenstein I, Pado S and Rudolph S 2012 LODifier: Generating linked data from unstructured text. *In Proc.Conf on Extended Semantic Web Conference*(Greece).

[2] Bizer C and Cyganiak R 2006 D2R server-publishing relational databases on the semantic web. *Proc.Conf on The 5thInternational Semantic Web*.

[3] Lehmann J, Isele R, Jakob M, Jentzsch A, Kontokostas D, Mendes P N, Hellmann S, Morsey M, Kleef P V, Auer S and Bizer C 2014 DBpedia-A large-scale multilingual knowledge base extracted from wikipedia *J. Sem. Web*.

[4] Paulheim H 2012 Browsing linked open data with auto complete *Proc Semantic Web Challenge* (Sydney).

[5] Knight S and Burn J 2005 Developing a framework for assessing information quality on the world wide web *J. Inf. Sci* **8** pp 159-172.

[6] Furber C and Hepp M 2010 Using semantic web resources for data quality management *Proc Int.Conf on The 17 the Knowledge Engineering and Knowledge Management* vol 6317 (Lisbon : Springer) pp 211-225.

[7] Furber C and Hepp M 2010 Using SPARQL and SPIN for data quality management on the semantic web *Proc.Int.Conf on The 13th Business Information Systems* vol LNBIP 47 (Berlin:Springer) pp 35-46.

[8] Hogan A, Harth A, Passant A, Decker S and Polleres A 2010 Weaving the pedantic web *Proc.Int.Works on The 3rd Linked Data on the Web*.

[9] Lei Y, Nikolov A, Uren V and Motta E 2007 Detecting quality problems in Semantic Metadata without the presence of a gold standard *Proc.Works on Evaluation of Ontologies for the Web at the WWW EON* pp 51-60.

[10] Guert C, Groth P, Stadler C and Lehmann L 2012 Assessing linked data mappings using network measures *Proc on The 9th Extended Semantic Web*(Crete).

[11] Furber C and Hepp M 2011 SWIGA: A semantic web information quality assessment framework. *Proc.Int on European Information Systems* (Helsinki).

[12] Singh M P and Huhns M N 2005 *Service-Oriented Computing: Semantics, Processes, Agents*. Wiley, 1st edition.

[13] Paulheim H and Pan J Z 2012 Why the semantic web should become more imprecise *What will the Semantic Web look like 10 years from now*.

[14] The Large Knowledge Collider. EU FP 7 Large-Scale Integrating Project 2008 (online: http://larkc.eu/ LarKC)

[15] Ji Q, Gao Z and Huang Z 2011 Reasoning with Noisy Semantic *Data Proc.Int on the 8th Extended Semantic Web ESWC* (Crete).

[16] Straccia U 2005 Towards a fuzzy description logic for the semantic web *Proc .Conf on The 2nd European Semantic Web: The Semantic Web: Research and Applications* vol LNCS 3532

pp. 167–181 (Springer).
[17] Lukasiewicz T. Probabilistic default reasoning with conditional constraints 2002 *J. Annals of Math. and Artifi. Intell* 34(1-3) pp 35-88
[18] d'Amato C and Fanizzi N, Esposito F 2010 Inductive learning for the semantic web: What does it buy? *Proc.Int on the 8th Extended Semantic Web ESWC* (Crete) vol1 (Springer ) pp 53-59.
[19] Giovanni A, Gangemi A, Presutti V and Ciancarini P 2012 Type Inference through the Analysis of Wikipedia Links *Proc. Works on Linked Data on the Web LDOW*(Lyon).
[20] Sleeman J and Finin T 2013 Type prediction for efficient co reference resolution in heterogeneous semantic graphs *Proc.Int on The 7th IEEE Semantic Computing* (California).
[21] Lavrac N and Dzeroski S 1994 *Inductive logic programming techniques and applications* (Ellis Horwood).
[22] Getoor L and Taskar B 2007 *Introduction to statistical relational learning* (The MIT Press).
[23] Stumme G, Hotho A and Berendt B 2006 Semantic web mining -state of the art and future directions *J. Web. Sem* vol 4 no 2 pp 124–143.
[24] Tresp V, Bundschus M, Rettingerand A and Huang Y 2008 Towards machine learning on the semantic web *Proc.Int.Works Uncertainty Reasoning for the Semantic Web URSW 2005-2007, Revised Selected and Invited Papers* (Springer) vol 5327 pp 282–314.
[25] Richardson M and Domingos P 2006 Markov logic network *J. Mach. Learn.*
[26] Kok S and Domingos P 2010 Learning markov logic network structure using structural motifs. *Proc.Int.Conf on the 27th Machine Learning* (Haifa : Omnipress).
[27] Maedche A and Staab S. 2001 Ontology learning for the semantic web 2001 J. IEEE Intell. Syst vol 16 no 2 pp 72-79.
[28] Zhu M, Gao Z, Pan J Z, Zhao Y, Xu Y and Quan Z 2013 Ontology learning from incomplete semantic web data by BelNet *Proc Int.Conf on The 25th IEEE Tools with Artificial Intelligence* (Herndon,USA).
[29] Zhu M 2011 DC proposal: Ontology learning from noisy linked data *Proc.Int.Conf on The 10th The Semantic Web* (Berlin: Springer-Verlag) part II pp 373-380.
[30] Buhmann L, Fleischacher D, Lehmann J, Melo A and Volker J 2014 Inductive lexical learning of class expressions *J. Knowl. Eng and Knowl. Man* **8676** 42-53.
[31] Zaveri A, Rulab A, Maurino A, Pietrobon R, Lehmann J, Auer S 2014 Quality Assessment for Linked Data: A Survey *J.Sem. Web.*
[32] Furber C and Hepp M 2011 SWIQA A semantic WEB information quality assessment framework. *Proc.Euro.Conf on Information Systems ECIS.*
[33] Bizer C and Cyganiak R 2009 Quality driven information filtering using the WIQA policy framework. *J. Web Semantics.*
[34] Tropper G, Knuth M and Sack H 2012 DBpedia ontology enrichment for inconsistency detection *Proc.Int on the 8th on Semantic Systems I-SEMANTICS.*
[35] Lehmann J and Buhmann L 2010 ORE- A tool for repairing and enriching knowledge bases. *Proc.Conf on the 9th International Semantic Web ISWC*(Shanghai).
[36] Paulheim H 2014 Identyfing wrong links between datasets by multi-dimensional outlier detection *Proc. Int. Works on Debugging Ontologies and Ontology Mappings* (USA).
[37] Kok S and Domingos P 2009 Learning markov logic network structure via hypergraph lifting *Proc.Int.Conf on the 26th Machine Learning* (Montreal :Omnipress) pp 505-512.

# Mapping of medicine data with k-means and apriori combinations based on patient diagnosis

**N P Dharshinni[1*], H Mawengkang[2*] and M K M Nasution[3*]**
[1]Department of Information Technology, Faculty of Computer Science and Information Technology,Universitas Sumatera Utara, Medan 20155, Indonesia
[2]Department of Mathematics, Faculty of Matematics and Natural Sciences,Universitas Sumatera Utara, Medan 20155,Indonesia
[3]Department of Information Technology, Faculty of Computer Science and Information Technology,Universitas Sumatera Utara, Medan 20155, Indonesia

*Email : [1]priyadharshinninaidu@gmail.com,[2]hmawengkang@yahoo.com, [3]mahyuddin@usu.ac.id

**Abstract**. Medicine is one of the items needed by sick society, the high influence of medicine on service and the economy in hospitals, requires mapping and planning the optimal need for medicines according to the conditions, because 50% -60% of hospital income is sourced from medicine sales. The purpose of this study was to find patterns of doctor's prescription medicine association with sales data using an apriori algorithm based on data grouping using a k-means algorithm. The results of the experiments show that medicine prescription data with medicine sales have significant differences so that the data can not be used as materials for medicine planning, this is due to some indication of one of the unavailability of medicine caused by mapping inaccuracy so that the planning of medicine requirements is not optimal. The results of this analysis can be used as input materials in decision making, so the planning needs of medicines can be in accordance with the development of patient disease patterns.
**Keywords** apriori, k-means, Medical Prescription, medicine

## 1. Introduction

Hospital pharmacy service is one of the support services and also serves as the center of the main income, because 90% of health care use of pharmaceuticals (medicine, chemicals, radiological materials, materials medical devices consumables, medical equipment, and medical gas ), and 50% of all income derived from the management of pharmaceutical supplies [1]. The high contribution to the service of the hospital pharmacy, requires a careful and full management responsibility in the medicine plan [2].Planning pharmaceutical care medicine is an activity which aims to get the type, amount of medication as needed and avoid the emptiness that is based on patterns of disease and the frequency of the number of visits patients [3]. Medicine is one of the items needed by a sick society [3]. The determination of medicine based on the results of the doctor's diagnosis that is written in the prescription, and all patient diagnostic data stored in the database, one method that can extract large amounts of data to obtain useful medicine information in planning based on patient's medical data is to apply data mining.

Data mining is a multidisciplinary field useful in finding and defining patterns from data sets in databases into information[4]. Data mining is widely applied in the medical field, including medicine prediction, heart disease and the discovery of clinical and pathological data relationships. Data mining is an efficient method for detecting some types of problems to pinpoint important patterns in a set of data, one of which is Frequent Itemsets Mining (FIM). Frequent Itemsets play an important role in many data mining problems that try to determine important patterns from the database[5]. Most data mining problems are grouped into two, namely (1) candidate generation using Breadth First Search (BFS) algorithm and (2) pattern growth with Depth First Approach (DPA) approach[6]. Depth First Approach (DPA) is considered more complicated because it uses data structures in its research[7], while Breadth First Search (BFS) is more widely used because it is more simple, one of them by using Algoritma Apriori[4].

Apriori algorithm is the most classic and quite important algorithm in Frequent Itemsets Mining (FIM). Although many similar algorithms are developed more efficiently, such as FP-Growth, LCM, and so on [8]. Apriori algorithm can reduce the number of candidates that must be calculated support by pruning, so it has good performance [6] but has a long period of computation due to the scanning process that must be done at every time iteration.

Clustering is one technique for grouping objects or patterns in a set (cluster) so that each cluster has similarities to each other [9]. FCM is one of the cluster algorithms that is quite optimal in the grouping but requires a long computation time [10]. FCM is very strong against noise [11] but is too sensitive to cluster initialization [12]. K-Means Clustering is an efficient algorithm for grouping large and fast datasets in computation [10] but is limited to numerical data [13].

In this research, an apriori method is proposed to conduct medicine mapping based on doctor's prescriptions sourced from diagnosis results, this doctor's prescribing as a source of mapping for the correlation of diagnosis result with medicine as useful information in medicine planning. This is because of the changing pattern of illnesses, there are still doctors who make prescriptions beyond standard set by the Pharmacy and Therapy Committee, so that the purchase of medicine to the outside pharmacy or unserviceable prescriptions, especially for inpatients and outpatients, due to the unavailability of the medicine. The analytical framework uses the data grouping method with the k-means algorithm to apriori, then we analyze the influence of grouping on the computation process and the accuracy of the mapping of medicine sales relation with doctor's prescription. This paper is presented as follows: Chapter 2 related research and Chapter 3 Proposed method. Chapter 4 discussion and Chapter 5 conclusion.

## 2. Related Research

Apriori data mining algorithm is the most commonly used method for extraction of data in large data with association rules in determining associational connectivity in item sets, each relationship can be measured using two different units of support value and confidence level. Any relationship that meets the minimum threshold is called the k-itemset, the level of processing in the apriori algorithm is determined the number of candidate sets with the set produced at each level so that the algorithm takes a high computation time, for optimization [14] proposes a combination of tree based frequent. Research of Li Wei et al [15], developed an apriori optimization algorithm for medical big documents based on MapReduce, WEN-JING ZHANG et al [12], apply an apriori algorithm for the health field with the purpose of breast cancer data classification so that doctors can be more efficient and accurate in diagnosing patients. V R Sadasivam et al [16], implemented an apriori algorithm for criminal pattern data mining, Xiaoyan Cui [17], applied an apriori algorithm for large data mining on medical and cloud computing with the Hadoop framework.

K. Zuhtuogullari [18], propose for the mining of health data to find the pattern of illness that often occurs in patients by seeking symptom relation with disorders in the medical database. Yan Yuguang et al [19], research developed a multi-model based on apriori algorithm at the hospital to extract data from the database to produce useful information in medical decision-making, Miao Wang et al [20], applies to the health field to discover the characteristics of headaches in traditional apriori-based treatments, making it easier for doctors to decide on prescriptions of various headache sufferers. Wang Renli [21], proposed the Apriori-BSO algorithm for the mining of health data, with the aim of finding data associations and symptoms of medical records, medicine combination data, and the relationship between symptoms and treatment.

## 3. Proposed Method

### *3.1. Apriori algorithm*

The apriori algorithm consists of several stages of iteration in the database[4], each iteration will generate a calculated frequency pattern by scanning the database to obtain the support for each item, items that have support above the minimum support selected into the frequency pattern high with one length or often called 1-itemset. K-itemset is a term for a set consisting of k items. In the second iteration process will produce 2-itemset which each set has two items[5]

### *3.2. K-Means Algorithm*

The K-Means algorithm determines the cluster of objects based on the attributes/features of the object into the cluster (K). The groups are distinguished by their *centroid*. This algorithm is very sensitive to placing the initial value of the *centroid*.The determination of cluster number (k) partition is very important in this algorithm, but there is no provision to determine the number of clusters (k) formed [22]. The steps in the K-means algorithm as follows:

1. Prepare training data.
2. Set the value of K cluster.
3. Set the initial value of centroids.
4. Calculate the distance between the data and the centroid using the formula (Euclidean Distance).
5. Partition data based on the minimum value.
6. Then do it iteration during the data partition is still moving (no more objects moving to another partition), if still moving to another partition then go back to point 3.
7. If the data group is now the same as the previous data group, then stop the iteration.
8. Data has been partitioned according to the final centroid value.

### *3.3. Dataset*

Dataset sourced from government hospitals and private hospitals with data of 2015 and 2016, these two sources of data will be used as a material for analysis and comparison.

### *3.4. Tool*

In this research, we apply apriori algorithm and k-means in WEKA open source application because it supports several data mining methods such as preprocessing data, clustering, regression classification, visualization and feature selection.

### *3.5. Research framework*

In this study, an apriori algorithm is applied to locate an item set of a set of data from the database of medicines in the hospital, but after the data are grouped by k-means algorithm. In Figure 1 is an illustration of the research framework we are proposing. Based on the results of our observation, mapping medicine use based on sales data of general pharmacy or pharmacy BPJS, because of that, we will make this as a reference in this case. We used two approaches, the first approach to apply grouping sales data and then applied to the apriori algorithm, while the second approach we apply grouping in medical data of the patient and then applied to the apriori algorithm.

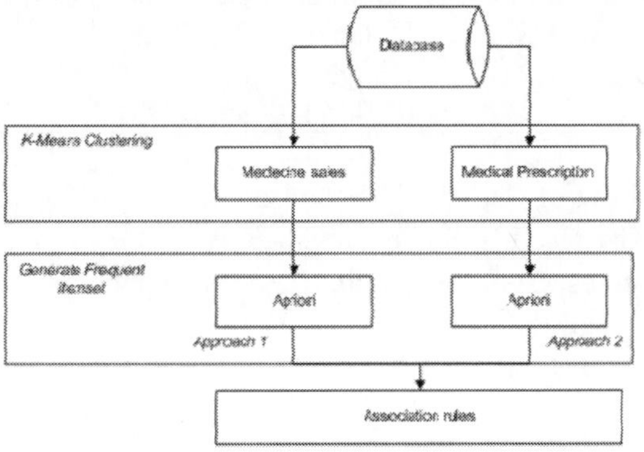


**Figure 1.** Research framework

IOP Conf. Series: Journal of Physics: Conf. Series **978** (2018) 012027 doi:10.1088/1742-6596/978/1/012027

In general, the process of medicine sales in hospitals sourced from the results of doctor's diagnosis written on the prescription and the patient brought the prescription to the pharmacy, so the doctor's prescription will be listed in pharmacies as medicine sales, but sometimes patients do not order at the hospital pharmacy because the availability of medicines at the pharmacy does not match the progression of the patient's illness. Our main objective of this study was to look for the relationship of patient prescription data to the sale of medicine, to serve as inputs in the medicine planning adjusted based on the patient's disease pattern.

## 4. Discussion

This study uses two different data sources, but there are linkages that need to be extracted to produce useful information in the decision making of medicine requirements planning. The first source is based on sales data with a total of 14,509 transactions consisting of 2,165 types of medicines and the second source of medical / prescription data totaling 14,509. Both of these sources will be applied to the apriori algorithm after the grouping of k-means data. in table 1 and table 2 is the result of grouping of two data sources.

**Table 1.** Cluster Sales Data

| | **#of Instance** | **ratio%** |
|---|---|---|
| $Cluster_0$ | 95 | 84% |
| $Cluster_1$ | 76 | 16% |

**Table 2.** Cluster patient medical data

| | **#of Instance** | **ratio%** |
|---|---|---|
| $Cluster_0$ | 101 | 32% |
| $Cluster_1$ | 76 | 30% |
| $Cluster_2$ | 54 | 13% |
| $Cluster_3$ | 45 | 20% |
| $Cluster_4$ | 39 | 5% |

In Table 1 The first approach uses a sales data source with four dominant attributes, $cluster_0$ is a medicine sale for inpatients, while emergency patients are the most sources of medicine sales in this cluster. $Cluster_1$ patient age is dominant between 50-60 years old who is sourced from an outpatient with lung disease and tuberculosis, while patient age between 0-7 year old comes from children's health with outpatient status with cough disease. In table 3 the result of association rules first approximation.

**Table 3.** Association rules related to Disease data (Min Sup: 20%, Min Conf: 65%)

| **Rules** | **Sup.** | **Conf.** | **Lift** | **Consequent** | **Antecedent** |
|---|---|---|---|---|---|
| R1 | 0.45 | 0.81 | 1.35 | Emergency Services | Inpatient |
| R2 | 0.31 | 0.79 | 1.47 | Diagnosis | Cataract |
| R3 | 0.23 | 0.85 | 1.23 | Diagnosis | Age = 50-60 |
| R4 | 0.26 | 0.77 | 1.17 | Children's Health | Cough |

In Table 3 the minimum limit of Support 20% and confidence 65% with four extraction results is called R1, R2, R3, and R4. The result of the analysis shows that emergency service is very dominant to conduct medicine sales transaction of pharmacies for inpatients, while outpatients who buy medicines are dominated from child health.

**Table 4.** Cluster General Model

| Attribute | Full Data (14.509) | $Cluster_0$ (95) | $Cluster_1$ (76) |
|---|---|---|---|
| Group | Emergency Services | Emergency Services | Children's Health |
| Diagnosis | Cataract | Cataract | Cough |
| Age | Age = 50-60 | Age = 50-60 | Age = 0-7 |
| Status of care | Inpatient | Inpatient | Outpatient |

The results of the analysis in Table 4 of the dataset were sourced from sales with 14.509 data, in $Cluster_0$ there were 95 data that were successfully grouped with the source of the patient from emergency services, the dominant diagnosis was a cataract, with patient age 50-60 and patient is an Inpatient. In $Cluster_1$, the dominant patient is an outpatient with a 0-7 year-old age who is sourced from Child's Health with a Cough disease diagnosis. Therefore, in medicine planning, consideration should be given to increasing the stock of medicines with categories for children age 0-7 year-old, because the highest sales are sourced from the Child Health for outpatients. The second approach uses a patient medical data source with four rules used in sales data, the results are shown in table 5.

**Table 5.** Associated rules related to patient medical data (Min Sup: 20% Min Conf: 65%)

| **Rules** | **Sup.** | **Conf.** | **Lift** | **Consequent** | **Antecedent** |
|---|---|---|---|---|---|
| R1 | 0.75 | 0.92 | 1.75 | Internal disease | Inpatient |
| R2 | 0.43 | 0.85 | 1.68 | Diagnosis | THT |
| R3 | 0.56 | 0.89 | 1.45 | Diagnosis | Age = 55-75 |
| R4 | 0.89 | 0.83 | 1.58 | Children's Health | Diarrhea |

In Table 5 the results of dataset extraction from doctor's prescriptions amounted to 14,509 prescriptions, a minimum limit of support 20% and 60% confidence. The results of this second approach analysis were dominated by inpatients from Internal Disease with age 55-75 years, the most dominant disease was THT, while outpatient was dominant from polyclinic of Child Health with a diagnosis of the dominant disease was Diarrhea.

**Table 6.** Cluster General Medical Model

| Attribute | Full Data (14.509) | $Cluster_0$ (101) | $Cluster_1$ (76) | $Cluster_2$ (54) | $Cluster_3$ (45) | $Cluster_4$ (39) |
|---|---|---|---|---|---|---|
| Group | Internal Disease | Internal Disease | Emergency Services | Children's Health | Internal Disease | Children's Health |
| Diagnosis | THT | THT | THT | Diarrhea | Cataract | Cough |
| Age | 50-60 | 50-60 | 50-60 | 0-7 | 60-75 | 0-7 |
| status of care | Inpatient | Inpatient | Outpatient | Inpatient | Inpatient | Outpatient |

The second approach to the data sourced from the medical records of patients there are differences in the association rules as shown in Table 6. In $Cluster_0$, the dominant disease group data is a group of internal diseases, with an THT diagnosis, the dominant age of the disease is between 50-60 and the status of inpatients. $Cluster_1$ is sourced from emergency services, $cluster_2$ child health is more dominant with age 0-7 is inpatient. $Cluster_3$ includes cataract disease with patient age 60-75 is inpatient, while the $cluster_4$ dominant health of children with cough disease is outpatient.

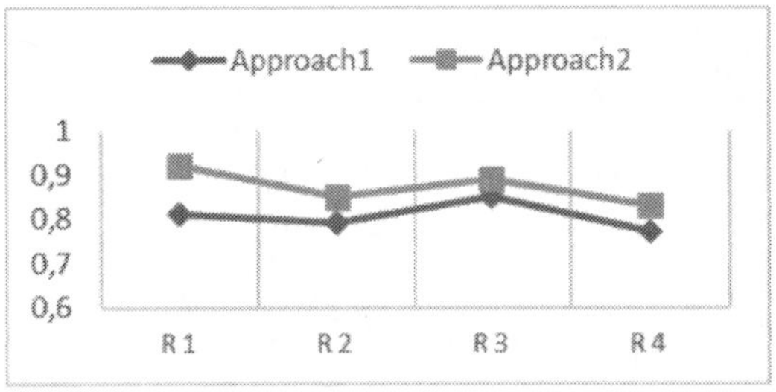


**Figure 2.** Confidence related to medicine data

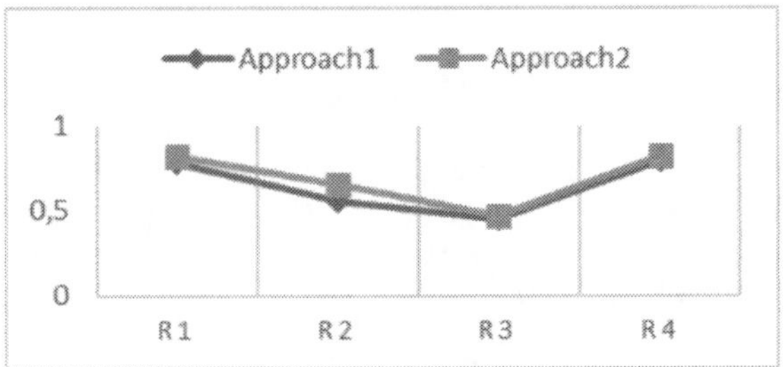


**Figure 3.**Confidence related to Patients treatment data

## 5. Conclusions

Optimal medicine planning based on the pattern of disease progression in the health sector is one of the things that greatly affect medical services and economics. The results of the mapping from the sales data and medical data of patients, we observe there are significant differences, where 35% of doctors research data on patients does not occur sales transactions, this is one good input for decision-making especially medicine planning. The application of k-means clustering method to apriori gives a significant influence in determining association rules and computation time. For further work In mapping data with data mining, it is necessary to compare with other methods to find more accurate methods in prediction, this will be useful in decision making.

## References

[1] Rohmani S, Fudholi A and Hakim L 2016 Analisis Faktor Internal-Eksternal terhadap Pengolahan Obat di Instalansi Farmasi RSUD dr.Moewardi Surakarta *Journal of Pharmaceutical Science and Clinical Research*, 01, 10-20

[2] Hamid T B 2005 Elemen Pelayanan Minimum Farmasi di Rumah Sakit , Direktorat Jendral Pelayanan Kefarmasian dan Alat Kesehatan, Departemen Kesehatan RI, Jakarta

[3] Satibi 2014 Managemen Obat di Rumah Sakit, Fakultas Farmasi, Universitas Gadjah Mada Yogyakarta

[4] Rao S and Gupta P 2012 Improved Algorithm. Over APRIORI Implementing Data Mining Association Rule Algorithm *IJCST* **Vol. 3**, Issue 1. 489-493

[5] Goswami D N , Anshu C and Raghuvanshi C S 2010 An Algorithm for Frequent Pattern Mining Based On Apriori *(IJCSE) International Journal on Computer Science and Engineering*, **Vol. 02**, No. 04, 942-947

[6] Patel B , et al 2011 Optimization of Association Rule Mining Apriori Algorithm Using ACO. *International Journal of Soft Computing and Engineering (IJSCE)* **Vol.1**, Issue 1. 24-26. ISSN: 2231-2307

[7] Kumar S S, et al 2012 Aprori M. Improved Algorithm Based on a bottom up approach using probability and Matrix *IJCSI International Journal of Computer Science* Issues, **Vol. 9**, Issue 2, No. 3. 242-246

[8] Kumar K S and Chezian R M 2012 A Survey on Association Rule Mining using Apriori Algorithm *International Journal of Computer Applications* (0975-8887). Volume 45- 5. 47-50

[9] Izakian H and Abraham A 2011 Fuzzy C-Means and fuzzy swarm for fuzzy clustering problem Expert Systems With Applications, 38 (3), 1835-1838

[10] Ghosh S and Dubey K S 2013 *Comparative Analysis of K-Means and Fuzzy C Means Algorithms. (IJACSA) International Journal of Advanced Computer Science and Applications*, **Vol. 4**, No. 4, 2013

[11] Hamasuna Y and Endo Y and Miyamoto S 2009 On tolerant fuzzy c-means clustering and tolerant possibilistic clustering. *Soft Computing*, 14(5), 487–494

[12] Zhang W- J, Ma D- L and Dong B, et al 2012 The Automatic Diagnosis System Of Breast Cancer Based On The Improved Apriori Algorithm *International Conference on Machine Learning and Cybernetics*, Xi'an, 15-17. 63- 66

[13] Bai L, Liang J and Dang C 2011 An initialization method to simultaneously find initial cluster centers and the number of clusters for clustering categorical data Knowledge-Based Systems, 24 (6), 785-795

[14] Yamuna N Y, Devi N and Devi J S 2016 A combinatorial tree-based frequent pattern mining *Journal of Theoretical and Applied Information Technology* 781- 789. ISSN: 1992-8645. E-ISSN: 1817-3195

[15] Wei L, et al 2016 Optimization and Application in Medical Big Document-Data of Apriori Algorithm based on MapReduce *International Conference on Computer Communication and Informatics* 978-1-4673-6680-9

[16] Sadasivam V R, Duraisamy K and Bharathi R M 2015 Association Rule Mining and Frequent Pattern Mining Applications on Crime Pattern Mining: A Comprehensive Survey *International Journal of Innovative Research in Science, Engineering and Technology* **Vol. 4**, Special Issue 6. 228- 236. ISSN (Online): 2319-8753

[17] Cui X, Yang S, and Wang D 2016 An Algorithm of Apriori based on Medical Big Data and Cloud Computing, Proceedings of CCIS 361-365

[18] Zuhtuogullari K and Allahverdi N 2011 An Improved Itemset Generation Approach for Mining Medical Databases, *IEEE*. 39-43. 978-1-61284-922-5

[19] Yuguang Y, Chunyan Wand Min L 2011 Application of the Model_Multi based on Apriori algorithm in a Supporting System of Medical Decision *International Conference on Measuring Technology and Mechatronics Automation* 566-569. 978-0-7695-4296-6, DOI 10.1109 / ICMTMA.2011.144

[20] Wang M, et al 2014 The Applicat ion Characteristics of Traditional Chinese Medical Science Headache Treatment on Data-Mining Based on Apriori Algorithm *IEEE International Conference on Bioinformatics and Biomedicine* 153-157

[21] Renli W, Yueming D and Liming D 2016 The Application Of Apriori-BSO algorithms in Medical Records Data Mining. 827-832

[22] Zhang C and Fang Z 2013 An Improved K-means Clustering Algorithm Traditional K-mean Algorithm *Journal of Information & Computational Science*, 1, 193-199

IOP Conf. Series: Journal of Physics: Conf. Series **978** (2018) 012028 doi:10.1088/1742-6596/978/1/012028

# E-learning process maturity level: a conceptual framework

**A Rahmah[1], H B Santoso[1] and Z A Hasibuan[1]**

[1]Faculty of Computer Science, Universitas Indonesia, Depok, Indonesia

**Abstract**. ICT advancement is a sure thing with the impact influencing many domains, including learning in both formal and informal situations. It leads to a new mindset that we should not only utilize the given ICT to support the learning process, but also improve it gradually involving a lot of factors. These phenomenon is called e-learning process evolution. Accordingly, this study attempts to explore maturity level concept to provide the improvement direction gradually and progression monitoring for the individual e-learning process. Extensive literature review, observation, and forming constructs are conducted to develop a conceptual framework for e-learning process maturity level. The conceptual framework consists of learner, e-learning process, continuous improvement, evolution of e-learning process, technology, and learning objectives. Whilst, evolution of e-learning process depicted as current versus expected conditions of e-learning process maturity level. The study concludes that from the e-learning process maturity level conceptual framework, it may guide the evolution roadmap for e-learning process, accelerate the evolution, and decrease the negative impact of ICT. The conceptual framework will be verified and tested in the future study.

## 1. Introduction

The rapid advancement of information and communication technology (ICT) from time to time is a sure thing. Indonesia has internet penetration number 50.4% [1] covering all population in 2016. However, the utilization for education domain is still low, only 7.8% for college students, and 6.3% for students % [1]. According to the ICT Development Index, a composite index measuring ICT usage to enhance country's growth a development, Indonesia ranks 115$^{th}$ from 175 countries in 2016 [2]. These phenomena may lead Indonesian citizen gain more negative rather than positive impacts from the technology advancement. Aside from the advantages of accessibility towards data, information, and knowledge on one hand, people are also perplexed by the information overload, explosion, hoax, misuse [3], cyber threats, and cyber crime [4].

ICT has the power to inspire us to transform the way we live, the way we teach, and the way we learn [5]. Learners become digital learners because of the ubiquity of technology that surrounds them [5]. In a broad sense, the way people use an electronic device (usually a computer) with learning technology is described as e-learning to develop new knowledge and skills individually or collaboratively [6]. Along with ICT advancement, e-learning implementation has evolved bringing us web-based learning, mobile learning, and the latest, ubiquitous learning. Those learning technologies increase the ease of the learning process regarding the recorded process, material accessibility, immediacy of access, interactivity, and integration of learning with daily activities [7]–[9]. However, this trend use technology dependency as the approach to define an e-learning state [8][6][7]. Those technologies are used to support mobility and ubiquitousness are expensive, hard to find on a regular basis, and not affordable for a particular type of learners. It relates to digital divide issues [5] where utilization level of e-learning is different regarding knowledge ability, comprehension of ICT literacy backgrounds, and ICT accessibility. How e-learning technologies can be effective to all types of learners regarding its advancement? A concept used to answer this problem is e-learning process evolution. It requires the learner to improve not only knowledge ability for a particular topic but also the learning process itself. This research uses context of e-learning process as the process of conducting learning supported by ICT, as used in [10].

By nature, people adopt ICT gradually, influenced by many factors. However in another side, there are government policies related to e-learning which apply globally and enforce simultaneously in all schools and universities in Indonesia. Enforcing a policy without a preliminary analysis of learner conditions may results in the ineffectiveness and inefficiency. The ineffectiveness and inefficiency bring a risk of country financial losses. Accordingly, there is a need for evolutionary path like a multi-level transformation from an initial to a target level that shows progression, called maturity model [11]. The target is to achieve maturity, the state of being complete, perfect, or ready [12]. This study explores related works about e-learning process evolution from the individual point of view using maturity level concept. Ideal conditions expected is achieving the learning objective using accessible ICT advancement to conduct the e-learning process in both formal and informal situations. Furthermore, this approach attempts to reduce potential losses regarding the government program in ICT utilization for education because of the improper use of the ICT. The following sections explain the study as follows: Section 2 discusses about related works to define the e-learning process influencing factors and e-learning maturity level. Section 3 discusses the resulting conceptual framework. The last section describes the conclusion of this study.

## 2. Related Works

This section discusses factors affecting the maturity level progression of e-learning process through the extensive literature review by comparing the similar ideas, contrasting the differences, and synthesizing the relevant ideas from these related works. It makes sense to use a maturity model when there are requirements for a structured approach, a measurable result based on the established body of knowledge, a defined roadmap, and ability to monitor and measure progress particularly in the presence of change [13]. Different domains have utilized a maturity model concept, so it may also be applied in ICT based learning development domain.

There are previous research using capability approach to explain the maturity level of learning, e-learning maturity model (eMM) with e-learning process capability over an achieved objective [14] and CMM in e-learning [15]. The level consists of delivery, planning, definition, management, and optimization. The process areas consist of learning, development, support, evaluation, and organization. Another capability related work is learning maturity model [16], which uses Bloom's Taxonomy Revised by Marzano [17] to define maturity levels as knowledge ability development assessment. The levels consist of motivation and knowledge retrieval development, synthesis development, analysis development, usage, and learning. The factors affecting the learning process are motivation, objective, and knowledge ability [16]. Learning objective is a baseline for knowledge ability development progression. Knowledge ability, based on Marzano's new taxonomy, consists of retrieve, synthesis, analysis, and application [17]. Learning performance is monitored through learning result tracking. These maturity models do not represent a progression roadmap, but to what extent the process in e-learning is adequate

A discussion of technology trend to support e-learning determines progression maturity levels suitable with this study context, in which commonly uses three stages: e-learning, mobile learning (m-learning), and ubiquitous learning (u-learning), as stated in [9], [6] and [18]. Related works in ICT-based learning determine u-learning as the highest trend in e-learning. Ogata defines u-learning characteristics as follows: permanency, accessibility, immediacy, interactivity, and situatedness [7][19]. Meanwhile, Hwang defines u-learning environment characteristics as context-aware environment that provides personalized support in the right way, time, and place, enables seamless learning, and adapts with the various mobile devices [8]. Both studies conducted by Ogata and Yano [7] along with Hwang [8] emphasize that u-learning is about the context-aware environment to support learning. Another point of view delivered by Cope and Kalantzis, who stated that u-learning exploits ubiquitous computing as a product of social needs bringing social effects such as situated, interactive, participatory, spatial, temporal, cognitive, and intuitive computing [5]

Besides defining u-learning characteristics as a single object, there are also works characterizing u-learning as a learning trend along with e-learning and m-learning. In the following research, Hwang and Liu explain u-learning as paradigm shifts in e-learning [6] requiring a particular type of ICT advancement. Yahya, et al. [9] describe these trends by adopting characteristics from Ogata and [7], Chen et al. [20], Curtis et al. [21], Hwang et al. [22], and Chiu et al. [23]. In another research, Ogata and Uosaki explain this trend using ICT advancement time chart in research on computers in education [19] covering three kinds of learners: personal, group, and social learners. Each improvement enables a particular learner to learn with specific ICT advancement like fixed/PC, multimedia like game-based learning, internet allowing the emergence of web-based learning/e-learning and computer-supported collaborative learning, and last, mobile and surround technology supporting m-learning and u-learning. In further technology-driven research, Ogata and Uosaki [19] offer seamless learning, which is the situations where students can learn whenever they want in a variety of scenarios. These progressions of e-learning discussed above are affected by certain factors.

Sfenrianto, et.al. discuss e-learning process factors for learning material personalization consisting of learning styles, motivation, and knowledge ability [10]. According to Hwang, the motivation includes learning needs, a learning style with preferences, and also considers schedules [24]. Meanwhile, Alkis et al. group factors related with e-learning acceptance into dimensions consisting of social environment, acceptance of e-learning application, individual technical and non-technical skills, desired system features to support e-learning, institutional factors defining learning motivation, and e-learning popularity [25]. Bonuauti et.al determine three factors used to describe the level of ubiquity in a learning environment which are the type of learning, learning methodologies, and technological mediation [26].

## 3. E-Learning Process Maturity Level Conceptual Framework

In design science, an approach to propose solution is by forming construct as research artefacts [27]. Schon in [27] states that constructs can provide the language in which problems and solutions are defined and communicated. Constructs are used to understand components for the to-be-built conceptual framework. The proposed conceptual framework adopting those constructs is illustrated in Figure 1 with elements consist of: learner, learning objectives, e-learning process, continuous improvement, evolution path, and factors influencing the maturity grouped as learner-related factors and technology-related factors.

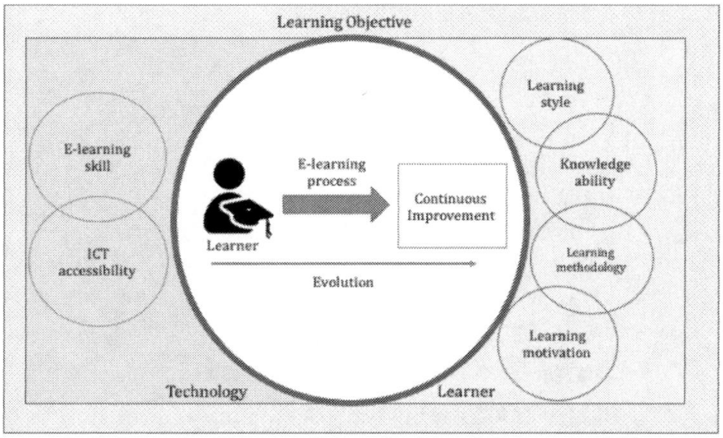


**Figure 1.** E-Learning Process Maturity Level Conceptual Framework

Learner is someone, a group, or a community who conduct the e-learning process. In conducting the process, learner has certain learning objective. Level of knowledge in Bloom Taxonomy revised [28] with creation as the highest goal, describes these learning objectives. The further meaning of reaching this state is that a learner can accomplish his/her job immediately after learning.

E-learning process is defined as conducting learning using ICT which can evolve to achieve continuous improvement in context of higher learning objective. E-learning process as a system with continuous improvement can be described with the I-P-O model. Input can be described as learner status (represented by motivation, knowledge ability, motivation, and learning objective). Output represents how far the e-learning process can improve the learning outcome (knowledge ability and e-learning skill) and achieve the learning objective. Through the output, the learner status and achievement can be evaluated. In addition, feedback can be formulated to improve it. This I-P-O and feedback model form a cycle of which a learner can conduct over and over following certain path of improvement, the maturity level for e-learning process.

The evolution path, as illustrated in Figure 2, describes ICT utilization progression from time to time. It uses the progression type of maturity model which has a levelling mechanism representing a transformation of the object, the e-learning process, into a more advanced state. The proposed maturity levels for the e-learning process adopts technology trend in supporting e-learning, which are web based learning, m-learning, and u-learning. As known, learning is possible conducted without ICT so that there should be a level 0. It is called 'traditional learning'. In addition, maturity level needs a maturity goal as an ideal state. This study proposes seamless learning to answer it.

Seamless learning, ability in doing learning anytime, anywhere, curiosity/inquiry based, and it generates applicable knowledge to get a job done immediately as the learning outcome. The analogy to explain this condition is that self-directed learners know what to learn and what material is needed, know how to access and use it. The approach used in seamless learning utilizes the learning material as a standalone entity. It relates to another material according to learners' needs to gain the capability to create something or get the job done. Meanwhile in conventional learning, they are grouped into topics and hierarchical following a curriculum.

Figure 2 describes the e-learning process evolution, explained as maturity levels in the chart of two dimensions (x, y). The x-axis represents the progression stage of maturity level. The y-axis represents particular e-learning advance's frequency of utilization. A variable represents each influencing factor. Those variables will then together contribute to achieve a maturity level. The notation of this equation can be written as follows. If each factor is represented by $\underset{a}{\rightarrow}$, $\underset{b}{\rightarrow}$, $\underset{c}{\rightarrow}$, $\underset{d}{\rightarrow}$, $\underset{e}{\rightarrow}$, $\underset{f}{\rightarrow}$, then the achievement of a maturity level is expressed by $\underset{v}{\rightarrow}$ (a,b,c,d,e,f).

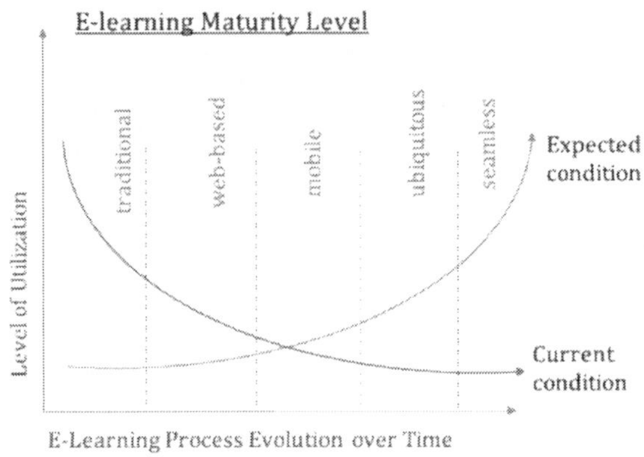


**Figure 2.** E-Learning Process Maturity Level

Moving up from one level to another follows an established improvement roadmap. It covers the levelling mechanism, the current condition measurement, and the expected condition. A learning objective becomes determinant to what extent the learning process utilizes particular ICT. This principle allows learners to improve the e-learning process, considering the condition of each factor affecting the process.

Several factors related to learners and ICT influence the individual e-learning process evolution and measured regarding its contribution in supporting the e-learning process to move up to a higher maturity level. Learner-related factors include learning styles, knowledge ability, methodology, and motivation. On the other hand, technology-related factors include ICT accessibility and e-learning skills. Each factor has its stages to represent the state of a learner in using e-learning with a defined goal as the highest stage, illustrated as a vector $\underset{a}{\rightarrow}, \underset{b}{\rightarrow}, \underset{c}{\rightarrow}, \underset{d}{\rightarrow}, \underset{e}{\rightarrow}, \underset{f}{\rightarrow}$. The ideal is achieving seamless learning with curiosity as the driver for learning methodology, using effort as motivation, the consciousness of learning styles, also capability in choosing and using specific accessible ICT as learning tools.

Learning style is adopted as a factor affecting the e-learning process based on previous research about influential factors of the e-learning process [10] and factors in mobile learning adoption [29]. In this factor, the highest goal is not related to a certain style, but the ability of a learner identifies which one is suitable for him/her. Learning styles and e-learning skills represent the ability of a learner to choose and use the most appropriate learning tools.

As mentioned in related research that effort is a fundamental indicator of a student's motivation [10], it becomes the highest goal of motivation to reach in learning in this research context. Another factor is learning methodology including directed learning, problem-based learning, project-based learning, and inquiry-based learning. Inquiry-based learning uses curiosity for the driver to do learning. It acts as proposed highest goal of learning methodology to be achieved to reach seamless learning.

Improving each learner's factors is not necessarily enough. There are also factors related to ICT to gain a higher level of maturity for the e-learning process. The first factor is e-learning skills, a proposed factor to measure capability to identify and use a particular learning tool. E-learning skill is part of comprehension in ICT literacy or mentioned as e-literacy [30] or digital literacy [3], specifically applied for learning tools. This skill covers the ability to identify, choose, and use appropriate ICT to support learning, and know-how to use basic functions of this technology to support learning (upload, download, save, retrieve, and share). This skill requires alignment with accessibility towards ICT itself.

Last factor proposed in this research is accessibility to ICT advances to support learning. The ability to access various technology types in which learners can experience web-based learning, m-learning, u-learning, and seamless learning is a prerequisite to conduct learning. In this context, ability to access particular technology is varied. Therefore, learners need a maturity roadmap to guide which level they are at related to technology accessibility and what technology should be used to get to the next level. It can be explained more with a theory about types and levels of technology use [31].

## 4. Conclusion

E-learning process is a form of self-directed learning in which its gradual improvement may be conducted by aligning learners' needs and conditions. This study proposes A Conceptual Framework of E-learning Process Maturity Level as an evolution roadmap for this improvement using maturity level concepts. The goal is achieving learning driven by curiosity, anytime, anywhere, using accessible ICT to accomplish learners' work immediately – called seamless learning. In order to achieve it, a learner should pass these levels: traditional learning, web-based learning, m-learning, u-learning. The framework has elements consists of learner factors, e-learning process, continuous improvement, evolution path, and technology-related factors. It serves as ICT based learning evolution roadmap, a guideline to accelerate the evolution, and a way to decrease the number of ICT utilization negative impacts by optimizing the use of ICT for learning. For future study, this research will be verified and tested in the future study.

## 5. Acknowledgement

This publication is within the research program funded by Universitas Indonesia – International Indexed Publications for Students Final Project Grant with number: 403/UN2.R3.1/HKP.05.00/2017 and Doctoral Program of Computer Science funded by Indonesia Endowment for Education.

## References

[1] APJII 2016 Survei internet APJII 2016
[2] International Telecommunication Union 2016 ICT development index
[3] Rahmah A 2015 Digital literacy learning system for indonesian citizen *Procedia Computer Science* **72** 94–101
[4] Setiadi F, Sucahyo Y G and Hasibuan ZA 2012 An overview of the development indonesia national cyber security *Int. J. of Information & Computer Science* **6** 108
[5] Cope B and Kalantzis M 2009 Ubiquitous learning *Encyclopedia of the Sciences of Learning*
[6] Liu GZ and Hwang GJ 2010 A key step to understanding paradigm shifts in e-learning: towards context-aware ubiquitous learning *British Journal of Educational Technology* **41** 1–9
[7] Ogata H and Yano Y 2004 Context-aware support for computer supported ubiquitous learning *Wireless and Mobile Technologies in Education* 27–34
[8] Hwang G 2006 Criteria and strategies of ubiquitous learning *Proc. IEEE Int. Conf. on Sensor Networks, Ubiquitous, and Trustworthy Computing* **1** 72–77.
[9] Yahya S, Ahmad EA and Jalil KA 2010 The definition and characteristics of ubiquitous learning: a discussion *Int. Journal of Education and Development using Information and Communication Technology* **6** 1
[10] Sfenrianto, Hasibuan ZA and Suhartanto H 2011 The influence factors of inherent structure in e-learning process *Int. Journal of e-Education, e-Business, e-Management and e-Learning* **1** 217
[11] Fraser P, Moultrie J and Gregory M 2002 The use of maturity models/grids as a tool in assessing product development capability **1** 244–49.
[12] Simpson J A and Weiner ESC 1989 *The Oxford English dictionary* 2nd ed New York: Clarendon Press ; Oxford University Press
[13] Mehravari N 2011 Everything you always wanted to know about maturity models Carnegie-Mellon University
[14] Marshall S 2010 A quality framework for continuous improvement of e-learning: the e-learning maturity model *Int. Journal of E-Learning & Distance Education* **24** 143–66

[15] Alrasheedi M and Capretz LF 2013 Applying CMM towards an m-learning context Int. Conf. i-Society 146–51.
[16] Marcinka J, Mirzianov O and Mitasiunas A 2014 Learning process maturity model *Software Process Improvement and Capability Determination* **477** 261–67.
[17] Marzano R J and Kendall J S 2007 *The new taxonomy of educational objectives*, 2nd ed. Thousand Oaks, CA: Corwin Press
[18] Priyogi B 2014 Analisis Kualitas Layanan Open Education (OE) dengan Pendekatan Deksriptif Kuantitatif *Thesis Report* Universitas Indonesia
[19] Ogata H and Uosaki M 2012 A new trend of mobile and ubiquitous learning research: towards enhancing ubiquitous learning experiences *International Journal of Mobile Learning and Organisation* **6** 64–78
[20] Chen Y, Kao T, Sheu J and Chiang J Y 2002 A Mobile Scaffolding-Aid-Based Bird-Watching Learning System *Proc. of the IEEE Int. Workshop on Wireless and Mobile Technologies in Education*
[21] Curtis M, Luchini K, Bobrowsky W, Quintana C and Soloway E 2002 Handheld use in K-12: A descriptive account *Proceedings - IEEE Int. Workshop on Wireless and Mobile Technologies in Education* 23–30
[22] Hwang GJ, Chin-Chung T and Yang S J 2008 Criteria, strategies and research issues of context-aware ubiquitous learning *Journal of Educational Technology & Society* **11** 2
[23] P.-S. Chiu, Y.-H. Kuo, Y.-M. Huang and T.-S. Chen 2008 A Meaningful Learning Based u-Learning Evaluation Model 77–81
[24] Hwang G J 2014 Definition, framework and research issues of smart learning environments-a context-aware ubiquitous learning perspective *Smart Learning Environments*, **1** 4
[25] Alkis N, Coskunçay D F and Yildirim S O 2014 A systematic review of technology acceptance model in e-learning context 1–5
[26] Bonaiuti *et al.* 2011 TAG introduction to an ubiquitous learning model to assess the ubiquity level in higher education institutions *Ubiquitous Learning: An International Journal* **5** 2
[27] Hevner AR, March ST, Jinsoo P and Ram S 2004 Design Science in Information System Research *MIS Quarterly*
[28] Krathwohl D R 2002 A Revision of Bloom's Taxonomy: An Overview *Theory Into Practice* **41, no. 4** 212–218
[29] Karimi S 2016 Do learners' characteristics matter? An exploration of mobile-learning adoption in self-directed learning *Computers in Human Behavior* **63** 769–76
[30] Indrajit R E 2005 Strategi dan kiat meningkatkan e-literacy masyarakat indonesia *Sistem Informasi: Berbagai Makalah tentang Sistem Informasi yang disampaikan dalam Konferensi Nasional Sistem Informasi* 37–45
[31] Sedek M, Mahmud R, Jalil H A and Daud S M 2012 Types and levels of ubiquitous technology use among ict undergraduates *Procedia - Social and Behavioral Sciences* **64** 255–64

IOP Conf. Series: Journal of Physics: Conf. Series **978** (2018) 012029 doi:10.1088/1742-6596/978/1/012029

# Develop applications based on android: Teacher Engagement Control of Health (TECH)

**Sasmoko[1,2], S R Manalu[1,3], S A Widhoyoko[1,4], Y Indrianti[1] and Suparto[5]**
[1]Research Interest Group in Educational Technology, Bina Nusantara University, Jakarta, Indonesia
[2]Faculty of Humaniora, Bina Nusantara University, Jakarta, Indonesia
[3]Mobile Application & Technology Department, School of Computer Science, Bina Nusantara University, Jakarta, Indonesia
[4]Podomoro University, Jakarta, Indonesia
[5]Universitas Kristen Krida Wacana, Jakarta, Indonesia

**Email :** {sasmoko, smanalu}@binus.edu, samuel.anidyo@podomorouniversity.ac.id, yasintaindrianti@gmail.com, suparto@ukrida.ac.id

**Abstract.** Physical and psychological condition of teachers is very important because it helped determine the realization of a positive school climate and productive so that they can run their profession optimally. This research is an advanced research on the design of ITEI application that able to see the profile of teacher's engagement in Indonesia and to optimize the condition is needed an application that can detect the health of teachers both physically and psychologically. The research method used is the neuroresearch method combined with the development of IT system design for TECH which includes server design, database and android TECH application display. The study yielded 1) mental health benchmarks, 2) physical health benchmarks, and 3) the design of Android Application for Teacher Engagement Control of Health (TECH).

## 1. Introduction

Teacher excellence is not merely an example of the knowledge, attitude and actions, but also exemplary in terms of keeping yourself both physically and psychologically. The idea that a healthy teacher will produce a healthy and successful student has emerged for a long time. Because teachers who are interested in taking care of their health will also tend to pay attention to the health of their students and become role models in generating healthy behaviours [1]. In addition to the physical, teachers must also be able to maintain the condition of psychic to stay healthy, because the psychological health of teachers is closely related to the source of stressors that affect their profession. This condition contributes negatively to the psychological climate in school [2].

School climate can be created positively and productively when teachers are able to run their profession optimally. Conditions in which teachers interpret and run the profession is called the teacher engagement. The teacher who is engage is a teacher who in his existence is always professional because of his ability to keep himself physically and psychologically [3]. Behavioural and emotional engagement proved to be an indicator of the internal dynamics of students' conditions that contributed to the success and success of students in schools because teachers who engage will experience a high level of connectivity with their job duties. Teachers will strive to achieve task-related goals with definitions and roles that lead to high performance achievement [4], [5]. Therefore, there is an effort to build the condition of teachers to engage with their profession by maintaining health condition both physically and psychologically, especially in Indonesia as a country having various demography condition.

This effort has been carried out throughout 2015 - 2017 by building an instrument of teacher engagement in accordance with the Indonesian context called the Indonesian Teacher Engagement Index (ITEI). ITEI is an instrument used to detect the condition of teachers' engagement in Indonesia. However, seeing the importance of physical and psychological health roles for teachers in carrying out their profession, ITEI needs to be equipped with instruments capable of detecting these conditions. ITEI as early detection has been realized in android application, for it required a familiar self-control in the daily activities of teachers through their gadgets.

According to statistics 'we are social' most of global users utilize in smartphone than personal computers, and it has been increased 5% in January 2016. Indonesia itself in January 2017 smartphone users dominated by Android reached 71.6% [6]. Regarding to this data and problems that have been explained before, this research proposed to develop an android application to control ITEI's teacher health. By this research, the teachers will be able to handle their concepts.

The basic concept of self-control centres on the hearts of each teacher that is related to the wellbeing problem. The more teachers can keep the heart, the joy will be able to improve the overall health of teachers [7]. No health is realized without mental health. Mental disorders will have an impact on physical disorders that will ultimately affect the quality of the work [8]. Although the teacher is professional, but if his ability is weak in controlling both physical and psychological, it will result in the teacher personally become not engage. To maximize the engagement of teachers in Indonesia required an intervention medium that served as an electronic supplement that can control themselves through android based applications. The application is then called Indonesian teacher engagement which is then named TEACHER ENGAGEMENT CONTROL OF HEALTH (TECH). The formulation of the problem is how the TECH application system?.

Media intervention is a new breakthrough for teachers in Indonesia as a complement of the approach of the Ministry of Education and Culture of the Republic of Indonesia is to assess the performance and competence of each teacher periodically. And the two assessments are not measuring what should be measured. Through this TECH, then the teacher will have the ability to assess himself, improve himself and progress because of himself. And this all in the end become a benchmark in the form of Electronic Master Teacher Engagement Indonesia.

## 2. Research Methods

The combination of qualitative, quantitative research summarized in the neuroresearch method became the main basis of this study. This method is also combined with the development of IT system design for TECH which includes server design, database and android TECH application display. The neuroresearch method is performed through three main stages: exploratory, explanatory and confirmatory stages with the result of theoretical basic concepts and instruments used in the application [9] - [11]. Neuroresearch methods are also performed to find the positioning of applications between similar applications so that these applications have more comprehensive and appropriate benefits in detecting the condition of teachers.

## 3. Result

### *3.1. Measuring mental health*

WHO explained that there is no healthy individual if it does not have a healthy mental [12]. Mental health is related to the condition of individuals who are psychologically positive. The positive concept of psychology is a general term that shadows the emotional condition of a person, the character is in a healthy psychological condition [13]. Someone is said to be in a healthy psychological condition when they feel wellbeing, have quality life and work and do not experience boredom and burnout conditions [14] - [16].

Subjective wellbeing has a diverse range of categories that include a variety of phenomena that can have a positive impact on one's health even the life span of a person [17]. Subjective wellbeing is more dominated by the psychological condition of the individual subjectively So is the quality of working life. As one of the concepts in positive psychology, quality of work life is a process that enables individuals to participate actively and effectively in the work environment and organization so that the expression can describe the environmental and humanistic values associated with one's work experience [15]. Conversely, one's psychological health also needs to be viewed from a different perspective one of them is burnout and boredom of work.

Burnout is associated with physical and psychological fatigue that affects health problems and productivity declines. Burnout needs to be seen Because it is a metaphor that generally describes a person's mental state in a depressed condition [16].

### *3.2. Measuring physical health*

**Table 1.** Vital Sign By Age [18]

| | Pulse (per minutes) | Blood Pressure | Respirations (breaths per minute) |
|---|---|---|---|
| Adult | 60 to 100 | 90- 140mmHg | 12 to 20 |
| Children (1 to 8 yrs) | 80 to 100 | 80- 110mmHg | 15 to 30 |
| Infants (1 to 12 mth) | 100 to 120 | 70-95mmHg | 25 to 50 |
| Neonates (1 to 28 days) | 120 to 160 | >60mmHg | 40 to 60 |

### *3.3. Android Application for Teacher Engagement Control of Health*

Teacher Engagement Control of Health (TECH) is one of supporting menu in android ITEI. This will help teachers to control their health, by measuring the status of their vision, hearing, blood pressure, heart rate, lung capacity, respiratory rate and Oxygen. Android ITEI is client-server application, as shown in Figure 1a, that implementing load balancer in server side to divide all incoming traffic to cloning server equally. By implementing load balancer will help to minimize the failure at server side[19], [20].

This research does not only measure the health status but also psychological status for subjective wellbeing, quality work life, burnout, and job boredom. Before developing the application, we conduct a research at google play store that compare the application TECH blue print to 3 of similar application about health and psychological measurement as shown in Table 2. iCare Health Monitor control the same health aspect in TECH, but not in psychological aspects. iCare Heart rate only control heart rate in health aspect, and iCare in respiration rate. Both applications do not control psychological aspects.

**Table 2.** Comparison of Similar Applications

| | **iCare Monitor Kesehatan** | **iCare Denyut Jantung** | **iCare Laju Pernafasan** | **TECH** |
|---|---|---|---|---|
| Psychological - Subjective well being | | | | √ |
| Psychological - Quality work life | | | | √ |
| Psychological - | | | | √ |

| | | | | |
|---|---|---|---|---|
| Burnout | | | | |
| Psychological - Job Boredom | | | | √ |
| Health - Vision | √ | | | √ |
| Health - Hearing | √ | | | √ |
| Health - Blood Pressure | √ | | | √ |
| Health - Heart Rate | √ | √ | | √ |
| Health - Lung Capacity | √ | | | √ |
| Health - Respiratory Rate | √ | | √ | √ |
| Health - Oxygen | √ | | | √ |

Teachers are required to login first before using the TECH, and filling the biodata form. Login uses google account data. This data is used as identity of the teacher. Biodata and identity record in database, to make sure that we can still use this data in future. TECH application process teachers' health status while they input data by using camera and screen smartphone. The process doing by mapping the user input into data in Table 1 and view the result ad the teacher's health profiling as shown in Figure 2b.

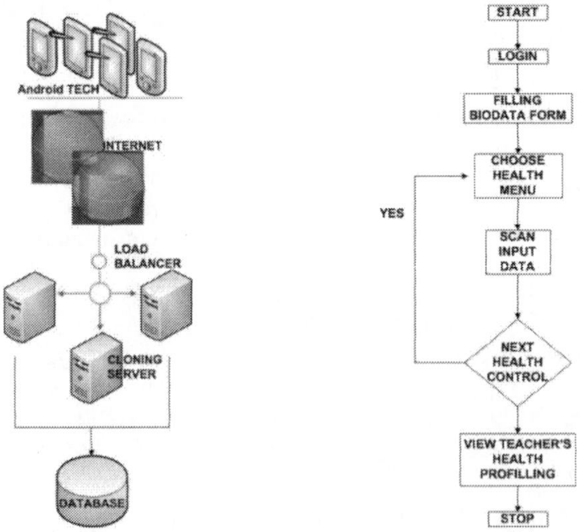


(a) (b)

**Figure 1. (a)** TECH Architecture; **(b)** 'Teacher Engagement Control of Health' Work Flow

Android TECH application is an application designed with the main purpose to improve the psychological condition of patients positively. Psychological conditions are first measured by answering many questions, and using measurements translated into specific algorithms will display results as well as related things to do. The main page display of Android TECH app is shown in Figure 2a, for the initial design there are eight main menus namely ITEI, Vision, Blood Pressure, Heart Rate, Insulin Schedule, Medicine Schedule, Visit Doctor Scheduler, and History. ITEI is the main menu of this application that serves to measure the psychological condition. This psychological condition is measured by answering many questions where the initial design. Users will be asked to answer questions whose answers are in the range 1 - 4. The answers will be reprocessed to show what needs to be done to improve the psychological condition of the patient positively.

Support menu is no less important is Vision. This menu serves to assess the condition of the patient's eye now. Includes a myopia (minus eye) Figure 2b is performed by checking the introduction of a number of shapes from a certain distance, and colour blind Figure 2c is performed by digitally recognizing the identification of numbers in a multi-coloured form. This application is also supported by blood pressure measurements that can be done on the Blood Pressure menu, the user only need to press the palm of the thumb at the point specified on the screen, the blood pressure measurement will automatically be performed and display the results as shown in Figure 2d. Based on the results shown Figure 2e will be given appropriate recommendations.

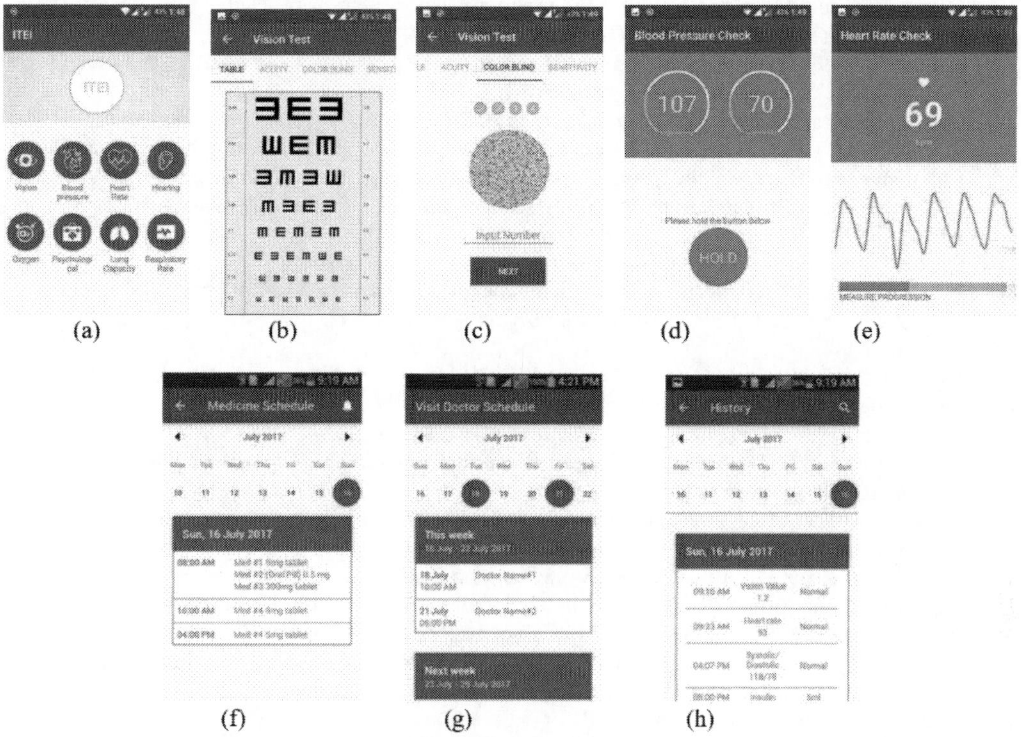


**Figure 2.** Support Menu Interface

TECH Android app also has a function to measure heart rate called Heart Rate. This measurement mechanism is the same as Blood Pressure, but the results shown are the number of heart rate and current health condition if you see the heart rate. Will also displayed recommendations in accordance with the results shown.

To support the health improvement of Android TECH applications also support the scheduling of drug consumption - Medicine Schedule and consultation time to doctor- Visit Doctor Schedule as depicted in Figure 2f-h. As well as one feature that is not less important is the History, this menu will record all the activities that have been done with applications ranging from assessment results have been obtained until the schedule of doctor's visit ever scheduled. Android TECH is also planned to manage features to provide dietary recommendations and exercises / activities that can be taken to improve the physical condition and psychology of patients.

## 4. Discussion

TECH android application is very important because the realization of teachers who engage with the profession is strongly influenced by physical and mental health. After TECH compared with existing applications, it turns out that existing applications have not been equipped with psychic test kea rah mental health measurement. Based on the comparison of existing applications, TECH proved to be more comprehensive and can be implemented to Indonesian teachers in self-assessment, developing itself based on expert system.

**5. Acknowledgment**
This work is partially supported by Ministry of Research, Technology and Higher Education of the Republic of Indonesia and The Ministry of Education and Culture as a partner of research. The authors also gratefully acknowledge the helpful comments and suggestions of the reviewers, which have improved the presentation.

**References**

[1] Schee C V and Gard M 2014 Healthy, happy and ready to teach, or why kids can't learn from fat teachers: the discursive politics of school reform and teacher health *Crit. Public Health* vol. 24 **2** pp 210–225
[2] Grayson J L and Alvarez H K 2008 School climate factors relating to teacher burnout: A mediator model *Teach. Teach. Educ* vol. 24 **5** pp 1349–1363
[3] Kahn W A 1990 Psychological Conditions of Personal Engagement and Disengagement at Work *Acad. Manag. J. VO - 33* **4** p 692
[4] Skinner E, Furrer C, Marchand G and Kindermann T 2008 Engagement and disaffection in the classroom: Part of a larger motivational dynamic? *J. Educ. Psychol* vol. 100 **4** pp 765–781
[5] Christian M S, Garza A S and Slaughter J 2011Work Engagement: a Meta-Analytic Review and Directions for Research in an Emerging Area *Acad. Manag.* no. September 2016
[6] KEMP S 2017 *Digital in 2017: Global Overview - We Are Social*
[7] Diener *Diener-Subjective_Well-Being.pdf*
[8] Prince M *et al 2007* No health without mental health *Lancet* vol 370 **9590** pp 859–877
[9] Sasmoko Y and Ying 2015 Construct Validity in NeuroResearch *Adv. Sci. Lett* vol 21 **7** pp 2438–2441
[10] Sasmoko 2012 *Metode Penelitian Eksplanatori-Konfirmatori, Edisi ke 11* (Jakarta)
[11] Fios F, Sasmoko and Gea A A 2016 Neuro-Research Method: A Synthesis Between Hermeneutics and Positivism *Adv. Sci. Lett* vol. 22 **9** pp 2202–2206
[12] Creswell J W 2012 *Educational research: Planning, conducting, and evaluating quantitative and qualitative research* vol. 4
[13] Seligman M E P, Steen T A, Park N and Peterson C 2005 Positive psychology progress: Empirical validation of interventions *Am. Psychol* vol 60 **5** pp 410–21
[14] Tov W and Diener E 1991 *Subjective well-being* pp 0–8
[15] Srivastava S and Kanpur R 2014 A Study On Quality Of Work Life: Key Elements & It's Implications *IOSR J. Bus. Manag. Ver. I* vol 16 **3** pp 2319–7668
[16] Schaufeli W B and Bakker A B 2004 Job demands, job resources, and their relationship with burnout and engagement: A multi-sample study *J. Organ. Behav.* vol 25 **3** pp 293–315
[17] Diener E and Chan M Y 2011 Happy People Live Longer: Subjective Well-Being Contributes to Health and Longevity *Appl. Psychol. Heal. Well-Being* vol 3 **1** pp 1–43
[18] ProHealthSys *Vital Signs Table*
[19] Schaelicke L *et al.* 2011 A 'cool' load balancer for parallel applications *Sc11* pp 1–11
[20] Handigol N and Seetharaman S 2009 Plug-n-Serve: Load-balancing web traffic using OpenFlow *Acm Sigcomm*

IOP Conf. Series: Journal of Physics: Conf. Series **978** (2018) 012030 doi:10.1088/1742-6596/978/1/012030

# An analysis of absorbing image on the Indonesian text by using color matching

**G A Hutagalung[1], Tulus[2], Iryanto[2], Y F A Lubis[3], M Khairani[3] and Suriati[3]**

[1]Department of Computer Science, Faculty of Computer Science and Information Technology, Universitas Sumatera Utara, Medan-20155, Indonesia
[2]Department of Mathematics, Universitas Sumatera Utara, Medan-20155, Indonesia
[3]Department of Informatics, Universitas Harapan Medan, Medan-20217, Indonesia

Email : tulus@usu.ac.id

**Abstract.** The insertion of messages in an image is performed by inserting per character message in some pixels. One way of inserting a message into an image is by inserting the ASCII decimal value of a character to the decimal value of the primary color of the image. Messages that use characters in letters, numbers or symbols, where the use of letters of each word is different in number and frequency of use, as well as the use of letters in various messages within each language. In Indonesian language, the use of the letter *A* to be the most widely used, and the use of other letters greatly affect the clarity of a message or text presented in the language. This study aims to determine the capacity to absorb the message in Indonesian language from an image and what are the things that affect the difference. The data used in this study consists of several images in JPG or JPEG format can be obtained from the image drawing software or hardware of the image makers at different image sizes. The results of testing on four samples of a color image have been obtained by using an image size of 1200 X 1920.

## 1. Introduction

Messages are information that can have different levels of privacy. Not all messages are publicly available. Some messages have some levels of privacy that must be maintained from unauthorized persons to know the content of messages intended for the recipient. One of message concealment method is steganography. This method hides the messages in a media either in the form of images, sounds or other media that can hide the message to the wrong recipients [1]. Digital image is one of the most commonly media used in steganography. This medium can be an option in message insertion, where messages are inserted into some pixels of digital images either in one bit or directly in the pixel decimal value. [2]

Digital images have the absorption capacity of messages to be inserted based on several things. Color values of the digital image can be used to be inserted in the images. The differences of color values will affect the number of words or letters of a sentence to be inserted into the image. Messages have different frequencies of letter use in each language used for the messages. For example, the English language has the letter *E* that is the highest value in use. On the other hand, the letter *A* is the most widely used letter reached 19.81% in every word that is used in Indonesian. [3] This study aimed to analyze how the absorption of the digital image to the message in Indonesian language. The analysis is performed to the number of words that can be inserted in the digital image and something that influence differences in absorption image of the messages.

## 2. Previous Work

Some previous researches on image insertion are performed by using different image types, i.e. color image and grayscale image, to produce a better steganography. In these works there are different ways of insertion that are done because of different color values of the selected image type as a sample [3]. The research on the frequency of use in Indonesian language, letters is carried out by taking samples of formal texts from official speeches, newspapers and through stories written in both fiction and nonfiction. In this research, the study yields an average of 6.6 letters in each word in Indonesian texts [4]. The research on steganography used in Arabic has been done using Novel Approach which used a data input to produce Arabic text using algorithm approach which is done after Encrypt Compressed Secret Message (ECSM) [5].

Research on calculating the frequency of using letters of the text in Malay language has also been done based on the average use of letters in some media in both print and online media, where the letter A in Malay language reaches 19% of its use in every word. Malay and Indonesian language have some similarities in the writing of words, but the difference that often occurs in the meaning and pronunciation of the word used [6]. The authors argue that for the use of text in Malay and Indonesian language have similarities. The previous research in the calculation of the frequency of letters in Malay can be used as a reference in the technique of calculating the frequency of the use of letters in Indonesian language. The N-gram method can be used to detect whether the language used in the text is Indonesian, that does not use English, and is better with the speed of search is also quickly completed with [7].

## 3. The Proposed Method

To obtain the results of the research objectives, the authors conducted research implementation measures or the research design, in outline, the authors carried out research which consists of the following steps :

1. Analysis and sampling, where the sample or object of this research are some digital images having different size and color values consisting of different image types (monochrome, grayscale, color).
2. Introduction of values on each sample image tested, because each image has a different byte value that is tailored to the constituent color of the image.
3. Testing by adjusting the color values in digital images with the value of the occurrence of letters in the Indonesian text.
4. The number of letters that appear adjusted to the average use in the word to find the number of words that can be accommodated in the image.
5. Results and conclusions from testing.

The data used in this study consists of several images in JPG or JPEG format can be obtained from the image drawing software or hardware of the image makers at different image sizes. For a sample image of the author using 9 pieces following image:

**Table 1.** Sample image

| NUMBER | TYPE | SIZE |
|---|---|---|
| 1 | Monochrome | 256 x 256 |
| 2 | Monochrome | 450 x 250 |
| 3 | Monochrome | 480 x 640 |
| 4 | Grayscale | 512 x 512 |
| 5 | Grayscale | 360 x 500 |
| 6 | Grayscale | 1080 x 1920 |
| 7 | Color | 512 x 512 |
| 8 | Color | 2835 x 4472 |
| 9 | Color | 459 x 736 |
| 10 | Color | 1200 x 1920 |

After the image is selected, the next process is the reading of values from pixels one by one. The values of a pixel which consists of red, green and blue, are separated into three values that are created in each color frame values. To make more data, the function of dividing frames is not just for one value for one pixel.

There are three values that can be made for the message insertion media. Frame division is based on three basic colors in each pixel that is Red, Green, and Blue. Next, the decimal values of color are changed in to the character based on the ASCII code. After getting the values, then the next process is the calculation of the number of characters that can be input into the image that is distinguished from the colors obtained. The color values of the character are adjusted based on color matching with the value of the character that has been converted into a decimal. The number of characters are calculated to fit an image to a message in Indonesian text. The steps to adjust the values of the image with the distribution of Indonesian letter are as follows:

1. Take the greatest number of characters in the image. The distribution of *A* in Indonesian language reached 19.81%.
2. Once the number of characters *A* is held, the next process is to count how many number of characters that can be inserted by dividing the number of characters *A* to the percentage of the letter *A*.
3. Adjust the other characters with the percentage of each individual character. It should be in accordance with the distribution table of the letter. There should be no character that in number is less than the number of minimum letters based on the percentage in the distribution table. If one of the characters is not suitable, then the length of the character is adapted to the number of letters of the characters that are not suitable.
4. If the number of letter *b* is only 5 and that is not enough to meet the number of characters that previously obtained, then the length of characters that can be inputted that is adjusted with the letter *b* namely 5 / 2.64 * 100 = 189 characters. Further adjustments to the other letters in the same way.
5. Once the total of the letter percentage number is corresponding to a character length, further adjustments to the number of words which, according to research conducted in [4], the average use of the letters in each word is 6.6 letters.

Some calculation is performed to find out the number of characters that can be used as the basis for calculating the amount of code numbers of each letter that must be in the image as follows:

$$Amount\ of\ Characters = Numbers\ of\ Character\ A\ /\ 19.81\ *\ 10)$$

If there is a smaller number of letters the percentage multiplied by the number of characters, the number of characters must be adapted to the number of letters in the letter.

$$Amount\ of\ Characters = Amount\ of\ Letters\ /\ Percentage\ of\ Letters *\ 100$$

The calculation of the number of words is:

$$Amount\ of\ Letters = Amount\ of\ Characters\ /\ 6.6$$

## 4. Result

Here are the results of testing on four samples of a color image using an image size of 1200 X 1920. From the data in table 1, the results of testing from samples of 4 color images shows that these images have the letter *A* with the total number of 51,989. From this data, the percentage of letters that must be met by the next letter is as described in table 2.

**Table 2.** Results of Sample 4 Image Color Adjustment

| Letter | Amount | % Letter Dis | Number of Character | % Number Of Letter |
|---|---|---|---|---|
| A/a | 51989 | 19,81% | 262437 | 100,00% |
| B/b | 50889 | 2,97% | 262437 | 97,88% |
| C/c | 51345 | 0,67% | 262437 | 98,76% |
| D/d | 51303 | 3,72% | 262437 | 98,68% |
| E/e | 54282 | 8,18% | 262437 | 104,41% |
| F/f | 53018 | 0,25% | 262437 | 101,98% |
| G/g | 51544 | 3,96% | 262437 | 99,14% |
| H/h | 52545 | 1,97% | 262437 | 101,07% |
| I/i | 52000 | 8,08% | 262437 | 100,02% |
| J/j | 53685 | 0,85% | 262437 | 103,26% |
| K/k | 52892 | 5,35% | 262437 | 101,74% |
| L/l | 52439 | 3,50% | 262437 | 100,87% |
| M/m | 50332 | 4,85% | 262437 | 96,81% |
| N/n | 49996 | 9,56% | 262437 | 96,17% |

| | | | | |
|---|---|---|---|---|
| O/o | 47763 | 1,60% | 262437 | 91,87% |
| P/p | 44655 | 2,77% | 262437 | 85,89% |
| Q/q | 44165 | 0,01% | 262437 | 84,95% |
| R/r | 47107 | 5% | 262437 | 90,61% |
| S/s | 45530 | 4,11% | 262437 | 87,58% |
| T/t | 48313 | 5,12% | 262437 | 92,93% |
| U/u | 47082 | 5,57% | 262437 | 90,56% |
| V/v | 48670 | 0,08% | 262437 | 93,62% |
| W/w | 48470 | 0,41% | 262437 | 93,23% |
| X/x | 46814 | 0,01% | 262437 | 90,05% |
| Y/y | 47754 | 1,60% | 262437 | 91,85% |
| Z/z | 46887 | 0,02% | 262437 | 90,19% |

In the letter *A* with a percentage of 19.81% of the total number of letters 51.989 it is obtained the number of characters as much as 42.336 characters. By adjusting to the next letter, the number of through the letters in the sample 2 of color images is greater than the number of characters from the letters with the greatest percentage. So the number of characters that can be accommodated by the image is 262.437 characters and also 39.762 words. The results of the implementation for all images are obtained as described in table 3.

**Table 3**. Entire Sample Test Result

| Number | Type | Size | Character | Letter |
|---|---|---|---|---|
| 1 | Monochrome | 256 x 256 | 0 | 0 |
| 2 | Monochrome | 450 x 250 | 0 | 0 |
| 3 | Monochrome | 480 x 640 | 0 | 0 |
| 4 | Grayscale | 512 x 512 | 0 | 0 |
| 5 | Grayscale | 360 x 500 | 24.426 | 3700 |
| 6 | Grayscale | 1080 x 1920 | 314.386 | 46.278 |
| 7 | Color | 512 x 512 | 0 | 0 |
| 8 | Color | 2835 x 4472 | 1.357.485 | 205.679 |
| 9 | Color | 459 x 736 | 42.336 | 6.414 |
| 10 | Color | 1200 x 1920 | 262.437 | 39.762 |

The analysis of the research results with respect to the absorptive performance of the images to the messages using Indonesian languages is an important thing to close the discussion. The accommodation and fit rating of the letters to the image use the decimal value of the primary color (Red, Green, Blue) of each image pixel and the decimal value of the letter (ASCII) [8]. Data distribution of letters and the use of letters in the word and the text using Indonesian language for the adjustment are data from previously conducted research with 2.042.616 characters and 306.359 words (6.6 letters per word). The monochrome images cannot accommodate the character because the images only have values of 255 and 0. The digital images with greater resolution and use of many colors will accommodate more words in Indonesian language.

## 5. Conclusion

Based on the discussion and testing on the research, it can be concluded as follows:

1. The number of letters obtained from matching the color decimal value with a decimal value of the ASCII letters must meet the percentage of the distribution table of letters for all types of letters. If there is one letter that the number is less, then the maximum number of letters to the Indonesian language text to be readjusted based on the percentage of letters less.
2. The binary image cannot accommodate an Indonesian-language text with matching colors.

3. The results of the study are the result of matching the color values of samples with the distribution of letters in the Indonesian language.

**6. References**

[1] Khalil C and Hilmat F 2011 Combining Steganography and Cryptography *International Journal on New Computer Architectures and Their Application* **1**(1) pp 199-208
[2] Atawneh S, Almomani A and Sumari P 2014 Steganography in digital images: Common approaches and tools *IETE Technical Review* **30**(4) pp 344-358
[3] Muhammad K, Ahmad J, Farman , Jan Z, Sajjad M and Baik S W 2015 A secure method for color image steganography using gray-level modification and multi-level encryption *KSII Transactions on Internet and Information* 9(5) pp 1938-1962
[4] Andana G 2010 Analisis Frekuensi Pada Teks Berbahasa Indonesia Dan Modifikasi Algoritma kriptografi Klasik. *Makalah IF3058 Kriptografi Tahun 2010* pp 9-18
[5] Suhad M And Shareef 2017 A Novel Approach for Arabic Text Steganography Based on the “BloodGroup” Text Hiding Method Engineering *Technology & Applied Science Research* **7**(2) pp 1482-1485
[6] Promadi 2012 Perbedaan Semantik Antara Bahasa Indonesia Dan Bahasa Malaysia: Satu Kajian Awal Upaya Mengelak Kesalahpahaman Dan Perbedaan Budaya Antara Bangsa Serumpun Di Asia Tenggara *Jurnal Sosial Budaya* **9**(2) pp 261-282
[7] Hamzah A 2010 Deteksi Bahasa untuk dokumen teks berbahasa indonesia, *Seminar Nasional Informatika 2010 UPN "Veteran" Yogyakarta* p. A5-A13
[8] Zhang Q and Wei X 2013 RGB Color Image Encryption Method Based on Lorenz Chaotic System and DNA Computation *IETE Technical Review* **30**(5) pp 404-409

IOP Conf. Series: Journal of Physics: Conf. Series **978** (2018) 012031 doi:10.1088/1742-6596/978/1/012031

# Microcontroller based automatic temperature control for oyster mushroom plants

**P Sihombing[1], T P Astuti[1], Herriyance[1] and D Sitompul[1]**
[1]Faculty of Computer Science and Information Technology, Universitas Sumatera Utara, Jl. Dr. Mansyur No.9, Padang Bulan, Medan, Sumatera Utara 20155

Email: poltakhombing@yahoo.com[1]

**Abstract**. In the cultivation of Oyster Mushrooms need special treatment because oyster mushrooms are susceptible to disease. Mushroom growth will be inhibited if the temperature and humidity are not well controlled because temperature and inertia can affect mold growth. Oyster mushroom growth usually will be optimal at temperatures around 22-28°C and humidity around 70-90%. This problem is often encountered in the cultivation of oyster mushrooms. Therefore it is very important to control the temperature and humidity of the room of oyster mushroom cultivation. In this paper, we developed an automatic temperature monitoring tool in the cultivation of oyster mushroom-based Arduino Uno microcontroller. We have designed a tool that will control the temperature and humidity automatically by Android Smartphone. If the temperature increased more than 28°C in the room of mushroom plants, then this tool will turn on the pump automatically to run water in order to lower the room temperature. And if the room temperature of mushroom plants below of 22°C, then the light will be turned on in order to heat the room. Thus the temperature in the room oyster mushrooms will remain stable so that the growth of oyster mushrooms can grow with good quality.

## 1. Introduction

In the beginning, most mushrooms grow outdoors in some parts of the world. So that mushroom farmer only depends on the condition of the natural environment, as if wild mushrooms grow in a natural atmosphere. But now, the farmers in certain areas have been able to regulate the growth of mushrooms for commercial purposes as well as for personal use. The farmers today are no longer just dependent on the growth of mushrooms naturally but they are able to produce mushrooms by making their own plantings. The room for mold growth can be controlled by utilizing technology, which can control temperature and inertia automatically. They have been able to plant and harvest mushrooms only by utilizing the land as in the warehouse or garage around the house [1].

In mushroom plants, the most important consideration is to watch for the prevention of possible pests and pathogens in addition to understanding the relationship between temperature, humidity, and air exchange. In this paper, we focus on discussing of room conditions i.e. room temperature, humidity, and air movement circulation. In the room is closed, when the room temperature rises, then the humidity of the air is relatively down. When the outside air flows into the room, the temperature and humidity of the room is relatively changed according to the condition of the outside air. Furthermore, to regulate the temperature of the room to remain stable at temperatures suitable for mushroom cultivation, it is necessary for the air circulation to be controlled automatically. Automatic room control will greatly assist mushroom farmers to create optimum room conditions to keep them in accordance with the needs of mushroom plants.

Related to this project, we will develop Arduino Uno Microcontroller based automatic tool and Android smartphone to support mushroom crops in the small area. We use sensors to detect the air

IOP Conf. Series: Journal of Physics: Conf. Series **978** (2018) 012031 doi:10.1088/1742-6596/978/1/012031

temperature and humidity. Then the detection results of those sensors will be sent to the Microcontroller Arduino Uno which serves as the brain of this tool to monitor the temperature and humidity of mushroom plants. The heat and the humidity of the mushroom plant room will also be displayed on the monitor of the Arduino Uno microcontroller. All of the data will be sent to the Android smartphone using WIFI (Wireless Fidelity) network, and Android smartphone can also to control the Arduino Uno microcontroller.

## 2. Related Work

There are a number of proposed systems in the past few years. One of them is a temperature and humidity monitoring system using AT89S52 as the microcontroller, SHT10 as the temperature and humidity sensor and TC35i GSM (Global System for Mobile communication) Module for wireless communication. When the temperature and humidity exceeds the limit, the system will send short messages to the mobile phones to alert the user [1]. Another system is a greenhouse monitoring and control system based on GSM. This system is more complicated if we compared to the previous system, because as it does not only allow a data acquisition, but it also allows the user to control humidity, temperature and soil moisture via SMS (Short Message Service) [2].

Mushroom cultivation is significantly grown in a controlled environment Based on mushroom growth follows a very complex and sensitive procedure. Automatic environment control on mushroom cultivation has been introduced by using Trend IQ3 controllers. By using the IQ3 Trend controller, the system automatically controls and monitors the environment at every stage of the growth cycle to maximize the yield of oyster mushrooms with good quality [3]. The control system interfaces are continually improved to monitor the condition of the rooms and the operation of plants and to adjust the needs of oyster mushroom plants [4].

The other research related to this project is using a controls that can be used to monitor plants in the real- time have been done by Dan Wang, et al in 2015 [5]. This study conducted the data monitoring on aquaponic plants that can be sent to smartphones via an internet.

The next related study is about the addition of the water automatically made by Rajeev L.M and Preet Jain in 2015 [6] which use an electrode as a benchmark of the amount of water in the plants. They developed if the water is below of the amount of a plant needs, then the pump will do watering to add the water.

There is another related research that develops about automatic plant watering has been done by Devika et al [7]. They developed the watering of the plants grown on soil where moisture sensors will detect soil moisture levels. If the soil is not moist / dry, then the system will automatically do watering on the plant.

## 3. System Description

To develop automatically control of the temperature of an oyster mushroom system, we use an Arduino Uno microcontroller. The sensors are placed around the root of oyster mushroom and will transmit the detection result every time to the Arduino microcontroller. The Arduino Uno microcontroller is the main part of automatic temperature control for this oyster mushroom plants. At the previously, we have embedded a module program on this microprocessor to support this control system. The Arduino Uno microcontroller is also serves to set the real-time system to set the Alarm to the water pump. When the alarm is activated, the Relay will be activated and the water pump will send the water solution into the oyster mushroom area. And if the relay is off (off), then the water pump will stop to supply water.

To support this project, we first have designed a Virtuino application on Android smartphone that serves to see the water level and temperature around the oyster mushroom plant. This application will control the Arduino microcontroller. Before running the Virtuino application design, we first provide the data storage by using the thingspeak.com module. This thingpeak.com module serves the store of data sent by the sensor to the microcontroller and other necessary data.

In Figure 1, we can see that a simple flow of the system starts from the detection of proximity sensor and temperature sensor. Other relays are connected to the microcontroller port. When the port pin of relay is high, means the pump will run the water and when the port pin of relay is low, and then the water pump will stop running water. Thus the water circulation will be flow regularly. This pump will be controlled by a relay to turn on and off the water flow. We also use the LCD screen to display the current time and date of the system.

*3.1 General Architecture of Our System*
For more details, we will describe the general Architecture of the system that we have developed. The architecture of the temperature control system and the moisture flow of the mushroom plant are shown in figure 1 below.

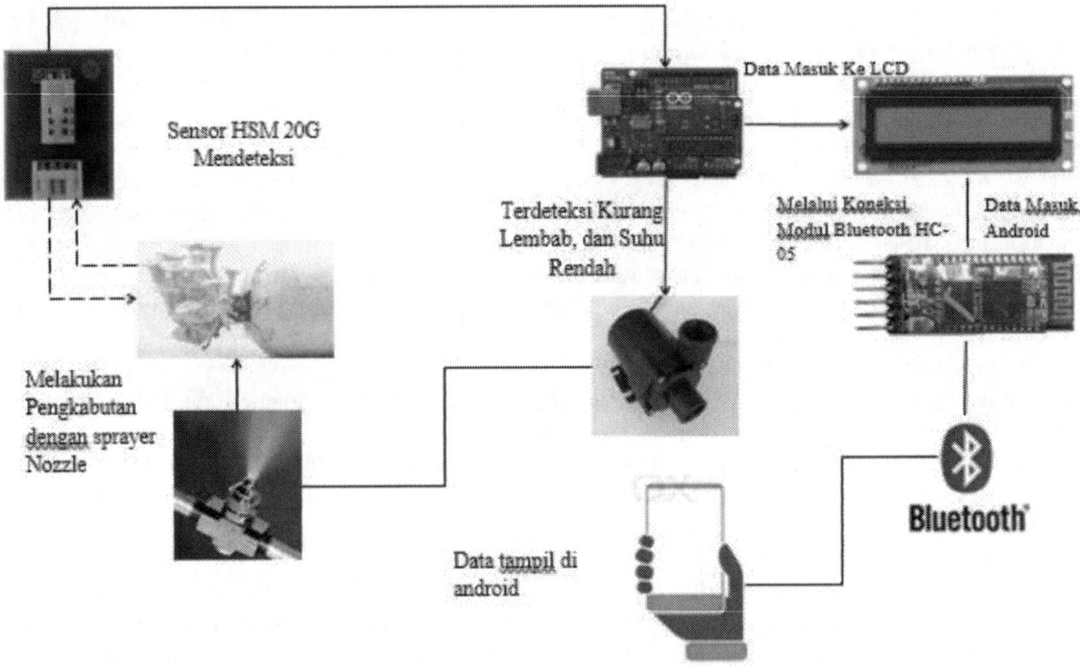


**Figure 1.** The General Architecture of Automated Mushroom Nutrition Plants Systems

The working process of the temperature control system and humidity of oyster mushroom plant are as follows; the temperature sensor (HSM - 250G) will detect the overheating value of the oyster mushroom area in units of the ° C range. If the temperature increased more than 28°C in the room of mushroom plants, then this tool will turn on the water pump automatically to run the water in order to lower the room temperature. And if the room temperature of mushroom plants below of 22°C, then the light will be turned on in order to heat the temperature room area.

*3.2 The Flowchart of Automatic Temperature Control for Oyster Mushroom plants Systems*
The system flow begins with the detection of the temperature sensor. In the temperature sensor, if heat and humidity are detected, the fan and pump will ignite and do the condensation. And if normal temperature and normal humidity (22-28°C) then the fan does not turn on, as well as cold temperatures and wet humidity then the fan also will not on. If any of the conditions are met then the results will show on the LCD and will connect to Android Smartphone with the help of Bluetooth. The system flowchart is presented in figure 2.

IOP Conf. Series: Journal of Physics: Conf. Series **978** (2018) 012031 doi:10.1088/1742-6596/978/1/012031

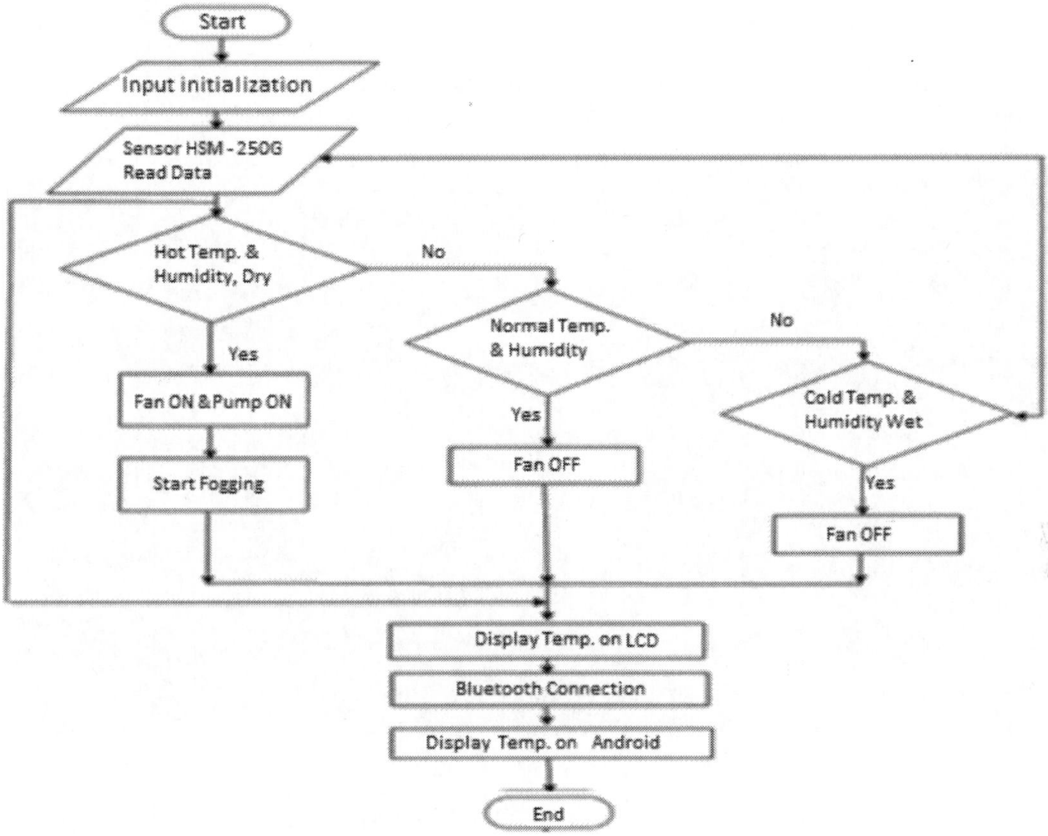


**Figure 2.** The Flowchart of Automatic Temperature Control.

*3.3 The Coding Program*

To support this project, we have embedded the module program in Arduino Uno microcontroller. A part of our coding program is presented in Figure 3 below.

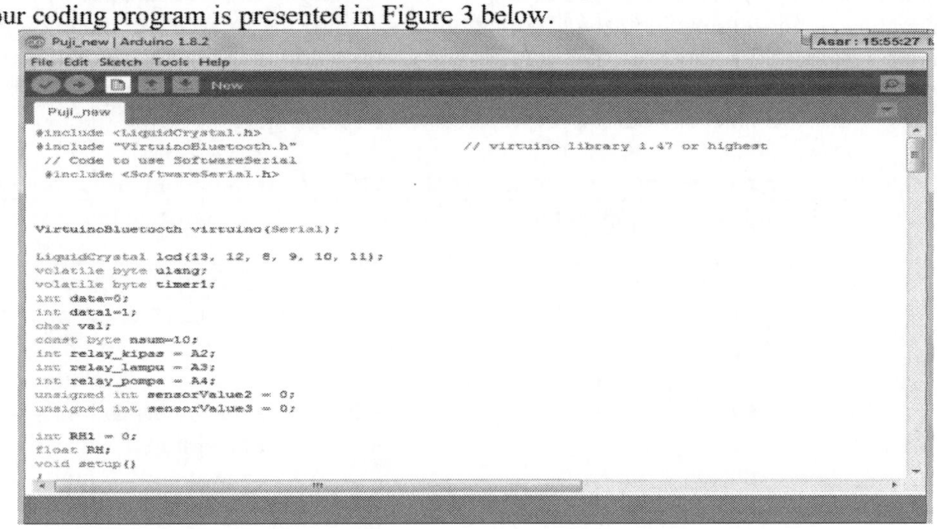


**Figure 3.** The program module

IOP Conf. Series: Journal of Physics: Conf. Series **978** (2018) 012031 doi:10.1088/1742-6596/978/1/012031

## 4. The Experimental setup and Result

The design and implementation of the Automatic Temperature Control for Oyster Mushroom plants Systems automatically using Arduino Uno microcontroller performed is shown in Figure 4 below.

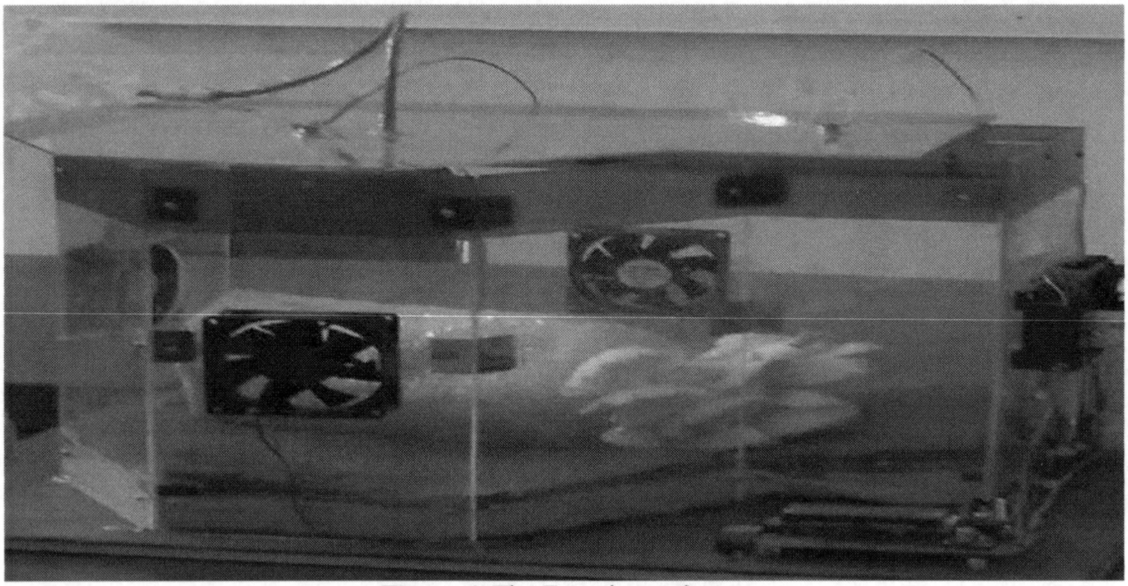

**Figure 4.** The Experimental setup

In table 1 below shows, the Wi-Fi module will send the humidity of water level and temperature values of to the mushroom area.

### *4.1. Automatic Temperature Control System*

In this experiment, automated and manual system testing has been performed. Testing is automatically done with control via smartphone. We have developed an application on a smartphone to control the temperature automatically. In applications that have been designed on this smartphone can be done by clicking the option of the automatic button. Through this system, if the room temperature has reached 30 °C, then the system will automatically order the pump relay to do the spraying to reduce the room temperature. Thus the room temperature in oyster mushrooms will be kept stable at the range 25-29 °C. The results of the tests on each hour are presented in table 1 and figure 5.

**Table 1.** The humidity, temperature and spraying data resulted by Arduino Uno Microcontroller.

| Time | Humidity (%) | Temperature (°C) | Spraying | Information |
|---|---|---|---|---|
| 8.00 | 72 | 27 | No | Normal Condition |
| 9.00 | 71 | 28 | No | Normal Condition |
| 10.00 | 70 | 29 | No | Normal Condition |
| 11.00 | 67 | 30 | Yes | Temperature conditions exceed normal limits |
| 12.00 | 65 | 31 | Yes | Temperature conditions exceed normal limits |
| 13.00 | 64 | 32 | Yes | Temperature conditions exceed normal limits |

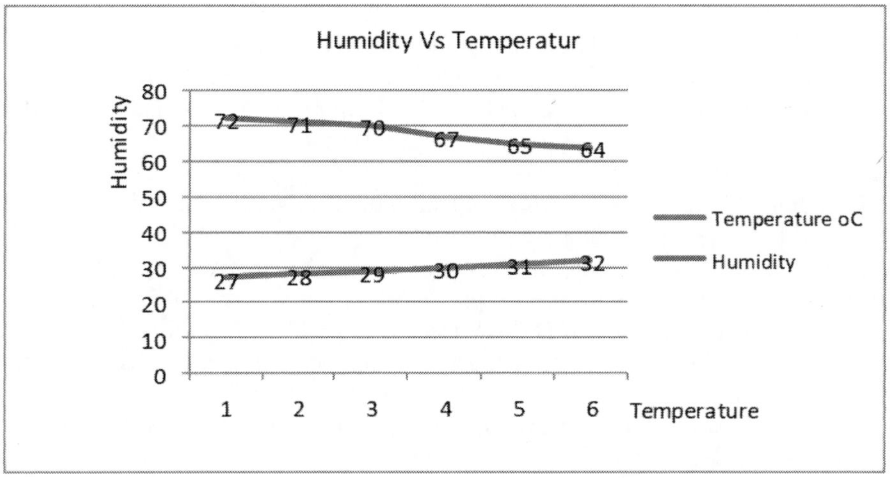


**Figure 5.** The Automatic Temperature Control System

*4.2. Manual Temperature Control System*

Manual testing has also been done by giving commands through applications that have been designed in the Android Smartphone. To run the system manually is done by clicking a selection of manual buttons on the smartphone. In this way, the system will be able to run the water pump by the user if the user wants to drain water or to heat the room temperature. By this manual system, the user can also adjust the system performance by directly activating or disabling the options button for a lamp, pump, or fan anytime. Users can also turn off the feature if needed to set the temperature and humidity in the room oyster mushrooms.

## 5. Conclusion

In this paper, we have described how to control the temperature and humidity of air in oyster mushrooms, whether done automatically by using a microcontroller run through a smartphone or to flow the water manually. All components of automatic temperature control device in the cultivation of oyster mushrooms can run as planned. An Arduino microcontroller tool can work well that is controlled via handphone. At temperatures greater than 29 $^0$C the pumps appear to live automatically to perform the fogging (condensing) temperatures in the room of the fungus. Data communications made to transmit temperatures or moisture to Android smartphones can be done via Bluetooth or with available Wi-Fi or using a Thing-Speak online cloud for monitoring and storage. The data could be accessed by the user anytime by using a computer that is connected to the Internet. The control system was able to control ac-powered humidifier, light, and fan based on the feedback of the sensors to maintain temperature, humidity, and light intensity at optimum growth condition in an actual mushroom farm. The experimental results show that at 67% humidity and the temperature is 30 $^0$C, the system starts to turn on the cooling fan and turn on the water pump to lower the room temperature and anticipate staying within the range of 25-29 $^0$C.

**References**

[1] Wang X 2014 Temperature and Humidity Monitoring System Based on GSM Module *International Journal of Computer, Consumer, and Control* Vol. 3 **1** pp 41–49

[2] Rahali A, Guerbaoui M, Ed-dahhak A, El A Y, Tannouche A, Lachhab A and Bouchikhi B 2011 Development of a data acquisition and greenhouse control system based on GSM *Int. J.Eng. Sci. Technol* Vol. 3 **8** pp 297–306

[3] Chang S, Chang S T and Miles P G 2004 Mushrooms, Cultivation, Nutritional Value, Medicinal Effect, and Environmental Impact *CRC Press*

[4] Baysal E 2003 Cultivation of oyster mushroom on waste paper with some added supplementary materials *Bioresour. Technol* vol. 89 **1** pp 95–97

[5] Wang D, Zhao J, Huang L and Xu D 2015 Design of A Smart Monitoring and Control System for Aquaponics Based on OpenWrt *Proceedings of Fifth International Conference on Information Engineering for Mechanics and Materials (ICIMM)*

[6] Mishra R L and Jain P 2015 Design and Implementation of Automatic Mushroom System using ARM Processor *International Journal of Advanced Research in Electrical, Electronics and Instrumentation Engineering* Vol. 4 I **9**

[7] Devika S V, Khamuruddeen S, Khamurunnisa S, Thota J and Shalk K 2014 Arduino Based Automatic Plant Watering System *International Journal of Advanced Research in Computer Science and Software Engineering* Vol. 4 I **10**

[8] Lokesh K K, Madhuri J & Anuradha K 2016 A ZigBee based Energy Efficient Environmental Monitoring Alerting and Controlling System *International Conference On Information Communication And Embedded Systems (ICICES2016)*

[9] Kalinin Y S, Velikov E K and Markova V I 2015 Design of Indoor Environment Monitoring System Using Arduino *Int. J. Innov. Sci. Mod. Eng.* Vol. 3 **7** pp 46–49, 20

[10] Grant J J 2002 An investigation of the airflow in mushroom growing structures, the development of an improved, three-dimensional solution technique for fluid flow and its evaluation for the modeling of mushroom growing structures. *Ph.D. thesis* (Dublin City University)

[11] Stamets P and Chilton J S 1983 The Mushroom Cultivator: A Practical Guide to Growing Mushrooms at Home *S. Cal. L. Rev.* p 416

[12] Van N, Bram and Oei P 2005 Small-scale mushroom cultivation oyster, shiitake and wood ear mushrooms (Agrodok;40. Agromisa/CTA, Wageningen, The Netherlands)

[13] Australian M G A Introduction to Mushroom Growing *AMGA* pp 1- 16

[14] Rosmiza M, Davies W, Aznie R C, Jabil M and Mazdi M 2016 Prospects for Increasing Commercial Mushroom Production in Malaysia: Challenges and Opportunities *Mediterr. J. Soc.Sci.* Vol. 7 **1** pp 406–415

[15] Aniket H, Hade and Sengupta M K 2014 Automatic Control of Drip Irrigation System & Monitoring Of Soil By Wireless *(IOSR-JAVS) E-ISSN: 2319-2380, P-ISSN: 2319-2372.*Vol 7 I **4** Ver. Iii

IOP Conf. Series: Journal of Physics: Conf. Series **978** (2018) 012032 doi:10.1088/1742-6596/978/1/012032

# Feature weighting using particle swarm optimization for learning vector quantization classifier

**A Dongoran*[1], S Rahmadani[1], M Zarlis[1] and Zakarias[1]**

[1]Magister Teknik Informatika, Fasilkom - TI , Universitas Sumatera Utara (USU), Medan - Indonesia

*Awaluddin.dongoran@gmail.com

**Abstract**. This paper discusses and proposes a method of feature weighting in classification assignments on competitive learning artificial neural network LVQ. The weighting feature method is the search for the weight of an attribute using the PSO so as to give effect to the resulting output. This method is then applied to the LVQ-Classifier and tested on the 3 datasets obtained from the UCI Machine Learning repository. Then an accuracy analysis will be generated by two approaches. The first approach using LVQ1, referred to as LVQ-Classifier and the second approach referred to as PSOFW-LVQ, is a proposed model. The result shows that the PSO algorithm is capable of finding attribute weights that increase LVQ-classifier accuracy.

## 1. Introduction

Classification is a process of learning by mapping a data into one or more predefined classes based on information discribed by its own features [6]. Learning that occurs in the classification process is also called an inductive-learning algorithm [1]. The purpose of the classification is to create a model that will be used to predict the value of the input data that is not yet known class. Simply split the classification process usually divide the sample data into two parts [2]. First referred to as training data used to train the data to get the expected classification model. The two are referred to as test data. This test data will be tested to determine the level of accuracy in predicting the value of new input data, or can also known error rate predictions.

A statistical approach can also be used to generate a classification model. In addition to statistical approaches, there are many more approaches used to generate classification models, such as decision tree, artificial neural network, genetic algorithm, evolutionary computing, swarm optimization, etc.

In this paper we will use competitive learning method based on competition on LVQ competitive network, and using particle swarm optimization (PSO) method to weighting each attribute of sample data. Both of these approaches will be compared with each other to find out the performance of both approaches in predicting the value of new mapped input data. The weighting method feature improves performance in this case increases accuracy or decreases error rate [5].

The weighting feature is part of the selection feature. The selection feature is a fairly complex job that can affect the performance of classification. Therefore approaches to feature selection continue to be developed. The most popular current approach is based on Evolutionary Computation (EC), Particle Swarm Optimization (PSO) [7] [8] based Swarm Intelligence. PSO is a simple and fast computing technique that produces convergent values. So PSO is an effective technique used for feature selection tasks [10].

## 2. Learning Vector Quantization (LVQ)

Learning vector quantization (LVQ) is a method that can be used to generate a classification model of trained sample data. LVQ is part of a supervised neural network method [9]. LVQ architecture is similar to Kohonen's SOM (Self Organizing Map) developed by Prof. Teuvo Kohonen in 1982.

Simply put, competitive layer on artificial neural networks have a single output layer, and each neuron is fully connected to the input node [1].

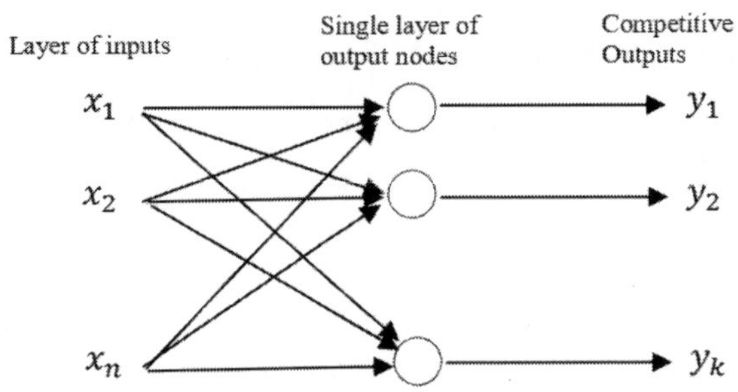


**Figure 1.** Architecture Simple LVQ

### *2.1. LVQ Algorithm*

The value of each attribute (information) training data is used to form a weight vector (prototype). The weight vector then iteratively becomes a model that will be used to classify the new sample data whose class is not yet known. The first stage of the LVQ algorithm is to determine the initial weights in the classification model search process. The initial weights used are derived from the training data samples. LVQ applies a winner-take-all strategy [5], choosing a winning vector as a winner. The winning weight vector is the vector that has the smallest distance or in other words, this weight has a resemblance to the input vector. Then the winning weights vector are updated or adjusted. While other weight vectors (which are not winners) are not updated. The weight vector update process is performed repeatedly until the termination conditions are met. The size of the similarity between the weight vector and the input vector is determined using the distance (Euclidean) formula in equation 1 below.

$$dist(x, w) = \sqrt{\sum_{i=1}^{n}(x_i - w_i)^2} \quad (1)$$

Where $n$ is the number of attributes, $x$ is the input vector, and $w$ is the weight vector.

In pattern recognition, the minimum Euclidean distance between the input vectors and the vector weights is the winner. This winning vector is then updated. If the value of input vector $x$ and weight vector $w$ (winner) are in the same class, then the weight vector $w$ is updated using equation 2.

$$w(t+1) = w(t) + \alpha\big(x(t) - w(t)\big) \quad (2)$$

And vice versa, if the input vector $x$ and weight vector $w$ (winner) are in different classes, and then the weight vector $w$ is updated using equation 3. Where $\alpha$ is called the learning rate.

IOP Conf. Series: Journal of Physics: Conf. Series **978** (2018) 012032 doi:10.1088/1742-6596/978/1/012032

$$w(t+1) = w(t) - \alpha\big(x(t) - w(t)\big) \tag{3}$$

The steps of LVQ learning process to generate the classification model, as follows:

1. Input: Input data, vector codebook, learning rate, maximum iteration
2. Initialization vector weights (taken from data input)
3. For each data input, calculate the Euclidean distance between each data input *x* and the weight vector *w* using equation 1
4. For each weighted vector *w*, select the smallest Euclidean distance as the winner.
5. Update the winning weight vector *w*:
   - If class *w* is equal to class *x*, update *w* using equation 2.
   - If class *w* is not the same as class *x*, update *w* uses equation 3.
6. If the termination condition is met, print the weighted vector *w* as the resulting classification model, otherwise return to step 4

## 3. Particle Swarm Optimization (PSO)

The swarm particle is a population-based optimization technique developed by James Kennedy and Russ Eberhart in 1995 [7] [8] inspired by the behavior of bird flocks. Although Kennedy and Eberhart started modeling with flocks, it turned out that in its development, the algorithm model is more similar to a model of swarming bees with collective intelligence [3].

### *3.1. PSO Algorithm*

The particle swarm-based optimization begins by initializing the positions of the particles randomly. Then each particle calculated its fitness value based on the evaluation function. This method will continue to iterate until the expected condition is reached (optimal condition). At each iteration, each particle is affected by two values [5]. The first is the best value each particle has achieved, called the *pbest*. The other value is the best overall value possessed by particles in the population, referred to as *gbest*. After obtaining the two best values, the particles renewed the velocity and their respective positions with equations 4 and 5.

$$v_i(t) = v_i(t-1) + c_1 r_1\big(x_{pi} - x_i\big) + c_2 r_2\big(x_g - x_i\big) \tag{4}$$

$$x_i(t) = x_i(t-1) + v_i(t) \tag{5}$$

Where *v* represents velocity, *x* is the particle position, *c* is the learning rate and *r* is a random number U (0,1).

$v_i(t-1)$ is also called inertia term [11] that forces particles to move in the same direction as the previous movement. $c_1 r_1\big(x_{pi} - x_i\big)$ Is referred to as the cognitive term [11] that forces the particle back to its best position. $c_2r_2(x_g - x_i)$ is referred to as social term [11] which forces particles to move to the best position of all particles.

The steps for the particle swarm-based optimization procedure are as follows:

1. Generate the initial population of the particles
2. Calculate the fitness value, if better than the fitness value during the run, set that value as *pbest*
3. Choose the particle with the best fitness value in the population, set that value as *gbest*
4. For each particles:
   a. Update *velocity* v using equation 4
   b. Update *position* x using equation 5
5. Stop if expected conditions are reached. If not, returned to step 2

## 4. Experiment

In this paper used two methods, namely LVQ and PSO. PSO as a weight seeker on feature (attribute) sample data, while LVQ is used to find the classification model (classifier). In the PSO algorithm, the datainput feature is used to search for the weight value, then the weight value is used as the LVQ input parameter to generate the classification model through the training process. The classification model is then validated in the testing process.Tthe performance level is then calculated for looking at the accuracy and error rate. The data used for training and testing is obtained from sample dataset using *Rotation* (n-*fold cross validation*).

### 4.1. Model PSO to find Feature Weighting

The particle swarm-based optimization begins with the initialization of the population of the particles by randomize way. The particles represent the sample dataset's attributes. The value of each attribute is randomized (from the minimum to the maximum value) to set the initial position of the particle. While the initial velocity of the particles is set to zero. The particle density must not exceed the maximum value of each attribute of the sample dataset. After initializing the value, each particle is then evaluated by an evaluation function to find its *pbest* value [5]. From the value obtained *pbest,* select one value with the best value as the *gbest* value [5]. Then update the *velocity* and *position* of each particle using equations 4 and 5. The fitness value evaluation process, density and position update, is done iteratively until the expected termination conditions are achieved. The value of the *learning rate* is updated iteratively to see the resulting convergence level. Convergence is met if the evaluation function tends to improve and stagnate, and the position of the particles has not changed significantly. The output of this PSO algorithm is the weighted value of each attribute of dataset.

### 4.2. Model LVQ to find Model Classifier

The weight of the attributes obtained from the PSO algorithm is then applied to the LVQ model. The weight of the attribute $a_j$, where *j* is the *j*'s attribute, is inserted into the euclidean distance formula, as shown in equation 6.

$$dist(x, w) = \sqrt{\sum_{i=1}^{n} \left(a_j * (x_i - w_k)\right)^2} \qquad (6)$$

Where *x* is the input data, $\boldsymbol{w_k}$ is vector weight for each class *k*, $\boldsymbol{a_j}$ is the feature weight generated by the PSO on each attribute *j*.

While the formula for the winner vector weighted update is modified due to the weight value of the resulting attribute. If the input *x* and the weight vector *w* are in the same class then the weight *w* is updated using equation 7, otherwise if *x* and *w* are not in the same class, the weight *w* is updated using equation 8.

$$w_{(t+1)} = a * w_{(t)} + \alpha\left(a * \left(x_{(t)} - w_{(t)}\right)\right) \quad (7)$$

$$w_{(t+1)} = a * w_{(t)} + \alpha\left(a * \left(x_{(t)} - w_{(t)}\right)\right) \quad (8)$$

Where $a$ is the attribute's weight, $\alpha$ is a learning rate.

## 5. Result

The dataset that used to train and test the above experiments was obtained from UCI Machine Learning Repository. The dataset has been classified before, each instance already has a certain class label. Table 1 shows a description of the 3 datasets with property A representing an attribute, I denoting the number of instances, C denoting the number of classes. Table 2 shows the optimal attribute weights generated by the PSO algorithm of each dataset. Table 3 shows the data validation results consisting of accurate predictive value, and error rate.

**Table 1.** Description of Datasets.

| | Attribute. (*A*) | Class. (C) | Instance. (I) |
|---|---|---|---|
| **Iris** | 4 | 3 | 150 |
| **Glass** | 9 | 7 | 214 |
| **Wine** | 13 | 3 | 178 |

**Table 2.** Attribute Weight of Datasets.

| **Iris** | | **Glass** | | **Wine** | | | |
|---|---|---|---|---|---|---|---|
| **Atrr** | **Weight** | **Attr** | **Weight** | **Attr** | **Weight** | **Attr** | **Weight** |
| A1 | 0.981 | A1 | 1 | A1 | 0.249 | A10 | 0.336 |
| A2 | 0.745 | A2 | 0.640 | A2 | 0 | A11 | 1 |
| A3 | 0.676 | A3 | 1 | A3 | 0.398 | A12 | 0 |
| A4 | 0.447 | A4 | 0.540 | A4 | 0 | A13 | 0 |
| | | A5 | 0.290 | A5 | 0 | | |
| | | A6 | 0.984 | A6 | 0.978 | | |
| | | A7 | 1 | A7 | 1 | | |
| | | A8 | 1 | A8 | 1 | | |
| | | A9 | 0 | A9 | 0 | | |

**Table 3.** Accuracy of Prediction.

| Dataset | Accuracy | | Classification Error | |
|---|---|---|---|---|
| | **LVQ-Classifier** | **PSOFW-LVQ** | **LVQ-Classifier** | **PSOFW-LVQ** |
| **Iris** | 96.00% | 97.33% | 4.00% | 3.27% |
| **Glass** | 72.55% | 77.06% | 27.45% | 22.94% |
| **Wine** | 90.98% | 98.33% | 9.02% | 1.67% |

## 6. Summary

In this paper, to form an optimal classification model used two methods namely, LVQ and PSO. The result of data processing and validation shows that there is an increase of classifier accuracy performance if it is weighted on attribute of each dataset. This improvement in accuracy shows that the PSO algorithm is capable of finding attribute weights that increase LVQ-classifier accuracy, although LVQ-Classifier itself has produced a fairly good classification model. Experiments conducted on this papaer is devoted to continuous values, future will be tried for non-continue values.

## References

[1] Mehmed K 2003 Data Mining: Concept, Models, Methods, And Algorithms *IEEE Press* (445 Hoes Lane, Piscataway, NJ 08854)
[2] Witten I H, Frank E and Hall M A 2011 Data Mining: Practical Machine Learning Tools and Techniques *Morgan Kaufmann Publisher. Elsevier.Inc* (Burlington, MA 01803, USA)
[3] Millonas M 1994 *Swarms, Phase Transition, and Collective Inteligence*
[4] Cardic C 1993 *Proc.Int.Conf. on Machine Learning*
[5] Brownlee J 2011 Clever Algorithms: Nature-Inspired programming *Recipes Creative Commons Attribution-Noncommercial-Share Alike 2.5* (Autralia License) ISBN: 978-1-4467-8506-5
[6] Xue B, Zhang M and Browne W N 2012 Particle Swarm Optimization for Feature Selection in Classification: A Multi Objective Approach *IEEE Transaction On Cybernetics Journal. IEEE Press*
[7] Kennedy J and Eberhart R 1995 *Proc. IEEE Int.Conf. Neural Network* vol. 4
[8] Shi Y and Eberhart R 1998 *Proc. IEEE Int.CEC* pp. 69-73.
[9] Kohonen T 1988 Self-Organization and Associative Memory *Springer-Verlag* (New York)
[10] Liu Y, Wang G, Chen H and Dong H 2011 *J.Bionic Eng.*,vol.8 **2** pp.191-200
[11] Alexandre S and Leandro N d C 2013 A Constructive Data Classification Version of the Particle Swarm Optimization Algorithm *Hindawi Publishing Corporation Mathematical Problems in Engineering*

IOP Conf. Series: Journal of Physics: Conf. Series **978** (2018) 012033 doi:10.1088/1742-6596/978/1/012033

# Social network extraction based on Web: 3. the integrated superficial method

**M K M Nasution[1*], O S Sitompul[1] and S A Noah[2]**

[1]Technical Information, Fasilkom-TI, Universitas Sumatera Utara, Padang Bulan 20155 USU, Medan, Indonesia
[2]Knowledge Technology Research Group, Faculty of Information Science & Technology, Universiti Kebangsaan Malaysia, Bangi 43600 UKM Selangor, Malaysia

E-mail: mahyuddin@usu.ac.id

**Abstract.** The Web as a source of information has become part of the social behavior information. Although, by involving only the limitation of information disclosed by search engines in the form of: hit counts, snippets, and URL addresses of web pages, the integrated extraction method produces a social network not only trusted but enriched. Unintegrated extraction methods may produce social networks without explanation, resulting in poor supplemental information, or resulting in a social network of durmise laden, consequently unrepresentative social structures. The integrated superficial method in addition to generating the core social network, also generates an expanded network so as to reach the scope of relation clues, or number of edges computationally almost similar to $n(n-1)/2$ for $n$ social actors.

## 1. Introduction

The Web as a source of information has a lot of potential usage in everyday life [1]. Information extraction relates to the structure of information to be generated [2, 3]. As with social networks, the extraction of social structures from the Web has a variety of different sources [4]. The sources can generally support each other, but there are fundamental differences in approach and outcomes. In general, supervised approaches always rely on more accurate explanations of the results of social network extractions, but are unable to grasp the meaning of changes that occur in information sources [5, 6, 7]. On the other hand, the unsupervised approach generally relies on the ability to capture change superficially, but can actually be enriched through the integration of different approaches to different sources of information from the Web.

So far, there are some superficial extraction methods. All of these methods have been developed differently by different researchers [8, 9, 10, 11]. One with other methods have been partially integrated, but the emphasis is to disclose the trusted information [6]. This paper intend to describe the integration of approaches into an enriched superficial method through different sources of information.

## 2. Related Works and Motivation

Although by involving the same queries, two steps for extracting social network from the Web, the queries produce different sources of information as an indicator of the relationship between social actors [12, 13]. For a pair of social actors $a_i$ and $a_j$, through query $q \Leftarrow t_{a_i}$ and $q \Leftarrow t$

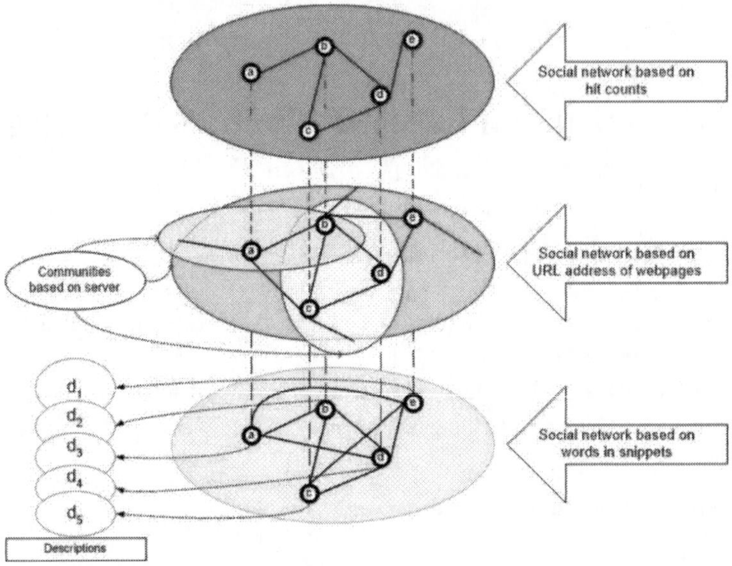


**Figure 1.** Levels of relationship in social networks

two singletons are generated: a hit count $\Omega_{a_i}$ and a hit count $\Omega_{a_j}$. Through the search engine, with which $t_{a_i}$ and $t_{a_j}$ are two different social actors names as query content. By involving the same search engine, the relation clue in semantic is generated by the query $q$ $t_{a_i}, t_{a_j}$ is a doubleton or hit count $\Omega_{a_i}$ $\Omega_{a_j}$. However, each query not only generates a hit count, but produces a snippet [5].

Each search engine produces a snippet as companion information from a hit count. Snippet serves as an explanation regarding query content. In each snippet there are a number of words and one identity uniquely for each source of information that is the URL address of the webpage [14]. Thus, for each query will result a set of $s_i$ snippets and a set of URL addresses $u_i$ if hit count $\Omega_a > 0$. For a collection of words from a non-empty snippet, it will contain the name of the social actor and also the words expressing an explanation of the activity of the social actor. For example, the affiliation of social actors, ideas or concepts contained in the paper title of social actor, performance targets to be planned by social actors, and so on [15, 16].

In general, the extraction of social networks from the Web uses a similarity measurement for a pair of social actors by involving three different hit counts [17]. This produces the strength relation. However, there is a very strong relationship up to the weakest relationship that can be generated through the similarity between the URL addresses [15]. This strongest relationship is based on the concept that one URL address as a representation of a webpage represents one event, and this results in social actors on the same webpage having a close relationship, which is generally described semantically as co-occurrence. However, the URL address of form is stratified as from directory to sub-directory, the lower part being the domain located at the base of the URL address. Furthermore, the similarity between URL addresses with each other is determined by the similarity between the parts separated by the slash from base to tip. While a collection of URL addresses are present via doubleton, it can be used as validation of the strength relation that arises from two sets of singleton-based URL addresses. In different cases, the number of URL addresses of doubleton becomes pure representation by co-occurrence[18].

In a different study, which also involves URL addresses, it has been revealed that a collection of words can express the relation clues of two social actors in addition to the description of the relationship. On different occasions, this is reinforced by involving a collection of words from singleton or doubleton-based snippets [19].

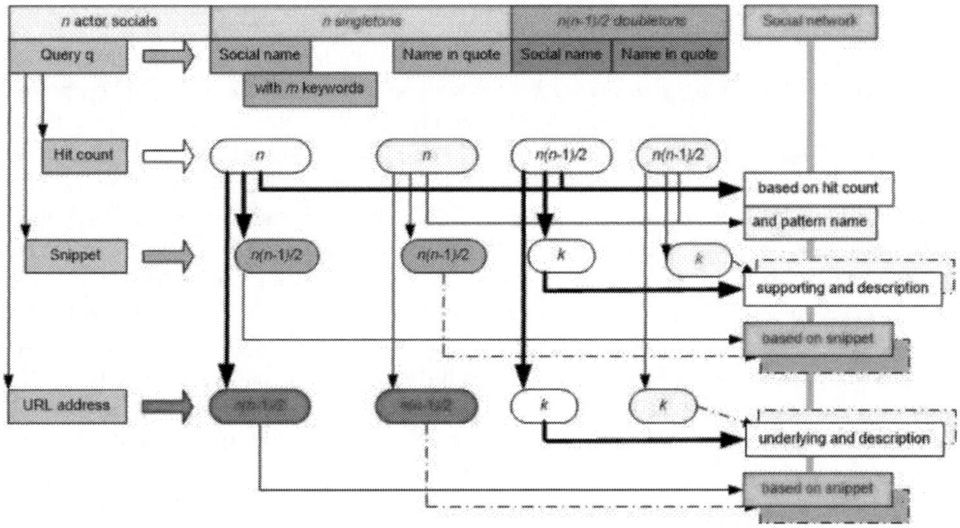


**Figure 2.** The cornerstone for integrating superficial methods

In general, the relationship between social actors derived from the same type of query will form layers of mutually supportive and explicit relationships, such as Fig. 1 for example.

## 3. The Proposed Approach

Based on the reviews have been made and the comparisons between different approaches in the extraction of social networks [20, 21], we construct an approach to the involvement of some superficial methods integratedly such as Fig. 2. Accordingly, different approaches are adopted to adjust to optimally implementation. That is adjustment to the limitations of search engine services for the number of queries and length of the query content [5].

When a query contains a social name in either singleton or doubleton, then the doubleton engagement in similarity is to show the function of reinforcing the existence of the relationship between two actors, since the doubleton arrangement of the two names for the two social actors causes a reduction of ambiguity, eventhough in singleton query processes by the search engine lifts all information based on similarity. For example, a social name in singleton such as $q$ = Mahyuddin K. M. Nasution, while two social names in a doubleton like $q$ = Mahyuddin K. M. Nasution,Opim Salim Sitompul. So if there is $n$ social actors, then this process requires the involvement of $n+n(n+1)/2$ queries and $n(n+1)/2$ computations. Normally, each search engine returns a maximum of 10 snippets in each page for the associated query. Please note most of search engines will feature snippets of webpages containing the most information about query content first in the top of the snippet list [22]. Therefore, if all snippets inside the first page of doubleton are used to correctly describe the strength relation (as supporting and description). It no longer involve a query, but it only spend $k$ computations for processing words in snippets. The next integration, involves to process a collection of URL addresses. For the same reason, the collection of URL addresses involved is a maximum of 10 first URL addresses based on doubleton (as underlying and description) [23]. Although the URL address uniquely represents the webpage, but it involves the URL address of the highest ranking webpage, so that it reveals a fundamental foundation for the establishment of a trusted social network in addition to being enriched with additional descriptions based on the webpage community as the origin of the server. Thus, this process also involve any additional queries, see Fig. 2. Significantly, this integrated approach involves $kn(n \quad 1)/2$ computations and follows $n + n(n + 1)/2$ queries [24].

By taking the name of social actors in the form of a pattern or exist in quotation marks, such as $q$ = Mahyuddin K. M. Nasution for singleton or $q$ = Mahyuddin K. M. Nasution,Opim Salim Sitompul for doubleton, the search engine produces the accurate information about hit count, snippet, and URL address. This is set on equality between the query content and the content of related webpages based on the pattern. However, the similar names of social actors can not be detected by search engines [25]. Thus, the second approach will automatically involve $kn(n \; 1)/2$ computations and follows $n + n(n \; 1)/2$ queries. If it involves also the keywords in query based on either singleton or doubleton, then we use the same process like above, i.e. $n(n \; 1)/2$ queries will be used and the number of computations is $kmn(n \; 1)/2$ where $m$ for processing keywords [23, 26], see Fig. 2.

In addition, social network extraction may also involve singleton from two different social actor descriptions. The descriptions derived from the snippet will form the relationship between two social actors through the similarity between the description details [27]. It only requires $n$ queries and $n(n \; 1)/2$ computations. Nevertheless, in general involvement of this description always further expanded the social network with the possibility that a non-existent relationship between two social actors will be formed from the description. Likewise, if it involves a collection of singleton-based URL addresses, it involves $n$ queries and $n(n \; 1)/2$ computations, and will expand social network differently between other social actors [19]. Therefore, the use of singleton both snippets and URL addresses are integrated to form social networks opinions that may be wider than actual social networks [28], see Fig. 2.

By expressing some opinions of integration approaches, generally superficial methods can be integrated as follows:

(i) Define a number of names of social actors.
(ii) Submit a query from each actors name to the search engine.
  (a) Record hit count based on singleton.
  (b) Record snippets of singleton in the first page.
(iii) Submit a query of a pair of actor names to a search engine.
  (a) Record hit count based on doubleton.
  (b) Record snippets of doubleton in the first page.
(iv) Calculate the strength relation for each pair of social actors.
  (a) Describe the description of the doubleton snippets to strengthen the evidence of the strength relation and calculate the comparison between the description weights and the hit count of doubleton.
  (b) Describe the URL address of the doubleton snippets to strengthen the proof of strength relation and calculate the number of URL addresses composition against the hit count of doubleton.
(v) Describe the resulting social network.

As for expanding social networks, an integrated approach may involve additional steps as follows:

(i) Describe the description of URL address of the singleton snippets of any social actor.
  (a) Calculate the clarity of description as additional strength relation between two social actors to expand the relations.
  (b) Calculate the URL addresses similarity between two social actors as additional strength relation to expand the relations.
(ii) Extend social networks through two additional strength relation

| | 1 | 2 | 3 | 4 | 5 |
|---|---|---|---|---|---|
| 1. DBLP | | | | | |
| 2. Microsoft | 0.8480 | | | | |
| 3. FTSM | 0.6937 | 0.6876 | | | |
| 4. Scribd | 0.6460 | 0.6802 | 0.6667 | | |
| 5. Myais | 0.2381 | 0.2613 | 0.2754 | 0.3074 | |
| 6. Research | 0.0814 | 0.0889 | 0.1956 | 0.1121 | 0.1863 |

**Figure 3.** Similarity of clusters based on URL addresses

**Table 1.** URL addresses as label of strength relation

| No. | URL address | id | Vertices | Edges |
|---|---|---|---|---|
| 1. | www.informatik.uni-trier.de | DBLP | 347 | 10,825 (85.79%) |
| 2. | academic.research.microsoft.com | Microsoft | 329 | 10,237 (81.13%) |
| 3. | www.ftsm.ukm.my | FTSM | 291 | 9,876 (78.27%) |
| 4. | www.scribd.com | Scribd | 279 | 8,623 (68.34%) |
| 5. | myais.fsktm.um.edu.my | Myais | 124 | 3,000 (23.78%) |
| 6. | research.ukm.my | Research | 63 | 1,043 (8.27%) |

## 4. Experiment and Discussion

We conduct an experiment by involving 462 social actors. The social network extraction by using a method that revealed hit count singleton and hit count doubleton yielded 31, 623 strength relations between social actors, or 29.70% of 462 461/2 potential relations. Through the same experiment, the doubleton snippet and the URL addresses doubleton can be processed the alternative relationship between two actors. Through the collection of snippets: words and URL addresses in doubleton we generate the descriptions of each strength relation.

Thus, in this experiment the usability of the URL addresses for each social actor can be obtained, then we use for grouping strength relations based on the URL addresses as the social network clusters, see Table 1. The similarity between clusters shows that, although number of vertices and edges between clusters and other clusters indicate the interdependence between one social network with another, but no social network sequence based on that cluster becomes purely a part of another social sequence, or we cannot found that a community becomes part of another community as a whole based on the servers addresses, see Fig. 3.

Generating the alternative relationships integratedly involves a singleton snippet containing the words and URL addresses. Although for 462 social actors, 12, 158 strength relations or 11, 41% of computation totality was done and the results of expansion was 11, 015 relations, while through the expanded URL addresses were 70, 604 relations. In an integrated manner, the social network expanded to 105, 168 edges or 98.76% of the potential relations.

## 5. Conclusion

The integrated superficial method is as an approach involving hit count, snippets or set of words and a collection of URL addresses either singleton or doubleton. This integration of information sources be an indicator of the relationships between social actors, not only provides the descriptions of strength relations but also results in the expansion of social networks in other forms. It is possible that the existence of built communities based on the underlying descriptions or the descriptions as explanation. With the information of clusters it is possible to see social behavior based on social structure, this be the target of subsequent research, as a follow-up in future.

IOP Conf. Series: Journal of Physics: Conf. Series **978** (2018) 012033 doi:10.1088/1742-6596/978/1/012033

## References


[1] Chakraborty T and Kearns M 2008 Bargaining solutions in a social network *WINE 2008* **LNCS 5385**.

[2] Boyd D M and Roychowdhury V P 2008 Social network sites: Definition, History, and Scholarship *Journal of Computer-Mediated Communication* **13**.

[3] Johnsen J W and Franke K 2017 Feasibility study of social network analysis on loosely structured communication networks *Procedia Computer Science* **108**.

[4] Tang J, Zhang D, Yao L and Li J 2008 Extraction and mining of an academic social network *Proceedings of WWW 2008*.

[5] Nasution M K M 2016 Social network mining (SNM): A definition of relation between the resources and SNA *International Journal on Advanced Science, Engineering and Information Technology* **6(6)**.

[6] Matsuo Y, Mori J, HamasakiM, Nishimura T, Takeda H, Hasida K and Ishizuka M 2006 POLYPHONET: An advanced social network extraction system *Proceedings of the 15th International Conference on World Wide Web* (WWW 2006).

[7] Nasution M K M, Sitompul O S, Sinulingga E P and Noah S A 2016 An extracted social network mining *Proceedings of 2016 SAI Computing Conference* (SAI 2016).

[8] Kautz H, Selman B and Shah M 1997 ReferralWeb: Combining social networks and collaborative filtering *Communication of the ACM* **40(3)**.

[9] Mika P 2005 Flink: Semantic Web technology for the extraction and analysis of social networks *Web Semantics: Science, Services and Agent on the World Wide Web* **3(2-3)**.

[10] Jin Y, Matsuo Y and Ishizuka M 2006 Extracting a social network among entities by web mining *Workshop on Web Content Mining with Human Language* (ISWC 2006).

[11] Jin Y, Matsuo Y, and Ishizuka M 2007 Extracting social networks among various entities on the Web *ESWC 2007* **LNCS 4519**.

[12] Matsuo Y, Mori J, Hamasaki M, Nishimura T, Takeda T, Hasida K and Ishizuka M 2007 POLYPHONET: An advanced social networks extraction system from the Web *Journal of Web Semantics: Science, Services and Agents on the World Wide Web* **5**: 262-278.

[13] Nasution M K M 2017 Social network extraction based on Web: 1. Related superficial methods *International Conference on Operational Research* (InteriOR).

[14] Nasution M K M and Noah S A 2017 Social network extraction based on Web: 2. Comparison of superficial methods *Procedia Computer Science*.

[15] Nasution M K M and Noah S A 2010 Superficial method for extracting social network for academics using web snipppets *Lecture Notes in Computer Science* **LNAI 6401**.

[16] Nasution M K M and Noah S A 2011 Extraction of academic social network from online database *2011 International Conference on Semantic Technology and Information Retrieval* (STAIR 2011).

[17] Nasution M K M, Sitompul O S, Nasution S, and Ambarita H 2017 New similarity *IOP Conference Series: Materials Science and Engineering* **180(1)**.

[18] Nasution M K M 2017 Semantic interpretation of search engine resultant *International Conference on Operational Research* (InteriOR).

[19] Nasution M K M and Sitompul O S 2017 Enhancing Extraction Method for Aggregating Strength Relation Between Social Actors *Artificial Intelligence Trends in Intelligent Systems* (AISC) **573**.

[20] Culotta A, Bekkerman R and McCallum A 2004 Extracting social networks and contact information from email and the Web *Computer Science Department Faculty Publication Series* **33**.

[21] Mitra A, Paul S, Panda S and Padhi P 2016 A study on the representation of the various models for dynamic social networks *Procedia Computer Science* **79**.

[22] Nasution M K M and Noah S A 2012 Information retrieval model: A social network extraction perspective *Proceedings - 2012 International Conference on Information Retrieval and Knowledge Management*.

[23] Nasution M K M 2014 New method for extracting keyword for the social actor." *Lecture Notes in Computer Science* **LNAI 8397 (PART 1)**.

[24] Nasution M K M 2017 Modelling and Simulation of Search Engine *Journal of Physics: Conference*

*Series* **801(1)**.
[25] Nasution M K M, Noah S A and Saad S 2011 Social network extraction: Superficial method and information retrieval *Proceedings of International Conference on Informatics for Development* (ICID11).
[26] Nasution M K M and Noah S A 2012 A methodology to extract social network form the Web Snippet *Cornell University Library* arXiv:1211.5877.
[27] Nasution M K M, Hardi M and Syah R 2017 Mining of the social network extraction *Journal of Physics: Conference Series* **801(1)**.
[28] Nasution M K M, Syah R and Elfida M 2018 Information retrieval based on the extracted social network *Applied Computational Intelligence and Mathematical Method, Advances in Intelligent Systems and Computing* **662**.

# Development of excavator training simulator using leap motion controller

**F Fahmi[1], F Nainggolan[2], U Andayani[2] and B Siregar[2]**
[1]Dept. of Electrical Engineering, Faculty of Engineering, Universitas Sumatera Utara, Jl. Dr. Mansur 9 Medan Indonesia
[2]Dept. of Information and Technology, Faculty of Computer Science and Information Technology, Universitas Sumatera Utara, Jl. Dr. Mansur 9 Medan Indonesia

Email. fahmimn@usu.ac.id

***Abstract***—Excavator is a heavy machinery that is used for many industries purposes. Controlling the excavator is not easy. Its operator has to be trained well in many skills to make sure it is safe, effective, and efficient while using the excavator. In this research, we proposed a virtual reality excavator simulator supported by a device called Leap Motion Controller that supports finger and hand motions as an input. This prototype will be developed than in the virtual reality environment to give a more real sensing to the user.

## 1. Introduction

Excavators are heavy equipment commonly used for excavation and flattening operations, material control, heavy lifting, and demolition work in the industry ranging from construction and forestry to agriculture and mining [1,2]. As one of a variety of heavy equipment, the use of excavators is not easy. Excavator operators must be adequately trained in a variety of capabilities to ensure safe, effective and efficient use of the equipment [1].

Excavator as heavy machinery is used for many industries purposes. Controlling the excavator is not easy. Its operator has to be trained well in many skills to make sure it is safe, effective, and efficient during using the excavator [1]. In this research, the author proposed a virtual reality excavator simulator supported by a device called Leap Motion Controller that supports finger and hand motions as an input.

Leap Motion Controller can improve the human and computer interaction [2,3]. By using this device, hands and fingers motions will be used to interact with the joystick and the lever inside the virtual simulator. The lever in front is used for moving the excavator forward and backward. And the joystick on the right side for boom and bucket, meanwhile the left joystick for an arm and swing the upper body. These movements are based on SAE Controls.

Leap Motion Controller is a computer hardware sensor tool that supports hand and finger movements as input, which can be equated with functions like a mouse, but does not require direct contact with the hands or touch. Leap Motion Controller is a small, Universal Serial Bus (USB) peripheral device designed to be placed on a physical desktop, facing up. Using two monochromatic IR cameras and three infrared LEDs, the tool observes a surface resembling a hemisphere on top with a range of about 1 meter.

Leap Motion Controller can enhance human interaction with computers [3,4]. Leap Motion Controller is designed to track hand and finger movements. At the time the system is run, the user can perform various activities with hand gestures. This hand movement is observed by sensors in the Leap Motion Controller as inputs to perform various things that can trigger the occurrence of events in the program. In the simulated excavator, this hand movement will directly come into contact with the excavator control which is a three-dimensional model.

The use of excavators requires learning in advance to understand the functions of excavator control [5,6]. Various methods of teaching the use of excavators can be done to facilitate the learning process, one of them by using simulation [7]. Therefore, simulation with hand gesture reading system is needed as an option of learning method that can help learning the use of an excavator.The purpose of this research is the use of Leap Motion Controller as a tool for system simulation system learning use excavator.

## 2. Material and methods

In order to develop the excavator learning system with LMC, we deploy a method that utilizes hand gestures and finger movement in controlling the visualization of a specific excavator machinery. The hand gesture is a static or dynamic motion that is used as a communication tool between human and machine or among human that using sign language [6].

By monitoring the changes of space coordinate of hand palms and each segment of a finger, the system will execute specific action affecting the 3D visualization of excavator machinery, including showing the affecting part of the machine.

The system was based on the input of the hand and finger tracking data provided by LMC. A 3D object is displayed following the hand movement, whether the direction is right or left, or up and down. The changes of the object orientation due to motions of hand and gesture are immediately shown in real time on the screen.

Every movement will be represented by a hand animation on the user interface. Each specific hand gesture triggers a specific response in the system. For example, the left-hand gesture can control the camera. And right-hand gesture can show the labels /name of excavator object. Labels of each excavator objects are the outputs of this system. The architecture of the developed system is shown in Figure 1.

### *A. Three Dimensional Excavator Objects*

A 3D excavator objects are used and will be displayed for the first time when user open the application. The 3D excavator objects are basically a file containing 3D models with the extension of *.obj. We developed 3D excavator objects created by specific software.

### *B. Leap Motion Controller (LMC)*

Leap Motion Controller (www.leapmotion.com) can potentially revolutionize the human-computer interactions [7]. With two IR monochromatic camera and three LED infrared, this device tracks a hemisphere surface above it within a 1-meter radius.

Basically, LMC can detect hand motion and image as a whole, ignoring the detailed parts of the hand like finger joints and knuckles. Nowadays, LMC is capable of observing some detailed characteristic parts of the hand like knuckles, right or left, hand grabbing factor, etc. This development is called V2Tracking. The new version of LMC was used in this study.

### *C. Motion Tracking*

Motion tracking process translated hand and finger motion as main input. These data are collected and processed by a specific part called Unity engine. Some of the possible inputs we developed in this system are described in detail in Figure 2.

The series of steps used starting from the input data in the form of observations of hand and finger movements by Leap Motion Controller. A three-dimensional object will be displayed following hand movements, be it right-handed, left-handed, or both. All hand gestures will be displayed as lightly as possible in the user interface. When the hand of usage comes into contact with the control joystick inside the system, the excavator moves according to its function.The input of this system is a hand gesture that is read by Leap Motion Controller. The process in this system is the display of three-dimensional.

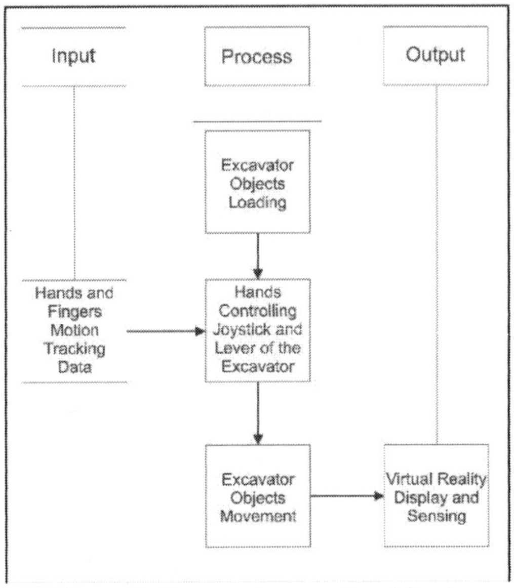


**Figure 1**. Proposed general architecture

Joystick model that when driven by a three-dimensional hand model will move the excavator in accordance with its function. The output of the system is the movement of the excavator in accordance with the control system.

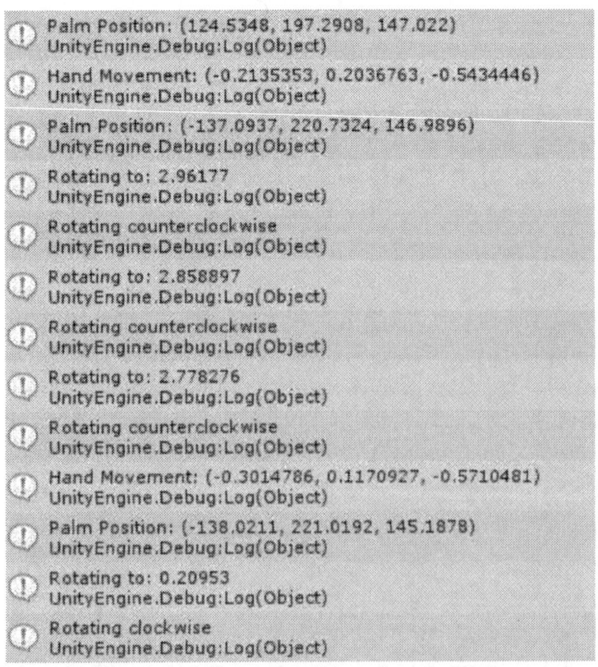


**Figure 2**. Hand and finger tracking data in Unity engine

These motion tracking data are being examined by the system, whether it is a specific gesture or simply general hand movement. In general, hand gestures corresponding to specific responses, that we use as the control of navigation system can be categorized as below:

*D. Levers and joysticks movement as a trigger for moving the excavator.*

1. Lever movement to move the excavator forward and backward.
2. Joystick movement to rotate the boom, arm, bucket, and upper body of the excavator.

The lever movement is used to determine the speed and the turn value of the excavator as shown in the equation (1) and equation (2).

$$x = (L_1 + L_2) \times C \quad (1)$$

$$\alpha = (L_1 - L_2) \times C \quad (2)$$

Which x is translation value of the excavator and $\alpha$ is rotation value of the excavator. $L_1$is the left lever rotation value and $L_2$ is the right lever. $C$is a constants value for the translation speed and rotation speed. When only the right lever is rotated forward by the virtual hand movement, the excavator will be moving forward and turn to the left simultaneously. And if it is the left lever, the excavator will be moving forward and turn to the rightsimultaneously.If both levers are rotated forward, the excavator will be moving straight forward.

Meanwhile, to rotate the boom, arm, bucket, and upper body of the excavator, equation (3) is used.

$$\theta_t = \theta_0 - \beta \times C \quad (3)$$

The function rotation of the left and right joysticks is determined by the axis the rotation. The left and right joysticks can only move is x and z-axis.There are 4 functions can be done by rotating the joystick in x and z-axis.

a) The arm rotation is determined by the rotation of the left joystick in the x-axis.
b) The upper body rotation is determined by the rotation of the left joystick in the z-axis.
c) The boom rotation is determined by the rotation of the right joystick in the x-axis.
d) The bucket rotation is determined by the rotation of the right joystick in the z-axis.

The equation (3) shows that $\theta_t$ is the final value of the angle of the part of the excavator (either arm, upper body, boom, or bucket) and $\theta_0$ is the initial value of the angle. Meanwhile $\beta$ is value of the angle of each joystick in eachcorrenponding axis and, $C$ is constants value for speed.

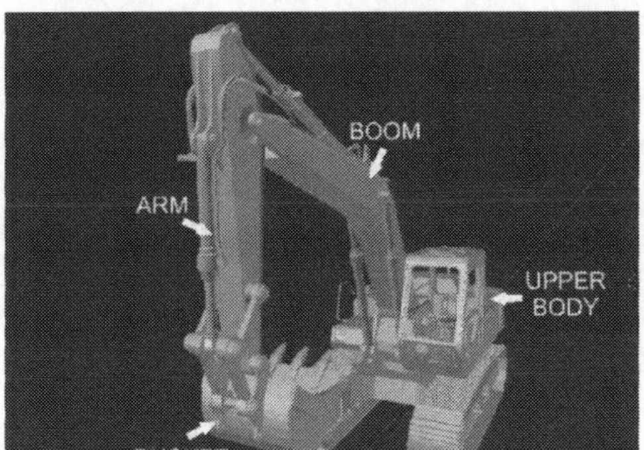


**Figure 3**. Parts of the excavator

IOP Conf. Series: Journal of Physics: Conf. Series **978** (2018) 012034 doi:10.1088/1742-6596/978/1/012034

## 3. Results

The result of this research reached a point where the user can learn about the basic of operating an excavator. Although the real excavator still has another button or lever that is customized and different for each own particular purpose. The result of the system is shown below :

**Figure 4**. Display of controlling scheme in ExcaVR

Compare to other techniques that currently available [2,3], excavator learning system using LMC has an advantage in its level of interactivity with users. LMC provides a fully interactive system that enables user gain more experiences and sense of adventure. This is important in the learning process that the more interactive the learning system, the more effective the learning process will be expected.

In the future, a study to increase the precision of LMC is needed. It is good to synchronize the hand movement showed on the screen compatible with the real hand movement. This compatibility will increase the easiness of. Furthermore, more detailed part of the excavator can be developed using more detail atlas involving more excavator objects in order to provide complete database of the excavator.

The benefits of this research are:
1. Create an interactive excavator simulation system with the user.
2. Understand how Leap Motion Controller works and implementation of a system.
3. Reduced cost utilization compared to some excavator simulations.

## 4. Conclusion

We have successfully developed a 3D excavator navigation using Leap Motion Controller for learning process in medical education. Based on the results, users were satisfied when using the system and agreed that it could be used as an effective supportive device in a 3D excavator learning process.
Leap Motion Controller can be used as a controlling device for excavator simulator in virtual reality. The users can see their virtual hand directly inside. This will help the human and computer interaction in the system.To avoid unnecessary deviations and extensions, the author uses a system for excavator control only on the joystick.

IOP Conf. Series: Journal of Physics: Conf. Series **978** (2018) 012034 doi:10.1088/1742-6596/978/1/012034

**References**

[1] Akyeampong J , Udoka S and Park E 2012 A Hydraulic Excavator Augmented Reality Simulator for Operator Training *Proceedings of the 2012 International Conference on Industrial Engineering and Operations Management Istanbul, Turkey, July 3 – 6*

[2] Haddock K 2007 The Earthmover Encyclopedia Motorbooks-*MBI Publishing Company: Minneapolis, Minnesota, USA*

[3] Nainggolan F L , Siregar B and Fahmi F 2016 Anatomy learning system on human skeleton using Leap Motion Controller In Computer and Information Sciences (ICCOINS), *2016 3rd International Conference on (pp 465-470) IEEE*

[4] Nowicki M , Pilarczyk O , Wąsikowski J and Zjawin K 2014 Gesture Recognition Library for Leap Motion controller *Poznan University of Technology: Poznań*

[5] Tao N , DingXuan Z , Yamada H and Shui N 2008 A low-cost solution for excavator simulation with realistic visual effect In Robotics, *Automation and Mechatronics, 2008 IEEE Conference on (pp 889-894) IEEE*

[6] Ni T , Zhao D and Ni S 2009 Visual system design for excavator simulator with deformable terrain In Mechatronics and Automation, *2009 ICMA 2009 International Conference on (pp 724-728) IEEE*

[7] Lim T H and Yang S Y 2006 Development and application of simulator for hydraulic excavator *Journal of the Korean Society for Precision Engineering*, **23(9)** pp 142-148

IOP Conf. Series: Journal of Physics: Conf. Series **978** (2018) 012035 doi:10.1088/1742-6596/978/1/012035

# Developing cloud-based Business Process Management (BPM): a survey

**Mercia[1], W Gunawan[1,3], A N Fajar[1] ,H Alianto[2] and Inayatulloh[2]**
[1]Information Systems Management Department, BINUS Graduate Program-Master of Information Systems Management, Bina Nusantara University, Jakarta, Indonesia 11480.
[2]Information Systems Department, School of Information System, Jakarta,Indonesia 11480.
[3]Faculty member, Post Graduate Program.
e-mail: mercia@binus.ac.id; gwang@binus.edu; afajar@binus.edu; hendraalianto@binus.ac.id; inay@binus.ac.id

**Abstract.** In today's highly competitive business environment, modern enterprises are dealing difficulties to cut unnecessary costs, eliminate wastes and delivery huge benefits for the organization. Companies are increasingly turning to a more flexible IT environment to help them realize this goal. For this reason, the article applies cloud based Business Process Management (BPM) that enables to focus on modeling, monitoring and process management. Cloud based BPM consists of business processes, business information and IT resources, which help build real-time intelligence systems, based on business management and cloud technology. Cloud computing is a paradigm that involves procuring dynamically measurable resources over the internet as an IT resource service. Cloud based BPM service enables to address common problems faced by traditional BPM, especially in promoting flexibility, event-driven business process to exploit opportunities in the marketplace.

## 1. Introduction

Since its introduction more than 20 years ago, Michael Hammer (1) and Thomas Davenport (2) initiated the term business process management (BPM) to illustrate the entire stream of activity that every organisation carries out. BPM has substantially matured over the last two decades and provided an effective way to monitor and improve business efficiency (1). With the increase of globalization, business management and the process becomes big challenges for all enterprises. Modern enterprises are dealing difficulties to cut unnecessary costs, eliminate wastes, streamline and automate their business processes to deliver huge gains (3) (4). BPM is expected to help the enterprise be both competitively agile and cost efficient, promote flexibility, event-driven business processes to exploit opportunities in the marketplace (5). To address these challenges, IT is used to manage business processes and eventually evolved into what is known as BPM today (6). BPM acts as a "business process advocate" using methods, techniques & software to design, implement, control & analyze operational processes involving people, organizations, applications, documents & other sources of information (7).

BPM takes data from various business enterprise applications in the form of data presentation, and then does the following two things (8): (1) to track how information is used to complete a business, so as to accurately locate and understand existing business processes; (2) to track the flow of information in various operations, to ensure business processes run well. BPM software was introduced along with the aim to provide solution that enables to integrate existing applications without additional technology investment. Even with limited investment, companies can further improve their work efficiency.

The article applies BPM method in cloud computing (CC) environment. Nowadays, CC has emerged as an attractive, high-performing multitenant environment that promises aggregator systems and business process delivery, business services and business content in an environment that drives innovation. Therefore, combining BPM with the cloud will produce a flexible and affordable environment for development of modern enterprise applications (3). With BPM software and applications connected with the cloud, the company receives all the benefits of internet applications along with the power and flexibility of the BPM software ecosystem. BPM located as Software as a Service (SaaS) bidding in the cloud that changes the way businesses view of cost structure of application creation and maintenance (9). By connecting BPM applications with the cloud, companies can handle large number of business processes at the same time and software development efforts become easier. The application of BPM in CC environment is expected to provide insight of BPM application in modern enterprises.

## 2. Theoretical Foundation

### *2.1. Business Process Management (BPM)*

BPM refers to discipline that enables to combine knowledge from IT and knowledge from management sciences, and applies them into operational business processes (7). BPM focuses on improving organization, integration, optimization, implementation, monitoring and process management. With the increasing needs of customers and dealing with competitors have made business process become more complex, heavily rely on information systems and may span multiple organizations (10).

Business processes in general can be classified into: human centric and system centric with combination of: person-to-person (P2P), person-to-application (P2A), and application-to-application (A2A) (11). The article examines the A2A processes in software systems that cover: transaction processing systems, enterprise architecture integration (EAI) platforms, and web-based integration servers. This software architecture enables all different business processes can be accessed dynamically through a protocol, known as service oriented architecture (SOA) and tailored to business needs of the company. The use of web-based SOA in BPM includes (12): (1) developing tools for users to personally define models with basic components; (2) business performance management tools to manage all functions as different processes and to monitor IT systems and business process operations. The architecture of web services technology that enables to link BPM are illustrated in the following:

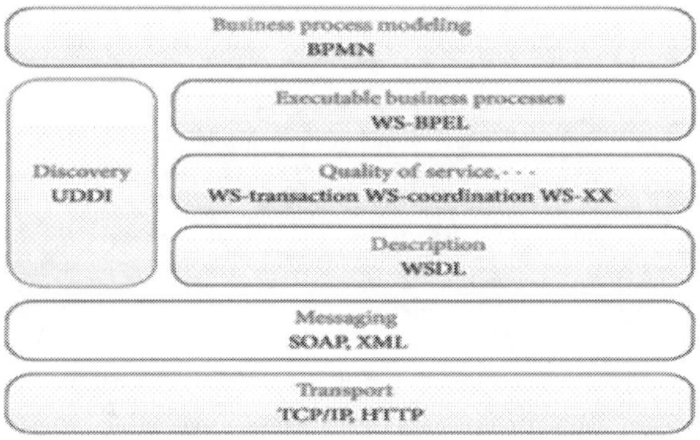


**Figure 1**. Web services technology that enables to link business process modeling.

### *2.2 Definition of Cloud Computing (CC)*

According to National Institute and Standard of Technology NIST (13), CC is a model that allows easy on-demand network access to shared containers of configurable computing resources (eg, networks, servers, storage, applications, and services) that can be set up and released quickly with little or no management effort or service provider interaction. CC has emerged to provide computing services based on demand, measurable, 'pay-per-us' and virtual centrally via the internet to improve the company's ability to cope with a flexible and highly competitive business environment.

IOP Conf. Series: Journal of Physics: Conf. Series **978** (2018) 012035 doi:10.1088/1742-6596/978/1/012035

The cloud technology has evolved through combining the advantages of SOA, virtualization, grid computing, and management automation with the following features (8): (1) virtualization, reuses hardware equipment to provide an expandable system environment with extra flexibility such as the use VMware and Xen that act as a demand-based virtualization IT equipment. Users can configure their personal network and system environment through virtualization network known as VPN; (2) service flow and workflow, provides a complete set of service environment as per demand; (3) web service and SOA, through standard of WSDL, SOAP, and UDDI, and other cloud services can be delivered in the web service; (4) web 2.0, can strengthen information share and interactive cooperation of users; (5) large-scale distributed systems, requires large-scale distributed memory system and computing ability to realize the rental of computing resources and memory spaces by users; (6) programming model, allows user to write application program under cloud environment.

CC involves dynamically measurable resource procurement through internet services. Through elastic and virtualized infrastructures, CC enables to lease and release the needed resources in an on-demand, utility-like fashion, with billing according to use and to scale the computing infrastructure up and down rapidly (rapid elasticity) (14). Cloud Computing also presents a significant technological trend, and has clear that it is reshaping the information technology process and IT market (15).

### *2.3. BPM in the Cloud Computing.*

The cloud based BPM refers to the use of BPM tools that are delivered as software service, Software as a Service (SaaS) over networks. According to Gartner by 2016, there will be more than 20% of all business processes around the world will be supported by cloud based BPM platforms. Cloud based BPM provides cloud users with valuable opportunities to use cloud software based in a pay-per-user manner. The application of cloud service model in organization can be summarized as:

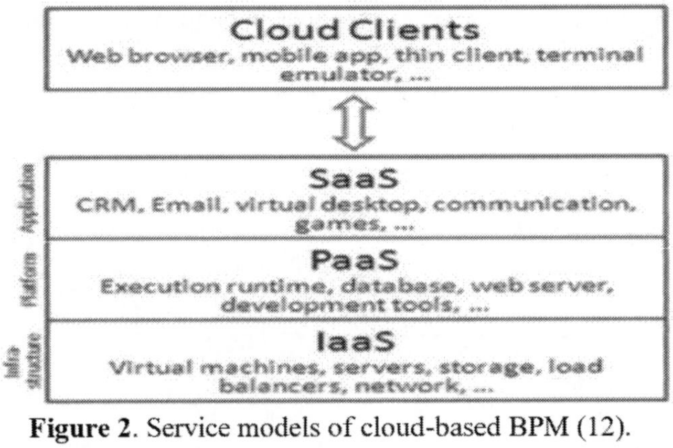


**Figure 2**. Service models of cloud-based BPM (12).

The figure 2 above shows service models of cloud-based BPM comprised of: (1) Infrastructure as a Service (IaaS), is a CC service that provides IT infrastructure of CPU, RAM, storage, bandwidth and other configurations. These components are used to build virtual computers. Virtual computer can be installed operating system and application as needed. The advantage of IaaS service is that it is not necessary to purchase computers so it allows creating cost-effective solution. Configuring a virtual computer can also be changed as needed. Suppose that when the storage is almost full, storage can be added immediately. Common IaaS providers such as: Amazon EC2, TelkomCloud and BizNetCloud; (2) Platform as a Service (PaaS), is a service that provides computing platform. Usually there are operating systems, databases, web servers and application framework for running applications. PaaS service enables developers to focus on the applications they create without thinking about the maintenance of the computing platform itself; (3) Software as a Service (SaaS), is a cloud computing service where applications are provided. The service provider manages the infrastructure and platform that runs the application. Examples of email application services are: Gmail, Yahoo and Outlook,

IOP Conf. Series: Journal of Physics: Conf. Series **978** (2018) 012035 doi:10.1088/1742-6596/978/1/012035

while examples of social media applications are Twitter, Facebook and Google +. These service enables users to utilize application without must purchase the license. Users only need a CC client device connected to the internet. There are also applications that require users to subscribe in order to access applications such as: Office 365 and Adobe Creative Cloud; (4) cloud clients, cover all important stakeholders of the company. They can be staffs, suppliers, business partners, and customers.

## 3. Theoretical Model

The article applies literature study method. The results are based on the information reviewed from selected relevant articles. The first step is to search the relevant articles from the last 10 years using a database of journals such as: Google Scholar, EBSCOhost, ProQuest and IEEE. By using the keyword "Business Process Management and Cloud Computing" then got some relevant articles.

The article examines the application of BPM in CC environment (see figure 3) with the following components (16) (8): (1) infrastructure service layer, consists of virtual resource environments, network and file storage systems, and service buses. Above the hardware layer supports dynamic configuration of virtual hardware facilities and enables to create a distributed file storage system. This storage system enables managing virtual resources and forms as separate file system on various physical machines distributed over LAN, including load balancing, fault tolerance treatment, dynamic node configurations, and more; (2) platform service layer, covers business process machines and business process pre-built libraries and other middle-wares. The service bus is in this layer with the function to uniformly manage, request and manage all service through Web Service, WSDL, SOAP, and UDDI services. The PaaS service function enables easy maintenance and improving transparency; (3) BPM as platform layer, has a BPM system built in it. BPM system has full life cycle management and specialized process services such as process modeling with BPM notation (BPMN) and business activity monitoring (BAM) monitoring. This layer provides the added benefits for helping companies to build and visualize their business needs on a cloud-based BPM; (4) software and service layer, sit on the top layer of CC that contains application services and software through internet services.

Those 3 layers of CC and BPM layer enable companies to visualize their BPM needs through BPMS. The core system of BPM is in PaaS, that enables companies to model their business processes. It enables to collect information, develop, optimize and monitor information easier. The use of BPM in CC enables to complement application pooling, allows easy system integration, managing sharing of such information and services with convenient tools. This allows the process architects to create and modify processes to solve business problems and adjust the processes needed, as business changes. In addition, cloud-based BPM also creates opportunities for SMEs with ease by simplifying complexity.

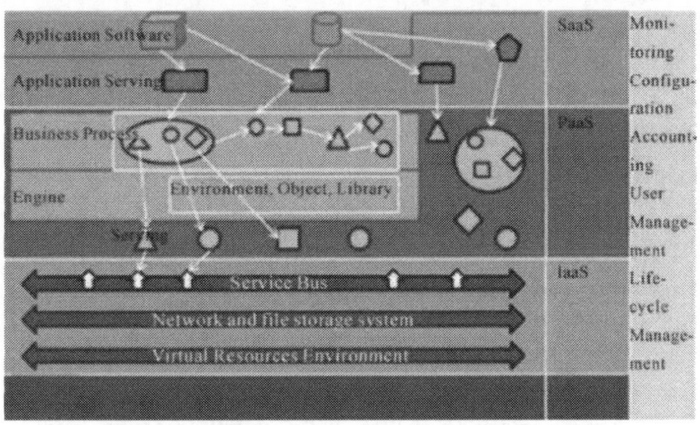


**Figure 3**. Framework Cloud-based BPM (8).

To maximize the BPM output, the cloud-based BPM framework needs to be supported with the use of data mining and DSS technology. It comprised of components of (17): (1) information sources, consists of current ERP and legacy system, point of sales (POS), other OLTP/web access, and external data that are fed as input data; (2) data management, consists of metadata, enterprise data-warehouse and replication; (3) information management, processes data marts of each departments into routine business reporting, OLAP, dashboards, intranet search for content; (4) operation management, carries out data and text mining, optimizing and simulating the data, and also automating decision system for board of directors. Data, information and analytical services are enabled through cloud services (see figure 4).

The figure 4 shows internal service oriented BPM enables to assist developing good BPM culture in organization, such as (18): (1) BPM enables important key activities to be managed and continuously improved to ensure consistent ability to deliver high quality standards of products/services; (2) business processes are critical and all-encompassing activities of design, manufacture, marketing, innovation, sales and others which deliver quality to the end customers; (3) process management constantly strive for excellence and how they stimulate innovation and creativity for process improvement and optimization; (4) BPM includes activities which refer to developing supplier quality management issues; (5) the management processes is conducted through performance measurement for setting targets for improvement and also for measuring product/service capability, process capability, supplier capability and efficiency/effectiveness aspects in terms of cycle time, quality standards, costs, etc.; (6) BPM, through continuous measurement and improvement will determine effectiveness of process design for streamlining and simplification. It ensures introducing best practices through benchmarking information and is based on valuable inputs from customers; (7) process management challenges practices (i.e. the dynamic aspects of each process and its behaviour) as much as the performance of each process (its output/metrics). Process management seeks to continuously strengthen all activities through the introduction of best practice, and to ensure that internal standards of performance are competitively acceptable; (8) BPM enables creating a systematic methodology supported by a problem-solving methodology to strengthen newly-designed processes, to reinforce the linkages between various functions and to ensure that optimum performance can be achieved.

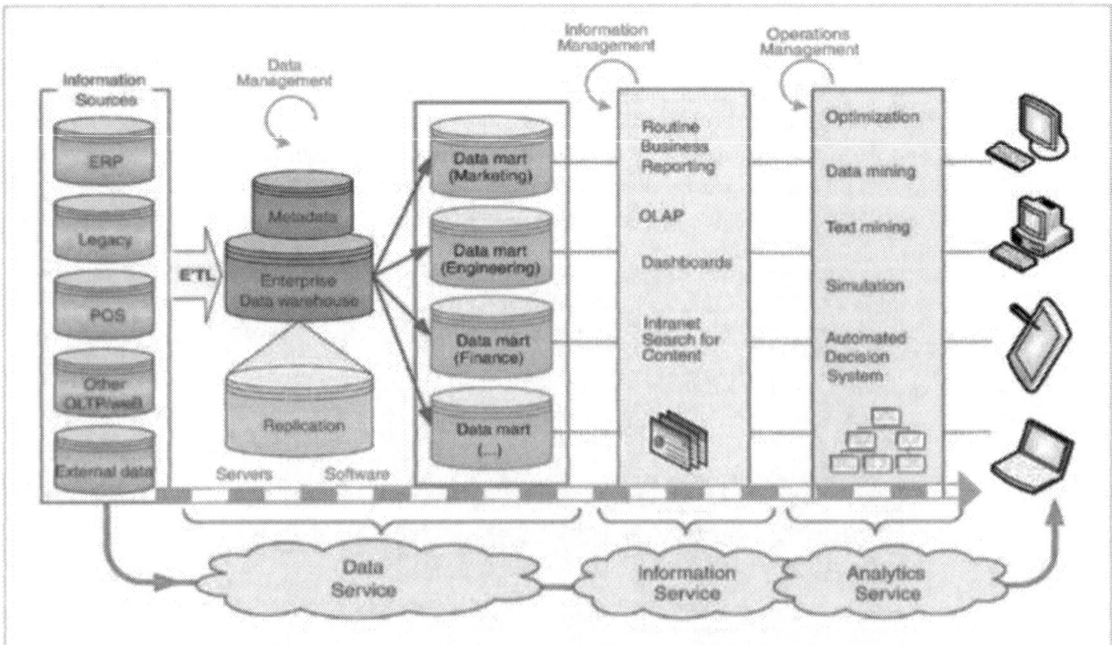


**Figure 4.** Conceptual architecture of service oriented BPM diagram (adapted from (17)).

## 4. Discussion.

Service oriented cloud based BPM enables to address data process inside organizations in the area of (19): (1) low degree of automation in implementation stage: the actual setup of business processes according to managerial needs is mainly done manually, often involving numerous consultants. The use of cloud BPM enables full degree automation in entire business process; (2) implementation delay: The dynamic composition of business processes is mostly impossible, increasing the time to market and reducing an organization's agility. Combining with mining technology has enabled to address specific problem and collaboration process to solve the problems; (3) cognitively inadequate complexity: The lack of a clear separation between business goals and implementation details makes the management of business processes overly complex. Cloud BPM enables to simplify the complexity in business processes and easy access for key stakeholders; (4) process blindness: Managers and other business experts cannot quickly determine whether a specific process can be composed out of existing atomic processes, nor can those stakeholders query the process space within their organization by logical expressions. Thus, checks for process feasibility (e.g. prior to the launch of new products or services) or compliance (e.g. ISO, etc.) are still to be done manually by business analysts. Cloud BPM enables fast query process and check process feasibility/compliance;

## 5. Conclusion.

This article discusses the basic concepts of CC and BPM, and how the cloud-based BPM framework is proposed. Cloud BPM enables easy and affordable cost management according to the needs. Along with the current development of business processes that are constantly changing and improving due to changing business conditions, combining BPM with the cloud is a promising approach to enable enterprises in all sizes to stay competitive and focus to their core business.

Cloud based BPM and supported with mining technology enables to create flexible business process, faster in delivery and provide better service for key stakeholders. CC is an emerging paradigm that enables to develop a better architecture for managing high complexity of business processes in modern enterprises. It involves easy data/information retrieval and processing, guaranteed data availability, fast and flexible information and analytical services.

## References

[1] Hammer M 1990 Reengineering work: don't automate, obliterate *Harv Bus Rev.* **68** (4) pp 104–12
[2] Davenport T H 1998 Putting the enterprise into the enterprise system *Harv Bus Rev.* **76** (4)
[3] Baeyens T 2013 BPM in the cloud. In: Business Process Management *Springer* pp 10–6
[4] Bae H, Cho E and Bae J 2007 A version management of business process models in bpms *Adv Web Netw Technol Inf Manag* pp 534–9
[5] Noel J 2005 BPM and SOA: Better together *White Pap IBM.*
[6] Karagiannis D, Ronaghi F and Fill H-G 2007 Business-oriented IT management: developing e-business applications with E-BPMS *Proceedings of the ninth international conference on Electronic commerce* pp 97–100
[7] Karagiannis D 1995 BPMS: business process management systems *ACM SIGOIS Bull.* **16** (1) pp 10–3
[8] Fang Z and Yin C 2010 BPM architecture design based on cloud computing *Intell Inf Manag.* **2** (5) p 329
[9] Duipmans E F, Pires L F and da Silva S L 2012 Towards a BPM cloud architecture with data and activity distribution. *Enterprise Distributed Object Computing Conference Workshops (EDOCW), 2012 IEEE 16th International.* pp 165–71
[10] Van D A W 2013 Business process management: a comprehensive survey *ISRN Softw Eng.*
[11] Van D A W, Dumas M and Ter H A 2005 Process-aware information systems: bridging people and software through process technology. *Wiley-Interscience*
[12] Tilwani H 2015 Cloud Business Process Management & Data Security. *Int J Comput Sci Technol Inf Technol.* **6** (3) pp 2089–93
[13] Mell P and Grance T et. al 2011 *The NIST definition of cloud computing*
[14] Schulte S, Janiesch C, Venugopal S, Weber I and Hoenisch P 2015 Elastic Business Process

Management: State of the art and open challenges for BPM in the cloud. *Futur Gener Comput Syst.* **46** pp 36–50

[15] Silver B 2009 *BPMS watch: Ten tips for effective process modeling.*

[16] Megersa B T and Zhu W 2012 Cloud-Enabled Business Process Management. *Int J Comput Theory Eng.* **4** (5) p 690

[17] Demirkan H and Delen D 2013 Leveraging the capabilities of service-oriented decision support systems: Putting analytics and big data in cloud. *Decis Support Syst.* **55** (1) pp 412–21

[18] Zairi M 1997 Business process management: a boundaryless approach to modern competitiveness. *Bus Process Manag J.* **3** (1) pp 64–80

[19] Hepp M, Leymann F, Domingue J, Wahler A and Fensel D 2005 Semantic business process management: A vision towards using semantic web services for business process management. *e-Business Engineering, 2005 ICEBE 2005 IEEE International Conference* pp 535–40

# Analysis of flood vulnerability in urban area; a case study in deli watershed

**I Indrawan[1] and R I Siregar[1*]**

[1]Civil Engineering Departement, Faculty of Engineering, Universitas Sumatera Utara, Jl. Dr. Mansur Kampus USU, Padang Bulan, Medan 20155, Indonesia

E-mail: ivan.indrawan@usu.ac.id

**Abstract**. Based on the National Disaster Management Agency of Indonesia, the distribution of disasters and victims died until the year 2016 is the largest flood disaster. Deli River is a river that has the greatest flood potential through Medan City. In Deli Watershed, flow discharge affected by the discharge from its tributaries, the high rainfall intensity and human activity. We should anticipate reducing and preventing the occurrence of losses due to flood damage. One of the ways to anticipate flood disaster is to predict which part of urban area is would flood. The objective of this study is to analyze the flood inundation areas due to overflow of Deli River through Medan city. Two-dimensional modeling by HEC-RAS 5.0.3 is a widely used hydraulic software tool developed by the U.S Army Corps of Engineers, which combined with the HEC-HMS for hydrological modeling. The result shows flood vulnerability in Medan by a map to present the spot that vulnerable about flood. The flooded area due to the overflowing of Deli River consists of seven sub districts, namely Medan Johor, Medan Selayang, Medan Kota, Medan Petisah, Medan Maimun, Medan Perjuangan and Medan Barat.

## 1. Introduction

Flood disaster is one of natural hazard that affected human safety and economic growth in the world. This disaster has a big risk in the future, it supported to climate change, high rainfall intensity, and flood frequency in many area of the world [1]. The system of urban development, land cover and land use change along watershed caused the loss of ecosystem services in urban area and an increase in urban flooding vulnerability. This problem combines the interaction among exposure, sensitivity, and adaptive capacity for studying flood vulnerability. There is construction for flood vulnerable concepts and developing flood emergency that investigated the spatiality of urban flooding vulnerability in a watershed by using geographic information system [2]. The study with high developing urbanization is potential about flood. In addition, the urban area that has bigger potential effect is not always has the bigger flood vulnerable in extreme climate condition. The system can effectively shows urban flooding vulnerability [3].

One of the factors is economic and social development that derived land cover and land use changes. It is potential to effect on water flow in a watershed. Land cover and land use change influence several parameters, they are infiltration, the evapotranspiration; surface runoff and groundwater recharge [4]. The effects in urbanization and population growth are catchment area responses especially to the hydraulic parameters. This condition will influence land cover change that related to landscape modifications. This problem is potential to increase catchments hydrologic response. Meanwhile, based on the observation data, this urbanization is not always tends to rose of flood vulnerability [5]. Some efforts to control the flood that happened in urban area, especially about flood infrastructure, but in modern cities over the world are still vulnerable to flood hazards. Even though, management of water resources focused on non-structural effort to face the flood hazard, the investigation will stay in a statement to control the flood. This case will consider as a disaster and make damage in urban area [6].

Those reviewed studies show that the study of flood vulnerability has to pay attention in urban area. The objective of this study is to analyze the flood inundation areas due to overflow of Deli River through the city of Medan. Flood inundation area will show flood vulnerability area in Medan by using a map. Results suggest this modeling is a feature of urban responses to Deli River overflow across different part regions in Medan. This flood map vulnerability will describe which part of Medan that vulnerable about flood. Limitations of the data obtained and the accuracy of selection of flood modeling methods will reduce the accuracy of the modeling results in predicting flooding. Therefore, the suitability of methods in predicting floods influences the flood control analysis to perform the best result.

## 2. Study Area and Materials

### *2.1. Study Area*

The study area is located in Medan, North Sumatera of Indonesia. It is located in the Deli Watershed that flows from Karo Regency, Deli Serdang Regency and passes Medan city then empties into Melaka Strait. Deli watershed administratively passes several cities and districts; it defines, as one area that has total area is 472 kilometers square and long river main is 55 kilometers. This study focused on Deli River that through Medan city.

The upper forest in Deli watershed is not able to absorb water with maximum, the rainfall that is not absorbed by tree roots in the forest, making the river water flow exceeds the normal limit. If the water seep into the soil so that water reserves continue to be available, if the rainy season is not flooded and the dry season is not short of water, but what happens is flooding in the main river and the river watersheds. Based on the study of JICA (Japan International Cooperation Agency), in 2015 there will be a water crisis in Medan. Then, some areas of Deli Serdang are if the Deli River condition cannot recover.

### *2.2. Data Sources*

In this study, the rainfall data used is the record data from 2006 to 2015. The data of rainfall sourced from the data of Major River Basin Organization of Sumatera, Ministry of Public Works. The data of debit recording in this research obtained from Research Centre and Development of Water Resources, Ministry of Public Works. The data of the discharge serves as the calibration and verification reference of the modeling created. Deli watershed utilizes remote sensing technology to provide information on land coverage by interpretation of objects in the image. The information obtained to show areas that have the most critical and most sensitive land capability. Digital Elevation Model data is used for two-dimensional modeling. This data interprets of topographic data that implemented flood inundation area.

## 3. Methods

### *3.1. Flood Modeling*

The data processing of the land cover and its classification system uses the help the geographic information data. Based on the data obtained there is ten classifications of land cover that will be determined the area of each for year 2010. Then the data will compare into the form of bar charts [7]. The method defines the flood vulnerability by a quantitative magnitude called relative vulnerability score and results in the vulnerability zoning of the area under assessment. The method applied in watershed and the obtained results showed that. The First, indexing method leads to a relative sense and an overall picture of the urban vulnerability toward floods. Secondly, the urban zones surrounded by two rivers, are in danger of a high level of the vulnerability, especially the first floors of the buildings. Then, it is necessary to review the location of the parking garage under construction, with regard to the flood vulnerability zoning [8]. A simple modeling framework provided based on accessible input data and a freely available and widely used hydrological model (HEC-RAS) to check the possible effect of land cover changes at a particular sub-catchment on the hydrograph at the basin outlet [9].

HEC-RAS is a widely used hydraulic software tool developed by the U.S Army Corps of Engineers, which combined with the HEC-HMS platform for hydrological simulations. HEC-RAS employs 1d flood routing in both steady and unsteady flow conditions by applying an implicit-forward finite difference scheme between sequent sections of flexible geometry. In all the above models, two boundary conditions required, which usually set at the upstream end of the channel through an imposed inflow as well as the assumption of uniform water depths at the upstream and downstream end (kinematic wave condition). Although an imposed depth would result in more stable solutions than the uniform flow, we choose the latter since, in practice, it is rare to know the temporal evolution of the water depth at a particular location. The models compute the appropriate time step based on the Courant number stability criteria [10].

This method applied the Hydrologic Engineering Center River Analysis System (HEC-RAS) model to estimate the potential catastrophes for different peak outflow scenarios with conclusions and recommendations [11]. Flooding damage appraisal can been obtained by interpolating real damage data caused by historical flooding events or accounting the effects of a flood in terms of the depreciation of assets. Most often, the expected damage evaluated by means of damage functions describing the relationship occurring between the damage and hydraulic characteristics of flood [12].

### *3.2. Flood Mapping Method*

After incorporating resampling and vertical errors, all resampled raster datasets used to create a 1D HEC-RAS model. HEC-RAS is the most commonly used flood model tool in Indonesia. The HEC-RAS parameters kept unchanged for different DEMs because the goal is to investigate the sensitivity of inundation maps to topographic errors instead of trying to create newly calibrated model for each topographic dataset [13]. Two types of tools, hydrodynamic models and geographic information systems, commonly employed for these evaluations. For this assessment, a coastal digital elevation model developed using the same data on hydrometric and elevation. These models used to project the present and future impact of storm surges on flood extent at Pigeon Point [14].

After obtaining the result of flood potency analysis in the form of modeling showing high flood and flood plain with HEC-RAS software, it can show areas of flood inundation and flood area that occur along the river area, with geographic information system predicted using software. Geographic information systems will work with HEC - Geo RAS extensions in importing data from HEC-RAS into geographic information systems.

### *3.3. Calibration and Validation Method*

The method compares the observed discharge, and then analyzed the parameters that influence the modeling results. Optimum parameter values in calibration found by statistical measures. The result of comparison then determined the strength in predicting the modeling by Nash-Sutcliffe model of efficiency coefficient (E). The Nash–Sutcliffe efficiency (NI) is the two criteria most widely used for calibration and evaluation of hydrological models with observed data [15]. Nash–Sutcliffe Index can range from $-\infty$ to 1. The value of 1 (NI=1) corresponds to a perfect match of modeled discharge to the observed data. The value of 0 (NI = 0) indicates that the model predictions are as accurate as the mean of the observed data, whereas it is less than 0; it occurs when the observed is the better predictor than the model [16]. The verification process based on the observation flow verified parameter is the runoff.

## 4. Results and Discussions

### *4.1. Flood Model*

In this paper, the model predictions of the flood are accurate and outperform the results of a simple rule based approach. Figure Two shows the elevation flood inundation while the water level at the time of the flood reached the banks. The area of flood inundation due to flood discharge with return period 25-years is about 3.69 Kilometers square.

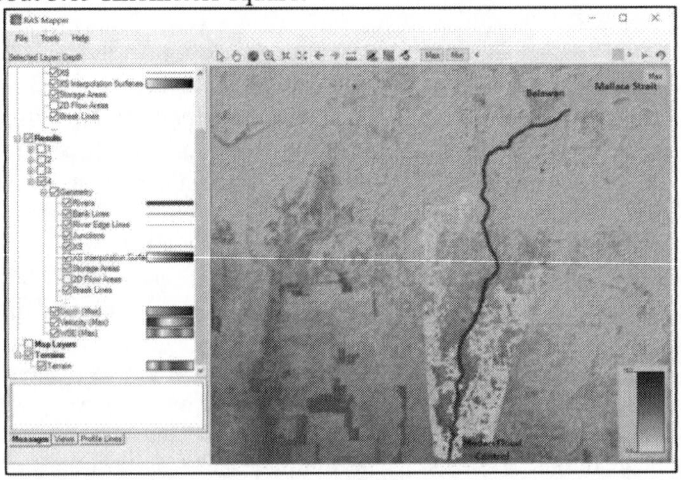

**Figure 1.** Deli River flood inundation when return period is 25 years

### *4.2. Flood Map Vulnerability*

Based on the result of the flood modeling with the help of HEC-RAS and with DEMs obtained flood inundation with Deli watershed, analysis of flooded areas using topographic data shows potential flooding for some areas as shown in the following figure. Figure 3 shows the area that potentially about flood in Deli watershed. Based on this pictures the flooded area due to the overflowing of Deli River consists of seven sub districts, Medan Johor, Medan Selayang, Medan Kota, Medan Petisah, Medan Maimun, Medan Perjuangan and Medan Barat.

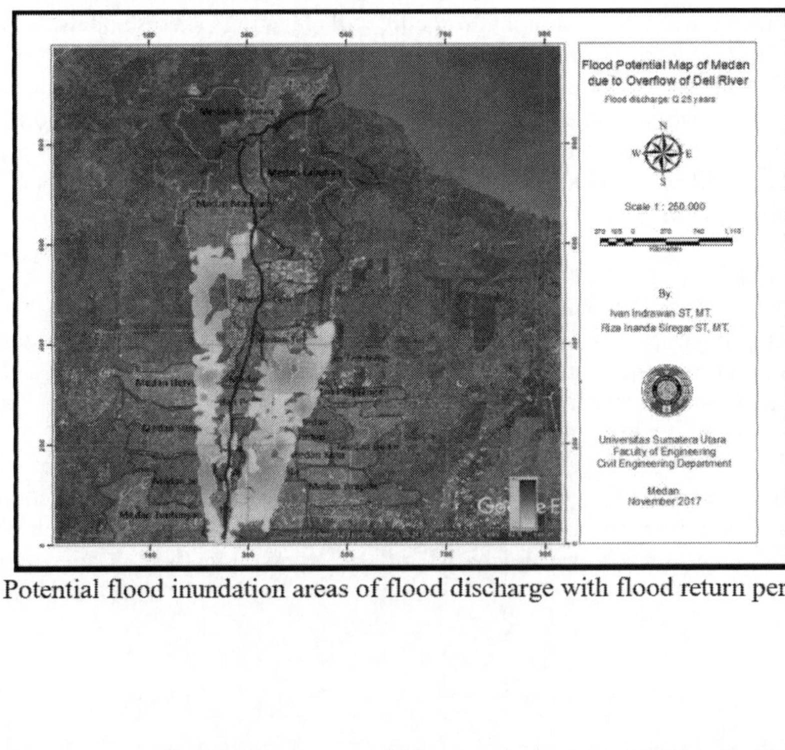


**Figure 3.** Potential flood inundation areas of flood discharge with flood return period 25-year

### *4.3. Validation*

The calibrated model applied for the discharge scenarios to assess the potential flood area impacts on the Nakayasu synthesized hydrographs obtained based on the parameters input into the Deli watershed, they analyzed for calibration and verification. The rain that occurs is the total daily rain that distributed evenly in the watershed. Calibration and verification of Deli Watershed carried out at each of the maximum floods in the study sites for three years, 2007, 2008 and 2010.

## 5. Conclusions

This study resulted in the deluge of the Deli River with a 10-year return period is 537 $m^3/s$; 25 years is 575 $m^3/s$; 50 years is 602 $m^3/s$; and a 100-year return period is 627 $m^3/s$. The inundation area is about 3.69 Kilometers square. The flooded area due to the overflowing of Deli River consists of 7 sub districts, Medan Johor, Medan Selayang, Medan Kota, Medan Petisah, Medan Maimun, Medan Perjuangan and Medan Barat.

**Acknowledgments**

The author gratefully acknowledged that the research institution of Universitas Sumatera Utara, Indonesia supported for the financial. Thanks to Major River Basin Organization of Sumatera, Public Work Ministry, for providing the data. Big thanks to Ahmad Perwira Mulia Tarigan and Makmur Ginting for giving information and knowledge to the authors about this research.

## References

[1] Ouma Y O and Tateishi R 2014 Urban Flood Vulnerability and Risk Mapping Using Integrated Multi-Parametric AHP and GIS: Methodological Overview and Case Study Assessment Water, *Elsevier* **6** 1515-1545

[2] Chang L F and Huang S L 2015 Assessing urban flooding vulnerability with an emergy approach *Elsevier* **143** 11-24

[3] Ajibade I, Mcbean G and Kerr R B 2013 Urban flooding in Lagos, Nigeria: Patterns of vulnerability and resilience among women *Elsevier* **23** 1714-1725

[4] Yira Y, Diekkruger B, Steup G and Bossa A Y 2016 Modeling Land Use Change Impacts on Water Resources in A Tropical West African Catchment (Dano, Burkina Faso) *Elsevier* **537** 187-199

[5] Salavati B, Oudin L, Percot C F and Ribstein P 2016 Modeling Approaches to Detect Land-Use changes: Urbanization Analyzed on a Set of 43 US Catchments *Elsevier* **538** 138-151

[6] Liao K H, Le T A and Nguyen K V 2016 Urban design principles for flood resilience: Learning from the ecological wisdom of living with floods in the Vietnamese Mekong Delta *Elsevier* **155** 69-78

[7] Lamboni M, Koeble R and Leip A 2016 Multi-Scale Land-Use Disaggregation Modelling: Concept and Application to EU Countries *Elsevier* **82** 183-217

[8] Fernandez P, Mourato S, Moreira M and Pereira L 2016 A new approach for computing a flood vulnerability index using cluster analysis *Elsevier* **94** 47-55

[9] Sanyal J, Densmore A L and Carbonneau P 2014 Analysing the Effect Of Land-Use/Cover Changes at Sub-Catchment Levels on Downstream Flood Peaks: A Semi-Distributed Modelling Approach with Sparse Data *Elsevier* **118** 28-40

[10] Dimitriadis P, Tegos A, Oikonomou A, Pagana V, Koukouvinos A, Mamassis N, Koutsoyiannis D and Efstratiadis A 2016 Comparative Evaluation of 1d and Quasi-2d Hydraulic Models Based on Benchmark and Real-World Applications for Uncertainty Assessment in Flood Mapping *Elsevier* **534** 478-492

[11] Butt M J, Umar M and Qamar R 2013 Landslide Dam and Subsequent Dam-Break Flood Estimation Using Hec-Ras Model in Northern Pakistan *Springer* **65** 241-254

[12] Notaro V, Marchis M D, Ffontanazza C M, Loggia G L, Puleo V and Freni G 2014 The Effect of Damage Functions on Urban Flood Damage Appraisal *Elsevier* **70** 1251-1260

[13] Saksena S and Merwade V 2015 Incorporating the Effect of Dem Resolution and Accuracy for Improved Flood Inundation Mapping *Elsevier* **530** 180-194
[14] Seenath A, Wilson M and Miller K 2016 Hydrodynamic versus GIS modelling for coastal flood vulnerability assessment: Which is better for guiding coastal management? *Elsevier* **120** 99-109
[15] Yucel I, Onen A, Yilmas K K and Gochis D J 2015 Calibration and evaluation of a flood forecasting system: Utility of numerical weather prediction model, data assimilation and satellite-based rainfall *Elsevier* **523** 49-66
[16] Gupta H V, Kling H, Yilmaz K and Martinez G F 2009 Decomposition of the Mean Squared Error And NSE Performance Criteria: Implications for Improving Hydrological modeling *Elsevier* **377** 80-91

IOP Conf. Series: Journal of Physics: Conf. Series **978** (2018) 012037 doi:10.1088/1742-6596/978/1/012037

# Study to the current protection of personal data in the educational sector in Indonesia

**E Rosmaini[1*], T F Kusumasari[2], M Lubis[2] and A R Lubis[3]**

[1]Universitas Sumatera Utara, Medan, 20155
[2]Telkom University, Jalan Telekomunikasi No. 1, Bandung, 40257
[3]Politeknik Negeri Medan, Jalan Almamater No. 1, Medan 20155

Email: muharmanlubis@telkomuniversity.ac.id

**Abstract**. This study examines how legal expert interpret UU ITE to protect personal data based on privacy principle by using content analysis. This act has importance in order to govern the process of collection, use, transfer, disclose and store personal data for profit or other commercial purposes. By recognizing both the right of individual for privacy and the need of organization to utilize the customer data, the Act, which was amended by Parliament at October, 27th 2016 have critical role for protection guideline in Indonesia. Increasingly, with the use of advanced technology, data protection became one of the main issues on various sectors, especially in the educational sector. Educational institutions require large amount of personal data to run their business process to support learning, teaching, research and administration process. It involves wide range of personal data from institution, agencies, colleges, lecturers, students and parents, which might consist of several sensitive and confidential data such as historical, health, financial, academic and experience background. Underestimating and ignoring these issues can lead to disaster such as blackmailing, stalking, bullying or improper use of personal data. In aggregation, they might deliver huge loss to institution either financially or trust. Thus, this study analyse the privacy principle of UU ITE through 21 coders from legal expert to obtain more understanding of appropriate approach to implement privacy policy in the educational sector.

## 1. Introduction

There is some changes in UU ITE for the purpose to prevent multi interpretation, to add new right, to decrease criminal threat, to execute the verdict of constitutional court, to synchronize with procedural law, to strengthen the role of investigators and government. The improvement by parliament to meet people's demands should well appreciated. Currently, privacy has obtained substantial concern from educational institution with regards to students' and staff' personal data information in aligning with the government's initiative for a paperless-based administration and information management. Besides that, there are many activities in school and university that involve storing, collecting, transferring and manipulating student and staff data for the purpose of certification, examination, information sharing, etc. Unfortunately, most of individuals do not consider privacy as the necessary matter; when the government collect personal information for certain people, many people will mentioned that they are not worried by said that I have got nothing to hide. Then they will declare that only if you are doing something wrong should you worry and then you have no right to keep it private, which undervalue the right of people to be left alone [1].

In terms of legislation, since April 2016, 108 countries have applied national data protection laws. Ninety-five of these laws, among others, enacted in Singapore, Malaysia and the Philippines are regulations that specifically and comprehensively govern the individual data protection of its citizens except in Indonesia that still use several laws to cater the needs. Although UU ITE has been enacted to protect privacy in general but to provide fundamental law on privacy right and to improve the interaction and communication with other nation economically, Indonesia must have one codified Act on Personal Data Protection [2]. Privacy as individual right has been protected by government fundamentally but not absolute, mostly based on specific context such as information, physical, opinion, group, territorial, communication, association, identity determination and property [3]. Meanwhile, certain privacy aspects has been implemented in various Acts [4] such as UU No. 7/1971 (Basic Provision in Archiving) or UU No. 36/1999 (Telecommunication), which discuss several aspects such as public, access, security, confidential and safety. However, due to the rapid advancement of technology, the current law often not effective to control and regulate personal data protection. This study has objective to analyze specific verse in UU ITE to identify the role of regulation in providing privacy principle for the purpose of developing effective guideline in educational institution.

## 2. Principle of PDP ACT

Several international convention acknowledge and declare privacy as essential human right to be protected such as article 12 of UDHR, article 21 of ASEAN HRD, article 16 of Arab Charter HR, article of article 14 of UNCMW, article 16 of the UNCRC, article 17 of ICCPR, article 4 of the African Union Principles on FE, article 8 of European Convention on HR, article 11 of American Convention of HR; all of them mentioned the importance of privacy. The development of protection measures of personal data based on UU ITE in the online context should accommodates several principles [5, 6, 17] such as:

### *2.1. General Principle*

It sets out the rights and obligations of the data user when processing personal data by virtue of this principle, thus, the processing of personal can only be done for a lawful purpose directly related to data user's activity. It also requires that the data processed must not be excessive (imagine if a bank requires from its customer to declare the history of his illnesses, a data which is not directly related and is likely excessive). UU ITE discussed the general principle under section I article 4 (electronic document), article 9 (electronic certificate) and article 12 (electronic signature). Meanwhile, section III explained the right and responsibilities under article 9 (complete information), article 11 (legal requirement), article 12 (security measures). However, the educational institution is recommended to develop guidelines that accommodate the using of social media platforms in specific activity of students either teaching learning, communication or sharing relevant information about their educational institution.

### *2.2. Notice and Choice Principle*

It prescribes that when collecting personal data, data user shall properly notify the data subjects as to the purpose of that collection or processing, as well as the related rights of data subject with regards to those. This could refer to when the data subject is first asked to provide the information or when the data controller first collects the personal information while at the same time, the data subject has chance to decide either agree with the process or decline it. UU ITE mentioned the implementation of electronic certification and electronic system under section IV article 14 (notification to data subject), verse 15 (proper implementation) and verse 16 (minimum responsibilities). Thus, the educational institution should assure that obvious choice have been given to student and staff in regard to opt-in consent in conducting their primary activities such as submitting report, competition, performance indicator, etc.

### *2.3. Disclosure Principle*

Another way of usage of personal data would be when due to several circumstances institution either school or university discloses personal data for their interest and benefit. Under section VI which lays out the disclosure principle, now prohibits such disclosure without the consent of the data subject, as well as this principle puts forward the restriction on disclosure of the personal data. Specifically, article 26 mentioned that data subject have right to file suit for damages under the provision of law. Meanwhile, the institution should provide clear understanding of several privacy attributes into their privacy policy such as retention, availability, integrity, accuracy, confidentiality, accessibility, operational, procedural and mechanical. Meanwhile, it is better for the educational institution to store personal data in isolated and protected location either physical and digital in which, it has been encrypted against possibility of attackers or illegal disclosure.

### *2.4. Security Principle*

Under section III and IV of the Act, which embodies the security principle, the data controller must take practical steps to protect and safeguard personal data from any loss, misuse, modification, unauthorized or accidental access or disclosure, alteration or destruction. Indeed, where data processor is to be used, there must be sufficient guarantees regarding its security features and measures. On the other hand, section VII (prohibited action) under article 30 discussed in specific the list of actions involve accessing, obtaining or penetrate illegally without proper authority from data controller or data processor while under article 32 stated that modified, added, reduced, transmitted, break, deleted, transferred and hide illegally with intention, which lead confidential data to be accessed may be subjected to the punishment under section XI (criminal provisions) articles 46 and 48. Therefore, educational institution should create clear guidelines for physical and digital data by developing privacy policies and procedure to set reference to show relevant person on their roles of responsibility, set appropriate activity process of data use and determine the acceptable workflows.

### *2.5. Retention Principle*

Regarding the period for retaining personal data, the section IV, which represents the retention principle, does not state the time frame allowed to store personal data but general criteria that instruct the data controller to destroy the data after all related business has ended. The development of guideline supposedly provide specific criteria in which terminate the right for data owner to store or use the data after the purpose for students or staff has been fulfilled except for reference only, so those data should be deleted permanently. Even though, the Act appears to make it optional but this principle can be aligned with general and notice principle that data controller dispose the data after the data no longer require to preserve the trust among them. It is better if the criteria specify the exact time for educational institution to keep data to make it clear for staffs after they change job or student after they graduate. The assessment should be done to consider the god and bad impact that might be influence the result in the educational institution such as certification, grading, health record, examination result, delinquency, parental fees paid, salary, reference, acknowledgement, award, gift, reunion, etc.to balance the right and responsibility by updating the terms and condition that allow relevant person decide for the own benefit.

### *2.6. Data Integrity Principle*

Educational institution in Indonesia, which hold and use data should provide suitable approach to assure that the personal data is accurate, complete, not deceptive and recent to avoid loss to the data owner and improve everyday decision in the institution. Thus, designing the system for data protection should be integrated into daily routine such as removing duplication wherever possible, stop collecting data that the school does not need and centralise as much data as possible under management of information system. Under section IV mentioned the duty of data user to ensure the accuracy and completeness of the personal data they collect, also, data integrity that avoid manipulation and illegal access. It is crucial to train and raise awareness among all staff members and organize continual professional development on data protection, at least once a year.

### *2.7. Access Principle*

PDP law is encouraging for more accountability that expect data users to create specific procedure where data owners or relevant authorized person have privilege to access related personal data and be able to update the information in which they think suit directly or indirectly. Thus, the educational institution should conduct data audit that classify and categorize personal data that they hold through verification and validation to identify what data easily extract, manipulate or delete on a per-person basis or through permission data owners. Under section X article 43 set the process to investigate the violation in the form of illegal access by considering privacy, confidentiality, public service, data integrity and accuracy. The access to the personal data must be differentiated through the restriction or limitation by creating several roles or position to avoid data leaked out of responsibilities.

## 3. Personal Data Practice in Educational Institution

In the last years, organizations have become more decentralized and often deal with unexpected situations whereby, increasingly, with greater interaction of physical and digital aspects but lack of security mechanism. Meanwhile, use outsourcing or part time staff to conduct several internal activities becomes a common practice to reduce costs and increase the performance. On the other hand, it has strong implication on the data protection requirements of organization and policy compliance, which personal data are often disclosed to an external supplier over whom the organization may not have direct control [8]. As a result, the system have difficulty to evaluate whether the objective has been met or the improvement should be made. Hassan [5] explained that the academic activities of educational institution are basically involves the education process (teaching & learning), research works and additional services. In conducting education activities, the various online materials are often stored in separate database according to respective department and transmitted electronically for certain purpose periodically. The most popular reason on why people reluctance to complete online registration form on the website is that there is no information available about how the data will be used (62%) and they do not trust the data collector (60%) [10]. In Canada, a survey showed that 85% of respondents received unsolicited ads, 74% expressed their concerns on misappropriate service and 82% believed they should get their approval before sending any type of ads. Meanwhile, 61% assumed that all types of telemarketing should be rejected despite the possibility they are leaving out on a good opportunity [7]. In educational sector, personal data has been used in various activities, not limited for the academic purpose only, in fact it has been displayed in the blog or social network for the purpose of promotion or communication. The data user usually maintains the data extracted from research with labelling as confidential and secret to assure specific objective that might be protected under intellectual property law. Thus, the data management requires strict security measure to assure personal data could not be disclose to unauthorized person, in which educational institution should be more proactive to assist relevant process upon request from the data subject. Importantly, the strategy need to be synergized between social and ethics approach to develop the level of understanding in the environment that in which using technology approach to prove the capability to do the in encountering cybercrime or preventing violation [9].

## 4. Research Methodology

The use of content analysis has to address prior questions on the origin of the available texts, what they mean and to whom, how they mediate between antecedent and consequent condition, and ultimately, whether they enable the analysts to select valid answer to questions concerning their contexts or not [12]. Percent agreement measure uses the point-by-point reliability or moment-by-moment reliability based on proportion of all observation occasion for which the two observers agree regarding whether or not target behavior occurred [13]. There is never 100% certainty about research results, even when statistical significance is achieved [14].

| | C1 | C2 | C3 | C4 | C5 | C6 | C7 | C8 | C9 | C10 |
|---|---|---|---|---|---|---|---|---|---|---|
| C1 | | 77.42 | 43.01 | 30.11 | 40.86 | 17.20 | 55.91 | 68.81 | 6.45 | 6.45 |
| C2 | 0.405 | | 36.56 | 24.73 | 37.63 | 35.48 | 58.07 | 64.52 | 3.27 | 3.27 |
| C3 | 0.178 | 0.108 | | 51.61 | 63.44 | 27.96 | 37.63 | 33.33 | 16.13 | 13.98 |
| C4 | 0.111 | 0.035 | 0.408 | | 80.65 | 32.26 | 25.81 | 26.88 | 18.28 | 18.28 |
| C5 | 0.162 | 0.131 | 0.541 | 0.763 | | 29.03 | 34.41 | 34.41 | 12.90 | 15.05 |
| C6 | -0.018 | 0.154 | 0.122 | 0.147 | 0.131 | | 25.81 | 37.63 | 21.51 | 19.36 |
| C7 | 0.101 | 0.219 | 0.157 | 0.064 | 0.116 | 0.126 | | 55.91 | 5.38 | 4.30 |
| C8 | 0.131 | 0.12 | 0.053 | 0.054 | 0.074 | 0.169 | 0.146 | | 4.30 | 12.90 |
| C9 | 0.023 | -0.014 | 0.013 | 0.02 | -0.014 | 0.008 | 0.008 | 0.008 | | 49.46 |
| C10 | 0.014 | -0.024 | -0.014 | 0.017 | 0.011 | -0.042 | -0.006 | 0.088 | 0.01 | |

| | C11 | C12 | C13 | C14 | C15 | C16 | C17 | C18 | C19 | C20 | C21 |
|---|---|---|---|---|---|---|---|---|---|---|---|
| C1 | 29.03 | 45.16 | 35.48 | 59.14 | 30.11 | 90.32 | 52.69 | 62.37 | 35.48 | 26.88 | 33.33 |
| C2 | 32.26 | 49.46 | 40.86 | 51.61 | 27.96 | 77.42 | 43.01 | 62.37 | 29.03 | 24.73 | 34.41 |
| C3 | 36.56 | 21.51 | 23.66 | 38.71 | 26.88 | 46.24 | 18.28 | 34.40 | 25.80 | 26.88 | 32.26 |
| C4 | 26.88 | 18.28 | 29.03 | 34.41 | 23.66 | 31.18 | 20.43 | 25.81 | 27.96 | 24.73 | 24.73 |
| C5 | 35.48 | 21.51 | 27.96 | 36.60 | 21.51 | 40.86 | 16.13 | 34.41 | 21.51 | 25.81 | 33.33 |
| C6 | 25.81 | 34.41 | 25.81 | 41.96 | 30.11 | 21.51 | 27.96 | 25.81 | 26.88 | 29.03 | 30.11 |
| C7 | 46.24 | 38.71 | 38.71 | 39.79 | 21.51 | 58.07 | 38.71 | 54.84 | 26.88 | 27.96 | 32.26 |
| C8 | 41.94 | 54.84 | 40.86 | 67.74 | 31.18 | 72.04 | 53.76 | 60.22 | 23.66 | 30.11 | 35.48 |
| C9 | 16.13 | 12.90 | 12.90 | 4.30 | 32.26 | 2.15 | 27.96 | 12.90 | 48.39 | 51.61 | 46.24 |
| C10 | 17.20 | 8.60 | 8.60 | 10.75 | 32.26 | 3.23 | 22.58 | 9.68 | 47.31 | 48.39 | 45.16 |

| | C11 | C12 | C13 | C14 | C15 | C16 | C17 | C18 | C19 | C20 | C21 |
|---|---|---|---|---|---|---|---|---|---|---|---|
| C11 | | 33.33 | 26.88 | 30.10 | 12.90 | 36.56 | 29.03 | 40.86 | 31.18 | 40.86 | 48.39 |
| C12 | 0.147 | | 41.94 | 43.01 | 34.40 | 49.46 | 64.52 | 44.09 | 19.36 | 20.43 | 24.73 |
| C13 | 0.061 | 0.156 | | 34.41 | 27.96 | 40.86 | 36.56 | 47.31 | 12.90 | 16.13 | 20.43 |
| C14 | 0.166 | 0.062 | 0.042 | | 38.71 | 62.37 | 41.96 | 46.24 | 25.81 | 19.36 | 25.81 |
| C15 | -0.055 | 0.104 | 0.074 | 0.14 | | 26.88 | 37.63 | 30.10 | 31.18 | 31.18 | 31.18 |
| C16 | 0.078 | 0.106 | 0.087 | 0.25 | 0.014 | | 52.69 | 66.67 | 34.41 | 24.73 | 32.29 |
| C17 | 0.036 | 0.472 | 0.118 | 0.1 | 0.155 | 0.095 | | 47.31 | 36.56 | 27.96 | 31.18 |
| C18 | 0.146 | 0.123 | 0.233 | 0117 | 0.085 | 0.193 | 0.12 | | 27.96 | 38.71 | 39.79 |
| C19 | 0.11 | 0.041 | -0.017 | 0.097 | 0.066 | 0.119 | 0.145 | 0.029 | | 60.22 | 60.22 |
| C20 | 0.225 | 0.065 | 0.021 | 0.045 | 0.072 | 0.031 | 0.049 | 0.178 | 0.324 | | 72.04 |
| C21 | 0.304 | 0.093 | 0.052 | 0.094 | 0.077 | 0.05 | 0.053 | 0.137 | 0.319 | 0.505 | |

| | C1 | C2 | C3 | C4 | C5 | C6 | C7 | C8 | C9 | C10 |
|---|---|---|---|---|---|---|---|---|---|---|
| C11 | -0.025 | 0.057 | 0.195 | 0.102 | 0.18 | 0.15 | 0.231 | 0.178 | -0.009 | 0.01 |
| C12 | 0.057 | 0.125 | -0.031 | -0.042 | -0.034 | 0.075 | 0.075 | 0.187 | 0.055 | -0.003 |
| C13 | 0.023 | 0.096 | 0.024 | 0.108 | 0.072 | 0.024 | 0.118 | 0.068 | 0.047 | -0.004 |
| C14 | 0.214 | 0.088 | 0.182 | 0.164 | 0.157 | 0.168 | 0.071 | 0.369 | -0.03 | 0.023 |
| C15 | 0.058 | 0.012 | 0.086 | 0.045 | 0.024 | -0.009 | 0.019 | 0.046 | 0.074 | 0.055 |
| C16 | 0.647 | 0.367 | 0.218 | 0.122 | 0.155 | 0.032 | 0.111 | 0.163 | 0 | 0.002 |
| C17 | 0.115 | 0.001 | -0.092 | -0.004 | -0.111 | 0.08 | 0.014 | 0.166 | 0.151 | 0.078 |
| C18 | 0.126 | 0.226 | 0.091 | 0.062 | 0.097 | 0.101 | 0.192 | 0.142 | 0.037 | 0.001 |
| C19 | 0.122 | 0.086 | 0.068 | 0.128 | 0.032 | 0.121 | 0.094 | 0.023 | 0.153 | 0.129 |
| C20 | 0.045 | 0.058 | 0.085 | 0.087 | 0.085 | 0.149 | 0.128 | 0.128 | 0.177 | 0.122 |
| C21 | 0.054 | 0.126 | 0.136 | 0.084 | 0.163 | 0.178 | 0.133 | 0.14 | 0.123 | 0.108 |

**Table 1.** Pairwise Percent Agreement (Orange) & Cohen Cappa (Blue) of UU ITE

When kappa values are below 0.60, the confidence intervals about the obtained kappa are sufficiently wide that one can surmise that about half the data may be incorrect [15]. Both percent agreement and kappa have strengths and limitations. The percent agreement statistic is easily calculated and directly interpretable [15]. The highest score for the percent agreement is 90.32% (coder1 & coder16), followed by 80.65% (coder5 & coder4) and 77.42% (coder16 & coder2). Meanwhile, the lowest score is 2.15% (coder9 & coder16) followed by 3.23% (coder10 and coder16) and the same value for both coder9 and coder10 over coder2 with 3.27%. There is 63 out of 210 of percent agreement, which has score above 40% with the exception of both closed-limit score of 39.79% (coder7 & coder14, coder18 & coder21). Therefore, the low average score occurred because three type of coding rules combined. It can also be interpreted as general picture on the quality of the current regulation has weak support based on coders' perspective and has no clarity in the context of PDP. It can be concluded that there is the range of 9.68%, 19.35% and 22.58% of the amount of data, which misrepresents the research data or incorrect data. The highest score in Cohen's kappa is 0.763 (coder4 & coder5), was followed by 0.647 (coder1 & coder16) and 0.472 (coder12 & coder17). For its reliability, it means that 24%, 35% and 53% of the data being analysed are erroneous. In the end, there is 21 out of 210 value of kappa, which below zero value – one have an exact zero, while there is three values in the percent agreement, which has score below 5% as the boundaries. Meanwhile, there is 23 out of 210 kappa value, which above 0.2 as the boundaries to be considered as the depiction of moderate level agreement among coders, though in the average, the agreement level was low. Despite the lack of agreement between coder, this act has the vital role as the basis for the direction of PDP in Indonesia in which the development of privacy policy can consider other perspectives to increase the concern on privacy [11].

## 5. Conclusion

To maintain the enforcement of personal data protection, educational institution is encouraged to create the periodical auditory, proper compliance and privacy impact assessment towards the implementation of the policy. Thus, the policy should measure the attribute of the process, data subject behavior and characteristics. Educational institution must ensure that all component; the system consisted of human, technology, task and structure; culture and policy aligned with relevant and related personal data protection act. It will determine whether the effort to protect the personal data will be effective or not in supporting the learning and teaching activities. The intention of having security policy was not to persuade users but to convince them, by letting the users reflect, on their own terms, on why information security is important and on how to react in certain circumstances [16].

## References

[1] Solove D J 2011 Why Privacy Matters even if you have nothing to hide *The Chronicle of Higher Education The Chronicle Review* May 15$^{th}$

[2] Latumahina R E 2014 Aspek Hukum Perlindungan Data Pribady di Dunia Maya *Jurnal Gema Aktualita* Vol 3 **2**

[3] Budhijanto D 2003 The Present and Future of Communication and Information Privacy in Indonesia *Jurnal Hukum Internasional University Padjadjaran* Vol 2 **2** pp 140

[4] Makarim E 2005 Pengantar Hukum Telematika (Suatu Kompilasi Kajian) *Jakarta: Raja Grafindo Persada*

[5] Hassan K M 2011 Personal Data Protection in the Business of Higher Education: Malaysian Law *International Conference on Sociality and Economics Development.* IPEDR Vol 10 (IACSIT Press, Singapore) pp 53-57

[6] Zulhuda S and Delpisheh M 2011 Personal Data "Up in the Air" *International Conference on Social Science and Humanity* IPEDR Vol 5 (IACSIT Press, Singapore) pp V1-300-V1303

[7] Sarabdeen J 2008 Laws on Right to Privacy in Mobile Marketing *WSEAS TRANSACTION on Business and Economic* Vol 5 ISSN: 1109-9526 **5** pp.280-290..

[8] Zannone N, Petkov M and Etalle S 2010 Towards Data Protection Compliance *Proceedings of the 5$^{th}$ International Conference on Security and Cryptography (SECRYPT)* pp 213-216

[9] Lubis M and Maulana F A 2010 Information and Electronic Transaction Law Effectiveness (UU-ITE) in Indonesia *IEEE ICT4M* pp C-13-19

[10] Noor R, Elena and Jawahitha 2003 Security and Privacy Issues as Barriers to E-Commerce Growth: A Consumer Perspective *Proceedings the International Business Information Management Conference* (Cairo December 16-18) p 114

[11] Lubis M, Kartiwi M and Zulhuda S 2017 Privacy and Personal Data Protection in Electronic Voting: Factors and Measures *Telkomnika* March vol. **15**(1) pp 512-521

IOP Conf. Series: Journal of Physics: Conf. Series **978** (2018) 012037 doi:10.1088/1742-6596/978/1/012037

[12] Krippendorff K 2004 Content Analysis: An Introduction to its Methodology *London, England: Sage Publication*

[13] Birkimer J C and Brown J H 1979 Back to Basics: percentage agreement measures are adequare, but there are easier ways *Journal of Applied Behavior Analysis* **12**(4) pp 535-543

[14] Marusteri M and Bacarea V 2010 Comparing groups for statistical differences: how to choose the right statistical test? *The Journal of Croatian Society of Medical Biochemistry and Laboratory Medicine* **20**(1) pp 15-32

[15] McHugh M L 2012 Interrater Reliability: the Kappa Statistic *The Journal of Croatioan Society of Medical Biochemestry and Laboratory Medicine* **22**(3) pp 276-282

[16] Ahlan A, Lubis M and Lubis A R 2015 Information Security Awareness at the Knowledge-Based Institution: Its Antecedents and Measures *Procedia Computer Science* **72** 361–373

[17] governorsagenda 2017 Data Protection law to be strengthened in 2018: Impact on Schools *https://governorsagenda.co.uk/2017/04/18/data-protection-law-to-be-strengthened-in-2018-impact-on-schools/*

IOP Conf. Series: Journal of Physics: Conf. Series **978** (2018) 012038 doi:10.1088/1742-6596/978/1/012038

# A real time mobile-based face recognition with fisherface methods

**D Arisandi[1], M F Syahputra[1], I L Putri[1], S Purnamawati[1], R F Rahmat[1] and P P Sari[1]**

[1]Department of Information Technology, Faculty of Computer Science and Information Technology, Universitas Sumatera Utara, Medan, Indonesia

Email: dedyarisandi@usu.ac.id | nca.fadly@usu.ac.id | iinleo096@gmail.com | sarahpurnamawati@usu.ac.id | romi.fadillah@usu.ac.id | perdana131402029@gmail.com

**Abstract**. Face Recognition is a field research in Computer Vision that study about learning face and determine the identity of the face from a picture sent to the system. By utilizing this face recognition technology, learning process about people's identity between students in a university will become simpler. With this technology, student won't need to browse student directory in university's server site and look for the person with certain face trait. To obtain this goal, face recognition application use image processing methods consist of two phase, pre-processing phase and recognition phase. In pre-processing phase, system will process input image into the best image for recognition phase. Purpose of this pre-processing phase is to reduce noise and increase signal in image. Next, to recognize face phase,we use Fisherface Methods. This methods is chosen because of its advantage that would help system of its limited data. Therefore from experiment the accuracy of face recognition using fisherface is 90%.

## 1. Introduction

At recent time, it cannot be denied that society benefits from ease of facility in cyberspace. Technology cannot be separated by society; that society could affects the progress of technology; and a thesis from determinist group could be dropped is one of many examples where technology deviates from its purpose originally, or its whole different effect comes from same technology [1], so technology can be utilized for a lot of things to obtain necessary information.

Face Perception or Face Recognition is a computational process in learning feature of face on image. As technology develops, face recognition is no longer able to utilized only in a desktop, but also in a smartphone, so it is more practical and able to be utilized anywhere in real time.

However, a problem arises when applying face recognition with data from university. Websites for student directories only have one image for every students in the university. This could give a significant effect due to lack of training data which will give impacts on accuracy of face recognition. By utilizing Fisherface Methods, this problem is expected to be solved.

## 2. Identification of Problems

Problem in this research is student identification process in a university using data from its public university websites. However, data from that website directory only have one image per person.

Therefore author will explain a methods where face recognition system could identify from data in directory from students directory and display the result to an application in Android operating system.

## 3. Previous research

Eigenfaces vs fisherfaces recognition using class specific linear projection [2] discuss the accuracy result in face recognition between Fisherface Methods and Eigenface Methods with specific linear projection on image objects in varying lighting and facial expressions. In this research, author concludes that Fisherface Methods is better at identifying face with varying facial expressions and lighting.

Pseudo-Fisherface Method for Single Image Per Person Face Recognition [3] discuss a modified method of Fisherface by researchers where the sample image for every identity is one. With this methods called Pseudo-Fisherface, every person could have several sample but only one dataset is saved and used in the system.

XFace: A Face Recognition System for Android Mobile Phones [4] discuss a face recognition with XFace Methods that developed by the researcher in mobile operating system, Android. XFace Methods is a face recognition with several pre-processing such as ROI (Region of Interest), LBP (Local Binary Pattern) feature extraction, reduction of feature extraction dimension with PCA (Principal Component Analysis) and LDA (Linear Discriminant Analysis) and classification of minimum distance.

## 4. Research methods

The general architecture which describes these methods which will be executed by the system can be seen in Figure 1.

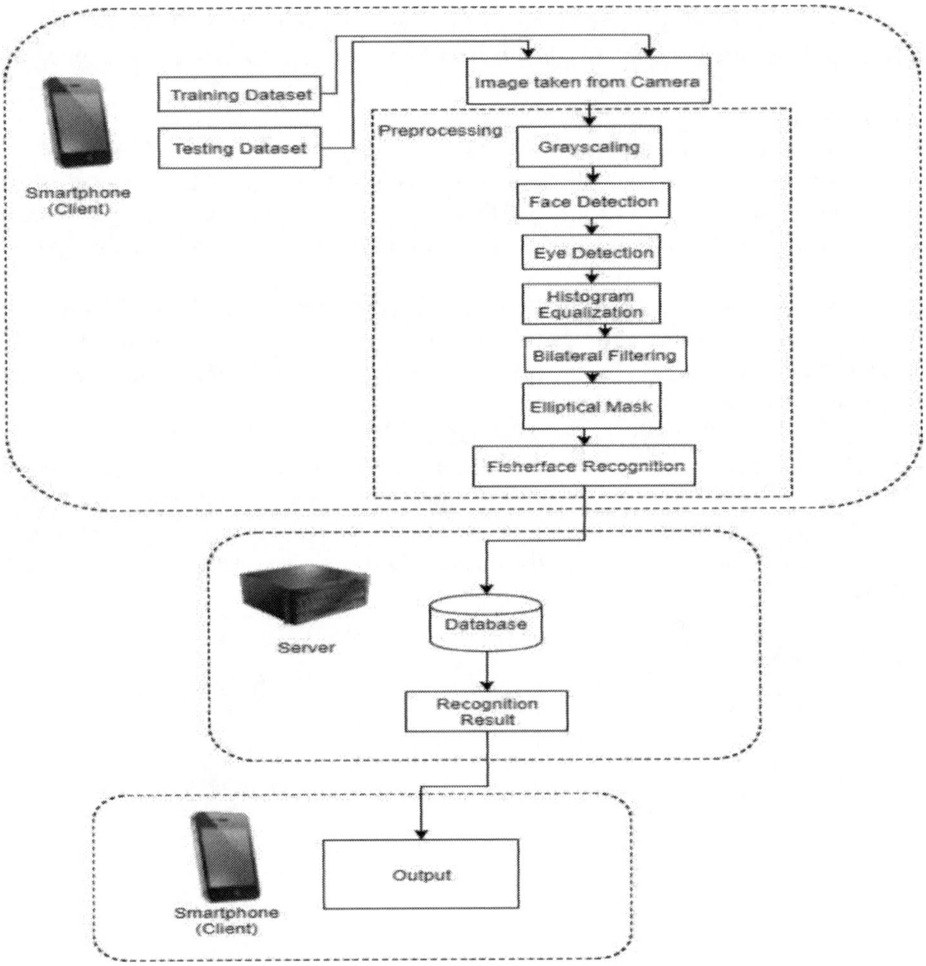


**Figure 1.** General architecture

*4.1. Grayscaling*

In this section, system will execute grayscaling on image. Grayscaling is a process to transform color image into grayscale image. Pixels image which originally consist value of RGB (Red Green Blue) and changed into Gray value in Grayscaling process[8]. The purpose of this grayscaling process is to reduce parameters which will be processed in next stage so the process would be faster.

*4.2. Face Detection*

In this section, the gray image will be processed to detect largest object in the image which probably is a face. This face detection process utilizes LBP (Local Binary Pattern) Methods Frontal Face. The reason for this is because LBP Methods uses minimum resource while reducing the accuracy about 10% to 20% when compared to Haar Classification [5].

*4.3. Eye Detection*

In this section, system will detect if there are eyes in the image. Once the system detect both eyes, the image will be rotated, resized, and translated so the result image will be an image with two eyes arranged to be horizontal from one with another.

In this system, two cascade classifier file will be used, one is Haar Cascade Eye and Haar Cascade Eye Tree Eyeglasses. Haar Cascade Eye is a cascade classifier which contains trained data for basic eye in image, while Haar Cascade Eye Tree Eyeglasses is a cascade classifier which contains trained data for eye obstructed by eyeglasses in image

### *4.4. Histogram Equalization*
In this section, system will execute a leveling on a histogram to leveling its lighting on an image. Usually, an image of face have a brighter side in one side when compared with its other side. Histogram equalization usually increases global contrast of many images, especially when the usable data of the image is represented by close contrast values [6]. Therefore, to reduce noise in that lighting, system will execute both side separately and entirely. The result of those process will be combined to make a whole new image consist of processed value from those images.

### *4.5. Bilateral Filtering*
In this section, system will execute a process called Filtering, In image processing, this is one of most crucial process. The reason why it is crucial is because it reduce noises in image which affects greatly on processing. In this system, Filtering process which will be executed is called Bilateral Filtering. Bilateral Filtering is a filtering called blurring the image while keeping the edge of image [7].

### *4.6. Elliptical Mask*
In this section, another filtering will be applied to processed image with elliptical shape to crop background and unnecessary image. This process is called Elliptical Mask since a layer with elliptical shape is used to mask the image. The purpose of this process is to minimalize noise in background and hair.

### *4.7. Fisherface Recognition*
Fisherface Recognition used a feature extraction method to look for class-specific linear equations, Linear Discriminant Analysis (LDA) or also called as Fisher's Linear Discriminant Analysis (FLDA) which is improved for Fisherface Recognition.

Fisherface Methods in OpenCV library require data training and data testing are in grayscale image. This Fisherface Methods also assumes that training data has same size with testing data. The disadvantage of Fisherface Method in OpenCV library is it doesn't support update.

Recognition result from Fisherface is the label data in traindata.yml file for its feature has the less difference in data testing. This result will be generated by the system if the current data has similarity in training data which is called confidence in this system. The less confidence between the data, system will assume that the data is similar. For this research, threshold value for confidence used in the system is 0.7.

### *4.8. Database*
Database in server has 2 tables, Face_Trait and sqlite_sequence. Face_Trait table is table where student data from directory is saved. This table has 4 fields which are id, image_name, nim, and name. Field id is an integer data with autoincrement, while image_name, nim, and name is a text-type of data. Sqlite_sequence table is a table generated by sqlite to save data for autoincrement for id in Face_Trait table.

### *4.9. Recognition Result*
After recognition result is matched in the database, server will send information about student based on recognition in JSON form to Client. In user's smartphone, system will shows data about student's name which is recognized by system.

## 5. Result and Discussion
Before testing the application, system will be trained with 40 training data. Then testing is executed with 20 testing data. There's two phase in testing data, Detection Testing and Recognition Testing.

IOP Conf. Series: Journal of Physics: Conf. Series **978** (2018) 012038 doi:10.1088/1742-6596/978/1/012038

### *5.1. Detection Testing*

Data is given value "True" if face detection result from system is the same as expected from author, and is given value "False" if the result isn't the same as what to be expected. Each left or right eye is given value "Detected" if the eye could be detected with more than 50% of the time, and is given value "Hard to Detect" if the eye is detected for less or 50% of the time. Eye detection for each side is important because with detected eyes, then the system will assume that the face faces front and will execute recognition process. For face and eye detection result can be viewed in Table 1.

**Table 1.** Result of Detection Testing

| No | Image | Face Detected | Left Eye Detected | Right Eye Detected |
|---|---|---|---|---|
| 1 | 151402001.jpg | True | Detected | Detected |
| 2 | 151402002.jpg | True | Detected | Detected |
| 3 | 151402003.jpg | True | Detected | Detected |
| 4 | 151402006.jpg | True | Detected | Detected |
| 5 | 151402007.jpg | True | Detected | Detected |
| 6 | 151402009.jpg | True | Detected | Detected |
| 7 | 151402010.jpg | True | Detected | Detected |
| 8 | 151402013.jpg | True | Detected | Detected |
| 9 | 151402014.jpg | True | Detected | Detected |
| 10 | 151402015.jpg | True | Detected | Detected |

Based on testing result, face detection in system is calculated as follows:

$$Accuracy = \frac{sum\ of\ data\ detected\ correctly}{sum\ of\ testing\ data} \times 100\%$$

$$Accuracy = \frac{20}{20} \times 100\%$$

$$Accuracy = 100\%$$

IOP Conf. Series: Journal of Physics: Conf. Series **978** (2018) 012038 doi:10.1088/1742-6596/978/1/012038

### *5.2. Recognition Testing*

For recognition testing, system is tested by the person directly as seen in Figure 2.

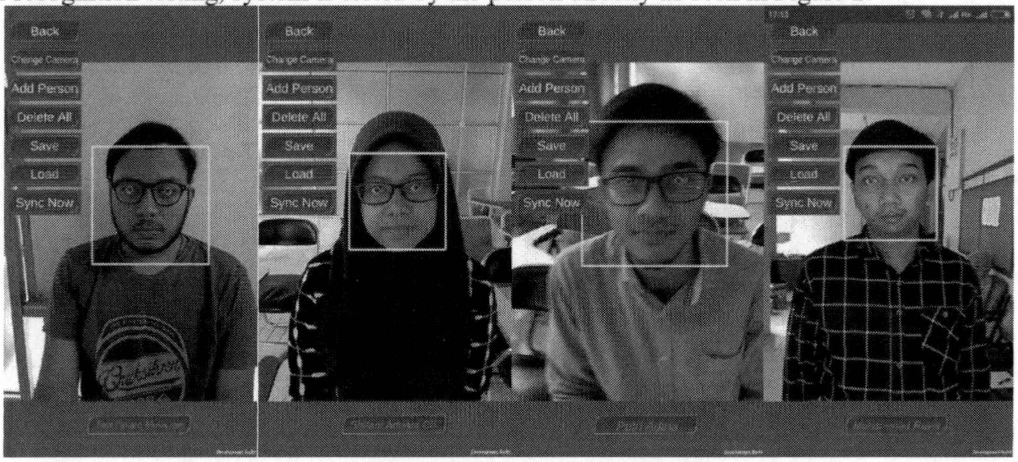


**Figure 2.** Direct testing to students

For result of this testing can be viewed in Table 2 as follows:

**Table 2.** Result of Recognition with Data Testing

| No. | Expected Result | Generated Result | Result |
|---|---|---|---|
| 1 | Surya Permana Putra | Surya Permana Putra | True |
| 2 | Shifani Adriani Ch. | Shifani Adriani Ch. | True |
| 3 | Sandy Pradana | Sandy Pradana | True |
| 4 | Ayu Dwi Rizky | Ayu Dwi Rizky | True |
| 5 | Muhammad Ravie | Muhammad Ravie | True |
| 6 | Grace Lusiana Siregar | Grace Lusiana Siregar | True |
| 7 | Riska Amaliyah | Riska Amaliyah | True |
| 8 | Zuhroh Nilakandi Maulida | Zuhroh Nilakandi Maulida | True |
| 9 | Muhammad Ilham Rizky | Muhammad Ilham Rizky | True |
| 10 | Hizkia Tarigan | Hizkia Tarigan | True |
| 11 | Riyandi Syahputra | Riyandi Syahputra | True |
| 12 | Muhammad Ari Winata | Muhammad Ari Winata | True |
| 13 | Putri Ramadannia | Putri Ramadannia | True |
| 14 | Rizki Sari Dewi | Rizki Sari Dewi | True |
| 15 | Winda Indriani Tanjung | Winda Indriani Tanjung | True |
| 16 | Reno Maulina | No Result | False |
| 17 | Faris Zharfan Alif | Faris Zharfan Alif | True |
| 18 | Tata Feraro Mukarram | Tata Feraro Mukarram | True |
| 19 | Rhama Perdani Ahmad | Putri Adina | False |
| 20 | Rany Ervina Gultom | Rany Ervina Gultom | True |

From 20 data testing which is tested, on 18 data result is not what expected because confidence which is generated by system between data testing and other system data is lower than supposed system data. This would cause system to give wrong identification to Client. Based on testing result, recognition testing from data testing is calculated as follows:

$$Accuracy = \frac{sum\ of\ data\ recognized\ correctly}{sum\ of\ testing\ data} \times 100\%$$

$$Accuracy = \frac{18}{20} \times 100\%$$

$$Accuracy = 90\%$$

Based on calculated result, the accuracy with Fisherface methods reached 90%. Overall testing result can be viewed in Table 3:

**Table 3.** Overall Testing Result

| Testing | Sum of Correct Data | Sum of Total Data | Accuracy |
|---|---|---|---|
| Face Detection | 20 | 20 | 100% |
| Eye Detection | (20+20) / 2 | 20 | 100% |
| Fisherface Recognition | 18 | 20 | 90% |

## 6. Conclusions

Based on testing and research on collected data, several conclusions can be drawn: Student directory data from Universitas Sumatera utara can be used as data training. While Face recognition with data directory is successful with 90% accuracy. Preprocessing in system raise accuracy in face recognition.

## Acknowledgment

The authors gratefully acknowledge that the present research is supported by Ministry of Research and Technology and Higher Education Republic of Indonesia. The support is under the research grant TALENTA USU scheme PENELITIAN DASAR of Year 2017 Contract Number 194/UN5.2.3.1/PPM/KP-TALENTA USU/2017.

## References

[1] MacKenzie D and Wajcman J 1985 The social shaping of technology *Prometheus Critical Studies in Innovation* vol 3 issue 2

[2] Belhumeur P N, Hespanha J P and Kriegman D J 1996 Eigenfaces vs fisherfaces recognition using class specific linear projection *European Conf. Computer Vision* pp 45–58

[3] Majumdar A and Ward A K 2008 Pseudo-fisherface method for single image per person face recognition *IEEE International conference on acoustics, speech and signal processing* pp 989-992

[4] Hu J, Peng L and Zheng L 2015 XFace: A face recognition system for android mobile phones *IEEE 3rd international conference on cyber-physicial systems, network, and applications* pp 13-18

[5] Chen D-S and Liu Z-K 2007 Generalized haar-like feature for fast face detection *IEEE Proceedings of the sixth international conference on machine learning and cybernetics* pp 2131-2135

[6] Zhihong W and Xiaohong X 2011 Study on histogram equalization *IEEE International Symposium on Intelligence Information Processing and Trusted Computing* pp 177-179

[7] Paris S, Kornprobst P, Tumblin J and Durand F 2008 Bilateral filtering: Theory and applications *IEEE Foundation and trends in computer graphics and vision* vol 4 no 1

[8] Rahmat R F, Chairunnisa T, Gunawan D and Sitompul O S 2016 Skin color segmentation using multi-color space threshold *International Conference on Computer and Information Sciences, ICCOINS 2016 – Proceedings* pp 391-396

[9] Rahmat R F, Pulungan A F, Faza S and Budiarto R 2017 Image classification of ribbed smoked sheet using Learning Vector Quantization *Journal of Physics: Conference Series* 801, p 12050

# Analysis of factors that inhibiting implementation of Information Security Management System (ISMS) based on ISO 27001

**R Tatiara[1], A N Fajar[1], B Siregar[2] and W Gunawan[1]**

[1]Information Systems Management Department, BINUS Graduate Program-Master of Information Systems Management, Bina Nusantara University Jakarta, Indonesia
[2]Department of information technology, Universitas Sumatera Utara, Medan, Indonesia,

Email: rullytatiara@yahoo.com , afajar@binus.edu , gwang@binus.edu , baihaqi@usu.ac.id

**Abstract**. The purpose of this research is to determine multi factors that inhibiting the implementation of the ISMS based on ISO 2700. It is also to propose a follow-up recommendation on the factors that inhibit the implementation of the ISMS. Data collection is derived from questionnaires to 182 respondents from users in data center operation (DCO) at bca, Indonesian telecommunication international (telin), and data centre division at Indonesian Ministry of Health. We analysing data collection with multiple linear regression analysis and paired t-test. The results are multiple factors which inhibiting the implementation of the ISMS from the three organizations which has implement and operate the ISMS, ISMS documentation management, and continual improvement. From this research, we concluded that the processes of implementation in ISMS is the necessity of the role of all parties in succeeding the implementation of the ISMS continuously.

## 1. Introduction

According to [1], information is one of the most valuable assets of an organization. According to [2], information security is a protection against the confidentiality, integrity and availability of information assets, whether in storage, processing, or transmission. According to [3] information security is the preservation of confidentiality, integrity, and availability of information. Confidentiality indicates that an information is not available or disclosed to individuals, entities, and processes that are not authorized for access the information. Integrity indicates that an information is protected against accuracy and completeness. Availability indicates that information can be accessed and used as requested by the entity which has having authority over the information, relate to the important asset to the organization. Decision makers in organization can make the important decisions with accurate and timely information while managing the organization. The information that has belonging to sensitive and critical organizations must also be secured. Meanwhile, the purposes is for protected accessed by unauthorized parties. Besides that, Information leakage is very detrimental because it can reduce the competitiveness of the organization and can also reduce the reputation of the organization. To secure information in an integrated, effective and efficient manner requires a good management framework. One example of a management framework created specifically for the management of information security is the ISO 27001 standard on Information Security Management Systems that called the ISMS. According to [4], there are related research that has been done, such as critical success factors in implementing the ISMS based on ISO 27001 that will be used by the author to conducting this research.

## 2. Related Work

There are a large number of researches that are focused on information security and information security management system in many organizations. Measuring the level of information security awareness using Multiple Criteria Decision Analysis (MDCA) in government organizations [4]. The results are showed that the level of awareness of information security is at "moderate" level, so it needs to be monitored for possible correction. Besides that, conduct an analysis to know problems of information security implementation in the organization using technical and operational requirement Combining ISO 27001: 2005 that has standard with maturity level [5]. According to [6], it is find out whether the security system on academic information system used is in accordance with the standards and the extent of the readiness of academic information systems in the application of information security standards by combining BS-7799 Standard with SSE-CMM. The results are the level of information security maturity on the average academic information system is still at the first level (Initial / ad hoc) in on Security Policy clause, Information Security Organization, Asset and Control Classification, Personnel / Human Resources Security, Information Security Incident Management, Aspect Business Sustainability Information Security [6]. Business processes in dynamic environment should be managed with [7]. According to [4], the critical success factors assessment of ISO 27001 certification in computer organization by test-retest reliability. This research has produced critical success factors that can be used for organizations while implementing the ISMS. Despite of [4], it believe that this methodology is useful for evaluating other business organizations with specific CSFs.

## 3. Results and Discussion

Stages in this research are can be described in the figure 1 below:

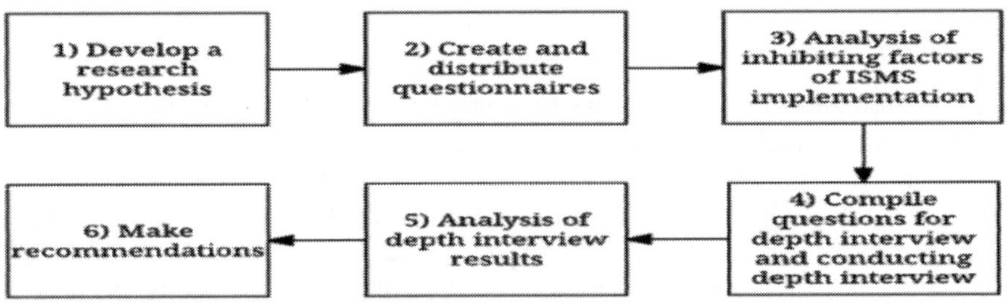


**Figure 1.** Research stages

The stages of this research consists of six step, such as : (1).develop a research hypothesis, (2). Create and distribute questionnaires, (3).analysis of inhibiting factors of ISMS implementation, (4). Compile questions for depth interview and conducting depth interview, (5). Analysis of depth interview results, and (6). make recommendations.

According to [4], the research variables in this study are described in table 1 below:

IOP Conf. Series: Journal of Physics: Conf. Series **978** (2018) 012039 doi:10.1088/1742-6596/978/1/012039

**Table 1.** The Research Variable

| Variable Y name | Variable X name | |
|---|---|---|
| Implementation of ISMS | X1: Commitment and leadership from top management | X8:ISMS documentation management |
| | X2: Effective information security policies and objectives | X9:Management of incidents, events, and weaknesses of information security |
| | X3 : An effective process approach | X10 :An effective internal audit |
| | X4 : An effective information security risk management | X11 :Continual improvement |
| | X5 : Implement and operate the ISMS | X12: Companies wide involving |
| | X6 : Assign roles, responsibilities, and authorities for the ISMS | X13:An effective management reviews |
| | X7:Training, awareness, and competence of effective human resources | X14:An effective motivation management |
| | | X15 : Knowledge management |

Based on these variables, the author determine the hypothesis in this study are:

1. H0: Variable X has no significant effect on Variable Y (implementation of ISMS). This means that the implementation of these factors within the organization is still not significant and a barrier to the implementation of the ISMS.
2. H1: Variable X has significant effect on Variable Y (implementation of ISMS). This means that the implementation of these factors can be quite significant.

The depiction of the relationship between variables X and Y is as follows in figure 2 below:

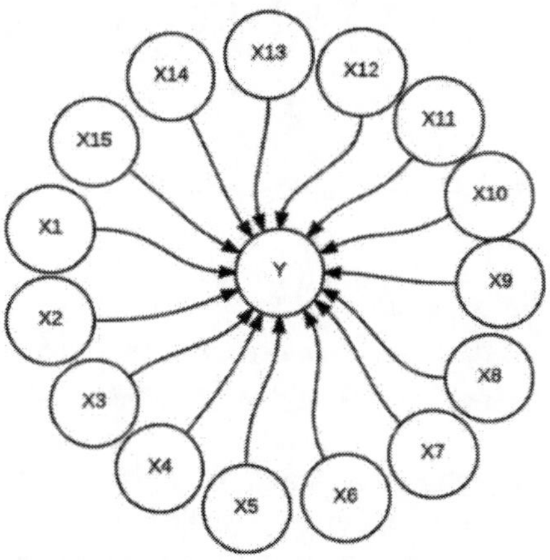


**Figure 2.** Conceptual Framework

### *3.1. Analysis of Inhibiting Factors of ISMS Implementation*

The authors conducted a test of correlation and different test using multiple linear regression analysis method and paired t test [8]. The authors are assisted by using SPSS application version 23. The statistical chart below shows the order of contribution of variable X to variable Y (implementation of ISMS) in each organization, such as figure 3,4, ,and 5 below:

IOP Conf. Series: Journal of Physics: Conf. Series **978** (2018) 012039 doi:10.1088/1742-6596/978/1/012039

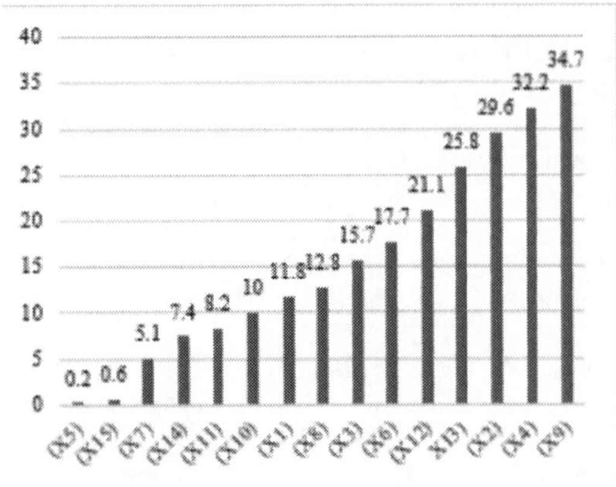


**Figure 3.** Contribution Variable X to Y (BCA)

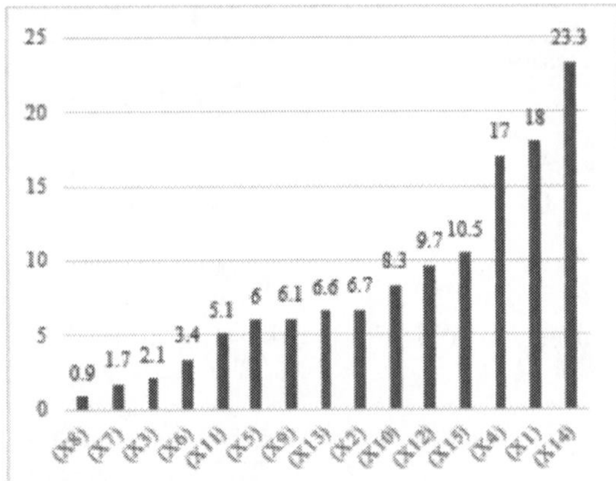


**Figure 4.** Contribution Variable X to Y (telin)

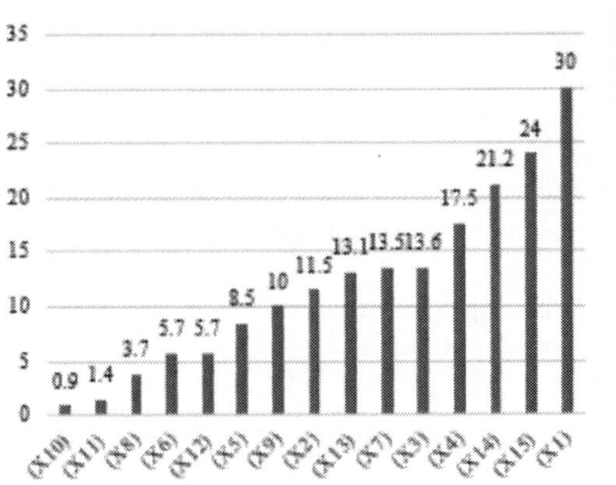


**Figure 5.** Contribution Variable X to Y (DATIN)

### *3.2. Factors that inhibiting the implementation of ISMS*

According to figure 3, figure 4, and figure 5, it can be seen that there is a range of values. The author decides to give the middle value of the range of values by using the median formula of odd numbers :

$$Me = X_{(n+1)/2} \tag{8}$$

With $Me$ is the order, and $n$ is the number of data. With 15 variables then the result is which the middle value of the range of values is the sequence to 8. The authors decides for values that fall on the middle value and below the middle value it will be considered as a factor inhibiting the implementation of the ISMS to the organization while the values fall over the middle value will be considered sufficiently implemented in the organization but it needs continuous improvement. The following is a table of author's conclusions:

**Table 2.** Factors That Inhibit the ISMS

| Org | Inhibiting Factors | | | | | | | |
|---|---|---|---|---|---|---|---|---|
| DCO BCA, | X5 | X15 | X7 | X14 | X11 | X10 | X1 | X8 |
| | 0.2 | 0.6 | 5.1 | 7.4 | 8.2 | 10 | 11.8 | 12.8 |
| NIT TELIN | X8 | X7 | X3 | X6 | X11 | X5 | X9 | X13 |
| | 0.9 | 1.7 | 2.1 | 3.4 | 5.1 | 6 | 6.1 | 6.6 |
| DATIN Centre | X10 | X11 | X8 | X6 | X12 | X5 | X9 | X2 |
| | 0.9 | 1.4 | 3.7 | 5.7 | 5.7 | 8.5 | 10 | 11.5 |

**Table 3.** Factors That Need Improvement

| Org | Need Improvement Factors | | | | | | |
|---|---|---|---|---|---|---|---|
| DCO BCA, | X3 | X6 | X12 | X13 | X2 | X4 | X9 |
| | 15.7 | 17.7 | 21.1 | 25.8 | 29.6 | 32.2 | 34.7 |
| NIT TELIN | X2 | X10 | X12 | X15 | X4 | X1 | X14 |
| | 6.7 | 8.3 | 9.7 | 10.5 | 17 | 18 | 23.3 |
| DATIN Centre | X13 | X7 | X3 | X4 | X14 | X15 | X1 |
| | 13.1 | 13.5 | 13.6 | 17.5 | 21.2 | 24 | 30 |

Based on the table 2and 3 above, the multi factors that inhibiting the implementation of the ISMS from each organization are as explain in table 3 below:

**Table 4.** factors that inhibiting the implementation of the ISMS from each organization

| DCO unit at BCA | NIT division at TELIN | DATIN centre of Ministry of Health RI |
|---|---|---|
| ISMS documentation management (X8), Commitment and leadership from top management (X1), An effective internal audit (X10), Continual improvement (X11), An effective motivation management (X14), Training, awareness, and competence of effective human resources (X7), Knowledge management (X15), Implement and operate the ISMS (X5) | An effective management review (X13), Information security incident, event, and security management (X9), Implement and operate the ISMS (X5), Continual improvement (X11), Assign roles, responsibilities, and authorities for the ISMS (X6), The effective process approach (X3), Training, awareness, and competence of effective human resources (X7),ISMS documentation management (X8) | Effective information security policies and targets (X2), Information security incident, event, and security management (X9), Implement and operate the ISMS (X5), Company wide involving (X12), Assign roles, responsibilities, and authorities for the ISMS (X6), ISMS documentation management (X8),Continual improvement (X11),An effective internal audit (X10) |

The authors also gave highlights on several variables to describe which variables are already quite well implemented in the three organizations as well as the X4 (An effective Information security risk management), and the variables that are the inhibiting factors in the three organizations at once are: Implement and operate the ISMS (X5), ISMS documentation management (X8), and Continual improvement (X11). In order to obtain qualitative data supporting for interpreting and the purpose of statistical data which has obtained by the authors, the authors conducted depth interviews with two experts in the field of ISMS and ISO 27001.The experts are have been experienced in assisting organizations in implementing the ISMS based on ISO 27001 from 2002. Both ISMS experts have accompanied the three organizations that have been sampled in this research in implementing the ISMS. In compiling a list of questions to do depth interview, the authors trying to dig up information based on personal experience of experts in assisting the implementation of ISMS in each organization and also by comparing the existing statistical data. The authors also made a presentation to the experts related to the results of statistical calculations that have been done by the author. Based on interviews with experts, an ISMS implementation is not easy to run in a short time. Based on expert experience, an organization can be said to experience significant improvement of ISMS implementation after passing the fifth year of running implementation. This condition should be supported by the consistency of the process running and balanced with the competence of human resources and updating technology that continues to increase.

## 4. Conclusion

In general, the multi factors that support the implementation of the ISMS has not been significantly implemented. Some things that must be followed up in order to support the implementation of the ISMS in the future are as follows: (1).Top management is more active in conducting management reviews by providing feedback for future ISMS improvement, (2). Socialize policies and procedures related to the management of information security incidents to all employees on a regular basis, (3). Conduct periodic reviews regarding implementation of the ISMS in accordance with the ISMS annual program, information security risk management plan, information security control plan implemented in the organization, (4). Communicate to all employees related to the existing improvement plan program every year so that all employees always make improvements related to the implementation of the ISMS in the organization, (5).Communicate related roles, responsibilities, and related authority of the ISMS to all employees on a regular basis, (6). Make a work program in the implementation of the ISMS every year and then communicate the work program on a regular basis to all personnel to be implemented optimally, (7). Communicate the information security policies and objectives that have been assigned to all employees on a regular basis,

**References**

[1] Pavlov G and Karakneva J 2011 Information Security Managemnet System In Organization *Trakia Journal of Sciences* ISSN: 1313-3551.2011 **9**(4) pp 20-25
[2] Whitman M E and Mattord H J 2011 Principles of Information Security (Fourth Edition) *United States of America: Cengage Learning*
[3] 2012 *International Standard ISO/IEC 27000 Second Edition 2012-12-0* (Switzerland)
[4] Hai, Hui-Lin and Kuei-Min W 2014 The critical success factors assessment of ISO 27001 certification in computer organization by test-retest reliability *African Journal of Business Management* **8**(27) pp 705-716
[5] SORA and Daniel 2012 Securing IT Networks with ISMS Family of Standards *Defense Resources Management In The 21st Century* (ISO 27001 Series)
[6] 2013 *International Standard ISO/IEC 27001 Second Edition 2013-10-01* (Switzerland)
[7] Fajar A N, Budiardjo E K and Hasibuan Z A 2012 System architecture in the dynamic environment based on commonality and variability business processes *ICCM*
[8] Landau, Sabine, Everitt and Brian S 2004 A Handbook of Statistical Analyses using SPSS *London: Chapman & Hall/CRC Press LLC*

IOP Conf. Series: Journal of Physics: Conf. Series **978** (2018) 012040 doi:10.1088/1742-6596/978/1/012040

# Augmented reality social story for autism spectrum disorder

**M F Syahputra[1], D Arisandi[1], A F Lumbanbatu[1], L F Kemit[1], E B Nababan[1] , O Sheta[2]**

[1]Department of Information Technology, Faculty of Computer Science and Information Technology, Universitas Sumatera Utara, Medan, Indonesia
[2]Department of Computer Science, Zagazig University, Egypt.

nca.fadly@usu.ac.id | dedyarisandi@usu.ac.id | anggif05@gmail.com | liviafyonikemit@gmail.com | ernastp@yahoo.com

**Abstract**. Augmented Reality is a technique that can bring social story therapy into virtual world to increase intrinsic motivation of children with Autism Spectrum Disorder(ASD). By looking at the behaviour of ASD who will be difficult to get the focus, the lack of sensory and motor nerves in the use of loads on the hands or other organs will be very distressing children with ASD in doing the right activities, and interpret and understand the social situation in determining a response appropriately. Required method to be able to apply social story on therapy of children with ASD that is implemented with Augmented Reality. The output resulting from this method is 3D animation (three-dimensional animation) of social story by detecting marker located in special book and some simple game which done by using leap motion controller which is useful in reading hand movement in real-time.

## 1. Introduction

Autism Spectrum Disorder (ASD) is a pervasive developmental disorder that affects the children in terms of interpersonal communication, social interaction, and imaginatively level in play. Many children with autism tend to self-stimulatory, such as hand-flapping, gloved, swaying or repetitive vocalizations [1]. Difficulty in realizing emotional change, lack of empathy with others and difficulty in responding appropriately to a disorder in social interaction for people with autism [2]. One therapy that can be done at the initial level of healing is social story.

A social story is a short story that can be used to help children with ASD in interpreting and understanding social situations and in determining a response appropriately in specially designated styles and formats [3]. Social story can be written or visual that can describe various social interactions, situations, skills, concepts and behaviour common to the community with the purpose of sharing social information that is accurate, convincing and easy to understand by children with ASD. Usually, the social story as therapy is done directly by the therapist to the autism children. This is done by read the story directly without visual support or with visual support but still with the visualization of two- dimensional (2D) which is considered still less accurate. A less interesting look can distract so that the message conveyed by the story will not be up and received by the children.

As the times progressed, a technology called Augmented Reality (AR) was established that can combine virtual world with its real state, which can be implemented in cultural preservation [7][8][9], historical virtual game [11], medicine and psychology. With the technology it also bring the side of medical therapy into the virtual world in the form and actual use.

## 2. Identification of Problems

Augmented Reality is a technique that can bring social story therapy into virtual world to increase intrinsic motivation of ASD children. By looking at the behaviour of children with ASD who will be difficult to get the focus, the lack of sensory and motor nerves in the use of loads on the hands or other organs will be very distressing children with ASD in doing the right activities, and interpret and understand the social situation in determining a response appropriately. Therefore it is necessary the method to be able to apply social story on therapy of children with ASD which is implemented with Augmented Reality.

## 3. Previous research

Zhen Bai, et al [4] has conducted a study using Augmented Reality method by applying target children with ASC (Autism Spectrum Conditions) at age 4 - 7 years and have been diagnosed positively with ASC. They make markers in the form of toys that are reflected through a screen that has an effect as a mirror.

Lizbeth Escobedo and Monica Tentori [5] in 2014 conducted research to help teachers or autistic child therapists provide additional visual support through the use of mobile-based augmented reality methods. The study was conducted over 8 weeks in 3 classrooms, each class consisting of 7 teachers or therapists and 14 children with autism. They are discussed how to train ASD children using virtual systems and augmented reality.

In 2014, Lakshmiprabha et al [1] studied the uses of Augmented and Virtual Reality (AR / VR) systems to train children with autism based on Applied Behaviour Analysis (ABA) techniques in viewing the behaviour of the child. The system helps in teaching children with asd in recognizing new images and objects and the use of appropriate words and sentences.

Qingguo Xu, et al [6] conducted a study using augmented-reality glass platform method for ASD by installing cameras and microphones as inputs used for interviews to be conducted on ASD children.

## 4. Research methods

### *4.1. Augmented reality*

Augmented Reality is the incorporation of existing objects in the real world with virtual worlds in certain places and circumstances that run interactively in real time and there is integration between objects in three dimensions (3D objects), virtual objects integrated with the real world.

Augmented reality (AR) technology enables technologies that combine virtual worlds generated by computer sensors in the form of graphics, sound and video with real world in real-time. The purpose of AR is to add information and meaning to an object or real space.

### *4.2. Social Story*

Social story is a short story that can be used to help children with asd in interpreting and understanding social situations and in determining a response appropriately in specially designated styles and formats. Social story can be written or visual that can describe various social interaction, situation, skill, concept and the behaviour common to society with the purpose of sharing social information that is accurate, convincing and easy to understand by children with autism.

### *4.3. Leap Motion*

Leap Motion Controller is a hardware sensor tool that supports hand and finger movements as input, which can be equated with functions like a mouse, but does not require direct contact with the hands or touch. Leap Motion Controller is a tool that aims to track hand and finger movements to perform a brief game on this research.

### *4.4. General Architecture*

The general architecture of Augmented Reality Social Story can be seen in Figure 1.

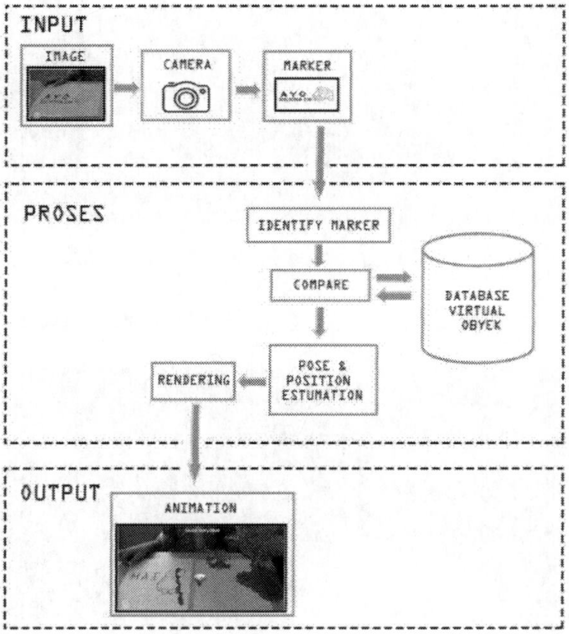


**Figure 1.** General Architecture Augmented Reality Social Story

This stage describes the general architecture of the application as shown in Figure 1:

- Input
  Markers are placed on sheet books that used as a place to show the 3D animations.
- Process
  The markers on the book sheets and Leap Motion Controller are placed between the camera and the screen. The use of Leap Motion helps the hand to move virtual 3D objects that will be detected and recognized by the camera. After that compare marker book with marker that available on database. If the information is appropriate then proceed to the estimate of laying the virtual objects on the marker contained at the book sheets. Then, done rendering process.
- Output
  The resulting output is an Augmented Reality object and 3D animation in the form of a social story on the book marker.

## 5. Result and Discussion

After testing the application by following the test plan, the results are stored as evidence that the application has successfully passed the test phase.

Test results obtained include:

- Implementation of AR Camera on marker detection done with a distance and certain condition.
- Augmented Reality and hand motion reading with leap motion goes well.

**Figure 2.** Augmented Reality Social Story

In Figure 2. can be seen that augmented reality shows a social story in the 3D animation.

**Figure 3.** Augmented Reality Social Story using Leap Motion Controller

In Figure 3. shows augmented reality with social story view integrated with leap motion controller. In this scene the user is invited to play with some simple games using leap motion controller.

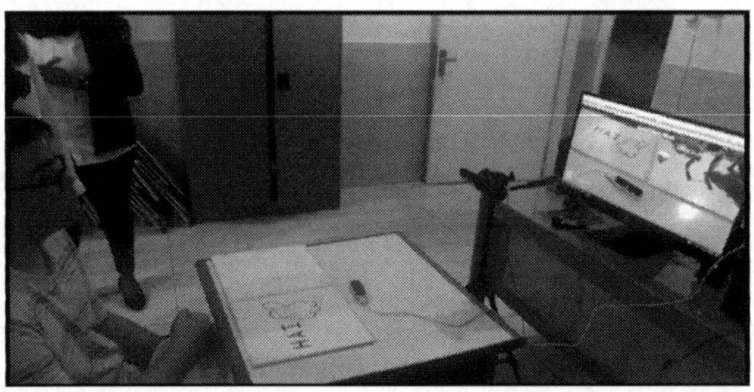

**Figure 4.** Child with autism using Augmented Reality Social Story

To examine the level of system use approval, the system was tested for three children with autism who were in therapy. When using the system, they are accompanied and guided by the therapist. Testing of children can be seen in Figure 4.

After that, the therapist is allowed to fill out a questionnaire containing some questions and statements about the level of conformity therapy to the system. All questions and statements are paired with Likert scale to get the results. Where on the scale do not agree is worth 1, less agree is worth 2, agree is worth 3 and strongly agree is worth 4.

$$\text{Average Value} = \frac{\sum (R \times AV)}{MAV \times R_{Total}} \times 100$$

Keterangan:
R = Respondents
AV = Answer Value
MAV = Maximum Answer Value

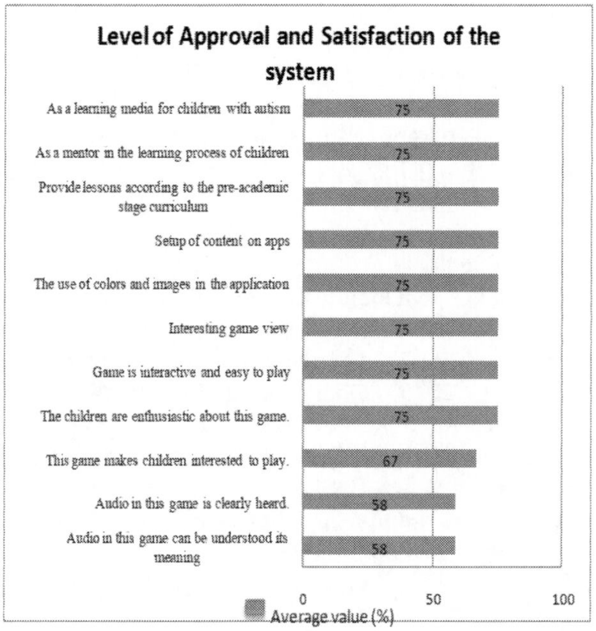


**Figure 5.** Level of System Use Approval

Tests of the system performed on three people with autism who was accompanied by each therapist (respondent). Based on the calculation with Likert scale, the average value of each statement at the level of system use approval can be seen in Figure 5.

Based on the results of questionnaires that have been done to three respondents, all respondents chose to agree that the first statement to the eighth states that if the social story application can be used as a medium of learning for children with autism, social story application helps mentors in the process of learning, this application can provide appropriate learning in the pre-academic stage curriculum, the arrangement of content on the application in accordance with the therapy, the use of colors and images in the application in accordance with the therapy, the display game interesting, the game is interactive and easy to play and children are enthusiastic with this game, where the average percentage of this statement is 75%. In the ninth statement, two respondents chose to agree and one respondent chose less agree if the game on this system to make children interested in doing the process of learning, where the average percentage of the statement is 67%. As for the tenth and eleventh statements, one respondent chose to agree and two other respondents chose less agree that the audio in this game is clear to be heard and the audio in this game can be understood its meaning, where the average percentage of this statement is 58%.

## 6. Conclusions

Some conclusion based on testing of the augmented reality social story therapy system for children with Autism Spectrum Disorder (ASD), are: Accuracy rate is higher when the number of therapy more. The more the amount of therapy can result more accurate the healing of children with ASD.NMarker distance and marker type changes as well as scenes given on visualization will increase the accuracy level when the therapy is done.Applications can run well when detecting markers, hand interaction using leap motion controller and displaying 3D objects. Based on the Likert scale with an average percentage of 71.18%, respondents agree that the system can support visualizing social story therapy for children with Autism Spectrum Disorder (ASD) in interpreting and understanding the surrounding social situations.

**Acknowledgment**

The authors gratefully acknowledge that the present research is supported by Ministry of Research and Technology and Higher Education Republic of Indonesia. The support is under the research grant TALENTA USU scheme PENELITIAN TERAPAN of 2017 Contract Number 81/UN5.2.3.1/PPM/KP-TALENTA USU/2017.

## References

[1] Lakshmiprabha N S, Santos A, Mladenov D and Beltramello O 2014 [Poster] An Augmented and Virtual Reality System for Training Autistic Children *IEEE International Symposium on Mixed and Augmented Reality 2014 Science and Technology Proceedings* (Germany)

[2] Kuriakose S and Lahiri U 2015 Understanding the Psycho-Physiological Implications of Interaction with a Virtual Reality based System in Adolescents with Autism: A Feasibility Study *IEEE Transactions on Neural Systems and Rehabilitation Engineering* pp 1-10

[3] Lorimer P A, Simpson R L, Myles B S and Ganz J B 2002 The use of social stories as a preventative behavioral intervention in a home setting with a child with autism *Journal of Positive Behavior Interventions* pp 53–60

[4] Bai Z, Blackwell F and Coulouris G 2013 Through the Looking Glass: Pretend Play for Children with Autism *IEEE International Symposium on Mixed and Augmented Reality 2013 Science and Technology Proceedings Australia*

[5] Escobedo, Lizbeth, Monica and Tentori 2014 Mobile Augmented Reality to Support Teachers of Children with Autism *International Journal on IEEE* pp 60-67

[6] Xu Q, Cheung S and Soares N 2015 Littlehelper: An Augmented Reality Glass Application To Assist Individuals with Autism in Job Interview *Proceedings of APSIPA Annual Summit and Conference 2015*

[7] Syahputra M F, Annisa T, Rahmat R F and Muchtar M A 2017 Virtual Application of Darul Arif Palace from Serdang Sultanate using Virtual Reality *Journal of Physics: Conference Series* **801** p. 12009

[8] Syahputra M F, Siregar R K and Rahmat R F 2017 Finger recognition as interaction media in Augmented Reality for historical buildings in Matsum and Kesawan regions of Medan City *Lecture Notes in Computer Science* **10325**

[9] Syahputra M F, I M Rizki, Fatimah S and Rahmat R F 2017 Implementation of player position monitoring for Tanjung Pura Palace Virtual Environment *Lecture Notes in Computer Science* **10325**

[10] Syahputra M F, Widhianto M and Rahmat R F 2017 Historical game of Majapahit Kingdom based on tactical role-playing game *Proceedings of the 12th International Joint Conference on Computer Vision, Imaging and Computer Graphics Theory and Applications (VISIGRAPP 2017)*, pp 327-332

# Hybrid cryptosystem RSA – CRT optimization and VMPC

**R Rahmadani[1], H Mawengkang[2] and Sutarman[2]**

[1]Master Program (S2) of Informatics Engineering, Faculty of Computer Science and Information Technology, Universitas Sumatera Utara, Medan, Indonesia.
[2]Department of Mathematics, Faculty of Mathematics and Natural Sciences, Universitas Sumatera Utara, Medan, Indonesia.

[1]reni_rahmadani@students.usu.ac.id, [2]hmawengkang@yahoo.com, [3]sutarman@usu.ac.id

**Abstract.** Hybrid cryptosystem combines symmetric algorithms and asymmetric algorithms. This combination utilizes speeds on encryption/decryption processes of symmetric algorithms and asymmetric algorithms to secure symmetric keys. In this paper we propose hybrid cryptosystem that combine symmetric algorithms VMPC and asymmetric algorithms RSA – CRT optimization. RSA – CRT optimization speeds up the decryption process by obtaining plaintext with $d_p$ and $p$ key only, so there is no need to perform CRT processes. The VMPC algorithm is more efficient in software implementation and reduces known weaknesses in RC4 key generation. The results show hybrid cryptosystem RSA – CRT optimization and VMPC is faster than hybrid cryptosystem RSA – VMPC and hybrid cryptosystem RSA – CRT – VMPC.
Keyword : *Cryptography, RSA , RSA – CRT, VMPC, Hybrid Cryptosystem.*

## 1. Introduction

Hybrid cryptosystem is a combination of symmetric and asymmetric key cryptography [1]. Ramaraj et al. [1] formed a security protocol using hybrid encryption on online transactions. This protocol is made using symmetric cipher (AESRijndael) and asymmetric cipher (RSA) with hash function. They have successfully fulfilled three cryptography objectives: integrity, confidentiality and authentication. Rasmi and Paul [2] combine circle symmetric key algorithm and RSA – CRT for e - commerce applications. The symmetric circle symmetric algorithm is based on 2-d geometry using the circle's nature and the center corner of the circle. This algorithm is a block cipher that has advantages in producing cipher text with fixed size. This symmetric algorithm does not require large computations. The time for the encryption / decryption process is not based on the key size. Kuppuswamy and Al-Khalidi [3] conducted research by creating a system that has two different cryptography algorithms. The public key cryptography algorithm used is based on linear block cipher and secret key algorithm based on simple symmetric algorithm. This algorithm fulfills the cryptography objectives of integrity and confidentiality.

VMPC (Variably Modified Permutation Composition) [4] is one of the symmetric cryptography algorithms included in the stream cipher found by Bartosz Zoltak. This algorithm is the modified version of RC4 algorithm. The author claims that the cipher text was designed to be efficient in software implementations and to resist attacks against the cipher's Key Scheduling Algorithm (KSA). As a symmetric cryptographic algorithm that uses the same key for encryption and decryption, it requires the sender to find a secure way to deliver the keys to the receiver [1].

RSA is a public key cryptography algorithm found by Rivest et al. in 1978 [5]. Although it is quite old, RSA still used for securing web traffic, email, and some wireless devices [6]. This is because RSA is a secure cryptographic algorithm for using a large encryption and decryption keys. Therefore, the process in RSA takes a considerable time. Many improvements have been made to speed up the process without reducing the level of security. One of them is RSA - CRT from Quisquater and Couvreur [7]. In this modification, the decryption process is based on the CRT (Chinese Remainder Theorem) method and improvisation on the modulus multiplication. So the whole process is accelerated by reducing the time for the decryption process using smaller exponents and moduli than the RSA algorithm [8].

In this paper we concentrated on RSA – CRT how to make it faster. In doing so, we reduced the decryption time by optimizing RSA – CRT to speed up the overall processing time and perform hybrid cryptosystems on RSA – CRT optimization with VMPC to produce faster algorithms.

## 2. Study Of Literature

### *2.1. Hybrid Cryptosystem*

One of the approaches of hybrid cryptosystem is to generate a secret key from symmetric cryptography algorithm then encrypt the key with asymmetric cryptography algorithm using receiver public key, and then encrypt the message using the private key of symmetric cryptography algorithm [2].

Hybrid cryptosystem scheme combines the convenience of an asymmetric cryptography algorithm scheme with the effectiveness of symmetric cryptography algorithm scheme. This scheme added security of the delivery process enhancing overall system performance [3].

### *2.2. RSA*

RSA stands for Rhivest, Shamir, and Adleman is a cryptography system which security rested on the difficulties of factoring large numbers [5].

RSA Key Generation

- Generate two large prime numbers $p \neq q$
- Find $n = p*q$ and $\phi(n) = (p\text{-}1)*(q\text{-}1)$, integer $n$ is called modulus (RSA)
- Find $e \in N$ where $e < 1 < n$ and GCD $(e, \phi\ (n)) = 1$
- Find $d \in N$ and $d < 1 < n$ where $ed \equiv 1 \pmod{\phi\ (n)}$
- Publish $(n, e)$ and $(d, p, q, \phi\ (n))$ remain private.

RSA encryption process

- Public key $(n, e)$ received
- Perform encryption process with the formula
  Encryption $C \equiv M^e \pmod{n}$
- Cipher text $C$ is sent to the owner of the public key
  RSA Decryption Process
  After the cipher text is received, decrypt the cipher text to get the message $M$ with formula
  Decryption $M \equiv C^d \pmod{n}$

$\phi(n)$ is read totien n is an Euler function where $n \geq 1$ and denotes the number of positive integers $< n$ relatively prime with $n$.

### *2.3. RSA – CRT*

Using the standard double and square method for modular exponentiation, the cost of RSA process is almost identical to the bit length $e$ (public key) and $d$ (private key) [9].

Research conducted by Quisquater and Couvreur [7] aims to reduce the length of decryption process in RSA. They use CRT (Chinese Remainder Theorem) to reduce the large size of modulus on decryption. In their research the decryption process is accelerated by decreasing the value of $d$ hence the calculation to get the plaintext becomes smaller. Additional savings in memory space are also possible due to reduced modulus size [8].

In this research the generation of keys and encryption process is the same as RSA, the difference is in the decryption process such as [7].

- Calculate $d_p = d \bmod p - 1$ and $d_q = d \bmod q - 1$
- Calculate $M_p = C^{dp} \bmod p$ and $M_q = C^{dq} \bmod q$.
- Calculate $M$ from $M_p$ and $M_q$ using CRT.

*2.4. VMPC (Variably Modified Permutation Combination)*

VMPC is a combination of basic operations on permutations and integers. The function in this algorithm can be calculated by three basic operations on permutations per byte. The resulting cipher text is designed to be more efficient in software implementation and reduce known weaknesses in RC4 key generation without reducing its speed and simplicity [4].

In VMPC algorithm there are two processes viz, key scheduling algorithm (KSA) and VMPC stream cipher [4]. VMPC stream cipher performs XOR process on cipher text and cryptography keys derived from KSA. VMPC stream cipher

$n = 0$
Repeat step 3 – 6 $L$ times
  $s = P[(s + P[n]) \bmod 256]$
  Output ^= $(P[(P[P[s]] + 1) \bmod 256])$
  Swap $(P[n], P[s])$
  $n = (n + 1) \bmod 256$
where: $P$ = 256-byte permutation table initialize on KSA VMPC
  $s$ = 8-bit variable initialize KSA VMPC
  $n$ = 8-bit variable
  $L$ = keystream length in bytes

KSA VMPC converts the cryptographic key into 256 permutation elements $P$ and initializes the variable $s$.

$s = 0$
for $n$ from 0 to 255: $P[n] = n$
  for $m$ from 0 to 767: execute step 4-6
  $n = m \bmod 256$
  $s = P[(s + P[n] + K[m \bmod c]) \bmod 256]$
  Swap ($P[n]$ and $P[s]$)
where : $c$ = length of cryptographic keys in bytes, $16 \le c \le 64$
  $K$ = table storing of $c$ – elements from cryptographic keys
  $m$ = 16 bit variable

## 3. Findings And Discussions

*3.1. RSA – CRT Optimization*

The proposed scheme to speed up the decryption process at RSA is by optimizing the RSA – CRT (ORSA – CRT). In the proposed scheme the encryption process is the same as RSA, while the decryption process:

- Calculate $d_p = d \bmod p - 1$
- Calculate $M = C^{dp} \bmod p$

The optimization performed on RSA - CRT is by eliminating the process to find the value of $M_p$, so the message is obtained only by calculating the cipher text with $d_p$ and $p$ that the CRT process is not necessary. In this scheme the $p$ and $q$ keys must be greater than 2 digit number.

### *3.2. Hybrid Cryptosystem RSA – CRT Optimization and VMPC*

In this study we propose a hybrid algorithm by utilizing the symmetric VMPC and RSA – CRT *optimization* algorithms to produce an efficient and secure cryptography algorithms. The proposed algorithm scheme can be seen in figure 1

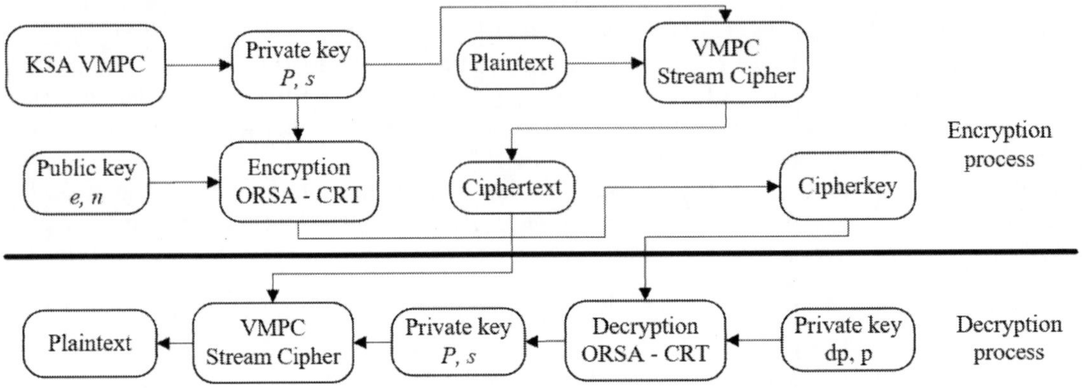


**Figure 1.** Encryption and Decryption Process

Encrypt the plaintext with VMPC stream cipher using $P$ and $s$ keys derived from KSA VMPC. $P$ and $s$ are encrypted with RSA – CRT optimization algorithms using previously generated public keys. Cipher text and cipher key are used as input on the decryption process. Then decrypt the cipher key using RSA – CRT optimization private key and obtained $P$, $s$. Decrypt cipher text with VMPC stream cipher using $P$ and $s$ keys to get the plaintext.

### *3.3. Result of Decryption Process*

Differences of RSA, RSA - CRT, and RSA - CRT optimization algorithms are in the decryption process. In the RSA - CRT optimization algorithm $p$ and $q$ value must be greater than 2 digit number to obtain the correct plaintext.

**Table 1.** Comparison of Plaintext from Decryption Process

| *p* | *q* | *e* | *d* | Decryption Result | | |
|---|---|---|---|---|---|---|
| | | | | RSA | RSA - CRT | RSA – CRT Optimization |
| 71 | 59 | 53 | 3677 | Experiment (*123#) | Experiment (*123#) | E1)+"& '- (*123#) |
| 337 | 509 | 881 | 44561 | Experiment (*123#) | Experiment (*123#) | Experiment (*123#) |
| 7901 | 2531 | 6651 | 18135851 | Experiment (*123#) | Experiment (*123#) | Experiment (*123#) |
| 85717 | 13829 | 62959 | 989130319 | Experiment (*123#) | Experiment (*123#) | Experiment (*123#) |
| 59393447 | 31229371 | 73990999 | 1846979500600279 | Experiment (*123#) | Experiment (*123#) | Experiment (*123#) |
| 612087229 3 | 260888650 7 | 7812862283 | 315964251945534483 5 | Attribute certificate | Attribute certificate | Attribute certificate |
| 779658128 | 397625195 | 930989909 | 171991145629106972 | Attribute | Attribute | Attribute |

IOP Conf. Series: Journal of Physics: Conf. Series **978** (2018) 012041 doi:10.1088/1742-6596/978/1/012041

| | | | | | | |
|---|---|---|---|---|---|---|
| 9641 | 8673 | 7409 | 59996609 | certificate | certificate | certificate |
| 830535911<br>436631 | 466963146<br>317137 | 123580510<br>5423229 | 298179093898968774<br>0694285909 | Attribute certificate | Attribute certificate | Attribute certificate |
| 716305387<br>111246346<br>857 | 264321434<br>037657934<br>747 | 4959594258<br>8047930834<br>3 | 663672772400511674<br>557000755109562584<br>42567 | Attribute certificate | Attribute certificate | Attribute certificate |
| 473992671<br>340071646<br>22983459 | 256611130<br>965415688<br>31790453 | 8471630557<br>5485889277<br>743037 | 849002681728117362<br>468465087160312075<br>28380926181429 | Attribute certificate | Attribute certificate | Attribute certificate |

Referring from table 1 RSA – CRT optimization algorithms obtain the correct plaintext as long as the $p$ and $q$ keys are larger than 2 digit number.

### *3.4. Results of Hybrid Cryptosystem*

Encryption and decryption process of all hybrid algorithms compared to see which hybrid algorithm is faster in processing. In table 2 we can see the difference in execution time of each hybrid algorithm with different key sizes $n$ and $e$. The size of $n$ and $e$ keys are in bytes.

**Table** 2. Execution Time with Different Keys

| $n$ length | $e$ length | RSA – VMPC | RSA – CRT – VMPC | RSA – CRT Optimization and VMPC |
|---|---|---|---|---|
| 256 | 128 | 26.275 sec | 14.841 sec | 11.653 sec |
| 512 | 256 | 176.1 sec | 88.44 sec | 70.08 sec |
| 1024 | 512 | 1260.78 sec | 620.04 sec | 491.82 sec |
| 2048 | 1024 | 9335.4 sec | 4501.92 sec | 3599.1 sec |

**Table 3.** Data Execution Time

| Char. Length | RSA – VMPC | RSA – CRT – VMPC | RSA – CRT Optimization and VMPC |
|---|---|---|---|
| 1151 | 23.609 sec | 13.249 sec | 10.531 sec |
| 4563 | 24.013 sec | 13.59 sec | 10.826 sec |
| 8283 | 24.267 sec | 13.925 sec | 11.146 sec |
| 15788 | 25.118 sec | 14.602 sec | 11.811 sec |
| 30656 | 26.274 sec | 15.709 sec | 13.104 sec |

Table 3 shows the execution time of each hybrid algorithm with different message length. The length of $n$ used is 248 bytes.

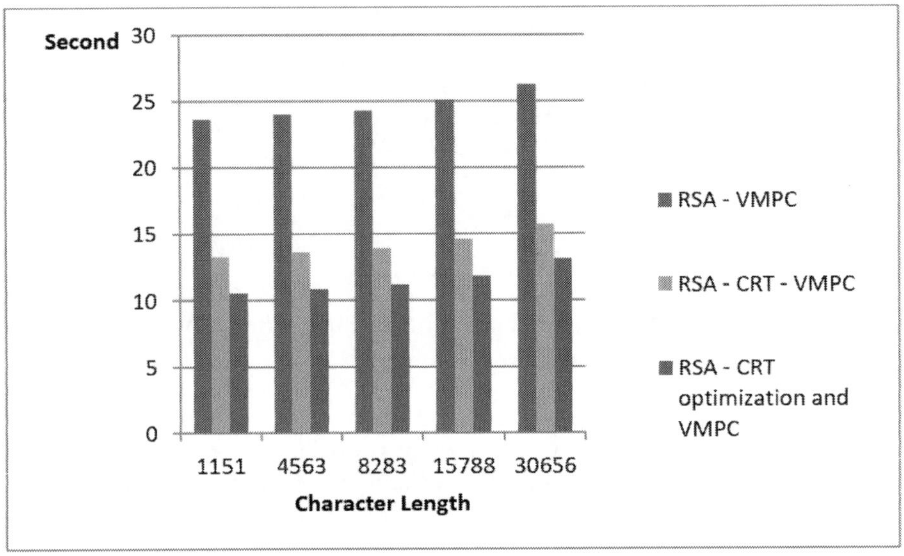


**Figure 2.** Comparison of Data Execution Time

Based on figure 2, hybrid cryptosystem RSA – CRT optimization and VMPC are faster than other hybrid cryptosystems. Hybrid cryptosystem RSA – CRT optimization and VMPC completed the process 53.49% faster than RSA – VMPC and 19.3% faster than RSA – CRT – VMPC.

## 4. Conclusion

Based on the findings and discussions described in the previous section, it can be concluded that

- RSA – CRT optimization speed up the process by calculating the cipher text only with dp and p that the CRT process is not necessary.
- RSA – CRT optimization can obtain plaintext with p and q key larger than 2 digit number.
- Hybrid RSA – CRT optimization and VMPC are 53.49 % faster than RSA – VMPC and 19.3 % faster than RSA – CRT – VMPC.

## References

[1] Ramaraj E, Karthikeyan S and Hemalatha M 2009 A Design of Security Protocol using Hybrid Encryp. Tech *AES- Rijndael and RSA* **17** 7886.
[2] Rasmi P S and Paul V 2011 A Hybrid Crypto System based on a new Circle – Symmetric key Algorithm and RSA with CRT Asymmetric key Algorithm for E-commerce Application *Proc. of Int. Conf. on VLSI, Comm. & Inst* (ICVCI) pp 14–18
[3] Kuppuswamy P and Al-Khalidi S Q Y 2014 Hybrid Encryp./Decryp Tech. Using New Public Key and Symm. Key Algo **19** 0113
[4] Zoltak B 2004 VMPC One-Way Function and Stream Cipher *Proc. of Fast Soft. Encryp* pp 210 –255.
[5] Rivest R, Shamir A and Adleman L 1978 A Method for Obt. Digital Sign. and Public-Key Crypto **21** 126.
[6] Boneh D and Shachan H 2002 Fast Var. of RSA **5** 52431
[7] Quisquater J. and Couvreur C 1982 Fast Decipher. Algo. for RSA Public-Key Crypto **18** 90507
[8] Seungkwang L, Dooho C. and Yongje C 2014 Improved Shamir's CRT-RSA Algo.: Rev. with the Mod. Chain. Method **36** 46978.
[9] Verma, Seema, Garg and Deepak 2015 Improvement in Rebalanced CRT RSA **12** 52431

IOP Conf. Series: Journal of Physics: Conf. Series **978** (2018) 012042 doi:10.1088/1742-6596/978/1/012042

# Insights to develop privacy policy for organization in Indonesia

**E Rosmaini[1*], T F Kusumasari[2], M Lubis[2] and A R Lubis[3]**

[1]Universitas Sumatera Utara, Medan, 20155, Indonesia
[2]Telkom University, Jalan Telekomunikasi No. 1, Bandung, 40257, Indonesia
[3]Politeknik Negeri Medan, Jalan Almamater No. 1, Medan 20155, Indonesia

muharmanlubis@telkomuniversity.ac.id

**Abstract**. Nowadays, the increased utilization of shared application in the network needs not only dictate to have enhanced security but also emphasize the need to balance its privacy protection and ease of use. Meanwhile, its accessibility and availability as the demand from organization service put privacy obligations become more complex process to be handled and controlled. Nonetheless, the underlying principles for privacy policy exist in Indonesian current laws, even though they spread across various article regulations. Religions, constitutions, statutes, regulations, custom and culture requirements still become the reference model to control the activity process for data collection and information sharing accordingly. Moreover, as the customer and organization often misinterpret their responsibilities and rights in the business function, process and level, the essential thing to be considered for professionals on how to articulate clearly the rules that manage their information gathering and distribution in a manner that translates into information system specification and requirements for developers and managers. This study focus on providing suggestion and recommendation to develop privacy policy based on descriptive analysis of 791 respondents on personal data protection in accordance with political and economic factor in Indonesia.

## 1. Introduction

Currently, 13 jurisdictions in Asia have comprehensive privacy laws namely Australia, Hong Kong, India, Japan, Kazakhstan, Kyrgyzstan, Macao, Malaysia, the Philippines, Singapore, South Korea, Taiwan and New Zealand. The latter is the only jurisdiction in the region that has been recognized by the European Commission as providing adequate protection. Notably absent from this list are countries such as China, Thailand, Vietnam and Indonesia, in which taking a piecemeal or sectorial approach [1]. In Indonesia, communication and information ministry is designed to carry out the law and the president's policies by performing number of different task and assignment, including regulation drafting for parliament consideration. Besides that, it also have legitimation through its administrative agencies mandated by constitution to assign certain bill or legislation called as PERPU to replace current regulation in conjuncture condition temporary. It will apply in certain period until next parliament meeting to discuss permanent regulation law. There are certain issues in the regulation form of language that is intrinsically vague and cannot translate to every factual situation in which it is applied, as well as cause by the political factors. Thus, it dictates the relevant agencies to interpret and decide their own strategy in enforcing regulation. However, any changes to the regulation might lead to drawback in accommodating society caused significant time frame to make changes even the minor one. The right to privacy limit other rights and freedoms such as association, belief and expression, in which the communication skill have critical impact in protecting freedom and assuring political accountability without fear of reprisal [2].

## 2. Literature Review

There are debates between participation and transparency with privacy to assure the legitimacy of certain activity especially in the governmental initiatives by overly limiting the rights of citizens in order to protect personal data [2]. Indeed, personal data of employee only disclose by organization through data subject request prior notice, in which under such circumstances, anonymity can be as option unless the organization have higher importance and priority [5]. The scope definition to explain certain terminology in practice often do not translate well in the regulation due to trend changes frequently, thus there is a need to have new mechanism concept to set specific consideration context and taxonomies, in which optimizing the interaction and communication [6]. In developing the privacy policy, it is critical that organizations address information quality in relate to current privacy issues such as accuracy, completeness, currency and reliability of information connected to data subject due to its public availability [7]. Consumers are often placed in a position where they are have to submit their personal data over service offered by organization, where those data will only be used based on agreement, it may be backed up by other promises such as trust mark, membership, discount, etc. Unfortunately, organization tend to break those promise to obtain more profits by sharing to the other or use them for other deceptive purpose [8]. Thus, data subjects must be given control and access to their personal data held by a data controller of course with their recognition, notification and agreement [5]. Meanwhile, the organization should define and develop the domain of consumer privacy, know its ethical dimensions, decide the firm-level drivers of customer privacy protection and more importantly train the employee to appreciate the value of consumers' data more than the others [9].

## 3. Research Methodology

Based on previous study, it showed that legal framework was inadequate to effectively regulate the development and use of voting machines, especially regarding security safeguards, the certification process and tabulation software [3][10][11]. The order list of PDP items as follows quick result of tabulation (1), accuracy (2), condition notification (3), reporting mechanism (4), cost reduction (5), *user permission* (6), physical authentication card (7), data encryption (8), *surveillance* (9), *anonymity value* (10), machine credibility (11), interface simplicity (12), *lack of cross validated* (13), nation proudness (14) and *physical attachment removal* (15).

**Table 1.** Inter-Item Correlation Matrix

| | PDP1 | PDP2 | PDP3 | PDP4 | PDP5 | PDP6 | PDP7 | PDP8 | PDP9 | PDP10 | PDP11 | PDP12 | PDP13 | PDP14 | PDP15 |
|---|---|---|---|---|---|---|---|---|---|---|---|---|---|---|---|
| **PDP1** | 1.000 | .616 | .374 | .227 | .429 | .058 | .346 | .252 | -.119 | .051 | .390 | .019 | .066 | .231 | .019 |
| **PDP2** | .616 | 1.000 | .268 | .243 | .498 | .106 | .549 | .404 | -.217 | .060 | .452 | .059 | -.061 | .396 | -.061 |
| **PDP3** | .374 | .268 | 1.000 | .432 | .423 | .317 | .483 | .442 | -.080 | .053 | .445 | .169 | .222 | .304 | .211 |
| **PDP4** | .227 | .243 | .432 | 1.000 | .442 | .306 | .511 | .480 | -.008 | .221 | .443 | .235 | .161 | .489 | -.084 |
| **PDP5** | .429 | .498 | .423 | .442 | 1.000 | .294 | .649 | .248 | -.130 | .123 | .416 | .200 | -.006 | .565 | .093 |
| **PDP6** | .058 | .106 | .317 | .306 | .294 | 1.000 | .394 | .505 | -.271 | .028 | .132 | -.057 | -.004 | .296 | .030 |
| **PDP7** | .346 | .549 | .483 | .511 | .649 | .394 | 1.000 | .531 | -.078 | .370 | .335 | .302 | .111 | .442 | -.007 |
| **PDP8** | .252 | .404 | .442 | .480 | .248 | .505 | .531 | 1.000 | .117 | .208 | .538 | .149 | -.268 | .325 | -.066 |
| **PDP9** | -.119 | -.217 | -.080 | -.008 | -.130 | -.271 | -.078 | .117 | 1.000 | .297 | .061 | .278 | -.060 | .009 | -.005 |
| **PDP10** | .051 | .060 | .053 | .221 | .123 | .028 | .370 | .208 | .297 | 1.000 | .153 | .377 | .001 | .344 | -.330 |
| **PDP11** | .390 | .452 | .445 | .443 | .416 | .132 | .335 | .538 | .061 | .153 | 1.000 | .239 | .101 | .425 | -.050 |
| **PDP12** | .019 | .059 | .169 | .235 | .200 | -.057 | .302 | .149 | .278 | .377 | .239 | 1.000 | .388 | .321 | -.155 |
| **PDP13** | .066 | -.061 | .222 | .161 | -.006 | -.004 | .111 | .268 | -.060 | .001 | .101 | .388 | 1.000 | -.016 | .253 |
| **PDP14** | .231 | .396 | .304 | .489 | .565 | .296 | .442 | .325 | .009 | .344 | .425 | .321 | -.016 | 1.000 | -.068 |
| **PDP15** | .019 | -.061 | .211 | -.084 | .093 | .030 | -.007 | -.066 | -.005 | -.330 | -.050 | -.155 | .253 | -.068 | 1.000 |

Internal consistency has a meaning as a psychometric property of test that describes whether the items in the test that are proposed to measure the same construct produce consisten score, which is associated with the degree of general factor saturation and degree of inter-item correlations but irrelevant to test length or simply all the items are parallel in a test or subtest [12]. Here, all item must be positive value, which indicate that all items in one same characteristic. Based on inter-item correlation, the cross-checking between PDP item indicated that PDP6, PDP9, PDP13 and PDP15 have the most negative correlation with the other items. Moreover, another reliability analysis will be conducted to determine the further step in strengthen the decision. Another method to check the internal consistency through a small corrected item-total correlation with value less than 0.2 or 0.3 that indicated the item is not measuring the same construct measured by the other items included. It has a meaning that the corresponding item does not correlate very well with the scale overall, thus, it may be dropped or removed. Therefore, when the item has limited number (less than 10), inter item correlation value will be high within 0.4 to 0.76 [13]. Interestingly, through corrected item-total correlation measurement, the indication has been supported that PDP9 and PDP15 obtain negative values. By using 0.3 as the upper boundaries, the study found PDP10 has 0.226, which shows there is no correlation with other criteria. It also have value under 0.1 in inter-item correlation matrix with PDP1 (0.051), PDP2 (0.06), PDP3 (0.053) and PDP13 (0.001). Thus, those five items is eliminated as they do not show reliability and consistency to measure privacy protection as well redundant with other items.

**Table 2.** Inter-Total Correlation Matrix

| | Scale Mean if Item Deleted | Scale Variance if Item Deleted | Corrected Item-Total Correlation | Square Multiple Correlation | Cronbach's Alpha if Item Deleted |
|---|---|---|---|---|---|
| **PDP1** | 64.98 | 49.046 | .363 | .495 | .699 |
| **PDP2** | 65.02 | 49.418 | .394 | .707 | .698 |
| **PDP3** | 65.36 | 44.981 | .571 | .469 | .674 |
| **PDP4** | 65.34 | 47.672 | .560 | .471 | .685 |
| **PDP5** | 65.02 | 46.441 | .568 | .650 | .679 |
| **PDP6** | 66.07 | 45.693 | .258 | .581 | .718 |
| **PDP7** | 65.20 | 44.213 | .695 | .745 | .663 |
| **PDP8** | 65.55 | 46.114 | .640 | .761 | .675 |
| **PDP9** | 66.07 | 52.902 | -.042 | .509 | .756 |
| **PDP10** | 66.09 | 47.666 | .226 | .457 | .718 |
| **PDP11** | 65.07 | 48.251 | .532 | .550 | .688 |
| **PDP12** | 65.20 | 50.213 | .352 | .524 | .702 |
| **PDP13** | 65.68 | 49.757 | .190 | .532 | .717 |
| **PDP14** | 64.95 | 47.579 | .551 | .544 | .685 |
| **PDP15** | 66.30 | 52.166 | -.026 | .375 | .760 |

A total of 790 out of 1000 questionnaires were completed and 79% of the respondents returned their completed survey of which 11 or approximately 1.39% was subsequently eliminated. The central tendency is important in data screening to replace missing or invalid value. A measure of central tendency is a single value that describes a set of data by identifying the central position within, in which called measures of central location under summary statistics through mean as the average value, median for middle value and the mode as frequent value. Those are all valid measures of central tendency but some measures become more appropriate to use than others [14]. As this study using likert scale, it is clearly show the average of the squared differences from the mean and how spread out the number are, in which 1 represent very disagreement while 6 as very agreement. Meanwhile, negative values for the skewness indicate data that are skewed left, which mean that the left tail is long relative to the right tail and vice versa. For positive kurtosis indicates a light tailed distribution relative to a normal distribution or lack of outliers. In short, there is no an observation point that is distant from other observation.

**Table 3.** Central Tendency, Measures Spread and Shape

| | PDP1 | PDP2 | PDP3 | PDP4 | PDP5 | PDP7 | PDP8 | PDP11 | PDP12 | PDP14 |
|---|---|---|---|---|---|---|---|---|---|---|
| **N Valid** | 779 | 779 | 779 | 779 | 779 | 779 | 779 | 779 | 779 | 779 |
| **Missing** | 0 | 0 | 0 | 0 | 0 | 0 | 0 | 0 | 0 | 0 |
| **Mean** | 5.19 | 4.89 | 5.05 | 5.03 | 4.96 | 4.91 | 4.85 | 5.10 | 4.81 | 5.27 |
| **Median** | 5.00 | 5.00 | 5.00 | 5.00 | 5.00 | 5.00 | 5.00 | 5.00 | 5.00 | 6.00 |
| **Mode** | 6 | 5 | 5 | 5 | 5 | 5 | 5 | 5 | 5 | 6 |
| **Sum** | 4045 | 3806 | 3937 | 3916 | 3864 | 3828 | 3780 | 3973 | 3745 | 4103 |
| **Percentile 25** | 5.00 | 5.00 | 5.00 | 5.00 | 5.00 | 5.00 | 4.00 | 5.00 | 4.00 | 5.00 |
| **50** | 5.00 | 5.00 | 5.00 | 5.00 | 5.00 | 5.00 | 5.00 | 5.00 | 5.00 | 6.00 |
| **75** | 6.00 | 6.00 | 6.00 | 6.00 | 6.00 | 6.00 | 6.00 | 6.00 | 6.00 | 6.00 |
| **Std. Error of Mean** | .036 | .039 | .036 | .036 | .041 | .040 | .041 | .035 | .039 | .036 |
| **Std. Deviation** | 1.006 | 1.101 | 1.013 | 1.000 | 1.131 | 1.122 | 1.138 | .974 | 1.083 | .999 |
| **Variance** | 1.012 | 1.212 | 1.025 | 1.001 | 1.280 | 1.259 | 1.296 | .949 | 1.174 | .998 |
| **Range (Min – Max)** | 5 (1-6) | 5 (1-6) | 5 (1-6) | 5 (1-6) | 5 (1-6) | 5 (1-6) | 5 (1-6) | 5 (1-6) | 5 (1-6) | 5 (1-6) |
| **Skewness** | -1.950 | -1.367 | -1.516 | -1.553 | -1.438 | -1.440 | -1.449 | -1.532 | -1.308 | -2.035 |
| **Std. Error of Skewness** | .088 | .088 | .088 | .088 | .088 | .088 | .088 | .088 | .088 | .088 |
| **Kurtosis** | 4.812 | 2.051 | 2.968 | 3.222 | 2.196 | 2.318 | 2.311 | 3.344 | 2.265 | 5.204 |
| **Std. Error of Kurtosis** | .175 | .175 | .175 | .175 | .175 | .175 | .175 | .175 | .175 | .175 |

## 4. Privacy Policy Development

Based on statistic result, it shows that user permission is not relevant for privacy policy, in which it might overlap with special circumstances for notification and the agreement for service use. It also shows that mostly the customer do not want to be attached with physical entity or responsibilities in regarding the security procedure as they expect the data controllers handle relevant activities. As originally enacted, the law imposed requirements for what must be included in a privacy policy [7] such as categories of identified personal data of consumers, third parties with whom the operator may share the personal data, description of activity process for users to update their personal data, user features in control theirs personal data and notification process and effective date of the privacy policy.

### *4.1. Registration Policy*

For organization, the registration is the critical phase to gather the personal data where potential customers, employees or projects provide almost all their sensitive information to fulfil the requirement or attract the attention. The website was common used to allow the online registration besides conventional through counter. The security by encryption of the website is the main concern in regards that sensitive data of the potential consumers is transferred and shared. Therefore, data subjects have the right to be told whether the data held on them and the purpose for which the information is used and to obtain a copy of the information about them by making a subject access request. It occurred in various types such as recruitment, admission, enquiries, research, examinations, graduation, accommodation and career. Occasionally, information is passed between different parts of the phase for operational reasons such as for notification, archive or announcement. Partial or full information disclosure can be done through the data subject request or the representative or data controller under authority. In addition, disclosing information about customers to the relevant parties are appropriate until the decision made on their accepted registration, of course with necessary information only. The personal data to be included in the registration must relate to an individual directly (personal request) or indirectly (business matters) [5].

### *4.2. Collection Policy*

In collecting personal data from participants or vice versa, the method used should be in accordance with the relevant Act. It means that the following methods should be made clear to anyone who is providing personal data. In regards for the role of organization as data controller or data user and that will hold the data, it should also be clear that the reason why the data is being collected. The data subject can be informed about the purpose at the times when the data that has been collected. The data user must be careful to ensure that there is no disclosure of information to third parties that are not known by the data subject, when the collection take part. Disclosure to a third party must be stated in a written notice to the data subject, to comply with a statutory obligation. Only after the data subject is aware of the existence of the third party and given the relevant approvals. Organization might collect data for the sole purpose of specifically planned, agreed and necessary such as performance report, service and statistic. It is clear that the collection and dissemination of private information will almost always raise issues of relevance to private life [2]. They can retain that personal data as long as those purposes remain valid, especially when relevant data subject is not part of organization business activity anymore. To avoid data leaked without proper control, individuals who have access to personal data should not leave computer screens open. Also, passwords should not be disclosed and e-mail should be used properly.

### *4.3. Data Processing Policy*
To conduct the processing of personal data, it should be clear what kind of data would be processed. Data user must provide the consent form to the data subject to be completed and signed as the evidence for approval. It is obligatory to the organization to remind that the relevant data can only be used for those specific reasons, which be signed by the data user with certain terms and condition stated in the document based on organization recognition. An "opt out" the sentence may need to be included so the consumers can have the opportunities to consider whether they want to reject or not. If the organization wants to maintain the relationship to the data subject, so portal, group community or report by the data user can be used to manually realize the type of processing relate to their personal data. Therefore, to put simplicity in the process, the frequent data processing can be categorized differently with the special case, whereby data subject just be informed once in the beginning. Furthermore, by having data user records anonymous are recommended, especially when they shared outside. The employee should be easily gain access to their friends' personal data, some restriction should be considered for certain personal data. Organization must explain exactly the approach they use to avoid personal data being hacked or exposed. The core challenge in IT service out sourcing would be the fact that data user loses a complete control over personal data of their data subjects to whom they are initially answerable [15].

### *4.4. Access Policy*
Commonly, data subject is willing to reveal or display their own personal data in order to promote their expertise, capacity and achievement. It will not cause any danger as long as it is used for good purposes and right manner. However, the administrations that keep and administer the data subject's data own a legal duty not to disclose them to unauthorized party. The bottom line is that data users are responsible to the security, integrity and reliability of the personal data that they process. In addition, the relevant data subjects should have full access to their own records, with the ability to correct false, incomplete or inaccurate information to prevent information misleading. Meanwhile, request for access by different party can be made in writing and data controllers are allowed to charge a fee for such requests. The circumstances where the controller may refuse to comply with the data access request especially when data controller is not supplied with relevant information on the identity of the requestor. Therefore, access to data stored on a computer to be controlled by passwords and where appropriate, there are the restriction or limitation to access certain file in regards to the authority. The policy need to cover the standard requirement to create the password while only authorized person know his/her own password whereby it is updated periodically.

### *4.5. Exemption Policy*
Personnel files or report that consist personal data should be stored responsibly in a specific place, so any illegal and unauthorized attempts to view and access them will be noticed immediately. They should have default of locked condition when not in use (require password) and could not be removed without authority. An exception in which personal data may be disclosed to third parties, if necessary for the prevention of crime, public interest or as authorized by law or court order. The protection by data user only through a reasonable assurance related to legal right and the consent of the data subject. The different between registration and collection policy stand on the data subject and user, where in registration data subjects submit their own personal data to gain such recognition by data user while in collection, data user ask the permission from data subject to give their personal data to specific purpose.

### *4.6. Deletion Policy*
Private data must be erased if the retention period has expired, it has served its purpose, or it is requested by the data owner. The deletion of private data must also be in accordance with prevailing laws and regulations and not be relevant to any case proceeding. Meanwhile, to ensure the awareness amongst the staff member, organization can conduct the campaign, training and special talk about code of practice. In case of any private data leak, the private data administrator must notify the data owner regarding the private data which was revealed, the time and sequence of events that led to the private data leak, efforts by the private data administrator to address the private data leak and contact information of the private data administrator [4]. Any breaches of security should be treated as a disciplinary issue. On the other hand, destroying or disposing of personal data also counts as processing. Such good care should be taken in the disposal of any personal data to ensure that it is appropriate.

## 5. Conclusion
To maintain the enforcement of personal data protection, organization is encouraged to create the periodical discussion of privacy policy evaluation among staff members. In general, regulation has many purposes like to provide activity guide to agency or committee body, to grant certain privilege to execute national policy, to declare the principle or strategy, to authorize bureaucracy in performing certain action or measures and many more. It is contrasted with a non-legislative act in the form of government regulation, which is adopted by administrative body under the authority of specific regulation act for implementing certain policy and ensuring uniformity of law application. It should take a note the utilization of personal data might be occurred in various activities such as application, complaint, suggestion, storing, transfer and sharing. Such control by the data user must be strictly enough in anticipating personal data to be improper use.

## References
[1] Rich C 2016 Privacy Laws in East, Central and South Asia and the Pacific *World Data Protection Report* Vol 16 **6**

[2] Mendel T, Puddephatt A, Wagner B, Hawtin D and Torres N 2012 Global Survey on Internet Privacy and Freedom of Expression *Unesco Series on Internet Freedom* ISBN 978-92-3-104241-6

[3] Lubis M, Kartiwi M and Zulhuda S 2017 Privacy and Personal Data Protection in Electronic Voting: Factors and Measures *Telkomnika March 2017* vol **15**(1) p 512-521

[4] Rahman A A 2016 Indonesia introduces a comprehensive privacy bill *Privacy Laws & Business International Report*

[5] Hassan K M 2011 Personal Data Protection in the Business of Higher Education: Malaysian Law *International Conference on Sociality and Economics Development* IPEDR Vol 10 (IACSIT Press, Singapore) pp 53-57

[6] Wisniewski P, Islam AKMN, Lipford HR and Wilson DC 2016 Faming and Measuring Multi-dimensional Interpersonal Privacy Preferences of Social Networking Site Users *Communications of the Association for Information Systems* Vol. 38 A **10**

[7] US Department of Justice 2006 Privacy Policy Development Guide and Implementation Templates *Global Justice Information Sharing Initiative*

[8] UNCTAD 2016 Data protection regulations and international data flows: Implications for trade and development (United Nations)

[9] Lanier C D and Saini A 2008 Understanding Consumer Privacy: A Review and Future Direction *Academy of Marketing Science Review* Vol 12 No 2 pp 1-44

[10] Lubis M and Kartiwi M 2014 Privacy Concern and Perceived Benefits towards Privacy and Election in Indonesia *IEEE ICT4M Proceedings*.

[11] Lubis M, Kartiwi M and Zulhuda S 2016 Current State of Personal Data Protection in Electronic Voting: Demand on Legislature's Bill *International Conference on Informatics and Computing (ICIC)* pp 233-238

[12] Tang W and Cui Y 2013 Internal Consistency: Do We Really Know, What It Is and How to Assess It? *AERA CRAME*

[13] Pallant J 2005 A step by step guide to data analysis using SPSS for Windows *(2nd Edition ed.). Crows Nest, New South Wales, Australia: Allen & Unwin.*

[14] Laerd Statistics. (n.d.). Descriptive and Inferential Statistics. *Retrieved 7th September 2016 from: https://statistics.laerd.com*

[15] Zulhuda S, Azmi IM and Hakiem N 2015 Big Data, Cloud and Bring Your Own Device: How the Data Protection Law Addresses the Impact of Datafication *Advanced Science Letters* Vol 21 pp 3347-3351

# Identification tibia and fibula bone fracture location using scanline algorithm

**M A Muchtar[1], S E Simanjuntak[1], R F Rahmat[1], H Mawengkang[2], M Zarlis[1], O S Sitompul[1], I D Winanto[3], U Andayani[1], M F Syahputra[1], I Siregar[4] and T H Nasution[5]**

[1]Department Information Technology, Universitas Sumatera Utara, Indonesia
[2]Department of Mathematics, Universitas Sumatera Utara, Indonesia
[3] Department of Medicine, Universitas Sumatera Utara, Indonesia
[4]Department of Industrial Engineering, Universitas Sumatera Utara, Indonesia
[5]Department of Electrical Engineering, Universitas Sumatera Utara, Indonesia

anggi.muchtar@usu.ac.id, ulfi.andayani@usu.ac.id

**Abstract.** Fracture is a condition that there is a damage in the continuity of the bone, usually caused by stress, trauma or weak bones. The tibia and fibula are two separated-long bones in the lower leg, closely linked at the knee and ankle. Tibia/fibula fracture often happen when there is too much force applied to the bone that it can withstand. One of the way to identify the location of tibia/fibula fracture is to read X-ray image manually. Visual examination requires more time and allows for errors in identification due to the noise in image. In addition, reading X-ray needs highlighting background to make the objects in X-ray image appear more clearly. Therefore, a method is required to help radiologist to identify the location of tibia/fibula fracture. We propose some image-processing techniques for processing cruris image and Scan line algorithm for the identification of fracture location. The result shows that our proposed method is able to identify it and reach up to 87.5% of accuracy.

## 1. Introduction

Fracture or broken bone is caused by trauma, high force impact or stress, or a certain medical condition that weaken bones. The soft tissue around the bone will determine whether the happened fracture is complete or not. A complete fracture happens when the parts of the fractured bone are completely separated from each other, while incomplete fracture extends partly across the bone. Tibia/fibula fracture happens more often than the other bone fracture because of the thin periosteum lining the tibia, especially in front area covered by skin, so that the bone often fractures because it is directly located under the skin [1]. The way to diagnose fracture is clinical examination and then followed by radiological examination (X-ray) [2]. Generally, radiologist analyze the bone fracture from the X-ray image manually and it takes longer time to diagnose it. The X-ray images also contain much noise, that possibly cause errors while diagnosing. Moreover, reading X-ray images requires highlighting to make the objects in it appear more clearly. Therefore, many previous researches based image processing to identify the fracture using various method to get higher accuracy [3].

In 2011, Mahendran and Baboo detected the fracture from X-ray image using Wavelet and morphological operations to segment the image. They used gray level method to extract features based texture and concluded that their proposed method can run exceedingly fast and can generate results in seconds [4].

Another method proposed by Chai, et al. to detect bones fracture. They used Gray Level Co-occurrence Matrix (GLCM) to analyze the texture of fracture by calculating features of homogeneity, contrast, energy, correlation. In the conclusion, the accuracy of their proposed method is achieved at least 86.67% which can be further improved [5].

Bone fracture detection through x-ray images was also done by Al-Ayyoub et al. in 2013. They used Sobel edge detector to detect the edges of the hand bone, used the combination of Wavelets, Curvelets and GLCM to extract features and compared some classifier methods, namely Neural Network (NN), Naïve Bayes (NB), Bayesian Network (BN) and Decision Tree (DT). They concluded that 91.8% of accuracy was obtained by applying boosting and bagging on the Bayesian Network classifier [6].

Next research in 2014, Kurniawan used OpenCV combined with canny edge detection for fracture detection. They stated that canny edge detection has optimal advantage for identifying the location of bone fracture from X-ray image [7].

In this research, we proposed some image processing techniques to enhance the image, canny edge detection to extract feature and scan line algorithm to identify the location of bone fracture. This paper is organized as follows; Section II describes the proposed method. Experimental results are presented on Section III. Section IV describes the summary and suggestions for future research.

## 2. Concept Development

In this research, there are several steps to identify the location of tibia and fibula fracture, namely collecting data of tibia and fibula image, cropping to get the region of interest (ROI), resizing to get the same image resolution of cruris image (image of detected fracture), gray scaling, sharpening to enhance the image, and then using canny edge detection and scan line algorithm to identify the location of tibia and fibula fracture. General architecture of our proposed method can be shown in Figure 1.

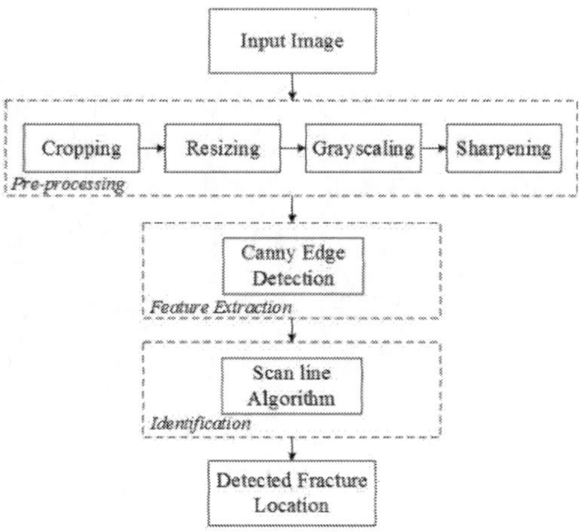


**Figure 1.** General Architecture

### *2.1. Image Acquisition*

The used data in this research is acquired cruris image from Haji Adam Malik Hospital and also from web. The digital images are fractured tibia/fibula, totaled 40 images in JPG (Joint Photographic Group) format and then will be used for the next step of the proposed method. A sample cruris image is shown in figure 2(a).

### *2.2. Pre-processing*

This step consist of four steps, namely cropping, resizing, grayscaling and sharpening, which will be clearly explained in the next sub subsection.

- *Cropping.* It is conducted to get the region of interest (ROI) and separate it from unnecessary area of cruris image. Figure 2(b) shows the result of cropping cruris image.
- *Resizing.* This process is conducted to make all images have the same size. In this research, we resize all the used image data into smaller size, 100 x 400 pixels, to make the process run quickly. The sample result from resizing can be shown in figure 2(c).
- *Grayscaling.* After resizing the colour image (has RGB components), it will change into grayscale. The conversion from RGB into grayscale can be done using equation (1). Figure 2 (d) shows the result of grayscaling process.

$$G = \frac{R + G + B}{3} \quad (1)$$

- *Sharpening.* As the principle of High Pass Filter, sharpening is done to enhance the image quality and emphasize the edge of image object. Figure 2(e) shows the result of sharpening.

### *2.3. Feature Extraction*

Feature extraction is to get unique feature from the image so that it will be an input for process of identification or classification. In this research, we use canny edge detection to extract the feature then the detected edges will be an input for scan line algorithm to identify the location of tibia/fibula fracture. There are several steps in canny edge detection, namely:

- Removing noise using Gaussian filter to remove unnecessary pixel so that it is undetected as an edge.
- Setting upper threshold 3f and lower threshold 50.5f. If the pixel value is greater that upper threshold, then it is detected as an edge. Figure 2(f) shows the result of canny edge detection. The binary image obtained from thresholding will be an input for scan line algorithm.

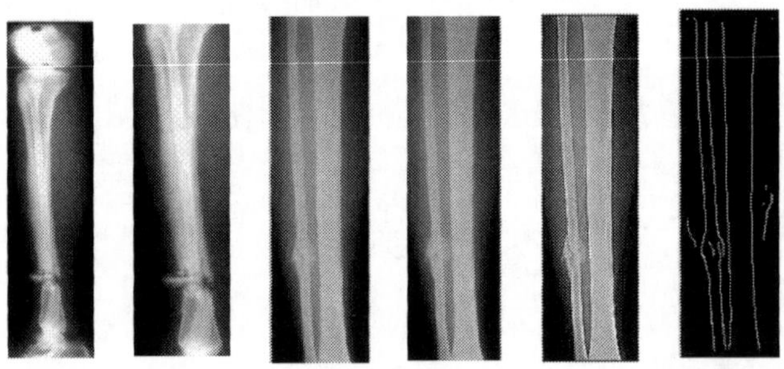

(a) (b) (c) (d) (e) (f)

**Figure 2.** (a) Cruris Image. (b) Cropping. (c) Resizing. (d) Grayscaling. (e) Sharpening. (f) Canny edge detection.

### *2.4. dentification using Scan line Algorithm*

Figure 3 shows the illustration of scan line algorithm. There are several steps of scan line algorithm for identifying bone fracture, namely:

- Calculate the number of white pixel (1) in a row using equation (2)

$$Rn = C1 + C2 + \cdots + Cn \quad (2)$$

- Calculate total number of white pixel for three rows around, formulated as in equation (3).

$$Sn = B(n-1) + Bn + B(n+1) \tag{3}$$

- Determine the rows which contains the highest number, then it will be the location of fracture.

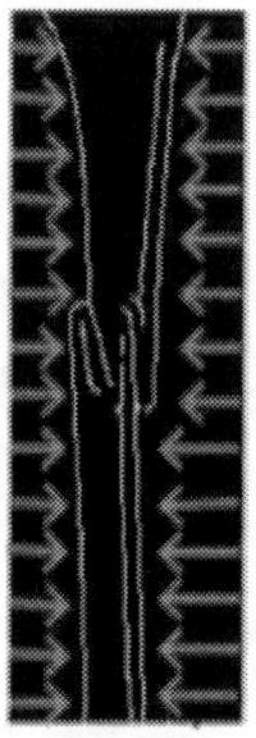

| | | | | | | | | | | |
|---|---|---|---|---|---|---|---|---|---|---|
| 0 | 0 | 1 | 0 | 0 | 0 | 0 | 1 | 1 | 3 | 6 |
| 0 | 0 | 1 | 0 | 0 | 0 | 0 | 1 | 1 | 3 | 9 |
| 0 | 0 | 1 | 0 | 0 | 0 | 0 | 1 | 1 | 3 | 9 |
| 0 | 0 | 1 | 0 | 0 | 0 | 0 | 1 | 1 | 3 | 9 |
| 0 | 0 | 1 | 0 | 0 | 0 | 0 | 1 | 1 | 3 | 11 |
| 0 | 1 | 1 | 0 | 0 | 0 | 1 | 1 | 1 | 5 | 15 |
| 0 | 1 | 1 | 1 | 1 | 1 | 1 | 0 | 1 | 7 | 19 |
| 0 | 1 | 1 | 1 | 1 | 1 | 1 | 0 | 1 | 7 | 17 |
| 0 | 1 | 0 | 0 | 0 | 1 | 0 | 1 | 0 | 3 | 13 |
| 0 | 1 | 0 | 0 | 0 | 1 | 0 | 1 | 0 | 3 | 9 |
| 0 | 1 | 0 | 0 | 0 | 1 | 0 | 1 | 0 | 3 | 9 |
| 0 | 1 | 0 | 0 | 0 | 1 | 0 | 1 | 0 | 3 | 9 |
| 0 | 1 | 0 | 0 | 0 | 1 | 0 | 1 | 0 | 3 | 6 |

(a) (b)

**Figure 3**. (a) Scan line Algorithm. (b) Illustration of scan line algorithm calculation

## 3. System Design

In the main page view of the application, there are several facilities such as cruris image selection, cruris image processing, and the result of identification of cruris image fracture location. The design of the main page view is shown in figure

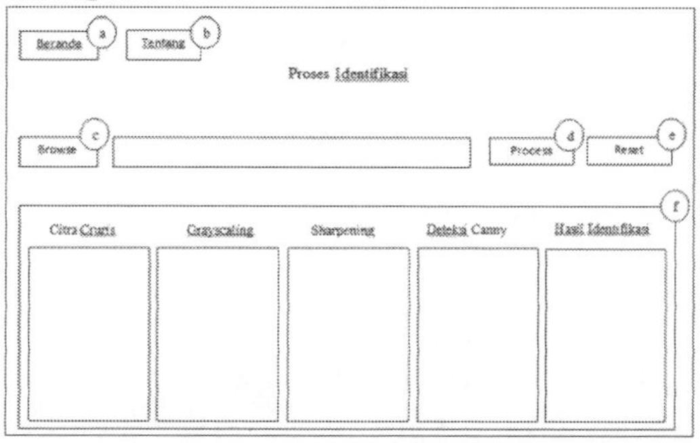


**Figure 4.** The Main Page Application

Information:

- The "Home" button allows the user to go back to the beginning.
- The "About" button allows the user to see a brief description of the system.
- The "Browse" button allows the user to select a bone cruris image on the image store. After the image is selected, the image will be displayed in the "Cruris Image".
- The "Process" button allows the user to perform the process of identifying the location of the fracture. After the identification process is complete, the pre-processing result image of the grayscaling, sharpening, and canny detection process will be displayed, and the result of identification of the location of the fracture will be displayed in the "identification result" section.
- The "Reset" button allows the user to delete the input image along with all the processing results.
- Represents each process that will be shown after the user selects the image and selects the "Browse" button.

## 4. System Prototyping

The methods used in identifying the location of tibia and fibula bone fractures are implemented into the system using Java programming language in accordance with the design that has been done.
In designing applications to identify the location of bone fractures of the tibia and fibula using Scanline algorithm requires hardware and software support include:

### *4.1. Specification of hardware and software used*

The hardware and software specifications used to construct a bone fracture identification system are as follows.

- Processor Intel Core i3-820M CPU 1.9GHz
- Memory (RAM) 4.00 GB DDR3
- 500 GB Hard Drive Capacity
- The operating system used is Microsoft Windows 8.1 Enterprise 64 Bit
- NetBeans IDE 7.0
- JDK 1.7

## 5. System Testing

The images of the selected tibia and fibula will then be displayed on the "Cruris Image" panel as shown in figure below. After the fracture image of the tibia and fibula bones is displayed on the "Cruris Image" panel, the "Process" button and the "Reset" button can be used. The "Process" button is used to initiate the process of identifying the location of the fracture from the selected bone fracture of the tibia and fibula, starting from pre-processing, feature extraction, and identification using the Scanline Algorithm. The results of the pre-processing process will be displayed in the grayscaling panel of grayscaling images, the sharpened image with the "Sharpening" panel, the feature extraction using canny edge detection will be displayed on the "Canny Detection" panel, and the results will be displayed on the panel "Identification Result" as shown in Figure below:

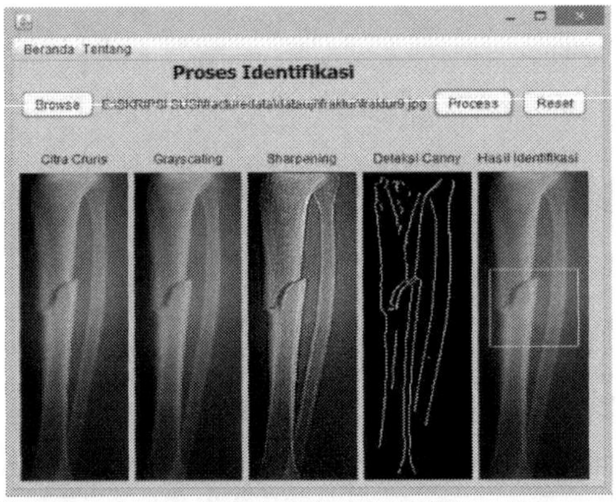


**Figure 5.** Application in Identifying Process

In the sharpening process, the edge of the image is sharper and clearer. The result is obtained by applying the principle of High Pass Filter (HPF) in sharpening process.

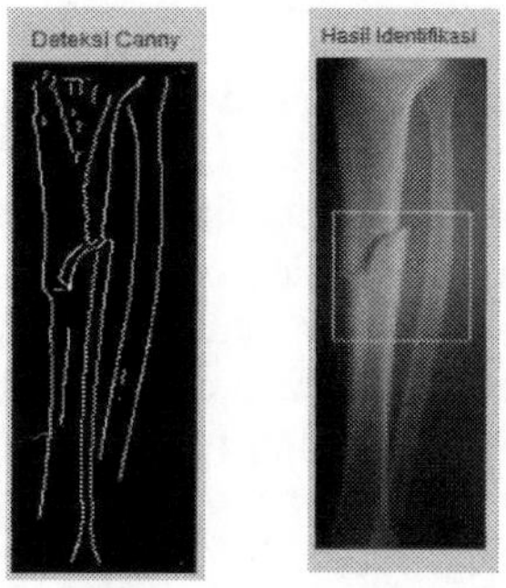


**Figure 6.** (a) Canny Detection (b) Identification Result

The results of the identification process using the Scanline Algorithm are displayed in the "Results Identification" panel. Display of result of process Identification of location of fracture of tibia and fibula bone image is shown in Figure 4.9. The value of the features obtained in the canny process will be calculated to get the highest score. The highest value will be the central point in determining the location of fractures of the tibia and fibula bone.

## 6. Result And Evaluation

In this section, we analyze the ability of our proposed method for identifying the location of tibia/fibula fracture. Therefore, system testing is conducted using 40 acquired images of tibia/fibula fracture (crusis image).The testing result is shown in Table 1.

**Table 1.** Result

| | | | | | |
|---|---|---|---|---|---|
| 1 | Fraktur1.jpg | True | 21 | Fraktur21.jpg | True |
| 2 | Fraktur2.jpg | True | 22 | Fraktur22.jpg | True |
| 3 | Fraktur3.jpg | True | 23 | Fraktur23.jpg | True |
| 4 | Fraktur4.jpg | True | 24 | Fraktur24.jpg | True |
| 5 | Fraktur5.jpg | True | 25 | Fraktur25.jpg | True |
| 6 | Fraktur6.jpg | True | 26 | Fraktur26.jpg | True |
| 7 | Fraktur7.jpg | True | 27 | Fraktur27.jpg | False |
| 8 | Fraktur8.jpg | True | 28 | Fraktur28.jpg | True |
| 9 | Fraktur9.jpg | True | 29 | Fraktur29.jpg | True |
| 10 | Fraktur10.jpg | False | 30 | Fraktur30.jpg | True |
| 11 | Fraktur11.jpg | True | 31 | Fraktur31.jpg | True |
| 12 | Fraktur12.jpg | True | 32 | Fraktur32.jpg | True |
| 13 | Fraktur13.jpg | True | 33 | Fraktur33.jpg | True |
| 14 | Fraktur14.jpg | True | 34 | Fraktur34.jpg | True |
| 15 | Fraktur15.jpg | True | 35 | Fraktur35.jpg | False |
| 16 | Fraktur16.jpg | False | 36 | Fraktur36.jpg | True |
| 17 | Fraktur17.jpg | True | 37 | Fraktur37.jpg | True |
| 18 | Fraktur18.jpg | True | 38 | Fraktur38.jpg | True |
| 19 | Fraktur19.jpg | True | 39 | Fraktur39.jpg | True |
| 20 | Fraktur20.jpg | False | 40 | Fraktur40.jpg | True |

Based on Table 1, from 40 images of testing data, there are 5 testing data detected error while identifying the location of tibia/fibula fracture, namely data to 10, 16, 20, 27 and 35. So, from all the testing result, we can conclude that the obtained accuracy for identifying the location of tibia/fibula fracture using scan line algorithm is 87.5%.

$$Percentage\ of\ Accuracy\ = \frac{35}{40} x 100\% \ = 87.5\%$$

## 7. Conclusion
Based on all testing results for identifying the location of tibia/fibula fracture, scan line algorithm is able to help the radiologist to identify the fracture location. The obtained accuracy also depends on the number of acquired data. In this research, our proposed method can give 87.5% of accuracy for testing 40 images. The identification result much depends on the image quality generated from pre-processing and feature extraction. The better the image quality, the higher the generated accuracy.

Future work can be done using another method for detecting more than one location of fracture. Another research possibly done is classifying two conditions of bone, namely normal and fracture, or identifying the type of fractures.

## Acknowledgement
This research was supported by University of Sumatera Utara. All the faculty, staff members and lab technicians of Information Technology Department, whose services turned my research a success.

## References
[1] Mahyudin L 2010 *Fraktur Diafsis Tibia* http://www.Belibis17.tk

[2] E-Radiography *Tibia & Fibula AP* http://www.e-radiography.net/technique/ankle-tib/tibia fibula.htm

[3] Mostafa S 2004 AdAgen : Adaptive Interface Agent For X-Ray Fracture Detection. *Int. J. Computing & Information Sciences* **2** 143-148

[4] Mahendran S K and Baboo S S 2011 Enhanced Automatic X-Ray Bone Image Segmentation using Wavelets and Morphological Operators *Int. Conf. on Information and Electronics Engineering* pp 125-129

[5] Chai H Y and Wee L K 2011 Gray level co-occurrence matrix bone fracture detection. *American J. Applied Sciences* **8** 26-32

[6] Al-Ayyoub M, Hmeidi I and Rababah, H 2013 Detecting Hand Bone Fractures in X-Ray Images *JMPT* **4** 155-168

[7] Kurniawan S F 2014 Deteksi fraktur tulang menggunakan open CV *J. Teoritical Indonesia and Information Technology* **64** 249-254

# Bioinformatics approach of three partial polyprenol reductase genes in *Kandelia obovata*

**M Basyuni[1*], R Wati[1], H Sagami[2], H Oku[3], S Baba[4]**

[1]Department of Forestry, Faculty of Forestry, Universitas Sumatera Utara, Jl. Tri Dharma Ujung No. 1 Medan, North Sumatera 20155, Indonesia
[2]Institute of Multidisciplinary Research for Advanced Material, Tohoku University, Sendai, 980-8577, Japan
[3] Molecular Biotechnology Group, Tropical Biosphere Research Center, University of the Ryukyus, 1 Senbaru, Nishihara, Okinawa 903-0213, Japan
[4]International Society for Mangrove Ecosystems, Faculty of Agriculture, University of the Ryukyus, 1 Senbaru, Nishihara, Okinawa 903-0213, Japan

[*]Email: m.basyuni@usu.ac.id

**Abstract**. This present study describesthe bioinformatics approach to analyze three partial polyprenol reductase genes from mangrove plant, *Kandeliaobovata*as well aspredictedphysical and chemical properties, potential peptide, subcellular localization, and phylogenetic. The diversity was noted in the physical and chemical properties of three partial polyprenol reductase genes. The values of chloroplast were relatively high, showed that chloroplast transit peptide occurred in mangrove polyprenol reductase. The target peptide value of mitochondria varied from 0.088 to 0.198 indicated it was possible to be present. These results suggested the importance of understanding the diversity of physicochemical properties of the different amino acids in polyprenol reductase. The subcellular localization of two partial genes located in the plasma membrane. To confirm the homology among the polyprenol reductase in the database, a dendrogram was drawn. The phylogenetic tree depicts that there are three clusters, the partial genes of *K. obovata* joined the largest one: C23157 was close to *Ricinus communis* polyprenol reductase.Whereas, C23901 and C24171 were grouped with *Ipomoea nil* polyprenol reductase, suggested that these polyprenol reductase genes form distinct separation into tropical habitat plants.

## 1. Introduction

Mangrove plants are well known a source of secondary metabolites mostly derived from isoprenoids [1].Polyisoprenoid alcohols are linear five unit polymers that are present in almost all living organisms. Long chain polyisoprenoids have occurred in various plant tissues [2]. These studies revealed the presence two types of polyisoprenoids regarding the stereochemistry: polyprenol (α-unsaturated isoprenoid alcohols) and dolichol (α-saturated isoprenoid alcohols). In plant photosynthetic tissues, polyprenols are usually abundantly detected in comparison to dolichols [2-4]. On the other hand, the sources for dolichols mainly are derived from animals (livers) [5], plant roots [6-8], and yeast cells [9].

Recently, dolichols but not polyprenols have been reported as the dominant polyisoprenoids alcohols of mangrove and coastal plants [6-8]. The occurrence of prosperous dolichols in the leaves of mangroves and coastal plants indicate that the enzyme of polyprenol reductase might be active to converse the polyprenols to dolichols [6-8]. In this context, it is essential to get more insight into the polyprenol reductase genes in mangrove plants. Nonetheless, the information on the polyprenol reductase has not been previously available in mangroves. The present study, therefore, aimed to analyze three predicted polyprenols genes in mangrove plant, *Kandelia obovata* using the bioinformatics approach.

## 2. Materials and method

### *2.1. Materials*

A total three partial polyprenol reductase genes from *Kandelia obovata* namely c23157, c23901, and c24171 were studied. These genes mainly were derived from genome sequence of *K. obvata* using a Blast search. These genes showed high homology to the plant polyprenol reductase genes.

### *2.2. Physicochemical properties of the polyprenol reductase gene*

Protparam online (web.expasy.org/protparam/) was used to analyze the composition, physical and chemical properties of three polyprenol reductase genes. The computed parameters describe the molecular weight, theoretical isoelectric point values, amino acid composition, atomic composition, extinction coefficient, estimated half-life, instability index, fat coefficient, and average hydrophilicity as previously reported [10].

### *2.3. Potential transit of peptide and subcellular localization of OSC gene*

The target P1.1 server online (www.cbs.dtu.dk/services/targetp/) was used to predict transit peptide. The location is based on the predicted occurrence of any of the *N*-terminal pre-sequences chloroplast transit peptide (cTP), mitochondrial targeting peptide (mTP) and secretory pathway signal peptide (SP). Furthermore, PSORT Prediction online (psort.hgc.jp/form.html) was used to determine the subcellular localization of polyprenol reductase genes as previously described [11].

### *2.4. Phylogenetic analysis of three polyprenol reductase*

The amino acid sequences were aligned using the FASTA version 3.4t26 [12] of the DNA Data Bank of Japan (Mishima, Shizuoka, Japan). Phylogenetic analysis of deduced amino acid alignment from three partial genes in addition to twenty one predicted polyprenol reductase was analyzedusing CLUSTAL W ver. 1.83 [13] of the DNA Data Bank of Japan followed by drawing with TreeView ver. 1.6.6 [14] based on a neighbour-joining method. Bootstrap analysis with 1000 replications was used to measure the strength of the nodes in the tree [15]. The GenBank accession numbers of the DNA and amino acid sequence of plant polyprenol reductase genes used this analysis as follows, *Arachis duranensis* polyprenol reductase 2 (XM_016081788), *A. ipaensis* polyprenol reductase 2 (XM_021109944), *Cajanus cajan* polyprenol reductase 2 (XM_020359734), *Elaeis guineensis* polyprenol reductase 1 (XM_019849699), *Eucalyptus grandis* polyprenol reductase 2 (XM_018861897), *Glycine max* polyprenol reductase 2 (XM_003537780), *Gossypium arboreum* polyprenol reductase 2-like (XM_017765134), *G. hirsutum* polyprenol reductase 2-like (XM_016848239), *G. raimondii* polyprenol reductase 2-like (XM_012601789), *Ipomoea nil* polyprenol reductase 2-like (XM_019302562), *Jatropha curcas* polyprenol reductase 2 (XM_012210100), *Juglans regia* polyprenol reductase 2-like (XM_018985852), *Malus x domestica* polyprenol reductase 2-like (XM_017330714), *Phoenix dactylifera* polyprenol reductase 1 (XM_017840866), *Populus euphratica* polyprenol reductase 2-like (XM_011038928), *Prunus persica* polyprenol reductase 2 (XM_007200320), *Pyrus x bretschneideri* polyprenol reductase 2-like (XM_018642260), *Ricinus communis* polyprenol reductase 2 (XM_015715302), *Theobroma cacao* polyprenol reductase 2 (XM_018128982), *Vigna radiata* var. *radiata* polyprenol reductase 2 (XM_014668834), and *Ziziphus jujuba*polyprenolreductase 2 (XM_016036804).

## 3. Results and Discussions

### *3.1. Physical and chemical properties of the polyprenol reductase gene*

Table 1 shows the start codon of the partial genes was not detected, where the stop codons were found only in C2390 (917TGA) and C24171 (504TGA). The open reading frame length was not identified due to the partial genes. Some encoded amino acids were 166 to 306. The more range of genes had the more values of the physicochemical parameters such as in molecular mass, the total number of atoms, extinction coefficient, Fat coefficient, and overall average hydrophilicity (Table 1). No stable proteins were found in predicted polyprenol reductase genes. The stability protein coefficient was generally below 40 as previously reported in *BgBAS* and *RsCAS* [10] of oxidosqualene cyclase genes and *KcAct1, BgAct1,* and *RsAct1* as members of plant actin genes [11]. The diversity was noted in the physicochemical properties of three partial polyprenol reductase genes.

**Table 1**. Physical and chemical properties of the polyprenol reductase in *K. obovata*

| Nucleotide name | C23157 | C23901 | C24171 |
|---|---|---|---|
| Length of genes/bp | 700 | 960 | 531 |
| Open reading frame length/bp | nd | nd | nd |
| Start site and codon | nd | nd | nd |
| Stop site and codon | nd | 917TGA | 504TGA |
| Number of encoded amino acids | 228 | 306 | 166 |
| Relative molecular mass | 59214.59 | 79063.22 | 44281.16 |
| Theoretical isoelectric point values | 5.15 | 4.98 | 5.17 |
| Total number of atoms | 7618 | 10023 | 5672 |
| Extinction coefficient | 0.148 | 0.183 | 0.175 |
| Half-life period | 1.2h | 7.2h | 30h |
| Instability coefficient | 48.67 | 47.77 | 56.94 |
| Fat coefficient | 22.46 | 28.81 | 34.40 |
| Overall average hydrophilicity | 0.576 | 0.870 | 0.962 |

nd= not detected

### *3.2. Potential transit of peptide and subcellular localization of polyprenol reductase gene*

Table 2 depicts the possibility of the possible transit peptide in *K. obovata* polyprenol reductase. There are four reliabilities concerning the possibility of the transit peptide namely chloroplast transit peptide, mitochondrial target peptide, signal peptide of secretory pathway, and reliability prediction (Table 2). The values of chloroplast were relatively high, showed the occurrence of chloroplast transit peptide. The contents of polyprenols and dolichols were enhanced during the life-course of a tissue or organ [6, 16].The target peptide value of mitochondria varied from 0.088 to 0.198 indicated it was possible to be present. The low signal peptide of the secretory pathway has been described in mangrove OSC genes and actin genes [10-11], indicating no signal peptide in the polyprenol reductase of *K. obovata.*

**Table 2**. Possibility of the potential transit peptide in *K. obovata* polyprenol reductase

| Nucleotide ID | Reliability | | | |
|---|---|---|---|---|
| | Chloroplast transit peptide | Mitochondrial target peptide | Signal peptide of secretory pathway | Reliability prediction |
| C23157 | 0.236 | 0.182 | 0.031 | 4 |
| C23901 | 0.204 | 0.198 | 0.009 | 4 |
| C24171 | 0.214 | 0.088 | 0.076 | 5 |

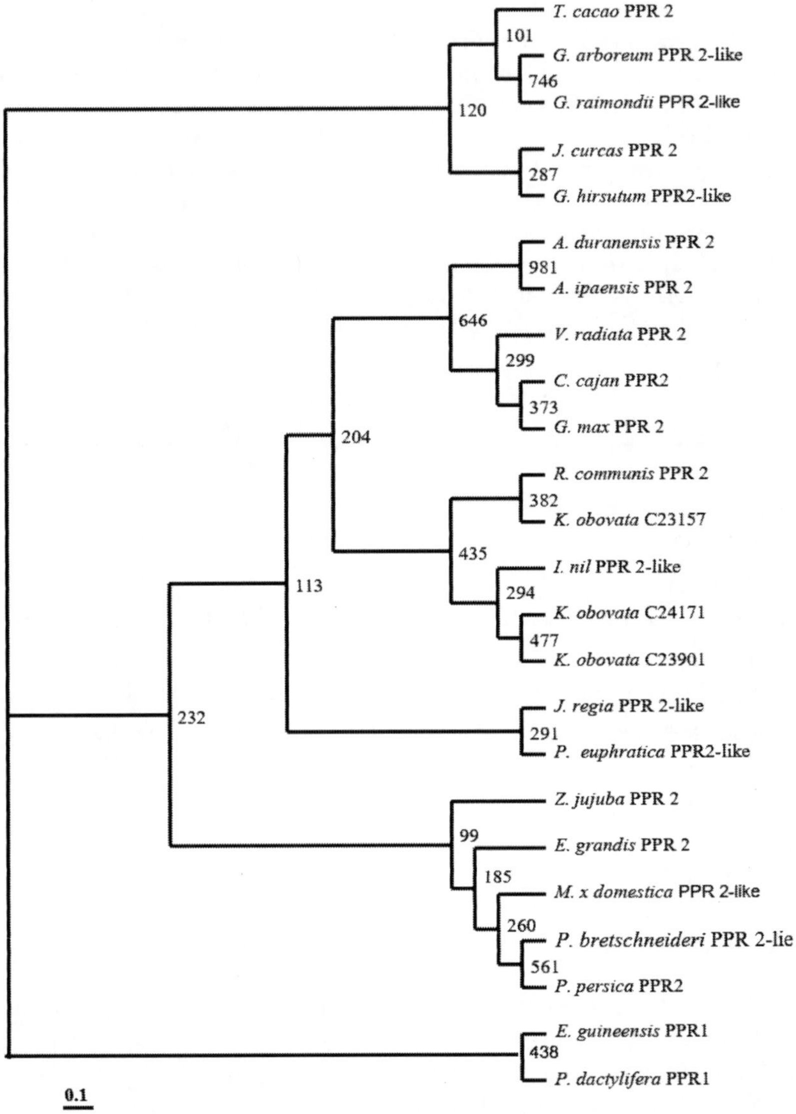


**Figure 1.** Phylogenetic tree of plant polyprenol reductase including three partial polyprenols reductase from *K. obovata*. Phylogenetic tree of deduced amino acid sequences was constructed with the neighbour-joining method of the CLUSTAL W [13]. The indicated scale represents 0.1 amino acid substitutions per site. Numbers indicate bootstrap value from 1000 replicates.The GenBank accession numbers of the amino acid sequence of using this analysis areshown in the Materials subsection.

Table 3 displays subcellular localization of polyprenol reductase genes in *K. obovata*. The subcellular localization of these genes was mostly in the endoplasmic reticulum (membrane and lumen). The C23901 has also located in Golgi bodies and plasma membrane, the C23157 was found in outside and lysosome. The C24171 was detected in the plasma membrane and outside. Recently, it has been shown that the gene expression of two triterpenoid synthases, *BgbAS* and *RsM1* enhanced the triterpenoid content of plasma membrane fractions as well as found in the plasma membrane [9, 16]. Furthermore, it has been established views that the plasma membrane is the first defence against changes in physicochemical variables of the altered environment [18-19]. Several salt tolerance genes from *Rhizophora stylosa* located in the plasma membrane supported the previous study that on the importance of layer for abiotic stress tolerance including salt stress [20].

**Table 3**. Subcellular localization of predicted polyprenol reductase

| Nucleotide ID | Golgi bodies | Plasma membrane | Endoplasmic reticulum (membrane) | Endoplasmic reticulum (lumen) | Outside | Lysosome |
|---|---|---|---|---|---|---|
| C23901 | 0.460 | 0.640 | 0.370 | 0.100 | nd | nd |
| C23157 | nd | nd | 0.550 | 0.100 | 0.100 | 0.317 |
| C24171 | nd | 0.717 | 0.100 | 0.100 | 0.437 | nd |

### *3.3. Phylogenetic analysis of polyprenol reductase gene*

To confirm the homology among the polyprenol reductase gene in *K. obovata* with plant polyprenol reductases, a phylogenetic tree was constructed by their amino acid sequences (Figure 1). The phylogenetic tree forms three clusters; the first branch consists of three genes: *T. cacao, G. arboretum,* and *G. raimondii.* The second group comprises the largest members where thee partial polyprenol genes joined in the tree. The C23157 was close to *Ricinus communis* polyprenol reductase. Whereas, C23901 and C24171 were grouped with *Ipomoea nil* polyprenol reductase, suggested that these polyprenol reductase genes form distinct separation into tropical habitat plants.

## 4. Conclusions

The present study confirmed the chloroplast transit peptide occurred in *K. obovata* polyprenol reductase. The cluster analysis indicated that polyprenol reductase from *K. obovata, R. communis,* and *I. nil* grouped distinct separation into tropical habitat plants.

**Acknowledgment**

This study was partly supported by an International Research Collaboration and Scientific Publication Grant (No. 003/SP2H/LT/DRPM/IV/2017 to MB) from the Directorate for Research and Community Service, Ministry of Research, Technology and Higher Education, Republic of Indonesia.

**References**

[1] Basyuni M, Oku H, Baba S, Takara K and Iwasaki H 2007 Isoprenoids of Okinawan mangroves as lipid input into estuarine ecosystem *J. Oceanography* **63** 601–608.

[2] Swiezewska E and Danikiewicz W 2005 Polyisoprenoids: Structure, biosynthesis, and function. *Prog. Lipid Res.* **44** 235–258.

[3] Swiezewska E, Sasak W, Mankowski T, Jankowski W, Vogtman T, Krajewska I, Hertel J, Skoczylas E and Chojnacki T 1994 The search for plant polyprenols *Acta Biochim. Pol.* **41** 221–260.

[4] Arifiyanto D, Basyuni M, Sumardi, Putri LAP, Siregar ES, Risnasari I and Syahputra I 2017 Occurrence and cluster analysis of palm oil (*Elaeis guineensis*) fruit type using two-dimensional thin layer chromatography *Biodiversitas* **18** 1487–1492.

[5] Ishiguro T, Morita-Fujimira Y, Shidoji Y and Sagami H 2014 Dolichol biosynthesis: The occurrence of epoxy dolichol in skipjack tuna liver *Biochem. Biophys. Res. Comm.* **451** 277–281.

[6] Basyuni M, Sagami H, Baba S, Iwasaki H and Oku H 2016 Diversity of polyisoprenoids in ten Okinawan mangroves *Dendrobiology* **75** 167–175.

[7] Basyuni M, Sagami H, Baba S and Oku H 2017 Distribution, occurrence, and cluster analysis of new polyprenyl acetones and other polyisoprenoids from North Sumatran mangroves *Dendrobiology* **78** 18–31.

IOP Conf. Series: Journal of Physics: Conf. Series **978** (2018) 012044 doi:10.1088/1742-6596/978/1/012044


[8] Basyuni M, Wati R, Sagami H, Sumardi, Baba S and Oku H 2018 Diversity and abundance of polyisoprenoid composition in coastal plant species from North Sumatra, Indonesia *Biodiversitas* **19** 1–11.

[9] Grabinska K and Palamarczyk G 2002 Dolichol biosynthesis in the yeast *Saccharomyces cerevisiae*: an insight into the regulatory role of farnesyl diphosphate synthase *FEMS YeastRes.* **2** 259–265.

[10] Basyuni M and R Wati 2017 Bioinformatics analysis of the oxidosqualene gene and the amino acid sequence in mangrove plants *J. Phys. : Conf. Ser.* **801** 012011.

[11] Basyuni M, Wasilah M and Sumardi 2017 Bioinformatics study of the mangrove actin genes *J. Phys. : Conf. Ser.* **801** 012013.

[12] Pearson WR and Lipman DJ 1988 Improved tools for biological sequence comparison *Proc. Natl. Acad. Sci. USA***85** 2444–2448.

[13] Thompson JD, Higgins DG and Gibson TJ 1994 CLUSTAL W: Improving the sensitivity of progressive multiple sequence alignment through sequence weighting, position-specific gap penalties, and weight matrix choice *Nucleic Acid Res.* **22** 4673–4680.

[14] Page RD 1996 TreeView: an application to display phylogenetic trees on personal computers *Comput. Appl. Biosci.* **12** 357–358.

[15] Felsenstein J 1985 Confidence limits on phylogenies: an approach using the bootstrap *Evolution***39** 783–791.

[16] Tateyama S, Wititsuwannakul R, Wititsuwannakul D, Sagami H and Ogura K 1999 Dolichols of rubber plant, ginkgo and pine *Phytochemistry* **51** 11–15.

[17] Inafuku M, Basyuni M and Oku H 2016 Triterpenoid modulates the salt tolerance of lanosterol synthase deficient *Saccharomyces cerevisiae,* GIL77 *Saudi J. Biol. Sci.* http://dx.doi.org/10.1016/j.sjbj.2016.10.009.

[18] Liu J, Zhu Y, Du G, Zhou J and Chen J 2013 Exogenousergosterol protects *Saccharomyces cerevisiae* from D-limonenestress *J. Appl. Microbiol.* **114** 482–491.

[19] Turk M, Plemenitas A and Gunde-Cimerman N 2011 Extremophilicyeasts: plasma-membrane fluidity as determinant of stress tolerance *Fungal Biol.* **115** 950–958.

[20] Basyuni M and Sumardi 2017 Bioinformatics approach of salt tolerance gene in mangrove plant *Rhizophora stylosa J. Phys. : Conf. Ser.* **801** 012012.

# An optimal generic model for multi-parameters and big data optimizing: a laboratory experimental study

**D N Utama[1], N Ani[1], M M Iqbal[2]**

[1]Computer Science, Universitas Mercu Buana, Jakarta, Indonesia
[2]Information Systems Department, UIN Syarif Hidayatullah, Jakarta, Indonesia

ditdit.nugeraha@mercubuana.ac.id

**Abstract.** Optimization is a process for finding parameter (parameters) that is (are) able to deliver an optimal value for an objective function. Seeking an optimal generic model for optimizing is a computer science study that has been being practically conducted by numerous researchers. Generic model is a model that can be technically operated to solve any varieties of optimization problem. By using an object-oriented method, the generic model for optimizing was constructed. Moreover, two types of optimization method, simulated-annealing and hill-climbing, were functioned in constructing the model and compared to find the most optimal one then. The result said that both methods gave the same result for a value of objective function and the hill-climbing based model consumed the shortest running time.

## 1. Introduction

Efficiency is one factor considered to choose an optimization model. A model's run time consumption is one of some variables that is able to be taken into account to see the efficiency of a model including the optimization model. Where, to develop a generic optimization model which is efficient is a challenging study. Besides efficient, it is going to be valuable to solve any types of optimization problem without changing any things to become suitable to the case. Also, it can be operated to find the precise decision that something problematic to get [1].

Numerous researchers have conducted research in optimizing field that implemented in several cases. [2] developed an optimization model to minimize the maximum index of the used frequency slots. The model technically constructed based on a combination of heuristic and genetic algorithms. The methods were used to sort connection requests and look for an optimal routing and spectrum assignment scheme. [3] developed two types of signal optimization model for addressing issues in asymmetric continuous flow intersection (CFI). The first model uses two step procedures, and the second one takes mixed-integer-linear-programming. The result revealed that the second model is more flexible in designing phase, however the first one performs better in reducing link queue length. Moreover, [4] constructed an optimization model that realized in decision support model to optimize a placement of stations in urban areas. Two parameters considered in this study; the number of reachable household and the cost of overall e-transportation energy. And [5] used the fuzzy evaluation method to create an optimization model for rationalizing and optimizing the structure of passenger transportation. The model can evaluate the condition of passenger transportation structure of both conditions; in model and real domain.

Furthermore, [6] constructed an optimization model to investigate a performance of underground gas drainage system. Several factors were considered in the study, e.g. pure gas drainage flow, gas concentration, and energy efficiency. The model was practically applied to a typical modern coal mine.

Then, [7] [8] have developed a decision support system based on optimization model. The model was developed based on the method water flow algorithm (WFA) and a combination between fuzzy-logic with WFA (called FWFA). It used to test the value of objective function of each decision alternatives that considers several parameters, e.g. traffic volume, density, etc.

This paper portrays the development of generic optimization model that is realized in solving the multi parameters and big data optimization case. Two types of optimization method, simulated-annealing (SA) and hill-climbing (HC), are a main method of the model that have been tested to see their efficiency of running time. The introduction section of the paper is followed by following sections; related works, research methodology, results and discussion, and conclusion and further works.

## 2. Related works

A generic method or model, as a main topic, has been studied by several researchers in many fields. In the field computers and operations research [9] proposed a generic Pareto based local search metaheuristic. It was used for marketing campaign. Here, a Greedy Randomized Neighborhood Structure was designed which was embedded in a multi-objective local search metaheuristic. In the field microprocessors and microsystems, [10] constructed a generic model to optimize the energy consumption of mobile platform. The parameters that taken into account in this study are power, response time, and thermal. Furthermore, in the field computer aided chemical engineering, [11] proposed a generic computer-aided methodology for synthesizing different processing networks. In this study, a superstructure optimization was operated. The methodology was able to handle different network optimization problems in many application fields. And also, in the environmental management field, [12] developed a generic methodology for the design of a self-optimizing controller in sewer systems, to keep the systems close to optimal performance. The result said that the proposed methodology has been successfully implemented and it is promising to contribute to performance increase of sewer systems.

## 3. Research methodology

In this study, two methods were technically functioned in developing a generic optimization model. They are methods simulated-annealing [13] [14] [15] [16]and hill-climbing [17] [15]. A class diagram, that is a part tool of object oriented method [17-18], was operated to design a schematic view of the constructed model. The laboratory experiments were practically conducted by involving a big data (a hundred thousand records with forty specified parameters, or similar to four million fields). They were done in Laboratory Optimization Models and Systems for Decision Support (UIN Syarif Hidayatullah, Jakarta, Indonesia) by using machine with processor Intel® Celeron® CPU B815 @ 1.60GHz (2 CPUs), memory 2048MB RAM, BIOS InsydeH2O V 1.06, and operating system Windows 8.1 Pro 64-bit (6.3, Build 9600). Those models also were tested in specific case (in a decision support model for renovating and rehabilitating mosques [19]).

The objective function used in the experiments is generically formulated in equation (1), where $n$ represents a number of parameter (in the experiments, 40 used to define $n$), $c_i$ is a $i$th parameter coefficient, and $P$ is a value of $i$th parameter. It means that the objective function is to obtain a maximum value of a summation function of forty coefficients and parameters multiplication.

$$Max\, f(x) = \sum_{i=1}^{n} c_i P_i \qquad (1)$$

## 4. Results and discussion

### *4.1. The constructed model*

IOP Conf. Series: Journal of Physics: Conf. Series **978** (2018) 012045 doi:10.1088/1742-6596/978/1/012045

Schematically, the constructed model is configured by Figure 1. This configuration is technically depicted by class diagram. Here, the constructed model fundamentally consists of two main classes; `OptimizationModel` and `SelectedCase`. The class `OptimizationModel` has two types of model, `SAOptimizationModel` and `HCOptimizationModel`; where they have attributes and operations that commonly belong to class `OptimizationModel`; only several attributes and operations that specifically belong to them. The class `SelectedCase` indicates that the model is generic, it can be operated in any types of case. The case totally depends on parameters and parameters' coefficient going to be considered. The objective function of optimization itself is defined in operation `definingOF` (defining objective function) in class `OptimizationModel`. For the explanations of other part can be seen in Figure 1 clearly.

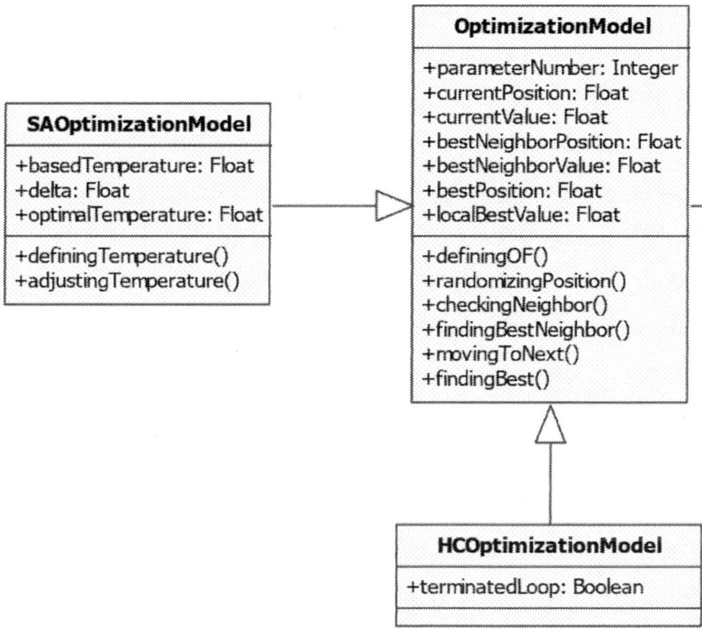


**Figure 1.** The generic class diagram of constructed model.

The generic pseudocode of procedure for SA and HC methods are showed in Code 1 and 2 respectively. Basically, SA was proposed by [13] and also by [14]. It is a probabilistic method for finding the global minimum of an objective function. In principle, it can possibly deliver several local minimum [20] after optimizing process. The idea of SA itself firstly came from the metallurgy field. The objective function is treated as the energy function of a molten metal exposed to an artificial temperature which is gradually cooled [21]. The temperature controls the probability that the optimizing process moves to a neighbor with a worse value of the objective function (as opposed to the HC method when such a move can never occur). The process will finish when the adjusted temperature reaches the optimal temperature. This scenario can still deliver a local optimum which is not the global optimum, but the algorithm has a chance to escape from this trap as long as the temperature is still high enough.

The variable `delta` is a basic conception in the SA method. As, when `curVal` (current value) is better than `bestNeighbor`, the `delta` is realistically operated to check the probability to move or to stop by calculating "`exp(-delta/baseTemp)`" and then comparing it to a value of random (0...1). This condition makes SA more flexible than HC. Can be concluded here that the other specific parameters used in simulated annealing models are `baseTemp` and `optTemp`.

```
Procedure SimulatedAnnealing(parameters)
Begin
      <...other variables definition...>
      baseTemp <-- 100; //for example
      optTemp <-- 40; // for example
      //randomizing new parameter combination and getting value of
      //objective function
      para <-- random()
      curVal <-- objFunction(para)

      //looping until local optimum is found
      While(baseTemp>=optTemp)
            //finding the best neighbor
            bestNeighbor <-- getBestNeigh()
            //finding local optimum
            If(bestNeighbor>=curVal)
                  curVal <-- bestNeighbor
                  bestNeighbor <-- 0
            Else
                  delta <-- bestNeighbor - curVal
                  //Probability
                  If(exp(-delta/baseTemp) > random(0..1))
                        curVal <-- bestNeighbor
                        bestNeighbor <-- 0
                  End if
            End if

            //adjusting baseTemp
            adjust(baseTemp)
      End while
End
```

Code 1 Generic Pseudocode of Simulated-Annealing Optimizing Procedure [15]

Furthermore, HC method is a local search method. It was offered by [15]. Straightforwardly stated, it contains a loop that continually moves in one direction of an increasing value (uphill), and never goes downhill. It is terminated when it reaches a 'peak', where there is no neighbor having a higher value. The probability to be trapped in one local maximum (one hill) is high. A local maximum is a peak that is higher than its neighbors but lower than the global maximum [22].

In Code 2, it is obviously mentioned that a looping process for seeking the optimum value can be terminated when the local optimum value is discovered. This occurs when the variable `curVal` is better than the value of the variable `bestNeighbor`. This means, no parameter combination that belongs to its neighbors can give a better value of the objective function than the variable `curVal`.

Several specific parameters required by HC method can be listed here. The parameters `curVal`, `bestNeighbor`, and `bestVal` are parameters necessitated in this method. They are respectively current value, best value of neighbor, and best value (should be as a local or global best value).

Figure 2 describes distributed data of execution experiment result for each models. The blue dots are depicting distributed data coming from SA based model's running time, while the read ones illustrate data of HC based model's running time. The experiments were conducted by manipulating a number of searching looping, 1 – 1000 looping times. Based on a number of searching looping, the running time increases linearly.

```
Procedure HillClimbing(parameters)
Begin
```

```
      <...variables definition...>
      //randomizing new parameter combination
      para <-- random()
      curVal <-- objFunction(para)

      //looping until local optimum is found
      While(search is not terminated)
            //finding the best neighbor
            bestNeighbor <-- getBestNeigh()
            //finding local optimum
            If(curVal>=bestNeighbor)
                  Local optimum is found
                  Search is terminated
            Else //move to next position
                  curVal <-- bestNeighbor
            End if
      End while
End
```

Code 2 Generic Pseudocode of Hill-Climbing Optimizing Procedure [15]

$$f(x) = 0.0007x - 0.0259 \quad (2)$$

$$f(x) = 0.0003x - 0.0205 \quad (3)$$

SA based model gives a linear increase pattern (trend-line) of running time in equation (2), where $x$ represents a number of searching looping, with $R^2$ is 0.8085; while, HC based model produces a pattern with equation (3) with $R^2$ is 0.6293. The equation (2) indicates that SA based model consumes longer running time than HC based model. It naturally happens, since in SA method, the optimal value searching process is still conducted when it has not reached an optimal temperature yet; even though the search starting process in the peak of hill. Conversely, in HC based model, the searching process is directly terminated when it gets the optimal value (peak of hill).

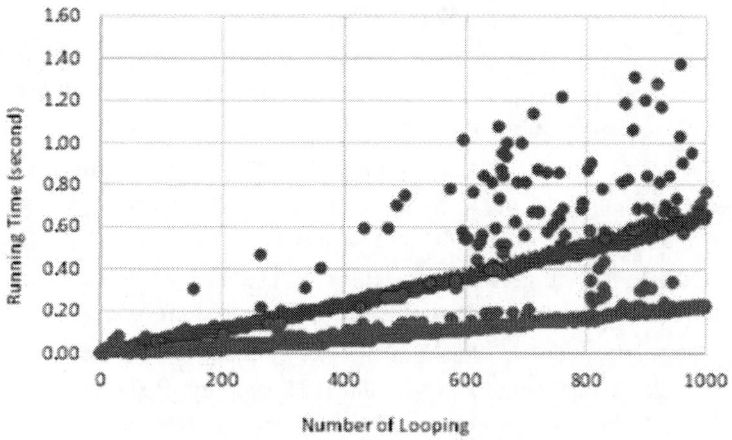


**Figure 2.** Running time of both methods.

*4.2. Discussion*

The model practically was implemented to tackle the specific problem [19]. However, in that study, only 196 records were used and the numbers of searching looping used are 100, 500, 1,000, 5,000,

10,000, 50,000, and 100,000 times. On the other hand, in this study, we generated 100,000 records (with forty valued parameters) by testing sequentially a number of searching looping until 1,000 times. Based on the result of laboratory experiments for the running time of model execution, both studies exposed a similar pattern for both methods.

## 5. Conclusion and further works

The generic optimization model to solve a big data problem was successfully constructed. The model consists of two types of optimizing method; SA and HC. The laboratory experiments were practically conducted to test the running time of 1 – 1,000 searching loops for both model. The final conclusion revealed that both models produced a linear pattern of running time, and the model based on HC has shorter running time. This it is more optimal than SA based model with similar capability in searching. Both models also have been implemented in specific case to recommend the best value of decision.

In addition, for the further study, the models essentially can be operated to solve several types of problem of decision making case. The models can be rationally implemented in bigger records as well, e.g. one million or billion records with more than forty parameters taken into account.

## References

[1] Noor N M M, Retnowardhani A, Triana Y S and Helawani S M 2013 The use of fuzzy alpha cut to improve the forecasting range results of the ARIMA model: assist decision making process *IJIPM: International Journal of Information Processing and Management* **4(7)** pp 59-71

[2] Xuan H, Wang Y, Xu Z, Hao S and Wang X 2017 New optimization model for routing and spectrum assignment with nodes insecurity *Optics Communications* **389** pp 42-50

[3] Yang X and Cheng Y 2017 Development of signal optimization models for asymmetric two-leg continuous flow intersections *Transportation Research Part C: Emerging Technologies* **74** pp 306-326

[4] Yi Z and Bauer P H 2016 Optimization models for placement of an energy-aware electric vehicle charging infrastructure *Transportation Research Part E: Logistics and Transportation Review* **91** pp 227-244

[5] Ma Y and Gao Y 2016 Passenger Transportation Structure Optimization Model Based on User Optimum *Procedia Engineering* **137** pp 202-209

[6] Wang X, Zhou F, Xia T and Xu M 2016 A multi-objective optimization model to enhance the comprehensive performance of underground gas drainage system *Journal of Natural Gas Science and Engineering* **36** pp 852-864

[7] Utama D N, Zaki F A, Munjeri I J and Putri N U 2016 FWFA Optimization based Decision Support System for Road Traffic Engineering *1st International Conference on Computing and Applied Informatics (ICCAI)* Medan Indonesia

[8] Utama D N, Zaki F A, Munjeri I J dan Putri N U 2016 A Water Flow Algorithm based Optimization Model for Road Traffic Engineering *International Conference on Advanced Computer Science and Information Systems (ICACSIS 2016)* Malang Indonesia

[9] Coelho V N, Oliveira T A, Coelho I M, Coelho B N, Fleming P J, Guimarães F G, Ramalhinho H, Souza M J, Talbi E G and Lust T 2017 Generic Pareto local search metaheuristic for optimization of targeted offers in a bi-objective direct marketing campaign *Computers & Operations Research* **78** pp 578-587

[10] Gupta U, Korrapati S, Matturu N and Ogras U Y 2016 A generic energy optimization framework for heterogeneous platforms using scaling models *Microprocessors and*

IOP Conf. Series: Journal of Physics: Conf. Series **978** (2018) 012045 doi:10.1088/1742-6596/978/1/012045

*Microsystems* **40** pp 74-87

[11] Bertran M O, Frauzem R, Zhang L and Gani R 2016 A Generic Methodology for Superstructure Optimization of Different Processing Networks *Computer Aided Chemical Engineering* **38** pp 685-690

[12] Mauricio-Iglesias M, Montero-Castro I, Mollerup A L and Sin G 2015 A generic methodology for the optimisation of sewer systems using stochastic programming and self-optimizing control *Journal of Environmental Management* 155 pp 193-203

[13] Kirkpatrick S, Gelat C D and Vecchi M P 1983 Optimization by Simulated Annealing *Science* **220** 4598 pp 671-680

[14] Černý V 1985 Thermodynamical approach to the traveling salesman problem: An efficient simulation algorithm *Journal of Optimization Theory and Applications* **45** 1 pp 41–51

[15] Utama D N 2015 The Optimization of the 3-d Structure of Plants, Using Functional-Structural Plant Models. Case Study of Rice (Oryza sativa L.) in Indonesia PHD Thesis Göttingen University

[16] Alaydrus M and Eibert T F 2007 Optimizing multiband antennas using simulated annealing *International ITG Conference on Antennas* pp 86-90

[17] Michalewicz Z 1992 *Genetic algorithms + data structures = evolution programs* Third Revised and Extended Edition Verlag: Springer

[18] Mathiassen L, Munk-Madsen A, Nielsen P A and Stage J 2000 *Object-oriented analysis and design* Aalborg: Marko

[19] Utama D N, Iqbal M M, Iksal M and Dharmawan T 2017 Decision Support Model for Mosque Renovation and Rehabilitation (Case Study: Ten Mosques in Jakarta Barat, Indonesia) *in press*.

[20] Bertsimas D and Tsitsiklis J 1993 Simulated Annealing *Statistical Science* **8** 1 pp 10-15

[21] Xiang Y, Gubian S, Suomela B and Hoeng J 2013 Generalized Simulated Annealing for Global Optimization: The GenSA Package.An Application to Non-Convex Optimization in Finance and Physics *The R Journal* **5** 1 pp 13-27

[22] Russell S and Norvig P 1995 Artificial Intelligence: A Modern Approach Pearson Education Inc.

IOP Conf. Series: Journal of Physics: Conf. Series **978** (2018) 012046 doi:10.1088/1742-6596/978/1/012046

# Readiness factors for information system strategic planning among universities in developing countries: a systematic review

**M Irfan[1], S J Putra[2], C N Alam[3], A Subiyakto[4] and A Wahana[5]**
[1,3,5]Department of Informatics, Universitas Islam Negeri Sunan Gunung Djati Bandung, Indonesia
[2,4]Department of Information System, Universitas Islam Negeri Syarif Hidayatullah Jakarta, Indonesia

E-mail: irfan.bahaf@uinsgd.ac.id

**Abstract.** The implementation of information system strategic planning (ISSP) in higher education institutions is to improve work efficiency, management effectiveness in order to improve organizational competitive advantage. However, the question of whether all universities are ready to implement ISSP as a way to achieve organizational goals has not been answered. This study aims to investigate the readiness phenomena through literature study. The method used is by using the Systematic Literature Review (SLR) instrument to identify readiness factors on the implementation of ISSP, especially among the institutions of higher education in developing countries. This study has identified 10 readiness measurement. There are three categories of measurement, namely people, processes and technologies that represent 11 factors of ISSP readiness measurement in universities.

## 1. Introduction

Every organization, either universities, governments or companies, has a vision to be achieved. There is a challenge to achieve the vision of the organizations because the vision is a future that has not happened. The vision can be achieved by looking at the current conditions and predicting future conditions. Vision can be realized and become a reality if every function has a good understanding of the system, the initiative to grow, a commitment to achieve the vision and able to work hard [1]–[3].

The universities in developing countries that have a vision and purposes, consciously need to grow and have competitive advantages. The role of information technology (IT) to achieve the vision and the objectives of the organizations is very significant. The existing conditions of IT at universities in developing countries have not run well because of high cost, IT literacy issues, and the IT is used only as a tool of the organizational population, not as a strategy to achieve the vision. On the other hand, the previous studies [4]–[8] show that the success of IT development projects still tends to be unsatisfactory. Based on these issues, the attention to ISSPs needs to be a concern of stakeholders in higher education institutions [9]–[12].

According to John Ward and Joe Peppard [13], there are three main goals of the implementation of information system or IT in an organization. The first goal is to improve work efficiency by automating various information processes. The second is to improve the effectiveness of management by satisfying the information needs for decision making. Lastly, the goal of IT is to improve the competitive advantages of organizations by changing the business way and style. By using IT, universities in developing countries are expected to achieve their vision and goals, be able to develop education, research and community services, be able to improve competitiveness, have effective and efficient management [9], [10], [12], [14].

The goals of using IT are to gain competitive advantages from business opportunities generated by IT, save costs for future infrastructure development, develop the resources and competencies to make IT successful in the organizations. The steps that must be taken to identify the readiness of the system is to integrate management and business knowledge with technical knowledge, develop business strategies, determine IT strategy, stakeholder interaction with IT and stakeholder business collaboration vision [15]–[17].

To achieve those goals, it is necessary to measure the readiness for implementing ISSP in the universities of developing countries. To determine the success of its implementation, E-Readiness is one of the most important aspects of the tool used [1], [17]–[21]. Therefore, the purpose of this study is to identify the existing mechanisms for measuring readiness, and to know the factors of readiness instruments for ISSP implementation at universities.

This paper is organized into four sections. First, the introduction section that explains the background, problems, and objectives of the study. Second, the research methods section that describes the review methods used for LSR. Third, finding and discussions. The last section is conclusion.

## 2. Research Methods

The method used in literature review is SLR. The SLR is a method to conduct the process of identifying, evaluating and interpreting relevant research and according to the topic area, phenomenon, and relevant interest. The SLR will perform an analysis of how data is obtained and generated [22]. The development of the review literature is done to summarize the results of the study and identify the readiness for the ISSP implementation. The process is carried out based on the instructions and guidelines for performing the SLR. Regarding with the LSR guidelines, to conduct LR there are three steps to be done: formulating LR questions, determining and selecting research and evaluation [18], [22]–[24].

The research questions were based on Population, Intervention, Comparison, Outcome, and Context (PICOC) [22], [23]. The criteria and scope of research questions are shown in Table I. Based on table I, the SLR question is as follows:

RQ1: How to explore the study of the readiness of ISSP implementation in higher education? RQ2: How to know the readiness factors used for ISSP implementation in higher education?

**Table 1.** Criteria and scope

| No | Criteria | Scope |
|---|---|---|
| 1 | Population | Any organizations |
| 2 | Intervention | Implementation of ISSP, issues and factors of readiness and measurement mechanism |
| 3 | Comparison | Mechanism of the readiness measurement of industry and academic |
| 4 | Outcome | Mechanism of the readiness measurement and factos of ISSP implementation |
| 5 | Context | Review of each study on ISSP implementation and preparedness factors and measurement mechanism |

Search online is done on a publisher database or digital library such as IEEEXplore, Google Scholar, Emerald, EBSCO, Elsavier, Science Direct, Proquest and digital campus library or professional associations. The results are in the form of research from Journal, Proceedings, thesis reports and scientific books. The basic keywords for search are "Information System", "Information Technology", "Strategic Planning", "Readiness", and "Higher Education". For optimization and accuracy of database search results (title, abstract and metadata), Boolean logic "AND" and "OR" are added. Year of search is started from 2010 to 2017 on articles in English.

At the beginning, around 500 relevant articles were found. After investigating the titles and abstracts, approximately 25 articles that were considered relevant to the topic of ISSP readiness were found. The results of article review are then grouped according to related criteria. After reading and reviewing the suitable and relevant articles, 10 research articles are accepted for the synthesis of SLR.

## 3. Finding and Discussion

Based on search results, papers, journals, books and research dissertations on the readiness of ICT application have been done by researchers and practitioners. This is reinforced by the number of literature studies on the readiness of the application of ICT to organizations. Here are the results of research based on the development of Research Questions. The result is a synthesis of the SLR that is the factors affecting the readiness of ISSP implementation.

### *3.1 Mechanism to measure readiness*

According to [25], there are several frameworks and instruments to assess the readiness of ICT implementation. Through SLR, 10 mechanisms used to measure ICT readiness can be identified. Mechanisms can measure the readiness of ICT initiatives in organizations. Therefore, this mechanism can be used to measure the readiness of the ISSP implementation in the organizations.

**Table 2.** List of Mechanisms to Measure Readiness

| Mechanisms to Measure Readiness | Reference | Factor | Purpose |
|---|---|---|---|
| A Conceptual Green-ICT Implementation Model Based-on ZEN and G-Readiness Framework | [26] | i. Attitude ii. Policy iii. Practice iv. Technology v. Governance | To develop a conceptual implementation model to adopt green-based ICT readiness |
| Influence of Social Technical Factors on ICT Readiness for Primary Schools in Bungoma County, Kenya | [19] | i. Professional Development ii. Accessibility of ICT iii. Ease of Use iv. Usefulness v. Technical Support vi. Leadership Support | To develop conceptual model for integrating technology in teaching |
| E – Readiness Assessment Of Large Organization In A Developing Country: The Case Of Iran | [27] | i. Culture & Resource ii. Leadership iii. Strategy and Policy iv. IT Security v. Process vi. Infrastructure | To measure an organization's ability to take advantage of the Internet as an engine of economic growth and human development |
| A Model for Organizational Readiness In Information Technology (IT) Project Implementation In The Malaysian Construction Industry | [28] | i. Infrastructure ii. People iii. Process iv. Work Environment | To develop readiness model In Information Technology (IT) Project Implementation |

| | | | |
|---|---|---|---|
| Green ICT Readiness Model for Developing Economies: Case of Kenya | [29] | i. Attitude ii. Policy iii. Practice iv. Technology v. Governance | To develop a conceptual implementation model to adopt green-based ICT readiness |
| ICT Readiness Assessment Model for Public and Private Organizations in Developing Country | [25] | i. Software and IS ii. ICT Hardware iii. ICT Infrastructure iv. People and Human Resource | Deliver model and assess readiness ICT for Public and Private |
| Measuring Organizational Readiness in Information Systems Adoption | [21] | i. Attributes of change ii. Leadership support iii. Internal context iv. Attributes of change targets v. It support | To measure the amount of readiness before implementing IS adoption |
| IT Readiness Assessment for Government Organizations | [30] | i. Architecture ii. Infrastructure iii. Processes | To Assess IT readiness and Evaluation Framework |
| E-Readiness to G-Readiness: Developing a Green Information | [31] | i. Attitude ii. Policy | To develop a conceptual implementation model to |
| Technology Readiness Framework | | iii. Practice iv. Technology v. Governance | adopt green-based ICT readiness |
| E-readiness assessment of nonprofit ICT SMEs in a developing country: The case of Iran | [32] | i. Awareness ii. Resources iii. Commitment iv. Governance | To assess the e-readiness of non-profit SMEs in a developing country. |

### *3.2 Factors for Readiness in Implementing ICT*

Universities in developing countries need to reorganize their business processes, people and technology. These three elements can be aligned by changing business strategy and relationship between them. The main factors of ICT implementation in developing country universities are process, people and technology. The three factors will be explained and their related attributes in the following table:

**Table 3.** Factors Related to Readiness in ISSP Implementation

| ICT Readiness Mechanism | Factors | | | | | | | | | | |
|---|---|---|---|---|---|---|---|---|---|---|---|
| | People | | | | | Process | | | | Technology | |
| | WC | L | Cp | R | CM | CI | G | A | S | I | Sc |
| 1. A Conceptual Green-ICT Implementation Model Based-on ZEN and G-Readiness Framework [26] | | √ | | | | | √ | √ | | √ | √ |
| 2. Influence of Social Technical Factors on ICT Readiness for Primary Schools in Bungoma Country, Kenya [19] | √ | √ | √ | √ | | √ | √ | √ | √ | | |
| 3. E – Readiness Assessment Of Large Organization In A Developing | | | | | | √ | √ | √ | √ | √ | √ |

| | | | | | | | | | | | |
|---|---|---|---|---|---|---|---|---|---|---|---|
| Country: The Case Of Iran [27] | | | | | | | | | | | |
| 4. A Model For Organizational Readiness In Information Technology (IT) Project Implementation In The Malaysian Construction Industry [28] | √ | √ | | | | √ | √ | √ | √ | √ | |
| 5. Green ICT Readiness Model for Developing Economies: Case of Kenya [33] | | | | | √ | | √ | √ | | √ | √ |
| 6. ICT Readiness Assessment Model for Public and Private Organizations in Developing Country[25] | | | √ | √ | | | | | | √ | |
| 7. Measuring Organizational Readiness in Information Systems Adoption [21] | √ | √ | | | √ | | | | | | |
| 8. IT Readiness Assessment for Government Organizations [30] | | | | | | √ | √ | √ | √ | √ | |
| 9. E-Readiness to G-Readiness: Developing a Green Information Technology Readiness Framework [31] | | | | | | √ | √ | | √ | √ | |
| 10.E-readiness assessment of non-profit ICT SMEs in a developing country: The case of Iran [32] | | | | √ | √ | √ | √ | | | | |

**Legend:**

WC – Workforce Capability, L – Leadership, Cp- Competency, R – Resources, CM – Change Management, Cl – Culture, G – Governance, A - Awareness, S – Strategy, I – Infrastructure, Sc – Security

**Table 4.** Definition of Factors

| Factor | Definition | Reference |
|---|---|---|
| Workplace Capability | Refers to the effectiveness of ICT training and capability of human resources | [19] [28] [21] |
| Leadership | Refers to highest hierarchy in organisation or stakeholders | [19] [21] [26] |
| Competency | Refers to skills, experience and knowledge | [25] |
| Resources | Refers to human resources, ICT resources and budget resources | [32] |
| Change Management | Refers to change commitment and change efficacy | [33] |
| Culture | Refers to activities in the environment of an organization | [19] [27] [31] |
| Governance | Refers to structure, procedures and routines, and communications involving business and IT | [26] [30] |
| Awareness | Refers to the understanding of the concept, sharing of experience and raising the level of knowledge | [27] [30] |
| Strategy | Refers to business and ICT strategy | [27] [31] |
| Infrastructure | Refers to technology, including software and hardware | [25] [26] [27] |
| Security | Refers to policy, information safety, and the legal and regulatory environment. | [26] [27] [33] |

## 4. Conclusion

Research on ISSP which is mostly done by academics and practitioners especially about the readiness of ISSP implementation in universities or companies is still limited. This shows that research on this topic will contribute to higher education to measure the readiness of IT/SI implementation in universities. Based on factors related to readiness in ISSP implementation, process and technology are the most dominant factors to measure the readiness of ISSP implementation on organizations, while some research articles do not include the people factor as the influencing factor for the ISSP implementation on the organizations. Therefore, in order to produce a readiness model for the implementation of ISSP at universities comprehensively, the future research needs to add a readiness factor that can improve the role of the people factor by adding the factor of policy (ZEN Framework) to measure the readiness of universities to implement and adopt ISSP [26].

## 5. References

[1] Kurniawan N B and Suhardi 2013 Enterprise architecture design for ensuring strategic business IT alignment (integrating SAMM with TOGAF 9.1) *Proc. 2013 Jt. Int. Conf. Rural Inf. Commun. Technol. Electr. Technol. rICT ICEV-T 2013*.

[2] Kearns G S and Sabherwal R 2006 Strategic alignment between business and information technology: a knowledge-based view of behaviors, outcome, and consequences *J. Manag. Inf. Syst.* vol. **23** no. 3 pp. 129–162.

[3] Curry J and Ferguson J 2000 Increasing the success of the global information technology strategic planning process process university of western sydney *Proc.33rd Hawaii Int. Conf. Syst. Sci. - 2000 Increasing* vol. **0** pp. 1–10.

[4] Subiyakto A, Ahlan A R, Kartiwi M and Putra S J 2016 Measurement of the information system project success of the higher education institutions in Indonesia: A pilot study *Int. J. Bus. Inf. Syst* vol. **23** no. 2.

[5] Putra S J, Subiyakto A, Ahlan A R and Kartiwi M 2016 A coherent framework for understanding the success of an information system project vol **14** no. 1 pp. 302–308.

[6] Subiyakto A, Ahlan A R, Putra S J and Kartiwi M 2015 Validation of information system project success model: a focus group study *SAGE Open* vol. **5** no. 2 p. 14.

[7] Subiyakto A, Ahlan A R and Sukmana H T 2014 An alternative method for determining critical success factors of information system project *TELKOMNIKA (Telecommunication Comput. Electron. Control* vol. **12** no. 3 p. 665.

[8] Subiyakto A and Bin Ahlan A R 2013 A coherent framework for understanding critical success factors of ICT project environment *Int. Conf. Res. Innov. Inf. Syst. ICRIIS* pp. 342–347.

[9] Semiawan T and Middleton M 1999 Strategic information systems planning and campus wide information systems in indonesia *Campus-Wide Inf. Syst* vol. **16** no. 2 pp. 70–76.

[10] Semiawan T and Middleton M 1999 Strategic information planning and campus information systems development in Indonesia *Campus-Wide Inf. Syst* vol. **16** no. 2 pp. 70–76.

[11] Adamk A 2014 A System Model and Applications for Intelligent Campusesv pp. 193–198.

[12] Pintari N 2008 ICT Strategy of Universities *Proc. 19th Cent. Eur. Conf. Inf. Intell. Syst. Sept. 24-26, 2008, Varaždin, Croat. Publ. Fac. Organ. Informatics, Ed. Boris Aurer, Miroslav Bača, Kornelije Rabuzin* pp.587-592 pp. 2–6.

[13] Ward J and Peppard J O E *AM Strategic Planning for Information Systems*. .

[14] Semiawan T and Middleton M 1999 Strategic information systems planning and campus-wide information systems in indonesia *Campus-Wide Inf. Syst.*, vol. **16** no. 2 pp. 70–76.

[15] Wahyudin A and Hasibuan Z 2015 Research classification in strategic information system planning development : a critical review," *Int. Conf. Sci. Inf. Technol* pp. 287–292.

[16] Yuhetty H 2004 ICT and Education in Indonesia *Proc. Asia Pacific Semin* no. 50.
[17] Koen R, Solms R V and Gerber M 2016 ICT Readiness for Business Continuity in local government *IST-Africa Conf. IST-Africa 2016*, pp. 1–11.
[18] Rohayani A H H, Kurniabudi and Sharipuddin 2015A literature review: readiness factors to measuring e-learning readiness in higher education *Procedia Comput. Sci* vol. **59** no. Iccsci pp. 230–234.
[19] Batoya I B, Wabwoba F and Kilwake J 2015 Influence of social technical factors on ict readiness for primary schools in bungoma county , Kenya vol. **1** no. 1 pp. 1–7.
[20] Iskandar A R 2009 Readiness For Implementing Data Warehouse of Higher Education System With Sodano â€™ s Perspective.
[21] Aziz K A and Yusof M M 2012 Measuring organizational readiness in information systems adoption *18th Am. Conf. Inf. Syst. 2012, AMCIS 2012.*
[22] Hussein S S, Ismail Z, Mahrin M N R, Abdullah M S, Rahim N Z A, Samy G N and Taib M Z M 2017 Towards designing an EA readiness instrument: A systematic review *Colloq. Inf. Sci. Technol. Cist*, pp. 158–163.
[23] Wahono R S 2015 A Systematic Literature Review of Software Defect Prediction : Research Trends , Datasets , Methods and Frameworks *J. Softw. Eng.* vol. **1** no. 1 pp. 1–16.
[24] Đurek V 2016 Review on e-readiness assessment tools pp. 161–168.
[25] Chanyagorn P and Kungwannarongkun B 2011 ICT Readiness Assessment Model for Public and Private Organizations in Developing Country *Int. J. Infromation Educ. Technol* vol. **1** no. 2 p 2017.
[26] Marcel 2017 A conceptual green-ICT implementation model based-on ZEN and G-readiness framework *2016 Int. Conf. Informatics Comput. ICIC 2016*, no. Icic, pp. 99–104.
[27] Sanaei M R 2014 E – Readiness Assessment of Large Organization in a Developing Country : the Case of Iran vol. **4** pp. 250–258.
[28] Aziz N M 2013 A model for organisational readiness in information technology (it) project implementation in the malaysian construction industry.
[29] Wabwoba F, Omuterema S, Wanyembi G and Omieno K. 2013 Green ICT Readiness Model for Developing Economies: Case of Kenya *Int. J. Adv. Comput. Sci. Appl* vol. **4** no. 1 pp. 51–65.
[30] IT Readiness Assessment for Government Organizations 2007 pp. 1–30.
[31] Molla A, Cooper V, Corbitt B, Deng H, Peszynski K, Pittayachawan S and Teoh S. Y 2008 EReadiness to G-Readiness: Developing a Green Information Technology Readiness Framework *19th Australas. Conf. Inf. Syst.*, pp. 669–678.
[32] Fathian M, Akhavan P and Hoorali M 2008 E-readiness assessment of non-profit ICT SMEs in a developing country: The case of Iran *Technovation* vol. **28** no. 9 pp. 578–590.
[33] Wabwoba F, Omuterema S and Wanyembi G W 2013 Green ICT Readiness Model for Developing Economies : Case of Kenya vol. **4** no. 1 pp. 51–65.

# Improving the accuracy of k-nearest neighbor using local mean based and distance weight

**K U Syaliman[1], E B Nababan[2*], and O S Sitompul[2]**

[1]Graduate School of Computer Science
[2]Department of Information Technology, Faculty of Computer Science and Information Technology, Universitas Sumatera Utara, Medan, Indonesia

ernabrn@usu.ac.id

**Abstract**. In k-nearest neighbor (kNN), the determination of classes for new data is normally performed by a simple majority vote system, which may ignore the similarities among data, as well as allowing the occurrence of a double majority class that can lead to misclassification. In this research, we propose an approach to resolve the majority vote issues by calculating the distance weight using a combination of local mean based k-nearest neighbor (LMKNN) and distance weight k-nearest neighbor (DWKNN). The accuracy of results is compared to the accuracy acquired from the original k-NN method using several datasets from the UCI Machine Learning repository, Kaggle and Keel, such as ionosphare, iris, voice genre, lower back pain, and thyroid. In addition, the proposed method is also tested using real data from a public senior high school in city of Tualang, Indonesia. Results shows that the combination of LMKNN and DWKNN was able to increase the classification accuracy of kNN, whereby the average accuracy on test data is 2.45% with the highest increase in accuracy of 3.71% occurring on the lower back pain symptoms dataset. For the real data, the increase in accuracy is obtained as high as 5.16%.

## 1. Introduction

The k-NN method is among the most widely used method in data mining and machine learning research, such as text categorization, pattern recognition, and classification (see for example the works of [1-8]). The kNN is recognized as an attractive, easy to apply, intuitive, simple and could be exploited in various application domains. However, it is found that the accuracy of kNN is still relatively low, especially when compared to other classification algorithms. In [9] the authors measured the accuracy of support vector machine (SVM) and kNN to obtain an average accuracy of kNN and SVM equal to 71.28% and 92.40%, respectively. Another study by [10] compared nearest centroid classifier (NCC) and kNN method. The result of their research revealed that NCC reach a highest accuracy of 96.67% and a lowest accuracy of 33.33%, whereas the kNN method was only capable to produce a highest accuracy of 26.7% and a lowest accuracy of 22.5%. A recent study by [11] found that the kNN method gave a best result of 48.78% with k = 8 when applied on a dataset which has 395 records, 30 attributes, and 4 classes.

The relatively low accuracy of kNN is caused by several factors. One of them is that every characteristic of the method has the same result on calculating distance. The solution of this problem is to give weight to each data characteristic [12]. Another factor that cause to the low accuracy of the kNN is the determination of new data classes which is based on a simple vote majority system [13-14], where the majority vote system ignores the closeness between data [15], this is unacceptable when the distance of each nearest neighbor differs greatly against the distance of the test data [16]. Furthermore, there will be a possibility of double majority class caused by the class determination system for new data based on the vote majority and determination of the number of nearest neighbors, where the number of nearest neighbors are chosen according to the desired level of success [17]. However, as suggested in [18], problems in determining new data classes with a vote majority system that ignores the closeness between data, which resulting in a misclassification, can be overcome by using distance weight. By using the method of distance weight the class determination for new data is based on the weights obtained through the distance between data.

In this research, we propose a replacement of the vote majority system in kNN using distance weight method, whereby in order to find the weight between data we combine the local mean based k-nearest neighbor (LMKNN) and distance weight k-nearest neighbor (DWKNN) methods, it is expected that the merging of these two methods is able to improve the accuracy result in the classification process. The rest of this paper is structured as follows. Section 2 will summarize previous studies on the theoretical foundation regarding the topic. In Section 3 we will provide the result and discussions and in Section 4 will provide with conclusions.

## 2. Local Mean Based K-Nearest Neighbor

In [18] a method called local mean based k-nearest neighbor (LMKNN) had been proposed. This method is a simple, effective and resilient nonparametric classification. This LMKNN has been proven to improve classification performance and also reduce the effect of existing outliers, especially in small data size [16]. The of LMKNN process could be describe as follows :

Step 1: Determination of Value $k$

Step 2: Compute of the distance between test data to each all training data using the Euclidean distance using the equation:

$$D(x,y) = ||x-y||_2 = \sqrt{\sum_{j=1}^{N} |x-y|^2} \quad (1)$$

Step 3: Sort distance of data from the smallest to the largest as much as $k$ for each data class

Step 4: Calculate local mean vector of each class with the equation [16] :

$$m_{w_j}^k = \frac{i}{k}\sum_{i=1}^{k} y_{i,j}^{NN} \quad (2)$$

Step 5: Define the test data class by calculating the closest distance to the local mean vector of each data class using the equation :

$$w_c = argmin_{w_j} d\left(x, m_{w_j}^k\right), j = 1, 2, \ldots, M \quad (3)$$

The LMKNN classification is equal to 1-NN if the value of $k = 1$ [18]. The value of $k$ in LMKNN is different from the original kNN, in original kNN value of $k$ is the number of nearest neighbors from all training data, whereas in LMKNN the value of $k$ is the number of nearest neighbors from each class in the training data [16].

In the class determination of the test data, LMKNN uses the closest distance measurements to each local mean vector of each data class, which is considered effective in overcoming the negative effects of outliers [16].

## 3. Distance Weight K-Nearest Neighbor

In [19] a method called distance weight k-nearest neighbor (DWKNN) had been proposed. This method specifies a new data class based on the weight value obtained from the distance between data, so that misclassification occurs due to ignoring the proximity between data can be overcome. This weighting method had a good performance because it can reduce the influence of outliers and distribution of unbalanced data sets [15]. The of DWKNN process could be describe as follows:

Step 1: Determination the value of $k$.

Step 2: Calculate the test data distance for each data in each class using the Euclidean distance

Step 3: Sort the distance between data from the smallest to the largest according to number of $k$.

Step 4: Calculate the weights from the distances between the ordered data.

In [19] solutions to count weight based on the distance between data, one of which may be use equation :

$$w_i = \frac{1}{d(x_q, x_i)} \tag{4}$$

Or

$$w_i = 1 - d(x_q, x_i) \tag{5}$$

In [15] give another option to calculate weight based on the distance between data, the weights given by using the equation :

$$w_i = \begin{cases} \frac{d_k^{NN} - d_i^{NN}}{d_k^{NN} - d_1^{NN}} \times \frac{1}{i}, & d_k^{NN} \neq d_i^{NN}, \\ 1 \quad , & d_k^{NN} = d_i^{NN} \end{cases} \tag{6}$$

Where :
$w_i$ is the weight of $i$ from the nearest neighbor.
$d(x_q, x_i)$ is the distance between test data and training data.
$d_k^{NN}$ is distance of $k$ nearest neighbor.
$d_i^{NN}$ is distance of $i$ from $k$ nearest neighbor.

## 4. Proposed Method

To further describe the combination of local mean based k-nearest neighbor (LMKNN) and distance weight k-nearest neighbor (DWKNN) methods, it will be explained step by step in this sub-chapter. The stages in outline can be seen in figure 1:

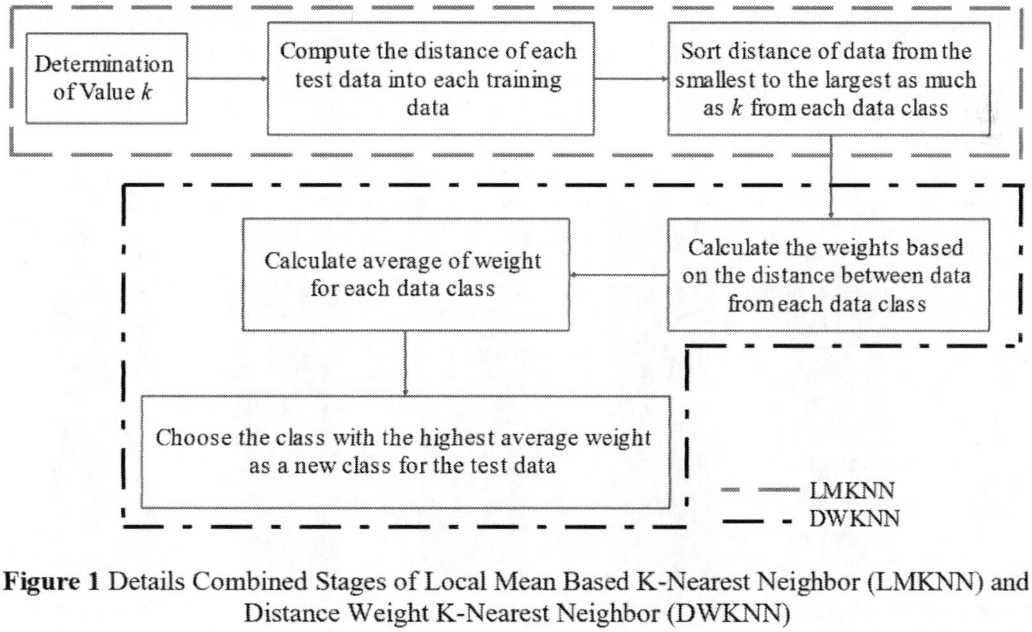


**Figure 1** Details Combined Stages of Local Mean Based K-Nearest Neighbor (LMKNN) and Distance Weight K-Nearest Neighbor (DWKNN)

From figure 1 it can be explained that the combined method of Local Mean Based K-Nearest Neighbor (LMKNN) and Distance Weight K-Nearest Neighbor (DWKNN) has several stages, among others :

Step 1 : Determination of value $k$, which is the number of nearest neighbors.

Step 2 : Compute the distance from each test data to each training data using the Euclidean using equation (1)

Step 3 : Sort the distance data from smallest to largest as many as k of each class.
Step 4 : Calculate the weights using equation (4)
Step 5 : Calculate average weight for each data class.
Step 6 : Choose class with the highest average weight as a new class for the test data.

Step 1 to step 3 is a contribution from LMKNN, and step 4 to step 6 is a contribution from DWKNN. The value of $k$ in this proposed method follows the rules of LMKNN, where the value of $k$ is the number of nearest neighbors from each class.

## 5. Result and Discussion

### *5.1. Result of the classification from data set*

To knowing the proposed method is better than original kNN, it will be tested using several datasets from the UCI Machine Learning repository, Kaggle and Keel, such as ionosphare, iris, voice genre, lower back pain, and thyroid (new thyroid). In addition, the proposed method is also tested using real data from a public senior high school in city of Tualang, Indonesia. Details of data can be seen in the table 1 :

**Table 1** Detail of Data

| Data | Attributes | Class | Total Data | Data | Attributes | Class | Total Data |
|---|---|---|---|---|---|---|---|
| Ionosphere | 34 | 2 | 351 | Lower Back Pain Symptomps | 13 | 2 | 310 |
| Iris | 4 | 3 | 150 | Thyroid | 5 | 3 | 215 |
| Voice Genre | 21 | 2 | 3168 | Majors of Student | 9 | 2 | 167 |

Data will be divided randomly into two category, 80% of the data will be used as training data and 20% of the data will be used as test data. Then it will be calculated using original kNN and proposed method. In this study, K is only worth 1 to 10. The average accuracy of each data can be seen in the figure 2 :

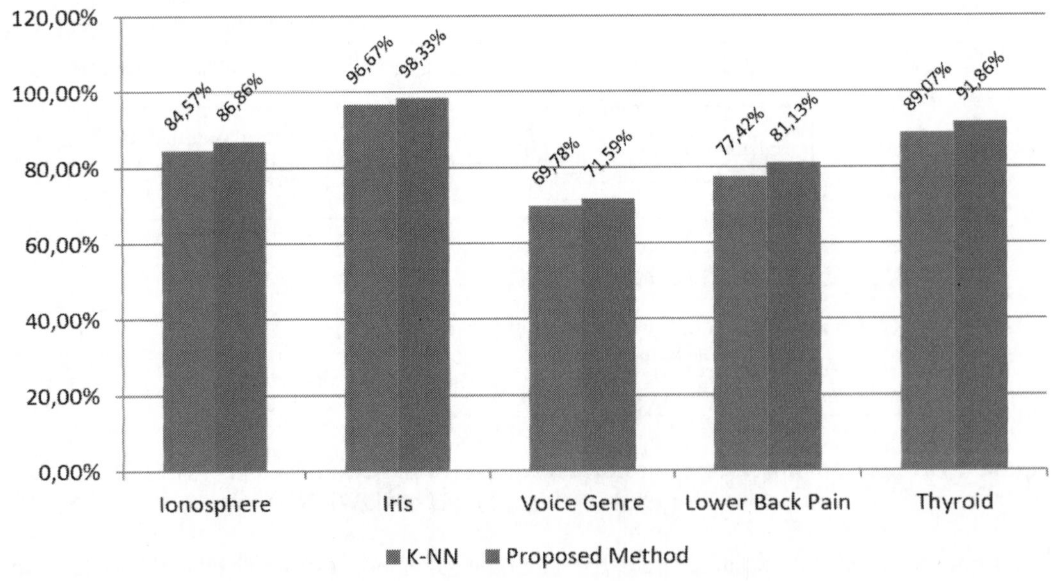


**Figure 2** Average Accuracy from All Data

Referring to figure 2 it can be seen that the proposed method has a higher accuracy value than the original kNN, where the highest difference of accuracy obtained in Lower Back Pain Symptoms data is worth 3.71%, and the lowest difference of accuracy obtained in the iris data set of 1.66%. Of all tested data the average increase the accuracy is 2.452%.

### 5.2. *Result of the classification from real data*

From the results of the tests performed in the previous sub-section was knowing, the proposed method is better than original kNN. To know with certainty whether the proposed method is better to make predictions in the real data from public high school in city of Tualang, Indonesia, it will be recomparison with original kNN. Details from devide of real data can be seen in table 2 and details of the test results can be seen in table 3 :

**Table 2** Details of Real Data

| No | Category | Total Data for Each Class | | Total of Data |
|---|---|---|---|---|
| | | Science | Social | |
| 1 | Training | 64 | 70 | 134 |
| 2 | Test | 15 | 18 | 33 |

**Table 3** Comparison of Accuracy from Real Data

| No | K | K-NN Conventional | | | Combine LMKNN+DWKNN | | | Method With Highest Accuracy |
|---|---|---|---|---|---|---|---|---|
| | | True Prediction | | Accuracy | True Prediction | | Accuracy | |
| | | Science | Sosial | | Science | Sosial | | |
| 1 | 1 | 14 | 14 | 84.85% | 14 | 14 | 84.85% | Both |
| 2 | 2 | 14 | 14 | 84.85% | 15 | 14 | 87.88% | Proposed Method |
| 3 | 3 | 14 | 13 | 81.82% | 15 | 13 | 84.85% | Proposed Method |
| 4 | 4 | 14 | 13 | 81.82% | 15 | 14 | 87.88% | Proposed Method |
| 5 | 5 | 13 | 13 | 78.79% | 15 | 14 | 87.88% | Proposed Method |
| 6 | 6 | 14 | 13 | 81.82% | 15 | 14 | 87.88% | Proposed Method |
| 7 | 7 | 14 | 13 | 81.82% | 15 | 14 | 87.88% | Proposed Method |
| 8 | 8 | 14 | 13 | 81.82% | 15 | 14 | 87.88% | Proposed Method |
| 9 | 9 | 14 | 13 | 81.82% | 15 | 14 | 87.88% | Proposed Method |
| 10 | 10 | 14 | 14 | 84.85% | 15 | 15 | 90.91% | Proposed Method |
| Average | | | | 82.42% | | | 87.58% | Proposed Method |

Table 3 shows that the proposed method gives the best prediction results in determining of majors public senior high school in city Tualang, Indonesia. The highest score reaching 90.91% when k = 10 and the lowest being 84.85% when k = 1. The proposed method are able to increase the average accuracy by 5.16% against accuracy value of original kNN.

## 6. Conclusion

Based on the findings and discussion in the previous section it can be concluded that:

- The proposed method is able to improve accuracy, where the average increase in accuracy of all test data is 2.492%. Highest average value of accuracy obtained in the data Lower Back Pain Symptoms of 3.91%, and the average value of the lowest accuracy value found in iris data that is equal to 1.66%.
- In the case of predicting majors of student in public senior high school in city of Tualang, Indonesia, the researcher proposes using the proposed method, because the proposed method proved to be better than the original kNN with average difference of accuracy is 5.16%, whereby highest accuracy value is 90.91% when k = 10.

**Acknowledgement**
High appreciation should be given to Lembaga Penelitian Universitas Sumatera Utara (LP USU), Graduate School of Computer Science Fasilkom-TI USU and Rector of Universitas Sumatera Utara for their support on the dissemination of this research work as well as the facilities provided.

**References**
[1] Nitin B and Vandana 2010 Survey of nearest neighbor techniques *Int. J. of Computer Science and Information Security* **8**(2) 302-5
[2] M Akhil j, Bulusu L D and Priti C 2013 Classification of heart disease using k-nearest neighbor and genetic algorithm Procedia Technology **10** 85-94
[3] A S Sánchez, F J Iglesias-Rodríguez, P R Fernándes & Francisco J de C J 2015 Applying the k-nearest neighbor technique to the classification of workers according to their risk of suffering musculoskeletal disorders Int. J. of Industrial Ergonomics **52** 92-9
[4] Zhibin P, Yidi W and Weiping K 2017 A new general nearest neighbor classification based on the mutual neighborhood information *Knowledge-Based System* **121** 142-52
[5] Nicolás G-P and Domingi O-B 2009 Boosting k-nearest neighbor classifier by means of input space projection *Expert Systems with Application* **36** 10570-82
[6] Jigang W, Predrag N and Leon N C 2007 Improving nearest neighbor rule with a simple adaptive distance measure *Pattern Recognition Letters* **28** 207–213
[7] Stefanos O and Georgios E 2012 Fast and accuratek-nearest neighbor classification using prototype selection by clustering The 16th Panhellenic Conference on Informatics (PCI)
[8] Yunsheng S, Jiye L, Jing L and Xingwang Z 2017 An efficient instance selection algorithm for k nearest neighbor regression *Neurocomputing* **251** 26-34
[9] Amri D, Devie P S, Dian A and Dini A 2016 Comparison of accuracy level k-nearest neighbor algorithm and support vector machine algorithm in classification water quality status *The 6th Int. Conf. on System Engineering and Technology (ICSET)* 137-41
[10] Elizabeth N T and Aditya W M Comparison of music genre classification using nearest centroid classifier and k-nearest neighbours *Int. Conf. on Information Management and Technology (ICIMTech)* 118-23
[11] Jessica M B 2017 Math test scores using k-nearest neighbor *IEEE Integrated STEM Conference (ISEC)*
[12] Maryam K 2016 A method to improve the accuracy of k-nearest neighbor algorithm *Int. J. of Computer Engineering and Information Technology* **8**(6) 90-5
[13] Syahfitri K L, Opim S S and Syahril E 2015 Sentiment analysis on Indonesian language text using support vector machine (svm) and k-nearest neighbor (kNN) (in Bahasa) *Seminar Nasional Teknologi Informasi dan Komunikasi* (SENTIKA) pp. 1-8
[14] Aditya D and Thendral P 2017 Enhancing classification accuracy of k-nearest neighbours algorithm using gain ratio *International Research Journal of Engineering and Technology (IRJET)* **4**(9) pp 1385-1388
[15] Jianping G and Taisong X and Yin K 2011 A novel weighted voting for k-nearest neighbor rule *J. of Computers* **6**(5) 833-40
[16] Zhibin P, Yidi W and Weiping K 2016 A new k-harmonic nearest neighbor classifier based on the multi-local means *Expert Systems with Applications* **67** 115-25
[17] Ö F Ertuğrul & M E Tağluk 2017 A novel version of k nearest neighbor: dependent nearest neighbor *Applied Soft Computing Journal* **55** 480-490
[18] Y Mitani and Y Hamamoto 2006 A local mean-based nonparametric classifier *Patern Recognition Letter* pp 1151-1159
[19] Gustavo E A P A B and Diego F S 2009 How k-nearest neighbor parameters affect it's performance *38° JAIIO - Simposio Argentino de Inteligencia Artificial* (ASAI) pp 95-106

IOP Conf. Series: Journal of Physics: Conf. Series **978** (2018) 012048 doi:10.1088/1742-6596/978/1/012048

# Mobile/android application for QRS detection using zero cross method

**M I Rizqyawan[1], A I Simbolon[1], M A Suhendra[1], M F Amri[1], D E Kusumandari[1]**

[1]UPT BPI, Indonesian Institute of Sciences, Bandung 40135, Indonesia

muha176@lipi.go.id, arth001@lipi.go.id, agung.fq@gmail.com, mfai001@lipi.go.id, esti_dek@yahoo.com

**Abstract.** In automatic ECG signal processing, one of the main topics of research is QRS complex detection. Detecting correct QRS complex or R peak is important since it is used to measure several other ECG metrics. One of the robust methods for QRS detection is Zero Cross method. This method uses an addition of high-frequency signal and zero crossing count to detect QRS complex which has a low-frequency oscillation. This paper presents an application of QRS detection using Zero Cross algorithm in the Android-based system. The performance of the algorithm in the mobile environment is measured. The result shows that this method is suitable for real-time QRS detection in a mobile application.

## 1. Introduction

Electrocardiogram (ECG) signal is a signal produced by measuring electrical activities from heart's muscle. It is a non-invasive instrument and due to this characteristic of ECG, it is widely used for cardiovascular diseases. It is also one of most popular research field in biomedical signal processing, with applications ranging from automatic detection of arrhythmia [1-3], sleep apnea [4, 5], to psychological stress [6, 7].

In automatic ECG signal processing, one of the main topics of research is QRS complex detection. This area of research has been explored for more than three decades [8]. Detecting correct QRS complex or R peak is important since it is used to measure several other ECG metrics, such as RR Interval, QRS Width, and Heart Rate Variability (HRV). These metrics are often used as features for ECG classification. The most common method for QRS detection is Pan Tomkins method [8] which use an adaptive threshold after the signal is processed with a bandpass filter, derivative, squaring, and moving average. Zero Cross method [9] is another QRS detection method which employed an adaptive filtering in the steps of its processing. This method uses an addition of high-frequency signal and zero crossing count to detect QRS complex which has a low-frequency oscillation. By employing the zero-crossing count, this method is robust against noise and performs well in the noisy ECG signal.

On the other hand, the other topic of research in ECG is the application of automatic ECG signal processing. Most of the study is intended to be applied in workstation/desktop computer. Some of the research is exploring the application of the techniques in hardware [10] (such as embedded system), and also mobile phones [11-14]. The exploration of ECG analysis in mobile phones is increasing due to the increase in availability of smartphones. Smartphones become a ubiquitous computer, with increasing capabilities in the terms of computation power. Although the computation power of today's smartphone is clearly higher than computer several decades ago, however, there's always a limitation to the mobile phone, which cannot have a full processing power and memory of a desktop computer in the same generation. Therefore, further study needed in the implementation of mobile ECG analysis to make the processing fast enough and feasible in real-time use-case.

## 2. Methodology

This research aims at developing a smart classifier of arrhythmia for android device. The research consists of two sub-research focuses, the classifier model development and development and implementation of the smart system in android device. The block diagram of our research can be seen in figure 1. As can be seen in figure 1, this paper falls on the development of the android smart system app, with the focus on QRS detection module. In this paper, our aim is to explore whether the algorithm fits to be applied in the mobile environment or not. The other sub-research focus such as classification and system architecture is discussed in our other paper [15, 16].

In this experiment, we use MIT-BIH arrhythmia database [17] as our testing data. Single channel (MLII channel) is used for the experiment. In this experiment, we use selected records of the database denotes as dataset 1 (ds1) in protocol proposed by Chazal et. al. [1]. The raw data is exported to CSV file, which will be loaded into the Android app later. In order to examine the computation load required by the algorithm, processing time of QRS detection is measured. The processing time is measured by setting up a timer just before and after the process of QRS detection using Java's *nanoTime*. The processing time is calculated by subtracting the end timer with start timer. The flowchart of the process is shown in figure 2.

## 3. Zero Cross Algorithm

Zero Crossing Count for QRS detection is proposed by Kohler et. al. at 2003 [9] with the aim to develop a QRS detection algorithm with low computational load and high detection performance. The algorithm can be divided into three main blocks; feature extraction, event detection, and temporal localization. The feature extraction aims to generate feature signal, which would be used as the input of event detection process. The event detection block searches the possible location of QRS complex by comparing the feature signal with the threshold. Temporal localization searches the actual peak location in the block of event detected in the previous process.

### *3.1 Feature Extraction*

This process aims to prepare the signal and process it all the way to the counting of the zero crossing of the signal to the x-axis. The first process of the feature extraction is filtering. Bandpass filter with cut-off frequency of 18 Hz and 35 Hz are used. Then the next process is the nonlinear transform, by squaring the signal multiplied by its signum. The process can be written as:

$$y(n) = sign(x_f(n)) \cdot x_f^2(n) \quad (1)$$

where $y(n)$ is the transformed signal and $x_f(n)$ denotes the bandpass filtered signal. Using this process, the important features of the signal can be distinguished from the rest of other features.

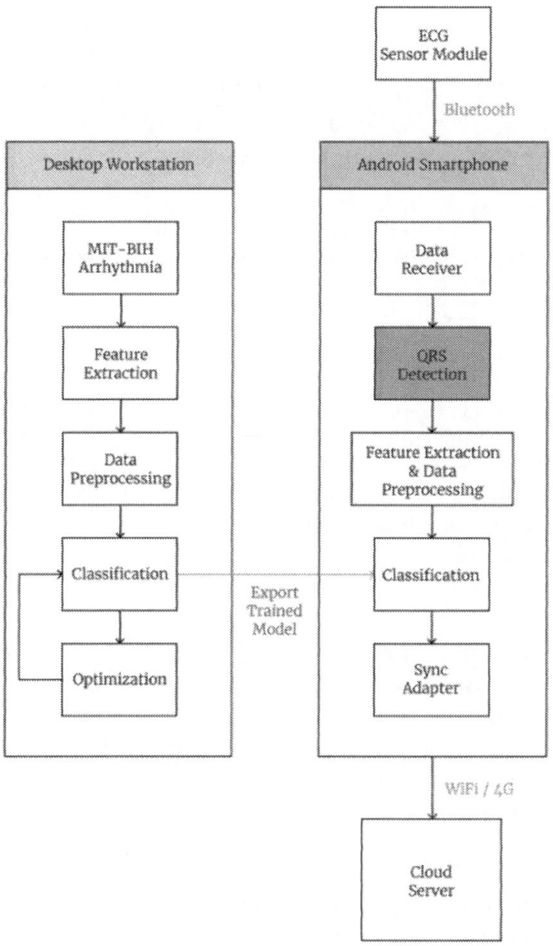


**Figure 1.** The big picture of the research project. This paper focuses on Android application development, especially on QRS Detection algorithm application.

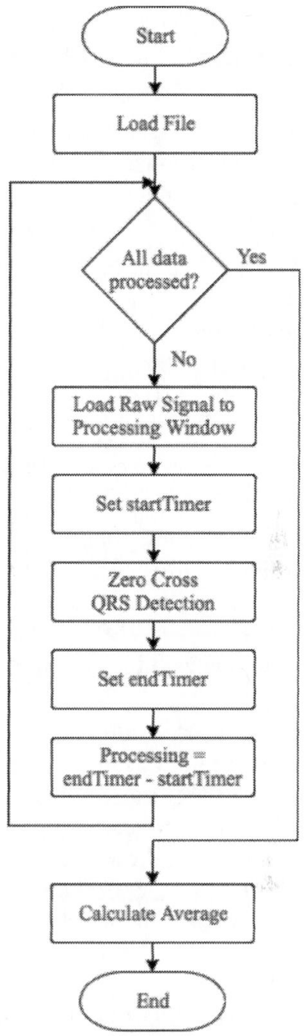


**Figure 2.** Flowchart of the Methodology

The next process is to add the high-frequency sequence to the nonlinear transformed signal. To do so, the amplitude of the high-frequency sequence must be calculated first. The estimated amplitude of the high-frequency sequence is determined by:

$$K(n) = \lambda_K K(n-1) + (1-\lambda_K)|y(n).c| \tag{2}$$

where $K(n)$ is the estimated amplitude, $\lambda_K$ is the forgetting factor, $y(n)$ is the nonlinear transformed signal, and $c$ denotes the constant gain. After the amplitude is estimated, then we can calculate the high-frequency sequence by:

$$b(n) = (-1)^n \cdot K(n) \tag{3}$$

where $b(n)$ is the high-frequency sequence and $K(n)$ is the estimated amplitude. The high-frequency sequence is added to the nonlinear transformed signal using:

$$z(n) = y(n) + b(n) \tag{4}$$

where $z(n)$ is the signal after high-frequency addition, $y(n)$ is the nonlinear transformed signal, and $b(n)$ is the high-frequency sequence. At this point, the signal is ready to be processed at the next step, which is zero crossing count. The zero crossing is counted by using autoregressive low pass filter:

$$D(n) = \lambda_D D(n-1) + (1-\lambda_D)d(n) \tag{5}$$

where $d(n)$ is the count of zero-crosses at the segment and $\lambda_D$is forgetting factor.

*3.2 Event Detection*

This process provides the suspected area of QRS complex by comparing the feature signal with an adaptive threshold. The adaptive threshold is computed by:

$$\Theta(n) = \lambda_\Theta \Theta(n-1) + (1-\lambda_\Theta)D(n) \tag{6}$$

where is $\lambda_\Theta$ forgetting factor, $D(n)$ is the feature signal and $\Theta(n)$ is the adaptive threshold. When feature signal is lower than the adaptive threshold, the event is marked as the start. The algorithm continued the scanning process, and when the feature signal is higher than the threshold, it is marked as the end. However, if the occurrence of start event is lower than time-out value, the two events are merged into one as it likely comes from one QRS complex.

*3.3 Temporal Localization*

This process searches the location of the R-wave in the interval provided by the previous process. To locate the R-wave, the algorithm search for the maximum and minimum value of the signal in the suspected interval of QRS complex. If the value of the maximum is lower than the absolute value of the minimum, then the R-wave is the temporal location of minimum value. But if not, the R-wave is determined by the temporal location of maximum value. To point the actual location of the peak in the raw signal, please note that the delay caused by the filtering needs to be considered.

## 4. Result

The zero cross algorithm is implemented using Java as the programming language, which is the official development language for android application development. The screenshot of the app can be seen in figure 3. To test the processing time of the algorithm, three phones are used which represents each category/class of smartphone. Motorola Moto E4, Oppo F1s, and Xiaomi Mi5 which represents low-end, middle-end, and high-end class Android phone respectively. Table 1 shows the result of processing time test using three different smartphones mentioned above.

**Table 1.** Processing time test on different devices.

| | Min (ms) | Max (ms) | Average (ms) |
|---|---|---|---|
| **Motorola Moto E4** | 14.85 | 33.5 | 15.7 |
| **Oppo F1s** | 270.0 | 426.1 | 299.8 |
| **Xiaomi Mi5** | 120.4 | 131.3 | 124.3 |

To measure the real-time capability of the method, the result from table 1 is converted to real-time factor, as proposed by [11] using:

$$R_f = \frac{P_t}{MeasurementWindow} \tag{7}$$

where $R_f$ is real-time factor value, and Pt is processing time acquired from table 1. Figure 4 shows the minimum, maximum, and average real-time factor required for the algorithm to run on the tested devices.

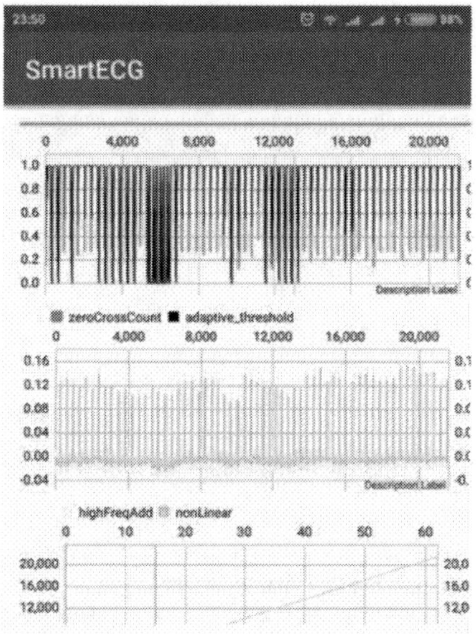


**Figure 3.** The screenshot of the application.

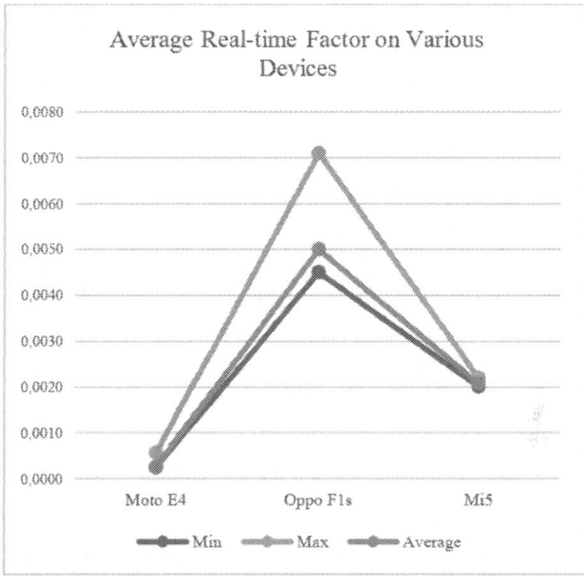


**Figure 4.** The minimum, maximum, and average real-time factor required for the algorithm to run on the tested devices.

## 5. Conclusion

This paper presents an application of QRS detection using Zero Cross algorithm in the Android-based system. The performance of the algorithm in the mobile environment is measured using three different devices as the test-bed.

The result shows that this method is suitable for real-time QRS detection in a mobile application. The average real-time factor of processing time in Xiaomi Mi5 is 0.0022, which is much less than 1. This result is much expected as the Xiaomi Mi5 presents the high-end class smartphone. The Oppo F1s shows the longest time required for processing. The maximum real-time factor of Oppo is highest than the other devices at 0.0071. But at this highest value, the result shows that the real-time factor is much less than 1. The surprising result comes from Motorola Moto E4 which obtained the lowest processing time and real-time factor required by the algorithm to run. This is surprising as the Moto E4 represents lower-end smartphone. Further study is required to investigate this result, but generally, the result shows that this method is feasible to be used in real-time QRS detection.

### Acknowledgments

This research was supported by the Insinas Ristek funded by Minister of Research, Technology, and Higher Education, Indonesia.

### References

[1] deChazal P, O'Dwyer M and Reilly R 2004 Automatic classification of heartbeats using ECG morphology and heartbeat interval features *IEEE Transactions on Biomedical Engineering*, **51**(7) pp 1196-1206

[2] Luz E, Schwartz W, Cámara-Chávez G and Menotti D 2016 ECG-based heartbeat classification for arrhythmia detection: a survey *Computer Methods and Programs in Biomedicine* **127** pp 144-164

[3] Escalona-Moran M, Soriano M, Fischer I and Mirasso C 2015 Electrocardiogram classification using reservoir computing with logistic regression *IEEE Journal of Biomedical and Health Informatics* **19**(3) pp 892-898

[4] Hassan A and Haque M 2016 Computer-aided obstructive sleep apnea screening from single-lead electrocardiogram using statistical and spectral features and bootstrap aggregating *Biocybernetics and Biomedical Engineering* **36**(1) pp 256-266

[5] Hassan A and Haque M 2017 An expert system for automated identification of obstructive sleep apnea from single-lead ECG using random under sampling boosting *Neurocomputing* **235** pp 122-130

[6] Lampert R 2015 ECG signatures of psychological stress *Journal of Electrocardiology* **48**(6) pp 1000-05.

[7] Han L, Zhang Q, Chen X, Zhan Q, Yang T and Zhao Z 2017 Detecting work-related stress with a wearable device *Computers in Industry* **90** pp 42-49

[8] Pan J and Tompkins W 1985 A real-time QRS detection algorithm *IEEE Transactions on Biomedical Engineering* **32**(3) pp 230-236

[9] Köhler B, Hennig C and Orglmeister R 2003 QRS detection using zero crossing counts *Progress in Biomedical Research* **8**(3)

[10] Zhang C and Bae T 2012 VLSI friendly ECG QRS complex detector for body sensor networks *IEEE Journal on Emerging and Selected Topics in Circuits and Systems* **2**(1) pp 52-59

[11] Sufi F, Fang Q and Cosic I 2007 ECG R-R peak detection on mobile phones *29th Annual International Conference of the IEEE Engineering in Medicine and Biology Society*

[12] Elgendi M, Eskofier B, Dokos S and Abbott D 2014 Revisiting QRS detection methodologies for portable, wearable, battery-operated, and wireless ECG systems *PLoS ONE* **9**(1) p. e84018

[13] Gradl S, Kugler P, Lohmuller C and Eskofier B 2012 Real-time ECG monitoring and arrhythmia detection using Android-based mobile devices *Annual International Conference of the IEEE Engineering in Medicine and Biology Society*

[14] Leutheuser H, Gradl S, Kugler P, Anneken L, Arnold M, Achenbach S and Eskofier B 2014 Comparison of real-time classification systems for arrhythmia detection on Android-based mobile devices *36th Annual International Conference of the IEEE Engineering in Medicine and Biology Society*

[15] Rizqyawan M, Simbolon A and Kusumandari D Weighted SVM with RR interval based features for Android-based arrhythmia classifier

[16] Rizqyawan M, Amri M, Pratama R and Turnip A 2016 Design and development of Android-based cloud ECG monitoring system *3rd International Conference on Information Technology, Computer, and Electrical Engineering (ICITACEE)*

[17] Moody G and Mark R 2001 The impact of the MIT-BIH arrhythmia database *IEEE Engineering in Medicine and Biology Magazine* **20**(3) pp 45-50

IOP Conf. Series: Journal of Physics: Conf. Series **978** (2018) 012049 doi:10.1088/1742-6596/978/1/012049

# Main factors in E-Learning for the Equivalency Education Program (E-LEEP)

**M B Yel[1] and Sfenrianto[2]**

[1]Faculty of Computer Science STIKOM CKI Jakarta, Indonesia
[2]Master in Information Systems Management, Jakarta, Indonesia

Betty.mby@stikomcki.ac.id, sfenrianto@binus.edu

**Abstract.** There is a tremendous learning gap between formal education and non-formal education. E-Learning can facilitate non-formal education learners in improving the learning process. In this study, we present the main factors behind the E-learning for the Equivalency Education Program (E-LEEP) initiative in Indonesia. There are four main factors proposed, namely: standardization, learning materials, learning process, and learners' characteristics. Each factor supports each other to achieve the learning process of E-LEEP in Indonesia. Although not yet proven, the E-learning should be developed followed the main factors for the non-formal education. This is because those factors can improve the quality of E-Learning for the Equivalency Education Program.

## 1. Introduction

In some countries, e-Learning has been implemented to support non-formal education. The e-Learning has become an international choice to improve the quality, access and equity between formal and non-formal education. A large amount of literature has addressed various e-Learning strategies to close the gap.

The study in Zambia by Richard and Foster states that distance education using e-Learning is a way of educating learners who are not getting formal education, and as a means of supporting and facilitating conventional education in developing countries [1]. E-Learning is also being developed in the project: The Distance Education Project for Rural Schools (DEPRS), to improve the quality of education in rural areas of China, especially in the poorer western provinces [2]. Then study in [3], proposes a Model to provide e-learning services to remote / rural areas of China in order to facilitate non-formal education. The Philippines develops eSkwela as an e-learning learning center for youth and adult outside school [4][5].

Meanwhile, Indonesia in 2016 has introduced e-Learning for equivalency education (see Figure 1) (https://kesetaraan.kemdikbud.go.id). E-Learning is built to facilitate learners to learn independently and to give broader access to learning. This is to accommodate the learning strategy of equivalency education programs that emphasize on independent learning 50%, face to face at least 20%, and tutorials at least 30% [6]. This learning approach is used as an alternative for students who are relatively difficult to meet directly with the tutor because 430386 learners and 12369 Community Learning Centre (CLC) at 2015/2016 spread across provinces, districts / cities and sub districts in Indonesia [7].

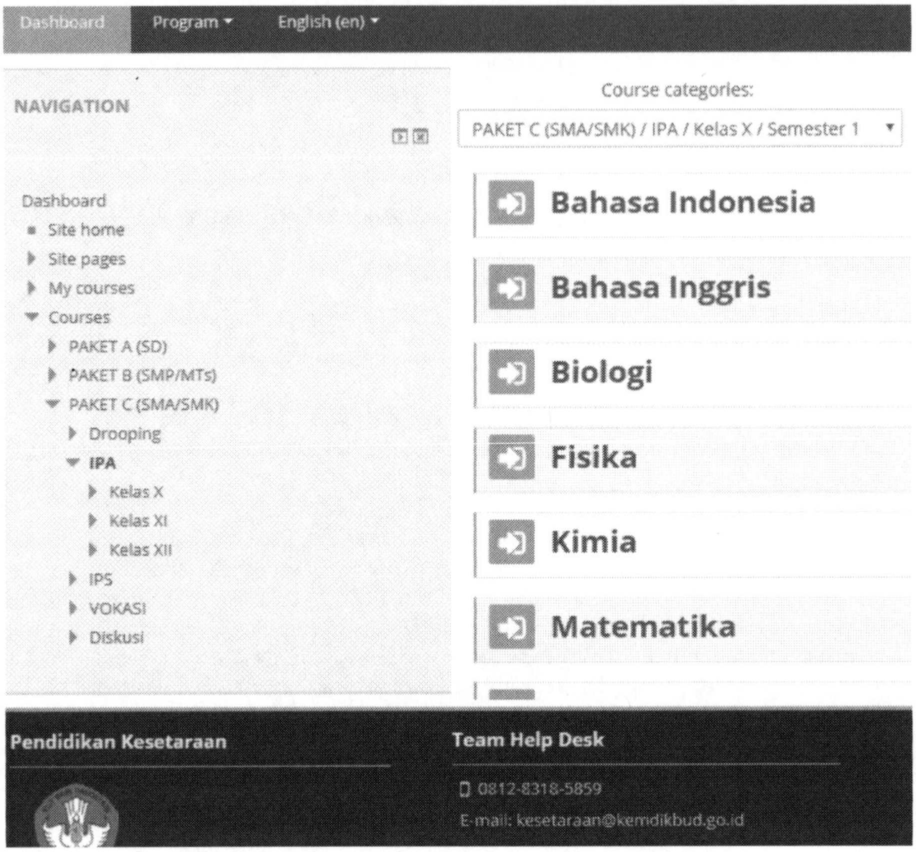


**Figure 1.** The e-Learning of Equivalency Education

Nevertheless, e-learning that has been applied had only change the content of learning into digital. However, many factors can be considered in accordance with the characteristics of e-Learning and equality education environment. With the proper application of e-learning, learners can optimize the learning process, so that they can achieve their educational goals.

This is the reason why, in our previous study [6], we have proposed e-Learning for equivalency education program (E-LEEP) model. The model consists of three main components: Users, programs and monitoring. Each component will support students in to achieve the goal of learning. In this study, main factors of E-LEEP initiatives (performance) are proposed.

## 2. Literature Review

In developing e-Learning, it is important to discern the success factors. Various important factors in e-learning have been established through research conducted by many researchers. Standardization is one of the important factors in the development of e-Learning [8][9]. The development of qualified e-Learning in an institution requires good standardization [10].

Learning content is also an important factor for the successful implementation of e-Learning. Learning content plays an important role in the performance of e-learning, which in turn affects learners' satisfaction and further their intentions to use e-learning [11]. Learners are very concerned about learning content that will be learned in the use of e-learning [12]. Learners should easily access learning content and get quality contents [13]. Learning content refers to teaching materials and how they are delivered to learners [11].

E-Learning can support and improve the learning process. Therefore, the learning process factor also affects the effectiveness of an e-Learning [14] [15]. The learning process in the context of e-learning is related to learning activities [15], discussion forums to encourage learners to be actively involved [16] and evaluation of learning outcomes [17].

In addition, various characteristics of learners such as: culture, background, speed of learning, learning style, motivation, and ability, required the implementation of learning in e-Learning. E-Learning should not only be able to transfer teaching materials from teachers to learners, but also facilitate the diversity of learners' characteristics as one of the influential factors in the e-Learning process. For example, how learners from different cultures can be actively involved in online learning [18]. There is an interrelationship between learning styles and motivation to knowledge skills in the e-Learning process[19]. One of the objectives of addressing the characteristics of learners in e-Learning is to adjust the need for teaching materials to suit their different characteristics.

## 3. Main Factors E-LEEPS

To determine the main factor of E-Learning for the Equivalency Education Program (E-LEEP), other than literature studies, one should also look at components related to the learning environment of equality. Therefore, the study proposes the main factor E-LEEP is the result of four factors interrelated: Standardization, Learning Content, Learning Process, and, Student Culture. Figure 2, explains the main factors in E-Learning for the Equivalency Education Program.

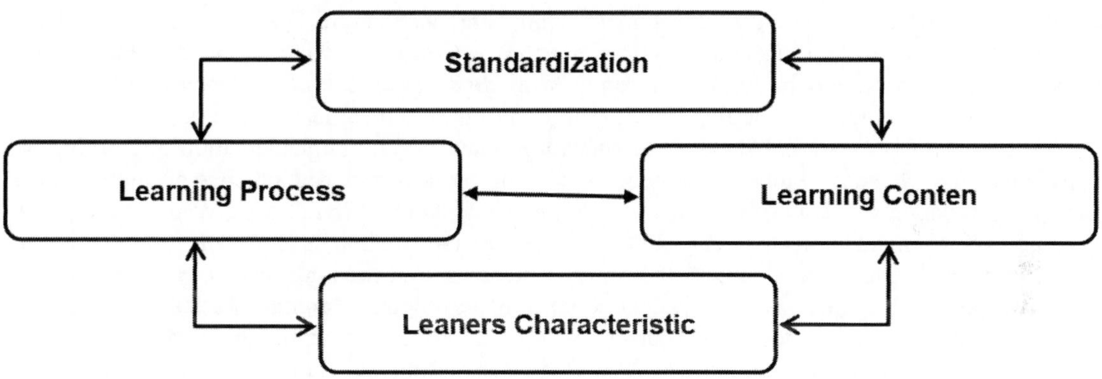


**Figure 2.** Main Factors in E-LEEP

### *3.1. Standardization Factor*

The implementation of e-Learning in education equality does not have a standard so that there is a misperception in its implementation. Therefore, one need to set up an appropriate standard service of the implementation of e-learning. E-LEEP must be organized based on Government Regulation (PP) 19 of the year 2005 on National Education Standards as adjusted in PP 32 of 2013 and PP 13 of the year 2015. Figure 3 describes the standardization component to support E-LEEP.

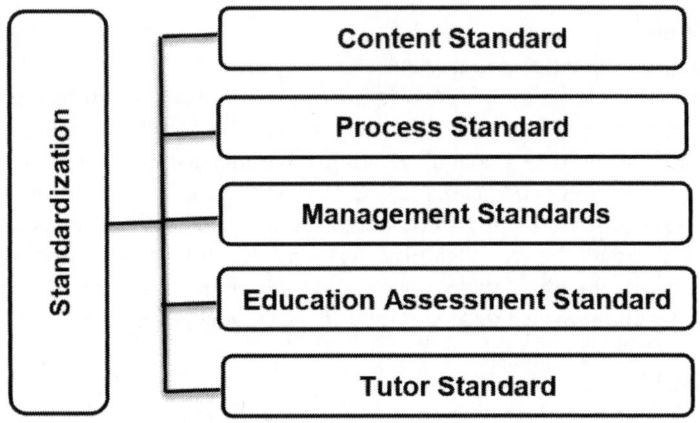


**Figure 3.** Standardization factors in E-LEEP

Content Standard for Package A Program, Package B Program, and Package C Program based on the regulation of national education minister no. 14 2007, covering the minimum material scope and the minimum competency level to achieve minimum graduate competency. The Content Standards for E-LEEP should take into account the educational calendar and curriculum. The curriculum of Package A, Package B, and Package C programs is implemented in a system of levels and degrees that are equivalent to a class system in formal education with their respective competencies. Educational calendar is the signs to organize learning activities that meet the needs of learners.

Process Standard is a national standard of education related to the implementation of learning on equality education. Standard Process Planning of learning process includes syllabus and lesson plan implementation which contains subject identity, competency standard (SK), basic competence (KD), competency achievement indicator, learning objectives, learning materials, time allocation, learning methods, learning activities, assessment of learning outcomes, and learning resources. Therefore, E-LEEP must apply in the process of equivalent equality of Package A Program, Package B Program, and Package C Program contained in the Minister of National Education regulation no. 3 year 2008.

Management Standards in accordance with the Minister of National Education's regulation no. 44 year 2009. Management Standards are the criteria of planning, implementing, and supervising educational activities at the educational, district / city, provincial, or national levels of education in order to achieve the efficiency and effectiveness of equality education. E-LEEP should pay attention to the management of each equality education at the central level undertaken by the Directorate General of Literacy and Equality. The provincial level is implemented by the Learning Activity Development Center (BPKB) or other similar names. The district / city district level is implemented by the Learning Activity Studio (SKB) or other names like that. At the sub-district level, there was an equivalency education supervisory unit (Package A, Package B, and Package C).

Education Assessment Standard refers to the regulation of the Minister of Education and Culture No. 66 year 2013. Assessment of learning outcomes of education equality learners in accordance with the basic and secondary education level is implemented based on applicable national education assessment standards. Therefore E-LEEP must pay attention to the mechanisms, procedures, and learning outcomes assessment instruments of learners nationally.

Successful implementation of equality education learning activities is largely determined by the competence of tutors or non-formal education personnel. Government Regulation No. 19 year 2005, has described the competence or standard of educators. Main task and tutor functions: Develop lesson plans; compile or develop learning materials; teach; and assess learning outcomes. Thus, the Tutor using E-LEEP can upload teaching materials, assignments, quizzes, exams and discussion forums, and participate in video conferences with learners according to schedule.

### *3.2. Learning Content Factor*

In general, the content of learning equality education in the form of learning module is used as a learning source. At each ends of discussion, there will be an Exercise Problem to be done alone or in groups. Nevertheless, Learning content E-LEEP must adapt the curriculum structure of Package A, Package B, and Package C programs implemented in a system of levels and degrees equivalent to the class system of formal education based on competence. Table 2 describes E-LEEP content learning based on the structure of the equivalency education curriculum.

Package A program includes: Level 1 with beginner (Awal) competency level equivalent to 3rd grade Elementary School, emphasizing literacy and numeracy skills (language and numeracy skills), so that learners are able to communicate via text in writing and oral, either in the form of letters or numbers. Level 2 with basic (Dasar) competency level equivalent to 6th grade Elementary School, emphasizes the gradual mastery of facts, concepts, and data, so that learners are able to communicate through text in writing and orally using simple natural and or social phenomena ethically, to have the skills basis in meeting the needs of daily living and continuing education to a higher level.

Package B program includes: Level 3 with competency level Proficient 1 (Terampil 1) is equivalent to 8th grade Junior High School, emphasizes the mastery and application of abstract concepts more widely and practicing improving thinking skills and acting logically and ethically, so that learners are able to communicate via text in writing and orally, and solving problems by using wider natural or social phenomena. Level 4 with competency level Proficient 2 (Terampil 2) is equivalent to 9th grade Junior High School, emphasizing the improvement of thinking skills and processing information and applying it to produce simple works that benefit themselves and society, so that learners are able to actively express themselves and communicate their work through text orally and written based on ethically accurate data and information, to meet the demands of simple job skills and to continue education to a higher level.

**Table 1.** Learning Contents For E-LEEP

| Equivalency Education | Level | Competency | Learning Contents |
|---|---|---|---|
| Package A Program | 1 | Beginner | Text based modul, exercise |
| | 2 | Basic | Text based modul and video, exercise |
| Package B Program | 3 | Proficient 1 | Text based modul, animation and video, exercise |
| | 4 | Proficient 2 | Text based modul, animation and video, ppt, exercise |
| Package C Program | 5 | Highly Proficient 1 | Text based modul, animation and video, ppt, reference link, exercise |
| | 6 | Highly Proficient 2 | Text based modul, animation and video, ppt, reference link, e-book, exercise |

Package C program includes: Level 5 with competency level Highly Proficient 1 (Mahir 1) equivalent to 10th grade Senior High School, directed to the achievement of the basics of academic competence and apply it to produce the work so that learners are able to communicate the concepts in a more scientific and ethical and prepare for able to work independently and develop a professional personality. Level 6 with competency level Highly Proficient 2 (Mahir 2) is equivalent to 12th grade Senior High School, directed to the achievement of academic ability and functional skills ethically, so that learners can work independently or entrepreneurship, be professional, active and productive participant in community life, and can continue education to a higher level.

### *3.3. Learning Process Factor*
The learning process in E-LEEP includes learning activities, discussion forums, and evaluations. Learning activities with e-learning system are conducted in accordance with the subjects held in equivalency education for each program scheduled in the coming semester. The lesson plan should be clear and adhered to. Teaching materials used in the learning process comes from various sources in various forms, formats, and media. Learning resources that have been packaged electronically and available to be accessed by participants anytime and from anywhere.

By utilizing E-LEEP, learners easily establish communication with the Tutor. Communication can be done through discussion forums with the purpose of learning. Discussion forums between learners and tutors can be used to discuss and to ask a lesson. In addition, discussion forums can also be conducted among fellow learners. Discussion forums should be able to attract students into the online learning process by encouraging them to take an active and central role in learning.

Meanwhile, evaluation of the learning process is done for the assessment of learning outcomes of learners. In equivalence education, evaluations carried out upon each subject ends in midterm and at the end of term, which will be graded in report cards. The evaluation features that E-LEEP provides are exercises, quizzes, personal tasks, and group tasks.

### *3.4. Students Characteristics Factor*
The learning process in e-Learning is very complex and many factors influence it, including the characteristics of learners. On the other hand, teaching materials should be flexible, so learners of different characteristics can get the teaching materials that fit the characteristics tersebu. E-LEEP should be able to identify differences in the characteristics of learners, so that learning materials are flexibly adjusted to the culture, learning style, motivation, ability and other characteristics.

## 4. Conclusion
The study has shown that there are four the main driven factors the E-learning for the Equivalency Education Program (E-LEEP) in Indonesia, namely: standardization factor, learning contents, learning process, and learners' characteristics. The standardization factor consists of five main standars: contents, processes, management, education assessment, and tutor. Learning contents based on the curriculum structure of Package A, Package B, and Package C programs that implemented in E-learning (E-LEEP). The Learning Process factor in E-LEEP includes learning activities, discussion forums, and evaluations. Meanwhile learners characteristic factor is an existence of the characteristics of learners in e-Learning (E-LEEP), such as culture, learning style, motivation, ability and other characteristics. Thus, each factor will dynamically support the learners to achieve the goal learning in E-LEEP. Our future research is analyzes an existence of the main factors in E-LEEP and test it on various courses. Preliminary studies aslo will be conducted, including analysis and discussion results will be shown in next paper.

## References
[1] Siaciwena R and Lubinda F 2008 The Role of Open and Distance Learning in the Implementation of the Right to Education in Zambia *The International Review of Research in Open and Distance Learning;: Special Issue - Role of Distance Learning in the Right to Education* **9**

[2] McQuaide S 2009 Making education equitable in rural china through distance learning *International Review of Research in Open and Distance Learning*

[3] Hussain S, Wang Z and Rahim S 2013 E-learning Services for Rural Communities *International Journal of Computer Applications* **68** 15–20

[4] Kamei M 2010 ESkwela project - ESchool for out-Of-school youths and adults, Philippines *2010 International Conference on Technology for Education, T4E 2010*

[5] Llantos O E 2017 Cloudification of my.eskwela for e-Governance in Philippine Education *Procedia Computer Science* **109** 680–5

[6] Yel, Mesra Betty; Sfenrianto S 2017 E-Learning Model for Equivalency Education Program in Indonesia *2017 4th International Conference on Electrical Engineering, Computer Science and Informatics (EECSI)* pp 683–7
[7] Bastari 2016 *Indonesia Education Statistics In Brief 2015/2016*
[8] Friesen N 2005 Interoperability and learning objects: An overview of e-learning standardization *Journal of E-Learning and Learning Objects*
[9] Richter T and Pawlowski J M 2007 The need for standardization of context metadata for e-learning environments *Proceedings ot the e-ASEM Conference 2007*
[10] Jung I 2011 The dimensions of e-learning quality: From the learner's perspective *Educational Technology Research and Development* **59** 445–64
[11] Uppal M A and Ali S 2004 Factors Determining E-Learning Service Quality 2 Understanding Quality in Higher Educational Service Delivery 165–75
[12] Goi C L and Ng P Y 2009 E - learning in Malaysia : Success Factors in Implementing E - learning Program **20** 237–46
[13] Cheawjindakarn B, Suwannatthachote P and Theeraroungchaisri A 2012 Critical Success Factors for Online Distance Learning in Higher Education: A Review of the Literature *Creative Education* **3** 61–6
[14] Sun P C, Tsai R J, Finger G, Chen Y Y and Yeh D 2008 What drives a successful e-Learning? An empirical investigation of the critical factors influencing learner satisfaction *Computers and Education* **50** 1183–202
[15] Chokri B 2012 Factors Influencing the Adoption of the E- Learning Technology in Teaching and Learning By Students of a University Class *European Scientific Journal (ESJ)* **8** 165–90
[16] Harris N and Sandor M 2007 Developing online discussion forums as student centred peer e-learning environments *Ascilite 2007 - ICT: Providing Choices for Learners and Learning* 383–7
[17] Zhang L, Zhang X, Duan Y, Fu Z and Wang Y 2010 Evaluation of Learning Performance of e-Learning in china: A methodology based on change of international mental model of learners *Turkish Online Journal of Educational Technology* **9** 70–82
[18] Hannon J and D'Netto B 2007 Cultural diversity online: student engagement with learning technologies *International Journal of Educational Management* **21** 418–32
[19] Sfenrianto, Hasibuan Z A, Suhartanto H and Selviandro N 2014 An approach to detect learning types based on triple-factor in e-learning process *Journal of Next Generation Information Technology* **5** 9–15

# Bioinformatics analysis of the predicted polyprenol reductase genes in higher plants

**M Basyuni[1*], R Wati[1]**

[1]Department of Forestry, Faculty of Forestry, Universitas Sumatera Utara, Jl. Tri Dharma Ujung No. 1 Medan, North Sumatera 20155, Indonesia

m.basyuni@usu.ac.id

**Abstract**. The present study evaluates the bioinformatics methods to analyze twenty-four predicted polyprenol reductase genes from higher plants on GenBank as well as predicted the structure, composition, similarity, subcellular localization, and phylogenetic. The physicochemical properties of plant polyprenol showed diversity among the observed genes. The percentage of the secondary structure of plant polyprenol genes followed the ratio order of α helix > random coil > extended chain structure. The values of chloroplast but not signal peptide were too low, indicated that few chloroplast transit peptide in plant polyprenol reductase genes. The possibility of the potential transit peptide showed variation among the plant polyprenol reductase, suggested the importance of understanding the variety of peptide components of plant polyprenol genes. To clarify this finding, a phylogenetic tree was drawn. The phylogenetic tree shows several branches in the tree, suggested that plant polyprenol reductase genes grouped into divergent clusters in the tree.

## 1. Introduction

Higher plants are known to produce long-chain polyisoprenoids, which is classified into polyprenol and dolichol concerning the α-isoprene unit [1]. The occurrence of polyisoprenoids has been reported in various plant tissues: leaves [2-3], roots [2-5], flowers [5], fruits [6], shoots [7], seeds [5-7], and needles [8]. These studies revealed the omnipresent distribution of polyisoprenoids in the plant kingdom. Polyprenols are predominant polyisoprenoids in plant photosynthetic tissues, and few dolichols occurred [1]. On the other hand, dolichols were found abundantly in plant roots, yeast, and animal (livers) tissues [2-5,9-10].

However, recently dolichols have been shown as the primary polyisoprenoid alcohols in mangrove and coastal plants [2-4]. The presence of abundant dolichols even in the mangroves leaves implies that the enzyme of polyprenol reductase might be energetic in the mangrove. This enzyme corresponding to the SRD5A3 protein in animals has been shown to catalyze the final step of polyprenol converted to dolichols in the biosynthesis of dolichol [11].

In this circumstance, it is crucial to further understand the polyprenol reductase genes in higher plants. Two polyprenol reductase genes from *Arabidopsis thaliana* have been reported [11]. Nonetheless, bioinformatics information on the plant polyprenol reductase is scarce. The present study aimed to analyze the predicted polyprenol reductase from higher plants using the bioinformatics method.

## 2. Materials and method

### *2.1. Materials*

A total of twenty-one predicted polyprenol reductase genes, were collected, along with predicted polyprenol reductase gene from *Kandelia obovata*: *KoPPRD1* (1192 bp, encoding for 382 amino acids). The Genebank accession numbers of the DNA and amino acid sequence used in this analysis are as follows: XM_016081788 (*Arachis duranensis* polyprenol reductase 2), XM_021109944 (*A. ipaensis* polyprenol reductase 2), XM_020359734 (*Cajanus cajan* polyprenol reductase 2), XM_019849699 (*Elaeis guineensis* polyprenol reductase 1), XM_018861897 (*Eucalyptus grandis* polyprenol reductase 2), XM_003537780 (*Glycine max* polyprenol reductase 2), XM_017765134 (*Gossypium arboreum* polyprenol reductase 2-like), XM_016848239 (*G. hirsutum* polyprenol reductase 2-like), XM_012601789 (*G. raimondii* polyprenol reductase 2-like), XM_019302562 (*Ipomoea nil* polyprenol reductase 2-like), XM_012210100 (*Jatropha curcas* polyprenol reductase 2), XM_018985852 (*Juglans regia* polyprenol reductase 2-like), XM_017330714 (*Malus x domestica* polyprenol reductase 2-like), XM_017840866 (*Phoenix dactylifera* polyprenol reductase 1), XM_011038928 (*Populus euphratica* polyprenol reductase 2-like), XM_007200320 (*Prunus persica* polyprenol reductase 2), XM_018642260 (*Pyrus x bretschneideri* polyprenol reductase 2-like), XM_015715302 (*Ricinus communis* polyprenol reductase 2), XM_018128982 (*Theobroma cacao* polyprenol reductase 2), XM_014668834 (*Vigna radiata* var. *radiata* polyprenol reductase 2), and XM_016036804 (*Ziziphus jujuba* polyprenol reductase 2).

### *2.2. Physicochemical properties of the polyprenol reductase*

The physical and chemical properties of DNA and amino acid sequences of polyprenol reductase were analyzed using Protparam online analysis (web.expasy.org/protparam/). The computed parameters consist of the molecular weight, theoretical isoelectric point values, composition of amino acid, atomic composition, extinction coefficient, estimated half-life, instability index, fat coefficient, and overall average hydrophilicity as previously described [12].

### *2.3. Potential transit of peptide and subcellular localization of predicted polyprenol reductase*

The targetP 1.1 server online (www.cbs.dtu.dk/services/targetp) was used to determine the prediction of transit peptide. The location assignment is according to the predicted presence of any of the *N*-terminal pre-sequences chloroplast transit peptide (cTP), mitochondrial targeting peptide (mTP) and secretory pathway signal peptide (SP). The subcellular localization of plant polyprenol reductase was predicted using PSORT Prediction online (psort.hgc.jp/form.html) as previously reported [13].

### *2.4. Phylogenetic analysis of predicted polyprenol reductase*

The amino acid sequences were aligned, and similarity scores were obtained using the FASTA ver. 3.4t26 [14] of the DNA Data Bank of Japan (Mishima, Shizuoka, Japan). Phylogenetic analysis of deduced amino acid alignment from polyprenol reductase genes was conducted with CLUSTAL W ver. 1.83 [15] of the DNA Data Bank of Japan followed by drawing with TreeView ver. 1.6.6 [16] based on a neighbour-joining method. Bootstrap analysis with 1000 replications was used to increase the strength of the nodes in the tree [17]. The GenBank accession numbers of the DNA and amino acid sequence of using this analysis is described in the Materials subsection.

## 3. Results and Discussions

### *3.1. Physical and chemical properties and secondary structure of the polyprenol reductase gene*

Table 1 shows the initiation codon of plant polyprenol reductase genes was ATG, where the stop codons consisted of TGA and TAG. The open reading frame length was varied among the genes observed. A number of encoded amino acids were 101 to 357. It is noteworthy to note the diversity of relative molecular mass, theoretical isoelectric point value, positively or negatively charged residues, extinction coefficient, Fat coefficient, and overall average hydrophilicity among the evaluated genes. *M. x domestica* was distinguishable from other plant polyprenol reductase genes in the total number of atoms, as there is at least one ambiguous position (B, Z or X) in the sequence considered, the atomic composition, therefore, cannot be computed.

The estimated half-life period was accurately same among the genes and was much longer that triterpenoid and cycloartenol synthase genes, the members of oxidoqualene cyclase (OSC) genes [13] and mangrove actin genes [19]. Based on stability coefficients, only three plant polyprenol reductase genes namely *G. max, G. arboretum,* and *G. raimondii* were stable proteins, the mostly remaining genes were non-stable proteins. These results suggested the importance of understanding the diversity and function of physicochemical properties of the different amino acids in plant polyprenol reductase genes [19].

**Table 1**. Physical and chemical properties of the plant polyprenol reductase in the database

| Plant species | *A. duranensis* | *A. ipaensis* | *C. cajan* | *E. grandis* | *E.guineensis* |
|---|---|---|---|---|---|
| Length of genes/bp | 2031 | 1556 | 1411 | 1320 | 1229 |
| Open reading frame length/bp | 1915 | 1415 | 1165 | 1057 | 909 |
| Start site and codon | 902ATG | 402ATG | 122ATG | 41ATG | 239ATG |
| Stop site and codon | 1915TGA | 1415TGA | 1165TGA | 1057TGA | 909TGA |
| Number of encoded amino acids | 337 | 337 | 347 | 338 | 223 |
| Relative molecular mass | 38916.89 | 38998.16 | 40210.04 | 39149.63 | 24760.07 |
| Theoretical isoelectric point values | 9.11 | 9.34 | 9.35 | 8.58 | 8.23 |
| Positively charged residues | 24 | 26 | 26 | 26 | 16 |
| Negatively charged residues | 18 | 17 | 18 | 23 | 14 |
| Total number of atoms | 5527 | 5550 | 5692 | 5509 | 3512 |
| Extinction coefficient | 2.456 | 2.310 | 2.448 | 2.480 | 2.334 |
| Half-life period | 30h | 30h | 30h | 30h | 30h |
| Instability coefficient | 48.48 | 48.13 | 42.65 | 51.95 | 57.23 |
| Fat coefficient | 109.35 | 110.21 | 105.07 | 99.79 | 112.06 |
| Overall average hydrophilicity | 0.487 | 0.487 | 0.349 | 0.326 | 0.537 |

Table continued

| | *G. arboreum* | *G. hirsutum* | *G. max* | *G. raimondii* | *I. nil* |
|---|---|---|---|---|---|
| Length of genes/bp | 1348 | 1310 | 1682 | 1438 | 2755 |
| Open reading frame length/bp | 1151 | 1036 | 1441 | 1229 | 2143 |
| Start site and codon | 81ATG | 128ATG | 413ATG | 180ATG | 1244ATG |
| Stop site and codon | 1151TGA | 1036TGA | 1441TGA | 1229TGA | 21436TAG |
| Number of encoded amino acids | 357 | 302 | 342 | 349 | 299 |
| Relative molecular mass | 40749.35 | 34808.15 | 39109.94 | 39842.36 | 34118.14 |
| Theoretical isoelectric point values | 7.25 | 9.45 | 9.23 | 8.17 | 9.30 |
| Positively charged residues | 24 | 27 | 24 | 24 | 28 |
| Negatively charged residues | 24 | 17 | 17 | 22 | 17 |
| Total number of atoms | 5737 | 4953 | 5550 | 5611 | 4830 |
| Extinction coefficient | 2.520 | 2.819 | 2.227 | 2.578 | 2.063 |
| Half-life period | 30h | 30h | 30h | 30h | 30h |
| Instability coefficient | 37.89 | 55.65 | 36.13 | 38.36 | 46.07 |
| Fat coefficient | 100.59 | 107.55 | 106.81 | 98.97 | 102.11 |
| Overall average hydrophilicity | 0.432 | 0.428 | 0.488 | 0.437 | 0.381 |

Table continued

| | *J. curcas* | *J. regia* | *Ko PPRD1* | *M.domestica* | *P. dactylifera* |
|---|---|---|---|---|---|
| Length of genes/bp | 1333 | 1727 | 1193 | 1464 | 928 |
| Open reading frame length/bp | 1175 | 1369 | nd | 1306 | 633 |
| Start site and codon | 261ATG | 353ATG | nd | 290ATG | 145ATG |
| Stop site and codon | 1175TGA | 1369TGA | nd | 1306TGA | 633TGA |
| Number of encoded amino acids | 304 | 338 | 357 | 338 | 162 |
| Relative molecular mass | 35067.13 | 38978.60 | 41140.93 | 38483.31 | 18250.43 |
| Theoretical isoelectric point values | 8.61 | 8.88 | 8.44 | 9.31 | 9.30 |
| Positively charged residues | 22 | 26 | 30 | 27 | 13 |
| Negatively charged residues | 19 | 21 | 25 | 18 | 9 |
| Total number of atoms | 4969 | 5507 | 5739 | - | 2600 |
| Extinction coefficient | 2.569 | 2.952 | 2.350 | 2.523 | 2.688 |
| Half-life period | 30h | 30h | 1.3h | 30h | 30h |
| Instability coefficient | 40.65 | 57.59 | 49.02 | 47.79 | 53.92 |
| Fat coefficient | 111.64 | 105.30 | 90.95 | 107.57 | 116.91 |
| Overall average hydrophilicity | 0.481 | 0.400 | 0.185 | 0.441 | 0.530 |

Table continued

| | *P. euphratica* | *P. persica* | *Pyrus x* | *R. communis* | *T. cacao* |
|---|---|---|---|---|---|
| Length of genes/bp | 1335 | 1569 | 472 | 1288 | 1317 |
| Open reading frame length/bp | 1117 | 1340 | 321 | 1133 | 1083 |
| Start site and codon | 197ATG | 339ATG | 16ATG | 258ATG | 58ATG |
| Stop site and codon | 1117TGA | 1340TGA | 321TGA | 1133TGA | 1083TGA |
| Number of encoded amino acids | 306 | 333 | 101 | 291 | 341 |
| Relative molecular mass | 35383.78 | 38143.66 | 11555.41 | 33656.41 | 38280.72 |
| Theoretical isoelectric point values | 9.33 | 9.41 | 5.88 | 9.33 | 8.09 |
| Positively charged residues | 26 | 27 | 7 | 26 | 23 |
| Negatively charged residues | 16 | 17 | 10 | 18 | 21 |
| Total number of atoms | 5009 | 5409 | 1639 | 4768 | 5395 |
| Extinction coefficient | 2.498 | 2.401 | 2.202 | 2.381 | 2.465 |
| Half-life period | 30h | 30h | 30h | 30h | 30h |
| Instability coefficient | 49.75 | 42.98 | 47.89 | 47.90 | 41.75 |
| Fat coefficient | 105.13 | 103.00 | 110.99 | 102.51 | 102.90 |
| Overall average hydrophilicity | 0.549 | 0.401 | 0.319 | 0.390 | 0.481 |

Table continued

| | *V. radiata* | *Z. jujuba* |
|---|---|---|
| Length of genes/bp | 1548 | 1724 |
| Open reading frame length/bp | 1309 | 1510 |
| Start site and codon | 287ATG | 494ATG |
| Stop site and codon | 1309TGA | 1510TGA |
| Number of encoded amino acids | 340 | 338 |
| Relative molecular mass | 39015.67 | 39208.84 |
| Theoretical isoelectric point values | 9.03 | 9.31 |
| Positively charged residues | 25 | 28 |
| Negatively charged residues | 20 | 18 |
| Total number of atoms | 5544 | 5539 |
| Extinction coefficient | 2.268 | 2.730 |
| Half-life period | 30h | 30h |
| Instability coefficient | 43.61 | 49.24 |
| Fat coefficient | 108.97 | 99.79 |
| Overall average hydrophilicity | 0.465 | 0.330 |

Table 2 depicts the percentage of the secondary structure of plant polyprenol genes followed the proportion order of α helix > random coil > extended chain structure. This similar pattern has been previously reported in the oxidosqualene cyclase genes [13] and mangrove actin genes [18]. The variation was notable among the genes observed, implied that the important function of secondary structure analysis of plant polyprenol reductase in the plant kingdom. Polyprenol may play a key role in early steps of protein N-linked glycosylation, the enzyme was required for the alteration of polyprenol into dolichol [10].

**Table 2**. Secondary structure analysis of plant polyprenol reductase

| Protein accession number | α helix number | α helix ratio (%) | Extended chain structure number | Extended chain structure ratio (%) | Random coil number | Random coil ratio (%) |
|---|---|---|---|---|---|---|
| XP_003537828 | 159 | 46.49 | 63 | 18.42 | 120 | 35.09 |
| XP_007200382 | 159 | 47.75 | 63 | 18.92 | 111 | 33.33 |
| XP_011037230 | 167 | 54.58 | 52 | 16.99 | 87 | 28.43 |
| XP_012065490 | 181 | 59.54 | 31 | 10.20 | 92 | 30.26 |
| XP_012457243 | 190 | 54.44 | 45 | 12.89 | 114 | 32.66 |
| XP_014524320 | 173 | 50.88 | 59 | 17.35 | 108 | 31.76 |
| XP_015570788 | 154 | 52.92 | 43 | 14.78 | 94 | 32.30 |
| XP_015892290 | 186 | 55.03 | 49 | 14.50 | 103 | 30.47 |
| XP_015937274 | 177 | 52.52 | 59 | 17.51 | 101 | 29.97 |
| XP_016703728 | 173 | 57.28 | 31 | 10.26 | 98 | 32.45 |
| XP_017186203 | 175 | 51.78 | 50 | 14.79 | 113 | 33.43 |
| XP_017620623 | 190 | 53.37 | 49 | 13.76 | 117 | 32.87 |
| XP_017696355 | 99 | 61.11 | 12 | 7.41 | 51 | 31.48 |
| XP_017984471 | 147 | 43.11 | 67 | 19.65 | 127 | 37.24 |
| XP_018497776 | 30 | 29.70 | 32 | 31.68 | 39 | 38.61 |
| XP_018717442 | 175 | 51.78 | 48 | 14.20 | 115 | 34.02 |
| XP_018841397 | 175 | 51.78 | 52 | 15.38 | 111 | 32.84 |
| XP_019158107 | 163 | 54.52 | 42 | 14.05 | 94 | 31.44 |
| XP_019705258 | 122 | 54.71 | 21 | 9.42 | 80 | 35.87 |
| XP_020215323 | 181 | 52.16 | 58 | 16.71 | 108 | 31.12 |
| XP_020965603 | 180 | 53.41 | 53 | 15.73 | 104 | 30.86 |
| KoPPRD1 | 157 | 43.98 | 51 | 14.29 | 149 | 41.74 |

*3.2. Potential transit of peptide and subcellular localization of OSC gene*

Table 3 shows the possibility of the possible transit peptide in plant polyprenol reductase genes. There are four reliabilities analyzed: chloroplast transit peptide, mitochondrial target peptide, the signal peptide of the secretory pathway, and the prediction probability. The values of chloroplast but not signal peptide were too low, indicated that few chloroplast transit peptide or relatively high signal peptide of secretion pathway in twenty-four plant polyprenol reductase genes. It is interesting to note that target peptide value of mitochondria diversed from 0.042 to 0.540, indicating that it likely existed. The highest mitochondrial target peptide was *KoPPRD 1,* a candidate of *K. obovata* polyprenol reductase gene. This result supported the previous report on the highest mitochondrial target peptide among the OSC genes from mangrove plants, was found in *KcCAS, K. candel* cycloartenol synthase gene with the product of 100% cycloartenol [13, 20].

**Table 3**. Possibility of the potential transit peptide in polyprenol reductase genes

| | Reliability | | | |
|---|---|---|---|---|
| Protein accession number | Chloroplast transit peptide | Mitochondrial target peptide | Signal peptide of secretory pathway | Reliability prediction |
| XP_003537828 | 0.011 | 0.056 | 0.399 | 4 |
| XP_007200382 | 0.005 | 0.376 | 0.364 | 5 |
| XP_011037230 | 0.025 | 0.094 | 0.242 | 5 |
| XP_012065490 | 0.037 | 0.308 | 0.024 | 5 |
| XP_012457243 | 0.027 | 0.075 | 0.197 | 5 |
| XP_014524320 | 0.036 | 0.042 | 0.286 | 5 |
| XP_015570788 | 0.004 | 0.181 | 0.371 | 5 |
| XP_015892290 | 0.006 | 0.264 | 0.296 | 5 |
| XP_015937274 | 0.021 | 0.059 | 0.098 | 4 |
| XP_016703728 | 0.002 | 0.217 | 0.313 | 5 |
| XP_017186203 | 0.004 | 0.225 | 0.380 | 5 |
| XP_017620623 | 0.015 | 0.025 | 0.184 | 3 |
| XP_017696355 | 0.024 | 0.276 | 0.806 | 3 |
| XP_017984471 | 0.072 | 0.069 | 0.287 | 5 |
| XP_018497776 | 0.014 | 0.100 | 0.129 | 1 |
| XP_018717442 | 0.006 | 0.477 | 0.118 | 4 |
| XP_018841397 | 0.023 | 0.422 | 0.084 | 4 |
| XP_019158107 | 0.002 | 0.105 | 0.687 | 3 |
| XP_019705258 | 0.026 | 0.393 | 0.308 | 5 |
| XP_020215323 | 0.037 | 0.435 | 0.025 | 4 |
| XP_020965603 | 0.013 | 0.084 | 0.128 | 4 |
| KoPPRD1 | 0.000 | 0.540 | 0.922 | 4 |

Table 4 shows subcellular localization of plant polyprenol reductase genes in higher plants. The subcellular localization of these genes mostly resided in Golgi bodies, plasma membrane, and endoplasmic reticulum. A few genes were chloroplast thylakoid membrane and microbody. Recently, it has been reported that the expression of two triterpenoid synthases, *BgbAS* and *RsM1* under a *GAL1* promoter in GIL77 enhanced the triterpenoid content of plasma membrane portions [13]. In the present data (Table 4), both gene *BgbAS* and *RsM1* were located at the plasma membrane, supported previous results on their subcellular localization situated in the plasma membrane [13].

### *3.3. Phylogenetic analysis of polyprenol reductase gene*

To confirm the similarity among the plant polyprenol reductase gene, a phylogenetic tree was drawn on the source of their amino acid sequences (Figure 1). The phylogenetic tree shows that there are five clusters in the tree. The first branch was *Z. juja;* the second cluster consisted *G. hirsutum*. The third branch comprised five members of polyprenol reductase genes. The fourth cluster was the largest members consisting of 13 genes. This cluster was the center of the phylogenetic tree in term of evolution.

A new member of *KoPPRD1 K. obovata* polyprenol reductase joined in the center branch. This gene joined with other plant polyprenol reductase in higher plants: *R. communis, I. nil, P. euphratica,* and *E. grandis*. Our present result, therefore, suggested that *KoPPRD1*gene is a new member of plant polyprenol reductase.

The close relationship between *KoPPRD1*gene with plant as mentioned earlier polyprenol reductase can be explained by their habitat in the tropical forests [2,3]. The apparent similarity and joining these five genes in the tree may due to the result of tropical or subtropical climatic conditions [2,21]. Mangrove forests are distributed in the tropical and sub-tropical area as well as *R. communis, I. nil, P. euphratica,* and *E. grandis.*species [21].

IOP Conf. Series: Journal of Physics: Conf. Series **978** (2018) 012050 doi:10.1088/1742-6596/978/1/012050

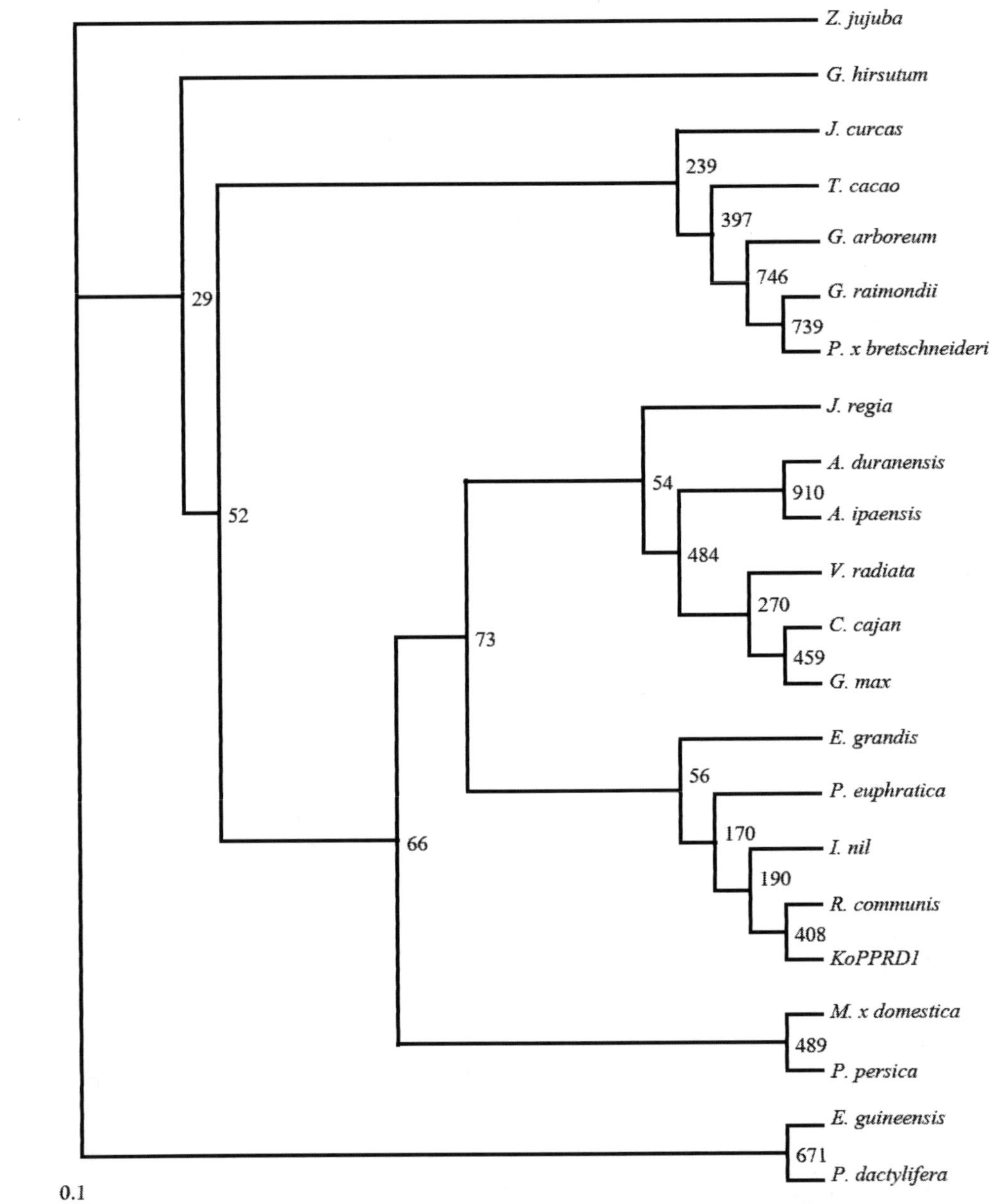


**Figure 1.** Phylogenetic tree of plant polyprenol reductase. Phylogenetic tree of deduced amino acid sequences was constructed with the neighbour-joining method of the CLUSTAL W [15]. The indicated scale represents 0.1 amino acid substitutions per site. Numbers point out bootstrap value from 1000 replicates. The GenBank accession numbers of the amino acid sequence of using this analysis are mentioned in the Materials subsection

.

**Table 4**. Subcellular localization of plant polyprenol reductase

| | Degree of certainty | | | | | |
|---|---|---|---|---|---|---|
| Protein accession number | Chloroplast thylakoid membrane | Golgi bodies | Plasma membrane | Endoplasmic reticulum | Microbody | Mitochondrial inner membrane |
| XP_003537828 | 0.426 | 0.400 | 0.600 | 0.300 | nd | nd |
| XP_007200382 | nd | 0.460 | 0.640 | 0.685 | nd | nd |
| XP_011037230 | nd | 0.460 | 0.640 | 0.685 | nd | 0.822 |
| XP_012065490 | nd | 0.460 | 0.640 | 0.685 | nd | nd |
| XP_012457243 | nd | 0.400 | 0.600 | nd | nd | 0.340 |
| XP_014524320 | 0.375 | 0.400 | 0.600 | nd | nd | nd |
| XP_015570788 | nd | 0.460 | 0.640 | 0.685 | nd | nd |
| XP_015892290 | nd | nd | 0.730 | 0.640 | 0.388 | nd |
| XP_015937274 | 0.438 | 0.400 | 0.600 | 0.300 | nd | nd |
| XP_016703728 | nd | 0.460 | 0.640 | 0.685 | nd | nd |
| XP_017186203 | nd | 0.460 | 0.640 | 0.685 | nd | nd |
| XP_017620623 | nd | 0.400 | 0.600 | 0.300 | 0.300 | nd |
| XP_017696355 | nd | 0.460 | 0.640 | 0.685 | nd | nd |
| XP_017984471 | 0.354 | 0.400 | 0.600 | 0.300 | nd | nd |
| XP_018497776 | 0.249 | nd | 0.700 | 0.200 | 0.640 | nd |
| XP_018717442 | nd | 0.460 | 0.640 | 0.685 | nd | nd |
| XP_018841397 | nd | nd | 0.460 | 0.100 | 0.388 | nd |
| XP_019158107 | nd | 0.460 | 0.640 | 0.685 | nd | nd |
| XP_019705258 | nd | 0.460 | 0.640 | 0.685 | nd | nd |
| XP_020215323 | 0.426 | 0.400 | 0.600 | 0.300 | nd | nd |
| XP_020965603 | 0.438 | 0.400 | 0.600 | 0.300 | nd | nd |
| KoPPRD1 | nd | 0.460 | 0.640 | 0.685 | nd | nd |

nd= not detected

## 4. Conclusions

The present study confirmed the diversity and function of physicochemical properties of the plant polyprenol reductase genes. Furthermore, our data supported previous results on the subcellular localization of triterpenoid synthase genes located in the plasma membrane.

## Acknowledgment

A part of this study was supported by an International Research Collaboration and Scientific Publication Grant 2017 (No. 003/SP2H/LT/DRPM/IV/2017 to MB) from the Directorate for Research and Community Service, Ministry of Research, Technology and Higher Education, Republic of Indonesia.

## References

[1] Surmacz L and Swiezewska E 2011 Polyisoprenoids–Secondary metabolites or physiologically important superlipids? *Biochem. Biophys. Res. Commun.* **407** 627–632.

[2] Basyuni M, Sagami H, Baba S, Iwasaki H and Oku H 2016 Diversity of polyisoprenoids in ten Okinawan mangroves *Dendrobiology* **75** 167–175.

[3] Basyuni M, Sagami H, Baba S and Oku H 2017 Distribution, occurrence, and cluster analysis of new polyprenyl acetones and other polyisoprenoids from North Sumatran mangroves *Dendrobiology* **78** 18–31.

[4] Basyuni M, Wati R, Sagami H, Sumardi, Baba S and Oku H 2018 Diversity and abundance of polyisoprenoid composition in coastal plant species from North Sumatra, Indonesia *Biodiversitas* **19** 1–11.

[5] Tateyama S, Wititsuwannakul R, Wititsuwannakul D, Sagami H and Ogura K 1999 Dolichols of rubber plant, ginkgo and pine *Phytochemistry* **51** 11–15.

[6] Arifiyanto D, Basyuni M, Sumardi, Putri LAP, Siregar ES, Risnasari I and Syahputra I 2017 Occurrence and cluster analysis of palm oil (*Elaeis guineensis*) fruit type using two-dimensional thin layer chromatography *Biodiversitas* **18** 1487–1492.

[7] Kurisaki A, Sagami H and Ogura K 1997 Distribution of polyprenols and dolichols in soybean plant *Phytochemistry* **44** 45–50.

[8] Yu J, Wang Y, Qian H, Zhao Y, Liu B and C Fu 2012 Polyprenols from the needles of *Taxus chinensis* var. *mairei Fitoterapia* **83** 831–837.

[9] Sagami H, Kurisaki O, Oguro K and Chojnacki T 1992. Separation of dolichol from dehydrodolichol by a simple two-plate thin layer chromatography. *J. Lipid Res.* **33**: 1857–1862.

[10] Grabinska K and Palamarczyk G 2002 Dolichol biosynthesis in the yeast *Saccharomyces cerevisiae*:an insight into the regulatory role of farnesyl diphosphate synthase *FEMS Yeast Res.* **2** 259–265.

[11] Jozwiak A, Gutkowska M, Gawarecka K, Surmacz L, Buckzkowska A, Lichocka M, Nowakowska J and Swiezewska E 2015 Polyprenol reductase2 deficiency is lethal in *Arabidopsis* due to male sterility *Plant Cell* **27**: 3336–3353.

[12] Basyuni M and Sumardi 2017 Bioinformatics approach of salt tolerance gene in mangrove plant *Rhizophora stylosa J. Phys. : Conf. Ser.* **801** 012012.

[13] Basyuni M and R Wati 2017 Bioinformatics analysis of the oxidosqualene gene and the amino acid sequence in mangrove plants *J. Phys. : Conf. Ser.* **801** 012011.

[14] Pearson WR and Lipman DJ 1988 Improved tools for biological sequence comparison *Proc. Natl. Acad. Sci. USA* **85** 2444–2448.

[15] Thompson JD, Higgins DG and Gibson TJ 1994 CLUSTAL W: Improving the sensitivity of progressive multiple sequence alignment through sequence weighting, position-specific gap penalties, and weight matrix choice *Nucleic Acid Res.* **22** 4673–4680.

[16] Page RD 1996 TreeView: an application to display phylogenetic trees on personal computers *Comput. Appl. Biosci.* **12** 357–358.

[17] Felsenstein J 1985 Confidence limits on phylogenies: an approach using the bootstrap *Evolution* **39** 783–791.

[18] Basyuni M, Wasilah M and Sumardi 2017 Bioinformatics study of the mangrove actin genes *J. Phys. : Conf. Ser.* **801** 012013.

[19] Swiezewska E and Danikiewicz W 2005. Polyisoprenoids: Structure, biosynthesis, and function. *Prog. Lipid Res.* **44** 235–258.

[20] Basyuni M, Oku H, Tsujimoto E and Baba S 2007 Cloning and functional expression of cycloartenol synthases from mangrove species *Rhizophora stylosa* Griff. and *Kandelia candel* (L.) Druce *Biosci. Biotechnol. Biochem.* **71** 1788–1792.

[21] Basyuni M, Baba S, Kinjo Y, Putri LAP, Hakim L and Oku H 2012 Salt-dependent increase in triterpenoids is reversible upon transfer to fresh water in mangrove plants Kandelia candel and Bruguiera gymnorrhiza *J. Plant Physiol.* **169** 1903–1908.

IOP Conf. Series: Journal of Physics: Conf. Series **978** (2018) 012051 doi:10.1088/1742-6596/978/1/012051

# The recognition of female voice based on voice registers in singing techniques in real-time using hankel transform method and macdonald function

**R Meiyanti[1], A Subandi[1], N Fuqara[2], M A Budiman[1] and A P U Siahaan[3,4]**

[1] Universitas Sumatera Utara, Medan, Indonesia
[2] Universitas Malikussaleh, Aceh Utara, Indonesia
[3] Universitas Pembangunan Panca Budi, Medan, Indonesia
[4] School of Communication and Computer Engineering, Universiti Malaysia Perlis

rinimeiyanti522@gmail.com , aminsubandi@yahoo.com , aransq92@gmail.com , mandrib@usu.ac.id and andiesiahaan@gmail.com

**Abstract**. A singer doesn't just recite the lyrics of a song, but also with the use of particular sound techniques to make it more beautiful. In the singing technique, more female have a diverse sound registers than male. There are so many registers of the human voice, but the voice registers used while singing, among others, Chest Voice, Head Voice, Falsetto, and Vocal fry. Research of speech recognition based on the female's voice registers in singing technique is built using Borland Delphi 7.0. Speech recognition process performed by the input recorded voice samples and also in real time. Voice input will result in weight energy values based on calculations using Hankel Transformation method and Macdonald Functions. The results showed that the accuracy of the system depends on the accuracy of sound engineering that trained and tested, and obtained an average percentage of the successful introduction of the voice registers record reached 48.75 percent, while the average percentage of the successful introduction of the voice registers in real time to reach 57 percent.

## 1. Introduction

Voice recognition technique consists of two main stages, training and recognition. In the training stage, an identified voice is inserted into a database. In the recognition stage, an undefined voice signal will be identified. Voice recognition method is used for both verifications as well as identification. [1] The voice register is the division of the human voice area by sound source, the sensation of resonance space, shape, color, sound timbre, and the resulting low tone. Along with audio signal separation, audio classification is important due to following reasons:

- Different audio types should be processed differently;
- The searching space after classification is reduced into a particular subclass during the retrieval process. [2]

Maybe we are often confused and wondering about terms in the distribution of voice registers like falsetto, head voice, chest voice, vocal fry and some other terms. But simply we can tell the difference. In this system, the author uses the Hankel Transformation which aims to recognize the input sound as a sample and can normalize sound waves. In addition, the author also uses Macdonald's function in

IOP Conf. Series: Journal of Physics: Conf. Series **978** (2018) 012051 doi:10.1088/1742-6596/978/1/012051

this voice recognition system as a combination to produce better and more accurate system performance in voice recognition process.

## 2. Theories

### *2.1. Voice Classification*

There are various approaches that can be used such as Artificial Neural Network (JST), DCT, Gabor Filter, Statistics, Wavelet, and others, to classify the sound. Various information that can be obtained from speech sounds coming from the mouth of humans, among others: Speech Identity, Expression, Dialect/Slog/Tribe, Gender Type, Distance of sound sources, Voice rate, Age, Word Recognition, Loud level, Saturation rate, Health condition and language quality.

### *2.2. WAV*

WAV is Microsoft's standard audio format and IBM for personal computer (PC), usually using PCM coding (Pulse Code Modulation). WAV is uncompressed data so that all audio samples are stored all on the hard disk. Software that can create WAV from analog sound, for example, is Windows Sound Recorder.

### *2.3. Hankel Transform*

The Hankel Transform is an integral transform developed by Hermann Hankel. Further computational investigation reveals a number of instances in which one member of a pair of sequences with the same Hankel transform is the Binomial or Invert transform of the other. [3] General Equation of Hankel Transform:

$$F_v(k) = \sum_{r=0}^{N-1} f(r) \times r \times K_v(kr) \tag{1}$$

For $r = 0, 1, 2, 3, \ldots, N\text{-}1$

### *2.4. Macdonald Function*

Certain problems of mathematical physics arising in spheroidal or cylindrical domains, e.g. Laplace's equation, have solutions that involve Macdonald's functions. [4] The Macdonald function is a modification of the Bessel function and the Cylinder function, which is made by H. M. Macdonald. The general equation of the Macdonald functions are:

$$K_v(kr) = \frac{\pi}{2} \frac{I_{-v}(kr) - I_v(kr)}{\sin v\,\pi} \tag{2}$$

Where $v$ is any non-integral real number

$$I_v(kr) = \sum_{m=0}^{\infty} \frac{(\frac{kr}{2})^{v+2m}}{m!\ \Gamma(v+m+1)} \tag{3}$$

$$I_{-v}(kr) = \sum_{m=0}^{\infty} \frac{(\frac{kr}{2})^{-v+2m}}{m!\ \Gamma(-v+m+1)} \tag{4}$$

*2.5. Gamma Function*

In mathematics, the gamma function (presented by the Greek capital letter Γ) is an extension or extension of the factorial function, with its argument shifted down by 1, to real and complex numbers. That is, if n is a positive integer, then:

$$\Gamma(n) = (n-1)! \tag{5}$$

Gamma function is defined for all complex numbers, except for negative integers and zeros. For complex numbers whose real part is positive, the gamma function is defined by an incorrectly converging integral.

In the previous research, the calculations using Hankel transform and Macdonald function can recognize the types of sounds such as soprano, alto, tenor and bass with different accuracy for both records sound and the type of voice in real time, where the average for recorded sounds reaches accuracy up to 66.87 %, while for real-time voice has accuracy up to 72%. [5]

## 3. The Scheme of System

The scheme of this study is divided into several stages, which are described in the following:

*3.1. Overall Research Scheme for Macdonald Function and Hankel Transformation*

The scheme of the overall method developed in this study is illustrated in Figure 1:

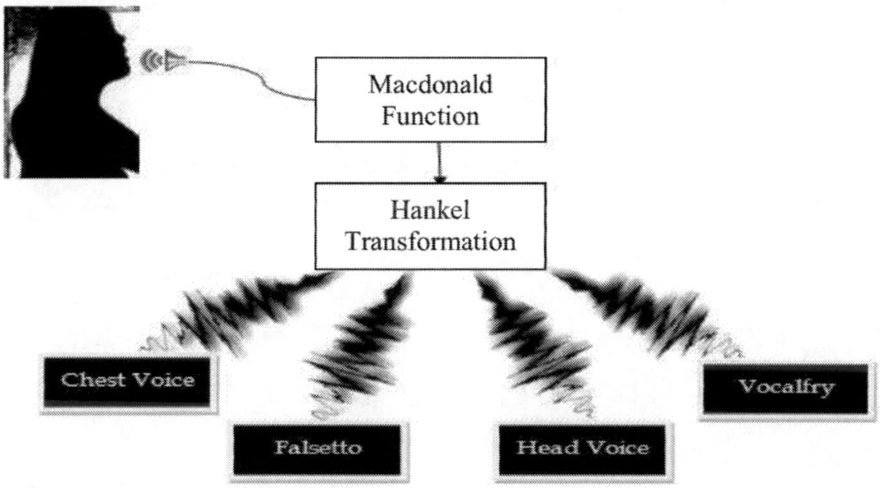


**Figure 1.** Overall system scheme.

*3.2. Voice coaching scheme using Macdonald function and Hankel's transform*

Flowchart for voice coaching processes are developed based on Figure 2:

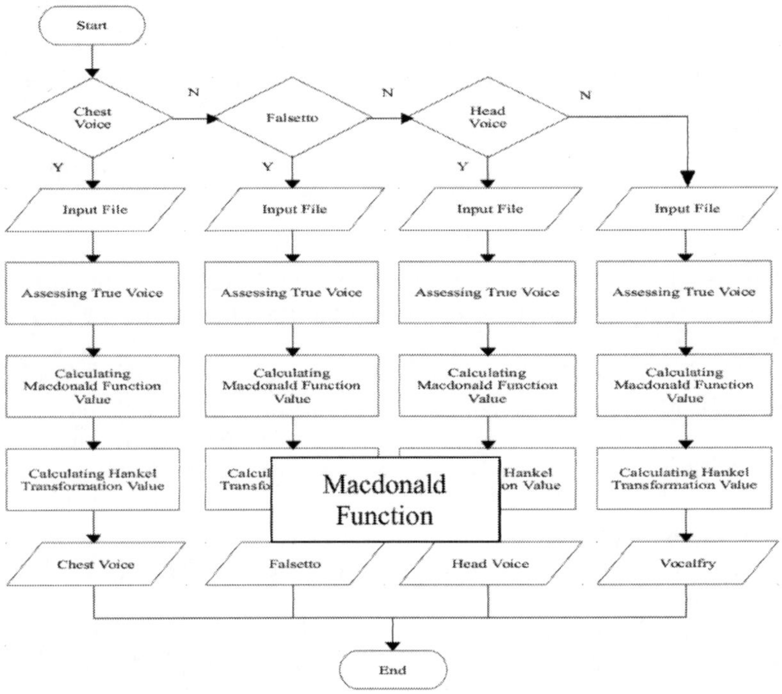


**Figure 2.** Voice coaching scheme using Macdonald function and Hankel's transform.

*3.3. Voice Testing scheme using Macdonald function and Hankel's transform*

Flowchart for voice testing processes are developed based on Figure 3:

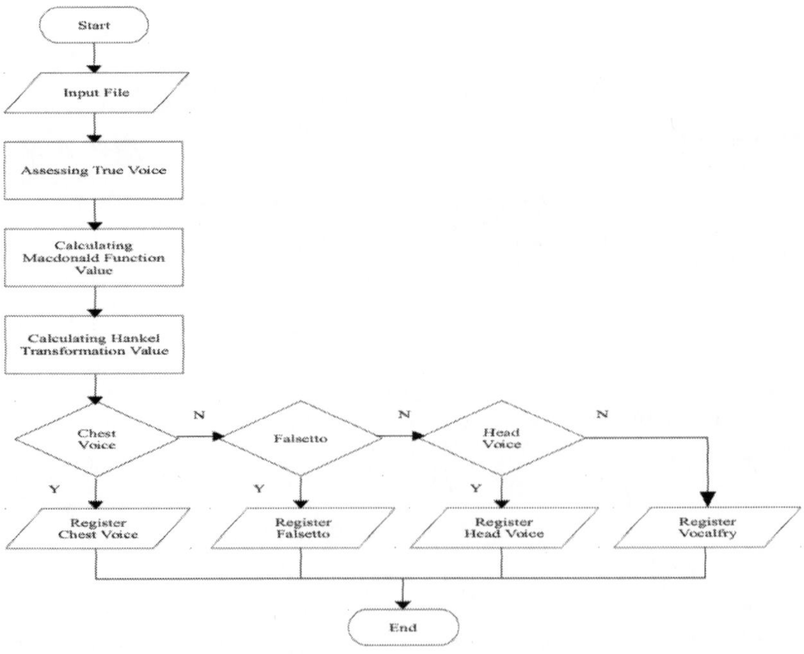


**Figure 3.** Voice testing scheme using Macdonald function and Hankel's transform.

## 4. User Interface

*4.1. View of The Main Form*
Before the operation, the program will display a page like this:

**Figure 4.** View of the main form.

*4.2. The Test Result*
This form is intended for testing of Head Voice register in real-time, here is a view of test results:

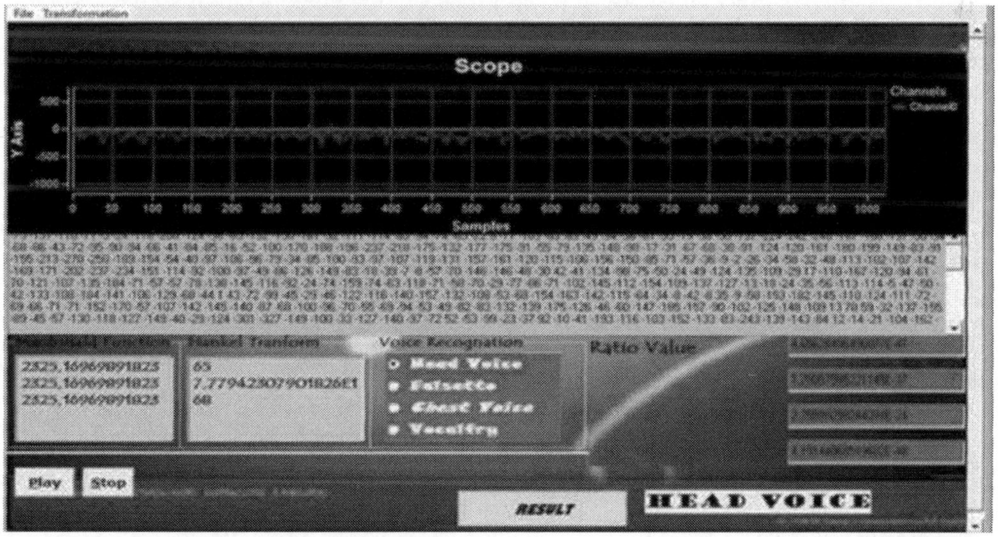


**Figure 5.** Sample testing process of head voice registers in real-time.

The picture shows that when the user inputs one of the voice register samples, the program will make an introduction by calculating the energy value in the sound sample and then get the end result of the introduction process, namely the voice register type.

IOP Conf. Series: Journal of Physics: Conf. Series **978** (2018) 012051 doi:10.1088/1742-6596/978/1/012051

## 5. Conclusions

The conclusions of this study are as follows:

- This application has a weakness that is only able to recognize the sound based on the energy value ratio between the training process and the testing process. The program can not directly detect the origin of the sound source being pronounced because the real voice register is not only seen from the high tone generated but also seen from the sound source, the sensation of resonance space and the color of the sound;
- Based on system performance, voice recognition process in real-time is more effective than voice recognition process by a record;
- The method applied in this application achieves the average percentage of the successful voice record recognition record reached 48.75%, while the average percentage of the successful introduction of voice registers in real time reached 57%.

## References

[1] Ayad R R 2013 Security of Iris Recognition and Voice Recognition Techniques *Ibn AL-Haitham Jour. for Pure & Appl. Sci.* **26** 326-57.

[2] Kumar, Choudhary, Tembhare, Pote 2014 Audio Signal Separation and Classification: A Review Paper *International Journal of Innovative Research in Computer and Communication Engineering* **2** Issue 11.

[3] John W L 2001 The Hankel Transform and Some of its Properties *Journal of Integer Sequences* **4** 0115.

[4] Passian A, Simpson H, Kouchekian S, and Yakubovich S B 2009 On the orthogonality of the MacDonald's functions. *Journal of Mathematical Analysis and Applications*, 360(2), 380–390.

[5] Fuadi W, Ula M., and Sadli M 2015 The Introduction Types of Vocal Sound in Choir in Realtime Using Hankel Transformation and Macdonald Function, 6(January), 1–14.

[6] Eric S W 2015 Applications of Incomplete Gamma Functions to the Incomplete Normal Distribution *Open Journal of Applied & Theoretical Mathematics (OJATM)* **1** No. 1 pp. 01-06.

# Mobile technology for expansion of service range medan public library

**A R Siregar and H Dewiyana**

Department of Library and Information Science, Faculty of Humanities, Universitas Sumatera Utara, Jl. Perpustakaan No. 3 Kampus USU, Medan, Indonesia

aridwansiregar@gmail.com; himmadewiyana.usu@gmail.com

**Abstract.** The information society, especially library users, expects an easier and faster way to get information and share knowledge. The library also implements the best way to provide the best service and the easiest way to disseminate and share information with its users. One technology that allows libraries more closely with their users is mobile technology. This paper aims to determine the basic conditions of mobile technology services and mobile services most needed library users, as well as constraints faced by libraries and users in the application of mobile technology. This paper will help libraries to develop better M-Library services in expanding the reach of library services. Methods of data collection were conducted through surveys through questionnaires, interviews, and direct observation. The results of the use of mobile technology applications in libraries bring great advantages to provide better library services and to improve accessibility of library resources in the easiest way, but due to lack of funds, lack of skills and lack of staff, many libraries are unable to provide this service with very successful.

## 1. Introduction

The information culture has become the lifestyle of today's society. Every aspect of life makes use of and uses information. One technology that provides the easiest and fastest way to communicate and share information and reduce the communication gap between two individuals and two communities is mobile technology. Mobile technology offers the ease and speed in utilizing information especially for quick reference and conducting extensive research. Adopting mobile technology in the library will assist in performing better service by communicating with users openly and sharing information and knowledge in an easy way. Mobile technology becomes a means for libraries to communicate better, service and share information and knowledge with library users anytime and anywhere. In this paper we put forward and develop a mobile library application for smart phones [1][2][3].

To know the basic service condition of cellular technology, to ensure the most common mobile service offerings made by libraries to their users, to find out which mobile services most users need, the benefits derived from the use of mobile technology, to find out the problems facing the library to provide m-library services, and to know which mobile services should be added for the development of better M-Library services in expanding the reach of library services?

## 2. Literature Review

The application of mobile technology in the library has been thoroughly studied and studied [1]. Liu and Briggs have examined the current state of mobile services among academic libraries of the

IOP Conf. Series: Journal of Physics: Conf. Series **978** (2018) 012052 doi:10.1088/1742-6596/978/1/012052

country's top 100 accredited mobile services at the university level [4]. This paper shows that all libraries offer at least one mobile service, and the majority offers multiple services. Some of the most common mobile services offered by the library are mobile sites, text messaging services, e-books, and mobile access to databases and catalogs. Sharma and Sahoo describe mobile technology applications in libraries and information services and explain new technologies, to provide effective and efficient services to library users as well as appropriate access to the required access and quick access to relevant information so that mobile applications to provide library and information services is an important step [5]. Kumbar and Bright, discuss about mobile technology and focus on finding that mobile device implementations help generate new services and provide faster access to their collections and also create opportunities and challenges for academic libraries [6]. Malaty and Kantha, outlines the application of mobile technology and development in mobile telecommunication systems to provide up-to-date, user-friendly, personal and dynamic information services to up-to-date information to library users. This paper reveals that the implementation of mobile communication systems helps provide and improve library services to users [7][8].

Conducted research on the utilization of cellular services in the educational environment [9]. This paper explores the nature of mobile use among students, and investigates student perceptions about the use of mobile phones in libraries and information services. This paper shows that the provision of telecommunication services in the library allows the effective use and access of information services and libraries using mobile phones. Nalluri and Gaddam discussed the main strategies used by libraries to provide services through mobile devices [10]. The paper also discusses an overview of mobile device trends and their applications for libraries, including mobile internet, mobile multimedia, and SMS, etc. and describes the various activities the device supports for library use, focusing on how libraries and librarians respond with services tailored for this device and what the library can do in the future in this regard. This study further explores the utilization of cellular services in the academic environment [11][12][13][14][15][16].

### *2.1. Mobile Technology*

Different types of mobile technology are widely used in today's society such as cell phones, multimedia phones, laptops, net books, e-book readers, cameras, and so on. This technology performs various activities including voice and sound. video calling, e-mail transfer, text messaging, clicking photos, playing videos, making videos, setting alarm clocks, searching the internet, searching school database information, accessing course management systems, reading or listening to books and articles, newspapers, shopping online, internet banking, using GPS navigation systems, games, uploading and downloading content from the internet [1].

### *2.2. Mobile Technology Service Application in Library*

Mobile technology provides the most flexible, fastest and easiest way to communicate access and share information by anyone from anywhere and anytime. Mobile technology services that libraries can adopt include:

**Table 1**. Mobile Technology Services

| | |
|---|---|
| SMS Notification Services: | To provide up-to-date information, news and event information remains current, SMS notification service to users. Moreover, users can get notifications of new book arrival, overdue, loan request, etc. |
| Access Library Website: | To provide users with dynamic information services through mobile devices, libraries should design library sites as interactive and participatory interactive library websites. |
| Allow accessing OPAC: | Libraries can provide access to OPAC libraries and resource varieties and databases via mobile technology. This will help to get library resource availability anytime anywhere. |

| | |
|---|---|
| Providing Mobile Library Instruction: | Implementation of library services through mobile technology will help support formal education, assist research activities in the e-learning environment |
| Free access of Library E-resources: | Some publishers have already sent e-books (both text and audio), e-Journals, Web databases, dissertations, etc. that can be accessed via mobile users. |
| Mobile Library Databases: | Provides access to various library and database resources via mobile technology. The library should also provide a Download link to the user's mobile technology. |
| SMS Refrence Service: | Through this library the service can provide instant answers, and links to articles and references in real time, to user communities, and more. |

## 3. Method
Research location is the Medan Public Library. Data collection was conducted from April 1st to May 10th 2017, using questionnaires and interviews with librarians and library users. For supporting data direct observation is made to find out the services provided by the library.

## 4. Results and Discussions

### *4.1. Current Condition of Mobile Technology Service Provided by Medan Public Library*
The purpose of implementing mobile technology services in libraries is to provide better library services in an easy way and to increase the use of library resources wherever and whenever. In this case the Medan Public Library still provides mobile technology services for certain services for users, such as ordering books, and storing detailed information on catalog search. For the library itself, this technology can be one of the new and more innovative ways to promote owned collections. Implementation of mobile technology services is still the development stage and still trying to give the best to library users according to their needs [11][12].

### *4.2. Mobile Basic Services Offered*
The Mobile Library Basic Service adopts new technologies and services to provide and improve accessibility to library resources. In this case the mobile technology services offered are E-book / E-articles, E-news, E-journals, downloadable, OPAC, book reservations, mobile reference service, personal service through mobile technology, chat / instant messaging, social media, and others.

**Table 2.** Basic Mobile Technology Services Library

| Library Services |
|---|
| E-book / E-articles |
| Downloadable e-resources |
| Mobile OPAC |
| Book reservation |
| Mobile Reference Service |
| SMS notification Services |
| Chat/IM |
| Social Media |
| Personal services |

### *4.3. Benefits of Mobile Technology Services for users and librarians Medan Public Library*
From the results of questionnaires to users and interviews to librarians is known that the benefits of mobile services are perceived, among others:

- Help improve access to library resources and services

- Help improve the use of information resources available in the library
- Facilitate the use of library resources
- Improve the quality and efficiency of the library
- Improve the quality of library resources
- Reduce the communication gap between the library and its users
- Giving easy communication with library users anytime
- Stay connected between libraries and users, and more.

### *4.4. Constraints and Challenges in Mobile Technology Implementation in Medan Public Library*
Evaluation of the application of mobile technology in the Medan Public Library shows many problems in handling new devices, namely maintenance and updating of content. In this case, the challenges facing libraries are lack of human resources, lack of skills; lack of funding, lack of support from government and institutions, communication problems and lack of content, and content management.

### *4.5. The Future Service of Mobile Technology*
Library has done its best to improve library services and increase the use of library resources by adopting new technologies and adding many services. So to improve the accessibility of library resources and to provide better service to users, based on survey results will add the following mobile technology services [1][3]:
M-library instruction to support learning process, research activity, formal education, and others
- Text messaging services to provide information, news and events, keep up with library and institutional information
- E-book access: providing electronic access through their mobile technology, anywhere
- Augmented reality
- M-OPAC: provides online catalog access via their mobile on or off campus, accessing databases, etc.
- Book renewal: online book extension from off-campus via their mobile phone
- Mobile interlibrary loans: book loans and reservation services
- Applications for accessing e-resources / databases: providing e-resources access from e-resources applications such as My Library, Pub Med, Springer, Science Direct, E-Library, etc.

## 5. Conclusions and Suggestions
The need for library user information is increasing every day following their expectation of technology, ease and speed and accuracy. So, adopting new technology in libraries is the only way to meet the needs of users precisely and on time. In this context, the use of mobile technology applications in libraries brings great advantages to providing better library services and to improve accessibility of library resources in the easiest way, but due to lack of funds, lack of skills and lack of staff, many libraries are unable to provide services this is very successful [12][13]. Therefore, providing sufficient funds and training staff on new technologies is essential to understanding the progress of these technologies and providing library services with the simplest for library users. Moreover, the provision of such services is not only a major requirement but also a big challenge [13][14][15][16][17]. Once accomplished, library users will be able to easily use library resources and services via their mobile phones. Libraries will be able to communicate better and easily share information and knowledge with their library users wherever they fit their demands [18][19][20].

## References
[1] Joan K L 2010 A Mobile future for academic libraries. *Ref. Services Rev.* **38** (2) pp 205-213
[2] Hui L and Zhao Q C 2016 Design and implementation of the mobile library app based on smart phone *Proc. Int. Conf. Machine Learning and Cybernetics (ICMLC)* 10-13 July 2016 vol **1**

pp 318-322 (Jeju: South Korea)

[3] Gattan A M and & Razek M A 2014 Towards efficient mobile library services. *5th Int. Conf. on Intelligent Systems, Modelling and Simulation* pp 562 – 567

[4] Liu Y Q and Briggs S 2015 A Library in the palm of your hand: Mobile services in top 100 university libraries *Information technology and libraries* pp 133-148

[5] Sharma D and Sahoo D R 2014 Application of Mobile Technology in Library Services: An Overview *Int. J. of Inf. Tech. and Lib. Sci.* vol **3** (1) pp 17 -24

[6] Kumbhar S S and Pawar R R 2014 *Mobile based services application and challenges* https://www.researchgate.net/publication/271906602

[7] Malathy S and Kantha P 2013 Application of Mobile Technologies to Libraries DESIDOC *J. of Lib. & Inf. Tech.* vol **33** (5), pp 361-366

[8] AL-Faresi S and Patel N 2012 The design of an adoption and acceptance framework for mobile digital library services *Int. Conf. on Inf. Soc. (i-Society 2012)* pp 216 – 224

[9] Karim N S A etal 2006 Mobile phone applications in academic library services: a student feedback survey *Campus Wide Information System* vol **23** (1) pp 35-51

[10] Nalluri S R and Gadddam B 2016 Mobile library services and technologies: A study. *Int. J. of Research in Lib. Sci.*vol **2** (2) pp 59-66

[11] Chaveesuk S, Vongjaturapat S and Chotikakamthorn N 2013 Analysis of factors influencing the mobile technology acceptance for library information services: Conceptual model. *Int. Conf. on Inf. Tech. and Electrical Eng. (ICITEE)* pp 18 – 24

[12] Robin Yeates 2014 M-Libraries 4: from Margin to Mainstream – Mobile Technologies Transforming Lives and Libraries *Program* vol **48** Issue: 4 pp 417-418 https://doi.org/10.1108/PROG-10-2013-0021

[13] Richard Palmer 2010 M-Libraries 2: A Virtual Library in Everyone's Pocket *Library Review*, vol **59** Issue: 9 pp725-726

[14] Madely du Preez 2012 M-libraries 3: Transforming Libraries with Mobile Technology *Online Information Review* vol **36** Issue: 5 pp 770-771

[15] Jan Weaver 2011 M-libraries 2: A Virtual Library in Everyone's Pocket *Library Management* vol **32** Issue: 3 pp 230-231

[16] Ina Fourie 2011 M-libraries 2: A Virtual Library in Everyone's Pocket, *The Electronic Library*, vol **29** Issue: 6 pp 851-852

[17] Tingting Z and Lili Z 2014 IT applications to mobile services in university library *2014 9th Int. Conf.e on Comp.r Sci. & Edu.* pp 654 – 658

[18] Vongjaturapat S and Chaveesuk S 2013 Mobile technology acceptance for library information service: A theoretical model. *Int. Conf. on Inf. Soci. (i-Society 2013)* pp 290 – 292

[19] Negi D S 2014 Using mobile technologies in libraries and information centers *Library Hi Tech News* vol **31** (5) pp 14-16

[20] Dewiyana H and Siregar A R 2017 V-Reference: Innovation in Developing Services According To Changing Needs of Users at The Public Library of Medan City *Int. J. of Trend in Research & Dev.* vol 4 pp71-75

**Acknowledgement**

Sincere thanks are conveyed to the Public Library of Medan City, The Society of Medan City, and Universitas Sumatera Utara for Research assistance in Grant DRPM Kemenristekdikti 2017.

IOP Conf. Series: Journal of Physics: Conf. Series **978** (2018) 012053 doi:10.1088/1742-6596/978/1/012053

# Multi-documents summarization based on clustering of learning object using hierarchical clustering

**M Mustamiin[1], I Budi[2] and H B Santoso[3]**

Faculty of Computer Science, Universitas Indonesia, Depok, Indonesia

m.mustamiin@gmail.com[1], indra@ui.ac.id[2], harrybs@cs.ui.ac.id[3]

**Abstract**. The Open Educational Resources (OER) is a portal of teaching, learning and research resources that is available in public domain and freely accessible. Learning contents or Learning Objects (LO) are granular and can be reused for constructing new learning materials. LO ontology-based searching techniques can be used to search for LO in the Indonesia OER. In this research, LO from search results are used as an ingredient to create new learning materials according to the topic searched by users. Summarizing-based grouping of LO use Hierarchical Agglomerative Clustering (HAC) with the dependency context to the user's query which has an average value F-Measure of 0.487, while summarizing by K-Means F-Measure only has an average value of 0.336.

## 1. Introduction

Open Educational Resources (OER) is a portal of teaching, learning, and research resources available in the public domain or have been released under an intellectual property license that allows a freely accessible [1]. The use of technology in education, including OER increases the number and variety of forms of learning content. The number of instructional content that causes the need for a convenient search system to obtain the desired learning content.

A previous study revealed that Learning Object (LO) search engine-based ontology has an average performance of 29.7%. It is better than using a search engine index-based document [2]. LO Ontology is designed to explain the concept and the relationship between learning content to facilitate the process of finding the learning content [2].

LO is granular more easily reused by other learning-related topics as well as the topic of learning itself [3]. Figure 1 shows that LO is a representation in the form of granules that can be reused as a material for the lesson topic that is still associated or used as components of the manufacture of new learning materials. LO in the form of granules and the search results on ontology-based LO search engines, in particular on the documents top of the search results, are often found materials that are related or even similar. It shows that between LO in the field of material related with LO in other materials, especially for materials that contain the same keywords, with these conditions it is possible to create an empowerment of the LO that is digital so they can be reused as material learning material a new one in accordance with the desired topic.

A study conducted by Paramartha [2] in relation to LO traceability techniques was limited to designing LO ontologies and search engines, while for LO use itself does not exist. The research also provides suggestions for creating a system that can create new materials based on LO ontologies that have been constructed by reusing sub-material that has been mapped.

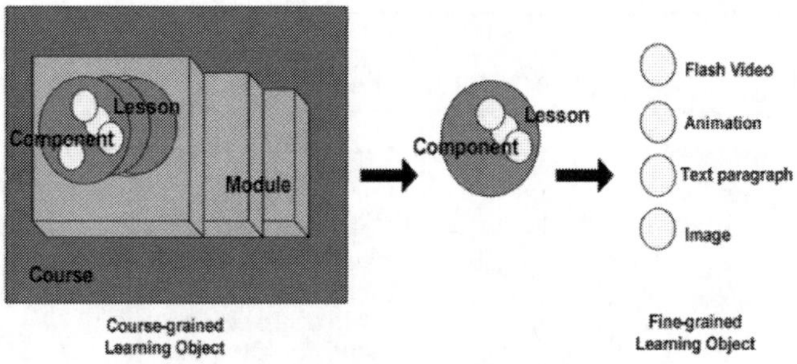


**Figure 1.** Grained LO [3].

LO utilization as a new learning materials that can be done by summarizing LO or sub-materials. Summarization is one of the topics in the acquisition of information retrieval to make information shorter and denser than the original document. Multi-documents summarization used to summarize more than one document, there are several methods in multi-documents summarization, ranging from methods based on graph-based and cluster-based [4]. Multi-documents summarization based graph using a specific query [5] and there is also multi-documents summarization using clustering sentence that produces good results [6].

To get the LO summarization diverse and do not overlap and can produce summarization accordance with the desires of user is required appropriate grouping method in the LO grouping of search results. According to Shepitsen [7] the context dependence on Hierarchical Agglomerative Clustering (HAC) can improve the result of recommendations in accordance with the user profile. In the context of search, user profiles can be compared with the wishes of the user in finding information. Therefore, reliance on HAC context that has been proven to improve on as the user desires can be applied to multi-documents summarization based on the results of the query entered by the user.

In this study, the authors conducted a development of the search engine LO-based ontologies created by Paramartha [2] by adding summarization features against LOs search results and representing it becomes a collection of information that can be a reference for the user to create learning materials related topics of interest. Document summarization process search results using multi-documents summarization based grouping HAC with dependency context of the user's query.

## 2. Material and Method

### *2.1. Research Data*

Data were taken from the research data tracking techniques ontology-based LOs by Paramartha [2]. The research data is in the form of ontology files, based on the information in the file, the file was created using OWL ontology API (version 3.4.2). where the ontology files is obtained from extraction of 115 lecture slides of five courses, including computer network (CN), web programming (WP), enterprise application integration (EAI), supply chain management (SCM), and e-commerce (EC) [2].

Figure 2 shows example in identifying titles, descriptions, and media of presentation slide. The format of presentation slides document used as test data is *.pptx. *.pptx is basically a collection of XML files compressed into a zip file. Each XML file represents a slide in a presentation slide in *.pptx format. XML files can be extracted to be identified further and can be mapped into ontology [2].

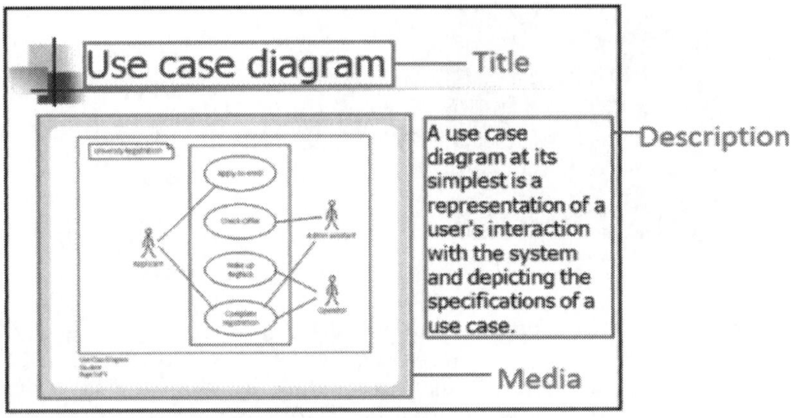


**Figure 2.** Title, description, and picture identification example [2].

### *2.2. The Method*

This study has several stages in the implementation include: (1) adding of summarization feature on search engines; (2) testing and evaluation of summarization experimental results; and (3) reporting research results. This part will focus on three research stages as shown on Figure 3.

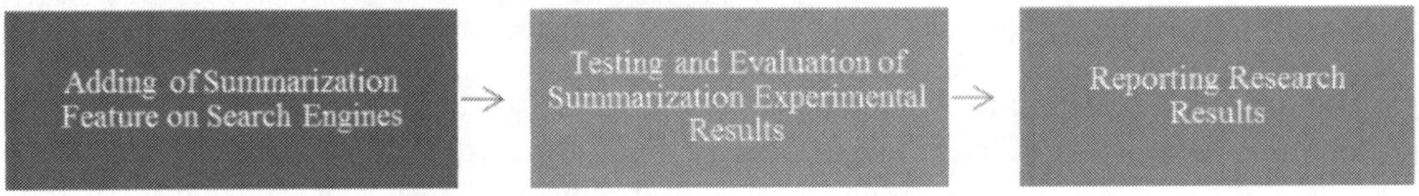


**Figure 3.** Additional Features on Search Engines.

#### *2.2.1. Adding of Summarization Feature on Search Engines*

At this stage, the addition of summarization features on ontology-based search engine LO, the documents that are used as ingredients of summarization are the top 10 search results. Automatic summarization based on user-query the basis of summarizing, for the purpose of summarization done is seeking information related to the user's wishes are taken from the queries that users enter into search engines LO. Adding of Summarization Feature on Search Engines can be seen in Figure 4.

After searching process with LO search engine is completed, there are four core steps: 1) indexing LO from top 10 search result documents, 2) selection of LO summary, 3) sorting LO result of summary process based on its original position in document , and 4) show the summary. The flow of the summary process can be seen in Figure 5.

Each LO have distance from the other based on the maximum value of similarity, i.e., 1 minus the value of the resemblance. LO can take the form of text, images, audio, video or animation, but in calculating the distance between LO based only on the content of the text is in it, for example if in one LO has an image of a computer and the components in it and a text image name "computer image" it is used as a determinant in the calculation of the distance with the other LO is the name of the picture only "computer image".

Figure 6 shows a branch of a tree cutting techniques (dendogram). Cutting a branch of a tree is determined by the division coefficient values as shown on the left. In this figure also shows an example of cutting dendogram based on context-dependent of users [7].

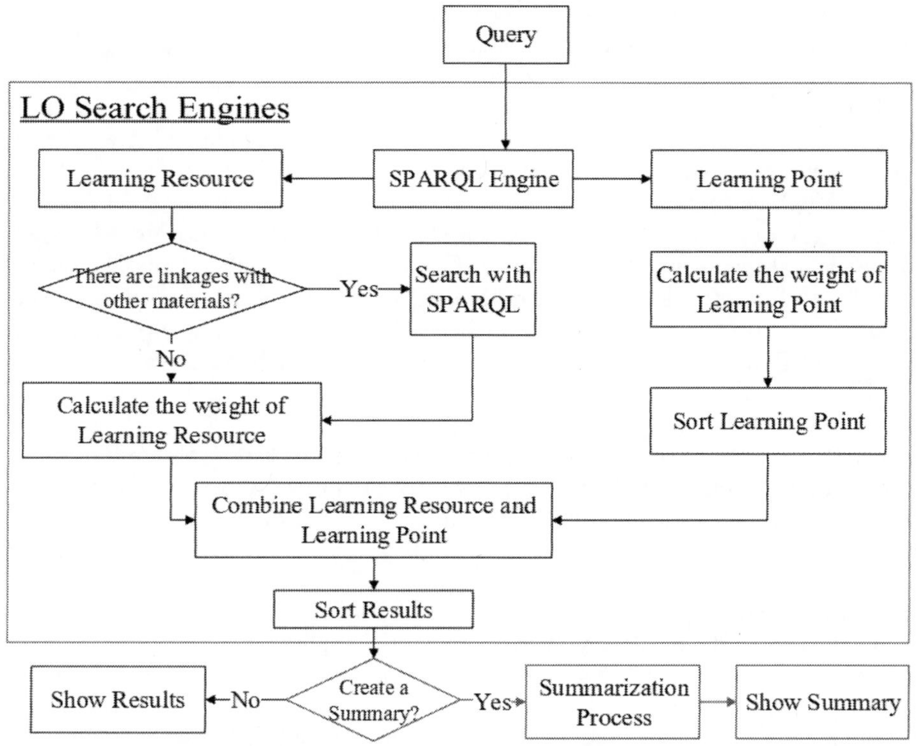


**Figure 4.** Additional Features on Search Engines.

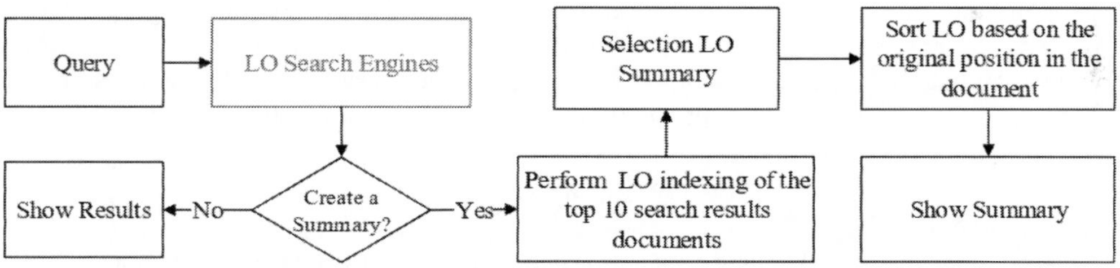


**Figure 5.** Flow Diagram of Summarization.

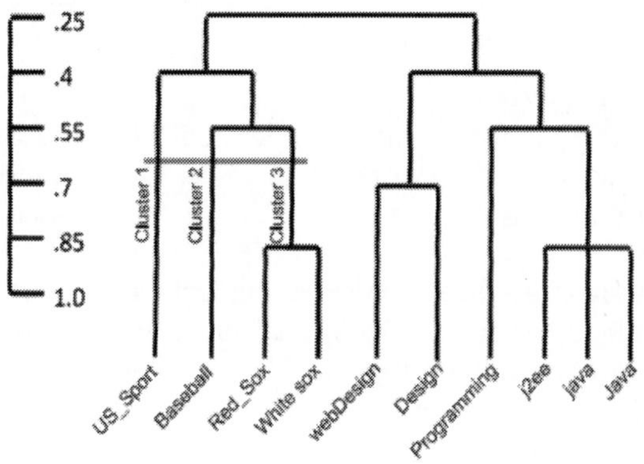


**Figure 6.** Context-Dependent Hierarchical Agglomerative Clustering Technique [7].

*2.2.2. Testing and Evaluation of Summarization Experimental Results*

Summarization experimental evaluation process is done manually by comparing the experimental results with a summary of the summary made by the human user as a comparison. Summarization manually made by people who have competence in the field of computer science for the test data in this study were associated data learning materials in the field of computer science. The result of the manual summarization is a collection of LO which is a collection of slides contained in the powerpoint or document which will be summarized.

To assess the summarization experimental results based on comparison summarized results with a summary of the manual then calculate precision and recall value of each summarization experimental results. After that, a summarization experiment to compare one with the other used F-Measure value calculation. Based on consideration of the importance of the accuracy of the election LO summary F-Measure value calculation in this study involves assigning weights doubled in value precision.

*2.2.3. Reporting Research Results*

At the reporting stage, researchers put the entire process from the initial research to the final stage. Components that are reported in the form of the activities performed during the study, namely: adding of summarization feature on search engines, testing and evaluation of summarization results as well as conclusions.

## 3. Results and Discussion

This section explains about the results and discussions of the study.

*3.1. Adding Summarization Feature on Search Engine*

Documents which will be summarized in this study are the top 10 search results document. A summary consists of a collection of Learning Object (LO) found on top of the document by using extractive summarization method.

In this study, Learning Point (LP) results in this study was defined as LO and implemented into the code in the Java programming language into a node are then grouped and be part of the summary. The stages in the manufacture of a summary of this research, namely: LO indexing, determining the distance between LO, LO election summaries by several methods, sequencing LO based document id and LO position on the original document.

*3.1.1. Random LO Selection*

After all of the top LO then the indexed documents are drawn at random as the value of the compression rate summary specified. In this research, three summarizations with random LO selecting which compression rate of 10%, 15%, and 20% of the total number LO obtained from the search results.

*3.1.2. HAC Method LO Selection*

Selection is based on the grouping LO using methods Hierarchical Agglomerative Clustering (HAC) with cutting dendogram only on the part corresponding to the user's query and the context of the electoral process is cutting dendogram LO performed on five coefficient distances between LO, namely 0.3, 0, 4, 0.5, 0.6, and 0.7.

LO grouping process with HAC method using hierarchical clustering support library-java with some modifications and adjustments as well as additional functions for the determination group and cutting dendogram. In the process of classifying the HAC, the method used is the average linkage method, namely the determination of the group based on the average distance from members of the other group members.

### *3.1.3. K-Means Method LO Selection*

In this study, summarization using LO election with K-Means method performed by three variations of the number of groups according to the degree of compression of the summary by 10%, 15%, and 20% of the total LO indexing results. Centroid determination early on K-Means clustering method is determined randomly and the maximum number of iterations in the process of classification is 10 iterations.

### *3.2. Summarization Experiment*

Summarization results will appear on the user's query and LO resulting from the summarization (see Table 1). Each LO displays the title and the contents in the form of text or images in accordance with the contents of the LO.

**Table 1.** Query List.

| No. Query | Query | Symbol |
|---|---|---|
| 1 | Supply Chain Management | Q1 |
| 2 | Enterprise Application Integration | Q2 |
| 3 | E-Commerce | Q3 |
| 4 | Perancangan dan Pemrograman Web | Q4 |
| 5 | Computer Networks | Q5 |

After the experimental results compared with the gold standard summarization then calculated the value of precision and recall summarization results. After that, an experiment to compare one with the other summarization used F-Measure value calculation. Based on consideration of the importance of precision in the manufacture of the summary calculation F-Measure value in this study involves assigning weights doubled in value precision. Figure 7 shows example of summarization results.

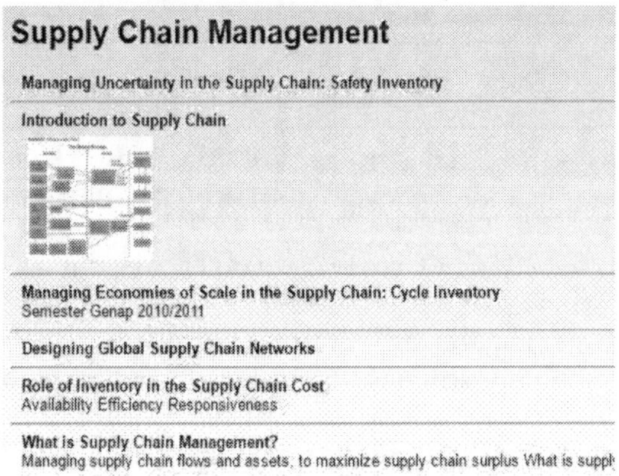


**Figure 7.** Interface Example of Summarization Results.

Figures 8, 9, 10 shows F-Measure value from summarization with Random LO Selection, HAC LO Selection and K-Means LO Selection. Fig. 11 shows that summarization with HAC method decisively superior to every query number. When viewed from their mean, the average value of the experiment HAC is 0.487 with the highest value at query experiment number 1 which has a value of F-Measure reached 0.768.

Based on analysis of queries and documents related to the search results that are used as ingredients in summarization can be seen that the search results heavily influence result of summarization. Good search results will get results LO considerable and are related to the search context.

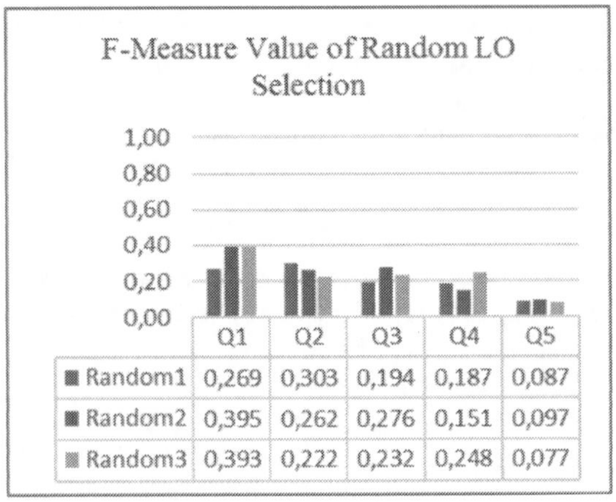


| | Q1 | Q2 | Q3 | Q4 | Q5 |
|---|---|---|---|---|---|
| Random1 | 0,269 | 0,303 | 0,194 | 0,187 | 0,087 |
| Random2 | 0,395 | 0,262 | 0,276 | 0,151 | 0,097 |
| Random3 | 0,393 | 0,222 | 0,232 | 0,248 | 0,077 |

**Figure 8.** *F-Measure* Value of Random LO Selection.

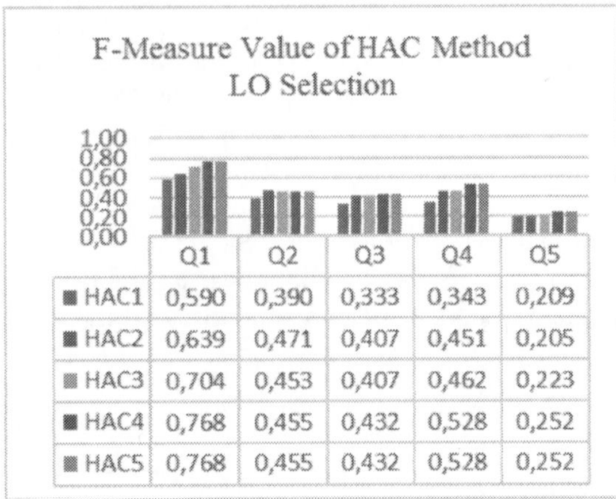


| | Q1 | Q2 | Q3 | Q4 | Q5 |
|---|---|---|---|---|---|
| HAC1 | 0,590 | 0,390 | 0,333 | 0,343 | 0,209 |
| HAC2 | 0,639 | 0,471 | 0,407 | 0,451 | 0,205 |
| HAC3 | 0,704 | 0,453 | 0,407 | 0,462 | 0,223 |
| HAC4 | 0,768 | 0,455 | 0,432 | 0,528 | 0,252 |
| HAC5 | 0,768 | 0,455 | 0,432 | 0,528 | 0,252 |

**Figure 9.** F-Measure Value of HAC Method LO Selection.

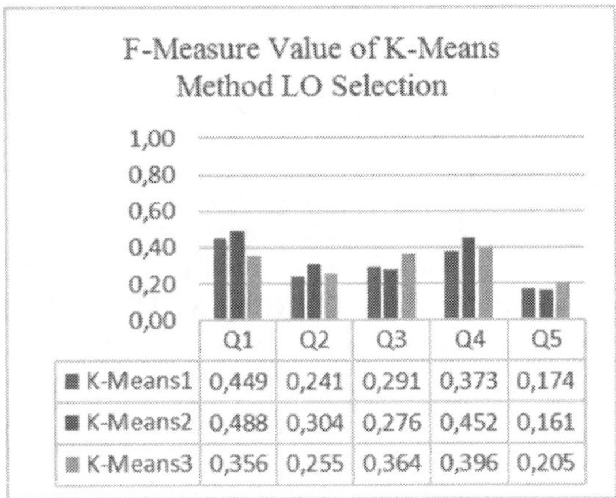


| | Q1 | Q2 | Q3 | Q4 | Q5 |
|---|---|---|---|---|---|
| K-Means1 | 0,449 | 0,241 | 0,291 | 0,373 | 0,174 |
| K-Means2 | 0,488 | 0,304 | 0,276 | 0,452 | 0,161 |
| K-Means3 | 0,356 | 0,255 | 0,364 | 0,396 | 0,205 |

**Figure 10.** F-Measure Value of K-Means Method LO Selection.

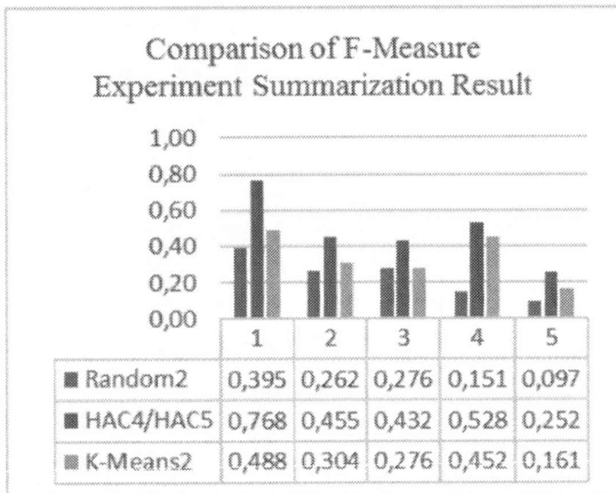


| | 1 | 2 | 3 | 4 | 5 |
|---|---|---|---|---|---|
| Random2 | 0,395 | 0,262 | 0,276 | 0,151 | 0,097 |
| HAC4/HAC5 | 0,768 | 0,455 | 0,432 | 0,528 | 0,252 |
| K-Means2 | 0,488 | 0,304 | 0,276 | 0,452 | 0,161 |

**Figure 11.** Comparison of F-Measure Experiment Summarization Result.

In the search results Q3 and Q4, LO coverage results are still lacking so much so that the summarization result is not optimal. Based on the analysis of the outcome document number from search results in Q3 and Q4, the number of search results that are slightly due to the material "e-commerce" and "design and web programming" has a lot of LO that do not contain the word search query.

Meanwhile, LO from Q1 and Q5 search result despite quite a lot but the LO many are outside the context of search. This occurs because the query that is used, the words "enterprise application integration" and "computer networks" contains a common word in which the word is also found in many other LO course material.

## 4. Conclusion

Learning content or LO are granular and can be reused for constructing new learning materials. LO from ontology-based search results can be used as an ingredient to create new learning materials according to the topic searched by users. The experimental results of this study indicate that summarization method using context dependency with HAC is better than summarization using the K-

Means. This is indicated by the F-Measure mean value of the summarization using HAC with context dependency better than a mean value of F-Measure with K-Means method.

**References**

[1] Hewlett Foundation. OER. [Online]. Available: http://www.hewlett.org/Programs/Education/OER
[2] Paramartha A G 2014 *Teknik Penelusuran Learning Object Berbasis Ontologi* (Depok: Tesis, Universitas Indonesia).
[3] Noor S F, Yusof N and Hashim S Z 2011 Creating granular learning object towards reusability of learning object in e-learning context. *Bandung: International Conference Electrical Engineering and Informatics (ICEEI)* pp. 1-5.
[4] Meena Y K, Jain A and Gopalani D 2014 Survey on graph and cluster based approaches in multi-document text summarization. *Jaipur: International Conference Recent Advances and Innovations in Engineering (ICRAIE)* pp. 1-5.
[5] Pandit S R and Potey M A 2013 A query specific graph based approach to multi-document text summarization: simultaneous cluster and sentence ranking. *Katra: Machine Intelligence and Research Advancement (ICMIRA)* pp. 213-217.
[6] Gupta V K and Siddiqui T J 2012 Multi-document summarization using sentence clustering. *Kharagpur: Intelligent Human Computer Interaction (IHCI)* pp. 1-5.
[7] Shepitsen A, Gemmell J, Mobasher B and Burke R 2008 Personalized recommendation in social tagging systems using hierarchical clustering. *New York: ACM* pp. 259-266.
[8] Ooi J M 2015 A survey of query expansion, query suggestion and query refinement techniques. *Kuantan: Software Engineering and Computer Systems (ICSECS)* pp. 112-117.
[9] Chauhan R, Goudar R, Sharma R and Chauhan A 2013 Domain ontology based semantic search for efficient information retrieval through automatic query expansion. *Gujarat: Intelligent Systems and Signal Processing (ISSP) International Conference* pp. 397-402.
[10] Grimm S 2010 Knowledge representation and ontologies. *Berlin: Scientific Data Mining and Knowledge Discovery* pp. 111-137.
[11] Hasibuan Z A and Cenka B A 2012 Toward indonesia open educational resources. *Bali: International Symposium on Open, Distance, and E-Learning (ISODEL)*
[12] Learning Technology Standards Committee. *Draft Standard for Learning Object Metadata.* New York: Institute of Electrical and Electronics Engineers (IEEE), 2002.
[13] Manning C D, Raghavan P and Schütze H 2009. *An Introduction To Information Retrieval.* (Cambridge: Cambridge University Press )
[14] PadmaLahari E, Kumar D V and Prasad S 2014 Automatic text Summarization with statistical and linguistic features using successive thresholds. *Ramanathapuram: IEEE* pp 1519-24.
[15] Princeton University. 2010. *About WordNet.* (Princeton University)
[16] Rehak D R and Mason R 2003 *Engaging with the Learning Object Economy, in Littlejohn, Allison, Reusing Online Resources: A Sustainable Approach to E-Learning* ( London: Kogan Page) pp. 22-30.
[17] Wang S, Li W, Wang F and Deng H 2010 *A Survey on Automatic Summarization*. Kunming: IEEE pp. 193-196.
[18] Wiley D A 2000 *Connecting Learning Objects to Instructional Design Theory: A Definition, A Metaphor, and A Taxonomy.*
[19] Santoso H B, Cenka B A, Priyogi B, Paramartha Y and Hasibuan Z A 2015 Platform development and design evaluation of the indonesia open educational resources. *Hangzhou: The 23rd International Conference on Computers in Education (ICCE)*

IOP Conf. Series: Journal of Physics: Conf. Series **978** (2018) 012054 doi:10.1088/1742-6596/978/1/012054

# Skipping Strategy (SS) for Initial Population of Job-Shop Scheduling Problem

**M Abdolrazzagh-Nezhad[1], E B Nababan[2*], and H M Sarim[3]**

[1]Department of Computer Engineering, Faculty of Engineering,
Bozorgmehr University of Qaenat, Qaen, Iran.
[2] Department of Information Technology, Faculty of Computer Science and Information Technology
Universitas Sumatera Utara, Medan, Indonesia
[3]Data Mining and Optimization Research Group, Center for Artificial Intelligence Technology
Universiti Kebangsaan Malaysia, 43600 UKM Bangi, Selangor, Malaysia

[*]Corresponding author: ernabrn@usu.ac.id

**Abstract**. Initial population in job-shop scheduling problem (JSSP) is an essential step to obtain near optimal solution. Techniques used to solve JSSP are computationally demanding. Skipping strategy (SS) is employed to acquire initial population after sequence of job on machine and sequence of operations (expressed in *Plates-jobs* and *mPlates-jobs*) are determined. The proposed technique is applied to benchmark datasets and the results are compared to that of other initialization techniques. It is shown that the initial population obtained from the SS approach could generate optimal solution.

## 1. Introduction

Researchers have been trying for many years to find effective algorithms for optimal feasible solutions in job-shop scheduling problem. Branch and bound algorithm [1], [2], were acceptable for small size instances. For other instances, meta-heuristic algorithms, such as tabu search [3], [4], [5] genetic algorithms [6], simulated annealing [7], ant colony optimization [8], particle swarm optimization [9], artificial immune system [10], DNA computing [12], and others[11], [13], [14], are mostly population-based algorithms that require an initial population to start exploring and exploiting the optimal schedule.

In this paper, a Skip Strategy (SS) is proposed. We first determine sequence of job on machine and sequence of operations followed by designing a set of rules to activate the *mPlates-Jobs*. Then skipping strategy based on the rules generated on *mPlates* is employed to obtain initial population.

## 2. Previous Research

Several random techniques were employed to get initial population of job shop scheduling problem [5],[7],[9],12],[15]. However it demands more computational time to get to near optimal solution. A new representation and schedule deduction method using a heuristic technique based on genetic algorithms (GA) to initialize the population was proposed in [17]. Additionally, integer linear programming [18] was also proposed, and the results were compared with some other proposed initialization procedures. However, there is still a need to consider different procedure to generate a population of initialization points to get a better solution.

## 3. Method

The job-shop scheduling problem formulated as follows: given set of jobs $J = \{J_1, J_2, \ldots, J_n\}$ on a set of machines $M = \{M_1, M_2, \ldots, M_m\}$. The operations of the jobs have to be processed on the machines. $O = \{O_{ij}, i = 1,2, \ldots, n \wedge j = 1,2, \ldots, m\}$. The objective function is:

$$min \quad C_{max} = \min\left(\max_{1\le i\le n}(\,C(O_{im})\,)\right) \tag{1}$$

where $C_{max}$ maximum amount of time required for any particular task to be completed
$C(O_{im})$ is the completion time of $O_{im}$
$O_{im}$ is the last operation of $J_i$ and $i = 1,2, \ldots, n$.

### 3.1 Pre-Processing

#### *3.1.1 Determining Sequence of Job on Machine (SJM)*

Matrix of a sequence of jobs on machines ($SJM$) is first determined to represent a solution space. Each row of the matrix represents jobs to be processed on a given machine and the machine has to be visited just once.

#### *3.1.2 Constructing Active Schedule*

We utilize an instance of six jobs six machines (Ft06 benchmark data set) to build an active schedule of a given $SJM$.

| | Operation | | | | | |
|---|---|---|---|---|---|---|
| | Machine (Processing Time) | | | | | |
| Job | 1 | 2 | 3 | 4 | 5 | 6 |
| 1 | 3(1) | 1(3) | 2(6) | 4(7) | 6(3) | 5(6) |
| 2 | 2(8) | 3(5) | 5(10) | 6(10) | 1(10) | 4(4) |
| 3 | 3(5) | 4(4) | 6(8) | 1(9) | 2(1) | 5(7) |
| 4 | 2(5) | 1(5) | 3(5) | 4(3) | 5(8) | 6(9) |
| 5 | 3(9) | 2(3) | 5(5) | 6(4) | 1(3) | 4(1) |
| 6 | 2(3) | 4(3) | 6(9) | 1(10) | 5(4) | 3(1) |

The sequence of operations ($SOJ$) of the instance is:

$$SOJ = \begin{bmatrix} 3 & 1 & 2 & 4 & 6 & 5 \\ 2 & 3 & 5 & 6 & 1 & 4 \\ 3 & 4 & 6 & 1 & 2 & 5 \\ 2 & 1 & 3 & 4 & 5 & 6 \\ 3 & 2 & 5 & 6 & 1 & 4 \\ 2 & 4 & 6 & 1 & 5 & 3 \end{bmatrix} = \begin{matrix} J_1 \rightarrow \\ J_2 \rightarrow \\ J_3 \rightarrow \\ J_4 \rightarrow \\ J_5 \rightarrow \\ J_6 \rightarrow \end{matrix} \begin{bmatrix} M_3 & M_1 & M_2 & M_4 & M_6 & M_5 \\ M_2 & M_3 & M_5 & M_6 & M_1 & M_4 \\ M_3 & M_4 & M_6 & M_1 & M_2 & M_5 \\ M_2 & M_1 & M_3 & M_4 & M_5 & M_6 \\ M_3 & M_2 & M_5 & M_6 & M_1 & M_4 \\ M_2 & M_4 & M_6 & M_1 & M_5 & M_3 \end{bmatrix}$$

$$= \begin{bmatrix} O_{11} & O_{12} & O_{13} & O_{14} & O_{15} & O_{16} \\ O_{21} & O_{22} & O_{23} & O_{24} & O_{25} & O_{26} \\ O_{31} & O_{32} & O_{33} & O_{34} & O_{35} & O_{36} \\ O_{41} & O_{42} & O_{43} & O_{44} & O_{45} & O_{46} \\ O_{51} & O_{52} & O_{53} & O_{54} & O_{55} & O_{56} \\ O_{61} & O_{62} & O_{63} & O_{64} & O_{65} & O_{66} \end{bmatrix} \tag{2}$$

### 3.2 Generating Initial Population

#### *3.2.1 Plates-Jobs and mPlates-Jobs*

After determining $SOJ$ we then generate *Plates-jobs* and *mPlates-Jobs* respectively. The first column/operation ($O_{i1}$) of $SOJ$ has two machines ($M_2$,$M_3$), the second one ($O_{i2}$) has four machines ($M_1$, $M_2$, $M_3$, $M_4$) and so on. For example $O_{i1}$ has 2 *Plates-Jobs*, a *Plate-Jobs* indexes with $M_2$ has three jobs ($J_2$, $J_4$ and $J_6$) and a *Plate-Jobs* indexes with $M_3$ with three other jobs ($J_1$, $J_3$ and $J_5$). Another example, $O_{i5}$ has four *Plates-Jobs* of which the first includes just one job ($J_1$), the second consists of two other jobs ($J_2$ and $J_5$), the third has one job $J_3$ and the last one contains two jobs, $J_4$ and $J_6$. The remaining operations/columns of $SOJ$, other *Plates-Jobs* is illustrated in Figure 1.

| | | | | |
|---|---|---|---|---|
| Plates-Jobs related to $O_{i1}$ → | $M_3$: $J_1$, $J_3$, $J_5$ | | $M_2$: $J_2$, $J_4$, $J_6$ | |
| Plates-Jobs related to $O_{i2}$ → | $M_1$: $J_1$, $J_4$ | $M_3$: $J_2$ | $M_4$: $J_3$, $J_6$ | $M_2$: $J_5$ |
| Plates-Jobs related to $O_{i3}$ → | $M_2$: $J_1$ | $M_5$ :$J_2$, $J_5$ | $M_6$ :$J_3$, $J_6$ | $M_3$ : $J_4$ |
| Plates-Jobs related to $O_{i4}$ → | $M_4$: $J_1$, $J_4$ | $M_6$ :$J_2$, $J_5$ | $M_1$ :$J_3$, $J_6$ | |
| Plates-Jobs related to $O_{i5}$ → | $M_6$: $J_1$ | $M_1$: $J_2$, $J_5$ | $M_2$: $J_3$ | $M_5$: $J_4$, $J_6$ |
| Plates-Jobs related to $O_{i6}$ → | $M_5$: $J_1$,$J_3$ | $M_4$: $J_2$, $J_5$ | $M_6$: $J_4$ | $M_3$: $J_6$ |

**Figure 1** *Plates-Jobs* based on *SOJ*

Furthermore, the *Plates-Jobs* of Ft06 are reorganized one-by-one, based on the numbers of the machines, to create *mPlates-Jobs*. For example, $M_1$ has three Plates-Jobs related to $O_{i2}$ ($J_1$ and $J_4$), $O_{i4}$ ($J_3$ and $J_6$) and $O_{i5}$ ($J_2$ and $J_5$) respectively. These operations are belongs to the same machine $M_1$. The same process will be repeated to the remaining machines $M_2$ to $M_6$. The *mPlates-Jobs* is illustrated in Figure 2.

| | | | | | | |
|---|---|---|---|---|---|---|
| *mPlates-Jobs* related to $M_1$ → | $O_{i2}$: $J_1$, $J_4$ | | $O_{i4}$: $J_3$, $J_6$ | | $O_{i5}$: $J_2$, $J_5$ | |
| available processing orders → | - | - | - | - | - | - |
| *mPlates-Jobs* related to $M_2$ → | $O_{i1}$: $J_2$, $J_4$, $J_6$ | | | $O_{i2}$: $J_5$ | $O_{i3}$: $J_1$ | $O_{i5}$: $J_3$ |
| available processing orders → | - | - | - | - | - | - |
| *mPlates-Jobs* related to $M_3$ → | $O_{i1}$: $J_1$, $J_3$, $J_5$ | | | $O_{i2}$: $J_2$ | $O_{i3}$: $J_4$ | $O_{i6}$: $J_6$ |
| available processing orders → | - | - | - | - | - | - |
| *mPlates-Jobs* related to $M_4$ → | $O_{i2}$: $J_3$, $J_6$ | | $O_{i4}$: $J_1$, $J_4$ | | $O_{i6}$: $J_2$, $J_5$ | |
| available processing orders → | - | - | - | - | - | - |
| *mPlates-Jobs* related to $M_5$ → | $O_{i3}$: $J_2$, $J_5$ | | $O_{i5}$: $J_4$, $J_6$ | | $O_{i6}$: $J_1$,$J_3$ | |
| available processing orders → | - | - | - | - | - | - |
| *mPlates-Jobs* related to $M_6$ → | $O_{i3}$: $J_3$, $J_6$ | | $O_{i4}$: $J_2$, $J_5$ | | $O_{i5}$: $J_1$ | $O_{i6}$: $J_4$ |
| available processing orders → | - | - | - | - | - | - |

**Figure 2** *mPlates-Jobs* obtained after reorganizing the *Plates-Jobs*

Individually *mPlate-Jobs* is split into two parts where the top part indicates the available jobs of the *mPlate-Jobs* and indexed operation, and the lower part shows an available process orders of the *mPlatet-Job*. The number of rows and its elements on the lower part of *mPlates-Jobs* are similar to the number of rows and its elements on $SJM$. The empty *mPlates-Jobs* will be filled up with the available jobs on each *mPlate* as shown in Figure 3.

| | | | | | | |
|---|---|---|---|---|---|---|
| mPlates-Jobs related to $M_1$ → | $O_{i2}: J_1, J_4$ | | $O_{i4}: J_3, J_6$ | | $O_{i5}: J_2, J_5$ | |
| The first row of $SJM$ → | $J_1$ | $J_4$ | $J_6$ | $J_3$ | $J_2$ | $J_5$ |
| mPlates-Jobs related to $M_2$ → | $O_{i1}: J_2, J_4, J_6$ | | | $O_{i2}: J_5$ | $O_{i3}: J_1$ | $O_{i5}: J_3$ |
| The second row of $SJM$ → | $J_2$ | $J_6$ | $J_4$ | $J_5$ | $J_1$ | $J_3$ |
| mPlates-Jobs related to $M_3$ → | $O_{i1}: J_1, J_3, J_5$ | | | $O_{i2}: J_2$ | $O_{i3}: J_4$ | $O_{i6}: J_6$ |
| The third row of $SJM$ → | $J_3$ | $J_1$ | $J_5$ | $J_2$ | $J_4$ | $J_6$ |
| mPlates-Jobs related to $M_4$ → | $O_{i2}: J_3, J_6$ | | $O_{i4}: J_1, J_4$ | | $O_{i6}: J_2, J_5$ | |
| The fourth row of $SJM$ → | $J_3$ | $J_6$ | $J_4$ | $J_1$ | $J_2$ | $J_5$ |
| mPlates-Jobs related to $M_5$ → | $O_{i3}: J_2, J_5$ | | $O_{i5}: J_4, J_6$ | | $O_{i6}: J_1, J_3$ | |
| The fifth row of $SJM$ → | $J_2$ | $J_5$ | $J_6$ | $J_4$ | $J_3$ | $J_1$ |
| mPlates-Jobs related to $M_6$ → | $O_{i3}: J_3, J_6$ | | $O_{i4}: J_2, J_5$ | | $O_{i5}: J_1$ | $O_{i6}: J_4$ |
| The sixth row of $SJM$ → | $J_6$ | $J_3$ | $J_2$ | $J_5$ | $J_1$ | $J_4$ |

**Figure 3** *mPlates-Jobs* filled up with available jobs on each *mPlate-Jobs*

Therefore, the $SJM$ *mPlates-Jobs* is:

$$SJM = \begin{matrix} M_1 \rightarrow \\ M_2 \rightarrow \\ M_3 \rightarrow \\ M_4 \rightarrow \\ M_5 \rightarrow \\ M_6 \rightarrow \end{matrix} \begin{bmatrix} J_1 & J_4 & J_6 & J_3 & J_2 & J_5 \\ J_2 & J_6 & J_4 & J_5 & J_1 & J_3 \\ J_3 & J_1 & J_5 & J_2 & J_4 & J_6 \\ J_3 & J_6 & J_4 & J_1 & J_2 & J_5 \\ J_2 & J_5 & J_6 & J_4 & J_3 & J_1 \\ J_6 & J_3 & J_2 & J_5 & J_1 & J_4 \end{bmatrix} = \begin{bmatrix} 1 & 4 & 6 & 3 & 2 & 5 \\ 2 & 6 & 4 & 5 & 1 & 3 \\ 3 & 1 & 5 & 2 & 4 & 6 \\ 3 & 6 & 4 & 1 & 2 & 5 \\ 2 & 5 & 6 & 4 & 3 & 1 \\ 6 & 3 & 2 & 5 & 1 & 4 \end{bmatrix} \quad (3)$$

Based on (4), the makespan (completion time) of the $SJM$ obtained is 60 while the optimal makespan for FT06 is 55. It shows the ability and efficiency of the proposed classification of *mPlates-Jobs.*

### *3.2.2 Generating Rules*

Furthermore to continue the process in producing initial population, we create rules to conform:

**1:** *Two consecutive operations of mPlates on a given machine can join together.*

**2:** *If $SJM$ on a given machine do not match their own mPlates or consecutive mPlates, move these jobs towards their own mPlates*

**3:** *Sorted the jobs on their mPlates based on the maximum of the remainder processing times (RPT).*

**4:** *If two consecutive mPlates on a given machine are joined together and one job is formed, the first mPlate becomes a candidate to move to the next mPlate, and the processing order of the candidate job on the operation, should be recalculated.*

### *3.2.3 Skipping Strategy (SS)*

By considering the conditions and following the generated rules we then employ a skipping strategy (SS) to obtain initial population. The brief outline of SS is as follows:

**Step 1:** *Define mPlates, the real processing orders ($r_{order}$) of jobs related to their mPlates and their current processing orders ($c_{order}$) on $M_\alpha$/row $\alpha$ of the $SJM$.*

**Step 2:** *Consider mPlate on $M_\alpha$, if it satisfy (4), sort its $c_{order}$ on $SJM$ follow the third rule of mPlates-Jobs, recalculate makespan of the exchanged primal $SJM$.*

$$\begin{cases} CT(O_{ij}) + PT(O_{rs}) \leq ST(O_{ij+1}) \\ \text{and} \\ \left( \begin{array}{c} \left(CT(O_{rs-1}) \leq ST(O_{ij})\right) \\ \text{or} \\ \left(ST(O_{ij}) + PT(O_{rs}) \leq ST(O_{rs+1})\right) \end{array} \right) \end{cases} \quad (4)$$

**Step 3:** *Match the $c\_orders$ of jobs on the primal $SJM$ with their $r\_orders$ predefined on the mPlate or consecutive mPlates (the first and second rule of mPlates-Jobs), by moving the $c\_orders$ of jobs toward their $r\_orders$ one-by-one, if condition (5) or (6) is satisfied, and then recalculate the makespan of the exchanged primal $SJM$.*

$$\begin{cases} \theta \leq CT(O_{rs+1}) \\ \text{and} \\ max\left(CT(O_{ij-1}), \theta\right) + PT(O_{ij}) \leq CT(O_{ab}) \end{cases} \quad (5)$$

*where $\theta = max\left(CT(O_{rs-1}), CT(O_{pq})\right) + PT(O_{rs})$, if there is no $J_a$ then $CT(O_{ab})$ equals the makespan of the primal $SJM$, and if $J_p$ is not available then $CT(O_{pq})$ equals zero. Conversely, if $\rho < \tau$, the jobs with c_orders $\rho - 1$, $\rho + 1$ and $\rho + 2$ are $J_p$ with operation $O_{pq}$, $J_i$ with operation $O_{ij}$ and $J_a$ with operation $O_{ab}$, respectively. If they satisfy condition (6), then swap the $c\_orders$ of $J_r$ and $J_i$, recalculate the makespan of the exchanged primal $SJM$, and repeat the procedure until the current order of $J_r$ reaches $\tau$.*

$$\begin{cases} \delta \leq CT(O_{ij+1}) \\ \text{and} \\ max(CT(O_{rs-1}), \delta) + PT(O_{rs}) \leq CT(O_{ab}) \end{cases} \quad (6)$$

*where $\delta = max\left(CT(O_{ij-1}), CT(O_{pq})\right) + PT(O_{ij})$, if there is no $J_a$ then $CT(O_{ab})$ equals the makespan of the primal $SJM$, and if $J_p$ is not available then $CT(O_{pq})$ equals zero.*

**Step 4:** *Any job in the mPlate next to the chosen mPlate with smaller RPT than the maximum of the RPTs on the chosen mPlate can get the order of the chosen mPlate (the fourth rule of mPlates-Jobs).*

**Step 5:** *At the same time as moving jobs from their current order to a new order (Steps 2-4).*

## 4. Results and Discussion

We implemented our proposed method to some of benchmark data and compared the result to those of some other algorithms in term of relative percentage of error (RPE) for the completion time.

$$RPE = \frac{bf - BKS}{BKS} \times 100 \quad (7)$$

where $bf$ is the best quality of initial population produced,

BKS is the Best Known Solution

$bf^{Target}$ is the best quality of generated points by the target algorithm.

Result comparison of our experiment is shown in Table 1, where the deviation from the BKS of the best point in the initial population produced by SS is on average 3.6% for Ft06, Ft10, and Ft20 datasets, 4.3% for La01–40 datasets, and so on. The negative RPE for Willem's dataset [18] by SS means that SS has succeeded in generating a point in the initial population 2.1% better than the BKS for Willem's dataset.

**Table 1** Result comparison with BKS

| | Ft | La | Abz | Ord | Swv | Willem | Yn |
|---|---|---|---|---|---|---|---|
| RPE (SS) | 3.6% | 4.3% | 7.9% | 9.6% | 23.7% | -2.1% | 18.3% |

## 5. Conclusion

In this paper, a new strategy (SS) to generate an initial population to near optimal solution has been proposed. The SS was designed based on *mPlates-Jobs*, with respect to $SOJ$. Then we create rules by considering the solution space, monitoring the processing order of jobs over machines, and checking their effects on the quality of the solutions and on gaps created in their schedules.

From the result obtained it is shown that SS has the ability to produce any size of initial population compare to those of available valid methods. The important point is the observation of the best known solution among the initial population of FT06, FT10, La04 and La15.

## 6. Acknowledgements

We would like to thank Prof. Dr. Runtung Sitepu, Rector of Universitas Sumatera Utara, and Lembaga Penelitian Universitas Sumatera Utara, Medan, Indonesia for supporting this research and to The Ministry of Higher Education of Malaysia under the Centre for Research and Instrumentation Management, Universiti Kebangsaan Malaysia (ERGS Code: ERGS/1/2013/ICT07/UKM /02/5). We would also like to thank Prof. Dr. Salwani Abdullah of the Data Mining and Optimization Research Group Universiti Kebangsaan Malaysia for her contributions to this research.

## References

[1] Brucker P, Jurisch B and Sievers B 1994 A branch and bound algorithm for the job-shop scheduling problem *Discrete Applied Mathematics* 49(**1**) 107-127

[2] Nababan E B, Abdullah S, Hamdan A R, Zakaria M S 2008 A branch and bound algorithm in optimizing job shop scheduling problems *IEEE Int. Symp. on Information Technology ITSIM'08*

[3] Vilcot G and Billaut J -C 2011 A tabu search algorithm for solving a multicriteria flexible job shop scheduling problem *Int. Journal of Production Research* 49(**23**) p 6963-80

[4] Nababan E B, Sitompul O S 2011 Manipulating tabu list to handle machine breakdowns in Job shop scheduling problem *American Institute of Physic (AIP) Conf.* 1337(**1**) p 224-9.

[5] Shaheed I M, Syukor S A, Nababan E B 2016 An empirical analysis of the relationship between the initialization method performance and the convergence speed of a meta-heuristic for Fuzzy Job-Shop scheduling problems *Journal of Theoretical and Applied Information Technology* 93(**2**) pp 297-311

[6] Moon I, Lee S,and Bae H 2008 Genetic algorithms for job shop scheduling problems with alternative routings *Int. Journal of Production Research* 46(10) p 2695-2705

[7] Zhang R, and Wu C 2011 A simulated annealing algorithm based on block properties for the job shop scheduling problem with total weighted tardinessobjective *Computers & Operations Research* 38(5) p 854-867

[8] Seo M and Kim D 2010 Ant colony optimisation with parameterised search space for the job shop scheduling problem *Int.l Journal of Production Research* 48(4) p 1143-54

[9] Hong-Wei G, Liang S, Yan-Chun L, and Feng Q 2008 An effective PSO and AIS-based hybrid intelligent algorithm for job-shop scheduling *IEEE Transactions*

[10] Qiu X and Lau H 2012 An AIS-based hybrid algorithm for static job shop scheduling problem *Journal of Intelligent Manufacturing* p 1-15

[11] Abdolrazzagh-Nezhad M, Nababan E B, Jaddi N S, Sarim H M 2016 Electromagnetic-like mechanism for job-shop scheduling by novel heuristic initialization *Journal of Theoretical and Applied Information Technology* 93(**1**) pp 174-184

[12] Zhixiang Y, Jianzhong C, Yan Y and Ying M 2006 Job shop scheduling problem based on DNA computing *Journal of Systems Engineering and Electronics* 17(3) p 654-9

[13] Shi-Jin W, Bing-Hai Z and Li-Feng X 2008 A filtered-beam-search-based heuristic algorithm for flexible job-shop scheduling problem *Int. Journal of Production Research* 46(11) p 3027-58.

[14] Wisittipanich W and Kachitvichyanukul V 2012 Two enhanced differential evolution algorithms for job shop scheduling problems *Int. Journal of Production Research* 50(10) p 2757-73

[15] Sha D Y and Lin H H 2010 A multi-objective PSO for job shop scheduling problems *Expert Systems with Applications* 37 p 1065-70.

[16] Amirthagadeswaran K S and Arunachalam V P 2006 Improved solutions for job shop scheduling problmes through genetic algorithm with a different method of schedule deduction *Int. Journal of Advanced Manufacturing Technology* 28 p 532-540

[17] Yahyaoui A, Fnaiech N and Fnaiech F 2011 A suitable initialization procedure for speeding a neural network job-shop scheduling *Industrial Electronics IEEE Transactions* 58(3) p 1052 - 60.

[18] Willems T M and Brandts L W 1995 Implementing heuristics as an optimation criterion in neural networks for job shop scheduling *Journal of Intelligent Manufacturing* 6 p 377-387.

IOP Conf. Series: Journal of Physics: Conf. Series **978** (2018) 012055 doi:10.1088/1742-6596/978/1/012055

# Hierarchical Rhetorical Sentence Categorization for Scientific Papers

**G H Rachman[1], M L Khodra[2], D H Widyantoro[3]**

[1,2,3]School of Electrical Engineering and Informatics, Institut Teknologi Bandung, Bandung, Indonesia

[1]ghoziyahaitan@gmail.com, [2]masayu@stei.itb.ac.id, [3]dwi@stei.itb.ac.id

**Abstract.** Important information in scientific papers can be composed of rhetorical sentences that is structured from certain categories. To get this information, text categorization should be conducted. Actually, some works in this task have been completed by employing word frequency, semantic similarity words, hierarchical classification, and the others. Therefore, this paper aims to present the rhetorical sentence categorization from scientific paper by employing TF-IDF and Word2Vec to capture word frequency and semantic similarity words and employing hierarchical classification. Every experiment is tested in two classifiers, namely Naïve Bayes and SVM Linear. This paper shows that hierarchical classifier is better than flat classifier employing either TF-IDF or Word2Vec, although it increases only almost 2% from 27.82% when using flat classifier until 29.61% when using hierarchical classifier. It shows also different learning model for child-category can be built by hierarchical classifier.

## 1. Introduction

Actually domain of scientific papers become one kind of document that is often needed by academics and researchers. Then they use it as reference or comparison for their research by collecting relevant information [2]. This information can be composed of rhetorical sentences to be structured from some categories as needed by readers [3]. Rhetorical sentence itself is each sentences in document which is assigned with label of defined rhetorical categories based on its text features and rhetorical structure is a connected structure of texts, in which every segment of a text (i.e. sentence) has a meaningful category in the body of each section in the document [1].

Teufel [4] initially conducted rhetorical sentence categorization by assigning one of seven rhetorical categories to every sentence. These seven are "aim", "textual", "own", "background", "contrast", "basis", and "other". After that, these seven is improved into 15 rhetorical categories [5], namely "aim", "nov_adv", "co_gro", "othr", "prev_own", "own_mthd", "own_fail", "own_res", "use", "own_conc", "codi", "gap_weak", "antisupp", "support", and "fut". For seven rhetorical categories, Merity et al. [6] has improved its F-measure performance become 92.93% by using Maximum Entropy classifier from Teufel's result [7]. Besides that, Widyantoro, et al [8] implemented 15 categories added with "textual" and employed multi-heterogeneous classifier. Its average of F-measure result is about 25%. They still used the technique of flat classification that only considers leaf nodes (child-category) without parent-category, as shown in Figure 1, in which no structure defines relationship between the existing categories [9] [10]. So based on our best knowledge, there is no hierarchical classification yet that is conducted for rhetorical sentence categorization in scientific papers. Due to this condition, in this paper, we develop hierarchical classifier for rhetorical sentence categorization by employing rhetorical categories from Teufel, et al [5]. This paper is actually next results of our research in rhetorical sentence categorization.

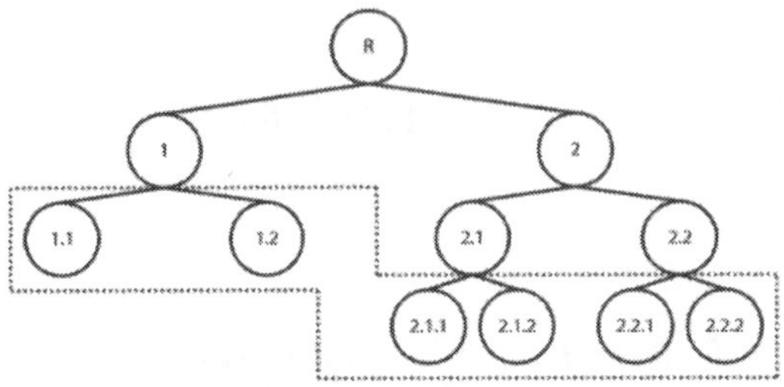


**Figure 1.** Flat classification approach using a flat multi-class classification [10]

The rest of this paper is organized as follows. The next section provides the related works on rhetorical sentence categorization from Teufel and implementation of Word2Vec semantic representation and hierarchical classification. Our method is explained in section 3. We also define various rhetorical categories and feature sets for classifying sentences. We describe our experiments in section 4. The results are analyzed in section 5. Finally, the conclusion and further work are described in section VI.

## 2. Related Work

Rhetorical sentence categorization is commonly used for scientific papers. Teufel and Moens [7] conducted this task by assigning one of seven rhetorical categories to every sentence. These seven are "aim", "textual", "own", "background", "contrast", "basis", and "other". They employed Naïve Bayes classifier and got the F-measure result from 26% to 86%. Then, its result has risen up until 92.93% after employing Maximum Entropy [6]. These seven categories next become 15 rhetorical categories [5], namely "aim", "nov_adv", "co_gro", "othr", "prev_own", "own_mthd", "own_fail", "own_res", "use", "own_conc", "codi", "gap_weak", "antisupp", "support", and "fut". The distribution detail of 15 categories from 7 rhetorical categories is shown in Table 1.

**Table 1.** Distribution of 15 category from 7 rhetoric category

| 7 Rhetoric Categories [4] | 15 Rhetoric Categories [5] |
|---|---|
| Aim | Aim |
| Background | Co_gro |
| Other | Othr |
| | Prev_own |
| Own | Own_mthd |
| | Own_fail |
| | Own_res |
| | Own_conc |
| Basis | Use |
| | Support |
| | Antisupp |
| Contrast | Codi |
| | Gap_weak |
| Textual* | |
| | Nov_adv** |
| | Fut** |

*This category is not continued in 15 rhetorical categories
**Teufel did not involve these categories to any 7 rhetorical categories

Widyantoro, et al. [8] implemented these 15 categories added with "textual" by adapting Teufel's features and employing multi-heterogeneous classifier. Its average of F-measure result is about 25%. Actually, they used different corpus from Teufel's [4]. This condition could affect the final rhetorical categorization model, because Teufel employed meta-discourse feature, in which their existing grammars depend on word-patterns that always appear in sentences from corpus.

For the second, Word2Vec is used to group words which have similar meaning into vector representation. This model is proposed by Mikolov [12] and released by Google in 2013. There are two architectures of Word2Vec, namely the continuous bag-of-words (CBOW) and skip-gram. The CBOW predicts the current word based on the context and the skip-gram predicts surrounding words given the current word [13]. Heffernan and Teufel [14] employed Word2Vec representation to identifying problem statement in scientific text. They used 18,753,472 sentences from a biomedical corpus based on all full-text Pubmed articles and then built model from the 200 semantically similar words to only "problem". Its result showed that Word2Vec model caused significant performance increase, because Word2Vec attributes have the greatest information gain compared the other features.

At last, flat classification is commonly conducted for text categorization. Actually, there is another technique named as hierarchical classification. The difference between them is flat classification only considers leaf nodes (child-category) without parent-category [9] [10], but hierarchical classification considers parent-category first then child-category from each parent. It assigns more suitable categories from the defined hierarchical category in document [15]. Tegegnie, et al [16] classified a heterogeneous collection of Amharic News Text by employing hierarchical structure. It shows that hierarchical classifier is better than flat classifier. The accuracy when using level three classifier with the top 15 features is 89.06%. It significantly improves 29.42% from the accuracy when using flat classifier.

## 3. Methodology

In this research, we use the data set of scientific paper from Khodra, et al. [3] added with 50 new scientific papers from ACL Anthology Reference Corpus (ACL-ARC). This data has 75 annotated scientific papers that have been split into 10,880 sentences. It is added with 50 new scientific papers, then our data set has 16,046 sentences. Training data is 100 and testing is 25 scientific papers. The number of training data set is 12,594 and testing data set is 3,452.

Every sentence in these papers has been annotated into one of 16 rhetorical categories from Teufel, et al. [5], namely "aim", "nov_adv", "co_gro", "othr", "prev_own", "own_mthd", "own_fail", "own_res", "own_conc", "codi", "gap_weak", "antisupp", "support", "use", "fut", and "textual". The description of these categories is explained in [8]. We adapt features from Teufel [4] and [3], namely content, absolute location, explicit structure, sentence length, syntax, citation, and formulaic & agentivity (meta-discourse). Its explanation can be seen in Table 2.

**Table 2.** Features and its description

| Type | Name | Description |
|---|---|---|
| | Cont-1 | Occurrence of 10 significant terms of document using TF-IDF |
| **Content** | Cont-2a | Incidence of words occurring in document title |
| | Cont-3 | Occurrence of 10 significant terms of abstract using TF-IDF |
| **Absolute location** | Loc | Sentence position in document to 10 segments |
| | Struct-1 | Sentence position within section |
| **Explicit structure** | Struct-2 | Sentence position within paragraph |
| | Struct-3 | Prototypical type of section title |
| **Sentence length** | Length | Whether sentence has longer than 15 words or not |
| **Syntax** | Syn | Occurrence of 1st finite verb and auxiliary modal |
| | Adj | Occurrence of adjective |
| **Citations** | Cit-1 | Occurrence of citation or self-citation |
| | Cit-2 | Citation location in sentence |
| **Formulaic** | Formu | Occurrence of formulaic expression |
| | Ag-1 | Occurrence of agent type |
| **Agentivity** | Ag-2 | Occurrence of action type |
| | Negation | Occurrence negation in sentence |

Our method is generally divided into four main processes. The first is pre-processing, and then constructing the vector representations of the vocabularies of data set by using TF-IDF and Word2Vec. The second is producing all the features. The third is building classification model. This will employ technique of hierarchical classification. The last is we conduct testing to get the F-measure by using some classifiers, namely Naïve Bayes [17] and SVM Linear utilizing LibSVM [18] library. The general process is shown in Figure 2.

Technique of hierarchical classification is conducted to the child category after doing the first classification in parent-category. So, there are two step of flat classification for parent-category and child-category. In this research, we use this technique for category "contrast", "own", "basis", and "other" because only these categories are divided into some child-categories [4] [5]. This hierarchical rhetorical category schema that will be done is shown in Figure 3. As shown in Table 1, "nov_adv" and "fut" are not derivative from any category, but in this paper we assume that these two are elaborated from category "own". It is due to the definition of "nov_adv" and "fut" that is mostly related to the definition of "own".

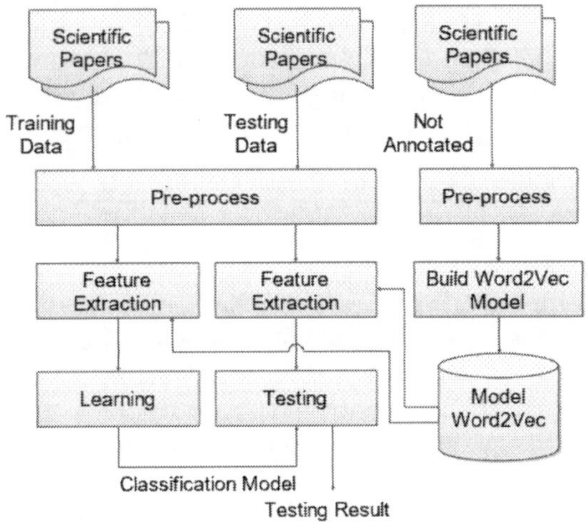


**Figure 2.** General process of our methodology

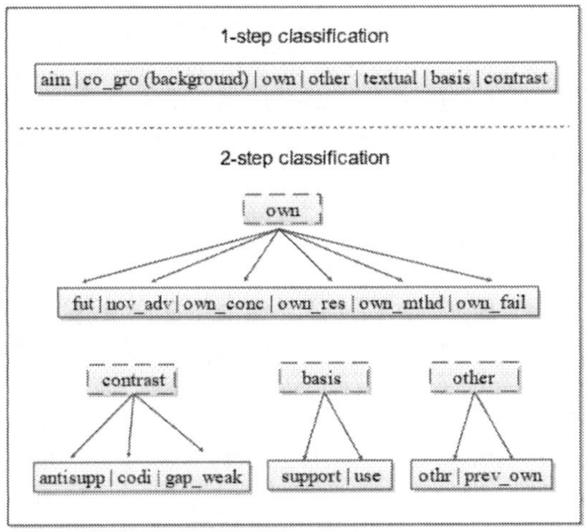


**Figure 3.** Hierarchical rhetorical category

We consider four pre-processing that consist of case folding, tokenization, stemming, and stop word elimination. These are done using Apache Lucene library and Weka [19]. All of them will always be applied in all experiments. Then this paper employs Word2Vec representation algorithm from Medallia[1] library. We use 20,157 scientific papers taken from ACL Anthology Reference Corpus (ACL-ARC) to build the word vector learning model itself. After that, we use our annotated pre-processed sentences to output word vectors as feature of words. The weight of word is calculated through TF-IDF and model from Word2Vec representation (Skip-Gram). We compare the usage of them. For Word2Vec, we set its layer size into the vector length of 300. In the end, all classification models are produced from the last process.

Especially for hierarchical classification, we employ SMOTE [20] and SpreadSubsample from Weka to try dealing with imbalance data set. Actually, the difference of this paper from our previous work [11], namely: employing meta-discourse feature from Teufel's; not using previous label; using Naïve Bayes and SVM; doing hierarchical classification technique; and adding data set for 16 rhetorical categories and data set for Word2Vec.

### *3.1. Experiment*

[1] https://github.com/medallia/Word2VecJava

In addition, there are four experiments that is conducted. It is related to implementation of TF-IDF, Word2Vec of Skip-Gram, and hierarchical classification. These are as follows:

- Baseline. The features to use are content, absolute location, explicit structure, sentence length, syntax, formulaic & agentivity (meta-discourse), and citation. These are adapted from [4] and [3];
- Scenario 1. In this experiment, all the features from baseline are used. Then, we employ TF-IDF to give score for bag of words from data set;
- Scenario 2. This experiment will employ all the features and involve Word2Vec to give score of semantic similarity words by using Skip-Gram algorithm. This scenario is conducted to compare the result between using Word2Vec and TF-IDF;
- Scenario 3. This experiment will employ all features from baseline and Word2Vec representation, then we conduct the hierarchical classification from seven rhetorical categories until 16 categories. In second step of hierarchical classification, we also employ SpreadSubsample and SMOTE.

## 4. Result and Analysis

First, we built the word vector learning model itself for the 20,157 scientific papers taken from ACL Anthology Reference Corpus (ACL-ARC). Besides that, we use 100 scientific papers for training and 25 for testing. We extracted all the features regarding scenario we planned. This was evaluated by two classifiers, namely Naïve Bayes and SVM Linear.

**Table 3.** Comparison of f-measure result from the first three experiments

| No | Rhetorical Category | Baseline | | Scenario 1 (TF-IDF) | | Scenario 2 (Word2Vec) | |
|---|---|---|---|---|---|---|---|
| | | Naïve Bayes | SVM Linear | Naïve Bayes | SVM Linear | Naïve Bayes | SVM Linear |
| 1 | othr | 13.40 | 3.50 | 10.50 | 23.20 | 4.60 | 15.30 |
| 2 | prev_own | 16.00 | 22.00 | 21.80 | 9.70 | 27.50 | 22.00 |
| 3 | fut | 32.30 | 35.30 | 17.80 | 9.10 | 13.70 | 34.60 |
| 4 | nov_adv | 15.50 | 0.00 | 4.70 | 0.00 | 2.80 | 10.50 |
| 5 | own_conc | 25.30 | 26.40 | 15.50 | 27.30 | 11.70 | 33.80 |
| 6 | own_fail | 4.50 | 0.00 | 2.90 | 0.00 | 0.00 | 0.00 |
| 7 | own_mthd | 68.10 | 68.00 | 32.80 | 69.20 | 17.20 | 71.50 |
| 8 | own_res | 10.90 | 0.00 | 29.50 | 13.60 | 9.40 | 24.60 |
| 9 | antisupp | 0.00 | 0.00 | 0.00 | 0.00 | 1.40 | 0.00 |
| 10 | codi | 0.00 | 0.00 | 1.80 | 0.00 | 2.50 | 0.00 |
| 11 | gap_weak | 23.40 | 7.70 | 15.80 | 7.50 | 12.60 | 31.20 |
| 12 | aim | 46.00 | 49.00 | 29.70 | 46.40 | 15.70 | 58.70 |
| 13 | textual | 17.90 | 1.00 | 36.00 | 51.70 | 23.30 | 63.20 |
| 14 | support | 21.40 | 18.80 | 12.00 | 19.60 | 8.00 | 23.40 |
| 15 | use | 16.80 | 17.70 | 12.60 | 14.20 | 16.30 | 17.30 |
| 16 | co_gro | 30.90 | 33.70 | 18.60 | 28.20 | 17.10 | 39.00 |
| Total | | 21.40 | 17.69 | 16.38 | 19.98 | 11.49 | 27.82 |

Table 3 shows the result of first three experiments. After employing bag of words TF-IDF feature in Scenario 1, the performance of baseline experiment using only Teufel's features is still higher than adding bag of words TF-IDF. It can be because co-occurrence of significant words using TF-IDF has been applied for feature of content (Cont-1 and Cont-3). This feature applies TF-IDF more specifically, in which the significant words produced for each document are different. Meanwhile bag of words TF-IDF is more general for all documents. So that it does not have significant impact for the performance.

In baseline, Naïve Bayes produced the highest F-measure result that is 21.40%, but in Scenario 1 and 2 it keeps decreasing when using bag of words. It is proven that Naïve Bayes is not suitable to handle high dimensional feature caused by bag of words, either employing TF-IDF or Word2Vec. This result is different when using SVM Linear. From baseline until Scenario 2, the F-measure result has risen up from 17.69% to 27.82%. This indicates that the classification method is suitable for high dimensional feature though.

Besides that, the increasing performance is because either TF-IDF or Word2Vec can capture some important words that often appear in sentences but do not exist in the dictionary of meta-discourse (formulaic & agentivity) from Teufel's [4]. Meanwhile only Word2Vec can capture words that is similar to those important words. Co-occurrence of these similar important words can affect the decision of assigning category to sentences. Some of these words are align, annotate, argument, aspect, concept, consist, context, data, depend, differ, direct, distribute, document, domain, embed, express, extract, feature, function, generate, grammar, include, inform, knowledge, label, language, learn, and many others.

Actually, as shown in Table 3, some categories do not improve at all after applying either TF-IDF or Word2Vec representation. These are "own_fail", "antisupp", and "codi". It can be because these categories have low number of sentences and no related important words appear often in sentences from these categories. This result is different from category "textual". This category significantly improves from experiment of baseline until Scenario 2. It means there are some important words that almost always appear in sentence from this category. Scenario 2 can capture these words, while baseline experiment cannot. Besides that, some categories have fairly stable performance since baseline experiment, just like "own_mthd", "aim", "fut", and "co_gro". It means that sentences from these categories have patterns score of baseline features that often appear. So it has been good enough to define these categories without TF-IDF or Word2Vec.

**Table 4.** Comparison of f-measure result from scenario 3

| No | Rhetorical Category | Hierarchical Classifier Step 1 SVM Linear | Rhetorical Category | Hierarchical Classifier Step 2 SVM Linear |
|---|---|---|---|---|
| 1 | other | 18.20 | othr | 14.20 |
| 2 | | | prev_own | 21.10 |
| 3 | own | 80.60 | fut | 40.00 |
| 4 | | | nov_adv | 20.10 |
| 5 | | | own_conc | 23.10 |
| 6 | | | own_fail | 8.30 |
| 7 | | | own_mthd | 71.30 |
| 8 | | | own_res | 31.50 |
| 9 | contrast | 19.90 | antisupp | 9.50 |
| 10 | | | codi | 4.00 |
| 11 | | | gap_weak | 20.00 |
| 12 | aim | 59.50 | aim | 59.50 |
| 13 | textual | 62.50 | textual | 62.50 |
| 14 | basis | 35.80 | support | 32.30 |
| 15 | | | use | 17.30 |
| 16 | background | 39.10 | co_gro | 39.10 |
| Total | | 45.09 | Total | 29.61 |

In Scenario 3, we only employ SVM Linear because it has been proven in most previous experiment that this is the best classifier so far for high dimensional feature. Then we conduct hierarchical classification for category "contrast", "own", "basis", and "other" because only these categories are divided into some child-categories [4] [5]. We build new learning model for child-category from these categories. Its total is four model. The advantage of applying hierarchical classification, we can build learning model for child-category by using the features we select. Every model can be different. Actually, for child-category from "own", "basis", and "other", we employ all features from Teufel [4] and [3], Word2Vec representation, and SMOTE. We employ SMOTE to know that whether it can improve some categories that have F-measure 0.00% in Scenario 2 with SVM Linear or not. Then learning model of child-category from "contrast" employs all features from Teufel, Word2Vec representation, and SpreadSubsample.

The result of Scenario 3 is shown in Table 4. There is no F-measure 0.00% for "own_fail", "antisupp", and "codi". The F-measure result for "own_fail" is 8.30%, "antisupp" 9.50%, and "codi"

4.00%. It is proven that SMOTE and SpreadSubsample can improve performance of categories with the low number sentences. Meanwhile, as shown in Fig 4, the performance of some categories has decreased quite highly, just like "own_conc" and "gap_weak". They decrease about 10% until 11%. It can happen because employing SMOTE in parent-category "own" and SpreadSubsample in parent-category "contrast". In category "own", we only employ SMOTE for "own_fail", "nov_adv", and "own_res" that have the low number of sentences. Besides that, in category "contrast", we only employ SpreadSubsample for "antisupp" and "codi". This means that SMOTE and SpreadSubsample can decrease the performance of categories that is not set.

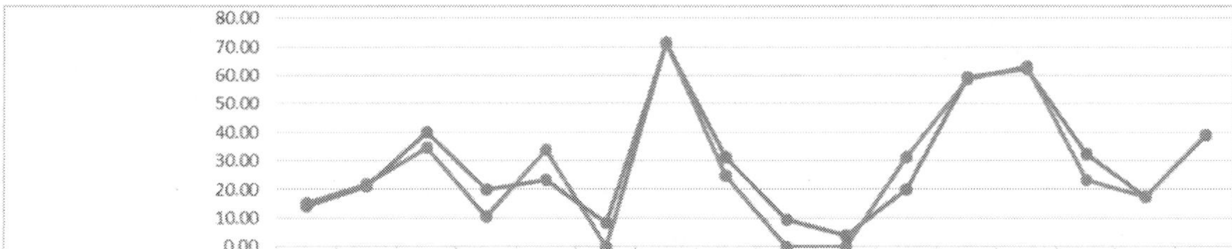

| | othr | prev_own | fut | nov_adv | own_conc | own_fail | own_mthd | own_res | antisupp | codi | gap_weak | aim | textual | support | use | co_gro |
|---|---|---|---|---|---|---|---|---|---|---|---|---|---|---|---|---|
| Hierarchical (Scenario 3) | 14.20 | 21.10 | 40.00 | 20.10 | 23.10 | 8.30 | 71.30 | 31.50 | 9.50 | 4.00 | 20.00 | 59.50 | 62.50 | 32.30 | 17.30 | 39.10 |
| Flat (Scenario 2) | 15.30 | 22.00 | 34.60 | 10.50 | 33.80 | 0.00 | 71.50 | 24.60 | 0.00 | 0.00 | 31.20 | 58.70 | 63.20 | 23.40 | 17.30 | 39.00 |

**Figure 4.** Comparison between f-measure result from scenario 2 and 3

Actually, the result between Scenario 2 using flat classification and Scenario 3 using hierarchical classification in Figure 4 shows that hierarchical classifier is better than flat classifier, although its difference is not significant. It increases only almost 2% from 27.82% until 29.61%. Hierarchical classifier can be better because we can build different learning model for child-category by using the features we select. For example, we employ SpreadSubsample only for parent-category "contrast" which is divided into "antisupp", "gap_weak", and "codi". We can select features as needed and suitable for group of child-category. In flat classification, we cannot conduct this technique, because we only build one learning model for all categories. Therefore, this hierarchical classification model is considered.

## 5. Conclusion and Further Work

The implementation of Word2Vec can improve the performance of some rhetorical sentence classification from baseline scenario. It can be because Word2Vec model can capture some related important words that do not exist in meta-discourse dictionary from [4]. The result is different from implementation of TF-IDF. This feature does not significantly affect the performance, because co-occurrence of significant words using TF-IDF has been applied for feature of content (Cont-1 and Cont-3). This feature applies TF-IDF more specifically, in which the significant words produced for each document are different. Besides that Word2Vec is also better than TF-IDF, because TF-IDF only capture co-occurrence of important words not related words.

Furthermore our work shows that the implementation of hierarchical rhetorical sentence categorization improves the performance of rhetorical sentence categorization, although is not significant. Hierarchical classifier is better than flat classifier because we can build different learning model for child-category. We can select features as needed and suitable for group of child-category. In flat classification, we cannot conduct this technique, because we only build one learning model for all categories. In addition, we do not use previous label in this paper, because in second step of hierarchical classification, we have to build model for group of child-category, in which its label is not structured any more like 16 rhetorical categories. So it is difficult to predict previous label and define "start" label like in [11].

For the future work, we plan to conduct research for every child-category in 16 rhetorical categories from Teufel [5]. It is to know that whether every child-category from same parent-category have high similar category or not.

**References**

[1] Taboada M and Mann W C 2006 *Rhetorical structure theory: Looking back and moving ahead* Discourse studies vol **8** no 3 pp 423-59

[2] Schwegler R A and Shamoon L K 1982 *The Aims and Process of the Research Paper College* English vol **44** no 8 pp 817-24

[3] Khodra M L, Widyantoro D H, Aziz E A and Trilaksono B R 2012 Automatic Tailored Multi-Paper Summarization based on Rhetorical Document Profile and Summary Specification *Journal of ICT Research and Applications* vol **6** no 3 pp 220-239

[4] Teufel S 1999 *Argumentative Zoning: Information Extraction from Scientific Text* (Edinburgh: University of Edinburgh)

[5] Teufel S, Siddharthan A and Batchelor C 2009 Towards discipline-independent argumentative zoning: evidence from chemistry and computational linguistics *Proc. of the 2009 Conf. on Empirical Methods in Natural Language Processing* (Association for Computational Linguistics) pp 1493-1502

[6] Merity S, Murphy T, and Curran J 2009 Accurate Argumentative Zoning with Maximum Entropy models, *Proc. of the 2009 Workshop on Text and Citation Analysis for Scholarly Digital Libraries* (Suntec: World Scientific Publishing Co Pte Ltd) pp 19-26

[7] Teufel S and Moens M 2002 *Summarizing scientific articles: experiments with relevance and rhetorical status* (Computational Linguistics - Summarization vol **28**) issue 4 pp 409-445

[8] Widyantoro D H, Khodra M L, Trilaksono B R and Aziz E A 2013 A multiclass-based classification strategy sentence categorization from scientific papers *Journal of ICT Research and Applications* vol **7** no 3 pp 235-49

[9] D'Alessio S, Murray K, Schiaffino R and Kershenbaum A 2000 *The effect of using hierarchical classifiers in text categorization* (Content-Based Multimedia Information Access-Volume **1**) pp 302-13

[10] Silla Jr C N and Freitas A A 2011 *A survey of hierarchical classification across different application domains* (Data Mining and Knowledge Discovery vol **22**) no 1-2 pp 31-72

[11] Rachman G H, Khodra M L, Widyantoro D H 2017 Rhetorical sentence categorization for scientific paper using word2vec semantic representation *IOP Publishing, Journal of Physics: Conference Series* vol **801** no 1

[12] Rong X 2014 *word2vec Parameter Learning Explained* (Cornell University Library) *Preprint* cs/1411.2738

[13] Mikolov T, Chen K, Corrado G and Dean J 2013 *Efficient estimation of word representations in vector space* (Cornell University Library) *Preprint* cs/1301.3781

[14] Heffernan K and Teufel S 2016 Identifying problem statements in scientific text, foundations of the language of argumentation, *Proc. of COMMA 2016 Workshop* (University of Potsdam) p 18

[15] Sun A and Lim E P 2001 Hierarchical text classification and evaluation, *Proc. IEEE International Conference on Data Mining, ICDM 2001* (IEEE) pp 521-528

[16] Tegegnie A K, Tarekegn A N and Alemu T A 2017 A comparative study of flat and hierarchical classification for amharic news text using svm *Int. Journal of Information Engineering and Electronic Business (IJIEEB)* vol **9** no 3 pp 36-42

[17] Rish I 2001 An empirical study of the naive Bayes classifier, *IJCAI 2001 workshop on*

*empirical methods in artificial intelligence* (IBM New York)
[18] Chang C C and Lin C J 2011 LIBSVM: a library for support vector machines *ACM Transactions on Intelligent Systems and Technology (TIST)* vol **2** no 3
[19] Hall M, Frank E, Holmes G, Pfahringer B, Reutemann P and Witten I H 2009 The WEKA data mining software: an update *ACM SIGKDD explorations newsletter* vol **11** no 1 pp 10-18
[20] Chawla N V, Bowyer K W, Hall L O and Kegelmeyer W P 2002 SMOTE: synthetic minority over-sampling technique *Journal of artificial intelligence research* vol **16** pp 321-57.

**Acknowledgement**
This research was partially supported by the Master's Program toward Doctoral Degree for Excellent Graduate (*Program Pendidikan Magister Menuju Doktor untuk Sarjana Unggul*/PMDSU) from Kemenristekdikti Indonesia within research entitled "Automatic Guided Summarization".

IOP Conf. Series: Journal of Physics: Conf. Series **978** (2018) 012056 doi:10.1088/1742-6596/978/1/012056

# Application of the fuzzy topsis multi-attribute decision making method to determine scholarship recipients

**I Irvanizam[1]**

[1]Department of Informatics, Syiah Kuala University, 5 Syech Abdurrauf Street, Darussalam Banda Aceh 23111, Aceh, Indonesia

Email: irvanizam.zamanhuri@unsyiah.ac.id

**Abstract**. Some scholarships have been routinely offered by Ministry of Research, Technology and Higher Education of the Republic of Indonesia for students at Syiah Kuala University. In reality, the scholarship selection process is becoming subjective and highly complex problem. Multi-Attribute Decision Making (MADM) techniques can be a solution in order to solve scholarship selection problem. In this study, we demonstrated the application of a fuzzy TOPSIS as an MADM technique by using a numerical example in order to calculate a triangular fuzzy number for the fuzzy data onto a normalized weight. We then use this normalized value to construct the normalized fuzzy decision matrix. We finally use the fuzzy TOPSIS to rank alternatives in descending order based on the relative closeness to the ideal solution. The result in terms of final ranking shows slightly different from the previous work.

## 1. Introduction

The national education system has a crucial role in creating a good learning and learning process [1]. Therefore, learners or students have to improve their potentials by doing some efforts and plans. The students are expected to become human beings who have strong religious faith, strong personality, healthy, knowledgeable, responsible, capable, creative, independent, and become citizens of a democratic. This is in line with what meantioned in the Act, Law No. 20 in Article 3, 2003 about The National Education System of Indonesia (Undang-Undang No. 20 Tahun 2003). In order to support this system, Ministry of Research, Technology and Higher Education of the Republic of Indonesia has been routinely offering some scholarships for students at Syiah Kuala University. One of those scholarships is including the Academic Achievement (AA) Scholarship.

The scholarship selection process is highly complex problem [2]. Typically, several eligible documents submitted by many scholarship recipicient candidates or students have to be individually checked and evaluated by a scholarship committee. The committee then assign and select a list of the scholarship recipicients based on their prejudices and personal preferences. As a result, the review process of assessing candidate becomes subjective and among vagueness [2].

Some researchers have developed decision making systems using Multi-Attribute Decision Making (MADM) techniques in order to solve problems in decision making. In their application systems, the MADM techniques were focused on how the experts or the decision makers assigned the weighting value of criteria based on their references. The experts gave numeric values in order to make the computation easier.

There are several methods that can be used to solve the MADM problem. One of those methods is Technique for order preference by similarity to an ideal solution (TOPSIS). This method has been developed by Hwang and Yoon [3]. The method selects the chosen alternative which is the shortest distance from the positive ideal solution and the farthest distance from the negative ideal solution [4]. As the decision makers give assessment values of criteria, the results have subjectivity and amount of vagueness [2,4]. In this case, this method cannot always be used. However, many years ago, several researchers have attempted to minimalize this subjectivity and vagueness by applying fuzzy set theory to multiple attribute evaluations [5]. Typically, the overall alternative values are presented by a fuzzy number which is later called the fuzzy utility. Then, the alternatives are ranked based on the comparison of their fuzzy utilities. In [6], a fuzzy MADM problem was transformed into a crisp value using centroid defuzzification. Tsaur et al. then use TOPSIS to solve the non-fuzzy MADM problem. In [7], a fuzzy MADM problem was also converted to a crisp and solved it using TOPSIS. Meanwhile, Chen and Tzeng in [8] used fuzzy integral to solve MADM problem after they converted a fuzzy MADM into a crisp MADM.

Additionally, TOPSIS method is suitable to solve a decision making problem using quantity multiplication operation of triangular fuzzy number [1]. Yuan-guang in [9] has proved that the TOPSIS method can be implemented effectively with less information and the result is reasonable and objective. Therefore, in this study, a TOPSIS method is used in fuzzy multiple-attribute decision making to determine the right scholarship recipicients by demonstrating a numeric example.

The rest of this paper is structurally organized as follows: in the second section, The previous research is described in brief. In section 3, some basic concepts of the TOPSIS are briefly described. Section 4 demonstrates the calculation of TOPSIS through a numerical example. In the last section, some conclusions are presented by showing comparison results of the TOPSIS and the previous research in [2].

## 2. The previous research

Some researches related to MADM problems for scholarship selection have been conducted by scholars. For instance, the decision making system, developed by Uyun and Riadi [1], used Unified Modeling Language (UML) to select the candidates for academic and non-academic scholarships through three requirement criteria. The paper was focused on how to build the system using UML and showed a comparison between TOPSIS and Weighted Product methods. However, this our study has four criteria, uses triangular number for fuzzy data, and emphasizes on how to determine the scholarship recipients using a numerical example.

## 3. Methodology

This section describes the concept of fuzzy set theory and fuzzy TOPSIS. The fuzzy set theory is used to handle the fuzziness of decision maker's judgment whereas the fuzzy TOPSIS is used to rank the best alternative and to select the scholarship recipients.

### *3.1. Fuzzy set theory*

Fuzzy set theory is one of the most preferred theories in decision making problem. Basically, this theory is an extended theory of ordinary set theory that was introduced by Zadeh [10] for handling uncertainty and imprecise information correlated with another. In some literatures, triangular fuzzy numbers (TFNs) are one of the fuzzy number forms that can be used in order to capture the vagueness of the criteria related to the study case. In this study, a TFN is used. A triangular fuzzy number $\tilde{a}$ $(a_1, a_2, a_3)$ will be used to consider the fuzziness of the parent income criteria. The membership function $\mu(x, a_1, a_2, a_3)$ of the triangular fuzzy number may be defined as in equation (1) [11].

IOP Conf. Series: Journal of Physics: Conf. Series **978** (2018) 012056 doi:10.1088/1742-6596/978/1/012056

$$\mu_{\tilde{a}}(x, a_1, a_2, a_3) = \begin{cases} 0, & x \leq a_1 \\ \frac{x-a_1}{a_2-a_1}, & a_1 < x \leq a_2 \\ \frac{a_3-x}{a_3-a_2}, & a_2 < x \leq a_3 \\ 0, & a_3 < x \end{cases} \quad (1)$$

*3.2. Technique for order preference by similarity to ideal solution (TOPSIS)*

The TOPSIS method is an MADM approach that was firstly introduced by Hwang and Yoon, and later Chen and Hwan developed a fuzzy TOPSIS. The TOPSIS attempts to find the alternative, which is the shortest distance from the positive ideal solution (FIS) and longest distance from negative ideal solution (FIS). This method will rank the alternatives in descending order based on the closeness coefficient representing the distances to the PIS and NIS.

In general, the TOPSIS method procedure follows these steps:

**Step 1:** Construct a fuzzy decision matrix ***D*** with *m* alternatives and *n* criteria that can be concisely presented as in equation (2).

$$D = \begin{matrix} \\ A_1 \\ A_2 \\ A_3 \\ \vdots \\ A_m \end{matrix} \begin{matrix} C_1 & C_1 & C_1 & \cdots & C_n \\ \end{matrix} \begin{pmatrix} x_{11} & x_{12} & x_{13} & \cdots & x_{1n} \\ x_{21} & x_{22} & x_{23} & \cdots & x_{2n} \\ x_{31} & x_{32} & x_{33} & \cdots & x_{3n} \\ \vdots & \vdots & \vdots & \ddots & \vdots \\ x_{m1} & x_{m2} & x_{m3} & \cdots & x_{mn} \end{pmatrix} \quad (2)$$

where $A_i$ $(i=1,2,3,..,m)$ is possible alternative, $C_j$ $(j=1,2,3,..,n)$ is possible criterion, and $x_{11}$ is the performance rating of the *i*-th alternative, $A_i$, with respect to the *j*-th criterion, $C_j$.

**Step 2:** Build normalized decision matrix ***R***. Each element of the matrix ***D*** can be normalized using equation (4).

$$R = \begin{matrix} \\ A_1 \\ A_2 \\ A_3 \\ \vdots \\ A_m \end{matrix} \begin{matrix} C_1 & C_1 & C_1 & \cdots & C_n \\ \end{matrix} \begin{pmatrix} r_{11} & r_{12} & r_{13} & \cdots & r_{1n} \\ r_{21} & r_{22} & r_{23} & \cdots & r_{2n} \\ r_{31} & r_{32} & r_{33} & \cdots & r_{3n} \\ \vdots & \vdots & \vdots & \ddots & \vdots \\ r_{m1} & r_{m2} & r_{m3} & \cdots & r_{mn} \end{pmatrix} \quad (3)$$

$$r_{ij} = x_{ij} * \left(\sum_{i=1}^{m} x_{ij}^2\right)^{-1/2} \quad (4)$$

**Step 3:** Construct weighted normalized matrix ***V***. In this study, the matrix ***V*** has given weight ***W*** = $[w_1, w_2, w_3, \ldots, w_n]$ which is either crisp or a triangular fuzzy number. Each element of the matrix ***V*** can be calculated using equation (6).

$$V = \begin{matrix} \\ A_1 \\ A_2 \\ A_3 \\ \vdots \\ A_m \end{matrix} \begin{matrix} C_1 & C_1 & C_1 & \cdots & C_n \\ \end{matrix} \begin{pmatrix} v_{11} & v_{12} & v_{13} & \cdots & v_{1n} \\ v_{21} & v_{22} & v_{23} & \cdots & v_{2n} \\ v_{31} & v_{32} & v_{33} & \cdots & v_{3n} \\ \vdots & \vdots & \vdots & \ddots & \vdots \\ v_{m1} & v_{m2} & v_{m3} & \cdots & v_{mn} \end{pmatrix} \quad (5)$$

$$v_{ij} = w_j * r_{ij} \quad (6)$$

**Step 4:** Determine the positive ideal solution (PIS), $A^+$ and the negative ideal solution (NIS), $A^-$. The PIS and NIS can be computed based on normalized weighted rating using equation (7) and (8) respectively.

$$PIS = A^+ = \left\{ \underset{i}{Max} v_{ij}; j \in J \right\} = \{v_1^+, v_2^+, v_3^+, \ldots, v_m^+\} \quad (7)$$

$$NIS = A^- = \left\{ \underset{i}{Min} v_{ij}; j \in J \right\} = \{v_1^-, v_2^-, v_3^-, \ldots, v_m^-\} \quad (8)$$

where $J$ is associated with benefit criteria.

**Step 5:** Calculate the separation measures using the Euclidean distance. The separation of each alternative from PIS, $D^+$ can be calculated using equation (9). Similarly, the separation of each alternative from NIS, $D^-$ can be calculated using equation (10).

$$D_i^+ = \left\{ \sum_{j=1}^{n} \left( v_{ij} - v_j^+ \right)^2 \right\}^{1/2}, 1 \leq i \leq m \quad (9)$$

$$D_i^- = \left\{ \sum_{j=1}^{n} \left( v_{ij} - v_j^- \right)^2 \right\}^{1/2}, 1 \leq i \leq m \quad (10)$$

**Step 6:** Compute the relative closeness to the ideal solution. The closeness of alternative $A_i$ with respect to $A^+$ can be calculated using equation (11).

$$RC_i^+ = \frac{D_i^-}{D_i^- + D_i^+}, \quad 1 \leq i \leq m \quad (11)$$

since $D_i^- \geq 0$ and $D_i^+ \geq 0$, then clearly, $RC_i^+ \in [0, 1]$.

**Step 7:** Rank the preference order. We can rank alternatives in descending order based on $RC_i^+$ value.

## 4. Numerical example

In this section, we demonstrate a numerical example, taken from [2], to illustrate the fuzzy TOPSIS algorithm for solving the scholarship selection problem with a triangular fuzzy number. We assume that Syiah Kuala University desires to select seven of ten students in order to distribute the AA scholarship. After having the requirement documents, the decision maker has identified four benefit criteria which are: Grade Point Average ($C_1$), the number of credit ($C_2$), the number of dependents parent ($C_3$) and parent incomes ($C_4$). The decision maker used the linguistic variable as shown in table 1 to assess the importance of each criterion. This linguistic variable is converted into triangular fuzzy numbers to construct the fuzzy decision matrix and determine the weight fuzzy number of each criterion, as shown in table 2.

**Table 1.** Linguistic variables for the importance weight of each criterion.

| Variable | A triangular fuzzy number |
|---|---|
| Very low (VL) | (0.0; 0.0; 0.1) |
| Low (L) | (0.1; 0.2; 0.2) |
| Medium low (ML) | (0.2; 0.3; 0.4) |
| Medium (M) | (0.4; 0.5; 0.6) |
| Medium high (MH) | (0.5; 0.6; 0.7) |
| High (H) | (0.6; 0.8; 0.8) |
| Very high (VH) | (0.8; 0.9; 1.0) |

**Table 2.** Importance weight of the criteria from a decision maker for the Academic Achievement (AA) Scholarship.

| | Fuzzy number | | | |
|---|---|---|---|---|
| Criteria | Linguistic Variable | Weight | Defuzzied value | Normalized weight |
| $C_1$ | VH | (0.8; 0.9; 1.0) | 0.9 | 0.45 |
| $C_2$ | M | (0.4; 0.5; 0.6) | 0.5 | 0.25 |
| $C_3$ | ML | (0.2; 0.3; 0.4) | 0.3 | 0.15 |
| $C_4$ | ML | (0.2; 0.3; 0.4) | 0.3 | 0.15 |

All criteria used crisp data except for the *4*-th criterion. The parent income criterion, $C_4$ had suitability degree with several alternatives decision: T(suitability) = {S,M,RG,B,VG}. The membership function of each element was represented by using triangular fuzzy numbers as shown in table 3.

**Table 3.** Fuzzy linguistic variables and fuzzy number for the criterion $C_4$

| Parent Income (IDR) | Fuzzy Set | | |
|---|---|---|---|
| | Linguistic Variable | A triangular fuzzy number | Defuzzied value |
| 0 - 1000000 | SMALL (S) | (0.0; 0.1; 0.4) | 0.15 |
| 1000001 - 2000000 | MODERATE (M) | (0.1; 0.4; 0.5) | 0.35 |
| 2000001 - 3000000 | RATHER BIG (RG) | (0.4; 0.5; 0.6) | 0.5 |
| 3000001 - 4000000 | BIG (B) | (0.5; 0.6; 0.8) | 0.625 |
| 4000001 - 5000000 | VERY BIG (VG) | (0.6; 0.8; 1.0) | 0.8 |

**Table 4.** The fuzzy decision matrix

| Alternative | $C_1$ | $C_2$ | $C_3$ | $C_4$ | |
|---|---|---|---|---|---|
| | | | | Income | Fuzzy Number |
| $S_1$ | 3.68 | 98 | 4 | 1500000 | 0.35 |
| $S_2$ | 3.9 | 98 | 4 | 2500000 | 0.5 |
| $S_3$ | 3.9 | 98 | 3 | 1000000 | 0.15 |
| $S_4$ | 3.7 | 54 | 2 | 2000000 | 0.35 |
| $S_5$ | 3.59 | 54 | 1 | 2500000 | 0.5 |
| $S_6$ | 3.43 | 54 | 4 | 2000000 | 0.35 |
| $S_7$ | 3.19 | 85 | 4 | 1000000 | 0.15 |
| $S_8$ | 3.28 | 41 | 4 | 1000000 | 0.15 |
| $S_9$ | 3.8 | 41 | 8 | 1000000 | 0.15 |
| $S_{10}$ | 3 | 20 | 6 | 5000000 | 0.8 |

The data were first constructed into a fuzzy decision matrix, that shown as in table 4. The algorithm then built a normalized fuzzy decision matrix using equation (3). A weighted normalized matrix was then computed using equation (5) and shown as table 6. The next steps are to compute the PIS ($A^+$) and NIS ($A^-$), respectively and shown in table 5.

Next, the separation measure of PIS ($D^+$) and NIS ($D^-$) are computed using equation (9) and (10), respectively. The closeness of each alternative $A_i$ is then calculated using equation (11). A larger value of the closeness shows that alternative $A_i$ is prefered to the AA scholarship. Finally, the result of the fuzzy calculation is shown in table 6. The first student who will receive the AA scholarship is $S_2$ and then $S_1$, $S_3$, $S_{10}$, $S_7$, $S_9$, and $S_6$ will be respectively. Meanwhile, the result from the previous work done by Irvanizam [2] is $S_3$, $S_2$, $S_1$, $S_7$, $S_9$, $S_4$ and $S_6$. This indicates that by using this method, minor changes in first and sixth ranking take place. Therefore, the result of this method is slightly different to the one using Simple Additive Weighting (SAW).

**Table 5.** The positive ideal solution (PIS) and the negative ideal solution (NIS) for the AA scholarship

| Alternative | The AA scholarship | | | |
|---|---|---|---|---|
| | $C_1$ | $C_2$ | $C_3$ | $C_4$ |
| Positive Ideal Solution ($A^+$) | 0.1559 | 0.1112 | 0.0862 | 0.0949 |
| Negative Ideal Solution ($A^-$) | 0.1199 | 0.0227 | 0.0108 | 0.0178 |

**Table 6.** The weighted normalized fuzzy decision matrix for the AA scholarship and The result of fuzzy TOPSIS for the AA scholarship

| Alternative | The AA scholarship | | | | | | | |
|---|---|---|---|---|---|---|---|---|
| | $C_1$ | $C_2$ | $C_3$ | $C_4$ | $D^+$ | $D^-$ | $RC^+$ | Rank |
| $S_1$ | 0.1471 | 0.1112 | 0.0431 | 0.0415 | 0.0692 | 0.1009 | 0.5932 | 2 |
| $S_2$ | 0.1559 | 0.1112 | 0.0431 | 0.0593 | 0.0559 | 0.1091 | 0.6612 | 1 |
| $S_3$ | 0.1559 | 0.1112 | 0.0323 | 0.0178 | 0.0941 | 0.0979 | 0.5101 | 3 |
| $S_4$ | 0.1479 | 0.0613 | 0.0215 | 0.0415 | 0.0979 | 0.0543 | 0.3569 | 9 |
| $S_5$ | 0.1435 | 0.0613 | 0.0108 | 0.0593 | 0.0980 | 0.0614 | 0.3853 | 8 |
| $S_6$ | 0.1371 | 0.0613 | 0.0431 | 0.0415 | 0.0869 | 0.0582 | 0.4012 | 7 |
| $S_7$ | 0.1275 | 0.0964 | 0.0431 | 0.0178 | 0.0940 | 0.0809 | 0.4626 | 5 |
| $S_8$ | 0.1311 | 0.0465 | 0.0431 | 0.0178 | 0.1123 | 0.0417 | 0.2707 | 10 |
| $S_9$ | 0.1519 | 0.0465 | 0.0862 | 0.0178 | 0.1007 | 0.0853 | 0.4585 | 6 |
| $S_{10}$ | 0.1199 | 0.0227 | 0.0646 | 0.0949 | 0.0979 | 0.0941 | 0.4899 | 4 |

## 5. Conclusion

In this study, we have demonstrated the fuzzy TOPSIS based on linguistic and crisp values for determining the scholarship recipients. The result in terms of final ranking shows slightly different from the previous work, done by Irvanizam [2]. However, this study still used a decision maker to assign the criteria weight so that this becomes the limitation of this study. The research still should be expanded on some study cases such as the use of triangular and trapezoidal fuzzy numbers for all criteria, the use of fuzzy multi-attribute group decision making (FMAGDM) and the use of a linear programming method.

**Acknowledgments**
This research was supported by the Department of Informatics, Syiah Kuala University and Database & Data Mining Information Retrieval Research Group (DMIR).

**References**
[1] Uyun S and Riadi I 2011 A fuzzy topsis multiple-attribute decision making for scholarship selection *J. TELKOMNIKA* **9** 37-46
[2] Irvanizam I 2017 Multiple attribute decision making with simple additive weighting approach for selecting the scholarship recipients at syiah kuala university *Int. Conf. on Electrical Engineering and Informatics (ICELTICs 2017)* vol 1 (Banda Aceh: October 18-20, 2017) pp 262-267
[3] Hwang CL and Yoon K 1981 *Multiple attribute decision making* (Berlin: Springer)
[4] Mahdavi I, Heidarzade A, Sadeghpour-Gildeh B and Mahdevi-Amiri N 2009 A general fuzzy topsis model in multiple criteria decision making *Int. J. Adv. Manuf. Technol* **45** 406
[5] Chen SJ and Hwang CL 1992 *Fuzzy multiple attribute decision making methods and applications* (Berlin: Springer)
[6] Tsaur SH, Chang TY and Yen CH 2002 The evaluation of airline service quality by fuzzy mcdm *Tourism Management* **23** pp 107-115
[7] Chu TC 2002 A survey on multi criteria decision making methods and its applications *Int. J. Uncertain Fuzziness Knowl-Based Syst* **10** pp 687-701
[8] Chen MF and Tzeng GH 2004 Combining grey relation and topsis concepts for selecting an expatriate host country *Math. Comput. Model* **40** pp 1474-1490
[9] Yuan-guang Y 2008 The topsis method of multiple attribute decision making problem with triangular-fuzzy-valued weight *Int. Conf. on Modelling, Simulation and Optimization* vol 1 (Hongkong: December 28, 2008) pp 11-14
[10] Zadeh LA 1975 The concept of a linguistic variable and its application to approximate reasoning *Information Sciences* **8** 199-249
[11] Önüt S, Soner S 2008 Transshipment site selection using the AHP and TOPSIS approaches under fuzzy environment *Waste Manag* **28** 1552-1559

# Decision Support Model for Mosque Renovation and Rehabilitation (Case Study: Ten Mosques in Jakarta Barat, Indonesia)

**[1]D N Utama, [1]Y S Triana, [2]M M Iqbal, [2]M Iksal, [2]I Fikri, [2]T Dharmawan**

[1]Computer Science, Universitas Mercu Buana, Jakarta, Indonesia
[2]Information Systems Department, UIN Syarif Hidayatullah, Jakarta, Indonesia

ditdit.nugeraha@mercubuana.ac.id

**Abstract.** Mosque, for Muslim, is not only a place for daily worshipping, however as a center of culture as well. It is an important and valuable building to be well managed. For a responsible department or institution (such as Religion or Plan Department in Indonesia), to practically manage a lot of mosques is not simple task to handle. The challenge is in relation to data number and characteristic problems tackled. Specifically for renovating and rehabilitating the damaged mosques, a decision to determine the first damaged mosque priority to be renovated and rehabilitated is problematic. Through two types of optimization method, simulated-annealing and hill-climbing, a decision support model for mosque renovation and rehabilitation was systematically constructed. The method fuzzy-logic was also operated to establish the priority of eleven selected parameters. The constructed model is able to simulate an efficiency comparison between two optimization methods used and suggest the most objective decision coming from 196 generated alternatives.

## 1. Introduction

Mosque is an important building for worshipping and multi-functional community place [1]. Millions people use the mosque daily. Particularly in Indonesia, there are more than 200 million Muslims with 200 thousand big mosques (masjid in Bahasa) and 190 thousand little mosques (mushala in Bahasa) [2]. To manage those mosques are realistically challenging effort to do.

Two activities of mosque management are renovation and rehabilitation. They are activities that an appropriate approach, such as decision support model (DSM), is reasonably required. DSM is an approach alternative that is possibly able to support the management activity. Via a combination between complete data and precise model, DSM can be used to manage the mosques, especially in accurate decision making related to renovation and rehabilitation; as an accurate decision is something difficult to reach [3].

Regarding DSM, several researchers have conducted a number of researches, where DSM as a main topic. [4] combined three types of method interpretive structural modeling, analytical network processing, and fuzzy logic in constructing a DSM for evaluating a performance of partnerships. The ranking of partnership performance index was produce by model as a basic reason to select the decision. [5] constructed a DSM for providing a robust decision making for automated railway level crossing. By using fuzzy logic control, the model was used to save the operation time, to avoid any accidental facilities, and to eliminate human errors. [6] studied a DSM ordinarily used in the solid management area. Where, several aspects are usually considered; environmental, economic, and also social.

Moreover, [7] developed a DSM to select the best facility. The best here is related to energy saving and facility improvement program. The research was conducted in elementary school in South Korea. [8] have developed a DSM to get a higher productivity on maritime automobile terminals. The productivity measured here was correlating to internal transport, exit truck gate operation, and wagon manipulation. [9] developed a DSM for renovation. It is able to suggest a reasonable and economic policy for renovating and rebuilding organizations' facilities.

Thus, the paper presents a constructed DSM that implemented in managing the mosques, mainly for their renovation and rehabilitation intention. Where, ten mosques in Jakarta Barat (Indonesia) were taken as an object of the research, as a basis of data behavior to be generated for simulation purpose. The section of introduction of the paper is followed by sections related works, research methodology, results and discussion, and conclusion and further works.

## 2. Related works

The studies relating the topic "mosque" that has been academically selected as a main research object were conducted by many researchers. The mosque is a unique phenomenon, where the researchers need to learn more. [10], in engineering field, investigated implications of stereotype mosque architecture on three dimensions of sustainability. The dimensions considered are social, economic, and social problem. [11] evaluated the effect of climatic factors on construction and local architecture in hot and arid regions with special focus on Varzaneh city. The study concluded that Varzaneh Jame mosque (as an object of the research) was climatically designed, it was harmonized with the economic and social context of religion. Moreover, to answer the problem of the uncontrolled and unnecessary mosque ornamenting, [12] studied the importance of Islamic art in mosque interior. They concluded that the mosque ornamentation should be considered as a part of mosque components. As it will methodically fulfil the human beings' psychological needs.

In social and behavioral sciences, [13] examined the role of Sheikh Zayed Grand Mosque (in Abu Dhabi) to identify the aspects that the mosque intensely requires to do to be more activated. The study was conducted by using theoretical framework as a criterion. [14] studied regarding design functionality of the communal mosques to satisfy the Muslim community's needs and to create a sustainable environment. The research itself was practically conducted through comparative study based on qualitative research, text interpretation, and observation. It suggested novel insights by formulating new design approach and guideline in designing future communal mosques.

## 3. Research methodology

In this research, ten mosques in area Jakarta Barat (Indonesia) were taken as an object of the research. These ten mosque empirical data were also used as a basic data to technically generate 196 experimental data for simulation.

Primarily, four stages were followed to conduct the research (Figure 1), where each stage is related to specific methods. In preliminary analyzing, method literature study was done. Several journal papers and other resources were systematically reviewed. And then, the detailed analyzing was conducted. Here, several methods were operated; i.e. literature study, interview of expert in justifying eleven selected parameters, and observation.

Furthermore, method of fuzzy-logic [15] concept was operated. It was benefitted to parameterize in obtaining a parameter priority value of each parameter that has been rationalized by expert. Finally, the DSM was constructed. The model was created based on principle of object-oriented method [16], specifically in analysis and design of model. To optimize the result of decision, two types of method were utilized and compared. The model is not only able to propose the most objective decision, but present the efficiency comparison between two methods simulated-annealing [17][18][19][20] and hill-climbing [19][21] also.

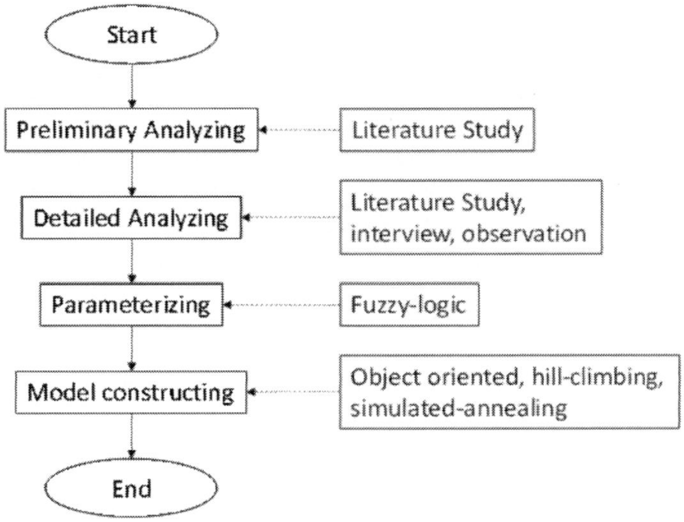


**Figure 1.** Stages of the research.

## 4. Results and discussion

### *4.1. The constructed model*

Eleven parameters are taken into account in developing the model. Those parameters are categorized into two groups, parameters mosque (with seven parameters) and damage (with four parameters). To justify those parameters' urgency, one expert involved. The justification result that is in language value and then converted (based on fuzzy-logic concept) to crisp output value (COV) to become priority value (PV) is presented in Table 1. The type of fuzzy membership function itself that was used is triangular type with five types of language variable: very unimportant (0, 0, 25), unimportant (10, 30, 50), rather important (35, 55, 75), important (65, 80, 95), and very important (85, 100, 100).

**Table 1.** Parameters' value.

| No. | Parameter | Language Value | COV | PV |
|---|---|---|---|---|
| Mosque Parameter | | | | |
| 1. | Mosque age | Very unimportant | 11.54 | 0.02 |
| 2. | Mosque type | Unimportant | 38.33 | 0.06 |
| 3. | Mosque capacity | Important | 82.09 | 0.14 |
| 4. | Mosque activity type | Unimportant | 35.00 | 0.06 |
| 5. | Land ownership status | Unimportant | 31.67 | 0.05 |
| 6. | Distance to other nearest mosque | Rather important | 55.00 | 0.09 |
| 7. | Mosque width | Unimportant | 30.00 | 0.05 |
| Damage Parameter | | | | |
| 8. | Damage age | Important | 80.00 | 0.13 |
| 9. | Rehabilitation intensity | Rather important | 69.22 | 0.12 |
| 10. | Rehabilitation cost | Important | 90.00 | 0.15 |
| 11. | Damage type | Rather important | 78.08 | 0.13 |

The model is interconnected with four actors, three human actors and a system actor. The human actors are `Expert`, `Religion Dept.` (religion department), and `Plan Bureau`. Ant the system actor is `Religion Dept. MIS` (Religion Department Management Information Systems). This interconnection is depicted by Figure 2 (in usecase diagram).

Five usecases are operated in the model. The usecase `Parameterizing` is related to actor `Expert`. Here, an expert justified parameters' urgency, where they are directly converted to priority values based on concept of fuzzy-logic. The data used in the model coming from MIS of Religion Department. The process of data extraction is done in the usecase `Extracting Data`. Optimization process (in the usecase `Optimizing`) is based on two types of methods, simulated-annealing and hill-climbing. The decision (in the usecase `Generating Decision`) can be identified by actors `Religion Dept.` and `Plan Bureau`. And the usecase `Reporting` is only interconnected to actor `Religion Dept.`

$$\max(P) = \sum_{i=1}^{n} c_i P_i \qquad (1)$$

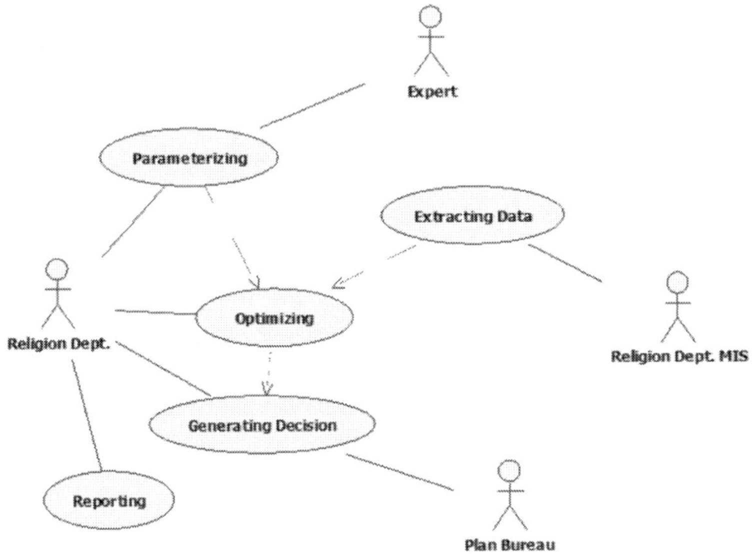


**Figure 2.** Usecase diagram for the constructed model.

One screenshot example of the constructed model is presented in Figure 3. The model recommends the most objective decision. The decision maker is able to decide which the mosque has to be renovated and rehabilitated first based on the model's suggestion. Principally, two types of optimization methods give a consistent decision, where the model is experimentally executed until a hundred thousand times. The objective function of the optimization process is generically delivered in equation (1), where $c$ represents a value of parameter priority and $P$ is an optimal value of parameter that is obtained by calculating the relative optimal value of it. The optimal value of parameter should be maximum or minimum. For example, to consider the parameter mosque capacity, the higher value of capacity will be more optimal than others.

Numerous laboratory experiments have been done to see the correlation between model execution number and running time. The correlation is represented by the graph in Figure 4. In second unit, the method simulated-annealing consumed 0.0142, 0.0779, 0.1663, 0.9036, 1.9469, 11.5169, and 23.7527 seconds, while the method hill-climbing consumed 0.0064, 0.0356, 0.0790, 0.4385, 0.9844, 5.5252, and 9.9542 seconds for number of model execution 100, 500, 1000, 5000, 10000, 50000, and 100000 times respectively. Here, can be concluded that the method hill-climbing consumes a shorter running time than the simulated-annealing does. Practically, the experiments were conducted in laboratory optimization models and systems for decision support (UIN Syarif Hidayatullah, Jakarta, Indonesia) by using machine with processor Intel® Celeron® CPU B815 @ 1.60GHz (2 CPUs), memory 2048MB RAM, BIOS InsydeH2O V 1.06, and operating system Windows 8.1 Pro 64-bit (6.3, Build 9600).

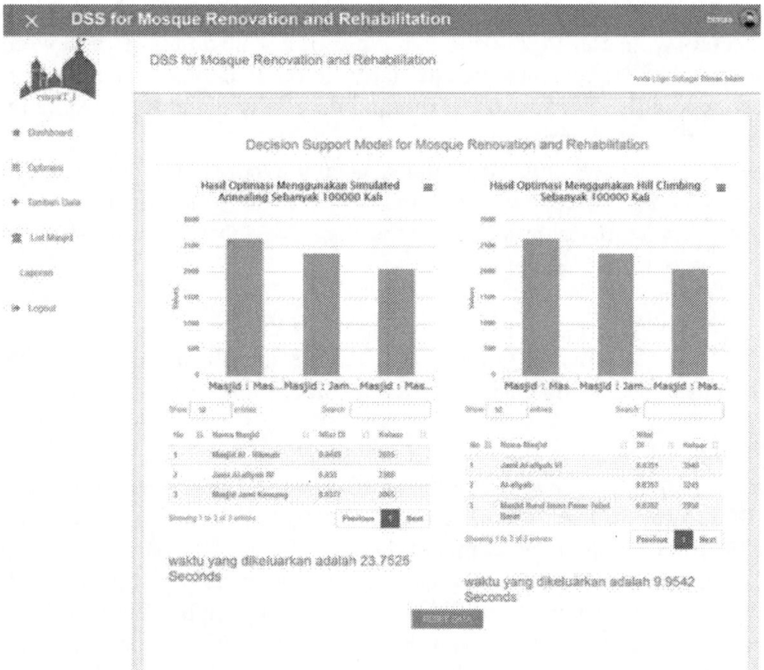


**Figure 3.** Screenshot example of the constructed model.

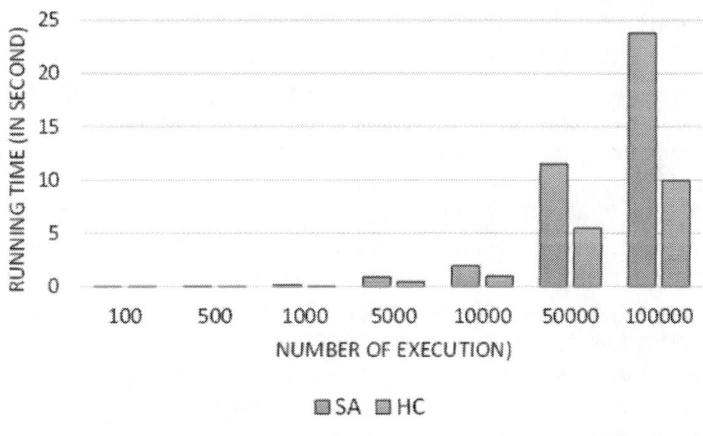


**Figure 4.** Running Time Comparison of Two Methods Simulated-Annealing (SA) and Hill-Climbing (HC).

### *4.2. Discussion*

Particularly in engineering field, [10], [11], and [12] conducted their research with "mosque" as a main topic. However, no one of them talked regarding renovation and rehabilitation. Differently, in our study, we focused on renovation and rehabilitation aspects, where we constructed the computer based model that can support the decision maker to decide the most objective decision for selecting which mosque has to be renovated and rehabilitated first among a lot of decision alternatives.

## 5. Conclusion and further works

A decision support model for mosque renovation and rehabilitation was technologically constructed. Eleven selected parameters were considered and two types of optimization method were operated. The constructed model can propose the best mosque alternative to renovated and rehabilitated primarily. The model also can clearly expose the comparison of running time of two methods, where the hill-climbing is more efficient than the simulated-annealing.

If several parameters related to environmental aspect are considered in next study, it will be interesting. They are such as a distance to water source, green condition of the mosque, etc. And also, it will be more fruitful if other heuristic optimization methods are considered in the research, e.g. water flow algorithm, ant colony optimization, etc.

## References

[1] Abdullah F H, Majid N H A and Othman R 2016 Defining issue of thermal comfort control through urban mosque Façade design *Procedia – Social and Behavioral Sciences* **234** pp 416-423

[2] DBI – Ditjen Bimas Islam 2017 *Data masjid dan mushala* Direktorat Urusan Agama Islam dan Pembinaan Syariah, http://simas.kemenag.go.id [access date: 22 January 2017]

[3] Noor N M M, Retnowardhani A, Triana Y S and Helawani S M 2013 The use of fuzzy alpha cut to improve the forecasting range results of the ARIMA model: assist decision making process *IJIPM: International Journal of Information Processing and Management* **4(7)** pp 59-71

[4] Piltan M and Sowlati T 2016 A multi-criteria decision support model for evaluating the performance of partnerships *Expert Systems with Applications* **45** pp 373-384

[5] Pattanaik L N and Yadav G 2015 Decision support model for automated railway level crossing system using fuzzy logic control *Procedia Computer Science* **48** pp 73-76

[6] Kamperis A C, Aravossis K, Tatsiopoulos I P and Sotirchos A 2013 Decision support models for solid waste management: Review and game-theoretic approaches *Waste Management* **33(5)** pp 1290-1301

[7] Hong T, Koo C and Jeong K 2012 A decision support model for reducing electric energy consumption in elementary school facilities *Applied Energy* **95** pp 253-266

[8] Beškovnik B and Twrdy E 2011 Managing maritime automobile terminals: an approach toward decision-support model for higher productivity *International Journal of Naval Architecture and Ocean Engineering* **3(4)** pp 233-241

[9] Rosenfeld Y and Shohet I M 1999 Decision support model for semi-automated selection of renovation alternatives *Automation in Construction* **8(4)** pp 503-510

[10] Ahmad T, Thaheem M J, Anwar A and Din Z 2016 Implications of stereotype mosque architecture on sustainability *Procedia Engineering* **145** pp 96-103

[11] Varzaneh E H, Amini M and Bemanian M R 2014 Impact of hot and arid climate on architecture (case study: Varzaneh Jame Mosque) *Procedia Engineering* **94** pp 25-32

[12] Othman R and Zainal-Abidin ZJ 2011 The importance of Islamic art in mosque interior *Procedia Engineering* **20** pp 105-109

[13] Qaraya B A 2015 The civic and culture role of the Sheikh Zayed Grand Mosque *Procedia – Social and Behavioral Sciences* **211** pp 488-497

[14] Baharudin N A and Ismail AS 2014 Communal mosques: Design functionality towards the development of sustainability for community *Procedia – Social and Behavioral Sciences* **153** pp 106-120

[15] Zadeh L A 1996 Fuzzy logic = computing with words *IEEE Transactions of Fuzzy Systems* **4(2)** pp 103-111

[16] Mathiassen L, Munk-Madsen A, Nielsen P A and Stage J 2000 *Object-Oriented Analysis and Design* Marko Publishing ApS, Aalborg Denmark

[17] Kirkpatrick S, Gelatt C D and Vecchi M P 1983 Optimization by simulated annealing *Science* **220(4598)** pp 671-680

IOP Conf. Series: Journal of Physics: Conf. Series **978** (2018) 012057 doi:10.1088/1742-6596/978/1/012057

[18] Černý V 1985 Thermodynamical approach to the traveling salesman problem: An efficient simulation algorithm *Journal of optimization theory and applications* **45(1)** pp 41-51

[19] Utama D N 2015 *The Optimization of the 3-d Structure of Plants, Using Functional-Structural Plant Models. Case Study of Rice (Oryza sativa L.) in Indonesia* PhD Thesis Georg-August Universität Göttingen

[20] Alaydrus M and Eibert T F 2007 Optimizing multiband antennas using simulated annealing *International ITG Conference on Antennas* pp 86-90

[21] Michalewicz Z 1992 *Genetic algorithms + data structures = evolution programs* Springer Science & Business Media

# PCA based feature reduction to improve the accuracy of decision tree c4.5 classification

**M Z F Nasution [1], O S Sitompul [2*] and M Ramli [3]**

[1]Graduate School of Computer Science, Faculty of Computer Science and Information Technology, Universitas Sumatera Utara, Medan, Indonesia
[2]Department of Information Technology, Faculty of Computer Science and Information Technology, Universitas Sumatera Utara, Medan, Indonesia
[3]Department of Mathematics, Syiah Kuala University, Aceh, Indonesia

[*]Corresponding author: opim@usu.ac.id

**Abstract**. Splitting attribute is a major process in Decision Tree C4.5 classification. However, this process does not give a significant impact on the establishment of the decision tree in terms of removing irrelevant features. It is a major problem in decision tree classification process called over-fitting resulting from noisy data and irrelevant features. In turns, over-fitting creates misclassification and data imbalance. Many algorithms have been proposed to overcome misclassification and overfitting on classifications Decision Tree C4.5. Feature reduction is one of important issues in classification model which is intended to remove irrelevant data in order to improve accuracy. The feature reduction framework is used to simplify high dimensional data to low dimensional data with non-correlated attributes. In this research, we proposed a framework for selecting relevant and non-correlated feature subsets. We consider principal component analysis (PCA) for feature reduction to perform non-correlated feature selection and Decision Tree C4.5 algorithm for the classification. From the experiments conducted using available data sets from UCI Cervical cancer data set repository with 858 instances and 36 attributes, we evaluated the performance of our framework based on accuracy, specificity and precision. Experimental results show that our proposed framework is robust to enhance classification accuracy with 90.70% accuracy rates.

## 1. Introduction

C4.5 algorithm is a decision tree classification approach based on information entropy. The algorithm uses modified splitting criterion called gain ratio [1]. On decision tree approach, pruning is a step to eliminate several branches which actually does not required [2] and create a large scale of decision tree process called over-fitting. Over-fitting for cost-sensitive learning method produces classification model by improving accuracy in training data. Nevertheless, it does not improved classification model when applied to unseen data in real data set [3]. Over-fitting caused by noisy-data and irrelevant feature [4] and creates misclassification resulting in overfitting and data imbalance. In turns, overfitting reduces accuracy of decision tree in classification model [5]. In order to reduce high data dimesions, a common approach is to use feature reduction to obtain a lower dimensional data, which depends on the feature criteria of the problem setting [6]. Feature reduction is followed by feature selection by correlation and gain ratio methods. Unsupervised method minimizes the information loss whereas the supervised reduction methods maximize the class discrimination [7].

In this research, we propose a framework for feature reduction using principal component analysis (PCA) to eliminate irrelevant features, selecting relevant and non-correlated features without affecting the information contained in the original data and then predictor is developed using decision tree C4.5 algorithm to classify data set used. The aim is to reduce some features of data by PCA and then build a prediction of classification model by decision tree C4.5 to obtain relevant features and to improve the accuracy of decision tree C4.5. To evaluate the proposed approach, we use cervical cancer data set from the UCI Machine Learning repository.

## 2. Principal Component Analysis (PCA)

PCA gives good result when applied to correlated attributes. In this research, PCA is applied in both training and testing attributes of the cervical cancer data set. The PCA will identify patterns in the data set, and finds their similarities and differences between each attributes. It acts as a powerful model to analyse the data. The original data and the average of the original data are chosen. The covariance matrix is computed whereby the result is used in calculating the eigen vectors and eigen values [8] and the eigen vector with the highest eigen value is chosen as the principle component of the cervical cancer data set as it exhibits the most significant relationship between the data set attributes. The Eigen values are sorted in ascending order to choose the most significant data and discard the least significant one. By this means data with higher dimensions is reduced to lower dimensions [9].

The variance is calculated to find the spread of data in the UCI Cervical cancer data set using equation (1) to determine the deviation of data in the sample data set [10].

$$\text{Var(x)} = \sigma^2 = \frac{1}{n}\sum_{i=1}^{n}(\tilde{z}_{ij} - \mu_j)^2 \quad (1)$$

Afterwards, covariance is calculated to find the relation between two classes, in which zero value indicates that there is no relation between two dimensions. The covariance is computed using (2) as suggested in [11].

$$\text{Cov (x,y)} = \frac{1}{n-1}\sum_{i=1}^{n}(x_{ij} - \mu_{xj})\,(y_{ij} - \mu_{yj}) \quad (2)$$

Finally, the Eigenvalues and Eigenvectors for the covariance matrix are calculated. The computed eigenvalues is then transformed (varimax orthogonal rotation) using equation (3) [11].

$$\text{Det}(A - \lambda I) = 0 \quad (3)$$

## 3. C4.5 Algorithm

The next step is classifying attributes based on information gain ratio using decision tree C4.5. It is a measure on the number of bits of information obtained for the class prediction. The expected information needed to classify a given sample data set is given by [1].

$$Entropy\ (S) = -(+)\ log2\ P(+) - P(-)\ log2\ P(-) \quad (4)$$

Where P is the probability of sample. Information Gain is computed using equation (5) as suggested in [4].

$$Gain(S, A) = Entropy\ (S) - \sum_{i=1}^{n}\frac{|S_i|}{|S|} \times Entropy\ (S_i) \quad (5)$$

The entropy of the expected information based on the splitting of subset to data set is given by (6)

$$SplitInfo_A(D) = -\sum_{j=1}^{v}\frac{|D_j|}{|D|} \times \log_2\left(\frac{D_j}{D}\right) \quad (6)$$

and the gain ratio is computed using (7) [4].

$$\text{Gain Ratio (A)} = \frac{Gain\ (A)}{SplitInfo\ (A)} \quad (7)$$

## 4. Previous Research

An hybrid PCA feature reduction and selection subset by using wrapper filter and information gain-ratio decision tree C4.5 had been proposed in [9]. Feature reduction resulted by the hybrid model did not calculate the performance accuracy of classification decision tree C4.5. A calculation on the classification performance accuracy was proposed by [12] using a prediction system to develop business processes in predictive diagnosis of hypertension by adding functionality prognosis. The prediction is comducted on health information of hypertension data set using decision tree C4.5 algorithms. A 93% output of classification performance accuracy is obtained for the level of health risk patients.

Research work by [13] proposed orthogonal local preserving projection (OLPP) and artificial neural network (ANN) classifier. Here, OLPP was used to reduce feature dimensions. Once the feature reduction is formed, the prediction was subsequently performed based on the classifier. The authors claimed a classification performance accuracy of 90% was achieved. Furthermore, an approach for efficient imputation and classification using imputation based on class based clustering (IMCBC) was developed in [5]. The classification accuracies calculated using the proposed imputation approach was deployed on the existing classifiers such as ID3, J48, Naïve Bayes, KNN and SVM. Naïve Bayes had a significant accuracy of 80%.

In addition, by sharing the same coefficient signs, research work by [14] used a regularization-based transfer learning strategy to enhance both source and target models. Cross-modality individual risk and cross-expert subjective was used to asses colposcopy cervical cancer images quality. A combination of logistic regression and SVM were claimed to obtained positive positive results. Another classification model approach based on cost-selectivity was proposed in [15]. Pre-processing was performed by calculating correlation value for each attributes and target attributes on UCI Cervical cancer data set. In the final step, features reduction method eliminates the highest values attributes correlation. The results achieved a true positive rate (TP rate) by 0.429.

Finally, [16] conducted a principal component analysis and artificial neural network (ANN) classifier to perform non-correlated feature selection. The analysis was performed on traumatic brain injury (TBI) patients, including intracranial hypertension. The predictivity of the components were tested and the results showed a promising accuracy with a mean absolute error of 0.025.

## 5. The Proposed Method

In this proposed framework, feature reduction method using PCA is conducted as an initial step towards reducing the number of attributes without losing the main purpose and objective information from the original data. The next step is developing a predictor with an improved accuracy in order to classify data set. With the objective to find the decision attributes that have more dependent information from data, the proposed framework consists of the several phases.

- Data acquisition. An experimental data set on Cervical cancer from UCI Machine Learning repository was used to predict cervical cancer indicators and diagnosis.
- Data pre-processing. New attributes should be constructed in order to replace missing values and inconsistencies. Hinselman, schiller, cytology and biopsy are among the targets that are reduced into two values target, namely yes and no when the value of constructed target is 1 and 0, respectively. Missing values of numerical attributes are replaced with average amount of the subset, whereas missing values of nominal attributes are replaced with the most commonly occurring value of the subset.
- Standardizing each attributes. Calculated using variance and standard deviation formula.
- Calculate correlation of each attribute based on covariance formula.
- Calculate eigenvalues of each attributes to produce principal components (PC).
- Find principal component by proportion of variance, these collectively account for 95% of principal component (PC) from eigenvalues.
- Calculate eigenvectors. Performed by transposing and multiplying matrices. Each element of the eigenvectors represents the contribution of a given attributes to principal component (PC).

- Find new attributes based on correlations between the original attributes and the principal component, which is called loading. The loading considered as the cut-off is 0.5.
- Building classification model of decision tree C4.5 classifier. A number of new attributes are classified and evaluated based on performance accuracy, particularity and precision by dividing data into 90% and 10% dor training and testing, respectively.

## 6. Result and Discussion

Cervical cancer data set from UCI Machine Learning repository contains 858 instances and 36 attributes including 4 medical test attributes constituted as the target of this data set. The proposed framework is implemented in Rapid Miner® version 5.3. The detail description of each attribute as described in the UCI repository before attribute reduction using PCA is shown in table 1.

**Table 1**. Cervical cancer data set description befor attribute reduction

| Attributes | Type | Attributes | Type | Attributes | Type | Attributes | Type |
|---|---|---|---|---|---|---|---|
| Age | Int | Hormonal Contraceptives (years) | Int | STDs:vulvo-perineal condylomatosis | Bool | STDs: HPV | Bool |
| Number of sexual partnes | Int | Intra Uterine Device (IUD) | Bool | STDs: syphilis | Bool | STDs: number of diagnosis | Int |
| Age of first sexual Intercourse | Int | IUD (years) | Int | STDs: pelvic inflammatory disease | Bool | STDs: time since first diagnosis | Int |
| Number of pregnancies | Int | Sexually Transmitted Diseases (STDs) | Bool | STDs: genital herpes | Bool | STDs: time since last diagnosis | Int |
| Smokes | Bool | STDs (number) | Int | STDs: molluscum contagiosum | Bool | Dx: Cancer | Bool |
| Smokes (years) | Bool | STDs: condylomatosis | Bool | STDs: AIDS | Bool | Dx:CIN (cervical intraephitelial neoplasia) | Bool |
| Smokes (packs/year) | Bool | STDs: cervical condylomatosis | Bool | STDs: HIV | Bool | Dx: HPV (Human Papillomavirus) | Bool |
| Hormonal Contraceptives | Bool | STDs: vaginal condylomatosis | Bool | STDs: Hepatitis B | Bool | Dx (diagnosis diseases) | Bool |

In order to select relevant attributes, a number of principal components (PC1 to PC12) is performed, which contribute eigenvalues of 99.80% and absolute value of eigenvectors above 0.5. The PCA has reduced 36 attributes to 12 relevant attributes. The detail of each attribute after attribute reduction using PCA is shown in table 2.

**Table 2**. Result of attributes reduction on cervical cancer data set

| Attributes | Eigenvector | Attributes | eigenvector |
|---|---|---|---|
| Age | 0.969 | Hormonal Contraceptives | 0.906 |
| Number of sexual partners | 0.93 | Hormonal Contraceptives (years) | 0.966 |
| First sexual Intercourse | 0.912 | IUD (years) | 0.953 |
| Number of pregnancies | 0.973 | STDs (number) | 0.712 |
| Smokes (years) | 0.886 | STDs: time since first diagnosis | 0.715 |
| Smokes (packs/year) | 0.889 | STDs: time since last diagnosis | 0.695 |

The new set of attributes is then classified using decision tree C4.5 classifier, which in turns are evaluated based on performance accuracy of the classifier. The data set is divided into 90% of data training and 10% of data testing. The accuracy, particularity and precision obtained can be seen in table 3. The accuracy was enhanced to a negligible margin from 86.05% to 90.70%, whereas for the same data set the number of attributes are decreased from 36 to 12 attributes. The comparison performance of both C4.5-Non-PCA and PCA+C4.5, shows in table 3.

**Table 3**. The performance metrics of model C4.5-Non-PCA and PCA+C4.5

| | Accuracy | Particularity | Precision |
|---|---|---|---|
| C4.5 Non PCA | 86.05% | 94.74% | 33.33% |
| PCA+C4.5 | 90.70% | 100% | 100% |

As seen in table 3, the evaluation metrics are analyzed for two machine learning model, by which we had obtained the objective of the proposed cervical cancer data set classification system. The performance of model C4.5-Non-PCA based on accuracy, particularity and precision values are 86.05%, 94.74% and 33.33%, respectively, whereas the performance of model C4.5+PCA based on accuracy, particularity and precision values are 90.70%, 100% and 100%.

## 7. Conclusion

The proposed framework has achieved a robust feature reduction approach based on PCA to enhance the accuracy of decision tree C4.5 classifier. The proposed framework was implemented in Rapid Miner® version 5.3. The experiments is conducted using the available data set from UCI Cervical cancer data set repository, which are evaluated based on the accuracy performance by applying feature reduction using PCA to reduce attributes and then classifying attributes using decision tree C4.5. The results obtained baed on performance accuracy shows a value as high as 90.70%. Between C4.5-Non-PCA and PCA+C4.5 there are differences in precision as high as 4.65% accuracy, 5.26% particularity and 66.67%, respectively. This result concludes that feature reduction is an important issue in the classification model.

## References

[1] John D K, Brian M N, Aoife D 2015 Fundamentals of machine learning for predictive data analytics (Cambridge: The MIT Press) chapter 4 pp 128-190

[2] Abellan J and Castellano J 2016. A comparative study on base classifiers in ensemble methods for credit scoring. *Expert Systems with Applications* 73 1-10

[3] Zhang S 2012 Decision tree classifiers sensitive to heterogeneous costs *Journal of Systems and Software* 85(4) pp 771–779

[4] Tao Wang, Zhenxing Q, Shicao Z and Chengqi Z 2012 Cost-sensitive classification with inadequate labeled data *Information Systems* 37 (2012) pp 508–516

[5] UshaRani Y and Sammulal P 2016 An Efficient Disease Prediction and Classification Using Feature Reduction Based Imputation Technique *IEEE Conference on Engineering & MIS* (ICEMIS) Morocco pp 1-5

[6] Ali H S, Guftar M, Qamar U and Muzaffar W A 2015 A feature reduction framework based on rough set for biomedical data sets *IEEE International Conference on SAI Intelligent Systems Conference (IntelliSys)* London pp 343-9

[7] Sivapriya T R and Nadira B K AR 2013 Hybrid feature reduction and selection for enhanced classification of high dimensional medical data *IEEE Int. Conf. on Computational Intelligence and Computing Research* (ICCIC) pp 1-4

[8] Kotu V and Deshpande B 2015 *Predictive Analytics and Data Mining* (Waltham: Morgan Kaufmann)

[9] Kavitha R and Kannan E 2016 An efficient framework for heart disease classification using feature extraction and feature selection technique in data mining *IEEE Int. Conf. on Emerging Trends in Engineering Technology and Science* (ICETETS) pp 1-5

[10] Jolliffe I T 2002 *Principal Component Analysis* 2nd Ed. (New York: Springer-Verlag)

[11] Johnson R A and Wichern D W 2007 *Applied Multivariate Statistical Analysis* 6th Ed. (New Jersey: Pearson Prentice Hall)

[12] Sandi G, Supangkat H S and Slamet C 2016 Health risk prediction for treatment of hypertension *IEEE Int. Conf. on Cyber and IT Service Management* (CITSM) pp 1-6

[13] Tarle B and Jena S 2016 Improved artificial neural network for dimension reduction in medical

data classification *IEEE Int. Conf. on Computing Communication Control and Automation* (ICCUBEA) pp 1-6

[14] Fernandes K, Cardoso J S and Fernandes J 2017 Transfer Learning with Partial Observability Applied to Cervical Cancer Screening *Iberian Conference on Pattern Recognition & Image Analysis* pp. 243-250

[15] Fatlawi H K 2017 Enhanced classification model for cervical cancer data set based on cost sensitive classifier *Int. J. of Computer Techniques* 4(4) pp. 115-120

[16] Naraei P and Sadeghian A 2017 A PCA based feature reduction in intracranial hypertension analysis *IEEE Int. Conf. on 30th Canadian Conference on Electrical and Computer Engineering* (CCECE) pp 1-6

**Acknowledgement**
High appreciation should be given to Lembaga Penelitian Universitas Sumatera Utara (LP USU), Graduate School of Computer Science Fasilkom-TI USU and Rector of Universitas Sumatera Utara for their support on the dissemination of this research work as well as the facilities provided.

# Specific acoustic models for spontaneous and dictated style in indonesian speech recognition

**C B Vista[1], C H Satriawan[2], D P Lestari[3] and D H Widyantoro[4]**

[1, 2, 3,4]School of Electrical Engineering and Informatics, Institut Teknologi Bandung, Bandung, Indonesia

[1]candrabellavista@gmail.com, [2]23515053@std.stei.itb.ac.id, [3]dessipuji@stei.itb.ac.id, [4]dwi@stei.itb.ac.id

**Abstract**. The performance of an automatic speech recognition system is affected by differences in speech style between the data the model is originally trained upon and incoming speech to be recognized. In this paper, the usage of GMM-HMM acoustic models for specific speech styles is investigated. We develop two systems for the experiments; the first employs a speech style classifier to predict the speech style of incoming speech, either spontaneous or dictated, then decodes this speech using an acoustic model specifically trained for that speech style. The second system uses both acoustic models to recognise incoming speech and decides upon a final result by calculating a confidence score of decoding. Results show that training specific acoustic models for spontaneous and dictated speech styles confers a slight recognition advantage as compared to a baseline model trained on a mixture of spontaneous and dictated training data. In addition, the speech style classifier approach of the first system produced slightly more accurate results than the confidence scoring employed in the second system.
**Keywords**. Specific model; speech recognition; spontaneous speech; Indonesian language

## 1. Introduction

A speech recognition system identifies spoken words or sentences from a speech audio stream through a series of computations [1]. Previous GMM-HMM based automatic speech recognition systems for the Indonesian language have produced accurate results [2,3]. However, the accuracy of such systems declines dramatically when recognizing spontaneous speech, such as during conversations, meetings, and interviews, compared to dictated or prepared speech. This could be due to the differences in acoustical and lexical characteristics between spontaneous speech and dictated speech [4]. Spontaneous speech contains repetition, revision, mispronunciation, filled pauses, and hesitation [5]. In addition, spontaneous speech often contains foreign words, typically English, which complicates recognition by incurring errors expressed as specified terms, affixation, and new phonemes influenced by the speaker's dialect, a common occurence in Indonesian conversation [3].

Compatibility between the trained acoustic model and language model with input speech greatly affects the accuracy of speech recognition. Several previous studies have attempted to build models that are more robust towards the differing characteristics of input speech through model adaptation [6]. To investigate the effect of model compatibility with speech input, some experiments to build specific models for spontaneous speech and dictated style were conducted.

## 2. Related Works

In previous research, models trained for specific speaker age groups led to a rise in recognition accuracy for Italian language spontaneous and dictated speech [7]. The usage of separate age-group-specific models led to a 0.4% and 0.7% decrease in word error rate (WER) when tested against adult and child speech data, respectively, over a baseline system utilizing a single model.

In [8], a Czech language automatic speech recognition and subtitling system was built using separate speech style models. This study used spontaneous speech training data from 64 male and 36 female speakers. The training of separate models for each gender led to a decline in the WER rate by 2%.

Based on previous research, the training of separate acoustic models for specific groupings promises improved speech recognition performance. In this study, we build specific acoustic models for two speech styles, namely spontaneous and dictated speech, then test this approach by decoding spontaneous and dictated speech with the corresponding model. The results of this approach are then compared to the baseline system consisting of a single acoustic model for both speech styles.

The differences in spontaneous and dictated speech vary by language. A previous study [9] analyzed the acoustic characteristics of Indonesian language phonemes in spontaneous speech by phoneme duration, log energy, and frequency of occurence. From a total of 33 analyzed phonemes in the Indonesian language, the occurrence rate of the @ phoneme, or 'ə' in the International Phonetic Alphabet (IPA), is significantly higher in spontaneous speech as it is commonly used as a filled pause in the Indonesian language. The average duration of phonemes towards the beginning of words is higher than towards the end of words in spontaneous speech as compared to dictated speech, indicating that during spontaneous speech speakers tend to slow down towards the end of words. The phonemes 'a' and 'e' are more frequently followed by certain strain in spontaneous speech thus phoneme's log energy of a and e (e in the word of 'enak') is higher compared to another phoneme.

## 3. Speech Corpus Description

For acoustic model training, a speech corpus containing both spontaneous and dictated speech were used. The speech corpora has been previously used in [6]. 300 speakers were recorded for the corpus, with each speaker uttering 300 dictated sentences from a prepared text and a variable number of spontaneous sentences from topical prompts. For the spontaneous portion, utterances were derived from longer statements by manual segmentation. The corpus was recorded in a studio acoustical environment at 16 kHz sampling rate and 16 bit resolution. In addition to speaker identity, each utterance further contained metadata for speaker gender, locale (assumed to influence dialect), and age group. Table 1 summarizes the statistics of the speech corpus used in this work.

**Table 1.** Statistics of the Spontaneous and Dictated Corpus

| **Type** | **Spontaneous** | **Dictated** |
|---|---|---|
| **Speakers** | 299 | 300 |
| **Semtences** | 18806 | 83310 |
| **Duration** | 74 hours | 150 hours |

## 4. Experiment Set Up

The acoustic models for the experiments are GMM-HMM-based. Feature extraction, training, and decoding was handled using the Kaldi toolkit [10]. The language model and lexicon from [6] throughout training and testing. The language model utilizes the standard 3-gram form. The text corpus used to build the language model consists of 613.054 sentences and spontaneous and dictated speech transcriptions. The lexicon was built from words contained in the text corpus, taking into account the frequency of words. Each word listed in the lexicon may have one or more alternative pronunciations due to differences in dialect contained in the results of spontaneous and dictated speech transcription.

Two systems were designed and built for the experiment. The systems share a common speech recognition system consisting of three separate acoustic models, namely a spontaneous-specific acoustic model, a dictation-specific acoustic model, and a baseline acoustic model that uses a mixture of both spontaneous and dictated training data. In addition to the common speech recogntion system, the first system utilizes a speech style random forest classifier, the details of which can be found in [9]. Based on the prediction results of the speech style classifier, each input testing utterance is decoded using the appropriate acoustic model. In contrast, the second system decodes each input testing utterance with each model in the common speech recognition system. It subsequently calculates a confidence score to predict which decoding process best predicted the input utterance. If the confidence score for decoding using the speech style specific models is higher than the score for decoding using the integrated model, than the prediction of the speech style specific models is used, and vice versa. Figure 1 and figure 2 shows the two architecture systems.

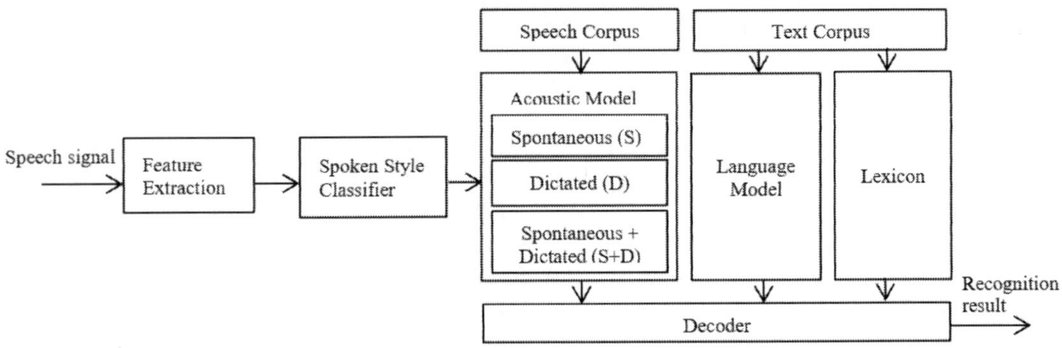


**Figure 1.** Architecture system of automatic speech recognition system using spoken style classifier

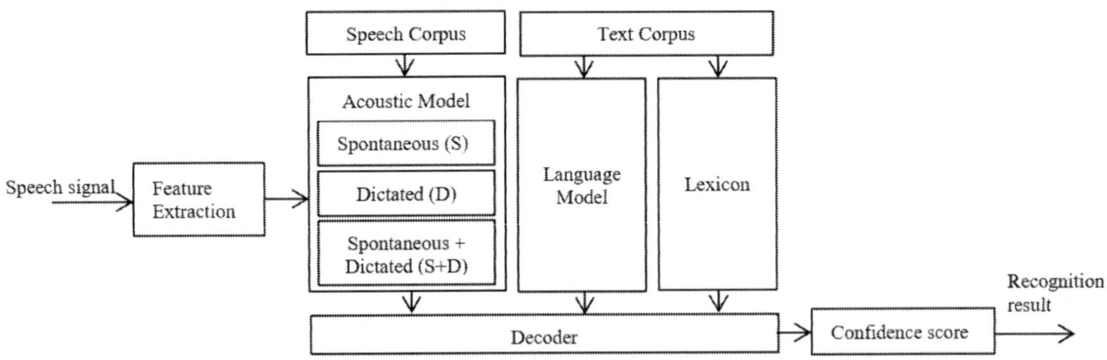


**Figure 2.** Architecture system of automatic speech recognition system using confidence score

During experimentation, an initial set of results are obtained in which each separate style-specific model is tested against speech data of the corresponding type. By comparing the recognition results of the speech style-specific models against the baseline (mixture of spontaneous and dictated) model, we ascertain whether such separate modeling is beneficial. These results also determine the theoretical upper bound of style-specific recognition, which occur when the predictions of the preceeding speech style classifier or of the subsequent confidence score calculation, as in the systems above, are perfectly accurate. This is under the assumption that all spontaneous speech is more accurately recognized using a spontaneous-specific model, and vice versa for dictated speech. Table 2 summarises the relevant combinations. Each separate style-specific model is tested against the appropriate speech style, while the integrated model is tested against all data.

**Table 2.** Experimental Design of Specific Acoustic Model

| Acoustic Models | Test Data |
|---|---|

| | |
|---|---|
| **Dictated (D)** | Dictated |
| **Spontaneous (S)** | Spontaneous |
| **Mixture of spontaneous and dictated (S+D)** | Dictated<br>Spontaneous |

## 5. Result

Initially, the speech recongition system is tested without the preceeding speech style classifier nor the subsequent confidence scoring as in the systems described previously. In this initial experiment and in all subsequent ones, the speech data from the 300 speakers in the dictated portion of the corpus and the 299 speakers of the spontaneous portion of the corpus are divided into ten subsets of data for 10-fold cross validation. Tests were performed ten times for each set of experiments, and the average value of the Word Error Rate (WER) as a percentage was calculated from these results. The experiment results are presented in Table 3.

**Table 3.** Result of Experiments on Specific Acoustic Model

| Test Data | Model | Monophone | Triphone |
|---|---|---|---|
| **Spontaneous (S)** | S | 49.5 | 34.4 |
| | S+D | 51.2 | 34.9 |
| **Dictated (D)** | D | 15.0 | 8.2 |
| | S+D | 16.2 | 8.6 |

When spontaneous speech data is decoded using the spontaneous monophone model, a WER of 49.5% is achieved, a 1.7% improvement over the baseline monophone model whereby spontaneous speech is decoded using the integrated S+D model. The same scenario conducted using the triphone model instead of the monophone results in a WER of 34.4%, a slight 0.5% improvement over the corresponding S+D model.

In the case of dictated speech, on average recognition error rates were lower than for spontanteous speech. Decoding dictated speech using the dictation-specific triphone and monophone model produced WERs of 15.0% and 8.2%, respectively. Decoding dictated speech using the mixed S+D triphone and monophone models produced WERs of 16.2% and 8.6%, respectively, a 1.2% and 0.4% decline against the dictation-specific triphone and monophone models, respectively.

By comparing the recognition results of style-specific models against their mixed model counterparts, we may conclude that on average employing style-specific models leads to an improvement in recognition results, although this improvement is usually slight. In addition, it must be considered that these results assume a perfect method to identify the speech style or determine the "better" result, and inaccuracies in the identification of speech style may offset any improvements in recognition gained by employing style-specific acoustic models. In the proceeding experiments we test the speech style classifier and the confidence scoring algorithm separately, and subsequently, in tandem with the speech recognition system.

The speech style classifier in [8] employs a random forest classifier to categorize the speech input as spontaneous or dictated. Before the decoding process, testing data will be recognized in advance by the spoken style classifier, which is labeled as either spontaneous or dictated speech. The test data that has been labeled as spontaneous or dictated is then decoded using the appropriate acoustic model. Table 4 shows the classification results of the dictated test data and spontaneous test data. The classification results of the dictated test data shows that 14.7% of the test utterances were misclassifieded as spontaneous utterances. In comparison, 12.3% of spontaneous test utterances were misclassified as being dictated utterences.

**Table 4.** Classification Result

| Test Data | Total Test Data | Misclassified | %Misclassified |
|---|---|---|---|

| | | | |
|---|---|---|---|
| **Dictated (D)** | 83310 | 12337 | 14.8 |
| **Spontaneous (S)** | 18806 | 2324 | 12.3 |

Evaluation of the confidence scoring method was conducted by decoding utterances of a certain style with both its style-specific model and the baseline S+D model, calculating confidence scores, and counting the number of utterances whereby the confidence score of decoding with the baseline model was higher than with the style-specific model. The results of the confidence score evaluation are displayed in Table 5 below. Classification using confidence scores results in a 37.7% and 52.2% misclassification rate for dictated and spontaneous speech, respectively. This "misclassification rate" signifies the number of utterances whereby the confidence score of utilizing the style-specific model for decoding was lower than utilizing the baseline S+D model.

**Table 5.** Classification Result

| Test Data | Total Test Data | Misclassified | %Misclassified |
|---|---|---|---|
| **Dictated (D)** | 83310 | 2961 | 37.7 |
| **Spontaneous (S)** | 18806 | 986 | 52.2 |

Finally, we experiment on the full systems as described in the previous section; the first system prepended by the speech style classifier and the second system appended by the confidence scoring algorithm. By comparing the initial baseline model experiments with the system evaluation results using speech style classification and confidence scores, we obtain the results shown in Table 6 below.

**Table 6.** Classification Result

| Acoustic Models | Spontaneous Test Data | Dictated Test Data | Average |
|---|---|---|---|
| **Mixture of spontaneous and dictated (S+D)** | 34.9 | 8.6 | 21.8 |
| **Spoken Style Classifier** | 34.0 | 8.0 | 21.0 |
| **Confidence Score** | 34.6 | 8.3 | 21.5 |

As compared to the mixed S+D model, the utilization of style-specific models in the two systems are shown to improve speech recognition results. The first system, employing the speech style classifier, produced WERs of 34.0% and 8.0%, a 0.9% and 0.6% improvement over the baseline model for spontaneous and dictated test data, respectively. The second system, utilizing confidence scoring, produced WER values of 34.6% and 8.3%, an improvement of 0.3% in both cases over the baseline S+D model for spontaneous and dictated speech, respectively.

These results show that the usage of style-specific models leads to increased recognition performance. On average, a 0.8% and 0.3% improvement in WER over recognition using a sole baseline model trained on both styles of speech was achieved with the usage of a speech style classifier and confidence scoring, respectively, in conjunction with separate style-specific models.

It can be seen that the speech style classifer appears to achieve recognition rates exceeding the theoretical upper bound determined through our initial results; the speech style classifier achieves 34.0% and 8.0% WER for spontaneous and dictated speech, respectively, whereas our intial results suggest 34.4% and 8.2% WER as the upper bound for spontaneous and dictated speech, respectively. This perceived discrepancy warrants further research, but can be explained by the fact that some spontaneous utterances were better recognized utilizing the dictation-specific model, and vice versa. With this in mind, we may come to the conclusion that parts of the training data were "mislabeled", in the sense that some "spontaneous" utterances were spoken in a manner more strongly resembling dictated speech, and vice versa. Further research is needed to determine whether this occurs at the utterance level or at the speaker or other group level.

## 6. Conclusion and Further Work

The results show that the usage of style-specific acoustic models for spontaneous and dictated speech improves speech recognition performance, as compared to a baseline model trained on a combination of spontaneous and dictated speech data. Through initial experimentation, it was shown that decoding spontaneous and dictated speech using a corresponding style-specific triphone model led to a decline in Word Error Rate (WER) of 0.6% and 0.2%, respectively.

The usage of a speech style classifier in conjuction with these separate models resulted in a 0.9% and 0.6% decline in WERs for spontaneous and dictated speech, respectively, over the baseline system. The usage of confidence scoring following the decoding of incoming speech through the separate models led to a 0.8% and 0.3% decline in WERs for spontaneous and dictated speech, respectively, over the baseline system.

Further research is needed to determine whether the use of style-specific models improves recognition for other style groupings aside from binary spontaneous/dictated groups. In addition, the use of hybrid or ensemble approaches to classifying speech styles, for instance employing both speech style classification and confidence scoring, improves recognition results.

## References

[1] Jurafsky D and Martin J H 2008 Speech and Language Processing: An Introduction to Natural Language Processing, Computational Linguistics, and Speech Recognition, Second Edition. Upper Saddle River, NJ: Prentice-Hall.

[2] Sakti S , Kelana E, Riza H, Sakai S, Markov K, and Nakamura S 2008 Development of Indonesian Large Vocabulary Continuous Speech Recognition System within A-STAR Project. *Procedings of the Workshop on TCAST* pp 19-24).

[3] Lestari D and Furui S 2010 Adaptation to Pronunciation Variations in Indonesian Spoken Query-Based Information Retrieval. *IEICE Transactions on Information and System* 93-D(9):2388-2396.

[4] Furui S, Nakamura M, Ichiba T and Iwano K 2005 Why is the Recognition of Spontaneous Speech so Hard. *8th International Conference Text, Speech, and Dialogue,* (Karlovy Vary, Czech Republic) pp 9-22.

[5] Nakamura M, Iwano K, and Furui S 2008 Differences between Acoustic Characteristics of Spontaneous and Read Speech and Their Effects on Speech Recognition Performance. *Computer Speech and Language* pp 171-184).

[6] Hoesen D, Lestari, D P and Khodra M 2016 Adaptation of Acoustic Model for Indonesian Using Varying Ratios of Spontaneous Speech Data. *2016 Conference of The Oriental Chapter of International Commitee for Coordination and Standardization of Speech Databases and Assessment Technique (O-COCOSDA)*. Bali, Indonesia.

[7] Gerosa M 2009 Speaker Adaptive Acoustic Modeling with Mixture of Adult and Children's Speech. 2193-96.

[8] Vanek J and Psutka J 2010 Gender-Dependent Acoustic Models Fusion Developed for Automatic Subtitling of Parliament Mettings Broadcasted by the Czech TV. *International Conference on Text, Speech, and Dialogue, Springer Berlin Heidelberg* pp 431-38.

*[9]* Satriawan C and Lestari D 2016 Analyzing Differences in Indonesian Spontaneous and Dictation Speech. *Bali, Indonesia.: 2016 Conference of The Oriental Chapter of International Commitee for Coordination and Standardization of Speech Databases and Assessment*

[10] Povey D, Ghoshal A, Boulianne G, Burget L, Glembek O, Goel N, Vesely K, Hannemann M, Motlicek P, Qian Y, Schwarz P,Silovsky J, Stemmer G 2011 The Kaldi Speech Recognition Toolkit. *IEEE 2011 Workshop on Automatic Speech Recognition and Understanding.*

IOP Conf. Series: Journal of Physics: Conf. Series **978** (2018) 012059 doi:10.1088/1742-6596/978/1/012059

**Acknowledgement**
This research was partially supported by the Master's Program toward Doctoral Degree for Excellent Graduate (*Program Pendidikan Magister Menuju Doktor untuk Sarjana Unggul*/PMDSU) from Kemenristekdikti Indonesia within research entitled "Indonesian Automatic Speech Recognition System".

IOP Conf. Series: Journal of Physics: Conf. Series **978** (2018) 012060 doi:10.1088/1742-6596/978/1/012060

# Attendance fingerprint identification system using arduino and single board computer

**M A Muchtar[1], Seniman[2], D Arisandi[3], S Hasanah[4]**

Department of Information Technology, Faculty of Computer Science and Information Technology, Universitas Sumatera Utara, Jl. Alumni No. 3 Kampus USU Medan 20155

Email: [1] anggi.muchtar@usu.ac.id,[2]seniman@usu.ac.id,[3]dedyarisandi@usu.ac.id, [4]anaesn16@gmail.com

**Abstract.** Fingerprint is one of the most unique parts of the human body that distinguishes one person from others and is easily accessed. This uniqueness is supported by technology that can automatically identify or recognize a person called fingerprint sensor. Yet, the existing Fingerprint Sensor can only do fingerprint identification on one machine. For the mentioned reason, we need a method to be able to recognize each user in a different fingerprint sensor. The purpose of this research is to build fingerprint sensor system for fingerprint data management to be centralized so identification can be done in each Fingerprint Sensor. The result of this research shows that by using Arduino and Raspberry Pi, data processing can be centralized so that fingerprint identification can be done in each fingerprint sensor with 98.5 % success rate of centralized server recording.

## 1. Introduction

Human beings have something unique that only belongs to them. This gives the idea of making the uniqueness of humanity as identity. Fingerprint is different to each of people, even for the twin. This makes fingerprints be appropriate used for biometrics technology. Another advantage of fingerprints is its practicality and durability [3]. Biometric fingerprint is a technology with physical characteristics of fingerprints. This type of biometrics is popularly used because it has a good performance and uniqueness. Moreover, fingerprints are permanent in one's lifetime, quite immune to aging and biological changes [1].

Biometric technology has great potential in various technical fields such as network security, public security, and financial industry systems [5]. At this time, the most developed technology is fingerprint recognition by using fingerprint sensor. Fingerprint sensor is a sensor used to detect fingerprint using optical system, where the detection is done by reading the contours (high level of the surface) fingerprint and static electricity body. The match rate of False Acceptance Rate (FAR) is up to <0.001% and the match rate limit of the Average Rejection Rate (FRR) is up to <1% [2].

However, the fingerprint sensor can only identify the user's fingerprint registered in the sensor. Therefore, we need a method that makes fingerprint recognition using a fingerprint sensor to be centralized. With the Arduino and Raspberry Pi, supporting data can be centralized to identify the user's fingerprint entirely and can perform fingerprint recognition in different fingerprint sensors. Using WebSocket makes it possible to provide a variety of interactions between browsers and websites. In this way, two-way continuous conversations can be made between browsers and servers [4].

## 2. Research Methodology

The identification of fingerprints for the attendance system using Arduino and Raspberry Pi which is built on this research consisted of several stages. The first stage is done with the initialization of the Fingerprint sensor, Arduino, Ethernet Shield and Raspberry Pi. The next stage is Fingerprint Sensor. At this stage the Fingerprint Sensor will take the user's fingerprint extraction thrice in one time. Then change the fingerprint capture results into data form as much as 8 bytes. Fingerprint data collection will be converted into hexadecimal form which will then send fingerprint extraction result to Arduino. Then Arduino will forward the results of fingerprint extraction to the server via Ethernet Shield. The stored data will be displayed in the form of monitoring that can be viewed on a web page. The general architecture of the system to be built can be seen in Figure 1.

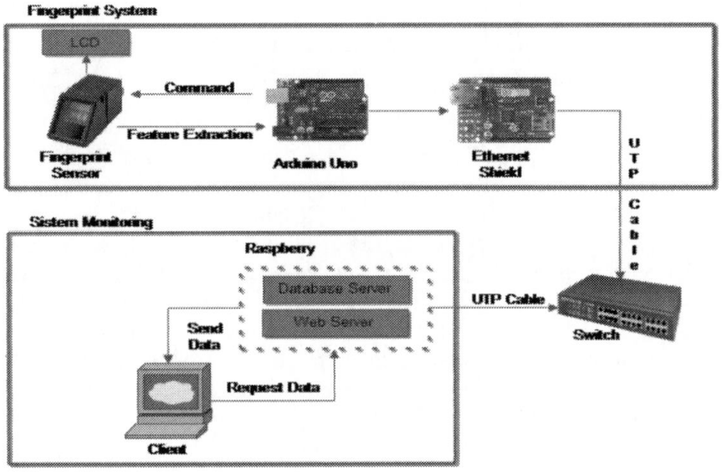


**Figure 1.** General Architecture

### *2.1. Sensor Design*

The steps that will be done to design the fingerprint sensor system are as follows:

1. Communication Fingerprint Sensor

The pins on the fingerprint sensor are used to communicate between the Arduino and the Biosec OK-300 Fingerprint Sensor using serial communication. This fingerprint sensor scans user's fingerprint. This sensor will work if user's finger is detected as well as the green LED lights on the sensor, further details, see Figure 2.

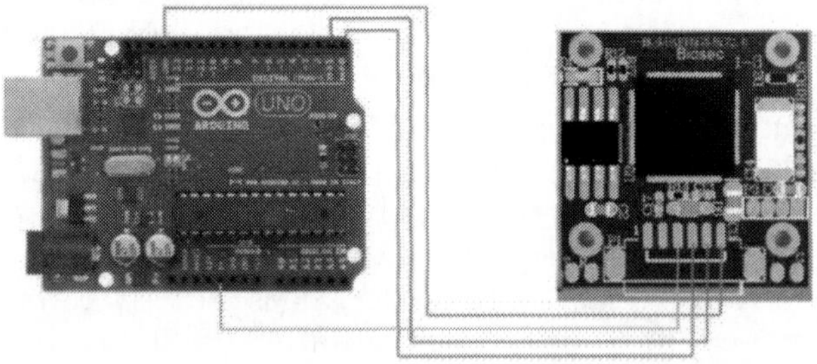

**Figure 2.** Communication between Arduino Uno and Fingerprint Module OK-300

2. Arduino Communication with LCD

LCD, in the system which will be built, serves to display the circumstances that occur within the system. When Arduino sends fingerprint extraction result to the server, the extraction data will be matched then

IOP Conf. Series: Journal of Physics: Conf. Series **978** (2018) 012060 doi:10.1088/1742-6596/978/1/012060

the server will send 'return' to Arduino and Arduino will give command to LCD to show that command succeed. To see clearly LCD display can be seen in Figure 3.

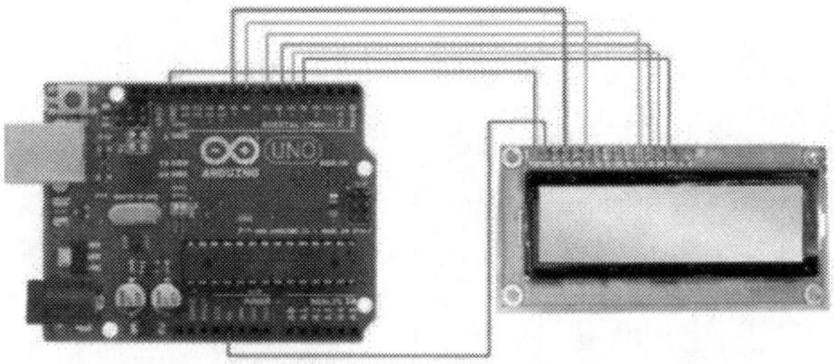


**Figure 3.** Arduino Communication with LCD

After all the sensors can communicate well to the arduino, then a circuit will be made into a single unit by using PCB (Printed Circuit Board) as seen in Figure 4.

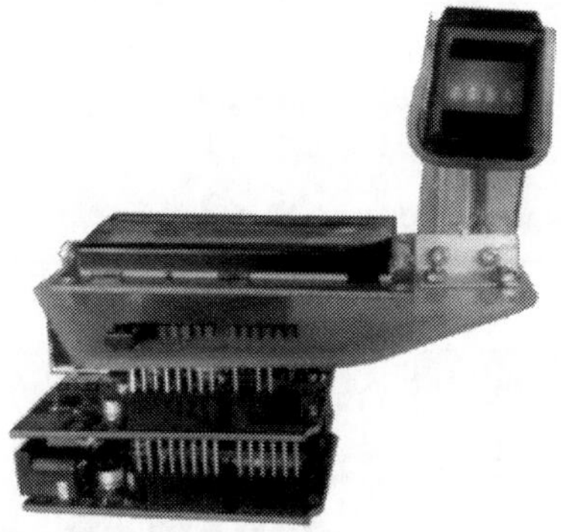

**Figure 4.** Fingerprint Sensor Series

After all the hardware has been installed, attendance system design has been integrated, then the user's fingerprint will be sent from arduino fingerprint sensor data via Local Area Network (LAN) to Raspberry server as designed in general architecture.

## 3. Result and Discussion

### *3.1. Fingerprint Identification Performance*

This stage will display the results of fingerprint sensor testing during registration. In the designed system, registration mode is selected manually in the fingerprint sensor by pressing the button on the sensor and test results can be seen through Table 1.

**Table 1.** Fingerprint Sensor Testing on Registration Process

| Steps of Registration Process | Image | | | |
|---|---|---|---|---|
| Display the current running mode | 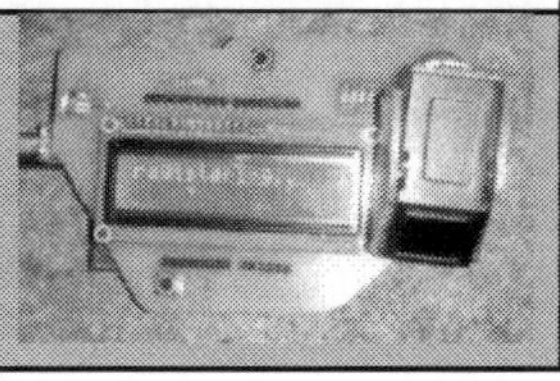 | | Wait for the fingerprint to be pressed | 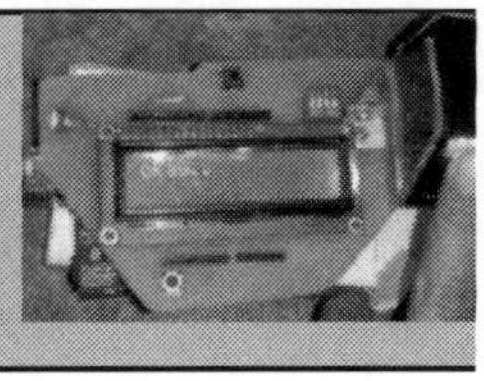 |

| | | | |
|---|---|---|---|
| **Take fingerprint extraction** | 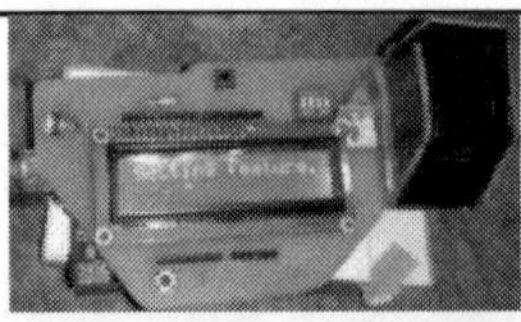 | **Fingerprint extraction completed** | 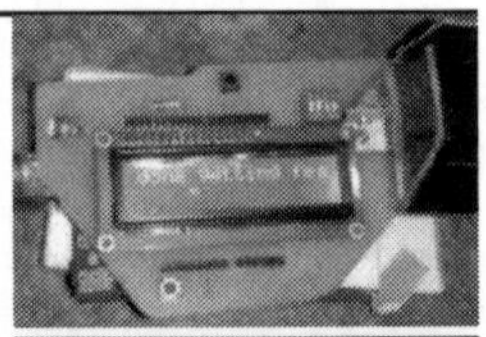 |
| **Sending the extraction results to the database** | 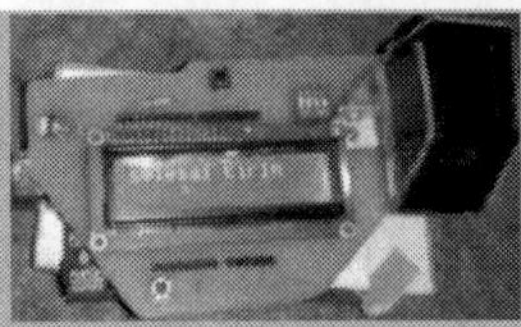 | **Sending completed** | 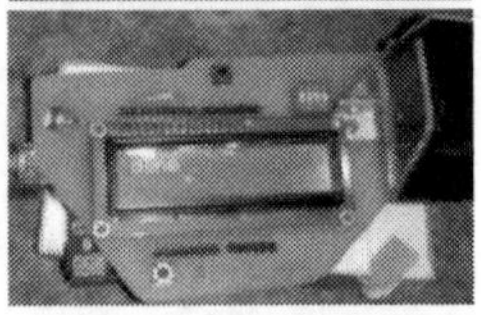 |
| **Extraction results go into the database** | 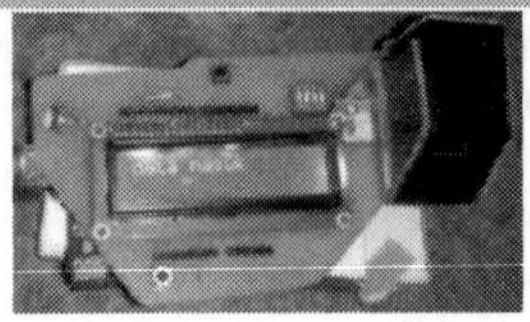 | **Broadcast a new fingerprint to all sensors** |  |

At this stage, it will display the sample results of fingerprint sensor testing while matching. Further details of the test results can be seen through Table 2.

**Table 2.** Fingerprint Sensor Matching Test

| **ID User** | **Fingerprint Match Testing** | | | | | | | | | | | | | | | | | | | |
|---|---|---|---|---|---|---|---|---|---|---|---|---|---|---|---|---|---|---|---|---|
| | **1** | **2** | **3** | **4** | **5** | **6** | **7** | **8** | **9** | **10** | **11** | **12** | **13** | **14** | **15** | **16** | **17** | **18** | **19** | **20** |
| **1** | ✓ | ✓ | ✓ | ✓ | ✓ | ✓ | ✓ | ✓ | ✓ | ✓ | ✓ | ✓ | ✓ | ✓ | ✓ | ✓ | ✓ | ✓ | ✓ | ✓ |
| **2** | ✓ | ✓ | ✓ | ✓ | ✓ | ✓ | ✓ | ✓ | ✓ | ✓ | ✓ | ✓ | ✓ | ✓ | ✓ | ✓ | ✓ | ✓ | ✓ | ✓ |
| **3** | ✓ | ✓ | ✓ | ✓ | ✓ | ✓ | ✓ | ✓ | ✓ | ✓ | ✓ | ✓ | ✓ | ✓ | ✓ | ✓ | ✓ | ✓ | ✓ | ✓ |
| **4** | ✓ | ✓ | ✓ | ✓ | ✓ | ✓ | ✓ | ✓ | ✓ | ✓ | ✓ | ✓ | ✓ | ✓ | ✓ | ✓ | ✓ | ✓ | ✓ | ✓ |
| **5** | ✓ | ✓ | ✓ | x | ✓ | ✓ | ✓ | ✓ | ✓ | ✓ | ✓ | ✓ | ✓ | x | ✓ | ✓ | ✓ | ✓ | ✓ | ✓ |
| **6** | ✓ | ✓ | ✓ | ✓ | ✓ | ✓ | ✓ | ✓ | ✓ | ✓ | ✓ | ✓ | ✓ | ✓ | ✓ | ✓ | ✓ | ✓ | ✓ | ✓ |
| **7** | ✓ | ✓ | ✓ | ✓ | ✓ | ✓ | ✓ | ✓ | ✓ | ✓ | ✓ | ✓ | ✓ | ✓ | ✓ | ✓ | ✓ | ✓ | ✓ | ✓ |
| **8** | ✓ | ✓ | ✓ | ✓ | ✓ | ✓ | ✓ | ✓ | ✓ | ✓ | ✓ | ✓ | ✓ | ✓ | ✓ | ✓ | ✓ | ✓ | ✓ | ✓ |
| **9** | ✓ | ✓ | ✓ | ✓ | ✓ | ✓ | ✓ | ✓ | ✓ | ✓ | ✓ | ✓ | ✓ | ✓ | ✓ | ✓ | ✓ | ✓ | ✓ | ✓ |
| **10** | ✓ | ✓ | ✓ | ✓ | ✓ | ✓ | ✓ | ✓ | ✓ | ✓ | ✓ | ✓ | ✓ | ✓ | ✓ | ✓ | ✓ | ✓ | ✓ | ✓ |
| **11** | ✓ | ✓ | ✓ | ✓ | ✓ | ✓ | ✓ | ✓ | ✓ | ✓ | ✓ | ✓ | ✓ | ✓ | ✓ | ✓ | ✓ | ✓ | ✓ | ✓ |
| **12** | ✓ | ✓ | ✓ | ✓ | ✓ | ✓ | ✓ | ✓ | ✓ | ✓ | ✓ | ✓ | ✓ | ✓ | ✓ | ✓ | ✓ | ✓ | ✓ | ✓ |
| **13** | ✓ | ✓ | ✓ | ✓ | ✓ | ✓ | ✓ | ✓ | ✓ | ✓ | ✓ | ✓ | ✓ | ✓ | ✓ | ✓ | ✓ | ✓ | ✓ | ✓ |
| **14** | ✓ | ✓ | ✓ | ✓ | ✓ | ✓ | ✓ | ✓ | ✓ | ✓ | ✓ | ✓ | ✓ | ✓ | ✓ | ✓ | ✓ | ✓ | ✓ | ✓ |
| **15** | ✓ | ✓ | ✓ | ✓ | ✓ | x | ✓ | ✓ | ✓ | ✓ | ✓ | ✓ | ✓ | ✓ | ✓ | ✓ | ✓ | ✓ | ✓ | ✓ |
| **16** | ✓ | ✓ | ✓ | ✓ | ✓ | ✓ | ✓ | ✓ | ✓ | ✓ | ✓ | ✓ | ✓ | ✓ | ✓ | ✓ | ✓ | ✓ | ✓ | ✓ |
| **17** | ✓ | ✓ | ✓ | ✓ | ✓ | ✓ | ✓ | ✓ | ✓ | ✓ | ✓ | ✓ | ✓ | ✓ | ✓ | ✓ | ✓ | ✓ | ✓ | ✓ |
| **18** | ✓ | ✓ | ✓ | ✓ | ✓ | ✓ | ✓ | ✓ | ✓ | ✓ | ✓ | ✓ | ✓ | ✓ | ✓ | ✓ | ✓ | ✓ | ✓ | ✓ |
| **19** | ✓ | ✓ | ✓ | ✓ | ✓ | ✓ | ✓ | ✓ | ✓ | ✓ | ✓ | ✓ | ✓ | ✓ | ✓ | ✓ | ✓ | ✓ | ✓ | ✓ |
| **20** | ✓ | ✓ | ✓ | ✓ | ✓ | ✓ | ✓ | ✓ | ✓ | ✓ | ✓ | ✓ | ✓ | ✓ | ✓ | ✓ | ✓ | ✓ | ✓ | ✓ |
| **21** | ✓ | ✓ | ✓ | ✓ | ✓ | ✓ | ✓ | ✓ | ✓ | ✓ | ✓ | ✓ | ✓ | ✓ | ✓ | ✓ | ✓ | ✓ | ✓ | ✓ |
| **22** | ✓ | ✓ | ✓ | ✓ | ✓ | ✓ | ✓ | ✓ | ✓ | ✓ | ✓ | ✓ | ✓ | ✓ | ✓ | ✓ | ✓ | ✓ | ✓ | ✓ |
| **23** | ✓ | ✓ | ✓ | ✓ | ✓ | ✓ | ✓ | ✓ | ✓ | ✓ | ✓ | ✓ | ✓ | ✓ | ✓ | ✓ | ✓ | ✓ | ✓ | ✓ |
| **24** | ✓ | ✓ | ✓ | ✓ | ✓ | ✓ | ✓ | ✓ | ✓ | ✓ | ✓ | ✓ | ✓ | ✓ | ✓ | ✓ | ✓ | ✓ | ✓ | ✓ |
| **25** | ✓ | ✓ | ✓ | ✓ | ✓ | ✓ | ✓ | ✓ | ✓ | ✓ | ✓ | ✓ | ✓ | ✓ | ✓ | ✓ | ✓ | ✓ | ✓ | ✓ |
| **26** | ✓ | ✓ | ✓ | ✓ | ✓ | ✓ | ✓ | ✓ | ✓ | ✓ | ✓ | ✓ | ✓ | ✓ | ✓ | ✓ | ✓ | ✓ | ✓ | ✓ |
| **27** | ✓ | ✓ | x | ✓ | ✓ | ✓ | ✓ | ✓ | ✓ | ✓ | ✓ | ✓ | x | ✓ | ✓ | ✓ | ✓ | ✓ | ✓ | ✓ |
| **28** | ✓ | ✓ | ✓ | ✓ | ✓ | ✓ | ✓ | ✓ | ✓ | ✓ | ✓ | ✓ | ✓ | ✓ | ✓ | ✓ | ✓ | ✓ | ✓ | ✓ |
| **29** | ✓ | ✓ | ✓ | ✓ | ✓ | ✓ | ✓ | ✓ | ✓ | ✓ | ✓ | ✓ | ✓ | ✓ | ✓ | ✓ | ✓ | ✓ | ✓ | ✓ |
| **30** | ✓ | ✓ | ✓ | ✓ | ✓ | ✓ | ✓ | ✓ | ✓ | ✓ | ✓ | ✓ | ✓ | ✓ | ✓ | ✓ | ✓ | ✓ | ✓ | ✓ |

| | | | | | | | | | | | | | | | | | | | | |
|---|---|---|---|---|---|---|---|---|---|---|---|---|---|---|---|---|---|---|---|---|
| **31** | ✓ | ✓ | ✓ | ✓ | ✓ | ✓ | ✓ | ✓ | ✓ | ✓ | ✓ | ✓ | ✓ | ✓ | ✓ | ✓ | ✓ | ✓ | ✓ | ✓ |
| **32** | ✓ | ✓ | ✓ | ✓ | ✓ | ✓ | ✓ | ✓ | ✓ | ✓ | ✓ | ✓ | ✓ | ✓ | ✓ | ✓ | ✓ | ✓ | ✓ | ✓ |
| **33** | ✓ | ✓ | ✓ | ✓ | ✓ | ✓ | ✓ | x | ✓ | ✓ | ✓ | ✓ | ✓ | ✓ | ✓ | ✓ | ✓ | x | ✓ | ✓ |
| **34** | ✓ | ✓ | ✓ | ✓ | ✓ | ✓ | ✓ | ✓ | ✓ | ✓ | ✓ | ✓ | ✓ | ✓ | ✓ | ✓ | ✓ | ✓ | ✓ | ✓ |
| **35** | ✓ | ✓ | x | ✓ | ✓ | ✓ | ✓ | ✓ | ✓ | ✓ | ✓ | ✓ | ✓ | ✓ | ✓ | ✓ | ✓ | ✓ | ✓ | ✓ |
| **36** | ✓ | ✓ | ✓ | ✓ | ✓ | ✓ | ✓ | ✓ | ✓ | ✓ | ✓ | ✓ | ✓ | ✓ | ✓ | ✓ | ✓ | ✓ | ✓ | ✓ |
| **37** | ✓ | ✓ | ✓ | ✓ | ✓ | ✓ | ✓ | ✓ | ✓ | ✓ | ✓ | ✓ | ✓ | ✓ | ✓ | ✓ | ✓ | ✓ | ✓ | ✓ |
| **38** | ✓ | ✓ | ✓ | ✓ | ✓ | ✓ | ✓ | ✓ | ✓ | ✓ | ✓ | ✓ | ✓ | ✓ | ✓ | ✓ | ✓ | ✓ | ✓ | ✓ |
| **39** | ✓ | ✓ | ✓ | ✓ | ✓ | ✓ | ✓ | ✓ | ✓ | ✓ | ✓ | ✓ | ✓ | ✓ | ✓ | ✓ | ✓ | ✓ | ✓ | ✓ |
| **40** | ✓ | ✓ | ✓ | ✓ | ✓ | x | ✓ | ✓ | ✓ | ✓ | ✓ | ✓ | ✓ | ✓ | ✓ | ✓ | ✓ | ✓ | ✓ | ✓ |

Based on the testing done in matching process, nine of the fingerprints cannot be identified which are marked with "x". This case is because of various problem such as fingerprint is not in the correct position, wet or oily finger and thin finger surface. While the successful ones are marked with "√". From the matching fingerprint data in Table 2, it can be seen that the accuracy of the designed system is 98.875%. The accuracy value is obtained based on equation 1.

$$\text{Percentage of Accuracy} = \frac{\text{number of successful fingerprint matching}}{\text{total number of fingerprint test data}}\ X\ 100\% \qquad (1)$$

$$\text{Percentage of Accuracy} = \frac{791}{800}\ X\ 100\ \%$$

$$= 98.875\ \%$$

### *3.2. Fingerprint Server Application*

At this stage, it will show the results of the interface design implementation of the built system. Here is a built-in interface pageview:

1. Add User Page

On this page, admin does not need to fill the member ID and fingerprint data because the data is obtained when the user scans his fingerprint on the fingerprint sensor. The added user pageviews can be seen in Figure 5.

**Figure 5.** Add User Pageviews

2. User Data Pageviews

On this page, admin can view or search for information from users like identity number, name, gender, address and mobile phone number. Admin can also edit user data in case of errors while filling out data and delete users. On this page also provides a print menu that works to print user data into PDF form. Views of user data pages can be seen in Figure 6.

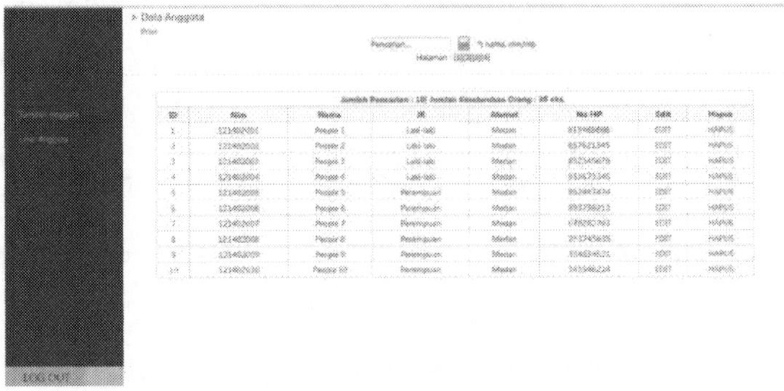

**Figure 6.** User Data Pageviews

3. User Attendance History Pageviews

On the user attendance history page, the admin must select the user first. To view user history can be done when an admin clicks on one of the user on the user data pageviews.

The provided information from this page is the description of the user's attendance. This page shows processed fingerprint mathing which have been processed and transmitted by the client arduino fingerprint system to the single board computer server. The pageviews of user's history can be seen in figure 7.

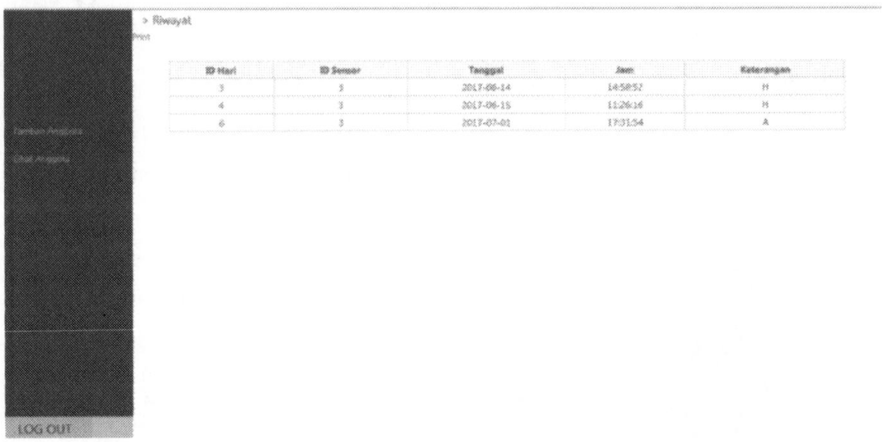


**Figure 7.** User Attendance History Pageviews

## 4. Conclusion

The conclusions can be obtained from fingerprint identification research using Arduino and Raspberry for attendance system are as follows:

- Research of fingerprint identification in attendance system using Arduino and Raspberry is successfully implemented with 98.875% success rate of centralized server recording.
- Based on the results of this study, it can be concluded that fingerprint data processing becomes centralized.
- Attendance list can be done in any fingerprint sensor connected in one network.

## 5. Acknowledgement

This submitted research paper is outcome of "Penelitian Produk Terapan" research funded by DRPM RISTEKDIKTI based on contract number 003/SP2H/LT/DRPM/IV/2017 date of 20th April 2017.

## References

[1] Amioy K, Tanvir S M and Ajay K 2009 Anatomy of Hand. Encyclopedia of Biometrics vol 1 pp 28-35.

[2] Biosec. 2015. OK-300 Data Sheet. (Online) http://biosec.com.cn/r/cms/www/biosec/OK300.pdf (15 Oktober 2016).

[3] Jain A K, Prabhakar S, and Pankanti S 2002 On the similarity of identical twin fingerprints. *Pattern Recognition* 35 pp 2653–63.

[4] Shuang, K., Shan, X., Sheng, Z., & Zhu, C. 2014. An Efficient ZigBee-WebSocket based M2M Environmental Monitoring System. IEEE 12th International Conference on Dependable, Autonomic and Secure Computing , 322-326.

[5] Zhu, Z., & Chen, F. 2011. Fingerprint Recognition-Based Access Controlling System for Automobiles. IEEE International Congress on Image and Signal Proce , 1899-1902.

# The gap values in the profile matching method by fuzzy logic

**S A Sitepu[1], S Efendi[2] and Z Situmorang[3]**

[1]Master Program (S2) of Informatics Engineering, Faculty of Computer Science and Information Technology, Universitas Sumatera Utara, Medan, Indonesia.
[2]Master Program of Informatics Engineering, Faculty of Computer Science and Information Technology, Universitas of Sumatera Utara, Medan, Indonesia.
[3]Department of Computer Science, Faculty of Computer Science and Information Technology, Universitas St. Thomas Medan, Indonesia. Email : [1]syafrina.amanah@gmail.com

**Abstract.** In this research, the determination of the appropriate values of Gap for the assessment of promotion criteria of position in an institution / company. In this study the authors use Fuzzy Sugeno logic on the determination of Gap values used in Profile Matching method. Test results of 5 employees obtained the eligibility of promotion with the position of Z* values between in 3.20 to 4.11.
Keywords: *Decision Supporting System, Fuzzy Profile Matching Algorithm, ranking, Gap values.*

## 1. Introduction

Calculation process in *Profile Matching* method is started by defining the minimum values for each variable of valuation. The Gap of each values of test data against the minimum values of each variable is became the Gap that should be given a weight. The weight of a variable will be calculated based on the average values of group of *Core Factor* (CF) and *Secondary Factor* (SF) variable. The CF composition added by SF is 100%, depend on the interest of user of this method. The last stage of the method is accumulation process of CF and SF values based on data *testing variable values*.

Weighting on method of Profile Matching, is the definite values of a certain values because the existing values is a member of the *crisp set*. In a crisp set, membership of an element in the set of stated explicitly, whether the object is a member of the set or not by using the function characteristics. If this is not done on the eligibility determination process, it will result in the classification of the inflexible and rigid. [4].

The final values obtained from the method of Profile Matching, is a firm, definite values because the values is a member of the *crisp set*. In the crisp set, membership is stated expressly. If it is associated with the process of determining the eligibility of employees who occupy a position in the newly promoted, this will result in rigidity for the *PDAM Tirtanadi Sumatera Utara* in the promotion of the employee. Whereas, in the process of the promotion of the position of the employee it is required to be more just and wise.

## 2. Study of literature

### *2.1. Fuzzy Logic*

Fuzzy logic is a branch of artificial intelligence systems (Artificial Intelligent) that emulates human thinking ability into the form of algorithms which then executed by the machine. This algorithm is

used in a variety of data processing applications that cannot be represented in binary form. Fuzzy logic interprets the vague statements into a logical sense [2].
Fuzzy logic is a branch of computer science that studies about the values of truth that worth a lot. In contrast to the values of truth in classical logic of values 0 (false) or 1 (true), Fuzzy logic has a values of truth in the real interval [0.1]. Fuzzy logic was first developed by Lotfi A. Zadeh a United States-born scientist of Iran nationality from the University of California at Berkeley. Even so, the fuzzy logic more developed by Japan practitioners [3].
In Fuzzy theory, membership of an element in the set is expressed with the degree of membership (membership values) whose values lies on [0.1]. Research on Fuzzy logic is used in the process of weighting variables feasibility assessment of a particular position promotion. This is done in the hope of accumulation of testing data values on the feasibility assessment process will are normally grouped with the predicate less, fair and good. So the company's management can apply the policy in determining the eligibility of a person to occupy a position in the office. The predicate of this classification, based on Fuzzy logic that is capable of bridging the expressive machine language with human language that emphasized the meaning based on natural language.

*2.2. Decision Support System*
Decision Support System abbreviated by DSS is a system that aimed for managerial decisions to support the semi structured decision. Herbert. A. Simon determining decision-making in three stages, namely Intelligent, Design Choice, Implementation. According to Whitten (2004) Decision Support System/DSS can be defined as a system of information that helps identify opportunities to make decisions or provide information to assist decision making. With a variety of special characters such as expressed above, decision support system can deliver profits or values in order for the users.
In the early 1970 's, Scott Morton is the first in articulating the concept of the crucial decision support systems. He defines a decision support system as "interactive computer-based systems, which help decision-makers to use data and models to solve unstructured problems." Other classic definition of "Decision support system combines the intellectual resource of individuals with computer capabilities to improve the quality of decisions" [5].

*2.3. Gap Values*
In the research of Kansil and Sutapa, conducted surveys of library service user satisfaction at Petra University (UK Petra). The UK Petra Library is one of the libraries that is very concerned about the quality of its services. This is demonstrated by the user satisfaction survey for service quality in 2011 and 2012. Data show CSI (Customer Satisfaction Index) in 2011 and 2012 high values, while CDI (Customer Dissatisfaction Index) is very low. This study not only examines the Gap between user satisfaction and user interest in the quality of the UK Petra library service (Gap 5), but also the interests of users with management interests (Gap 1), user interests with staff interests (Gap 6), and interests management and staff interests (Gap 7). The research is done by measuring the values of Gap (Gap) on the quality of library service with Servqual method [6].
The result of the research is the smallest Gap values 5 found in the service quality of the library staff is neat and polite, with the score is -0.46. The greatest Gap values is found in the quality of the library staff's liveliness offering assistance to users, with a score of -1.47. The most positive Gap 1 values is found in the quality of updated data collection service on iSPEKTRA online log (0.81). Quality of service that has a values of Gap of 1 close to 0 is the library staff is believed to maintain the serenity of the library (0,08), the ambiguity of room signs (0,08), the comfort of the room furniture (-0.06). The values of the 1 pal-ing negative Gap is the quality of the Digital Theses collection conformity service to user requirements (-1.00). The most positive values of Gap 6 is found in the quality of interior room service (0.56). The values of Gap 6 approaching 0 is found in the quality of the collection service arranged neatly on the shelf (0.06). The values of Gap 6 which has the most negative values is found in the quality of DiVo's presence service as a means of information and promotion (-0.15). The most positive Gap values found in the quality of the collection service is neatly arranged on the shelf (0.44). The values of Gap 7 approaching 0 is in 3 service quality. The quality of this services is the speed of

the library staff in helping users obtain the required information (0.06), the completeness of the iSPEKTRA online catalog feature [6].

### *2.4. Profile Matching Method*

In the research of Sopianti and Bahtiar, conducted the determination of specialization on mathematics and Natural Sciences, Social Sciences, and Language and Culture Sciences. The learners conducted since class X. in this study applied Profile Matching method to calculate the competence of each individual based on the criteria given. Implementation of Profile Matching method is optimized by placing core and secondary factor dynamically in each department so as to obtain ideal calculation result. The results of experiments on this system can determine the interest of students based on the data criteria that have been determined by the school and generate data on the determination of interest with a match rate of 94.73%. Supporting decision support system of learners can give consideration in the determination of student interest in accordance with the priority field of alternative majors [7].
The goals of decision support system are:

1. Assist managers in decision-making over issues of semi structured.
2. Provide support for the consideration of the Manager and not intended to replace the function of the Manager.
3. Increase the effectiveness of the decisions taken the Manager over on the efficiency improvements.
4. Speed of computation. The computer allows the decision makers to do a lot of computing rapidly and costs are low.
5. Increased productivity. Build a group decision makers, especially the experts, can be very expensive. Supporters of the group could reduce the size of the computerized and allows its members to be in various different locations (save travel costs). In addition, support staff productivity (e.g. financial and legal analysis) could be improved.
6. Productivity can also be upgraded using the optimization tools to determine the best way to run a business.
7. Quality support. Computers can enhance the quality of the decisions made. For example, the more data is accessed more alternative can also be evaluated. Risk analysis can be done quickly and the views of the experts (some of them are in what is a much) can be collected quickly and with lower costs. Expertise could even be taken directly from a computer system via a method of artificial intelligence. With computers, decision-makers can perform complex simulations, check out the many scenarios that are possible, and assess the range of influences in a fast and economical. All of these capabilities lead to better decisions.
8. Competitive power. Management and empowerment of company resources. The pressure of competition led to the decision maker to be difficult task. The competition is based not only on price but also on the quality, speed, kustomasi products, and customer support. Organizations must be able to frequently and rapidly changing operating mode, reverse engineer restarted the process and structure, empowering employees, and innovate. Technology decision making can create significant empowerment by way of allowing someone to make good decisions quickly, even if they have a less knowledge.[i]
9. Overcoming limitations in cognitive processing and storage. According to Simon (1977), the human brain has a limited ability to process and store information. People are sometimes difficult to remember and use information in a manner that is free from errors.

## 3. METHODS OF ANALYSIS

The assessment process for promotion is done by the method of Profile Matching standard and with the method of Fuzzy Profile Matching. On the methods of Profile Matching, it is known the existence of the level difference between the values of test data with a minimum values called the Gap. On the assessment process using the standard method of Profile Matching, Gap values have been determined by decision makers.
In Fuzzy Profile Matching method, employee classification process for promotion of position, weighting of Gap values using Fuzzy logic where Gap values is given weight, so it can be determined average values of composition of the comparison values in the form of fuzzy set based on idea to

extend the range of function characteristics such that the function will include real numbers at intervals [0.1]. The membership values indicates that an item in the universe of speech is not only 0 or 1, but also the values that lies between them. In other words, the truth values of an item is not only true or false or 0 indicates false, values 1 indicates true. That there are still more values lies between right and wrong. By using Fuzzy Profile Matching method, it is expected that promotion of appraised positions can be categorized not only feasible and not feasible, but can be categorized in detail that the category "less" is not feasible and the categories "adequate" and "good" are considered feasible.
Step-by-step method of Profile Matching is:

1. Determine the variable data that is needed.
2. Determine the aspects used for assessment.
3. Mapping the Gap profile.
   Gap *Profile Matching* = (Minimal profile – Profile of test data)
   Gap *Fuzzy Profile Matching* calculated with the approach of Fuzzy Logic.
4. After the next Gap values obtained are given weights for each values of the Gap.
5. Calculation and classification of Core Factor and Secondary Factor.
6. After determining the weighting values of Gap, then grouped into 2 groups, the Core Factor (a major factor) and a Secondary Factor (factor endowments).
7. The Calculation Of The Total Values.

Determination of the ranking calculation where the end result of the process of Profile Matching is a ranking that can be made by the values of the threshold in a classification of promotion of the employee. Weights on the values is given to each criteria.

**Table 1.** Weighting of Assessment

| Values | Weight |
|---|---|
| 0 – 39 | 1 = very bad |
| 40 – 59 | 2 = bad |
| 60 – 69 | 3 = less |
| 70 – 89 | 4 = good |
| 90 – 100 | 5 = very good |

**The calculation of values by method of Profile Matching**
Table Values obtained in accordance with weighting of assessment.

**Table 2.** Values of Work Attitude Aspect

| No | Employee id | 1 | 2 | 3 | 4 | 5 | 6 | 7 | 8 | 9 | 10 |
|---|---|---|---|---|---|---|---|---|---|---|---|
| 1 | K001 | 3 | 4 | 4 | 4 | 4 | 4 | 4 | 4 | 4 | 4 |
| 2 | K002 | 4 | 4 | 4 | 5 | 5 | 5 | 5 | 5 | 5 | 5 |
| 3 | K003 | 4 | 4 | 4 | 4 | 4 | 4 | 4 | 4 | 4 | 4 |
| 4 | K004 | 4 | 4 | 4 | 4 | 4 | 4 | 4 | 4 | 3 | 3 |
| 5 | K005 | 5 | 5 | 5 | 5 | 5 | 4 | 5 | 5 | 4 | 5 |

**Table 3.** Values of Intelligence Aspects

| No | Employee id | 1 | 2 | 3 | 4 | 5 | 6 | 7 | 8 |
|---|---|---|---|---|---|---|---|---|---|
| 1 | K001 | 4 | 4 | 4 | 4 | 4 | 4 | 4 | 4 |
| 2 | K002 | 5 | 5 | 5 | 5 | 5 | 5 | 5 | 5 |
| 3 | K003 | 4 | 4 | 4 | 4 | 4 | 4 | 4 | 4 |
| 4 | K004 | 4 | 4 | 4 | 4 | 4 | 4 | 4 | 4 |
| 5 | K005 | 5 | 5 | 5 | 5 | 5 | 5 | 5 | 5 |

**Table 4.** The Values of Competence Aspect

| No | Employee id | 1 | 2 | 3 | 4 | 5 | 6 | 7 | 8 |
|---|---|---|---|---|---|---|---|---|---|
| 1 | K001 | 4 | 4 | 4 | 4 | 4 | 4 | 4 | 4 |
| 2 | K002 | 5 | 5 | 5 | 5 | 5 | 5 | 5 | 5 |
| 3 | K003 | 4 | 4 | 4 | 4 | 4 | 4 | 4 | 4 |
| 4 | K004 | 4 | 4 | 4 | 4 | 4 | 4 | 4 | 4 |
| 5 | K005 | 5 | 5 | 5 | 5 | 5 | 5 | 5 | 5 |

The calculation of the values of the Gap at any values derived
**Gap = Values Attribut – Values Target**
Description:
Attribute Values: The values obtained by officers of the appropriate weighting.
Target Values: The values of a standard that has been set.

**Table 5.** The Final Values of Profile Matching

| No | Employee_Id | Ns | Nkc | Nkp | Final Values |
|---|---|---|---|---|---|
| 1 | K001 | 4,4 | 4,75 | 2,74 | 3,68 |
| 2 | K002 | 4,2 | 4,25 | 3,00 | 3,62 |
| 3 | K003 | 4,5 | 4,75 | 2,74 | 3,70 |
| 4 | K004 | 4,45 | 4,75 | 2,74 | 3,69 |
| 5 | K005 | 3,8 | 4,25 | 2,36 | 3,22 |

The results of the ranking of promotion on table 3.5 above can be seen as in Table 3.6.

**Table 6.** Results of the Ranking by Method of Profile Matching

| No | Employee_Id | Ranking Result |
|---|---|---|
| 1 | K003 | 3,70 |
| 2 | K004 | 3,69 |
| 3 | K001 | 3,68 |
| 4 | K002 | 3,62 |
| 5 | K005 | 3,22 |

**Calculation method of Fuzzy values with Profile Matching**

Table interval of eligibility category of promotion with the method of Fuzzy Profile Matching as in Table 7 below:

**Table 7.** Interval Values of Eligibility of Promotion Category

| No | Z* | Category | Final Values |
|---|---|---|---|
| 1 | $Z \leq 2$ | Less | Not Viable |
| 2 | $2.1 \leq Z \leq 3.5$ | Less | Viable |
| 3 | $2.5 \leq Z \leq 4.5$ | Less Good | Viable |
| 4 | $3.6 \leq Z \leq 4.99$ | Good | Viable |
| 5 | $Z \geq 5$ | Good | Viable |

The results of the determination of position of promotion with the method of Fuzzy Profile Matching as follows:

**Table 8.** Result of *Fuzzy Profile Matching Method*

| No | Employee_Id | Result Z* | Final Values |
|---|---|---|---|
| 1 | K003 | 3.25 | Viable |
| 2 | K001 | 3.55 | Viable |
| 3 | K004 | 2.45 | Viable |
| 4 | K002 | 3.20 | Viable |
| 5 | K005 | 4.11 | Viable |

## 4. Conclusion

Based on the results of the analysis of the calculation of values of the variables used in the promotion of Office have some conclusions as follows:

a. The process of classification using profile matching method will have the final result of the definite values (crisp).
b. It is recommended to use classification method in the case of the decision support system, because this method can do a ranking of test data/test in large quantities.
c. The results of the classification by using the methods of fuzzy logic can categorize data in more than one category depending on the set are used. So this method contains elements of policies on the promotion.

## 5. REFERENCES

[1] Angkasa S 2016 Application of The Method of Profile Matching to Determine Feasibility of Granting Credit In The PNPM Mandiri City of Banjarmasin. *Journal of information technology* Vol. **11**.

[2] Purba S and Sudiarso A 2013 Leather Handicraft Products Price Determination Using Fuzzy Logic Approach Taking Into Account The Projected Profits, Consumer and Competitor Price Perception.

[3] Li Y, Zhuang Y, Lan H, Niu X and Sheimy N E 2016 A Profile-Matching Method for Wireless Positioning. *IEEE Communications Letters,* **20** no 10.

[4] Joshi A and Amity J S S 2013 Attributes Similarities Supports Profile Matching In Social Network. *International Journal of Engineering Sciences & Emerging Technologies* University, Noida. ISSN: 2231 – 6604 vol **6**, Issue 3, pp: 368-72.

[5] Indriantoro S C and Utami A W 2016 Decision Support System Of The Supersemar Scholarship Using Web-Based Promethee Method (Case Study: Faculty Of Engineering State University Of Surabaya).

[6] Kansil A A K P and Sutapa I N 2013 Measurement Gap (Gap) Quality of Service Library UK Petra with Servqual Method. *Journal Titra* Vol.**1**.

[7] Sopianti L and Bahtiar N 2015 Students Major Determination Decision Support Systems using Profile Matching Method with SMS Gateway Implementation. *Journal of Science and Mathematics* Vol. **23**.

IOP Conf. Series: Journal of Physics: Conf. Series **978** (2018) 012062 doi:10.1088/1742-6596/978/1/012062

# Simulation evaluation of capacitor bank impact on increasing supply current for alumunium production

**S Hasan, K Badra, R Dinzi and Suherman***
Department of electrical engineering Univesitas Sumatera Utara 20155, Indonesia

*E-mail: suherman@usu.ac.id

**Abstract**. DC current supply to power the electrolysis process in producing aluminium at PT Indonesia Asahan Aluminium (Persero) is about 193 kA. At this condition, the load voltage regulator (LVR) transformer generates 0.89 lagging power factor. By adding the capacitor bank to reduce the harmonic distortion, it is expected that the supply current will increase. This paper evaluates capacitor bank installation impact on the system by using ETAP 12.0 simulation. It has been obtained that by installing 90 MVAR capacitor bank in the secondary part of LVR, the power factor is corrected about 8% and DC current increases about 13.5%.

## 1. Introduction

Capacitor bank is widely used in electrical power system to improve the power factor. Capacitor is acting as a reactive load generator and connected in paralel to load compensating the lagging current [1-4]. Capacitor reduces reactive current and reactive power so that the power loss and dropped voltage decrease, and load voltage increases. Capacitor bank topology consits of grounded and ungrounded wye as shown in Figure 1.

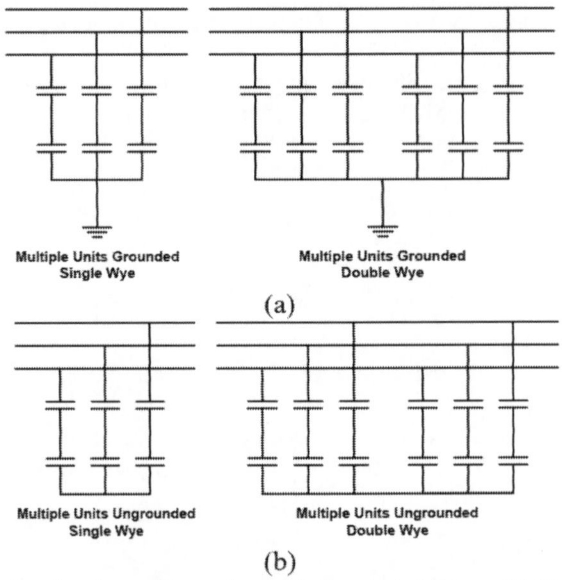


**Figure 1**. Capacitor bank configuration: (a) Grounded Wye, (b) Ungrounded Wye

This paper reports the simulation conducted when planning capacitor bank integration to PT Indonesia Asahan Aluminium (Persero) or PT. Inalum, to increse electrolysis current.

## 2. Research Method

Power flow calculation in existing condition is performed by modelling the electrical network within PT Inalum starting from generator, transmission system, smelter loads and the simplified interconnection with the public electrical operator (PLN) by using ETAP 12.0. The system states are based on the measured parameters using SCADA of PT Inalum [5]. Manual calculation is performed to determine the size of the capacitor bank.

The main station (Gardu Induk) of PT. INALUM has 4 units LVR. Each has 182 MVA capasity. The LVR adjusts and stabilizes output voltage for rectifier transformer [6]. Each electrolysis potline is served by a single unit LVR with 3 taps of NVTC (No Voltage Tap Changer). Each NVTC tap has 27 OLTC (On Load Tap Changer) taps. Tap works as a voltage regulator to anticipate load fluctuation. Figure 2 shows the model of the electrical system within PT. Inalum.

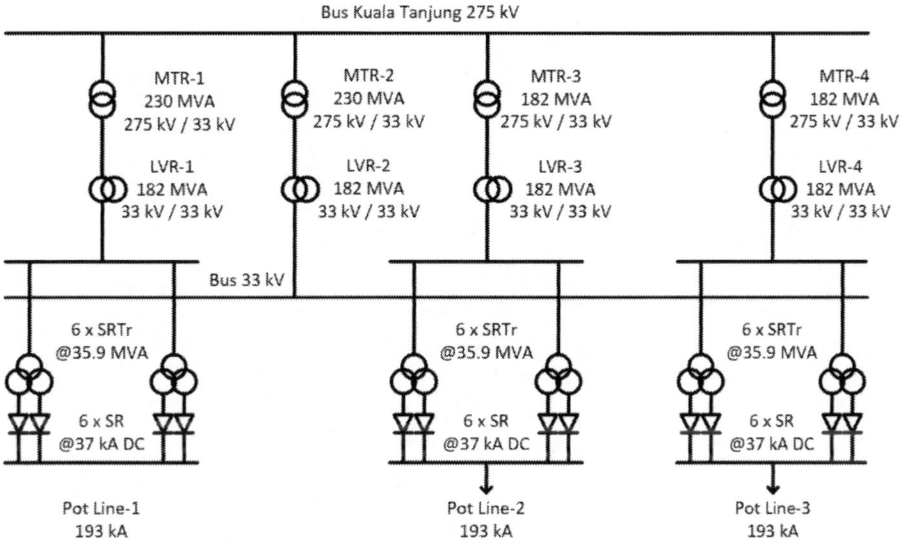


**Figure 2**. Electrical network model of PT. Inalum

## 3. Evaluation Results

### *3.1. Capacitor Bank Calculation*

The reactive power generation in some conditions are approximated by using Equation1 and 2.

$$Potline\ AC\ Power = Potline\ DC\ Power \times Conversion\ Constant \quad (1)$$

$$Potline\ DC\ Power = I_{dc} \times V_{dc} \times PIO \quad (2)$$

The minimum current, the existing current, upgrade alternative 1 and 2 are plotted in Table 1. The alternative capacitor bank implementations are based on maximum and partly compensated capacitor bank values. The maximum current (6 x 37 kA = 222 kA) requires 89.9 MVAR reactive power. This research uses the capacitor bank of 2x45 MVAR as plotted in the maximum current of 222 kA. It is referred to as upgrade 1, while 45 MVAR is upgrade 2.

IOP Conf. Series: Journal of Physics: Conf. Series **978** (2018) 012062 doi:10.1088/1742-6596/978/1/012062

**Table 1**.Reactive power requirement

| Condition | I dc (kA) | V dc | PIO | P dc P/L (MW) | P ac P/L (MW) | Q P/L (MVAR) | S P/L (MVA) |
|---|---|---|---|---|---|---|---|
| Min.Current | 130 | 4.4 V | 165 pot | 94.4 | 97.5 | 52.6 | 110.8 |
| Normal | 193 | 4.4 V | 165 pot | 140.1 | 144.7 | 78.1 | 164.5 |
| upgrade 1 | 210 | 4.4 V | 165 pot | 152.5 | 157.5 | 85.0 | 179.0 |
| upgrade 2 | 222 | 4.4 V | 165 pot | 161.2 | 166.5 | 89.9 | 189.2 |

### *3.2. Power system model with capacitor addition*

By integrating capcitor bank to model as in Figure 2, the integration model is shown in Figure 3. Capacitor bank is in grounded Wye model.

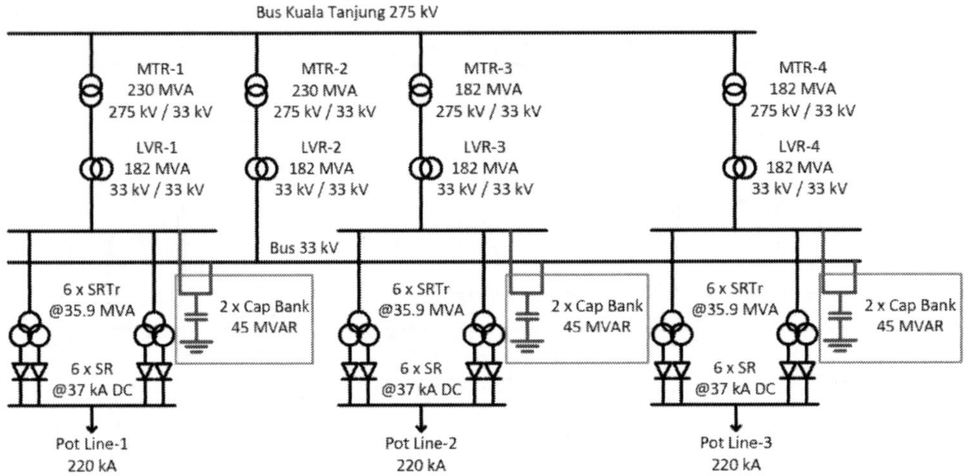


**Figure 3**. Electrical network model of PT. Inalum with capacitor bank integration

There are 3 pot lines shown in Figure 3, with 182 MVA MTR and 182 MVA LVR. The capacitor bank is parallel to STR and SR.

### *3.3. Impact on power factor and current improvements*

Figure 4 shows impacts of capacitor bank integration. At normal condition, the power factor is slightly above 0.89. By adding upgrade 1 45 MVAR, power factor moves upward to about 0.96. By fully connecting all capacitors, power factor is stable in 0.99.

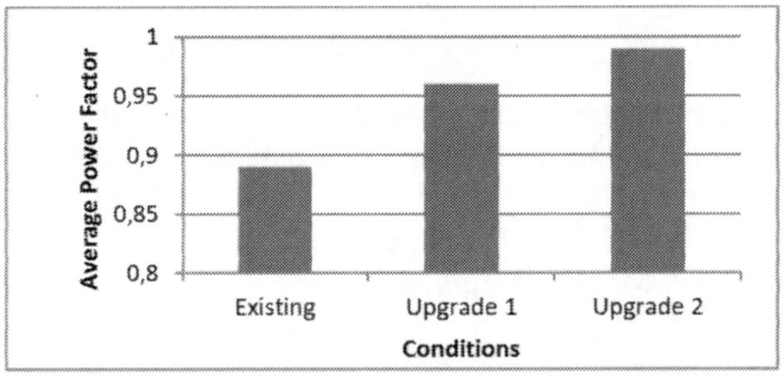


**Figure 4.** Impact on power factor

The reactive power decrement reduces secunder currents of LVR from 2930 A to 2727 A (upgrade 1) and 2587 A (upgrade 2). The LVR secunder current reductions increase the rating of electrolysis potline, so that the current can achieve 219 kA.

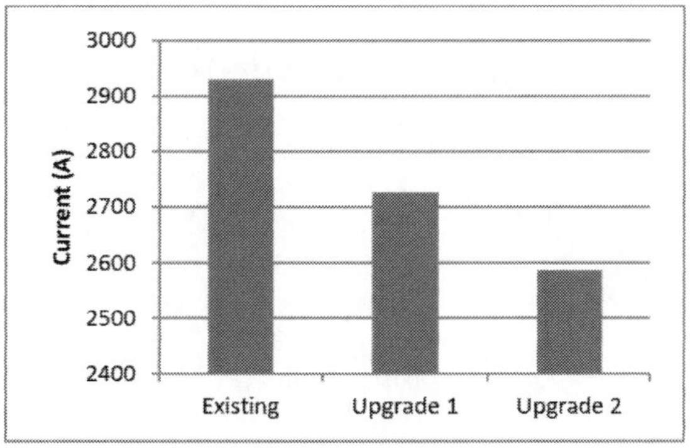


**Figure 5**. LVR current decrement

## 4. Conclusions

Based on the results of ETAP 12.0 simulation, a single 45 MVAR capacitor bank installation on the system increases the power factor from 0.89 to 0.95. The 2x45 MVAR capacitor bank installations improve power factor to 0.99.

Further, the LVR secondary current reduces from 2930 A to 2727 A (for 45 MVAR capacitor bank) and 2587 A (for 2x45 MVAR capacitor bank).This reductions enable the potline current capacity increases from 193 kA to maximum 219 kA. This increment is useful for increasing alumnium production.

## References

[1] Maswood A I and Fangrui L 2009 A Unity PF Rectifier-Inverter Under Unbalanced Supply. *Power & Energy Society*
[2] Pontt J, Rodriguez J, and Martin S J 2005 Power Electronics Specialists Conference (PESC), *IEEE 36th.*
[3] Sheldon P K 2002 IEEE Transactions on Industry Applications, **38** (4).
[4] Malaviya A K and Bundell G A 2001 *IEEE Transactions on Industry Applications*, **37** (3).
[5] Load Dispatching Center Hand Book,( PT Indonesia Asahan Aluminium (Persero), Indonesia).
[6] Yang J 2012 *IET Electric Power Applications* **7**(6).

# No research without publication: early mining

M K M Nasution[1*]

[1]Computer Science, Fasilkom-TI, Universitas Sumatera Utara, Padang Bulan 20155 USU, Medan, Indonesia

E-mail: mahyuddin@usu.ac.id

**Abstract.** In every college in Indonesia, Tridharma with three pillars: education, research and community service is the task of institutions and individuals within the educational institution. In this paper, we want to prove the implementation of research either by the institution or personally lecturers in Indonesia. Based on the data of scientific publications as research evidence derived from the indexing database, Scopus, it is found that less than 11% of colleges are proven to conduct research, not until of 17.08% lecturers who proved to conduct research.

## 1. Introduction

Education, research, and community service have become an integral part of every college in Indonesia [1]. This concept is known as the Tridharma of college, or in short Tridharma, which is implemented in an integrated way through higher education institutions academics either individually or in groups [2]. In this case, research not only supports the transformation of knowledge (the implementation of education) but also proves that there is knowledge development in every college, with which community service becomes the cornerstone of the implementation of the downstream of the other two dharma: It enables research to be applied and improves social welfare, while alumni are well accepted by the community and get a place in the world of work [3].

Globally through information technology, evidence of research can be collected from information sources such as the Web and the Internet [4, 5]. Sources of information contain documents from the results of scientific activities such as scientific papers and documents of intelectural property organized in the publisher database or indexer and / or stand alone in the Web, and this requires management procedures [6, 7, 8, 9]. Although some of Indonesias knowledge management systems have been launched such as Garuda[1] and Sinta[2], but the study of research evidence in Indonesia has received little attention. Therefore, this paper intends to describe the phenomenon of Indonesian research seen from its presence in the database of Scopus indexers.

## 2. Problem Definition

A research is stated to have been going well and correctly if it meets the scientific method, and this always produces the outcome [10]. Outcomes are not always tangible innovations, but they may be proof of unrighteousness. This is possible: A research is based on prejudices and logical

[1] http://id.portalgaruda.org/
[2] http://sinta.ristekdikti.go.id/

assumptions that may be true or false [11]. However, a research must produce an outcome[3], and this is minimally expressed in a scientific paper [12]. A research course must be based on recent reviews about the field of research or it is recognized as the latest developments of what will be studied [13]. Thus, the outcome of the research can be a review article [14, 15].

Regardless, a research started from a research proposal [16]. The proposal contains a problem statement, objectives or interests, method, studies, or research schedule [17]. Ideas have been disclosed in the proposal and for the idea to be guaranteed renewable in addition to being based on a review must also be claimed more first [12, 18]. Therefore, a research proposal may also be an article or scientific work, at least presented in a scientific meeting to become a scientific publication paper [19]. This is to avoid any claims from others about the same idea. In addition, many of researcher present papers or articles related to the methods used in research, sometimes expressed in special forms such as frameworks [20, 21] or methodologies [22].

A strong and quality research not only produces scientific publications, but also has an outcome of innovation [23]. Innovation consists of two parts: Discovery is an innovation that is to introduce to many people, something that already exists but is not well recognized. While the invention, is really the result of a new creation, it is generally recognized as a research that has novelty, although it comes from something already existing but there are additional values [2]. Therefore, innovation always gives birth to intellectual property right or patent. They have documents and is registered in the database and can be viewed online.

Every research, other than having a report must also have an outcome: scientific publication (published in journals, scientific meeting proceedings, or books), intellectual property right and patents. All outputs are literally documented and can be detected online. The placement of all research outsourced documents online is the first protection against possible duplication [24], and this prevents the same thing from being developed in many places by different researchers [24]. This principle is at the heart of knowledge management. Thus, research behaviour can be proven through the existence of the documents.

## 3. An adaptive approach

The execution of the research has proceeded well. It expressed by its outcome. Research outcomes can be scientific publications such as scientific papers or books, but all outcomes must be well tested[4] [2]. Research, as the ongoing existence of knowledge development in a college, has need of a process of measurement as performance to be tested. Of course, research outputs such as scientific publications also meet test standards. Scientific publications must pass through expert reviews: each article is reviewed by reviewers/editors/executives in each journal, likewise also other papers such as conference papers reviewed by experts by enforcement of the peer reviews [25]. The reviewers also must recognized generally in their field through their expertise by other expert: not only through their measurable works, but also acknowledged in the proved categorization of the H-index. The work of reviewers and H-index is based on a recognized database of indexers [26]. Therefore, the proof that the research carried out by either someone individually or the related college is approached by involving reputable indexed scholarly publications. In other words, there is no research without scientific publication.

In the adaptive approach, the distribution of authors by the number of documents gives its own meaning from the point of comparison, with which the probability gives an understanding that when the number of documents is proportional to the number of documents means the distribution of research is evenly distributed [27]. Likewise, if the author is replaced by the lecturers.

[3] Peraturan Menteri Keuangan Republik Indonesia Nomor 106/PMK.02/2016 tentang Standar Biaya Keluaran Tahun Anggaran 2017

[4] Undang-Undang Nomor 12 tahun 2012 tentang Pendidikan Tinggi and Peraturan Menteri Riset, Teknologi, dan Pendidikan Tinggi Republk Indonesia Nomor 44 Tahun 2015 tentang Standar Nasional Pendidikan Tinggi

**Table 1.** Number of colleges in Scopus

| College (C) categories | Number of C in Indonesia | Number of C in North Sumatra |
|---|---|---|
| PTN-BH | 11 | 1 |
| PTN | 26 | |
| UN/IKIP | 9 | |
| UIN/IAIN | 13 | |
| PTS | 83 | 2 |
| Politeknik | 12 | |
| High School/Academic | 13 | |

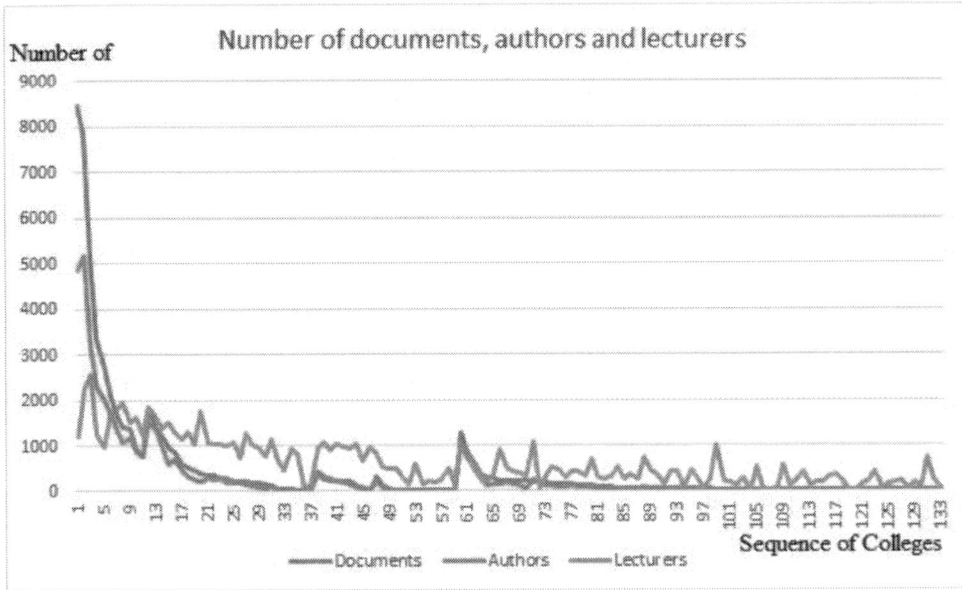


**Figure 1.** Information from Scopus for Indonesia colleges

## 4. Some Comparisons

According to the database from the minister of research, technology and higher education (Kemenristekdikti Republik Indonesia), Indonesia has 4, 539 colleges with 25, 876 studies program and 245, 019 lecturers (among of them are 13% Ph.D)[5]. On the other hand, since every college and lecturer has long been obliged to organize and implement the college tridharma, where research becomes one of its pillars. Assuming this, the evidence for the conduct of the research is only directly measurable from scientific publications, through the issuance of articles in reputable international journals and/or through reputable conference paper claims, then one reputable indexer database of scientific publications may serve as evidence of research already undertaken, and this shows the indirect behaviour of knowledge development. This database contains track record of research evidence that has long been recording the selected and best documents.

At present, the number of institutions in Indonesia as scientific publishing affiliations in Scopus[6] reputable indexer is only 252 units, of which there are 167 college or only 3.68% are

[5] https://forlap.ristekdikti.go.id/
[6] https://www.scopus.com

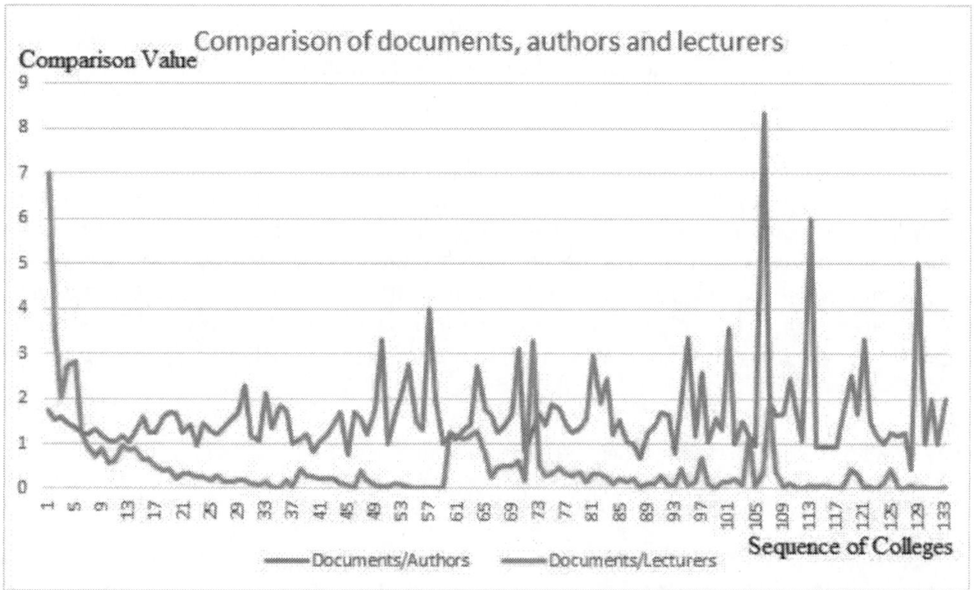


**Figure 2.** Comparison in % constant between Docoments, Authors, and Lecturers

evident from the number of college should conduct research. The list of Indonesian colleges registered in the Scopus indexer is shown in Table 1 briefly in categories. Meanwhile, the number of reputable paper/articles was generated until this data was collected as many as 58, 810 documents with 41, 853 authors. Thus, there is only a maximum of 17.08% of the total number of lecturers in Indonesia who make reputable scientific publications as evidence of research implementation.

To disclose evidence of research based on information available online, we collected documents and authors for a number of existing colleges, outside polytechnics, from Scopus index database, whereas lecturer data for each college is taken from each colleges webpage or obtained from the information available on other webpages as a comparison. Based on this, the research behaviour for each college is not same. Three data items: documents, authors, and lecturers, aligned in the order of college under the categories: PTN-BH (state-owned universities), PTN (state colleges), UN/IKIP (state colleges derived from education or education-based), UIN/IAIN (state-based religious colleges), and PTS (private colleges). Fig. 1 shows that the presence of a document of scientific publication as research evidence depends on the author. Some colleges, especially PTN-BH have different researches activities than other colleges. Although the enhancement of number of documents on the PTN-BH is followed by the number of authors, but at other colleges it is more likely to be lower or the authors are the same people. In PTN-BH the author does not as a whole comes from lecturers, but at other colleges are more likely the writer is a lecturer. This is also expressed by Fig. 2: Although the comparison between the documents with the authors in general is greater than or equal to one, but the comparison between documents and lecturers is lower than 1 so it can be said that many lecturers are not proven to conduct research.

In Indonesian colleges, most of lecturers arenot proven to conduct research, but it is possible only to carry out simple education (do learning with the old lecture materials). Since it is not proven to conduct research, the knowledge development does not occur in the first dharma: education. This behaviour is shown in Fig. 2, i.e. a few colleges in Indonesia only have a probability ratio (%) of the number of documents and lecturers equal to one or more. In other words, less than 11% of colleges are proven to conduct research, or a few of lecturers are proven to conduct research, or not many the lecturers that fully develop their knowledge on an ongoing basis.

## 5. Conclusion

By involving technology and information sources, research implementation can be proved in every college or by the lecturer concerned. There is no research without scientific publication, explaining that indexed documents by reputable indexers are important outcomes for evidence of research implementation. So, from the number of universities in Indonesia and the number of lecturers, few colleges and lecturers are proven to conduct research. The next study is to look at other behaviors in research implementation in Indonesia, such as collaborative behavior.

## References

[1] Nasution M K M 2016 Fenomena riset (Opini - Harian Analisa.)
[2] Nasution M K M 2016 Hilirisasi penelitian berbasis teknologi pada perguruan tinggi (Opini - Harian Analisa.)
[3] Bernett R 1994 The Limits of Competence: Knowledge, Higher Education and Society( Open University Press).
[4] Boyd D M and Roychowdhury V P 2008 Social network sites: Definition, History, and Scholarship *Journal of Computer-Mediated Communication* 13.
[5] Nasution M K M and Noah S A 2010 Superficial method for extracting social network for academics using web snipppets *Lecture Notes in Computer Science* LNAI 6401.
[6] Nasution M K M and Noah S A 2011 Extraction of academic social network from online database *International Conference on Semantic Technology and Information Retrieval* (STAIR 2011).
[7] Nasution M K M, Noah S A M and Saad S 2011 Social network extraction: Superficial method and information retrieval *Proceedings of International Conference on Informatics for Development* (ICID11).
[8] Nasution M K M, Hardi M and Syah R 2017 Mining of the social network extraction *Journal of Physics: Conference Series* 801(1).
[9] Nasution M K M 2016 Social network mining (SNM): A definition of relation between the resources and SNA *International Journal on Advanced Science, Engineering and Information Technology* 6(6).
[10] Cox R, Hyde M, Gatehouse S, Noble W, Dillon H, Bentler R, Stephens D, Arlinger S, Beck L, Wilkerson D, Kramer S, Kricos P, Gagné J P, Bess F, Halberg and Lillemor 2000 Optimal outcome measures, research priorities, and internal cooperation *Ear & Hearing* 21(4).
[11] Wilson T D 1999 Models in information behaviour research *The Journal of Documentation* 55(3).
[12] Nasution M K M 2016 Carut marut menulis karya ilmiah *Opini - Harian Waspada.*
[13] Nasution M K M 2016 Karya ilmiah dosen & mahasiswa *Opini - Harian Waspada.*
[14] Aha D W 1997 Editorial: Lazy learning *Artificial Intelligence Review* 11.
[15] Nasution M K M 2017 Social network extraction based on Web: 1. Related superficial methods *Interior.*
[16] Kirchhoff L , Stanoevska-S K, Nicolai T, Fleck M 2008 Using social network analysis to enhance information retrieval systems *Social Networks Applications Conference.*
[17] Nasution M K M 2017 Cara menulis karya ilmiah *Teknik Penulisan Karya Ilmiah* Bagian 2.
[18] Nasution M K M (to print) Theory of sharing knowledge: An introduction *Bulletin of Mathematics*
[19] Besnard P and Hunter A 2001 A logic-based theory of deductive arguments *Artificial Intelligence* 128(1-2).
[20] Nasution M K M, Nuradi T E and Syah R 2017 SumutSiana: A framework for applying ICT to preserve the cultural heritage of Sumatera Utara Indonesia *Journal of Telecommunication, Electronic and Computer Engineering* 9(2-4).
[21] Syah R, Nuradi T E, and Nasution M K M 2017 A framework to apply ICT for bequeathing the cultural heritage to next generation *Journal of Physics: Conference Series* 801(1).
[22] Nasution M K M and Noah S A 2012 A methodology to extract social network from the Web snippet *Cornell University Library* arXiv:1211.5877.
[23] Chaudhuri A 1994 The diffusion of an innovation in Indonesia *Journal of Product & Brand Management* 3(3).
[24] Simmonds N W 1985 Duplicatin of research: A good or a bad thing? *The Journal of the Operational Research Society* 361.
[25] Bird J E and Bird M D 1999 Do peer-reviewed journal papers result from meeting abstracts of the Biennial Conference on the Biology of Marine Mammals? *Scientometrics* 46(2).
[26] Egghe L 2011 The single publication H-index and the indirect H-index of a researcher *Scientometrics* 88.
[27] Gupta B M, Kumar S and Rousseau R 1998 Applicability of selected probability distributions to the number of authors per article in theoretical population genetics *Scientometrics* 42(3).

IOP Conf. Series: Journal of Physics: Conf. Series **978** (2018) 012064 doi:10.1088/1742-6596/978/1/012064

# Image processing analysis of geospatial uav orthophotos for palm oil plantation monitoring

**F Fahmi [1], D Trianda[1], U Andayani[2] and B Siregar[2]**
[1]Department Electrical Engg, Faculty of Engineering, University of Sumatera Utara, Jl. Dr. Mansur 9 Medan Indonesia
[2]Department Information and Technology, Faculty of Computer Science and Information Technology, University of Sumatera Utara, Jl. Dr. Mansur 9 Medan Indonesia

Email. fahmimn@usu.ac.id

***Abstract*—** Unmanned Aerial Vehicle (UAV) is one of the tools that can be used to monitor palm oil plantation remotely. With the geospatial orthophotos, it is possible to identify which part of the plantation land is fertile for planted crops, means to grow perfectly. It is also possible furthermore to identify less fertile in terms of growth but not perfect, and also part of plantation field that is not growing at all. This information can be easily known quickly with the use of UAV photos. In this study, we utilized image processing algorithm to process the orthophotos for more accurate and faster analysis. The resulting orthophotos image were processed using Matlab including classification of fertile, infertile, and dead palm oil plants by using Gray Level Co-Occurrence Matrix (GLCM) method. The GLCM method was developed based on four direction parameters with specific degrees 0°, 45°, 90°, and 135°. From the results of research conducted with 30 image samples, it was found that the accuracy of the system can be reached by using the features extracted from the matrix as parameters Contras, Correlation, Energy, and Homogeneity.

## 1. INTRODUCTION

Image processing today has a vast spectrum of applications in various fields of life including in the fields of biomedicine [1], astronomy, archeology, biometrics [2], daily life [3,4] image and document archives, industry and remote sensing using satellite imagery technology[5] .

Utilization of satellite imagery has been done mainly to identify changes in shape, extent or other conditions of a specific region. In this study, we utilized Unmanned Aerial Vehicle (UAV) to get geospatial orthophotos for monitoring oil palm plantations. The aim is to check the growth of oil palm plants from the beginning, in order to identify which part of the plantation land is fertile for planted crops which mean to grow perfectly, less fertile in terms of growth but not perfect, or not growing at all. This information can be known quickly by the using of UAV geospatial orthophotos.

In previous research conducted by Rizatus Shofiyanti (2011) [6], they utilized the image of UAV for indraja application for monitoring agricultural land. Muhammad Rendana et al. have established a system to detect palm oil nutrient stress, and Salim Malek (2014) examines the UAV image to detect palm oil rods [7,8].

A UAV or an unmanned aircraft is an unmanned flight or a glyph which can be controlled by a remote control or even with autonomous self-control. Unmanned drone controls are two main variants, the first variation being controlled via remote control and a second variation is a plane that flies independently according to the program included before flying [9].

An orthophoto is a kind of photographic map, which is quite possible to perform measurements as if it were a standard map. It is part of the photogrammetry field and is generally performed by Unmanned Aerial Vehicles (UAV). The resulting orthoimage can be synchronized with other maps containing other urban or technical elements such as agriculture or plantations, power grids, dams, etc. Example of orthophotos can be seen in Figure 1.

Geospatial or geo-space is a spatial aspect which indicates the location and position of an object or event that is under or on the surface of the earth expressed in a particular coordinate system. Geospatial data is data about geographic location, dimension or size, and / or characteristic of natural and / or man-made objects located below or on the surface of the earth [10].

Image or photo interpretation is an action to examine a photographic image for the purpose of identifying the object and assessing the significance of the image or photograph. Image interpretation is a process of recognizing objects in the form of images (images) for use in certain disciplines such as geology, geography, and other disciplines.

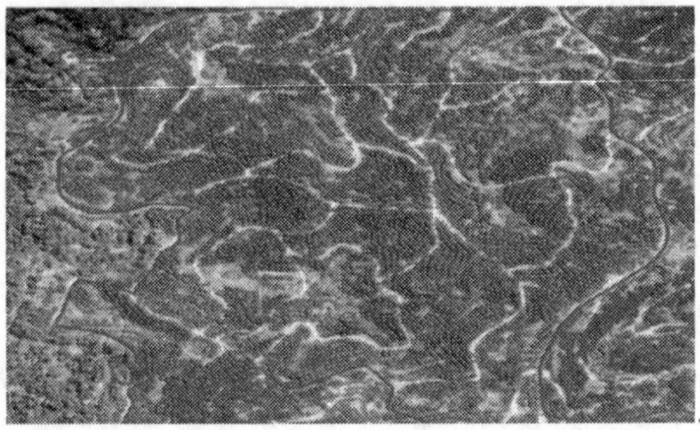

**Figure 1**. Example of Orthophotos mapped by UAV

The purpose of this study was to develop an orthophotos image analysis for monitoring vegetation within this palm oil plantation. The aim is to automatically measure the area of palm oil plantation and classify it into three categorize : fertile, nonfertile and no grow area.

## 2. Material and Methods

Digital image processing is work to process and analyze a digital image that involves a lot of image data in the form of pixel representation. The stage of pre-processing is part of the preparation for digital image processing to be analyzed later on. Several methods can be done as part of the preparation of the image in accordance with the needs of the next process.

### *2.1 Image Segmentation and Extraction*

Image segmentation is an important step in the pattern recognition process. Once the object is successfully segmented, then we can perform the image character extraction process. Characteristic extraction is a stage that aims to extract the characteristics of an object in which the trait is used to distinguish between one object to another object.

This characteristic extraction process relates to the quantization of image characteristics into a set of corresponding feature values. One technique for obtaining statistical features is to calculate the probability of an adjacency relationship between two pixels at a certain distance and angular orientation. This approach works by forming a co-occurrence matrix from the image data, followed by defining the feature as a function of the intermediate matrix.

### *2.2 Gray Level Co-Occurrence Matrix (GLCM)*

GLCM is a matrix whose elements are the number of pixel pairs that have a certain degree of brightness, in which the pixel pair is separated by a distance d, and with an inclination angle θ [11]. In other words, the co-occurrence matrix is the probability of the appearance of gray level i and j of two separate pixels at distance d and angle θ. Orientation is formed in four angular directions with 45 ° angle intervals, i.e., 0 °, 45 °, 90 °, and 135 °. While the distance between pixels is usually set by 1 pixel. These four directions can be represented in Figure 2.

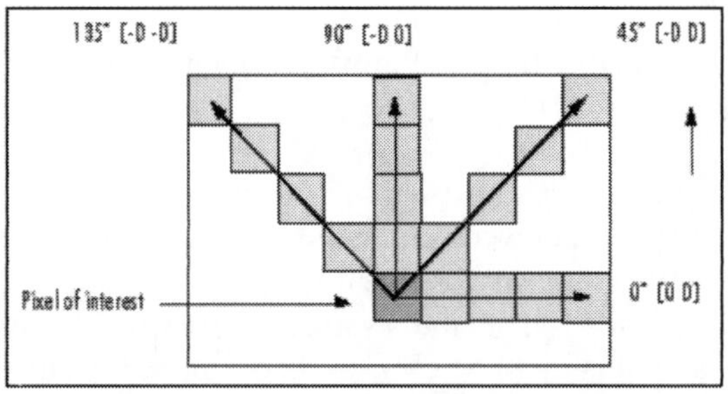


**Figure 2**. Direction of Gray level Orientation

Co-occurrence means a common occurrence, i.e., the number of occurrences of one level of pixel value adjacent to one level of another pixel value in the distance (d) and a certain angular orientation (θ). Distance is expressed in pixels and orientation expressed in degrees.

Haralick et al. proposed various types of textural characteristics that can be extracted from the co-occurrence matrix, namely Angular Second Moment, Contrast, Correlation, Variance, Inverse Difference Moment, and Entropy (Haralick, K. Shanmugam, and I. Dinstein, 1973).

$$ASM = \sum_i \sum_j \{p(i,j)\}^2 \text{ ...................... } (1)$$
$$CON = \sum_k k^2 \left[\sum_i \sum_j p(i,j)\right] \text{ .............. } (2)$$
$$COR = \frac{\sum_i \sum_j (i,j) . p(i,j) - \mu_x \mu_y}{\sigma_x \sigma_y} \text{ .................. } (3)$$
$$VAR = \sum_i \sum_j (i - \mu_x)(j - \mu_y) p(i,j) \quad (4)$$
$$IDM = \sum_i \sum_j \frac{1}{1+(i-j)^2} p(i,j) \text{ .............. } (5)$$
$$ENT_2 = - \sum_i \sum_j p(i,j) . 2_{\log p(i,j)} \text{ ...... } (6)$$

In the above equation, p (i, j) denotes the probability, which starts from zero to one, i.e., the element value in the co-occurrence matrix. While i and j, denote the adjacent pair of intensities, which in each co-occurrence matrix becomes the row number and column number.

*2.3 Classification*

At the classification stage, palm oil will be classified based on three (3) categorize fertile, infertile and non-growing categories using the distance method, i.e., the Euclidean method as a classification method.

$$d_{Euc} = \sqrt{\sum_{i=1}^{n} (P_i - Q_i)^2} \text{ ........... } (7)$$

Where Pi = Test image pattern and Qi = standard image pattern

*2.4 Research Method*

The research method is done with the following stages:

a. Preparation of Orthophotos Imagery

Image data retrieval is done by UAV / Drone type DJI 3 type Phantom (Figure 3)

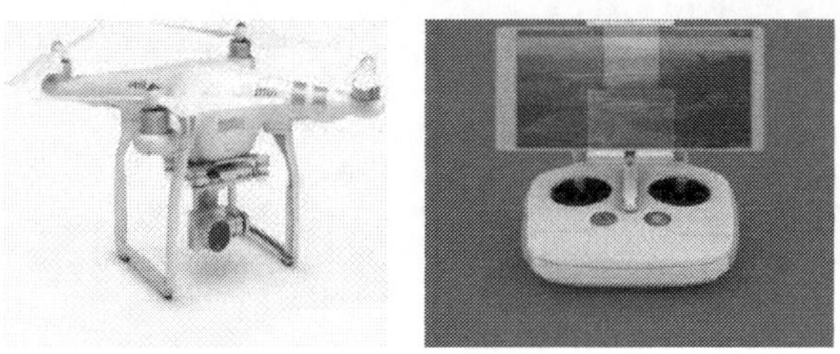

**Figure 3**. UAV type Drone DJI 3 Phantom and its remote control

b. Calculation of the Area
The wide of specific area will automatically be determined when UAV / Drone is flown to retrieve image data from oil palm plantations [12].

c. Image Segmentation and Feature Extraction
Image segmentation and feature extraction were conducted to divide an image into objects into homogeneous regions based on the similarity between the gray level of a pixel to the gray level of its neighboring pixels. This step is included in the Feature Extraction process by entering scale level segmentation to determine what objects will be extracted. The image of the segmentation result is then extracted by the Gray Level Co-occurrence Matrix (GLCM) method, for each class of the object / segment that has been formed from the segmentation.

d. Classification
Classification is done by first input sample data for palm oil image in fertile category 10 sample, less fertile 10 sample and not grow 10 sample as the training set. The threshold value was obtained from the training session and then used as a parameter to classified the unknown area.

e. Data validation and testing
From the sample tested, it will be calculated are based on four features: Energy, Correlation, Contrast, and Homogeneity.

## 3. Results
The palm oil image data obtained from the drone type DJI phantom type is shown in Figure 4.

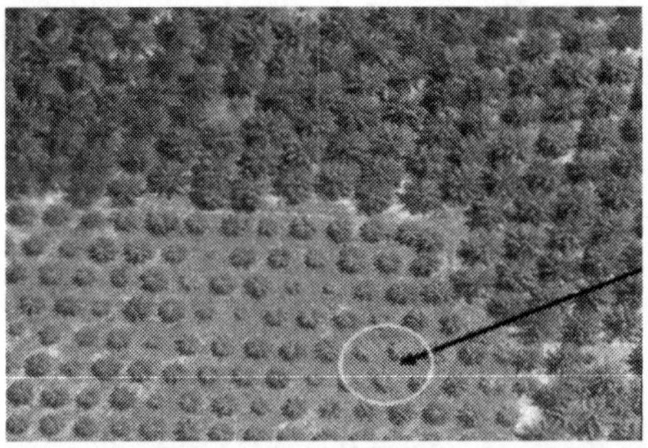

**Figure 4**. Palm oil Plantation captured by Drone

Table 1 shows the data of oil palm image used in this study.

**Table 1.** Palm Oil Image dataset

| Palm Oil | Sample | Baseline | Format |
|---|---|---|---|
| Fertile | 10 | 1 | .jpg |
| Less Fertile | 10 | 1 | .jpg |
| Not Growing | 10 | 1 | .jpg |

Examples of palm oil image for fertile, less fertile and not growing are shown in Figures 5 (a),(b), and (c).

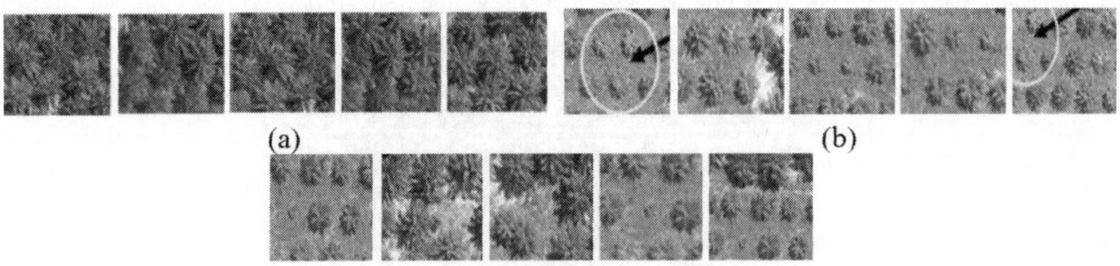


(c)

**Figure 5**. Palm oil Area (a) Fertile, (b) Less Fertile and (c) Not Growing

IOP Conf. Series: Journal of Physics: Conf. Series **978** (2018) 012064 doi:10.1088/1742-6596/978/1/012064

After the image data was cropped, the image will be changed into a matrix form that collects the pixel color of the gray image data and the direction value of the degree of image grayscale with 4 categories.

The results of testing for oil palm samples for fertile, infertile and non-growing areas can be seen in Table 2.

**Table 2**. Result of GLCM feature parameters

| Image Dataset | Contrast | | | | Correlation | | | | Energy | | | | Homogeneity | | | |
|---|---|---|---|---|---|---|---|---|---|---|---|---|---|---|---|---|
| **Fertile Samples** | | | | | | | | | | | | | | | | |
| | 0° | 45° | 90° | 135° | 0° | 45° | 90° | 135° | 0° | 45° | 90° | 135° | 0° | 45° | 90° | 135° |
| Sample1 | 4,2118 | 7,6933 | 4,1154 | 6,1307 | 0,8692 | 0,7609 | 0,8723 | 0,8095 | 0,0113 | 0,0085 | 0,0113 | 0,0096 | 0,5223 | 0,4502 | 0,5242 | 0,4792 |
| Sample2 | 3,4975 | 6,7414 | 3,6888 | 5,2860 | 0,8560 | 0,7228 | 0,8483 | 0,7827 | 0,0140 | 0,0105 | 0,0135 | 0,0116 | 0,5434 | 0,4635 | 0,5332 | 0,4908 |
| Sample3 | 3,7022 | 6,9786 | 3,6815 | 5,3527 | 0,8327 | 0,6827 | 0,8322 | 0,7566 | 0,0140 | 0,0104 | 0,0135 | 0,0116 | 0,5389 | 0,4574 | 0,5274 | 0,4856 |
| Sample4 | 4,0719 | 7,2346 | 3,7178 | 5,7002 | 0,8130 | 0,6653 | 0,8285 | 0,7363 | 0,0135 | 0,0103 | 0,0135 | 0,0112 | 0,5287 | 0,4560 | 0,5310 | 0,4770 |
| Sample5 | 4,3881 | 7,3142 | 3,7905 | 6,1264 | 0,8179 | 0,6958 | 0,8425 | 0,7451 | 0,0126 | 0,0098 | 0,0133 | 0,0108 | 0,5261 | 0,4552 | 0,5397 | 0,4854 |
| Sample6 | 3,5207 | 6,1008 | 3,3594 | 5,1056 | 0,8226 | 0,7004 | 0,8354 | 0,7492 | 0,0155 | 0,0119 | 0,0156 | 0,0130 | 0,5560 | 0,4832 | 0,5576 | 0,5088 |
| Sample7 | 5,9551 | 11,2707 | 6,2329 | 8,5564 | 0,7969 | 0,6155 | 0,7883 | 0,7081 | 0,0094 | 0,0073 | 0,0092 | 0,0080 | 0,4794 | 0,4090 | 0,4750 | 0,4331 |
| Sample8 | 5,7921 | 11,0510 | 5,9909 | 8,3755 | 0,8075 | 0,6321 | 0,8008 | 0,7212 | 0,0092 | 0,0071 | 0,0091 | 0,0078 | 0,4802 | 0,4083 | 0,4771 | 0,4325 |
| Sample9 | 5,4995 | 10,8631 | 5,9411 | 7,8689 | 0,8095 | 0,6216 | 0,7925 | 0,7260 | 0,0101 | 0,0075 | 0,0096 | 0,0085 | 0,4896 | 0,4069 | 0,4721 | 0,4423 |
| Sample10 | 6,3933 | 12,5300 | 6,9843 | 9,3393 | 0,8137 | 0,6348 | 0,7971 | 0,7278 | 0,0086 | 0,0064 | 0,0081 | 0,0071 | 0,4789 | 0,3984 | 0,4641 | 0,4264 |
| **Less Fertile Samples** | | | | | | | | | | | | | | | | |
| Image Dataset | Contrast | | | | Correlation | | | | Energy | | | | Homogeneity | | | |
| | 0° | 45° | 90° | 135° | 0° | 45° | 90° | 135° | 0° | 45° | 90° | 135° | 0° | 45° | 90° | 135° |
| Sample1 | 2,9808 | 5,9216 | 3,3563 | 4,3456 | 0,8190 | 0,6419 | 0,7969 | 0,7370 | 0,0300 | 0,0234 | 0,0281 | 0,0253 | 0,5761 | 0,4992 | 0,5658 | 0,5319 |
| Sample2 | 3,2708 | 6,4362 | 3,6381 | 4,8599 | 0,8149 | 0,6359 | 0,7948 | 0,7252 | 0,0261 | 0,0203 | 0,0247 | 0,0226 | 0,5654 | 0,4849 | 0,5489 | 0,5178 |
| Sample3 | 5,4231 | 9,7542 | 5,4277 | 8,0278 | 0,8477 | 0,7278 | 0,8486 | 0,7759 | 0,0124 | 0,0097 | 0,0120 | 0,0106 | 0,5106 | 0,4417 | 0,5055 | 0,4659 |
| Sample4 | 4,9593 | 8,9281 | 5,1490 | 7,7714 | 0,8482 | 0,7264 | 0,8425 | 0,7617 | 0,0184 | 0,0150 | 0,0180 | 0,0159 | 0,5396 | 0,4751 | 0,5321 | 0,4926 |
| Sample5 | 3,4958 | 6,9842 | 3,8866 | 5,2175 | 0,8363 | 0,6727 | 0,8180 | 0,7555 | 0,0207 | 0,0160 | 0,0195 | 0,0179 | 0,5557 | 0,4784 | 0,5401 | 0,5118 |
| Sample6 | 2,3976 | 4,8563 | 3,1086 | 4,4408 | 0,8783 | 0,7539 | 0,8419 | 0,7748 | 0,0241 | 0,0178 | 0,0215 | 0,0184 | 0,6068 | 0,5212 | 0,5732 | 0,5290 |
| Sample7 | 4,2730 | 7,8296 | 4,4025 | 6,2405 | 0,8129 | 0,6592 | 0,8083 | 0,7285 | 0,0184 | 0,0144 | 0,0178 | 0,0158 | 0,5338 | 0,4608 | 0,5203 | 0,4844 |
| Sample8 | 2,1865 | 4,2861 | 2,6293 | 3,5826 | 0,8695 | 0,7435 | 0,8423 | 0,7856 | 0,0294 | 0,0221 | 0,0267 | 0,0234 | 0,6128 | 0,5263 | 0,5816 | 0,5435 |
| Sample9 | 3,7980 | 7,1455 | 4,0638 | 5,6712 | 0,8206 | 0,6634 | 0,8090 | 0,7328 | 0,0213 | 0,0166 | 0,0203 | 0,0179 | 0,5520 | 0,4805 | 0,5437 | 0,5019 |
| Sample10 | 5,3022 | 9,5309 | 5,6343 | 8,5540 | 0,8400 | 0,7118 | 0,8294 | 0,7413 | 0,0151 | 0,0120 | 0,0148 | 0,0130 | 0,5164 | 0,4481 | 0,5098 | 0,4690 |
| **Not Growing Samples** | | | | | | | | | | | | | | | | |
| Image Dataset | Contrast | | | | Correlation | | | | Energy | | | | Homogeneity | | | |
| | 0° | 45° | 90° | 135° | 0° | 45° | 90° | 135° | 0° | 45° | 90° | 135° | 0° | 45° | 90° | 135° |
| Sample1 | 2,5597 | 4,9242 | 2,9423 | 4,0216 | 0,8484 | 0,7088 | 0,8264 | 0,7621 | 0,0276 | 0,0212 | 0,0255 | 0,0227 | 0,5956 | 0,5162 | 0,5763 | 0,5366 |
| Sample2 | 2,9963 | 5,9502 | 3,2870 | 4,4742 | 0,8577 | 0,7179 | 0,8439 | 0,7878 | 0,0154 | 0,0111 | 0,0145 | 0,0125 | 0,5676 | 0,4775 | 0,5509 | 0,5087 |
| Sample3 | 5,6832 | 11,0656 | 6,2809 | 8,7417 | 0,8737 | 0,7541 | 0,8605 | 0,8056 | 0,0079 | 0,0059 | 0,0077 | 0,0066 | 0,4884 | 0,4138 | 0,4836 | 0,4406 |
| Sample4 | 5,0802 | 9,8605 | 6,4220 | 9,2017 | 0,8872 | 0,7802 | 0,8569 | 0,7950 | 0,0093 | 0,0071 | 0,0087 | 0,0075 | 0,5078 | 0,4349 | 0,4912 | 0,4539 |
| Sample5 | 7,5858 | 14,1140 | 8,0362 | 11,5997 | 0,8453 | 0,7115 | 0,8364 | 0,7629 | 0,0065 | 0,0048 | 0,0061 | 0,0052 | 0,4658 | 0,3931 | 0,4468 | 0,4091 |
| Sample6 | 7,4790 | 14,0363 | 7,9481 | 11,2842 | 0,8514 | 0,7209 | 0,8423 | 0,7758 | 0,0064 | 0,0047 | 0,0061 | 0,0053 | 0,4663 | 0,3890 | 0,4478 | 0,4146 |
| Sample7 | 2,7960 | 5,3915 | 3,1939 | 4,4329 | 0,8426 | 0,6967 | 0,8211 | 0,7508 | 0,0269 | 0,0207 | 0,0253 | 0,0225 | 0,5878 | 0,5075 | 0,5712 | 0,5328 |
| Sample8 | 7,4931 | 14,4286 | 8,3382 | 11,5500 | 0,8471 | 0,7057 | 0,8299 | 0,7644 | 0,0063 | 0,0047 | 0,0060 | 0,0052 | 0,4630 | 0,3888 | 0,4443 | 0,4079 |
| Sample9 | 5,3408 | 10,2140 | 5,8298 | 8,2820 | 0,8739 | 0,7593 | 0,8624 | 0,8048 | 0,0082 | 0,0061 | 0,0079 | 0,0068 | 0,4963 | 0,4199 | 0,4868 | 0,4470 |
| Sample10 | 3,1493 | 6,6105 | 4,0308 | 5,3978 | 0,8967 | 0,7831 | 0,8677 | 0,8229 | 0,0163 | 0,0121 | 0,0149 | 0,0130 | 0,5660 | 0,4821 | 0,5413 | 0,4996 |

Of the three samples tested it can be seen that contrast for fertile samples is higher than the other two samples. As for the other three categories, the fertile sample has more of the same value than the two samples with less fertile and non-growing categories.

For more details, the difference can be seen in Figure 6 with direction 0° for four parameters Contrast, correlation, energy, and homogeneity. From Figure 6 it can be seen that the average value of the fertile area is higher than other categories.

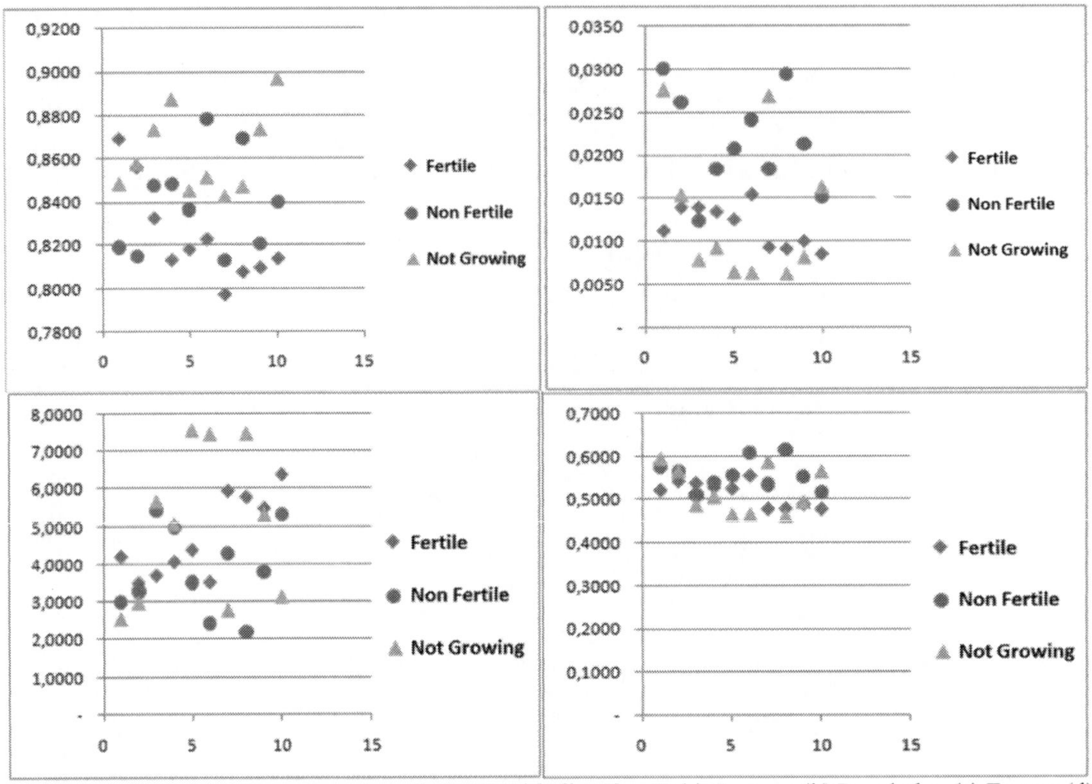


**Figure 6**. Feature Parameter value of GLCM with direction 0° (a) Contrast (b)Correlation (c) Energy (d) Homogeneity

## 4. CONCLUSION

From the research that has been done, it can be concluded that the use of UAV for monitoring of oil palm can identify the fertile area, less fertile and not grow automatically. The application can be made using GLCM method. From the results of research conducted with 30 image samples, it was found that the accuracy of the system can be reached by using the features extracted from the matrix as parameters Contras, Correlation, Energy, and Homogeneity.

## REFERENCES

[1] Fahmi F 2014 3D movement correction of CT brain perfusion image data of patients with acute ischemic stroke *Neuroradiology* **56:6** pp 445-452.

[2] Siswanto A, Tarigan P and Fahmi F 2013 Design of contactless hand biometric system with relative geometric parameters *Instrumentation, Communications, Information Technology, and Biomedical Engineering (ICICI-BME), 2013 3rd International Conference on* IEEE

[3] Triantoro T, Batubara R F and Fahmi F 2014 Image based water gauge reading developed with ANN Kohonen *Electrical Engineering and Computer Science (ICEECS), 2014 International Conference on* IEEE

[4] Muhammadiy W and Fahmi F 2016 Mobile application to differentiate flesh meat between beef and pork *Computational Intelligence and Cybernetics (CYBERNETICSCOM), 2016 International Conference on* IEEE

[5]Tarigan A P M and Fahmi F 2017 Mapping a Volcano Hazard Area of Mount Sinabung Using Drone: Preliminary Results *IOP Conference Series: Materials Science and Engineering* Vol **180** No 1 IOP Publishing

[6] Shofiyanti R 2011 Teknologi Pesawat Tanpa Awak untuk Pemetaan dan Pemantauan Tanaman dan Lahan Pertanian, *Inform Pertan* , vol **20**, no 2, pp 58–64

[7] Guldogan O, Rotola-pukkila J, Balasundaram U, Le T and K Mannar 2016 Automated Tree Detection and Density Calculation using Unmanned Aerial Vehicles *Vcip2016* pp 4–7

[8] Moranduzzo T and Melgani F 2014 Monitoring Structural Damages in Big Industrial Plants With Uav Images pp 4950–4953

[9] Emirul B 2015 Drone 1 pp 1–19

[10] Mitchell J J, Glenn N F, Anderson M O , Hruska R C and Charlie A H 2012 Unmanned Aerial Vehicle

( UAV ) Hyperspectral Remote Sensing for Dryland Vegetation Monitoring Hyperspectral Image and Signal Sensing *4th Work Hyperspectral Image Signal Process*
[11] Saifudin and Abdul F 2013 Sistem Identifikasi Citra Kayu Berdasarkan Tekstur Menggunakan *Gray Level Co-occurrence Matrix* (Glcm) Dengan Klasifikasi Jarak Euclidean ISSN: 1410-233
[12] Hepi H and MariaRegina Caeli 2016 Studi Perhituungan Jumlah Pohon Kelapa Sawit Menggunakan Metode Klasifikasi Berbasis Obyek (Tugas Akhir, ITS)

# Simulating 3D deformation using connected polygons

**J T Tarigan, I Jaya, S M Hardi and E M Zamzami**

Faculty of Computer Science and Information Technology,
Universitas Sumatera Utara

jostarigan@usu.ac.id

**Abstract**. In modern 3D application, interaction between user and the virtual world is one of an important factor to increase the realism. This interaction can be visualized in many forms; one of them is object deformation. There are many ways to simulate object deformation in virtual 3D world; each comes with different level of realism and performance. Our objective is to present a new method to simulate object deformation by using a graph-connected polygon. In this solution, each object contains multiple level of polygons in different level of volume. The proposed solution focusses on performance rather while maintaining the acceptable level of realism. In this paper, we present the design and implementation of our solution and show that this solution is usable in performance sensitive 3D application such as games and virtual reality.

## 1. Introduction

Deformation effects such as fractured walls, shattering glass, or destructions of objects are becoming common feature in today's computer games. These effects may increase the realism of a scene and gives a better experience to the player by allowing a more immersive interaction between the player and the 3D objects. However, based on the approach used in the system, simulating deformation of a 3D object can be very expensive performance-wise. Creating a deformation with a high level of realism is complicated and requires complex calculation. Moreover, the deformation may be vary based on many properties such as point of impact, direction, and object rigidness. Examining all these properties may increase the level of realism but may significantly hit the performance.

Real time systems such as games and virtual reality, requires a certain amount of performance efficiency that does not allow complex calculation to be performed in-game. Hence, computer graphic researchers have developed algortihms that able to simulate object deformation in real time. Most of the methode proposed to simulate deformation in real time system requires pre-calculated information regarding the deformation process. This concept allows the system to simply go through the pre-calculated data to render the result. Other even proposed a pre-calculated model to simply replace the collided object with a deformed version of the object and add smoke and mirrors effects to hide the replacement process.

In our paper, we propose a method to implement deformation by using pre-calculated connected polygons. Each object contains multiple level of polygons that are connected to a polygon in a different level. These objects are then stored in a modified obj file format. Later, we use this obj file to simulate the object and its deformation in a 3D application. Our objective is to show that our algorithm can successfully simulate object deformation with the least amount of processing time. Moreover, we would also like to implement our method in a 3D computer graphic application to show the usability of our system. This paper is structured as follow: in the next section we will go through some works that has been done previously related to our case. Section 3 will describe the algorithm of our method followed by the thougrough description on how we implemented the system. The later section will discuss the test and results followed by conclusion.

## 2. Related Works

Object deformation is not new in computer graphic research. One of the first and pioneer in object deformation simulation was presented by Terzopoulos et al. [7] where they presented a formulation of deformable models by using reference component. There are a lot of methods algortihm introduced to simulate object deformation in 3D world. Usually, the algortihm is categorized into two groups; real-time and pre-calculated. The real time methods focus on simulating the deformation in-game. Algortihm that falls into this category mostly considers user interaction such as the point of impact. Hence, the result may be more accurate but there will be a performance hit. Parker et al. introduces a real-time deformation and fracture suitable for games [1]. The deformation process is based around a corotational tetrahedral finite element method. The authors also provide an example of the process implemented in published console game. An intresting work from Glodu et al. shows how fracture patterns in a photograph can be used as samples to create cracs and fractures in 3D objects [2]. While the authors argue that the algortihm can be performed in real time, it requires sampling/learning processes that may take 15 to 25 minutes based on the complexity of the sample. Brown et al. presented a new method to simulate object deformation in surgical training 3D application [3]. The algorithm developed is complex due to the accuracy required in detecting collision and performing deformation. Yet, the paper shows that the deformation process can be done in real time. Bao et al. proposed a hybrid approach to simulate fracture on rigid materials [5]. The system uses rigid body simulation for efficient dynamic evolution and fully linearized, small displacement, finite elemets to determine rich and realistic fracture pattern. Tang et al. developed a system that allows mobile users to interact with 3D computer graphic application and perform object deformation [6]. However, due to the nature of complexity of the deformation process, the calculation is done by server and streamed to user's mobile phone.

The second category is pre-calculated method where the deformation polygons are built before the application starts. This method is common in games where performence in critical and deformation details are less required. Clothier et al. introduces a method to create a destructible object using 3D Voronoi Subdivision tree [4]. This method divides a 3D object into two Voronoi cell regions and store the subdivisions in a tree. Based on the impact on the object, the system will move downward to the lower level of the tree. This process is repeated until it reached the lowest level of the tree which means the object has been completely removed from the scene. Another precomputed method is presented by Muller et al. [8]. The method perform a computation to find a linear stiffness matrix to be used during runtime.

## 3. Deformation Based on Connected Polygons

The basic idea of our method is simple: each polygon is connected to other polygon based on their connection to each other. There are two relation between polygon in an object. The first relation is group. Polygons connected in the same group will act equally. If one of the polygon is deformed, the other polygons that are connected in the same group are also deformed. Each group has a starting polygon that starts the deformation process. The same polygon will also activate the starting polygon in the lower leve. The second relation is level. Each polygon contains a level. This level value defines the layer of the polygon where polygon at outmost layer will have level 0.

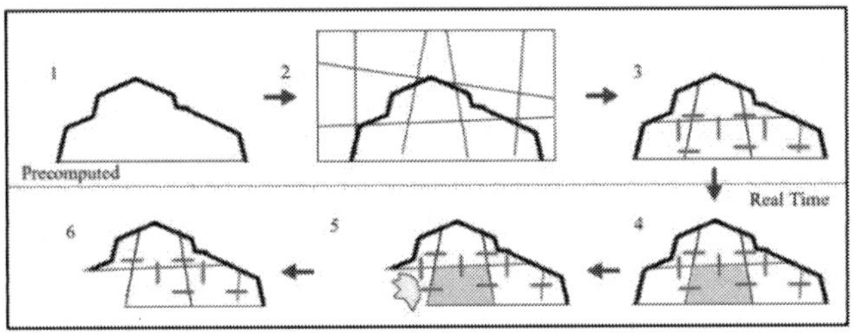


**Figure 1**. The flow of creating and simulating deformation

The deformation process goes as follows: a collision triggered a group and deformation process is activated. The group's starting polygon will set itself invisible and trigger the next polyon in the same group deactivate. This process is continued until it reached the starting polygon. At the same time, the starting polygon will trigger the starting polygon from the next level to activate itself and other polyon in the the same group. The deformation process is shown in the next figure.

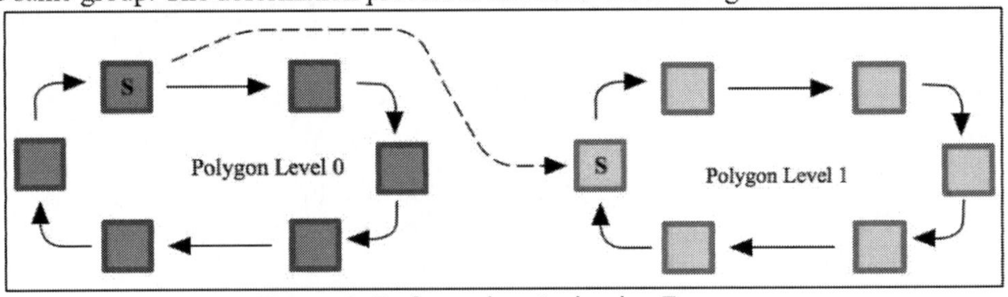


**Figure 2**. Deformation Activation Process

## 4. Implementation

### *4.1 Building the Object*

The first step of the implementation is to build a 3D objects with multiple layer of polygons. We use a common 3D content generation application for this process. The creation process starts with the non deformed shape of the objects, which contains polygons with the lowest level 0. The polygons are grouped based on their position. Later we deformed the polygon and increase the level by 1. This process is repeated until we've reached the smallest deformed shape of the object. To simplify this process, we treat each group as a single object since we have to manually merge the polygon in every level manually. Figure 2 shows the non-deformed object grouped by colour.

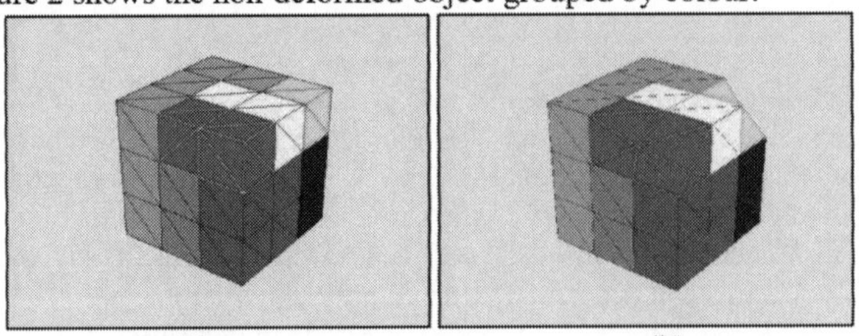

(a) (b)

**Figure 3**. Polygon colored based on group
(a) Non-deformed shape,(b) the deformed shape

### *4.2 Exporting to obj File Format*

Since we have to export each state of the object as a single obj file, it is important to merge all the obj file into one file. Unfortunately, there is no easier way to do this, we have to go through each obj, find the corresponding polygons and group them. We also have to enter the group and level information to the obj file manually. This information is attached in the end of each line of face data. Figure 3 shows how the modification is made to add an information that the corresponding face is group 4 and level 2.

**Original obj Format**

```
f 62/66/80 63/67/81 57/69/82
```

**Modified obj Format**

```
f 62/66/80 63/67/81 57/69/82 4 2
```

**Figure 4**. Modified obj format that contains group and level data

*4.3 Testing the Object in Application*

To test wether the method can sucessfully simulate deformation in real time, we develop a 3D computer graphic application. The application is built using C++ and OpenGL library. To help the windowing and I/O, we use GLUT library. At our current test, we do not collect performance data (such as frame per second) since the 3D object used in the test is too simple to affect the performance. Figure 4 shows the application running as a desktop application.

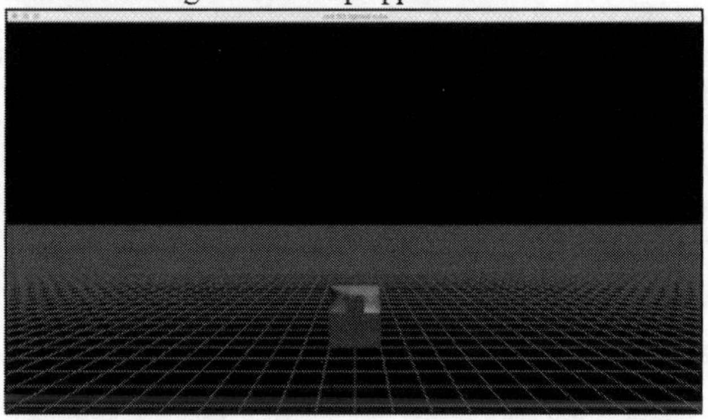

**Figure 5**. 3D Application to test the deformation process

Based on our observation during the test, the proposed method is capable to simulate deformation. There is no performance or memory issue arise during the test.

## 5. Conclusion and Future Works

We have shown that our connected polygon method is capable in simulating object deformation in real time. Our proposed system uses predefined 3D object with deformation data. Each polygons of the 3D object is connected to other polygon. When a deformation is triggered, the system will go trhough connected polygon to find which polygon is activated and which is deactivated.

There are a lot of improvement that needed to be added in the future. The most significant problem is the grouping and leveling process is done manually by modifying the obj file. The process is exhaustive consindering an object may contain thousands of polygon. Modifying thousand lines of faces information is exhaustive and time consuming. In the future, we would like to do these process automatically by creating a plug-in in the 3D content generation application. Hence, the deformation modelling and object exporting can automatically creates a file based on the method we proposed.

## 6. Acknowledgement

The authors would like to thank the head and staff of "Lembaga Penelitian USU (Research Center of Universitas Sumatera Utara)", Prof. Dr. Erman Munir, MSc, and staff for all the support during this research. The author would also like to thank the Dean of Faculty of Computer Science and

Information Technology Universitas Sumatera Utara, Prof. Dr. Opim Salim Sitompul, M.Sc for their widest to support this work.

**References**

[1] Parker E G and O'Brien J F 2009 Real Time Deformation and Fracture in a Game Environment *Proceedings of Eurographics/ACM SIGGRAPH Symposium on Computer Animation 2009* p 165-175

[2] Glondu L, Muguercia L, Marchal M, Bosch C, Rushmeier H, Dumont G and Drettakis G 2012 *Computer Graphics Forum* vol **31** no. 4 p 1547-1556

[3] Brown J, Sorkin S, Bruyns C, Latombe J C, Montgomery K, Stephanides M 2001 Real-Time Simulation of Deformable Objects: Tools and Application *Proceedings Computer Animation 2001. Fourteenth Conference on Computer Animation* p 228-258

[4] Clothier M and Bailey M 2015 Creating Destructible Objects Using a 3D Voronoi Subdivision Tree *Proceedings of 15th International Conference on Computational Science and Its Applications* p 42-46

[5] Bao Z, Hong J, Teran J, and Fedkiw R 2007 Fracturing Rigid Materials *IEEE Transactions on Visualization and Computer Graphics* vol. **13** no. 2 p 370-378

[6] Tang Z, Ozbek O, and Guo X 2011 Rea-Time 3D interaction with deformable model on mobile devices *Proceedings of the 19th ACM International Conference on Multimedia* p 1009-1012

[7] Terzopoulos D and Witkin A 1988 Physically Based Models with Rigid and Deformable Components *Proceedings of Graphics Interface Conference* p 146-154

[8] Muller M, Dorset J, McMillan L, Jagnow R and Cutler B 2002 Stable real-time deformations *Proceedings of the 2002 ACM SIGGRAPH/Eurographics Symposium on Computer Animation* p 49-55

IOP Conf. Series: Journal of Physics: Conf. Series **978** (2018) 012066 doi:10.1088/1742-6596/978/1/012066

# Control of motion stability of the line tracer robot using fuzzy logic and kalman filter

**M S Novelan [1], Tulus [2] and E M Zamzami [3]**

[1]Department of Information Technology, Faculty of Computer Science and Information Technology, Universitas Sumatera Utara, Medan 20155, Indonesia
[2]Department of Mathematics, Faculty of Math and Science, Universitas Sumatera Utara, Medan 20155, Indonesia
[3]Department of Information Technology, Faculty of Computer Science and Information Technology, Universitas Sumatera Utara, Medan 20155, Indonesia

**Abstract**. Setting of motion and balance line tracer robot two wheels is actually a combination of a two-wheeled robot balance concept and the concept of line follower robot. The main objective of this research is to maintain the robot in an upright and can move to follow the line of the Wizard while maintaining balance. In this study the motion balance system on line tracer robot by considering the presence of a noise, so that it takes the estimator is used to mengestimasi the line tracer robot motion. The estimation is done by the method of Kalman Filter and the combination of Fuzzy logic-Fuzzy Kalman Filter called Kalman Filter, as well as optimal smooting. Based on the results of the study, the value of the output of the fuzzy results obtained from the sensor input value has been filtered before entering the calculation of the fuzzy. The results of the output of the fuzzy logic hasn't been able to control dc motors are well balanced at the moment to be able to run. The results of the fuzzy logic by using membership function of triangular membership function or yet can control with good dc motor movement in order to be balanced

**Keywords:** Line Tracer, Robot Balance, Logika Fuzzy, Kalman Filter

## 1. Introduction

Robots are all something that can be programmed and reprogrammed , by having manipulator mechanical / propulsion designed to moving objects , components or special tools with various a flexible program / easy adapted to do various duties. The one example is the robot tracer a line. In general , tracer a line robot made by using a wheel three or four, two rear wheels connected to a DC motor serve as a drive and one or two front wheels acting as a guide but a fuzzy logic line robot will be created with the expected smooth movement compared to the PID method. This is because the line sensor that still no noise. The noise will be filled by using kalman filter method so that the noise from the line sensor output will be lost

## 2. Theoritical Basis

### *2.1 Robot Line Follower / Line Tracer Robot*

Line follower or Line tracer robot is a robot that can move following the line guide path. The wave guide line used in this case is a white line placed on a dark surface, or vice versa, the black line placed on the surface is white. As for robot electronics that is a robot that includes the main controller circuit (main controller), the sensor circuit, and the driver circuit. And the important system in making the follower robot is the programmer language (software). There are two kinds of line follower robot is a regular line follower without using program and line follower with microcontroller program.

### 2.2 *Definition of Robot*
The word robot comes from the Czech language, meaning robota worker, began to become popular when a Czech national writer, Karl Capek, making a comedy show in 1921 entitled RUR (Rossum's Universal Robot). Robot can be interpreted as a machine that can work continuously either automatically or in control. Robots that are used to help human tasks do things that are difficult or do humans directly. For example handling radioactive materials, assembling cars in the car assembly industry, exploring planet mars, as a medium of defense or war, and so on.

### 2.3 *Fuzzy Logic*
Fuzzy logic was first introduced by Jan Lukasiewicz in the 1920s as a probability theory. This logic extends the range of truth values to all real numbers at intervals between 0 and 1. Selanjutnya further investigated by Max Black in the 1930s in his research on vagueness: an exercise on logical analysis. In 1965, Professor and head of the department of electrical engineering at the University of California at Berkeley, Lotfi Zadeh, rediscovered, identified, explored, promoted, and fought for fuzzy logic. Professor Zadeh expanded the work space of probability theory into a formal mathematical logic system, and a new concept for applying natural language terms to his research, the 'Fuzzy sets'. This new logic is called fuzzy logic. Fuzzy logic is widely used because fuzzy logic is similar to human thinking. Fuzzy logic systems can represent human knowledge in mathematical form by resembling human thinking. Fuzzy logic has membership degrees in the range 0 (zero) to 1 (one), in contrast to digital or discrete logic that has only two values, namely 1 (one) or 0 (zero). Fuzzy logic is used to translate a quantity expressed using language (linguistic), suppose the rate of speed of the vehicle is expressed slowly, so fast, fast, and very fas. Fuzzy logic can process uncertain values of limits, such as "very," "a little", "more or less."

### 2.4 *Kalman Filter*
Kalman filter is a digital filter that is able to estimate any process very effectively. In the process state, the filter calm algorithm is believed to provide an estimated condition. Kalman filters can also minimize square errors in noise-sensitive control systems and can reduce the noise-exposed measurements of sensors before entering the control system.

$$x_k = F_k x_k - 1 + G_k u_k + w_k,$$

$$y_k = H_k x_k - 1 + v_k,$$

Where $x_k$ expressed as a state or state of the system, $y_k$ is the measurement of output, $u_k$ is an input that is known by the system, $w$ is the process of noise, $v$ is the measurement noise, $_k$ is an index of time and $F,G$ and $W$ is a matrix. Each number is a vector containing more than one element. On algorithm kalman filter writer will make $u$ as input result acceleration, $z$ as the output corner and $x$ as actuator speed and position. Noise and noise measurements can be presented oleh matriks konvarian $Q_w$ and $R_v$. value Qw and Rv plays an important role to determine the ability or output performance of the filter calm. To get the value of matrix authors use advanced statistic equations or manually tuned so that will get the desired output.

## 3. The Scheme Of System
This research has several stages in the form of a system scheme to create a model that will be used for robot line tracer. To get optimal result in stability of robot wheeled two researcher do some stages before giving input value which will be used for control of robot of tracer line of two wheeled that is by testing robot using original input data of sensor used. After the sensor data input value is obtained, the researcher will implement the kalman filter to smooth the noise on the sensor, Then the researchers will get the appropriate values to minimize noise on the sensor, From the test results will then get the value to be combined with the method to be studied and will get the robot response value to stay

balanced in the event of noise and disturbances that occur in the robot. The system scheme of the stages in this study is illustrated in Figure 1.

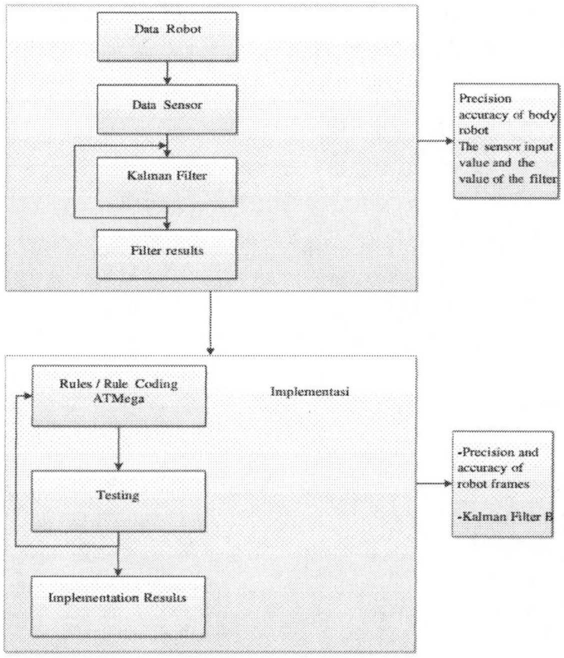


**Figure 1.** Stages of Line Tracer Robot Research Using Fuzzy Logic with Kalman Filter

### 3.1 *Flowchart Program*

In the flowchart this program describes the stages of how the system works the program used in starting the robot through the black line by taking into account the speed and stability of the robot in fuzzy logic algorithm. Flowchart program to run Line Tracer Robot is illustrated in Figure 2:

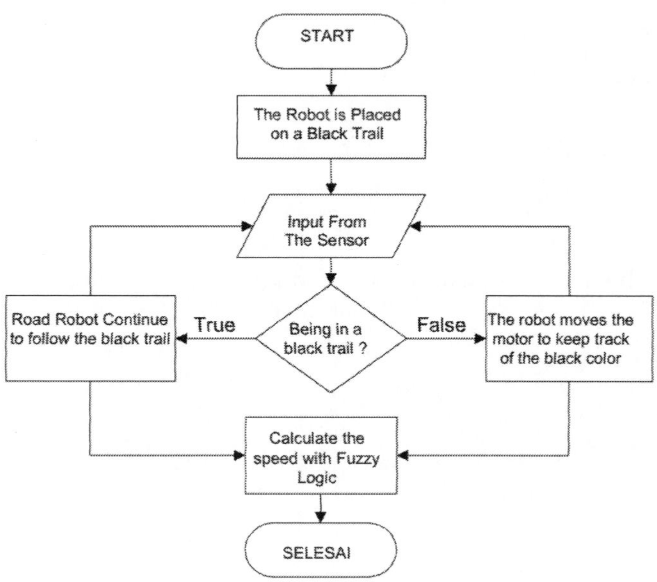


**Figure 2.** Flowchart program BASCOM AVR Line Tracer

IOP Conf. Series: Journal of Physics: Conf. Series **978** (2018) 012066 doi:10.1088/1742-6596/978/1/012066

### 3.2 *Flowchart System*

In order to see the structure of the way the system will be created and in carefully made flowchart (flowchart). Flowcharts are used as the basic reference in making a system or program flow designed. The implementation flowchart of the line tracer robot system is illustrated in Figure 3:

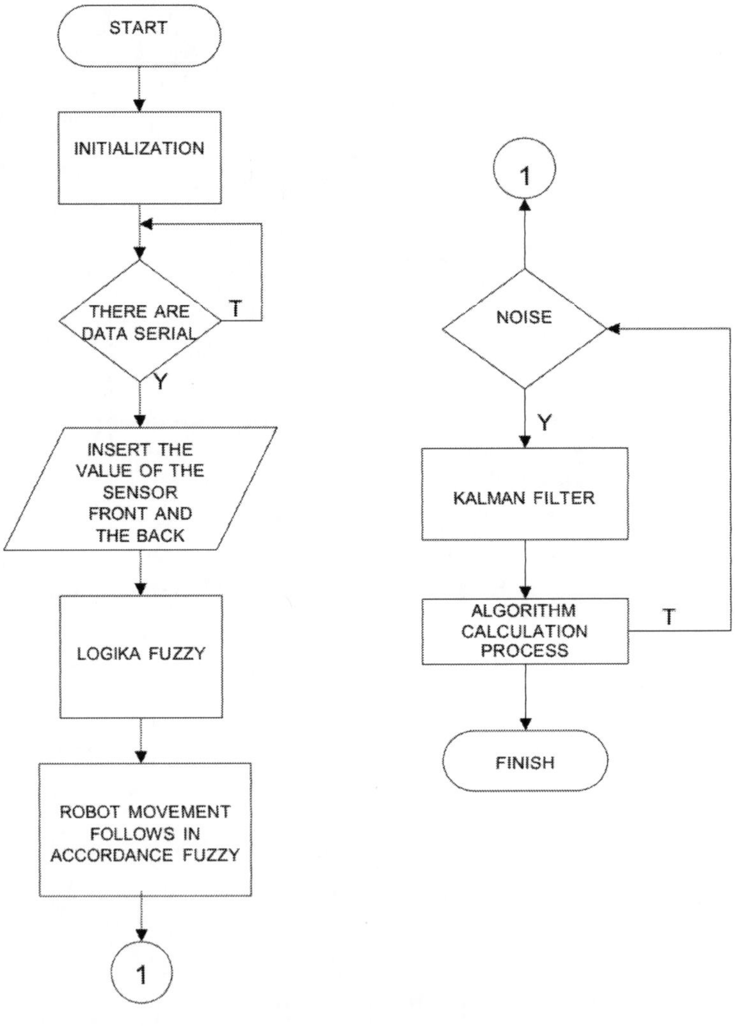


**Figure 3.** Flowchart Implementation of robot line tracer algorithm using fuzzy logic with kalman filter

## 4. *User Interfaces*

*4.1 See the preview*

The graph of the difference between the raw sensor values and the filtered sensors values of the filter cylinders is illustrated in the Figure 4 :

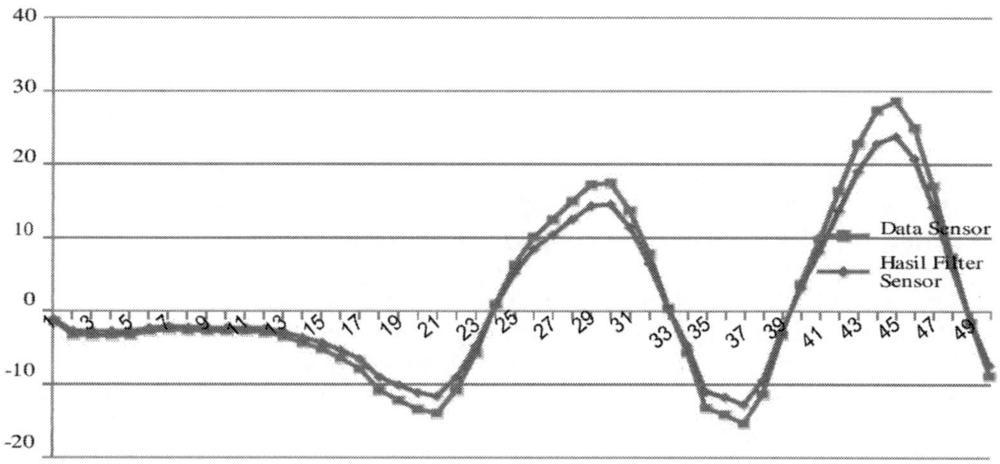


**Figure 4.** graph of Motion Management System and Line Tracer Robot Balance

*4.2 Process Control Using Fuzzy Logic*

Graph of dc motor response in reading input and output data

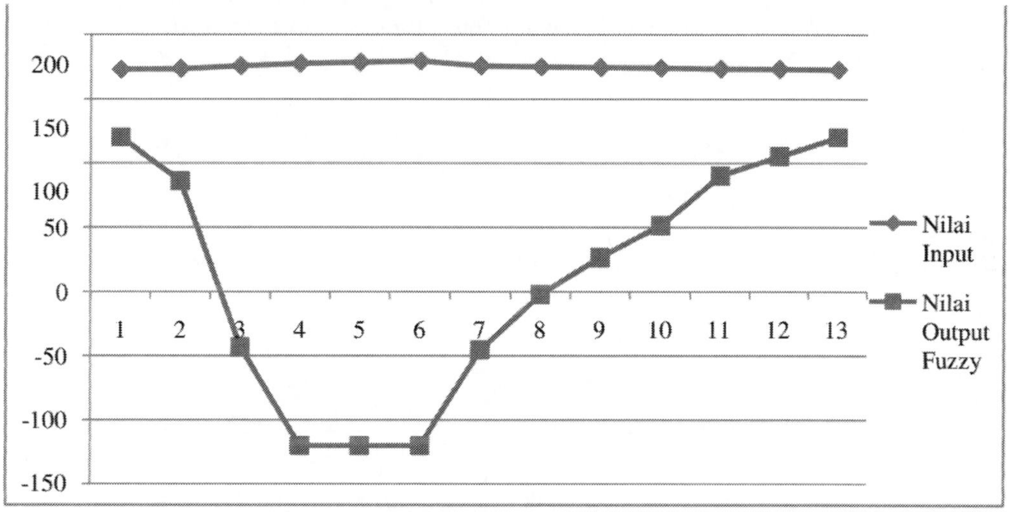


**Figure 5.** Graph of dc motor response

**5. Conclusions**

From the results of research and discussion undertaken conclusions can be drawn as follows:

1. Test results from the fuzzy logic method is not too good for the balance control system on the two-wheeled robot when passing through the interference.
2. Test results from the Kalman filter method can reduce the original value or the raw value of the sensor.
3. Test results from a combination of fuzzy logic and PID methods can pass through disturbances as high as 1 cm.
4. Better rule-based is required for a two-wheel balance control system that is able to maintain balance while passing through interference

**References**

[1] Bakdi A and Hentout 2017 Optimal Path Planning and Execution for Mobile Robots using Genetic Algorithm and Adaptive FuzzyLogic Control *International Journal of Fuzzy Systems, 16(3)*

[2] Dwisaputra I, Sulistijono I A and Nugraha M I 2015 Two Wheels Balancing Line Tracer Robot Using Fuzzy Logic Control, *The 13th Industrial Electronics Seminar 2011 (IES 2015), Electronic Engineering Polytechnic Institute of Surabaya (EEPIS)*, Surabaya.

[3] Apriliani E, Subchan F, Yunaini and Hartini S Estimation and Control Design of Mobile Robot Position *Far East Journal of Mathematical Sciences (FJMS)* (Puspha Publisher, Surabaya ).

[4] Farooq U, Amar M, Asad M U, Abbas G and Hanif A 2014, Fuzzy Logic Reasoning System for Line Following Robot *IACSIT International Journal of Engineering and Technology*, No. 4, Vol. **6**, Hal. 244-248

[5] Holland J M 2015 Designing Autonomous Mobile Robot, Inside the Mind of an Intelligent Machine (Elsevier Inc, US America)

[6] Nguyen H-N, Zhou J, Kang H-J 1990 A calibration method for enhancing robot accuracy through integration of an extended Kalman filter algorithm and an artificial neural network *IEEE International Conference on Robotics and Automation, Proceedings,* vol. **2**, 1990, pp. 1032–1037

# Developing and pilot testing M-health care application for pregnant and toddlers based on user experience

**I D Lestantri[1], Putrima[2], A Sabiq[3] and E Suherlan[4]**

Universitas Yarsi, Jakarta, Indonesia

1)indahdl2005@gmail.com, 2)putrima@gmail.com, 3)sabiq.net@gmail.com, 4)elansuherlan70@gmail.com

**Abstract**. Indonesia is the country number 4 as the largest smartphone user. In addition to functioning as a social media, the use of smartphones for other purposes has begun to develop in order to facilitate the work. Similarly in the field of maternal and child health. Most of the public, pregnant mothers and checking the development of children under five, were conducted to the Public Health Center. As a service center, Public Health Center need to improve its services. Among other things, it needs to be supported with technology. The high number of queues, the length of queue time and the traffic jam, prompted the need for the development of health applications, based on mobile. The role of the user is crucial to the successful implementation of a technology. This study aims to measure whether the application of M-Health Care application, as a prototype of the application developed to meet the needs of users. M-Health Care Application was developed with a prototyping approach. While for the measurement of M-Health Care Application done by using approach of USE Questionnaire. From result of data processing got result, usability measurement value equal to 88,3%. This indicates the M-Health Care application prototype is considered satisfactory for the user.The results of this measurement can be used as a reference for the next M-Health Care application development model based on user experience.

Keywords—m-health, care, mobile, application, pregnant, todddlers

## 1. Introduction

Mobile applications are developed and used as utilization to process and present information. Indonesia will become the country with the largest smartphone active users, after China, India, and America. The population of Indonesia is about 250 million people is a big market. Smartphone users in Indonesia are growing rapidly. Digital marketing research institute E-Marketer estimates that by 2018, the number of active smartphone users in Indonesia will reach more than 100 million people (Wahyudi, 2015)[1].

GfK is a leading market research institute in Indonesia and one of the largest market research companies in the world. According to GfK data, smartphone penetration and digitalization process happen very rapidly in Indonesia. Recorded more than 88 million Internet users in Indonesia. Nine out of 10 internet users or (93%) access the internet via smartphone. There are more 51% dominated by young people with age ranges between 13 and 27 years old. Male users (56%) and female users (44%). The online population in Indonesia is very active. The average usage of 5.5 hours a day is spent accessing about 46 applications and web domains through smartphones. Online use is done via desktop at 11% and through tablets by 5% (GfK, 2016) [2].

The Global Positioning System (GPS) is a satellite-based navigation system developed by the US Department of Defense in the early 1970s. Initially, GPS was developed as a military system to meet military needs. However, it was then made available to civilians, and is now a dual use system accessible to military and civilian users. Google currently provides programming functionality provided by Google Maps, the Google Maps API. With the Google Maps API, Google Maps can be integrated into application that will be created to display maps, markers and more. By using these tools, we can develop mobile applications for various purposes, including the improvement of health services.

Innovative M-Health applications have the potential to transform healthcare in both the developing and the developed world. They can contribute to bringing healthcare to unserved or underserved populations; increasing the effectiveness and reducing the costs of healthcare delivery; improving the effectiveness of public health programs and research; preventing illness (including through behaviour change); managing and treating chronic diseases; and keeping people out of hospital [3].

Mobile communications can greatly increase public awareness of health issues. An initiative in Uganda sent out an SMS quiz on HIV awareness, which resulted in a 40% increase in people requesting free HIV/AIDS testing. In the UK, the NHS could take this a step further by adding SMS to the National Pandemic Flu Service's existing online and telephone resources. Similarly, the NHS consultation phone line (NHS Direct) could send an SMS after a phone consultation, directing patients to the nearest clinic or surgery based on their responses [4].

Maternal and child health is the first priority. Maternal Mortality has already declined, but is still far from the MDGs target by 2015, although the number of deliveries helped by health workers has increased. This condition is probably caused by among others the quality of maternal health services that are not adequate, the condition of pregnant women are not healthy and other determinant factors. The main causes of maternal death are hypertension in pregnancy and post partum bleeding. This cause can be minimized if the quality of Antenatal Care implemented properly. Some conditions that can cause unhealthy pregnant women conditions include handling complications, anemia, pregnant women who have diabetes, hypertension, malaria, and other pregnancy factors. Other pregnancy factors include getting pregnant at too young, less than 20 years; pregnancy in old age, more than 35 years old; consecutive pregnancies, less than 2 years; and too many pregnancies. As many as 54.2 per 1000 women under the age of 20 have given birth, while women who gave birth to age over 40 years were 207[5].

One of the services in Public Health Centers is services for mothers and children. Maternal and child services include services for pregnant women. Child services include services for newborns, through neonatal medical records and immunization services. As a health service, Public Health Center need to improve health services quickly and appropriately. One of them by reducing the queue time for patients. The number of patients in Public Health Center on average can reach 100 people per day. With a registration time of 5-7 minutes per person. Administrative records are still manual, and some cases of registration time, such as losing or forgetting to bring a medical card, adding to the length of registration time. While the waiting time for treatment can reach 10-15 minutes per person. The condition of the registration time and the long waiting time of accumulation become the problem of most of the public for treatment to the Public Health Center. The distance is quite far and traffic jams, often a supporting factor in addition to the problem of waiting a long time.

One way to solve the problem is the need for the development of information technology. With reference to the number of smarphone users in Indonesia, the development of android based health applications is one of the right solutions. M-Health Care application is a mobile application. This application was developed with various features in it, one of them is registration to the Public Health Center through mobile application. This application was developed with prototyping approach. Prototype has been made communicated to the public to get input, modeling, and design appropriate, in accordance with public needs.

Research related to mobile application development for health has been done. Becker (2016) in his research on the implementation of mobile applications for mental health care. This mobile mental health care application is quite acceptable to the public. This application is quite supportive and is a tool that can be used anytime by the user [6]. Other research done by Dwivedi et all (2015). Mobile application in the health field are applied to the use of antibiotic preciption. Once implemented for one year, data shows that among users, 62% use the OS system while 38% use the Android system [7]. Mobile health applications are also implemented in Spain. Based on the research Garcia et all (2017) presented a mobile application to help training activities. This application is tested for the purpose of measuring the quality of the application as well as the acceptance of the mobile application. The results of these studies show that mobile mental and physical health applications provide high utility values to the public [8]. The application has great potential for use [8]. On the other hand mobile applications of this mental and physical health, there is a relationship between application users, usage time, age, nationality and education level [8]. Research on the implementation of m-Health is also done by Kim et all (2017). Research conducted on m-Health interventions aimed at improving immunization programs, can effectively provide greater access to vaccine services [9]. The results of the research show that m-Health is very important. m-Health is a tool that can be used to improve the range and impact of the program vaccine [9].

One of the features that will be developed in M-Health Care application is the electronic medical records of the owners. This note contains data that is very personal and become one of the important information that must accompany someone wherever he goes. Information available on this electronic will be beneficial for both patients and other institutions, among others, the Public Health Center or Hospital. Suppose the patient suddenly had an accident, the hospital will be greatly helped by the information available on M-Health.

Mobile health applications in Indonesia are already emerging. Some of these mobile applications are Daftar Pasien, Public Health Center Mobile, Infokes Mobile, e-Health Surabaya, Poli Umim Z, Public Health Center Kuta Utara, ImunInfo. Those applications are the similar application in playstore. Summary results on the eight applications can be seen in Table 1

**Table 1.** Summary of M-Health Application

| **Application Name** | **Description** | | | |
|---|---|---|---|---|
| | ***Rating*** | ***Number of User*** | ***Fitur*** | ***Weakness*** |
| Daftar pasien | 21 | 221 | User: Public<br>User registration<br>Registrastion online<br>Registration sort number<br>Outpatient List<br>New Patient List | The Service has not been in sync with real condtotion<br>The data base is not good<br>Application not working propoerly |
| Public Health Center Mobile | 4 | 29 | User : Nurse<br>Data Input Patient | Limited user<br>UI is not very friendly. The functions that can be used have not been thorough |
| Infokes Mobile | 4,5 | 35 | User : Public<br>Registration online<br>Check the queue number<br>Check the number of BPJS<br>Share artcles<br>Online Consultation | There are still some validationn bugs to fix |
| eHealth Surabaya | 3,5 | 279 | User : Public<br>Online Registration RS<br>Check the queue number<br>Online consultation<br>Online complains<br>Check the location of the hospital | |

| **Application Name** | | | **Description** | |
|---|---|---|---|---|
| | ***Rating*** | ***Number of User*** | ***Fitur*** | ***Weakness*** |
| | | | A special service application for Suarabaya | |
| Poli Umum Z | 5 | 3 | User: Nurse<br>Integartion with Pcase application<br>Record medic<br>Input drug to the pharmacy<br>Patient referral BPJS | Limited user |
| North Kuta Health | 5 | - | User : Public<br>Online registration<br>View history visit<br>View data health<br>Settiing | This application only be used for treatment to the Nort Kuta Health Center |
| ImunInfo | 3,8 | 5 | User : Public<br>User registration<br>Read self data<br>Add babydata<br>Get informasion of vaksin<br>Check schedule<br>Public Health Center and Integrated service post<br>Addimmunization data<br>Seemap of Public Health Center<br>And Integrate service post | |
| Loket | 3 | 4 | Registrasi patient | |

The above application includes several mobile applications developed to support health services. Existing mobile application have not been used optimally by the public with some constraints, including feature completeness, data integration, software stability, and socialization.

This research is different from previous research. M-Health Care application developed focused on pregnant women and toddlers. The M-Health Care application is similar to ImunInfo application. The difference between M-Health Care application and ImunInfo application lies in the features. The features developed in M-Health Care application include medical treatment number, Public Health Center information, schedule of doctor information, patient visit history, information of location ambulance, facility to show the fastest route to certain Public Health Center and integrate with medical record of pregnant mother, neonate and immunization. Medical record recording pregnant women and medical record of neonate and immunization, that is different from the patient's general medical record. Medical records of pregnant women should be recorded in detail the data of pregnancy control visits until childbitrh. Medical records in infants include neonates and immunizations. M-Health Care application integrated e-Neonate application and e-pregnancy application. Neonate application is a application to record neonate and immunization. And e-pregnancy is a application to recording pregnant woman.

Software quality is very important in a software [10]. The purpose of software engineering is to provide high quality software, and SPI is a recommendation approach to improve software processes and produce high quality software [11]. The role of user experience in measuring a software is necessary. Research related user experience in software one of them done by Fagan et all (2015). Fagan conducting quantitative studies on the use of messages software update messages to identify some user perceptions, experience, and beliefs about software update mechanisms. The results show that the user sometimes confused and annoyed with the appearance of update messages. This indicates a negative attitude associated with warning and update messages that appear on the software [12].

This study aims to measure, whether the M-Health Care application prototype has been developed is enough to meet the needs of users in the review of the four dimensions, including usefulness, ease of use, ease of learn, and satisfaction. To measure the four dimensions, questionare is used, using USE

Questionnaire. Prototype measurement is based on user experience, to generate input modeling and design according to user needs. Results from research to assist in developing Smart e-Health.

## 2. Methodology

### *2.1. Prototyping Approach*

M-Health Care applications that will be developed are expected to be implemented in 16 Public Health Centers and used by the public. The application development stage of M-Health Care application uses prototyping approach. The prototyping development stage can be seen in Figure 1. The prototyping approach taken is throw away prototyping. Throw away prototyping selected because it to is more suitable to the condition, used to reduce requirement risk, so we must validate or derive the system requirement [13].

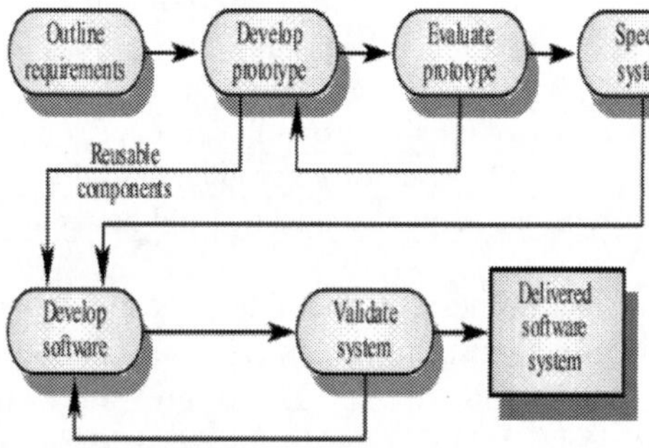


**Figure 1.** Prototyping Approach

Researchers use a prototyping approach. This prototyping approach is chosen because it makes it easier to explore user needs, reduces the risk of misunderstanding between users with developed software, enhanced software development, and reduces the risk of development errors.

### *2.2. Research Methodology*

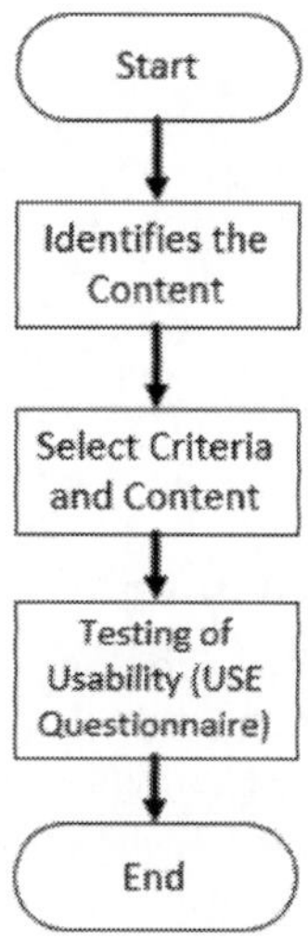


**Figure 2.** Research Flow Diagram

The M-Health Care application prototype results are presented to the public at the research site to get input, modeling, and design appropriate for the public. The location of data collection in Kemayoran area. The social background of Kemayoran people mostly traders, entrepreneur, and some are office workers. The research flow diagram can be seen in Figure 2. From Figure 2 can be seen, the stages of research conducted are: 1) Identities the Content. 2) Select Criteria and Content. 3) Testing of Usability. 4) Analisis of Results.

This type of research is a quantitative research. Quantitative approach is chosen because based on the data distributed through questionnaires to users offline. According to (Nielsen, 2012) usability is a quality variable used to know how the interface is used [14]. The usability test was conducted by using a questionnaire. USE Questionnaire published by Arnold M. Lund in 2001. According to Lund (2001), to evaluate the usability of a product is recommended to use four dimensions of the USE Questionnaire [15]. These four dimensions are usefulness, ease of use, ease of learning, and satisfaction dimensions. Description of each dimension can be seen in Table 2 [16].

**Table 2.** Description of Dimension

| Usability Dimension | ***Description*** |
|---|---|
| Usefulness | related to the usefulness of the product for the user. How much is the product useful and utilized by the user to achieve user goals. |
| Ease of use | related to the ease of the user using the product |
| Ease of learning | how fast a user can be operate the product until the user understands how to use it. |
| Satisfaction | with regard to user acceptance, feelings and opinions of the product. |

To meaure the four dimensions (usefulness, ease of use, ease of learning, and satisfaction) the author uses USE Questionnaire. USE Questionnare is a tools of questionnaires. USE stands for Usefullness, Satisfaction, and Ease of Use [17]. There are three main dimensions in initial development using USE Questionnaire[17]. USE Questionnaire is used to measure the usability of the four dimensions by including 30 questions. Assessments using USE Questionnaire were conducted on a graded scale of ratings. Instrument Use Questionnaire can be seen in Figure 3

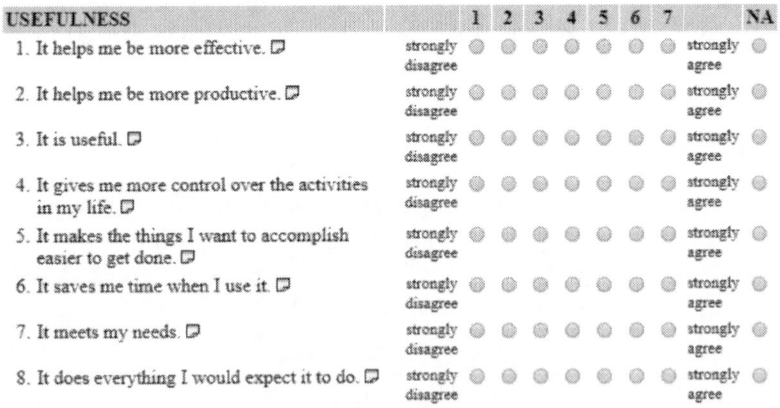


**Figure 3.** USE Questionnaire

After determining the four dimensions measured, then analyzed the results of the questionnaire that spread. Results of data analysis to determine the usability level of M-Health Care application prototype based on usefulness dimension, ease of use dimension, ease of learn dimension, and satisfaction dimension. Total score usability calculation is calculated from the average of all dimensions multiplied by 100 percent

### *2.3. Participant*

Data collection by distributing questionnaires, followed by 31 female respondents with an age range of 20-60 years. The age distribution of respondents is: age 20-30 years as many as 19%, age 31-40 years as much as 26%, age 41-50 years as much as 48%, and age 50-60 years as much as 7%. Number of respondents who have a toddler as much as 47%.

### *2.4. Research Model*

M The model of this research flow is illustrated in Figure 2. From the research model is made hypothesis as follows:

- Usefulness has a significant positive effect on Usability
- Ease of use has a significant positive effect on Usability
- Ease of learnings has a significant positive effect on Usability
- Satisfaction has a significant positive effect on Usability

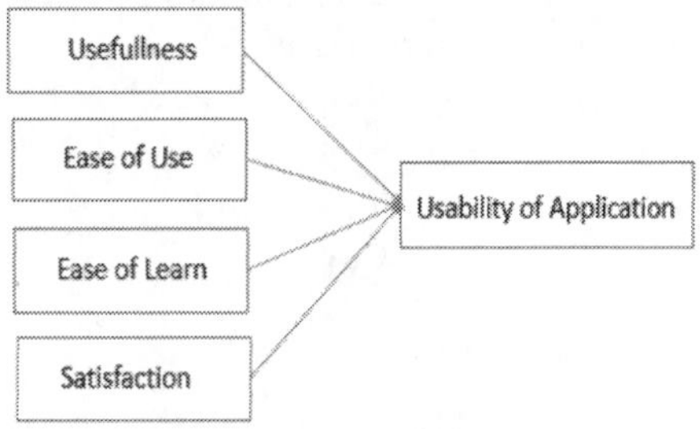


**Figure 4.** Research Model

## 3. Result And Discussion

### *3.1. Model*

M This M-Health Care application prototype has the following features:

- Public Health Center Services. In this service there is a feature for the user to choose the location of Public Health Center to be the destination, get information about the Public Health Center. Information on available Public Health Center includes the types of poly services available at Public Health Center, doctor's schedule, number of available rooms, address and contact number of Public Health Center, and location of ambulance existence. In this Public Health Center service, the patient can register the medication at the targeted health center, and get the sequence number of the patient queue.
- The Maps Service, available on this service features to see the location of Public Health Center around the user, the distance that must be taken from the user location and the estimated time required from the user location to the destination.
- Service History Visit, on this service the patient can see history of patient visit to Public Health Center and polyclinic ever go.
- This M-Health Care application integrates with e-pregnancy and e-neonate applications. E-pregnancy is an application to record medical records of pregnant women and e-neonates is an application for newborn baby recording and immunization. Display of M-Health Care application can be seen in Figure 5.

**Figure 5.** M-Health Care Application Prototype

The M-Health Care application to be developed has a single data base. This M-Health Care application will integrate with e-pregnancy application and e-neonate application. The e-pregnancy application is an application developed for the recording of maternity medical records, maternity record-keeping, and maternal postpartum. This e-pregnancy application also features to visualize maternal pregnancy data and prediction of mother's birth type. Is it normal or cesarean, so that should be referred to the hospital. The e-neonate application is an application for recording newborns and immunization medical records. M-Health Care applications, in addition to having the function to register, in the future will be added features for patient medical records, services to obtain information about maternal and child health, and as a media of consultation through mobile application. M-Health Care application prototype modeling can be seen in Figure 6.

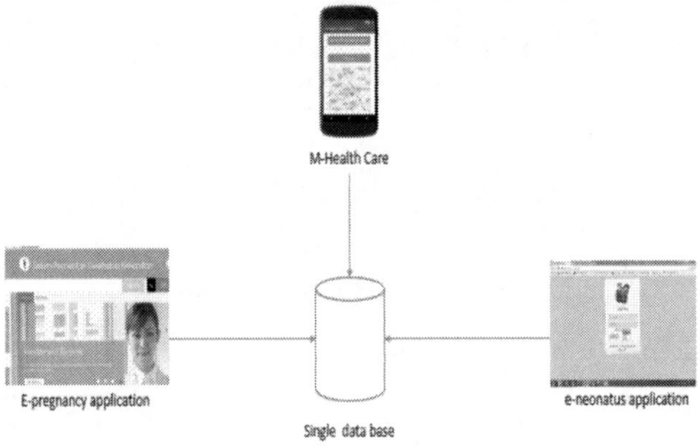


**Figure 6.** M-Health Care Model

### *3.2. Testing*

The result data from the questioner is tested. The calculation results can be seen in Table 3. Table 4 shows the usability level of each dimension. The highest dimension is Usefullness as big as 89,66% and the lowest dimension is Ease of Use as big as 85,85%. Usefullness dimension has the highest value, it indicates the need of the community to use M-Health Care application. While at the lowest score is on the dimensions of satisfaction. This happens, because based on user experience in mobile application usage. The test results of each dimension can be seen in Table 4.

**Table 3.** Calculation Results

| Usability Dimension | ***Score*** |
|---|---|
| Usefulness | 89,66% |
| Ease of use | 85,85% |
| Ease of learning | 87,83% |
| Satisfaction | 89,61% |

Overall ratings on each dimension get good results, with an average of over 85%. The assessment results in each dimension describe the condition of M-Health Care prototype as follows:

- Measurement results on Usefullness dimension of 89.66%, it shows has a high usability value for the user.
- The measurement result on Ease of Use dimension has value equal to 85,85%. This shows that this prototype of M-Health Care application is easy to use by user.
- The measurement result on Ease of Learning dimension has a value of 87.83%. This shows that the prototype of M-Health Care application is easy to learn by user. Users do not need to take too long u learn to use this application.
- The result of measurement on Satisfaction dimension has value equal to 89,61%. This shows the prototype M-Health Care is satisfactory for the user.

**Table 4.** Result of Each Usability Dimension

| Dimensions | ***Indi cators*** | ***Descriptions*** |
|---|---|---|
| Usefullness | U1 | 87,3% |
| | U2 | 93% |
| | U3 | 94% |
| | U4 | 91,3% |
| | U5 | 93% |
| | U6 | 91,3% |
| | U7 | 80,6% |
| | U8 | 84% |
| | U9 | 92% |
| | U10 | 91,3% |
| Ease of Use | E1 | 91,3% |
| | E5 | 86,6% |
| | E3 | 89,3% |
| | E4 | 84,6% |
| | E5 | 81,3% |
| | E6 | 81,3% |
| | E7 | 84% |
| | E8 | 84% |
| | E9 | 84% |
| Ease of Learn | L1 | 89,3% |
| | L2 | 88,6% |
| | L3 | 88,6% |
| | L4 | 82% |
| Satisfaction | S1 | 87,3% |
| | S2 | 82,6% |
| | S3 | 87,3% |
| | S4 | 87,3% |
| | S5 | 82,6% |
| | S6 | 90,6% |
| | S7 | 90,6% |

This study was conducted to determine the level of needs and usefulness of the application of M-Health Care. Of the 30 questions in USE Questionnaire., the third questioner item (It gives me more control over the activities in my life) in the Usefullness dimension has a value of 94%. This item is one indicator that represents the usefullness valuation indicator. This indicates that the prototype of M-Health Care application has usability that helps users. The smallest measurement value is on the Ease of Use dimension. In the Ease of Use dimension, the third and fourth indicator items (Is is user friendly and it requires the fewest steps poosible to accomplish what I want to to with it) have a value of 81.3%. This indicates that not all users are easy and familiar in using M-Health Care applications, it needs improvement from the user interface.

Total usability measurement results of 88.3%. This indicates that overall prototype M-Health Care application is quite satisfactory according to user. This indicates that the M-Health Care application prototype needs to be further developed. Although the overall value of usability is good, and each dimension has a good enough value as well, it still needs to be improved on some things. Among other things from the interface side. Therefore it is necessary to test methods to measure aspects of use from the system interface side.

## 4. Conclusion

Prototype M-Health Care application is tested in the public to get the right input and modeling for further M-Health Care application development. Tests were performed on five variables, including usefullness, ease of use, ease of learn., satisfaction, and usability

The test results in each dimension are: 89,66% for usefullness dimension, 85,85% for ease of use dimension, 87,83% for ease for learning dimension, and 89,61% for satisfaction dimension. This shows that from all sides of the dimension, the prototype of the M-Health Care application is quite good according to the user and its existence needs to exist.

Usability dimension of 88,3%. The usufullness, ease of use, ease of learning, and satisfaction dimensions, each has a positive effect on usability variables. This shows that the public needs this M-Health Care application and recommends its use.

The maturity of this application needs to be improved. This app needs to be added to other features. Some features that need to be developed include check the number of BPJS and integration with other application, example Family Information System applications.

## References

[1] Wahyudi and Ade 2015 Indonesia, Raksasa Teknologi Digital Asia (Tempo).

[2] Gfk, Tech Trends 2016 2016 2016, Accessed on 2nd September, 2016

[3] FREng I L and Sansom R 2011 Model Communication for Medical Care – Final Repor

[4] GSMA 2010 MobileTechnology's Promise for Healthcare

[5] Riskesdas. Laporan Hasil Riset Kesehatan Dasar (Kementerian Kesehatan RI 2010).

[6] Becker D 2016 Acceptance of Mobile Mental Health Treatment Applications *Procedia Computer Science* pp 220-227.

[7] Yogesh K. Dwivedi, Shareef M A, Simintras A C, Lal B, Weerakkody V 2015 A Generalised Adoption Model for Services: A Cross Country Comparison of Mobile Health (m-Health) (Governmnet Information Quartely) GOVINF-01102.

[8] Garc'ia I P, S'anchez C M, Espilez A S and G-Campayo I Development And Initial Evaluation of A Mobile Application To help With Mindfullness Training and Practise

[9] Kim S S, Patel M, Hinman A 2017 Use of m-Health in Polio Eradication and Other Immunization Activities in Developing Countries ,Vaccine.

[10] Martín C L, Nassif A B and Abran A A 2017 Training Process for Improving The Quality of Software Projects Developed by A practitioner *The Journal of System and Software*.

[11] Sanchez-G M L and O'Connor R V 2015 Understanding The Gap Between Software Process Practices And Actual Practice In Very Small Companies *Software Qualilty Journal, Springer,* In Press, DOI 10.1007/s11219- 015-9282-6

[12] Fagan M, Khan M H and Buck R 2015 A Study of Users Experiences and Beliefs About Software Update Messages *Computer in Human Behavior* pp 504-519

[13] Sommervile I 2011 Software Engineering, *9th Ed. Addison Wesley Publishing, New York*

[14] Nielsen J 2001 Success Rate: the Simplest Usability Metric. http://www.nngroup.com/articles/success-rate-the-simplest-usability-metric. Accessed on 30th August, 2017

[15] Lunda and Arnold, 2016 Measuring Usability with The USE Questionnaire. Https://www.researchgate.net/publication/230786746, Accessed on 2nd September, 2017

[16] Rubin, Jeffrey, Chrisnell and Dana 2008 Handbook of Usability Testing: How to Plan, Design and Conduct Effective Tests.( 2nd Ed. Wiley Publishing.)

# Hybrid cryptosystem for image file using elgamal and double playfair cipher algorithm

**S M Hardi, J T Tarigan and N Safrina**
Faculty of Computer Science and Information Technology, Universitas Sumatera Utara, Jl. Doktor Mansyur No.9, Padang Bulan, Medan Baru, Sumatera Utara 20155

Email: nanda.safrina@usu.ac.id

**Abstract.** In this paper, we present an implementation of an image file encryption using hybrid cryptography. We chose ElGamal algorithm to perform asymmetric encryption and Double Playfair for the symmetric encryption. Our objective is to show that these algorithms are capable to encrypt an image file with an acceptable running time and encrypted file size while maintaining the level of security. The application was built using C# programming language and ran as a stand alone desktop application under Windows Operating System. Our test shows that the system is capable to encrypt an image with a resolution of 500x500 to a size of 976 kilobytes with an acceptable running time.

## 1. Introduction

There are many algorithms that can be used to perform an encryption of a file. These algorithms can be categorized into two types: symmetric and asymmetric. In symmetric encryption, both sender and receiver hold identical key which can be used to encrypt and decrypt the data. This fact may cause a problem since both party have to communicate to share the key, and this process may be eavesdropped by the middleman. Therefore, sender and receiver must have a secured channel to share the key, which is not possible most of the time.

Asymmetric encryption solves this problem by using two different keys. First introduced by Diffie et al. [1], asymmetric encryption uses public keys which distributed widely and private keys which is known only by the owner. Hence, unlike symmetric encryption, both parties do not require a secure channel to distribute the key. These keys often rely on mathematical problems with no efficient way to solve such as integer factorization, discrete algorithm, and elliptic curve. However, due to it needs to perform mathematical operations to perform encryption, the time to perform encryption heavily depends on the file size. Hence, asymmetric encryption is often not suitable to encrypt large files.

Since both encryptions have their own weakness, it is not possible to a single type of encryption to secure a data. Hybrid cryptosystem tries to solve this problem by mixing both algorithm to increase the security while maintaining the performance. The system uses symmetric encryption to encrypt the plain text and asymmetric to encrypt the symmetric key. Since the size symmetric key is usually small, there wouldn't be a significant time added to encrypt the key.

In this paper, we propose a hybrid cryptosystem that combines asymmetric encryption Elgamal and symmetric encryption Double Playfair Cipher. We chose image file as our plain text to see the performance of the algortihm in both time and space. Our objective is to measure the usability of the system and to see whether the combination of these two algorithms are applicable. The paper is structured as follow: the next section will discuss related previous works that can be used as reference. Later, we will go shortly describe the algorithms used in this paper. Section 4 will describe the design and implementation of the system, followed by test and results in section 5. In the last section, we will go through the conclusions and future works.

## 2. Related Works

Hybrid cryptosystem has been used in most security. It capability to increase security while maintaining security level is one of the reason of it's use in modern cryptography. One of the most commonly used cryptosystem that built based on hybrid method is Pretty-Good-Privacy. Developed and introduced by Phillip Zimmerman [2], this method has become a backbone in internet security [3]. The encryption process of PGP contains symmetric encryption, usually AES [4] or IDEA [5], to encrypt the data and asymmetric encryption, usually RSA [6], to encrypt the key for symmetric encryption. Additionally, PGP also contains a process to validate message authentication by using certificates and digital signatures. Currently, PGP is used in most computer security system such as Transport Layer Security (TLS) [7] and its predecessor, Secure Socket Layer (SSL) [8]. First defined in 2007 by Internet Engineering Task Force, OpenPGP has become the standard part of TSL [9].

Other than PGP, there are many research focused on hybrid cryptosystem. Some of the work focus on using other encryption algorithm and test its usability. Hazra et al. developed a cryptosystem by using Blowfish and Diffie-Hellman algorithm [10]. Dahake et al. developed a fingerprint based cryptosystem where the biometric fingerprint data is used for key generation [11].

## 3. Encryption Algorithm

### *3.1 Elgamal Algorithm*

Elgamal is an asymmetric algorithm developed by Taher Elgamal and introduced firstly in 1985 [12]. The security of the algorithm relies on the difficulty of computing discrete logarithms over finite fields. It is based on key exchange algorithm proposed by Diffie-Hellman [1]. The implementation of Elgamal Algorithm is as follow: Let say Bob tries to send a message to Alice. Alice has to provide a public key to be used as encryption key. To generate this public key, Alice have to choose these values: a large prime number $P_A$ a primitive element $\alpha_A modulo\ P_A$, and a random integer $d_A$ where $2 \leq d_A \leq P_a - 2$. Using these values, Alice calculates $\beta_A$ where $\beta_A \equiv {\alpha_A}^{d_a}(mod\ P_a)$. Alice then publish the public key $(P_A, \alpha_A, \beta_A)$ and keep the private key $d_A$.

Bob as the sender of message $M$ where $M < P_A$ choses as random integer $k$. Bob calculates then computes the encrypted message $(r, t)$ where $r \equiv {\alpha_A}^k(mod\ P_A)$ and $t \equiv {\beta_A}^k M(mod\ P_A)$. The encrypted message $(r, t)$ is sent to Alice. Alice then decrypt the message by using the private key $d_A$ in the equation $tr^{-d_A} \equiv M\ mod\ P_A$. Should any eavesdropper intercept the message $(r, t)$, it will require him to know the value $d_A$. While it is possible to calculate the value by using the public key $(P_A, \alpha_A, \beta_A)$, it would be too difficult to be practical since the algorithm suggest the use of a very large prime number (200-300 digits integer).

### *3.2 Double Playfair Chiper*

Playfair Cipher is one of the earliest encryption algorithm and categorized as classical cipher. It was first introduced by Charles Wheatstone in 1854. The idea is to use a 5x5 table to substitute each character in the plain text. Each cell from the table contains one character from A to Z and is unique to each other. The characters are placed sequentially. However, the first set of the cell contains the character of the key used in the encryption. The table below shows the table for playfair cipher with the key "EXAMPLE". It is also important to notice that since there are 25 cell in the table, there will be one missing character on the table. In the table below, we assume the character 'i' and 'j' is interchangeable.

**Table 1.** Example of Playfair table with Key "EXAMPLE"

| | | | | |
|---|---|---|---|---|
| E | X | A | M | P |
| L | B | C | D | F |
| G | H | I | K | N |
| O | Q | R | S | T |
| U | V | W | Y | Z |

The message is then split into digraph and replaced by the character on the table by following these rules:

1. If both letters are the same, add a predefined character, eg. "X" after the first letter and encrypt the new pair.
2. If the letters appear on the same row of the table, replace them with the letters to their immediate right respectively.
3. If the letters appear on the same column, replace them with the letter immediately below respectively.
4. If the letters are not on the same row or column, replace them with the letters on the same roe respectively but at the other pair of corners of the rectangle defined by the original pair. Notice that in this step, the order is important.

However, the behavior of playfair chiper that does not allow a double letter diagram eg. EE, makes it easy to detect by using cryptanalysis. Moreoever, playfair digraph and its reverse will decrypt the same letter pattern in the plain text. This is a problem especially in English where reversed digraph (eg. RE and ER, DE and ED) is common (eg. REceivER, DEpartED). The use of double playfair cipher, or usually called two-square cipher, solves this problem. The idea is simple; to use two tables and two keys instead of one. In addition, the substitution rules are modified as follow:

1. For a vertical two-square, the first character of both plaintext and cipher text uses the top matrix, while the second use the bottom matrix.
2. For a horizontal two-square, the first character of both digraphs uses the left matrix, while the second character uses the right.
3. Find the first letter in the digraph in the upper/left text matrix.

**4. Implementation**

As previously stated, we will simulate the encryption process by building a stand alone desktop application. To simplify the process, this process is done in a single computer; data communication will be simulated by creating files. The application is built using C# and run under Windows Operating System. To simplify the process, we also use a small digit prime number as key for asymmetric encryption. While this may vague the running time, our concern is only to see the usability of the algorithm. Figure 1 shows the process of encryption that simulated in our program. There are two role of user in the program: sender and receiver. The process begins when the sender sends a request to send an encrypted message. The receiver will generate a public/private key using ElGamal algorithm. The receiver sends the public key to the sender. The sender will generate a 2 key for the symmetric algorithm (as required by the double playfair algorithm). The sender encrypts the message using these 2 keys, and encrypt the symmetric keys using receiver's public key.

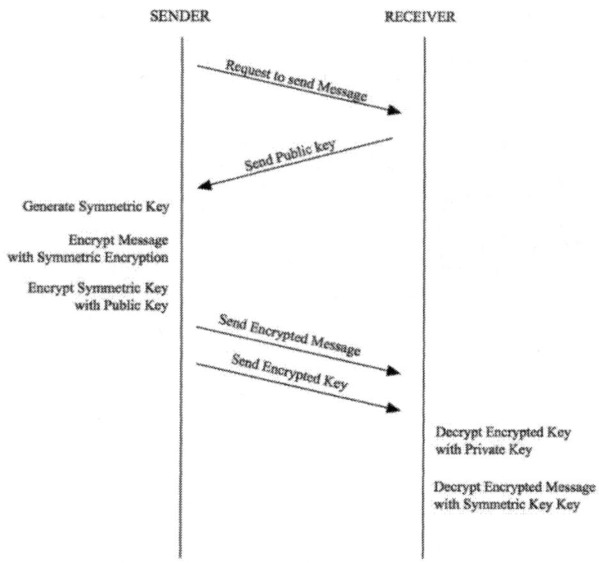


**Figure 1**. Hybrid Cryptosystem Simulated in the Program

Figure 2 below shows the first step of the encryption process. User is acting as a sender and required to enter 3 values as the parameter of the public and private key. The public key $(P_A, \alpha_A, \beta_A)$ will be kept saved as a file and loaded later upon decryption.

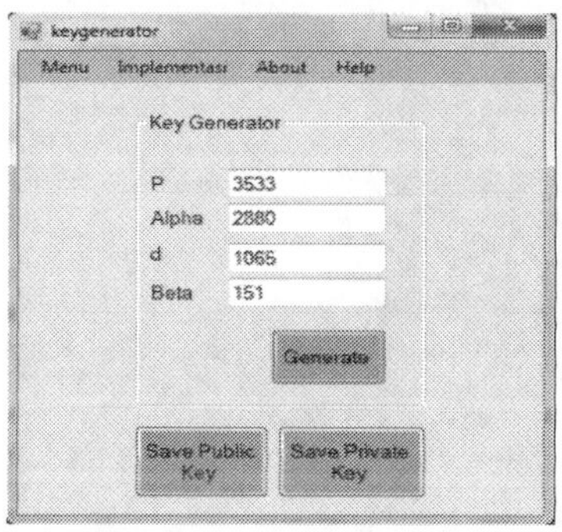


**Figure 2**. Input Form for Asymmetric Generator

In the next form, user will be asked to choose the file that will be encrypted. User will also need to enter two keys for the double play fair algorithm. When all the required fields are done, the program will encrypt the image and save it as an encrypted file format. Figure 3 shows the encryption form described previously.

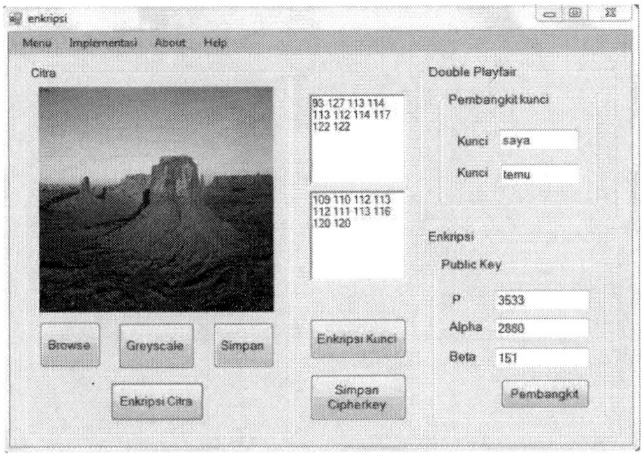


**Figure 3.** Input Form for Encryption Process

Acting as a receiver, the user will be presented with a different form, as shown as figure 4. Unlike the previous one, in this form the user does not have to enter the symmetric key. However, the user have to choose the file that contains the encrypted key and enter the receiver's private key in the input form. If the encryption is successfully performed, the same image will be rendered on the left part of the form as shown as the figure below.

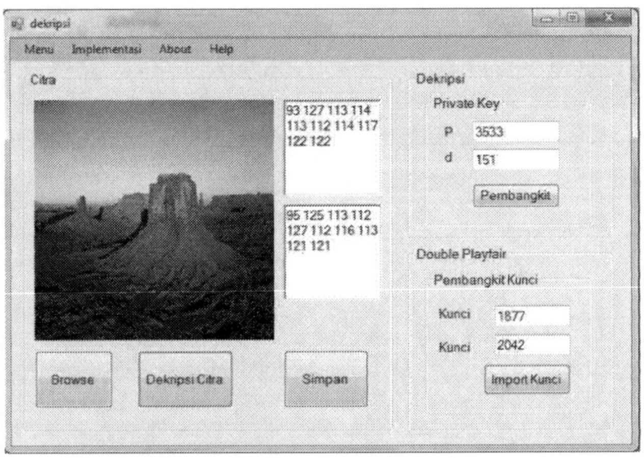


**Figure 4.** Input Form for Decryption Process

It is also important to know that the encrypted file size is significantly larger than the original file. Our tests show the size is increased averagely 6 to 8 times. As an example, an image file with a resolution 400x400 is 105 KB before the encryption and 625 KB after the decryption. This is however acceptable considering the image is rendered per pixel.

## 5. Conclusion and Future Work

We have successfully build a cryptosystem application that allows the use ElGamal and Double Playfair as the encryption algorithms. The implementation shows that the system has successfully perform an encryption and decryption process when the correct input has been entered. However, due to the nature of playfair as a classic encryption, it is better to perform the symmetric encryption using a more advanced symmetric algorithm such as AES.

IOP Conf. Series: Journal of Physics: Conf. Series **978** (2018) 012068 doi:10.1088/1742-6596/978/1/012068

**Acknowledgments**
The authors gratefully thank to all who have supported in the completion of this research. And to the Universitas Sumatera Utara that has given me the opportunity to publish this paper.

**References**
[1] Diffie W and Hellman M 1976 New Directions in Cryptography *IEEE Transactions on Information Theory* **22** 644
[2] Zimmermann P R 1996 The Official PGP Users's Guide (Massachusetts Ave: The MIT Press)
[3] Callas J, Donnerhacke L, Finney H, Shaw D and Thayer R 2007 OpenPGP Message Format https://tools.ietf.org/html/rfc4880
[4] NIST 2001 Advanced Encryption Standard *Federal Information Processing Standard Publications 197* http://nvlpubs.nist.gov/nistpubs/FIPS/NIST.FIPS.197.pdf
[5] Rivest R L, Shamir A and Adleman L 1977 A Method for Obtaining Digital Signatures and Public-Key Cryptosystems (Massachusetts Ave: USA) http://people.csail.mit.edu/rivest/Rsapaper.pdf
[6] Lai X 1992 On the Design and Security of Block Ciphers (Zurich: Hartung-Gorre)
[7] Dierks T and Rescorla E 2008 The Transport Layer Security (TLS) Protocol Version 1.2 https://tools.ietf.org/html/rfc5246
[8] Freier A, Karlton P and Kocher P 2011 The Secure Sockets Layer (SSL) Protocol Version 3.0 https://tools.ietf.org/html/rfc6101
[9] Mavrogiannopoulus N 2007 Using OpenPGP Keys for Transport Layer Security (TLS) Authentication https://tools.ietf.org/html/rfc5081
[10] Hazra T K, Mahato A, Mandal A and Chakraborty A K 2017 A Hybrid Cryptosystem of Image and Text Files Using Blowfish and Diffie-Hellman Techniques *Proc. Industrial Automation and Electromechanical Engineering Conf.* (Bangkok: Thailand) p 137
[11] Dahake P and Nimbhorkar S 2015 Hybrid Cryptosystem for Maintaining Image Integrity Using Biometric Fingerprint *Proc. Int. Conf. on Pervasive Computing* (Pune: India) p 1031

# Openlobby: an open game server for lobby and matchmaking

**E M Zamzami, J T Tarigan, I Jaya and S M Hardi**

Faculty of Computer Science and Information Technology,
Universitas Sumatera Utara

elvi_zamzami@usu.ac.id

**Abstract.** Online Multiplayer is one of the most essential feature in modern games. However, while developing a multiplayer feature can be done with a simple computer networking programming, creating a balanced multiplayer session requires more player management components such as game lobby and matchmaking system. Our objective is to develop OpenLobby, a server that available to be used by other developers to support their multiplayer application. The proposed system acts as a lobby and matchmaker where queueing players will be matched to other player according to a certain criteria defined by developer. The solution provides an application programing interface that can be used by developer to interact with the server. For testing purpose, we developed a game that uses the server as their multiplayer server.

## 1. Introduction

Modern video games industry is flooded with games with online multiplayer feature. The idea of multiplayer computer games has been around since 1961. The first real computer game that allows multiple player to play in the same session is Spacewar, an MIT product used to demonstrate a new PDP-1 computer. The game allows two players to shoot torpedo at one another. In the early 70s, University of Illinois' ILLIAC build a computer system called PLATO. This multi purpose computer has a server and running on a nearly a dozen computers. The system also contains two games; Empire and Airfight, which are considered as the first computer games to allow multiplayer game via computer network. One of the significant milestone in computer network games is DOOM in 1993. It features a four player cooperative mode using IPX protocol from Novell [1]. It uses peer-to-peer network architecture and can be used in via LAN or modem connection [2].

While multiplayer feature is considered capable to lengthen the lifespan of a game by allowing player to interact to other players, not all in-game interaction can enhance the game quality. Player satisfaction is highly depending on other players. Since it is impossible to fully control in-game interaction, multiplayer game usually group players according to their properties. Most games uses experience (or a value equivalent to it) to group players based on their skill. Other game requires a more complex matchmaking formula that uses additional properties such as region, latency, kill/death ratio, or even play style. To meet this need, matching players has become a complicated problem of how to match a group of available player to ensure the quality of a session. Instead of directly matching available players, most games use a lobby and a matchmaker algorithm to group optimal players. Lobby is where the players meet and wait to be matched with other player. Dhupelia et al. gives a thourough architecture of game lobby and its properties in common modern games [5].

As the player waiting in the lobby, a matchmaking system will work to find the best group of player based on some criteria. This matchmaking problem is one of the most discussed topic in game networking. There are a few researches that focus on developing an optimal matchmaking system.

IOP Conf. Series: Journal of Physics: Conf. Series **978** (2018) 012069 doi:10.1088/1742-6596/978/1/012069

Agarwal et al. developed Htrae, a latency prediction system that can be used in matchmaking application [4]. A similar system was proposed by Manweiler et al. with an application called Switchboard [7]. Similar to Htrae, this application is a matchmaking system for mobile games that heavily rely on latency between game player. Armitage et al. [6] proposed REED, a matchmaking system focused on optimizing host selection in a client-server model multiplayer game session. Another interesting research by Delalleau et al. shows how matchmaking process can be optimized to increase game quality by applying neural network to find the optimal team amongst selected player [3]. The system proposed uses a more complex data set to calculate a player's attribute such as number of matches played, kill/death ratio, firing accuracy, and headshot percentage.

However, developing an online multiplayer feature in a game is not a trivia task. It requires a specific skills and tools to develop the feature. While large game developers already have enough resources to develop this feature, (network programmer, game/database server), small game developers might not be able to have these privileges. Common game engines such as Unity and Unreal has also provide a higher level API to decrease the complexity in developing a multiplayer feature. However, this feature still requires networking programming skill and developed specifically to a certain game engine.

In this paper, we propose a solution that able to help small developers to add an online multiplayer feature in their product. We will start by discussing the architecture of the system in the next section. Section 3 will discuss the implementation of the system and the specification of hardware and software used to develop and run the system. In section 4, we will discuss the testing process which involves developing a multiplayer game application using the OpenLobby API. We conclude our work in section 5 and present the future work regarding the system.

## 2. Application Architecture

The overall architecture of our server is as follow:

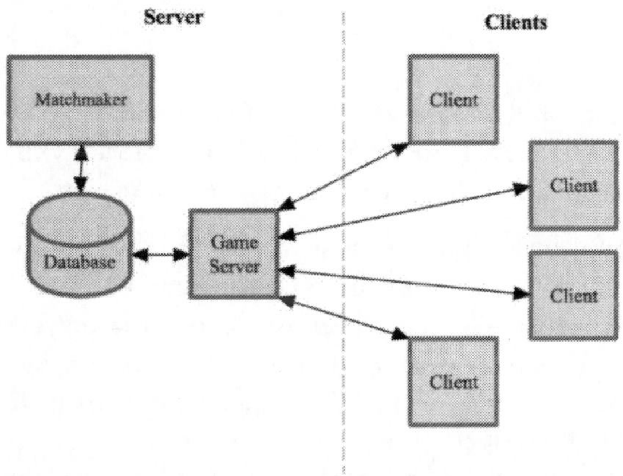


**Figure 1.** OpenLobby Application Architecture

The server contains 3 important items; web server, database, and daemon program. The web server is responsible in receiving and processing request from client and sending the appropriate response to client. Currently, our server is able to process basic functions of a game server. The first basic function of the server is to register a new player who would like to join a multiplayer game session. When a client is requesting to join a multiplayer game, it send a request to join to the server. The server is then kept the basic information related to the request such as request arrival time and client's IP address. The second basic function is to inform the client whether the client has been joined to a multiplayer session. Our system follows a pull-notification method to perform this function; clients send a request to get the update from the server. The third function is to update the player's information related to the game such as score, win/lose, rank, etc. This information is sent after a multiplayer session has completed.

In designing the database server, we have to consider the various requirement of games that may connect to our server. Different games have different requirements and criteria to create an ideal session. Most games, if not all, rely on player's rank and experience to define an ideal set of players. Some games use additional player's properties such as region, latency, and preference. To provide the different requirements of games, we have to develop a flexible concept to store player's properties. We can not define one player's property as a single column since each game may require different properties. We came up with a simple solution to store various properties; a text based values delimited with a specific unusable character. Hence, we only need a single column to store all properties defined by the game. The figure below shows the overall architecture of our database design:

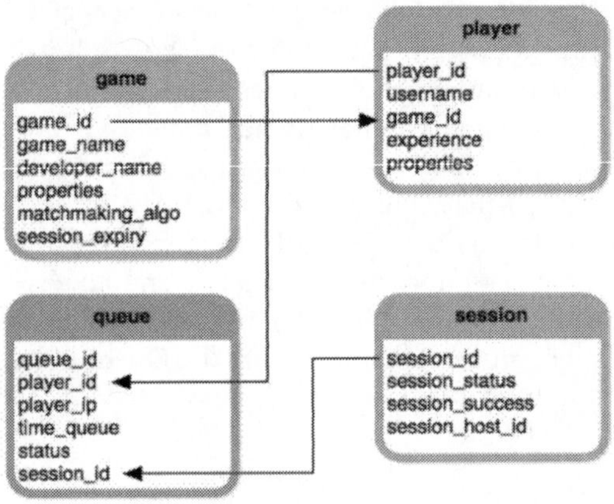


**Figure 2.** OpenLobby Database Desing

## 3. Implementation

In this section, we will go through the implementation of the architecture previously described. The web server is built using PHP. It uses HTTP protocol to communicate with the clients. In communicating with the server, application may use this 5 methods:

1. `queue`: this function is called when a user wants to be listed as queueing in a game queue. It requires player ID and game ID as its parameter. If the game ID is not registered in the system, the request will be rejected. Otherwise, the server will check the player's information based on the player ID. If the player's ID is not listed on the game database, the system will register the player ID with default experience. After both player and game are confirmed to be registered, the server will return the queue id to the player.
2. `get_status`: this function is called by user to check its queue status. The server is then return the user's status whether its waiting, found, or cancelled. When the user has been assigned to a session, the response will contain the host identification. Additionally, if a player has already queued for 5 minutes, the queue will be dropped and the player has to resend the request.
3. `get_session_info`: this function is called by user to find information regarding the session it is in. The information sent contains the host IP address, `session_id`.
4. `session_started`: this function is called by the host right after a session is started.
5. `update_result`: this function is called by the host after a session is over. This function will update the player's information based on the result of the session. Host can only update a session.
6. `remove_queue`: this function is called when a user wants to be removed from a queue. This action can not be performed if a player has already been matched to a session.

IOP Conf. Series: Journal of Physics: Conf. Series **978** (2018) 012069 doi:10.1088/1742-6596/978/1/012069

## 4. Test and Result

To test the usability of the server, we build a game that uses OpenLobby as player management in its multiplayer feature. The game developed on Android Platform and built using Java. The game is a simple math-based quiz game, Cogitare. In this game, two players will compete to answer a math question by pressing a larger number. In each session, both players will be given a limited amount of time to answer a series of questions. Each question will give an amount of point based on its difficulty. The player with more points win the session. The game also implements an adaptive difficulty method which makes the question becomes harder if the player able to answer an amount of correct answers consecutively. Harder question will grant more points. Figure 4 shows the screenshot of our game Cogitare that uses the OpenLobby to add lobby and matchmaking. To increase the quality of each session, the game also requires players with similar experience to compete in a session. The game holds two value as player's properties. The first property, experience, is linear to the amount of game played by the user. Each game completed by the user is calculated 10 experience point. The experience point is tripled if the player wins the game. The second property is correct/incorrect ratio which holds the comparison between correct and incorrect ratio of a player. However, to calculate the ratio, we need to keep two values; the amount of correct and incorrect value.

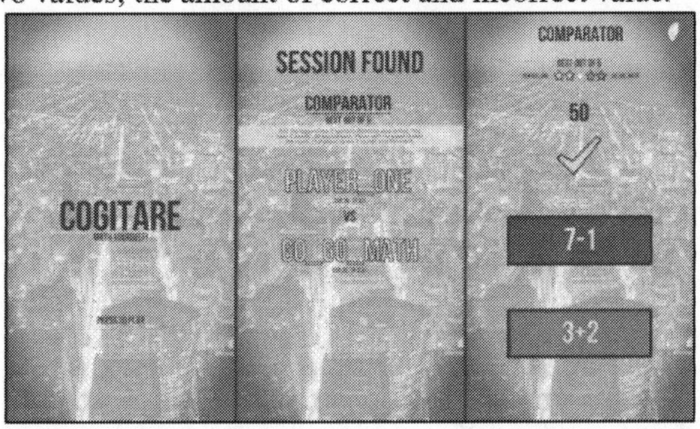


Figure 3. Screenshot of Cogitare: Menu (left), Session Found after matchmaking process (middle), and in-game (right)

The multiplayer session begins when a player sends a queue request to join a multiplayer game. The server will store the queue in the system to be processed later by the matchmaker. Every 5 second, the game checks the status of the queue; should a session has been found, it may request more information regarding its session such as players' id, name, and properties. The server will also decide the host's of each session and send the host's IP to all players. If the player has been chosen as a host, it will wait for other player to communicate with the host. When both players are ready, the multiplayer session starts. When the session is over, the host is responsible to calculate properties changes of both players and pass the information to the server. The server stores the adjusted data to the database and the OpenLobby's protocol ended.

## 5. Conclusion and Future Works

We have successfully build OpenLobby, an open player management service for games regarding their platform. The server acts as a lobby and matchmaking service and provide an API for communicating with games via HTTP Request. The server also provides player's properties database to keep in track of player's progress. To test the functionality of the server, we developed a game that uses OpenLobby's service.

While the system already works as intended, it does not yet applicable for commercial use due to lack of some important elements such as players and games registration process. Another essential feature that should be implemented in the near future is the security of communication process. Currently, we do not have any security feature implemented in our games. Additionally, add the social experience in the server, we would also like to implement leaderboards and friend-networking features in the next phase of the development.

**References**

[1] Novell *2001 Novell NetWare 6: Internet Packet Exchange* www.novell.com/documentation.

[2] Armitage G, Claypool M and Branch P 2006 *Networking and Online Games* (England: John Wiley & Sons, Ltd.) p 12

[3] Delalleau O, Contal E, Thibodeau-Laufer E, Ferrari R C, Bengio Y and Zhang F 2012 Beyond Skill Rating: Advanced Matchmaking in Ghost Recon Online *IEEE Transactions on Computational Intelligence and AI in Games* **4** p 167-177

[4] Agarwal S and Lorch J R 2009 Matchmaking for Online Games and Other Latency-Sensitive P2P Systems *Proceedings of SIGCOMM 2009* (Barcelona: Spain) p 315-326.

[5] Dhupelia S and Kirmse A 2004 General Lobby Design and Development *Game Programming Gems 4* (Charles River Media, Inc.) p 544.

[6] Armitage G and Heyde A 2012 REED: Optimizing First Person Shooter Game Server Discovery using Network Coordinates *ACM Transactions on Multimedia Computing, Communications, and Applications* **8**

[7] Manweiler J, Agarwal S, Zhang M, Choudhury R R, and Bahl P 2011 Switchboard: A Matchmaking System for Multiplayer Mobile *Proceeding of MobiSys 2011* p 71-84

[8] Ducheneaut N and Moore R J 2004 The Social Side of Gaming: A Study of Interaction Patterns in a Massively Multiplayer Online Game *Proc. of CSCW 2004* (New Orleans, LA: USA) p 360-369.

[9] Riegelsberger J, Counts S, Farnham S D and Philips B C 2007 Personality Matters: Incorporating Detailed User Attributes and Preferences into Matchmaking Process *Proc. of the 40th Hawaii Int. Conf. on System Sciences* (Waikoloa, Big Island, HI: USA) p 1-10.

[10] Siitonen M 2009 Exploring the Experiences Concerning Leadership Communication in Online Gaming Groups *Proceedings of MindTrek 2009* p. 90-93.

[11] Fristch T, Voigt B, Schiller J 2007 The next Generation of Competitive Online Game Organization *Proc. of Netgames* (Melbourne: ACM).

[12] Elo A E 2008 *The Rating of Chess Players, Past and Present* Ishi Press.

[13] Unity Technologies 2017 Matchmakers, Unity Documentation https://docs.unity3d.com/520/Documentation/Manual/UNetMatchMaker.html, 2015, retrieved October 25, 2017.

[14] Minka T and Zaykov Y 2005 *TrueSkill Ranking System* Microsoft Research, https://www.microsoft.com/en-us/research/project/trueskill-ranking-system/.

[15] Wang H, Yang H and Sun C 2015 Thinking Style and Team Competition Game Performance and Enjoyment *IEEE Transactions on Computational Intelligence and AI in Games* **7** p 243-254.

[16] Achterbosch L, Pierce R and Simmons G 2008 Massively Multiplayer Online Role-Playing Game: The Past, Present, and Future *ACM Computers in Entertainment* **5**.

IOP Conf. Series: Journal of Physics: Conf. Series **978** (2018) 012070 doi:10.1088/1742-6596/978/1/012070

# Design and architecture of retailapp: an application to support conventional retailers

**I Jaya, J T Tarigan, S M Hardi and M K M Nasution**

Faculty of Computer Science and Information Technology,
Universitas Sumatera Utara

ivanjaya@usu.ac.id

**Abstract.** Over the last few years, the preference on shopping over internet application, usually called online shopping or e-shopping, has increased significantly. One of the major advantage of online shopping is the capability to use digital content to support the marketing process. Prospective buyers can conveniently browse and look to a certain object and find the information needed with a few clicks. Moreover, the use of multimedia (such as images, sound, and movies) may help prospective buyers to make a decision faster than conventional shopping at retailers. However, shoppers still consider that conventional shopping is the best way to perform shopping due to the ability to make a physical contact to the object. Our objective is to merge these experiences by enabling user to find digital content that relate to nearby retailers. In this research, we built a mobile application that collects data from nearby retailers and shows it to its user. By using this application, it will be easier to users to find the location of the object and to find nearby promotion.

## 1. Introduction

### *1.1. Background*

In the past few years, there are tendencies that users prefer to perform shopping via online shop than conventional retailer and store. There are many factors that may trigger this tendency such as easier and lesser effort to perform the shopping and the capability to find more information regarding the desired product. User can easily search the desired product simply by using the search feature which is an obligatory feature in online shops. Moreover, user can easily find useful information regarding the product such as detailed specifications, reviews, and recommendations. Both of these features are unavailable in conventional shopping; all information regarding the product came from the seller who tend to give positive information to prospective buyer.

However, the inability to directly access the product can sometimes repel customer to perform online shopping. Some products, such as wearable items, may require direct contact to measure its quality. Moreover, easily expired product such as foods and drinks might never have a chance to be sold online. Hence, the role of conventional shopping still has an advantage against modern online shopping. To keep itself relevant, conventional store have to maximize these advantages and give a different shopping experience compared to online shopping to be able to attract modern shoppers. One of the many solutions available is to use computer and information technology to assist stores and retailers in creating a distinct experience.

In this paper, we present the design and architecture of RetailApp, an application that may help conventional store and retailer to add online shopping features to its shopping experience. The objective of this paper is to discuss the design of the proposed system and to analyze possible technologies that may be used to build the application in the future. While we understand that retailers contain massive and lengthy processes, our current system only covers marketing and customer experience. The paper will be structured as follows: the next section will present works from other researches that related to our system. Later, we will define the requirement of the system by analyzing the current state of conventional store and retailer. Based on these requirements, we will design the architecture of the system and explore possible technologies compatible to the proposed design. The last section will discuss the conclusion and future works of our research.

### *1.2. Related Works*

One of the major audit service provider in the world, KPMG, publish an article that discuss the global trends in retail industry. Among the top 5 trends listed in the article, the first two is directly related to the use of technology; the use of technology to increase the quality of customer experience (such as virtual reality and augmented reality) and the growth of mobile shopping [1]. The other 3, while not directly related to computer technology, also suggests the use of computer to create an advantage over competitors. Needless to say, the article suggest that the use of computer based system is vital in order to attract technology-aware customer. A similar suggestion is presented by Price Waterhouse Cooper, another major auditor service provider. It released a list of investment retailers can make to increase the competitiveness of its company [2]. Four of the 10 items in the list are related to computer based solution; mobile site, big data, secure platform, and amazon-based strategy. These two articles show the importance of conventional retailers and store to adopt information technology to survive the modern competition.

In accordance with the need of modern computer-based solutions to aid conventional retailers, some researches have proposed solutions related to this need. Adhikari et al. developed IntelligShop [3], an implementation of location based augmented reality that helps user to find information regarding a certain store. User can simply capture the front of a store using his/her smartphone camera and the application will search and show reviews related to the store. Rienzo et al. presents smart retailer environment [4], a concept that uses screens and RFID to increase the efficiency of marketing inside a conventional retailer shop. The system detects nearby prospective customer by using wireless technology such as bluetooth and warns nearby system to display a certain content based on user's profile. Mahoney et al. analyze the relation of retails location and the structure of the consumer community [5]. Early investigation shows that each community has their own characterization and it requires different approach and it is possible to gather this information using the data in social media. Meneweger et al. propose a solution to use mobile computer technology as sales assistants [6] to aid conventional human-based sales assistant in providing customer profile and products marketing. Karamshuk et al. developed Geo-Spotting, an application that may help in deciding store placement based on location-based data [7]. By using the data of customer attendance and purchase history, the system able to predict the optimal location of a store.

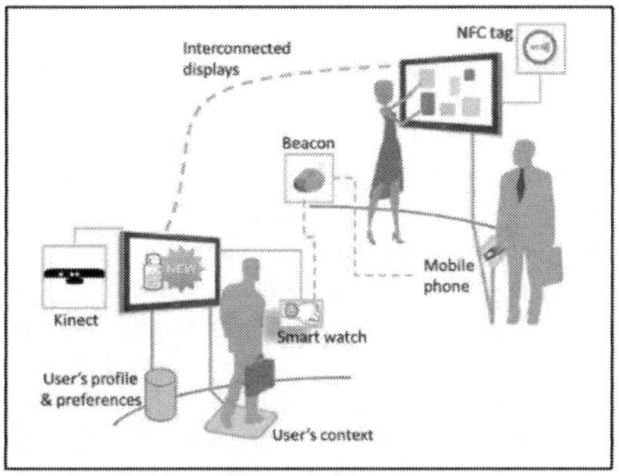


**Figure 1.** Smart Retailer Environment proposed by Rienzo et al. [4]

Some technologies proposed to help conventional retailers depend heavily on customer location. These researches suggest that the most prospective buyer is the one located nearby the store. Hence, it is important for conventional store to gather nearby user's profile and promote based on that information. In 2014, Apple proposed iBeacon [8], a system that detects nearby user by using Bluetooh Low Energy [9]. The system is directed to conventional store to create an on-the-go marketing technique to attract a certain customer. Moreover, the system allows store to send a notification to nearby prospective customers which gives a new method to perform a marketing.

## 2. System Requirements

Based on the previous background observation, we developed a list of requirements that can be used as a reference to build the architecture. This section will list the requirements of our future system capability.

### *2.1. Customer Localization*

The system depends heavily on customer localization. The localization is categorized into two process; macro-localization and micro-localization. The macro-localization is used to detect user's location based on its smartphone localization system (which mostly depends on information gathered from GPS and Wi-Fi Routers). By using this macro-localization, the server will gather information from nearby stores and send the information to the user.

The second localization is micro-localization, which is used to predict the exact location of the user. By using the exact location, the system gathers the user profile, characteristic, and behavior to be used by the store to create a more personalized marketing technique. Moreover, this information may assist store assistant to give a better service to the customers by giving a detailed and thorough information regarding a specific product. Store assistant ability to predict the user's preference may increase their efficiency by focusing on a certain case or knowledge that important to the user. For example, a user that visited a sport apparel store tend to have an interest in sport product. This information is important to other sport apparel stores since there is one or more prospective customer nearby.

### *2.2. Intelligent Shopping Assistant*

Another feature we would like to add to the system is AI powered shopping assistant. The purpose of this feature is to assist users to perform shopping such as suggesting similar product, gathering reviews, and planning shopping route. The use of artificial intelligence as shopping assistant is common in modern online shopping. Users often suggested products based on their recently viewed product or visited website. We would like to add similar experience to the proposed system. The system should be able to predict products suitable to customer based on three information:

IOP Conf. Series: Journal of Physics: Conf. Series **978** (2018) 012070 doi:10.1088/1742-6596/978/1/012070

1. *Purchased product*
   The information of recently purchased product is vital to predict the prospective product. Users tend to buy multiple similar products. Moreover, users tend to buy other product that may complement to the previous products. Users that purchased a computer tend to buy additional accessories such as display, mouse, and keyboard.
2. *Visited store*
   Visited store can give a slight information regarding user's purpose or interest. We argue that a user visiting a store without purchasing a product is currently looking for a better option of the product sold on the store. The system may help the user by providing similar products to the users.
3. *Shopping behavior*
   Users tend to have a repeat pattern of shopping. By using this information, the system must be able to predict the user's shopping list before they even perform the shopping. This information is also given to stores and retailers so that they may predict customer's behavior and create a more efficient strategy to attract prospective customers.

*2.3. Shopping Privacy*

While the system provides information regarding customers shopping behavior, the system must be able to kept the privacy of users. Some users prefer to keep their anonymity during the shopping process, yet others prefer a more personalize and efficient shopping experience. The system must allow the users to retrieve information from nearby stores without sending any information regarding to the users to the store. This security feature is important to keep the control of information in the hand of the customer.

## 3. Application Architecture

Our system is based on client-server architecture with both customer application and store application both act as clients. All data regarding customer information and store notification is stored in an online database server. To keep the communication protocol as simple as possible, we build the use HTTP for communication between the client and the server. Figure 2 below is the architecture of our system.

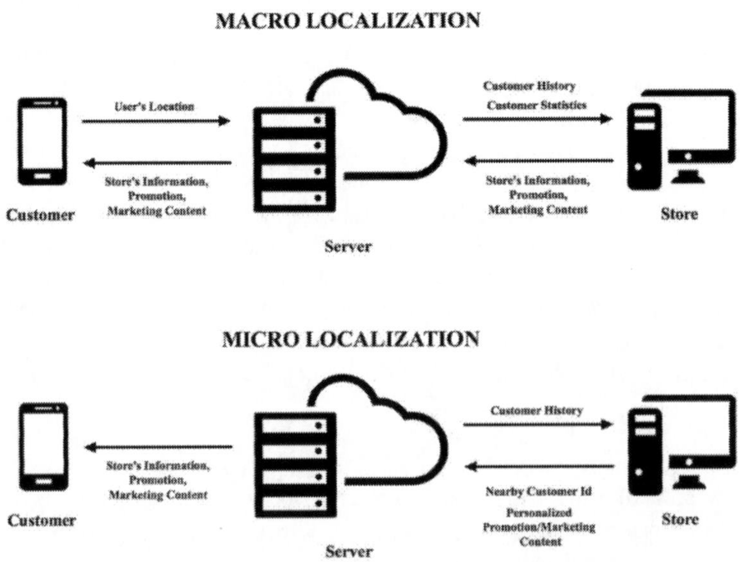


**Figure 2.** Data Flow Diagram for Macro-localization (top) and Micro-localization (bottom)

IOP Conf. Series: Journal of Physics: Conf. Series **978** (2018) 012070 doi:10.1088/1742-6596/978/1/012070

The system contains two sub-systems: the macro-localization which responsible in giving overall statistics of nearby customers to the store and nearby store information and promotion to the customer. The overall statistics contains a rough information regarding users demographic nearby the stores such as the amount of nearby customer and rough information regarding customer's average behavior. The micro-localization is responsible a detailed information regarding a customer to the store. The system detects nearby customer by using a short-ranged wireless system such as iBeacon or Bluetooh Low Energy technology and gather the customer's information from the server. This information is presented to the store owner, manager, or assistant so they can provide a personalized service to the customer.

## 4. Conclusion and Future Works

This paper presents the design and architecture of RetailersApp. It objectives is to assist conventional store to give a unique shopping experience in order to compete against online shop. We present the requirement and the thorough architecture of the system that can be used in future development process. The next step of this research is to build and implement the system in real life situation.

However, there are many things needed to be considered in the future regarding the feature of the system. We have yet not discussed the security aspect of the system which is important considering the system holds a detailed customer's shopping history and behavior. We also haven't design the implementation of the system, which requires intense coordination with retailers and stores.

## 5. Acknowledgement

The authors would like to thank the college teachers of Faculty of Computer Science and Information Technology Universitas Sumatera Utara, Indra Aulia and Dedy Arisandi for all the support during this work.

## References

[1] KMPG 2017 *Global Retail Trends* https://assets.kpmg.com/content/dam/kpmg/xx/pdf/2017/06/retail-trends.pdf

[2] PWC 2017 *10 Retailer Investments for an Uncertain Future* www.pwc.nl/en/publicaties/total-retail-2017.html

[3] Adhikari A, Zheng W V, Lin M, Fang Y, Cao H and Chang C K 2015 IntelligShop: Enabling Intelligent Shopping in Malls through Location-based Augmented Reality *Proc. Int. Conf. on Data Mining* (Atlantic City, NJ: USA) p 1604

[4] Rienzo D A, Fra C, Garzotto F, Valla M and Cremonesi P 2015 Towards a smart retail environment *Proc. Int. Joint Conf. on Pervasive and Ubiquitous Computing and Proc. Int. Symp. on Wearable Computers* (Osaka: Japan) p 779

[5] Mahoney J, Lawson S and Stone R 2014 "What do you think of the return of dungarees?": Social Media Interactions between Retail Locations and their Customers *Proc. Int. Conf. on Human Factors in Computing Systems* (Toronto, Ontario: Canada) p 1927

[6] Meneweger T, Zachhuber D, Wilfinger D, Tscheligi M and Asian I 2013 Towards the Counter Free Store: Requirements for Mobile Sales Assistants *ACM Conf. on Pervasive and Ubiquitous Computing Adjunct Publication* (Zurich: Switzerland) p 999

[7] Karamshuk D, Noulas A, Scellato S, Nicosia V and Mascolo C 2013 Geo-Spotting: Mining Online Location-based Services for Optimal Retail Store Placement *KDD*'13 (Chicago, Illionis: USA) p 793

[8] Chen BW and Ji W 2016 Intelligent Marketing in Smart Cities Crowdsourced Data for Geo-Conquesting *IT Professional* **18** p 18

[9] Aman S M, Jiang H, Quint C, Yelamarthi K and Abdelgawad A 2016 Reliability Evaluation of iBeacon for Micro-Localization *Ubiquitous Computing, Electronics & Mobile Communication Conf.* (New York: USA)

# Investigation of the ripeness of oil palm fresh fruit bunches using bio-speckle imaging

**R Salambue[1], A Adnan[2] and M Shiddiq[3]**

[1] Department of Information System, Universitas Riau
[2] Department of Statistic, Universitas Riau
[3] Department of Physics, Universitas Riau

roni.salambue@lecturer.unri.ac.id

**Abstract**. The ripeness of the oil palm Fresh Fruit Bunches (FFB) determines the yield of the oil produced. Traditionally there are two ways to determine FFB ripeness which are the number of loose fruits and the color changes. Nevertheless, one drawback of visual determination is subjective and qualitative judgment. In this study, the FFB ripeness was investigated using laser based image processing technique. The advantages of using this technique are non-destructive, simple and quantitative. The working principle of the investigation is that a FFB is inserted into a light tight box which contains a laser diode and a CMOS camera, the FFB is illuminated, and then an image is recorded. The FFB image recorder was performed on four FFB fractions i.e. F0, F3, F4 and F5 on the front and rear surfaces at three sections. The recorded images are speckled granules that have light intensity variation (bio-speckle imaging). The feature extracted from the specked image is the contrast value obtained from the average gray value intensity and the standard deviation. Based on the contrast values, the four fractions of FFB can be grouped into three levels of ripeness of unripe (F0), ripe (F3) and overripe (F4 and F5) on the front surface of base section of FFB by 75%.

## 1. Introduction

In the last few decades the problem of determining the maturity and ripeness level of oil palm fresh fruit bunches (FFB) has become a popular research topic in the palm oil industry. Indonesia, which is one of the largest Crude Palm Oil producing and exporting countries in the world, makes palm oil the main source of foreign exchange in Indonesia. Fruit maturity and ripeness determine the yield of oil produced [1], therefore the determination of FFB maturity and ripeness is one of the problems for farmers and oil palm companies in Indonesia. The determination of FFB ripeness before harvested or in sorting and grading process is still done manually or traditionally using the vision (human vision) by trained and experienced workers. Several studies on FFB ripeness have been conducted in the last decade, one of which is a review of several techniques for predicting maturity of oil palm fruit based on visual observation, oil content and image processing [2].

FFB ripeness is classified by several standards. In general, there are three categories of FFB ripeness those are unripe, ripe, and overripe. Traditionally there are two ways to determine the ripeness stages of FFB that are base on the number of fruits that detached and fell from a FFB and based on the color of FFB. However, one drawback of visual determination is subjective and qualitative judgment.

This research investigates FFB ripeness categories using a laser based image processing technique. In order to get the image, a FFB is inserted into a light tight box containing a 650 nm laser diodes and a CMOS camera. The diode laser illuminates the FFB and the FFB speckle image is recorded using the camera. The resulting image has a light and dark speckle pattern. In the field of biological science, the activity of biological matter such as a fruit which is irradiated by a laser beam is called Bio-speckle. Bio-speckle imaging and image processing techniques are potential combination to analyze and classified ripeness of oil palm FFBs.

## 2. Literature Review

The use of laser light in agriculture become intensive in last decade because it is non-destructive, simple, and fast. The optical method is potential to substitute or accompany laboratory test methods that are expensive, invasive, and time consuming. Laser Speckle Imaging (LSI) is an optical method that can be used to detect the level of ripeness and abnormalities in the fruit. This technique relies on the interaction of light and the fruit material and monochromatic light. When laser light illuminates the fruit surface, light can be absorbed, reflected, transmitted, and scattered by the surface of the material. Fruit has several physical properties which are represented by physical parameters such as density, porosity, surface roughness, diameter, and color. Light that concerns uneven surfaces will experience two types of reflections that are specular reflections and diffuse reflections. The reflected light beams will have different phases that will cause constructive and destructive interference resulting in some light and dark speckle granules. The method of speckling pattern caused by biological material such as fruit is called Bio-speckle. The bio-speckle phenomenon occurs when biological matter is irradiated by coherent light (laser light). The diffused light from surface roughness is optically recorded and produces the image with light and dark speckles of light [3].

There are two type of speckle pattern: static and dynamic. The static pattern is derived from the static part of the fruit network, while the dynamic pattern comes from particles or molecules that move in the tissues i.e in the process of maturity or ripeness. The dynamics of this speckle pattern can be statistically analyzed. This method is based on variations of laser light intensity or contrast dissipated by biological samples. Fruit is a complex element consisting of molecules of fruit tissue built of various sizes depending on the level of maturity and ripeness, These molecular sizes will affect the optical properties of the fruit associated with the skin's ability and the fruit tissue to absorb and scatter light so that it can be used to monitor fruit maturity levels [4].

The efforts of finding nondestructive methods for the detection of fruits for both pre- and post-harvest conditions have been widely practiced. The nondestructive method for estimating mango maturity uses color as one of the characteristics of maturity [5]. In addition, the use of image processing algorithm has also been developed for sorting and grading in fruits and vegetables [6].
A study of the determination of FFB maturity and ripeness has also been done and many methods have been developed. Identification of nondestructive maturity generally uses computer vision and image processing techniques. The characteristic of the extraction is the color of the image. The CCD camera recording is then processed using software to classify FFB maturity [7]. This technique provides classification accuracy between 83.5% - 94%. In addition, a grading system using a computer-based photogrammetric method, has been conducted in which the FFB is illuminated by a homogeneous light bulb in a box [8]. Furthermore, the ripeness category is determined from the Color Digital Number (DN). Automatic FFB grading systems have also been developed in estimating the ripeness fraction, oil content and free fatty acids of FFB using two statistical methods of forward stepwise multiple linear regression and multilayer-perceptron artificial neural network analysis. The estimation accuracy of both statistical methods is about 90%[9].

## 3. Material and Methodology

### *3.1 Tools and materials*

The tools used in this research are a diode laser, a CMOS sensor camera, biconcave lenses, image processing software, and data processing as well as other optical components. The sample material studied was FFB of varieties of Tenera (Lonsum clone). FFBs are selected by four maturity fractions (representing three levels of maturity) i.e. F0 (unripe), F3 (ripe), F4 and F5 (overripe). The tools and materials used are presented in the following table

**Table 1.** Tool and materials

| Tool and materials | Information |
|---|---|
| Oil palm FFB | Tenera Variety |
| Optical bench | Thorlabs |
| Diode laser 650 nm | Roithner |
| Monochrome CMOS camera | Imaging Source |
| Camera lens Kamera 1.4/12mm | Navitar |
| Biconcave lens f = -50 mm | Thorlabs |
| Laptop | Sony |
| IC Capture 2.4 | Imaging Source |
| Matlab | http://mathworks.com |

*3.2 Data*

The FFB samples used represented three levels of ripeness and were selected from trees under the age of eight years. The ripeness stages of harvested FFBs were determined by an experience harvester. The FFBs were then brought to Lab and cleaned. The measurements were completed in 24 hours after harvest. The image recording is performed on two FFB surfaces, the rear and front surfaces. The rear surface is the surface facing the palm tree while the front surface is facing sunlight. The surface of FFB is irradiated in three parts, namely the base, the middle and the tip, so that each FFB gets six specimen images. A specimen image of each FFB was analyzed using MATLAB software to obtain average gray value intensities and standard deviation values on all parts of each surface. Both values are computed to obtain the value of the contrast that represents the ripeness level of each FFB.

*3.3 Method*

To observe the bio-speckle activity at three levels of FFB ripeness, we have calculated the speckle modulation pattern expressed as contrast in the FFB image. The contrast equation is as follows:

$$C = \frac{\sigma I}{\langle I \rangle} \tag{1}$$

$$I = \frac{\Sigma f(x,y)}{n} \tag{2}$$

$$\sigma I = \sqrt{\frac{\Sigma (f(x,y) - I)^2}{n-1}} \tag{3}$$

where

| | |
|---|---|
| $C$ | = contrast |
| $\sigma I$ | = standard deviation |
| $I$ | = average of grey value intensity |
| $f(x, y)$ | = grey value on spatial coordinates |
| $n$ | = the number of spatial coordinates |

The average gray value intensity shows the intensity distribution of the gray level, the standard deviation shows the intensity of gray and the contrast intensity levels represents variations in gray level intensity distribution. Equation (1) becomes the characteristic of extraction algorithm in this study.

## 4. Results and Discussion

Image of speckled FFB captured uses a monochrome CMOS camera with 650 nm diode laser wavelength. Supporting components are camera lens, biconcave lens -50 mm. The camera lens is used to adjust the focus and camera aperture and bi-concave lenses are used to spread the laser beam. Biconcave lens arrangement focusehe laser beam only on 2-5 pieces on FFB fruits because illuminating the entire surfaces of FFB will produce dark dominant speckles. FFS objects are irradiated alternately in three parts of the base, middle and ends of the front and back surfaces of FFB. Each fraction is represented by three FFB, so that for each part on the surface 12 images are produced. CMOS cameras are installed as far as 44.3 cm and perpendicular to FFB. The laser distance to FFB is 58.6 cm, and the distance of the laser to the camera is 17.5 cm. The distances between these components is kept constant, so the width of the beam of laser beams on the surface of the FFB remains constant. The system is built in black box with size 60 cm x 60 cm x 90 cm. The black box function is to eliminate outside light interference during the recording process. Illustration of illumination and recording of FFB and FFB biospeckle results is presented in the following figure.

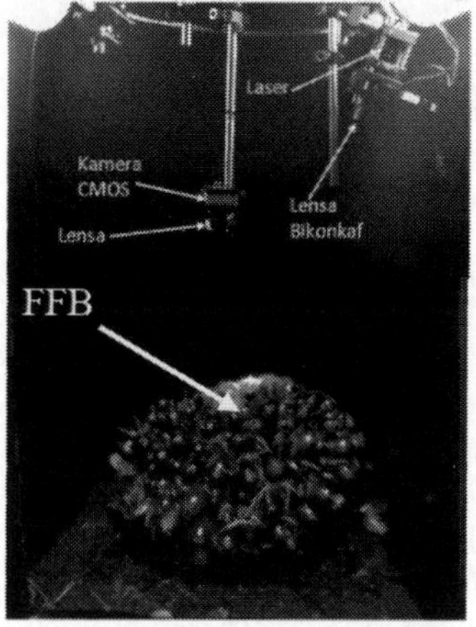


**Figure 1.** Radiation system and bio-speckle image recording

Image recording is done experimentally at different levels of FFB ripeness. The sample used was four fractions of FFB with three variations of ripeness level i.e. F0 (unripe), F3 (ripe), F4 and F5 (overripe). The speckled data retrieval is taken from the front and back surfaces of FFB. The fruits on the back surface of FFB have a variety of different shape and size. The backbone FFB base has many spikelet and many small fruits. Rear surface fruits have fewer gaps than the front surface. The front surface of FFB has uniform fruits on all parts. The front surface FFB tip has a smaller size than the rear end FFB tip.

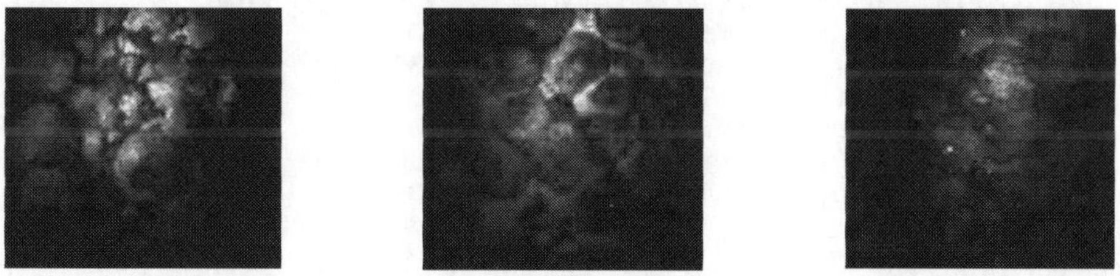

(a) (b) (c)

**Figure 2.** The pattern of the frontal specimen of FFB fraction F4 (a) base (b) middle (c) top

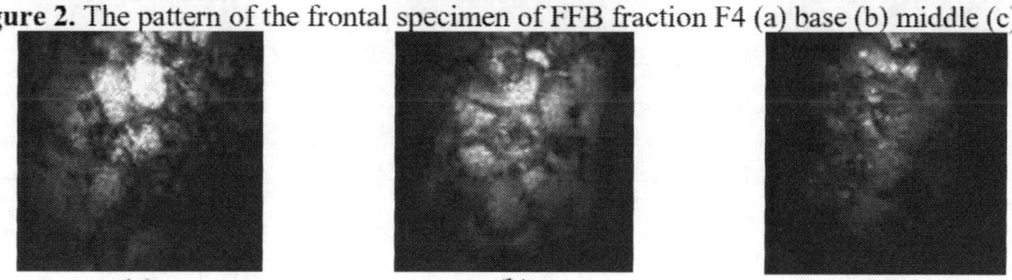

(a) (b) (c)

**Figure 3.** The pattern of the backside FFB specimen fraction F4 (a) base (b) middle (c) top

IOP Conf. Series: Journal of Physics: Conf. Series **978** (2018) 012071 doi:10.1088/1742-6596/978/1/012071

Speckle image data processing is done using MATLAB. Via this software, the average features of gray value and standard deviation intensity are extracted using equations (2) and (3). Then both values are entered into equation (1) to obtain a contrast value. Based on the observation of contrast value on 72 image data, it can be determined that the contrast value for F0 ≤ 1.4, F3 ≤ 1.5, F4 and F5 > 1.5. The results of the feature extraction, computation and category of FFB speckle patterns are presented in the following table.

**Table 2.** The results of feature extraction and category of front surface FFB

| No | Label | $(I)$ | $(\sigma I)$ | $(C)$ | Category | Results | Remarks |
|---|---|---|---|---|---|---|---|
| 1 | F0S1 DP | 9,2335 | 11,7414 | 1,2716 | **F0** | **True** | DP: Front surface of base section |
| 2 | F0S2 DP | 9,1673 | 12,3662 | 1,3489 | **F0** | **True** | |
| 3 | F0S3 DP | 7,6634 | 10,5114 | 1,3716 | **F0** | **True** | |
| 4 | F3S1 DP | 11,2172 | 16,1706 | 1,4416 | **F3** | **True** | |
| 5 | F3S2 DP | 11,1350 | 16,0150 | 1,4383 | **F3** | **True** | |
| 6 | F3S3 DP | 10,7062 | 15,9368 | 1,4886 | **F3** | **True** | |
| 7 | F4S1 DP | 16,0704 | 24,4794 | 1,5233 | **F4/F5** | **True** | |
| 8 | F4S2 DP | 12,0800 | 17,3159 | 1,4334 | F3 | False | |
| 9 | F4S3 DP | 9,8844 | 13,1788 | 1,3333 | F0 | False | |
| 10 | F5S1 DP | 12,0197 | 18,4556 | 1,5354 | **F4/F5** | **True** | |
| 11 | F5S2 DP | 12,0103 | 17,2111 | 1,4330 | F3 | False | |
| 12 | F5S3 DP | 17,3642 | 27,2451 | 1,5690 | **F4/F5** | **True** | |
| 13 | F0S1 DT | 7,9177 | 10,8452 | 1,3697 | **F0** | **True** | DT: Front surface of middle section |
| 14 | F0S2 DT | 7,6918 | 10,9207 | 1,4198 | F3 | False | |
| 15 | F0S3 DT | 7,3038 | 10,5891 | 1,4498 | F3 | False | |
| 16 | F3S1 DT | 11,8859 | 15,3337 | 1,2901 | F0 | False | |
| 17 | F3S2 DT | 9,8199 | 12,8930 | 1,3129 | F0 | False | |
| 18 | F3S3 DT | 10,9066 | 15,5500 | 1,4257 | **F3** | **True** | |
| 19 | F4S1 DT | 15,7896 | 21,9578 | 1,3907 | F3 | False | |
| 20 | F4S2 DT | 12,9032 | 18,9321 | 1,4672 | F3 | False | |
| 21 | F4S3 DT | 9,2836 | 14,1572 | 1,5250 | **F4/F5** | **True** | |
| 22 | F5S1 DT | 9,7572 | 14,0671 | 1,4417 | F3 | False | |
| 23 | F5S2 DT | 12,3533 | 18,9965 | 1,5378 | **F4/F5** | **True** | |
| 24 | F5S3 DT | 10,9792 | 17,1891 | 1,5656 | **F4/F5** | **True** | |
| 25 | F0S1 DU | 7,8667 | 11,6992 | 1,4872 | F3 | False | DU: Front surface of top section |
| 26 | F0S2 DU | 7,5373 | 11,1222 | 1,4756 | F4 | False | |
| 27 | F0S3 DU | 7,4140 | 11,1486 | 1,5037 | F4/F5 | False | |
| 28 | F3S1 DU | 13,8776 | 19,4875 | 1,4042 | **F3** | **True** | |
| 29 | F3S2 DU | 10,0506 | 13,5783 | 1,3510 | F0 | False | |
| 30 | F3S3 DU | 8,9027 | 12,2714 | 1,3784 | **F3** | **True** | |
| 31 | F4S1 DU | 12,0523 | 16,7271 | 1,3879 | F3 | False | |
| 32 | F4S2 DU | 11,2896 | 15,5545 | 1,3778 | F3 | False | |
| 33 | F4S3 DU | 8,8804 | 13,8123 | 1,5554 | **F4/F5** | **True** | |
| 34 | F5S1 DU | 8,2798 | 11,6025 | 1,4013 | F3 | False | |
| 35 | F5S2 DU | 8,9208 | 12,6784 | 1,4212 | F3 | False | |
| 36 | F5S3 DU | 8,4947 | 12,3973 | 1,4594 | F3 | False | |

**Table 3.** The results of feature extraction and category of back surface FFB

| No | Label | $(I)$ | $(\sigma I)$ | $(C)$ | Category | Results | Remarks |
|---|---|---|---|---|---|---|---|
| 1 | F0S1 BP | 17,8477 | 27,6307 | 1,5481 | F4/F5 | False | BP: Rear surface of base section |
| 2 | F0S2 BP | 20,6675 | 29,5511 | 1,4298 | F3 | False | |
| 3 | F0S3 BP | 15,7328 | 26,0196 | 1,6539 | F4/F5 | False | |
| 4 | F3S1 BP | 18,8325 | 28,3307 | 1,5043 | F4/F5 | False | |
| 5 | F3S2 BP | 15,0107 | 21,5276 | 1,4341 | **F3** | **True** | |
| 6 | F3S3 BP | 15,0268 | 22,9386 | 1,5265 | F4/F5 | False | |
| 7 | F4S1 BP | 22,7972 | 37,5626 | 1,6477 | **F4/F5** | **True** | |
| 8 | F4S2 BP | 22,2380 | 34,5376 | 1,5531 | **F4/F5** | **True** | |
| 9 | F4S3 BP | 15,0004 | 20,5883 | 1,3725 | F0 | False | |
| 10 | F5S1 BP | 20,2776 | 32,8684 | 1,6209 | **F4/F5** | **True** | |
| 11 | F5S2 BP | 16,9439 | 26,6742 | 1,5743 | **F4/F5** | **True** | |
| 12 | F5S3 BP | 24,0895 | 38,0983 | 1,5815 | **F4/F5** | **True** | |

| | | | | | | | |
|---|---|---|---|---|---|---|---|
| 13 | F0S1 BT | 14,1247 | 24,3208 | 1,7219 | F4/F5 | False | BT: Rear surface of middle section |
| 14 | F0S2 BT | 14,3170 | 22,0723 | 1,5417 | F4/F5 | False | |
| 15 | F0S3 BT | 8,2993 | 11,3388 | 1,3662 | **F0** | **True** | |
| 16 | F3S1 BT | 19,2050 | 28,7487 | 1,4969 | **F3** | **True** | |
| 17 | F3S2 BT | 13,2111 | 20,1834 | 1,5278 | F4/F5 | False | |
| 18 | F3S3 BT | 15,0398 | 20,6931 | 1,3759 | **F3** | **True** | |
| 19 | F4S1 BT | 21,7583 | 29,4354 | 1,3528 | F3 | False | |
| 20 | F4S2 BT | 17,2836 | 24,7171 | 1,4301 | F3 | False | |
| 21 | F4S3 BT | 12,0697 | 16,0695 | 1,3314 | F0 | False | |
| 22 | F5S1 BT | 12,9132 | 19,7892 | 1,5325 | **F4/F5** | **True** | |
| 23 | F5S2 BT | 13,3411 | 21,3701 | 1,6018 | **F4/F5** | **True** | |
| 24 | F5S3 BT | 14,4245 | 21,1381 | 1,4654 | F3 | False | |
| 25 | F0S1 BU | 8,2519 | 11,5079 | 1,3946 | F3 | False | BU: Rear surface of top section |
| 26 | F0S2 BU | 7,9346 | 12,8586 | 1,6206 | F4/F5 | False | |
| 27 | F0S3 BU | 7,6093 | 11,3591 | 1,4928 | F3 | False | |
| 28 | F3S1 BU | 9,5628 | 13,6629 | 1,4287 | **F3** | **True** | |
| 29 | F3S2 BU | 11,1023 | 15,4696 | 1,3934 | **F3** | **True** | |
| 30 | F3S3 BU | 9,7705 | 13,4804 | 1,3797 | **F3** | **True** | |
| 31 | F4S1 BU | 13,1232 | 19,3499 | 1,4745 | F3 | False | |
| 32 | F4S2 BU | 11,6777 | 17,0577 | 1,4607 | F3 | False | |
| 33 | F4S3 BU | 10,2687 | 14,3380 | 1,3963 | F3 | False | |
| 34 | F5S1 BU | 8,9288 | 12,9348 | 1,4487 | F3 | False | |
| 35 | F5S2 BU | 8,6688 | 12,7948 | 1,4760 | F3 | False | |
| 36 | F5S3 BU | 8,9799 | 13,6102 | 1,5156 | **F4/F5** | **True** | |

Based on the tables, the most category is found on the front surface of base section of FFB by 75%, which means that the correct image is classified is 9 of 12 images. On the middle section, the category is 42% (5 of 12) and for the top section of the category result is 25% (3 of 12). Tests performed on the back surface of base and middle section, the classification success rate is 50% (6 of 12) and at the top section of the classification results only 33.3% (4 of 12).

The highest classification results produced by the front surface of base section, because the front surface has a fruit similarity on all parts so that it has a higher scattering intensity than the other parts. This higher scattering intensity is due to a more homogeneous scattering surface than any other surface. However, the lowest classification results are produced on the front surface of top section because it has smaller fruit sizes than the rest so that small scattering areas produce low scattering intensity.

## 5. Conclusion

Based on the implementation of research conducted we may conclude that:

1. The highest category is obtained on the front surface of FFB at the base of 75% because the front surface has a fruit similarity on all parts so the scatter intensity is higher than the other parts.
2. The lowest category is obtained on the front surface of the ends of the FFB i.e. 25% because the size of the fruit in this section is smaller than the other parts resulting in low scattering intensity.

## References

[1] USAID 2009 Handbook of small-scale palm oil mills for the production of biofuel raw materials No. 497,.
[2] M A H and Razali S R H and Somad A 2012 A Review on Crop Plant Production and Ripeness Forecasting," *Int. J. Agric. Crop Sci.*, vol. **4**, no. 2, pp. 54–63.
[3] Zdunek A, Adamiak A and Pieczywek P M and Kurenda A 2013 The biospeckle method for the investigation of agricultural crops : A review *Opt. Lasers Eng.*, pp. 1–10.
[4] Ansari M Z and Nirala A K 2014 Assessment of fruits during shelf-life storage using biospeckle laser vol. **16**, no. 3, pp. 223–229.
[5] Slaughter D C 2009 Nondestructive Maturity Assessment Methods for Mango pp. 1–18.
[6] Mahendran R, Gc J and Alagusundaram K 2012 Application of Computer Vision Technique on Sorting and Grading of Fruits and Vegetables pp. 1–7.

[7] Fadilah N and Mohamad-Saleh J 2014 Color feature extraction of oil palm fresh fruit bunch image for ripeness classification *13th Int. Conf. Appl. Comput. Appl. Comput. Sci.*, pp. 51–55.

[8] Roseleena J, Nursuriati J, Ahmed J and Low C Y 2011 Assessment of palm oil fresh fruit bunches using photogrammetric grading system *Int. Food Res. J.*, vol. **18**, no. 3, pp. 999–1005.

[9] Makky M, Soni P and Salokhe V M 2014 Automatic Non-destructive Quality Inspection System for Oil Palm Fruits *Int. Agrophysics*, vol. **28**, no. 3, pp. 319–329.

IOP Conf. Series: Journal of Physics: Conf. Series **978** (2018) 012072 doi:10.1088/1742-6596/978/1/012072

# Molybdenum-99 production calculation analysis of SAMOP reactor based on thorium nitrate fuel

**Syarip[1], E Togatorop[2] and Yassar[2]**

[1] Center for Accelerator Science and Technology, National Nuclear Energy Agency
Jalan Babarsari, Yogyakarta, Indonesia
[2] Departements of Nuclear Engineering and Engineering Physics of Universitas Gajah Mada
Jalan Grafika 2 Yogyakarta 55281, Indonesia

Email : syarip@batan.go.id, estelita.togatorop@mail.ugm.ac.id, fadhil.yassar@mail.ugm.ac.id

**Abstract.** SAMOP (Subcritical Assembly for Molybdenum-99 Production) has the potential to use thorium as fuel to produce $^{99}Mo$ after modifying the design, but the production performance has not been discovered yet. A study needs to be done to obtain the correlation between $^{99}Mo$ production with the mixed fuel composition of uranium and with SAMOP power on the modified SAMOP design. The study aims to obtain the production of $^{99}Mo$ based thorium nitrate fuel on SAMOP's modified designs. Monte Carlo N-Particle eXtended (MCNPX) is required to simulate the operation of the assembly by varying the composition of the uranium-thorium nitrate mixed fuel, geometry and power fraction on the SAMOP modified designs. The burnup command on the MCNPX is used to confirm the $^{99}Mo$ production result. The assembly is simulated to operate for 6 days with subcritical neutron multiplication factor ($k_{eff}$ = 0.97-0.99). The neutron multiplication factor of the modified design ($k_{eff}$) is 0.97, the activity obtained from $^{99}Mo$ is 18.58 Ci at 1 kW power operation.

## 1. Introduction

Radioisotope $^{99}Mo$ is one of the most essential radioisotopes in the nuclear medical field. It is required as a parent radionuclide in a radioisotope generator system that produces $^{99m}Tc$ for diagnostic purposes. Radionuclides $^{99}Mo$ can be produced in several ways, i.e.: through a $^{235}U$ fission reaction fired with neutrons produced on the deuteron and proton accelerator reaction (D, n) and (p, n), through the activation of neutrons from $^{98}Mo$ (n, γ) $^{99}Mo$, through the process of photography, by utilizing $^{100}Mo$ isotope, which is $^{100}Mo$ (γ, n) $^{99}Mo$. [1]

The world's demand for $^{99}Mo$ is produced by some of the world's major reactors such as the National Research Universal (NRU) in Canada, High Flux Reactor (HFR) in the Netherlands, Belgian Reactor-2 (BR-2) in Belgium, South African Fundamental Atomic Research Installation-1 (Safari-1) in South Africa and most recently the Australian Open Water Pool Light Reactor (OPAL) in Australia [2]. However, the global crisis of $^{99}Mo$ has happened recently because of some reactors have been ceased. Two major reactors, HFR and SAFARI-1, were shut down in 2013-2014, this led to a major $^{99}Mo$ supply disruption. OSIRIS, a light-water-cooled research reactor located in France, was shut down in December 2015. The shutdown of BR-2 happened in March 2015 due to refurbishment, including the replacement of the Beryllium reflector. The operating license of NRU should be ended in 2016, but the Canadian government has announced that NRU will continue to operate until March 2018 in order to support the global demand crisis, but NRU will not be used for routine production of

$^{99}$Mo. This extended operating license was a settlement due to the unexpected circumstances of $^{99}$Mo shortage from 2016 to 2018. [3]

Based on those facts, new research toward reactors continue to develop to meet the world's needs for $^{99}$Mo. Nearly $^{99}$Mo of total production with reactor using HEU (High Enriched Uranium) as target. The development of other alternative fuels and targets is also continuously sought with respect to the RERTR (Reduced Enrichment for Research and Test Reactors) program linked to the use of HEU to LEU (Low Enriched Uranium) with less than 20% uranium enrichment [4]. The OPAL reactor is one of the nuclear facilities that uses LEU as nuclear fuel as well as nuclear target to produce $^{99}$Mo.

The half-life of radioisotope $^{99}$Mo is about 66 hours, and the half-life of its daughter, $^{99m}$Tc, is about 6 hours. Therefore, it is needed an assembly that can provide the $^{99}$Mo locally and regionally. In Indonesia, BATAN has proposed Subcritical Assembly for Molybdenum-99 Production (SAMOP) in order to supply the needs of $^{99}$Mo especially in Indonesia. The initial concept of SAMOP is to use a fuel containing slightly $^{235}$U with a neutron multiplication factor value ($k_{eff}$) smaller than one or subcritical ($0.97 \leq k_{eff} \leq 0.99$). With this concept, SAMOP does not require the same harsh reactor safety requirements. SAMOP is designed to use low enriched uranyl nitrate as the fuel. [5]

The development of design and production analysis is needed to complete the data towards SAMOP's compatibility. The flexibility of SAMOP using different fuel needs to be discovered, for that reason, the research to use thorium as fuel is conducted. Apart from the use of LEU, the use of thorium as nuclear fuel is also widely developed. Thorium ($^{232}$Th) is a fertile material that can be used to produce fissile material of $^{233}$U isotope inside the reactor. The use of thorium as a fuel has several advantages over the use of uranium, other than its abundance of 3 to 4 times more than uranium abundance, the thorium-$^{233}$U cycle also produces fewer minor actinides than uranium-plutonium cycles. Then, thorium is also superior in safety aspect because it does not require enrichment process, and it produces less waste than uranium. [6]

Studies to see the potential of SAMOP using thorium nitrate mixed fuel have been done, however, the neutron multiplication value ($k_{eff}$) obtained is much smaller than the expected range. Therefore, a modified design is considered to meet the range of subcritical neutron multiplication factor value ($k_{eff}$). SAMOP is able to operate with thorium nitrate after modifying the patent design, but, study of the $^{99}$Mo productivity characteristics from the modified design has not yet been analyzed [7]. Thus, this study is important to be done to determine the productivity characteristics of $^{99}$Mo if using the mixture of thorium nitrate as its fuel and target.

## 2. Monte Carlo N-Particle eXtended (MCNPX)

The Monte Carlo method is a method of calculation by applying probability and statistical laws to the natural sciences. This method uses various random number distributions in which each distribution is a representation of a process that occurs sequentially. The process, for example, is the diffusion of neutrons in various materials, Monte Carlo method is used to calculate samples that is close to the actual value of the neutron diffusion. The Monte Carlo method was later converted into a computational science system used to predict random events in nature.

MCNPX (Monte Carlo N-Particle Code eXtended) is developed by Los Alamos National Laboratory (Los Alamos National Laboratory). MCNPX contains radiation transfer codes. It is designed to track many types of particles over a wide energy range. MCNPX is used for simulating nuclear processes such as fission and other simulations involving neutrons, photons and electrons. Specifically, MCNPX can be used for simulation of radiation protection and dosimetry, radiation shield, radiography, medical physics, nuclear safety, detector design and analysis, target accelerator design, fission and fusion reactor design, decontamination and decommissioning. The MCNPX software is basically a newer version of its own MCNP software. An important additional feature of MCNPX is the burnup calculation used in this study to obtain fission product results after the reactor is operated. [8]

The basic structure of MCNPX code consists of cell card, surface card, and data card. Data entered in the form of cell card contains information in the form of cell number, material density and cell position based on certain reference point. The surface card contains the constituent form of the design, such as circles, rectangles and balls. The data card contains the desired command to be completed, such as the burnup code and also the type of reactor constituent material, including the composition fraction, the mass number and the atomic number of the material [8]. The MCNPX software comes with Visual Editor software used to see the design of reactor geometry that has been created in either two or three dimensions.

## 3. Method

A study towards the design of the assembly according to SAMOP patent number 34511 is conducted, then the related parameters that affect the design especially the neutronic characteristic is determined. The geometry design of SAMOP is created using program MCNPX. The patent SAMOP design is the reference to determining the thickness and height of the cores, cladding, and reflector. [5]

The fuel composition using thorium nitrate (Th $(NO_3)_4$) is calculated. SAMOP uses two main fuels, the TRIGA fuel, which is a component of the Kartini reactor [9], and the mixture of uranium-thorium nitrate. The mixture of uranium-thorium nitrate lies at the center and annulus fuel tube of the assembly. After incorporating the design dimension and fuel composition, the MCNPX program is run to determine the value of the reactor neutron multiplication factor. The desired value of the neutron multiplication factor is in the range 0.97 to 0.99 to be regarded as a subcritical reactor. Variations to obtain the desired criticality value can be done by changing the design of the reactor or changing the reactor fuel composition.

The burn code on MCNPX is then inserted to get the result of fission product especially to get the radioisotope $^{99}$Mo result. Burn-up is performed in one 6-day operation cycle for each variation of the fuel composition. The variation of SAMOP power is also performed. The results on each variation of the nuclear mixed fuel and the power variation are analyzed and then compared. The minimum production of $^{99}$Mo is expected to be worth 300 mCi of each operating cycle. The Kartini reactor in BATAN, Yogyakarta, currently operates at a power of 1 kW. Thus, the $^{99}$Mo production value obtained and can be applied in the near term is at a power of 1 kW. The flowchart of the research is described in Figure 1.

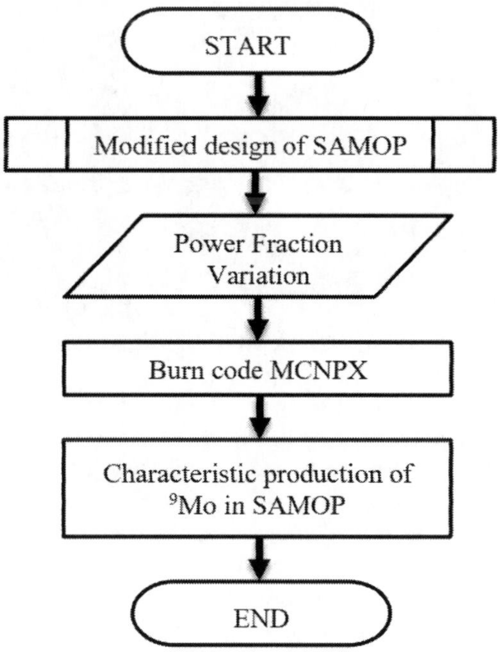


**Figure 1.** Flowchart of the Research

IOP Conf. Series: Journal of Physics: Conf. Series **978** (2018) 012072 doi:10.1088/1742-6596/978/1/012072

## 4. SAMOP Modified Design

SAMOP modified design has the following geometric dimension as shown in the Table 1.

**Table 1.** The Geometry of SAMOP Modified Designs

| Parameter | Mod-1 |
|---|---|
| Fuel | Thorium Nitrate |
| Center Fuel (Tube) Diameter | 5 cm |
| Annulus Fuel (Tube) Diameter | 31.4 cm |
| Center Fuel (Tube) Height | 46 cm |
| Annulus Fuel (Tube) Height | 40 cm |
| Number of Fuel Rods | 12 |
| Cladding Thickness | 3.6 cm |
| Reflector Thickness | 20 cm |
| Cladding | 0.1 cm |
| Reflector | Graphite |
| Coolant | Water |
| Framework | SS-316 |

The designs later inserted in the MCNPX cards. Both designs can be viewed using Visual Editor as shown in Figure 2. It can be seen that the color determines each materials used or contained in the design. The light blue color on the edge of the design is the framework of the assembly (SS316), while dark blue is the water coolant, yellow is the reflector (graphite), orange is the TRIGA fuel, red in the center and annulus reactor is the fuel in the form of mixed uranium-thorium nitrate, and light red above the fuel is the air.

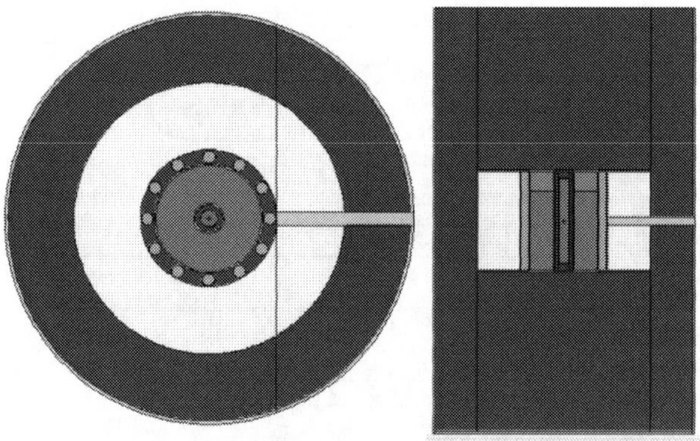

**Figure 2.** SAMOP Modified Design

## 5. Result and Analysis

The modified design is analyzed by varying the composition of uranium in the mixed uranium-thorium nitrate fuel. The uranium-thorium nitrate mixed fuel consists of 95% water and 5% uranium-thorium nitrate mixed fuel. The ratio between uranium and thorium in that 5% mixed fuel is varied, if the research uses 20% of $^{233}U$ in the mixed fuel, it means the fraction of uranyl nitrate fraction ($UO_2(NO_3)_2$) is 1% of the total fuel solution and the remaining 4% is thorium nitrate ($Th(NO_3)_4$).

The density of the uranium-thorium nitrate mixed fuel is calculated using the same formula to calculate the density of uranyl nitrate from the SAMOP patent [10]. The output result of the MCNPX will show information such as mass, atom density, fraction density and the activity of all the actinides and non-actinides on the first day and the sixth day operation in each burned materials. It also provide information such as the neutron multiplication factor ($k_{eff}$) of the assembly. Certain information needed to analyze the performance of SAMOP is chosen to be compared as shown in Table 2. The $^{99}$Mo of production is obtained with a power operation of 1 kW with operating time for 6 days.

**Table 2.** $K_{eff}$ Value and the $^{99}$Mo Based on the $^{233}$U Fraction Variations

| $^{233}$U fraction | $K_{eff}$ | Std. Deviation | $^{99}$Mo in the U-Th Fuel | | $^{99}$Mo in the TRIGA UZrH | |
|---|---|---|---|---|---|---|
| | | | Mass (gr) | Activity (Ci) | Mass (gr) | Activity (Ci) |
| 20% | 1.02795 | 0.00290 | 4.077E-05 | 19.59 | 3.244E-05 | 15.59 |
| 18% | 1.01366 | 0.00255 | 3.976E-05 | 19.11 | 1.911E+01 | 16.40 |
| 16% | 0.97542 | 0.00250 | 3.867E-05 | 18.58 | 3.541E-05 | 17.02 |
| 14% | 0.94605 | 0.00246 | 3.714E-05 | 17.85 | 3.670E-05 | 17.64 |
| 12% | 0.91305 | 0.00275 | 3.561E-05 | 17.11 | 3.915E-05 | 18.81 |
| 10% | 0.87221 | 0.00243 | 3.318E-05 | 15.94 | 4.207E-05 | 20.22 |

It can be seen from the Table 2 that 16% uranium fraction meets the qualification of neutron multiplication factors in the subcritical range, which is 0.975. The assembly operates at a 1 kW power operation, and it obtains about 18.58 Ci of the $^{99}$Mo activity.

SAMOP operates at the subcritical state, therefore, the composition with 16% of $^{233}$U is used as the control variable to obtain the $^{99}$Mo activity by varying the SAMOP power. The activity of $^{99}$Mo multiplies along the multiplication of the power. It can be seen in the Table 3 the activity resulted in different value of power. SAMOP Power is multiplied ten times from 1 kW to 100 kW.

**Table 3.** Activity of $^{99}$Mo based on the Power Variations

| Power (kW) | $^{99}$Mo Activity (Ci) |
|---|---|
| 1 | 18.58 |
| 10 | 184.4 |
| 100 | 1832 |

Besides $^{99}$Mo, there are more fission products (non-actinides) resulted in the assembly. Below is shown radioisotopes using 16% fraction of $^{233}$U as control variable and at 1 kW SAMOP power.

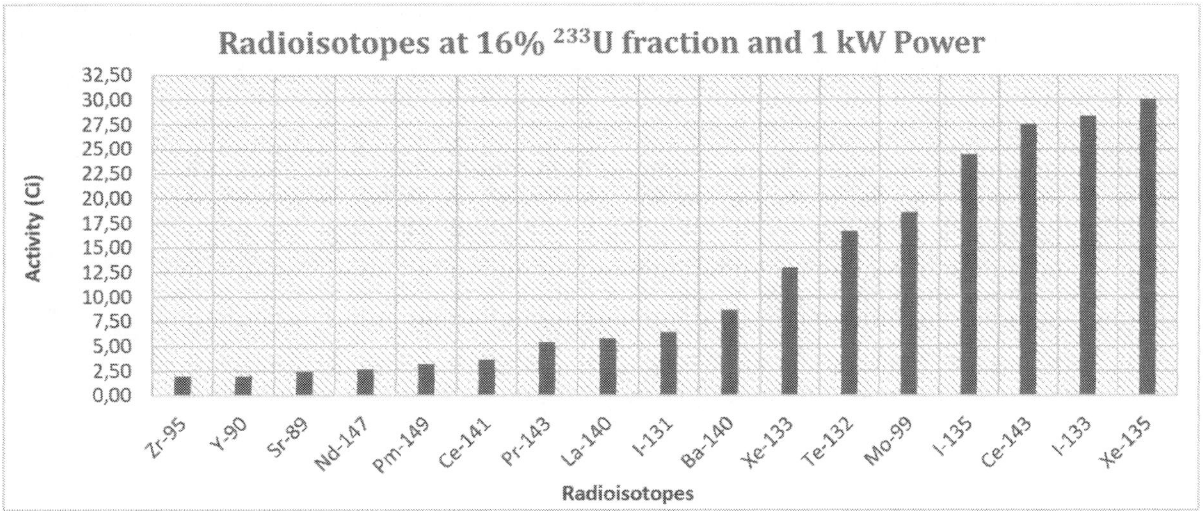


**Figure 3.** Radioisotopes Resulted in the Assembly

IOP Conf. Series: Journal of Physics: Conf. Series **978** (2018) 012072 doi:10.1088/1742-6596/978/1/012072

## 6. Conclusion

MCNPX can be used to simulate a nuclear assembly to obtain the neutronic and productivity characteristic. The characteristic that is obtained by analyzing the result can be used to complete the data regarding to the performance of SAMOP, particularly using thorium nitrate as fuel mixture. It is also provide the consideration for refinement and development of modified SAMOP design. The total activity of radioisotope $^{99}$Mo both in center and annulus fuel tubes on the modified design is 18.58 Ci with a 1 kW power operation, neutron multiplication factor = 0.975 (subcritical).

## 7. References


[1] IAEA Nuclear Energy Series 2013 *Non-HEU Production Technologies for Molybdenum-99 and Technetium-99m* Technical Document (Vienna: IAEA)

[2] Huisman M 2013 Medical Isotope Production Reactor Master Thesis (Netherlands: Delft University of Technology)

[3] National Academics of Sciences, Engineering, and Medicine 2016 *Molybdenum-99 for Medical Imaging* (Washington DC: The National Academies Press)

[4] A. Travelli. *The U. S. Reduced Enrichment Research and Test Reactor (RERTR) Program.* Technical Document. Argonne National Laboratory, Illinois.

[5] BATAN 2013 *Perangkat Reaktor Subkritik untuk Memproduksi 99Mo Nomor Paten 34511* Patent (Jakarta: DJKI)

[6] IAEA Nuclear Energy Series 2005 *Thorium Fuel Cycle; Potential Benefits and Challenge.* Technical Document (Vienna: IAEA) View at Google Scholar

[7] Yassar F 2017 Neutronic Analysis of Subcritical Assembly For $^{99}$Mo Isotope Production Based on Thorium Nitrate Fuel *Padjajaran International Physics Symposium* (Bandung: Universitas Padjajaran)

[8] Pelowitz D B 2008 *MCNPX User's Manual Version 2.6.0* Tech. Report LA-CP-07-1473 (USA: Los Alamos National Laboratory)

[9] Sutondo T and Syarip 2014 *Ganendra Majalah IPTEK Nuklir (Journal of Nuclear Science and Technology)* 17, 2, 83-90

[10] Syarip and Sutondo T 2017 Analytical Method of Atomic Density Determination of Uranyl Nitrate Solution *International Conference on Computation in Science and Engineering* (Bandung: ITB)

[11] Gholamzadeh Z, Feghhi S A H, Mirvakili S M, Joze-Vaziri A and Alizadeh M 2015 Computational Investigation of $^{99}$Mo, $^{89}$Sr, and $^{131}$I Production Accelerator, *Nuclear Engineering Technology* 47 875-883

[12] Budisantoso E T and Syarip 2002 Studi Produksi Radioisotop Mo-99 dengan Bahan Target Larutan Uranil Nitrat pada Reaktor Kartini "Ganendra" *Journal of Nuclear Science and Technology* 5 1 1-8

[13] Artem V ,Matyskin, Ridikas D, Skuridin V S, Sterba J and Steinhauser G 2012 Feasibilty Study for Production of 99mTc by Neutron Irradiation of MoO3 in a 250 kW TRIGA Mark II Reactor *(J Radioanal Nucl. Chem.* DOI 10.1007/s10967-012-2381-y)

[14] Hummel and Andrew J 2013 *Molybdenum-99 Production in the Oregon State TRIGA Reactor: Analysis of the Reactor Design Using a New LEU Target as Fuel* PhD Dissertation (USA: Oregon State University)

IOP Conf. Series: Journal of Physics: Conf. Series **978** (2018) 012073 doi:10.1088/1742-6596/978/1/012073

# Applying transpose matrix on advanced encryption standard (AES) for database content

**E B P Manurung[1], O S Sitompul[2] and Suherman[3*]**
[1,2] Department of Computer Science, Universitas Sumatera Utara, Indonesia
[3] Department of Electrical Engineering, Universitas Sumatera Utara, Indonesia

*Email: suherman@usu.ac.id

**Abstract.** Advanced Encryption Standard (AES) is a specification for the encryption of electronic data established by the U.S. National Institute of Standards and Technology (NIST) and has been adopted by the U.S. government and is now used worldwide. This paper reports the impact of transpose matrix integration to AES. Transpose matrix implementation on AES is aimed at first stage of chypertext modifications for text based database security so that the confidentiality improves. The matrix is also able to increase the avalanche effect of the cryptography algorithm 4% in average.

## 1. Introduction

The most common threat on databases is the security lack on integrity, availability, and confidentiality so that the 4 (four) database security mechanisms should be implemented, namely: access control, inference control, flow control, and data encryption [1].

The access control on database is performed by requesting user password, but its content is not protected [2], some unsecured informations may be exposed [3]. Data encryption on databases may increase the security; however, the databases performance may decrease as it introduces additional processing delay [4]. As a matter of fact the speed of computer system increases, processing delay on encryption may not be an issue.

The Advanced Encryption Standard (AES) algorithm uses symetrical keys for encryption and desciption [6]. Research on adding encryption has been conducted by many researchers. The AES is one of the methods used in protecting database. For instance, Vishal and Patel [7] applied AES for cloud database, while Dawle *et al.* uses intrusion detection [8].

This paper modifies AES by using transpose matrix to enhace security when the plain text is altered before encrypted. Further, the transpose matrix insertion is also aimed at improving the avalanche effect so that the small change on plaintext results significant differences in cyphertext.

## 2. Research Method

The proposed approach for securing database content using AES algorithm is initialized by transposing the cypher text within the AES algorithm. The database contents are protected by the original AES as well as the modified one so that transpose matrix impact can be identified. Afterwards, the descripted information is analyzed.

The main parameter to be extracted from both method is the number of altered bits on chyper text that changed when the plain text is slightly changed. The next parameter is the percentage of avalanche effect. The higher avalanche effect the higher the security. The research flow is depicted in Figure 1.

IOP Conf. Series: Journal of Physics: Conf. Series **978** (2018) 012073 doi:10.1088/1742-6596/978/1/012073

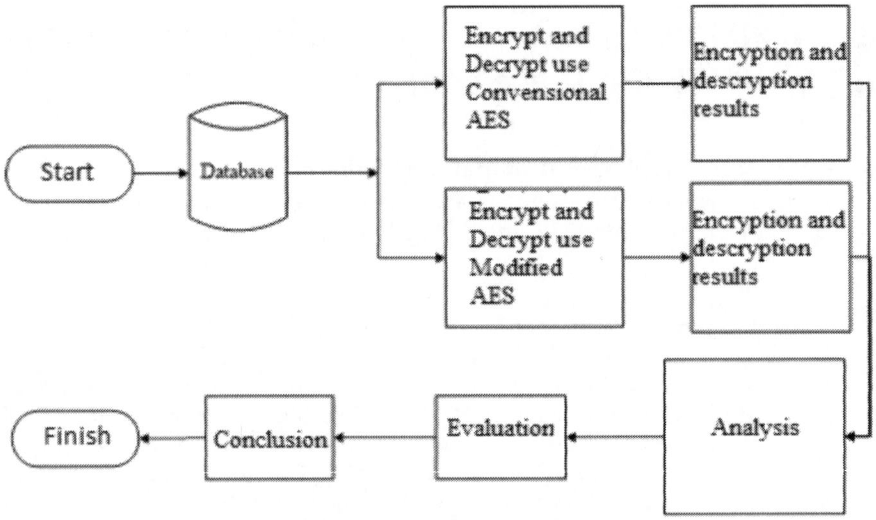


**Figure 1.** Research flow

In order to evaluate the method, My-SQL was employed and accessed by using php scripted web page. The network is simplified, and error correction due to network condition is guaranteed by the lower layer. The same evaluation runs for both AES and modified AES.

## 3. AES modification

Modification of AES is performed by implementing transpose matrix on MixColumns process on encryption and descryption process as shown in Figure 2.

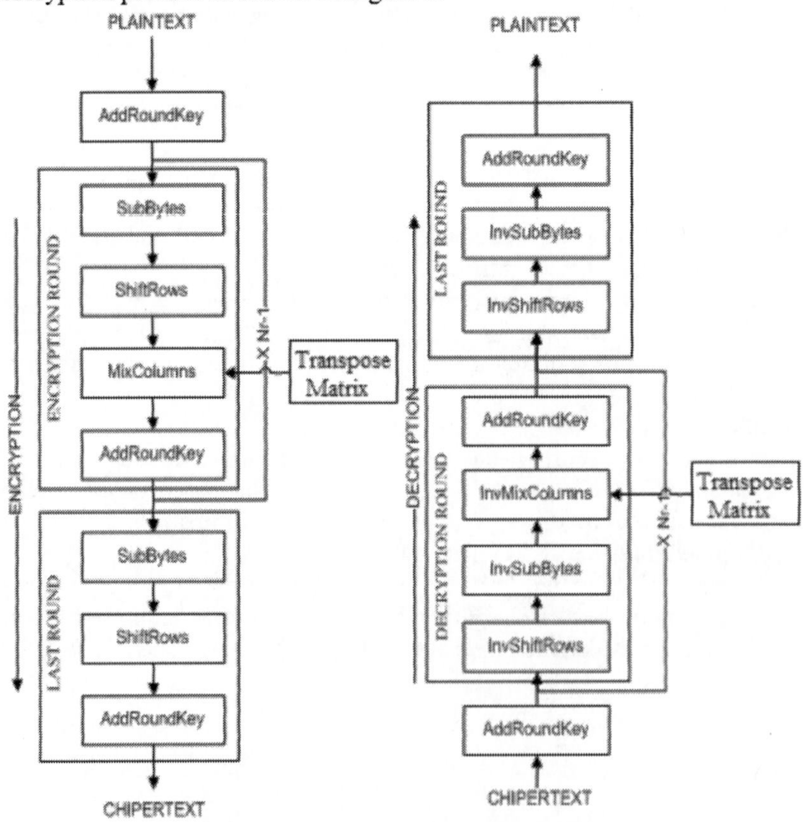


**Figure 2**. Transpose matrix insertion

In description process, transpose matrix is also applied to the opposite process which is InvMixColumn.

*3.1 Transforming MixColumns dan InvMixColumns*

MixColumns transformation on AES is performed by randomizing data in each array state column. Afterward, AddRoundKey performs XOR process between current state and round key. MixColumns process each state column. This operation converts column as polynomial on GF ($2^8$) [6].

**Table 1.** Multiplication matrix of MixColumns [5]

| | | | |
|---|---|---|---|
| 2 | 3 | 1 | 1 |
| 1 | 2 | 3 | 1 |
| 1 | 1 | 2 | 3 |
| 3 | 1 | 1 | 2 |

The values on Table 1 are calculated by using the following equations:

$$S_{0,C} = (\{02\} \bullet S_{0,C}) \oplus (\{03\} \bullet S_{1,C}) \oplus S_{2,C} \oplus S_{3,C}$$
$$S_{1,C} = S_{0,C} \oplus (\{02\} \bullet S_{1,C}) \oplus (\{03\} \bullet S_{2,C}) \oplus S_{3,C}$$
$$S_{2,C} = S_{0,C} \oplus S_{1,C} \oplus (\{02\} \bullet S_{2,C}) \oplus (\{03\} \bullet S_{3,C})$$
$$S_{3,C} = (\{03\} \bullet S_{0,C}) \oplus S_{1,C} \oplus S_{2,C} \oplus (\{02\} \bullet S_{3,C})$$

InvMixColumns treats each column as 4 variables of polynomials in the Galois field which are multiplied to c(x) as $(x^4+1)$ $c(x)=11x^3+13x^2+9x+14$ [6]. The InvMixColumns transformation employs the same process as MixColumns as shown in Table 2.

**Table 2.** Multiplication matrix of InvMixColumns [5]

| | | | |
|---|---|---|---|
| 0E | 0B | 0D | 09 |
| 09 | 0E | 0B | 0D |
| 0D | 09 | 0E | 0B |
| 0B | 0D | 09 | 0E |

The following formulas are employed.

$$S_{0,C} = (\{0E\} \bullet S_{0,C} \oplus (\{0B\} \bullet S_{1,C}) \oplus (\{0D\} \bullet S_{2,C}) \oplus (\{09\} \bullet S_{3,C})$$
$$S_{1,C} = (\{09\} \bullet S_{0,C} \oplus (\{0E\} \bullet S_{1,C}) \oplus (\{0B\} \bullet S_{2,C}) \oplus (\{0D\} \bullet S_{3,C})$$
$$S_{2,C} = (\{0D\} \bullet S_{0,C} \oplus (\{09\} \bullet S_{1,C}) \oplus (\{0E\} \bullet S_{2,C}) \oplus (\{0B\} \bullet S_{3,C})$$
$$S_{3,C} = (\{0B\} \bullet S_{0,C} \oplus (\{0D\} \bullet S_{1,C}) \oplus (\{09\} \bullet S_{2,C}) \oplus (\{0E\} \bullet S_{3,C})$$

*3.2 MixColumns and InvMixColumns modifications*

MixColumns randomizes data in each array state column based on multiplication matrix [5]. MixColumns and InvMixColumns on AES are modified as shown in Table 3 and 4.

IOP Conf. Series: Journal of Physics: Conf. Series **978** (2018) 012073 doi:10.1088/1742-6596/978/1/012073

**Table 3.** Multiplication matrix modification of MixColumns

| | | | |
|---|---|---|---|
| 02 | 01 | 01 | 03 |
| 03 | 02 | 01 | 01 |
| 01 | 03 | 02 | 01 |
| 01 | 01 | 02 | 03 |

**Table 4.** Multiplication matrix modification of InvMixColumns

| | | | |
|---|---|---|---|
| 0E | 09 | 0D | 0B |
| 0B | 0E | 09 | 0D |
| 0D | 0B | 0E | 09 |
| 09 | 0D | 0B | 0E |

## 4. Results and Discussion

A sample of encryption process to the field of database by using modified AES is shown in Figure 3.

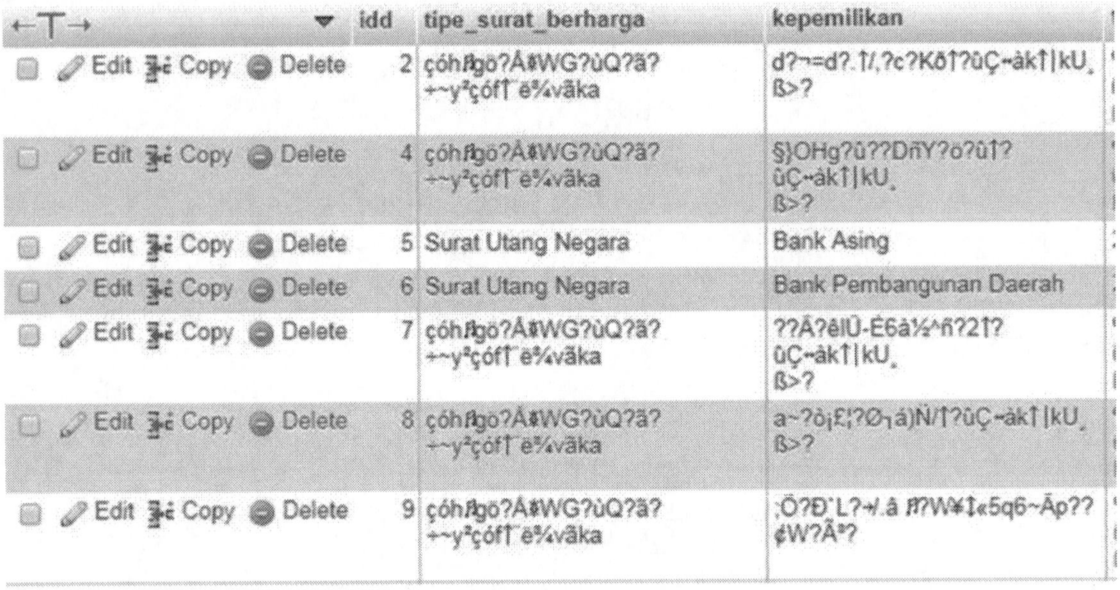

| ←T→ | idd | tipe_surat_berharga | kepemilikan |
|---|---|---|---|
| Edit Copy Delete | 2 | çóhﬁgö?Å‡WG?ùQ?ã? ÷~y²çófΤ¨ê¾vãka | d?¬=d?.Ť/,?c?Kő↑?ûÇ÷àk↑\|kU¸ ß>? |
| Edit Copy Delete | 4 | çóhﬁgö?Å‡WG?ùQ?ã? ÷~y²çófΤ¨ê¾vãka | §}OHg?û??DñY?ö?û↑? ûÇ÷àk↑\|kU¸ ß>? |
| Edit Copy Delete | 5 | Surat Utang Negara | Bank Asing |
| Edit Copy Delete | 6 | Surat Utang Negara | Bank Pembangunan Daerah |
| Edit Copy Delete | 7 | çóhﬁgö?Å‡WG?ùQ?ã? ÷~y²çófΤ¨ê¾vãka | ??Â?ëlÜ-É6à½^ñ?2↑? ûÇ÷àk↑\|kU¸ ß>? |
| Edit Copy Delete | 8 | çóhﬁgö?Å‡WG?ùQ?ã? ÷~y²çófΤ¨ê¾vãka | a~?ò¡£¦?Ø┐á)Ñ/↑?ûÇ÷àk↑\|kU¸ ß>? |
| Edit Copy Delete | 9 | çóhﬁgö?Å‡WG?ùQ?ã? ÷~y²çófΤ¨ê¾vãka | ;Ö?Ð˙L?÷/.â ﬁ?W¥‡«5q6~Ãp?? ¢W?Ã°? |

**Figure 3** MySQL database encryption

The analysis is performed by comparing encryption results of an original plain tex and a single character change plaint text for both original AES and the modified one. By performing 10 rounds of encryption, the bit changes are shown in Figure 4.

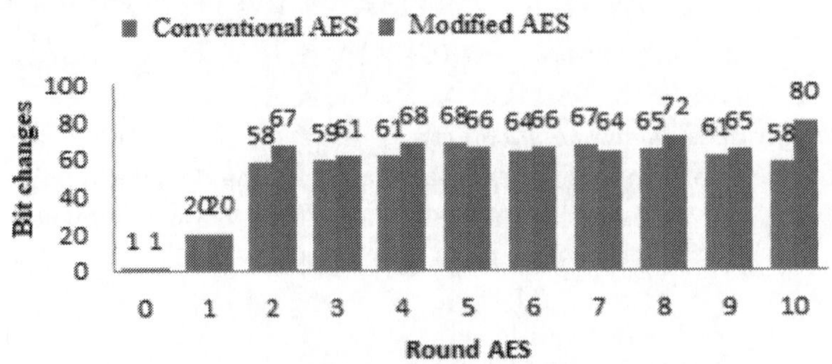


**Figure 4**. Bit changes as plaint text changes

Conventional AES results bit changes up to 68 bits while the modified AES results up to 80 bits. In average the modified AES produce about 5 bit higher than the existing one.

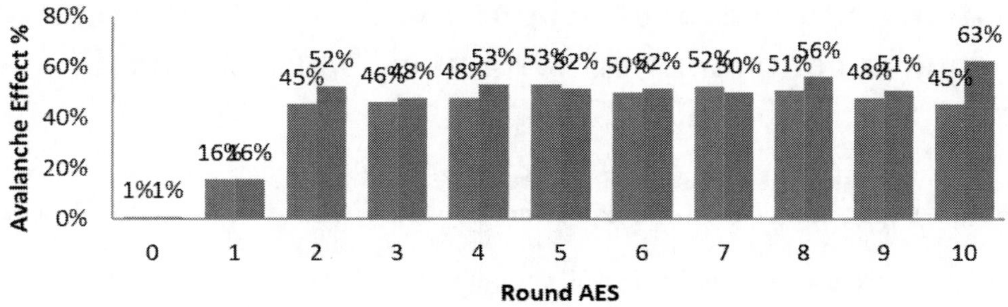


**Figure 5**. Avalanche effects

The avalanche effect of the modified AES are about 63% in average, while the standard one is 53% as shown in Figure 5. Avalanche effect increases as the transpose matrix is applied. It mean the security level is improved.

## 5. Conclusion

The application of transpose matrix to AES algorithm is successfully changing the plain text. When a single character in plain text is changed, significant bit changes occured. In average, the modification results in 5 bits higher and 4% higher avalanche rates compared to the original one.

## References

[1] Rezk A, Ali H A and Barakat SI 2012 *International Journal of Computer Applications* **49**(19).

[2] Basharat I, Azam F and Muzaffar A W 2012 *Int. Journal of Computer Applications* **12** (47): 28-34.

[3] Al Khalil F, Gabillon A and Capolsini P 2014 *Foundations and Practice of Security* pp. 189-208.

[4] Singh P and Kaur P. 2015 *Int. Conference on Futuristic trend in Computational Analysis and Knowledge Management* (12): 353-358.

[5] Mahajan P and Sachdeva A 2013 *Global Journal of Computer Science and Technology* **13**(15).

[6] Bacon J, Eyers D, Pasquier T F M, Singh J, Papagiannis I, and Pietzuch P 2014, *IEEE Transactions on Network and Service Management* **11**(1): 76-89.

[7] Pancholi VR and Patel B P 2016 *International Journal for Innovative Research in Science and Technology* **2**(9):18-21.

[8] Dawle Y, Naik M, Vande S, and Zarkar N. *IJLERA*. 2017 2(03):01-6.

IOP Conf. Series: Journal of Physics: Conf. Series **978** (2018) 012074 doi:10.1088/1742-6596/978/1/012074

# Simulation comparison of proportional integral derivative and fuzzy logic in controlling AC-DC buck boost converter

**A Faisal, S Hasan and Suherman**
Department of electrical engineering Univesitas Sumatera Utara 20155, Indonesia

*E-mail: ahmadfaisal.st@gmail.com,suherman@usu.ac.id

**Abstract** - AC-DC converter is widely used in the commercial industry even for daily purposes. The AC-DC converter is used to convert AC voltage into DC. In order to obtain the desired output voltage, the converter usually has a controllable regulator. This paper discusses buck boost regulator with a power MOSFET as switching component which is adjusted based on the duty cycle of pulse width modulation (PWM). The main problems of the buck boost converter at start up are the high overshoot, the long peak time and rise time. This paper compares the effectiveness of two control techniques: proportional integral derivative (PID) and fuzzy logic control in controlling the buck boost converter through simulations. The results show that the PID is more sensitive to voltage change than fuzzy logic. However, PID generates higher overshoot, long peak time and rise time. On the other hand, fuzzy logic generates no overshoot and shorter rise time.

## 1. Introduction

The higher level of welfare of the population is directly proportional to the increasing demand for electrical energy. Today many consumers of electric power use electronic power converters such as uninterruptible power supply (UPS), charger, and electric regulator in low voltage channels. Various types of converters have been widely used in those converters.

AC-DC converters generally have three steps of working: rectifying, filtering, and regulating. The first stage aims to rectify AC voltage into DC voltage. The second stage is to smooth the DC voltage. The third stage is to regulate the output voltage ($V_{out}$) [1]. There are several regulating techniques employed by AC-DC converters, one of them is buck converter. Buck converter works when the output voltage is smaller than the input voltage. If the output voltage is greater than the input voltage, the converter is called the boost converter. If the output voltage is larger or smaller than the input voltage, the converter is called buck boost converter [2].

Control techniques are increasingly important nowadays in the buck boost converters due to their rapid responses and high accuracies. In previous studies, the proportional integral (PI) control method was used on an AC-DC buck boost converter to improve the power factor [4]. The open loop sliding control (OLSC) method is used on buck and AC-DC buck boost converters to improve total harmonic distortion (THD) and power factor [5]. The fuzzy logic control method is used on boost converter to regulate the output voltage and to improve the power factor [6]. The hysteresis current controller (HCC) method is compared to the fuzzy logic method on boost converter to regulate the output voltage, to improve THD and power factor on the input side [7]. This paper compares the effectiveness of PID and fuzzy logic in controlling pulse width modulation (PWM) on MOSFET switch to regulate output voltage based on reference voltage or desired voltage.

## 2. Research Methodology

This paper evaluates the AC to DC converter with diode bridge rectifier with a power MOSFET as a switching component that controlled by PWM signal through simulation.The magnitude of the output voltage is set according to the duty cycle (D) of PWM. If D > 0.5, then the output voltage ($V_{out}$) will be greater than the input voltage ($V_{in}$). Whereas if D <0.5 then the output voltage will be smaller than input voltage, and $V_{in} = V_{out}$ at D = 0.5 [3]. This paper evaluates varying values of duty cycle.

The circuit is modeled as depicted in Figure 1. The PID and fuzzy logic on buck booster converter is applied to feedback output voltage to MOSFET.

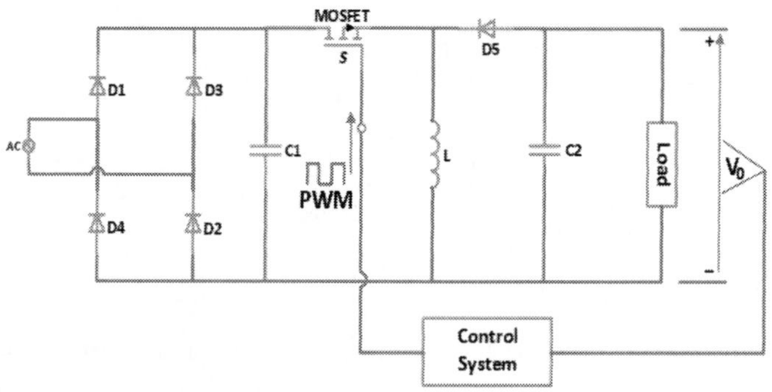


**Figure 1.** Circuit diagram of AC-DC buck boost converter with control system

### 2.1. *Parameters of Simulation*

Table 1 contains the parameters used in the simulation circuit. Source voltage is set 220 V with frequency of 50 Hz. The reference voltages are at 200, 225 and 250 Volts. Parameters of diode, MOSFET and capasitor are also shown in Table 1.

**Table 1.** The component parameters in simulation

| Parameter | Value | Unit |
|---|---|---|
| Source voltage ($V_{in}$) | 220 | Volt |
| Source Frequency ($F$) | 50 | Hz |
| $V_{referensi}$ | 200, 225 dan 250 | Volt |
| **Rectifier Diode** | | |
| Resistance Ron | 0,1 | Ω |
| Forward voltage | 0,8 | Volt |
| Snubber resistance | 500 | Ω |
| Snubber capacitance | $250 \times 10^{-3}$ | F |
| **Filter capacitor** | $2.3 \times 10^{-3}$ | Farad |
| **MOSFET** | | |
| Resistance Ron | 0,1 | Ω |
| Internal dioda resistance | 0,01 | Ω |
| Snubber resistance | $1 \times 10^{5}$ | Ω |
| Inductor | $1 \times 10^{-3}$ | Henry |
| **Diode regulator** | | |
| Resistance Ron | 0,1 | Ω |
| Forward voltage | 0,8 | Volt |
| Snubber resistance | 500 | Ω |
| Snubber capacitance | $250 \times 10^{-3}$ | F |
| Regulator capacitor | $12 \times 10^{-3}$ | Farad |
| Load R | 50 | Ω |

### *2.2. Simulated Circuit with PID Control*

The overall model of AC-DC buck boost converter circuit with PID control is shown in Figure 2.

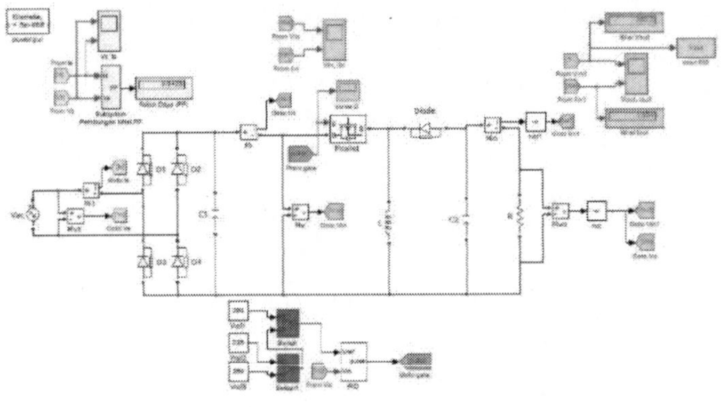

**Figure 2.** Simulated circuit of AC-DC buck boost converter with PID control

Simulation with PID control uses one input error (e) of voltage and produces an output voltage as expressed in Equation (1).

$$e(t) = V_{reff} - V_o \quad (1)$$

The input error (e) of the voltage is processed each by multiplying it by proportional constant ($K_p$), integral constant ($K_i$), and derivative constant ($K_d$) then the result is summed to produce the output voltage. The expressions for the voltage error, proportional constant, integral constant and derivative constant are shown in Equations (2), (3), and (4), respectively.

$$P_{out} = K_p . e(t) \quad (2)$$

$$I_{out} = K_i \int_0^t e(\tau) d\tau \quad (3)$$

$$D_{out} = K_d \frac{de(t)}{dt} \quad (4)$$

### *2.3. Simulated Circuit with Fuzzy Logic Control*

The model of the AC-DC buck boost converter with fuzzy logic control is shown in Figure 3.

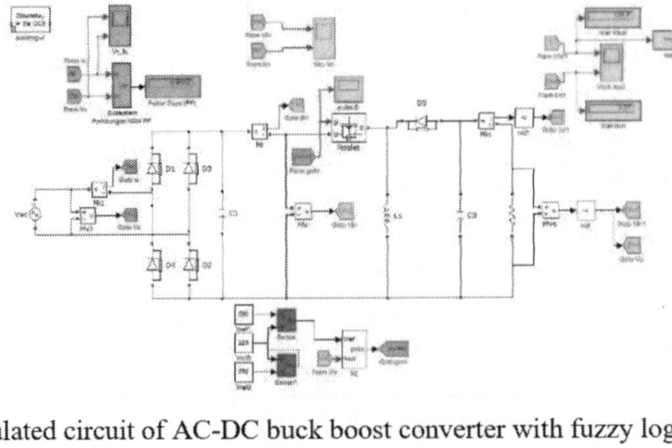

**Figure 3.** Simulated circuit of AC-DC buck boost converter with fuzzy logic control

The simulation with fuzzy logic control consists of two inputs: error e(k) and derror Δe(k). Error e(k) is the difference between the desired voltage ($V_{ref}$) and the output voltage ($V_{out}$), while the derror Δe(k) is the difference between the current error and the previous error shown in Equations (5) and (6), respectively .

$$e(k) = V_{ref} - V_o(k) \tag{5}$$

$$\Delta e(k) = e(k) - e(k-1) \tag{6}$$

## 3. Results of Simulation and Discussion

The results of the simulation using PID and fuzzy logic control systems on AC-DC buck boost converter with Vref of 200 V are shown in Figure 5 and Figure 6.

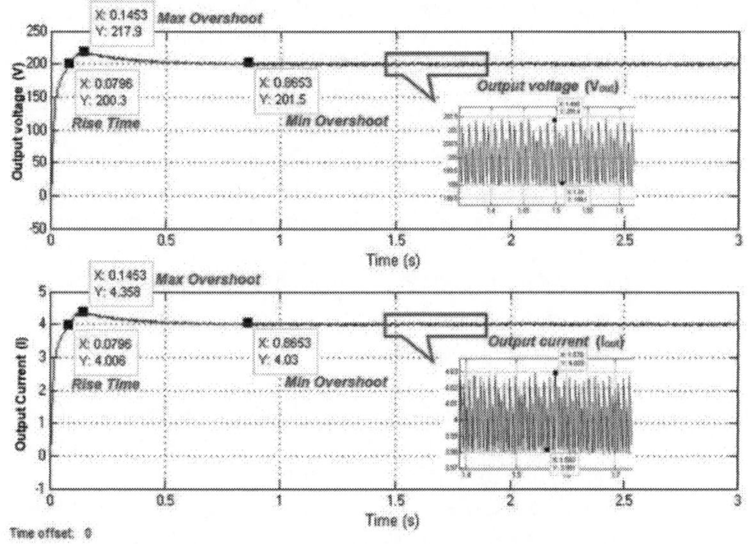


**Figure 4.** Voltage wave and output current for $V_{referensi}$ of 200 V with PID control

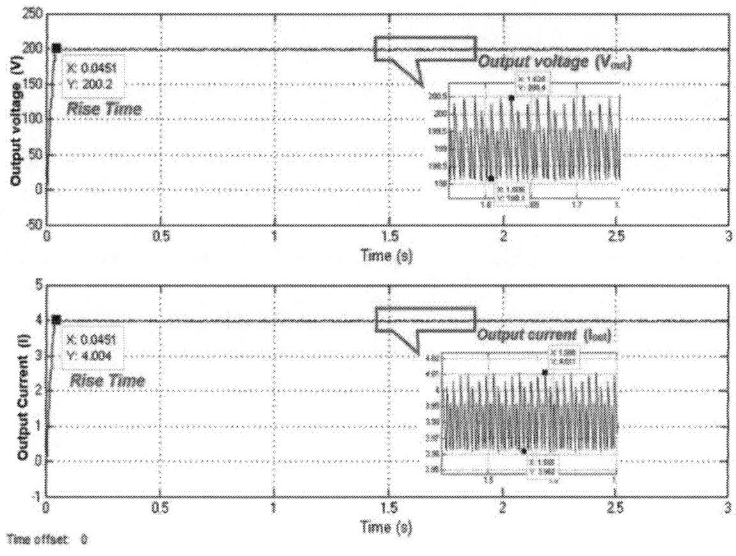


**Figure 5.** Voltage wave and output current for $V_{referensi}$ of 200 V with fuzzy logic control

Based on the results, each control system is able to maintain the power factor value (cos ϕ) on the input side and the output voltage based on the desired voltage Vref. However, the differences are apparent in the characteristics of output currents and output voltages, i.e. overshoot, peak time, and rise time which can be seen in Table 2.

**Table 2.** Comparison of PID control and fuzzy logic control.

| No | Characteristic | $V_{referensi}$ 200 V | | $V_{referensi}$ 225 V | | $V_{referensi}$ 250 V | |
|---|---|---|---|---|---|---|---|
| | | **PID** | **Fuzzy logic** | **PID** | **Fuzzy logic** | **PID** | **Fuzzy logic** |
| 1 | Rise Time | 0,079 ms | 0,04 ms | 0,14 ms | 0,04 ms | 0,23 ms | 0,04 ms |
| 2 | Peak Time | 0,145 ms | - | 0,24 ms | - | 0,41 ms | - |
| 3 | Max Overshoot $V_{out}$ | 217,9 V | - | 252,5 V | - | 288,3 V | - |
| 4 | Min Overshoot $V_{out}$ | 201,5 V | - | 226,1 V | - | 251,6 V | - |
| 5 | Max Overshoot $I_{out}$ | 4,35 A | - | 5,05 A | - | 5,73 A | - |
| 6 | Min Overshoot $I_{out}$ | 4,03 A | - | 4,52 A | - | 5,03 A | - |

Table 2 proves that by using PID control on AC-DC buck boost converter, the overshoot and peak time in voltage and current of output remain exist. The larger the Vref is given, the greater the values of overshoot and peak time are obtained. While using fuzzy logic control, overshoot does not exist, and the peak time does not occur.

The value of rise time generated by the PID is not constant and increasing to references voltage. Whereas by using fuzzy logic control, the value of rise time remains constant.

## 4. Conclusions

Based on the conducted simulation and parameters, fuzzy logic performs better than PID. PID control on the simulated AC-DC buck boost converter still generates the overshoot in the output voltage wave and output current. Its rise time and peak time increase to reference voltage.

On the other hand, the simulated AC-DC buck boost converter with fuzzy logic suppressed the overshoot on output voltage and output current. Peak time does not occur as there is no overshoot. Further, rise time is constant to reference voltage and shorter than the PID control.

## References

[1]. Helmy M H, Onny S and Hadi S 2016 *JNTETI*, **5** (1).
[2]. Rashid H M 2004 Power Electronics: Circuit, Devices, and Aplication.Third Edition, 2004.
[3]. Daniel W H 1997 Introduction to Power Electronics. (Valparaiso University, Indiana: Prentice-Hall International, Inc).
[4]. Rajalakshmi G, Sivaranjani K and Abinaya K. Microcontroller Implementation of Power Factor Improvement Strategy for a Buck-Boost AC-DC Converter. *IEEE*, 978-1-4799-3826-1
[5]. Mohammed M S K, Shamsul, Rubaiyat, Ashfanoor K, Amina H A and Choudhury M A 2012 Input Switched Single Phase Buck and Buck-Boost AC-DC Converter with Improved Power Quality. *IEEE*, 978-1-4673-1436-7, 2012.
[6]. Periyasamy K 2012 IJERA, **2** (5).
[7] Shobana P,Geetha N and Gnanavadivel J 2012 Enhancement of Power Quality of AC-DC Boost Converter with HCC and FLC-A Comparative Study. *IEEE* 978-1-4673-0210-4.

IOP Conf. Series: Journal of Physics: Conf. Series **978** (2018) 012075 doi:10.1088/1742-6596/978/1/012075

# Computer simulation and implementation of defected ground structure on a microstrip antenna

**H Adrian[1], A H Rambe[2*] and Suherman[2]**
[1]M.Eng student of Electrical Engineering, Universitas Sumatera Utara, Indonesia
[2]Department of Electrical Engineering, Universitas Sumatera Utara, Indonesia

*E-mail: ali3@usu.ac.id

**Abstract.** Defected Ground Structure (DGS) is a method reducing etching area on antenna ground to form desirable antenna's ground field. This paper reports the method impact on microstrip antennas working on 1800 and 2400 MHz. These frequencies are important as many radio network applications such mobile phones and wireless devices working on these channels. The assessments were performed by simulating and fabricating the evaluated antennas. Both simulation data and implementation measurements show that DGS successfully improves antenna performances by increasing bandwidth up to 19%, reducing return loss up to 109% and increasing gain up to 33%.

## 1. Introduction

Defected Ground Structure (DGS) is a method reducing surface wave by partially etching the ground structure of an antenna. DGS enables antena designers to adjust ground field shape so that antenna works better. Since microstrip antennas experience high surface wave losses [1], DGS may improve microstrip antenna performances.

DGS impact on microstrip antenna has been studied by Dash et al [2], Ayman and Adel [3], and Ajay et al [4] which results performance improvements. Dash applied complimentary DGS, Ayman and Adel used C-shape DGS, and Ajay et al employed U-slot. This paper discusses the squared-DGS implementation on a dualband microstrip antenna which is designed to work on channel 1800 MHz and 2400 MHz. The designed antenna is a two-elements-square-shape microstrip antenna which searches the working frequencies in between 1600 MHz and 2600 MHz. The antenna size is adjusted to get dualband characteristics. The defected area is applied in between the two patch elements.

As the advantages of microstrip antenna rely on its low profile characteristics: small, light, thin, easy to fabricate, cheap and frequency ranges [5, 6], DGS implementation is expected to enhance performance without reducing microstrip benefits.

## 2. Method

### *2.1 The designed antennas*

The designed antenna contains two elements. The two elements are fed by using the same transmission line. The DGS is performed by removing a small square-shaped part of conductors between the two elements with size changes from 7.5 mm to 23.5 mm with 2 mm space as shown in Figure 1.

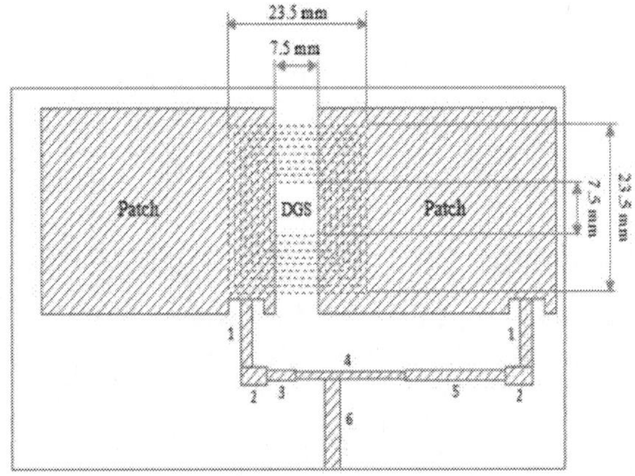


**Figure 1**. Microstrip antenna with DGS design

The designed antenna are realized by using FR-4 epoxy substrate with dielectric constant ($\varepsilon r$) 4.4 , 1.6 mm tickness and dielectric loss tangent (tan $\delta$) 0.02. The size of the designed antenna is shown in Table 1.

**Table 1**. Antenna size

| Component | Dimension (mm) |
|---|---|
| Ground plane | 94 × 54 |
| Patch | 40 × 29 |
| 1 | 2.0 × 9.5 |
| 2 | 4.5 × 2.5 |
| 3 | 5.0 × 1.5 |
| 4 | 18.5 × 1.0 |
| 5 | 17.0 × 1.5 |
| 6 | 2.5 × 13.0 |
| DGS 1 | 7.5 × 7.5 |
| DGS 2 | 9.5 × 9.5 |
| DGS 3 | 11.5 × 11.5 |
| DGS 4 | 13.5 × 13.5 |
| DGS 5 | 15.5 × 15.5 |
| DGS 6 | 17.5 × 17.5 |
| DGS 7 | 19.5 × 19.5 |
| DGS 8 | 21.5 × 21.5 |
| DGS 9 | 23.5 × 23.5 |

### *2.2 Antenna simulations*
Antenna modeling and simulation were performed by using the AWR simulator. The evaluated parameters include return loss (S11), S21 and gain.

## 3. Results

### *3.1 The fabricated antenna*
Antenna fabrication produces two antennas: with and without DGS. Figure 2 shows the front and back views of the realized antennas with DGS.

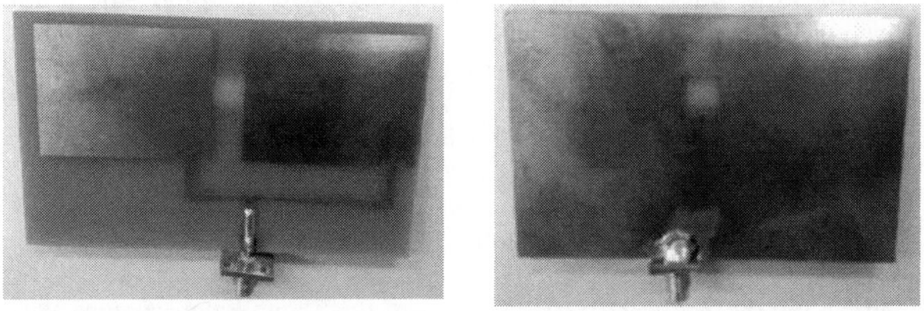

(a) (b)

**Figure 2**. The fabricated microstrip antenna with DGS (a) Front view,(b) Back view

*3.2 Simulation results*

The simulation results (Table 2) shows that for frequency of 1800 MHz, DGS causes bandwidth, return loss and gain reductions. However, the bandwidth is significantly improved for 2400 MHz. Antenna gain remains poorer.

**Table 2**. Simulation results

| Freq. *Dualband* (MHz) | Kind of Antenna | Freq. Resonant $f_r$ (MHz) | Band frequency (MHz) | | | RL ($S_{11}$) (dB) | $S_{21}$ (dB) | *Gain* (dB) |
|---|---|---|---|---|---|---|---|---|
| | | | $f_{up}$ | $f_{down}$ | BW | | | |
| 1800 | Non DGS | 1840 | 1877 | 1706 | 171 | -14.78 | -25.86 | 6.351 |
| | DGS | 1840 | 1879 | 1718 | 161 | -20.6 | -24.44 | 5.993 |
| 2400 | Non DGS | 2430 | 2557 | 2352 | 205 | -19.27 | -18.71 | 6.434 |
| | DGS | 2390 | 2474 | 2230 | 244 | -19.56 | -14.46 | 6.321 |

*3.3 Measurement results*

Measurement results (Table 3) show that for the frequency of 1800 MHz, DGS causes bandwidth widen, increasing about 5.4%. For the frequency of 2400 MHz, DGS results better return loss and gain. Gain is increasing about 33.3%.

.

**Table 3**. Measurement results

| Freq. *Dualband* (MHz) | Kind of Antenna | Freq. Resonant $f_r$ (MHz) | Frequency Band (MHz) | | | RL ($S_{11}$) (dB) | *Gain* (dB) |
|---|---|---|---|---|---|---|---|
| | | | $f_{up}$ | $f_{down}$ | BW | | |
| 1800 | Non DGS | 1862 | 1905 | 1664 | 241 | -25.46 | 4.87 |
| | DGS | 1862 | 1906 | 1652 | 254 | -22.35 | 4.70 |
| 2400 | Non DGS | 2440 | 2585 | 2384 | 201 | -18.75 | 6.16 |
| | DGS | 2415 | 2531 | 2331 | 200 | -39.17 | 8.21 |

## 4. Conclusions

DGS impact on microstrip antennas vary for different frequencies. Simulation results show that DGS reduces bandwidth for 1800 MHz, but enlarge 19% bandwidth for 2400 MHz. Return loss is successfully reduced by DGS for both frequencies.

On the other hand, return loss decreases in 2400 MHz experiment measurement. Gain increases 33% on this channel. Generally, for both simulation and experiment, DGS is potential for reducing the return loss, enlarging the bandwidth and increasing gain.

## References

[1] Garg R, Bhartia P, Bahl I and Ittipibon A 2001 *Microstrip Antenna Design Handbook* (London: Artech House).

[2] Dash JC, Mishra GP, Mangaraj BB 2016 Design of dual band patch antenna with bandwidth enhancement using complementary defected ground structure *2nd Int. Conf. Contemporary Comput. Inf.* pp 271-4

[3] Ayman MI and Adel BA 2015 Isolated Elements Dual-Band Microstrip Patch Antenna Array Using C-Shape DGS *IEEE 32nd NRSC.*

[4] Thatere A, Zade PL and Arya D 2015 Bandwidth enhancement of microstrip patch antenna using 'U'slot with modified ground plane *IEEE ICMOCE* pp 408-11.

[5] Raj NA and Dwivedi RP 2015 High gain antenna with DGS for wireless applications *IEEE Int. Conf. SPIN* pp 19-24.

[6] Varun V and Sunil A 2014 Bandwidth Optimization using Fractal Geometry on Rectangular Microstrip Patch Antenna with DGS for Wireless Applications *IEEE Int. Conf. MedCom.*

# Improved hybridization of Fuzzy Analytic Hierarchy Process (FAHP) algorithm with Fuzzy Multiple Attribute Decision Making - Simple Additive Weighting (FMADM-SAW)

**B E Zaiwani[1], M Zarlis[2] and S Efendi[3]**

Department of Information and Technology, Faculty of Computer Science and Information Technology, Universitas Sumatera Utara, Medan, Indonesia.

zaiwani.babyesly@students.usu.ac.id

**Abstract**. In this research, the improvement of hybridization algorithm of Fuzzy Analytic Hierarchy Process (FAHP) with Fuzzy Technique for Order Preference by Similarity to Ideal Solution (FTOPSIS) in selecting the best bank chief inspector based on several qualitative and quantitative criteria with various priorities. To improve the performance of the above research, FAHP algorithm hybridization with Fuzzy Multiple Attribute Decision Making - Simple Additive Weighting (FMADM-SAW) algorithm was adopted, which applied FAHP algorithm to the weighting process and SAW for the ranking process to determine the promotion of employee at a government institution. The result of improvement of the average value of Efficiency Rate (ER) is 85.24%, which means that this research has succeeded in improving the previous research that is equal to 77.82%. Keywords: Ranking and Selection, Fuzzy AHP, Fuzzy TOPSIS, FMADM-SAW.

## 1. Introduction

In the research [1] hybridization algorithm of Fuzzy Analytic Hierarchy Process (FAHP) with Fuzzy Technique for Order Preference by Similarity to Ideal Solution (FTOPSIS) was conducted. Hybridization is done by weighting with FAHP algorithm and ranking with FTOPSIS algorithm. The result of this research is the average value of Efficiency Rate (ER) of 77.82% which is still felt too low. The conclusion of this research suggest that in this method developed again by hybridization of FAHP algorithm with Multiple Attribute Decision Making (FMADM).

## 2. Study of Literature

Decision Support System (DSS) is an interactive computer-based system that helps decision makers utilize data and models to solve a problem. There are several methods including Analytical Hierarchy Process (AHP) and Technique For Order Preference by Similarity to Ideal Solution (TOPSIS) [2]. AHP is a method in a decision-making system that uses several variables with a multilevel analysis process [3]. The analysis is done by giving the priority value of each variable, then do the pairwise comparison of the variables and alternative. TOPSIS is a method based on the concept that the best-chosen alternative not only has the shortest distance from the ideal solution, but also has the longest distance from the ideal solution.

In [1] an integrated approach of Fuzzy Analytical Hierarchical Process (FAHP) and Fuzzy Technique for Fuzzy Technique for Order Preference by Similarity to Ideal Solution (FTOPSIS) is used to calculate the weight of criteria and the FTOPSIS method is applied to prioritize the optimal alternative according to the criteria.

## 3. Findings and Discussions

### *3.1. Flowchart Research*

The Flowchart hybridization algorithm Fuzzy Analytic Hierarchy Process (FAHP) with Fuzzy Multiple-Attribute Decision Making - Simple Additive Weighting (FMADM-SAW) can be seen as in Figure 1.

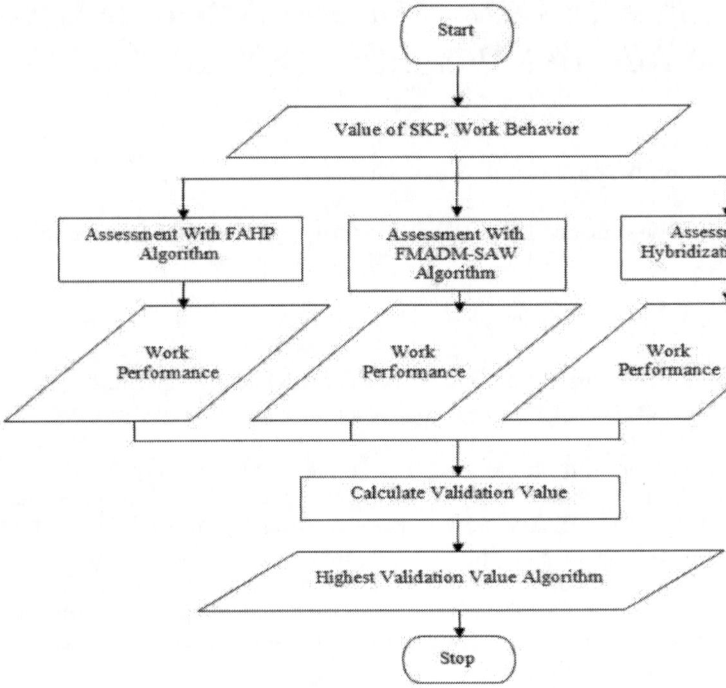


**Figure 1.** Flowchart Research

In the flowchart above, the input data in the form of SKP value and the value of Work Behavior taken 2 years. Furthermore, the data is processed by using FAHP, FMADM-SAW and hybridization algorithm is a combination of FAHP-FMADM-SAW algorithm. The result of the process is the Work Performance Value of each algorithm and validation value which is the determination of whether or not an employee is given promotion.

### *3.2. Data used*

The data used comes from Aparatur Sipil Negara (ASN) at Badan Kepegawaian Daerah dan Pengembangan Sumber Daya Manusia of Medan City (BKDPSDM) which is calculated according to the provisions of Badan Kepegawaian Negara as in Table 1.

**Table 1.** Employee Criteria Assessment Data

| Alternative | Name Employee | Value SKP-1 | Value SKP-2 | Value SKP-3 | Value Behavior-1 | Value Behavior-2 | Value Behavior-3 |
|---|---|---|---|---|---|---|---|
| A1 | Agus | 65 | 85 | 65 | 75 | 90 | 40 |
| A2 | Budi | 74 | 80 | 60 | 85 | 96 | 85 |
| A3 | Charli | 60 | 80 | 60 | 85 | 95 | 80 |
| A4 | Deni | 85 | 65 | 76 | 90 | 94 | 45 |
| A5 | Endang | 84 | 74 | 84 | 75 | 94 | 55 |
| A6 | Fajar | 90 | 84 | 90 | 75 | 80 | 96 |
| A7 | Gina | 50 | 56 | 60 | 70 | 70 | 87 |
| A8 | Harry | 74 | 55 | 50 | 60 | 70 | 75 |
| A9 | Indah | 75 | 71 | 45 | 62 | 90 | 65 |
| A10 | Jenny | 60 | 48 | 60 | 70 | 60 | 85 |

In table 1 contains the assessment results of each criterion for each employee scale 100 consisting of criteria SKP Value (Employee Objectives) 1 to 3 and Work Behavior 1 to 3

*3.3. FAHP algorithm*
The calculation steps of the FAHP algorithm are as in Table 3.

*3.3.1. Structure of hierarchy*
The hierarchical structure of the selection problem of promotion can be seen in Figure 2.

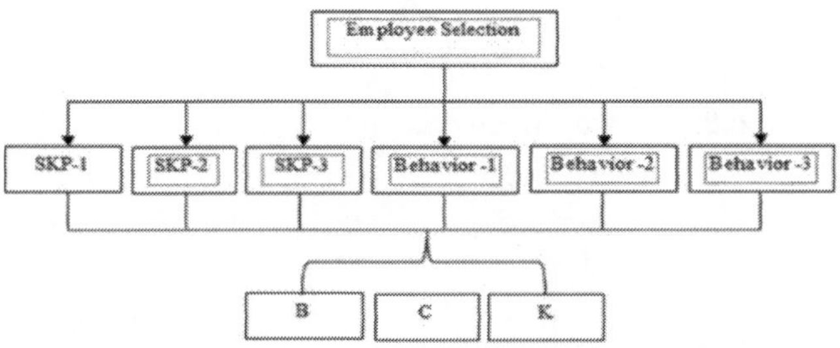


**Figure 2.** Structure of Employee Selection Hierarchy

Figure 2 above shows the hierarchical structure of the employee selection problems with selection criteria are SKP-1, SKP-2, SKP-3, Work Behavior-1, Work Behavior-2 and Work Behavior-3. The purpose of this assessment is to obtain the category of Good (B), Enough (C) or Less (K).

*3.3.2. Determination of synthesis value*

**Table 2.** Comparison of SKP matrix in pairs of Fuzzy AHP criteria

| | K1 = SKP-1 | | | K2 = SKP-2 | | | K3 = SKP-3 | | |
|---|---|---|---|---|---|---|---|---|---|
| | B | C | K | B | C | K | B | C | K |
| SKP-1 | 1 | 3 | 1 | 1 | 3 | 2 | 1 | 3 | 1 |
| SKP-2 | 0.33 | 1 | 0.5 | 2 | 1 | 2 | 1 | 0.55 | 0.33 |
| SKP-3 | 0.5 | 1 | 2 | 1 | 0.33 | 1 | 3 | 1 | 2 |

Value Table 2 above is obtained from the comparison between 1 element of SKP criteria with other SKP criteria elements.

**Table 3.** Comparison of Matrix Working Matched Behaviour of Fuzzy AHP criteria

| K1 = Value Behaviour-1 | K2 = Value Behaviour-2 | K3 = Value Behaviour-3 |
|---|---|---|

| | B | C | K | B | C | K | B | C | K |
|---|---|---|---|---|---|---|---|---|---|
| Behaviour-1 | 1 | 3 | 1 | 1 | 3 | 2 | 1 | 3 | 1 |
| Behaviour-2 | 0.33 | 1 | 0.5 | 2 | 1 | 2 | 1 | 0.55 | 0.33 |
| Behaviour-3 | 0.5 | 1 | 2 | 1 | 0.33 | 1 | 3 | 1 | 2 |

### *3.4. FMADM-SAW algorithm*

Step I: Evaluate the Fuzzy set of selected alternatives.

a. The linguistic variables that represent the weight of importance for each criterion are T (importance)

$$W = \{SR, R, C, T, ST\}$$

With:
SR = Very Low
R = Low
C = Enough
T = High
ST = Very High

Where each is represented by the triangle fuzzy number as follows:
SR = (0, 0, 0.25)
R = (0, 0.25, 0.5)
C = (0.25, 0.5, 0.75)
T = (0.5, 0.75, 1)
ST = (0.75, 1, 1)

b. The degree of suitability of alternatives to the decision criteria is:

$$T\ (match)S = \{SK, K, C, B, SB\}$$

with:
SK = Very Less
K = Less
C = Enough
B = Good
SB = Very Good

Where each is represented by the triangle fuzzy number as follows: SK = (0, 0, 0.25)
K = (0, 0.25, 0.5)
C = (0.25, 0.5, 0.75) B = (0.5, 0.75, 1)

c. Rating for each decision criterion
Ratings for each decision criterion can be seen as in Table 4.

**Table 4.** Rating of interest for each criterion

| **Criterion** | **Alias** | **Rating Interests** |
|---|---|---|
| Value SKP-1 | C1 | Enough (C) |
| Value SKP-2 | C2 | Very High (ST) |
| Value SKP-3 | C3 | Enough (C) |
| Value Behaviour-1 | C4 | High (T) |
| Value Behaviour-2 | C5 | Enough (C) |
| Value Behaviour-3 | C6 | Enough (C) |

### *3.5. Implementation of Hybridization Algorithm (FAHP-FMADW-SAW)*

Hybridization algorithm is the application of FAHP algorithm on weighting and FMADM-SAW for ranking based on input from FAHP to determine promotion of ASN with the case study of BKDPSDM. Ways of weighting and consistency ratio with FAHP algorithm and ranking with the FMADM-SAW algorithm can be seen as Chapter 3.3 and 3.4 above.

**Table 5.** Result Weighted FAHP Algorithm

| Criterion | Weight | Percentage |
|---|---|---|
| SKP-1 | 0.028 | 0.2% |
| SKP-2 | 0.085 | 0.8% |
| SKP-3 | 0.14 | 14.0% |
| Behaviour-1 | 0.20 | 2.0% |
| Behaviour-2 | 0.25 | 25.0% |
| Behaviour-3 | 0.28 | 28.0% |

**Table 6.** Result FMADM-SAW Algorithm Ranking

| No. | Alternative | End Value | Point |
|---|---|---|---|
| 1 | Agus | 0.366 | 100 |
| 2 | Budi | 0.342 | 85 |
| 3 | Fajar | 0.191 | 70 |
| 4 | Denny | 0.121 | 50 |
| 5 | Charli | 0.010 | 50 |
| 6 | Gina | 0.000 | 0 |

In Table 5 we can see the weighting results for each criterion (SKP or Work Behavior). In Table 6 shows the ranking for each employee (alternative)

Furthermore, the weight obtained with FAHP algorithm is calculated by the FMADM-SAW algorithm to step rank for determination of promotion in the employee. Flowchart Hybridization algorithm can be seen as in Figure 3.

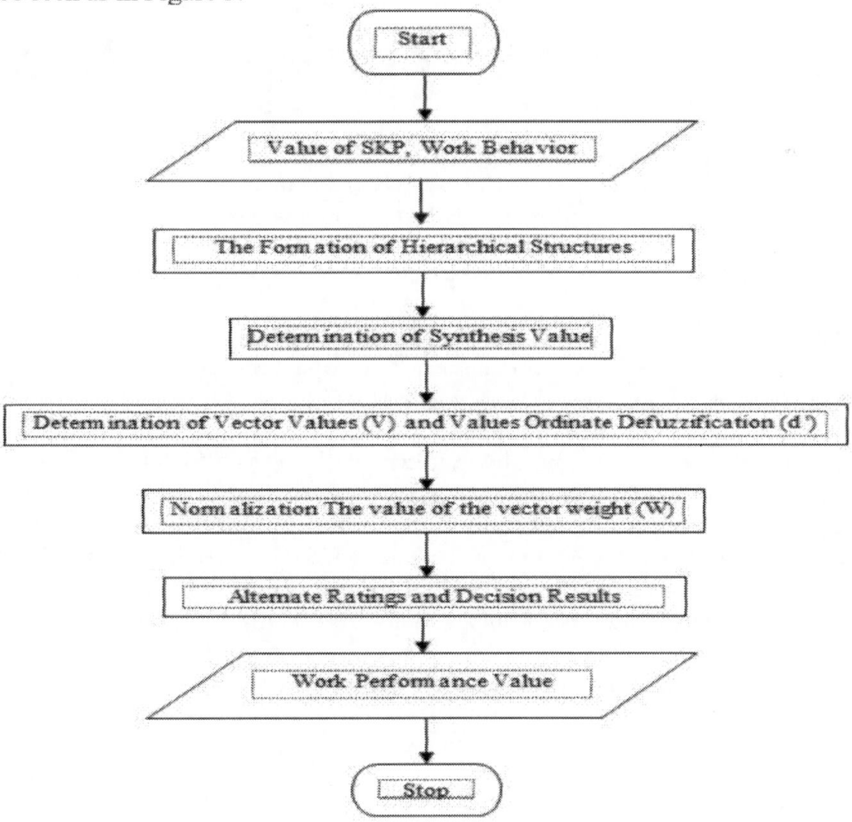


**Figure 3.** Flowchart Hybridization Algorithm

## 4. User interface

### *4.1. View of example alternative data*

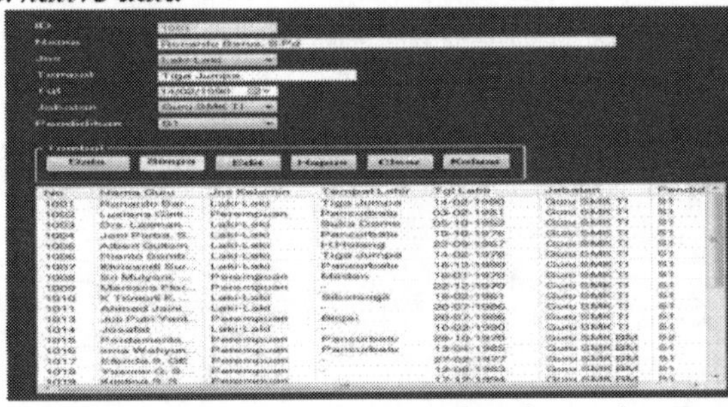

**Figure 4.** View of Example Alternative Data

In this form look the data input alternative the employee biodata that will be processed ranking for determination promotion.

### *4.2. View of ranking process*

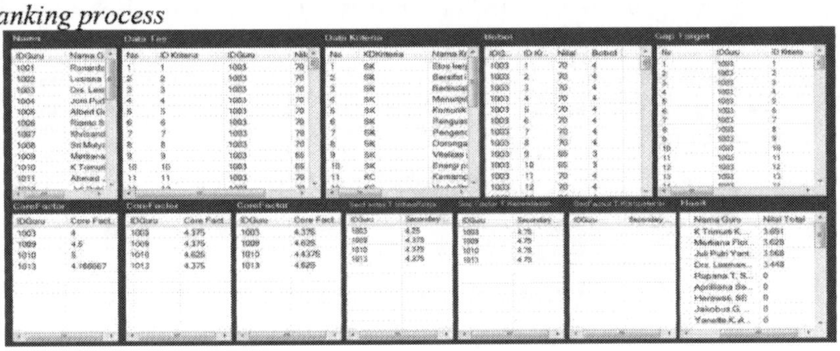

**Figure 5.** View of Ranking Process

In this form seen the ranking process of each alternative that can be seen the result on the bottom right.

## 5. Conclusion

There is an improvement of Dooki, A.E., Bolhasani, P. & Fallah, M. (2017) research, where the average value of Efficiency Rate (ER) is only 77.82% which is still too low. In Dooki, Bolhasani and Fallah's research, hybridization of Fuzzy Analytic Hierarchy Process (FAHP) algorithm was adopted by Fuzzy Technique for Order Preference by Similarity to Ideal Solution (FTOPSIS) to select the best bank chief inspector based on several qualitative and quantitative criteria with various priorities. The AHP and TOPSIS Fuzzy Methods are used to determine the criteria and ranking weight of each of the selected inspectors. In this research, we made an improvement of Dooki and Bolhasani and Fallah research by improving the FAHP algorithm with Simple Additive Weighting (SAW), which is done by applying FAHP Algorithm to the weighting process and SAW for the ranking process to determine the promotion of the employee. The result of improvement is the average value of ER is 85.24%, which means that this research has succeeded in improving Dooki and Bolhasani and Fallah research by 7.42%.

## References

[1] Dooki AE, Bolhasani P and Fallah M. 2017 An Integrated Fuzzy AHP and Fuzzy TOPSIS Approach for Ranking and Selecting the Chief Inspectors Of Bank: A Case Study *J. Appl. Res. Ind. Eng.* vol 4 **1** 8–23.

[2] Supraja S and Kousalya P 2016 A comparative study by AHP and TOPSIS for the selection of all round excellence award *Proc. IEEE Int. Conf. Electrical Electron. Optimization Tech.*

*(Hyderabad).*

[3] Xiulin SI and Dawei LI 2014 An improvement analytic hierarchy process and its application in teacher evaluation *Proc. Int. Conf. Intell. Syst. Des. Eng. Appl. (Liaoning).*

[4] Xie Z, Zhang F, Cheng J and Li L 2013 Fuzzy multi-attribute decision making methods based On improved set pair analysis *Int. Symp. Comput. Intell. Des. (Qingdao).*

[5] Lv ZY, Liang XN and Liang XZ 2015 A fuzzy multiple attribute decision making method based on possibility degree *Int. Conf. Fuzzy Syst. Knowledge Discovery.*

[6] Engel MM, Utomo WH and Purnomo HD 2017 Fuzzy Multi Attribute Decision Making–Simple Additive Weighting (MADM-SAW) for Information Retrieval (IR) in e-commerce recommendation *Int. J. Comp. Sci. Software Eng.* vol 6(6).

[7] Novian, D.2010, Sistem Pendukung Keputusan Mutasi, Enumerasi dan promosi pegawai menggunakan metode AHP (Analytic Hierarchy Process). Jurnal Ilmiah Media Elektrik,Vol 5, No.2, Desember 2010.

[8] Sutikno, Sistem Pendukung Keputusan Metode AHP Untuk Pemilihan Siswa Dalam Mengikuti Olimpiade Sains Di Sekolah Menengah Atas. Program Studi Ilmu Komputer. UNDIP.

[9] Purnomo, E. N. S., Sihwi, S. W. & Anggrainingsih, R. 2013. Analisis Perbandingan Menggunakan Metode AHP, TOPSIS, dan AHP-TOPSIS dalam Studi Kasus Sistem Pendukung Keputusan Penerimaan Siswa Program Akselerasi. Informatika, Fakultas MIPA, Universitas Sebelas Maret Surakarta. Jurnal ITSMART Vol 2. No 1. Juni 2013 ISSN: 2301–7201.

IOP Conf. Series: Journal of Physics: Conf. Series **978** (2018) 012077 doi:10.1088/1742-6596/978/1/012077

# Predicted cycloartenol synthase protein from *Kandelia obovata* and *Rhizophora stylosa* using online software of Phyre2 and Swiss-model

**M Basyuni[1]*, N Sulistiyono[1], R Wati[1], Sumardi[2], H Oku[3], S Baba[4], H Sagami[5]**

[1] Department of Forestry, Faculty of Forestry, Universitas Sumatera Utara, Jl. Tri Dharma Ujung No. 1 Medan, North Sumatera, 20155, Indonesia
[2] Faculty of Pharmacy, Universitas Tjut Nyak Dhien, Jl. Rasmi No. 28 Medan 20123, Indonesia
[3] Molecular Biotechnology Group, Tropical Biosphere Research Center, University of the Ryukyus, 1 Senbaru, Nishihara, Okinawa 903-0213, Japan
[4] International Society for Mangrove Ecosystems, Faculty of Agriculture, University of the Ryukyus, 1 Senbaru, Nishihara, Okinawa 903-0213, Japan
[5] Institute of Multidisciplinary Research for Advanced Material, Tohoku University, Sendai, 980-8577, Japan

*Email: m.basyuni@usu.ac.id

**Abstract**. Cloning of *Kandelia obovata KcCAS* gene (previously known as *Kandelia candel*) and *Rhizophora stylosa RsCAS* have already have been reported and encoded cycloartenol synthases. In this study, the predicted *KcCAS* and *RsCAS* protein were analyzed using online software of Phyre2 and Swiss-model. The protein modelling for *KcCAS* and *RsCAS* cycloartenol synthases was determined using Pyre2 had similar results with slightly different in sequence identity. By contrast, the Swiss-model for *KcCAS* slightly had higher sequence identity (47.31%) and Qmean (0.70) compared to *RsCAS*. No difference of ligands binding site which is considered as modulators for both cycloartenol synthases. The range of predicted protein derived from 91-757 amino acid residues with coverage sequence similarities 0.86, respectively from template model of lanosterol synthase from the human. Homology modelling revealed that 706 residues (93% of the amino acid sequence) had been modelled with 100.0% confidence by the single highest scoring template for both *KcCAS* and *RsCAS* using Phyre2. This coverage was more elevated than swiss-model predicted (86%). The present study suggested that both genes are responsible for the genesis of cycloartenol in these mangrove plants.

## 1. Introduction

Phytosterols and pentacyclic triterpenoids are widespread in higher plants including mangrove forests [1]. Molecular cloning of two mangrove plants namely *Kandelia obovata KcCAS* gene (previously known as *Kandelia candel*) and *Rhizophora stylosa RsCAS* had been cloned, characterized, and encoded cycloartenol synthases [2]. The *KcCAS* and *RsCAS*, therefore, are involved in phytosterol biosynthesis, a member of oxidosqualene cyclases (OSCs) gene. The open reading frame (ORF) of both genes consist of 2277 bp that encode 58 amino acid polypeptide, respectively [2]. Salt tolerance mechanism in two mangrove plants *K. obovata* and *R. stylosa* showed variation responses: mRNA level of *KcCAS* was not modulated by salinity concentration in the roots and was lowered after removal the salinity and transfer to freshwater [3-4]. By contrast, *KcCAS* decreased in the leaves of *K. obovata* [3]. In case of *RsCAS*, the mRNA level was enhanced with salinity only in the roots but not in the leaves [5].

Particular prominence was on the phytosterols because these compounds were significant membrane permeability for steroid hormone and biomarkers for marine organic matter [6-7] and might give more insight into the physiological roles of phytosterols in mangrove plant species. It has been reported that the highest mitochondrial target peptide in the possibility of the potential transit peptide was in *KcCAS* [8]. Regardless of the critical part of the protein in the plant kingdom, the predicted protein model of cycloartenol synthases from mangrove plants have rarely been studied. Thus, the present study aimed to analyze protein structure from two cycloartenol synthases of *K. obovata* and *R. stylosa*.

## 2. Materials and Method

### *2.1. Material*
Two cycloartenol synthase genes from NCBI database (https://www.ncbi.nlm.nih.gov/) were collected. The GenBank accession numbers of the DNA and amino acid sequences of using this analysis is as follows: AB292609, BAF73930 (*K. obovata KcCAS*) and AB292608, BAF73929 (*R. stylosa RsCAS*).

### *2.2. Template search and model building*
Template search and the modelling of the three-dimensional structure of the protein was performed using Phyre2 (http://www.sbg.bio.ic.ac.uk/~phyre2/html/page.cgi?id=index) [9] and Swiss-model (https://swissmodel.expasy.org/) [10]. The templates with the highest quality have then been selected for model building.

### *2.3. Prediction of ligand binding site*
The ligand binding site prediction in the modelled protein structure was carried out using 3DLigandSite server (http://www.sbg.bio.ic.ac.uk/3dligandsite/) [11].

### *2.4. Transmembrane helix prediction*
Transmembrane helices were predicted the sequence using phyre2 (http://www.sbg.bio.ic.ac.uk/~phyre2/html/page.cgi?id=index) [9]. The extracellular and cytoplasmic sides of the membrane were labelled [9]. The start and end of each transmembrane helix described a number indicating the residue index.

## 3. Results and Discussion

### *3.1. Protein modelling using Phyre2 and Swiss-model*
Table 1 showed protein modelling using Pyre2 of *RsCAS* and *KcCAS* cycloartenol synthases. Both genes had similar results with slightly different in sequence identity. The protein model used human lanosterol synthase with protein database (PDB) ID: c1w6kA as a template structure with confidence 100% and coverage 93%.

Similarly, results of protein modelling using Swiss-model was listed in Table 2. *KcCAS* slightly had higher in sequence identity (47.31%) and Qmean (0.70, estimated model reliability between 0-1) compared to *RsCAS*. The range of predicted protein derived from 91-757 amino acid residues with coverage sequence similarities 0.86, respectively from template model of lanosterol synthase from human (Table 2).

**Table 1**. Results of protein modelling using Phyre2

| Gene | Template ID | Sequence Identity | Residue | Confidence | Coverage (%) |
|---|---|---|---|---|---|
| KcCAS | c1w6kA | 46 | 706 | 100 | 93 |
| RsCAS | c1w6kA | 45 | 706 | 100 | 93 |

**Table 2**. Results of protein modelling using Swiss-model

| Gene | Template | Sequence Identity | Sequence similarity | Range | Coverage | QMEAN |
|---|---|---|---|---|---|---|
| KcCAS | 1w6k.1.A | 47.31 | 0.43 | 91-757 | 0.86 | 0.70 |
| RsCAS | 1w6k.1.A | 46.85 | 0.43 | 91-757 | 0.86 | 0.69 |

*3.2. Predicted binding site*

Ligands which are considered in the current study as modulators for both cycloartenol synthases are displayed in Table 3. The ligands bind particular residue, amino acid, contact, and average distance. *KcCAS* had seven residues from 677, 678, 679, 680, 683, and 74 which corresponded to amino acids of His, Val, Val, Asn, Trp, and Met. The longest distance belongs to Val678.

**Table 3**. Predicted binding site of two cycloartenol synthase

| Gene | Residue | Amino Acid | Contact | Average distance |
|---|---|---|---|---|
| KcCAS | 677 | HIS | 24 | 0.00 |
| | 678 | VAL | 11 | 0.69 |
| | 679 | VAL | 24 | 0.00 |
| | 680 | ASN | 21 | 0.29 |
| | 683 | TRP | 15 | 0.46 |
| | 724 | MET | 24 | 0.00 |
| RsCAS | 678 | ILE | 11 | 0.69 |
| | 679 | VAL | 24 | 0.00 |
| | 680 | ASN | 21 | 0.29 |
| | 683 | TRP | 15 | 0.46 |
| | 724 | MET | 24 | 0.00 |

In contrast to this observation, *RsCAS* possed five residues Ile678, Val679, Asn680, Trp683, and Met724 with contact value was 11-24. A similar result was obtained for average distance for Ile678 (0.69) of *RsCAS* cycloartenol synthase.

It is noteworthy that the phytosterol composition from leaves and roots of *K. obovata* and *R. stylosa* was diversity. β-sitosterol is the major phytosterol component of both species in leaves and roots, as could be expected for higher plants including mangrove species [1, 5]. The concentration of cycloartenol, a precursor for phytosterol biosynthesis, is usually low in the plant tissue [12]. Our previous study also found the low level of cycloartenol in the leaf of both of *R. stylosa* and *K. candel* [1], suggesting that both genes are responsible for the genesis of cycloartenol in these mangrove plants. It is yet to come to demonstrate the involvement of cycloartenol synthase gene in the phytosterol biosynthesis of the plant kingdom.

*3.3. Phyre2 and Swiss-protein model*

Homology modelling revealed that 706 residues (93% of the amino acid sequence) had been modelled with 100.0% confidence by the single highest scoring template for both *KcCAS* and *RsCAS* using Phyre2 (Figure 1 A and B). This coverage was higher than the Swiss-model predicted (86%) as showed in Figure 2A and B. Furthermore the number aligned proteins both genes were 93 with some matched PDB structures were 31. This results supported therefore by the predict molecular weight for both *KcCAS* and *RsCAS* were 83 kD, and approximately theoretical isoelectric residues were 4.91, respectively.

This slight difference model may be due to the difference zonation of mangrove between *K. obovata* and *R. stylosa*. *R. stylosa* usually grows close to the sea or coast, and *K. obovata* distributes more inland ward [5]. These results suggested the importance of understanding the diversity of phytosterol genes in mangroves. Therefore further investigation is needed to clarify this finding.

**A**

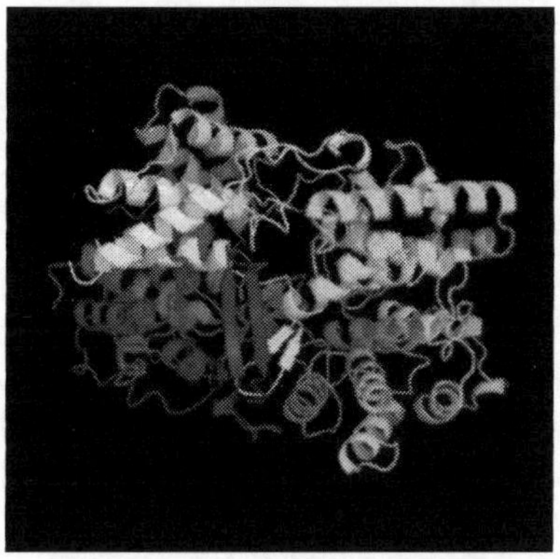

Image coloured by rainbow N → C terminus
Model dimensions (Å):
X:76.216 Y:72.109 Z:69.233

**B**

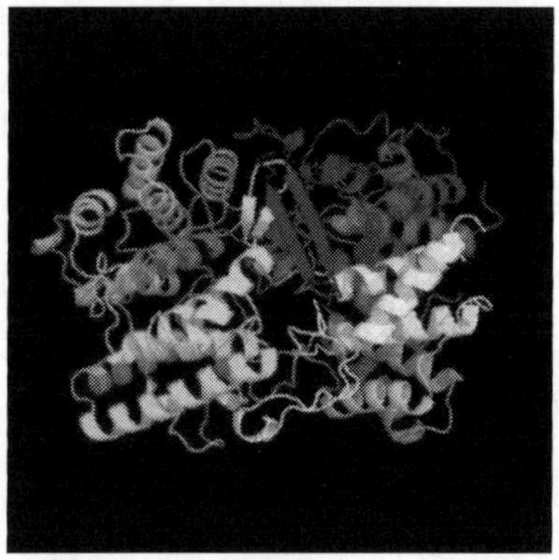

Image coloured by rainbow N → C terminus
Model dimensions (Å):
X:78.753 Y:72.109 Z:69.233

**Figure 1.** Phyre2 protein model for*KcCAS* (A) and *RsCAS* (B).

**A**

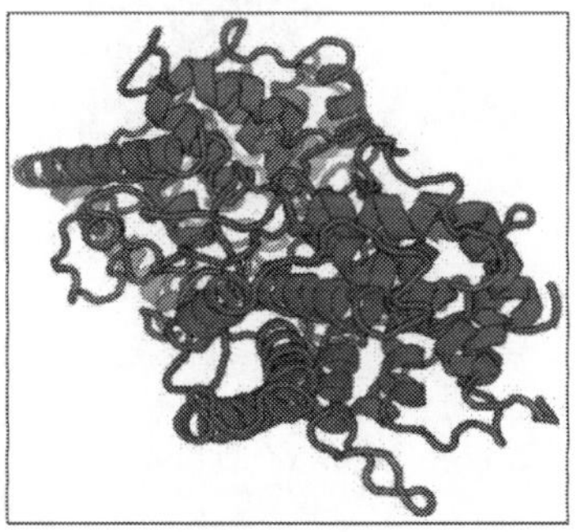

**B**

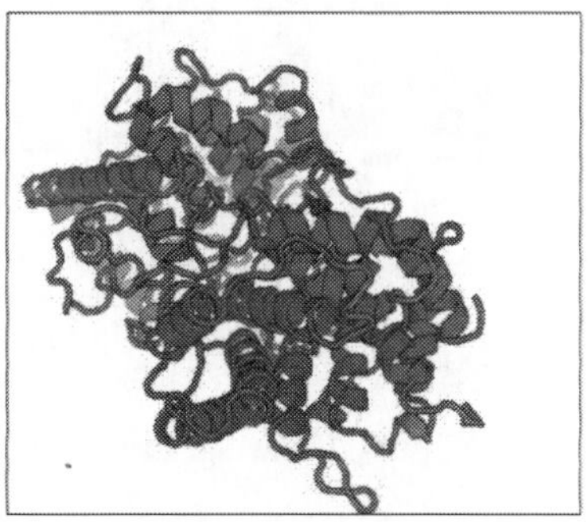

**Figure 2.** Swiss protein model for *KcCAS* (A) and *RsCAS* (B).

### *3.4. Transmembrane helices*

Figure 3 showed transmembrane (TM) helices in both *KcCAS* and *RsCAS*. Both genes had four TM helices (S1-S4). The different position between *KcCAS* and *RsCAS* were on S2 (257-272 and 256-271) and S3 (330-315 and 334-319) both in the extracellular and cytoplasmic. Several lines of the study demonstrated that the sterol is a modulator of membrane permeability [1, 12]. Triterpenoid alcohols have been structurally differentiated from phytosterols, although they share common biosynthetic pathway and hence the similar chemical structure [2]. However, there has been no scientific rationale to distinguish these compounds functionally as the structural membrane lipid. In this context, our *in vitro* study showed that triterpene, as well as phytosterols, was incorporated into the lipid bilayer of the liposome, suggesting that triterpenoid is also a modulator of membrane permeability [4-5].

**A**

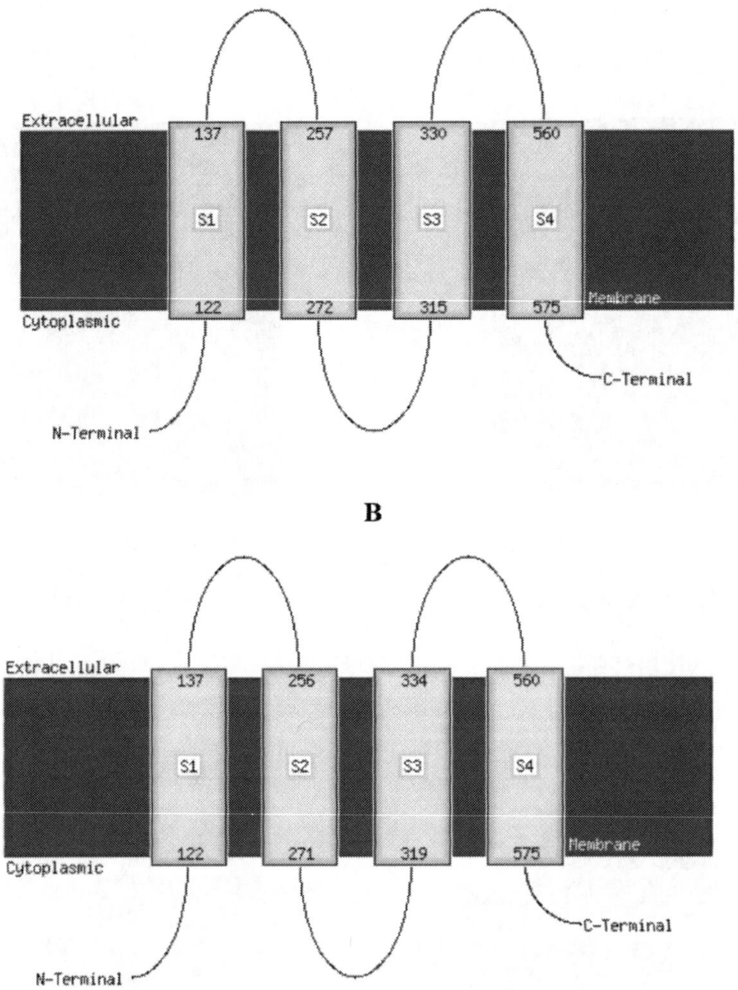


**Figure 3**.Transmembrane helices of *KcCAS* (A) and *RsCAS* (B)

## 4. Conclusions

The predict protein modelling for *KcCAS* and *RsCAS* cycloartenol synthases showed similar results with slightly different in sequence identity using Phyre2 and Swiss-model. In a case for the Swiss-model for *KcCAS* fairly had higher sequence identity and Qmean compared to *RsCAS*. The present study suggested that both genes are responsible for the origin of cycloartenol in these mangrove plants.

## Acknowledgment

A part of this work was supported by an International Research Collaboration and Scientific Publication Grant 2017 (No. 003/SP2H/LT/DRPM/IV/2017 to MB) from the Directorate for Research and Community Service, Ministry of Research, Technology and Higher Education, Republic of Indonesia.

**References**

[1] Basyuni M, Oku H, Baba S, Takara K, Iwasaki H 2007 Isoprenoids of Okinawan mangroves as lipid input into estuarine ecosystem *J Oceanography* **63** 601–608.

[2] Basyuni M, Oku H, Tsujimoto E, Baba S 2007 Cloning and functional expression of cycloartenol synthases from mangrove species *Rhizophora stylosa* Griff. and *Kandelia candel* (L.) Druce *Biosci. Biotechnol. Biochem.* **71** 1788–1792.

[3] Basyuni M, Baba S, Inafuku M, Iwasaki H, Kinjo K, Oku H 2009 Expression of terpenoid synthase mRNA and terpenoid content in salt-stressed mangrove *J. Plant Physiol.***166** 1786–1800.

[4] Basyuni M, Baba S, Kinjo Y, Putri LAP, Hakim L, Oku H 2012 Salt-dependent increase in triterpenoids is reversible upon transfer to fresh water in mangrove plants *Kandelia candel* and *Bruguiera gymnorrhiza J. Plant Physiol.* **169** 1903–1908.

[5] Basyuni M, Baba S, Kinjo Y, Oku H 2012 Salinity increases the triterpenoid content of a salt secretor and a non-salt secretor mangrove *Aquat. Bot.* **97** 17–23.

[6] Rontani JF, Vaultier F, Bonin P 2014 Biotic and abiotic degradation of marine and terrestrial higher plant material in intertidal surface sediments from Arcachon Bay (France): A lipid approach *Mar. Chem.* **158** 69–79.

[7] He D, Mead RN, Belicka L, Pisani O, Jaffé R 2014 Assessing source contributions to particulate organic matter in a subtropical estuary: a biomarker approach *Org. Geochem.* **75** 129–139.

[8] Basyuni M, R Wati 2017 Bioinformatics analysis of the oxidosqualene gene and the amino acid sequence in mangrove plants *J. Phys. : Conf. Ser.* **801** 012011.

[9] Kelley LA, Mezulis S,Yates CM, Wass MN, Sternberg, MJE 2015 The Phyre2 web portal for protein modeling, prediction and analysis *Nat. Protocols* **10** 845–858.

[10] Arnold, K., Bordoli, L., Kopp, J. and Schwede, T. 2006 The SWISS-MODEL workspace: a web-based environment for protein structure homology modelling *Bioinformatics* **22** 195–201.

[11] Wass M.N., Kelley LA, Sternberg MJ 2010 3DLigandSite: predicting ligand-binding sites using similar structures *Nucleid Acid Res.* **38** W469–473.

[12] Uddin M, Sarker M, Islam Z, Ferdosh S, Akanda M, Haque J, Easmin M, Shamsudin B, Hadijah S, Yunus KB 2015 Phytosterols and their extraction from various plant matrices using supercritical carbon dioxide: a review *J. Sci. Food Agric.* **95** 1385–1394.

# 5W1H Information Extraction with CNN-Bidirectional LSTM

**A Nurdin[1] and N U Maulidevi[2]**

[1,2]School of Electrical Engineering and Informatics, Institut Teknologi Bandung, Bandung, Indonesia

Email : [1]23515035@std.stei.itb.ac.id, [2]ulfa@stei.itb.ac.id

**Abstract.** In this work, information about *who, did what, when, where, why,* and *how* on Indonesian news articles were extracted by combining Convolutional Neural Network and Bidirectional Long Short-Term Memory. Convolutional Neural Network can learn semantically meaningful representations of sentences. Bidirectional LSTM can analyze the relations among words in the sequence. We also use word embedding word2vec for word representation. By combining these algorithms, we obtained F-measure 0.808. Our experiments show that CNN-BLSTM outperforms other shallow methods, namely IBk, C4.5, and Naïve Bayes with the F-measure 0.655, 0.645, and 0.595, respectively.

**Keywords**: information extraction; 5w1h; deep learning; Convolutional Neural Network; Bidirectional Long Short-Term Memory

## 1. Introduction

News articles provide information about event that consist of information *who did what, when, where, why*, and *how* (5W1H). Unfortunately, people often miss these main information since there are too much information about an issue, known as information overload. To overcome this problem, 5W1H extraction system can automatically present the main information in each news is proposed.

Information extraction is a task in Natural Language Processing (NLP) that aims to extract semantic content from a text [1]. This task can be solved by deep learning approach. In recent years, deep learning in machine learning field have succeeded show the best performance in classification task e.g. speech recognition [2], multimedia classification [3], and natural language processing [4-12]. To the best of our knowledge, there is no existing 5W1H information extraction research on Indonesian news articles using deep learning method.

In NLP task, convolutional neural network have been proven to be efficient for capturing meaningful representation of sentences e.g. classification and language modeling [4], sentiment analysis [5], named tagging and semantic role labeling [6], and event extraction [7,8]. Bidirectional LSTM is a type of RNN that popular for sequential modeling and achieving the best performance in machine translation [9], part-of-speech tagging [10], and language modeling [11,12].

In this paper, the task of 5W1H extraction was transformed into text categorization. Similar to previous research in 5W1H extraction on Indonesian news article [13-15], this task is performed as sequence tagging with BIO (Begin In Other) labelling scheme. Each token in an article will be classified into one of 13 classes. Label token depends on other token observations in the sequence and information of the entire sentence. We propose method that using CNN to extract information syntactic and semantic token within the sentence and using bidirectional LSTM to learn the relations among tokens in the sequence.

The rest of the paper is organized as follows. In section 2, related works in this area are presented. Section 3 describes our method. In section 4, experiments are discussed. Section 5 presents conclusion and future work.

## 2. Related Work

The research to extract 5W1H information on Indonesian news articles has been done by Khodra [13], Wicaksana [14], and Ilyas et.al. [15]. The three studies used machine learning to get key information in the news, what events occurred, the participants involved, the time and the location of the incident, why and how the incident occurred.

Khodra [13] extracted 5W1H information on 90 news articles using the C4.5 and AdaboostM1 algorithms with features set consist of the currentWord, bef1Word, bef2Word, currentTag, bef1Tag, bef2Tag, currentNETarg, bef1NETag, bef2NETag, bef1Label, Bef2Label, isTimePattern, and isTitleToken.

Another research was conducted by Ilyas et.al [15]. In their research, sequence labeling technique, such as previous research, was used on 190 news articles and overcame the imbalanced dataset by using SMOTE (Synthetic Minority Over-Sampling). IBk (Instance-based Learning) algorithms, SVM, and C4.5 were used for classifying. The features used in their research are token, token kind, contextual, morphological, post, ne, sentence_number, location, bef-one-token, bef-two-token, bef-one-post, bef-two-post, bef-one-ne, bef-two-ne, bef-one-class, bef-two-class, and current class.

Wicaksana [14] also conducted research to extract 5W1H information. The dataset used consisted of 80 news articles. Wicaksana [14] used bef1class, idxsentence, bef1pofe, and ne. The algorithms which used in the study are Naïve Bayes and IBk.

Unlike the previous researches [13-15] that used shallow learning algorithm, this research use deep learning CNN and BLSTM algorithm for modeling 5W1H information extraction in news articles. In addition for representation word, we use word2vec [16] which can represent the semantic distribution of words. In Khodra [13], Wicaksana [14], and Ilyas [15] research, the performance of the algorithm was evaluated by using 10-fold cross validation. The use of cross validation in Ilyas [15] and Wicaksana [14] was performed after resampling the dataset with the oversampling method of SMOTE. This could lead to overfitting as there may be replication of training data appearing in fold testing. Therefore, there is the same training data as the testing data. To avoid this phenomenon, cross validation is performed to split data into training and testing set then apply oversampling only on training set or by providing testing data for evaluation. In this research, we divide the dataset into training and testing data. Random oversampling is applied to train data to overcome the imbalanced dataset.

Chen, et.al. [7] and Nguyen, et.al. [8] employed CNN for event detection on ACE (Automatic Content Extraction) 2005 corpus with 529 training documents and 40 testing documents. Nguyen, et.al. [8] used CNN to detect events by transforming each token into real-valued vector via word embedding, position embedding, and entity type embedding. Chen, et.al. [7] used a CNN with dynamic multi-pooling to extract sentence and lexical level features so that it can detect trigger and argument of events in a sentence. To extract sentence level features, three types of inputs are used i.e. context-word features, position features, and event-type features. All inputs are transformed into real-valued vectors using word embedding, position embedding, and event type embedding and then combined into an input matrix on the convolutional layer. The output from the pooling layer is combined with lexical level features of trigger and argument candidate, one token before and one token afterwards.

## 3. Methods

We formalize the 5W1H information extraction as a multi-class classification problem. For each token in an article, we want to predict if the current token is an element of *what, who, where, when, why,* or *how.* We use BIO notation, as sequence tagging labelling scheme. Each token will be classified into one of 13 classes (*B_who, I_who, B_what, I_what, B_when, I_when, B_where, I_where, B_why, I_why, B_how, I_how, Other*).

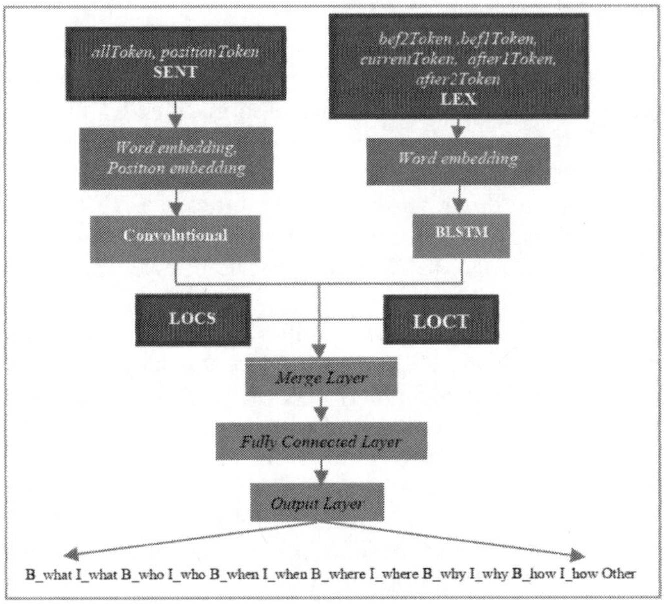


**Figure 1.** Proposed Network Architecture

Label of each token depends on other token observations in the sequence. Here, sequence tokens consist of current token and context token (token before and after current token). For example, consider the following sentences:

**S1: Indonesia** menjadi tuan rumah penyelenggara SEA GAMES XXVI.
(***Indonesia*** *hosts The 26$^{th}$ SEA GAMES.)*

**S2:** SEA GAMES XXVI akan diselenggarakan di **Indonesia**.
(*The 26$^{th}$ SEA Games will be held in* ***Indonesia****.)*

In S1, *Indonesia* is a subject that labeled as *who* element. In S2, *Indonesia* indicates a location name, so that labeled as *where* element. This label depends on the context token. In S1, *Indonesia* is followed by a verb, *menjadi*. *Indonesia* tells the subject of the event. As a general rule, a subject word usually followed by a predicate word or a verb. In S2, *Indonesia* is a location because it is preceded by preposition of place and position, *di*.

Beside information about current token and its context, we also need to observe token within a sentence. For example, consider the following sentence:

**S3:** *Pelaku* ***tabrak lari*** *yang menewaskan Siska* ***berhasil ditangkap*** *di jalan Monginsidi.*
(*The* ***hit-and-run*** *actor who killed Siska was captured on the Monginsidi road.*)

In S3, there are two possible phrases that can fill *what* slot, *tabrak lari* and *berhasil ditangkap*. Both phrases are preceded by a noun that can act as a subject. But, when we see the entire sentence, the phrase of *berhasil ditangkap* is *what* element and the phrase of *tabrak lari* describes the word of *pelaku* which is also the *who* element.

Since this 5W1H extraction is a sequence labelling task, we use bidirectional long short-term memory (BLSTM). BLSTM can incorporate past and future input information [17]. Network steps through the input sequence in both directions, backward and forward.

To induce sentence-level features, we use convolutional neural network, which has been successfully applied to event detection [7,8]. Output from BLSTM units and max pooling layer in convolutional neural network are combined with another additional features and then are fed into classifier. figure 1. illustrates the proposed network architecture.

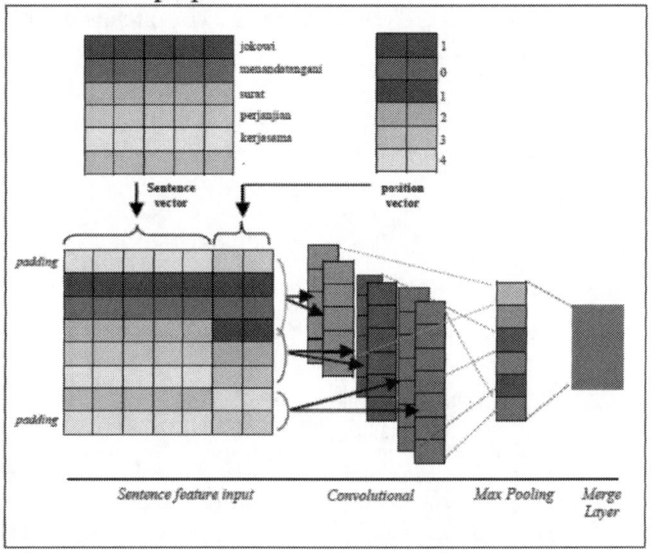


**Figure 2.** Convolutional Neural Network. It illustrates the processing of token *menandatangani* from sentence "*Jokowi menandatangani surat perjanjian kerjasama.*". We use two filter with window size in the set {2,3,4} to generate feature maps, and apply max pooling to each feature map.

*3.1. Extracting Sentence Features with Convolutional Neural Network*
Convolutional neural network can learn semantically meaningful representations of sentences. With max pooling layer, we can get the most useful information that represents the entire sentence. Figure 2. describes our convolutional neural network in details. This convolutional neural network is inspired by [7,8].

*3.1.1. Input.* In the CNN, an input must have fixed size. Therefore, we limit the context to a fixed window size by trimming longer sentence and padding shorter sentence with 0 when necessary. Before entering convolutional layer, each token is transformed into real-valued vector by using the following embedding:

- **Word Embedding.** We use word2vec [16] embedding trained on 10.956 Indonesian news articles. Each token in a sentence is transformed into 50-dimensional word vector. If token was not found in word embedding, we randomly initialized the word embedding vectors using uniform distribution with the interval from -0.25 to 0.25.
- **Position Embedding.** To specify the token that will be predicted, we use position embedding. This embedding randomly initialized. Position embedding provides information about relative distance each token within a sentence with current token.

We concatenate real-valued vector from word embedding and position embedding into one 2-dimensional matrix. We then fed this input matrix into convolutional layer.

*3.1.2. Convolutional Layer.* In this layer, vector matrix will be convoluted by filter with some window size *k*. This filter is composed by weight vectors and bias, and slides over the input matrix. After operating non-linear function, we obtained some valuable features with lower dimension in feature map.

*3.1.3. Max Pooling Layer.* The pooling units perform max operation to capture the most important features (max value) within each feature map. These features are concatenated with other features into single vector before they are fed into classifier.

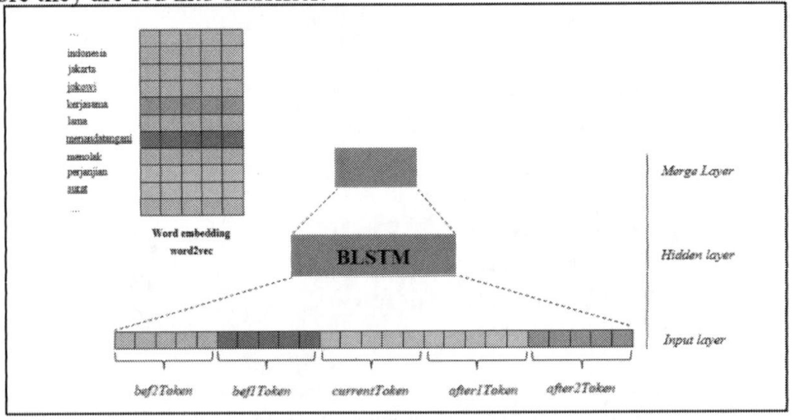


**Figure 3.** Bidirectional LSTM for sequence labelling. It illustrates the processing of one sequence that consist of current token *surat,* bef1token *menandatangani,* bef2token *jokowi,* after1token *perjanjian* and after2token *kerjasama.*

*3.2. Sequence Labelling with Bidirectional Long Short Term Memory.*
Given a sequence $x_{n-2}$, $x_{n-1}$, $x_n$, $x_{n+1}$, $x_{n+2}$ with label $y$. Bidirectional LSTM is used to model this sequence and to analyze the relationship among tokens in the sequence. Each token $x$ is transformed into word vector by using word embedding. Output BLSTM will be combined with other features and then be fed into classifier. Figure 3. illustrates this BLSTM unit.

*3.3. Additional Features*
Other than all tokens and their relative position within a sentence (**SENT**) and lexical sequence (**LEX**), we use the following additional features:

- **LOCT:** Word's position within a sentence has a positive meaning for determining its role.

$$\text{LOCT} = 1 - \frac{\log t}{\log n} \qquad (1)$$

Where $n$ is the number of tokens in the sentence and $t$ is token location within the sentence. For instance, if token $x$ is at beginning of the sentence, $t$=1.

- **LOCS:** Sentence token location within an article. *Who, what, where,* and *when* elements often appear at the beginning of a news article. And *why* and *how* elements often appear at the middle or end of the article.

$$\text{LOCS} = 1 - \frac{\log s}{\log n} \qquad (2)$$

Where $n$ is the number of sentences in an article and $s$ is sentence token location within the article.

## 4. Experiments

*4.1. Dataset*
We use 5W1H corpus of Indonesian news articles that have been collected by Wicaksana [14]. The corpus labeled manually by human annotator, consists of 484 news articles. We utilized test set with

145 news articles, and training set with the remaining 339 articles. Figure 4. shows an example of 5W1H information on Indonesian news article [13].

From 484 news articles, we built dataset of BIO-labeled token. After annotating, we obtained 123.586 instances. Table 1. shows the label distribution.

From the distribution labels, we can see that it is an imbalanced dataset. To overcome this problem, we apply random oversampling to minority classes in training set.

**Table 1.** Label Distribution Based on BIO Labelling Scheme

| **Label** | Count | |
|---|---|---|
| | ***Training Set*** | ***Testing Set*** |
| **B_who** | 340 | 145 |
| **I_who** | 1.586 | 713 |
| **B-what** | 340 | 145 |
| **I_what** | 3.924 | 1.418 |
| **B_where** | 403 | 241 |
| **I_where** | 921 | 509 |
| **B_when** | 324 | 142 |
| **I_when** | 1.300 | 624 |
| **B_why** | 302 | 120 |
| **I_why** | 3.863 | 1.583 |
| **B_how** | 324 | 122 |
| **I_how** | 5.164 | 2.060 |
| **Other** | 69.179 | 27.794 |

| **Rakyat Venezuela Mulai Pilih Pengganti Chavez** ***[Venezuelans Vote to Choose Chavez Successor]*** | **5W1H Information** |
|---|---|
| REPUBLIKA.CO.ID, KARAKAS – Transisi kekuasaan akan dimulai di Venezuela. Tidak kurang dari 19 juta pemilih, akan menentukan masa depan politik negara tersebut saat Ahad (14/4) waktu setempat. Beberapa hasil polling masih menempatkan, pejabat presiden sementara, Nicolas Maduro unggul dari pesaingnya, Hanrique Capriles. Tapi, prediksi tersebut dikhawatirkan berbalik. Reuters mengatakan Maduro, menurut polling, unggul dua dijit meninggalkan bekas Gubernur Negara Bagian Miranda tersebut. Namun, kesenjangan popularitas itu kian menyempit menjelang pemilihan. Dikatakan jarak popularitas hanya terpaut tujuh persen. Caprilles mengatakan, aksi manipulasi politik Maduro telah memecah belah masyarakat Venezuela. Dirinya yakin akan mengambil peran presiden, and akan menjungkalkan rezim sosialisme di negara itu. Dalam kampanye, dia mengatakan, pesaingnya itu culas memanfaatkan mendiang Presiden Hugo Chavez untuk kepentingan pribadinya. Padahal, kata dia, mendiang Chavez adalah milik semua masyarakat Venezuela. Tempat pemungutan suara akan di buka Ahad (14/4) pukul 06:30 waktu setempat, atau Senin (15/4) waktu Indonesia. TPS akan dibuka selama 10 jam, dengan voting secara elektronik. Pemenangnya akan dilantik 19 April mendatang, and akan menentukan pemerintahan sampai 2019 mendatang. | **What**: menentukan masa depan politik negara tersebut *[determine political future of the country]*<br><br>**Who:** 19 juta pemilih *[19 million voters]*<br><br>**Where**: Venezuela<br><br>**When:** Ahad (14/4) *[Sunday, 14 April]*<br><br>**Why**: Pemenangnya akan dilantik 19 April mendatang, and akan menentukan pemerintahan sampai 2019 mendatang *[The winner is due to be sworn in on 19 April and serve until January 2019]*<br><br>**How**: TPS akan dibuka selama 10 jam, dengan voting secara elektronik. *[Polls will open for 10 hours, by electronic voting]* |

**Figure 4.** Example of 5W1H information on Indonesian news article [13]

### *4.2. Hyperparameter*

We train the network with batch size=64 and epoch=10. We used Adadelta [18] optimizer (lr=1.0, rho=0.95, decay=0.0, epsilon=1e-08). ReLU activation function was used on convolutional layer and fully connected layer. Because we convert the output variable into one-hot encoding representation, we apply sigmoid function on output layer.

For convolutional neural network, we use multiple window size 2, 3, 4 and utilized two features maps for each window size. For bidirectional LSTM, we set 250 units LSTM. In order to avoid overfitting, we apply dropout [19]. Dropout is one of the regularization techniques that randomly turns off neurons during training phase. We set dropout rate 0.2 for input layer CNN, 0.3 for output layer CNN and fully connected layer, and 0.5 for BLSTM layer.

**Table 2.** Label Performance Based On Different Feature Set

| Features | P | R | F |
|---|---|---|---|
| LEX (BLSTM) | 0.698 | 0.641 | 0.667 |
| SENT (CNN) | 0.747 | 0.591 | 0.648 |
| SENT + LEX (CNN+BLSTM) | 0.783 | 0.831 | 0.8 |
| + additional features | | | |
| **SENT + LEX + LOCT + LOCS** | **0.798** | **0.828** | **0.808** |

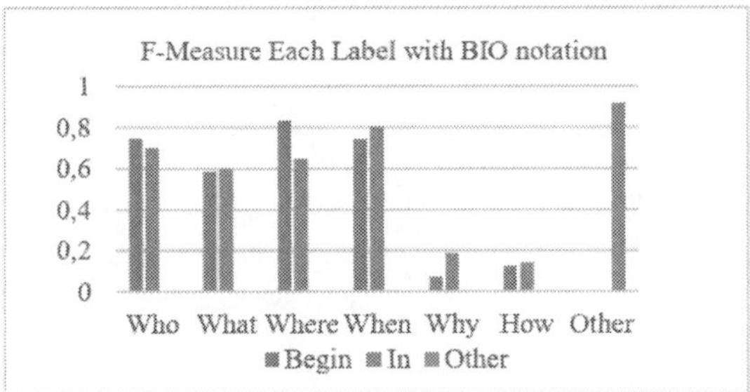


**Figure 5.** Performance based on BIO notation

*4.3. Performance*

In our experiments, we use precision (P), recall (R), and F-measure (F) to evaluate extraction performance. In the first experiment, we test the performance when introducing different feature set. Table 2. shows the result.

The results reveal that the best performance obtained by using all features, SENT+LEX+LOCT+LOCS, with F-measure 0.808, Precision 0.798, and Recall 0.828. From the experimental results, lexical sequence BLSTM with F-measure 0.667 and then sentence semantic features resulted by CNN contributed to performance with F-measure 0.648. The combination of these two features gave F-measure 0.8.

Deep learning has the ability in feature learning that learns about the features of raw data through its algorithm so it does not depend complicated features which are defined manually by involve experts or
feature engineering [20]. This can be seen from the F-measure performance of the CNN + BLSTM model without the additional features is good enough in doing token classification. In addition of token location features (LOCT and LOCS) increase F-measure by 0.008 thus achieve the best performance with F-measure 0.808. The best performance was obtained without using features from NLP tools like POStag and NEtag.

Figure 5. shows the model performance of the 13 labels. The highest value is shown by *when* element. Because the pattern of this element is most easily detected. When element, which denotes time, has a fixed pattern, consisting of a set of days and or numbers showing the date and time. These results are also shown in the previous researches [13,15], where *when* element is the element most easily detected and has the highest accuracy. *Why* and *how* elements have a low accuracy value.

The *why* and *how* elements, which provide an additional description of the event, are commonly contained in one sentence with the *who* and what elements, but more often found in separate sentences that are in the middle or end of the news. When the *why* and *how* elements are a separate sentence, the *why* sentence can be detected through the use of certain words e.g. the word "because", "for", "caused by" which indicates the *why* element. But sometimes there are sentences that contain the word but not the *why* element. These sentences have no semantic relation to *what* element.

While the *how* sentence has no special characteristics, the characteristics of this sentence have similarities with the sentence of *who* and *what*. There should be information about the semantic relationship between *why* and *how* with *what* (the element that states the event) so that it can distinguish the sentence with a sentence labeled *other*. The absence of features that represent semantic relation among sentences for these two elements causes low accuracy.

**Table 3.** Extraction Performance of Naïve Bayes, C4.5, IBk, and CNN-BLSTM

| **Method** | **P** | **R** | **F** |
|---|---|---|---|
| Naïve Bayes | 0.702 | 0.526 | 0.595 |
| C4.5 | 0.732 | 0.587 | 0.645 |
| IBk | 0.672 | 0.682 | 0.655 |
| **CNN + BLSTM** | **0.798** | **0.828** | **0.808** |

### 4.4. *Performance Comparison*

This experiment aims to compare the performance of CNN-BLSTM with other shallow methods i.e. IBk, C4.5, and Naïve Bayes on our dataset and features set. We cannot directly compare our results with the results in previous studies [13-15] because of differences in dataset. Table 3. shows the experimental results. We can conclude that the CNN-BLSTM method has the best performance with F-measure of 0.808. IBk, C4.5, and Naïve Bayes gave the average F-measure of 0.655, 0.645, and 0.595, respectively.

This demonstrates the reliability of deep learning in feature abstractions from high-dimensional data and the ability in generating feature functions from raw data thereby enhancing classification precision. Shallow learning is difficult to approach the functions of complex variables consequently resulting lower precision.

## 5. Conclusion and Future Works

5W1H extraction on Indonesian news article is a multiclass classification problem. By using BIO notation, each token in an article is classified into one of 13 classes, *B-who, I-who, B-what, I-what, B-when, I-when, B-where, I-where, B-why, I-why, B-how, I-how,* and *Other*. Label token depends on other token observations in the sequence and the information the entire sentence. We use CNN to extract information syntactic and semantic token within a sentence and bidirectional LSTM to learn the relations among tokens in the sequence. From six news elements, 5W1H, *when* elements are most easily detected, otherwise *why* and *how* elements are the most difficult ones.

Deep learning has the advantage of feature abstraction from high-dimensional data and the ability in generating feature functionality from the data thereby enhancing classification precision. While shallow learning cannot approach the function of complex variables. This is demonstrated by the results of tests comparing the performance of CNN-BLSTM with IBk, C4.5, and Naïve Bayes. The best performance was obtained by CNN-BLSTM with the F-measure 0.808. IBk, C4.5, and Naïve Bayes obtain the F-measure 0.655, 0.645, and 0.595, respectively.

For future works, we will improve model by adding number of training data. Deep learning algorithms perform better with more data. To sequential data modeling, we will employ development of LSTM unit, *Gated Recurrent Units* (GRU) and *Minimal Gate Unit* (MGU), to compare with the BLSTM in this task. Furthermore, we will explore more potential features and provide information that can represent semantic relation between *why* and *how* sentences with *what* elements and title/news headline so it can improve model performance in extracting *why* and *how* information.

**Acknowledgement**
This research receives grant from Superior Scholarship (*Beasiswa Unggulan*) by Bureau of Planning and Foreign Cooperation, Ministry of Education and Culture RI.

**References**

[1] Jurafsky D and Martin J H 2008 *Speech and Language Processing: An Introduction to Natural Language Processing, Computational Linguistics, and Speech Recognition, 2nd Edition* (Upper Saddle River: NJ: Prentice-Hall)

[2] Hinton G, Deng L, Yu D, George E D, Mohamed A-R, Jaitly N, Senior A, Vanhoucke V, Nguyen P, Sainath T N and Kingsbury B 2012 *Deep Neural Network for Acoustic Modeling in Speech Recognition: The Shared Views of Four Research Groups* (IEEE Signal Processing Magazine vol. 29) pp 82 – 97

[3] Yan Y, Chen M, Shyu M-L and Chen S-C 2015 *Deep Learning for Imbalanced Multimedia Data Classification, Int. Symp. on Multimedia* (Miami: IEEE)

[4] Kalchbrenner N, Grefenstette E and Blunsom P 2014 A Convolutional Neural Network for Modelling Sentences *Proc. 52nd Annu. Mtg. Assoc. Comput. Linguistics (Baltimore)* pp 655–65

[5] Kim Y 2014 Convolutional Neural Networks for Sentence Classification *Proc.Conf. EMNLP (Doha)* pp 1746–51

[6] Collobert R, Weston J, Bottou L, Karlen M, Kavukcuoglu K and Kuksa P 2011 Natural Language Processing (Almost) from Scratch *J. of Machine Learning Research* **12** pp 2493-2537

[7] Chen Y, Xu L, Liu K, Zeng D and Zhao J 2015 Event Extraction via Dynamic Multi-Pooling Convolutional Neural Network *Proc. 53rd Annu. Mtg. Assoc. Comput. Linguistics (Beijing: Association for Computational Linguistics)* pp 67–176

[8] Nguyen T H and Grishman R 2015 Event Detection and Domain Adaptation with Convolutional Neural Network *Proc. 53rd Annu. Mtg. Assoc. Comput. Linguistics (Beijing: Association for Computational Linguistics)* pp 365–71

[9] Sundermeyer M, Alkhouli T, Wuebker J and Ney H 2014 Translation Modeling with Bidirectional Recurrent Neural Network *Proc. Conf. EMNLP (Doha, Qatar)* pp 14–25

[10] Wang P, Qian Y, Soong F K, He L, Zhao H 2015 *Part-of-Speech Tagging with Bidirectional Long Short-Term Memory Recurrent Neural Network* (Cornell University Library) *Preprint* cs/1510.06168

[11] Sundermeyer M, Ney H and Schluter R 2015 From Feedforward to recurrent lstm neural networks for language modelling *J. IEEE/ACM Trans. Audio Speech Lang. Process.* **23** issue 3 pp 517–29

[12] Sundermeyer M, Schluter R and Ney H 2012 *LSTM Neural Networks for Language Modeling, Proc. ISCA Interspeech* (Portland) pp 194-7

[13] Khodra M L 2015 Event Extraction on Indonesian News Article Using Multiclass Categorization *ICAICTA (Chonburi: IEEE)*

[14] Wicaksana A P 2015 *Ekstraksi Informasi 5W1H Pada Berita Bahasa Indonesia* (Bandung: ITB)

[15] Ilyas R and Khodra M L 2015 Ekstraksi Informasi 5W1H pada Berita Online Bahasa Indonesia *J. Cybermatika* **3** JC03(2015)006

[16] Mikolov T, Chen K, Corrado G and Dean J 2013 Distributed Representations of Words and Phrases and their Compositionality *Proc. Int. Conf. Neural Inf. Process. Syst. (Lake Tahoe, Nevada)* pp 3111-9

[17] Graves A, Fernández S and Schmidhuber J 2005 Bidirectional LSTM networks for improved phoneme classification and recognition *Proc. 15th Int. Conf. Artificial Neural Networks: Formal Models Appl.* vol II (Warsaw, Poland :Springer Berlin Heidelberg) pp 799-804
[18] Zeiler M D 2012 *Adadelta: An Adaptive Learning Rate Method* (Cornell University Library) *Preprint* cs/1212.5701
[19] Srivastava N, Hinton, Geoffrey E, Krizhevsky, Alex, Sutskever, Ilya and Salakhutdinov R 2014 Dropout: a simple way to prevent neural networks from overfitting *J. of Machine Learning Research* **15** issue 1 pp 1929–1958
[20] Bengio Y 2012 *Deep Learning of Representations for Unsupervised and Transfer Learning, Proc. of the 2011 Int. Conf. on Unsupervised and Transfer Learning workshop vol 27* (Washington: JMLR) pp 17-37

# Implementation of Chaotic Gaussian Particle Swarm Optimization for Optimize Learning-to-Rank Software Defect Prediction Model Construction

**M A Buchari, S Mardiyanto and B Hendradjaya**

School of Electrical Engineering and Informatics, Institut Teknologi Bandung, Bandung, Indonesia

Email : 23514008@std.stei.itb.ac.id

**Keywords :** software metrics; defect modules ranking; software defect prediction; learning-to-rank; meta-heuristic

**Abstract**. Finding the existence of software defect as early as possible is the purpose of research about software defect prediction. Software defect prediction activity is required to not only state the existence of defects, but also to be able to give a list of priorities which modules require a more intensive test. Therefore, the allocation of test resources can be managed efficiently. Learning to rank is one of the approach that can provide defect module ranking data for the purposes of software testing. In this study, we propose a meta-heuristic chaotic Gaussian particle swarm optimization to improve the accuracy of learning to rank software defect prediction approach. We have used 11 public benchmark data sets as experimental data. Our overall results has demonstrated that the prediction models construct using Chaotic Gaussian Particle Swarm Optimization gets better accuracy on 5 data sets, ties in 5 data sets and gets worse in 1 data sets. Thus, we conclude that the application of Chaotic Gaussian Particle Swarm Optimization in Learning-to-Rank approach can improve the accuracy of the defect module ranking in data sets that have high-dimensional features.

## 1. Introduction

Software defect prediction helps us in identifying defect prone software modules. This process is really helpful for software testing activity [1]–[3]. There are 3 different techniques on software defect prediction, that are classification purposes, defect number prediction and construct the defect modules ranking.

Defect ranking prediction and defect classification are the most used technique in the software defect domain. Defect modules ranking is important for the software testing phase, as they can be used to ensure that test resources are allocated efficiently [4]. There are two approaches to construct a ranking defect prediction. The first one is predicting the module defect number, then order it and the second one is Learning-to-Rank approach that directly optimizes the ranking performance [5].

Learning-to-Rank approach uses simple linear model as a prediction model with software metrics [6] as a linear model variable and α as the corresponding parameter for software metrics. The value for

$\alpha$ is obtained by training the software module's data with known defect number. Learning-to-Rank approach utilizes Composite Differential Evolution (CoDE) to find the optimal value for $\alpha$ [5]. If $\alpha$ have already been trained, the model is ready to use for predicting the defect modules ranking.

Because $\alpha$ is a corresponding parameter for software metrics in learning-to-Rank linear model, the number of $\alpha$ increases as the number of usages of software metrics in the development of predictive models. CoDE is a meta-heuristic proposed by Wang et al [7], in their research CoDE was evaluated using 25 test function and each function run at 30 dimensional search space. The software defect prediction data sets used nowadays has more than 190 software metrics [3], [8].

Chaotic Gaussian Particle Swarm Optimization is meta-heuristic proposed by Jia et al [9] has good test results in solution search space with dimensions of 100 to 500. We propose Chaotic Gaussian Particle Swarm Optimization as an $\alpha$ determinant for the Learning-to-Rank approach by considering that improving the performance of prediction model construction can increase the prediction accuracy.

The remaining of this paper is organized as follows. In Section 2, we mention research that relate to our research. Section 3 describes the implementation of Chaotic Gaussian Particle Swarm Optimization in the Learning to Rank approach. We report and discuss our empirical evaluation in Section 4. We conclude our research in the last section.

## 2. Literature study

### *2.1. Meta-heuristic in software defect prediction*

Can et al [10] constructed software defect prediction model using SVM and utilize Particle Swarm Optimization for determine some parameter value in SVM. Their research showed better accuracy for classification purpose. Coelho et al [11] applied Particle Swarm Optimization for both ensemble constructor and consensus function for ensemble clustering technique. This technique showed that it could predict defects as good as or better than the other algorithms. Wahono and Suryana [12] applied Particle Swarm Optimization and Bagging for noise feature and data imbalance problem. Their result showed that the classification accuracy had significantly improvement in almost classification method after they implement PSO and Bagging for noise attribute and data imbalance problem.

### *2.2. Ranking approach software defect prediction*

Li et al [13] arranged top 4 subsystem that suspected to have most defect ratio for testing priority purposes. Zimmermann et al [8] used linear regression to predict defect number for further analysis correlation post and pre-release defect modules ranking. They used a combination of complexity metrics in constructing prediction models. The result showed that the complexity metrics were good indicators for software defect prediction. Li et al [14] classified the data to 3 classifications, defect prone modules, non-defect prone modules and deferment modules. They performed ranking method for each class separately. The result showed that the proposed method gave a good ranking prediction. Yang et al [5] showed to us that directly optimize the ranking performance can give good defect modules ranking. They used simple linear model with meta-heuristic algorithm to optimize the $\alpha$ for the model.

### *2.3. Chaotic Gaussian Particle Swarm Optimization*

Particle Swarm Optimization was proposed by Kennedy and Eberhart in 1995 [15]. PSO had simple and efficient characteristics when looking for solutions in the search space. However, Particle Swarm Optimization often experienced premature convergence when searching for solutions in high dimensional search space. Therefore Jia et al [9] performed a hybrid algorithm between Particle Swarm Optimization, Chaotic and Gaussian Local Search.

## 3. Methodology and implementation

In this section, we discuss the implementation of Learning-to-Rank approach with Chaotic Gaussian Particle Swarm Optimization. In this research we use Chaotic Gaussian Particle Swarm Optimization based on research conduct in [9].

### *3.1. Data sets*

In this research, we have used 2 public benchmark data sets. The first data sets is 3 version of ECLIPSE data sets [8] and the second data sets is 5 open-source project AEEEM data sets [3]. The characteristics of these data sets are shown in table 1.

The first data sets is public benchmark data that was submitted by D'Ambros et al consisted of 5 software projects Eclipse JDT Core, Eclipse PDE UI, Equinox Framework, Mylyn and Apache Lucene. In each software project D'Ambros extracted 6 categories of software metrics: process metrics, previous defect metrics, source code metrics, entropy of change metrics, metric source code and entropy (metric source code). This data can be used for defect prediction at the software class level.

The second data sets is ECLIPSE software project that was proposed by Zimmermann et al consisted of various source code metrics and divided into 3 versions of release i.e. version 2.0, 2.1 and 3.0. This data can be used to predict defects-prone at the file and software package level.

**Table 1.** Experimenal data

| Data Sets | Number of Module | Number of Software Metric | Defect Module Percentage | Number of Defect |
|---|---|---|---|---|
| Apache Lucene | 691 | 212 | 9.30% | 97 |
| Eclipse JDT Core | 997 | 212 | 20.70% | 374 |
| Equinox Framework | 324 | 212 | 39.80% | 244 |
| Eclipse PDE UI | 1497 | 212 | 14% | 341 |
| Mylyn | 1862 | 212 | 13.20% | 340 |
| Eclipse_File2.0 | 6729 | 198 | 14.5% | 1692 |
| Eclipse_File2.1 | 7888 | 198 | 10.8% | 1182 |
| Eclipse_File3.0 | 10593 | 198 | 14.8% | 2679 |
| Eclipse_Package2.0 | 377 | 207 | 50.4% | 917 |
| Eclipse_Package2.1 | 434 | 207 | 44.7% | 662 |
| Eclipse_Package2.0 | 661 | 207 | 47.4% | 1534 |

### *3.2. Chaotic Gaussian Particle Swarm Optimization Implementation for Learning-to-Rank Approach*

Learning-to-Rank [5] approach calculates relative defect number of a module that denote as f(x) use the following equation:

$$f(x) = \sum_{i=1}^{d} \alpha_i x_i \tag{1}$$

x_i is the i-th metric from software metrics vector that represent module x = (x1,x2, . . . . . , xd), when d is total software metrics and represent for the dimension of solution search space on Chaotic Gaussian PSO that is utilized to determine the value of vector α. For Chaotic Gaussian PSO fitness function, we have use Fault Percentile Average (FPA) [16] as suggested in [5]. FPA is also used for evaluating ranking prediction results. The Fault Percentile is defined as below

$$\frac{1}{k}\sum_{m=1}^{k} \frac{1}{n}\sum_{i=k-m+1}^{k} n_i \tag{2}$$

where n_i is actual defect number in module i, k is k-th module from top i actual defect modules ranking and m is top m relative defect modules ranking. Vector α is translated into the position of each particle or usually called solution candidate in the swarm in order to determine the optimal value of

IOP Conf. Series: Journal of Physics: Conf. Series **978** (2018) 012079 doi:10.1088/1742-6596/978/1/012079

vector α. Then Chaotic Gaussian PSO [9] begins the optimizing process for vector α using the parameter setting from table 2 and FPA as the fitness function.

### *3.3. Implementation scenario*

Our experiment scenario for Learning-to-Rank software defect prediction approach is based on [5]. The implementation of Learning-to-Rank software defect prediction is divided into 2 scenarios. This division is based on different treatments on the data that will be used for construct prediction model and data used to predict the ranking of modules. Five data sets are single-release software project data (AEEEM data sets) and 6 other data sets are multiple-release software project data (Eclipse V2.0, 2.1, 3.0 in file and package level). The detail scenario is shown in figure 1 and 2.

**Table 2**. Chaotic gaussian pso parameters setting

| Chaotic Gaussian PSO Parameter | Value of Parameter |
|---|---|
| W | 0.7289 |
| $c_1$ | 1.49618 |
| $c_2$ | 1.49618 |
| population size | 60 |
| Particle initial position upper boundary[a] | 20 |
| Particle initial position lower boundary[a] | -20 |
| Maximum Fes | Number of dimension × 5000 |

[a] only used in the initialization phase

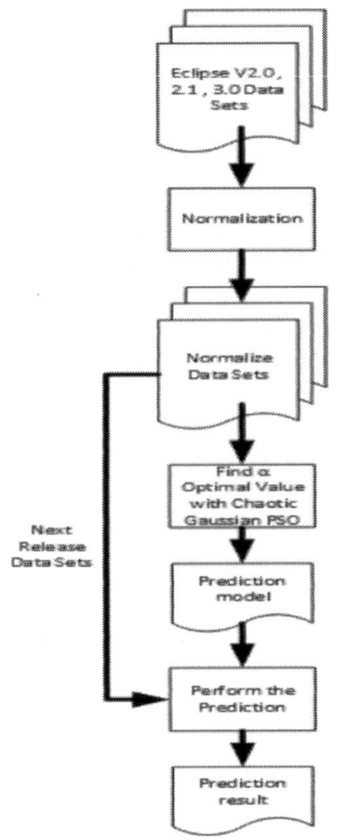


**Figure 1.** Implementation scenario for multiple-release data sets

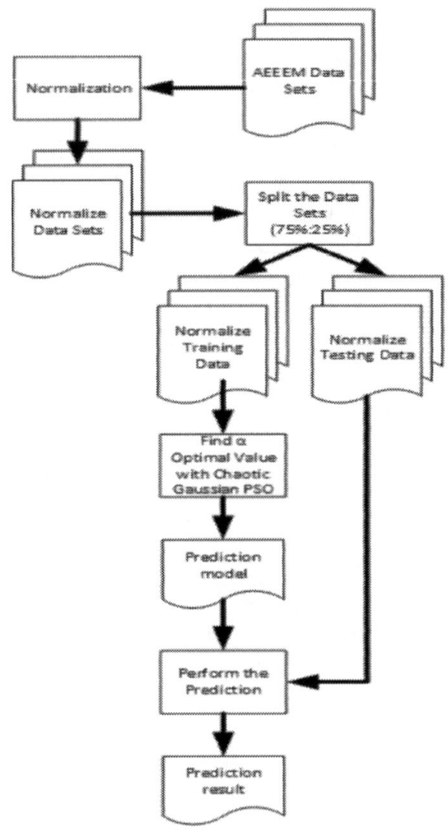


**Figure 2.** Implementation scenario for multiple-release data sets

The first step taken on the implementation of Learning-to-Rank is to normalize the data sets. It then transforms all values in each column to the range zero to one. After normalizing the data, we divide each AEEEM project data set based on 75% -25% ratio. 75% of the data is used to train prediction models, the remaining 25% is used to evaluate the prediction model as test data. We then perform the splitting, because the 5 data sets from the AEEEM data set had only 1 version of each data set.

The next step is to train prediction models using Chaotic Gaussian PSO. With the vector α that is produced by Chaotic Gaussian PSO, model is ready to be used to predict modules rank.We use the prediction model that constructs from former release data sets to predict next release data sets such as a prediction model construct from Eclipse_File2.0 data sets will predict the defect modules ranking for Eclipse_File2.1 and Eclipse3.0 data sets. For single release data sets we predict the defect modules ranking of data test with the prediction model construct by corresponding training data sets.

**Table 3.** Experiment result for multiple release data sets

| Data Sets | CGPSO (Train) | CoDE (Train) | CGPSO (predict) | CoDE (predict) | Ideal |
|---|---|---|---|---|---|
| Eclipse_File2.0-2.1 | **0.8470±.001** | 0.8362±.000(.00) | **0.7568±.004** | 0.7441±.004(.00) | 0.9586 |
| Eclipse_File2.0-3.0 | | | **0.7867±.001** | 0.7801±.003(.00) | 0.9506 |
| Eclipse_File2.1-3.0 | **0.8032±.002** | 0.7872±.001(.00) | 0.7452±.008 | 0.7490±.013(.73) | 0.9506 |
| Eclipse_Package2.0-2.1 | **0.8223±.001** | 0.8080±.001(.00) | **0.7763±.005** | 0.7706±.003(.02) | 0.8870 |
| Eclipse_Package2.0-3.0 | | | 0.7823±.007 | 0.7825±.009(.91) | 0.8944 |
| Eclipse_Package2.1-3.0 | **0.8123±.001** | 0.7962±.001(.00) | 0.7903±.005 | 0.7884±.009(.47) | 0.8944 |

**Table 4.** Experiment result for single release data sets

| Data Sets | CGPSO (train) | CoDE(train) | CGPSO (predict) | CoDE (predict) | Ideal |
|---|---|---|---|---|---|
| Eclipse | **0.8776±.001** | 0.8667±.000(.00) | 0.8396±.004 | **0.8460±.005**(.01) | 0.9323 |
| Equinox | **0.8299±.002** | 0.8204±.001(.00) | **0.7925±.008** | 0.7779±.019(.02) | 0.8620 |
| Apache Lucene | **0.9145±.002** | 0.8948±.001(.00) | 0.8203±.029 | 0.8455±.019(.08) | 0.9672 |
| Mylyn | **0.8525±.002** | 0.8300±.001(.00) | **0.7798±.010** | 0.7611±.009(.00) | 0.9451 |
| Eclipse PDE UI | **0.8362±.002** | 0.8160±.001(.00) | 0.8164±.011 | 0.8179±.016(.73) | 0.9453 |

## 4. Result and discussion

As we mentioned earlier, we have used FPA value for evaluating both training performance and prediction accuracy. We have performed Wilcoxon rank-sum test at 0.05 significance to determine whether models produce by two meta-heuristic has statistically different prediction accuracy. Wilcoxon rank-sum test has been performed for both training phase FPA and prediction accuracy (FPA value as prediction result). Experimental results for multiple-release data sets are shown in table 3.

Eclipse_File2.0-2.1 in table 3 means the prediction model is trained using Eclipse_File2.0 data sets and then the prediction model is used to predict defect modules ranking for Eclipse_File2.1 data sets and so on. We have trained 10 prediction models for each data set. Each value in table column represents the average of 10 FPA value from training process or average of 10 predicted result along with its deviation standard, while the number in the bracket is p-value of nonparametric test results of Wilcoxon rank-sum.

Table 3 shows that the prediction model trained using Chaotic Gaussian Particle Swarm Optimization has significantly better accuracy on 3 data sets and has an equivalent accuracy in 3 other data sets. This proves that Chaotic Gaussian Particle Swarm Optimization has a better ability than CoDE to find optimal solutions in data sets with large dimensions. It is confirmed by CGPSO FPA train value that is statistically better than CoDE FPA train value in all data sets.

Table 4 represents the experimental results of Learning-to-Rank software defect prediction implementation in single-release data sets. All prediction models in table 4 have been trained with 75% of the total data sets, and then the trained models have been used to predict the defect module ranking for 25% of unused data on the model training process.

In Table 4, Chaotic Gaussian Particle Swarm Optimization has significantly better accuracy in 2 data sets, has an equivalent accuracy on 2 data sets and has worse accuracy results in 1 data sets. The result is approximately the same as the result in table 3. However, in Table 4 Chaotic Gaussian PSO has worse accuracy than CoDE on eclipse data sets. However, we still have not been able to find the exact reason CoDE has a higher accuracy on predicted defect modules ranking on eclipse data sets. We suspect that the model that we have trained with eclipse data sets using Chaotic Gaussian Optimization is overfitting.

## 5. Conclusion

In this research, we have presented and evaluated the implementation of Chaotic Gaussian Particle Swarm Optimization on the Learning-to-Rank software defect prediction approach for train model parameter. The results show the prediction model construct using Chaotic Gaussian Particle Swarm Optimization has better FPA train value on all data sets used in the model training process. An improved prediction accuracy is also seen in some data sets. Although one of the prediction results on Eclipse data sets produces worse accuracy than the prediction results produces by CoDE.

The overall results show prediction models construct using Chaotic Gaussian Particle Swarm Optimization gets better accuracy on 5 data sets, more or less the same results in 5 data sets and worse results in 1 data sets. Finally, we conclude that the application of Chaotic Gaussian Particle Swarm Optimization in Learning-to-Rank approach can improve the accuracy of the defect module ranking in data sets that have high-dimensional features (over 190 software metrics).

In the future, we plan to modify the fitness function on meta-heuristic to increase the prediction accuracy and explore the impact of various software metrics we have not used in this research on learning to rank software prediction approach.

**References**

[1] Jiang Y, Cukic B and Ma Y 2008 Techniques for evaluating fault prediction models *Empir. Softw. Eng.* **13** 561–95

[2] Mende T and Koschke R 2011 Effort-aware defect prediction models *Proc. Eur. Conf. Softw. Maint. Reengineering, CSMR* 107–16

[3] D'Ambros M, Lanza M and Robbes R 2012 Evaluating defect prediction approaches: a benchmark and an extensive comparison *Empir. Softw. Eng.* **17** 531–77

[4] Yang X, Tang K and Yao X 2012 A Learning-to-Rank Algorithm for Constructing Defect Prediction Models *Intelligent Data Engineering and Automated Learning - IDEAL* ed H Yin, J A F Costa and G Barreto (Natal, Brazil: Springer Berlin Heidelberg) pp 167–75

[5] Yang X, Tang K and Yao X 2015 A Learning-to-Rank Approach to Software Defect Prediction *IEEE Trans. Reliab.* **64** 234–46

[6] Fenton N and Bieman J 2014 *Software Metrics: A Rigorous and Practical Approach, Third Edition* (CRC Press)

[7] Wang Y, Cai Z and Zhang Q 2011 Differential evolution with composite trial vector generation strategies and control parameters *IEEE Trans. Evol. Comput.* **15** 55–66

[8] Zimmermann T, Premraj R and Zeller A 2007 Predicting defects for eclipse *Third International Workshop on Predictor Models in Software Engineering, PROMISE'07*

[9] Jia D, Zheng G, Qu B and Khan M K 2011 A hybrid particle swarm optimization algorithm for high-dimensional problems *Comput. Ind. Eng.* **61** 1117–22

[10] Can H, Jianchun X, Ruide Z, Juelong L, Qiliang Y and Liqiang X 2013 A new model for software defect prediction using Particle Swarm Optimization and support vector machine *25th Chinese Control Decis. Conf.* 4106–10

[11] Coelho R A, Guimaraes F dos R N and Esmin A A A 2014 Applying Swarm Ensemble Clustering Technique for Fault Prediction Using Software Metrics *13th Int. Conf. Mach. Learn. Appl.* 356–61

[12] Wahono R S and Suryana N 2013 Combining particle swarm optimization based feature selection and bagging technique for software defect prediction *Int. J. Softw. Eng. its Appl.* **7** 153–66

[13] Li P L, Herbsleb J, Shaw M and Robinson B 2006 Experiences and results from initiating field defect prediction and product test prioritization efforts at ABB Inc. *Proceeding of the 28th international conference on Software engineering - ICSE '06* (New York, New York, USA: ACM Press) p 413

[14] Li W, Huang Z and Li Q 2016 Three-way decisions based software defect prediction *Knowledge-Based Syst.* **91** 263–74

[15] Kennedy J and Eberhart R 1995 Particle swarm optimization *IEEE Int. Conf. Neural Networks* **6** 1942–8

[16] Weyuker E J, Ostrand T J and Bell R M 2010 Comparing the effectiveness of several modeling methods for fault prediction *Empir. Softw. Eng.* **15** 277–95

IOP Conf. Series: Journal of Physics: Conf. Series **978** (2018) 012080 doi:10.1088/1742-6596/978/1/012080

# Modeling of Sedimentation Process in Water

**Tulus**[1], **I. Z. Sefnides**[1], **Sawaluddin**[1], **Suriati**[2], **M Dwiastuti**[3]

[1] Department of Mathematics, Universitas Sumatera Utara, Padang Bulan 20155 USU Medan Indonesia
[2] Department of Informatics, Sekolah Tinggi Teknik Harapan, Medan 20218 Indonesia
[3] Department of Language Science and Technology, Saarland University, Germany

E-mail: tulus@usu.ac.id

**Abstract.** Sedimentation process occurs in container of water reservoir. In the sedimentation process will be observed how the fluid flow occurs and how sedimentation occurs in water. The sedimentation is influenced by the density of the water type and the speed of water in the container of water reservoir. This study aims to find out how to model the mixture of solid and liquid particles (suspension) on the container of water by way of modeling it into a mathematical form that is by using differential equations. In this case, the factors affecting the distribution of the sludge concentration will be used as a reference for modeling the mathematical equations will then be solved using differential equations. The results of this study will show that the equation model of mixed solid and liquid particles (suspension), which then is called the momentum transport equation.

## 1. Introduction

The water supply company is a regional-owned company engaged in the processing and clean water industry for public society. Some facilities in clean water processing include: intake, raw water tower, clarifier, fast mixing building (Rapid Mixing), slow mixing building, filtration building, and reservoir. All of these equipments can be operated by computer system. The water supply company also use chemicals such as chlorine and alum in water processes [1]. There are several stages of clean water process: filtering and precipitation, coagulation, flocculation, sedimentation, filtration and disinfection. One of the water treatment process is sedimentation. From the problem of the sedimentation process in water, some researchers performed the analysis of the sedimentation process in water by modeling the sedimentation process into the model of mathematics [2, 3].

Model plays an important role in the field of science. Usually in terms of economics to save time and cost or other valuable commodities. Modeling can avoid the risk of real system damage. Thus a model is required when an experiment with a real system becomes blocked because it is expensive, dangerous or something that is impossible to do. Mathematical models are ideal representations of real systems that are expressed or expressed in terms of symbols and mathematical statements. In other words, the mathematical model represents a system in the form of quantitative and logical relations, in the form of a mathematical equation. In artificial mathematical models of events or natural events are described through a set of mathematical equations. In this case, modeling will change the sedimentation process in water into Mathematical model to make it easier to find solutions to the problem.

## 2. Materials and Methods

### Fluid Mechanics

Fluid mechanics is the discipline within the broad field of applied mechanics that is concerned with the behavior of liquids and gases at rest or in motion. It covers a vast array of phenomena that occur in nature (with or without human intervention), in biology, and in numerous engineered, invented, or manufactured situations. There are few aspects of our lives that do not involve fluids, either directly or indirectly [6]. A fluid is defined as a substance that deforms continuously when acted on by a shearing stress of any magnitude [4]. A shearing stress (force per unit area) is created whenever a tangential force acts on a surface as shown by the figure in the margin. Classification of fluid flows are uniform, one, two and three dimensional flows, viscous and inviscous flows, laminar and turbulent flows and incompressible and compressible flows [5].

Sedimentation is a solid-liquid separation using gravity precipitation to set aside suspended-solid The suspension is a fluid mixture containing solid particles. Or in other words, a heterogeneous mixture of liquids and solids dissolved in the liquid. Examples of suspensions are mud, flour, mist, paint and others. One example of such suspension is mud. Mud is a liquid, mixture between water and soil. Mud occurs when the soil is wet. Geologically, mud is a mixture of water and particles of sediment. Factors that influence of Sediment are concentration, pressure, density, surface tension, viscosity, and velocity.

Mathematical models can be defined as equations, equations that model some real situations. Research can help model the situation into the form of a differential equation. Technology can be used to help solve the equation (computer program gives answer). Answers based technology would be interpreted or communicated in accordance with real situations.

*The Differential Equation*

To solve a partial differential equation for a dependent variable, certain conditions are required which means that the independent variables must be determined at certain values of independent variables. If the independent variables are spatial coordinates such as speed, the conditions are called boundary conditions. If the independent variable is time, the conditions are called initial conditions. The boundary condition is the velocity component to the perpendicular in the non-condensed flow. In the non-viscous flow in which viscosity is neglected, the velocity vector has a tangential direction to the boundary. These differential equations will be derived using cartesian coordinates.

*Initial and Boundary Conditions*

For the problem sedimentation in this research is assumed as follows:

(i) Assume sedimentation is mud
(ii) Viscous effects are negligible
(iii) It is assumed that water flows and does not rotate, incompressible flow, and laminar flow
(iv) Cohesion force and adhesion force are not considered

## 3. Results and Discussion

*Two-Phase Flow Modeling of a Dense Suspension*

Liquid-solid mixtures (suspensions) are important in a variety of industrial fields, such as oil and gas refinement, paper manufacturing, food processing, slurry transport, and wastewater treatment. A suspension is a mixture of solid particles and a liquid. The dynamics of a suspension can be modeled by a momentum transport equation for the mixture. It uses the following equation of momentum transport

*The Differential Continuity Equation*
To derive the differential continuity equation, the infinitesimal element is utilized. It is a small control volume into and from which the fluid flows. It is shown in the $xy$-plane with depth $dz$. Let us assume that the flow is only in the $xy$-plane so that no fluid flows in the $z$-direction. Since mass could be changing inside the element, the mass that flows into the element minus that which flows out must equal the change in mass inside the element. This is expressed as

$$\rho udydz - (\rho u + \frac{\partial(\rho u)}{\partial x}dx)dydz + \rho vdxdz - (\rho v + \frac{\partial(\rho v)}{\partial y}dy)dxdz = \frac{\partial}{\partial t}(\rho dxdydz) \tag{1}$$

where the density $\rho$ is allowed to change across the element. Simplifying the above, recognizing that the elemental control volume is fixed, results in :

$$\frac{\partial(\rho u)}{\partial x} + \frac{\partial(\rho v)}{\partial y} = -\frac{\partial \rho}{\partial t} \tag{2}$$

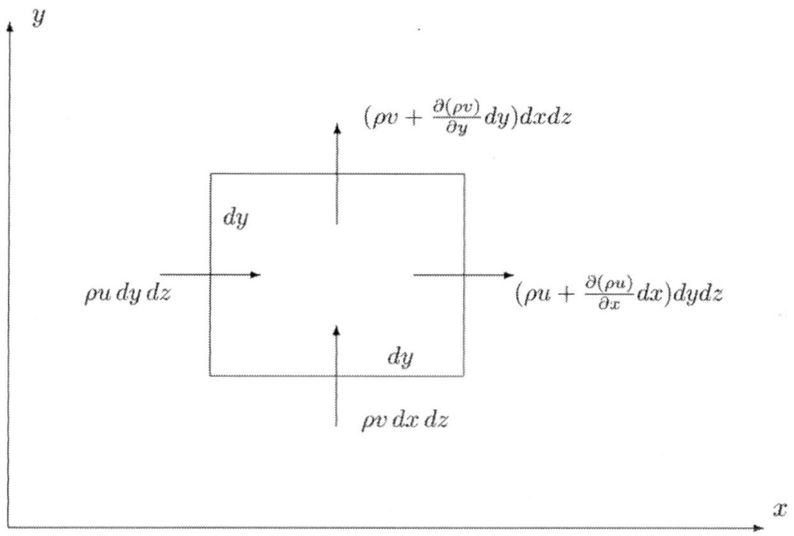


**Figure 1.** Infinitesimal Control Volume.

Differentiate the products and include the variation in the $z$-direction. The equation is described and the result is the differential continuity equation. The differential continuity equation for an incompressible flow is

$$\frac{D\rho}{Dt} + \rho\nabla \cdot V = 0 \tag{3}$$

*The Differential Momentum Equation*
The differential continuity equation derived in the differential continuity equation contains the three velocity components as the dependent variables for an incompressible flow. If there is a flow of interest in which the velocity field and pressure field are not known, such as the flow around a turbine blade or over a weir, the differential momentum equation provides three additional equations since it is a vector equation containing three component equations. The four unknowns are then $u$, $v$, $w$, and $p$ when using a rectangular coordinate system. The four

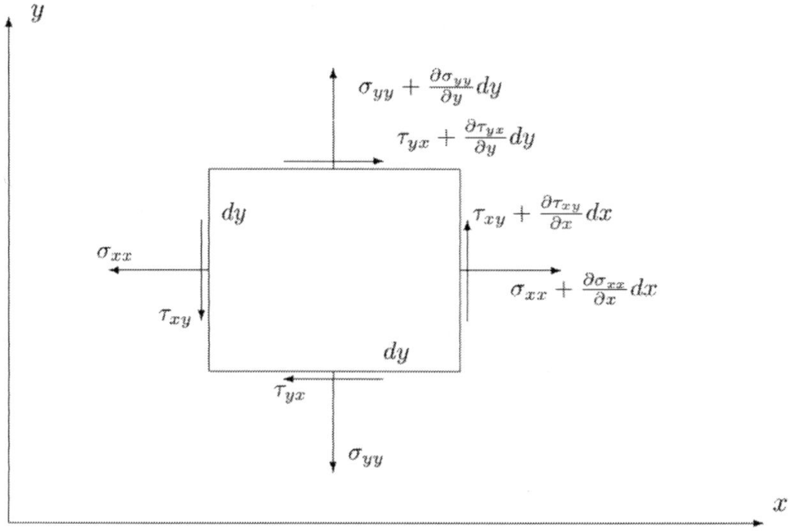


**Figure 2.** Rectangular stress components on a fluid element.

equations provide us with the necessary equations and then the initial and boundary conditions allow a tractable problem.

Apply Newtons second law to the element of Fig. 2, assuming that no shear stresses act in the $z$-direction (we will simply add those in later) and that gravity acts in the $z$-direction only:

$$(\sigma_{xx} + \frac{\partial \sigma_{xx}}{\partial x} dx)dydz - \sigma_{xx}dydz + (\tau_{xy} + \frac{\partial \tau_{xy}}{\partial y} dy)dxdz - \tau_{xy}dxdz = \rho dxdydz\frac{Du}{Dt} \tag{4}$$

$$(\sigma_{yy} + \frac{\partial \sigma_{yy}}{\partial y} dy)dxdz - \sigma_{yy}dxdz + (\tau_{xy} + \frac{\partial \tau_{xy}}{\partial x} dx)dydz - \tau_{xy}dydz = \rho dxdydz\frac{Dv}{Dt} \tag{5}$$

The equation above is described and the result is the Navier Stokes equations. The Navier Stokes equations can be written in vector form as

$$\rho\frac{Dv}{Dt} = -\nabla p + \mu\nabla^2 V + \rho g \tag{6}$$

*The Mixture Model Equation*
A mixed model (or an algebraic slip model) is a simplified formulation of a multiphase flow equation (two phases). The mixed model consists of continuity and momentum equations for mixed and continuity equations for dispersion or separate phases. If in the fluid flow there is a solid particle which is assumed to be a mud to form sedimentation, then there is a series of new equations found in the Navier-Stokes equation. The new equation is the effect of sedimentation on the flow of fluid flowing in a container. In the same way as deriving the differential continuity equation, we find an equation describing the solid particles present in a fluid stream.

This equation is called the momentum transport equation for two-phase flow of suspension that can solve and calculate how much sedimentation in a container of water reservoir.

The momentum transport equations can be written as

$$\rho\frac{DV}{Dt} = -\nabla p - \nabla(\rho m_p(1 - m_p)V_{slip}V'_{slip} + \mu\nabla^2 V + \rho g \tag{7}$$

where $V$ is the mass averaged mixture velocity $(m/s)$, $p$ denotes the pressure $(Pa)$, $g$ refers to the acceleration of gravity $(m/s^2)$, $m_p$ is the dimensionless particle mass fraction, $V_{slip}$ gives the relative velocity between the solid and the liquid phases $(m/s)$ and $\rho$ is mixture density.

This equation is called the momentum transport equation for two-phase flow of suspension. It can solve and calculate how many sedimentation in a container of water reservoir.

## 4. Conclusion

From the results and discussion, it shows that the problems in the real world can be modeled by using Mathematical Model, in this case is the problem of sedimentation on the container of water reservoir in the sedimentation container at the time of clean water process. The problem can be modeled into Mathematical form, which is to be an equation called the momentum transport equation. The momentum transport equation is $\rho\frac{DV}{Dt} = -\nabla p - \nabla(\rho m_p(1 - m_p)V_{slip}V'_{slip} + \mu\nabla^2 V + \rho g$. It also shows that mathematics can be applied in everyday life.

## Acknowledgments

The authors gratefully acknowledge that the present research is supported by Ministry of Research and Technology and Higher Education Republic of Indonesia. The support is under the research grant TALENTA USU of Year 2017.

## References

[1] Tulus, Suriati, M. Situmorang, and D. M. Zain 2017 Computational Analysis of Sedimentation Process in the Water Treatment Plant *Journal of Physics Conference Series* **890(1)** 012077.

[2] Tulus, and M. Situmorang 2017 Modeling of Sedimentation Process in the Irrigation Channel IOP Conference Series: Materials Science and Engineering **180(1)** 012021

[3] Tulus, Suriati, and M. Situmorang 2017 Computational Analysis of Suspended Particles in the Irrigation Channel *Journal of Physics: Conference Series* **801(1)** 012094

[4] Tulus, O. S. Sitompul, Sawaluddin, and Mardiningsih 2015 Mathematical Modeling and Simulation of Fluid Dynamic in The Continuous Stirred Tank *Bulletin of Mathematics* **7**(2) 1-12

[5] D.F. Young, B.R. Munson, T.H. Okiishi, and W.W. Huebsch 2010 A Brief Introduction to Fluid Mechanics, Fifth Edition. New York:John Wiley & Sons

[6] M. C. Potter and D. C. Wiggert 1997 Fluid Mechanics; New Jersey:Prentice-Hall Inc,

[7] B. Triatmodjo 1993 Hidraulika II. Yogyakarta:Beta Offset.

# Online thesis guidance management information system

**T H Nasution[1], F Pratama[1], K Tanjung[1], I Siregar[2] and A Amalia[3]**

[1]Department of Electrical Engineering, Universitas Sumatera Utara, Medan, Indonesia
[2]Department of Industrial Engineering, Universitas Sumatera Utara, Medan, Indonesia
[3]Department of Computer Science, Universitas Sumatera Utara, Medan, Indonesia

Email: tigor.nasution@usu.ac.id

**Abstract.** The development of internet technology in education is still not maximized, especially in the process of thesis guidance between students and lecturers. Difficulties met the lecturers to help students during thesis guidance is the limited communication time and the compatibility of schedule between students and lecturer. To solve this problem, we designed an online thesis guidance management information system that helps students and lecturers to do thesis tutoring process anytime, anywhere. The system consists of a web-based admin app for usage management and an android-based app for students and lecturers.

## 1. Introduction

The development of technology is very fast and plays an important role to support human activities in order to time optimization. In the development of internet technology in the world of education nowadays is still not maximized, especially for the guidance process of student's thesis to the lecturer. Thesis is a term used in Indonesia to illustrate a scientific paper in the form of exposure to the writings of undergraduate research that discusses a problem / phenomenon in a certain field of science by using applicable rules [1]. One of the problems is the limitation of communication due to limited meeting schedules. To overcome these problems, need to be designed a thesis guidance support system, which can solve the problems faced in the process of thesis guidance. The system will be built based on android and website.

There was several online management information system design research done previously in various fields such as finance [2], [3]. Some researchers have developed Information Management Systems in the health [4] - [6]. For the education field itself there have also been some research management information systems [7] - [9].

In contrast to previous research that had been done before, this application is expected to make students and supervisors may do the thesis guidance and mentoring easier, so as to minimize the final task duration that is caused by the bustle of students and lecturers. This application is also equipped with a Website that can be monitored by the Head of the Department, which final task guidance who have problems or the execution time of the final task is too long for the Head of the Department can evaluate the lecturer or student concerned.

## 2. Method

### *2.1 System requirement analysis*

The system consists of 3 groups of users, i.e. Admin, Student, and Lecturer. Admin is the department manager who will organize and monitor the process of guidance. Student is a student user who will perform thesis guidance. Lecturers are assigned to guide the thesis. The role of each user group can be seen in Use Case Diagram in figure 1.

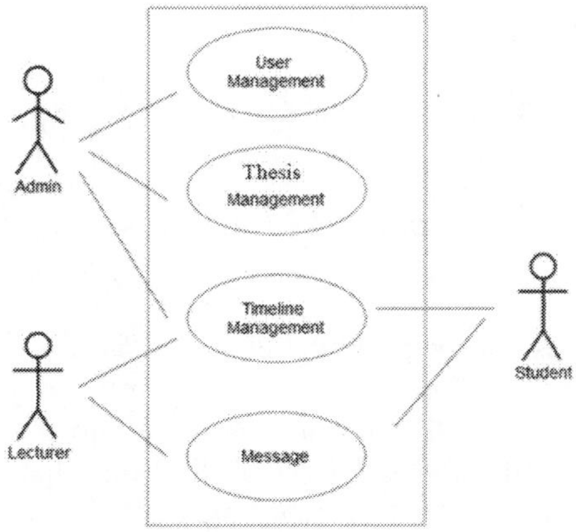


**Figure 1.** Use Case Diagram

Based on the Use Case Diagram in figure 1, it can be seen the interactivity between users on the system. Admin is assigned to register Student who register the submission of Thesis and also determine the mentor Lecturer for the student. Students will create a timeline of their activities in a system that can be monitored and discussed with lecturers from every stage of the activities in the timeline. Admins can also monitor the timeline to keep track of the student's thesis process. Lecturers will interact with students through a student-made timeline. Lecturers will set "Done" of each stage status in the timeline if the student has completed the current stage. Data Flow Diagram (DFD) of this system can be seen in figure 2.

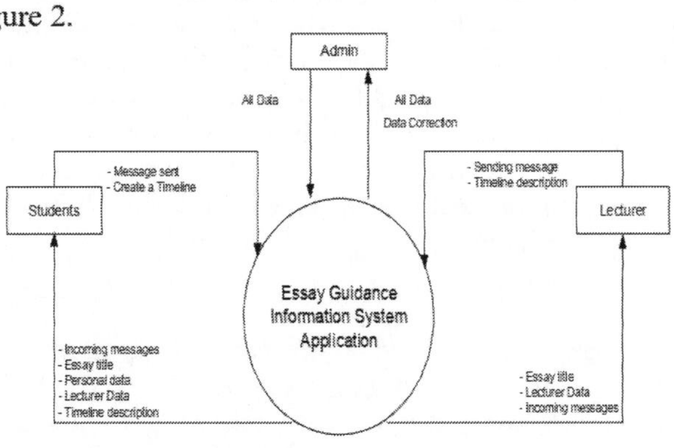


**Figure 2.** Data Flow Diagram

### *2.2. Database design*

Based on the needs of the existing analysis is done database design to store data transactions that exist on the system. After normalization, there are 6 tables used in this system. The tables are tbl_user, tbl_admin, tbl_lecture, tbl_student, tbl_timeline, and tbl_message. The relationships between tables displayed in the Entity Relationship Diagram (ERD) in figure 3.

After finished the database design, then continue to the MySQL database server design. MySQL was originally developed to handle large databases much faster than existing solutions and has been successfully applied in highly demanding production environments for several years. Connectivity, speed, and security make MySQL server very suitable for accessing databases on the Internet [10].

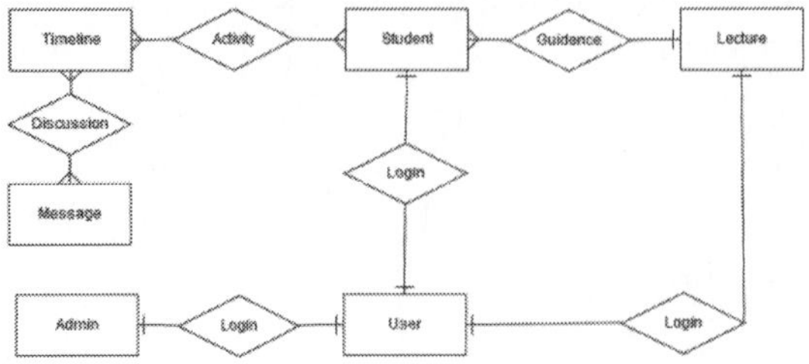


**Figure 3.** ERD of system

### *2.3. Interface design*

Interface that built on this system consists of three types based on existing user groups ie admin, students and lecturers. For admin interface designed web-based applications using PHP programming language. PHP is a server side script that can be inserted into html code [11]. For Students and Lecturers, it has been designed an android-based applications that can be used on android-based mobile devices. This is done so that the Students and Lecturers can interact anywhere with their mobile devices. The choice of mobile movements based on android because android mobile devices are popular mobile devices that are widely used because of many free apps provided [12], [13].

Admin application interface consists of Login interface, User Management interface, and also Management Guidance. The Login interface is used to validate the users who will be logged into the app. User Management is used to manage user data including students and lecturers. Management Guidance is used to determine the pair between students and mentor lecturers. Management guidance is also used to monitor the progress of the guidance process that occurs. All the interfaces are accessed through the web page by using the browser application.

For student application interface can be accessed through android application consisting of login interface, Timeline Management interface and Message interface. Same with the Admin app interface, the Login interface is used to validate the users who will be logged into the app. The Timeline Management interface is used by students to create timelines from the each work stages of the thesis. While the Message Management interface is used for student to interact with the lecturer.

Same with the student application interface, the lecturers' application can also be accessed via android application and consists of the Login interface, Timeline Management interface and also Message interface. The difference in the lecturer Management Timeline interface's status can only be changed if the current stage has been completed.

## 3. Results and Discussion

From the system designed, there are 3 applications that can be used by each user. For admin applications the process begins in the Login page on the website as shown in figure 4. Admin will login using the Employee Identity Number and password that have been registered before. To perform user management (lecturer and students), the admin will access the user management page shown in figure 5 and figure 6.

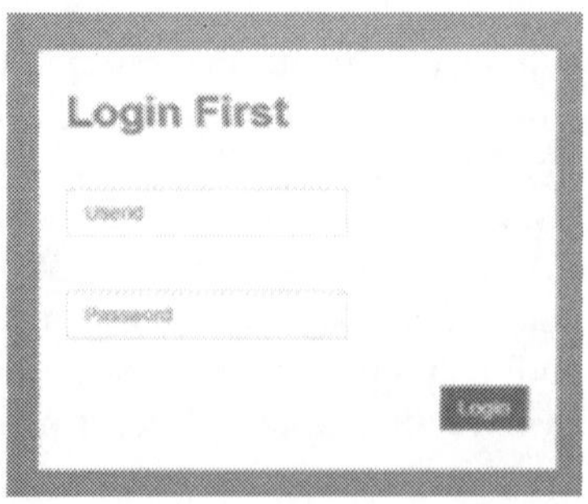


**Figure 4.** Login Page

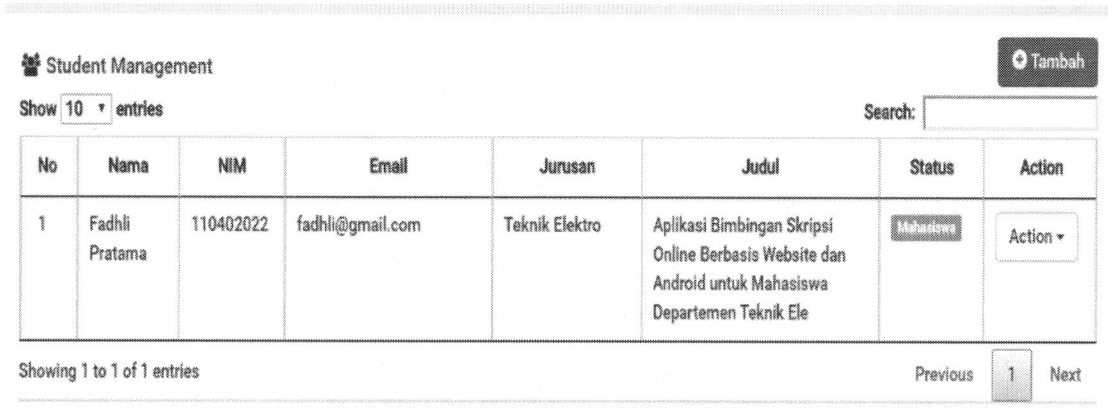


**Figure 5.** Student's Management Page

**Figure 6.** Lecturer's Management Page

Students and lecturers apps are accessed with an android application by doing Login at the application as shown in figure 7. For student login done by using Student's Number, while lecturer use Lecturer's Number and password that has been registered by admin. Furthermore, students will first make the stages of activity on the Timeline menu as shown in figure 8.

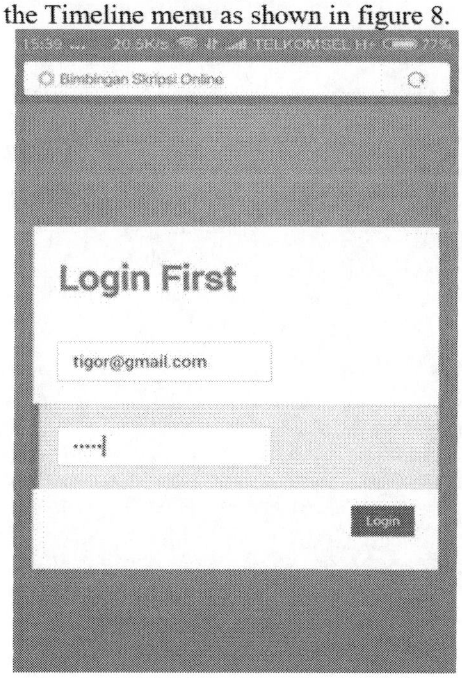


**Figure 7.** Login Interface in Android Application

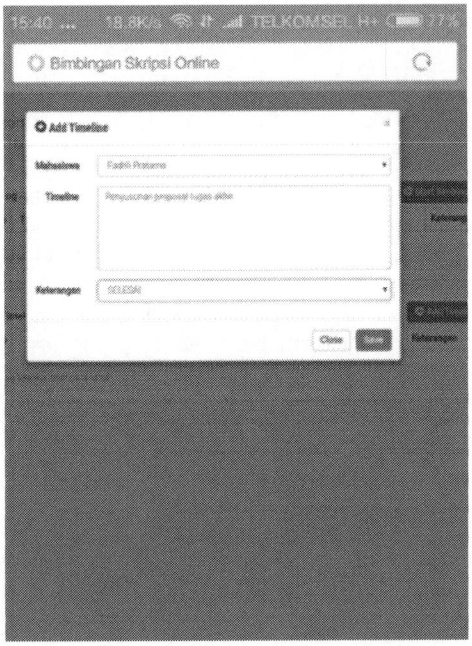


**Figure 8.** Management Timeline Interface

Lecturer will then interact with the students through the Message menu on the timeline as shown in figure 9 below. At each stage that has been completed by the student or the lecturer then the status of the stages will be changed into Done.

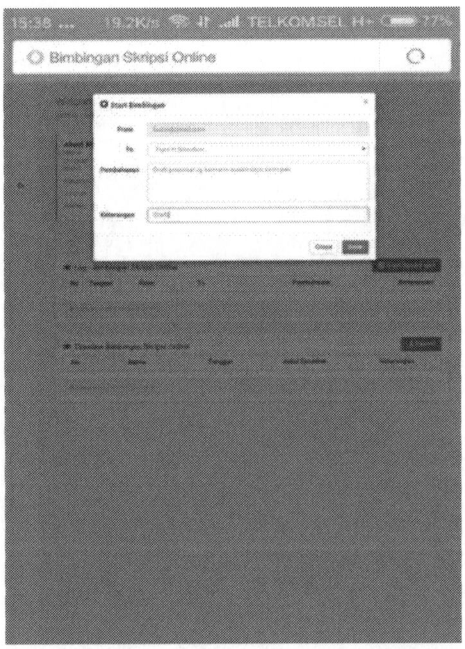


**Figure. 9.** Messages Menu Interface

## 4. Conclusion

Online Thesis Guidance System is designed to assist the student's thesis guidance process. The Design of Online Thesis Guidance System consists of a website for the Department to input both students and lecturers' data. And the android application design is used for students and lecturers to communicate by using smartphones. Based on application features that have been designed, Online Thesis Guidance Application could help the communication process between student and lecturer and make them being easier.

**Acknowledgments**
This research was supported by University of Sumatera Utara. We thank the Vice Rector for Research, Service, and Cooperation University of Sumatera Utara that have supported this research activity. All the faculty, staff members and lab technicians of Electrical Engineering Department, whose services turned my research a success. My Wife, Erna Kurnia and children, Zafran Ahmad Nasution and Zaura Aisya Nasution, family members and friends, without whom I was nothing; they not only assisted me financially but also extended their support morally and emotionally

**References**
[1] Noor D J 2011 *Metodologi Penelitian: Skripsi, Tesis, Disertasi, & Karya Ilmiah* (Jakarta: Prenada Media) pp 107–37.
[2] Huang Y 2016 Design and Implementation of Financial Supervision and Management Information System *Proc. ICSCSE* pp 214–7.
[3] Ying L 2016 Design and Implementation of the Management Information System for Chain Business Corporation *Proc. 8th ICMTMA* pp 195–8.
[4] Hapsari D A, Permanasari A E, Fauziati S, and Fitriana I 2016 Management information systems development for veterinary hospital patient registration using first in first out algorithm *1st Int. Conf. Biomed. Eng.* pp 1–5.
[5] Srivastava S, Khurana P S, Rai A, Cheema A S and Srivastava P K 2016 High performance and adaptive lab report generation in hospital management information systems *Proc. IEEE INDICON* pp 1–6.
[6] Fahmi F, Nasution T H and Anggreiny 2017 Smart cloud system with image processing server in diagnosing brain diseases dedicated for hospitals with limited resources *Technol. Heal. Care*, vol 25 no 3.
[7] El-Ebiary Y A B, Al-Sammarraie N A, Al Moaiad Y and Alzubi M M S 2016 The impact of Management Information System in educational organizations processes *Proc. IC3e* pp 166–9.
[8] Teimoornia M, Hamidi F, Jomeh S M R I and Foroozesh-nia S 2011 The implementation of information and communication technology (ICT) in extracurricular activities of education system in Iran *Proc. Comput. Sci.* vol 3 pp 617–22.
[9] Jain A and Chande S V 2013 Information portal system for a digital campus based on information architecture *Proc. IEEE Int. Conf. MOOC Innovation Technol. Edu.* pp 349–52.
[10] Di Giacomo M 2005 MySQL: Lessons learned on a digital library, *IEEE Softw.* vol 22 no 3 pp 10–3.
[11] Olsson M 2013 *PHP quick scripting reference* vol 9781430262848.
[12] Wu T, Liu J, Xu Z, Guo C, Zhang Y, Yan J and Zhang J 2016 Light-weight, inter-procedural and callback-aware resource leak detection for android apps *IEEE Trans. Softw. Eng. 42(11)* 1054-76.
[13] Ruiz I J M, Nagappan M, Adams B, Berger T, Dienst S and Hassan A E 2016 Analyzing Ad Library Updates in Android Apps *IEEE Softw.* 33(2) pp 74–80.

# Plagiarism Detection for Indonesian Language using Winnowing with Parallel Processing

**Y Arifin[1,3], S M Isa[2], L A Wulandhari[3] and E Abdurachman[2]**

[1]Computer Science Department, BINUS Graduate Program – Doctor of Computer Science, Bina Nusantara University, Jakarta, Indonesia 11480
[2]Computer Science Department, BINUS Graduate Program – Master of Computer Science, Bina Nusantara University, Jakarta, Indonesia 11480
[3]Computer Science Department, School of Computer Science, Bina Nusantara University, Jakarta, Indonesia 11480

Email: yarifin@binus.edu, sisa@binus.edu, lwulandhari@binus.edu, edia@binus.edu

**Abstract.**The plagiarism has many forms, not only copy paste but include changing passive become active voice, or paraphrasing without appropriate acknowledgment. It happens on all language include Indonesian Language. There are many previous research that related with plagiarism detection in Indonesian Language with different method. But there are still some part that still has opportunity to improve. This research proposed the solution that can improve the plagiarism detection technique that can detect not only copy paste form but more advance than that. The proposed solution is using Winnowing with some addition process in pre-processing stage. With stemming processing in Indonesian Language and generate fingerprint in parallel processing that can saving time processing and produce the plagiarism result on the suspected document.

## 1. Introduction

Searching technology is growing and increasing easily to get any information on the internet. Through a search website like Google makes it easy for anyone to get good information in the form of news, science, results of assignments and others. Computers and the Internet are the main cause of increased plagiarism.[1]. Plagiarism in general has several meanings: acknowledging the work of others or the ideas of others without rewarding or citation referring to the producer of the work, giving false information about the reference source, changing some words from a series of sentences from the work of others without give rewards [2]. The plagiarism can happen in any context such at the industry or at the academy, or can be in documents or in programming or other context [3]. According to the Pew Research Center study in 2011, there is an increasing in cyber plagiarism among students, it is around 55% of students admitted to plagiarism in the work of scientific tasks. Not only among the students, but also among researchers also do plagiarism in making scientific writing [4].

M. Alzahrani divides plagiarism into two type, the first one is literally plagiarism by adding or changing similar words and the second one is intelligently plagiarism by doing paraphrasing or translation [5]. Plagiarism detection is influenced by the type of plagiarism, it can be seen in previous research in this field. The previous researcher had been proposed technique to detect plagiarism including mono language or cross language, such as substring matching, keyword similarity, term sequence matching with digital digests that known as fingerprint [6]. The fingerprint can be compute by digital digest through hashing function. This function will generate a collection numbers as a fingerprint, for example of hashing function is MD-5 hashing, it is one of most popular to be used, introduced by Stein and Meyer [7], the others are the Rolling Hash or SHA1 [8].

In Indonesian Language, there are some research that used the fingerprint method in plagiarism detection either with winnowing or just the origin of fingerprint method [9][10]. The winnowing algorithm used by Kusnawan et al in-plagiarism detection applications in Indonesian uses only simple pre-processing text such as removing spaces, and unneeded characters. This kind of method only can detect plagiarism in simple way but it cannot detect the more complex form such as changing active to passive voice. With addition stemming process in pre-processing that can help to detect this kind of plagiarism form. The proposed approach uses the winnowing algorithm in addition to the pre-processing stage that affects time processing. To overcome processing time, in pre-processing and processing processes requiring large quantities of document checking can utilize CPU capabilities that currently can consist of multiple cores to perform multiple processes simultaneously known as parallel processing. According to Khatter, many current CPUs consist of 8 - 12 cores that can be upgraded for more efficient process [11]. Through some trials can be seen time that can be saved by using parallel processing that involves several cores in one process. By checking the plagiarism with the winnowing algorithm through parallel processing can provide better results with more optimal performance.

## 2. Plagiarism Detection Theory

### *2.1. A Winnowing Algorithm*

The methodology used in performing plagiarism detection using the fingerprint method with the Winnowing algorithm. The fingerprint method is a method that represents a document with a set of numbers obtained through a hash function [12]. This method was developed with winnowing algorithm [13]. How it works winnowing can be seen in the following stages:

a. A collection of text is read from a document.
   For example: "There are a lot of food"
b. A collection of texts is a cleaning process of unrelated characters such as spaces and dots.
   For example: after the cleaning is done the set of text becomes " Therearealotoffood ".
c. Furthermore, the word separation is based on N-Grams.
   For example: The N-Grams used are 5, then after the word separation becomes: There herea erear reare earea areal realo ealot aloto lotof otoff toffo offoo ffood
d. By using a certain hashing function, each word is determined by the number of hash fuction.
   For example : 656 333 1582 3172 4022 9914 1425 656 2355 3051 333 8395 1524 3181
e. From the sequence of numbers hereinafter referred to as the hash value will be formed into several windows. Each window consists of the same number of numbers. A window can consist of 4 hash keys or another one.
   For example : Each window only contains 4 hash values , then the results will be like these :
   [656 **333** 1582 3172] [333 1582 3172 4022] [1582 3172 **4022** 9914] [3172 4022 9914 **1425**] [4022 9914 1425 **656** ] [9914 1425 656 2355] [656 2355 3051 **333**] [2355 3051 333 8395] [3051 333 8395 1524 ] [333 8395 1524 3181]
f. From each window, the smallest hash key is chosen to form the fingerprint of a document. If there is the same smallest hash key in a different window, then it is taken that occupies the rightmost position.
   For example: 333 4022 1425 656 333 333

After do all the step above, the fingerprint that represent the example document are 333 4022 1425 656 333 333. It can be varied depend on the size of the document and the hashing method to be used to generate fingerprint

### *2.2. Jaccard Coeficient*

The detection of plagiarism can be seen from the degree of similarity between documents. Measuring the degree of similarity between documents using Jaccard Similarity Coeficient [14] :

$$Jaccard\ Similarity\ (A, B) = \frac{P\ (A \cap B\ )}{P\ (A \cup B)} \quad (1)$$

For examples A and B are two documents. Measures of similarity between documents A and B are derived from the same number of features between document A and document B then divided by the total of all the features present in document A and document B.

## 3. Methodology

The methodology that been used in this research briefly can be seen in figure 1. The detail explanation from each stage can be seen below:

1. Preparing the Dataset

   The dataset is collected from the online library of Bina Nusantara University that contain e-thesis file of students of Computer Science with Multimedia and Game topics. The data collected consists of 100 documents that have different file sizes (See Table 1) and using Indonesian Language. As is known thesis consists of several chapters, and the most likely most plagiarism occurs is in chapter 1 which is the introductory chapter and chapter 2 which consists of theoretical basis used in the study.

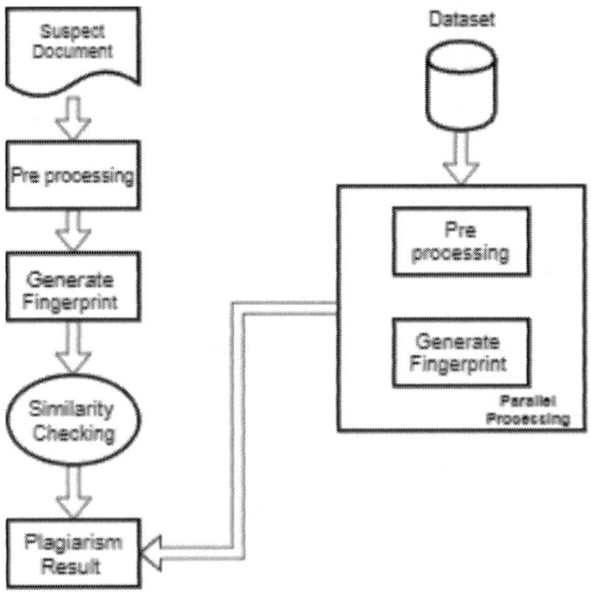


**Figure 1.** The stage of the Plagiarism Detection Technique

**Table 1.** Dataset Testing

| No | Category | Quantity |
|---|---|---|
| 1 | Short Document (5 -10 pages) | 49 |
| 2 | Long Document ( > 10 pages) | 51 |
| | Total Document | 100 |

IOP Conf. Series: Journal of Physics: Conf. Series **978** (2018) 012082 doi:10.1088/1742-6596/978/1/012082

Treatment of the dataset to be tested consists of several treatments, namely the addition of the exact same sentence with another document, the change of sentence intonation from active sentence to passive sentence and there is also a document that does not receive any treatment so according to the original result of the document obtained (See Table 2).

**Table 2.** Some Treatment in Dataset

| No | Treatment | Quantity |
|---|---|---|
| 1 | Original Document | 40 |
| 2. | Add plagiarism 20% in Short Document | 20 |
| 3. | Add plagiarism 40% in Long Document | 20 |
| 4. | Active – Passive Changed | 20 |
| | Total Document | 100 |

2. Generate Corpus
This activity generate corpus from the dataset that contain two processing: pre-processing and generate fingerprint for all documents in the dataset. In the pre-processing stage, the document must have docx extension and if the document has different extension then it converts first to docx extension. Then the all sentences in the document will be change in basic word through stemming process [15]. The next process is to generate fingerprint from the stemming result. Before generate fingerprint, all the white space or unrelated character will be removed from the stemming result. Separate the basic word using N-gram then generate number from the basic word using SHA1 hashing function that generate hash values Collect the hash value in each window then select the appropriate hash value. All the selected hash value knowns as the fingerprint from one document. The illustration from corpus generating can be seen in Figure 2. All the documents have been processed through parallel processing to saving processing time. This step to generate corpus only have to run one times as long the parameter in the plagiarism detection does not changed and can be used as reference corpus for any suspected document.

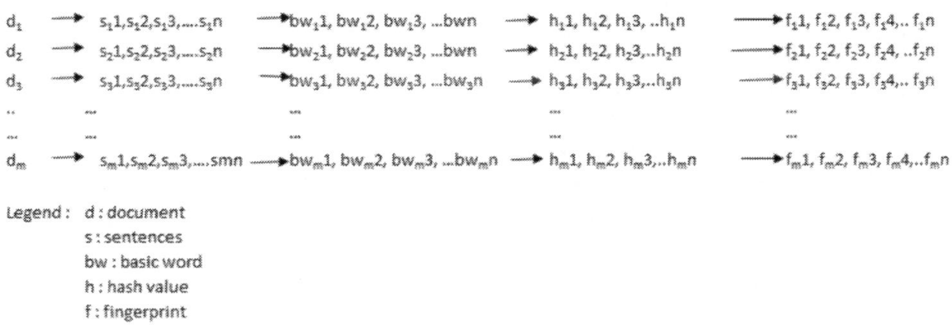


**Figure 2.** The illustration of each process in generating corpus

3. Checking of the Suspected Document
The suspected document that want to be checked must have docx extension and also do the preprocessing and generate fingerprint as same as generate corpus.
4. Similarity Checking
Identify plagiarism using the similarity checking with Jaccard Coeficient. The fingerprint from the suspected document checked with the fingerprint from the corpus through the similarity

checking and the result from this similarity checking represented in percentage to define how similar between both documents.

5. Plagiarism Result
Based on the result from similarity checking then compared with the threshold 0.45 [9]. If the similarity degree is more than 45% , the suspected document has plagiarism, and if less than equal with 45% ,the suspected document does not have plagiarism.

## 4. Result and Discussion

### *4.1. Experiment Result*

The experiment consists of two parts: test on various windows and 5 N-Gram, then test on process done without multiprocessing and with multiprocessing. Here are the test results by trying different window variations that are used to determine the level of similarity (Table 3):

**Table 3**. Similarity Degree related with the Window Size and 5 N-Grams

| [Gram,Window Size] | [5,5] | [5,10] | [5,20] | [5,30] |
|---|---|---|---|---|
| Document 1 | **47.66%** | **46.58%** | 43.39% | 42.36% |
| Document 2 | **47.32%** | **45.02%** | 41.95% | 37.96% |
| Document 3 | **46.60%** | 44.37% | 40.80% | 37.86% |
| .... | ... | ... | ... | ... |
| .... | ..... | .... | .... | ... |
| Document 100 | 26.79% | 25.87% | 23.17% | 19.89% |

From the test results seen with the same N-Gram that is 5 N-Gram with various window variations show the greater the window size, the value of similarity is decreasing. Therefore, it can be concluded that the larger window size decreases the similarity level between the corpus document and the document suspected. In the winnowing algorithm if the window size is 10 means from all the fingerprint in one document will be grouped by ten in one window then the fingerprint with the smallest number be selected in each window. It can decrease the chance to be the selected fingerprint as compare with the size of window is 5. The more bigger size of the window has less selected fingerprints than the smallest size of the window. With more selected fingerprint can increase the chance to be compare with the fingerprint of suspected document.

In table 3 document 1 has the highest level of similarity compared to other documents when compared with suspected documents. According to threshold 0.45 then it can be concluded that the suspected document is a plagiarism document because the similarity degree with document 1,2 and document 3 more than 45% with window size is 5 and 5 N-Gram.

For the second stage of testing comparing processing time between parallel processing processes and without parallel processing (See Table 4). The process tested in the second stage is the stage of generating a fingerprint from Corpus that has 100 documents. The core process that is owned on the laptop used for testing consists of 4 suspected cores.

**Table 4.** Comparison of Processing Time

| | **Without Parallel Process** | **With Parallel Process** | **Saving Time** |
|---|---|---|---|
| **Time Processing** | 208.7894 Minutes | 52.9052 Minutes | 74,66% |

Based on table 4, the process fingerprint that using parallel processing takes 52.9052 minutes and tit save processing time around 74.66%. If we cannot use the GPU's processor, this kind of technique is one of solution to process a lot of document more quickly.

## 5. Conclusion

The detection of plagiarism in Indonesian Texts can be done with Winnowing preceded by a pre-processing stemming stage. Although the addition of this stage affects time processing, but it can save time with parallel processing. The appropriate number of hash values for winnowing based on test results is to use 5 hash values on each winnow. With winnow the greater the effect on the level of similarity between documents.

For the further research, this plagiarism detection technique will be improved not only in Indonesian Language, but cross language between Indonesian with English. As we know there are more English text or documents that provide online in internet.

## References

[1] Parker K, Lenhart A,and Moore K 2011 The Digital Revolution and Higher Education *Pew Res.* **202** 29.

[2] Maurer HA, Kappe F, Zaka B 2006 Plagiarism-a survey. *J. UCS.* **12(8)** 1050-84.

[3] Joy M and Luck M 1999 Plagiarism in programming assignments *IEEE Trans. Edu.* **42(2)** pp 129–33.

[4] Pupovac V and Fanelli D 2015 Scientists Admitting to Plagiarism: A Meta-analysis of Surveys *Sci. Eng. Ethics* **21(5)** pp 1331–52.

[5] Alzahrani SM, Salim N, Abraham A and Member S 2012 Understanding Plagiarism Linguistic Patterns, Textual Features, and Detection Methods *IEEE Trans. Syst. Man Cybernetics Part C (Applications and Reviews)* **42(2)** 133-49.

[6] Stein B and zu Eissen SM 2007 Fingerprint-based Similarity Search and its Applications *Forsch. und wissenschaftliches Rechn.* pp 85–98.

[7] Stein B and Meyer S 2006 Near Similarity Search and Plagiarism Analysis *Data Inf. Anal. Knowl. Eng.* 430-7.

[8] Wang J, Shen HT, Song J and Ji J 2014 Hashing for similarity search: A survey *arXiv preprint arXiv:1408.2927.*

[9] Wibowo AT, Sudarmadi KW and Barmawi AM 2013 Comparison between fingerprint and winnowing algorithm to detect plagiarism fraud on Bahasa Indonesia documents *IEEE Proc. ICoICT* pp 128-33.

[10] Kusmawan PY, Yuhana UL and Purwitasari D 2009 Aplikasi Pendeteksi Penjiplakan pada File Teks dengan Algoritma Winnowing pp 1–11.

[11] Khatter H and Aggarwal V 2014 Efficient parallel processing by improved CPU-GPU interaction *Proc. ICICT* pp 159–61.

[12] Meuschke N and Gipp B 2013 State-of-the-art in detecting academic plagiarism *Int. J. Educ. Integr.* **9(1)** pp 50–71.

[13] Schleimer S, Wilkerson DS, Aiken A and Berkeley UC 2003 Winnowing : Local Algorithms for Document Fingerprinting *Proc. ACM SIGMOD Int. Conf. Manage. Data* pp 76-85.

[14] Niwattanakul S, Singthongchai J, Naenudorn E and Wanapu S 2013 Using of Jaccard Coefficient for Keywords Similarity *Int. MultiConference Eng. Comput. Sci.* vol I pp 380–4.

[15] Librarian A and Kukuh R 2017 Sastrawi [Online]. Available: https://github.com/sastrawi/sastrawi.

IOP Conf. Series: Journal of Physics: Conf. Series **978** (2018) 012083 doi:10.1088/1742-6596/978/1/012083

# Feature Detection of Curve Traffic Sign Image on The Bandung - Jakarta Highway

**M Naseer[1], I Supriadi[2] and S H Supangkat[3]**

[1,2]Departement of Informatics, Sekolah Tinggi Teknologi Bandung
[3]School of Electrical Engineering and Informatics, Institut Teknologi Bandung

Email : [1]naseer@sttbandung.ac.id, [2]irwin@sttbandung.ac.id, [3]suhono@stei.itb.ac.id

**Abstract**. Unsealed roadside and problems with the road surface are common causes of road crashes, particularly when those are combined with curves. Curve traffic sign is an important component for giving early warning to driver on traffic, especially on high-speed traffic like on the highway. Traffic sign detection has became a very interesting research now, and in this paper will be discussed about the detection of curve traffic sign. There are two types of curve signs are discussed, namely the curve turn to the left and the curve turn to the right and the all data sample used are the curves taken / recorded from some signs on the Bandung - Jakarta Highway. Feature detection of the curve signs use Speed Up Robust Feature (SURF) method, where the detected scene image is 800x450. From 45 curve turn to the right images, the system can detect the feature well to 35 images, where the success rate is 77,78%, while from the 45 curve turn to the left images, the system can detect the feature well to 34 images and the success rate is 75,56%, so the average accuracy in the detection process is 76,67%. While the average time for the detection process is 0.411 seconds.

## 1. Introduction

The ability to recognize or understand an object is one of the major components in the field of computer vision and machine learning that are widely implemented in many things, such as object matching, monitoring systems, video analysis, and information description existing in an image or video that produces a smart system. Another breakthrough done in the last decade is detector and identificator of various traffic symbols or signs that exist in the field of Transportation / Intelligent Transportation System (ITS), and also an interesting focus on development of autonomous driving [1] or Smart vehicle.

Traffic Signs are part of the Road fixtures in the form of symbols, letters, numbers, sentences, and / or combinations that serve as warnings, prohibitions, orders or directions for the Road User [2]. Although traffic signs have already owned its standards through/by/from the Transportation Ministry of the Republic of Indonesia in 2014 [2], but traffic signs have very wide characteristics for their visual appearance in real world environments, for example display may different when affected by the changes of global light, weather conditions, light reflection, discoloration and physical form of signs that may be worn out, so in practice it needs to be done with a very high accuracy. Traffic Signs have been designed to be easily read/recognized by humans who are certainly very good at recognizing the existing symbols, but for a computer system, the problem of detecting, recognizing, and classifying traffic signs still has been a challenge until now, where that is included into the focus pattern recognition. In the field of computer vision and learning machine, until now it is continuously conducted an improvement for

IOP Conf. Series: Journal of Physics: Conf. Series **978** (2018) 012083 doi:10.1088/1742-6596/978/1/012083

solving and refining the problems that occur in the process of detection and recognition of traffic signs[3].

The traffic signs discussed in this Paper are road curve traffic signs, where from the conducted searches, found the data that the single accident is often caused from driver negligence, which from conducted research at the Center of Automotive Safety Research, University of Adelaide - Australia , by [4] it was found that problems with the road surface were a common cause of road crashes, particularly when combined with curves. This inspired this research to focus on curve signs as a preparation of Smart Driver Assistance System in Indonesia, as had been done by [5]. The dataset used in this study is traffic signs data in Indonesia which refers to the Minister of Transportation Regulation no. 13 of 2014 on Traffic Signs in the Republic of Indonesia..

## 2. Related Work

Research about the detection and recognition of traffic signs or known as Traffic Sign Detection (TSD) and Traffic Sign Recognition (TSR) has been widely conducted in some countries, and each of these studies demonstrates different approaches, methods, and outcomes. The differences are certainly influenced by many factors, one of those is a different traffic sign model in each country. As had been conducted by [6][7], the study discusses the probability and measurement of detection rates to traffic signs in the United States, where the research is conducted as a form of development from the German Traffic Sign Detection Benchmark (GTSDB) competition which refers only to signs in Europe.

Research that was conducted on [8] mentions that using a region interest detector or feature detector such as SIFT (Scale Invariant Feature Transform) [9] or SURF[10] is known to be an improvement from SIFT, making the detection process of interest points to be better, this also applies to feature detection. Another study using SURF for traffic sign detection was conducted at [11], from that study, SURF was combined with FPGA hardware for realtime detection process, but it is conveyed that there was a limitation in realtime detection due to the complexity of the SURF algorithm that slowed the process.

By combining the color threshold, polygon approximation algorithm, and SURF, in the study [12] segmentation is conducted on potential of traffic signs by color threshold, and a polygon approximation algorithm s used to detect appropriate polygons. The potential signs are compared with the template signs in the database using SURF feature matching method. In the identification step, it was used SURF for a CPU system only and a CPU with GPGPU System, the result is there are an improvement in accuracy and efficiency towards some data. A recent study using SURF for traffic sign detection was conducted on [13], where a combination of SURF and Artificial Neural Network Classifier was conducted and the result of accuracy was 97%.

## 3. Method

Method used in The Study consisted of several steps as follows : Image Input using Mobile Phone Camera Placed on Car Dashboard, Pre Stage Image Processing Consist of Scaling and Gray Scaling, Interest Point Detection, Feature Detection by SURF detector[10] by using dataset of Curve Traffic Sign. As The result, the image input will contain suitable feature with the dataset. All images were captured along on the Bandung – Jakarta Highway. The general architecture of the research methodology can be seen in Figure 1.

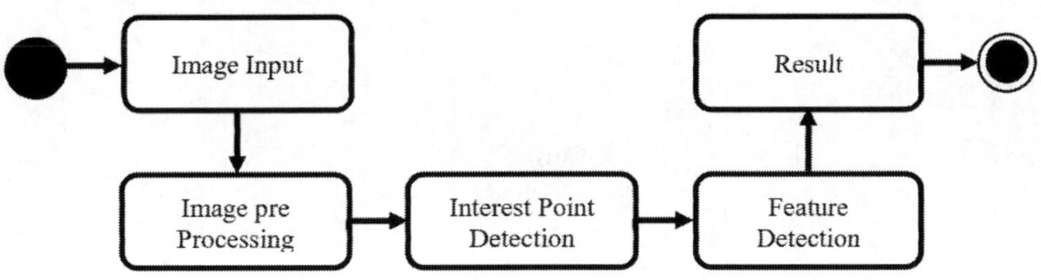


**Figure 1**. General Architecture

### *3.1. Image Input*

All input images used in this research were recorded by using camera of mobile phone that was placed into the car dashboard. Recorded image data were taken per frame that contains cure traffic sign, either that directed to the rights or directed do the left. The set of data can be seen from Figure 2 below:

**Figure 2**. Set of Image Scene

### *3.2. Pre-Processing*

Pre-processing image step was done to produce a better image to be processed to the next steps of detection phase. In this study, pre-processing the image taken was image scaling 800x450 and the conversion of gray level image (gray scaling).

### *3.3. SURF Detector*

Interest Point Detection was used to select points that contains a lot of information and at the same time is stable towards the local or global distortion changes in digital images. With the SURF of algorithm, in the scene image that contains a curve traffic sign was selected interest point detector that has the invariant character towards the scale, that was the blob detection. A blob is an area of a digital image that has properties that are constant or vary within a certain range. To compute this blob detection, the determinant of the Hessian (DoH) matrix from the image was used. In the SURF algorithm, the determinant of the Hessian matrix was calculated from the Haar wavelet using integral image optimally. The determinant of the Hessian matrix was used as the basis of the SURF algorithm because of its invariant characteristic and "to able to face" the difference in scale, stability and viewing the angle well.

### *3.4. Detection Process*

The detection results in the first step were scene image that had got an interest point or feature interest, as seen from Figure 3 as follows.

**Figure 3**. Detected Features on Image Scene

**Figure 4**. Matching Line on Image Scene

In Figure 3 can be seen small blue circles, these points were interest points or features detected by the system through the SURF Detector mechanism. These points were much existing in the Right Curve Traffic Sign, so it shaped a yellow line indicating that the area was match with the image in the dataset. In addition to the detected keypoints, there was also a matching line shown in Figure 4. Matching line in Figure 4 connects the keypoints detected in the scene image to the keypoints on the dataset located at the top of the road scene.

## 4. Result

The final result in this research is image detection based on match feature between the image contained in the road scene image and match it with the dataset image. Detected circuit images can be seen in Figure 5.

**Figure 5**. Detected image for Right Curve (top) and Left Curve (bottom)

Table 1 and Table 2 below are the test results when testing the image scene was conducted. Table 1 is the result of Right Curve Detection and Table 2 is the result of Left Curve Detection. Both tables also contain the duration of time required in the detection process, and time is presented in Second.

**Table 1**. Result for Right Curve Sign Detection

| No. | Image | Features Det. | Time |
|---|---|---|---|
| 1. | curveRight (1).jpg | Detected | 0,482 |
| 2. | curveRight (2).jpg | Detected | 0,361 |
| 3. | curveRight (3).jpg | Detected | 0,382 |
| 4. | curveRight (4).jpg | Detected | 0,445 |
| 5. | curveRight (5).jpg | Detected | 0,422 |
| 6. | curveRight (6).jpg | Detected | 0,335 |
| 7. | curveRight (7).jpg | Detected | 0,418 |
| 8. | curveRight (8).jpg | False Feature | 0,522 |
| 9. | curveRight (9).jpg | Detected | 0,424 |
| 10. | curveRight (10).jpg | Detected | 0,367 |
| 11. | curveRight (11).jpg | Detected | 0,399 |
| 12. | curveRight (12).jpg | False Feature | 0,490 |
| 13. | curveRight (13).jpg | False Feature | 0,426 |
| 14. | curveRight (14).jpg | Detected | 0,307 |
| 15. | curveRight (15).jpg | Detected | 0,376 |
| 16. | curveRight (16).jpg | Detected | 0,526 |
| 17. | curveRight (17).jpg | Detected | 0,497 |
| 18. | curveRight (18).jpg | Detected | 0,402 |
| 19. | curveRight (19).jpg | Detected | 0,384 |

**Table 2**. Result for Left Curve Sign Detection

| No. | Image | Features Det. | Time |
|---|---|---|---|
| 1. | curveLeft (1).jpg | Detected | 0,477 |
| 2. | curveLeft (2).jpg | Detected | 0,392 |
| 3. | curveLeft (3).jpg | Detected | 0,375 |
| 4. | curveLeft (4).jpg | Detected | 0,363 |
| 5. | curveLeft (5).jpg | False Feature | 0,422 |
| 6. | curveLeft (6).jpg | False Feature | 0,466 |
| 7. | curveLeft (7).jpg | Detected | 0,353 |
| 8. | curveLeft (8).jpg | Detected | 0,435 |
| 9. | curveLeft (9).jpg | Detected | 0,367 |
| 10. | curveLeft (10).jpg | Detected | 0,385 |
| 11. | curveLeft (11).jpg | Detected | 0,402 |
| 12. | curveLeft (12).jpg | Detected | 0,462 |
| 13. | curveLeft (13).jpg | Detected | 0,442 |
| 14. | curveLeft (14).jpg | False Feature | 0,484 |
| 15. | curveLeft (15).jpg | Detected | 0,317 |
| 16. | curveLeft (16).jpg | Detected | 0,364 |
| 17. | curveLeft (17).jpg | False Feature | 0,539 |
| 18. | curveLeft (18).jpg | Detected | 0,329 |
| 19. | curveLeft (19).jpg | Detected | 0,522 |

IOP Conf. Series: Journal of Physics: Conf. Series **978** (2018) 012083 doi:10.1088/1742-6596/978/1/012083

| | | | |
|---|---|---|---|
| 20. | curveRight (20).jpg | False Feature | 0,445 |
| 21. | curveRight (21).jpg | Detected | 0,307 |
| 22. | curveRight (22).jpg | False Feature | 0,409 |
| 23. | curveRight (23).jpg | Detected | 0,438 |
| 24. | curveRight (24).jpg | Detected | 0,466 |
| 25. | curveRight (25).jpg | False Feature | 0,504 |
| 26. | curveRight (26).jpg | Detected | 0,465 |
| 27. | curveRight (27).jpg | Detected | 0,532 |
| 28. | curveRight (28).jpg | Detected | 0,382 |
| 29. | curveRight (29).jpg | False Feature | 0,423 |
| 30. | curveRight (30).jpg | Detected | 0,497 |
| 31. | curveRight (31).jpg | False Feature | 0,533 |
| 32. | curveRight (32).jpg | Detected | 0,402 |
| 33. | curveRight (33).jpg | Detected | 0,423 |
| 34. | curveRight (34).jpg | Detected | 0,331 |
| 35. | curveRight (35).jpg | Detected | 0,371 |
| 36. | curveRight (36).jpg | Detected | 0,455 |
| 37. | curveRight (37).jpg | False Feature | 0,549 |
| 38. | curveRight (38).jpg | Detected | 0,346 |
| 39. | curveRight (39).jpg | Detected | 0,447 |
| 40. | curveRight (40).jpg | False Feature | 0,441 |
| 41. | curveRight (41).jpg | Detected | 0,483 |
| 42. | curveRight (42).jpg | Detected | 0,511 |
| 43. | curveRight (43).jpg | Detected | 0,344 |
| 44. | curveRight (44).jpg | Detected | 0,417 |
| 45. | curveRight (45).jpg | Detected | 0,495 |

| | | | |
|---|---|---|---|
| 20. | curveLeft (20).jpg | Detected | 0,352 |
| 21. | curveLeft (21).jpg | Detected | 0,316 |
| 22. | curveLeft (22).jpg | False Feature | 0,481 |
| 23. | curveLeft (23).jpg | Detected | 0,551 |
| 24. | curveLeft (24).jpg | Detected | 0,311 |
| 25. | curveLeft (25).jpg | Detected | 0,505 |
| 26. | curveLeft (26).jpg | Detected | 0,473 |
| 27. | curveLeft (27).jpg | False Feature | 0,454 |
| 28. | curveLeft (28).jpg | False Feature | 0,405 |
| 29. | curveLeft (29).jpg | Detected | 0,366 |
| 30. | curveLeft (30).jpg | Detected | 0,334 |
| 31. | curveLeft (31).jpg | Detected | 0,447 |
| 32. | curveLeft (32).jpg | False Feature | 0,487 |
| 33. | curveLeft (33).jpg | Detected | 0,544 |
| 34. | curveLeft (34).jpg | Detected | 0,406 |
| 35. | curveLeft (35).jpg | Detected | 0,428 |
| 36. | curveLeft (36).jpg | Detected | 0,548 |
| 37. | curveLeft (37).jpg | Detected | 0,471 |
| 38. | curveLeft (38).jpg | False Feature | 0,473 |
| 39. | curveLeft (39).jpg | Detected | 0,422 |
| 40. | curveLeft (40).jpg | False Feature | 0,434 |
| 41. | curveLeft (41).jpg | Detected | 0,453 |
| 42. | curveLeft (42).jpg | Detected | 0,379 |
| 43. | curveLeft (43).jpg | Detected | 0,523 |
| 44. | curveLeft (44).jpg | Detected | 0,446 |
| 45. | curveLeft (45).jpg | False Feature | 0,476 |

To calculate The accuracy of the test, we use the following equation :

$$\text{Accuracy} = \frac{\text{Detected Features}}{\text{All Tested Data}} \times 100\%$$

The accuracy of the test results in Table 1 & Table 2 can be seen in Table 3.

**Table 3**. Accuracy Result

| No. | Curve Sign Type | Number of Data | Number of Correct Detected | Accuracy | Detection Time (Second) |
|---|---|---|---|---|---|
| 1 | Curve to Right | 45 | 35 | 77,78% | 0,392 |
| 2 | Curve to Left | 45 | 34 | 75,56% | 0,431 |

$$\text{Average Accuracy} = \frac{(35 + 34)}{90} \times 100\% = 76{,}67\%$$

## 5. Conclusion and Future Work

Regarding to some related conducted researches, for case of traffic sign detection, SURF algorithm can not independently role in the detection process, it needs to be accompanied by other methods / algorithms as the initial process such as shape detection or region interest detection. By using only feature detection, detection towards the curve traffic sign produces a fairly low accuracy with an average of 76.67%, in order that, the next work is to add a shape detection method where the curve traffic signs have a distinctive and special shape, so that is expected to increase the accuracy of detection. In addition, will be conducted a Development of Mobile Apps that will detect the Curve Traffic Sign image to give an alert to the drivers.

## Acknowledgment

This paper is funded by Research Grant Award from the Directorate of Research and Community Service (DRPM), Ministries of Research, Technology, and Higher Education (Kemristekdikti), Indonesia, 2017. The Authors would like to thank to the Kemristekdikti, Sekolah Tinggi Teknologi Bandung, and Institut Teknologi Bandung for the opportunity for allowing authors to conduct research at the two universities. Special thanks to Bapak Suhono who has taken his time for directing the authors.

## References

[1] Levinson J, Askeland J, Becker J, Dolson J, Held D, Kammel S, Kolter J Z, Langer D, Pink O, Pratt V, Sokolsky M, Stanek G, Stavens D, Teichman A, Werling M and Thrun S 2011 Towards fully autonomous driving: Systems and algorithms *2011 IEEE Intell. Vehicles Symp. (IV)* (IEEE) pp 163–8

[2] Kementerian Perhubungan Republik Indonesia 2014 Peraturan Menteri Perhubungan Republik Indonesia Nomor PM 13 Tahun 2014 Tentang Rambu Lalu Lintas

[3] Stallkamp J, Schlipsing M, Salmen J and Igel C 2012 Man vs. computer: Benchmarking machine learning algorithms for traffic sign recognition *Neural Networks* **32** 323–32

[4] Mrj B, Cn K and Aj M 2008 *In-depth research into rural road crashes Report documentation*

[5] Soendoro D and Supriana I 2011 Traffic sign recognition with Color-based Method, shape-arc estimation and SVM *Proc. 2011 Int. Conf. on Electrical Eng. Inf.* (IEEE) pp 1–6

[6] Mogelmose A, Dongran Liu and Trivedi M M 2014 Traffic sign detection for U.S. roads: Remaining challenges and a case for tracking *17th Int. IEEE Conf. ITSC* pp 1394–9

[7] Mogelmose A, Liu D and Trivedi M M 2015 Detection of U.S. Traffic Signs *IEEE Trans. Intell. Transp. Syst.* **16** 3116–25

[8] Salti S, Petrelli A, Tombari F, Fioraio N and Di Stefano L 2015 Traffic sign detection via interest region extraction *Pattern Recognit.* **48** 1039–49

[9] Lowe D G 2004 Distinctive image features from scale-invariant keypoints *Int. J. Comput. Vis.* **60** 91–110

[10] Bay H, Ess A, Tuytelaars T and Van Gool L 2008 Speeded-Up Robust Features (SURF) *Comput. Vis. Image Underst.* **110** 346–59

[11] Zhao J, Zhu S and Huang X 2013 Real-Time Traffic Sign Detection Using SURF Features on FPGA

[12] Ding D and Yoon J 2015 Traffic sign detection and identification using SURF algorithm and GPGPU Traffic sign detection and identification using SURF algorithm and GPGPU pp 10–3

[13] Dhar P and Deb K 2016 Traffic Sign Recognition Using SURF : Speeded Up Robust Feature Descriptor and Artificial Neural Network Classifier 198–201

# Nearby Search Indekos Based Android Using A Star (A*) Algorithm

**B Siregar[1], EB Nababan[1], JA Rumahorbo[1], U Andayani[1] and F Fahmi[2]**

[1]Department of Information Technology, Faculty of Computer Science and Information Technology, University of Sumatera Utara, Indonesia.
[2]Department of Electrical Engineering, Faculty of Engineering, University of Sumatera Utara, Indonesia.

Email: baihaqi@usu.ac.id

**Abstract.** Indekos or rented room is a temporary residence for months or years. Society of academicians who come from out of town need a temporary residence, such as Indekos or rented room during their education, teaching, or duties. They are often found difficulty in finding a Indekos because lack of information about the Indekos. Besides, new society of academicians don't recognize the areas around the campus and desire the shortest path from Indekos to get to the campus. The problem can be solved by implementing A Star (A*) algorithm. This algorithm is one of the shortest path algorithm to a finding shortest path from campus to the Indekos application, where the faculties in the campus as the starting point of the finding. Determination of the starting point used in this study aims to allow students to determine the starting point in finding the Indekos. The mobile based application facilitates the finding anytime and anywhere. Based on the experimental results, A* algorithm can find the shortest path with 86,67% accuracy.

## 1. Introduction

Society of academicians who come from out of town need a temporary residence, such as Indekos or rented room during their education, teaching, or duties. They are often found difficulty in finding a Indekos because lack of information about the Indekos. Besides, new society of academicians don't recognize the areas around the campus and desire the shortest path from Indekos to get to the campus. The need of such information can be served if there is a place, system, or an application for collecting and broadcasting information about the place of the Indekos. Due to that case, the finding of Indekos can be done using Android smartphone with GPS and Google Maps. Then, to find the nearest Indekos can be done by using A Star algorithm. A Star algorithm is a best-first search that uses cost function. The cost function is used to count distance, area, or another factor to decide either a route is selected or not. The cost function checks the feasibility of the cost required to reach a node from another node [1]. In this study, the authors create an application using A Star. Google Maps is used in this application to determine the coordinate point. The finding of Indekos using each faculty as the starting point. To get the shortest path, the application uses heuristic function and A Star algorithm to find the nearest path. In addition, traffic congestion is set as one of the parameter to find the shortest path in this application of study. This journal is structed by the following sections: Section II describes the problem identification in this study. Section III describes some previous researches that have been done. Section IV describes how the application in this study is developed by the proposed method. The results of the study are discussed in Section V. Section VI contains research conclusions, as well as suggestions for further research.

## 2. Problem Identification

Students are often found difficulty in finding a Indekos because they don't recognize the areas around the campus very well. Moreover, the information of Indekos often less actual so students do not have any standards to find a Indekos. The finding shortest path to the Indekos based on Android application is expected to solve this problem.

## 3. Previous Research

Several previous researches have been conducted to solve the problem of finding the shortest path in some cases. Research conducted in 2011 by Tilawah applies A Star (A*) algorithm to solve maze problem. He found that A* algorithm can be applied to solve the maze problem by storing data the position of the maze wall as a non-walkable node [2]. Research conducted by Hannawati, Thiang, and Eleazhar is to find the shortest path using Genetic algorithm. They found that genetic algorithm is quite effective and easy to use, especially in terms of finding the shortest path and the shortest time based on path conditions [3]. Research conducted in 2015 by Hutabarat is to find the shortest path using the Floyd-Warshall algorithm in online taxi application. The shortest path is found by making points of intersection in the protocol streets in Medan city. He found that the Floyd-Warshall algorithm can calculate the distance of the nearest taxi from the client [4]. Research conducted in 2016 by Siregar is to find the shortest path of food delivery. He applies A Star algorithm and uses Geographical Information System (GIS) for vehicle usability. By applying A Star algorithm to find the shortest path, the food delivery system works well. However, it should also consider traffic congestion in finding the shortest path [5].

## 4. Methodology

The method proposed in finding the shortest path to get the nearest Indekos information is by using A Star algorithm.

### *4.1. General Architecture*

The application is developed to find the nearest Indekos. Users of this application are society of academicians who need a temporary residence, such as Indekos or rented room. The general architecture of the system can be seen in Figure 1.

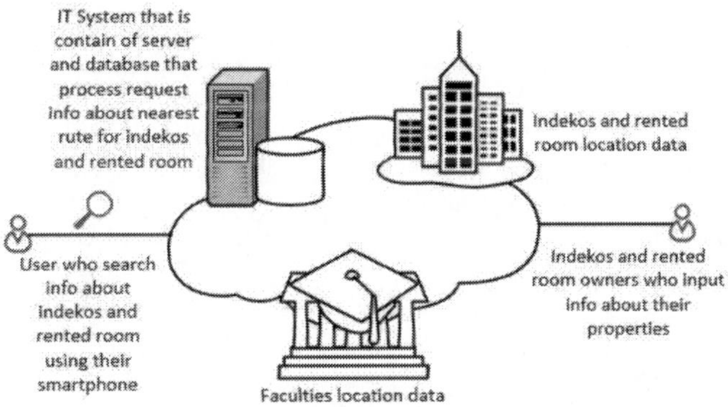


**Figure 1.** General Architecture

The data used in this application is data of Indekos, rented room, faculty, traffic congestion point, coordinate point, and relation point. An administrator inputs the data into the database so the data can be used to find the shortest path to the Indekos from the faculty. Users (society of academicians) who want to find Indekos can choose a faculty marked in the map. After that, the user can choose to find either Indekos or rented room. The system will look for the nearest Indekos or rented room from the faculty chosen by the user. Determination of the location of the nearest Indekos or rented room is done

IOP Conf. Series: Journal of Physics: Conf. Series **978** (2018) 012084 doi:10.1088/1742-6596/978/1/012084

by using the Euclidean Distance heuristic function. Then, the system will look for the shortest path using the A Star algorithm according to the path weight due to some traffic congestion points that may exist on the path. As the output, the map will show the shortest path that has been processed by the system. After finding the nearest Indekos or rented room, the user can view the more information about the Indekos by tapping on the Indekos point.

### 4.2. Floyd-Warshall Algorithm

We use 5 points to analyze A Star algorithm. These five points are arranged in a simple graph. The points can be seen in Figure 2.

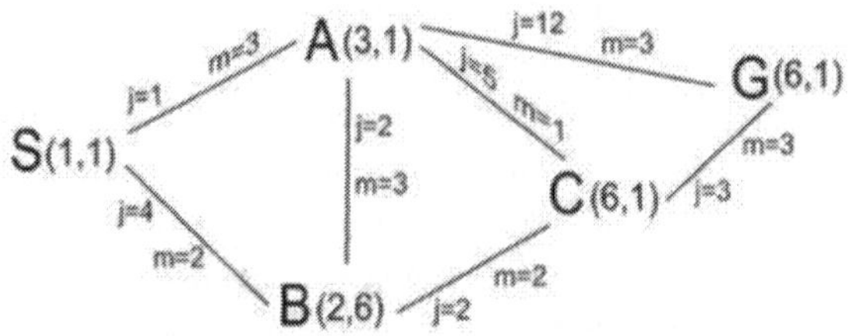


**Figure 2.** Sample Graph with 5 Points

In the figure, there are five points, such as S, A, B, C, and G. The starting point is S and the destination is G. The coordinates of each points are S (1,1), A (3,1), B (2,6), C (6,1), G (6.5). To get the value of h(n) we will use the calculation of heuristic function. The heuristic function used is Euclidean Distance. This function provides better results than other heuristic functions [6]. For the value of g(n) we will use the distance value (j) multiplied by the traffic congestion weight (m) to manipulate the weight of each points relation using Euclidean Distance formula, as shown in Equation 1.

$$d(x,y) = |\sqrt{(x_1-y_1)^2 + (x_2-y_2)^2 + \cdots + (x_n-y_n)^2} \quad (1)$$

Calculate the heuristic value from the related point: S-A, S-B, A-C, A-G, A-B, B-C, C-G by using Euclidean Distance formula.

$$S(1,1)\ to\ A(3,1)$$
$$d(x,y) = |\sqrt{(1-1)^2 + (3-1)^2} = 2$$
$$A(3,1)\ to\ C(6,1)$$
$$d(x,y) = |\sqrt{(3-1)^2 + (6-1)^2} = 5.38$$
$$A(3,1)\ to\ B(2,6)$$
$$d(x,y) = |\sqrt{(3-1)^2 + (2-6)^2} = 4.47$$

$$S(1,1)\ to\ B(2,6)$$
$$d(x,y) = |\sqrt{(1-1)^2 + (2-6)^2} = 4$$
$$A(3,1)\ to\ G(6,5)$$
$$d(x,y) = |\sqrt{(3-1)^2 + (6-5)^2} = 2.23$$
$$B(2,6)\ to\ C(6,1)$$
$$d(x,y) = |\sqrt{(2-6)^2 + (6-1)^2} = 6.4$$

For the value of $g(n)$ taken from the distance value multiplied by the traffic congestion weight, as show in Table 1.

**Table 1.** Value of $g(n)$

| | | | | |
|---|---|---|---|---|
| **S-A** | = | **1x3** | = | **3** |
| **S-B** | = | **4x2** | = | **8** |
| **A-C** | = | **5x1** | = | **5** |
| **A-G** | = | **12x3** | = | **36** |
| **A-B** | = | **2x3** | = | **6** |
| **B-C** | = | **2x2** | = | **4** |
| **C-G** | = | **3x3** | = | **9** |

After getting the value of $h(n)$ and $g(n)$, then we will look for the value $f(n)$ using A Star algorithm with the formula:

$$f(n) = h(n) + g(n) \quad (2)$$

S is set as the starting point and G as the destination, then the value of f(n) points related to the point S is S-A and S-B.

$$S - A = h(n) + g(n) = 2 + 3 = 5 \qquad S - B = 4 + 8 = 12$$

So, S-A route will be used because it provides the smallest f(n) value. Then proceed with the relation point A with other points are A-C, A-B, and A-G.

| | | |
|---|---|---|
| $A - C = 5.38 + 5 = 10.38$ | $A - G = 2.23 + 36 = 38.23$ | $A - B = 4.47 + 6 = 10.47$ |

The smallest f(n) value is 10,38 provided by A-C route. Next, the other C-point relation is C-G.

$$C - G = 5.09 + 9 = 14.09$$

From the calculation, the A Star algorithm found that the shortest path is S-A-C-G.

## 5. Study and Result

System performance test is performed to determine the performance of the system in monitoring whether the system is running well or not. Test is performed on the buttons in the application to show the output of the system. Design of the application in this system starts with home activity with six buttons, labeled as Indekos, Kontrakan, Peta, Tentang, Bantuan, dan Keluar. The home activity can be seen in Figure 3.

**Figure 3.** Home Activity of Application

Indekos and Kontrakan buttons show all the Indekos and rented room information from the database. The Indekos and rented room activity can be seen in Figure 4.

(a) (b)

**Figure 4.** Indekos Activity (a), Rented Room Activity (b)

Peta button show the map integrated with Google Maps. There are several types of points in this map, such as faculties, dormitories, rented rooms, and traffic congestion points. The map activity can be seen in Figure 5(a). To find the nearest Indekos or rented room, user taps on the faculty to show pop up window that gives three options: cancel, nearest Indekos, and nearest rented room. The activity can be seen in Figure 5(b).

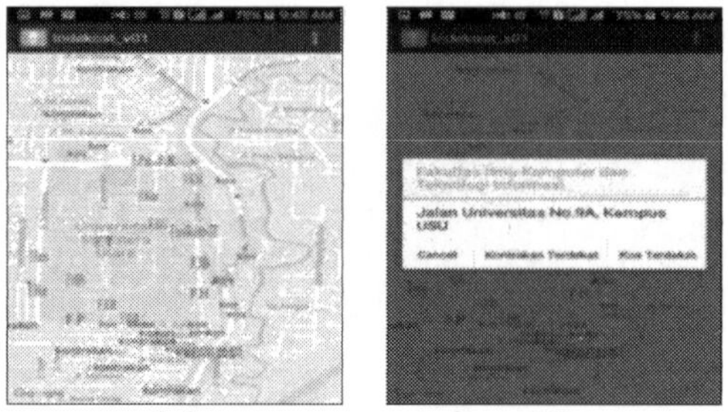

(a) (b)

**Figure 5.** Map Activity (a), Faculty Pop Up Activity (b)

Once the nearest Indekos or nearest rented room is selected, the application will show the path from the selected faculty to the selected Indekos or rented room. The path is shown as red, blue, and green lines. After the path appears, we can zoom in the map to get better details of the path. The Indekos or rented room shown by the path can be tapped to get pop up information about Indekos or rented room. The path and pop up information of the Indekos and rented room can be seen in Figure 6.

(a) (b)

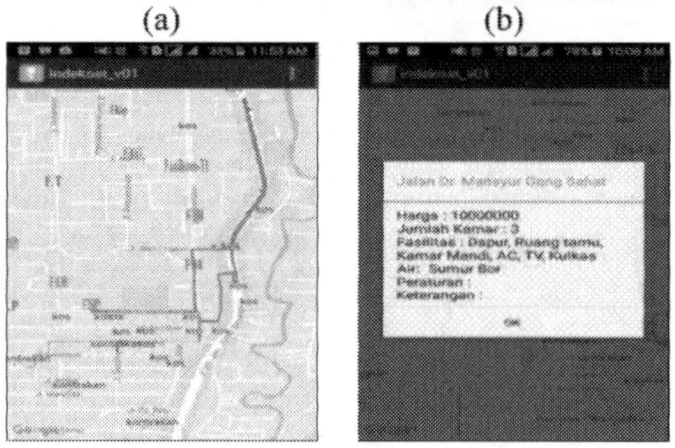

(c) (d)

**Figure 6.** Nearest Indekos Activity (a), Indekos Pop Up Activity (b), Nearest Rented Room Activity (c), Rented Room Pop Up Activity (d)

From the test results, we can see that the A Star algorithm can find the shortest path from faculty to the nearest Indekos or rented room. In this case, bad internet connection will cause failure in routing and some path are not found. The success rate of this research can be seen in Table 2.

**Table 2.** Application Test Results

| Faculty | Nearest Indekos | Nearest Rented Room |
|---|---|---|
| Kedokteran | Success | Success |
| Kesehatan Masyarakat | Fail | Success |
| Ilmu Komputer & Teknologi Informasi | Success | Fail |

| | | |
|---|---|---|
| Ilmu Budaya | Success | Success |
| Hukum | Success | Success |
| Ilmu Sosial & Ilmu Politik | Success | Success |
| Ekonomi & Bisnis | Success | Success |
| Kedokteran Gigi | Success | Success |
| Keperawatan | Success | Success |
| Teknik | Success | Fail |
| Psikologi | Success | Success |
| Pertanian | Success | Success |
| Matematika & Ilmu Pengetahuan Alam | Success | Fail |
| Farmasi | Success | Success |
| Hukum | Success | Success |

From the test results of Indekost Android based application using A Star algorithm to find the shortest path from University of Sumatera Utara, we obtain the accuracy in process of finding the shortest path with 86,67% accuracy. The results of the accuracy value can be seen as follows:

$$successful\ rate = \frac{number\ of\ success\ test}{number\ of\ test}\ x\ 100\% = 86.67\%$$

## 6. Conclusion

Based on the test results, the authors conclude that:

- A Star algorithm can be used to find the shortest path to Indekos from campus.
- Bad internet connection will cause failure in routing.
- Traffic congestion points are not really helpful and do not impress the path because almost every exit points from USU are traffic congestion points.

## 7. Suggestion

Here are some suggestions that can be used as considerations for further research:

- Further research is expected to find another location as the case so the impression of traffic congestion points in finding the path can be tested.
- Coordinate points must be determined accurately to get better results.
- Research implements different shortest path algorithm to compare the shortest path results.
- Further research can use standalone avoid internet connection problem when accessing map (Google Maps).

## 8. References

[1] Russel SJ and Norvig P 2003 Artificial Inteligence A Modern Approach (New Jersey: Prentice Hall).

[2] Tilawah H 2011 Penerapan Algoritma A-Star (A*) Untuk Menyelesaikan Masalah Maze *J. Teknik Informatika Institut Teknologi Bandung*.

[3] Hanawati, Thiang dan Eleazar 2002 Pencarian Rute Optimum Menggunakan Algoritma Genetika *J. Teknik Elektro* **2(2)** pp 78-83.

[4] Hutabarat NC, Gunawan D dan Purnamawati S 2015 Aplikasi Pencarian Taksi Terdekat Menggunakan Algoritma Floyd-Warsahll (Medan: Universitas Sumatera Utara).

[5] Siregar B, Gunawan D, Lubis ES, Andayani U dan Fahmi F 2016 Food Delivery System with the Utilization of Vehicle Using Geographical Information System and A Star Algorithm *J. Phys. Conf. Ser.* **801(1)** p 012038.

[6] Rayner DC, Bowling MH, Sturtevant NR 2011 Euclidean Heuristic Optimization *AAAI*.

IOP Conf. Series: Journal of Physics: Conf. Series **978** (2018) 012085 doi:10.1088/1742-6596/978/1/012085

# Analysis backpropagation methods with neural network for prediction of children's ability in psychomotoric

**F Izhari, H W Dhany, M Zarlis and Sutarman**
Informatics Engineering Study Program Master, Faculty of Computer Science and Information Technology, Universitas Sumatera Utara, Medan, Indonesia

E-mail: fahmiizhari14@gmail.com, dhany.hanna@gmail.com, m.zarlis@usu.ac.id, sutarman@usu.ac.id

**Abstract.** A good age in optimizing aspects of development is at the age of 4-6 years, namely with psychomotor development. Psychomotor is broader, more difficult to monitor but has a meaningful value for the child's life because it directly affects his behavior and deeds. Therefore, there is a problem to predict the child's ability level based on psychomotor. This analysis uses backpropagation method analysis with artificial neural network to predict the ability of the child on the psychomotor aspect by generating predictions of the child's ability on psychomotor and testing there is a mean squared error (MSE) value at the end of the training of 0.001. There are 30% of children aged 4-6 years have a good level of psychomotor ability, excellent, less good, and good enough.

*Keys - Artificial neural network, backpropagation, Java, Psychomotor.*

## 1. Introduction

The quality of education in Indonesia is still relatively low, it is seen from the average of very low learning achievement. To improve the condition of education in Indonesia is done through the reform of the learning curriculum. Government regulations explain that the competence of graduates includes attitudes, knowledge, and skills so as to describe the overall behavior as a result of child learning. With this it can be concluded that, learning with affective aspects (attitude), cognitive (knowledge), and psychomotor (skills). Assessment on the cognitive aspects includes sub-aspects of knowledge, understanding, analysis, and assessment. Psychomotor assessment is an assessment of the child's skills and appearance. In general, assessments reported and written by educators are usually cognitive judgments. Seen in the results of student learning reports, there are only values of cognitive aspects, while psychomotor only exist on some points. Though psychomotor aspect has criteria and detail specification. Hamalik (2005: 23) says "Although the greatest proportion of educational goals belongs to this cognitive goal. However, in actual practice these three aspects of assessment are closely related to one another. Objectives of the cognitive aspect can be achieved when it has mastered the objectives of the psychomotor aspect, and the purpose of the psychomotor aspect can be achieved when it has mastered the affective goal". It is necessary to assess and report on the psychomotor so that the parties can know in detail how the ability especially in the assessment psychomotor. So, to know the weaknesses and advantages of ability and must be followed up in order to achieve maximum assessment.

## 2. Research Methods

### *2.1. Psychomotor*

Psychomotor is derived from the word psyche and motor means psychological or mental, while the motor means motion. Thus, the development of psychomotor means development related to psychology or mental and movement.

### *2.2. Physical and motor development*

Motoric is anything that has to do with body movements. In motoric development, the elements that determine the muscles, nerves, and brain. The three elements that carry out each role in a "positive interaction", meaning that elements are one related to each other, mutual support, complementary with other elements to achieve a more perfect motor conditions. In addition to relying on muscle strength was the perfection of the brain also helped determine the state of the child whose brain growth experienced the disorder seemed less skill to move his body. The speed of motor development is influenced by nutrition, health and other physical environments such as the availability of games and opportunities given to children to train various movements. At the age of three, most children can walk on straight lines. At the age of four the child can walk along a circular line. After the age of five, they were able to run as fast as adults. Generally at the age of 3 years the child is able to make throwing movements without losing balance. At the age of 5 years old children jump by maintaining their balance. Skill development is rapidly evolving through physical play exercises through jumping, climbing, running, and riding a bicycle. At the age between 4-5 years usually children are able to create images, pictures of people. The shape of a person's image is usually shown with a great environment. Usually it is a head and added small round as the eyes, nose, mouth, and ears.

### *2.3. Characteristics of children aged 4-6 years*

Snowman (in Soemiarti Patmonodewo, 2000), suggests the traits of children aged 4-6 years who usually exist in Playgroup. Characteristics presented include the physical, social, emotional, and cognitive aspects of the child.

#### *2.3.1. Physical characteristics*

Appearance and movements are easily distinguished with children who are in the earlier stages, namely:

- Children aged 4-6 years are usually very active, because it has a mastery of the body that has the activities undertaken.
- After the children perform various activities, children need adequate rest.
- The big muscles in the child are more developed than the control of the fingers and hands. The cost of the child has not been able to perform complicated activities.
- Children still often have difficulty if they have to focus their views on small objects.
- Although it has a flexible body, but the skull is still soft. Should be careful and give explanation about the danger.
- Although boys are bigger than women, girls are more skilled in practical tasks, especially in fine motor tasks. Keep away from comparative attitudes in skill competencies.

#### *2.3.2. Social Features*

- In general, the child at the social stage has one or two friends, but friends of children quickly change. Children generally can quickly adjust themselves socially, children want to play with friends.
- The play groups tend to be small and not too well organized.
- Younger children often play next to older big children.

### *2.4. Artificial Neural Network Backpropagation*

Backpropagation is a supervised learning algorithm and is commonly used by perceptrons with multiple layers to change the weights associated with neurons in the hidden layer.
The backpropagation training algorithm basically consists of 3 stages, namely:

- Input the value of training data to obtain the output value.
- Back propagation of the error value obtained.
- Adjust the connection weight to minimize the error value.

All three stages are repeated continuously to get the desired error value. After the training is done, only the first stage is needed to utilize the artificial neural network. Then, testing of the neural network requires advanced propagation and followed by backward propagation. Both are done for all training

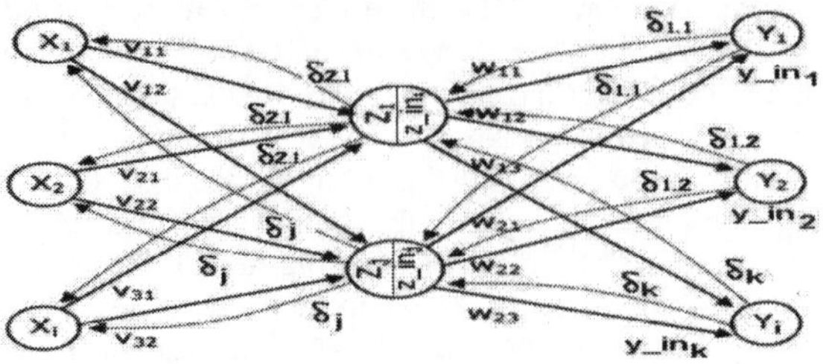

patterns.

**Figure 1.** Picture of Backpropagation Network Architecture (Kusumadewi, 2003 : 236)

### *2.5. Artificial Neural Network Design*

Artificial neural networks have multiple network architectures, single layer net, multiple layer networks, and competitive layer networks. Network architecture with multiple layers where one or more layers are located between the input layer and the output layer, there is a weighted layer that lies between two adjacent layers as shown.

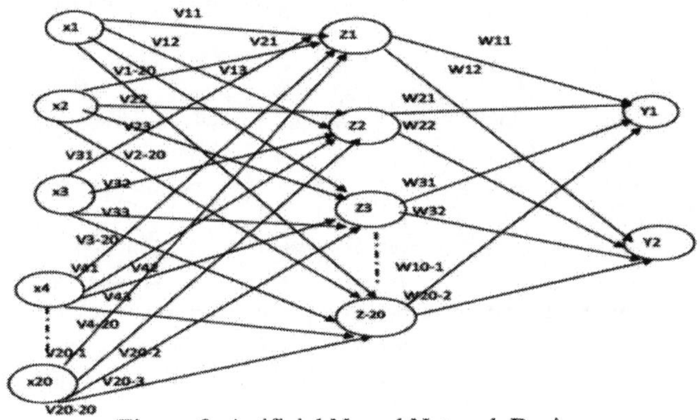


**Figure 2.** Artificial Neural Network Design

Artificial Neural Network Architecture to predict children's ability based on psychomotor aspect.
Descriptions:
X = input
V = weights on the hidden layer
Z = hidden layer
W = weights on the output layer
Y = output

Artificial neural network architecture used here consists of 3 layers, namely 1 layer input 1 hidden layer, and 1 output layer. The number of input neurons to artificial neural networks with backpropagation algorithms is 20 neurons. The number of neurons hidden layer as much as 20 neurons, and the number of neurons output layer as much as 2 pieces of neurons with the activation function used is a binary sigmoid function with a range of 0 to 1. Target test (target output) which is the output of the output layer neuron is a level prediction the child's ability on the psychomotor aspect. Where in the form of 4 categories is less or failed, enough, good and very good.

IOP Conf. Series: Journal of Physics: Conf. Series **978** (2018) 012085 doi:10.1088/1742-6596/978/1/012085

## 3. Results and Discussion

In the Backpropagation method to predict psychomotor ability by determining the number of input neurons, the number of hidden neurons, and the number of output neurons. The number input neurons used in the program 20 psychomotor neurons, 20 hidden neurons and 20 output neurons.

**Table 1.** List of neurons

| No. | Psychomotor Characteristics | Initialization of the input neuron in ANN |
|---|---|---|
| 1 | Segmental Reflex | X1 |
| 2 | Intersegmental Reflex | X2 |
| 3 | Suprasegmental Reflex | X3 |
| 4 | Locomotor Movement | X4 |
| 5 | Non-locomotor Movement | X5 |
| 6 | Manipulatif Movement | X6 |
| 7 | Kinesthetic Discrimination | X7 |
| 8 | Visual Discrimination | X8 |
| 9 | Auditory Discrimination | X9 |
| 10 | Tactile Discrimination | X10 |
| 11 | Coordinated Discrimination | X11 |
| 12 | Endurance | X12 |
| 13 | Power | X13 |
| 14 | Flexibility | X14 |
| 15 | Agility | X15 |
| 16 | Adaptive Skills | X16 |
| 17 | Integrated Adaptive Skills | X17 |
| 18 | Comprehensive Adaptive Skills | X18 |
| 19 | Expressive Movement | X19 |
| 20 | Interpretive Movement | X20 |

Then establishes training data as it is well known that backpropagation trains the network to gain a balance between the network's ability to recognize the patterns used in the training. Training data used in the program to study patterns, training data used to predict psychomotor abilities.

**Table 2.** Training data

| Training Data of Artificial Neural Network | | | | | | | | | | | | | | | | | | | | | |
|---|---|---|---|---|---|---|---|---|---|---|---|---|---|---|---|---|---|---|---|---|---|
| No | x1 | x2 | x3 | x4 | x5 | x6 | x7 | x8 | x9 | x10 | x11 | x12 | x13 | x14 | x15 | x16 | x17 | x18 | x19 | x20 | Target |
| 1 | 2 | 2 | 2 | 3 | 3 | 3 | 4 | 4 | 4 | 4 | 4 | 5 | 5 | 5 | 5 | 1 | 1 | 1 | 2 | 2 | 10 |
| 2 | 5 | 5 | 5 | 4 | 4 | 4 | 3 | 3 | 3 | 3 | 3 | 2 | 2 | 2 | 2 | 1 | 1 | 1 | 5 | 5 | 10 |
| 3 | 2 | 2 | 2 | 2 | 2 | 2 | 2 | 2 | 2 | 2 | 2 | 2 | 2 | 2 | 2 | 2 | 2 | 2 | 2 | 2 | 01 |
| 4 | 2 | 2 | 1 | 3 | 3 | 2 | 4 | 3 | 2 | 3 | 4 | 1 | 2 | 2 | 1 | 1 | 2 | 1 | 3 | 3 | 01 |
| 5 | 5 | 5 | 5 | 4 | 4 | 4 | 4 | 5 | 5 | 5 | 5 | 2 | 2 | 2 | 2 | 4 | 4 | 4 | 5 | 5 | 11 |
| 6 | 4 | 4 | 4 | 4 | 4 | 4 | 4 | 4 | 4 | 4 | 4 | 4 | 4 | 4 | 4 | 4 | 4 | 4 | 5 | 5 | 11 |
| 7 | 1 | 1 | 1 | 1 | 1 | 1 | 1 | 1 | 1 | 1 | 1 | 1 | 1 | 1 | 1 | 1 | 1 | 1 | 1 | 1 | 00 |
| 8 | 1 | 1 | 1 | 1 | 1 | 1 | 1 | 1 | 1 | 1 | 1 | 1 | 1 | 1 | 1 | 1 | 1 | 1 | 2 | 1 | 00 |
| 9 | 5 | 5 | 5 | 2 | 2 | 2 | 2 | 2 | 1 | 1 | 1 | 1 | 3 | 3 | 3 | 2 | 2 | 2 | 2 | 2 | 01 |
| 10 | 3 | 3 | 3 | 3 | 2 | 2 | 3 | 4 | 3 | 4 | 3 | 2 | 2 | 2 | 2 | 2 | 2 | 2 | 2 | 2 | 01 |
| 11 | 4 | 4 | 4 | 4 | 4 | 4 | 4 | 4 | 4 | 4 | 4 | 4 | 4 | 4 | 4 | 4 | 4 | 4 | 5 | 5 | 11 |
| 12 | 2 | 2 | 2 | 5 | 5 | 5 | 5 | 4 | 4 | 4 | 4 | 5 | 5 | 5 | 5 | 4 | 4 | 4 | 4 | 4 | 11 |
| 13 | 1 | 1 | 1 | 1 | 1 | 1 | 1 | 1 | 1 | 1 | 1 | 1 | 1 | 1 | 1 | 1 | 1 | 1 | 2 | 2 | 00 |
| 14 | 2 | 2 | 2 | 1 | 1 | 1 | 1 | 1 | 1 | 1 | 1 | 1 | 1 | 1 | 1 | 1 | 1 | 1 | 1 | 1 | 00 |
| 15 | 3 | 3 | 3 | 4 | 4 | 4 | 2 | 2 | 2 | 2 | 2 | 4 | 4 | 4 | 4 | 3 | 3 | 3 | 4 | 4 | 10 |
| 16 | 4 | 4 | 4 | 3 | 3 | 3 | 2 | 2 | 2 | 2 | 2 | 4 | 4 | 4 | 4 | 5 | 5 | 5 | 2 | 2 | 10 |
| 17 | 4 | 4 | 4 | 2 | 2 | 1 | 2 | 1 | 1 | 1 | 3 | 4 | 3 | 3 | 3 | 3 | 3 | 3 | 2 | 2 | 01 |
| 18 | 3 | 3 | 3 | 4 | 3 | 4 | 2 | 2 | 3 | 3 | 3 | 2 | 1 | 2 | 1 | 3 | 4 | 2 | 2 | 2 | 01 |
| 19 | 3 | 3 | 3 | 4 | 4 | 4 | 4 | 4 | 4 | 4 | 4 | 4 | 4 | 5 | 5 | 5 | 5 | 5 | 4 | 4 | 11 |
| 20 | 1 | 1 | 1 | 5 | 5 | 4 | 4 | 5 | 5 | 5 | 5 | 5 | 4 | 4 | 4 | 5 | 5 | 5 | 5 | 5 | 11 |
| 21 | 2 | 1 | 1 | 1 | 1 | 2 | 1 | 1 | 1 | 1 | 1 | 1 | 1 | 2 | 1 | 1 | 2 | 1 | 1 | 1 | 00 |
| 22 | 1 | 1 | 1 | 3 | 2 | 3 | 1 | 1 | 1 | 1 | 1 | 1 | 1 | 1 | 1 | 1 | 1 | 1 | 1 | 1 | 00 |
| 23 | 2 | 2 | 2 | 3 | 3 | 3 | 4 | 4 | 4 | 4 | 4 | 5 | 5 | 5 | 5 | 3 | 3 | 3 | 2 | 2 | 10 |
| 24 | 1 | 1 | 1 | 3 | 3 | 3 | 4 | 4 | 4 | 4 | 5 | 5 | 5 | 5 | 3 | 3 | 3 | 4 | 4 | 4 | 10 |

| | | | | | | | | | | | | | | | | | | | | | |
|---|---|---|---|---|---|---|---|---|---|---|---|---|---|---|---|---|---|---|---|---|---|
| 25 | 3 | 4 | 3 | 3 | 3 | 3 | 2 | 2 | 3 | 1 | 2 | 3 | 2 | 2 | 3 | 4 | 2 | 2 | 3 | 3 | 01 |
| 26 | 3 | 3 | 3 | 1 | 1 | 1 | 3 | 3 | 3 | 3 | 2 | 2 | 2 | 3 | 4 | 4 | 4 | 4 | 2 | 2 | 01 |
| 27 | 2 | 2 | 2 | 5 | 5 | 5 | 5 | 4 | 4 | 4 | 4 | 5 | 5 | 5 | 5 | 5 | 5 | 5 | 4 | 4 | 11 |
| 28 | 5 | 5 | 5 | 5 | 5 | 5 | 5 | 5 | 4 | 4 | 4 | 4 | 4 | 3 | 3 | 3 | 3 | 3 | 5 | 5 | 11 |
| 29 | 1 | 1 | 1 | 1 | 1 | 1 | 1 | 1 | 2 | 2 | 1 | 1 | 1 | 1 | 1 | 2 | 1 | 2 | 1 | 3 | 00 |
| 30 | 2 | 2 | 1 | 1 | 1 | 2 | 1 | 1 | 2 | 4 | 1 | 1 | 1 | 1 | 1 | 1 | 1 | 1 | 1 | 1 | 00 |
| 31 | 4 | 4 | 4 | 5 | 5 | 5 | 3 | 3 | 3 | 3 | 3 | 4 | 4 | 4 | 4 | 2 | 2 | 2 | 3 | 3 | 10 |
| 32 | 4 | 4 | 4 | 5 | 5 | 5 | 3 | 3 | 3 | 3 | 4 | 4 | 4 | 4 | 4 | 2 | 2 | 2 | 3 | 3 | 10 |
| 33 | 3 | 1 | 1 | 1 | 2 | 1 | 1 | 1 | 1 | 1 | 1 | 1 | 1 | 1 | 1 | 1 | 1 | 1 | 4 | 4 | 00 |
| 34 | 2 | 1 | 2 | 1 | 1 | 1 | 1 | 2 | 1 | 1 | 2 | 1 | 1 | 1 | 1 | 3 | 3 | 3 | 1 | 1 | 00 |
| 35 | 5 | 5 | 5 | 5 | 5 | 5 | 5 | 5 | 5 | 5 | 5 | 5 | 5 | 5 | 5 | 5 | 5 | 5 | 4 | 4 | 11 |
| 36 | 5 | 5 | 5 | 5 | 5 | 5 | 5 | 5 | 5 | 5 | 5 | 5 | 5 | 5 | 5 | 5 | 5 | 5 | 5 | 5 | 11 |
| 37 | 2 | 2 | 2 | 4 | 4 | 4 | 5 | 5 | 5 | 5 | 5 | 3 | 3 | 3 | 3 | 4 | 4 | 4 | 3 | 3 | 10 |
| 38 | 3 | 3 | 3 | 5 | 5 | 5 | 4 | 4 | 4 | 4 | 4 | 2 | 2 | 2 | 2 | 4 | 4 | 4 | 5 | 5 | 10 |
| 39 | 3 | 2 | 2 | 3 | 4 | 4 | 5 | 3 | 3 | 3 | 2 | 1 | 1 | 1 | 3 | 3 | 3 | 3 | 5 | 5 | 01 |
| 40 | 3 | 3 | 3 | 2 | 2 | 2 | 4 | 4 | 4 | 4 | 2 | 2 | 2 | 2 | 2 | 4 | 4 | 4 | 3 | 3 | 01 |

The third is the determination of initial weights and the proper learning rate for artificial neural networks to predict psychomotor abilities. In this experiment the initial weighting was performed by randomizing the numbers (random) between -0.5 to 0.5. The experiment is done by changing the learning rate between 0.01 to 0.4 and the maximum number of epochs achieved. Each learning rate is experimental for 3 times.

**Table 3.** Learning Rate

| Learning Rate | Test | Number of Epoch | Max Epoch | Target Error | % Convergence of Training Data |
|---|---|---|---|---|---|
| 0.4 | 1 | 6008 | 10000 | 0.001 | 100% |
| | 2 | 4965 | 10000 | 0.001 | 100% |
| | 3 | 5978 | 10000 | 0.001 | 100% |
| 0.3 | 1 | 7932 | 10000 | 0.001 | 100% |
| | 2 | 8150 | 10000 | 0.001 | 100% |
| | 3 | 7505 | 10000 | 0.001 | 100% |
| 0.2 | 1 | 12776 | 50000 | 0.001 | 100% |
| | 2 | 12430 | 50000 | 0.001 | 100% |
| | 3 | 13494 | 50000 | 0.001 | 100% |
| 0.1 | 1 | 24315 | 50000 | 0.001 | 100% |
| | 2 | 24473 | 50000 | 0.001 | 100% |
| | 3 | 25610 | 50000 | 0.001 | 100% |
| 0.09 | 1 | 29096 | 50000 | 0.001 | 100% |
| | 2 | 28845 | 50000 | 0.001 | 100% |
| | 3 | 30960 | 50000 | 0.001 | 100% |
| 0.08 | 1 | 30824 | 50000 | 0.001 | 100% |
| | 2 | 31449 | 50000 | 0.001 | 100% |
| | 3 | 34529 | 50000 | 0.001 | 100% |
| 0.07 | 1 | 38904 | 50000 | 0.001 | 100% |
| | 2 | 38884 | 50000 | 0.001 | 100% |
| | 3 | 38508 | 50000 | 0.001 | 100% |
| 0.06 | 1 | 49301 | 50000 | 0.001 | 100% |
| | 2 | 46934 | 50000 | 0.001 | 100% |
| | 3 | 47737 | 50000 | 0.001 | 100% |
| 0.05 | 1 | 57791 | 100000 | 0.001 | 100% |
| | 2 | 61462 | 100000 | 0.001 | 100% |
| | 3 | 53342 | 100000 | 0.001 | 100% |
| 0.04 | 1 | 77575 | 100000 | 0.001 | 100% |
| | 2 | 68549 | 100000 | 0.001 | 100% |
| | 3 | 66425 | 100000 | 0.001 | 100% |
| 0.03 | 1 | 89365 | 100000 | 0.001 | 100% |
| | 2 | 93779 | 100000 | 0.001 | 100% |
| | 3 | 92952 | 100000 | 0.001 | 100% |
| 0.02 | 1 | 140267 | 150000 | 0.001 | 100% |
| | 2 | 126733 | 150000 | 0.001 | 100% |
| | 3 | 150001 | 150000 | 0.001 | 100% |
| 0.01 | 1 | 200001 | 200000 | 0.001 | 100% |
| | 2 | 200001 | 200000 | 0.001 | 100% |
| | 3 | 200001 | 200000 | 0.001 | 100% |

IOP Conf. Series: Journal of Physics: Conf. Series **978** (2018) 012085 doi:10.1088/1742-6596/978/1/012085

The results obtained are the higher learning rate used then the number of epoch to achieve the target error is less, and the smaller the learning rate used, the more the number of epoch that reaches the target error. The initial weights used in the program are the initial weights at the learning rate of 0.4 in the 2nd experiment, with the number of epoch 4965.

## 4. Conclusion

In the trial used data derived from the results in the psychomotor aspects of children. The comparing results of manual calculations with the program. To calculate the percentage of truth results from the application created using the following formula:

$$Percentage = \frac{data\ successes}{the\ amount\ of\ data} \times 100\%$$

Known:
Data successes = 17
Amount of data = 20

The success of data obtained from one of the data reduced the amount of data so that:

$$Percentage = \frac{17}{20} \times 100\%$$

$$Precentage = 85\ \%$$

So the results have been obtained a success percentage of 85%.

## References


[1] Hamalik O 2005 *Teknik Pengukuran dan Evaluasi Pendidikan* (Bandung: Mandar Maju)

[2] Kohli M and Prasad TV 2000 Identifying dyslexic students by using artificial neural networks. *Proc. World Congr. Eng.* vol 1.

[3] Patmonodewo S 2000 *Pendidikan Anak Prasekolah* (Jakarta: Rineka Cipta).

[4] Kusumadewi S 2003 Artificial Intelligence (Teknik dan Aplikasinya) (Yogyakarta: Graha Ilmu).

[5] Cohen IL, Sudhalter V, Landon-Jimenez D, Keogh M 1993 A neural network approach to the classification of autism J. Autism Dev. Disord. 23(3) pp 443–66.

[6] Tiwari S, Naresh R and Jha R 2013 Comparative Study of Backpropagation Algorithms in Neural Network Based Identification of Power System Int. J. Comput. Sci. Inf. Technol. 5(4) pp 93-107.

[7] Imlahi H and Kissani I 2015 Intelligence quotient and its environmental factors in children Marocco: Al Akhawayn University vol 6 pp 697-704.

[8] Ting F, Wei JH, Kim CT, Tian Q 2003 Question classification for e-learning by artificial neural network Proc. Joint Conf. 4th Int. Conf..

# An Implementation of RC4$^+$ Algorithm and Zig-zag Algorithm in a Super Encryption Scheme for Text Security

**M A Budiman, Amalia and N I Chayanie**

Departemen Ilmu Komputer, Fakultas Ilmu Komputer dan Teknologi Informasi, Universitas Sumatera Utara, Jl. Universitas No. 9-A, Kampus USU, Medan 20155, Indonesia

Email: mandrib@usu.ac.id, amalia@usu.ac.id, noerindachayanie@student.usu.ac.id

**Abstract**. Cryptography is the art and science of using mathematical methods to preserve message security. There are two types of cryptography, namely classical and modern cryptography. Nowadays, most people would rather use modern cryptography than classical cryptography because it is harder to break than the classical one. One of classical algorithm is the Zig-zag algorithm that uses the transposition technique: the original message is unreadable unless the person has the key to decrypt the message. To improve the security, the Zig-zag Cipher is combined with RC4$^+$ Cipher which is one of the symmetric key algorithms in the form of stream cipher. The two algorithms are combined to make a super-encryption. By combining these two algorithms, the message will be harder to break by a cryptanalyst. The result showed that complexity of the combined algorithm is $\theta(n^2)$, while the complexity of Zig-zag Cipher and RC4$^+$ Cipher are $\theta(n^2)$ and $\theta(n)$, respectively.

## 1. Introduction

Communication is the process of transferring information from one party to another. One example of communication is sending letters. With more advanced developments, the letter is no longer done by handwriting but by means of electronic messaging. Sending massages electronically via insecure channels may allow unauthorized parties to read the messages. To maintain the messages confidentiality, one solution is to encrypt the messages using cryptographic algorithms. Cryptography is the science and art to preserve messages confidentiality. According to Request for Comments (RFC), cryptography is a mathematical science that deals with transforming data to make the meaning unintelligible (to hide the meaning of the data), to prevent it from unauthorized changes, or to prevent it from unauthorized use. A cryptographic system is a system that can be used to change a message in order to make the message not understood by anyone other than the recipient. This process is called encryption [1]. Two types of cryptography are classical cryptography and modern cryptography. To further improve security, a combination of classical and modern algorithms is performed. Classical cryptographic algorithm to be used for this data encoding process is Zig-zag Cipher and modern algorithm that will be used is RC4$^+$ Cipher.

Zig-zag Cipher is one of the classical cryptography that use transposition techniques that use character permutations, in which the original message unreadable unless the person has the key to decrypt the message [2]. The Zig-zag Cipher algorithm includes an unsafe classical cryptography algorithm. This is evidenced by research that proves that the more difficult algorithm of the Zig-zag

Cipher algorithm, that is Affine Cipher can already be solved by cryptanalysis [3]. $RC4^+$ Cipher is one type of RC4 algorithm, which uses a variable whose key length is from 1 to 256 bits used to initialize a 256 bit long table. The two algorithms combined are called Super Encryption. By combining these two algorithms, we can get a stronger cipher so it is not easy to break, and also to prevent of using a single cipher only which is comparatively weak [4].

## 2. Method

One of the key symmetric algorithms in the form of stream cipher is $RC4^+$ cipher algorithm. Undoubtedly, the importance of stream ciphers in computer applications cannot be ignored. Therefore, a standardized model for the stream cipher design is certainly today's requisite [5]. $RC4^+$ uses a structure like RC4 and adds some operations to amplify the cipher. This structure tries to exploit good points on RC4 then provides some additional features for better security limits [6]. The $RC4^+$ cipher algorithm has two main parts: Key Scheduling Algorithm (KSA) in Table 1 and Pseudo Random Generation Algorithm (PRGA) in Table 2.

**Table 1.** KSA of $RC4^+$ Cipher [6]

```
for i from 0 to 255
   S[i] := i
endfor
j := 0
for i from 0 to 255
   j := (j + S[i] + key[i mod keylength]) mod
256
   swap values of S[i] and S[j]
endfor
```

**Table 2.** PRGA of $RC4^+$ Cipher [6]

```
while |cipherkey| < |plaintext|:
   i := i + 1
   a := S[i]
   j := j + a

Swap S[i] and S[j]        (b := S[j]; S[i] := b; S[j]
:= a)

   c := S[i<<5 ⊕ j>>3] + S[j<<5 ⊕ i>>3]
   output (S[a+b] + S[c⊕0xAA]) ⊕ S[j+b]
endwhile
```

Where *i* and *j* are 8-bit array indexes, *S* is all possible 256 permutations, << and >> are left and right shift, ⊕ is exclusive OR.

The technique applied to the Zig-zag Cipher algorithm is the technique of transposing the encryption cipher and decrypting the message by altering the order of the letters in the plaintext to the ciphertext in such a way that the contents of the message are not understood except by certain people. How to make Zig-zag Cipher algorithm as powerful as modern algorithm that is by using ASCII characters and then convert it to binary by playing in its bit. Each symbol that appears on the computer screen has a different ASCII code, the length of each ASCII code in binary is 8 bits and there are 28 unique symbols contained in the ASCII table [7].

Zig-zag transpositions can be performed in succession by forming rows or columns arranged in matrix format. If the digit of the key is i, then the position of the transposition message matrix column is read as follows:

$$(1, i)\,(2, i + 1)\,(3, i)\,(4, i + 1)\,(5, i) \ldots (m, n) \quad (1)$$

After a transposition is processed with a lock on a symmetric encryption cipher, the same key is used for process of decryption. If the digit of the key is i, then the position of the transposition message matrix row is read as follows:

$$(i, 1)\,(i + 1, 2)\,(i, 3)\,(i + 1, 4)\,(i, 5) \ldots (m, n) \quad (2)$$

The number of digits in the key used in the zig-zag cipher algorithm depends on the number of rows when using row transpositions and depends on the number of columns when using transposition columns [8].

## 3. Result and Discussion

The result showed that complexity of the combined algorithm is $\theta(n^2)$, while the complexity of Zig-zag Cipher and RC4$^+$ Cipher are $\theta(n^2)$ and $\theta(n)$, respectively. The test results obtained for the .doc file extension are divided into several parts: plaintext text in Figure 1, plaintext symbols in the Figure 2 and the plaintext complex (text, symbols and images) in Figure 3.

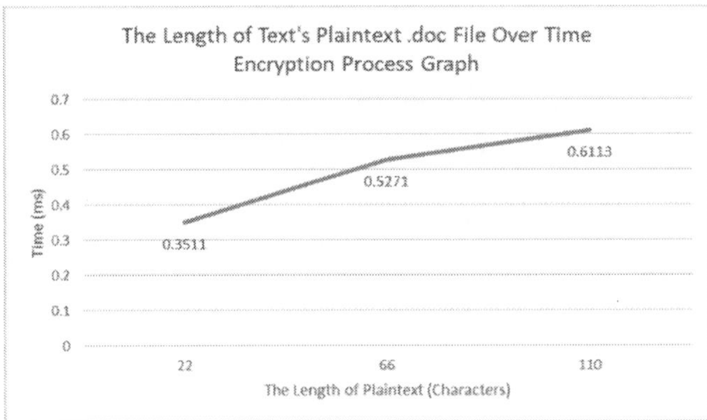


**Figure 1.** The Length of Text's Plaintext .doc File Over Time Encryption Process Graph

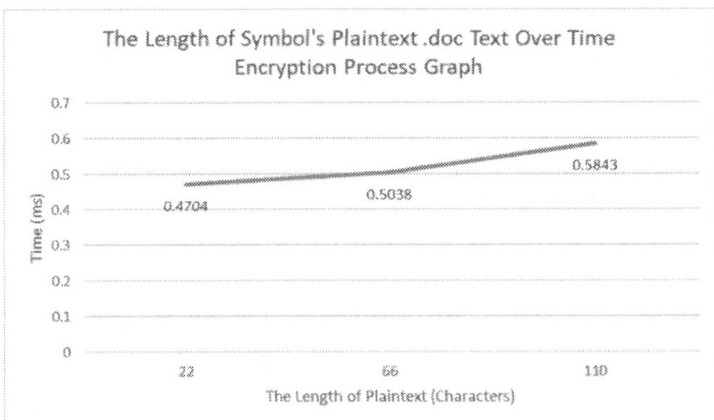


**Figure 2.** The Length of Symbol's Plaintext .doc File Text Over Time Encryption Process Graph

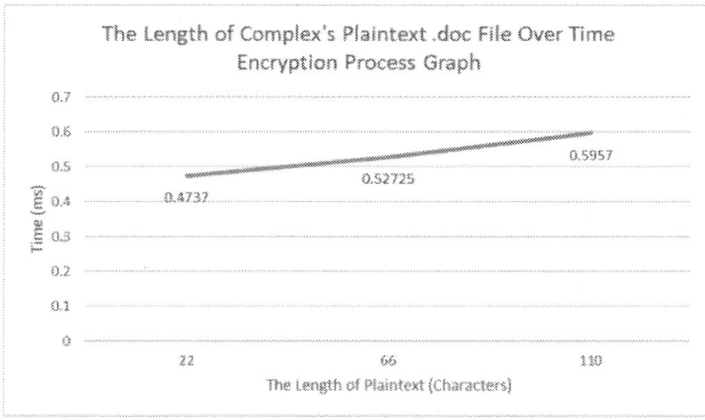


**Figure 3.** The Length of Complex's Plaintext .doc File Over Time Encryption Process Graph

The test results obtained for the pdf file extension are also divided into several parts: plaintext text in Figure 4, plaintext symbols in Figure 5 and plaintext complexes (text, symbols and drawings) in Figure 6.

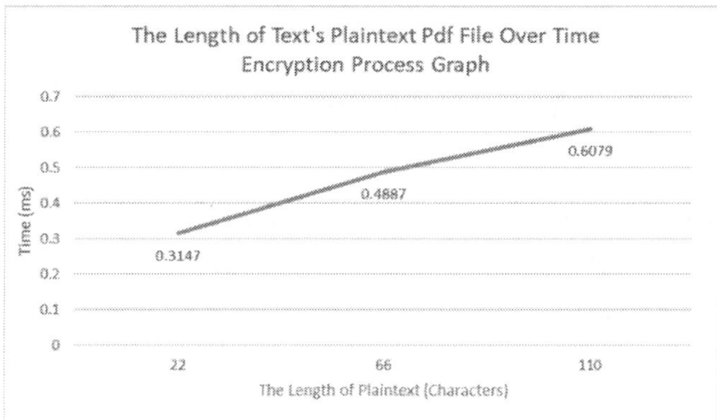


**Figure 4.** The Length of Text's Plaintext Pdf File Over Time Encryption Process Graph

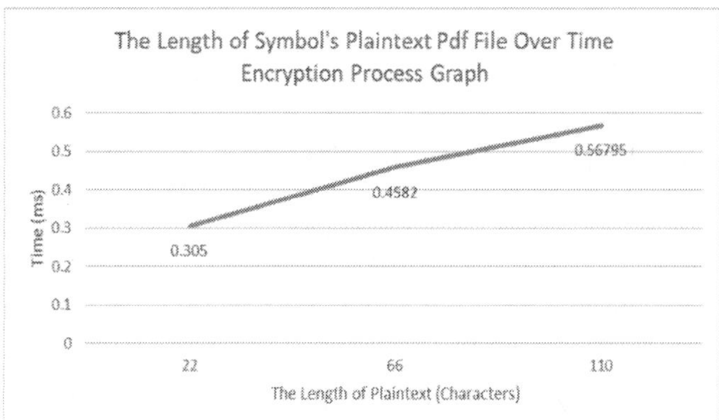


**Figure 5.** The Length of Symbol's Plaintext Pdf File Over Time Encryption Process Graph

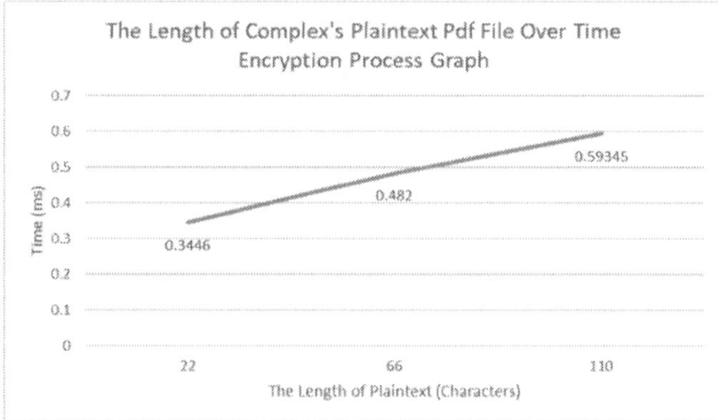


**Figure 6.** The Length of Complex's Plaintext Pdf File Over Time Encryption Process Graph

## 4. Conclusion

The conclusions of an implementation of $RC4^+$ Algorithm and Zig-zag Algorithm in Super Encryption Scheme for Text Security are:

- The transposition method of Zig-zag Cipher algorithm used at the time of encryption must be the same as the transposition method used at the time of decryption to obtain the same plaintext result.
- When encrypting more than once using Zig-zag Cipher algorithm and $RC4^+$ Cipher algorithm with plaintext and same key will produce different ciphertext. This is due to the addition of capital letters to the remaining Zig-zag Cipher matrix algorithms that have not been filled.
- The result of the decryption process of Zig-zag Cipher algorithm combination and $RC4^+$ Cipher algorithm is the same although the encryption obtained is different when encrypting more than once with the same plaintext and key.
- Based on the plaintext length graph of the process time, the plaintext length is proportional to the linear graph. That is, the longer plaintext then the time required to perform the encryption is longer too.
- Based on the complexity value of both algorithm used is Zig-zag Cipher algorithm and $RC4^+$ Cipher algorithm obtained by result $\Theta(n^2)$. This complexity sometimes has a faster time so that it's form a parabolic graph.

## References

[1] Churchhouse RF 2002 *Code and Ciphers: Julius Caesar, the Enigma and the Internet* (Cambridge: Cambridge University Press).

[2] Ariyus D 2008 *Pengantar ilmu Kriptografi, Teori, analisis dan Implementasi* (Yogyakarta: Penerbit Andi).

[3] Biryukov A, De Canniere C, Braeken A and Preneel B 2003 A toolbox for cryptanalysis: Linear and affine equivalence algorithms *Eurocrypt* **2656** pp 33-50.

[4] Iswahyudi C, Setyaningsih E and Widyastuti N 2012 Pengamanan kunci enkripsi citra pada algoritma super enkripsi menggunakan metode end of file *Proc. SNAST* **3** pp 278-85.

[5] Alghazzawi DM, Hasan SH and Trigui MS 2013 Stream Ciphers: A Comparative Study of Attacks and Structures *Int. J. Comput. Appl.* **83**(1) pp 27-9.

[6] Paul G and Maitra S 2012 *RC4 Stream Cipher and Its Variants* (London: CRC Press Taylor & Francis Group).

[7] Budiman MA and Rachmawati D 2017 On Using Goldbach G0 Codes and Even-Rodeh Codes for Text Compression *IOP Conf. Series: Mater. Sci. Eng.* **1** 1-5.

[8] Padmapriya, Ph.D, A. & Mu. Annalakshmi, 2013. Zigzag Ciphers: A Novel Transposition Method. *International Journal of Computer Applications (0975 – 8887),* 8-12. Annalakshmi M and Padmapriya A 2013 Zigzag Ciphers: A Novel Transposition Method *Proc. IJCA* **2** pp

IOP Conf. Series: Journal of Physics: Conf. Series **978** (2018) 012086 doi:10.1088/1742-6596/978/1/012086

8-12.

# Comparison of Naive Bayes and Decision Tree on Feature Selection Using Genetic Algorithm for Classification Problem

**S Rahmadani, A Dongoran, M Zarlis, Zakarias**

Magister Teknik Informatika, Fasilkom - TI , Universitas Sumatera Utara, Medan - Indonesia

Email: dhani.harahap@gmail.com

**Abstract**. This paper discusses the problem of feature selection using genetic algorithms on a dataset for classification problems. The classification model used is the decicion tree (DT), and Naive Bayes. In this paper we will discuss how the Naive Bayes and Decision Tree models to overcome the classification problem in the dataset, where the dataset feature is selectively selected using GA. Then both models compared their performance, whether there is an increase in accuracy or not. From the results obtained shows an increase in accuracy if the feature selection using GA. The proposed model is referred to as GADT (GA-Decision Tree) and GANB (GA-Naive Bayes). The data sets tested in this paper are taken from the UCI Machine Learning repository.

## 1. Introduction

The issue of the huge dataset feature is still an important issue in data mining task. The main problem is how to choose or even reduce one or more features on the dataset that may affect the quality of data. Features that may interfere with data, irrelevant, correlated, or even overlapping must be cleaned so that the quality and performance of the data can be increased [1].

There are two common tasks related to feature reduction, *Feature Selection* and *Feature Composition*. There are two feature selection methods , *feature-ranking algorithm* and *minimum subset algorithm*. Feature selection algorithm is very helpful data mining process to seek new knowledge in finding models of data patterns that have high accuracy. One of feature selection technique is to compare means and variance. The trivial technique is its unknown feature distribution. If the distribution is normal then the results will be better statistically but weak on the assumption. If the distribution is not known then means and variance are heuristic so the value is not too exact. In this paper, the technique used is non statistical technique but using heuristic method, Genetic Algorithm.

The feature selection is a fairly complex task that can affect the performance of classification. The most popular approach is based on Evolutionary Computation (EC) such as Genetic Algorithm [12] and Swarm Intelligence called Particle Swarm Optimization (PSO).

## 2. Naive Bayes

Naive Bayes is a classic model that has been studied since the 1950s. It became popular after being tested for text retrieval in the 1960s [8], combined with Support Vector Machine [9], and also applied to medical diagnosis [10]. Naive Bayes is one of the statistical approach of inductive-inferencing in the classification problem that uses bayes theorem.

IOP Conf. Series: Journal of Physics: Conf. Series **978** (2018) 012087 doi:10.1088/1742-6596/978/1/012087

Let H be a hypothesis and X is a data residing in a certain C class. Then P (H / X) is called the posterior probability that expresses our confidence level on a hypothesis H after X data is given. P (H) represents the H prior probability for all sample data. P (H / X) is certainly more informative than P (H). Bayes's theorem describes the relationship between P (H / X), P (H), and P (X) is shown on equation 1 as follow :

$$P(H/X) = P(X/H) * P(H)/P(X) \tag{1}$$

By using the bayes theorem, the Naive Bayes Classifier can be formulated as follows on equation 2.

$$P(C_i/X) = P(X/C_i) * \frac{P(C_i)}{P(X)} \tag{2}$$

P (X) is the constant probability of the X dataset for all classes. $P(C_i)$ is the number of training samples of the class Ci/m (m is the number of training data samples). In this case $P(X/C_i) * P(C_i)$ is a unit that can be optimized so that $P(C_i/X)$ becomes optimal, since P (X) and $P(C_i)$ are constant. Based on the aforementioned assumptions, $P(X/C_i)$ is formulated as follows on equation 3.

$$P(X/C_i) = \prod_{t=1}^{n} P(x_t/C_i) \tag{3}$$

Where $x_t$ is the value of the sample attribute X. The probability value $P\ (x_t/C_i)$ can be estimated from the training dataset.

### *2.1. Naive Bayes Algorithm*
The steps of Naive Bayes's Algorithm in the classification of data are as follows:
- Find $P(C_i)$, that is the class i probability by calculating total class i in total m training dataset.
- Calculate the probability $P(x_t/C_i)$ for each attribute value of the new data sample X using the training dataset.
- Calculate $P(X/C_i)$ Using equation 3 for each class.
- Calculate $P(C_i/X)$ Using equation 2 for each class.
  - $P(C_i/X) \approx P(X/C_i) * P(C_i)$.
- Choose a maximum probability value $P(C_i/X)$ Which is the result of class predictions of the new data.

## 3. Decision Tree
One of method that is quite reliable and efficient in generating a classifier model is the decision tree. The Decision Tree is called a predictive model that maps the observed data into the target value. In predicting the model, the Decision tree usually uses a *statistical approach*, *data mining* and *machine learning*. Decision tree is included into the inductive learning algorithm. Decision tree has a special feature in the learning process that is using a *top-down strategy*.

### *3.1. Decision Tree Algorithm*
Assumed that all features have finite discrete domains, and there is a target feature called classification. Each element in the classification is called a class. Decision Tree learns from source data that has been divided into two parts, data training and data testing. The algorithm for forming a decision tree typically uses a top-down method [3].

There are many decision tree algorithms, such as ID3, C4.5, CART, CHAID, MARS, Conditional Inference Trees. In this case, the algorithm used is C4.5.
The steps to form a tree using the C4.5 algorithm are as follows:
- Select the feature (attribute) as the root

- Create a branch for each value
- Divide the cases into branches
- Repeat the process for each branch until all the cases on the branch have the same class.

To select the attribute as the root is used the highest gain information of each attribute. The gain information formula in equation 4.

$$Gain\ (S, A) = Entropy(S) - \sum_{i=1}^{n} \frac{|S_i|}{|S|} * Entropy(S_i) \quad (4)$$

where $S$ is a set of cases. $A$ is a set of feature (atribute), $n$ represents number of atribute partitions, $|S_i|$ is number of case on partition to-i, and $|S|$ is number of cases in $S$. While to calculate the value of entropy used equation 5.

$$Entropy\ (S) = \sum_{i=1}^{n} -(p_i * \log_2 p_i) \quad (5)$$

where $S$ is a set of cases, n is number of partition, and $p_i$ is a proportion of $S_i$ of $S$.

## 4. Genetic Algorithm (GA)

Genetic Algorithm is one of the heuristic search algorithms that mimics the process of natural evolution. GA is used for the problem of optiamsi and search that is inspired by natural evolution such as inheritance, mutation, selection and marriage. GA was first proposed by Fraser [12], and then Bremerman [13].

### *4.1. GA for Feature Selection*

In GA, mutations are done by turning on and off attributes. Crossover is defined as a feature change that is used.

Selection feature stages are as follows:

- Generate initial population composed of individual *p*
- Each attribute is enabled with probability *pi*. The size of *p* and *pi* is adjusted by population size and initial *p*.
- All individuals in the population perform mutations, for example setting the attributes used to be unused attributes based on the probability of the mutation *pm* and vice versa. The probability of *pm* is obtained from the initial parameter of *pm*.
- Choose two individuals from the population to cross-crossover based on the probability of crossover *pc*. *Pc* is taken from the initialization of *pc* parameters. The type of crossover used can be selected from crossover type parameters.
- Mapping all individuals based on their fitness values for the selection process. Take individuals randomly according to their respective probabilities.
- As the fitness value increases, repeat step 2. Unless, the process is finished.

### *4.2. Crossover*

In binary representation, crossover is done in the sexual way of two parents [12]. Some of the crossover operators used in GA, among others:

- *One-point crossover* : segments of the genes are exchanged between parents who form offspring, and not single genes [14].
- *Two-point crossover* : two bit positions are randomly selected and the bitstring subsequently exchanged.
- *Uniform crossover* : n - dimensional masks are created randomly [10]. If px = 0.5 (bit-swaping probability) it means the chance of bits to be exchanged equally.

In this paper, the crossover operator used using Uniform Crossover.

### *4.3. Mutation*

For binary data, the usual mutation operator is as follows:

- *Uniform (Random) Mutation* [14]: the bits position is selected and the selected bits are reversed, from 0 to 1, and vice versa.
- *Inorder Mutation*: two points of mutation are randomly selected, bits between these two points behind the bits.
- *Gaussian Mutation*: for each chromosome, the random number of the Poisson is distributed to determine the gene that is mutated.

## 5. Design Experiment

The experiment design of this paper is shown in Figure 1.

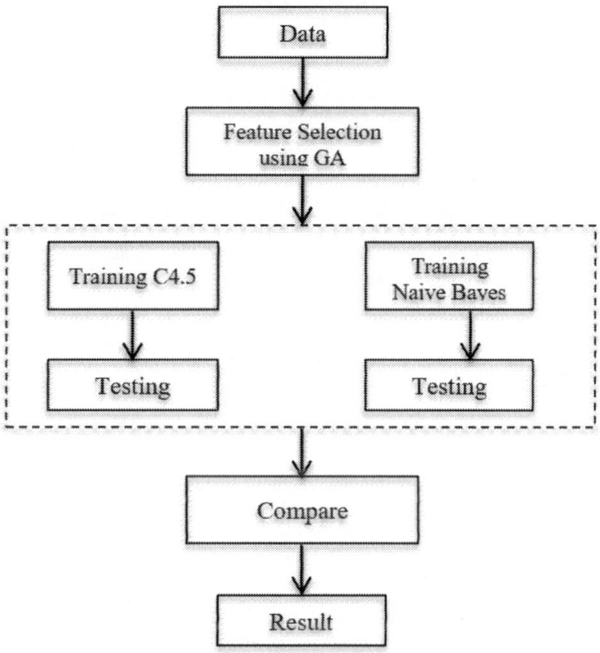


**Figure 1**. Methodology of Experiment

## 6. Result

The dataset used to train and test the above experiments was obtained from UCI Machine Learning Repository. Table 1 shows the dataset with property A represents the attribute, I denotes the number of instances, C denotes the number of classes. Tables 2, 3, and 4 show the performance vector of the three Iris, Glass, and Credit datasets. Table 5 shows the results of data validation by training using Naive Bayes, and Table 6 shows the Decision Tree result.

**Table 1.** Description of Datasets.

| | Attribute. (*A*) | Class. (C) | Instance. (I) |
|---|---|---|---|
| **Iris** | 4 | 4 | 150 |
| **Glass** | 9 | 7 | 214 |
| **Credit** | 20 | 2 | 1000 |

**Table 2.** Performance Vector Germany Credit GA-DT.

| | True GOOD | True BAD | Class Precision |
|---|---|---|---|
| Pred. GOOD | 178 | 52 | 77.39 % |
| Pred. BAD | 30 | 38 | 55.88 % |
| Class Recall | 85.58 % | 42.22 % | |

**Table 3.** Performance Vector Iris GA- DT.

| | True Iris-Setosa | True Iris-VersiColor | True Iris-Virginica | Class Precision |
|---|---|---|---|---|
| Pred. Iris-Setosa | 25 | 0 | 0 | 100.00 % |
| Pred. Iris-VersiColor | 0 | 25 | 1 | 96.15 % |
| Pred. Iris-Virginica | 0 | 0 | 24 | 100 .00% |
| Class Recall | 100.00 % | 100 % | 96.00 % | |

**Table 4.** Performance Vector Glass GA-DT.

| | True 1 | True 2 | True 3 | True 5 | True 6 | True 7 | Class Precision |
|---|---|---|---|---|---|---|---|
| Pred. 1 | **14** | 1 | 3 | 0 | 0 | 1 | 73.68 % |
| Pred. 2 | 4 | **19** | 0 | 0 | 0 | 0 | 82.61 % |
| Pred. 3 | 2 | 2 | **2** | 0 | 0 | 0 | 33.33 % |
| Pred. 5 | 0 | 1 | 0 | **4** | 0 | 0 | 80.00 % |
| Pred. 6 | 1 | 0 | 0 | 0 | **3** | 0 | 75.00 % |
| Pred. 7 | 0 | 1 | 0 | 1 | 0 | **8** | 100.00 % |
| Class Recall | 66.67 % | 82.61 % | 40.00 % | 100.00 % | 100.00 % | 88.89 % | |

**Table 5.** Accuracy of Prediction Naive Bayes

| | NB Naive Bayes | GA-NB | Acc. Improvement |
|---|---|---|---|
| **Iris** | 96.00 % | 100 % | 4.0 % |
| **Glass** | 56.92 % | 64.62 % | 7.7 % |
| **Credit** | 73.67 % | 80.00 % | 6.33 % |
| | | **Average** | 6.01 % |

**Table 6.** Accuracy of Prediction Decision Tree

| | DT Decision Tree | GA-DT | Acc. Improvement |
|---|---|---|---|
| **Iris** | 94.67% | 98.67 % | 4.0 % |

| | | | |
|---|---|---|---|
| **Glass** | 55.38 % | 76.92 % | 21.54 % |
| **Credit** | 70.00 % | 72.48 % | 2.48 % |
| | | **Average** | 9,34 % |

## 7. Summary

From the results obtained above, it is seen that the accuracy for each dataset is different. The highest accuracy is obtained by the Iris dataset, then Glass and last Data Germany Credit. However, when viewed from the level of increase in accuracy before and after done GA-Featured Selection, then the Glass dataset has the highest value. When compared between the Decision Tree and Naive Bayes models, for the three Iris, Glass, and Credit datasets the Decision tree performance is slightly better than Naive Bayes. Thus it can be concluded that Feature Selection using GA can improve the accuracy of both the Decision Tree and Naive Bayes models, and the Decision Tree shows slightly better accuracy than Naive Bayes.

## References

[1] Kantardzic M 2003 *Data Mining: Concept, Models, Methods, And Algorithms* (IEEE Press 445 Hoes Lane, Piscataway, NJ 08854).
[2] Witten IH, Frank E, Hall MA 2011 *Data Mining: Practical Machine Learning Tools and Techniques*. (Burlington: Morgan Kaufmann Publisher).
[3] Rokach L and Maimon O 2005 Top-down induction of decision trees classifiers-a survey *IEEE Trans. Syst. Man Cybernetics Part C (Appl. Rev.)* **35**(4) pp 476-87.
[4] Cardic C 1993 Proc. Int. Conf. Mach. Learn.
[5] Brownlee Jason 2011 *Clever Algorithms Nature-Inspired programming Recipes* (Australia: Creative Commons Attribution-Noncommercial-Share Alike 2.5 Autralia License).
[6] Strobl, C.; Malley, G.; Tutz 2009. An Introduction to Recursive Partitioning: Rationale, Application and Char- acteristics of Classification and Regression Trees, Bag- ging and Random Forests. Psychological Methods 14 (4): 323–348. Strobl C, Malley J and Tutz G 2009 An introduction to recursive partitioning: rationale, application, and characteristics of classification and regression trees, bagging, and random forests *Psy. Methods* **14**(4) pp 323.
[7] Breiman L, Friedman J, Stone CJ and Olshen RA 1984 *Classification and regression trees* (Monterey: CRC press).
[8] Rokach L, Maimon O 2008 *Data mining with decision trees: theory and applications*. (Singapore: World Scientific Pub Co Inc)
[9] Hebb D 1949 *The Organization of Behavior* (New York: Wiley).
[10] Farley BW, Clark W 1954 Simulation of self-organizing systems by digital computer *Trans. IRE Prof. Group Inf. Theory* **4**(4) pp 76-84.
[11] Goldberg DE 1989 *Genetic Algorithms in Search, Optimization, and Machine Learning*. (Boston: Addison-Wesley).
[12] Fraser AS 1957 Simulation of Genetic Systems by Automatic Digital Computers I Introduction *Aust. J. Biol. Sci.* **10** pp 484–91.
[13] Reed J, Toombs R and Barricelli NA 1967 Simulation of Biological Evolution and Machine Learning *J. Theor. Biol.* **17** pp 319–42.
[14] Holland JH 1975 *Adaptation in Natural and Artificial Systems* (Michigan: University of Michigan).

# Implementation of Super-Encryption with Trithemius Algorithm and Double Transposition Cipher in Securing PDF Files on Android Platform

**M A Budiman, D Rachmawati and Jessica**

Departemen Ilmu Komputer, Fakultas Ilmu Komputer dan Teknologi Informasi, Universitas Sumatera Utara, Jl. Universitas No. 9-A, Kampus USU, Medan 20155, Indonesia

Email: mandrib@usu.ac.id, dian.rachmawati@usu.ac.id, jessica13@student.usu.ac.id

**Abstract.** This study aims to combine the trithemus algorithm and double transposition cipher in file security that will be implemented to be an Android-based application. The parameters being examined are the real running time , and the complexity value. The type of file to be used is a file in PDF format. The overall result shows that the complexity of the two algorithms with duper encryption method is reported as $\Theta\ (n^2)$.However, the processing time required in the encryption process uses the Trithemius algorithm much faster than using the Double Transposition Cipher. With the length of plaintext and password linearly proportional to the processing time.

## 1. Introduction

The word 'cryptography' comes from Greek. According to the language, the word cryptography is divided into two, namely 'crypto' and 'graphia'. 'Crypto' means secret and 'graphia' means writing. According to its terminology, cryptography is the science and art of keeping messages safe when messages are sent from one place to another [1]. Thus, cryptography can help us in preventing the arrival of important information, into unauthorized hands. The purpose of cryptography is 4. ie confidentiality, integrity, authentication,andnon-repudation. Confidentiality focuses on data confidentiality, integrity focuses on received data is true and genuine data, authentication is done to prevent false data, and non-repudation aims to prevent the shielding of the data from the sender. Cryptography algorithm divided 2 that is, symmetric algorithm and asymmetric algorithm. However, in this study the algorithm used is a classical algorithm, namely Trithemius Algorithm and double transposition cipher.

Trithemius Cipher is one of the polyalphabetic codes designed to be easier to use. Instead of using a random combination of letters of the alphabet, Trithemius uses a special table of Trithemius Trithemius Cipher is one of the polyalphabetic codes designed to be easier to use. Instead of using a random combination of letters of the alphabet, Trithemius uses a special table Trithemius, which has 24 rows and columns 24 of 24 letters in the Latin alphabet except the letter j and v [2]. However, in order to make the system more useful, in this study we used a trithemius table consisting of 65 rows and columns. Where, all characters have entered into the table. Both lowercase and uppercase. Plus numbers 0 through 9, and some punctuation.

Double Transposition Cipher consists of two rounds of transition columns, using 2 different keys. After the key in the first round is determined, then each letter in the key is numbered in alphabetical order, and the plaintext is read per column based on the sequence of numbers. Where, the direction of the sequence should be determined first whether left-to-right, or right-to-left, to facilitate when there is the same letter on the key.Ciphertext generated from the first round, encrypted again in the second round in the same way, but using a different key [3].

File to be used in this research is a file with PDF format. PDF files are considered safe for static file formats . The advantage of a document created in PDF format is that it can not be changed directly by the recipient of a document unlike a document file with the format .doc, .txt or the other [4].

The purpose of this study are using Trithemius Algorithm and Double Transposition Cipher in super encryption scheme to avoid theft of files / data, calculates the super encryption complexity in Big Theta (Big-θ), and measure Real Running Time super encryption in millisecond (ms).

## 2. Method

On the encryption process, Trithemius algorithm will be used first. In the encryption process, the next letter on the plaintext will be replaced by the relevant letter of the next row in the table. After using the last line, one should move back to the first line. This means that all letters on the plaintext change according to the number of positions specified by the actual row. Therefore the first letter is encrypted without a shift, the second letter with the shift is determined by the second row (so by one position), the third letter with the shift determined by the third row (so by two positions) and so on. Or, can also be calculated by the formula:

$$C \equiv P + K(mod\ n) \tag{1}$$

$$P \equiv C - K(mod\ n) \tag{2}$$

Where :
C: Ciphertext
P: Plaintext
K: Key
N: number of letters used

This cipher consists of two rounds of transition columns, using 2 different keys. After the key in the first round is determined, then each letter in the key is numbered in alphabetical order, and the plaintext is read per column based on the sequence of numbers. Where, the direction of the sequence should be determined first whether left-to-right, or right-to-left, to facilitate when there is the same letter on the key.

Ciphertext generated from the first round, encrypted again in the second round in the same way, but using a different key. Ciphertext generated from the second round, then divided into blocks. Where, each block consists of 5 letters to facilitate the transmission process. One of the requirements of encryption process using this cipher is, the table we use must be fully loaded. In order for the table used to be fully charged, padding character is used.

While in the decryption process by using Double Transposition Cipher, start by using the second key first. Where, the number of rows in the table is determined by the division of ciphertext length by key. (Salomon, 2005). Then, the ciphertext is inserted into the table based on the number of fragments in alphabetical order in the key used.

## 3. Result and Discussion

The system was built using Android Studio 1.3.2. with the programming language is Java. This system is tested with Personal Computer with processor intel core i7 – 3632QM CPU @ 2.20 GHz, Memory 4 GB RAM.

**Table 1.** Complexity of Trithemius Algorithm Encryption Method

| No | Program Code | *C* | # | *C#* |
|---|---|---|---|---|
| 1. | public String enkripsi_ttm(String plaintext, String key){<br>char[]arraykey = createkey(key, plaintext); | *C1* | 1 | *C1* |
| 2. | char[]arrayplaintext = plaintext.toCharArray(); | *C1* | 1 | *C1* |
| 3. | char[]arrayciphertext = new<br>char[arrayplaintext.length]; | *C1* | 1 | *C1* |
| 4. | int p,k,c; | *C1* | 1 | *C1* |
| 5. | for (int i=0; i<arrayplaintext.length;i++){ | *C2* | *n* | *C2n* |
| 6. | if(getindex(arrayplaintext[i])==-1){ | *C3* | *n* | *C3n* |
| 7. | arrayciphertext[i]=arrayplaintext[i]; | *C1* | *n* | *C1n* |
| 8. | continue; | *C4* | | 1 *C4n* |
| 9. | } | | | |
| 10. | p= getindex(arrayplaintext[i]); | *C1* | *n* | *C1n* |
| 11. | k= getindex(arraykey[i]); | *C1* | *n* | *C1n* |
| 12. | c= (p+k)%index.length(); | *C1* | *n* | *C1n* |
| 13. | arrayciphertext[i]=getcharacter(c); | *C1* | *n* | *C1n* |
| 14. | } | | | |
| 15. | String hasil = new String(arrayciphertext); | *C1* | 1 | *C1* |
| 16. | return hasil; | *C5* | 1 | *C5* |
| 17. | } | | | |

$$T(n) = \sum C_i \#_i$$

$$= 5C_1 + 5C_1 n + C_2 n + C_3 n + C_4 n + C_5$$

$$= (5C_1 + C_5)n^0 + (5C_1 + C_2 + C_3 + C_4)n^1$$

$$= \Theta(n)$$

**Table 2.** Complexity of Trithemius Algorithm Decryption Method

| No | Program Code | C | # | C# |
|---|---|---|---|---|
| 1. | public String dekripsi_ttm(String ciphertext, String key){<br>char[]arraykey = createkey(key, ciphertext); | C1 | 1 | C1 |
| 2. | char[]arrayciphertext = ciphertext.toCharArray(); | C1 | 1 | C1 |
| 3. | char[]arrayplaintext = new char[arrayciphertext.length]; | C1 | 1 | C1 |
| 4. | int p,k,c; | C1 | 1 | C1 |

| | | | | |
|---|---|---|---|---|
| 5. | `for (int i=0; i<arrayciphertext.length;i++){` | C2 | n | C2n |
| 6. | `if(getindex(arrayciphertext[i])==-1){` | C3 | n | C3n |
| 7. | `arrayplaintext[i]=arrayciphertext[i];` | C1 | n | C1n |
| 8. | `continue;` | C4 | n | C4n |
| 9. | `}` | | | |
| 10. | `c= getindex(arrayciphertext[i]);` | C1 | n | C1n |
| 11. | `k= getindex(arraykey[i]);` | C1 | n | C1n |
| 12. | `p= (c-k);` | C1 | n | C1n |
| 13. | `if(p<0) {` | C3 | n | C3n |
| 14. | `p=p+index.length();` | C1 | n | C1n |
| 15. | `}` | | | |
| 16. | `else{` | | | |
| 17. | `p=p%index.length();` | C1 | n | C1n |
| 18. | `}` | | | |
| 19. | `arrayplaintext[i]=getcharacter(p);` | C1 | n | C1n |
| 20. | `}` | | | |
| 21. | `String hasil = new String(arrayplaintext);` | C1 | 1 | C1 |
| 22. | `return hasil;` | C5 | 1 | C5 |
| 23. | `}` | | | |

$$T(n) = \sum C_i \#_i$$

$$= 5C_1 + 7C_1n + C_2n + 2C_3n + C_4n + C_5$$

$$= (5C_1 + C_5)n^0 + (7C_1 + C_2 + 2C_3 + C_4)n^1$$

$$= \Theta(n)$$

From the above calculation, the result of complexity of Trithemius algorithm is Θ (n).

**Table 3.** Complexity of Double Transposition Cipher Encryption Method

| No | Program Code | *C* | # | *C#* |
|---|---|---|---|---|
| 1. | `public String enkripsi_double(String plain, String key1, String key2) {`<br>`String k1 = key1, k2 = key2, cipher1 = "", cipher2 = "";` | *C1* | 1 | *C1* |
| 2. | `Integer i1 = (int) Math.ceil(plain.length() * 1.0 / k1.length());` | *C1* | 1 | *C1* |

| | | | | |
|---|---|---|---|---|
| 3. | Integer j1 = k1.length(); | *C1* | 1 | *C1* |
| 4. | char[][] tabel1 = new char[i1][j1]; | *C1* | 1 | *C1* |
| 5. | int k = 0; | *C1* | 1 | *C1* |
| 6. | } | | | |
| 7. | for (int i = 0; i < i1; i++) | *C2* | $n$ | $C2n$ |
| 8. | for (int j = 0; j < j1; j++) { | *C2* | $n^2$ | $C2n^2$ |
| 9. | if (k < plain.length()) | *C3* | $n^2$ | $C3n^2$ |
| 10. | tabel1[i][j] = plain.charAt(k); | *C1* | $n^2$ | $C1n^2$ |
| 11. | else | | | |
| 12. | tabel1[i][j] = 'Ā'; | *C1* | $n^2$ | $C1n^2$ |
| 13. | k++; | *C1* | $n^2$ | $C1n^2$ |
| 14. | } | | | |
| 14. | Integer[] runtutan = urutan(k1); | *C1* | 1 | *C1* |
| 15. | for (int i = 0; i < runtutan.length; i++) | *C2* | $n$ | $C2n$ |
| 16. | for (int j = 0; j < i1; j++) | *C2* | $n^2$ | $C2n^2$ |
| 17. | cipher1 += tabel1[j][runtutan[i]]; | *C1* | $n^2$ | $C1n^2$ |
| 18. | Integer i2 = (int) Math.ceil(cipher1.length() * 1.0 / k2.length()); | *C1* | 1 | *C1* |
| 19. | Integer j2 = k2.length(); | *C1* | 1 | *C1* |
| 20. | char[][] tabel2 = new char[i2][j2]; | *C1* | 1 | *C1* |
| 21. | k = 0; | *C1* | 1 | *C1* |
| 22. | for (int i = 0; i < i2; i++) | *C2* | $n$ | $C2n$ |
| 23. | for (int j = 0; j < j2; j++) { | *C2* | $n^2$ | $C2n^2$ |
| 24. | if (k < cipher1.length()) | *C3* | $n^2$ | $C3n^2$ |
| 25. | tabel2[i][j] = cipher1.charAt(k); | *C1* | $n^2$ | $C1n^2$ |
| 26. | else | | | |
| 27. | tabel2[i][j] = 'Ī'; | *C1* | $n^2$ | $C1n^2$ |
| 28. | k++; | *C1* | $n^2$ | $C1n^2$ |
| 29. | } | | | |
| 30. | runtutan = urutan(k2); | *C1* | 1 | *C1* |
| 31. | for (int i = 0; i < runtutan.length; i++) | *C2* | $n$ | $C2n$ |
| 32. | for (int j = 0; j < i2; j++) | *C2* | $n^2$ | $C2n^2$ |
| 33. | cipher2 += tabel2[j][runtutan[i]]; | *C1* | $n^2$ | $C1n^2$ |
| 34. | return cipher2; | *C4* | 1 | *C4* |

35. }

---

$$T(n) = \sum C_i \#_i$$

$$= 11C_1 + 4C_2n + 8C_1n^2 + 4C_2n^2 + 2C_3n^2 + C_4$$

$$= (11C_1 + C_4)n^0 + 4C_2n + + (8C_1 + 4C_2 + 2C_3)n^2$$

$$= \Theta(n^2)$$

From the above calculation, the result of complexity for Double Transposition Cipher is $\Theta(n^2)$. Thus, it can be concluded that the complexity of the super-encryption method between the Trithemius algorithm and the Double Transposition Cipher is $\Theta(n^2)$. It means, the length of plaintext and password proportional to linear processing time. The longer the plaintext and password the longer the time it takes.

## 4. Conclusion

- The process of encryption and decryption using a combination of Trithemius algorithm and Double Transposition Cipher in super encryption method, successfully done.
- The value of super encryption complexity between the two algorithms used is Trithemius and Double Transposition Cipher algorithm yield value $\Theta$ ($n^2$)
- In the encryption process, the longer the plaintext and password the longer the time it takes. In other words, the length of the plaintext and the password is linearly proportional to time.

## 5. References

[1] Ariyus, Doni. 2008. *Pengantar ilmu Kriptografi, Teori, analisis dan Implementasi*. Yogyakarta: Penerbit Andi.

[2] Mollin, Richard A.2005.Codes: the guide to secrecy from ancient to modern times : United State of America.

[3] Salomon, David.2005. *Coding for Data and Computer Communications* : United State of America.

[4] Kurnanda, Rahman. 2016. Perbandingan Algoritma 3DES dan Rinjdael Dalam Perancangan Aplikasi Keamanan Data. Skrpsi. Medan: Universitas Sumatera Utara.Alghazzawi, Daniyal M., Hasan, Syed Hamid. & Trigui, Mohamed Salim. 2013. Stream Ciphers: A Comparative Study of Attacks and Structures. *International Journal of Computer Applications* **1** 27-29.

IOP Conf. Series: Journal of Physics: Conf. Series **978** (2018) 012089 doi:10.1088/1742-6596/978/1/012089

# Analysis Spectrum of ECG Signal and QRS Detection during Running on Treadmill

**M. Agung Suhendra*[1], M. Ilham R.[1], Artha I. Simbolon[1], M. Faizal A.[1], A. Munandar[1].**

[1]Technical Implementation Unit for Instrumentation Development, Indonesian Institute of Sciences, Indonesia.

*agung.fq@gmail.com

**Abstract**. The heart is an important organ in our metabolism in which it controls circulatory and oxygen. The heart exercise is needed one of them using the treadmill to prevent health. To analysis, it using electrocardiograph (ECG) to investigating and diagnosing anomalies of the heart. In this paper, we would like to analysis ECG signals during running on the treadmill with kinds of speeds. There are two analysis ECG signals i.e. QRS detection and power spectrum density (PSD). The result of PSD showed that subject 3 has highly for all subject and the result of QRS detection using pan Tomkins algorithm that a percentage of failed detection is an approaching to 0 % for all subject.

## 1. Introduction

The heart is primarily organ in humans, it pumps circulatory, supply oxygen and nutrients to body tissues. Many people were deaths caused by heart disease in which it suddenly attack us. We never know when it occurs to us however we can to prevent it such as sports or physical exercise. One of the tools to analyze the heart is Electrocardiograph (ECG). In general, ECG is applied to investigating and diagnosing anomalies of the heart based on electrical activity. The electrical conduction of the heart is generated by two main of the cell: cardiomyocytes in which it provides electrical potentials when contraction and cells particularized in the production and conduction of the action potentials.

The characteristic of ECG signal as shown in Fig. 1, which there are three elements in the ECG waveform i.e. isoelectric line, segments, and intervals. Isoelectric describes as the horizontal line during no electrical activity, Segments is defined as the time duration of the isoelectric among waves, and Intervals is the term for the time among in around the same segments [1].

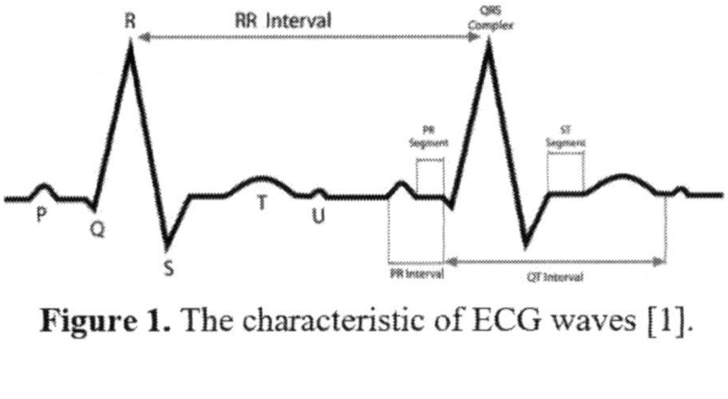


**Figure 1.** The characteristic of ECG waves [1].

IOP Conf. Series: Journal of Physics: Conf. Series **978** (2018) 012089 doi:10.1088/1742-6596/978/1/012089

In this paper, we would like to analyze ECG signal when the subjects using a treadmill with various speed to identify anomaly and quality of the heart. The researchers of ECG fields and cardiologist were proposed several methods and experiments to heart rate exercise such as physical activity consisted of two hours per week, daily swimming routine, fitness and others training then the whole is compared to heart rate each of them [2-3]. The ECG signal recorded by the device on wireless during running on the treadmill and the data will process using wavelet transform (WT) algorithms to removing artifact and it provides to analyze in the frequency domain and finally, we detect a QRS peak of ECG signal using pan Tomkins algorithm.

## 2. Material and Method

### 2.1. Data Acquisition

The five subject were recorded ECG signal with the mean of age from 25 - 28 years, each of them given instruction to running on treadmill to one minutes for one kinds of experiments in which there are five various based on treadmill speed (m/s): 0, 1.5, 3.0, 4.5, and 6.0. The device used has specification are frequency sampling (FS) 1000 Hz/s, wireless based, 1-lead (1-positive, 1-negative, 1-reference) and Bipolar differential measurement. The position of placement sensor and the experimental setup is given in Fig. 2.

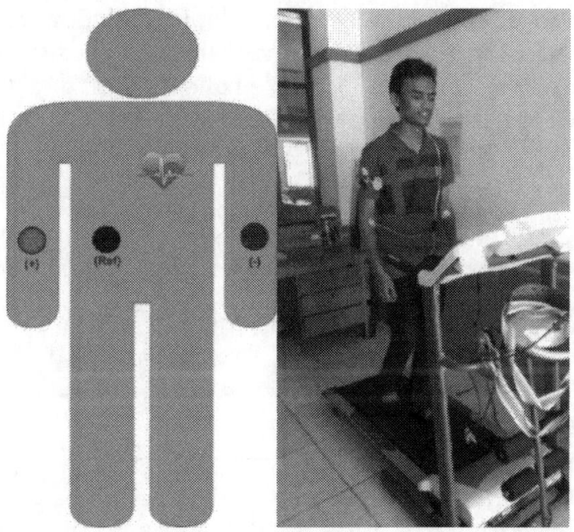


**Figure 2.** The placement of sensor (left) and experimental setup (right)

### 2.2. Preprocessing

The scheme of signal processing to removing noise and artifact during the experiment as shown in figure 3, which is the signal acquisition was described foregoing. In addition, there is three part of preprocessing i.e. detrending data, windowing, and finite impulse response (FIR) filter. The detrending data described to remove the baseline data by eliminating the trend of ECG data in which to calculate it follows:

$$Level_{Trending} = \left[ \frac{\log_2 2_t}{\log_2 N} \right] \tag{1}$$

where $t$ is the time duration of sampling and $N$ is the number of sampling. The windowing reduces the amplitude of the discontinuities in data and The FIR filter generally used to operate on discrete-time signals and occurs weighted of the data in which it describes as

$$y[n] = \sum_{k=0}^{M} b_k x[n-k] \tag{2}$$

where M is finite [4].

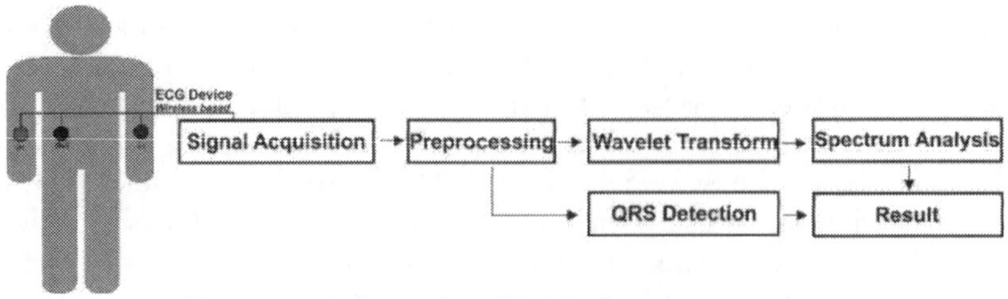


**Figure 3.** Schematic of ECG signal processing

### 2.3. *Wavelet Transform*

Wavelet Transform (WT) commonly used to signal processing for several application such as EEG an ECG, which it provides to analyze the signal in time domain, signal decomposition and signal compression. In mathematics, WT describe as

$$\psi_{a,b} = \frac{1}{\sqrt{a}}\psi(\frac{t-b}{a}) \tag{3}$$

where ($\psi_{a,b}$) as wavelet function and a, b $\in$R, a > 0, and R is the *wavelet space* [5,6]. Furthermore, WT has several of properties are can be represented on multiple resolution scales called Discrete wavelet transform (DWT). In particular, DWT used to analyze various signals to obtain good resolutions by decomposing them. Moreover, it can yield approximation (a) and detail coefficient (d), DWT decomposition can be defined as

$$a_{(i)} = x(k) * \varphi_{i,l}(k), \tag{4}$$

$$d_{(i)} = x(k) * \psi_{i,l}(k), \tag{5}$$

where $a_{(i)}(l)$ , $d_{(i)}(l)$ is the approximation and detail coefficient and $i$ is the resolution [7,8]. In figure 4 describe as scheme of DWT in which the filter bank implementation of wavelet transform for three-level wavelet decomposition.

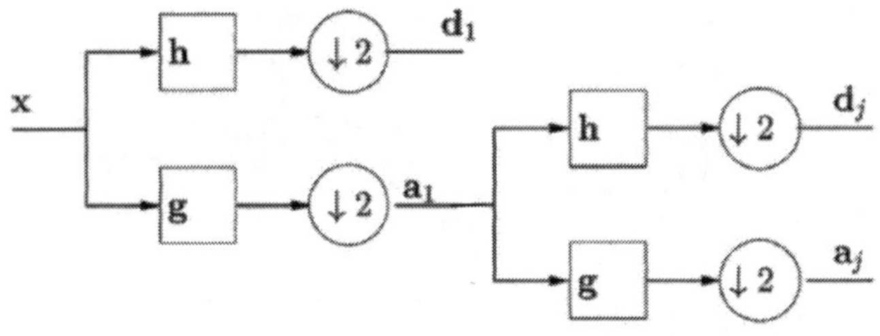


**Figure 4.** The properties of wavelet decomposition.

### 2.4. *QRS Detection using Pan Tomkins' alghorithm*

There are many of kinds method for QRS detection or beat per minutes (BPM) calculation such as Pan Thompkins' algorithm, Savitsky-Golay algorithm, and the simplest algorithm is applying three processing: linear high-pass filter, non-linear low-pass filter, and the decision making [9-10].

Pan Tompkins developed by J. Pan and W. J. Tompkins in which they are was built an algorithm for detection of the QRS on ECG signal in real time. In general, the scheme of the algorithm is given in figure 5. First is an integer coefficient bandpass filter composed of cascaded low-pass and high-pass filters, Its function is noise rejection. The signal was filtered next is a filter that approximates a derivative in which five-point with the transfer function (Eq. 6), the amplitude response (Eq. 7), and the difference equation (Eq. 8).

$$H(z) = (1/8T)(-z^{-2} - 2z^{-1} + 2z^{1} + z^{2}) \tag{6}$$

$$|H(wT)| = (1/4T)[\sin(2\omega T) + 2\sin(\omega T)] \tag{7}$$

IOP Conf. Series: Journal of Physics: Conf. Series **978** (2018) 012089 doi:10.1088/1742-6596/978/1/012089

$$y(nT) = (1/8T)[-x(nT-2T) - 2x(nT-T) + 2x(nT+T) + x(nT+2T)] \tag{8}$$

the results from derivation, do square process using mathematical operation $y(nT) = [x(nT)]^2$, the signal moves into a moving-window integrator. Finally, Adaptive thresholds then separate the locations of the QRS complexes [11].

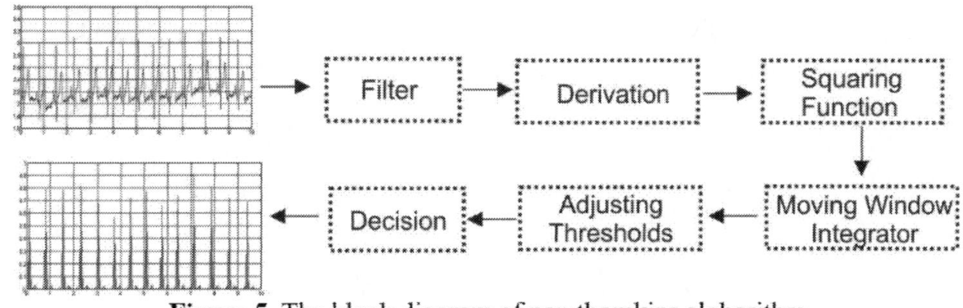


**Figure 5.** The block diagram of pan thomkins alghorithm

## 3. Results and Discussions

The raw signal was recorded as shown in figure 6 based on each of speed on the treadmill, their signal was included noise caused by error transmission device through wireless and artifact caused by hand or any body movement to disturb when ECG sensing. So, required to remove them and obtained the real ECG signal. To implementation, do signal processing. First, pre-processing step i.e. detrending data, windowing, and FIR filter and last, used discrete wavelet transform to analyze the signal in approximation and coefficient then using FFT to get power spectrum density (PSD) for the complete subject as shown in figure 7 in which for subject 3 has PSD highly than another subject.

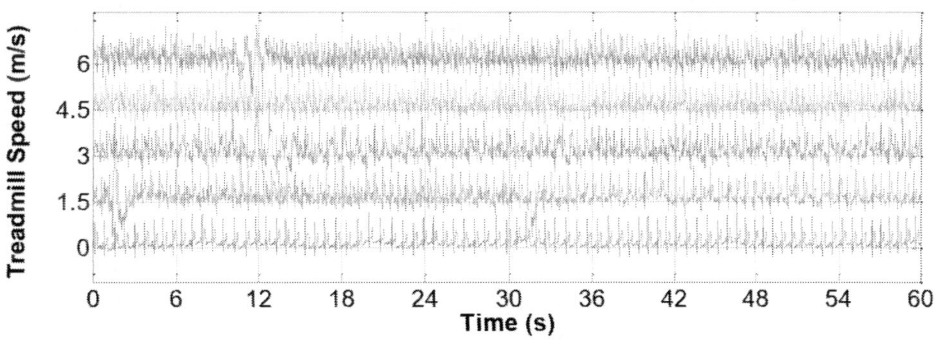


**Figure 6.** The raw data ECG signal from subject 1.

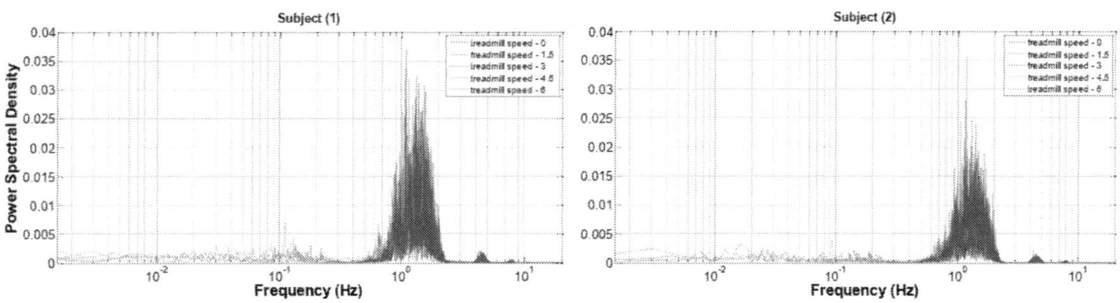

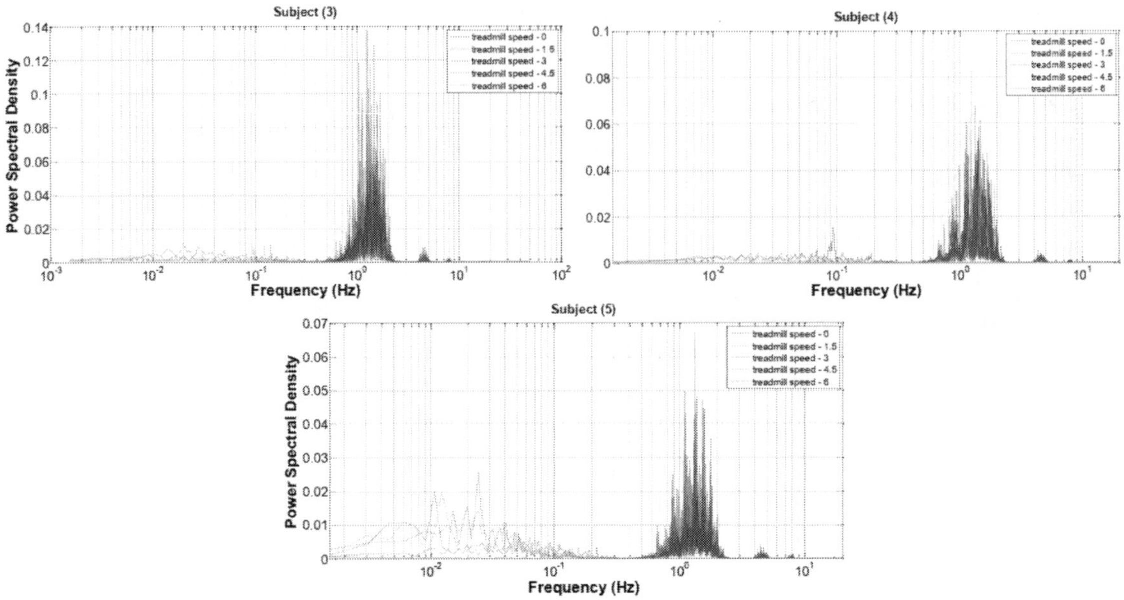


**Figure 7.** The result of PSD for all subject.

The results of QRS detection using pan Tompkins algorithm is given in table 1 in which it powerful to detects them, Moreover, a percentage of failed detection is an approaching to 0 % for all subject. The total peak calculated manually peak by peak in ECG signal and this table showed an increasing speed of treadmill it's so difficult to detect them but overall, this algorithm can detect it.

**Table 1.** The Result of QRS determinaton

| Speed (m/s) | Subject | Total Peak | QRS Detection | Failed detection (%) | Speed (m/s) | Subject | QRS Peak | QRS Detection | Failed detection (%) |
|---|---|---|---|---|---|---|---|---|---|
| **0** | 1 | 83 | 83 | **0** | **4,5** | 1 | 128 | 128 | **0** |
| | 2 | 87 | 87 | **0** | | 2 | 129 | 131 | **0.02** |
| | 3 | 97 | 97 | **0** | | 3 | 136 | 136 | **0** |
| | 4 | 135 | 135 | **0** | | 4 | 147 | 147 | **0** |
| | 5 | 80 | 80 | **0** | | 5 | 118 | 122 | **0.03** |
| **1,5** | 1 | 99 | 101 | **0.02** | **6** | 1 | 136 | 134 | **0.02** |
| | 2 | 111 | 111 | **0** | | 2 | 138 | 137 | **0.01** |
| | 3 | 111 | 111 | **0** | | 3 | 159 | 158 | **0.01** |
| | 4 | 135 | 135 | **0** | | 4 | 168 | 168 | **0** |
| | 5 | 95 | 97 | **0.02** | | 5 | 131 | 131 | **0** |
| **3** | 1 | 110 | 110 | **0** | | | | | |
| | 2 | 115 | 115 | **0** | | | | | |
| | 3 | 125 | 125 | **0** | | | | | |
| | 4 | 133 | 133 | **0** | | | | | |
| | 5 | 133 | 133 | **0** | | | | | |

## 4. Conclusions

In this paper, we would like to analyze ECG signal during running on the treadmill with power spectrum density and QRS detection using pan Tomkins. There is five subject for the participant in this experiment. The raw data was recorded that increasing peak for each of speeds on the treadmill as shown figures foregoing. The result of PSD shown that subject 3 has highly PSD than another subject. In addition, QRS detection using pan Tomkins shown that a percentage of failed detection is an approaching to 0 % for all subject.

**Acknowledgments**
This research was supported by Technical Management Unit for Instrumentation Development (Deputy for Scientific Services), Indonesian Institute of Sciences, Indonesia and INSINAS 2017 program from Ministry of Research, Technology and Higher Education, Indonesia.

**References**

[1] Gacek A 2011 *ECG signal processing, classification and interpretation: a comprehensive framework of computational intelligence* ed Pedrycz W (Berlin: Springer Science & Business Media).
[2] Furlan R, Piazza S, Dell'Orto S, Gentile E, Cerutti S, Pagani M and Malliani A 1993 Early and late effects of exercise and athletic training on neural mechanisms controlling heart rate *Cardiovasc. Res.* vol 27 pp 482-8.
[3] Aubert AE, Seps B and Beckers F 2003 Heart rate variability in Athletes *Sports Med* **33**(12) pp 889-919.
[4] Kumar KS, Yazdanpanah B and Raju GS 2014 Performance Comparison of Windowing Techniques for ECG Signal Enhancement *Int. J. Eng. Res.* **3**(12) pp 753-6.
[5] A. Prochazka, J. Kukal, and O. Vysata. "Wavelet transform use for feature extraction and EEG signal segments classification," In Proc. IEEE ISCCSP, pp. 719-722, Malta, 2008. Prochazka A, Kukal J and Vysata O 2008 Wavelet transform use for feature extraction and EEG signal segments classification *Proc. IEEE ISCCSP* pp 719-22.
[6] Murugappan M, Nagarajan R and Yaacob S 2010 Combining spatial filtering and wavelet transform for classifying human emotions using EEG signals *J. Med. Biol. Eng.* **31**(1) pp 45-51.
[7] Subasi A 2007 EEG signal classification using wavelet feature extraction and a mixture of expert model *Expert Syst. Appl.* **32** pp 1084-93.
[8] Ciupe A M and Roman NM 2015 Study of ECG signal processing using wavelet transfroms *Proc. IEEE ATEE* pp 27-30.
[9] Das S and Chakraborty M 2012 QRS detection algprithm using savitzky-golay filter *Int. J. Signal Image Process. vol 03.*
[10] Chen HC and Chen SW 2013 A moving average based filtering system with its application to real time QRS detection *Proc. IEEE Comput. Cardiology.*
[11] Pan J and Tompkins WJ 1985 A real-time QRS detection alghorithm *Proc. IEEE. Trans. Biomed. Eng.* vol BME-32.

# An implementation of super-encryption using RC4A and MDTM cipher algorithms for securing PDF Files on android

**M A Budiman, D Rachmawati and M R Parlindungan**

Department of Computer Science, Faculty of Computer Science and Information Technology, Universitas Sumatera Utara, Jl. Universitas No. 9-A, Kampus USU, Medan 20155, Indonesia

Email: mandrib@usu.ac.id, dian.rachmawati@usu.ac.id and mutiararizkyparlindungan2013@student.usu.ac.id

**Abstract.** MDTM is a classical symmetric cryptographic algorithm. As with other classical algorithms, the MDTM Cipher algorithm is easy to implement but it is less secure compared to modern symmetric algorithms. In order to make it more secure, a stream cipher RC4A is added and thus the cryptosystem becomes super encryption. In this process, plaintexts derived from PDFs are firstly encrypted with the MDTM Cipher algorithm and are encrypted once more with the RC4A algorithm. The test results show that the value of complexity is $\Theta(n^2)$ and the running time is linearly directly proportional to the length of plaintext characters and the keys entered.

## 1. Introduction

Nowadays, sensitive information may become very vulnerable to be known to others, as it can be captured and manipulated by unauthorized parties. In cryptography, there are some processes that can be done to ensure the security of the information transmitted; one of them is by encoding the information into data that can not be read or understood by other parties, but can be read by the sender and the recipient. PDF files are considered more secure than the other file types since they can be set to disallow editing. But technological developments are increasing over time, the security of confidential messages in PDF format is perceived to be unsafe, as more and more applications can change or destroy messages in PDF format [1].

The MDTM Cipher algorithm stands for 'Monograph-Digraph-Transposition-Monograph' wherein the relationship between individual letters of plain text and ciphers is quite complex because each original letter is replaced by two letters each of which separately combines with the letters of the other pair to form Digraph, which is then re-converted into a monograph. This algorithm is part of a character-based classical algorithm that is very vulnerable to attack.

To provide dual security to the data specially in PDF file, this time the MDTM Cipher algorithm will be combined with the RC4A algorithm. The RC4A algorithm is a modified cryptographic byte-oriented stream cipher algorithm of RC4 ciphers whose security level is stronger than the usual RC4. The principal design principle of RC4A is to reduce the correlation between byte output and internal variables by making the byte output dependent on more random variables. This algorithm is part of a modern symmetric algorithm whereby modern algorithms are more known to be safe than classical algorithms. The technique of combining the two algorithms above is referred to as a super encryption method in which a classical symmetry algorithm combined with modern symmetric algorithms that the modern symmetric algorithm is expected to reinforce the previous algorithm.

## 2. Method

The RC4A algorithm is a symmetric key algorithm proposed by the FSE in 2004 by Paul S. and Bart Preneel which is a stream cipher modified by its predecessor RC4 algorithm [2]. The RC4A algorithm tries to improve RC4 security by using two arrays, there are $S_1$and $S_2$ from 1 to 256 characters, two indexes $j_1$and $j_2$, and two keys $k_1$and $k_2$ corresponding to the Key Scheduling Algorithms (KSA) in Figure 1 and Pseudo Random Generation Algorithm (PRGA) process in Figure 2.

```
for i = 0 to 255 do
   s1[i] ← i
end for
for i = 0 to 255 do
   s2[i] ← i
end for
j1 ← 0
j2 ← 0
for i = 0 to 255 do
   j1 ← (j1 + s1[i] + k1[i mod l]) mod 256
   swap (s1[i],     s1[j1])
end for
for i = 0 to 255 do
        j2 ← ( j2 + s2[i] + k2[i mod l])mod 256
        swap (s2[i],     s2[j2])
end for
```

**Figure 1.** KSA of RC4A

```
Initialization:
   i ← 0
   j1 ← 0
   j2 ← 0
Loop:
        i ← i ++
        j1 ← S1( j1 + S1[i])
        swap (S1[i], S1[j1])
        Z(1) ← S2[S1[i] + S1[j1]]
        j2 ← S2(j2 + S2[i]
        swap (S2[i], S2[j2])
        Z(2) ← S1[S2[i] + S2[j2]]
```

**Figure 2.** PRGA of RCA

Whenever $i$ is added, two bytes are generated [3] with formula: $j_1 = S_1[j_1 + S[i] + key[i\ mod\ key - length]]\ mod\ 256$ where the key is the input value of the user and the "key length" is the byte number of The key and range of the key length is from 1 to 256. For each iteration, values of $j_1$and $j_2$are calculated, $S_1[i]$ and $S_1[j_1]$ in swap (exchanged), and the sum of $S_1[i]$ and $S_1[j_1]$ mod 256 After the result $S_1$ then done generate the first key and then for the calculation of $S_2$. These steps are repeated on $S_2$ and $j_2$, and $S[S_2[i] + S[j_2]]$ is the key stream output [4]. In the PRGA, the key generator (key stream) is XOR with plaintext to generate encryption text on the form based on the byte stream. The same operation applies to array $S_2$. Byte is returned, then XOR with one individual letter on plaintext to convert it to cipher text.

In MDTM Cipher algorithm formed a 16x16 square matrix. The sixteen lines are identified by the letters A, B, C, D, E, F, G, H, I, J, K, L, M, N, O, and P. Then arrange the first plaintext into the box. Fill in the remaining boxes with unused characters from characters. If the keywords included in it are repeated, then they are ignored [5]. Each letter in the plaintext (called monograph) is now replaced with rows and column letters such that, for example, M becomes EM (called digraph). The resulting cipher text will be twice as long as the original. There are two types of keys used, the second key is a transposition key by inputting a number from 1 to 10 at random. The transposition keys are then used for the transposition process by arranging them into transposition tables and arranging them in order. The result is changed again to monograph in accordance with the first monograph table.

## 3. Results and Discussion

The results showed that the algorithm complexity obtained was quadratic with the result Θ ($n^2$). For the results of the processing time can be seen in Figure 3 and Figure 4 below.

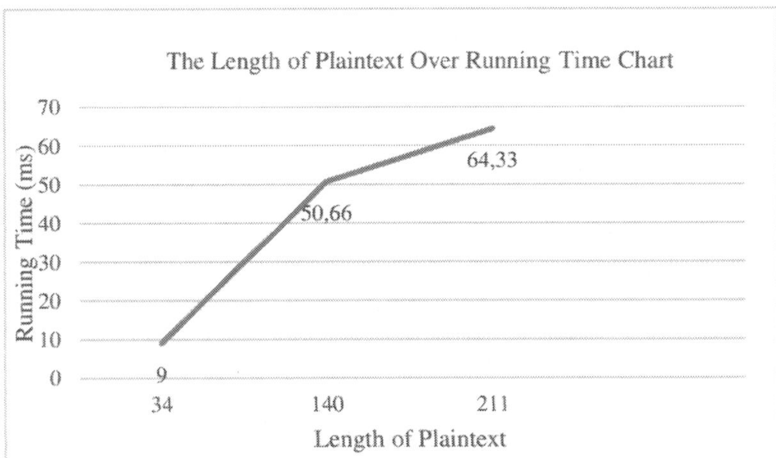


**Figure 3.** The Length of Plaintext Over Running Time Chart

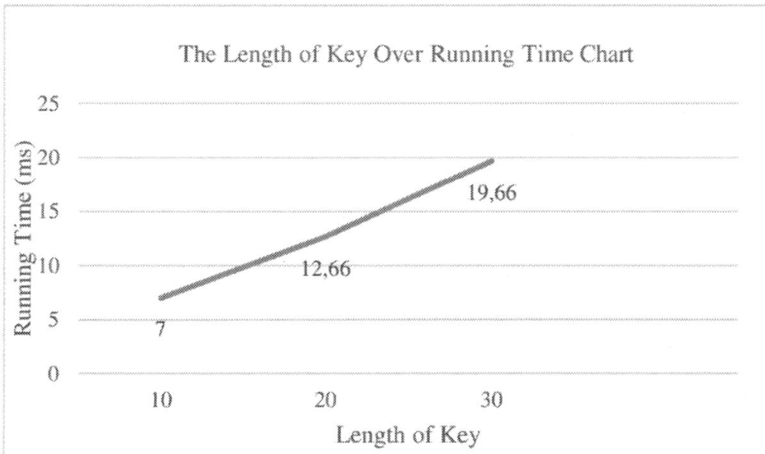


**Figure 4.** The Length of Key Over Running Time Chart

The picture above shows the plaintext and key plaintext graphs against the time of the encryption process, in which the length of the plaintext and the key is directly proportional to the processing time that produces a linear line. The linear line states that the longer the plaintext character and the key to be processed the longer the process time required.

## 4. Conclusion

As for the results of RC4A and MDTM Cipher algorithms implementation in the encryption of files on the Android platform, then the conclusion is:

1. The application designed in this research is capable of encryption and decryption using RC4A algorithm and MDTM Cipher shows that the final state array of the encryption and decryption process is the same.
2. The value of complexity of both algorithm used is RC4A and MDTM Cipher algorithm which consist of Key Scheduling Algorithm (KSA) and Pseudo-Random Generation Algorithm (PRGA) resulted $\Theta$ ($n^2$). This complexity is directly proportional to the user input so that the graph formed is a linear graph.
3. Test results at the time of the process obtained that the algorithm process time is directly proportional to the length of plaintext characters and keys. This means that the longer the plaintext used, the longer it will be The time required for encryption and decryption of plaintext and key.

## References

[1] Dibiyo M S and Aisyatul K 2016 *Teknik Penyembunyian Pesan PDF Terenkripsi Menggunakan Algoritma Kriptografi Vernam Cipher Dan Steganografi End Of File (Eof) Dalam Media Gambar* (Semarang : Universitas Dian Nuswantoro).

[2] Paul S & Bart P 2004 A New Weakness in the RC4 Keystream Generator and an Approach to Improve the Security of the Cipher* *Int. Workshop on Fast Software Encryption* pp 245-259.

[3] Maximov A 2004 *Two Linear Distinguishing Attacks on VMPC and RC4A and Weakness of RC4 Family of Stream Ciphers (Corrected)* Sweden.

[4] Sarkar S 2010 *Further Non-randomness in RC4, RC4A and VMPC* (Chennai, India: Chennai Mathematical Institute).

[5] Churchhouse R F 2004 *Codes and Ciphers Julius Caesar, the Enigma and the internet* : Cambridge.

IOP Conf. Series: Journal of Physics: Conf. Series **978** (2018) 012091 doi:10.1088/1742-6596/978/1/012091

# Application of Open Garden Sensor on Hydroponic Maintenance Management

**S Nasution[1], B Siregar[1], M Kurniawan[1], H Pranoto[2], U Andayani[1] and F Fahmi[3]**

[1]Department of Information Technology, Faculty of Computer Science and Information Technology, Universitas Sumatera Utara, Indonesia.
[2]Department of Electronic, Politeknik Negeri Medan, Indonesia.
[3]Department of Electrical Engineering, Faculty of Engineering, Universitas Sumatera Utara, Indonesia.

Email: baihaqi@usu.ac.id

**Abstract.** Hydroponic farming system is an agricultural system that uses direct water as a nutrient without using soil as a planting medium. This system allows smallholder farmers to have the opportunity to develop their crop production with less capital. In addition, hydroponic planting has also been widely adapted by individuals as a personal hobby. Application of technology has penetrated various fields including agricultural fields. One of the technologies that can be applied in a hydroponic farming system is the sensor. Sensors are devices that used to convert a physical quantity into a quantity of electricity so that it can be analyse with a certain electrical circuit. In this study, the technology to be applied is wireless sensor technology applied in human life to help get information quickly and accurately. Sensors to be used in this study are pH sensors, conductivity sensors, temperature sensors and humidity. In addition to sensors, the study also involved Arduino technology. Arduino is a microcontroller board that is used to interact with the environment based on programs that have been made. The final results of the application testing show that the system success to display diagram in real-time in an environment from Arduino board to database and web server.

## 1. Introduction

The role of wireless sensor technology can be applied in human life to help get information quickly and accurately. Open Garden consists of three different kits, each for a specific type of indoor growing plants (house and greenhouse), outside (gardens and fields) and hydroponics (plants in water installations). Open Garden has been designed to assist in caring for gardens by automating controls and organizing work that takes place in hydroponic pots. This platform allows to control the state of the plant by sensing some parameters such as soil moisture, temperature, humidity, light consisting of water sensors: pH, conductivity, and temperature. Hydroponic farming system is an agricultural system that uses direct water as a nutrient without using soil as a planting medium. The use of soil as a growing medium can be replaced with other growing media such as sand, broken bricks, coconut husk, foam, Rockwool, etc. [1]. The use of hydroponic farming systems in developed countries especially in countries with high population levels has been done by individuals as a hobby and also on a large scale for commercial purposes [2]. By looking at the above background, the authors lifted the title of this study as, "Application of Open Garden Sensor on Hydroponic Maintenance Management"

IOP Conf. Series: Journal of Physics: Conf. Series **978** (2018) 012091 doi:10.1088/1742-6596/978/1/012091

## 2. Problem Identification

Technology that can help the management of the treatment is an open garden sensor that can monitor the state of the plant at any time but it takes the system to organize information from open sensors to be displayed to the user especially when the user is away from the plant.

## 3. Previous Research

In 2014, Saputri undertook research for the manufacture of Voice Recognition Application System as the Arduino UNO Electrical Equipment Controller. Created a smart home control that allows humans to control home electrical devices using EasyVR as voice recognition [3]. In 2013, Saenz, et al conducted a study entitled Strawberries Collecting Robot Prototype in the Greenhouse Hydroponic System. In this study Saenz undertook the development of collecting robot strawberry prototypes in a greenhouse hydroponic system, using artificial vision in identifying the maturity and redness of strawberries [4]. By 2015, Ali. M, et al conducted a study on the title of Automated pH Controller System for Hydroponic Cultivation. In this study Ali automated pH levels in aqueous solutions administered by microcontrollers and measured by sensors [5]. In 2016, Siregar, et al. [6] publish research that using Wireless Sensor Network to monitor plant growing. Plant growth is highly depending on the sunlight, if the consumption of sunlight is enough, it will grow well. The plant will be green because of its chlorophyll and it can perform photosynthesis at maximum; but if the plants get less sunlight, it will make the plants be yellowing. In the same year, Siregar, et al. also publish research about air pollution [7], one of the indicators is the measurement of the level of environmental air humidity in a city where the study was conducted. This research also uses Wireless Sensor Network technology.

## 4. Methodology

In this study consists of several steps, which begins with the general architecture of the use of sensors and tools used and work processes of the built system can be seen in Figure 1.

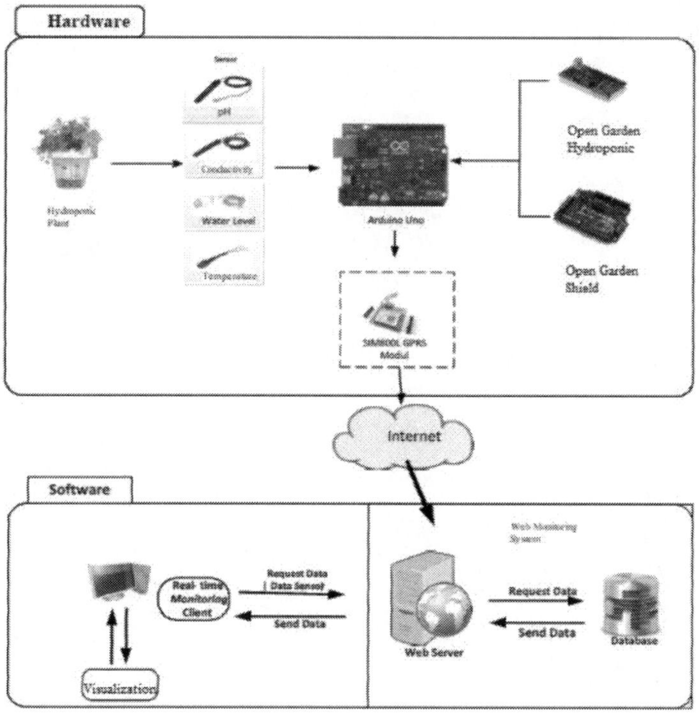


**Figure 1.** General Architecture.

### *4.1. Hardware and Sensor Monitoring System Design*

The elaboration of the stages of the system's general architecture is as follows: At hardware stage shows the process of taking data with the sensor to be sent to Arduino. Submission of data by Arduino to the monitoring application system using SIM module 800L GPRS GSM Module. This process starts from the pH sensor, conductivity, water level and temperature placed on the plant. The sensor will collect data from the plant and sent to the Arduino. Sensors will transmit data via digital pins. The Arduino then calculates the data received from the sensors placed in the plant. Data obtained from the pH sensor, conductivity, water level, air temperature, air humidity, and light level are sent to the database where data collection through the server to be carried out into a real-time monitoring system. Delivery of data to server using SIM800L GPRS GSM Module and connected to server. Hydroponic monitoring application system built using a web-based application system using PHP.

#### 4.1.1. *Web Server*

Web Server serves as a storage and processing data between Arduino, database, and client. The data received by the sensor is sent to the Arduino for processing and sent to the server where the data is collected. The data obtained will be stored into the database and ready to be displayed to the client in graphical form. Graphs are displayed in time with a certain interval and will always move updated automatically every second as long as the Arduino sends data to the server.

#### *4.1.2. Client*

The system is made based on the web to perform monitoring. Client dealing with a page containing the display of sensor data presented in the form of graphs. The data displayed are pH sensors, conductivity, water level, water temperature, air humidity, air temperature, and light level.

### *4.2. System Design*

In designing the system will be designed how the application will display data from the existing sensors in hydroponics plants.

#### *4.2.1. Use Case Diagram*

Use case describes a typical interaction between the user and the system itself with a member of a narrative about how the system is used [6]. System use case diagram to be viewed in the Figure 2.

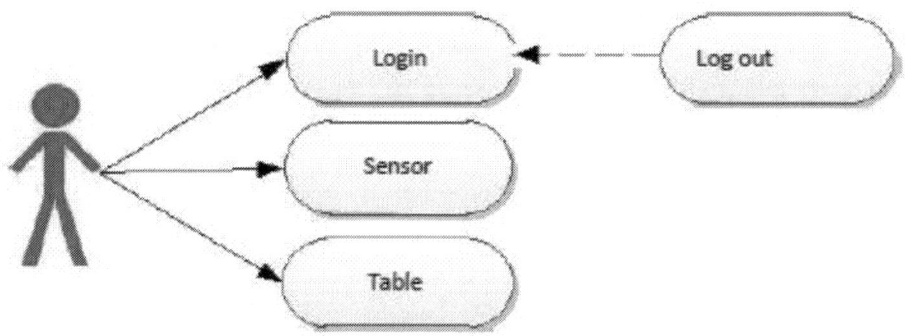


**Figure 2.** Use Case Diagram.

#### *4.2.2. Use Case Specification*

Use case specification is a description of the use case diagram, explaining how a use case it works. Specification of use case to display sensor data.

**Table 1.** Use Case Specification for Main Page.

| Use Case Name | Specification |
|---|---|
| Actor | Computer user. |
| Description | Used by user to view sensor data. |
| Pre-Condition | System turned-on. |
| Characteristic of activation | Could be done by anyone who has given authorization. |
| Basic flow | User sees sensor data at main page. |
| Post condition | User can view sensor data condition of the plant with periods of 15 seconds. |
| Limitations | User can only monitor phenomena that captured by appropriate sensors function. |

*4.2.3. Web Application Page Layout*

After the user successfully login, then the user directly into the main page. On the main page there are buttons labeled Home, Tanaman, Log Sensor, and Log Out. In the middle of the page is a description of the sensor graph for each sensor and indicator lights as a notification whether the value captured by the sensor reaches a low, medium, or high threshold. The design of this main page can be seen in Figure 3.

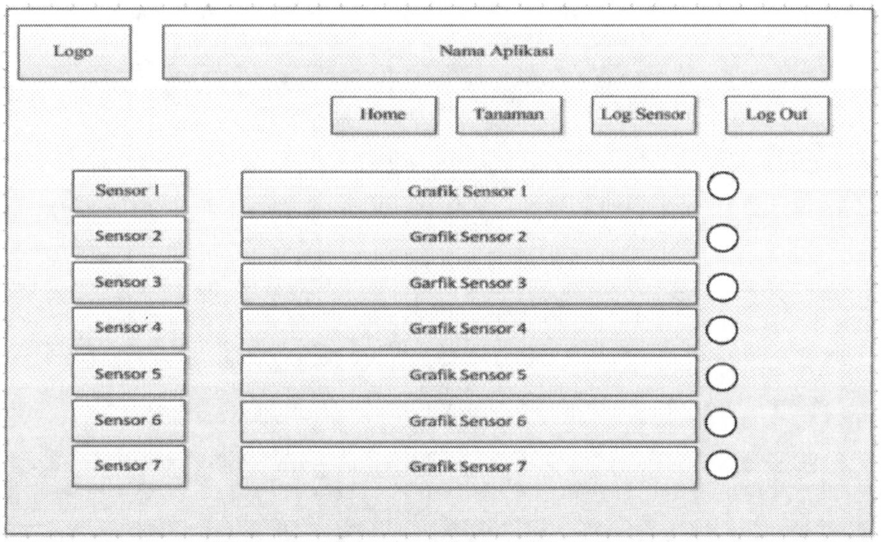


**Figure 3.** Main Page Design View Layout

## 5. Study and Result

Implementation of interface design that has been done previously on the system is the web dashboard that shown running graph periodically can be seen in Figure 4.

IOP Conf. Series: Journal of Physics: Conf. Series **978** (2018) 012091 doi:10.1088/1742-6596/978/1/012091

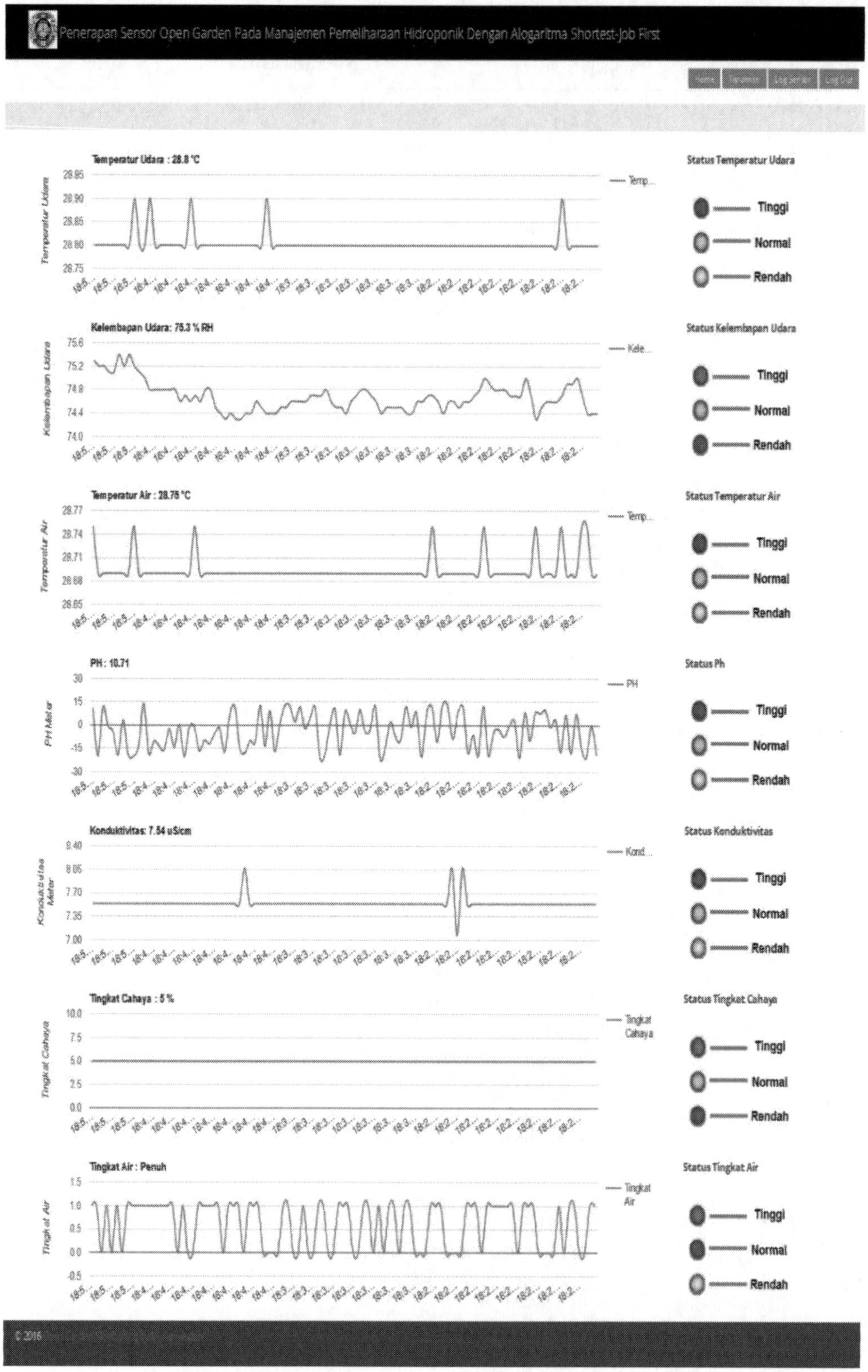


**Figure 4.** Web Dashboard that Shows Running Graph.

In this dashboard page users can see the value of the sensor in the form of a diagram of each sensor. On this page there is also a display of 7 diagrams, the first diagram is an air temperature diagram

IOP Conf. Series: Journal of Physics: Conf. Series **978** (2018) 012091 doi:10.1088/1742-6596/978/1/012091

showing data from the surrounding air. The second is the air humidity diagram showing the high humidity. The third is a water temperature diagram showing the water temperature. The fourth is a pH sensor diagram showing the water in an acid or base pot. The fifth is the sensor conductivity diagram showing the value of electrical conductivity in the solution. The sixth is a light level diagram that shows the darkness of a room. The seventh is a water level diagram showing the dose of water in a pot. On the next page, as shown in Figure 5, the value of numbers obtained from sensors attached to the plant is tabulated. The views on the page will be updated every 15 seconds.

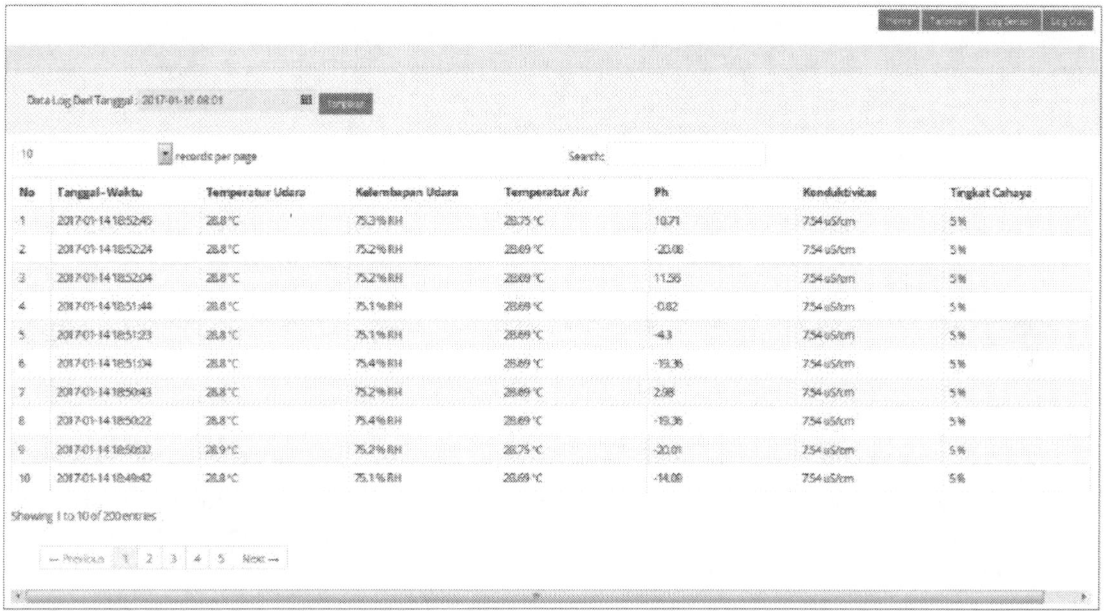


**Figure 5.** Sensor Log Page

Testing of integrity is done to ensure that the phenomenon captured by the plant is successfully delivered and received into the database without any data being changed or corrupt. The test results can be seen in Table 2.

**Table 2.** Data Sensing Integrity

| **Data Integrity of Sensing Results Received by Database Periodically** | **States** |
|---|---|
| Air Temperature | Whole |
| Air Humidity | Whole |
| Water Temperature | Whole |
| pH | Whole |
| Conductivity | Whole |
| Light Level | Whole |
| Water Level | Whole |

Some points on the graph can be highlighted to more details about the captured data of a particular sensor. As the example can be seen in Figure 6, which is more detailed information about the water level in the hydroponic plant container at certain times. For hydroponic plants, water level conditions in the container are necessary to get special attention. Plants in containers should not suffer from shortages or excess water.

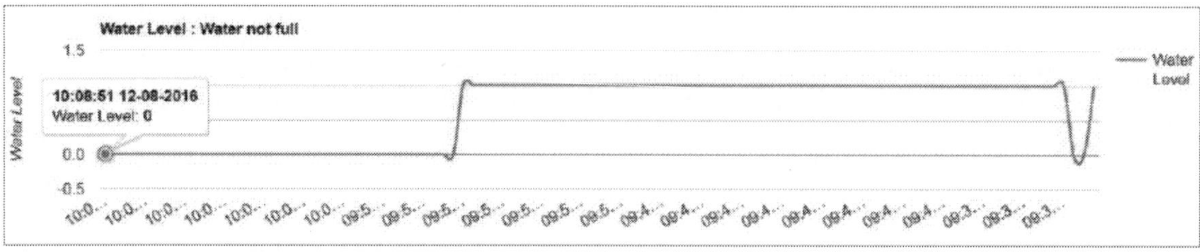


(a)

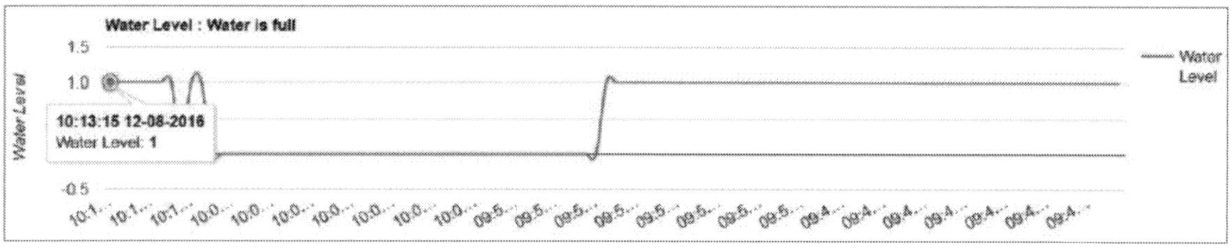


(b)

**Figure 6.** Water Level Indicator: (a) Not Full, (b) Full

## 6. Conclusion

The results of the application testing show that data from sensor readings success put into database and then shown in web dashboard in real time. The time range for data updates is 15 seconds, so the user can do a comparison to study the condition of the plant.

## References

[1] Lingga P 1984 *Pertanaman Ubi-Ubian* (Jakarta: Penebar Swadaya).

[2] Istiqomah S 2013 *Menanam Hidroponik* (Jakarta:Ganeca Exact).

[3] Saputri ZN 2014 *Aplikasi Pengenalan Suara Sebagai Pengendali Peralatan Listrik berbasis Arduino Uno* (Malang: Universitas Brawijaya).

[4] Saenz E, Ramirez M and Ramirez A 2013 Strawberries Collecting Robot Prototype in Greenhouse Hydroponic Systems *Proc. STSIVA*.

[5] Saaid MF, Sanuddin A, Ali M and Yassin MSAIM 2015 Automated pH Controller System for Hydroponic Cultivation in *Proc. IEEE ISCAIE*.

[6] Siregar B, Fadli F, Andayani U, Harahap LA and Fahmi F 2017 Monitoring of Solar Radiation Intensity using Wireless Sensor *J. Phys. Conf. Ser.* vol 801.

[7] Siregar B, Azmi AB and Fahmi F 2016 Integrated Pollution Monitoring System for Smart City *Int. Conf. ICT Smart Soc.*

# Classification of stroke disease using convolutional neural network

**J T Marbun[1], Seniman[2] and U Andayani[3]**

Department Information Technology, Universitas Sumatera Utara, Indonesia

Email: jodymarbun@gmail.com, seniman@usu.ac.id, ulfi.andayani@usu.ac.id

***Abstract*—** Stroke is a condition that occurs when the blood supply stop flowing to the brain because of a blockage or a broken blood vessel. A symptoms that happen when experiencing stroke, some of them is a dropped consciousness, disrupted vision and paralyzed body. The general examination is being done to get a picture of the brain part that have stroke using Computerized Tomography (CT) Scan. The image produced from CT will be manually checked and need a proper lighting by doctor to get a type of stroke. That is why it needs a method to classify stroke from CT image automatically. A method proposed in this research is Convolutional Neural Network. CT image of the brain is used as the input for image processing. The stage before classification are image processing (Grayscaling, Scaling, Contrast Limited Adaptive Histogram Equalization, then the image being classified with Convolutional Neural Network. The result then showed that the method significantly conducted was able to be used as a tool to classify stroke disease in order to distinguish the type of stroke from CT image.

## 1. Introduction

Stroke is a condition that occurs when the blood supply stop flowing to brain because of a blockage or a broken blood vessel. It can happen because there is a plaque in the brain or the patient suffers from hypertension. Stroke can have complications such dementia, afasia and disrupted vision [1]. Stroke can be diagnosed by clinical examination and then followed by Computerized Tomography (CT) Scan by a radiologist. CT Scan will give some image of the brain part that have stroke. In this research, we proposed several image processing techniques and Convolutional Neural Network to enhance the image and classify stroke from CT Scan image.

## 2. Problem Identification

Image that have been produced from CT Scan will be manually checked by a neurologist in order to classify stroke between hemorrhagic and ischaemic which is need a proper lighting to see the image precisely. Therefore, we need a method that can help doctor to classify stroke automatically.

## 3. Previous Research

Previous research by utilizing CT image had been done to identify stroke from CT image using Contrast Limited Adaptive Histogram Equalization (CLAHE) to raise contrast and Statistical Region Merging to segment the image. They used Statistical Feature Extraction to characterize histogram with 4 parameters such as mean, standard deviation, skewness, and kurtosis. It was concluded that their proposed method can clarify head CT scan image significantly and identify stroke well with the statistical measurement [2].

Another method proposed by Chawla to classify stroke. He used Wiener Filtering to remove noise in image, then using Contra-Later Symmetry to change unsymmetry object become symmetry. For the first level of classification, Histogram Features is used to count the histogram value and for the second level of classification, they use Wavelet-based features to count the result of histogram value from the first level of classification. The proposed method classify stroke well with accuracy 90% [3].

Next research is to classify brain hemorrhage from CT Scan image with neural network method like Extreme Learning Machine. Grayscale and Scaling method is used to be the image pre-processing before the image being segmented with Threshold to get the object. Then, Extreme Learning machine classify the image well with accuracy 90% [4].

## 4. Methodology

In this research, there are several steps to classify stroke, namely collecting data of Hemorrhagic and Ischaemic Stroke, Grayscaling to change the RGB image into grey image, Scaling to reduce the pixel of the image, Contrast Limited Adaptive Histogram Equalization (CLAHE) to raise image contrast, Threshold to change image into black and white image , and then using Convolutional Neural Network to classify stroke. General architecture of our proposed method can be shown in Figure 1.

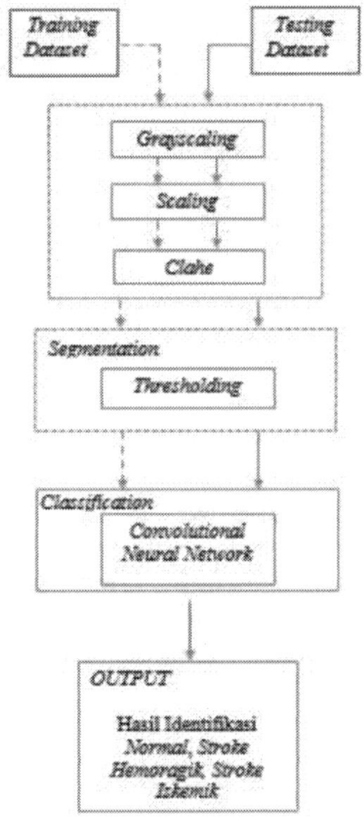


**Figure 1.** General Architecture

### *4.1. Image Acquisition*

Head CT Scan data is acquired from https://radiopaedia.org/ [8]. There are 45 digital images that used for Training and Testing Dataset, 10 images of Hemorrhagic Stroke, 10 images of Ischaemic Stroke, 10 images of Normal brain for Training Dataset and 5 images of Hemorrhagic Stroke, 5 images of Ischaemic Stroke and 5 images of Normal brain for Testing dataset in JPG (Joint Photographic Group) format and then will be used for the next step of the proposed method. A sample of head CT Scan image is shown in figure 2.

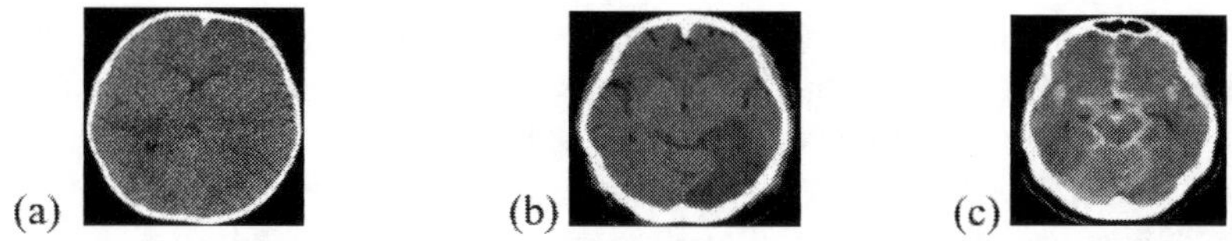

**Figure 2**. (a) Normal brain. (b) Stroke Ischaemic. (c) Stroke Hemorrhaegic.

### *4.2. Pre-processing*

This process consist of three steps, namely Grayscaling, Scaling and CLAHE, which will be clearly explained in the next sub subsection.

- *Grayscaling*. It's a process to convert RGB image into grayscale image. The conversion from RGB into grayscale can be done using equation (1). Figure 3 (b) shows the result of grayscaling process.

- *Scaling*. It's a process to reduce the pixel amount of the image that used as input to neural network. So the computation duration can be shorthened. In this research we reduce the 250x250 image into 30x30 image. Figure 3 (c) shows the result of scaling.

- *CLAHE*. It's a process to raise contrast in image, so the stroke object looks clearer. The conversion from RGB into grayscale can be done using equation (1). Figure 3 (d) shows the result of CLAHE process.

### *4.3. Segmentation*

Segmentation is implemented to distinguish the object and the background. We use Thresholding to segmented image. Thresholding will create a binary image. Each pixel in image will be used as input for Convolutional Neural Network to classify stroke. The Threshold value is set to 170, so:

- If the pixel colour $\leq$ 170, change the colour to black (0).
- If the pixel colour $\geq$ 170, change the colour to white (1). Figure 3(e) shows the result of Thresholding. The binary image obtained from thresholding will be an input for scan line algorithm.

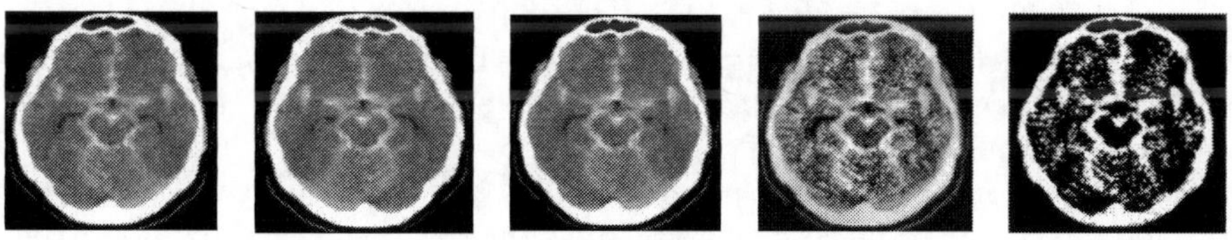

(a) (b) (c) (d) (e)

**Figure 3**. (a) CT Image. (b) Grayscaling. (c) Scaling. (d) CLAHE. (e) Thresholding.

### *4.4. Classification using Convolutional Neural Network*

There are two process of convolutional neural network to classify stroke, namely:

- Training

  Training process is where CNN being trained with 10 data training of each type classification. In training process, CNN consist of two process (feedforward and backpropagation). Feedforward count all the input neuron from input layer in hidden layer. Weights from Hidden

Layer will be sent to Output Layer. Backpropagation will trace the error by counting all the weight from Output Layer and then sent it back to Hidden Layer so the neural network obtained a new weights with minimum error. That two process is done for 1 EPOCH.

- Testing
  Testing process is where CNN being test with 5 data testing of each type classification and compare the weights from data testing with weights that has been obtained from data training. In testing process, CNN only have feedforward process.

We use Learning Rate = 0,2 , Hidden Node = 30 and maximum EPOCH = 1000 as the parameters in CNN for classifying stroke.

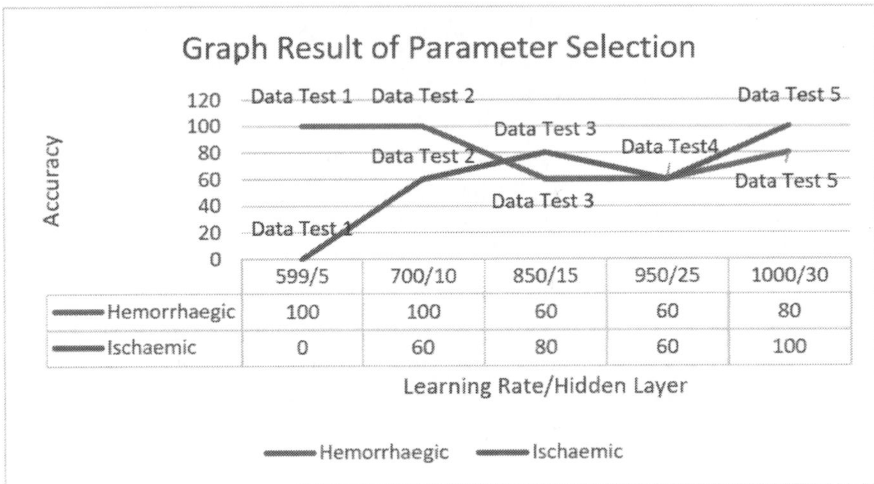


**Figure 4.** Graph Result of Parameter Selection

## 5. Development and Testing

The methods used in clasifying stroke are implemented into the system using C# programming language in accordance with the design that has been done. The images of the selected stroke will then be displayed on the left panel as shown in figure 5 below. After that, the "CLAHE" button is used to implement the CLAHE process to image, image which has been done CLAHE will be shown in the center panel of the system. Then use the “Proses” button to implement Threshold to the image, the threshold image will be shown in the center panel and replace the clahe image. After all that process, press the “Uji” button to get the result of the clasification, the result will be show in the right panel of the system. Figure 5 and figure 6 will show all of that process.

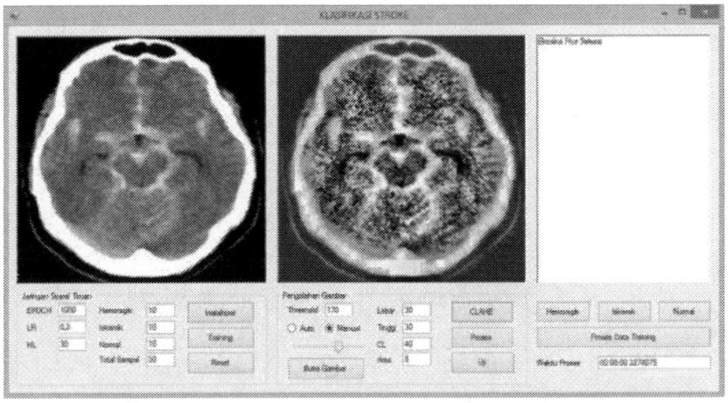

**Figure 5.** Application in Clasifying Process

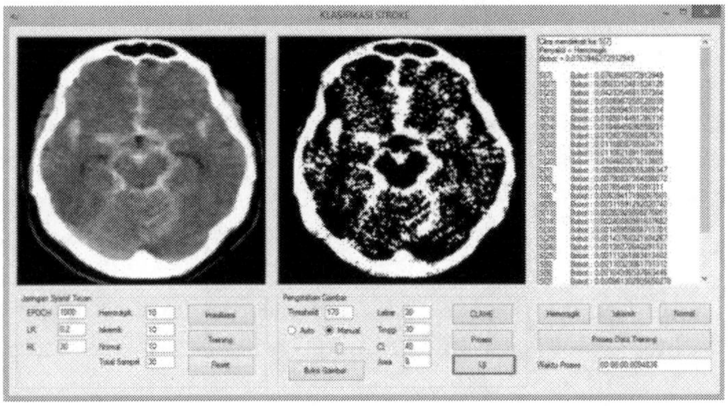

**Figure 6.** Application in Classifying Process (part 2)

The value of the testing process will be calculated in neural network to get the most similar value with the value from training process.

## 6. Result and Evaluation

In this section, we analyze the ability of our proposed method for classifying stroke. The testing result is shown in Table 1. We did 10 test for each image.

**Table 1.** Result

| No. | Name | Classification Result | | Accuracy |
|---|---|---|---|---|
| | | True | False | |
| 1 | Hemoragik1.jpg | 7 | 3 | 70% |
| 2 | Hemoragik2.jpg | 8 | 2 | 80% |
| 3 | Hemoragik3.jpg | 10 | 0 | 100% |
| 4 | Hemoragik4.jpg | 10 | 0 | 100% |
| 5 | Hemoragik5.jpg | 7 | 3 | 70% |
| 6 | Iskemik1.jpg | 10 | 0 | 100% |
| 7 | Iskemik2.jpg | 9 | 1 | 90% |
| 8 | Iskemik3.jpg | 9 | 1 | 90% |
| 9 | Iskemik4.jpg | 10 | 0 | 100% |
| 10 | Iskemik5.jpg | 0 | 10 | 0% |
| 11 | Normal1.jpg | 10 | 0 | 100% |
| 12 | Normal2.jpg | 10 | 0 | 100% |
| 13 | Normal3.jpg | 10 | 0 | 100% |

| | | | | |
|---|---|---|---|---|
| 14 | Normal4.jpg | 10 | 0 | 100% |
| 15 | Normal5.jpg | 10 | 0 | 100% |

Based on Table 1, from 15 images of testing data, there are 5 testing data detected error while classifying stroke, namely "Iskemik5.jpg". So, from all the testing result, we can conclude that the obtained accuracy for classifying stroke using convolutional neural network is 90%.

$$Percentage\ of\ Accuracy\ = \frac{14}{15} x100\%\ = 90\%$$

## 7. Conclusion

Based on the result for classifying stroke from CT head scan image, convolutional neural network is able to help neurologist to classify stroke. The obtained accuracy also depends on the number of acquired data for training dataset. In this research, our proposed method can give 90% of accuracy for testing 15 images of each type of stroke. The classification result much depends on how much images that being used in training process. More images used in training process, the higher the accuracy.

Future research can be done using other methods for classifying stroke so it can be compare with convolutional neural network. Other Feature Extraction methods also can be used to give specific feature of image to the neural network.

## ACKNOWLEDGEMENT

This research was supported by Universitas Sumatera Utara. All the lecturer and staff members of Information Technology Department, whose services turned our research a success.

## REFERENCES

[1] Setyopranoto I 2011 *Stroke: Gejala dan penatalaksanaan* (Yogyakarta: Cermin Dunia Kedokteran) **38**(4) p 247.

[2] Nurhayati OD and Windasari IP 2015 Stroke Identification System on the Mobile Based CT Scan Image *Proc. ICITACEE.*

[3] Chawla M, Sharma S, Sivaswamy J and Kishore LT 2009 A method for automatic detection and classification of stroke from brain CT images *Proc. Annu. IEEE Int. Conf. EMBS* pp. 3581-84.

[4] Rizki DA 2017 *Klasifikasi Pendarahan Otak Menggunakan Extreme Learning Machine* (Medan: Universitas Sumatera Utara).

[5] Faisal, A 1991 *Hasil Pemeriksaan CT Scanning pada Penderita Stroke* (Yogyakarta: Berkala Ilmu Kedokteran).

[6] de Freitas GR, Christoph DDH and Bogousslavsky J 2009 Topographic classification of ischemic stroke *Handbook of Clinical Neurology vol. 93(3)*, ed Fisher M. (Amsterdam: Elsevier).

[7] Koya CHM 2017 Brain Tumor Detection and Classification in MRI Images *Proc. NCACCES.*

[8] Free Educational Radiology Resource. (2017). http://radiopaedia.org (online)

[9] Sutoyo T 2009 *Teori Pengolahan Citra Digital* (Yogyakarta:Andi).

IOP Conf. Series: Journal of Physics: Conf. Series **978** (2018) 012093 doi:10.1088/1742-6596/978/1/012093

# Auditing Albaha University Network Security using in-house Developed Penetration Tool

**M E Alzahrani**

Department of Computer Science/Engineering, College of Computer Science & IT, Albaha University, Saudi Arabia

meid@bu.edu.sa

**Abstract**. Network security becomes very important aspect in any enterprise/organization computer network. If important information of the organization can be accessed by anyone it may be used against the organization for further own interest. Thus, network security comes into it roles. One of important aspect of security management is security audit. Security performance of Albaha university network is relatively low (in term of the total controls outlined in the ISO 27002 security control framework). This paper proposes network security audit tool to address issues in Albaha University network. The proposed penetration tool uses Nessus and Metasploit tool to find out the vulnerability of a site. A regular self-audit using in-house developed tool will increase the overall security and performance of Albaha university network. Important results of the penetration test are discussed.

## 1. Introduction

Network security management becomes a main concern of enterprise network recently. One of its main aims is to prevent unauthorized person accessing their classified information to be used against the organization for further own interest. The bigger the network size the more vulnerable the network from information stealing. The possibility that unauthorized personnel to access classified information should made to as minimum as possible. Thus, the network security should be prioritized and reinforced. It is the only way to guarantee smooth and productive operations.

Human interactions directly or indirectly cause network attacks. Employees themselves pose the biggest threat to enterprise. As an example, employees unintentionally install piracy software that is infected with viruses, worms or Trojans. Or, users may forget to secure their PCs, leaving them open as an easy target to potential attackers. Other users or employees may give sensitive information to outsiders, or even play a role in an important part of an attack.

User may extract classified information from the company's servers or storage devices by gaining physical access to those network devices. The gaining access to network resources from outside enterprise premise is done through a common way network security is compromised, the Internet. Nevertheless, many cases of attacks launched from internal network. Therefore, a security policy should include internal and external threats.

Penetration testing using commercially available automated tool scan help avoid such faults, however the analysis of several popular testing tools reveals significant drawbacks in the performance. Identifying the input vectors of a web application and checking the results of an attack are important parts of penetration testing. Then vulnerability assessments are necessary for discovering potential threats throughout the environment. In general, the scope of a normal vulnerability assessment arrangement is not necessary to be full exploitation of systems and services. Systems are typically enumerated and evaluated for vulnerabilities, and testing can often be done with or without authentication. Most vulnerability management and scanning solutions provide actionable reports that detail mitigation strategies such as applying missing patches, or correcting insecure system configurations.

According to the internal security audit conducted by a vendor on 7 September 2014, Albaha University network still having the following problems, in term of security: IT Department needs more skilled resources, single engineer cannot handle day to day operations

- Security requirements and business criticality for ALBAHA need to be defined and communicated
- More procedures and processes required to be applied for ensuring compliance to ISO27001
- More technical investment is required for Managing and Monitoring performance and capacity
- Incident management need more security controls
- Business continuity or disaster recovery plans need to be planned and tested
- There is no plans to conduct periodic vulnerability assessments or penetrations tests

This work attempts to improve the security of the university network, by developing a penetration test tool and use the tool to conduct self-assessment periodically.

## 2. Related Work

Works by [1-5], discuss steps and techniques commonly used to penetrate a system as follows.

1. Reconnaissance, which is a preparatory phase used to gather more information about the target.
2. Scanning, which is the pre-attack phase to find the basic information.
3. Gaining access, the first penetration phase, used to find holes in the system and get access to highly detailed information.
4. Maintaining access, in which the intruders establish control on an ownership level, leaving part of their presence in the system so as to gain at will access.
5. Clearing their tracks, in which the intruders hide traces of their activity by removing evidence from the system logs.

The methodology is used to gather information, scan for vulnerabilities and initiate penetration mechanisms (such as sniffing, password guessing, backdooring and flooding).

Mukhopadhyay et al. [6] surveyed existing available web penetration tools and introduced a penetration test architecture using Nessus and metasploit tool to perform vulnerabilities scan of a website.

In their survey, they identified a tool to discovery all the known vulnerabilities. They report the increase of number of web applications vulnerabilities. Research work in [7] concluded that existing approaches are based on the Tainted Mode vulnerability model that cannot handle inter-module vulnerabilities.

In another paper two important steps of penetration testing are identifying; the IVs of a web application and determining whether an attempted attack was successful. This paper proposed a new approach to penetration testing that improves both of these steps. The approach incorporates a conservative static analysis of the web application that identifies IVs directly from the application's code. The approach improves the response analysis by leveraging automated dynamic analyses that accurately detect when an attack has succeeded. The proposed steps' performance is compared against two state-of-the-art penetration testing tools and found the result is satisfactory [8].

## 3. Proposed Method

The architecture of the proposed method is shown in Figure 1. The proposed method consists of four main steps. In the first steps (information gathering), we use automated scanning to gain information related to the target application. This information is later used to launch the attack in the second steps

IOP Conf. Series: Journal of Physics: Conf. Series **978** (2018) 012093 doi:10.1088/1742-6596/978/1/012093

(Attack generation). We combine the recognised information and domain knowledge on possible vulnerabilities, to generate attacks. Combination of existing commercial as well as open-source tools are utilized to automate the attacks generation. In the third step (the response analysis), we do the checking whether an attack has succeeded. If so, the information about the attack is logged into a file. The final step is the visualization of the final result of the penetration testing process. The report contains details of the discovered vulnerabilities and corresponding attacks. Based on this information, we can recommend to the system/network manager steps or actions to eliminate the vulnerabilities and improve the security of their network [9].

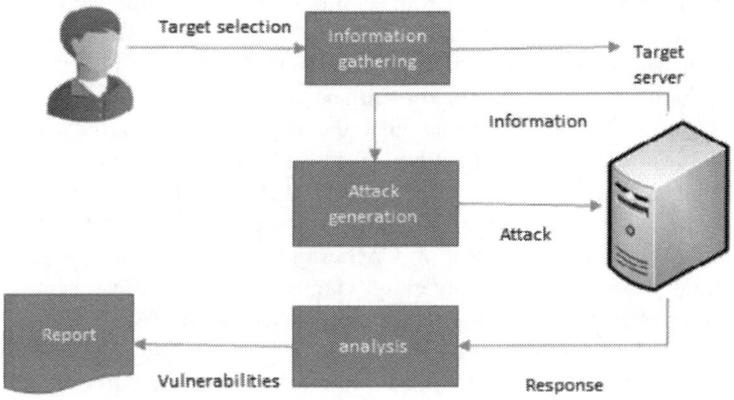


**Figure 1.** The proposed penetration test architecture (Adapted from [6])

### *3.1. Implementation of the Pentest Tool*

The main three techniques in performing penetration test are: ports scanning, gaining access, and maintaining the access. In this work, the scanning is done by using Nessus tool. Having done a vulnerability analysis using Nessus, a list of vulnerability is obtained. The list is used as a basis to launch the attacks using Metasploit tool. The ability to correlate information gained by different types of requests is equally important. Thus, the penetration test tool has to be based on consistent and standard procedure to provide integrated support for detecting a maximum number of vulnerabilities with a minimal false positive number. The visualization module is developed using JavaScript.

### *3.2. Nature of audit*

Following the guidelines setup by ISO27001, the study will include physical inspection, policy study, and a detailed assessment of security measures. The tasks on each element above however limited to data gathering (read-only, no changes), and if active testing is to be performed in the process of data gathering such as probing the university's network, the task is restricted to non-destructive.

Security measures are not limited to investment on tools or devices for protecting the university's assets. A good security planning allocates human resource aspect such as team building, clear set of responsibilities, staff training are some of the examples. Indirectly from this study, this aspect of security is included in the study's scope.

While external test is mainly focused on collection of data in broad and general sense, internal (onsite) audit is revolving around collecting specific and presumably accurate data. The internal auditing is limited to assessing security measures on university's selected data center -- a center that housed all important servers. The main tasks at this stage are divided into three elements, review of security policy, interview of key personnel, and the actual internal pen-test; and typically these tasks were accomplished for over a maximum period of two days.

The first step is to study and review security policy and other related documentations. This task is essential as the policy provides management decisions in regard to security and upon which the auditor can check and verify security practices. Other localized inside information such as network topologies, security management team, physical infrastructure and monitoring measures are often provided by the university.

Interview session guided by prepared checklists with IT Center staff is carried out just after the review of security policy. Typically, this task proceeds concurrently whereby each member of audit team will correspond face-to-face with respective IT Center technical staff. The checklist is service or device specific, and the questionnaires were customized from the standard practices stated in ISO27001 handbook of computer security. Some samples of the questions were extracted from SAN's GIAC program for Security Auditing (Track 5). The questions are of both objective where the answer is of either true/false or yes/no, and when there is question that requires subjective answer a scale from 1 – 10 is used to rate the likeliness of the respondent answer.

The final task of the onsite audit is the actual internal pen-test normally performed once internal information gathering tasks are completed (policy study and interview). The main aim of this task is to test and verify whether the guidelines set forth in the policy have been properly enforced. Common tasks performed at this stage include inspections of filtering devices (routers, firewalls, IDSs) configurations, running through a set of test on Operating Systems (OS) patches, testing on particular services or applications offered by the university, running tools to assess security vulnerability on internal network. Some of these tasks can be accomplished via the usage of automatic tools, while other may requires manual intervention assisted by the IT Center staff. Depending on the univeristy policy in regard to internal pen-test – deploying foreign tools on internal network could be against the university security policy, hence the usage of automated tools is only materialized upon the approval by IT Center management.

## 4. Results

On average Albaha University has an overall performance of up to 30% of the total controls outlined in the ISO 27002 security control framework (see Figure 2).

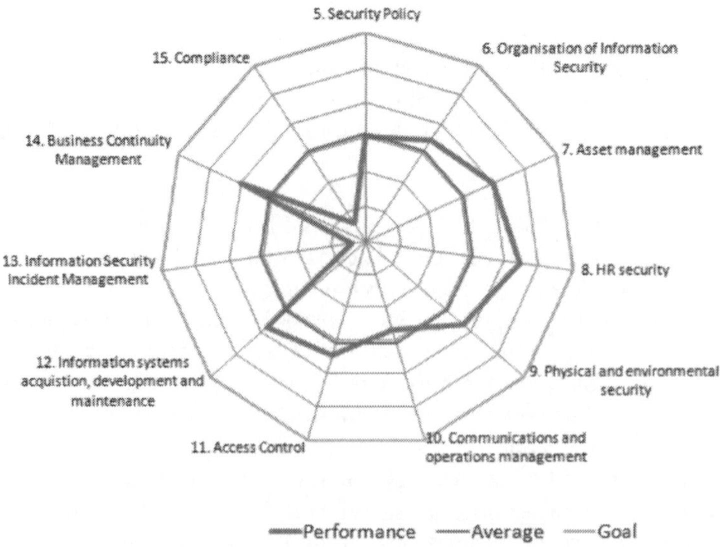


**Figure 2.** Overall performance of Albaha University in terms of ISO/IEC 27002 (Security controls) strengths and conformities.

Overall Vulnerability Assessment Results for Windows based servers and Linux/Unix based servers are shown in Figure 3a and Figure 3b, respectively.

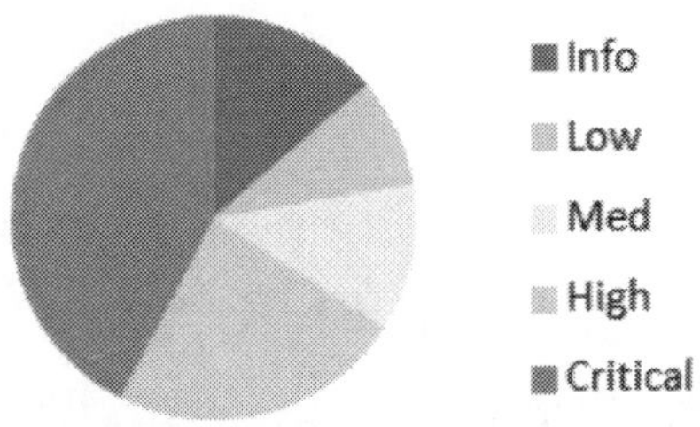


**Figure 3a.** Windows servers vulnerabilities

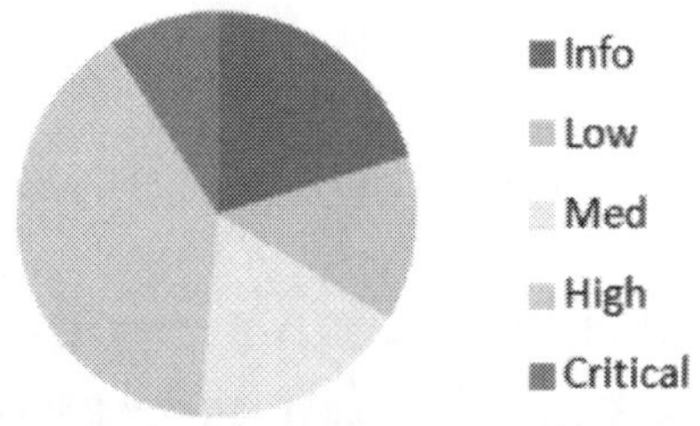


**Figure 3b.** Linux/Unix server vulnerabilities

Among the critical vulnerabilities, include:

- Microsoft Windows Administrator Default Password Detection.
- OpenSSL Heartbeat Information Disclosure (Heartbleed)
- The remote host is running Microsoft Windows XP. Support for this operating system by Microsoft ended April 8th, 2014.

## 5. Conclusion

The proposed penetration tool successfully provided a report on overall network security and performance according to the ISO/IEC 27002 (Security controls) strengths and conformities. The tool helps Albaha University IT Center staff in performing regular self-assessment of their security.

As a future work it is to consider combining Nessus and other metasploit as a holistic penetration tool that can accurately discover more vulnerabilities of a site.

**Acknowledgment**
The author would like to thank Albaha University for the research funding under research contract number 70/1438.

**References**

[1] Stiawan D, Idris MY, Abdullah AH, Aljaber F and Budiarto R 2017 Cyber-attack penetration test and vulnerability analysis *Int. J. Online Eng.* **13**(1) pp 125-32.
[2] Kim W, Jeong OR, Kim C and So J 2011 The dark side of the Internet: Attacks, costs and responses *Inf. Syst.* **36**(3) pp 675-705.
[3] Richard B and Integralis A 2001 Hacking Techniques: The tools that hackers use, and how they are evolving to become more sophisticated *Comput. Fraud Secur.* pp 9-12.
[4] Hansman S and Hunt R 2005 A taxonomy of network and computer attacks, Computers Security *Comput. Secur.* **24** pp 31-43.
[5] Zhijie L, Chongjun W and Shifu C 2008 Correlating multi-step attack and constructing attack scenarios based on attack pattern modeling *Int. Conf. Inf. Secur. Assur.* pp 214-9.
[6] Mukhopadhyay I, Goswami S and Mandal E 2014 Web penetration testing using nessus and metasploit tool *IOSR J. Comput. Eng.* **16**(3) ver IV pp 126-9
[7] Petukhov A and Kozlov D 2008 *Detecting security vulnerabilities in web applications using dynamic analysis with penetration testing* (Moscow: Moscow State University).
[8] Halfond WGJ, Choudhary SR and Orso A 2011 Improving penetration testing through static and dynamic Analysis *Softw. Test. Verif. Reliab.* **21**(3) pp 195-214
[9] Halfond WG, Choudhary SR and Orso A 2009 Penetration Testing with Improved Input Vector identification *IEEE ICST* pp 346-55.

# A Low-Complexity Subgroup Formation with QoS-Aware for Enhancing Multicast Services in LTE Networks

**M Algharem[1], M H Omar[1], R F Rahmat[2] and R Budiarto[3]**

[1]InterNetWorks Research Group, School of Computing, College of Arts and Sciences, Universiti Utara Malaysia, Malaysia
[2]Dept. of Information Technology, Faculty of Computer Science and Information Technology, Universitas Sumatera Utara, Medan, Indonesia
[3]Dept. of Computer Information System, College of Computer Science and Information Technology, Albaha University, Albaha, Saudi Arabia

{algharem,mhomar}@internetworks.my, romi.fadillah@usu.ac.id, rahmat@bu.edu.sa

***Abstract***— The high demand of Multimedia services on in Long Term Evolution (LTE) and beyond networks forces the networks operators to find a solution that can handle the huge traffic. Along with this, subgroup formation techniques are introduced to overcome the limitations of the Conventional Multicast Scheme (CMS) by splitting the multicast users into several subgroups based on the users' channels quality signal. However, finding the best subgroup configuration with low complexity is need more investigations. In this paper, an efficient and simple subgroup formation mechanisms are proposed. The proposed mechanisms take the transmitter MAC queue in account. The effectiveness of the proposed mechanisms is evaluated and compared with CMS in terms of throughput, fairness, delay, Block Error Rate (BLER).

## 1. Introduction

To handle the rapid growth of demand on multicast services such as Internet video streaming, video conferencing, news broadcast, and weather forecasts, Evolved Multimedia Broadcast Multicast Service (E-MBMS) as part of the Third-Generation Partnership Project (3GPP) LTE standard was proposed to deal with multicast services [1]. In E-MBMS, a multirate is used to server several users at the same time using same channel, which efficiently utilizes the network resources. Multicast services can be operated either in Conventional Multicast Scheme (CMS), Opportunistic Multicast Scheme (OMS), multirate (MR), or multilayers (ML). However, in CMS system, the network performance bounded by the user with the worst channel gain (WCG) in the multicast group. The fairness between users is increased on account of wasting network resources. In OMS, in each scheduling time, the best users are served. Then, in next transmission time, the best users among remain users are served. This process is repeated until all users get served. Even OMS utilizes the users diversity, it still waste the bandwidth by transmitting the same data several times. Moreover, it is difficult to use OMS in fast fading environment. To overcome the drawback of the CMS and OMS, the multirate system was used to improve the network resources utilization, as well as enhances the multicast throughput.

In contrary to CMS and OMS, multirate splits the users in each multicast group into several subgroups according to users channels gain. Each subgroup will receive the data at the worst user rate.

As result, different rates will be used to serve each multicast group. Frequently, users will join and leave the subgroup according to their instance channel gain.

In order to enhance the multicast performance, the number of subgroups and the number of users in each subgroup should be selected precisely. Finally, in multilayers, the video is splitted into several layers; base layer (BL) and one or more enhanced layers (ELs). The BL is transmit to all multicast group users, whereas the ELs layers are by the users with better channels gain. In fact, obtaining the best subgroup configuration or layer configuration, which results on maximizing the multicast performance is an NP-hard problem [2]. The complexity of finding the optimal subgroup configuration exponentially depends on the free network resources and on the number of the subgroups. The BL is sent to all subgroups whereas the ELs layers are sent to the subgroups which can decode the EL layers, which in turn, increase the network utilization. Authors in [3-6] proposed resources allocation approach by exploiting the frequency selectivity to schedule the EL to the subgroups that guarantees the best utilization of network resources. Consequently, determining the number of subgroups and the amount of radio resources that should be allocated to each subgroup is still an open issue and need to be properly selected.

In this paper, an innovative Radio Resource Management (RRM) mechanism is proposed to increase the E-MBMS performance by implementing the multirate scheme. The proposed RRM efficiently splits each multicast group into two or more subgroups according to the users channel quality, and allocated the resource to each subgroup in such a way that increases the network resources utilization. The proposed RRM uses two thresholds to split the multicast group into three subgroups (upper, lower, and medium subgroups). These thresholds are selected according to the Standard Division (StD), and the average of users channels quality.

This paper has been divided into following sections. Section 2 provides background of resource allocation in multicast. Section 3 introduces the proposed subgrouping mechanism. After that, simulation results are presented in Section 4. Finally, a conclusion is summarized in Section 5.

## 2. Long -Term Evolution (LTE) Network

Orthogonal Frequency Division Multiplexing (OFDM) is a promising technique for the future generation of wireless communication systems due to its tolerance to interference and fading. LTE networks use Orthogonal Frequency Division Multiplexing Access (OFDMA) which is based on OFDM [7]. However, in OFDMA, the full frequency bandwidth is divided into orthogonal subcarriers, where each subcarrier is allocated 15 kHz. The LTE frame consists of 12 consecutive subcarriers and 10ms duration. Each frame consists of 10 sub-frames; each sub-frame is 1ms, which is identical to the Transmission Time Interval (TTI); and then each sub-frame is equal to two-time slots, where each slot is 0.5 ms in the time domain and 12 subcarriers in the frequency domain. However, each slot is composed of a resource block (RB), which is the minimal radio resource allocation unit in the LTE. Each RB consists of seven symbols when the normal Cycle Prefix (CP) is used or six symbols when the extended CP is used. The extended CP is used in E-MBMS sub-frame [8]. Moreover, to take advantage of the frequency selectivity in the LTE network, each user periodically or non-periodically sends its Channel Quality Indicator (CQI) feedback to its base station which is called Evolved NodeB (eNodeB), as such the eNodeB dynamically selects the Modulation and Coding Scheme (MCS) level for that user which leads to high throughput while maintaining the block error rate (BLER) lower than 1% [9]. The high MCS means high transmission rate. Therefore, there are 15 levels of MCS supported by LTE network, as listed in Table 1.

Multicast data can be transmitted either in multirate or single rate mode. Each transmission mode has its own techniques, which are used to utilize the system resources and select the suitable transmission rate. Until recently, single rate transmission is more commonly used because of its low complexity and simplicity. On other hands, more attentions lately have been paid on multirate [10]. In the single rate mode, such as Conventional Multicast Scheme (CMS), a single data rate is selected to transmit the data to all users belong to the same multicast group. Usually, the data rate is selected according to the user with the worst channel gain (WCG) [11], [12], [13]. Moreover, the data rate can be selected either by using predetermined fixed rate (PFR) [11], [14], by using average multicast group rate (AVG) [12], [15], [16], weighted average rate [11], or target Spectrum Efficiency [13], [12].

However, transmitting the data using the AVG or PFR may result in a high packet loos especially for cell edge users as they cannot decode the AVG rate. Furthermore, for a reliable multicast delivery, the transmission rate should be selected equal to the WCG rate. The WCG mechanism provides high fairness among users belong to the same multicast group on account of wasting the network resources by forcing the users with high channel gain to receive the data with low rate.

Multirate Multicast Scheme (MMS) was proposed to utilize the merit of single rate transmission and the multiuser diversity of OFDMA for reliable data decoded [7]. In MMS, the data are transmitted to all multicast group users using several streams with different data rates. In compare to the CMS, multirate transmission techniques are more computational complexity, challenging in grouping users with high mobility [10]. MMS can be operated using Information Decomposition Techniques (IDT) [17-22], which splits the transmitted data into several sub-streams or layers. Each sub-stream or layer is transmitted using different MCS levels to ensure that each user can receive the data with bit rate equivalent to its MCS level. The base sub-stream or layer is transmitted with a MCS level which can be decoded by all users. The other sub-stream or layer (so-called enhanced substream or layer) will be received by users with good channel condition. There are two types of IDT: Multiple Description Coding (MDC) [17-19 and Hierarchical Layering (HL) [20-22].

The main different between MDC and HL is the sequence priority of received layers, in HL is required while in MDC is not. Another multirate technique is a cooperative multicast technique, in which a user with high gain channel can operate as a relay by forwarding the received packets to the rest users with bad channel condition who could not receive the packets. Cooperative multicast is a new technique, where the first was used in unicast network to improve the system performance. More details about cooperative multicast can be found in [23-26]. Furthermore, MMS can be operated using Multicast Subgroup Formation (MSF) techniques, or called as a group split, [16], [27], which overcome the limitation of CMS by separating users of a multicast group into two or more subgroups according to their channels gain. Thus, users with high channel gain can be formed as a subgroup and received a data with high bit rate, whereas, users with bad channel will receive the data with robust low bit rate. However, our proposed mechanisms are based on MSF techniques.

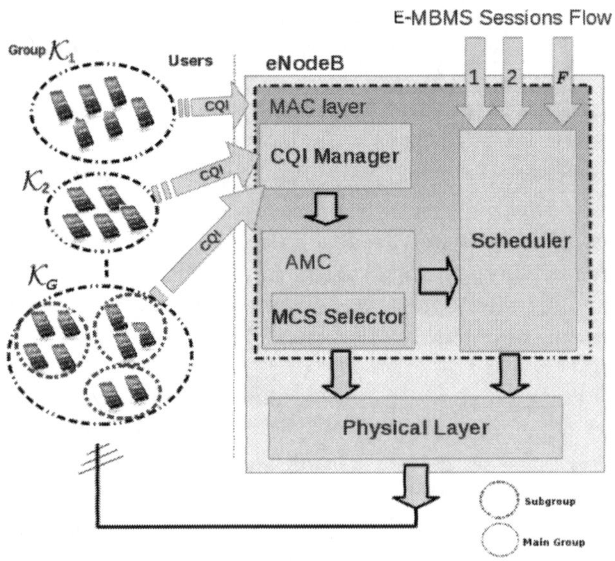


**Figure 1**. Subgrouping Formation Schema.

## 3. System Model and Problem Formulation

In LTE network, there are 15 levels of CQI, each level associated with an MCS level as listed in Table 1. Let $L=15$ is the CQI levels, then $CQI_l=l$, where $l=\{1,2,...,L\}$. Thus, $MCS_i$ is the MCS associated with the $CQI_l$. Any user has sent a $CQI_l$ feedback to its eNodeB, it can successfully receive and decode the transmitted data which is transmitted with the $MCS_i$ level where $i \leq l$.

**Table 1**. CQI indices and their interpretations.

| CQI Index | Modulation | Code rate x 1024 | SE [bit/s/Hz] |
|---|---|---|---|
| 1 | QPSK | 78 | 0.1523 |
| 2 | QPSK | 120 | 0.2344 |
| 3 | QPSK | 193 | 0.3770 |
| 4 | QPSK | 308 | 0.6016 |
| 5 | QPSK | 449 | 0.8770 |
| 6 | QPSK | 602 | 1.1758 |
| 7 | 16QAM | 378 | 1.4766 |
| 8 | 16QAM | 490 | 1.9141 |
| 9 | 16QAM | 616 | 2.4063 |
| 10 | 64QAM | 466 | 2.7305 |
| 11 | 64QAM | 567 | 3.3223 |
| 12 | 64QAM | 666 | 3.9023 |
| 13 | 64QAM | 772 | 4.5234 |
| 14 | 64QAM | 873 | 5.1152 |
| 15 | 64QAM | 948 | 5.5547 |

Let consider $K$ users belong to $G$ groups that receiving the data using $N$ subcarriers over a single eNodeB as illustrated in Figure 1. The set of users, groups, and subcarriers are represents by $\mathcal{K}, \mathcal{G}, \mathcal{N}$ respectively. The system bandwidth $W$ is equally shared between all subcarriers, so the bandwidth of subcarrier $n$ is $B_n = \frac{W}{N}$. In this research, for simplicity reason, we use one multicast group $G=g=1$. Thus, all users belong to one multicast group. Furthermore, we assume each subcarrier $n$ have equal power $P_n$. Assume a perfect orthogonality preserved, and perfect synchronization, so there is no inter symbol interference and no inter-carrier interference. All users receive transmitted data over one or more subcarriers without any interference. It is assumed that the eNodeB using reliable feedback channels to received CQI report from each user without any delay. Let $\mathcal{K}_g$ denotes the $g$ users group set, and the cardinality $\parallel \mathcal{K}_g \parallel$ denotes the number of users in group $g$, where $g = \{1,2,\dots,G\}$. Thus, for multicast $\parallel \mathcal{K}_g \parallel \geq 2$, and $\parallel \mathcal{K}_g \parallel = 1$ in unicast.

Moreover, $\sum_{i=1}^{G} \parallel \mathcal{K}_g \parallel = K$, $\mathcal{K} = \bigcup_{g=1}^{G} \mathcal{K}_g = \mathcal{K}_1 \cup \mathcal{K}_2 \cup \cdots \cup \mathcal{K}_G$. Let $MCS_g^v$ represents the MCS vector of users in group $g$, where $MCS_g^v = \{MCS_1, MCS_2, \cdots, MCS_M\}$, $M = \parallel \mathcal{K}_g \parallel$.

Let also consider $r_{k,n}$is the data rate of user $k$ on subcarrier $n$. The total multicast throughput depends on the number of groups, group rate, and the number of subcarrier allocated to each group. The RRM allocates the system subcarriers to all groups in a way that can maximize the system throughput while keep the fairness between users in acceptable level. According to [17], CMS throughput is bounded by the users with WCG, and will saturate when the users number increases in Rayleigh and Ricean fading environments. As aforementioned, the MMS techniques were emerged to overcome the limitations of the CMS. The MSF maximizes the multicast throughput by spliting the $G$ groups into $S$ subgroups with set $\mathcal{S}$, where $\mathcal{S} = \{1,2,\dots,S\} and\ G \leq S \leq K$. In details, each group $g$ can be splitted into $\mathcal{S}_g$ subgroups, where $\mathcal{S}_g = \mathcal{S}_g^1 \cup \mathcal{S}_g^2 \cup \dots \cup \mathcal{S}_g^l$, and $1 \leq l \leq L$. Then it transmits the data to each subgroup using a WCG rate. To efficiently utilize the multiuser diversity, the MG users are splitted according to their channel gains. Thus, each subgroup $S_g^i$ contains a set of users with same or close channel gains.

$$max \sum_{s=1}^{S} \sum_{n=1}^{N} \sum_{k=1}^{K} r_{s,n} \omega_{s,n} \alpha_{s,k} \tag{1}$$

$$\omega_{s,n}, \alpha_{s,k} \in \{o, 1\}, \forall s \in \mathcal{S}, \forall k \in \mathcal{K}, \forall n \in \mathcal{N} \tag{2}$$

$$\sum_{s=1}^{S} \omega_{s,n} = 1, \sum_{s=1}^{S}, \alpha_{s,k} = 1, \forall s \in \mathcal{S}, \forall n \in \mathcal{N}, \forall g \in \mathcal{G} \tag{3}$$

where $\omega_{s,n}, \alpha_{g,s,k}$ are binary indicators, which indicate either subcarrier $n$ and user $k$ belong to subgroup $s$ or not.

The RRM have to assign subcarriers and users to proper subgroups in order to maximize the total throughput. Thus, problem that is described in equations (1)-(3) is considered as NP-Hard (Non-deterministic Polynomial-time) Hard problem which does not have an optimal solution. Indeed, there is a solution for this problem by using exhaustive search algorithm, but it is usually a high complex and time-consuming computations [33]. Moreover, the complexity of NP-Hard increases exponentially with the number of subcarriers, subgroups, and groups which make it unrealistic to practical used. Furthermore, there are $S^N$ solutions for scheduling radio resources between all subgroups $S$ [2, 6]. Therefore, a lower complexity suboptimal mechanisms with an acceptable performance are usually preferable for practical systems. Nevertheless, this paper introduces simple but efficient suboptimal mechanisms which use the standard deviation (StD) of users channel gains to split each multicast group to three subgroups, worst, best, medium subgroups. The worst subgroup will contain all cell edge users whose MCSs are extremely low. The best subgroup will contain the cell center users whose MCSs are excessively high, whereas the medium subgroup will contain the remain users with medium MCSs levels. Another proposed mechanism split the multicast group users into two subgroups. First subgroup contains cell edge users, whereas the remain users will form the second subgroup.

## 4. The Multicast Packet Scheduling

To ensure that the proposed splitting mechanisms increase the multicast network, the eNodeB MAC queue status and utilized RBs increased as the number of subgroups increased. Moreover, the transmission delay increases as the number of subgroups increase. Therefore, the optimal subgroups number should be determined base on the MAC queue status and available RBs. In details, the Head of Line Delay (HOL), queue size, and available RBs are obtained to determine how many subgroups can be formed from a group g and can be served with the free RBs before the packets delay meet the predefined maximum video delay.

$$T_{free} = HOL - (\frac{qs}{r_g}) \tag{4}$$

Where $T_{free}$ is the free slot might remain after transmitting all packets which waiting in the MAC queue, qs represent the current size of all packets in the MAC queue, and $r_g$ is the selected transmission rate in multicast group $g$. Therefore, if $T_{free}>0$, then group $g$ can be splitted to $S_1$, and $S_2$ subgroups (or clusters). Thus, $S_2$ rate is selected to be able to transmit the data during the $(T_{free})_i$ slots. Then, it will contain users who can receive the data with $S_2$ rate. The remain users will join $S_1$. Moreover, if $T_{free}$ is more than enough to transmit the data to $S_1$, and $S_2$, then $S_1$ will be splitted to get $S_3$. These steps are repeated until $T_{free}=0$. In fact, the number of obtains subgroups is affected by the users channel gain, video rate, video type, available BW, and maximum video target rate. We used the $T_{free}$ in the account when implementing our subgrouping mechanisms.

## 5. The Proposed Subgrouping Mechanisms

In order to satisfy all users, each user should receive the data with its rate, hence; users with the same rate will form a subgroup. As consequently, the maximum number of subgroups is equal to the number of available rate offered by LTE network, which is equal to 15. In fact, the subgroups number will increase as well as the multicast group users increase, as shown in Figure 2a. The reason behind that is, when the number of user increases, the probability to use more different rate levels increased. Unfortunately, the benefit of multicast is reduced as the subgroups number increased (increasing the multiuser diversity on account of multicast gain), as illustrated in Figure 2b. As result, the multicast group throughput decreased as the subgroups number increased.

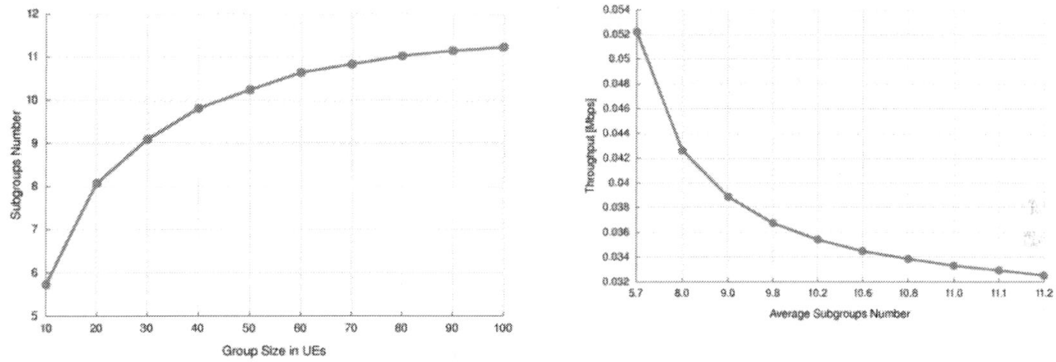


(a) (b)

**Figure 2**. Group-Based Evaluation Results (a) Subgroups Number. (b) Average User Throughput.

Therefore, a trade off between multicast gain and users deversity selection should be applied. In this paper, two subgrouping mechanisms are proposed. First one uses two thresholds (upper and lower thresholds), thus, three subgroups are formed. The second mechanism uses the lower threshold to split users into two subgroups.

### *5.1. StD-based formation mechanism*

This paper introduces a simple but efficient suboptimal mechanism referred as StD-Based, which uses the *StD* of users channel gains to split each multicast group to three subgroups, worst, best, and medium subgroups. The worst subgroup will contain all cell edge users whose MCSs are extremely low. The best subgroup will contain the cell center users whose MCSs are extremely high, whereas the medium subgroup will contain users with medium MCSs levels. As shown in Figure 3, the medium subgroup size is the biggest which means most users form it, whereas the smallest subgroup is the lower subgroup

which contains only few users (usually cell edge users). The proposed subgrouping mechanism uses the StD to show how the users MCS levels are distributed and deviated from the average value of all users MCS level values. The StD with small value mean MCSs of all users are closed together, whereas, the big StD value mean that all users are far from each other. However, in case of the users MCSs standard deviation value is enough small; it means the MCS level of the worst user case will be suitable for all users. The StD and average of users MCSs will be used to divide the multicast users into several categories or subgroups by using upper and lower thresholds. These thresholds will be used to split each multicast group g into three subgroups. The following equations are used to calculate the upper and lower MCS thresholds for each multicast group g.

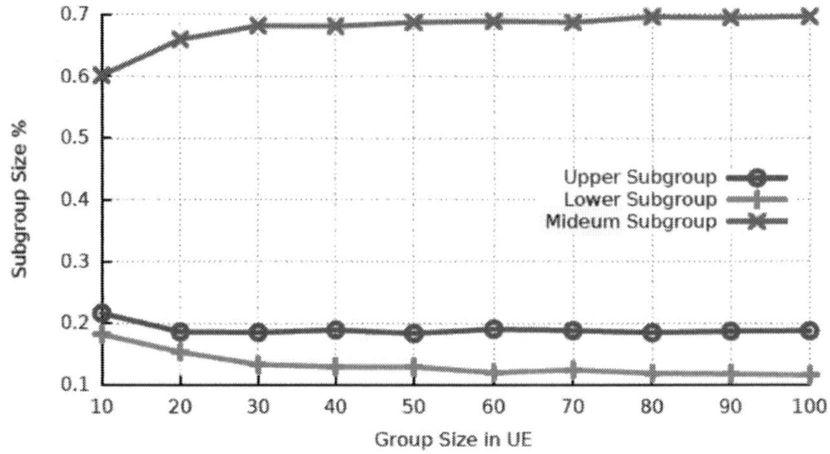


**Figure 3**. Rate Size of Upper, Medium, and Lower Subgroups

$$T_{up}^{g} = \overline{MCS_{g}^{v}} + \sigma_{g} \tag{5}$$

$$T_{low}^{g} = \overline{MCS_{g}^{v}} - \sigma_{g} \tag{6}$$

Then, each group $g$ in $\mathcal{K}$ set will be splitted into three subgroups ($S_{g}^{b}, S_{g}^{m}, S_{g}^{w}$), where

$$S_{w}^{g} = \{MCS_{i} | \ \forall\ MCS_{i} < T_{low}^{g}\},$$
$$S_{b}^{g} = \{MCS_{i} | \ \forall\ MCS_{i} > T_{up}^{g}\},$$
$$S_{m}^{g} = \{MCS_{i} | \ \forall\ MCS_{i} \leq T_{low}^{g}, MCS_{i} \leq T_{up}^{g}\},$$

*5.2. M-WCG-based formation mechanism*

As mentioned by work in [29], the Modified WCG mechanism (M-WCG) sacrifices users with low channel rate, which deviate from remain users channel rates and degrades the multicast performance. The threshold which is used to determine users with low channel gain is obtained from equation 6. In fact, in urinate system, these users with low channel gain cannot receive the data until they improve their channel gain. To overcome the limitation of M-WCG, we will use multirate technique (MR) with our M-WCG (so-called MR-M-WCG) to split the users into two subgroups. The worst users will form a subgroup; whereas, other users will form the second subgroup. The algorithm to select the MCS level is as follows.

**ALGORITHM**: MCS Selection Using Multirate Technique
1: **Start**
2: **Input:** $T_{up}^{i}, T_{low}^{i}, \ MCS_{i}^{v}, \ (i = 1,2,\ldots, g)$
3: **Output:** $MCS_{S}^{i}, S_{b}^{i}, S_{m}^{i}, S_{w}^{i}, \ (i = 1,2,\ldots, g)$
4: **Method:**
5: **Foreach** $g$ in $\mathcal{K}$
6: Calculate($\overline{MCS_{g}^{v}}$);
7: Calculate($\sigma_{g}$);
8: $T_{up} = \overline{MCS_{g}^{v}} + \sigma_{g}$;
9: $T_{low} = \overline{MCS_{g}^{v}} - \sigma_{g}$;
10: **Foreach** $MCS_{i}$ in $MCS_{g}^{v}$;
11: $S_{g}^{w} = \{MCS_{i} | \ \forall\ MCS_{i} < T_{low}^{g}\}$;

IOP Conf. Series: Journal of Physics: Conf. Series **978** (2018) 012094 doi:10.1088/1742-6596/978/1/012094

```
        S_g^b = {MCS_i | ∀ MCS_i > T_up^g};
        S_g^m = {MCS_i | ∀ MCS_i ≥ T_low^g, MCS_i ≤ T_up^g};
    EndForeach
    Foreach S in g
        MCS_g^s = argmin{S};, //select the smallest MCS in each subgroup
        Return MCSpw;
    EndForeach
EndForeach
End
```

## 6. Simulation of the model and results

In this section, the proposed mechanisms performance are evaluated and compared with M_WCG, WCG (or known as CMS), and MCS-Based mechanisms.

### *6.1. Simulation Scenario*

Since we use multirate mode, each multicast group may be splitted to one or more subgroups, several rates of Foreman video sequences (traced files) are used, which can be obtained from a Video Trace Library [30]. The trace file is coded by H.264/AVC coders at 30 fps. At first, we used the Foreman traced file with 242 kb/s rate as a based stream to serve the worst subgroup. Then, we generate several copies of the based stream with different rates by following the steps which recommended by Seeling et al. [31]. First, we scale the frame size to get a new trace file with higher rate than base traced file rate. Then, we wrap around the base trace file, if its end is reached. Therefore, the simulation continues transmits the data from the beginning of the traced file again once it reaches the end. By these steps, we generate a set of video traced files with different bit rates.

The proposed RRM mechanisms have been compared with the CMS in terms of average users throughput, fairness, delay, Block Error Rate, and resources utilization. For simplicity seek, only E-MBMS has been activated in the simulation. Thus, the whole bandwidth was assigned to the E-MBMS. For more accurate, each scenario was performed with a different number of users who uniformly distributed. Each scenario was repeated until obtain 95% confidence interval, then the average results were calculated and plotted. A realistic video trace files with 440 Kbit/s was used, which is available in [30]. The main simulation parameters are selected as defined in [32]. More details about simulation parameters are listed in Table 2.

**Table 2**. Simulation parameters.

| Parameter | Value |
|---|---|
| Carrier Frequency | 2GHz |
| Path loss | 100.54 + 34.1*$\log_{10}$(d), d in km |
| Thermal noise | –174 dBm /Hz |
| Downlink Bandwidth | 3 MHz |
| Symbols for TTI | 12 |
| Sub-Frame Length | 1 ms |
| Frame Type | FDD |
| eNodeB radius | 1 km |
| eNodeB Power transmission | 43 dBm |
| Modulation Schemes | QPSK, 16QAM, 64QAM (dynamic) |
| No. of users | 20-100 users |
| No. of E-MBMS group | 1 |
| User transmission power | 23 dBm |
| User distribution | Randomly and uniformly distributed |
| User speed | 30 km/h |
| User mobility model | Random direction model |

| | |
|---|---|
| CQI scheme | Full Bandwidth |
| Application flows | Video |
| Video rate | 440 kbps |
| Simulation time | 20 Second |
| Shadow fading | Log normal, μ= 0, σ = 8 (dB) |
| Scheduling correlation | 50 m [32] |
| Fast fading | ITU-R PedB (extended for OFDM [32]) |
| TTI | i ms |
| MIMO conclusions | 1 Tx, 2Rx |
| Penetration loss | 20 dB |

### *6.2. Performance metrics*

The proposed mechanisms are compared by using the following performance metrics.

- Resource Utilization Rate indicates the rate of utilized resources that consumed to support the multicast group to deliver its video stream. Thus, It is computed as the number of assigned RBs, during the session, divided by the total number of available RBs. Therefore, the Resource Utilization Rate for group g, referred as $U_g$ can be obtained as follows:

$$U_g = \frac{1}{T*N} \sum_{i=1}^{T} RB_g^i \tag{7}$$

  where $RB_g^i$ is the number of RBs which is assigned to group g in i scheduled time (TTI), and N is the number of RBs that associated with the selected bandwidth (for example, 15 RBs within 1.4 MHz system bandwidth).
- User Fairness Index (FI): this metric is used to measure the how the RBs assigned between the users within a multicast group [17]. The FI is defined as:

$$FI = \frac{(\sum_{k=1}^{K} r_k)^2}{K \sum_{k=1}^{K} (r_k)^2} \tag{8}$$

  where $r_k$ denotes the throughput of the *k-th* user, and FI value varies between (1/K) ≤FI ≤1. The maximum FI=1 can be obtained when all users are served with the same rate.
- Average User Throughput is the cumulative throughput of multicast group g divided by the number of users in that group.
- Delay is calculated as the time taken from transmit the packet until it is received at the target node.

### *6.3. Simulation Results*

The number of subgroups formed in each mechanism is shown in Figure 4. The proposed MR-M-WCG produced a lower BLER compared to other mechanism as shown in Figure 5.

The first evaluation is in term of average user throughput as shown in Figure 6. The proposed MR-M-WCG and StD-based subgrouping mechanisms outperform the CMS and MCS-Based in term of throughput. The StD-based mechanism provides the best throughput among other mechanisms especially, when the users are more than 20. The MCS-Based throughput degrades rapidly as the number of users increased. The rational is because the subgroups number increase as the users of multicast group increasedThus, the resource blocks are distributed between all subgroups, which increases the multiuser diversity in account of reducing the multicast gain.

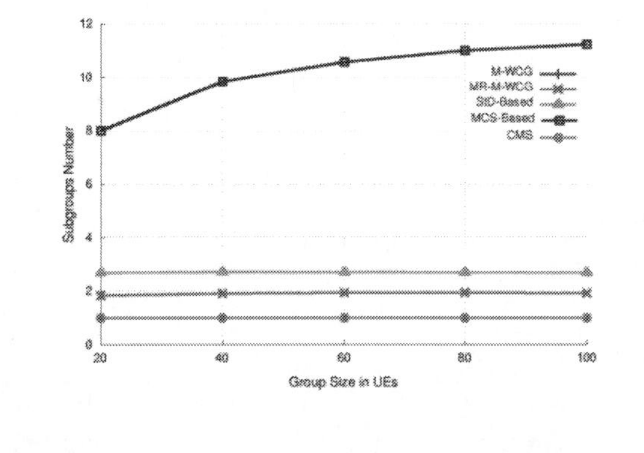


**Figure 4**. Number of subgroups formed in each mechanisms

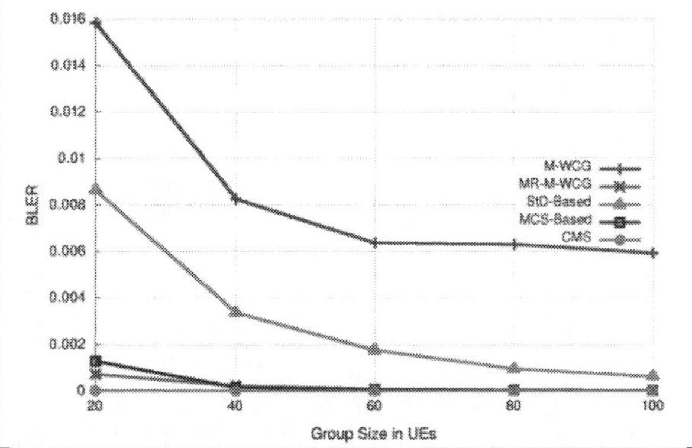


**Figure 5**. BLER between Proposed subgrouping Mechanisms and CMS

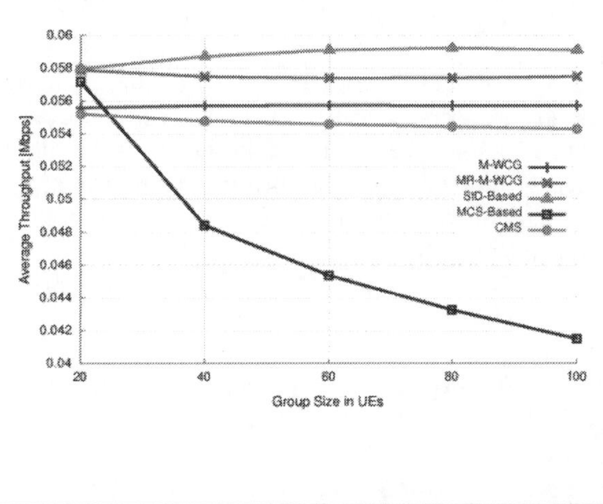


**Figure 6**. Average User Throughput between Proposed subgrouping Mechanisms and CMS

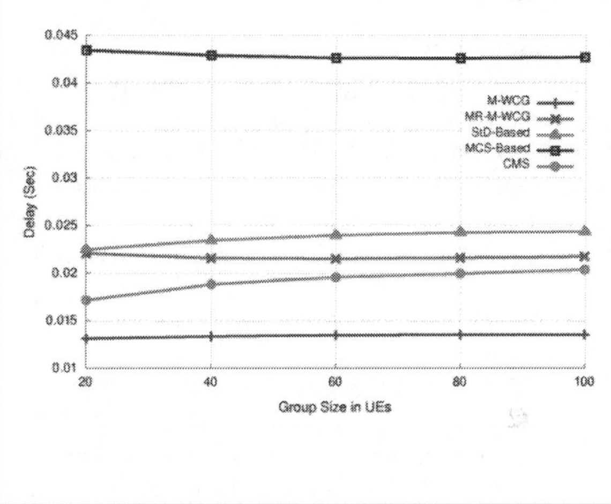


**Figure 7**. Delay of Proposed subgrouping Mechanisms and CMS

As it clearly appears in Figure 7, the lowest delay is achieved by M-WCG since it does not include the abnormal users who suffer very bad channel gain. The highest delay is performed by MCS-Based mechanism. The reasons behind this is because MCS-Based has more subgroups than other mechanisms, which take several TTIs until all its subgroups get served. The of the proposed subgrouping mechanism provide moderate delay.

As shown in Figure 8, the maximum fairness is achieved by CMS. The FI of the proposed subgrouping mechanism is less than FI of CMS. Therefore, the FI of the proposed subgrouping mechanism decreased as the number of users increased, because the subgroups size increases as the users increase. However, the FI of proposed mechanism is near the optimal value because each multicast group was splitted to only three subgroups, which mean three different rates available for all users in the multicast group.

In Figure 9 we can observe that the network resources utilized increases as the subgroup number increased. Therefore, the MCS-Based utilizes all available resources.

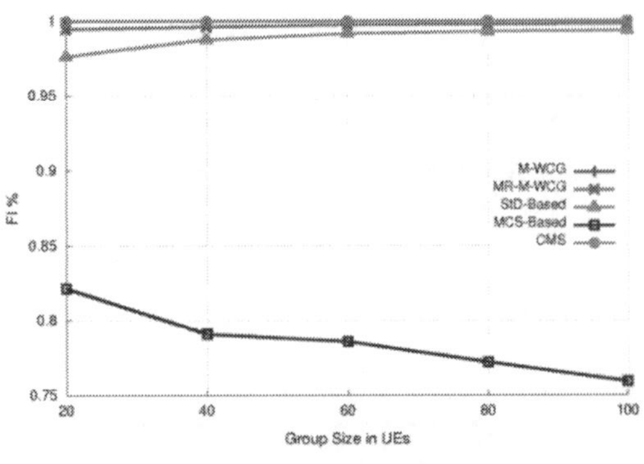


**Figure 8**. FI between Proposed subgrouping Mechanisms and CMS

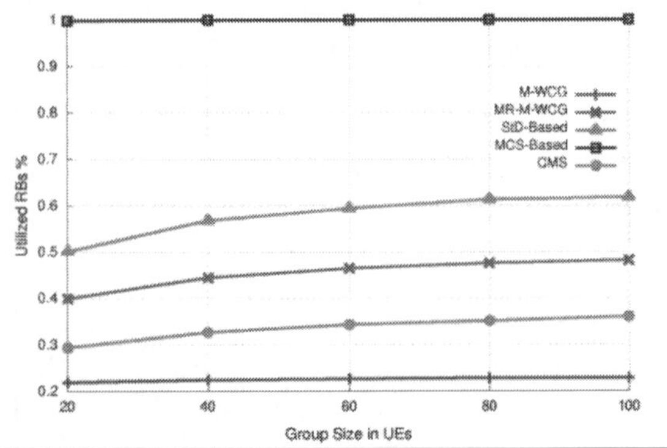


**Figure 9**. Network Resources Utilization Rate between Proposed subgrouping Mechanisms and CMS

## 7. Conclusion

In this paper, an efficient and low complex subgrouping mechanism was proposed to improve the performance of the EMBMS. The SINR of a multicast group users have been used to split the users into several subgroups. The Std and average of the users SINR were used as criteria to classify the users according to their SINR. The simulation results showed that the proposed subgrouping mechanism improves the throughput of the E-MBMS compared to the Conventional Multicast Scheme, while keeping the fairness between all users in acceptable level.

**References**

[1] 3GPP, Multimedia Broadcast/Multicast Service (MBMS); Architecture and functional description, TS 23.246, 3rd Generation Partnership Project (3GPP) (2009). URL http://www.3gpp.org/ftp/Specs/html-info/23246.htm.

[2] Afolabi R, Dadlani A and Kim K 2013 Multicast scheduling and resource allocation algorithms for ofdma-based systems: A survey, *IEEE Commun. Surveys Tutorials* **15**(1) 240–54.

[3] Eusebio P and Correia A 2005 *2nd Int. Symp. Wireless Commun. Syst.* pp 777-81.

[4] Araniti G, Scordamaglia V, Molinaro A, Iera A, Interdonato G and Spano F 2011 *IEEE Int. Symp. BMSB* pp 1-5.

[5] Militano L, Condoluci M, Araniti G and Iera A 2012 *IEEE VTC Fall* pp 1-5.

[6] Condoluci M, Araniti G, Molinaro A and Iera A 2016 Multicast resource allocation enhanced by channel state feedbacks for multiple scalable video coding streams LTE networks **65**(5), pp 2907–21.

[7] Tan C K, Chuah T C and Tan SW 2014 *Comput. Networks* **64** pp 180–194.

[8] Rinne M and Tirkkonen O 2010 *Comput. Commun.* **33**(16) pp 1894–1906.

[9] Mehlführer C, Wrulich M, Ikuno J C, Bosanska D and Rupp M 2009 Simulating the long term evolution physical layer *7th Eur. IEEE Signal Process. Conf.* pp 1471–78.

[10] Afolabi R O and Kim K 2012 *4th ICUFN* pp 255-60.

[11] Kim J and Cho DH 2005 *62nd IEEE VTC-2005-Fall* pp 725-9.

[12] Bochrini S, Bouras C and Kokkinos V 2013 *6th Joint IFIP WMNC* pp 1-8.

[13] Alexiou A, Bouras C, Kokkinos V, Papazois A and Tsichritzis G 2010 Spectral efficiency performance of mbsfn-enabled lte networks *IEEE 6th Int. Conf. WiMob* pp. 361–7.

[14] Agashe P, Rezaiifar R and Bender P 2004 *IEEE Communications Magazine* **42** 2 pp 83-89.

[15] Won H, Cai H, Eun D Y, Guo K, Netravali A, Rhee I and Sabnani K 2009 *Wireless Commun.* **8** 9 pp 4540-49.

[16] Koh C H and Kim YY 2006 *64th IEEE VTC-2006 Fall* pp 1-5.
[17] Suh C and Mo J 2006 *25th IEEE INFOCOM* pp 1-12.
[18] Yoon J, Zhang H, Banerjee S and Rangarajan S 2012 MuVi: a multicast video delivery scheme for 4G cellular networks *ACM Proc. 18th Annu. Int. Conf. Mobile Comput. Network.* pp 209–20.
[19] Shao M, Dumitrescu S and Wu X 2011 *IEEE Trans. Multimedia* **13** 2 pp 353-65.
[20] Deb S, Jaiswal S and Nagaraj K 2008 *27th IEEE Communications Society Conference on Computer Communications INFOCOM* pp 2252–60.
[21] Suh C and Hwang CS 2004 *15th IEEE Int. Symp. PIMRC* **3** pp 2102-6.
[22] Han T and Ansari N 2011 *IEEE Communications Letters* **15** 6 pp 620-2.
[23] Elrabiei S M and Habaebi MH 2010 *21st IEEE Int. Symp. PIMRC* pp 1719-24.
[24] Elrabiei S M and Habaebi MH, 2010 *5th IEEE Int. Symp. Wireless Pervasive Comput.* pp 600-5.
[25] Hou F, Cai I-L X, Ho P-H, Shen X and Zhang J 2009 *IEEE Trans. Wireless Commun.* **8** 3 pp 1508-19.
[26] Shrestha N, Saengudomlert P and Ji Y 2010 *Int. Conf. ECTICON* pp 410-414.
[27] 3GPP, 2013 Evolved Universal Terrestrial Radio Access (E-UTRA); physical layer procedures, TS, 36.213, 3rd Generation Partnership Project (3GPP).
[28] Video trace library. [Online]. Available: http://trace.eas.asu.edu/.
3GPP, 2006 Physical layer aspect for evolved Universal Terrestrial Radio Access (UTRA)," TR, 25.814 3rd Generation Partnership Project (3GPP), 2006.
[29] Algharem M, Omar M H, Alghamdi I, and Budiarto R 2015 An efficient modulation and coding scheme selection mechanism for single-cell mode e-MBMS *2nd Int. Conf. Electrical Eng. Comput. Sci. Inf.* **2**(1).
[30] Seeling P and Reisslein M 2005 Evaluating multimedia networking mechanisms using video traces *IEEE potentials* **24** (4) pp 21–5.
[31] 3GPP 2006 Physical layer aspect for evolved Universal Terrestrial Radio Access (UTRA), TR 25.814, 3rd Generation Partnership Project (3GPP) (Oct. 2006).
[32] Sharangi S, Krishnamurti R and Hefeeda M 2011 *IEEE Trans. on Multimedia* **13** 1 102-15.

IOP Conf. Series: Journal of Physics: Conf. Series **978** (2018) 012095 doi:10.1088/1742-6596/978/1/012095

# Protein modelling of triterpene synthase genes from mangrove plants using Phyre2 and Swiss-model

**M Basyuni[1]*, R Wati[1], N Sulistiyono[1], R Hayati[1], Sumardi[2], H Oku[3], S Baba[4] and H Sagami[5]**

[1] Department of Forestry, Faculty of Forestry, Universitas Sumatera Utara, Jl. Tri Dharma Ujung No. 1 Medan, North Sumatera, 20155, Indonesia
[2] Faculty of Pharmacy, Universitas Tjut Nyak Dhien, Jl. Rasmi No. 28 Medan 20123, Indonesia
[3] Molecular Biotechnology Group, Tropical Biosphere Research Center, University of the Ryukyus, 1 Senbaru, Nishihara, Okinawa 903-0213, Japan
[4]International Society for Mangrove Ecosystems, Faculty of Agriculture, University of the Ryukyus, 1 Senbaru, Nishihara, Okinawa 903-0213, Japan
[5] Institute of Multidisciplinary Research for Advanced Material, Tohoku University, Sendai, 980-8577, Japan

*Email: m.basyuni@usu.ac.id

**Abstract**. Molecular cloning of five oxidosqualene cyclases (OSC) genes from *Bruguiera gymnorrhiza, Kandelia candel,* and *Rhizophora stylosa* had previously been cloned, characterized, and encoded mono and -multi triterpene synthases. The present study analyzed protein modelling of triterpene synthase genes from mangrove using Phyre2 and Swiss-model. The diversity was noted within protein modelling of triterpene synthases using Phyre2 from sequence identity (38-43%) and residue (696-703). *RsM2* was distinguishable from others for template structure; it used lanosterol synthase as a template (PDB ID: w6j.1.A). By contrast, other genes used human lanosterol synthase (1w6k.1.A). The predicted bind sites were correlated with the product of triterpene synthase, the product of *BgbAS* was β-amyrin, while *RsM1* contained a significant amount of β-amyrin. Similarly *BgLUS* and *KcMS*, both main products was lupeol, on the other hand, *RsM2* with the outcome of taraxerol. Homology modelling revealed that 696 residues of *BgbAS, BgLUS, RsM1*, and *RsM2* (91-92% of the amino acid sequence) had been modelled with 100% confidence by the single highest scoring template using Phyre2. This coverage was higher than Swiss-model (85-90%). The present study suggested that molecular cloning of triterpene genes provides useful tools for studying the protein modelling related regulation of isoprenoids biosynthesis in mangrove forests.

## 1. Introduction

Secondary metabolites including triterpenesand phytosterols are common chemical constituents in mangrove forest [1]. A variety of triterpenes and phytosterols extensively distributed in plants are biosynthesized from a precursor 2,3-oxidosqualene, with the contribution of oxidosqualene cyclases (OSCs). 2,3-Oxidosqualene, therefore, situates at the branching point of isoprenoid pathway toward triterpenes or phytosterols biosynthesis [2]. Molecular cloning of OSC genes from mangrove forests namely *Bruguiera gymnorrhiza, Kandelia candel*, and *Rhizophora stylosa* had previously been cloned, characterized, and encoded mono and -multi triterpene synthases [3-4].

Recently it has been reported the bioinformatics analysis of the OSC genes and the amino acid sequence in mangrove plants [5]. Molecular mechanism of salinity tolerance in mangrove plants has been described: the level mRNA of triterpene synthase genes of mangrove leaves and roots increased with increasing salt concentration [6-8]. These studies suggested that triterpenes play an essential physiological role in the adaptation of mangroves trees as the self-protecting barrier against the external salt stress.

Furthermore, OSCs have been attracted our consideration because of their possibility to modify the chemical structures of triterpenoids, as well as, their significance as the primarily committed enzymes in the triterpene biosynthesis [3-4]. In higher plants, OSC family member of triterpenes are involved in triterpenes synthesis, and other OSCs are responsible for sterol biosynthesis. Given the diversity of the triterpene genes found in mangrove plants, it became interesting to understand the function of a protein of each triterpene synthase. However, the predict proteins from mangrove forests and their modelling have not been studied yet. Thus, the present study aimed to describe the protein modelling of triterpene synthase genes from mangrove forests using online software of Phyre2 and Swiss-model.

## 2. Materials and Method

### *2.1. Materials*

A total of five mangrove oxidosqualene cyclase genes deposited in NCBI (https://www.ncbi.nlm.nih.gov/) were analyzed. The GenBank accession numbers of the DNA sequence and amino acid sequence of used these analyses are as follows: AB289585, BAF80443 (*B. gymnorrhiza* BgbAS), AB289586, BAF80444 (*B. gymnorrhiza* BgLUS), AB257507, BAF35580 (*K. obovata* KcMS), AB263203, BAF80441 (*R. stylosa* RsM1), AB263204, BAF80442 (*R. stylosa* RsM2).

### *2.2. Template search, model building, and quality assessment*

Template search was performed using Phyre2 (http://www.sbg.bio.ic.ac.uk/~phyre2/html/page.cgi?id=index) [9] and against the Swiss-model template library (https://swissmodel.expasy.org/) [10]. The *BgbAS, BgLUS, KcMS, RsM1* and *RsM2* were searched with BLAST [12] against the amino acid sequence contained in the Protein Data Bank (PDB). The templates with the highest scoring crystal structure have then been selected for model building. Models are made according to the target-template alignment using ProMod3 [10]. The global and per-residue model quality was assessed using the QMEAN scoring function [12].

### *2.3. Prediction of ligand binding site*

The prediction of ligand binding site in the modelled protein structure was made using 3DLigandSite server (http://www.sbg.bio.ic.ac.uk/3dligandsite/) [13].

### *2.4. Transmembrane helix prediction*

Transmembrane helices were predicted using the sequence to adopt the topology using Phyre2 (http://www.sbg.bio.ic.ac.uk/~phyre2/html/page.cgi?id=index) [9]. The extracellular and cytoplasmic sides of the membrane are labelled and the beginning and stopping of each transmembrane helix illustrated with a number representing the residue index [9].

## 3. Results and Discussion

### *3.1.* Protein modelling using Phyre2 and Swiss-model

Table 1 depicts protein modelling using Phyre2 of five triterpene synthase genes: *B. gymnorrhiza* β-amyrin synthase (*BgbAS*), *B. gymnorrhiza* lupeol synthase (*BgLUS*), *R. stylosa* multifunctional triterpene synthase (*RsM1* produced germanicol (63%), β-amyrin (33%), and lupeol (4%), *R. stylosa* multifunctional triterpene synthase (*RsMS2* produced taraxerol, β-amyrin, and lupeol in the ratio of 70:17:13), and *K. obovata* multifunctional triterpene synthase with product of lupeol (50%), β-amyrin (50%), and α-amyrin (50%).

**Table 1**. Results of protein modelling using Phyre2

| Gene | Template | Sequence Identity | Residue | Confidence | Coverage (%) |
|---|---|---|---|---|---|
| *BgbAS* | c1w6kA | 43 | 696 | 100 | 92 |
| *BgLUS* | c1w6kA | 39 | 696 | 100 | 91 |
| *RsM1* | c1w6kA | 43 | 696 | 100 | 92 |
| *RsM2* | c1w6kA | 38 | 703 | 100 | 91 |
| *KcMS* | c1w6kA | 39 | 696 | 100 | 91 |

The diversity was noted within protein modelling of triterpene synthases using Phyre2 from sequence identity (38-43%) and residue (696-703). The human lanosterol synthase was used for protein model (PDB ID: c1w6kA as template structure with confidence 100% and coverage 91-92%.

Similar results were obtained using Swiss-model as displayed in Table 2. *RsM1* had the relatively highest value in sequence identity, sequence similarity, and range of protein sequence. *BgbAS, RsM2,* and *RsM2* were covered by 0.90, while *BgLUS* and *RsM1* was 0.85. It is noteworthy that Qmean was varied among the genes (0.67-0.70). *RsM2* was distinguishable from others for template structure; it used lanosterol synthase as a template (w6j.1.A). By contrast, other genes used human lanosterol synthase (1w6k.1.A). This result supported the previous result on the high relative molecular weight of *RsM2* (84.8 kDa) [5]. The remaining triterpene synthases had a molecular weight lower (83.4 to 83.7 kDa) than that *RsM2.*

**Table 2**. Results of protein modeling using Swiss-model

| Gene | Template | Sequence Identity | Sequence similarity | Range | Coverage | QMEAN |
|---|---|---|---|---|---|---|
| *BgbAS* | 1w6k.1.A | 43.50 | 0.42 | 23-756 | 0.90 | 0.70 |
| *BgLUS* | 1w6k.1.A | 39.75 | 0.40 | 90-765 | 0.85 | 0.67 |
| *RsM1* | 1w6K.1.A | 44.50 | 0.43 | 93-756 | 0.85 | 0.68 |
| *RsM2* | w6j.1.A | 39.16 | 0.40 | 32-753 | 0.90 | 0.68 |
| *KcMS* | 1w6k.1.A | 41.20 | 0.41 | 54-755 | 0.90 | 0.68 |

### *3.2. Predicted binding site*

The predicted ligands bind particular residue, amino acid, contact, and average distance. *BgbAS* and *RsM1* had the same parameter of the binding site, *BgLUS* and *KcMS* had four parameters, and *RsM2* had six parameters. *BgbAS*had seven residues from 679, 680, 681, 684, and 725 which corresponded to amino acids of Leu, Val, His, Trp, and Thr. The longest distance belongs to Leu679. The contact was varied from 12 to 22 (Table 3). *RsM1* possed five residues Leu679, Val680, His681, Trp684, and Thr725 with contact value was 13-22. The longest distance belongs to Leu679.

In contrast to this observation, *BgLUS* and *KcMS* had Leu678, Val679, Gln680, Ser724 and Leu678, Ile679, His680, and Ser724; respectively. Both had the longest distance of Leu678 and contact values 14-22. *RsM2* was found to have six binding sites: Asn680, Leu681, Val682, Gln683, Trp686, and Thr727. *RsM2* had the contact with 12-24 and longest average distance 0.67 (Table 3).

The predicted bind sites were correlated with a product of triterpene synthase, the product of *BgbAS* was β-amyrin, while *RsM1* contained the significant amount of β-amyrin [4], both genes sit together in the phylogenetic tree [4]. Similarly *BgLUS* and *KcMS,* both main products was lupeol and were close to the branch of lupeol synthase [3, 4]. On the other hand, *RsM2* with the outcome of taraxerol was scattered in the phylogenetic tree [4].

There are three motifs of critical residues of plant OSCs namely MW(L)CYCR, MQSFGSQ, and FIKKSQ motifs. These motifs added the established one: DCTAE and QW motifs in the plant OSCs. The *BgbAS, RSM1* and *RsM2* had the essential residues of Trp257, *BgLUS* and *KcMS* had Leu257 [4].

β-amyrin and lupeol have three critical residues SerPhe in the MQSFGSQ motifs as previously described [4]. The last amino acid residue was Lys449 in the FIKKSQ motifs has been reported to control triterpene synthase product [4].

**Table 3**. Predicted binding site of five triterpene synthases

| Clones | Residue | Amino Acid | Contact | Average distance |
|---|---|---|---|---|
| *BgbAS* | 679 | LEU | 13 | 0.69 |
| | 680 | VAL | 22 | 0.00 |
| | 681 | HIS | 19 | 0.30 |
| | 684 | TRP | 12 | 0.45 |
| | 725 | THR | 22 | 0.00 |
| *BgLUS* | 678 | LEU | 14 | 0.69 |
| | 679 | VAL | 22 | 0.00 |
| | 680 | GLN | 21 | 0.34 |
| | 724 | SER | 22 | 0.21 |
| *RsM1* | 679 | LEU | 14 | 0.69 |
| | 680 | VAL | 22 | 0.00 |
| | 681 | HIS | 21 | 0.34 |
| | 684 | TRP | 13 | 0.50 |
| | 725 | THR | 22 | 0.00 |
| *RsM2* | 680 | ASN | 24 | 0.00 |
| | 681 | LEU | 12 | 0.67 |
| | 682 | VAL | 24 | 0.00 |
| | 683 | GLN | 23 | 0.29 |
| | 686 | TRP | 14 | 0.47 |
| | 727 | THR | 24 | 0.00 |
| *KcMS* | 678 | LEU | 14 | 0.69 |
| | 679 | ILE | 22 | 0.00 |
| | 680 | HIS | 21 | 0.34 |
| | 724 | SER | 22 | 0.21 |

### *3.3. Phyre2 and Swiss-protein model*

Homology modelling revealed that 696 residues of *BgbAS, BgLUS, RsM1,* and *RsM2* (91-92% of the amino acid sequence) had been modelled with 100% confidence by the single highest scoring template using Phyre2 (Figure 1A-E). This coverage was higher than Swiss-model (85-90%) as depicted in Figure 2A-E. The slight different protein model may be due to distinct zonation of mangrove species [5].

It has been shown that the composition of triterpene may be a reflection of the distribution of triterpene synthase in the cells of mangrove plants [4]. Furthermore, a significant proportion of triterpene is found in the external parts of the root and may provide an advance indication for the shielding roles of triterpenoids in mangroves tree species [1]. The invention of several triterpene synthase genes from mangroves [3-4], whose molar ratio is defined by multifunctional OSCs, may be useful to the plant by depiction it more tolerant to the environmental stress, such as salinity stress [3-4,8]. Furthermore, previous works have demonstrated that the expression *PgTPS* terpene synthase increased by salt stress in *Panax ginseng* [14]. Our previous studies also showed salinity-dependent increases in the content of triterpenes and triterpene synthase gene expression in both secreting and non-secreting mangrove roots and leaves [6-8, 15-16].

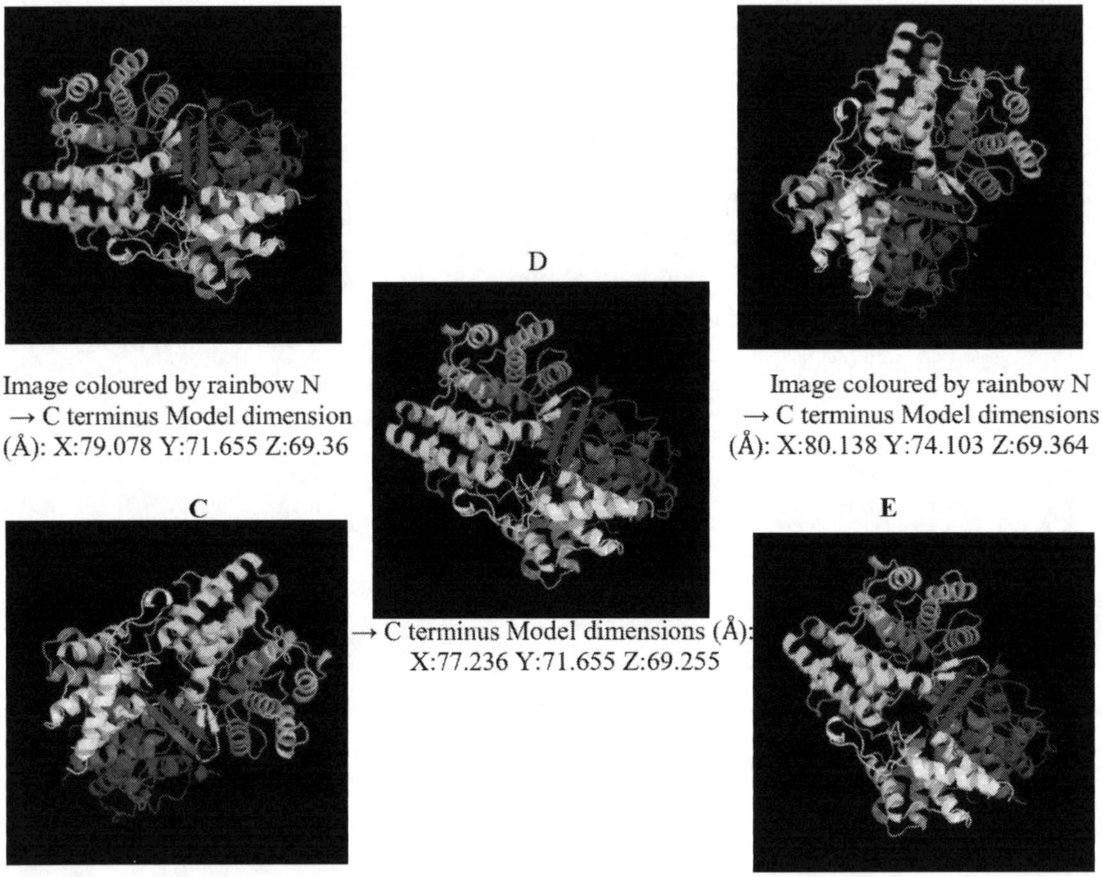

A

Image coloured by rainbow N → C terminus Model dimension (Å): X:79.078 Y:71.655 Z:69.36

B

Image coloured by rainbow N → C terminus Model dimensions (Å): X:80.138 Y:74.103 Z:69.364

C

Image coloured by rainbow N → C terminus Model dimensions (Å): X:79.078 Y:71.004 Z:69.364

D

→ C terminus Model dimensions (Å): X:77.236 Y:71.655 Z:69.255

E

Image coloured by rainbow N → C terminus Model dimensions (Å): X:80.138 Y:74.103 Z:68.968

**Figure 1.** Phyre protein model for *BgbAS* (A), *BgLUS* (B), *RsM1* (C), *RsM2* (D), and *KcMS* (E)

It has been disputed for a long time whether triterpene synthases ensuing different products are discrete proteins or if they are produced post-translational from one gene product to another [3]. This condition because the activities of OSCs are subject to changes in pH, detergents and electrolyte concentrations [3]. Some successful cloning of triterpene synthases has proved the presence of multiple OSCs as distinct proteins even in one plant species [3-4]. These changes may be because of differences in catalytic efficiency and specificity of multifunctional triterpene synthases such as *KcMS, RsM1,* and *RsM2* or other plant multifunctional triterpene synthases by the alteration in tertiary protein structure with different functions by transcriptional regulation [3-4].

*3.4. Transmembrane helices*

Figure 3 shows transmembrane (TM) helices in five triterpene synthases. *BgbAS*, *BgLUS*, *RsM1*, and *KcMS* had four TM helices (S1-S4), only *RsM2* had five TM helices (S1-S5) both in the extracellular and cytoplasmic. These results were parallel with protein model and binding sites of triterpene synthase genes from mangroves.

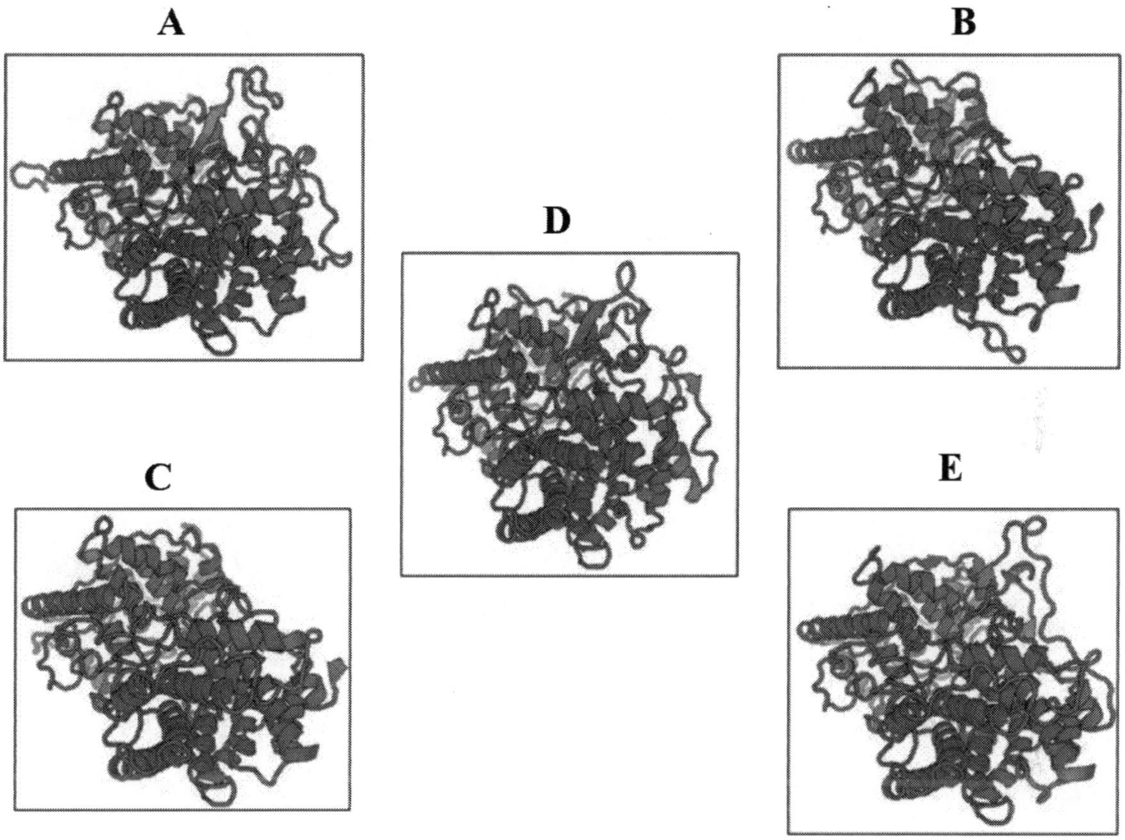


**Figure 2.** Swiss protein model for *BgBAS* (A), *BgLUS*, (B) *RsM1* (C), *RsM2* (D), and *KcMS* (E).

To gain more imminent into structural and functional roles of triterpenes, the well-defined and simple *in vitro* membrane models (liposomes) and *in vitro* study showed that triterpene, as well as phytosterol, was incorporated into the lipid bilayer of liposome [17], suggesting that triterpene was also a modulator of membrane permeability. Triterpene alcohols have been structurally distinguished from phytosterols, even though they share common biosynthetic pathway and hence the similar chemical structure. However, there has been no scientific basis to functionally identify these compounds as the structural membrane lipid [16-17]. Changes in the membrane lipid composition and properties represent an essential factor in the adaptation to high salt concentration [18]. These observations strongly suggested that the triterpene and triterpene synthase contribute to the salt tolerance of mangrove plants probably by changing the property of the cell membrane [6-8].

Recently, it has been revealed that the expression of two triterpene synthases, *BgbAS* and *RsM1* under a GAL1 promoter in GIL77 increased the triterpene content of both whole cell body and plasma membrane fractions [19]. In previous results [5], both gene *BgbAS* and *RsM1* were placed at the plasma membrane, supported earlier results on their subcellular localization situated in the plasma membrane [19].

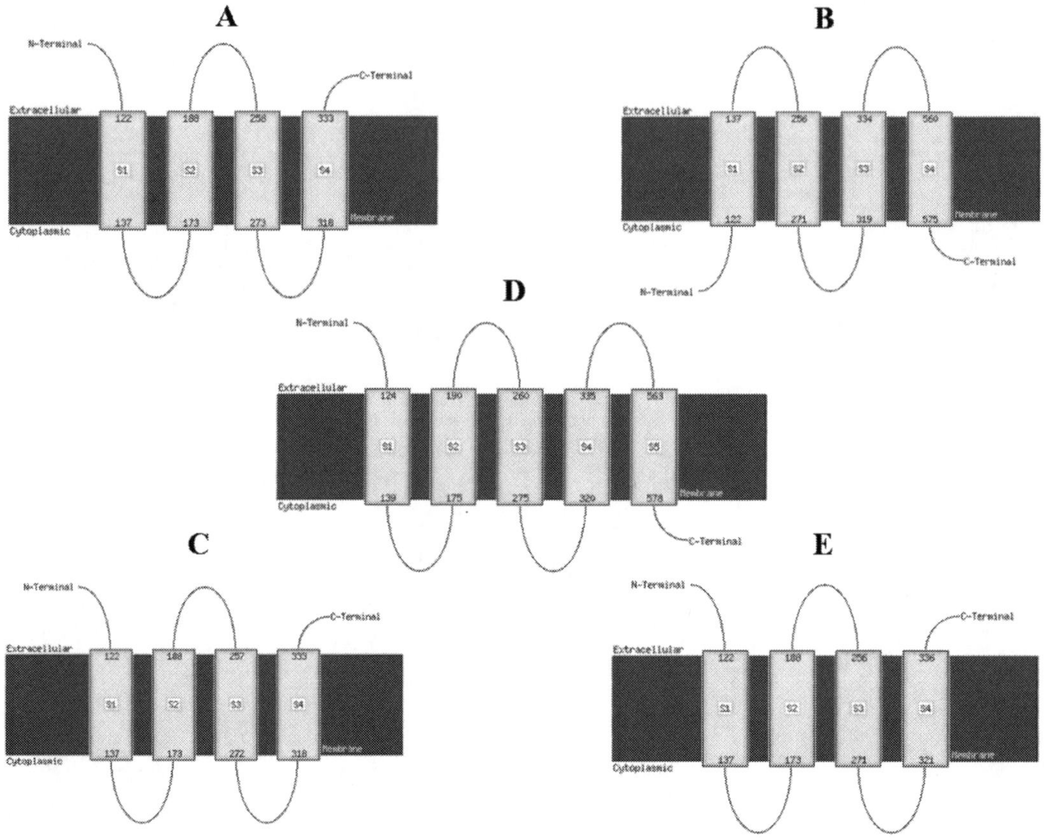


**Figure 3**. Transmembrane helices of *BgBAS* (A), *BgLUS* (B), *RsM1*(C), *RsM2* (D), and *KcMS* (E)

## 4. Conclusion

Molecular cloning of triterpene genes provides useful tools for studying the protein modelling related regulation of isoprenoids biosynthesis in mangrove plants. Some successful cloning of triterpene synthases has proved the presence of multiple OSCs as distinct proteins even in one plant species. The present study suggested that the triterpene and triterpene synthase contribute to the salt tolerance of mangrove forests probably by changing the property of the cell membrane.

## Acknowledgment

This study was supported by an International Research Collaboration and Scientific Publication Grant 2017 (No. 003/SP2H/LT/DRPM/IV/2017) from the Directorate for Research and Community Service, Ministry of Research, Technology and Higher Education, Republic of Indonesia.

## References

[1] Basyuni M, Oku H, Baba S, Takara K and Iwasaki H 2007 Isoprenoids of Okinawan mangroves as lipid input into estuarine ecosystem *J. Oceanography* **63** 601–8.

[2] AbeI, Rohmer M and Prestwich GD 1993 Enzymatic cyclization of squalene and oxidosqualene to sterols and triterpenes *Chem. Rev.* **93** 2189–2208.

[3] Basyuni M, Oku H, Inafuku M, Baba S, Iwasaki H, Oshiro K, Okabe T, Shibuya M and Ebizuka Y 2006 Molecular cloning and functional expression of a multifunctional triterpene synthase cDNA from a mangrove species *Kandelia candel* (L.) Druce *Phytochemistry* **67** 2517–2524.

[4] Basyuni M, Oku H, Tsujimoto E, Kinjo K, Baba S and Takara K 2007 Triterpene synthases from the Okinawan mangrove tribe, Rhizophoraceae *FEBS J.* **274** 5028–5042.

[5] Basyuni M, R Wati 2017 Bioinformatics analysis of the oxidosqualene gene and the amino acid sequence in mangrove plants *J. Phys. : Conf. Ser.* **801** 012011.

[6] Basyuni M, Baba S, Inafuku M, Iwasaki H, Kinjo K, Oku H 2009 Expression of terpenoid synthase mRNA and terpenoid content in salt-stressed mangrove *J. Plant Physiol.* **166** 1786–1800.

[7] Basyuni M, Baba S, Kinjo Y, Putri LAP, Hakim L, Oku H 2012 Salt-dependent increase in triterpenoids is reversible upon transfer to fresh water in mangrove plants *Kandelia candel* and *Bruguiera gymnorrhiza J. Plant Physiol.* **169** 1903–1908.

[8] Basyuni M, Baba S, Kinjo Y, Oku H 2012 Salinity increases the triterpenoid content of a salt secretor and a non-salt secretor mangrove *Aquat. Bot.* **97** 17–23.

[9] Kelley LA, Mezulis S, Yates CM, Wass MN, Sternberg, MJE 2015 The Phyre2 web portal for protein modeling, prediction and analysis *Nat. Protocols* **10** 845–858.

[10] Arnold, K., Bordoli, L., Kopp, J. and Schwede, T. 2006 The SWISS-MODEL workspace: a web-based environment for protein structure homology modelling *Bioinformatics* **22** 195–201.

[11] Altschul SF, Madden TL, Schaffer AA, Zhang J, Zhang Z, Miller W, Lipman DJ 1997 Gapped BLAST and PSI-BLAST: a new generation of protein database search programs *Nucleic Acids Res.* **25** 3389–3402.

[12] Benkert P, Biasini M, Schwede T 2011 Toward the estimation of the absolute quality of individual protein structure models *Bioinformatics* **27** 343–350.

[13] Wass M.N., Kelley LA, Sternberg MJ 2010 3DLigandSite: predicting ligand-binding sites using similar structures *Nucleic Acid Res.* **38** W469–473.

[14] Kim YJ, Ham AR, Shim JS, Lee JH, Jung DY, In JG, Lee BS, Yang DC 2008 Isolation and characterization of terpene synthase gene from *Panax ginseng J. Ginseng Res.* **32** 114–119.

[15] Basyuni M, Kinjo Y, Baba S, Shinzato N, Iwasaki H, Siregar EBM, Oku H 2011 Isolation of salt stress tolerance genes from roots of mangrove plant, *Rhizophora stylosa* Griff., using PCR-based suppression subtractive hybridization *Plant Mol. Biol. Rep.* **29** 533–543.

[16] Oku H, Baba S, Koga H, Takara K, Iwasaki H 2003 Lipid composition of mangrove and its relevance to salt tolerance *J. Plant Res.* **116** 37–45.

[17] Basyuni M, Oku H, Tsujimoto E, Baba S 2007 Cloning and functional expression of cycloartenol synthases from mangrove species *Rhizophora stylosa* Griff. and *Kandelia candel* (L.) Druce *Biosci. Biotechnol. Biochem.* **71** 1788–1792.

[18] Turk M, Mejanelle L, Šentjurc M, Grimalt JO, Gunde-Cimerman N, Plemenitaš A 2004 Salt-induced changes in lipid composition and membrane fluidity of halophilic yeast-like melanized fungi *Extremophiles* **8** 53–61.

[19] Inafuku M, Basyuni M, Oku H 2016 Triterpenoid modulates the salt tolerance of lanosterol synthase deficient *Saccharomyces cerevisiae,* GIL77 *Saudi J. Biol. Sci.* http://dx.doi.org/10.1016/j.sjbj.2016.10.009.

IOP Conf. Series: Journal of Physics: Conf. Series **978** (2018) 012096 doi:10.1088/1742-6596/978/1/012096

# Fluid flow and heat transfer characteristics of an enclosure with fin as a top cover of a solar collector

**H Ambarita[1*], A D Ronowikarto[1,2], R E T Siregar[1,2] and E Y Setyawan[3]**

[1]Sustainable Energy and Biomaterial Centre of Excellent, Universitas Sumatera Utara, Jl. Almamater Kampus USU, Medan 20155, Indonesia
[2]Mechanical Engineering Department, Faculty of Engineering, Universitas Sumatera Utara, Jl. Almamater Kampus USU, Medan 20155, Indonesia
[3]Mechanical Engineering, National Institute of Technology Malang, Malang 65145, Indonesia

*Email: himsar@usu.ac.id

**Abstract**. To reduce heat loses in a flat plate solar collector, double glasses cover is employed. Several studies show that the heat loss from the glass cover is still very significant in comparison with other losses. Here, double glasses cover with attached fins is proposed. In the present work, the fluid flow and heat transfer characteristics of the enclosure between the double glass cover are investigated numerically. The objective is to examine the effect of the fin to the heat transfer rate of the cover. Two-dimensional governing equations are developed. The governing equations and the boundary conditions are solved using commercial Computational Fluid Dynamics code. The fluid flow and heat transfer characteristics are plotted, and numerical results are compared with empirical correlation. The results show that the presence of the fin strongly affects the fluid flow and heat transfer characteristics. The fin can reduce the heat transfer rate up to 22.42% in comparison with double glasses cover without fins.

## 1. Introduction

Nowadays the world is facing energy problems includes global warming, lack of energy supplies and environmental concerns as air pollution. Boltzmann [1] concluded that the struggle for life is not a struggle for basic elements or energy, but a struggle for the availability of negative entropy transfer from the hot sun to the cold earth. World populations are expected to double by the middle of the 21st century [2]. To mitigate these problems, fossil fuel replacement with renewable energy is proposed. These problems require energy management as a method of energy use in an efficient and environmentally friendly way. Kabir et al. concluded [3], only three renewable energy sources (i.e., biomass, geothermal, and solar) can be utilized to yield sufficient heat energy for power generation. Of these three, solar energy exhibits the highest global potential since geothermal sources are limited to a few locations, and the supply of biomass is not ubiquitous [4].

There is a need to develop an ingenious method of solar energy conversion systems and then to substitute it where applications of fossil fuels are most vulnerable. One of the most widely developed today is solar energy. Solar energy has been widely investigated for water heating process [5,6], solar cooker [7,8], drying process [9], solar refrigeration [10,11], solar desalination [12,13,14] and so forth. To absorb the heat of the sun and convert it into a source of energy, a solar collector is needed. The main source of a solar collector is the collector itself. Fudholi et al. [15] had concluded that the solar collector

IOP Conf. Series: Journal of Physics: Conf. Series **978** (2018) 012096 doi:10.1088/1742-6596/978/1/012096

is the vital component of solar thermal systems. In the solar thermal system, the sun heats the collector, and the heat is then transferred to a working fluid. Numerous studies on solar collectors have been reported, and the focuses varied widely. Pinto et al. [16] had concluded, the natural convection study in cavities with inner bodies has been of great interest nowadays due to several engineering applications includes solar collector application. Maliska [17] concluded the advanced in knowledge computational in fluid dynamics (CFD) capability has contributed to more sophisticated equipment development with highest performance levels. Karim and Hawlader [18] studied the mathematical model based on the thermal performance of the finned and V-corrugated absorber plates. The study determined the factors that affect collector performance, such as the optimum angle of the triangular collector and the effect of changes in the shape of absorber plates. The analysis was based on Rayleigh number, inclination angle, AR(aspect ratio) and wavelength laminar. Numerical studies related to solar collector were reported by Beghein et al. [19] who analyzed the influence of buoyancy force on heat or mass transfer and Lewis number in a square cavity.

Utilization of Computational Fluid Dynamic (CFD) commercial code to optimize the design of a flat-plate type solar collector has come under scrutiny. Martinopoulos et al. [20] used CFD to investigate the behavior of a polymer solar collector. Selmi et al. [21] employed CFD commercial code to simulate heat transfer process in a flat-plate type solar collector. Recently, Ambarita [22] explore the characteristics of the enclosure of solar collector double glass cover. The above-reviewed studies show that heat transfer analysis in the enclosure of double glass cover plays an important role in designing highly efficient solar collector. One of the potential innovation to increase the performance of the solar collector is the finned enclosure. This paper focuses on numerical analysis in the finned enclosure between double glass cover of a solar collector. The objective is to explore the heat transfer and flow characteristics in the enclosure with fin. The results are expected to support the necessary information in developing high-performance flat-plate type solar collector.

## 2. Method

Our research group Sustainable Energy and Biomaterial Centre of Excellent, Faculty of Engineering, Universitas Sumatera Utara is developing high-performance flat-plate solar collector for several applications. Based on the previous studies, it was concluded that the main heat loss from the collector is the top cover. To decrease the loss double glass cover with fin is proposed. Typical of the flat-plate type solar collector is shown in figure 1. The focus of the present study is the comparison between double glass top cover non-finned and finned solar collector.

### *2.1. Numerical Method*

In this study, only the enclosure between the upper and bottom glasses is taken account into domain. The computational domain is shown in figure 1b (enclosure without fin) and figure 1c (enclosure with fin).

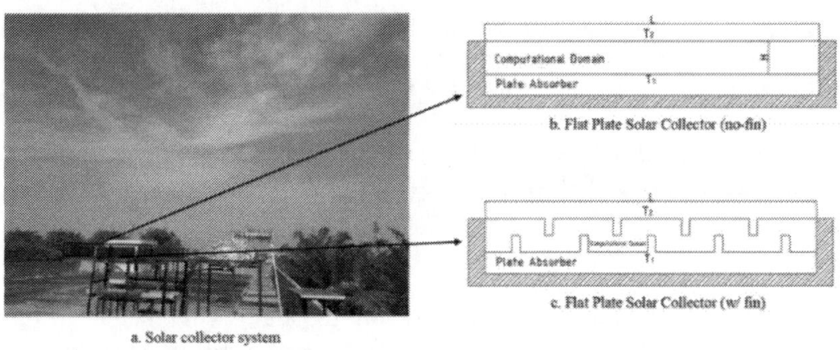


**Figure 1.** Photograph of the system and computational domain

In the present analysis, the domain is assumed to be the two-dimensional case. The flow in the domain is incompressible laminar flow and steady state condition. The additional assumptions are: there is no viscous dissipation, the gravity acts in the vertical direction, and fluid properties are constant except in density of the fluid. The Boussinesq approximation is employed to model the buoyancy force. By using those assumptions, the governing equations are presented below.

$$\frac{\partial u}{\partial x}+\frac{\partial v}{\partial y}=0 \tag{1}$$

$$u\frac{\partial u}{\partial x}+v\frac{\partial u}{\partial y}=-\frac{1}{\rho}\frac{\partial P}{\partial x}+\frac{\mu}{\rho}\left(\frac{\partial^2 u}{\partial x^2}+\frac{\partial^2 u}{\partial y^2}\right)+g\beta(T-T_r)\sin\phi \tag{2}$$

$$u\frac{\partial u}{\partial x}+v\frac{\partial u}{\partial y}=-\frac{1}{\rho}\frac{\partial P}{\partial y}+\frac{\mu}{\rho}\left(\frac{\partial^2 v}{\partial x^2}+\frac{\partial^2 v}{\partial y^2}\right)+g\beta(T-T_r)\cos\phi \tag{3}$$

$$u\frac{\partial T}{\partial x}+v\frac{\partial T}{\partial y}=\alpha\left(\frac{\partial^2 T}{\partial x^2}+\frac{\partial^2 T}{\partial y^2}\right) \tag{4}$$

Where $u$ and $v$ are the velocity vector in $x$- and $y$-directions, respectively. In the analysis, non-dimensional parameters are stated in the followings.

$$Ra=\frac{g\beta\Delta TH^3}{\nu^2}\Pr \tag{5}$$

Where Pr is Prandtl number and calculated by

$$\Pr=\frac{\nu}{\alpha} \tag{6}$$

Convective heat transfer coefficient ($h$) is calculated using the non-dimensional Nusselt number. It is given by

$$h=\frac{Nu\times k}{H} \tag{7}$$

Where $k$ [W/mK] is the conductive heat transfer coefficient of the air and $H$ [m] is the distance between the double glasses. The Nusselt number is presented using local and average Nusselt number. The local Nusselt number in the bottom and top surfaces are given by

$$Nu_x=\frac{H}{(T_1-T_2)}\left.\frac{\partial T}{\partial y}\right|_{y=0} \tag{8}$$

and

$$Nu_x=\frac{H}{(T_1-T_2)}\left.\frac{\partial T}{\partial y}\right|_{y=H} \tag{9}$$

, respectively. The average Nusselt number is calculated by

$$\bar{N}u=\frac{1}{L}\int_{x=0}^{L} Nu_x dx \tag{10}$$

All of those governing equations are converted into linear equations by employing finite volume method. The system of linear equations for all fields is coupled using SIMPLE algorithm. The commercial code of ANSYS FLUENT is used to carry out the simulation.

### *2.2. Empirical correlation*

The results of the present simulation will be compared with the analytical results. However, there is no correlation of finned surface is found in the literature. Thus only the smooth surface correlation will be used. There several empirical correlations are found in the literature. The first one is given by Jacob [23].

$$\bar{Nu} = 0.195 Ra_H^{0.25} \text{ for } 10^4 < Ra_H < 4\times10^5 \tag{11a}$$

$$\bar{Nu} = 0.068 Ra_H^{1/3} \text{ for } 4\times10^5 \le Ra_H < 10^7 \tag{11b}$$

In the more recent experimental study using air inside the enclosure, Holland et al. [24] proposed an empirical correlation between the average Nusselt number and Rayleigh number. The equation is given below.

$$Nu = 1 + 1.44\left[1 - \frac{1708(\sin 1.8\phi)^{1.6}}{Ra\times\cos\phi}\right]\left[1 - \frac{1708}{Ra\times\cos\phi}\right]^{+} + \left[\left(\frac{Ra\times\cos\phi}{5830}\right)^{1/3} - 1\right]^{+} \tag{12}$$

The above equation is claimed to be valid for tilt angle from 0° to 75°. In the equation the meaning of the "+" exponent is that only positive values of the terms in the square bracket are it will be zero if the value is negative. The additional condition for using equation (12) are the aspect ration (AR) of the enclosure and the Rayleigh number must be $AR \ge 12$ and $Ra_H < 10^5$.

## 3. Results and Discussion

### 3.1. Numerical Validation

The present numerical method has been validated in the previous work [22]. The selected numerical validation case is a laminar natural convection heat transfer from a square cavity heated left wall and cooled from right wall. While, the top and bottom walls are insulated. The results are compared for Rayleigh number $10^5$ and $10^6$. The results of the present method show a very good agreement with the previous results. Thus, the present method can be used to explore the problem.

### 3.2. Flow characteristic and heat transfer of smooth and finned collector

Figure 2 shows flow characteristics of the fluid inside the enclosure for smooth and finned flat-plate collector. The simulation was made for AR of 10. The dimension of the present enclosure is also 1 m × 10 cm with the dimension of fins are 5 cm × 2.5 cm. It is shown clearly that inside the enclosure without fin several circulation flows are generated. The mechanism can be explained in the following. The heated fluid flows upward and before reaching the top surface it will be divided into two flows. Each flow generates a circulation flow and it repeats. The velocity of the fluid follows the Benard cells [22]. The fluid flow characteristics in the enclosure with fin, the Bernard cells clearer and filled the areas between fins. Figure 2 shows this fact very clear.

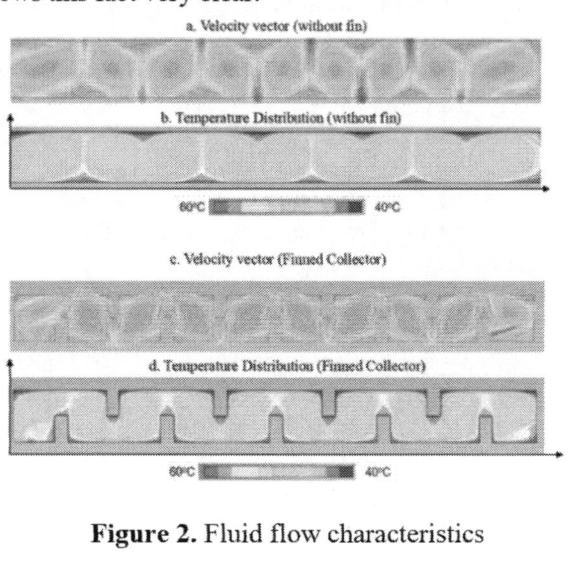


**Figure 2.** Fluid flow characteristics

The Benard cells affect local heat transfer characteristics. It is divided the bottom surface into several dead zones, where the velocity almost zero. Between two dead zones the velocity increases and makes temperature gradient higher. Heat transfer coefficient in the dead zone decreases due to lower temperature gradient. On the other hand, the area between dead zones the local heat transfer coefficient increases and reaching maximum value. This is because of higher temperature gradient. This fact is clearly shown in Figure 3. For the enclosure with fin, the local heat transfer coefficient shows a quite different trend. It still shows the sinusoidal pattern. However, on the dead zone areas there is a small wavy heat transfer.

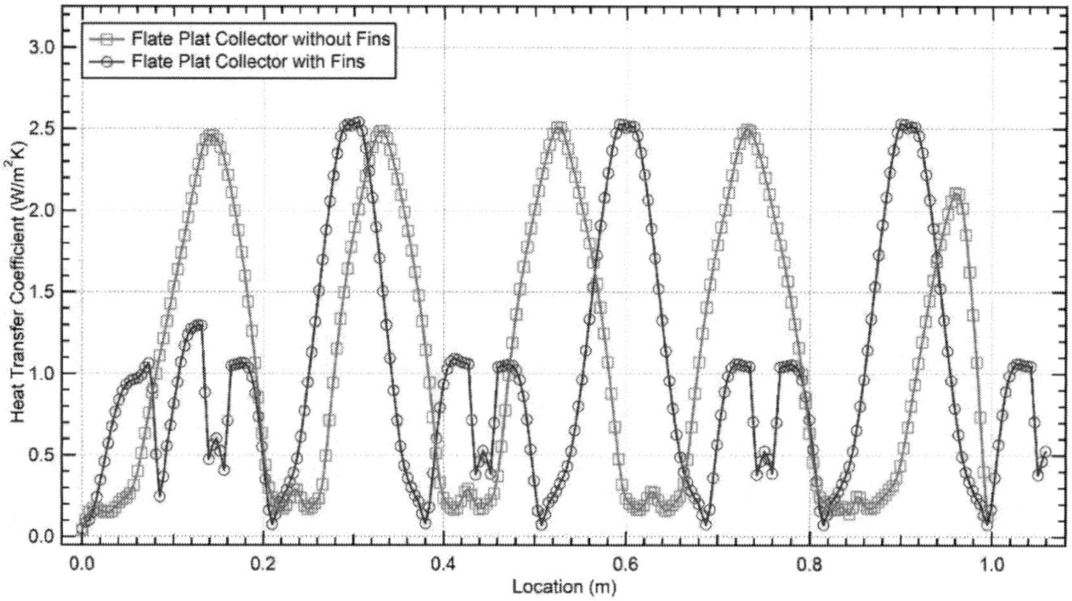


**Figure 3.** Locally heat transfer coefficient

### *3.3. Heat transfer coefficient*

The heat transfer coefficient from the present result and the estimation using equation (23) and equation (24) are shown in Table 1. The analyses are made at Rayleigh number $Ra_H = 1.90 \times 10^6$. The table shows that for the smooth enclosure, there is a small difference of average heat transfer coefficient among those methods. However, it is still in the range of less than 10%. For the surface with fin, the heat transfer coefficient decreases significantly. It is reduced 42.27% in comparison with smooth enclosure. This is because installing fin on the surface of glass cover will break some peak heat transfer coefficient as shown in the local heat transfer coefficient of figure 3. Since the peak heat transfer coefficients are broke down by the fins, the average heat transfer coefficient of the finned surface decrease significantly in comparison with the surface without a fin. This fact suggests that glass cover with fin will effective to reduce the heat loss from a solar collector to the ambient.

Table 1 Results comparison

| Aspect Ratio | Parameter | Present Results | | Equation (11) | Equation (12) |
|---|---|---|---|---|---|
| | | Finned | Smooth | Ref. [23] | Ref. [24] |
| $AR = 10$ | $\overline{Nu}$ | 4.63 | 8.02 | 7.241 | 8.321 |
| $L = 1m$ | $\bar{h}$ [W/m²K] | 1.19 | 1.94 | 1.752 | 2.014 |
| $H = 10.0cm$ | $Q''$ [W/m²] | 22.42 | 38.84 | 35.046 | 40.275 |

## 4. Conclusions

In this work, the heat transfer and fluid flow characteristics in the enclosure between double glasses cover of a smooth and finned flat-plate type solar collector have been numerically studied using commercial code CFD code. The numerical simulation shows that in the enclosure several Benard cells for two types of the solar collector are captured. The presence of the Benard cells strongly affects the local heat transfer coefficient. For the surface with fin, the heat transfer coefficient decreases significantly. It is reduced 42.27% in comparison with smooth enclosure. Since the peak heat transfer coefficients are broke down by the fins, the average heat transfer coefficient of the finned surface decrease significantly in comparison with the surface without the fin. This fact suggests that glass cover with fin will effective to reduce the heat loss from a solar collector to the ambient.

## References

[1] Boltzmann 2014 *Encyclopedia of Ecology,* Amsterdam, Elsevier
[2] Population Reference Bureau, 2017, Human Population: Population Growth.
[3] Kabir E, Kumar P, Kumar S, Adelodun A A and Kim H K 2018 *Renewable and Sustainable Energy Reviews* **(82)**, 894-900
[4] Ambarita H 2017 *Journal of Physic: Conference Series* **801(1)**, 012093
[5] Ambarita H and Sitepu T 2017 *IOP Conference Series: Materials Science and Engineering* **237** 012014
[6] Ambarita H, Abdullah I, Siregar C A, Siregar R E T and Ronowikarto A D 2017 *IOP Conference Series: Material Science and Engineering* **180(1)**, 012030
[7] Ambarita H 2017 *ARPN Journal of Engineering and Applied Sciences* **12(19)**, 5357-5365
[8] Sitepu T, Gunawan S, Nasution D M, Ambarita H, Siregar R E T and Ronowikarto A D 2016 *IOP Conference Series: Materials Science and Engineering* **180(1)** 012032
[9] Dina S F, Ambarita H, Napitupulu F H and Kawai H 2015 *Case Studies in Thermal Engineering* **5** 32-40
[10] Ambarita H and Kawai H 2016 *Case Studies in Thermal Engineering* **7** 36-46
[11] Sitorus T B, Napitupulu F H and Ambarita H 2016 *International Journal of Technology* **7(5)**, 910-922
[12] Ambarita H 2016 *Case Studies in Thermal Engineering* **8** 346-358
[13] Ambarita H 2016 *IOP Conference Series: Materials Science and Engineering* **180(1)** 012024
[14] Setyawan E Y, Napitupulu R A M, Siagian P and Ambarita H 2017 *IOP Conference Series: Materials Science and Engineering* **237** 012012
[15] Fudholi A, Sopian K, Othman Y M, Ruslan H M and Bakhtyar B 2013 *Energy Conversion and Management* **75**, 234-240
[16] Pinto J R, Guimaraes M P and Menon J G 2016 *Journal of Fluid Dynamics* **6**, 75-87
[17] Maliska C R 1995 *Transferencia de Calor e Mecanica dos Fluidos Computational,* Spain, LTC
[18] Karim M A and Hawlader M N A 2006 *Energy* **31**, 452-470
[19] Beghein C, Haghighat F and Allard F 1992 *International Journal Heat Mass Transfer* **35**, 883-846
[20] Martinopoulos G, Missirlis D, Tsilingiridis G, Yakinthos K and Kyriakis N 2010 *Renewable Energy* **35** 1499-1508
[21] Selmi M, Al-Khawaja M J and Marafia A 2008 *Renewable Energy* 33 383-387
[22] Ambarita H 2017 *IOP Conference Seris: Material Science and Engineering* (Accepted for publication)
[23] Jacob M 1949 *Heat Transfer*, New York, Wiley
[24] Hollands K G T, Unny T E, Raithby G D and Konicek L 1976 *Journal of Heat Transfer* **98**, 189-193

# Effects of the inclination angle on the performance of flat plate solar collector

**H Ambarita[1,2*], R E T Siregar[1,2], A D Ronowikarto[1,2] and E Y Setyawan[3]**

[1]Sustainable Energy and Biomaterial Centre of Excellent, Universitas Sumatera Utara, Jl. Almamater Kampus USU, Medan 20155, Indonesia
[2]Mechanical Engineering Department, Faculty of Engineering, Universitas Sumatera Utara, Jl. Almamater Kampus USU, Medan 20155, Indonesia
[3]Mechanical Engineering, National Institute of Technology Malang, Malang 65145, Indonesia

[*]Email: himsar@usu.ac.id

**Abstract**. Double glasses cover is typically used in a flat plate solar collector to decrease heat losses to ambient. The working principal of the cover is to allow the solar irradiation hit the plate absorber and blocks it using natural convection mechanism in the enclosure between the glasses. The performance of the enclosure to block the heat loss to the surrounding affected by the inclination angle of the collector. The objective of this study is to explore the effect of the inclination angle to the performance of the solar collector. Numerical simulation using commercial code Computational Fluid Dynamic (CFD) has been carried out to explore the fluid flow and heat transfer characteristics in the enclosure. In the result, streamline, vector velocity, and contour temperature are plotted. It was shown that the inclination angle strongly affects the performance of the collector. The average heat transfer coefficient decreases with increasing inclination angle. This fact suggests that too high inclination angle is not recommended for solar collector.

## 1. Introduction

As a fact, the world population has increased significantly. It is estimated that the present world population is 7.5 billion people and it will be increased more than 1.5 billion, in the last two decades. To support the people activities and life needs, the world's energy demand also increasing significantly. Preventing an energy crisis is one of the most common issues of the 21st century. Solar energy, among other renewable sources of energy, is a promising and freely available energy source for managing long term (sustainable) issues in energy crisis. Solar energy industry is developing steadily all over the world because of high demand for energy while major energy source, fossil fuel, is limited and other sources are expensive [1]. In particular, Indonesia has a big potency of solar energy [2] that can be harvested in solar thermal and photovoltaic. Solar energy thermal can be used as an energy source to power solar desalination [3,4,5], solar refrigeration [6,7], solar water heater [8,9], solar cooker [10,11], solar drying process [12], etc. One of the key component of solar thermal development is the solar collector. Many types of solar collector can be found in the literatures and the flat-plate type solar collector is the simplest solar collector and it is very popular [13].

In the flat-plate type solar collector, to decrease the heat loss from the top, a double glass cover is typically used. Studies on the free convective heat transfer mechanism in the enclosure between cover of flat-plate solar collector have been reported in literature. Varol and Oztop [14] reported their work on a comparative numerical study on natural convection in inclined wavy and flat-plate solar collectors. It was shown that flow field and thermal characteristic are affected by the shape of the enclosure and heat transfer rate increases in the case of wavy enclosure than that of flat enclosure. The effect of inclination to the wavy and flat-plate solar collectors has also explored [15]. Kumar [16] reported the study on natural convective heat transfer in trapezoidal enclosure of box-type solar cooker. It was shown that the values of convective heat transfer coefficient and top loss coefficient for rectangular enclosure are lower by 31-35% and 7 %, respectively.

Recently, Computational Fluid Dynamic (CFD) commercial code has been used to optimize the design of a flat-plate type solar collector. Martinopoulos et al. [17] studied the behavior of a polymer solar collector experimentally and numerically. Solar irradiation as well as convection and heat transfer in the circulating fluid and between the parts of the collectors is considered in their model. Selmi et al. [18] employed CFD commercial code to analyze heat transfer process in a flat-plate type solar collector. The above studies show that heat transfer analysis and flow characteristics within the enclosure of double glass cover play an important role in designing and developing a high performance flat-plate type solar collector. In the field, the flat-plate type solar collector usually installed with inclined position. Thus, the fluid flow and heat transfer coefficient within an inclined double glasses cover need to be explored. The objective of this paper is to explore the hat transfer and flow characteristics in the inclined enclosure of a double glass cover. The numerical results will be compared with empirical correlations found in the literature. The results are expected to support the necessary information in developing high performance flat-plate type solar collector.

## 2. Method

As a note, our research group known as Sustainable Energy and Biomaterial Centre of Excellent, Faculty of Engineering University of Sumatera Utara is developing high performance flat-plate solar collector. The collector is designed for several applications such as drying, adsorption cycle, solar desalination, and solar water heater. Typical of the flat-plate type solar collector is shown in Figure 1. The focus on the present study is the enclosure in the top cover of the solar collector, and the inclination angle of the flat-plate solar collector is from $30^0$, $45^0$, $60^0$, $70^0$ respectively.

### *2.1. Numerical Method*

The problem of the present works is solved using numerical approach. For simplicity, only the enclosure between the upper and bottom glasses is taken account into consideration. The computational domain is shown in Figure 1. In the analysis, the domain is assumed to be two-dimensional case. The flow in the computational domain is incompressible laminar flow and steady state. In addition, there is no viscous dissipation, the gravity acts in vertical direction, and fluid properties are constant except in density of the fluid. The Boussinesq approximation is employed to model the buoyancy force. By using the above assumptions, the governing equations are shown in the following.

$$\frac{\partial u}{\partial x}+\frac{\partial v}{\partial y}=0 \tag{1}$$

$$u\frac{\partial u}{\partial x}+v\frac{\partial u}{\partial y}=-\frac{1}{\rho}\frac{\partial P}{\partial x}+\frac{\mu}{\rho}\left(\frac{\partial^2 u}{\partial x^2}+\frac{\partial^2 u}{\partial y^2}\right)+g\beta\left(T-T_r\right)\sin\phi \tag{2}$$

$$u\frac{\partial u}{\partial x}+v\frac{\partial u}{\partial y}=-\frac{1}{\rho}\frac{\partial P}{\partial y}+\frac{\mu}{\rho}\left(\frac{\partial^2 v}{\partial x^2}+\frac{\partial^2 v}{\partial y^2}\right)+g\beta\left(T-T_r\right)\cos\phi \tag{3}$$

$$u\frac{\partial T}{\partial x}+v\frac{\partial T}{\partial y}=\alpha\left(\frac{\partial^2 T}{\partial x^2}+\frac{\partial^2 T}{\partial y^2}\right) \tag{4}$$

Where $u$ and $v$ are the velocity vector in $x$- and $y$-directions, respectively. In the analysis, the non-dimensional parameters are explained below.

$$Ra = \frac{g\beta\Delta TH^3}{\nu^2}\Pr \tag{5}$$

Where Pr is Prandtl number and calculated by

$$\Pr = \frac{\nu}{\alpha} \tag{6}$$

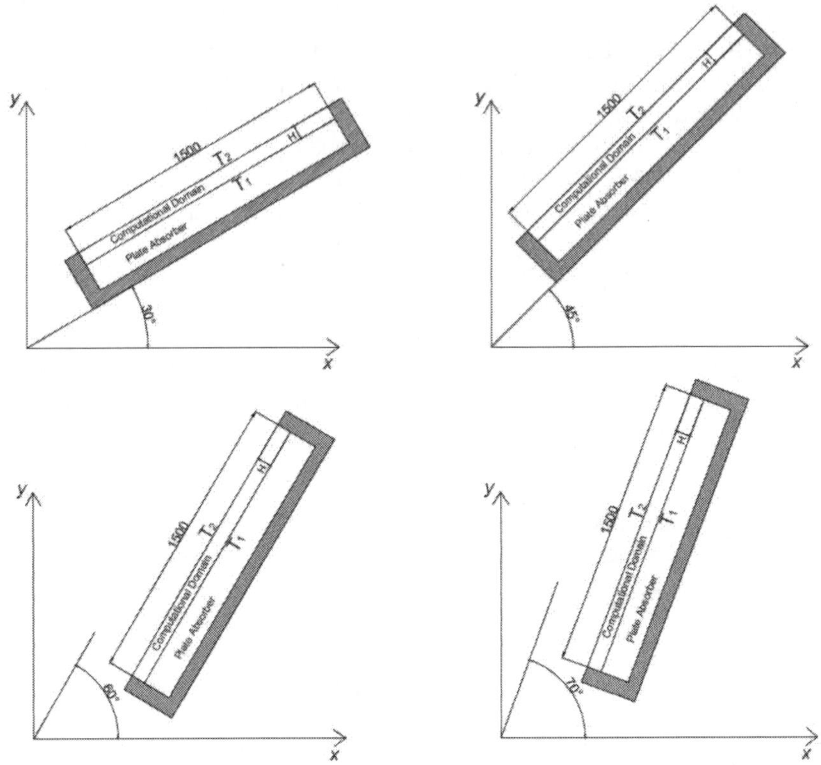


**Figure 1.** Computational domain for several analysis

The convective heat transfer coefficient ( $h$ ) is calculated using non-dimensional Nusselt number. It is calculated using the below equation.

$$h = \frac{Nu \times k}{H} \tag{7}$$

Where $k$ [W/mK] is the conductive heat transfer coefficient of the air in the enclosure and $H$ [m] is the distance between the double glasses. The Nusselt number in equation (7) is presented using local and average Nusselt number. The local Nusselt number in the bottom and top surfaces are formulated by

$$Nu_x = \frac{H}{\left(T_1 - T_2\right)} \frac{\partial T}{\partial y}\bigg|_{y=0} \tag{8}$$

and

$$Nu_x = \frac{H}{\left(T_1 - T_2\right)} \frac{\partial T}{\partial y}\bigg|_{y=H} \tag{9}$$

, respectively. The average Nusselt number is calculated by the below equation.

$$\bar{N}u = \frac{1}{L}\int_{x=0}^{L} Nu_x dx \tag{10}$$

All of those governings equations are converted into linier equations by employing finite volume discritization method. The system of liner equations for all fields are solved using SIMPLE algorithm. The commercial code of ANSYS FLUENT is employed to carry out the simulation.

### 2.2. *Empirical correlation*

The results of numerical simulation will be compared with the analytical results. In the analytical solution, the below empirical correlation will be used. There several empirical correlations are found in the literature. The first correlation is given by Jacob [19].

$$\bar{N}u = 0.195 Ra_H^{0.25} \text{ for } 10^4 < Ra_H < 4\times10^5 \tag{11a}$$

$$\bar{N}u = 0.068 Ra_H^{1/3} \text{ for } 4\times10^5 \leq Ra_H < 10^7 \tag{11b}$$

In the more recent experimental study using air inside the enclosure, Holland et al. [20] proposed an empirical correlation for inclined enclosure between average Nusselt number and Rayleigh number. The equation is given in the below equation.

$$Nu = 1 + 1.44\left[1 - \frac{1708(\sin 1.8\phi)^{1.6}}{Ra\times\cos\phi}\right]\left[1 - \frac{1708}{Ra\times\cos\phi}\right]^{+} + \left[\left(\frac{Ra\times\cos\phi}{5830}\right)^{1/3} - 1\right]^{+} \tag{12}$$

The above equation is claimed to be valid for tilt angle from 0° to 75°. In that equation the meaning of the "+" exponent is that only positive values of the terms in the square bracket are to be used and if the value is negative, the value will be converted into zero.

## 3. Results and Discussions

### 3.1. *Numerical Validation*

The present numerical method has been validated with previous work and reported in the previous work [21]. The selected numerical validation case is a laminar natural convection heat transfer from a square cavity heated left wall and cooled from right wall. While, the top and bottom walls are insulated. The results are compared for Rayleigh number $10^5$ and $10^6$. The results of the present method show a very good agreement with the previous results. Thus, the present method can be used to explore the problem.

### 3.2. *Flow and heat transfer characteristics*

Numerical simulation has been performed for a case with Rayleigh number $Ra_H = 1.90\times10^6$. The dimension of the enclosure are $L = 1m$ and $H = 10.0cm$. Thus, the aspect ratio of the enclosure is $AR = 10$. The inclination angle is varied. Figure 2 shows streamlines, temperature distributions, and vector velocity for the case with inclination angle of 30°. The streamlines show that the Bernard cells are generated in the enclosure. However, the dimension is not similar as shown in the horizontal enclosure [21]. The temperature distribution shows that on the hot surface of the glass, several hot spots (dead zones) are captured. On these areas the heat transfer coefficient will be decreased due to low temperature gradient. The plotted vector velocity shows that the pattern of Bernard cell follows the inclination surface. In other words, the form of Bernard cell is not horizontal rectangular but a rhombus form. These facts will affect the heat transfer characteristics.

In order to explore the effect of the inclination to the flow field characteristics, streamline, temperature distribution, and vector velocity for inclination angle of 45°, $60^0$, $70^0$ respectively are shown in figure 3, figure 4, and figure 5, respectively. The comparison of the figure shows when the inclination angle increased over than 30° the flow pattern become unpredictable. At inclination angle of 45°, only several Bernard cells presence in the enclosure. For inclination angle of $60^0$ and $70^0$, only small cells are captured and at the upper part a bigger circulation flow is captured.

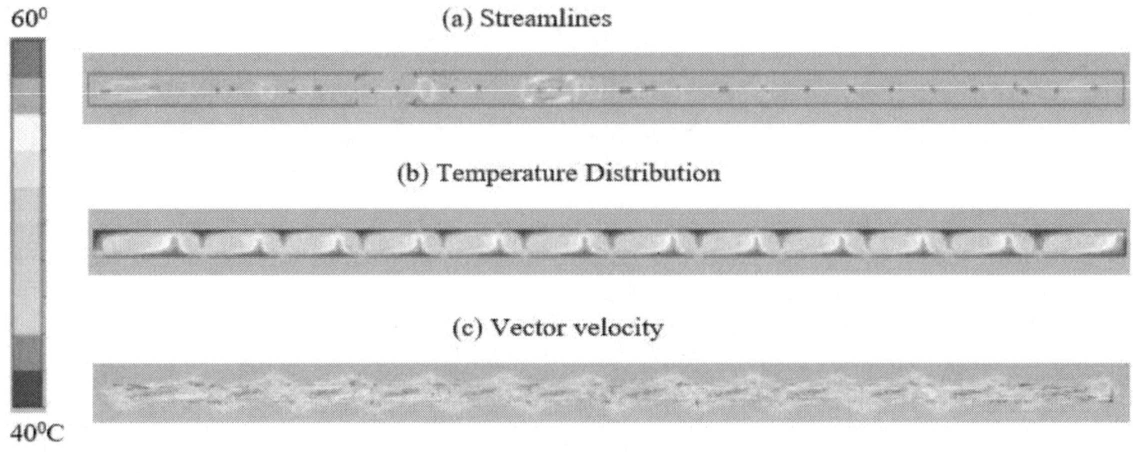


**Figure 2.** Flow characteristics with angle $30^0$

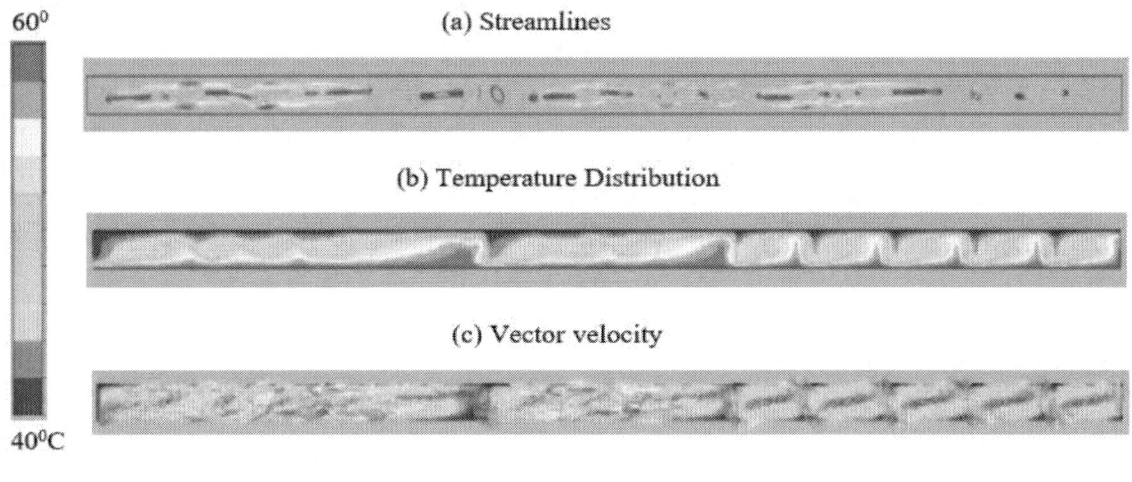


**Figure 3.** Flow characteristics with angle $45^0$

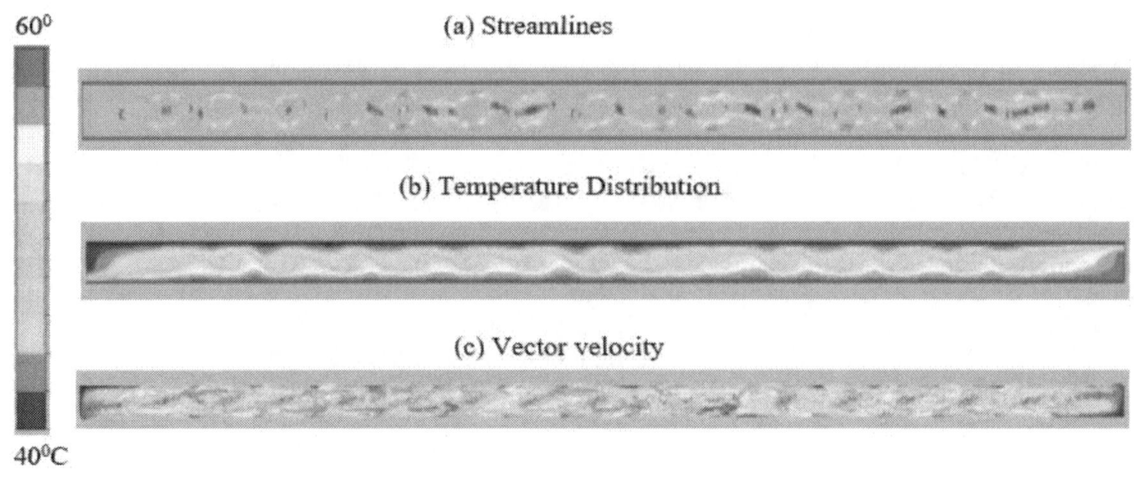


**Figure 4.** Flow characteristics with angle $60^0$

IOP Conf. Series: Journal of Physics: Conf. Series **978** (2018) 012097 doi:10.1088/1742-6596/978/1/012097

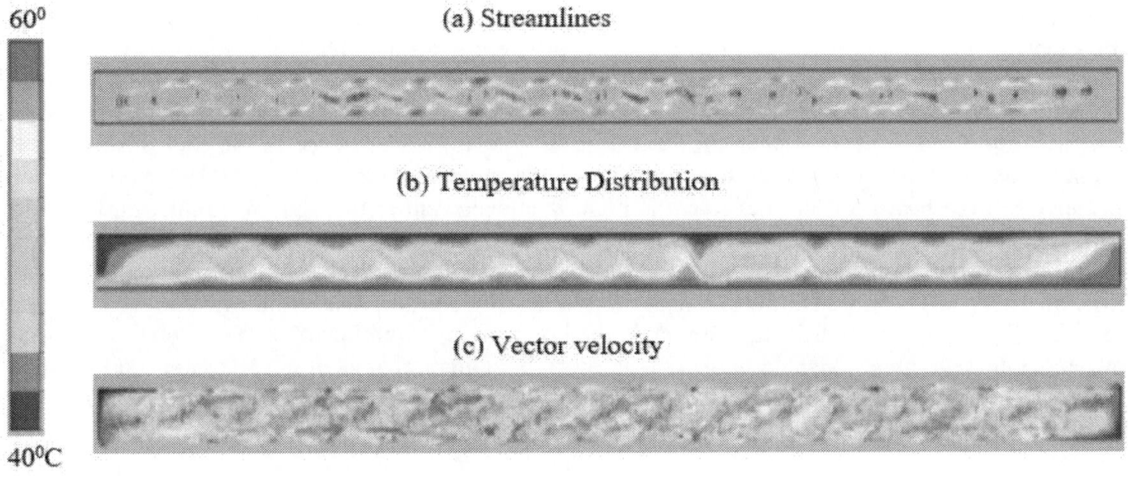


**Figure 5.** Flow characteristics with angle 70⁰

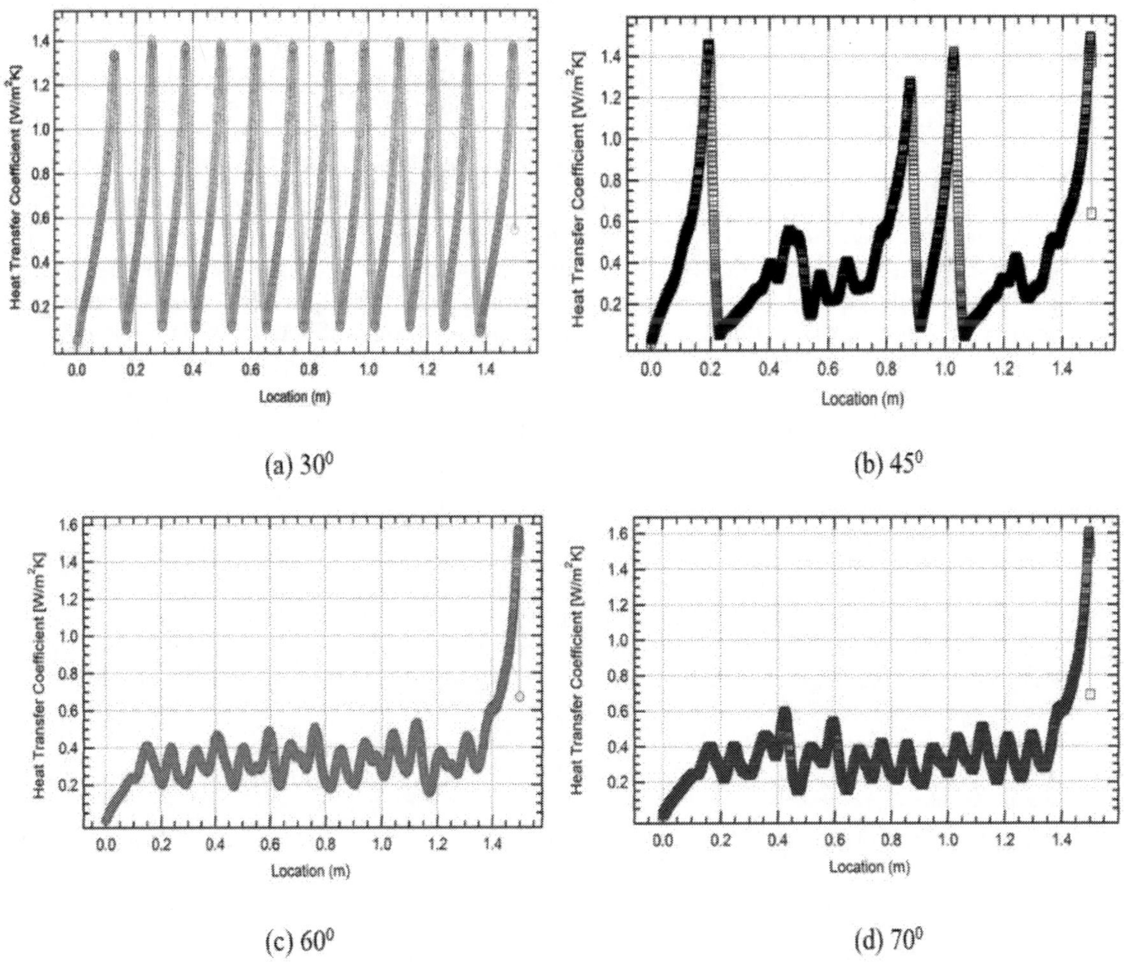


**Figure 6.** Locally heat transfer Coefficient

Figure 6 shows local heat transfer coefficient on the hotter surface of the glass cover. The figure shows that at 30°, the pattern of local heat transfer coefficient is similar. This is because the Bernard cell in the enclosure presence with similar size. This results in similar pattern, maximum on the areas with high temperature gradient and minimum at low temperature gradient. When the inclination angle increased to 45°, only several maximum heat transfer coefficients presence in the enclosure. For the inclination angle of $60^0$ and $70^0$, the heat transfer rate on the surface is mainly low. However, only on the upper part the heat transfer coefficient is high. It is even higher than the maximum value at low inclination angle.

The average heat transfer coefficient and average Nusselt number at different inclination angles are shown in Table 1. The table shows that heat transfer coefficient decreases with increasing inclination angle. This is because at high inclination angle the hot surface is dominated by dead zones which has low temperature gradient. This fact suggests that too high inclination angle is not recommended for solar collector. However, further research need to be carried out to combine the effect of solar radiation to the absorber plate.

**Table 1.** Average heat transfer coefficient and Nusselt number

| Collector Angle | $\bar{N}_u$ Number | $\bar{h}$ [W/m$^2$K] |
|---|---|---|
| $30^0$ | 4.35 | 2.38 |
| $45^0$ | 4.15 | 2.27 |
| $60^0$ | 3.84 | 2.09 |
| $70^0$ | 3.53 | 1.93 |

## 4. Conclusion

A numerical simulation using commercial code CFD has been carried out in order to explore the heat transfer and fluid flow characteristics in enclosure between double glasses cover of a flat-plate type solar collector. The method has been validated for a natural convection is a square enclosure heated and cooled from side walls, respectively. The validated method than used to explore the problem. The conclusions of this study are as follows. The inclination angle strongly affects the fluid flow and heat transfer characteristics of in the enclosure. The average heat transfer coefficient decreases with increasing inclination angle. This fact suggests that too high inclination angle is not recommended for solar collector. However, further research need to be carried out to combine the effect of solar radiation to the absorber plate.

## References

[1] Kannan N and Vakeesan D 2016 *Renewable and Sustainable Energy Reviews* **62**, 1092-1105
[2] Ambarita H 2017 *Journal of Physic: Conference Series* **801(1)**, 012093
[3] Ambarita H 2016 *Case Studies in Thermal Engineering* **8**, 346-358
[4] Ambarita H 2016 *IOP Conference Series: Materials Science and Engineering* **180(1)**, 012024
[5] Setyawan E Y, Napitupulu R A M, Siagian P and Ambarita H 2017 *IOP Conference Series: Materials Science and Engineering* **237** 012012
[6] Ambarita H and Kawai H 2016 *Case Studies in Thermal Engineering* **7** 36-46
[7] Sitorus T B, Napitupulu F H and Ambarita H 2016 *International Journal of Technology* **7(5)**, 910-922
[8] Ambarita H and Sitepu T 2017 *IOP Conference Series: Materials Science and Engineering* **237** 012014
[9] Ambarita H, Abdullah I, Siregar C A, Siregar R E T and Ronowikarto A D 2017 *IOP Conference Series: Material Science and Engineering* **180(1)**, 012030
[10] Ambarita H 2017 *ARPN Journal of Engineering and Applied Sciences* **12(19)**, 5357-5365
[11] Sitepu T, Gunawan S, Nasution D M, Ambarita H, Siregar R E T and Ronowikarto A D 2016 *IOP Conference Series: Materials Science and Engineering* **180(1)** 012032

[12] Dina S F, Ambarita H, Napitupulu F H and Kawai H 2015 *Case Studies in Thermal Engineering* **5** 32-40
[13] Pandey K M and Chaurasiya R 2017 *Renewable and Sustainable Energy Reviews* **67** 641-650
[14] Varol Y and Oztop H F 2008 *Building and Environment* **43** 1535-1544
[15] Varol Y and Oztop H F 2007 *Building and Environment* **42** 2062-2071
[16] Kumar S 2004 *Renewable Energy* **29** 211-222
[17] Martinopoulos G, Missirlis D, Tsilingiridis G, Yakinthos K and Kyriakis N 2010 *Renewable Energy* **35** 1499-1508
[18] Selmi M, Al-Khawaja M J and Marafia A 2008 *Renewable Energy* 33 383-387
[19] Jacob M 1949 *Heat Transfer*, New York, Wiley
[20] Hollands K G T, Unny T E, Raithby G D and Konicek L 1976 *Journal of Heat Transfer* **98**, 189-193
[21] Ambarita H 2017 *IOP Conference Seris: Material Science and Engineering* (Accepted for publication)

IOP Conf. Series: Journal of Physics: Conf. Series **978** (2018) 012098 doi:10.1088/1742-6596/978/1/012098

# The optimum intermediate pressure of two-stages vapor compression refrigeration cycle for Air-Conditioning unit

**H Ambarita[1,2*] and H V Sihombing[2]**

[1]Sustainable Energy and Biomaterial Centre of Excellent, Universitas Sumatera Utara, Jl. Almamater Kampus USU, Medan 20155, Indonesia
[2]Mechanical Engineering Department, Faculty of Engineering, Universitas Sumatera Utara, Jl. Almamater Kampus USU, Medan 20155, Indonesia

Email: himsar@usu.ac.id

**Abstract**. Vapor compression cycle is mainly employed as a refrigeration cycle in the Air-Conditioning (AC) unit. In order to save energy, the Coefficient of Performance (COP) of the need to be improved. One of the potential solutions is to modify the system into multi-stages vapor compression cycle. The suitable intermediate pressure between the high and low pressures is one of the design issues. The present work deals with the investigation of an optimum intermediate pressure of two-stages vapor compression refrigeration cycle. Typical vapor compression cycle that is used in AC unit is taken into consideration. The used refrigerants are R134a. The governing equations have been developed for the systems. An inhouse program has been developed to solve the problem. COP, mass flow rate of the refrigerant and compressor power as a function of intermediate pressure are plotted. It was shown that there exists an optimum intermediate pressure for maximum COP. For refrigerant R134a, the proposed correlations need to be revised.

## 1. Introduction

To avoid the globe from global warming, many countries have released the target on reducing Greenhouse gas (GHG) emissions. There are two strategies that can be used to mitigate GHG emissions. They are deforestation and sustainable energy development. In the sustainable energy development consists of enhancing renewable energy utilization and energy efficiency technology. Thus, increasing the energy efficient technology is extremely needed in the mitigation actions to reduce GHG emission. One of the potential application of energy efficient technology is vapor compression refrigeration cycle. As a note, vapor compression refrigeration cycle typically employed to provide cooling for air-conditioning unit. Considering the growth of economy and living standard, it is estimated that vapor compression cycle will consume huge amount of electricity. There are several works related to increasing energy efficiency of a vapor compression cycle such as using heat recovery technique [1,2], multi-stages cycle, etc.

The present paper focus in the multi-stages strategy. Bilge and Temir [3] reported the study on the optimum numbers of stages in the vapor compression refrigeration system. The cases of single-stage and multi-stages compression have been compared on a system with 600 kW cooling capacity and the working fluid is ammonia. Goodarzi et al. [4] studied the comparative analysis of the improved two-stage multi-inter-cooling ejector-expansion trans-critical $CO_2$ refrigeration cycle. In the study, an internal heat exchanger has been included within the system for improvement in the cooling capacity. Purohit et al. [5] reported the study on the effect of inter-stage pressure on the performance of a two stages refrigeration cycle using intercooler. Six common refrigerants are selected for the analyses include R22, R143a, propane, carbon dioxide and nitrous oxide. It was shown that the effect of inter stage pressure is expected to be an important consideration in warmer climate condition. Jian et al [6] investigated the role of optimum intermediate pressure in the design of two-stage vapor compression systems as a further investigation. Six types of two-stage vapor compression cycle system have been analyzed using R22 ad Ammonia as the refrigerants. Chopra et al. [7] performed energy, exergy and sustainability analysis of two-stage vapour compression refrigeration system. Eight types of refrigerants have been analyzed on the basis of energetic and exergetic performance. Xuan [8] developed a general staging model to optimize multistage exo-reversible refrigeration system affected solely by internal ir-reversibilities.

Those studies show that multi-stage technology to optimize the vapor compression refrigeration cycle has come under scrutiny. The present work focuses on the investigation of optimum intermediate pressure of two-stage vapor compression refrigeration cycle. The used refrigerant here is R134a which is a common type of refrigerant in the air-conditioning application. The literature shows that very limited studies reported the using of R134a in the optimization of multi-stage vapor compression refrigeration cycle. The objective is to explore the maximum coefficient of performance (COP) of the system at several different working pressures of evaporator and condenser. The results are expected to supply the necessary information on development highly efficient vapor compression refrigeration cycle.

## 2. Solution Method

The schematic diagram of two-stage vapor compression refrigeration cycle is shown in figure 1. In the figure, the *p-h* diagram is also presented. It can be seen that the component of the system consists of two compressors, condenser, flash cooler, and evaporator.

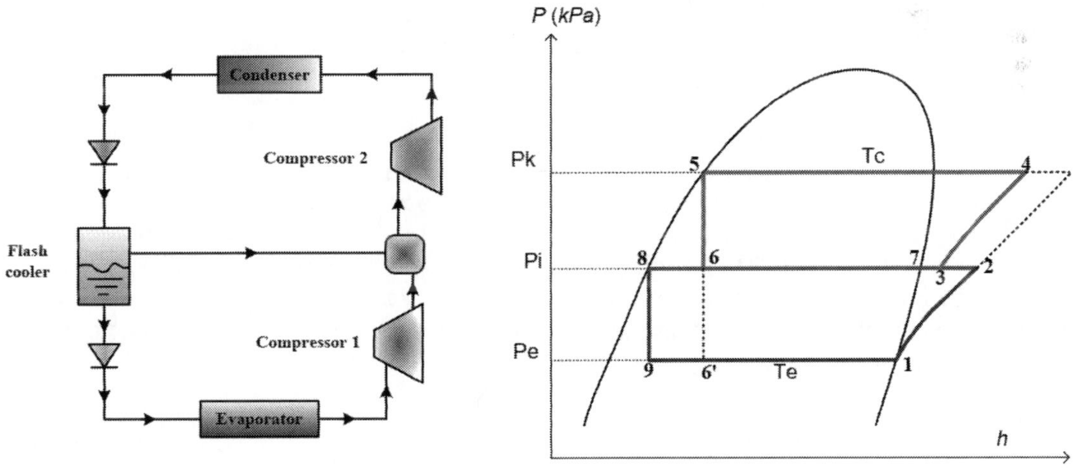


**Figure 1.** Schematic and *P-h* diagrams of the system

The governing equations of the systems are developed as follows. The power in the first ($\dot{W}_{c1}$) annd second compressors ($\dot{W}_{c2}$) are calculated by

$$\dot{W}_{c1} = \dot{m}_1 (h_2 - h_1) \quad (1)$$

$$\dot{W}_{c2} = \dot{m}_2 (h_4 - h_3) \quad (2)$$

where $h_1$, $h_2$, $h_3$, and $h_4$ are enthalpy of the refrigerant at the inlet and the exit of first compressor and at inlet and the exit of the second compressor, respectively. The total of the power to system is calculated by

$$\dot{W}_{tot} = \dot{W}_{c1} + \dot{W}_{c2} \tag{3}$$

The heat release by the system to ambient is given by equation (4).

$$\dot{Q}_c = \dot{m}_2(h_4 - h_5) \tag{4}$$

where $h_4$ and $h_5$ are enthalpy of the refrigerant at the inlet and the exit of the condenser. Here $\dot{m}_c$ is defined as the flow rate of the refrigerant entering the condenser. The heat absorbed by the refrigerant in the evaporator, $\dot{Q}_e$ is given by:

$$\dot{Q}_e = \dot{m}_1(h_1 - h_9) \tag{5}$$

Refrigeration effect (*ER*) of the system is calculated by equation (6).

$$ER = h_1 - h_9 \tag{6}$$

The coefficient of performance of the system (COP) is given by

$$COP = \frac{\dot{Q}_e}{\dot{W}_{tot}} \tag{7}$$

In the literature it is very limited equation can be used to estimate the optimum intermediate pressure ( $P_i$ ) in terms of suction and discharge pressures. To the best knowledge of the author, only the below equation is proposed to estimate the optimum intermediate pressure [12].

$$P_i = \sqrt{P_e \times P_c} \tag{8}$$

Where $P_e$ and $P_c$ are the pressure of the evaporator and the condenser. A computer program has been developed to solve the above equations. The properties of the refrigerant is modeled using the data provided by American Society of Heating, Refrigeration, and Air Conditioning Engineers (ASHRAE).

## 3. Results and Discussions

The analysis is carried out for refrigerant R134a. Three operating conditions, here after named as case, are analyzed, the first case is at temperature evaporation of -15$^0$C and temperature condensation of 35$^0$C, the second case is at temperature evaporation of -10$^0$C and temperature condensation of 40$^0$C, and the third case is at temperature evaporation of -5$^0$C and temperature condensation of 45$^0$C. For all cases the cooling load is assumed to be constant at 1000 Ton.

### *3.1. Mass flow rate of the refrigerant*

The mass flow rate of the refrigerant from in the system for case 1 and case 2 are shown in figure 2. In the figure, the mass flow rate from the evaporator is shown by green line, while mass flow rate from the flash cooler to the second compressor line is shown by the blue line. The total mass flow rate in the system shown by red line. The figure shows the same trend for all cases. The intermediate pressure has no effect to the mass flow rate from the evaporator. However, mass flow rate from the flash cooler increases with increasing intermediate pressure. As a result, the mass flow rate in the system increases with increasing intermediate pressure.

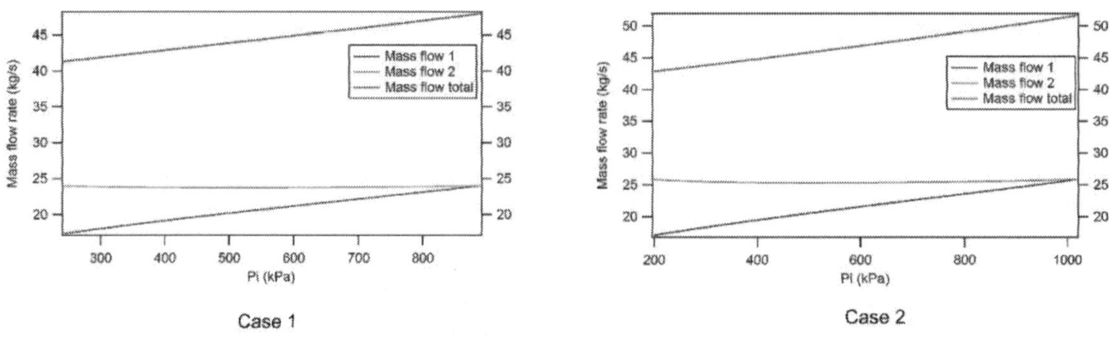


**Figure 2.** Mass flow rate in the system for case 1 and case 2

### *3.2. Compressor Power*

Figure 3 shows the compressor power of the system for all cases. The figure shows the almost similar trend. In the first compressor, the power increases with increasing intermediate pressure. This is because of increasing intermediate compressor will increase the pressure difference between the inlet and the exit of the compressor. The higher pressure difference results in higher compressor power. Different phenomenon is shown by the second compressor. The power of the second compressor decreases with increasing intermediate pressure. As a note, the top pressure and low pressure of the present system is fixed. This is because the evaporation temperature and condensing temperature are fixed. Thus, increasing the pressure intermediate means increasing pressure difference in the firs compressor. Since the top and low pressures are fixed, pressure difference in the second compressor will decrease. This fact is clearly shown in figure 3.

As a system the power of the system is calculated by using equation (3). In the figure, the total of the compressor power in the system is shown by red line. It was clearly shown that there exists a minimum compressor power for an optimum intermediate pressure. If the intermediate pressure lower that its optimum value, increasing the intermediate pressure will decrease the total compressor power. But, if the intermediate pressure over than optimum one, increasing intermediate pressure will increase the compressor power. This fact reveals that there will be exist an optimum performance of the system. This will be discussed in the next section.

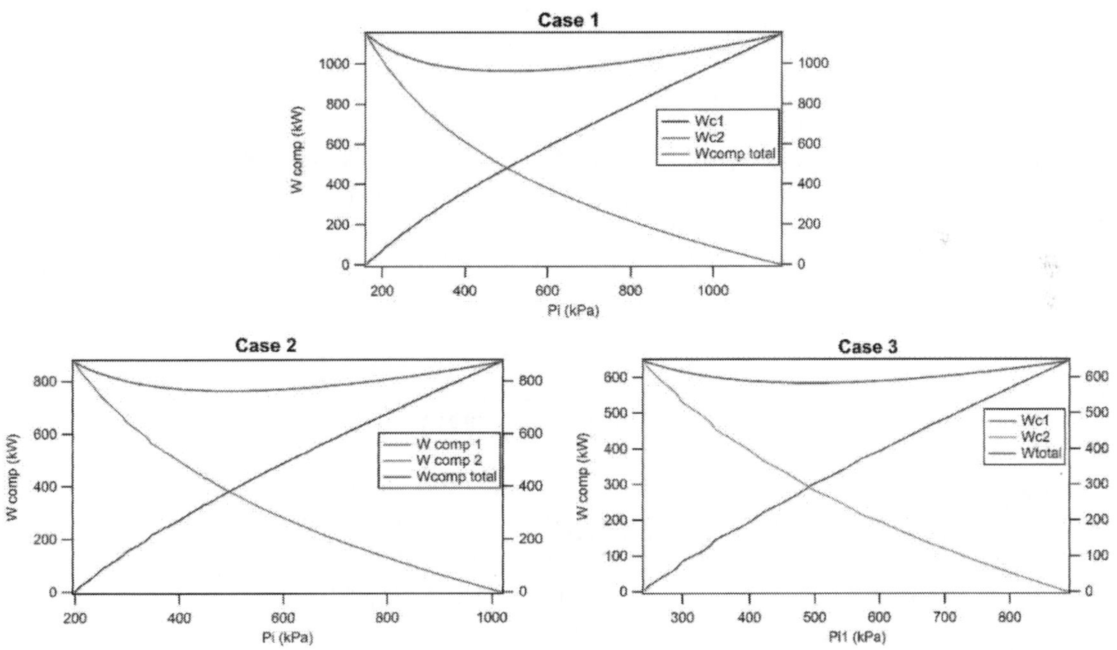


**Figure 3.** Compressor power for all cases

### *3.3. Coefficient of Performance*

Figure 4 shows the performance of the system for all cases. The figure shows that, in general, the COP of the system increases with decreasing pressure difference of the system. As a note, the highest pressure difference in the system is for case 1 and followed by case 2 and case 3. It can be seen that COP of case 3 is higher than case 2 and case 1. Also, the COP of case 2 is higher than case 1. As expected, there exists an optimum intermediate pressure for maximum COP. If the system is operated at its optimum intermediate pressure, increasing and decreasing the intermediate pressure will decrease the COP of the system.

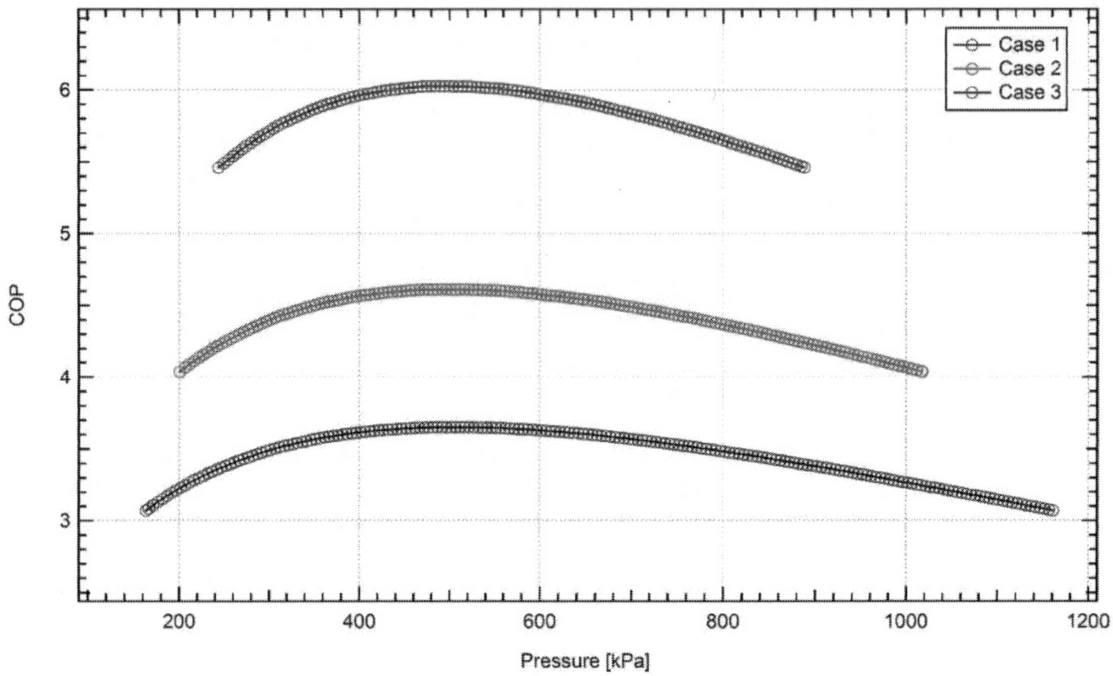


**Figure 4.** Coefficient of the performance

In order to provide a convenient discussion, the value of maximum COP and the optimum intermediate pressure are presented in Table 1. The theoretical optimum intermediate pressure suggested by Arora [12] is calculated using equation (8) is also presented in the table. The data reveals that there is a discrepancy of the present optimum intermediate pressure in comparison with theoretical one. The difference varies from 7.19% to 11.3%. This fact suggest that equation (8) need to be revised by proposing an correction factor.

**Table 1.** Optimum intermediate pressure

| | COPmax | Intermediate pressure [kPa] | | Discrepancy [%] |
|---|---|---|---|---|
| | | Present | Theory Equation (8) | |
| Case 1 | 3.65 | 491.80 | 436.22 | 11.30 |
| Case 2 | 4.61 | 486.80 | 451.81 | 7.19 |
| Case 3 | 6.02 | 509.80 | 464.87 | 8.81 |

## 4. Conclusions

In this work the optimum intermediate pressure of two-stage vapor compression refrigeration cycle has been explored. The conclusions are as follows.

a. The intermediate pressure has no effect to the mass flow rate from the evaporator. However, mass flow rate from the flash cooler increases with increasing intermediate pressure. As a result, the mass flow rate in the system increases with increasing intermediate pressure.
b. There exists a minimum compressor power for an optimum intermediate pressure. This fact reveals that there will be exist an optimum performance of the system.
c. There exists an optimum intermediate pressure for maximum COP.
d. There is a discrepancy of the present optimum intermediate pressure in comparison with theoretical one. The difference varies from 7.19% to 11.3%.

e. It is suggested to perform further analysis to propose a correction factor for the optimum intermediate pressure.

## References

[1] Ambarita H, Nasution A H, Siahaan N M and Kawai H 2016 *Case Studies in Thermal Engieering* **8**, 105-114.
[2] Ambarita H, Nasution D M, Gunawan S and Nasution A H 2017 *IOP Conf. Series: Material Science and Engineering* **180 (1),** 012027
[3] Purohit N, Gupta D K, and Dasgupta M S 2016 *Energy Procedia* **90**, 171-178
[4] Gadalla M and Saghafifar M 2016 *Energy Conversion and Management* **119**, 187-202
[5] Jiang S, Wang S, Jin X and Yu Y 2016 *International Journal of Refrigeration* **67**, 77-89
[6] Chopra K, Sahni V and Mishra R S 2015 *Journal of Thermal Engineering* **4**, 440-445
[7] Lee H, Hwang Y, Radermacher R and Chun H H 2015 *International Journal of Refrigeration* **51**, 103-111
[8] Goodarzi M, Gheibi A, Motamedian M 2015 *Applied Thermal Engineering* **81**, 58-65
[9] Manjili F E and Yavari M A 2012 *Applied Thermal Engineering* **40**, 202-209
[10] Bilge D and Temir G 2004 *American Journal of Applied Science* **1(2)**: 71-75
[11] Xuan C C 2003 *Cryogenics* **43**, 117-124
[12] Arora R C 2010 *Refrigeration and Air Conditioning*, PHI, New Delhi

IOP Conf. Series: Journal of Physics: Conf. Series **978** (2018) 012099 doi:10.1088/1742-6596/978/1/012099

# Numerical study on the effects of absorptivity on performance of flat plate solar collector of a water heater

**D R S. Tambunan[1], Y P Sibagariang[2], H Ambarita[1,2,*], F H Napitupulu[1] and H Kawai[3]**

[1]Mechanical Engineering Department, Faculty of Engineering, Universitas Sumatera Utara, Jl. AlmamaterKampus USU, Medan 20155, Indonesia
[2]Sustainable Energy and Biomaterial Centre of Excellent, Universitas Sumatera Utara, Jl. AlmamaterKampus USU, Medan 20155, Indonesia
[3]Mechanical Engineering Department, Muroran Institute of Technology, 27-1 Mizumoto-cho, Muroran 8585, Hokkaido, Japan

Corresponding author: himsar@usu.ac.id

**Abstract.** The characteristics of absorber plate of a flat plate solar collector play an important role in the improvement of the performance. In this work, a numerical analysis is carried out to explore the effect of absorptivity and emissivity of absorber plate to the performance of the solar collector of a solar water heater. For a results comparison, a simple a simple solar box cooker with absorber area of 0.835 m × 0.835 m is designed and fabricated. It is employed to heat water in a container by exposing to the solar radiation in Medan city of Indonesia. The transient governing equations are developed. The governing equations are discretized and solved using the forward time step marching technique. The results reveal that the experimental and numerical results show good agreement. The absorptivity of the plate absorber and emissivity of the glass cover strongly affect the performance of the solar collector.

## 1. Introduction

In order to avoid the catastrophe of the globe due to global warming, emission of Greenhouse gas gases to the atmosphere must be reduced. One of the potential solutions is to enhance the use of renewable energy resources. Solar energy is known as a potential renewable energy resource. It is convinced, it will play an important role in the future energy demand. In a year, the globe is exposed by 3,400,000 Exa-Joule of solar energy. This is a vast amount of energy. It can fill the present world energy consumption by only 1 hour 15 minutes of irradiation [1]. However, at the present time, only a very little portion of the human energy consumption is filled with the solar energy. According to REN21, since 2030 solar energy application will increase significantly. Thus, studies on solar energy have increased significantly [2]. Industry related to solar energy application is developing constantly in all over the world. This is because of twofold, increasing energy demand and limitation on major energy source the fossil fuel. Indonesian islands have a big potency of solar energy. It is also constant in every year [3]. The solar energy can be harvested by solar photovoltaic and solar thermal methods. In the solar thermal method, the solar energy can be collected and used as a heat energy for several applications such as solar drying process [4], solar adsorption refrigeration [5], solar desalination [6,7,8], solar water heater [9], solar cooker [10], etc.

There are many types of solar collectors have been studied and reported in literature. The most popular one is flat-plate type of solar collector due to simplicity and easy to use [11]. However, the thermal performance of the flat-plate type of solar collector is still low. Many researchers have been reported their study on the heat transfer characteristics and heat loss from a flat-plate type solar collector. Varol and Oztop [12] reported a comparative numerical study on natural convection in inclined wavy and flat-plate of solar collectors. It was shown that flow and thermal fields were affected by the shape of the air space (cavity). The heat transfer rate increases in the case of the wavy cavity in comparison with the flat surface cavity. The influence of inclination to the flat-plate and wavy solar collectors has also investigated [13]. It was shown that heat transfer is increased with increasing Rayleigh number and aspect ratio, and it is decreased with increasing wavelength. Kumar [14] studied natural convective heat transfer in the trapezoidal profile of an enclosure of a box-type solar cooker. It was revealed that in comparison with the rectangular profile, the values of convective heat transfer and top loss coefficient for rectangular enclosure are lower by 31-35% and 7 %, respectively.

Those reviewed studies showed that investigation on the heat transfer characteristics of flat-plate type solar collector has come under scrutiny. The focuses are mainly on the profile and operating parameters such as Rayleigh number. To the best knowledge of the authors, the only limited study focused on thermal characteristics of the flat-plate solar collector materials. In this study, a flat-plate type solar collector that is used in a simple solar box cooker is studied numerically. The objective is to explore the effect of the absorptivity and transmissivity of the solar collector material to the performance. The results are expected to supply the necessary information in developing an effective and efficient solar box cooker.

## 2. Method

In this study, two approaches have been carried out, they are numerical and experimental methods. The experimental work is used to validate the numerical results. The validated numerical method will be used to investigate the effect of absorptivity and transmissivity of the solar collector. Here, an experimental apparatus has been designed and fabricated. The experimental apparatus is depicted in figure 1.

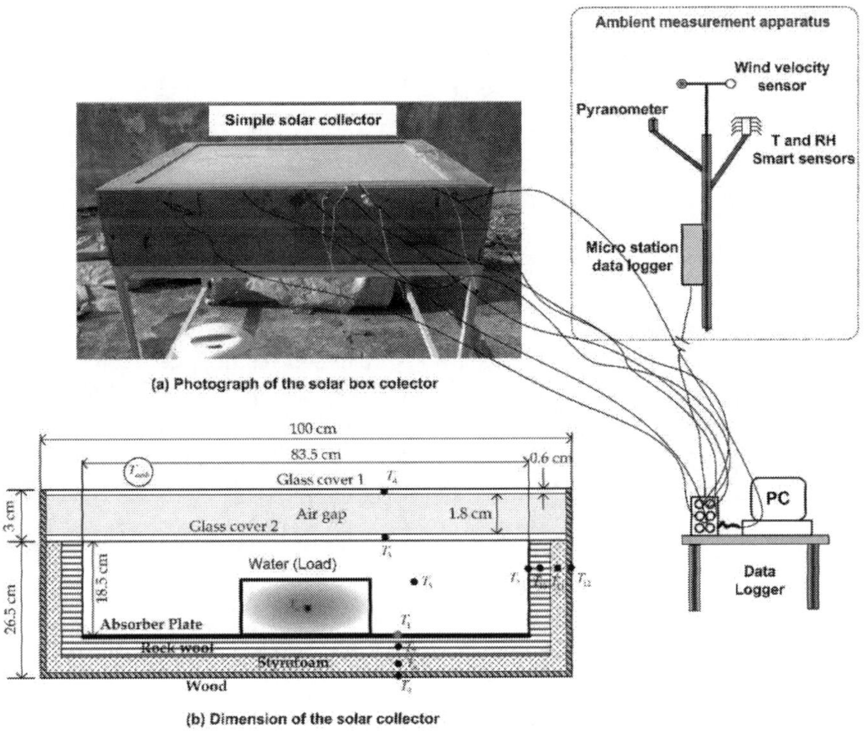

**Figure 1.** (a) Photograph of the solar collector and (b) computational domain

As shown in figure 1, the solar collector is categorized as a flat-plate type solar collector with double glasses cover in the top with absorber area of 0.835 m × 0.835 m. In order to decrease the heat loss, the bottom and wall of the solar collector are made of a series of insulation materials. They are rock wool, styrofoam, and wood, respectively. The dimension and configuration of the material are shown in the figure. The load to the solar collector is water which is placed in a cylindrical cooking container and placed in the middle of the collector. The container is made of aluminium with diameter and height of 30 cm and 12 cm, respectively.

*2.1 Numerical Method*

In the numerical method, the solar collector is divided into 12 components. The lumped capacity assumption is employed for all divided components. Thus, the temperature of all materials is uniform. The properties of each component are assumed to be uniform. Figure 2(b) shows the dimensions and temperature notation of every component. The governing equation is developed by employing energy conservation to each material. The energy conservation for absorber plate yield:

$$m_p c_p \frac{\partial T_1}{\partial t} = \tau^2 \alpha I A_p - FC_{15}(T_1 - T_5) - FR_{13}(T_1 - T_3) - FR_{12}(T_1 - T_2) - \frac{1}{R_{17}}(T_1 - T_7) \tag{1}$$

where $m_p$ [kg] and $c_p$ [J/kg K] are the mass and specific heat of the absorber plate. The parameters of $I$ [W/m$^2$], $\alpha$, and $\tau$ are solar irradiance, absorptivity of the absorber plate and transmissivity coefficients of the glass cover, respectively. The air inside the solar collector is assumed to be one substance with uniform temperature of $T_5$. Thus, energy conservation to the air gives:

$$(\rho V_{ol} c_p)_{air} \frac{\partial T_5}{\partial t} = FC_{15}(T_1 - T_5) - FC_{52}(T_5 - T_2) - FC_{53}(T_5 - T_3) - FC_{56t}(T_5 - T_6) - FC_{56w}(T_5 - T_6) \tag{2}$$

Where $FC$ [W/K] is a factor to represent convective heat transfer coefficient from any surface. The subscript 1, 5, 2, 3, 6 refer to surfaces shown in figure 1. In addition, 6t and 6w refer to the top surface and the wall surface of the container, respectively. The double glasses cover is made of two glasses with similar emissivity ($\varepsilon$) and thickness. The energy conservation to the glass gives:

$$m_g c_g \frac{\partial T_4}{\partial t} = (1-\tau) I A_g + FR_{34}(T_3 - T_4) + FC_{34}(T_3 - T_4) - FR_{4s}(T_4 - T_{amb}) - h_w A_g (T_4 - T_{amb}) \tag{3}$$

$$m_g c_g \frac{\partial T_3}{\partial t} = (1-\tau)\tau I A_g + FC_{53}(T_5 - T_3) + FR_{13}(T_1 - T_3) + FR_{23}(T_2 - T_3) - FC_{34}(T_3 - T_4) - FR_{34}(T_3 - T_4) \tag{4}$$

Where $h_w$ [W/m$^2$ K], $A_g$ and $T_{amb}$ [°C] are convective heat transfer coefficient from the top glass to the ambient air and temperature of the ambient air, respectively.

As mentioned above that the load for solar collector is water placed in a cooing vessel. In the process, the cooking vessel and water receive the heat from the air inside the solar collector through side wall and top wall of the vessel. In addition, the top surface of the cooking vessel also acting as a solar collector and it receives solar irradiance. Heat transfer from the absorber plate to the bottom and radiative heat transfer from the wall of the vessel are neglected. The temperatures of the vessel and the water are assumed to be homogenous. Thus, energy conservation in the container yields to:

$$(m_f c_f + m_c c_c)\frac{\partial T_6}{\partial t} = FC_{56w}(T_5 - T_6) + FC_{56t}(T_5 - T_6) + \tau^2 \alpha I A_{ct} \tag{5}$$

Where the subscript $f$ and $c$ represent the water and the vessel, respectively.

Employing the energy conservation to the rock wool, styrofoam, and wood result in the following equation.

$$m_{rw} c_{rw} \frac{\partial T_7}{\partial t} = \frac{1}{R_{17}}(T_1 - T_7) - \frac{1}{R_{78}}(T_7 - T_8) \tag{6}$$

$$m_{st}c_{st}\frac{\partial T_8}{\partial t}=\frac{1}{R_{78}}(T_7-T_8)-\frac{1}{R_{89}}(T_8-T_9) \quad (7)$$

$$m_{wd}c_{wd}\frac{\partial T_9}{\partial t}=\frac{1}{R_{89}}(T_8-T_9)-FC_{9a}(T_9-T_a) \quad (8)$$

Where the subscript $rw$, $st$, $wd$, and $a$ refer to the rock wool, styrofoam, wood, and ambient, respectively. Employing the energy conservation in every material of the wall of the solar box solar cooker (inner plate, rock wool, Styrofoam, and wood) will give the following equations:

$$m_pc_p\frac{\partial T_2}{\partial t}=FR_{12}(T_1-T_2)+FC_{52}(T_5-T_2)-FR_{23}(T_2-T_3)-\frac{1}{R_{210}}(T_2-T_{10}) \quad (9)$$

$$m_{rw}c_{rw}\frac{\partial T_{10}}{\partial t}=\frac{1}{R_{210}}(T_2-T_{10})-\frac{1}{R_{1011}}(T_{10}-T_{11}) \quad (10)$$

$$m_{st}c_{st}\frac{\partial T_{11}}{\partial t}=\frac{1}{R_{1011}}(T_{10}-T_{11})-\frac{1}{R_{1112}}(T_{11}-T_{12}) \quad (11)$$

$$m_{wd}c_{wd}\frac{\partial T_{12}}{\partial t}=\frac{1}{R_{1112}}(T_{11}-T_{12})-\frac{1}{R_{12a}}(T_{12}-T_a) \quad (12)$$

All of those governing equations, equation (1) to equation (12), are converted into linier equation system by using discretization technique. In this method, the used technique is forward time step marching. After several numerical experiments, the value of $\Delta t = 1\text{sec}$ is used due to stability consideration.

## 3. Results and Discussions

### 3.1. Solar Irradiance and Numerical validation

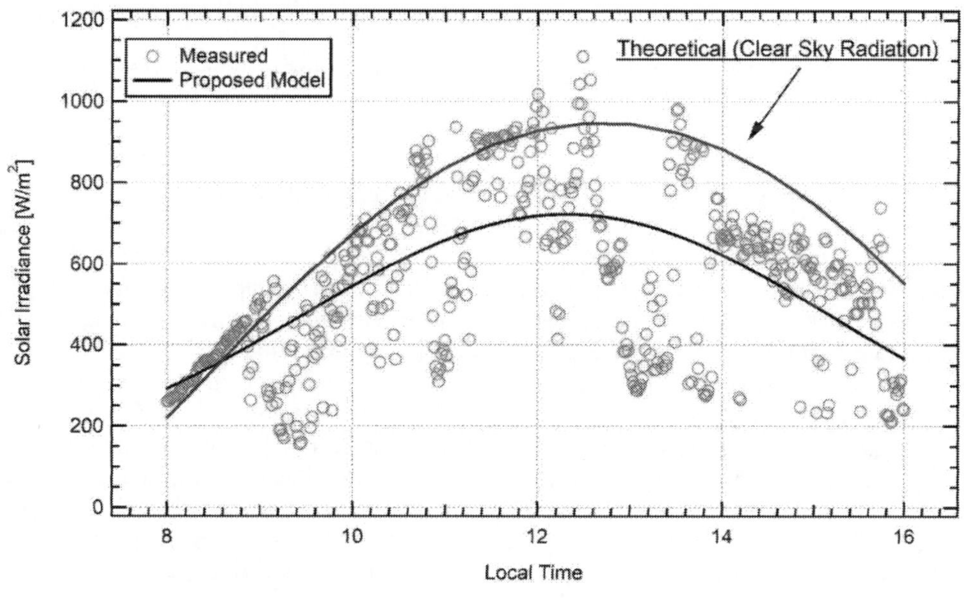


**Figure 2**. Measured, theoretical and modelled solar irradiance

As a note, the source of energy for the solar collector is solar irradiance. The solar irradiance during the experiment is measured and recorded. The measured data will be used to model energy input in the numerical simulation. The theoretical clear sky irradiance measured solar irradiance, and the modelled one is depicted in figure 2. In the modelled solar irradiance, the maximum solar irradiance is 722 W/m2, and it occurred at 12.19 of local time. The total solar energy resulted from the solar radiation and by the proposed model is the same. This model will be used in the numerical simulation as heat flux to the absorber plate of the collector.

A numerical validation test has been performed to the method before it is employed to analyze the problem. Two experiments have been carried out. Temperature history of water vessel at loads of 3 kg and 6 kg water are selected for comparison. Numerical simulations using the developed method are also carried out at the same load as experiments. Temperature history resulted from numerical simulation, and experimental one is shown in Figure 3. The figure reveals that for all loads the temperature history from the experimental and numerical results agree very well. It can be said that the numerical simulation accurately predicts the value of the temperature of the water. Based on this numerical validation, the developed numerical method can be employed to analyze the problem.

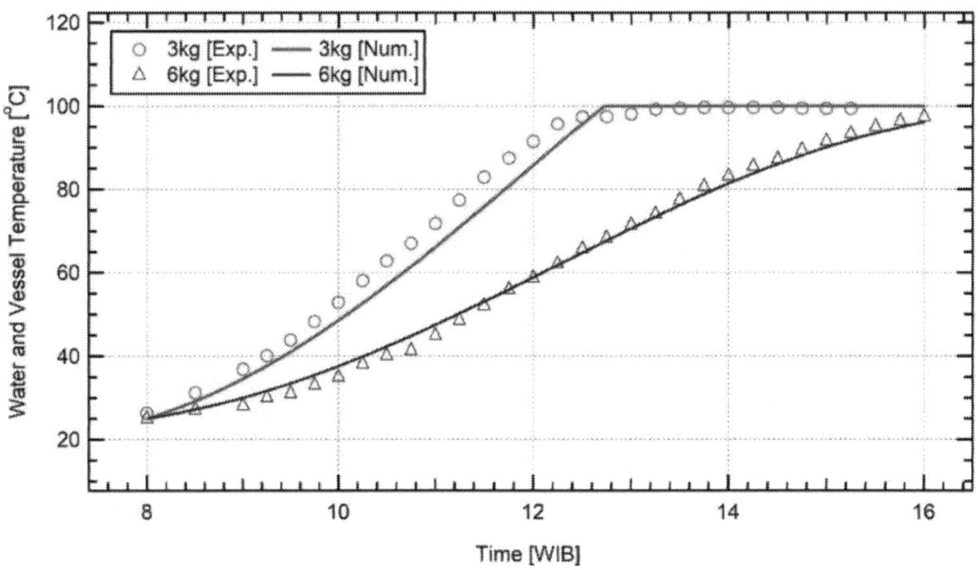


**Figure 3**. Numerical validation of the proposed model

### *3.2. Effect of glass transmissivity*

The simulation is made for water load of 5 kg. Figure 4 shows the effect of glass transmission coefficient to the characteristics of the solar collector. Here the characteristics are examined by using temperature of the absorber plate, temperature of the glass cover 1 and temperature of the glass cover 2. In this study, the coefficient of glass transmission is varied from 0.70 to 0.95. The figure shows that temperature of the absorber plate increases with increasing glass transmission coefficient. This is because the higher transmission coefficient transfers more solar irradiance to the plate absorber. The increasing temperature of absorber plate is proportional to the increasing glass transmission. It can be seen that at transmission coefficient of 0.7, the maximum temperature of the absorber plate was only 102°C. If the transmission coefficient increased to 0.95 the maximum increased up to 122°C. However, the increasing maximum temperature is not followed by glass cover temperature. The figure shows that temperature of bot glasses cover decreases with increasing transmission coefficient. This is because the higher transmission coefficient of the glass means less solar irradiance absorbed by the glass. As a result, there is no sufficient energy to increase its temperature. It can be seen that transmissivity of the glass shows the stronger effect to the top glass in comparison with the lower glass. This is because of the lower glass closer to the plat absorber. This makes it receives more energy from the absorber plate.

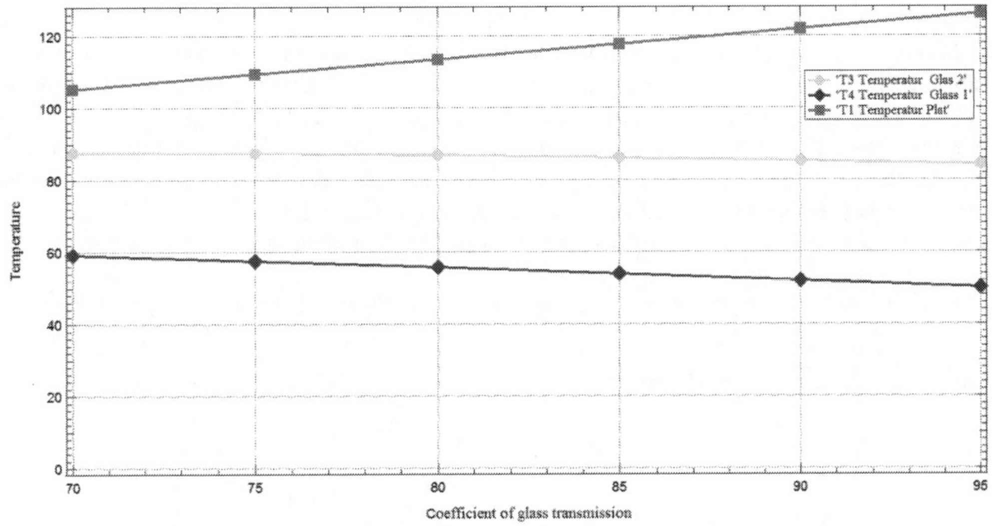


**Figure 4.** Effect of glass transmissivity

### 3.3. *Effect of absorptivity*

Figure 5 shows the effect of the absorptivity to the characteristic of the solar collector. In the figure, the simulation is made for water load of 5 kg, and transmissivity is fixed at 0.95. The absorptivity is varied from 0.7 to 0.95. The figure shows that temperature maximum of the plate absorber increases linearly with increasing absorptivity. This is because at higher absorptivity more solar energy converted into heat and made the temperature of the absorber plate increases. Furthermore, the temperature of the glass cover also increases with increasing absorptivity. However, the rate of increasing glass temperature is lower than the rate of absorber plate. This is because the absorptivity has no direct impact on the glass cover. The increasing temperature of the glass cover is caused by convective heat transfer from the plat absorber to the glass cover.

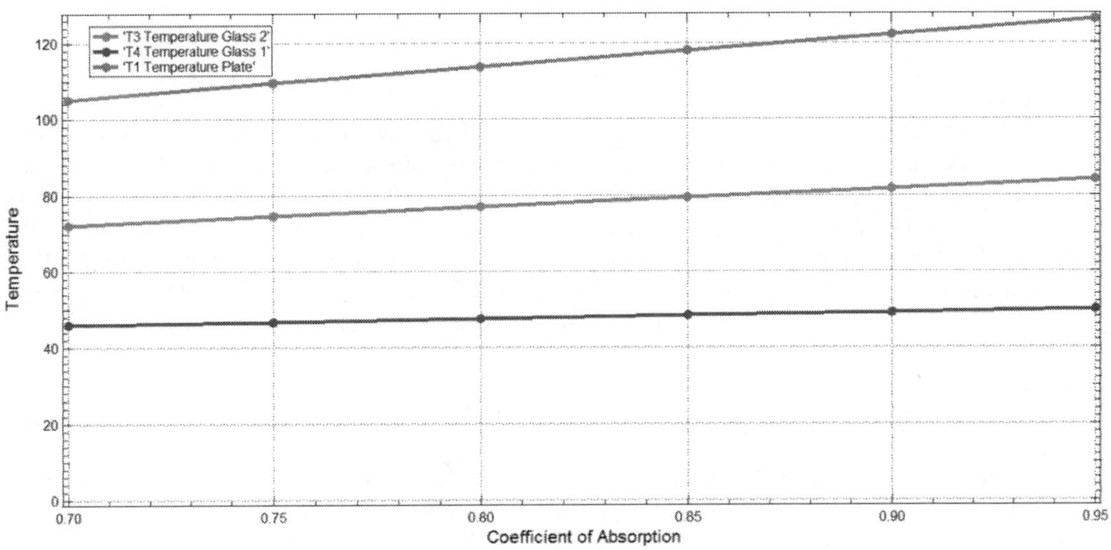


**Figure 5**. Effect of absorber absorptivity

### *3.4. Performance*

The effect of the transmissivity and absorptivity to the performance is shown in figure 6. In the figure, the simulation is made for the load of 5 kg. For the case of variation absorptivity, the transmissivity is fixed at 0.95. On the other hand, for the case with the variation of emissivity the absorptivity is fixed at 0.95. The figure shows that absorptivity and emissivity show the strong effect to the thermal efficiency of the solar collector. Increasing absorptivity will increase the thermal efficiency of the collector. Furthermore, thermal efficiency increases with increasing transmissivity. The comparison of both coefficients shows that the effect of transmissivity is stronger than the effect of absorptivity.

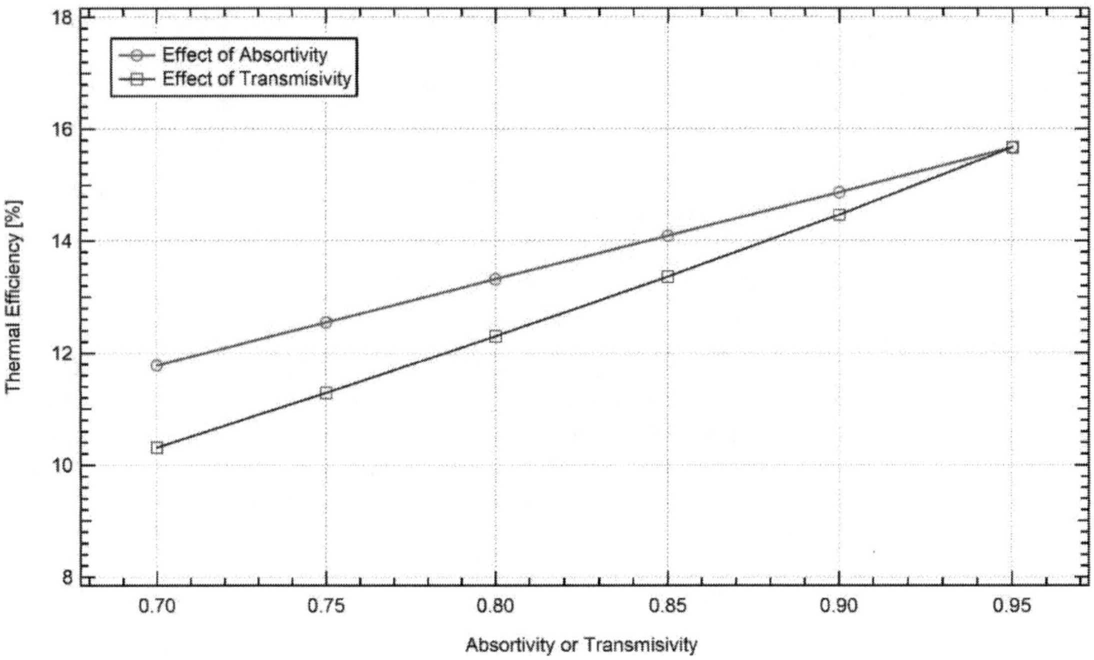


**Figure 6.** Effects of Absorptivity and Transmissivity to the performance

## 4. Conclusions

A numerical study has been conducted to study the effect of glass cover transmission coefficient and absorption coefficient. The governing equations are developed and solved numerically. The conclusion of this study is as follows. The increase of glass transmission coefficient will increase the temperature of the absorber plate. Increasing absorptivity coefficient will also increase the thermal efficiency of the solar cooker. The comparison of both coefficients shows that the effect of transmissivity is stronger than the effect of absorptivity. This fact shows that in the design of solar cooker it is recommended to use appropriate paint with high absorption coefficient and high coefficient of glass transmission.

## References

[1] Thirugnanasambandam M, Iniyan S and Goic R 2010 *Renewable and Sustainable Energy Reviews* **14** 312-322.
[2] Kannan N and Vakeesan D 2016 *Renewable and Sustainable Energy Reviews* **62** 1092-1105
[3] Ambarita H 2017 *Journal of Physics: Conference Series* **801(1)**, 012093
[4] Dina S F, Ambarita H, Napitupulu F H and Kawai H 2015 *Case Studies in Thermal Engineering* **5** 32-40
[5] Ambarita H and Kawai H 2016 *Case Studies in Thermal Engineering* **7** 36-46
[6] Ambarita H 2016 *Case Studies in Thermal Engineering* **8** 346-358

IOP Conf. Series: Journal of Physics: Conf. Series **978** (2018) 012099 doi:10.1088/1742-6596/978/1/012099

[7] Ambarita H 2016 *IOP Conference Series: Materials Science and Engineering* **180(1)** 012024
[8] Setyawan E Y, Napitupulu R A M, Siagian P and Ambarita H 2017 *IOP Conference Series: Materials Science and Engineering* **237** 012012
[9] Ambarita H and Sitepu T 2017 *IOP Conference Series: Materials Science and Engineering* **237** 012014
[10] Sitepu T, Gunawan S, Nasution D M, Ambarita H, Siregar R E T and Ronowikarto A D 2016 *IOP Conference Series: Materials Science and Engineering* **180(1)** 012032
[11] Pandey K M and Chaurasiya R 2017 *Renewable and Sustainable Energy Reviews* **67** 641-650
[12] Varol Y and Oztop H F 2008 *Building and Environment* **43** 1535-1544
[13] Varol Y and Oztop H F 2007 *Building and Environment* **42** 2062-2071
[14] Kumar S 2004 *Renewable Energy* **29** 211-222

# Designing a holistic end-to-end intelligent network analysis and security platform

**M Alzahrani**

Department of Information Technology, College of Computer Science & IT, Albaha University, Saudi Arabia

msawileh@bu.edu.sa

**Abstract**. Firewall protects a network from outside attacks, however, once an attack entering a network, it is difficult to detect. Recent significance accidents happened. i.e.: millions of Yahoo email account were stolen and crucial data from institutions are held for ransom. Within two year Yahoo's system administrators were not aware that there are intruder inside the network. This happened due to the lack of intelligent tools to monitor user behaviour in internal network. This paper discusses a design of an intelligent anomaly/malware detection system with proper proactive actions. The aim is to equip the system administrator with a proper tool to battle the insider attackers. The proposed system adopts machine learning to analyse user's behaviour through the runtime behaviour of each node in the network. The machine learning techniques include: deep learning, evolving machine learning perceptron, hybrid of Neural Network and Fuzzy, as well as predictive memory techniques. The proposed system is expanded to deal with larger network using agent techniques.

## 1. Introduction

Access control, firewalls and anti-virus software in information security have successfully helped us to stop people from doing wrong things in a computer network system. However, it still have limitation in providing insights into human behavior and advanced malware because the rules are set up statically in the traditional reliance. Thus, the rules are hard to update dynamically when needed due to the changes in the network. Security information and event management (SIEM), an approach to security management, provides us with a better insight into patterns of user behavior.

A technique called User Behavior Analytics has started to become particularly useful in giving solutions that have a level of flexible pattern recognition which rules are unsuitable, and which humans simply cannot achieve due to the excruciatingly large amount of data involved. The user behavior is considered a hard problem due to the fact that human behavior is erratic and hard to predict

A focus area of information security is cybercrime. It is difficult to prevent or fight this cybercrime due to the fact that attacks techniques are also getting sophisticated as more and more tools are available in the Internet that try to make the attacks very difficult if not impossible to detect. The attacker can launch the attacks form a different geographical location or continent or just from our next door neighbor. Government, businesses or even individuals may be targeted by this crime, regardless of size.

In fact System Administrators are often based only on ad hoc methods developed from years of experience in spotting and finding anomalous behaviors in their networks. Many commercial and open source tools have been developed to assist the System Administrators to guard their networks from the attacks. Nevertheless, these tools are not dynamic to adapt any traffic changing that requires the System Administrators to redefine policies from time to time with the aim to trigger alerts. The accuracy of the detection depends on how good the description of the anomalous behavior.

## 2. Related Works

### *2.1. User behaviour analytics in traffic anomaly/malware detection*

Detecting insider attacks against a network is really difficult due to the fact that the defense system may consider it was a normal activity. Works by Wang et al. [1] and W. Kim et al. [2] showed that emotions affected user superficial trust in new developed applications with either positive or negative emotion.

Sun et al. [3] found out "the effect of behavior" and presented "typical access uses of social web site". The profile of user behavior is generated from user's habit activity, and it has to be updated periodically to include the most recent changes. Yin et al. [4] described "a method for anomaly/intrusion detection on linear prediction and Markov chain model then are combined with signature verification to detect attacks more efficiently". The introduced method is based on the temporal behavior of application, and used data from University of New Mexico. Qiao et al. [5] proposed "a behavior analysis-based learning framework". A cluster-based outlier detection algorithm to detect anomalous data is introduced. The authors also described the set of factor of user behavior profile, program profile and system resource usage. "The proposed framework is sufficient to analyze behavior in host-based intrusion without explaining about how to capturing data in stream network and result of the alerts to be forwarded to response mechanism after identifying attack". Lim & Jones [6], presented "taxonomy of anomaly detection techniques in network-based intrusion". The authors confirmed the extent of the use of misuse-based detection system has become the dominant strategy for countermeasure from suspicious attacks and identified two approaches for building the behavior model; learning-based and specification-based models. Authors in [7] used fast-learning method with neural network based on pre-processed component. Prior to this work, Galassi et al. [8] through series of experiments investigated the possibility of automatically constructing a user profile from the logs users' activities, and introduced a standard formalism for regular expressions adopted for describing episodes and profiles.

Work of [2], presented "user behavior in using social network". The authors elaborated methods on how to attempt organizing the status, uses, and issues of social network into a comprehensive framework for the future of social networks/websites. The authors mapped the users' (government, business and individual) behavior onto uses of social website.

"A taxonomy of the most relevant users' behavior features of current solutions" [9]. The behavior features and models were proposed to classify existing techniques. "Users' behavior is very difficult to measure and is related with concepts such as social class, geographical culture, technical knowledge and many others" [10].

### *2.2. Deep Learning*

The existing network policies are not adequate to adapt to the continually changing network conditions arising from the explosive traffic growth. Thus, there is a need to rethink how the network traffic control can be improved. The incorporating intelligence into network traffic control systems can play a significant role in guaranteeing security and Quality of Service (QoS) in Internet Protocol (IP)-based networks [11]. Over the past few decades, machine learning (ML) has been exploited to intelligently dictate traffic control in wired/wireless networks [12-14]. Authors in [15] mentioned: "Since the early excitement stirred by machine intelligence in the 1950s, smaller subsets of machine intelligence have been impacting a myriad of applications over the last three decades". Recently, an even smaller subset of Machine Intelligence and ML techniques, known as deep learning, has emerged with the potential of creating even larger disruptions [16-18].

### *2.3. Automatic run-time extraction of network protocol specification*

A network protocol is a set of rules that describe the messages format and the way they are exchanged (i.e., the protocol state-machine) between two or more network devices. The process of analyzing and discovering Application-level specification of unknown/proprietary network protocols is known as reverse engineering.

Security applications require knowledge on protocol specification. For example a network management software depends on the knowledge to correctly recognize and classify monitored network traffic; penetration testing that generates network inputs to an application to uncover potential vulnerabilities can take advantage of such knowledge to improve the testing process by generating "malicious" inputs more efficiently; a network-based intrusion prevention and detection applications require the knowledge to perform deep packet inspection. The standard protocols specification such as SMTP and HTTP, can be obtained easily because they are available openly. However, for a proprietary protocols such as SMB or Skype, the specification has to be re-engineered. Unfortunately, the manual application of reverse engineering is a time-consuming, tedious, and error-prone task due to the following problems: (1) A message field may not be static; (2) A single protocol message may contain a large number of fields; and (3) complicated relationships (e.g., sequential, parallel, and hierarchical) or dependencies among the fields may exist. This work consider a number of real-world protocols and server applications such as HTTP and SMTP.

### *2.4. Runtime behaviour Analytics*

Traffic classification/intrusion detection system requires specific logics to describe attack signatures, and specific model-checking algorithms. Compared to standard misuse detection tools, a temporal logic allows one to describe behavior involving several events over time: standard misuse detection tools (e.g., anti-virus software or simple network intrusion detection systems) match a library of patterns against single events, and emit an alert once single so-called dangerous events occur. These days, more attacks involving complex, correlated sequences of events, which are usually individually kind.

Instamon is equipped by future traffic classification and intrusion detection tools which capable of analyzing and correlating events over time, in real time. Its purpose is to detect, report, and take countermeasures against intruders. The core of the engine is based on a defined language and algorithm. There are many other formalisms attempting to detect complex intrusion detection scenarios, using means as diverse as Petri nets, parsing schemata, continuous data streams, etc.

### *2.5. Sequential Hierarchical Superset (SHS) Neocortex Memory System*

As described earlier, human neocortex memory system forms the basis of intelligence, perception, creativity, and even consciousness. Our brain learns and understands knowledge by utilizing its neocortex. As described in the memory prediction framework theory [19], neocortex stores sequences of data in a hierarchy. It also recalls data auto-associatively in a form of prediction and match it with the actual input data.

It is constructed of 6 layers of cortical sheet. In the lower hierarchy, the stored data will be more specific features/details and fast changing. So it can be said that long term memory is stored in the top layer and short term memory is stored in the bottom layer [20].

## 3. The proposed design and discussion

The open-ended challenge in securing enterprise network are the driver to many innovative network monitoring and security solutions. High speed data capture, real-time analysis, exponential growth of captured data size, and complex data are among the common problems in existing network monitoring and security solutions. This paper presents a high-level proposal to extend the existing work on network monitoring platform and develops a holistic end-to-end intelligent network analysis and security cloud platform not only leveraging *pfsense*, a wildly adapted, robust, and proven open source security library for traffic data analysis and protection, but also adopting as well as proposing various machine learning techniques for intelligent data representation and security analysis. Figure 1 depicts the high-level design of the proposed system named as *Instamon*.

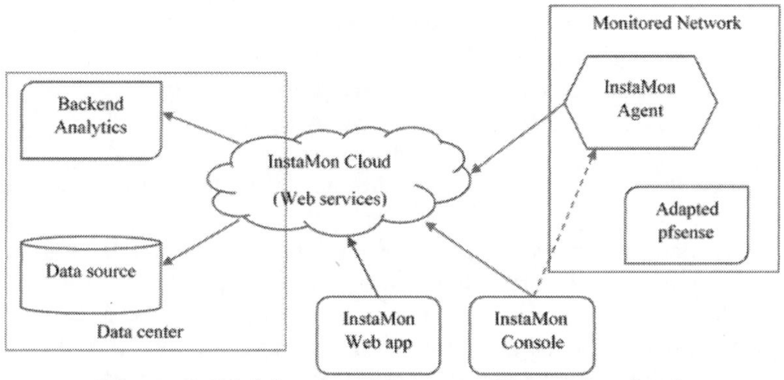


**Figure 1.** High level architecture of *InstaMon* cloud.

The main idea is to have cloud-based network monitoring platform where organization can monitor their network with multiple segments or locations from anywhere anytime through desktop, mobile apps, or web browser. As a passive monitoring solution, the InstaMon agent will sit at the network edge and work together with the adapted pfsense module to collect and analyse network traffic. The proposed InstaMon cloud is designed to be modular to allow organizations to subscribe or active only certain type of information that they intend to monitor such as the commonly monitored network usage statistics, web usage, server health monitoring, etc. as well as the analytics and intelligent monitoring modules.

### *3.1. The backend engine*

At the back of InstaMon cloud lies the backend analytics module which regularly pulls data from various tables in the databases for certain analytics as well as intelligent analysis. This backend analytics module will adopt extensible design to allow new analytics plugin or new intelligent analysis plugin be added in the future. On the other hand, the data source will receive compact data from multiple monitoring points where InstaMon agent is installed. These compact data will not be as detailed as the raw data but instead will have different compact structures according to the purpose of the data source collected at various network points. Figure 2 shows the design of the backend analytics module with proposed initial default plugins.

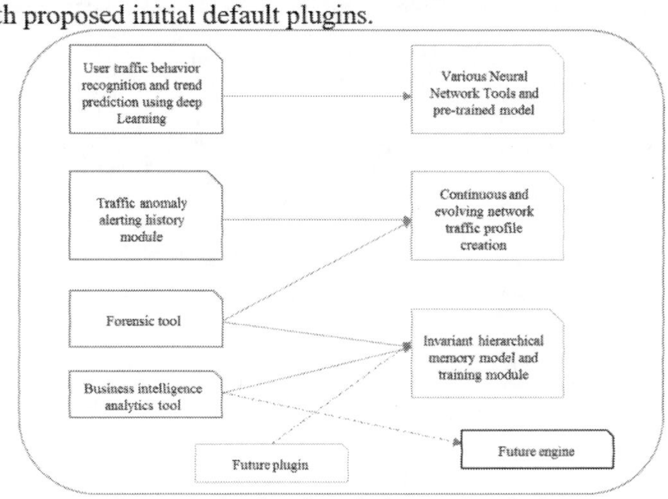


**Figure 2.** Backend analytics engine design.

IOP Conf. Series: Journal of Physics: Conf. Series **978** (2018) 012100 doi:10.1088/1742-6596/978/1/012100

### *3.2. Human brain's neocortex inspired invariant hierarchical memory model*

There will be 3 machine learnings as well as brain inspired engine used to support the analytics and intelligent monitoring module. The first one will be a collection of various neural network tools and pre-trained neural network model, which will be used to run deep learning analysis of user traffic behavior and predict future trends. Depending on the configuration, some output of this analysis would be list of unusual user activities, user upload/download pattern, user web usage pattern, etc. The second engine is an adaptation of the existing traffic profiling engine implemented using evolving connectionist technique enabling a real-time traffic anomaly detection at InstaMon agent by pulling the appropriate daily profile. This work also develops alerting archive module to give network administrator the power to manually adjust how the profile would evolve over previous anomalous activities detected.

Lastly, this work includes an implementation of human brain's neocortex inspired invariant hierarchical memory model as a powerful tool to learn network traffic data from as low as every byte inside a network packet hierarchically into high level network traffic pattern. This brain model will be exported into InstaMon agent and be used for various monitoring activities from as low as packet decoding process and creating compact format of traffic data to be passed back to InstaMon cloud data source. Combined with the traffic profile, the brain model will also be used to perform forensic analysis at a later stage inspecting traffic from 5 minutes interval to hourly, daily, and monthl

Figure 3 depicts the idea of multi-segment monitoring architecture of InstaMon cloud solution. The idea is to adapt pfsense to act as border control firewall that regulates internet traffic and as high speed capturing and analysis box. This pfsense based capturing and analysis box will tap on the edge of every network segment will periodically pass both collected packets and statistical analysis data to InstaMon agent. InstaMon agent will perform real-time traffic analysis and use various profiles, rules and engines from the backend analytics created to detect anomalous traffic, abnormal user usage behavior, as well as matching the traffic prediction trend with the actual traffic.

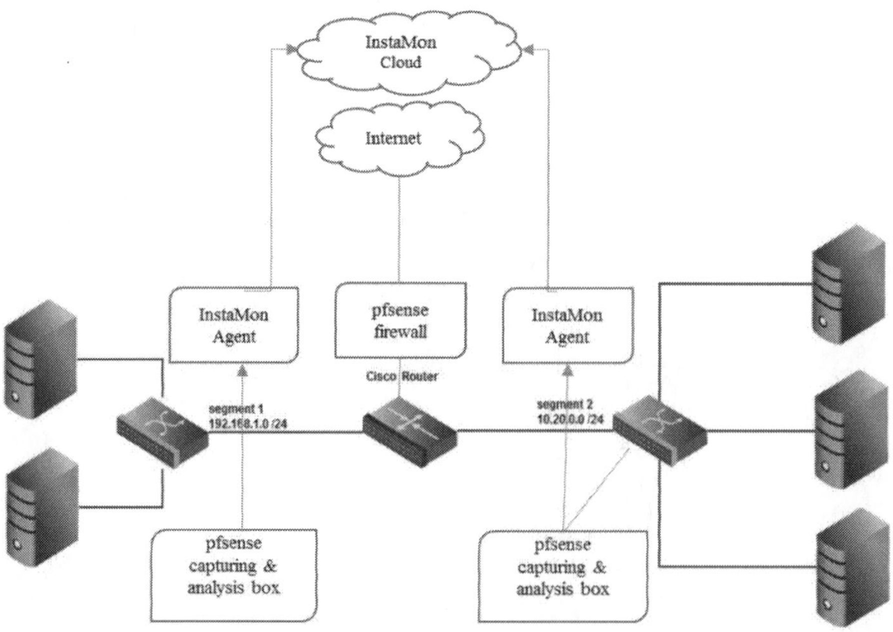


**Figure 3.** The architecture of multi-segment monitoring.

### *3.3. The proposed Sequential Hierarchical Superset (SHS) model*

The biggest problem lies in the fact that neural networks have little similarity with the actual processes in the brain other than both being constructed of many interconnected units. To address this issue, a more biologically plausible model is needed that reflects the actual human intelligence. In fact, it is a need to first understand how intelligence developed in human brain from neuroscience perspective.

Our knowledge on the detailed biological process of the human gene interaction, human brain evolution, and other biological processes is still in its infancy. But two recent important discoveries in neuroscience have shed some lights on different perspectives on modeling the brain. The first one is the fact that the evolution of human brain structure is significantly affected by its genome structure and the second one is a theory called memory-prediction framework which explains the high-level architecture of the human neocortex and how it works. The memory-prediction framework states that human's neocortex memory system forms the basis of intelligence.

The task of network forensic requires human intelligence to analyze and understand what is happening to the network. The Sequential Hierarchical Superset (SHS) model have the advantage of being able to process data in high speed and able to perform basic analysis on the data in real-time fashion. The proposed approach of Sequential Hierarchical Superset (SHS) implementation is translating the biological model of neocortex memory system into a simple working model based on the memory prediction framework [20] and with their smaller part in distributed neuro-gen adaptive engine [21][22][23]. Figure 4 shows the framework, adopted from [20].

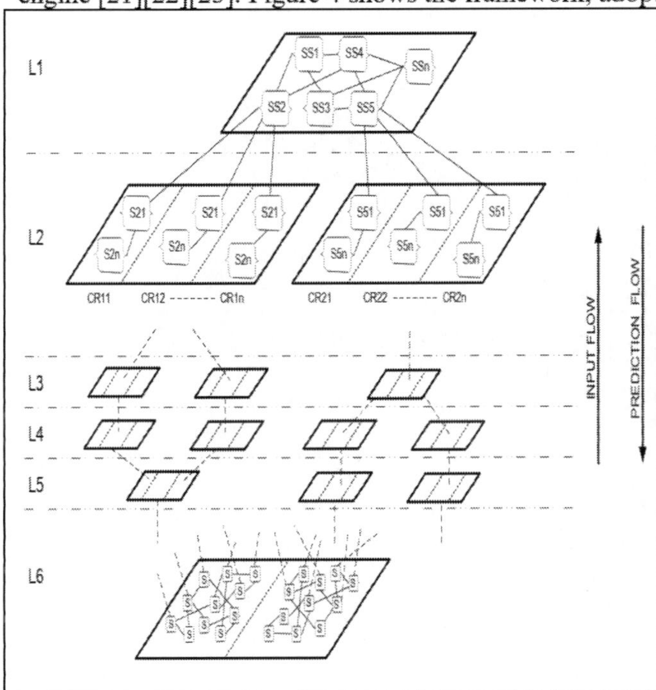


**Figure 4.** Neocortex memory system framework

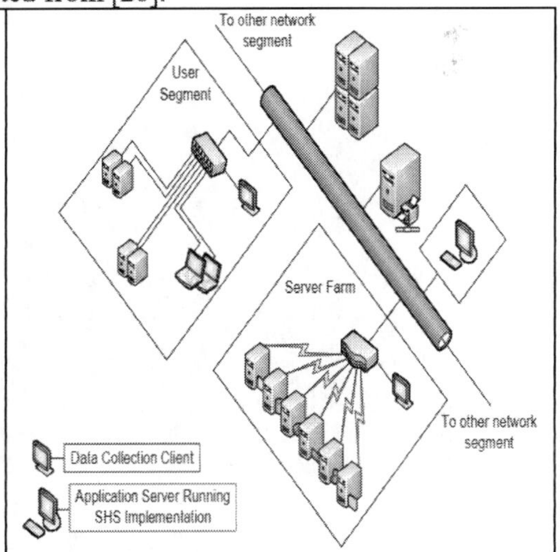


**Figure 5.** Typical implementation in enterprise networks.

## 4. Initial implementation & results

The first phase of the implementation is undergoing. The system is developed using Java programming under Eclipse technology platform. The typical implementation of the proposed system in an enterprise system is shown in Figure 5. Some intelligent features are shown in Figure 6, Figure 7 and Figure 8.

The preliminary results from experiments show a promising performance of SHS in terms of the way it manages and relates information about the network. The results were analyzed after running the experiment for around one month. The first and second week can be considered as the SHS learning period. Figure 8 shows the number of sets inside SHS memory after 1 month of experiment.

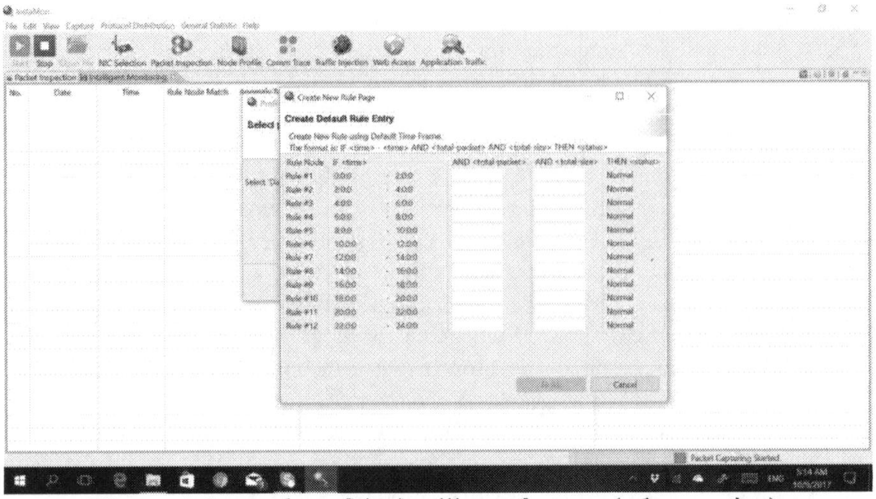

**Figure 6.** Screenshot of the intelligent features (rules creation).

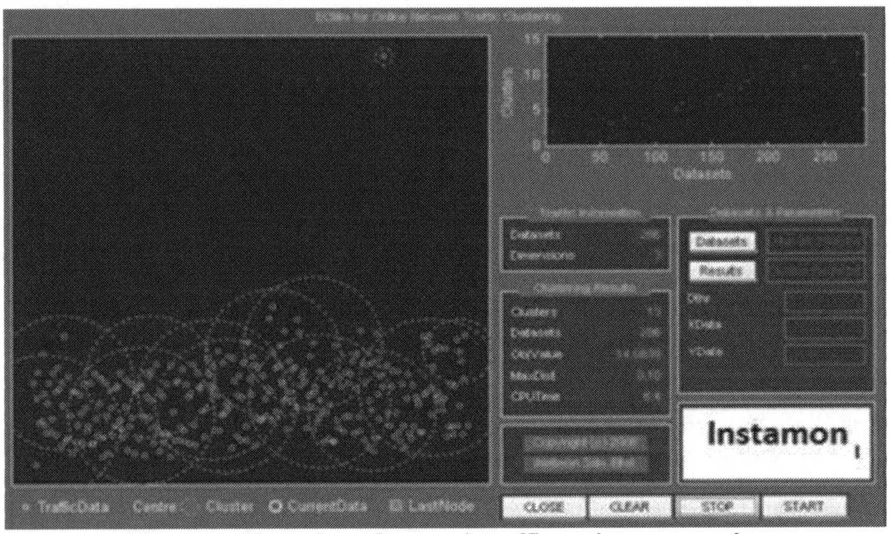


**Figure 7.** Clustering of network traffic to detect anomaly.

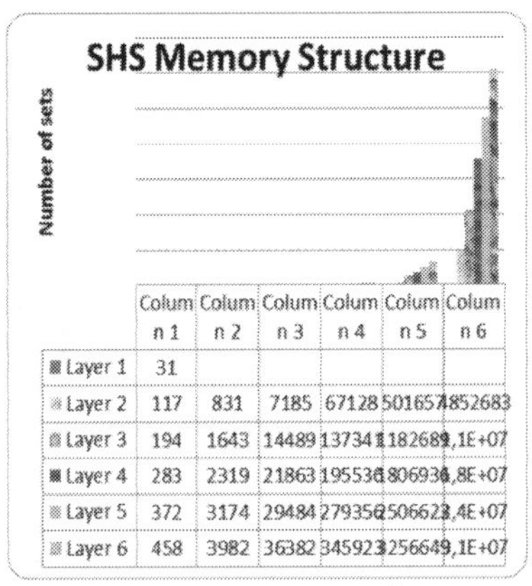


**Figure 8.** The SHS memory structure after 1 month of running.

IOP Conf. Series: Journal of Physics: Conf. Series **978** (2018) 012100 doi:10.1088/1742-6596/978/1/012100

## 5. Conclusion

In conclusion, it is believed the proposed InstaMon cloud solution could provide holistic end-to-end network monitoring and protection to assist network administrator to manage and protect their organization's network.

## References

[1] Wang, C.Y., Chou, S.-cho T. & Chang, H.-ching, 2009. Emotion and motivation: understanding user behavior of web 2 . 0 application, IEEE Computer Society Seventh Annual Communications Networks and Services Research Conference, pp.1341-1346.

[2] Kim, W., Jeong, O.K. & Lee, S.W., 2010. On social Web sites. Journal of Information Systems, 35, pp.215-236.

[3] Sun, Y. et al., 2010. Hybrid Regular Expression Matching for Deep Packet Inspection on Multi-Core Architecture. IEEE Proceeding Computer Communications and Networks (ICCCN), pp.1-7.

[4] Yin, Q. et al., 2004. Based on Behavioral Method Based on Behavioral Model. AnalysisIEEE Proceeding of the 5th World Congress on Inteligent Control and Automation, pp.4370-4374.

[5] Qiao, H. et al., 2007. Behavior Analysis-Based Learning Framework for Host Level Intrusion Detection. Proceedings of the 14th Annual IEEE International Conference and Workshops on the Engineering of Computer-Based Systems (ECBS'07).

[6] Lim, S.Y. & Jones, A., 2008. Network Anomaly Detection System : The State of Art of Network Behaviour Analysis British Telecommunications plc ., Security Research Centre , Ipswich , United Kingdom. International Conference on Convergence and Hybrid Information Technology 2008, pp.459-465.

[7] Koch, R., 2009. Changing Network Behavior. 2009 Third International Conference on Network and System Security, pp.60-66.

[8] Galassi, U., Giordana, A. & Mendola, D., 2005. Learning User Profile from Traces. 2005 Symposium on Applications and the Internet Workshops (SAINT 2005 Workshops), pp.166-169. Available at: http://ieeexplore.ieee.org/lpdocs/epic03/wrapper.htm?arnumber=1620003.

[9] Oh, S.H. & Lee, W.K., 2003. An anomaly intrusion detection method by clustering normal user behavior. Computers & Security, 22(7), pp.596-612.

[10] Pais, I. & Almeida, M., 2009. End User Behavior and Performance Feedback for Service Analysis. Intelligence in Next Generation Networks, 2009. ICIN 2009, pp.1-6.

[11] Barabas M, Boanea G, and Dobrota V, 2011. Multipath routing management using neural networks-based traffic prediction, in The 3rd International Conference on Emerging Network Intelligence, Lisbon, Portugal, Nov. 2011.

[12] Zhang, J. et al., 2009. A Distributed IPS Model Based On Neighbor Distance 1. IEEE Proceeding Ubiquitous, Autonomic and Trusted Computing, (09), pp.465-469.

[13] Ali M.K.M. and Kamoun F, 1993. Neural networks for shortest path computation and routing in computer networks, IEEE Transactions on Neural Networks, vol. 4, no. 6, pp. 941–954, Nov. 1993.

[14] Boanea G, Barabas M, Rus A.B., Dobrota V, and Domingo-Pascual J, 2011, Performance evaluation of a situation aware multipath routing solution, in 2011 RoEduNet International Conference 10th Edition: Networking in Education and Research, Iasi, Romania, Jun. 2011, pp. 1–6.

[15] Fadlullah Z, Tang F, Mao B, Kato N, Akashi O, Inoue T, Mizutani K, 2017. State-of-the-Art deep learning: evolving machine intelligence toward tomorrow's intelligent network traffic control systems, IEEE Communications Surveys & Tutorials. 2017, Issue: 99, DOI: 10.1109/COMST.2017.2707140

[16] Liu, F., Findlay, R. D. and Song, Q., "A Neural Network Based Electrical Loss Prediction of Bare Overhead ACSR Conductors" Proceeding of the First International Conference on Innovative Computing, Information and Control - Volume II (ICICIC'06), IEEE Press, 2006,

pp. 392-395.

[17] Gwon Y. L and Kung H. T, 2014, Inferring origin flow patterns in wi-fi with deep learning," in Proceedings of the 11th International Conference on Autonomic Computing (ICAC '14), pp. 18-20, Philadelphia, PA, USA, Jun. 2014.

[18] Ngia J, Khosla A, Kim M, Nam J, Lee H, and Ng A.Y ,"Multimodal Deep Learning," in International Conference on Machine Learning (ICML), Bellevue, USA, Jun. 2011.

[19] Hawkins, J., and Blakeslee, S., "On Intelligence", Owl Book, New York, 2005.

[20] Pasha M.F, Budiarto R, Ramadass S, and Syukur M, 2008. A sequential hierarchical superset implementation of neocortex memory system and its case study of automated network forensic analysis, Proc. of Int. Conf. on Artificial Intelligence 2008, Las Vegas, 14-16 July 2008, pp. 490-495.

[21] Rahmat R, Pasha M F, Syukur M, Budiarto R 2015 A gene-regulated nested neural network *International Arab Journal of Information Technology* ***12*** pp. 532–539

[22] Pasha M F, Rahmat R F, Budiarto R, and Syukur M 2010 A distributed autonomous neuro-gen learning engine and its application to the lattice analysis of cubic structure identification problem *International Journal of Innovative Computing, Information and Control* ***6*** pp. 1005–1022

[23] Hasibuan Z A, Rahmat R F, Pasha M F and Budiarto R 2009 Adaptive Nested Neural Network (ANNN) based on Human Gene Regulatory Network (GRN) for Gene Knowledge Discovery Engine *IJCSNS International Journal of Computer Science and Network Security* **9** pp. 43–54

IOP Conf. Series: Journal of Physics: Conf. Series **978** (2018) 012101 doi:10.1088/1742-6596/978/1/012101

# Virtual reality interactive media for universitas sumatera utara – a campus introduction and simulation

**R F Rahmat, Anthonius, M A Muchtar, A Hizriadi and M F Syahputra**

Department of Information Technology, Faculty of Computer Science and Information Technology, Universitas Sumatera Utara, Medan, Indonesia

Email: romifadillahrahmat@gmail.com

**Abstract**. Universitas Sumatera Utara is one of the public universities that have over 100 buildings with total area of more than 133.141 square meters. Information delivery on the location of the institutional buildings becomes challenging since the university land reaches 93.4 Ha. The information usually delivers orally, in video presentation and in the form of two-dimensional such as maps, posters, and brochures. These three techniques of information delivery have their advantages and disadvantages. Thus, we know that virtual reality has come to existence, touching every domain of knowledge. In this paper we study and implement virtual reality as a new approach to distribute the information to cover all of the deficiencies. The utilization of virtual reality technology combined with 3D modeling is aims to introduce and inform the location of USU institutional buildings in interactive and innovative ways. With the application existence, the campus introduction is expected to be more convenient so that all the USU students will be able to find the exact location of the building they are headed for.

## 1. Introduction

Information delivery about one location is done with oral, video and in the form of two-dimensional media such as maps, posters, and brochure. The two-dimensional shape is one of the most common ways used to since it is easy to create and distribute. Although the conventional information delivery is more economical, the information contained is less interactive and attractive.

Information of the location, to be more appealing, can be delivered using video. Information delivered using video is more effective than using two-dimensional media since the information presented on video more detailed and clearer, such as the similarity of the color and height of the actual building. However, it costs a lot of time and money in making it, and the result is less interactive because we cannot receive any feedback from the user.

Another method to deliver the information is oral technique. This technique is considered as the most interactive approach since it offers two-way communication with the speakers, so it is easier to get and understand the given information. However, the drawback of this method is the lack of speaker's availability to be ready 24/7 to deliver the information. Therefore, virtualization of the location can be considered as an efficient solution because it is able to overcome all the shortcomings of the above methods and accommodate all the advantages.

Virtual reality technology can be implemented in the virtualization of a location. The term of virtual reality (VR) first mentioned in several magazines and newspapers owned by Jaron Lanier, founder of Visual Programming Language (VPL) research company, in 1989. VR is a technology that uses computers and electronic technologies to produce a realistic three-dimensional atmosphere so that users can feel through sight, hearing, touching and forming a virtual world [1].

According to Zhang & Zheng [2], VR technology is a type of interface technology between humans and machines that can simulate people virtually being in a natural environment through sight, hearing, movement and other actions. Not only can illustrate the real environment, VR also allows users to observe virtual environments and to feel like they are in that location. Therefore, VR technology creates a new approach for visualizing spatial information. The technology has three characteristics, namely immersion, interactivity, and imagination [3].

Today, VR keeps growing in the field of urban planning, urban development, and urban management [4]. The virtual world has also been used by many educational institutions, organizations, medical academies for education purposes, business and marketing, etc. In the medical sector, for instance, virtual reality provides various assets for rehabilitation that exceeds the reach of the traditional methods[5]. Our previous research was connecting virtual reality with digital cultural preservation [14][15][16] and historical game [17]. The technology also can be put to use to help the rehabilitation process of stroke patients [6]. As for educational institutions, virtual reality can be implemented in the development of virtual campus so that people are able to explore the campus virtually. There are currently 29 universities in the world that utilize the technology for campus tour [7].

University of Sumatera Utara (USU) has 151 departments in its land area of 93.4 Ha. It has over 100 building with the total area of more than 133.141 square meters. Although it holds orientation week and campus introduction for freshman year students every year, the students' knowledge on campus area is considered below average. Therefore, a virtual visualization for USU campus is required to illustrate a clear, attractive visual representation and resembling the real campus area. In Indonesia, virtual campus tours are still limited to using combined photographs. In this research, the author will create a virtual tour using 3D model and integrate it with Oculus Rift so the experience will be more attractive and compelling.

In 2011, Didik Dwi Prasetya conducted research entitled Web-Based Virtual Tour Application as Tourism Promotion Media [8]. In the study, Didik used a set of images supported with music, narration, and text to simulate a tourist attraction location. The key to success which also serves the main component of a virtual tour is panoramic images. Moreover, the virtual tour is also combined with geolocation service for location mapping. The primary objective of the research is an innovative and interactive medium of tourism promotion and marketing.

In 2013, Nurul Mizaanatul Abror did a research entitled Virtual Tour Museum Prototype of E-Super Museum to introduce East Java and DIY Batik Culture [9]. In the research, Nurul did not use either audio or video, and only used a set of images and text to provide information. The images and texts were then compiled and processed using Macromedia Flash which would eventually be uploaded to the website.

In 2013, Fatchur Rohman Alhabso, Muhtadin, and Ahmad Zaini conducted research entitled 3D Object Visualization of Panorama 360 Virtual Touring [10]. In the web-based research, they use 3D scanners to reconstruct three-dimensional objects. Before implemented in the virtual tour, mesh decimation process and texture mapping modeling on the result of the reconstruction process. The main purpose of the research is to visualize museum collections that are not good enough to be visualized only in a two-dimensional form.

## 2. Methodology

The proposed method for this research consists of four stages, namely data acquisition, modeling, materializing-texturing and virtual reality environment. In data acquisition, USU maps that have been obtained from various sources are evaluated and perfected by conducting field inspections and interviews with experts. The initial 3D model design is performed based on USU maps in modeling stage. After conducting field inspections and interviews with experts, the initial 3D model is perfected into the final 3D model. The next step would be materializing & texturing. In this stage, the final 3D model is given the color and material for the model to look more real. The final stage is to create a virtual reality environment. In this stage, the support of game engine is used to generate an environment that resembles the reality.

The general architecture of the proposed method is shown in Figure 1.

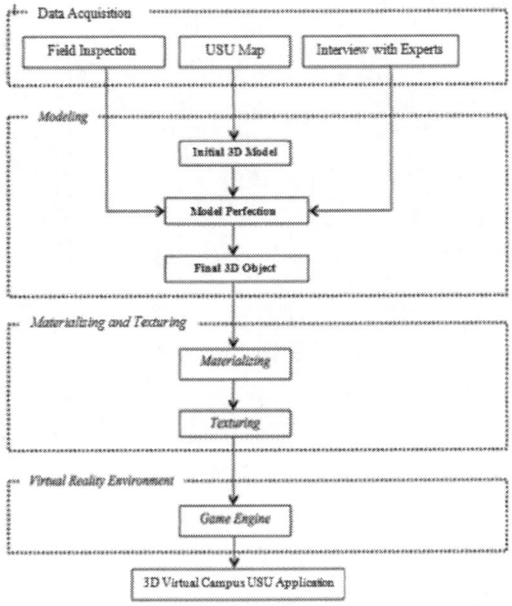


**Figure 1.** General architecture of the proposed method

### *2.1. Data acquisition*

There are three data resources for this research, they are USU maps, field inspection and interview with experts. Here we described every resources below.

#### *2.1.1. USU Maps*

The USU map is used as the base for the development of 3D virtual USU campus. The USU map is divided into two types, namely 2D map, and 3D map. 2D map is a map printed in 2011 by the USU Press. This map can be seen in Figure 2.

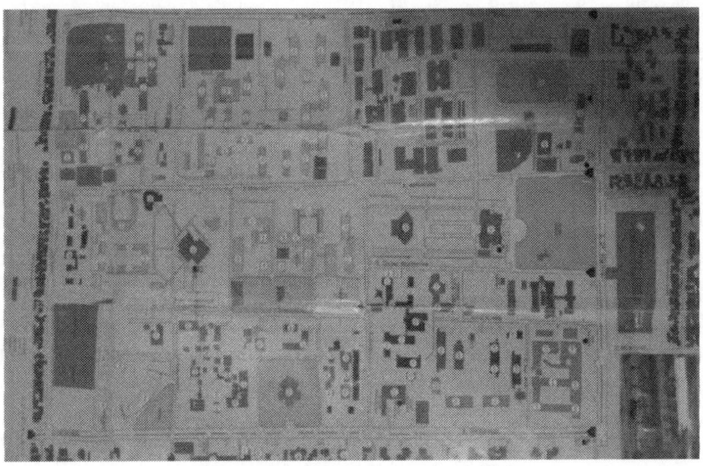

**Figure 2.** USU maps in 2011

In addition to the map printed by the USU Press, 2D map is also available from Google Maps. 2D USU map obtained from Google Maps in 2015 can be seen in Figure 3 below.

IOP Conf. Series: Journal of Physics: Conf. Series **978** (2018) 012101 doi:10.1088/1742-6596/978/1/012101

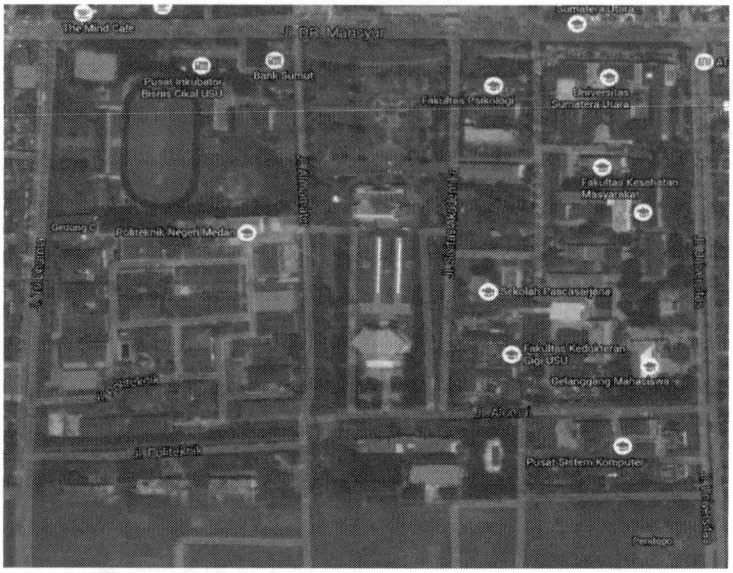


**Figure 3.** USU map based on Google Maps in 2015

The 3D USU maps created by the Architecture Studies Program have been used as references in this research. It can be seen in Figure 4.

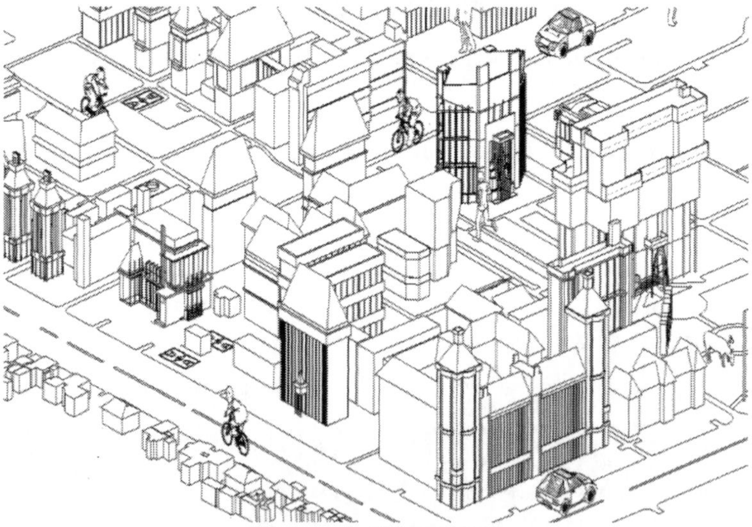

**Figure 4.** 3D USU Map

*2.1.2. Field inspection*

Field inspections are required in this research because the information presented on maps is not sufficient. Moreover, field inspection is also useful to match previously-obtained data with the actual form of the field so that the final result can become information sources that closest to the actual data. We collect mostly every main building in every faculty and department, including its infrastructure, road and garden. The sample of field inspection result can be seen in Figure 5.

**Figure 5.** Field inspection of faculty of Computer Science and Information Technology

Figure 5 was taken at faculty of Computer Science and Information Technology. In order to obtain good modeling, the images of every building were taken from various angles.

### *2.1.3. Interview with experts*

The mentioned expert for this research is architect experts. The purpose of this stage is to check whether the modeling results are in accordance with the actual form seen from architecture. If not, then the 3D model will be revised.

### *2.2. Modelling*

There are three phases of the modelling process. The first phase is initial modeling process. Data used in this process is based solely on USU map. Once the rough draft is complete, the field inspections result is used to improve the initial modeling results. Afterwards, interview with expert is conducted to obtain the final 3D model.

3D model built in this research is a solid 3D model to generate more realistic results. There are two techniques used in modeling, namely subdivision modeling and NURBS (spline modeling).

Subdivision modeling is one of the polygon modeling techniques. The stages in subdivision modeling can be divided into planning, rough draft, rough detail draft, final detail, and clean up.

The planning stage begins with measuring and calculating the model size to be built. After the measurement is completed, the rough draft stage can be started by selecting primitive geometry objects, such as cube, cylinder, sphere, cone, pyramid, and torus. The objects are then placed in a position in accordance with the planning result. After the rough draft is complete, the basic shape of the model can be seen even though it is not detailed. Therefore, it is necessary to perform refinement to improve and refine the model. Refinement process is conducted several times to generate final model details. Sample result of the model details final stage is shown in Figure 6a. The final model is then cleaned at the clean-up stage so the unnecessary geometry objects can be removed.

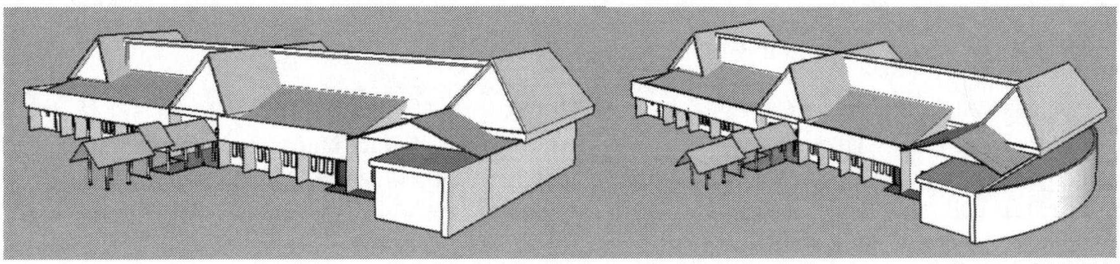

**Figure 6 (a).** The result of the model details final stage **(b).** Mixed subdivision modeling and NURBS model

For the curved model, a non-uniform rational B-spline (NURBS) technique is required. NURBS has stable floating points error, requires less memory and is able to represent any type of curve or surface. In contrast to the subdivision modeling, a polygonal geometry, the NURBS model has no face, edge, or vertices. The NURBS model is built using two or more Bezier curves or also called splines. An example of a mixed subdivision modeling and NURBS model can be seen in Figure 6b.

### *2.3. Materializing and Texturing*

Texturing or texture mapping is a shading technique to synthesize images where texture images are mapped to the surface in a three-dimensional scene, much like wallpaper affixed to a wall. The advantage of texture mapping is that it can add numerous details to an object even though it takes a little extra time when rendered [11].

Slightly different from textures containing bitmap data, essentially, the material is a shader or surface characteristic of an object, such as slick, solid or transparent. The material contains parameters that determine how light interacts with the model. The most commonly used materials are translucent, metal, stone and wood. This stage aims to make the final model look more real. Sample of texturing and materializing result is shown in Figure 9

**Figure 7.** Result of texturing and materializing

### *2.4. Virtual Reality Environment*

The environment is a very important aspect of virtual reality and must be built with caution, in order to provide an immersive experience to users. An immersive virtual environment can give the users a sensation as if physically in the virtual world [12]. About 80% of the information that humans feel comes from the eye, making visual perception plays an important part [13]. Therefore, the virtual environment is the key so that users can interact with the world entirely and allow the users to be unaware of the surrounding environment in the real world. In this research, game engine Unity 3D version 5 was implemented in the development of a virtual environment.

## 3. Experimental Result and Analysis

### *3.1. Virtual Tour Facilities*

In 3D Virtual Campus USU application, there are several features to facilitate the user while improving the quality of virtual tour, which are Faculty building marker, Street marker, Minimap to mark the user location, USU full map, USU Bus, Teleportation to selected location, Choice of virtual tour duration, Guidelines in using the application.

### *3.2. Application Initial Display*

There are several displays showed when application launches, namely the splash screen display, the welcoming title display and the main menu display. Splash screen display is shown in Figure 8a.

**Figure 8 (a).** Splash screen display **(b).** Welcoming title display

On the splash screen, three different time differences are available in the application, i.e., night, morning and dawn. The splash screen will appear for eight seconds. The time is used to load the required assets for the next scene.

The welcoming title view can be seen in Figure 8b. The welcoming title display aims to greet users while adding aesthetic value. The user can proceed to the main menu by pressing the Virtual Reality button.

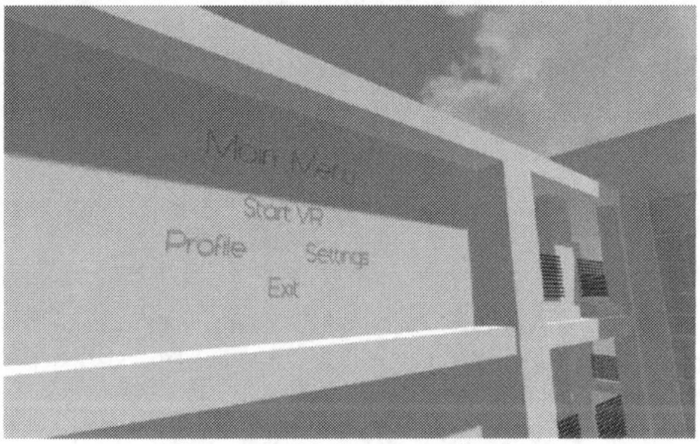


**Figure 9.** Main menu display

The main menu display can be seen in Figure 9. In the main menu, there are four buttons, namely Start VR, Profile, Settings, and Exit. The Start VR button will bring the users into USU and conduct a virtual tour. The Profile button will display the names of all people who play a role in this research completion. The Settings button will display the settings for time duration and sound. The Exit button takes the users out of the app.

**Figure 10 (a).** Application usage guidelines display **(b).** Virtual reality environment

### *3.3. VR Display*
Shortly after the user presses the Start VR button and enters the USU environment, the user will be told how to access the application usage guideline, by pressing Tab key. The application usage guideline is shown in Figure 10a. In Figure 10a, there is a building presented behind the application usage guidelines. The building is one of building modeling and is the lecture building of Information Technology department. A sample of virtual environments that have been built can be seen in Figure 10b. The built environment includes objects such as trees, grass, cars, street lights, etc.

### *3.4. Application Testing*
Application testing is required to test all components that have been designed and implemented. The test aims to ensure each part has functioned properly and has been in accordance with the expectation. There are two methods used in testing, namely black box testing and pluralistic walkthrough.

In the black box testing method, testing was performed on the interface regardless of coding. Based on black box testing, it shows that the interface has been functioning correctly.

In pluralistic walkthrough method, the test was conducted with end-user involvements. The purpose of pluralistic walkthrough is to acknowledge the visualization results of the institutional building that have been made. Numerous questions related to modelling results were given in the form of questionnaires. There are 20 end-users from various backgrounds invited to the test. Based on the questionnaires, it shows that more than 80% of end users stated that the modeling buildings could represent the actual ones and 100% of end users stated that 3D Virtual Campus USU is able to help inform the location of the college building in USU and benefit the students especially for new students every year.

## 4. Conclusion and Future Research

### *4.1. Conclusion*
In accordance with the testing of 3D Virtual Campus USU application, it can be concluded as follows:
1. The implementation result of virtual reality to the system is still depended on computers with high memory capacities.
2. Virtual reality implementation in 3D Virtual Campus USU application can be a breakthrough in providing information on lecture building locations to USU students excitingly and innovatively.

### *4.2. Future Research*
In the development of 3D Virtual Campus USU application, still, there are many drawbacks. Therefore, numerous improvements should be done in future research. Here are suggestions that can be used for further research:
1. Making 3D objects lighter, because the process of building rendering in large quantities will require high-capacitated memory.
2. The addition of the institutional building interiors and other buildings. For example, public structures such as a pavilion, gathering building and library.
3. More detailed activities on campus, such as walking and running students.
4. Artificial intelligence implementation to determine the shortest path to a location in the app.
5. The addition of multiplayer and chat features for the app to be used by more than one person at a time and can exchange information in the virtual world.

## References
[1] Liu Q 2012 The virtual reality technology in art design *2nd Int. Conf. Consumer Electron. Commun. Networks* 2226-28.

[2] Zhang L and Zheng G 2011 The virtual campus scene based on vrml *Int. Conf. Multimedia Technol.* 912-15.

[3] Feng L D, Ying H and Qin Z K 2010 Virtual reality technology and its application in digital campus system *World Congr. Software Eng* 356-60.

[4] Su P and Wang S 2012 Virtual reality practice in architecture design *IEEE Symp. Electrical Electron. Eng.* 98-10.

[5] Rizzo A A, Cohen I, Weiss P L, Kim JG, Yeh S C, Zali B and Hwang J 2004 Design and development of virtual reality based perceptual-motor rehabilitation scenarios *Proc. 26th Annu. Int. Conf. IEEE Eng. Med. Biol. Soc.* 4852-55.

[6] Sen S L, Xiang Y B, Ming E S L, Xiang K K, Fai Y C and Khan Q I 2015 Enhancing effectiveness of virtual reality rehabilitation system: Durian Runtuh. *10th Asian Control Conf.* 1-6.

[7] Chen Z, Qin H, Huang H L, Liu S J, Ni W C and Mo W J 2015 The construction of digital campus with virtual reality technology—taking xinhua college of sun yat-sen university as an example. *Int. Conf. Edu. Manage. Commerce Soc.* 224-28.

[8] Prasetya D D 2011 Web-based virtual tour application as tourism promotional medium *Seminar Electrical Inf. Edu.* 58-63.

[9] Abror N M 2013 *Virtual Tour Museum Prototype of E-super museum To Introduce East Java And DIY Batik Culture* (Universitas Dian Nuswantoro: Thesis).

[10] Alhabso F R, Muhtadin and Zaini A 2013 *3D Object Visualization of Panorama 360□ Virtual Touring* (Institut Teknologi Sepuluh November: Thesis).

[11] Heckbert P S 1989 *Fundamentals of Texture Mapping and Image Warping* (University of California: Thesis).

[12] Sharma S L, Rajeev S P and Devearux P 2015 An immersive collaborative virtual environment of a university campus for performing virtual campus education drills and tours for campus safety *Int. Conf. Collaboration Technol. Syst.* 84-89.

[13] Ying L W 2010 The design of campus roaming system based on virtual reality technology *Int. Conf. Audio Lang. Image Process.* 274-77.

[14] Syahputra M F, Annisa T, Rahmat R F and Muchtar M A 2017 Virtual Application of Darul Arif Palace from Serdang Sultanate using Virtual Reality *J Phys. Conf. Ser.* **801** 12009.

[15] Syahputra M F, Siregar R K and Rahmat R F 2017 Finger recognition as interaction media in Augmented Reality for historical buildings in Matsum and Kesawan regions of Medan City *Lect. Notes Comput. Scie* **10325.**

[16] Syahputra M F, I M Rizki, Fatimah S and Rahmat R F 2017 Implementation of player position monitoring for Tanjung Pura Palace Virtual Environment *Lect. Notes Comput. Scie.* **10325.**

[17] Syahputra M F, Widhianto M and Rahmat R F 2017 Historical game of Majapahit Kingdom based on tactical role-playing game *Proc. 12th Int. Joint Conf. Comput. Vision Imaging Comput. Graphics Theor. Appl. (VISIGRAPP 2017),* 327-332.

# An Interactive Augmented Reality Implementation of Hijaiyah Alphabet for Children Education

**R F Rahmat[1], F Akbar[1], M F Syahputra[1], M A Budiman[1] and A Hizriadi[2]**

[1]Department of Information Technology, Faculty of Computer Science and Information Technology, Universitas Sumatera Utara, Medan, Indonesia

Email: romi.fadillah@usu.ac.id

**Abstract**. Hijaiyah alphabet is letters used in the Qur'an. An attractive and exciting learning process of Hijaiyah alphabet is necessary for the children. One of the alternatives to create attractive and interesting learning process of Hijaiyah alphabet is to develop it into a mobile application using augmented reality technology. Augmented reality is a technology that combines two-dimensional or three-dimensional virtual objects into actual three-dimensional circles and projects them in real time. The purpose of application aims to foster the children interest in learning Hijaiyah alphabet. This application is using Smartphone and marker as the medium. It was built using Unity and augmented reality library, namely Vuforia, then using Blender as the 3D object modeling software. The output generated from this research is the learning application of Hijaiyah letters using augmented reality. How to use it is as follows: first, place marker that has been registered and printed; second, the smartphone camera will track the marker. If the marker is invalid, the user should repeat the tracking process. If the marker is valid and identified, the marker will have projected the objects of Hijaiyah alphabet in three-dimensional form. Lastly, the user can learn and understand the shape and pronunciation of Hijaiyah alphabet by touching the virtual button on the marker

## 1. Introduction

Technology is evolving very rapidly; this development makes technology help humans in all fields. One of the technological development examples is augmented reality. Augmented reality is a technology that combines two-dimensional or three-dimensional virtual objects into actual three-dimensional circles and projects them in real time. Augmented reality can be applied to the sense of hearing and touch. Besides being utilized in areas such as healthcare, military, and manufacturing industries augmented reality can be implemented to learning media, one of which is learning the Qur'an.

Augmented reality (AR) technology is an impressive multimedia technology development since it is not only impressing the users but also making them feel entertained and enjoy the technology as well as gain some useful information. Augmented reality (AR) technology is a variation of virtual environments that thoroughly immerse users in a synthetic environment [1].

Childhood is the time to learn and play. Lessons delivered by educators or parents should include many playful and fascinating activities. If the fun element is not on the table, the children will simply get bored and reluctant to follow the lessons to be given by educators or parents. Furthermore, to learn Hijaiyah alphabet, it is vital to have the fun element so that the children can enjoy learning Hijaiyah alphabet at an early stage. By utilizing augmented reality technology, the unpleasant and boring lessons of Hijaiyah alphabet can be minimized, so that it would be more fun and exciting.

The augmented reality which usually runs on Android platform has the advantages that can provide experience and knowledge of the subject easy to be understood by the learners. It also can be implemented to give information to the user [6]. With these advantages, augmented reality can be used as an engaging and interactive method of learning. Based on these matters, we were motivated to develop augmented reality application to give children a different learning process on understanding Arabic letters in the Holy Quran. It also can stimulate the children to be more active in learning Hijaiyah letters through Android in the future. In learning Hijaiyah alphabet, a child tends to experience tiredness and boredom. Therefore, a solution is required to tackle this problem so children do not easily get bored and saturated during the learning period. Other studies related to augmented reality has been conducted earlier, among which is Pratama (2014). In his research, they implemented augmented reality in the development of Android-based bataknese musical instrument Taganing introduction [2]. In the study, He used a marker with five 3D objects, and each object produced a different sound.

Yudiantika et al. (2013) conducted research to implement augmented reality on museum by developing an educational application for museum visitors [3]. In 2014. Nugraha, in his research, utilized augmented reality in Piano basic learning. [4]. Nugraha developed a useful application for theory learning of Piano to assist the users on Piano chord lessons.

## 2. Methodology

System modeling aims to simplify the development and comprehension of the system. System modeling consists of general architecture, general design flowchart, 3D object creation flowchart, marker creation flowchart, application development flowchart and use case diagram.

### 2.1. General Architecture

The general architecture of the proposed method for the learning process of Hijaiyah alphabet is shown in Figure 1. The proposed steps of the application development using augmented reality are as follows: User is going to place the marker that has been registered; smartphone camera will be tracking the marker. If the marker tracking is invalid, user should repeat the tracking process. If the marker is valid or identified, the marker will have projected the objects of Hijaiyah alphabet in three-dimensional form. The object of Hijaiyah alphabet will be placed on the identified marker; the user can listen to Hijaiyah letters pronunciation by touching the marker on the virtual button; system generates the sound of the selected Hijaiyah letter through the smart phone.

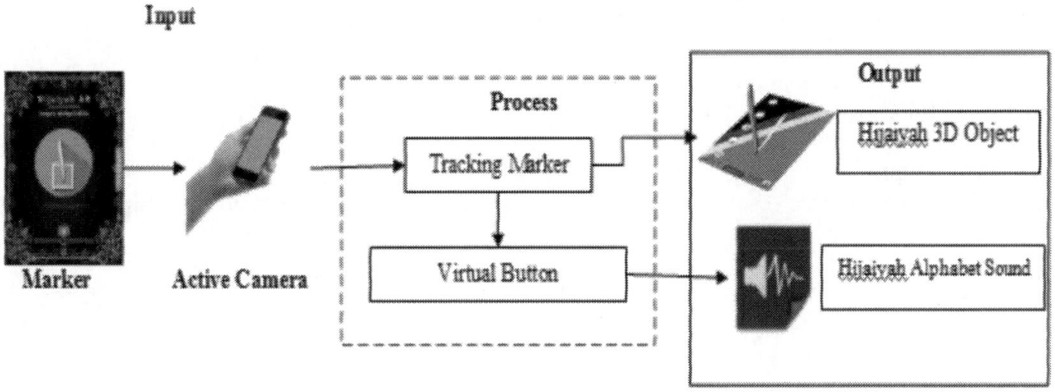


**Figure 1.** General architecture of the proposed method

IOP Conf. Series: Journal of Physics: Conf. Series **978** (2018) 012102 doi:10.1088/1742-6596/978/1/012102

*2.1.1. Input*
In this stage, the camera will identify the selected images, after the image pattern is saved in the library.

*2.1.2. Process*
Before the application can project 3D objects and Hijaiyah letters sound, it must be subjected to some process.

1. *Tracking marker*

The process of target registration is using Vuforia. The user will upload images for the target to be tracked and registered, then download the target resource, which bundled with the application. In this process, the target resource is downloaded in the form of libraries that have been bundled with Unity 3D application.

The working process of Vuforia AR SDK is as follows [5]: the camera captures images of the real world to track the marker then register the marker. The captured image serves as marker converted from YUV 12 to RG565 format for OpenGL ES and then set the lighting for marker tracking; the marker is turned into multiple frames, using a computer vision algorithm to detect and track real objects taken from the camera. The object is evaluated, and the result will be stored and accessed by the application; after getting the perfect position of the camera, the captured object will be rendered and visualized in the form of video in real-time.

2. *Virtual button*

The targeted image is used to place the virtual button, so when the marker is touched, the app will generate the sound of Hijaiyah alphabet in accordance with the selected visual button. The workflow is as follows: the virtual button on targeted images, each targeted image has five virtual buttons. Alter the name of each virtual button to simplify the configuration process; then add the audio file of Hijaiyah alphabet on each virtual button by dragging audio file on window asset into virtual button; on inspector audio source window, Bypass Effects and Play on Awake are unchecked and positioned the Pan Level at zero so the sound can be heard; reset the existing script of the virtual button, to declare the variables of Audio Source data so that the audio source variables appear on the target inspector image window and to set the virtual button for the audio to be heard when the hand touches the coordinates of the virtual button [4].

3. *Output*

   Outputs generated are the 3D objects and sounds of Hijaiyah alphabet.

## 3. Experimental Results And Analysis

3.1. Marker Design
There are 28 markers in this research, divided into one marker as the cover and the others as Hijaiyah letters.

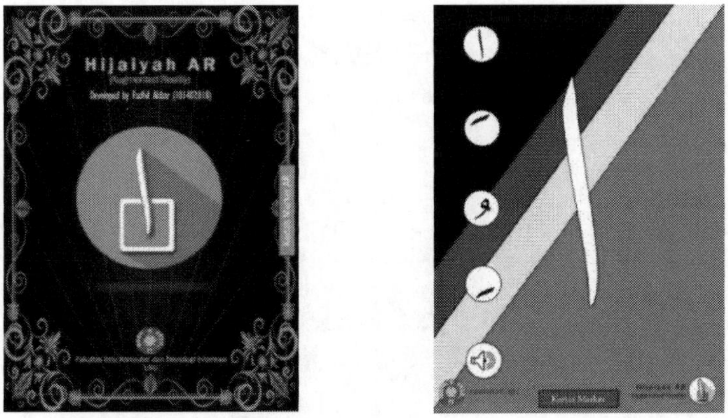


(a) (b)

**Figure 2.** Design marker (a) cover and (b) letter Alif

## 3.2. Application Layout

### *3.2.1. Main menu layout*
Main menu page consists of a start button, information button, help button and exit button. It is shown in Figure 3.

**Figure 3.** Main menu display

### *3.2.2. Start submenu layout*
Start submenu shows the function of smartphones camera to track the augmented reality in order to project 3D objects of Hijaiyah augmented reality.

On the left side of the marker, there is a virtual button that consists of five circles and each circle has a certain code that serves as a button that will sound when touched or covered. For more detailed information, it can be seen in Figure 4.

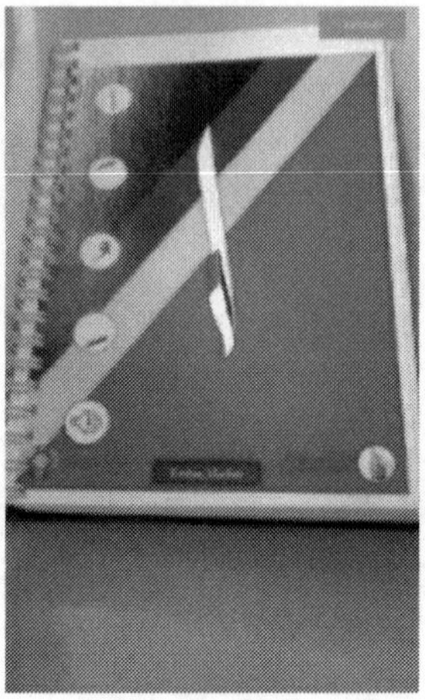

**Figure 4.** Start submenu layout

*3.2.3. Information submenu layout*

Information submenu displays brief information about Hijaiyah letters and the application. The layout design of submenu information can be seen in Figure 5 below:

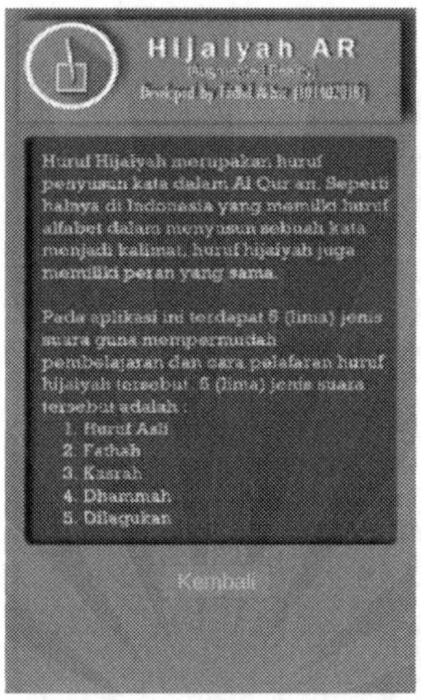


**Figure 5.** Information submenu layout

3.3. Marker Testing

There are several tests to detect the markers,i.e., marker trial, distance trial, Android camera resolution test, and lighting test.

*3.3.1. Marker trial*

In this research, authors limit the detection system to a single marker. So if there is more than one marker, the application is going to show one 3D object.

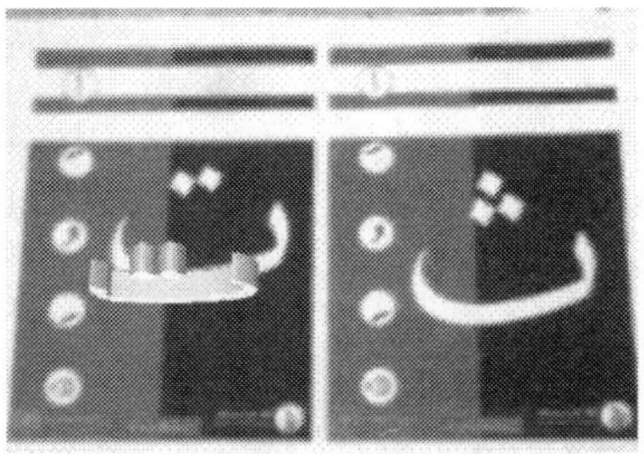

**Figure 6.** Detected 'Ba' letter

*3.3.2. Distance test*

In distance test, the closer the distance between the camera, the marker will result in the bigger marker being recognized by the camera so that the marker can be adequately captured.

Table 1. Marker distance test

| *No* | *Distance* | ***Tracking Marker Result*** |
|---|---|---|
| 1 | 15 cm | Well detected |
| 2 | 30 cm | Well detected |
| 3 | 50 cm | Well detected |
| 4 | 100 cm | Well detected |

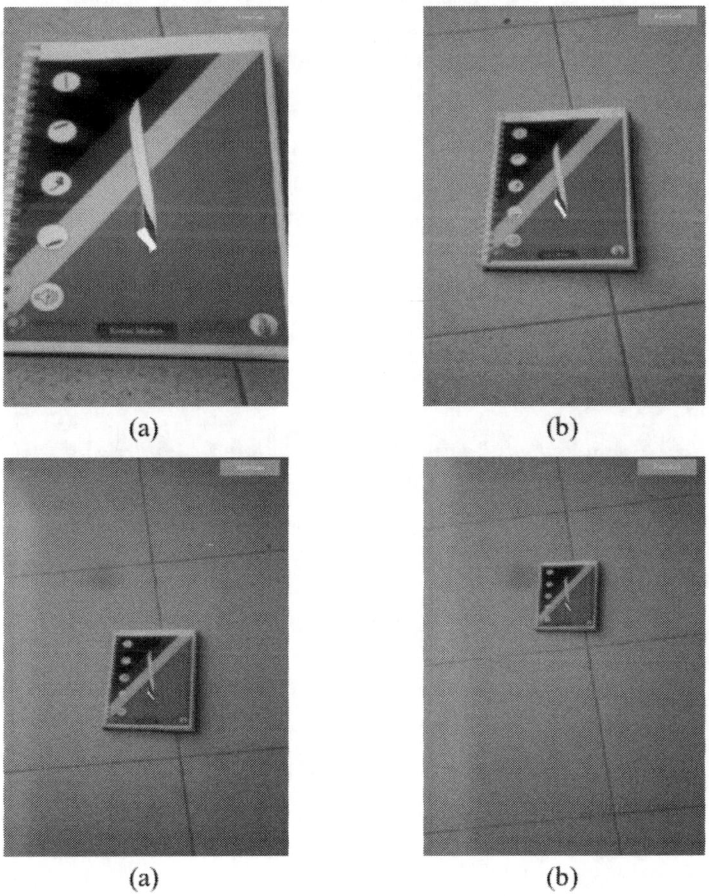

(a) (b)

(a) (b)

**Figure 7.** Distance test (a) 15 cm, (b) 30 cm, (c) 50 cm dan (d) 100 cm

### *3.3.3. Lighting test*

In lighting test, to detect the marker, lighting has great influence on 3D object quality. The trial was conducted in dark room without any lighting, the only source of lighting came from the brightness of the laptop screen. A marker is placed on keyboard, and laptop screen is lowered to allow light to illuminate the marker. Lighting test is shown in Figure 8.

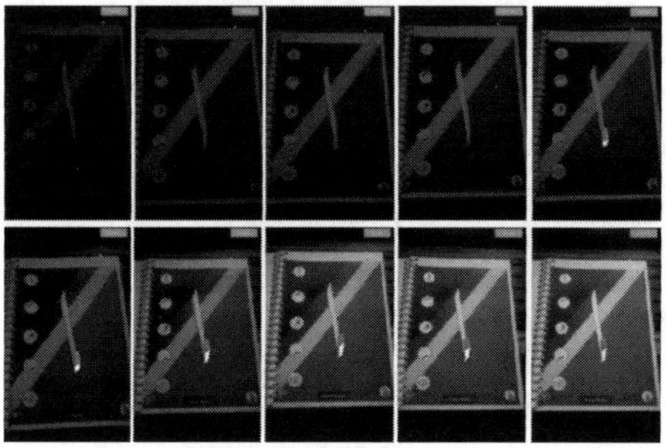

**Figure 8.** Lighting test

## 4. Conclusion and Future Research

Based on the results of analysis and testing of android-based learning application of Hijaiyah letters using AR, it can be concluded as follows; AR applications are still a new thing for children, so it looks more appealing to try the application of learning hijaiyah letters using AR; the marker detection on the application has functioned well. 3D objects Hijaiyah can appear on the marker; each virtual button can make a sound based on the button selected by the hand; in the application usage, must be considered the distance and focus of the camera with the marker; the application cannot run in a place that has less light (dark), because the camera requires light to track AR marker.

For future research, to be able to create a more responsive virtual button, to make it easier to use; in building applications, it is wise for researchers to master 3D Blender application not as a modeling but also animation, in order to create changes effects of 3D objects when touched

## 5. References

[1] Azuma T R 1997 *A Survey of Augmented Reality* (Presence: Teleoperators and Virtual Environments)

[2] Pratama M A 2014 *Augmented Reality Implementation In Android-based Batak Taganing Application Design* (Universitas Sumatera Utara: Medan)

[3] Yudiantika A R, Pasinggi E S, Sari I P, Hartono B S 201 *Augmented Reality Implementation At Museum: Study case Educational Application for Museum Visitors* (Universitas Gajah Mada: Yogyakarta)

[4] Nugraha I 2014 *Augmented Reality Utilization For Piano Instrument Learning* (Universitas Diponegoro: Semarang)

[5] Siregar A S 2011 Application Development Of Interrelationship Determinant Between Student Data And Length Of Study Using Multiple Linear Regression Algorithm STT Harapan Medan Students Case Study (Universitas Sumatera Utara: Medan)

[6] Muchtar M A, Syahputra M F, Syahputra N, Ashrafia S and Rahmat R F 2017 Augmented Reality for Searching Potential Assets in Medan using GPS based Tracking *Journal of Physics: Conference Series* **801** p. 12010

IOP Conf. Series: Journal of Physics: Conf. Series **978** (2018) 012103 doi:10.1088/1742-6596/978/1/012103

# Rangku Alu – A Traditional East Nusa Tenggara Game in Android Platform

**R F Rahmat[1], R Ramadhan[1], D Arisandi[1], M F Syahputra[1] and O Sheta[2]**

[1]Department of Information Technology, Faculty of Computer Science and Information Technology, Universitas Sumatera Utara, Medan, Indonesia
[2]Department of Computer Science, Zagazig University, Egypt

Email: romi.fadillah@usu.ac.id

**Abstract**. Rangku Alu is a traditional Indonesian game originated from Manggarai, East Nusa Tenggara, which is played using two pairs of bamboos or sticks in motion until the opponent's foot is wedged by the bamboos. However, nowadays the game is rarely played, as the rapid development of technology, the game can be played individually by anyone through an online game using media devices such as mobile or PC. Rangku Alu is a game where the moves of a dancer or player varied in each dance. In this research, Fisher-Yates Shuffle algorithm was used as a randomization method to determine the next moves to prevent the tap areas to appear at the same place more than once in a row. From the results, it shows that the tap areas have never been appeared at the same place in succession twice or more.

## 1. Introduction

Digital cultural perseveration has become a new field in the area of multimedia. The digitalization of culture has become a trend nowadays. Virtual Reality [1], Augmented Reality[2], Game[3], Computer Vision, and Mobile Games all of the areas has been related to preserving our culture. The traditional game as part of cultural heritage has become unknown right now, especially for our children's generation. The traditional game is a simple game played by children in the past to direct and lead them to be part of social events and community activities. However, nowadays the game is rarely played, as the rapid development of technology; the game can be played individually by anyone through an online game using media devices such as mobile or PC. This research positively affects the traditional game which begins to lose its existence as a medium of entertainment in the past

Rangku Alu is a traditional Indonesian game originated from Manggarai, East Nusa Tenggara, which is played using two pairs of bamboos or sticks in motion until the opponent's foot is wedged by the bamboos [4]. How to play the Rangku Alu Game begins by dividing the players into two teams, Player / Jumper team, which consists of 1–4 players, and Guard Team consisting of 4 players. The guard team squatted square and hold two bamboos in charge of moving them while singing the dance song, or other traditional songs. Jumper group whose turn to play will jump on the sidelines of bamboo according to the rhythm of the bamboo motions. The Jumpers should avoid their foot to get wedged by bamboos [4].

The property or tools used in Rangku Alu is four pieces of 2 meter-long bamboo; due to its difficulty of finding bamboo nowadays, it is usually replaced with scout sticks of the same size. The playground area should also be considered when playing this game. The field with the hard and non-grassy ground is the perfect place for this game because the grass can trigger the risk of slipping

IOP Conf. Series: Journal of Physics: Conf. Series **978** (2018) 012103 doi:10.1088/1742-6596/978/1/012103

during the play. Rangku alu is a game where the moves of a dancer/player varied in each dance. In determining the next possible moves is required a randomization method to build Rangku Alu game.

Juardi et al described some of the cultural preservation on Android-based game. In 2013 who built Karapan Sapi game using waterfall method model that is Requirement Analysis and Definition, System and Software Design, Implementation and Unit Testing, Integration and System Testing, Operation and Maintenance. They successfully created an educational game of "Karapan Sapi Race" with an attractive, entertaining and easy to play gameplay [5].

In 2013, Farizi et al. developed an Angklung simulator game on Android-based Smartphone through Game Design and Technical Design stages. These stages are then implemented and developed using AndEngine in Java. Based on the unit test results and integration, it can be concluded that the module unit of the game program met the functional requirements of the design. The performance test result shows an optimum performance when the game runs on a Smartphone that has a minimum processor of 600 MHz [6]. In 2015, Ovy Rizki performed a test on ball moves to determine the sensitivity level of an object and a test on a random function to determine the percentage of the random events occurrence possibility in Bekel adaptation game [7]. Fisher-Yates algorithm implementation has been applied by Ryan et al. (2015) to a 3D Unity based insect's species identification in The Lost Insect application with user satisfaction percentage at 80% and the rate of easy to use application is at 66.25% [8].

In this development, authors intended to build a game application that is accustomed from the traditional game of Rangku Alu by applying Fisher-Yates Shuffle as a method in determining the next possible moves in Android-based Rangku Alu game which will be performed particularly in its Random Mode.

## 2. Methodology

The proposed method for this research will provide an overview of the general architectural plan to be used in this development. Starting from users accessing the mobile to users facing the user interface that contains detailed information about the system that has game logic connected to the database to receive data from the system usage which returned to users. Details of the general architecture will be explained further in subsequent chapters. General architecture implemented by authors to build the Android-based game of Rangku Alu can be seen in Figure 1.

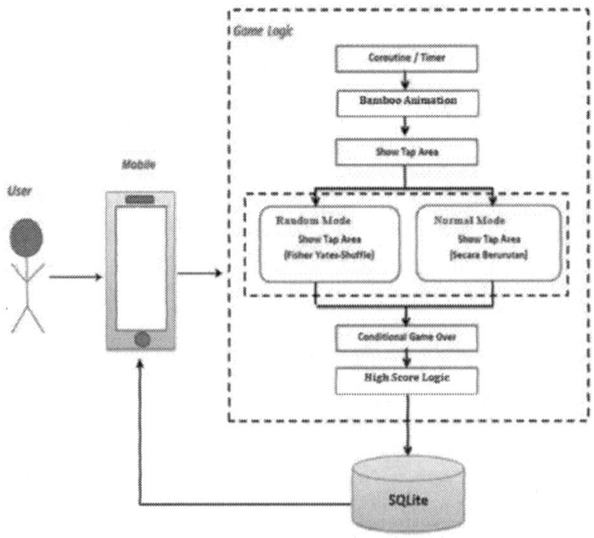


**Figure 1.** General architecture of the proposed method

In Figure 1, to access the game, it begins with selecting the Start menu which consists of "Play," "Score," and "Quit." In menu "play" the user will be presented with a selection of modes for game variation, and a choice of songs before entering the game area. This game has two types of modes that are normal mode and random mode. If random mode is chosen, then the tap area will be processed using Fisher-Yates Shuffle algorithm as in Figure 2. Otherwise, the tap area will be concocted manually. Then the system will receive input from the user as the user touches the same object with the tap area. An additional 100 points will be added to the user score. Otherwise, the game ends, and the high score will be stored in the database.

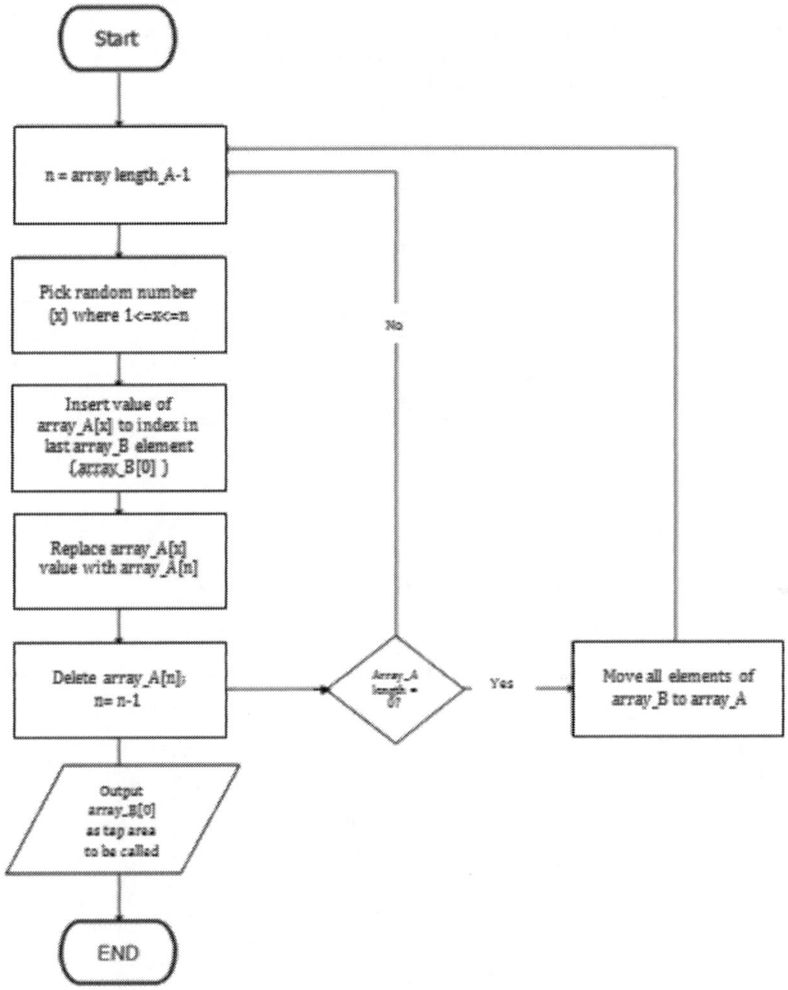


**Figure 2.** Flowchart of Fisher-Yates Shuffle Algorithm

## 2.1. Game Logic

Game Logic is the primary flow of Rangku alu game. Game Logic uses the randomization function to determine every next possible move during the game. For more detailed explanation can be seen as follows.

### *2.1.1. Coroutines /timer*

Coroutines is a subprogram that has multiple entry points, which are controlled by coroutines itself [9]. It is also known as the asymmetric control because of the mechanisms of the called and caller coroutines. Each coroutine has the same degree. Calling coroutine is called resume. The first resume is when the coroutine begins, while the next resume will start after the last statement is executed on coroutine, so coroutines can mutually resume as long as possible. Coroutines allow quasi-concurrent execution on other coroutines where coroutines are executed simultaneously yet not overlapping.

Coroutines function is IEnumerator function that authors use as time control function. Coroutines itself is a component of computer program commonly used for multitasking. Coroutines can have more than one entry point which is useful for suspending and resuming during program execution. A coroutine is like a function that can stop the execution and restore control to Unity but then continue where it left off in the next frame. In C#, the coroutine is revealed as in Figure 3.

```
IEnumerator Fade() {
    for (float f = 1f; f >= 0; f -= 0.1f) {
        Color c = renderer.material.color;
        c.a = f;
        renderer.material.color = c;
        yield return null;
    }
}
```

**Figure 3.** Sample of coroutine in C#

The pseudo code of coroutine function in this research is as follows:

```
IEnumerator Fade() {
    for (float f = 1f; f >= 0; f -= 0.1f) {
        Color c = renderer.material.color;
        c.a = f;
        renderer.material.color = c;
        yield return null;
    }
}
```

**Figure 4.** Pseudo code of coroutine as timer

2.2. Bamboo Animation

Animation of bamboo motions is adjusted using the time calculation given by coroutine function. The bamboo animation is the change of bamboo location in times. The pseudo code of bamboo animation can be seen in Figure 5.

```
class GameManager : Monobehaviour {

    Get GameObject Bamboos

    void MoveBamboo(bool closePosition, bool downPosition){

        Get Bamboos position when
        openPosition, closePosition, downPosition

        if(closePosition == false AND downPosition == false){
            Bamboos transform position to openPosition
        }
        else if(closePosition == true AND downPosition == false){
            Bamboos transform position to closePosition
        }
        else if(closePosition == false AND downPosition == true){
            Bamboos transform position to downPosition
        }
        else{
            output "there is no such position"
        }
    }

    // called every frame
    void Update(){

        //call coroutine so Bamboo won't moved every frame
        //but it get to wait for Coroutine to finish instead

        MoveBamboo(closePosition, downPosition)
        StartCoroutine(IEnumerator TapPlaying(delayTime))
    }
}
```

**Figure 5.** Pseudo code of bamboo animation

### 2.3. Appearing Tap Area

The game Logic mechanism on this system has two game modes which are random mode and a normal mode that focused on the tap area. Calling the tap area in normal mode is done manually and continuously or monotonically, while in random mode is done by randomization using the Fisher-Yates shuffle algorithm.

### 2.4. Fisher-Yates shuffle

Fisher-Yates shuffle is a randomization algorithm which generates a random permutation of a finite set, or in other words, Fisher-Yates shuffle can be used to randomize a set, e.g., a set of arrays containing integers into a set of integers which positions have been randomized [10].

In the random mode of Rangku Alu game, the sequence of tap area appearances is randomized to variations of the game. In Unity, two libraries can be used to get random numbers, which are UnityEngine.Random and System.Random. However, in Rangku Alu, the result of calling the random numbers is expected to be different every time the call is made, so randomization of numbers using two libraries is not enough.

Fisher-Yates shuffle algorithm can be used to prevent the tap area to appear at the same place more than once. The pseudo code of Fisher-Yates shuffle algorithm implemented in Rangku Alu mobile game can be seen in Figure 6.

```
-- To randomize possible move from
-- current tapArea to next possible tapArea

List TapArea =  list of other tapArea excluding
                itself

List Result as new List

public int RandPossible(){

    if(TapArea.Length equals to 0){
        move all elements in Result to TapArea
    }

    i = Random number between 0 to tapArea.Length-1

    insert tapArea[i] to beginning of result[]
    replace tapArea[i] with last element in TapArea[]

    remove last element in tapArea[] so the
    length reduced by 1

    return result[0] as possible move
}
```

**Figure 6.** Pseudo code of bamboo animation

In Figure 6, the next possible moves to attain from the current area are saved in a single list named "TapArea." For the result of shuffle or tap area randomization, will be stored in a new list named "Result." Fisher-Yates algorithm in the system contains the randpossible function. The steps performed in the function are:

1. Pick a random number, from 0 to length of TapArea -1 list repeatedly.
2. Take an element of TapArea with index [i] and insert it to the Result list in the earliest order or on the index
3. Replace the contents of TapArea [i] with the last element in TapArea.
4. Delete the last element in TapArea. Then the length reduced by 1.
5. This process will be repeated until TapArea list is empty.
6. Then check whether the TapArea list is empty. If it is already empty, then move all the contents of the Result List to the TapArea list.
7. Return to the first step.

This function will return Result [0] value as a possible move.

### 2.5. Game Over

In system development, the authors built the system design for conditions such as whether the user is not able to continue the game (game over). With this condition, the author hopes that the system can check and call the game over function when the player touches the bamboo or when the player is too fast / too late to hit the tap area.

### 2.6. The Highest Score Logic

The logic for the highest score is called at the beginning of the game initiation to display the highest score on the user interface, and when the game is over, the logic is able to store the scores to the database automatically.

## 3. Result and Analysis

### 3.1. Main Menu Display

The main menu display is the initial display of the game before entering the game display. The main menu display consists of the displays of a start menu, select the mode, select the song, and how to play. Start display (a) is an opening display when the application runs consisting of the menus "Start Play," "Score," and "Exit." Select the mode display (b) is display of selection mode chosen by the players. The mode in Rangku Alu is the tap variations when playing the game. There are two modes in Rangku Alu which are Normal mode and Random mode. Select the song mode (c) is a display where the player is expected to pick a song to be played in the game area as background music. There are three choices of songs, namely "Ampar-Ampar Pisang," "Rasa Sayange," and "Tari Tongkat." How to play display is made to guide the players about the game before it starts. The main menu displays can be seen in Figure 7.

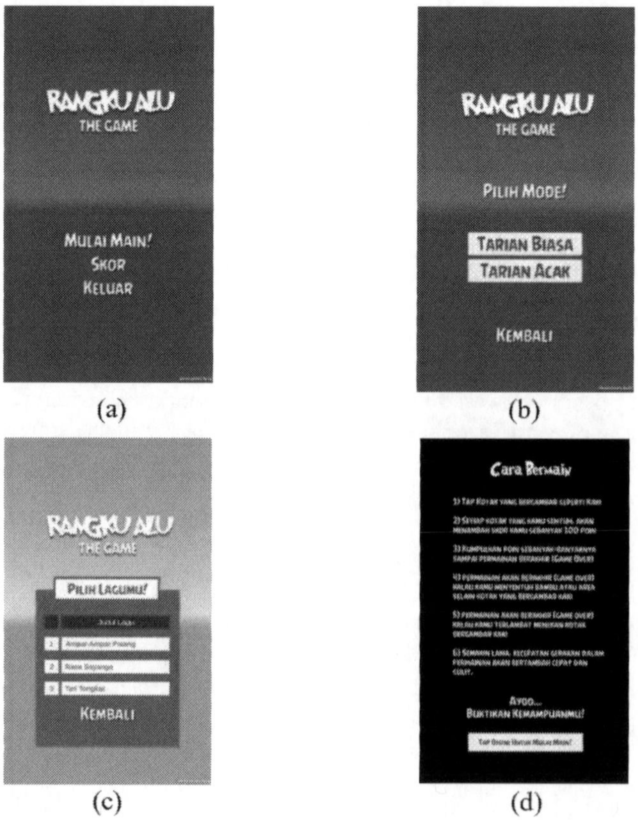


(a) (b) (c) (d)

**Figure 7.** Main menu display; (a)Start menu; (b) Select the mode; (c) Select the song; (d) How to play

### 3.2. Game Arena Display

Game area display is the game display of Rangku Alu after selecting a song to be played in the game and after reading the How to Play. Game area display can be seen in Figure 8.

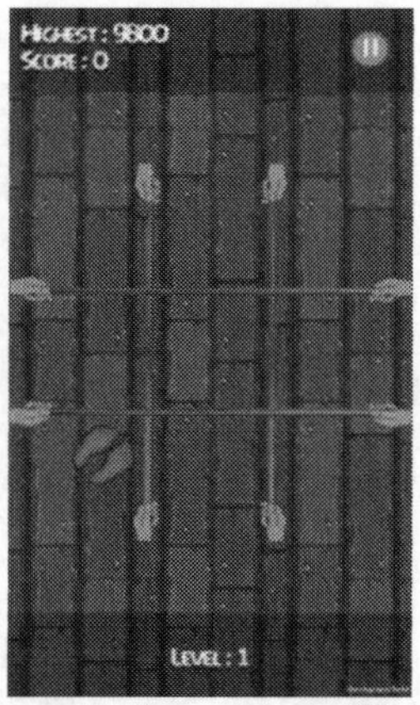


**Figure 8.** Game arena display

### 3.3. High score display

High score display is a display to show the obtained score when the game ends. In this screen, ten people with the highest score will be put on the scoreboard. High score display can be seen in Figure 9.

**Figure 9.** High score display

### 3.4. Notification Display

Notification display is a display serves as a notification to the player about the conditions that occur when playing. Notification display can be seen in Figure 10 and Figure 11.

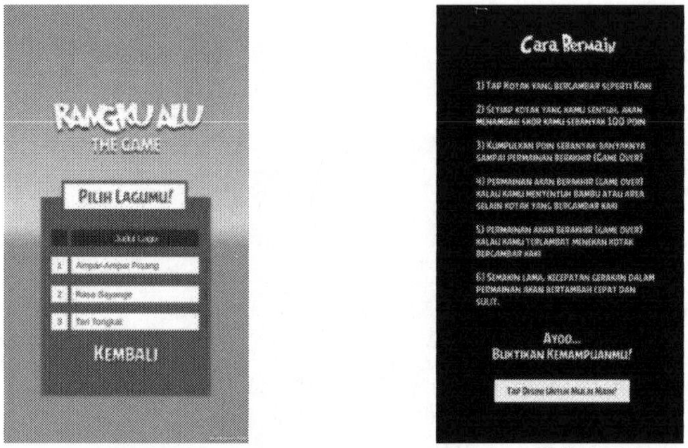


(a) (b)

**Figure 10.** Game pause notification display; (a) Options; (b) Exit confirmation

(a) (b)

**Figure 11.** Game over notification display; (a) Game over of miss jump; (b) Game over of miss click

Figure 10 (a) is notification display when button "pause" is clicked, options given by the notification are to continue the game, to repeat the game and to exit the game. Figure 10 (b) is a display showing confirmation question when button "exit" in Figure 10 (a) display is chosen.

Figure 11 (a) is the display when the player makes a mistake by jumping too late, and Figure 11 (b) is a display showing the game over notification when the wrong tap area is clicked.

## 3.5. Operational Procedures

Main menu displays as shown in Figure 7, is an initial stage in Rangku Alu game consisting of the displays of start menu, select the mode, select the song, and how to play. In selecting the mode display, there are two modes in Rangku Alu which are Normal mode and Random mode, while in select the song, three choices of songs are provided namely "ampar-ampar pisang," "rasa sayange" and "Tari Tongkat" for the background music. In rangku Alu, there are variations of footstep movements in both modes. Fisher-Yates shuffle was applied for the footstep movements to determine the next possible moves during the play.

### *3.5.1. Normal mode*

Normal mode in this application is a mode with regular and monotonous tap movements in the game. This mode is aligned with the simplest rules of the real Rangku Alu game, where the pattern of footstep movements has been defined from the beginning.

In normal mode, the sequence of footstep movements is stored in an array. The footstep movements are executed by calling the array elements, one in every few seconds depending on the level of the game.

This footstep movement mechanism will continue and repeat from the beginning when all the array elements have been called. The footstep movements in normal mode can be seen in Figure 12.

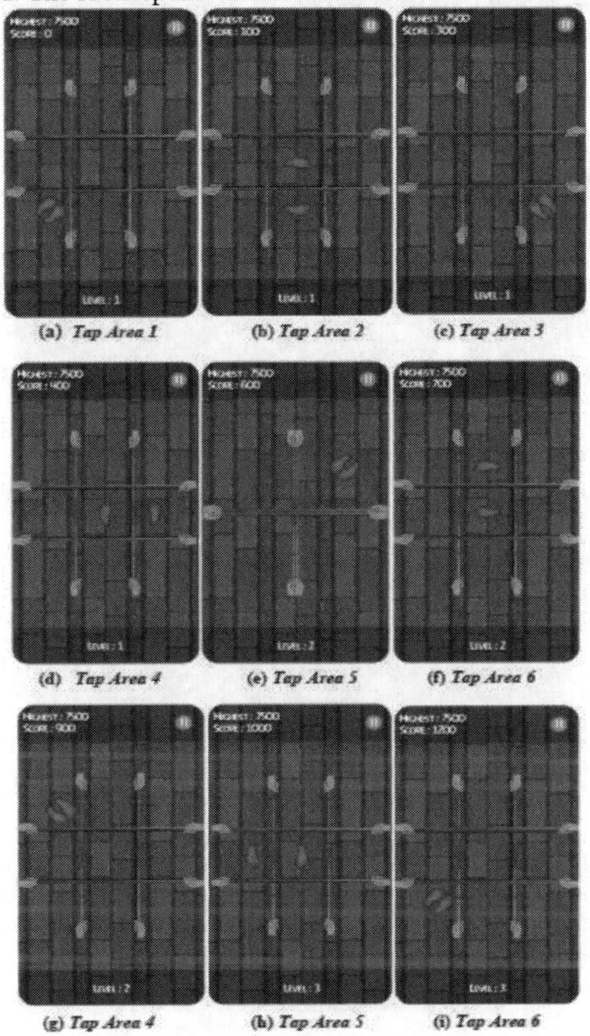


**Figure 12.** Footstep movements in normal mode

### *3.5.2. Random mode*

The random mode in Rangku Alu game is more varied, and this is because authors use moves randomization algorithm so that players do not jump with a monotonous rhythm. The possible algorithm for moves randomization is Fisher-Yates shuffle. This algorithm was used to prevent the tap area to appear at the same time more than once.

Figure 13 and Figure 14 are the results of one round of random mode game in two trials, where the positions of tap area which appear consecutively in Figure 13 are not the same as the positions of tap area shown in Figure. 13. The positions of tap area in random mode can be seen as follows.

IOP Conf. Series: Journal of Physics: Conf. Series **978** (2018) 012103 doi:10.1088/1742-6596/978/1/012103

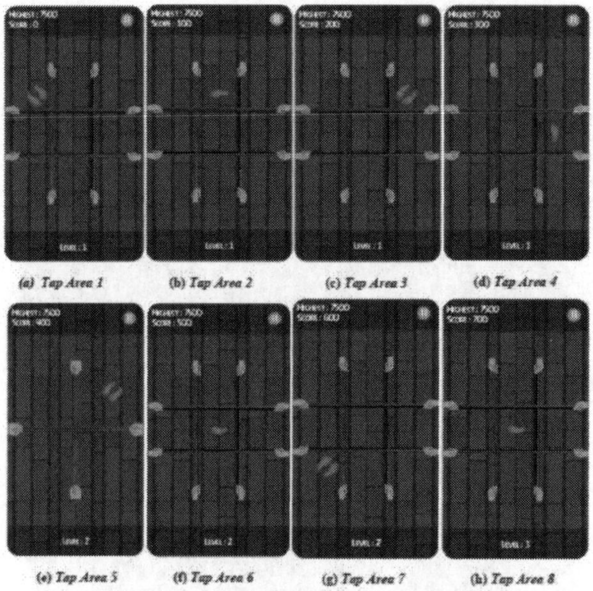


**Figure 13.** Footstep movement in random mode - first round

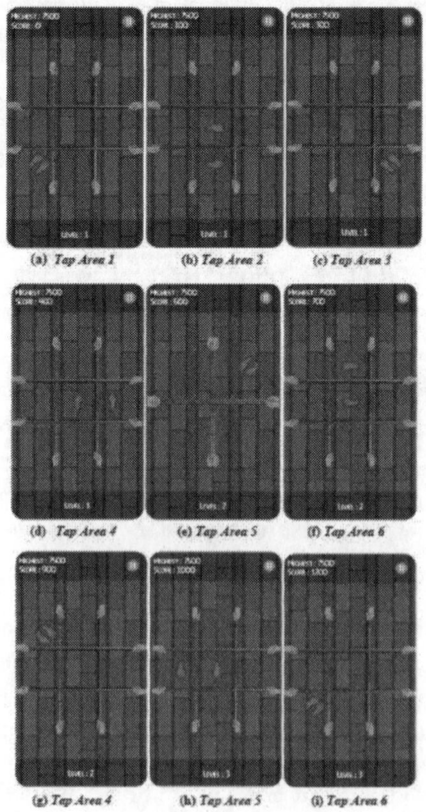


**Figure 14.** Footstep movement in random mode – second round

3.6. Application Testing

Authors performed a test on system to measure the game difficulties. The test was conducted on 20 individuals who play the game application directly. The test result can be seen in Table 1.

Table 1. The Result of Players Difficulty Level Test

| NO | Responder | Game Score | |
|---|---|---|---|
| | | Mode | |
| | | Normal | Random |
| 1 | R1 | 6700 | 8600 |
| 2 | R2 | 2300 | 8300 |
| 3 | R3 | 12300 | 8000 |
| 4 | R4 | 12300 | 8500 |
| 5 | R5 | 12300 | 1800 |
| 6 | R6 | 10700 | 11800 |
| 7 | R7 | 4300 | 2800 |
| 8 | R8 | 7300 | 4300 |
| 9 | R9 | 8000 | 4000 |
| 10 | R10 | 5600 | 7800 |
| 11 | R11 | 6000 | 8500 |
| 12 | R12 | 5200 | 4000 |
| 13 | R13 | 1100 | 5900 |
| 14 | R14 | 4700 | 3000 |
| 15 | R15 | 3100 | 8200 |
| 16 | R16 | 12000 | 9000 |
| 17 | R17 | 9000 | 7300 |
| 18 | R18 | 8300 | 5400 |
| 19 | R19 | 6500 | 5000 |
| 20 | R20 | 7800 | 8000 |

Table 1 presents the scores obtained by the player in both game modes which are normal mode and random mode. Normal mode has monitors tap changes while random mode does not. From the observation, the average score in each level can be calculated as follows:

$$\text{Average score} = \frac{Total\ Score}{Number\ of\ Players}$$

1. Average Score in Normal Mode = 7275
2. Average Score in Random Mode = 6510

The score range in each level on the application is as follows:

- Level 1 = 100-2800
- Level 2 = 2900-3700
- Level 3 = 3800-5300
- Level 4 = 5300 -15000
- Level 5 = > 15000

From the calculation result above, the level that possibly reached by 20 individuals who play the game in both normal and random mode is in level 4, thus it can be concluded that Rangku Alu game has a medium high level of difficulties.

## 4. Conclusion and Future Research

The conclusions that can be drawn from this research are as follows:

1. The implementation of the randomization algorithm Fisher-Yates Shuffle is able to determine the next moves so that this Rangku Alu game has tap variations with two game modes. From the research, it can be concluded that Rangku Alu game has a medium-high level of difficulties based on the average score.

2. In general, the footstep movements in Rangku Alu dance have many variations depending on the dancers. Authors can only adopt the simplest movements and add random footstep movements as variation.
3. Developing mobile game with an attractive interface requires a long and challenging process, so it should be done as a team so that more optimal results obtained.
4. Rangku Alu traditional game in groups or team will become our future study and development.

## 5. References

[1] Syahputra M F, Annisa T, Rahmat R F, and Muchtar M A 2017 Virtual application of Darul Arif palace from Serdang sultanate using virtual reality *Journal of Physics: Conference Series* **801** 12009

[2] Muchtar M A, Syahputra M F, Syahputra N, Ashrafia S, and Rahmat R F 2017 Augmented reality for searching potential assets in medan using GPS based tracking *Journal of Physics: Conference Series* **801** 12010

[3] Syahputra M, Widhianto M and Rahmat R F 2017 Historical game of Majapahit kingdom based on tactical role-playing game *Proc. The 12th International Joint Conference on Computer Vision* 327-32

[4] Ajim N 2016 *Rangku Alu Traditional Games* Retrieved from http://www.mikirbae.com/2016/05/permainan-tradisional-rangku-alu.html

[5] Juardi A, Tommy, Salim W 2011 Racing game application "karapan sapi racing" on Android-based smartphone using libgdx engine *Jurnal Bina Nusantara University*

[6] Farizi H R, Aryadita H, and Adam J E M 2013 Simulator mobile game development of angklung on Android-based system. *Repository of PTIIK UB Student Journal* **1** 5

[7] Rahmat R F, Fahrani O R, Purnamawati S and Pasha M F 2018 The development of indonesian traditional bekel game in android platform *Journal of Physics: Conference Series*

[8] Nugraha R et al. 2015 *Implementation of Algoritma Fisher-Yates on Unity 3D-Based The Lost Insect Application For Insect Species Identification.* (Palembang: Informatics)

[9] Unity Technologies 2016 Coroutines. Retrieved from https://docs.unity3d.com/Manual/Coroutines

[10] Ade I and Abejide O 2012 A simulated enhancement of fisher-yates algorithm for shuffling in virtual card games using domain-specific data structures *International Journal of Computer Applications*

IOP Conf. Series: Journal of Physics: Conf. Series **978** (2018) 012104 doi:10.1088/1742-6596/978/1/012104

# Measuring the power consumption of social media applications on a mobile device

**A I M Dunia, Suherman, A H Rambe and R Fauzi**
Department of electrical engineering Univesitas Sumatera Utara 20155, Indonesia

*E-mail: suherman@usu.ac.id

**Abstract**. As fully connected social media applications become popular and require all time connection, the power consumption on mobile device battery increases significantly. As power supplied by a battery is limited, social media application should be designed to be less power consuming. This paper reports the power consumption measurement of social media running on a mobile device. Experimental circuit was developed by using a microcontroller measuring an android smartphone on a 802.11 controlled network. The experiment results show that whatsapp consumes the power less than others in stand by and chat. While other states are dominated by line. The blackberry consumes the power the worst.

## 1. Introduction

Mobile device application progresses very rapidly. Fully connected applications such as social media become daily needs. As battery supplying the power on a mobile device is limited, social media should minimize its signaling to maintain low power consumption. As an initial study, this paper reports the measurement schema to determine the power consumption by the social media towards a green social media application.

Previous work measures TCP energy consumption on a mobile device [1]. Details on mobile device power consumptions have also been explored in [2, 3]. Some efforts on decreasing power consumption have also been performed, such as power saving mode [4], routing [5] and transport protocol [6-8]. This paper reports the measurement of power consumption of a mobile device when running a social media application.

The assessed social media are line, whatsapp and blackberry messenger. Whatsapp is a messaging application that enables users to exchange text, image, video, voice call and video call. Initially, it was designed for IPhone but now is available for BlackBerry, Android, Windows and Symbian phones, as well as desktop application [3]. Line is also a messaging application that works for various platforms, designed by a Japanese company. It also works for Mac and Windows. Blackberry messenger is initially for blackberry phones but now is available for various platform.

## 2. Method

In order to measure the social media power consumption, a circuit controlled by arduino has been developed. The circuits periodically measure voltage and current absorbed by a mobile device (Advan Vandroid). The circuit diagram is shown in Figure 1 and the implementation is in Figure 2.

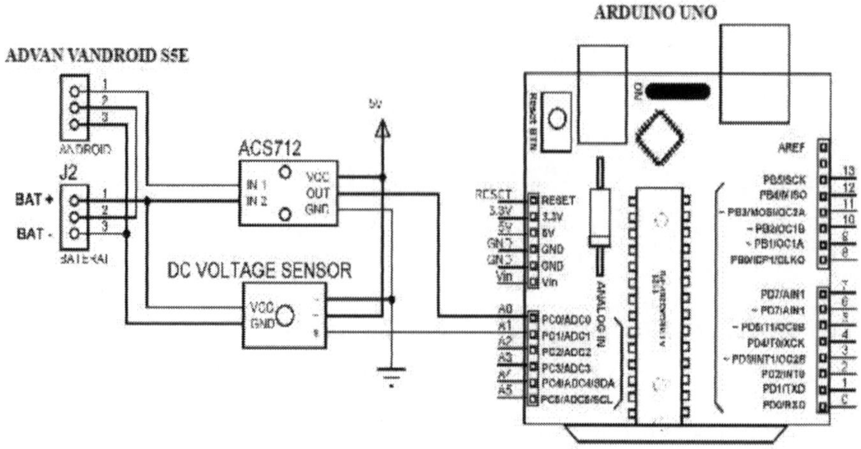


**Figure 1**. The employed circuit

DC voltage sensor measure the voltage across the mobile phone's terminal, while the ACS712 current sensor measures current flows to mobile phone. The assessed mobile phone is Advan Vandroid. The arduino measure the voltage and current, multiply them to get power in Watt (W). The measurement results are sent to computer by the arduino though USB connection.

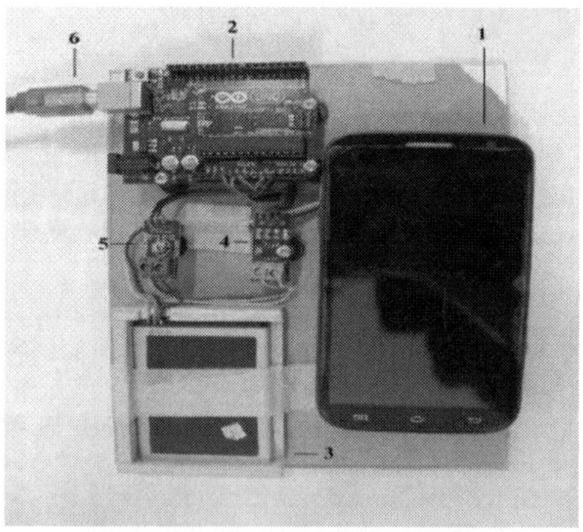


**Figure 2.** Measurement device

The power consumption is evaluated for stand by, chat, voice call, video call and image transmissions. In order to get stable connection, the 802.11 access point provides the internet connection.

The evaluated social media applications are line, whatsapp, and blackberry messenger. The measurement is last for 60 second. Stand by state is conditioned to have minimum background process.

The mobile device is connected to the internet through a 802.11 access point, but only for the social media application. The 802.11 network is set to work alone, so that no other access point interfere the frequency. However, the other node is connected to internet via mobile data network which has unknown network traffic and queue. As the other node is out of the supervised network, this is the limitation of the research.

Every social media is treated similarly. Battery is ensured to be filled at the same percentage, so the energy consumption is fair enough.

## 3. Measurement results

Stand by state is a condition where no transmission occurred. Chatting or text transmission uses the same text to be send. So does voice call and other services. The measurement results are plotted in Table 1.

**Table 1**. Power consumption of social media application

| No. | States | Line | WhatsApp | Blackberry |
|---|---|---|---|---|
| 1 | Standby | 0.91 W | 0.88 W | 0.97 W |
| 2 | Chatting | 1.05 W | 0.98 W | 1.19 W |
| 3 | Voice Call | 1.03 W | 1.06 W | 1.07 W |
| 4 | Video Call | 1.21 W | 1.39 W | 1.42 W |
| 5 | Image transmission | 1.04 W | 1.12 W | 1.13 W |

In stand by state, whatsapp has the smallest average power consumption about 0.88 W. While the line and blackberry messenger are 0.91 W and 0.97 W subsequently.

In chatting states or text transmission, whatsapp also has the smallest power consumption, about 0.98 W, followed by line 1.05 W and blackberry 1.19 W.

Voice call produces different rank where line generates the lower power consumption, 1.03 W, followed by whatsapp 1.06 W, and blackberry 1.07 W.

Video call energy saving is dominated by line, 1.21 W, followed by whatsapp, 1.39 W and blackberry 1.42 W.

Image transmission power consumption is also dominated by line, 1.04 W, followed bya whatsapp 1.12 W and blackberry 1.13 W.

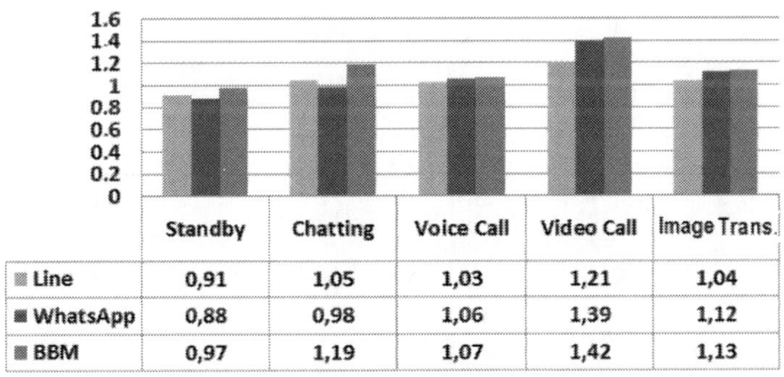


**Figure 3**. Power consumption comparisons

## 4. Conclusions

In stand by state and chatting, whatsapp has the lowest power consumption. However, when operating actively for voice cal, video call and image transmission, line is the most in conserving power. Blackberry performs the worst as it consumes the highest power in all states.

The measurement is conducted without extracting the details of the transmission signal. Future works may consider capturing the signal flows in the internet and determine what signal consumes the most and the less power, so that future application may consider it to achieve a greener application.

## References

[1] Suherman 2016 Impact of Window and Datagram Sizing on Mobile Device Power Consumption *JiTEK* **5 (2).**

[2] Carroll A and Heiser 2010 G *Proc. USENIXAT (Boston-USA)* **10** 21-2

[3] Longhao Z, Ali J and Muntean G M 2017 *IEEE Broadband Multimedia Systems and Broadcasting* 1-6.
[4] Jang S B and Kim YG 2010 *IEEE Transactions on Consumer Electronics* **56** (4).
[5] Zhu J and Wang X 2007 Model and Protocol for Energy Efficient Routing over Mobile Ad Hoc Networks, IEEE.
[6] Khalifa T 2014 *IEEE Transactions on Smart Grid* **5** (1), 381 –91.
[7] Chen S 2013 *IEEE Conf. on Wireless Communications and Networking (WCNC),* 1291 – 96.
[8] N Mubarakah and Suherman 2016 *IEEE Isemantic*, (**1**).

IOP Conf. Series: Journal of Physics: Conf. Series **978** (2018) 012105 doi:10.1088/1742-6596/978/1/012105

# Development of building security integration system using sensors, microcontroller and GPS (Global Positioning System) based android smartphone

**P Sihombing[1] , Y M Siregar[1], J T Tarigan[1] , I Jaya[1] and A Turnip[2]**

[1]Faculty of Computer Science and Information Technology, Universitas Sumatera Utara, Jl. Dr. Mansyur No.9, Padang Bulan, Medan, Sumatera Utara 20155.
[2]Lembaga Ilmu Pengetahuan Indonesia (LIPI), Bandung.

Email: poltakhombing@yahoo.com

**Abstract**. Security system is one of the common problems to protect an environment such as personal house or a warehouse. There are numerous methods and technologies that can be used as part of a security system. In this paper, we present a security system that offers a better efficiency. The purpose of this study is to build a system that can monitor home security at any time in particular fire and theft. Through sensors, the system will be able to provide warning information of hazard conditions via LCD monitor, sound, and alarm. This information will be sent automatically to the home owner's smartphone as well as to the corresponding to the security agency. Thus the prevention of theft and fire hazards can be immediately anticipated by the police and firefighters. The system will also notify the position of the coordinates of the location of the building (the house) by a link to the Google map in order to make it easier to get the location quickly.

## 1. Introduction

The security system is one of the most important things right now. The availability of integrated security systems with multiple agencies has become a necessity nowadays. Using only labor for security is no longer considered an effective method. Therefore, the role of technology needs to be involved to support the security system.

The cost to hire adequate person to cover a certain area and the inability to guarantee an effective 24-hour security are two of the biggest problem in security system. Hence, the role of technology has replaced man based security as the vital part of the security. There are many methods and technology available to develop a solid security system. However, the use these technologies to achieve a better security should be in line with efficiency. For example, while recording a 24-hour event is considered as the most effective system, it is still considered inefficient due to redundant amount of recording that contains no crucial information. Moreover, the cost of maintaining the system to work in 24-hour can be extremely expensive.

The objective of this work is to introduce a method that solves the efficiency problem without jeopardizing the effectiveness of the security system. This proposed system uses the sensors, such as gas sensor, infrared and vibration sensor. The sensors will detect if any occurring intrusion event or fire accident. In addition, the system also able to perform a one-way communication to registered user in order to notify the user should any event is occurred. We build this system using Raspberry Pi as its main board. Moreover, the system will be connected to a PC that will operate as an SMS gateway.

## 2. Related Work

Due to increasing crime, the current security system is also a lot of modifications by using sensors that are connected directly to the microcontroller and Raspberry Pi. Raspberry Pi [1] is basically a single board computer mini hardware that can be used as a computer server. This minicomputer was developed and introduced first by the Raspberry Pi Foundation in 2012, with the aim of this device is to simplify computer hardware while reducing costs. Raspberry Pi is made simpler but its ability to work as computer hardware is the same as any other computer. In addition, with Raspberry Pi also provides an opportunity for students and researchers to create specific computers for specific tasks such as for building security systems in this study. Raspberry size is made relatively small compared to ordinary computers, but Raspberry Pi functions like a common computer hardware that uses processor, memory, storage, it also has the ability to run basic operating systems like Linux and Windows. There are many research-based uses Raspberry Pi as its platform. Russell Barnes wrote several projects using raspberry pi [2]. Shaijupaul, et al wrote Home Based Android Automation Using Raspberry Pi [3]. Pawan Singh, et al has conducted a review paper on GSM Smart Home Automation System. [4].

Basically, the sensor can detect an event in a building that can be connected directly to the microcontroller and Raspberry Pi. For example, a PIR (Passive Infra-Red) sensor is an electronic sensor that detects events based on the infrared light changes emitted from the object in view. The principle of the infrared sensor operation is that all of the objects radiate heat energy in the form of radiation. This radiation is largely invisible because it emits infrared wavelengths. Infrared sensors are able to detect the radiation emitted by objects. Normally, the sensor does not need to expend any energy to perform the detection, then the term passively pinned to the sensor. Infrared vision is widely used to detect objects in dark environments. In the case of security, where disturbances mostly occur at night to avoid crowds, the use of infrared vision may be very important for recording objects [5]. One of the PIR sensor models can be seen in Figure 1.

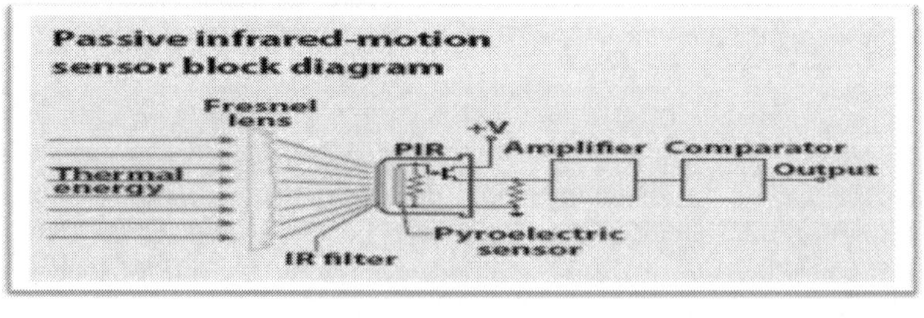


**Figure 1.** The main part of the PIR sensor

The PIR sensor used in this project only detects infrared transmitters with wavelengths of 8-14 micrometers. Beyond that wavelength, the sensor will not detect it. Generally, humans have a body temperature that can produce infrared emissions with a wavelength of about 9-10 micrometers (standard value 9.4 micrometers), the wavelength can be detected by the PIR sensor. (Generally, PIR sensors are designed to detect humans).

## 3. System description

### *3.1. The Block Diagram*

In the Block Diagram described a system configuration used. The system uses electricity from PLN 220 Volt AC 50/60 Hz, then transmitted to 2 output DC voltage (Direct Current) which is 12 Volt DC is used to supply Arduino ATMEGA 2560 and 5 Volt DC power microcontroller for Active Power

Speaker. All sensors and modules use power from a microcontroller that is converted by a microcontroller from 12 Volts to 5 Volts DC. All gas, smoke and PIR sensors will report their respective output values each time to the microcontroller. The microcontroller will identify the detection value of the sensor. If the microcontroller gets the sensor value at a predetermined threshold, then the microcontroller will perform the reporting function and the neutralizing function.

The detection resulted will send to the smartphone in the form of SMS via phone to a specified number (e.g. to the building owner, security, and police) by utilizing SIM900 Module. May also report in the form of sounds using VS1053 MP3 Module and forward it to the active speaker. Then use the 16X2 LCD monitor as a writing indicator, light alarm as alarm warning alarm. Then there is this Neutralization function like fan blower (Exhaust Fan) by using the Opto-Triac driver. Control on/off channel 1-5, This alarm light uses direct power from PLN 220 Volt AC with SMS command received by SIM900 module then executed according to SMS command by the microcontroller to enable Opto-Triac as a driver from channel 1- 5 and Light Alarm. The PIR sensor will report if there is theft at some point that has been installed sensors on the doors (windows) building. We can see the Block diagram in Figure 2.

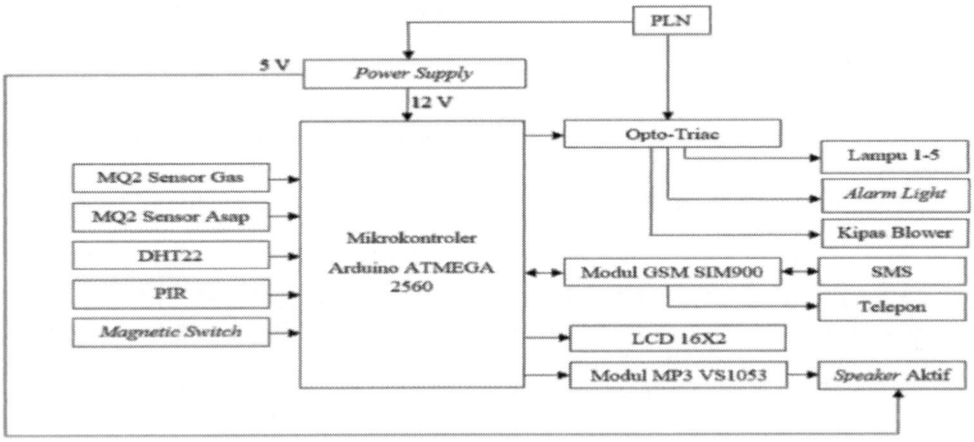


**Figure 2.** The block diagram of Building Security system.

The main components system is: MQ2 sensor gas, DHT22 sensor, PIR (infrared sensor), vibration sensor, and a personal computer as a SMS Gateway. The system workflow can be summarized as follows: the two sensors, infrared and vibration are continuously sensing the targeted area. If one of the sensor senses an event, it will send a signal to Raspberry Pi. Raspberry Pi, acting as the main controller, sends the signal to a connected SMS Gateway. The SMS Gateway will then build a message based on the detected event and send the message to a registered user.

## 4. The experimental set up and testing result

The first part of the implementation process is to connect all of the devices described above. In this section, we will go through the process of connecting the devices based on the architecture.

### *4.1. The Experimental Set Up*

After compiling all the wiring neatly and sequentially, then re-check the wiring that has been installed to ensure that the wiring matches the schematic. All modules and sensors are installed in miniature of the house as shown in Figure 3.

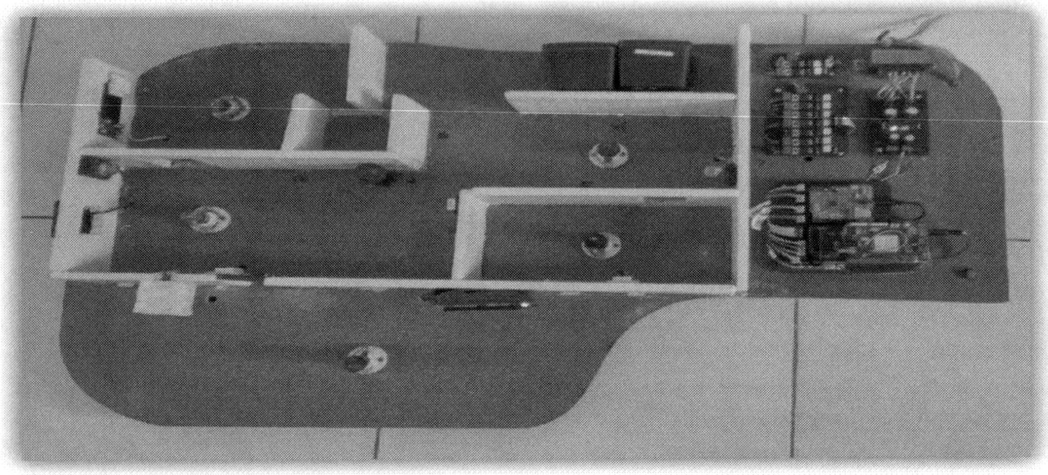

**Figure 3.** The miniature of the Building

*4.2. Testing system sensors*
In a testing of the performance of the sensor functions on the system carried out aims to test the sensors used by the system.
The test consists of 7 sensor testing i.e. PIR sensor, MQ-2 gas, MQ-2 smoke, DHT-22 temperature sensor and 2 magnetic switch sensor:

*4.2.1. PIR sensor testing.* PIR sensor testing aims to determine the ability of the sensor in detecting human presence at a distance of 500 Cm from the sensor. This test by simulating the existence of human and microcontroller will send data to SIM900 Shield which will send SMS message "There is People In Front of House!!". This experiment was conducted as much as 9 times and resulted in the test seen in Table 1

**Table 1**. PIR Sensor Test Results

| No. | Distance (Cm) | SMS status | System Performance Status |
|---|---|---|---|
| 1. | 50 | Sent | Works |
| 2. | 100 | Sent | Works |
| 3. | 150 | Sent | Works |
| 4. | 200 | Sent | Works |
| 5. | 250 | Sent | Works |
| 6. | 350 | Sent | Works |
| 7. | 400 | Sent | Works |
| 8. | 450 | Sent | Works |
| 9. | 500 | Sent | Works |

By the table 1 it can be concluded that the PIR sensor used to detect a distance of 50 to 500 Cm and work well as expected. Long delivery of SMS depending on operators Provider used. In this test the old SMS sending is 4-8 seconds.

4.2.2. *The MQ-2 gas and MQ-2 smoke sensors testing.* The MQ-2 gas and MQ-2 smoke sensor test aims to find out whether the sensor can detect the presence of gas and smoke and send a report of threshold conditions via SMS. The MQ-2 gas test will be tested using a gas lighter and the smoke MQ-

2 is tested using cigarette smoke. Lighters and cigarette smoke near the sensor for fast detection. This experiment was conducted 7 times and produced test data which can be seen in the Table 2.

**Table 2.** Sensor Test Results of MQ-2 Gas and MQ-2 Smoke

| No. | Range MQ-2 Gas (%) | Gas Detected (%) | Range MQ-2 Smoke (%) | Smoke Detected (%) | SMS Status | System Performance Status |
|---|---|---|---|---|---|---|
| 1. | 30-100 | 32 | 10-100 | 13 | Sent | Works |
| 2. | 30-100 | 42 | 10-100 | 16 | Sent | Works |
| 3. | 30-100 | 34 | 10-100 | 18 | Sent | Works |
| 4. | 30-100 | 45 | 10-100 | 15 | Sent | Works |
| 5. | 30-100 | 37 | 10-100 | 17 | Sent | Works |
| 6. | 30-100 | 33 | 10-100 | 12 | Sent | Works |
| 7. | 30-100 | 46 | 10-100 | 15 | Sent | Works |

If there is a gas leak detected by the MQ-2 gas sensor in the house above of the threshold value of 10%, then the system will report the condition in the form of LCD with the text format as follows: "Ada Gas bocor!! Kadar (detected)%". The gas leak monitor is shown in Figure 4.

**Figure 4.** The Text LCD Indicator Gas Leak

When the leaking gas is detected by the sensor above 40% then the system will report the condition via a short phone and SMS to the homeowner, firefighters, and Police or another number that has been set. The text format "FIREFIGHT: There's Gas Leaking, Triggering Fire, Google Map Locations: https://www.google.com/maps/place/3.55468,98.63993".

*4.2.3. The DHT-22 temperature sensor testing.* The DHT-22 temperature sensor used to detect the ability of the sensor in detecting heat of temperature. This test simulates will show the presence of hot temperatures by using matches as a trigger of hot temperatures. The microcontroller will send the data to SIM900 Shield which will send SMS message "Detected heat with Temperature: (temperature detected), Humidity (moisture detected)". This experiment was conducted 7 times and resulted in Table 3

**Table 3.** Temperature Sensor DHT-22 Test Result

| No. | Range DHT-22 | Gas Detected | System Performance Status | System Performance Status | SMS Status |
|---|---|---|---|---|---|
| 1. | 35-100 $^0$C | 40 $^0$C | Works | Works | Sent |
| 2. | 35-100 $^0$C | 37 $^0$C | Works | Works | Sent |
| 3. | 35-100 $^0$C | 48 $^0$C | Works | Works | Sent |
| 4. | 35-100 $^0$C | 43 $^0$C | Works | Works | Sent |
| 5. | 35-100 $^0$C | 36 $^0$C | Works | Works | Sent |
| 6. | 35-100 $^0$C | 44 $^0$C | Works | Works | Sent |
| 7. | 35-100 $^0$C | 47 $^0$C | Works | Works | Sent |

The DHT-22 sensor testing and SMS Status Condition of DHT-22 as shown in Figure 5 (a) and (b).

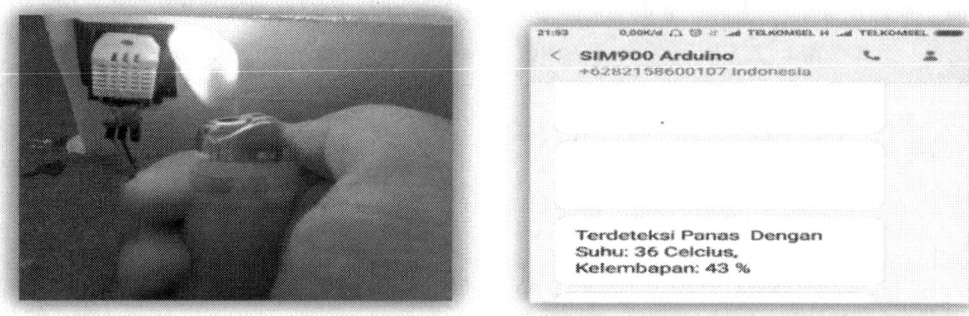


(a) (b)

**Figure 5.** (a) DHT-22 Sensor Testing, (b) SMS Status Condition of DHT-22

*4.2.4. Testing the magnetic switch sensor.* Sensor magnetic sensor testing aims to determine the ability of sensors in this system to detect whether the door/window is open or closed. This magnetic switch sensor has the same conditions as the PIR sensor that provides input to the microcontroller in the form of 2 conditions i.e. open or closed conditions (logic 1 or 0). This test is done by simulating a miniature door/window that opened and/or closed intentionally. Then Microcontroller will send data to SIM900 Shield which will send SMS message "Door / Open window!!". This test is done 7 times and the result is shown in Table 4.

**Table 4.** The Magnetic Sensor Switch Test Results

| No. | Magnetic *Switch* Status | SMS Status | System Performance Status |
|---|---|---|---|
| 1. | Open | Sent | Works |
| 2. | Open | Sent | Works |
| 3. | Open | Sent | Works |
| 4. | Open | Sent | Works |
| 5. | Open | Sent | Works |
| 6. | Open | Sent | Works |
| 7. | Open | Sent | Works |

## 5. Conclusions

The design and development of integrated Home Security systems Using SMS Android-Based Microcontroller ATMEGA 2560 has been described in this paper, and all components work properly. In this Project, we can see that the sensors automatically detect the state of the building and send the detection resulted to the ATMEGA 2560 Arduino microcontroller and controlled by the smartphone. We have tested for any sensor, e.g. PIR sensor testing aims to determine the ability of the sensor in detecting human presence at a distance up to 500 Cm from the sensor. In the gas sensor, when the leaking gas is detected by the sensor above 40% then the system will report the condition via a short phone and SMS to the homeowner, firefighters, and Police or another number that has been set. The text format "FIREFIGHT: There's Gas Leaking, Triggering Fire, Google Map Locations: https://www.google.com/maps/place/3.55468,98.63993". In the gas sensor, we have tested that the DHT-22 temperature sensor testing aims to determine the ability of the sensor in detecting heat temperature. And the last testing is magnetic sensor testing aims to determine the ability of the sensors in this system to detect the door/window is open or closed. This magnetic switch sensor has the same conditions as the PIR sensor that provides input to the microcontroller in the form of 2 conditions i.e. open or closed conditions (logic 1 or 0). This test is done by simulating a miniature door/window that opened and/or closed intentionally. Then microcontroller will send the data to the SIM900 Shield which will send by SMS message.

IOP Conf. Series: Journal of Physics: Conf. Series **978** (2018) 012105 doi:10.1088/1742-6596/978/1/012105

**Acknowledgment**
The authors would like to thank the Rector of Universitas Sumatera Utara (USU), Prof. Dr. Runtung Sitepu, SH, MHum, for the excellent services to support this research. The head and staff of "Lembaga Penelitian USU (Research Centre of Universitas Sumatera Utara), Prof. Dr. Erman Munir, MSc and staff. To the Dean and staff of Faculty of Computer Science and Information Technology (Fasilkom-TI) USU, Prof. Dr. Opim Salim Sitompul, MSc for their widest to support this work.

**References**
[1] Raspberry pi.[Online].Available:http://www.raspberrypi.org/
[2] Russel Barnes, The Official Raspberry Pi Project Book. Available: https://www.raspberrypi.org/magpi-issues/Projects_Book_v1.pdf
[3] Paul S, Antony A, Aswathy B 2014 Android Based Home Automation Using Raspberry Pi *Int. J. Comput Technol* **1**(1) 143-147.
[4] Singh P, Chotalia K, Pingale S, Kadam S 2016 A Review Paper on Smart GSM Based Home Automation System. *Int. Res. J. Eng. Technol.* **3**(4) 1838-1843.
[5] Gupta, A 2015 Intelligent Home security using GSM communication module.
[6] Asif O, Hossain MB, Hasan M, Rahman MT, Chowdhury ME 2014 Fire-Detectors Review and Design of an Automated, Quick Responsive Fire-Alarm System Based on SMS *Int. J. Commun. Network Syst. Scie.* vol. 7 386-395
[7] Gurav MR, Jagtap MR 2015 Wireless digital notice board using GSM technology *Int. Res. J. Eng. Technol.* vol. 2 1-3.
[8] Sadagopan VK, Rajendran U, Francis AJ 2011 Anti theft control system design using embedded system. *Proc IEEE ICVES* vol.85 239- 242.
[9] Pany JK, Choudhury RD 2011 Embedded Automobile Engine Locking System, Using GSM Technology *Int. J. Instrum. Control Automa.* **1**(2).
[10] Sebastian S 2013 Literature survey on automated person identification techniques *Int. J. Comput. Scie. Mobile Comput.* **2**(5) 232-237.
[11] Sehgal VK, Singhal M, Mangla B, Singh S, Kulshrestha S 2012 An embedded interface for GSM based car security system *Proc. 4th CICSyN*
[12] Bajaj R, Ranaweera SL, Agrawal DP 2002 GPS: location-tracking technology *Comput. Scie. Eng.* 92-94.
[13] Kumar BP, Raj VD, Harika P, Mallikarjun P, Kumar DK 2013 Wireless Unmanned all-Terrain Vehicle With Gsm and Gps *Proc. IJMCA* 1(3)
[14] Mazidi MA, Mazidi JG, McKinlay RD 2006 the 8051 microcontroller and embedded systems *Department of Computer Science and Information Engineering National Cheng Kung University of Taiwan*

IOP Conf. Series: Journal of Physics: Conf. Series **978** (2018) 012106 doi:10.1088/1742-6596/978/1/012106

# Hypertensive retinopathy identification through retinal fundus image using backpropagation neural network

**M F Syahputra[1], C Amalia[2], R F Rahmat[3], D Abdullah[4], D Napitupulu[5], M I Setiawan[6], W Albra[7], Nurdin[8] and U Andayani[9]**

[1,2,3,9]Department of Information Technology, Faculty of Computer Science and Information Technology, Universitas Sumatera Utara, Medan, Indonesia
[4,8] Department of Informatics, Universitas Malikussaleh, Aceh, Indonesia
[5]Research Center for Quality System and Testing Technology - Indonesian Institute of Sciences
[6]Department of Civil Engineering, Narotama University, Surabaya, Indonesia
[7]Department of Management, Universitas Malikussaleh, Aceh, Indonesia

nca.fadly@usu.ac.id | cut_amalia@students.usu.ac.id | romi.fadillah@usu.ac.id | dahlan@unimal.ac.id | darwan.na70@gmail.com | ikhsan.setiawan@narotama.ac.id | wahyuddin@unimal.ac.id | nurdin@unimal.ac.id

**Abstract.** Hypertension or high blood pressure can cause damage of blood vessels in the retina of eye called hypertensive retinopathy (HR). In the event Hypertension, it will cause swelling blood vessels and a decrese in retina performance. To detect HR in patients body, it is usually performed through physical examination of opthalmoscope which is still conducted manually by an ophthalmologist. Certainly, in such a manual manner, takes a ong time for a doctor to detetct HR on aa patient based on retina fundus iamge. To overcome ths problem, a method is needed to identify the image of retinal fundus automatically. In this research, backpropagation neural network was used as a method for retinal fundus identification. The steps performed prior to identification were pre-processing (green channel, contrast limited adapative histogram qualization (CLAHE), morphological close, background exclusion, thresholding and connected component analysis), feature extraction using zoning. The results show that the proposed method is able to identify retinal fundus with an accuracy of 95% with maximum epoch of 1500.

## 1. Introduction

Hypertension is a disease that spreads through out the human body in various forms. This happens because the blood pressure exceeds the normal limit and could cause damage in retina. Impaired retina due to hypertension is known as hypertensive retinopathy (HR) [1]. HR is long-term damage to the retina because high blood pressure has accumulated in the eye for several years even with patients who control their blood pressure with drugs [2].

Based on the results of the Basic Health Research in 2013, Number of people with Hypertension were declining from 31.7 percent to 25.8 percent in the span of 6 years. Existed research showed 25.8 percent of total victims are people with age of over 18 and that men have higher possibility to have hypertension in productive years. However, this do not apply to people in their 50s or above, since in this range of age, woman tend to have higher rates of possibility in experiencing this disease.

HR causes damage to blood vessels in the eye. The severity of retinal damage (retinopathy) is described on a scale of I to IV. In stage I, usually not found symptoms. In stage IV, it is marked with swelling of the optic nerve (papilledema) and macula (the center of the vision on the retina), which cause a decrease eye vision function. At an advanced stage, blood can enter the retina. Retinal spots are damaged by lack of blood supply and fat will accumulate in the retina. In addition to experiencing impaired vision, the patient will feel a headache.

Usually the examination of this disease is performed with physical examination and examination using funduscopy. It is also used a brightly shining ophthalmoscope device used to see the narrowing of blood vessels and excess fluid entering the blood vessels. Inspection is done manually and takes a long time. In addition, the analysis of the funduscopic is also still manual it allows for errors in consist of further action on the patient.

## 2. Identification of Problems

Hypertensive retinopathy (HR) occurs due to high blood pressure and causes damage to retina. In the event of Hypertension, the blood vessel will swell and decrease the performance of retina. In general, identifying hypertensive retinopathy disease through fundoscopy is still done manually by ophthalmologists. Therefore, a method is needed to help the ophthalmologist in diagnosing hypertensive retinopathy automatically so as to obtain better examination results than manual diagnosis.

## 3. Previous research

Quinn and Krishnan aims to segmentation of blood vessels in diabetic patients retinopathy and hypertensive retinopathy. In the preprocessing stage, use the green channel because it can show the vessels brighter than the background image. Then the equalization histogram improves the image contrast by changing the intensity of the image. To display an image that consists of edges (edges) using curvelet transform. Edge detection uses a modified Top-Hat transform with MSE Morphology. To eliminate the wrong edges that are not part of the blood vessels using morphological opening by reconstruction. Then CCA and Length Filtering to eliminate the wrong edge remnants accurately. Up to the detection of blood vessels can be done less than a minute [3].

Agurto et al through analysis of digital color fundus image obtained characteristic of hypertensive retinopathy (HR) such as AVR, silver/copper wiring, tortuosity and vessel abnormalities. These characteristics are used as inputs for classification using linear regression classifier to detect HR. The results of this research achieved an accuracy rate of 80% [2].

Nugroho et al to detect exudates on colored fundus images in diabetic retinopathy, the early stage of improving image quality by using contrast-limited adaptive histogram equalization (CLAHE) on green channels. Red channels are also used to detect and remove optical disks in imagery using median filtering and thresholding. In the segmentation stage using K-means clustering. The feature extraction stage uses GLCM and Lacunarity. For the classification using the Naive Bayes method achieved an accuracy of 92.13% [4].

Purandare and Noronha through fundus image analysis to classify diabetic retinopathy. In the preprocessing stage, use adaptive histogram equalization (AHE) to eliminate inappropriate sections of the background. Uses 2-D Gabor Wavelets for segmentation of blood vessels. The feature extraction stage uses gray level co-occurrence. SVM is used to classify normal retina or diabetic retinopathy. The accuracy achieved is 92.55% [5].

Jadhav, A., D'Cruz, J., Chavan, V., Dighe, A. and Chaudhari, J for detecting lung cancer using a CT Scan (Computer Tomography Scanner) image. In the preprocessing stage, the RGB image in the transformation becomes the grayscale and dark image, then the image enhancement process is done. A Genetic algorithm is used for image character extraction. Backpropagation Neural Network is used to classify images of lung cancer or not based on tumor size and stage of cancer [6].

Nurrahmadayeni is the identification of hypertensive retinopathy disease through retinal fundus image. Stages performed prior to identification is the process of image extraction feature extraction using two methods namely fractal dimension and invariant moments. Fractal dimension using Box

Counting algorithm. Then use the Probabilistic Neural Network method for normal retinal classification or hypertensive retinopathy. This research resulted in an excellent identification rate of 100% accuracy [7].

Sandri is the identification of retinoblastoma through retinal fundus image. Stages performed are image processing, image quality improvement and image characteristic extraction using gray level co-occurrence (GLCM). Backpropagation neural network method is used for normal retinal classification or retinoblastoma. This research resulted in an identification rate of 90% accuracy [8].

## 4. Research methods

The identification of hypertensive retinopathy in this research consists of several steps that begin with retinal fundus images are consist of normal image and hypertensive retinopathy image that is obtained through Structure Analysis of the Retina (STARE). This data will be used as training data and test data, followed by green channel process on the retinal fundus image to show the blood vessels and retinal structures more clearly. CLAHE is used to obtain better contrast results so that the blood vessels and retinal structures can be seen more clearly and reducing noise. Morphological close eliminates optical disk and background. Background exclusion is the process of subtraction between the result of CLAHE and morphological close. Furthermore, segmentation by using the thresholding process converts image to binary image. when thresholding process, there are still the objects size less than 70

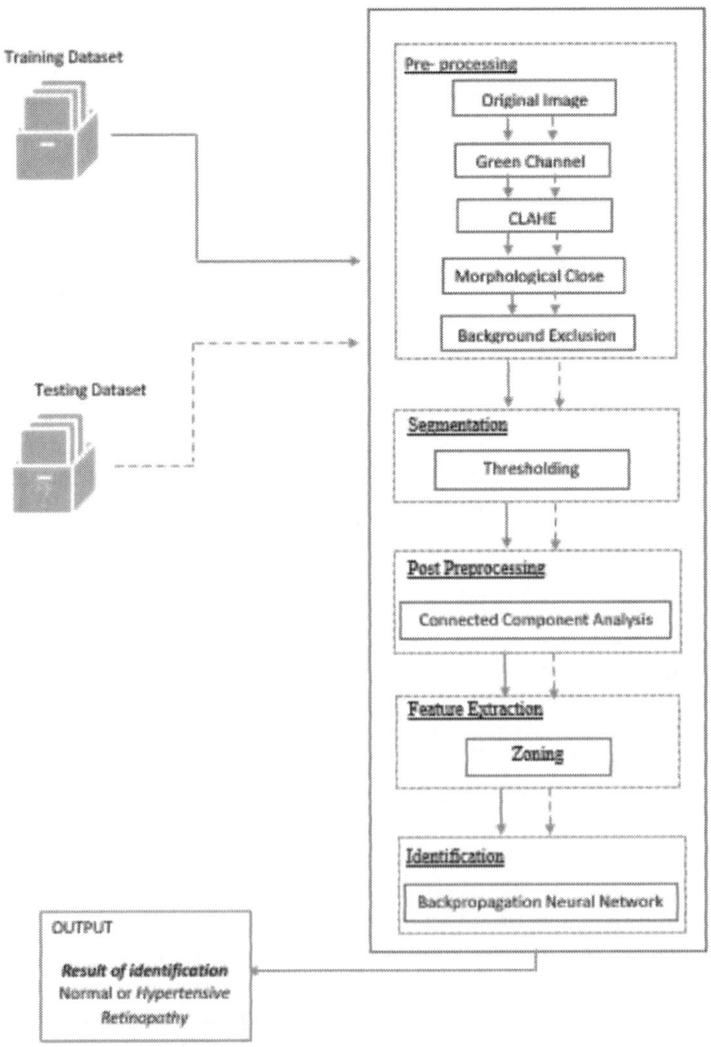


**Figure 1.** General Architecture

piksel, using connected component analysis eliminates them. The feature extraction using zoning and backpropagation neural network was used as a method for retinal fundus identification

Each step will be described in detail in the following sections. The methodology of this research can be seen in the general architecture of Figure 1.

*4.1. Prepocessing*

*4.1.1. Green Channel*. The early stages use the green channel because it has the best light reflections that can produce significant image information about blood vessels and retinal structures more clearly than red channels and blue channels. Green channel is a composition of red (red), green (Green) and blue (Blue). Can be calculated by the following equation:

$$I(x,y) = 0.R + 1.G + 0.B = G$$

*4.1.2. CLAHE.* The next stage of image contrast enhancement is using the Contrast Limited Adaptive Histogram Equalization (CLAHE) technique. CLAHE is an improved version of the previous method of Adaptive Histogram Equalization (AHE). CLAHE is used to obtain better contrast results so that the blood vessels and retinal structures can be seen more clearly, reducing noise problems and assigning a histogram boundary value

*4.1.3. Morphological Close.* The next stage extracts (splits) the background and optical disk by using morphological close because the object does not include the characteristics to be extracted. Morphological close used is dilation and erosion. Dilation process (image object thickening) then continued erosion process (separation of image object). Dilation process can be calculated by the following equation:

$$D(A,B) = A \oplus B = \{x : B_x \cap A \neq \emptyset\}$$

The erosion process can be calculated by the following equation:

$$E(A,B) = A \ominus B = \{x : B_x \cap X\}$$

*4.1.4. Background Exclusion.* The next stage of separation of blood vessels and retinal structures from the background is called background exclusion. The stages apply subtract operations between CLAHE image results and morphological close images to obtain blood vessels and retinal structures separate from the background.

*4.2. Segmentation*

*4.2.1. Thresholding.* The next stage is thresholding to obtain a binary image that is worth 0 and 1 (Black and White). Can be calculated the equation:

$$g(x,y)\begin{cases}1 \; if \; f(x,y) > T \\ 0 \; if \; f(x,y) \leq T\end{cases}$$

*4.3. Post Prepocessing*

*4.3.1. Connected Component Analysis.* The next stage is to remove objects that are not blood vessels that is by using a connected component analysis. Objects that are eliminated are objects that have a size of fewer than 70 pixels and are considered not blood vessels. Connected component analysis is a technique that can be used to classify regions or objects in digital images. This technique utilizes connectivity pixel theory in the image. The pixels in the region are called connected (there is connectivity or connectivity) when obeying the pixel rules. This pixel rule exploits the properties of the pixel sizes, the existing pixels linked by neighboring relationships. There is two connectivity that can be used that is 4-connectivity (4-connected neighbors), and 8-connectivity (8-connected neighbors). 4

connectivity (4-connected neighbors) can be seen in Table 1. 8-connectivity (8-*connected neighbors*) can be seen in Table 2.

**Table 1.** 4 connectivity (4-*connected neighbors*)

| | | |
|---|---|---|
| | P (x-1, y) | |
| P (x-1, y) | P (x, y) | P (x, y+1) |
| | P (x, y+1) | |

**Table 2.** 8 connectivity (8-*connected neighbors*)

| | | |
|---|---|---|
| P (x-1, y-1) | P (x, y-1) | P (x+1, y-1) |
| P (x-1, y) | P (x, y) | P (x+1, y) |
| P (x+1, y+1) | P (x, y+1) | P (x+1, y+1) |

### *4.4. Feature Extraction*

*4.4.1. Zoning.* Zoning is a method that borders the region into a MxN size. Image is divided into multiple zones, where each zone will generate a feature value by counting the highest white pixel count. At this stage, the image size of 700 x 605 pixels will be divided into 10 columns and 10 rows to get 100 zones representing 100 features. In each zone, the pixel size used for this zoning method is 520 pixels in size. the process of zoning method in feature extraction process on retinal fundus image include:

- Calculate the number of white pixels of each zone from zone Z1-Z100
- Specify the zone that has the highest white pixel count
- Calculate the feature value of each zone of Z1-Z100 with the Equation

$$Feature\ Value\ Zn = \frac{Z_n}{Z_{highest}}$$

range $1 \leq n \leq 100$

Zoning calculations include:

- The number of white pixels each zone among others Z1 = 90, Z15 = 140, Z40 = 200, Z80 = 180
- The zone that has the highest white pixel count is Z40 = 200
- Value Feature of each zone include:

Z1 = 90/200 = 0.45
Z15 = 140/200 = 0.7
Z40 = 200/200 = 1
Z80 = 180/200 = 0.9

Rounding on the feature value of each zone with the provision if:

$$Z_n < 0.5,\ Z_n = 0$$
$$Z_n \geq 0.5,\ Z_n = 1$$

This rounding is done so that the resulting feature value will be in the form of binary value that will be used as an input value is the process of classification in the next stage. So from the zone calculation is obtained the feature value of Z1 = 0, Z15 = 1, Z 40 = 1, Z80 = 1.

From the calculation of zoning, the feature extraction stage will generate 100 features that will be used as the input value in the classification process in accordance with figure 2.

| | | | | | | | | | | | | | | | | | |
|---|---|---|---|---|---|---|---|---|---|---|---|---|---|---|---|---|---|
| 0 | 0 | 1 | 1 | 0 | 0 | 1 | 1 | 0 | 0 | 0 | 1 | 0 | 1 | 1 | 0 | 0 | 1 |
| Z1 | Z2 | Z3 | Z4 | Z5 | | | | | | | Z40 | | Z80 | | | | Z100 |

**Figure 2.** Feature Extraction Value using Zoning

### *4.5. Classification*

*4.5.1. Backrpopagation Neural Network.* After getting the feature value on feature extraction process using Zoning, the next step is image classification process using Backpropagation Neural Network method. The stages are as follows: design phase backpropagation Neural Network architecture, backpropagation training stage and backpropagation testing phase. Can be calculated by the following equation [9]:

1. initialization

$$\left(-\frac{2{,}4}{F_i}, +\frac{2{,}4}{F_i}\right)$$

2. Activation (Feed Forward)

$$y_j(p) = sigmoid\left[\sum_{i=1}^{n} x_i\,(p).\, w_{ij}\,(p)\right]$$
$$y_k(p) = sigmoid\left[\sum_{i=1}^{m} x_{jk}\,(p).\, w_{jk}\,(p)\right]$$

3. Training Weights

$$\delta_k(p) = y_k\,(p) - y_{dk}\,(p)$$
$$\Delta w_{jk}(p) = \alpha.\, y_j\,(p).\, \delta_k\,(p) - \mu.\, \Delta w_{jk}\,(p-1)$$
$$w_{jk}\,(p+1) = w_{jk}\,(p) - \Delta w_{jk}\,(p)$$
$$\delta_j(p) = \left[\sum_{k=1}^{m} \delta_k\,(p).\, w_{jk}\,(p)\right].\, y_j\,(p).\left(1 - y_j\,(p)\right)$$
$$\Delta w_{ij}(p) = \alpha.\, x_i\,(p).\, \delta_j\,(p) - \mu.\, \Delta w_{ij}(p-1)$$
$$w_{ij}(p+1) = w_{ij}\,(p) - \Delta w_{ij}(p)$$

4. initialization

## 5. Conclusions and Recommendations

The conclusions that can be taken based on testing the identification system of hypertensive retinopathy disease through retinal fundus image using Backpropagation Neural Network are as follows:

- Identification of retinal fundus image can be done by using backpropagation neural network as classification method according to the specified target with accuracy value reaching 95%..
- The selection of backpropagation neural network parameter values have an effect on the result accuracy. In parameter testing at backpropagation neural network parameter, the test provides an accuracy of 95% at the maximum value of epoch 1500.

## References

[1] Faheem M R and Muizzuddin 2015 Diagnosing Hypertensive Retinopathy through Retinal Images *Biomed. Res. Therapy* **2**(10): 385-388.

[2] Agurto C, Joshi V, Nemeth S, Soliz P, Barriga S 2014 Detection of Hypertensive Retinopathy Using Vessel Measurements and Textural Features *2014 36th Annu. Int. Conf. IEEE Eng Med. Biol.* pp. 5406–5409.

[3] Quinn EA and Krishnan KG 2013 Retinal Blood Vessel Segmentation using Curvelet Transform and Morphological Reconstruction *Int. Conf. Emerging Trends Comput. Commun. Nanotechnol. (ICECCN)* pp. 570-575.

[4] Nugroho HA, Oktoeberza KW, Adji TB and Najamuddin F 2015 Detection Of Exudates On Color Fundus Images Using Texture-Based Feature Extraction. *Int. J. Technol.* **2**: 121-129.

[5] Purandare M and Noronha K 2016. Hybrid System for Automatic Classification of Diabetic Retinopathy using Fundus Images. *Online Int. Conf. Green Eng. Technol. (IC-GET)* pp. 1-5.

[6] D'Cruz MJ, Jadhav MA, Dighe MA, Chavan MV and Chaudhari JL 2016 Detection of Lung Cancer Using Backpropagation Neural Networks and Genetic Algorithm. *Int. J. Adv. Res. Comput. Commun. Eng.* 5(4): 963-967.

[7] Syahputra M F, Nurrahmadayeni, Aulia I, Rahmat R F 2017 Hypertensive retinopathy identification from retinal fundus image using probabilistic neural network *International Conference on Advanced Informatics, Concepts, Theory, and Application*

[8] Sandri W E 2017 *Identifikasi Retinoblastoma Menggunakan Backpropagation Neural Network.* Skripsi. Universitas Sumatera Utara.

[9] Negnevitsky M 2005 *Artificial Intelligence: A Guide to Intelligent System. 2nd Edition. Pearson Education Limited: Upper Saddle River.*

[10] Dataset Structured Analysis of the Retina http://www.ces.clemson.edu/~ahoover/stare/.

IOP Conf. Series: Journal of Physics: Conf. Series **978** (2018) 012107 doi:10.1088/1742-6596/978/1/012107

# Identification Male Fertility Through Abnormalities Sperm Based Morphology (Teratospermia) using Invariant Moment Method

**M F Syahputra[1], R Chairani, Seniman[2], R F Rahmat[3], D Abdullah, D Napitupulu[5], M I Setiawan[6], W Albra[7], C I Erliana[8] and U Andayani[9]**

[1,2,3,9]Department of Information Technology, Faculty of Computer Science and Information Technology, Universitas Sumatera Utara, Medan, Indonesia
[4]Department of Informatics, Universitas Malikussaleh, Aceh, Indonesia
[5]Research Center for Quality System and Testing Technology - Indonesian Institute of Sciences
[6]Department of Civil Engineering, Narotama University, Surabaya, Indonesia
[7]Department of Management, Universitas Malikussaleh, Aceh, Indonesia
[8]Department of Industrial Engineering, Universitas Malikussaleh, Aceh, Indonesia

nca.fadly@usu.ac.id | rouva@students.usu.ac.id | seniman@usu.ac.id | dahlan@unimal.ac.id | darwan.na70@gmail.com | ikhsan.setiawan@narotama.ac.id | wahyuddin@unimal.ac.id | cutitha@gmail.com

**Abstract**. Sperm morphology is still a standard laboratory analysis in diagnosing infertility in men. Manually identification of sperm form is still not accurate, the difficulty in seeing the form of the invisible sperm from the digital microscope image is often a weakness in the process of identification and takes a long time. Therefore, male fertility identification application system is needed Through sperm abnormalities based on sperm morphology (teratospermia). The method used is invariant moment method. This study uses 15 data testing and 20 data training sperm image. That the process of male fertility identification through sperm abnormalities based on sperm morphology (teratospermia) has an accuracy rate of 80.77%. Use of time to process Identification of male fertility through sperm abnormalities Based on sperm morphology (teratospermia) during 0.4369 seconds.

## 1. Introduction

The process of combining sperm cells and eggs along with the chromosomes of both cells is also called the fertilization process (fertilization). The process of fertilization can occur when the sperm cells can penetrate the egg. Sperm cells and eggs play an important role in the process of fertilization. In the process of fertilization required a good level of fertility so that sperm cells and eggs can melt into one. The level of fertility of a woman can be seen from the eggs that have been ripe and ready to be fertilized.

The level of fertility of a man can be seen from the number of sperm contained in semen (sperm fluid), sperm motion activeness, and normal sperm form.

The level of fertility depends not only on women, but male fertility also plays an important role in this fertilization process [1]. The frequency of infertility of about 40% is caused by abnormalities in men. Half of male infertility cases are caused by low sperm motility (asthenozoospermia), sperm coun (oligozoospermia), and morphological abnormalities of sperm (teratozoospermia) [2]. According to data from WHO the normal and perfect form of sperm cells is to have head, body, and tail. The number of factors causing the decline in productivity of a man like today's lifestyle is not supporting a healthy sperm environment, food type, diet, and unhealthy daily habits make sperm quality worse. Laboratory standards to see sperm fertility are still done based on sperm morphology [3]. Currently the identification of sperm form is still done manually and is subjective [4]. For that the researchers intend to conduct research that serves to identify a man's fertility through the deformity of sperm.

Fertility is the most important part in the process of fertilization. Eggs and sperm that have been melted and cooked will become a zygote and then become an embryo. The eggs produced should be good in order to get a good embryo, as well as the quality of sperm should be the best. Good sperm quality can be seen based on the morphology of sperm, sperm form with the best quality consists of regular oval head, body, and long straight tail. Sperm morphology is still a standard laboratory analysis in diagnosing infertility in men. Currently the identification of sperm form is still done manually and is subjective. Manually identification of sperm form is still not accurate, the difficulty in seeing the form of the invisible sperm from the digital microscope image is often a weakness in the process of identification and takes a long time. Therefore, male fertility identification application system is needed through sperm disorder based on sperm morphology (teratospermia).

## 2. Previous research

This study describes the algorithm used to find sperm in low contrast images. First, an algorithm for image repair is applied to remove additional particles from the image. Then, the foreground particles (including sperm and round cell) are segmented to separate from the background. Finally, sperm can be detected and separated from other cells [5].

This study implements image processing algorithms In computer aids for objective analysis of morphology of human Sperm, known as Automated Sperm Morphology Analyzer (ASMA). The first stage of Explanation is the detection and extraction of spermatozoon Individuals from an image that contains some Spermatozoa and biology as well. A new method for acrosom, core and mid segmentation A piece of spermatozoon A method called n-fusion In this study, the segmentation algorithm Implemented in this computer device, the algorithm has been tested with a 250 image database of Spermatozoon and the results are quite accurate [6].

This study focuses on the classification of good spermatozoa morphology that is normal or abnormal using matlab. The first stage is the pre stage Image processing that involves converting RGB images to gray scale images and thennoise images removed using median filter. The second stage is the detection and extraction of individual spermatozoon involving the extraction of sperm objects from images using a sobel edge detection algorithm. The third stage of spermatozoon is divided into various areas of interest such as aperm head, body and tail. The fourth stage involves statistical measurements of the spermatozoon classify normal or abnormal spermatozoa [7].

For segmentation of spermAcrosome, Nucleus, Midsepotong and identification of sperm through some points placed on the sperm utilizing expectation-entropy based on expectation maximization (EM) algorithm and Markov Random Field (MRF) Model for obtaining and upgrading the classconditional probability density function (CCPDF) and a priori probability of each class [8].

## 3. Methodology

The identification of male fertility in this study consists of several steps that begin with the normal sperm microscope image collection and abnormal sperm digital microscope images that will be used as training data and test data, followed by a grayscale process on the sperm digital microscope image to obtain Gray level of image. Furthermore, gaussian process is done to eliminate the noise in the image of digital microscope sperm. After the gaussian process is completed with segmentation by using the thresholding process, the labeling process is done for the numbering (labeling) of the object and counting the objects present in one sperm microscope image, the bounding box process is done to recognize the object in one image of the sperm microscope, and the classification using the value approach is euclidean distance.

After these steps are done it will be obtained from the identification of male fertility through sperm abnormalities based on sperm morphology (teratospermia). The methodology of this research can be seen in the general architecture at Figure 1.

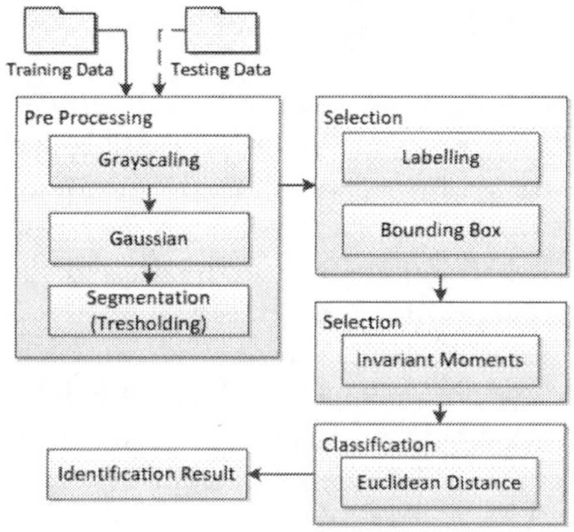


**Figure 1.** General Architecture

*3.1 Graphic Image Formation*

At this stage the RGB image is converted into a gray image to get the gray value of each pixel present in the image[9]. To change the colored image that has the value of each r, g, and b be grayscale with the value of s. Conversion can be done by taking the average value of r, g, and b Can be calculated by the following equation:

$$G = (R+G+B)/3$$

*3.2 Image Smoothing (Gaussian)*

This process is done after the grayscale process, this image smoothing process is done to reduce the noise in the image. In addition, the value obtained from this gaussian will be used for the process of object recognition is the feature extraction by using invariant moment. Gaussian filtering is obtained from convolution operation. The multiplication operation performed is the multiplication of the kernel matrix with the original image matrix. The gauss kernel matrix derived from the computational function of the gaussian distribution can be calculated using the follow :

$$G(i,j)=c.e\text{-}((i-u)^{\wedge}2+(j-v)^{\wedge}2)/(2\sigma^{\wedge}2)$$

The 5 x 5 gauss kernel matrix with $\sigma$= 1.0 is indicated In Table 1.

**Table 1.** The 5 x 5 gauss kernel matrix

| | | | | |
|---|---|---|---|---|
| 1 | 4 | 7 | 4 | 1 |
| 4 | 16 | 26 | 16 | 4 |
| 7 | 26 | 41 | 26 | 7 |
| 4 | 16 | 26 | 16 | 4 |
| 1 | 4 | 7 | 4 | 1 |

Multiplication of the weight of the original image matrix with the weight of the gauss kernel matrix is shown in equation 2.3.

*3.3 Image Segmentation (Thresholding)*
In the next stage the image that has undergone the process of image smoothing (gaussian) will experience the process of image segmentation. Image segmentation is done to separate objects from the background. In this process the image will be divided into two parts namely the object and the background. This thresholding process will generate binary images that provide information about the location of the object and the background. Where the image will only be black and white only. The thresholding process is shown in the following equation:

*3.4 Labeling of Objects In Labels Imagery (Labeling)*
Labeling is done to label the objects contained in the image. This process is done if in one image has many objects. The label will be given according to the result object of the image segmentation (thresholding). Labeling is one form of image processing operation for digital image segmentation. Connected component labeling is a technique that can be used to classify regions or Objects in digital imagery. This technique utilizes connectivity pixel theory in image. The pixels in the region are called connected (there is connectivity or connectivity) when obeying the pixel rules. This pixel rule exploits the nature of the pixel sizes, the existing pixels linked by the neighboring relationship. There are two connectivities that can be used which are 4-connectivity (4-connected neighbors), and 8-connectivity (8-connected neighbors). 4 connectivity (4 connected neighbors) can be seen in Table 2.

**Table 2.** 4 connectivity (4-connected neighbors)

| | | |
|---|---|---|
| | P (x-1, y) | |
| P (x-1, y) | P (x, y) | P (x, y+1) |
| | P (x, y+1) | |

8-connectivity (8-connected neighbors) can be seen in Table 3.

**Table 3.** 8 connectivity (8-connected neighbors)

| | | |
|---|---|---|
| P (x-1, y-1) | P (x, y-1) | P (x+1, y-1) |
| P (x-1, y) | P (x, y) | P (x+1, y) |
| P (x+1, y+1) | P (x, y+1) | P (x+1, y+1) |

*3.5. Typing Objects In Sperm Images (Bounding Box)*
In this process the object contained in the image will be boxed that serves to facilitate the introduction of objects and minimize the object area. The bounding box process is done by giving the box on the object according to the size of the object and in accordance with the labeling results. Where the object consists of head, body, and tail only. The bounding box technique is very easy by checking the pixels first, after that by comparing the maximum and minimum values in x and y coordinates. Bounding box can be done after labeling process.

*3.6. Invariant Moment*
Invariant moment is a method that describes geometry characteristic in object identification system and character recognition. After the image objects obtained, then calculated the seven values invariant momentnya. Invariant moment calculation process with sample gaussian image file size 100 x 100 pixel. The first step to find the invariant moment value can be started by calculating the moment value of each gaussian process object. The result of the gaussian process is calculated by the equation with the following equation:

The results of the Gaussian process can be seen in Table 4.

**Table 4.** Gaussian process results

| x/y | 0 | 1 | 2 | 3 | 4 | 5 | ... | 99 |
|---|---|---|---|---|---|---|---|---|
| 0 | 85 | 127 | 141 | 137 | 142 | 145 | ... | 52 |
| 1 | 130 | 169 | 188 | 189 | 190 | 192 | ... | 118 |
| 2 | 145 | 189 | 214 | 217 | 216 | 216 | ... | 141 |
| 3 | 141 | 190 | 218 | 220 | 219 | 218 | ... | 141 |
| 4 | 144 | 191 | 217 | 219 | 218 | 219 | ... | 142 |
| 5 | 145 | 192 | 216 | 127 | 218 | 221 | ... | 144 |
| ... | ... | ... | ... | ... | ... | ... | ... | ... |
| 99 | 71 | 129 | 148 | 142 | 141 | 145 | ... | 86 |

After the results of the gaussian process are obtained, the value The moment can be calculated by the following equation:

The value of moments obtained using the above equation is as follows:

1. μ_00 = 160650
2. μ_10= 8479770
3. μ_01= 7304730

After the value of m00, m10 and m01 obtained, then the calculation is continued by calculating the center moment value with the following equation:

Where: and

Central moment values obtained Using the above equation is as follows.

x ̅= 52.78413

y ̅= 45.46984

After the value of x and value y is obtained, then:

1. μ_11= 38224.09
2. μ_20= 4966543.52
3. μ_02= 1.78
4. μ_30= 655804.31
5. μ_03 = 1.40
6. μ_12= -2579968.51
7. μ_21= -1.21

After value μ11, μ20, μ02, μ30, μ03, μ12 and μ21 obtained for each object, then proceed with normalization of the center moment value with the equation as the following:

ηpq = μ_pq/(μ_(〚00〛^γ ) )

Where: γ = (p+q)/2 +1

μ_00=m_00

Then the value of normalization of the central moment is obtained, namely:

1. η11 = 1.4810705783391106e-6
2. η20 = 1.924388646307049e-4
3. η02 = 6.916438376251929e-4
4. η30 = 6.339754566407415e-8
5. η03 = 1.3541176602811373e-5
6. η12 = -2.4940926429292113e-7
7. η21 = -1.1717710666820208e-6

The final stage to obtain seven invariant values moment for each object is done in a way calculate the value of ϕ . After the ϕ value is obtained, thencthe value is directly defined in | log (|ϕ |) |.

ϕ1 = η20 + η02

$\phi 2 = (\eta 20 - \eta 02)2 + 4\eta 112$

$\phi 3 = (\eta 30 - 3\eta 12)2 + (3\eta 21 + \eta 03)2$

$\phi 4 = (\eta 30 + \eta 12)2 + (\eta 21 + \eta 03)2$

$\phi 5 = (\eta 30 - 3\eta 12)(\eta 30 + \eta 12)[(\eta 30 + \eta 12)2 - 3(\eta 21 + \eta 03)2] + (3\eta 21 - \eta 03)(\eta 21 + \eta 03)[3(\eta 30 + \eta 12)2 - (\eta 21 + \eta 30)2]$

$\phi 6 = (\eta 20 - \eta 02)[(\eta 30 + \eta 12)2 - (\eta 21 + \eta 03)2] + 4\eta 11(\eta 30 + \eta 12)(\eta 21 + \eta 03)$

$\phi 7 = (3\eta 21 - \eta 03)(\eta 30 + \eta 12)[(\eta 30 + \eta 12)2 - 3(\eta 21 + \eta 03)2] - (\eta 30 - 3\eta 12)(\eta 21 + \eta 03)[3(\eta 30 + \eta 12)2 - (\eta 21 + \eta 03)2]$

Then the value of the seven invariant moments of the object Contained in the image is as follows:
1. $\phi 1$ = 7.030959945114469
2. $\phi 2$ = 15.20484673197635
3. $\phi 3$ = 21.914878625296456
4. $\phi 4$ = 22.60034274190014
5. $\phi 5$ = 44.86779365123266
6. $\phi 6$ = 30.203467326090035
7. $\phi 7$ = 50.89496763743394

*3.7. Euclidean Distance*

Classification is done by using euclidean distance is a value approach. Value is obtained from training data will be stored into database, then the value of data testing obtained will look for the smallest distance with the existing value in the database. The smaller the value spacing obtained then the object will be more similar, vice versa if the distance value obtained greater then objects increasingly unlike. Distance values can be calculated use the following equation:

$$P = (p1, p2, \dots, pn)$$

$$Q = (q1, q2, \dots, qn)$$

$$d = \sqrt{(p_1-q_1)^{2} + (p_2-q_2)^2 + \cdots + (p_n-q_n)^2} = \sqrt{\sum_{i=1}^{n} (p_1-q_1)}$$

## 4. Conclusions and Recommendations

The conclusions that can be taken based on testing the male fertility identification system using invariant moment are as follows:
- Invariant moment method can be used to identify male fertility through sperm abnormalities based on sperm morphology (teratospermia).
- The process of male fertility identification through sperm abnormalities based on sperm morphology (teratospermia) has an accuracy rate of 80.77%.
- The use of time for the process of male fertility identification through sperm abnormalities based on sperm morphology (teratospermia) for 0.4369 seconds.

## References

[1] Venkatesh S, Deecaraman M, Kumar R, Shamsi MB, Dada R 2009 Role of reactive oxygen species in the pathogenesis of mitochondrial DNA (mtDNA) mutation in male infertilit *Indian J. Med. Res.* **129**:127 – 37.

[2] Shamsi MB, Kumar R, Bhatt A, Bamezai RN, Kumar R, Gupta NP, Das TK, Dada R 2008 Mitochondrial DNA mutations in etiopathogenesis of male infertility *Indian J. Urol.* **24**:150-4.

[3] World Health Organization 2010 WHO laboratory manual for the Examination and processing of human semen fifth edition.

[4] Johny B F 2011 Fertility Checkup (Fertility) in Men and Women Http://www.medicinesia.com. Retrieved 7 July 2015.

[5] Nafisi VR, Moradi MH, Nasr-Esfahani MH 2005 Sperm Identification Using Elliptic Model and Tail Detection *World Academy of Science, Engineering and Technology*.

[6] Carrillo H, Villarreal J, Sotaquirá M, Goelkel A, Gutierrez R 2007 Spermatozoon Segmentation Towards an Objective Analysis of Human Sperm Morphology. *IEEE* (1845-5921).

[7] Abbiramy V S and Shanthy V 2010 Spermatozoa Segmentation and Morphological Parameter Analysis Based Detection of Teratozoospermia. *Int. J. Comput. Appl.* **3** (7).

[8] Bijar A, Benavent AP, Mikaeili M 2012. Fully automatic identification and discrimination of sperm's parts in microscopic images of stained human semen smear *J. Biomed. Scie. Eng.* pp. 384-395.

[9] Rahmat R F, Chairunnisa T, Gunawan D and Sitompul O S 2016 Skin color segmentation using multi-color space threshold *Int. Conf. Comput. Information Scie. ICCOINS 2016 – Proceedings* pp. 391-396

IOP Conf. Series: Journal of Physics: Conf. Series **978** (2018) 012108 doi:10.1088/1742-6596/978/1/012108

# Implementation of augmented reality to models sultan deli

**M F Syahputra, N P Lumbantobing, B Siregar, Rahmat R F, U Andayani**

Department of Information Technology, Faculty of Computer Science and Information Technology, Universitas Sumatera Utara, Medan, Indonesia

nca.fadly@usu.ac.id | nathan_p_l@students.usu.ac.id | baihaqi@usu.ac.id | romi.fadillah@usu.ac.id

**Abstract**. Augmented reality is a technology that can provide visualization in the form of 3D virtual model. With the utilization of augmented reality technology hence image-based modeling to produce 3D model of Sultan Deli Istana Maimun can be applied to restore photo of Sultan of Deli into three dimension model. This is due to the Sultan of Deli which is one of the important figures in the history of the development of the city of Medan is less known by the public because the image of the Sultanate of Deli is less clear and has been very long. To achieve this goal, augmented reality applications are used with image processing methodologies into 3D models through several toolkits. The output generated from this method is the visitor's photos Maimun Palace with 3D model of Sultan Deli with the detection of markers 20-60 cm apart so as to provide convenience for the public to recognize the Sultan Deli who had ruled in Maimun Palace.

## 1. Introduction

Sultan Deli is one of the important figures in the history of Medan. But the people of Medan themselves are still less familiar with the Sultan of Deli who had ruled in Maimun Palace. This is based on the lack of documentation on Deli Sultanate which has been very old and less clear so that the introduction of Sultan Deli to the community is still minimal. Therefore it is necessary a technology that supports to be able to display the model of Sultan Deli and his royal bodyguard with details so that people can recognize the Sultan of Deli who ever ruled. The technology is one of them is by using augmented reality technology.

Multimedia in a technology can complement the technology [1]. Multimedia has an important role in all aspects, because multimedia is a trigger (trigger) to get something 'more' from the topic being studied. One of the multimedia that will provide the convenience is modeling the object. Modeling the object changes the two-dimensional shape into three dimensions so that the object will appear more real. Augmented Reality can combine systems of three-dimensional and two-dimensional objects with the real world. Augmented Reality makes it easy for users to understand and recognize the existing cultural heritage. With a very interesting Augmented Reality function, the application of Augmented Reality technology to the introduction of Sultan Deli who once ruled in Maimun Palace is needed.

The implementation of augmented reality for cultural preservation especially in modelling historical ground such as palace has been conducted before. They area Darul Arif Palace in Serdang Sultanate [9], Tanjung Pura Palace [11], Historical Building in Medan City [10] and Historical Game [12]. Augmented Reality The Sultan of Deli can display three dimensional models from Sultan Deli along with his royal bodyguards coupled with camera features that allow visitors to Maimun Palace to be able to take pictures with the three-dimensional model. Sultan Deli is loaded in three dimensions through this application provides an overview for users to know the history of the Deli Sultanate in the Palace Maimun. The Deli Sultanate in Maimun Palace is one of history in the city of Medan so that the augmented reality application of Sultan Deli can provide optimal function for its visitors.

Augmented Reality Technology Sultan Deli is expected to provide optimal results and in accordance with the design of the Augmented Reality.

## 2. Identification of Problems

The documentation of the photo of the Sultan of Deli in Maimun Palace is still very minimal, unclear and very old. The available images are still black and white. By utilizing one of the existing photographs as a reference of this research then formed the model of Sultan Deli and royal bodyguards including clothing with detail using 3D model applications so that visitors Maimun Palace can know more clearly than the photos in the Palace Maimun it. Models formed into three dimensions will be displayed via smarthphone with augmented reality technology. Augmented reality technology will be built using the camera feature so that visitors can take pictures with the model of Sultan Deli. Through this application is expected to provide optimal function for visitors Maimun Palace.

## 3. Previous research

Prince et al [2] conducted an interactive augmented reality research with users viewing markers as a reference for displaying 3D camera screen. The result of this augmented reality gives subjects a real part of the 3D scene.

Zollner et al [3] developed augmented reality to present cultural heritage. In each of these cultural heritages, interactive information is provided and raises videos related to the cultural heritage located at the study site. The creation of augmented reality uses mixed reality.

Croatia DSP Studio and Momentum Studio [4] designed augmented reality by creating a 3D animation of a famous figure in Croatia precisely in the city of Sibenik. Juraj Dalmatinac who is a famous sculptor in the city. This application presents the history of the city and allows to take pictures with Juraj Dalmatinac.

## 4. Research methods

### *4.1. Augmented reality*

Augmented Reality is a technology that combines two-dimensional and three-dimensional objects into a real environment that is then projected in real-time. Augmented Reality is a combination of real and virtual objects in a real environment, running interactively and real-time and there is integration between objects in three dimensions of virtual objects that are integrated in the real world. Merging of real and virtual objects is possible with the appropriate display technology. Interactivity is possible through certain input devices, and good integration requires effective tracking [5]

### *4.2. Maimun Palace*

In 1890, Sultan Ma'mun Al Rashid moved the central government of Deli Deli back to Medan. In 1876, Sultan Ma'moen Al-Rashid established a magnificent palace called the Maimun Palace. The first stone building of Maimun palace was laid by Sultan Makmoen Al Rashid on August 26, 1888, the construction of this palace took almost 3 years, and began to occupy on May 18, 1891 [6]. At that time the capital Deli was officially moved to Medan. Maimun Palace is the Palace of Greatness of Deli Sultanate with its yellow color (yellow is the color of the Malay kingdom) and typical Malay architectural style on the east coast. It is one of the most famous landmarks in Medan, the capital of North Sumatra. It has an area of 2,772 square meters, and has 30. Satisfied with Van Erp's work, he later returned Van Erp to design Masjid Raya Al-Mashun [7].

### *4.3. General Architecture*

The general architecture of Augmented Reality Sultan Deli can be seen in Figure 1.

IOP Conf. Series: Journal of Physics: Conf. Series **978** (2018) 012108 doi:10.1088/1742-6596/978/1/012108

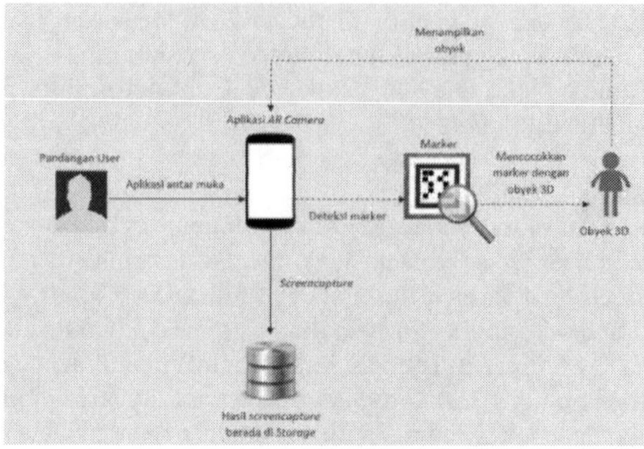


**Figure 1.** General architecture

This stage describes the general architecture of the application as shown in Figure 1:

- User
  This stage is the use of augmented reality applications on an android-based smartphone.
- System
  This application system begins with an augmented reality application that implements the camera feature and will detect a marker with a tracking system. After tracking the marker, it will connect to the 3D model that will be displayed. The 3D model will adjust the position and orientation of the marker so that it appears on the user's smartphone screen.
- Output
  The resulting output is a photo file with the format. JPG (JPEG, JPE). This file will be saved to the storage inside the smartphone.

*4.4. Methodology*
The methodology of Augmented Reality Sultan Deli application can be seen in Figure 2.

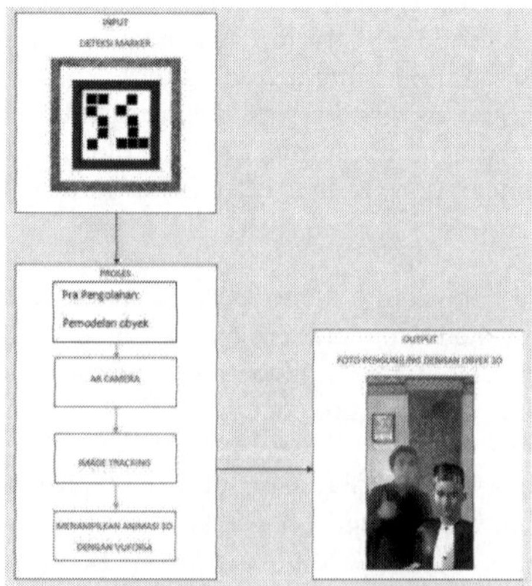


**Figure 2.** Application methodology of Augmented Reality Sultan Deli.

This stage describes the methodology in the application as in Figure 2, namely:

- Input
  This stage uploads the marker into the app as a detector to show 3D models that have been created.
- Process
  This stage is a three-dimensional model processing process created through an image of Sultan Deli and will be incorporated into an augmented reality application. Models will be created using 3D applications such as Blender, Zbrush, Facegen, and DAZ. Application Augmented Reality Sultan Deli will be shaped AR camera that is augmented reality with photo capture feature. 3D model processing results will be inserted into the AR camera so that the smartphone screen will display the 3D model.
- Output
  The resulting output is a photo file with the format. JPG (JPEG, JPE). The photo file is a visitor's photo with the 3D model of Sultan Deli.

*4.5. Making Marker*

Making markers is done by creating an image that will be used as image tracking. The image will be uploaded to the Qualcomm Developer website. Files that have been uploaded will be rated quality by the system. All markers that have been uploaded through vuforia will generate source code in the form of an XML file. This XML file is a configuration of Vuforia against markers that have been uploaded. In making marker required a JPG file to be uploaded to Vuforia.

The marker that has been made will be placed in the Maimun Palace at the appropriate location as shown by Figure 3.

**Figure 3.** Marker location.

The marker will be placed on seven points and at any point will only show one 3D model of Sultan Deli. The map of where the marker is placed is shown by Figure 4. In Figure 4. shows the plot of the marker location to be placed inside the Maimun Palace. As seen in Figure 4., then the number 1 is a marker of Sultan Lamanjiji which is located just under the photo of Sultan Mahmud Lamanjiji Perkasa Alamsyah. Number 2 is a marker of Sultan Mahmud Al Rashid who sits on a couch of relics of the previous Deli Sultanate. Number 3 is the marker of Sultan Ma'moen Al Rashid just above the collection of kris of the Sultan of Deli. Number 4 is a marker of Sultan Osman Al Sani Perkasa Alamsyah beside a cabinet relics Sultan Mai'moen. Number 5,6,7 is a marker of the royal guards.

IOP Conf. Series: Journal of Physics: Conf. Series **978** (2018) 012108 doi:10.1088/1742-6596/978/1/012108

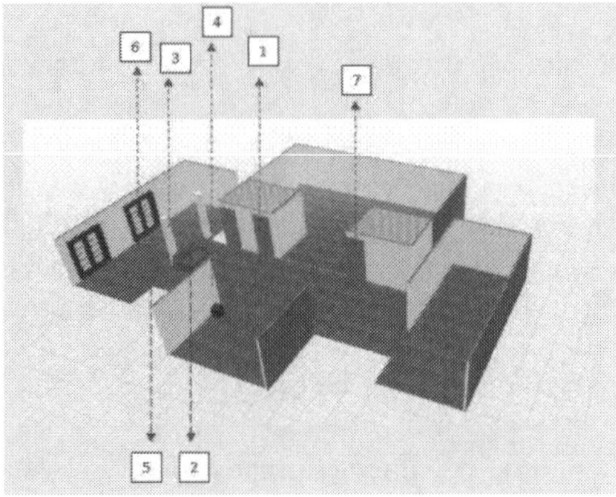


**Figure 4.** Marker location plan.

*4.6. Objects Creation*
3D modeling is done with some 3D tools. Modeling is done by utilizing photos of Sultan Deli as a reference in this study. Modeling the face of the Sultan deli is formed by adjusting the photo of Sultan Deli as in the picture.

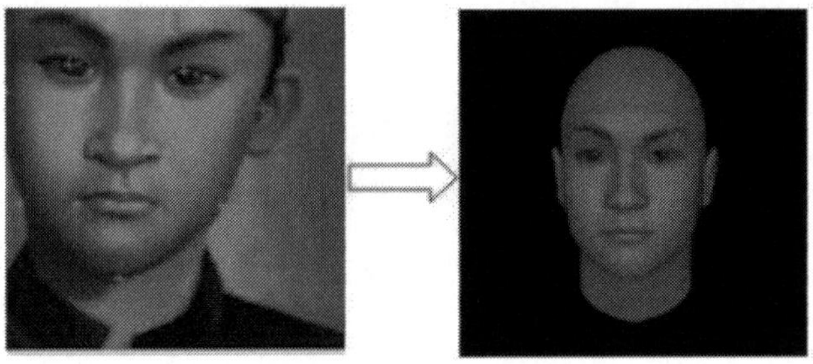

**Figure 5.** The making of the face of Sultan Mahmud Lamanjiji.

The technique used in making this face model is fitting and generating. By determining the points on the neutral image then the image will be generated into a 3D object. The next step is modeling the body and clothing in detail. The body is formed by adjusting the character to the Sultan Deli model as shown in Figure 6.

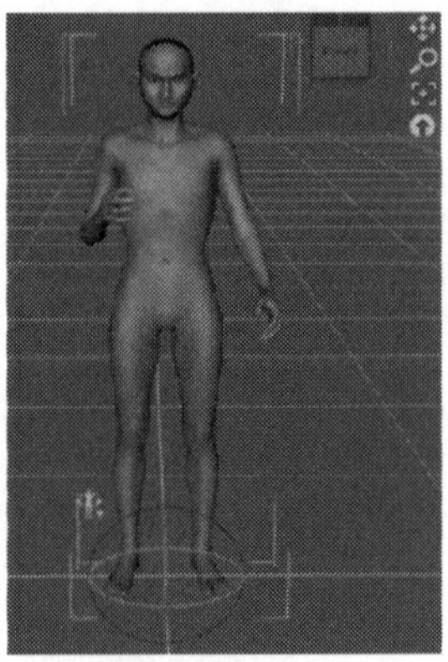

**Figure 6.** DAZ Studio 3D Models.

IOP Conf. Series: Journal of Physics: Conf. Series **978** (2018) 012108 doi:10.1088/1742-6596/978/1/012108

The model is then adjusted to the original form. This stage is done morph and fit process to make adjustments to the original character.

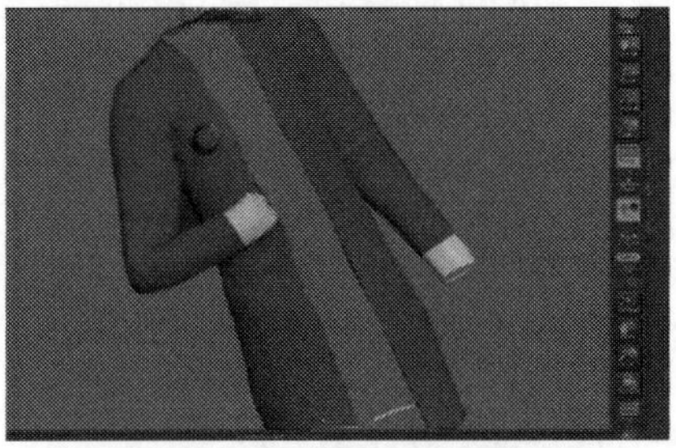

**Figure 7.** The Sultan's 3D Apparel Model.

In Figure 7, the Sultan's clothing is made by arranging the shape of the object to form a clothing object. With mask and hide point techniques can remove unneeded geomotor so that the clothing model will be formed more realistically. The mask technique is used to mark the part of the object for change, and the hide point is to remove the geometry portion marked by the mask. Details and colors adjusted with the photo of Sultan Deli so as to produce a real object.

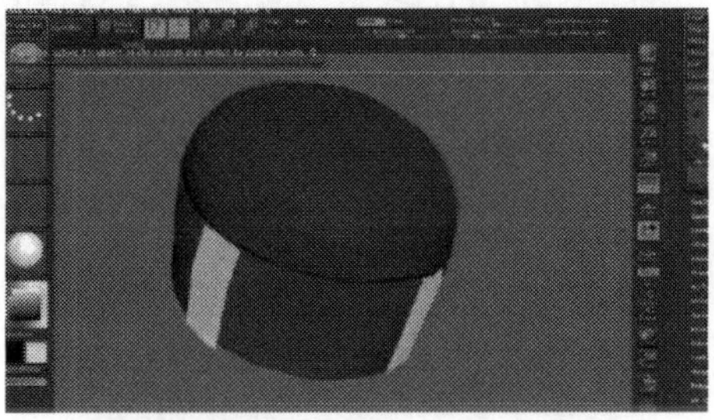

**Figure 8.** Sultan cap model.

At the stage of making hats or caps of the Sultan performed texturing process is to make the details of the cap that will be made so that the Sultan's hat will be worn more realist look.

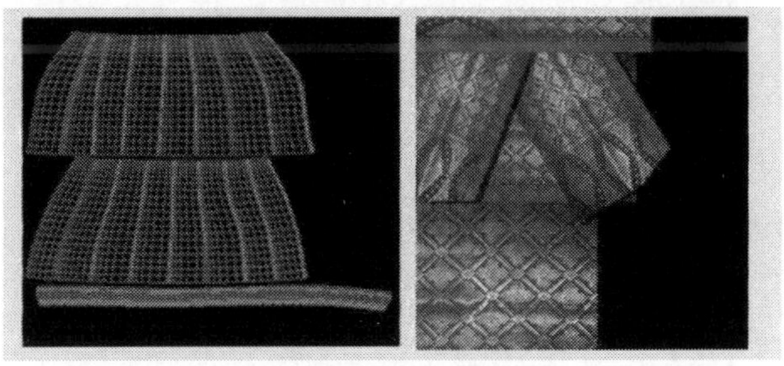

**Figure 9.** Detail sarong with texture atlas.

Detailing of sarong is done with texture mapping. The previous texture was made into a bitmap that was edited in such a way that the final model looks more real. The next stage is to do mapping which means to mapping texture (bitmap) to a three-dimensional object. [8]

**Figure 10.** Model of Sultan Lamanjiji.

The last stage is the incorporation of all three-dimensional objects by arranging the position and orientation of objects within a scene. The result of the 3D model is then exported into the FBX form so that it can be inserted into the Unity 3D application so that the three-dimensional object created can be merged into the application AR Camera made in Unity 3D.

### *4.7. Interface Creation and Augmented Reality*

After making 3D model of Sultan Deli hence required designing interface by using Unity 3D. The design is made by using tools and scripts that have been provided by unity.

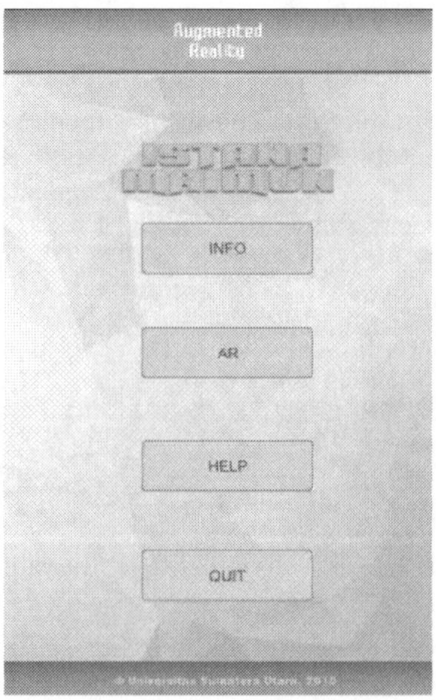


**Figure 11.** Display interface.

The built-in interface page is an info page aimed at displaying the information of the Maimun Palace and the ever-ruling Sultan of Deli, the Ar page aims to showcase augmented reality apps, and the help page aims to guide application usage. On the page Ar formed an augmented reality application created using package vuforia, which then in the innovation into augmented reality with the camera feature (screencapture).

## 5. Result and Discussion

After testing the application by following the test plan, the results of the trial are stored as evidence that the application has successfully passed the test phase.

Test results obtained include:

- Application of AR Camera in marker detection done with distance and certain condition.
- Augmented Virtual Stage application can be run well with menu display and AR Camera with AR Camera Application results in form in JPG format like.

**Figure 12.** Detection of markers.

The explanation of Figure 12. is the result of a test on the detection of a marker in which the marker can only be detected at a distance of 20 -60 cm apart from that the markable surface area can be detected only until the marker 25% marker is closed. The rest of the smartphone can not detect the marker attached.

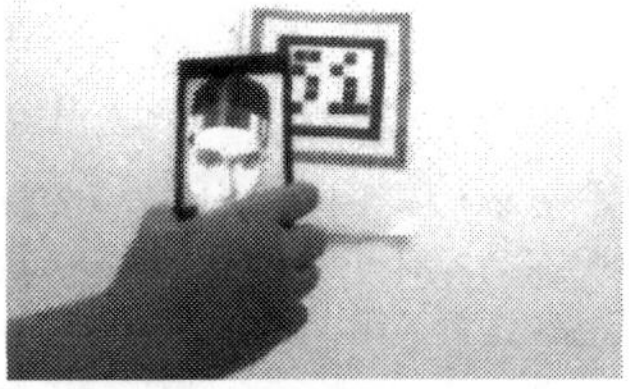

**Figure 13.** Result of marker detection.

In Figure 13. It appears that the smartphone managed to detect markers so that it displays the model of Sultan Deli Mahmud Lamanjiji Perkasa Alam.

**Figure 14.** Application test results.

In figure 14, it has been tested against the application in Maimun Palace so that the application can display the model of Sultan Deli and perform the screencapture process in accordance with the research design.

## 6. Conclusions and Recommendations

### *6.1. Conclusion*

After passing the stage of implementation and testing of the system, then there are some conclusions found in this study, namely:

- Application of augmented reality Sultan Deli can display the model of Sultan Deli who has ruled in the Palace of Maimun like Sultan Lamanjiji Perkasa Alam, Sultan Mahmud Al Rashid, Sultan Ma'moen Al Rashid, Sultan Osman Al Sani Perkasa Alamsyah and bodyguard of the kingdom in three dimensions complete with Clothing details.
- By detecting markers with a distance of 20 cm - 60 cm then the device can read marker and show Sultan Deli 3D well.
- Application Augmented Reality Sultan Deli able to display system AR Camera for screencapture.

### *6.2. Suggestion*

The author wants to give some suggestions for research in the field of Augmented Reality next, namely:

- Using hardware that is very supportive due to rendering and manufacturing process that requires large RAM and VGA.
- Using the official and paid DAZ Studio plugins to generate more interactive three-dimensional objects such as the addition of more realistic animations or details.

## References

[1] Vaughan W 2011 Digital Modeling *New Riders : USA.*

[2] Prince S, Cheok AD, Farbiz F, Williamson T, Johnson N, Billinghurst M, Kato H 2002 3D live: real time captured content for mixed reality *ISMAR 2002*

[3] Zoellner M, Keil J, Wuest H, Pletinckx D 2009 An Augmented Reality Presentation For Remote Cultural Heritage Sites *Fraunhover IGD Jerman.*

[4] Davies J 2014 Ones-Site Digital Heritage Interpretation: Current uses and future possibilities at World Heritage Sites *Durham University* 48-50

[5] Azuma R, Baillot Y, Behringer R, Feiner S, Julier S, MacIntyre B 2001 Resent advances in Augmented Reality.

[6] Teruna T A A 2006 Sultan Makmoen Al Rasyid dan Berdirinya Pemerintahan Kota Medan serta Istana Maimoon. Bandung: Melajoe Marie Meladjoe.

[7] Luthfi, Friza. 2014. Transformasi Gaya Arsitektur, Studi Kasus: Istana Maimun, Medan. Universitas Sumatera Utara

[8] Aditya 2007 50 Kreasi Modeling & Animasi 3D Spektakuler dengan 3DS MAX 8. PT Elex Media Komputindo, Jakarta.

[9] Syahputra M F, Annisa T, Rahmat R F and Muchtar M A 2017 Virtual Application of Darul Arif Palace from Serdang Sultanate using Virtual Reality *J. Phys. Conf. Ser.* **801** p. 12009.

[10] Syahputra M F, Siregar R K and Rahmat R F 2017 Finger recognition as interaction media in Augmented Reality for historical buildings in Matsum and Kesawan regions of Medan City . *Lecture Notes in Computer Science* **10325**

[11] Syahputra M F, I M Rizki, Fatimah S, Rahmat R F 2017 Implementation of player position monitoring for Tanjung Pura Palace Virtual Environment *Lecture Notes in Computer Science* **10325**

[12] Syahputra M F, Widhianto M, Rahmat R F 2017 Historical game of Majapahit Kingdom based on tactical role-playing game *Proc.12th Int. Joint Conf. Comput. Vision Imaging Comput. Graphics Theo. Appl. (VISIGRAPP 2017)* pp 327-332

IOP Conf. Series: Journal of Physics: Conf. Series **978** (2018) 012109 doi:10.1088/1742-6596/978/1/012109

# Implementation of augmented reality to train focus on children with special needs

**M F Syahputra[1], P P Sari[2], D Arisandi[3], D Abdullah[4], D Napitupulu[5], M I Setiawan[6], W Albra[7], Asnawi[8] and U Andayani[9]**

[1,2,3,9]Department of Information Technology, Faculty of Computer Science and Information Technology, Universitas Sumatera Utara, Medan, Indonesia
[4]Department of Informatics, Universitas Malikussaleh, Aceh, Indonesia
[5]Research Center for Quality System and Testing Technology - Indonesian Institute of Sciences
[6]Department of Civil Engineering, Narotama University, Surabaya, Indonesia
[7]Department of Management, Universitas Malikussaleh, Aceh, Indonesia
[8]Department of Elementary, Teacher Education, Universitas Samudra, Aceh, Indonesia

nca.fadly@usu.ac.id | perdana131402029@gmail.com | dedyarisandi@usu.ac.id | dahlan@unimal.ac.id | darwan.na70@gmail.com | ikhsan.setiawan@narotama.ac.id | wahyuddin@unimal.ac.id | asnawi@unsam.ac.id

**Abstract.** Differences possessed by children with special needs such as differences in mental traits, sensory abilities, social and emotional behavior etc. of normal children causing learning and developmental barriers on children, so that requiring more specific services. Autism is one category of children with special needs who experience delays in responding to the things around him that caused by chaos in the brain work system. The tendency of autistic children who only focus on themselves makes it difficult to concentrate on one thing and easily switch when being talked. Training focus of the child can be done by giving them basic commands presented with visually appealing from the utilization of augmented reality and leap motion controller.

## 1. Introduction

Every children often experiences obstacles in their learning and development. However, some children with special needs require special attention and treatment from the people around them to overcome these obstacles[1]. Children with special needs or exceptional children are children who differ in mental traits, sensory, physical and neuromascular abilities, social and emotional behavior, communication skills, or a combination of two or more of them from other normal children.

Autism is one category of children with special needs who experience delays in responding to the things around him that caused by chaos in the brain work system. The tendency of autistic children who only focus on themselves makes it difficult to concentrate on one thing and easily switch when being talked[2]. Therefore, it takes an application that can help people with autism to be more focused and concentrated so as to reduce barriers in the learning process and its development.

In this study, the authors will create a system that can help train the focus of people with autism with augmented reality using leap motion controller. Focus level of the child is seen from the ability of his hands to move the object that is read by leap motion according to the instruction shown to the user. The use of augmented reality in this system is seen when user seems to hold the object to move it but the fact is the user does not hold the object.

## 2. Identification of Problems

Due to the frequent occurrence of obstacles in the learning process as well as the development of autism sufferers due to the difficulty of the child's ability to focus and concentration, it is necessary an application that can train focus and concentration for children with autism.

## 3. Previous research

Research on children with special needs and more autism has been done before as in 2013, Zhen Bai and his colleagues conducted research on autism. In the study Zhen bai made augmented reality with a pretend play model as a treatment process for people with autism. This pretend play-shaped game can demonstrate the communication and social interaction skills of autistic children. Zhen Bai interprets marked beams as a means of transportation and displays them in the form of augmented reality. On the screen will be shown the child is running the train. Whereas in reality, the child is pretending to move the beam as if it were a train. This study received a positive response from parents of people with autism. They really liked the augmented reality in the form of pretend play in helping the healing of his son. And 9 out of 10 users also prefer to play pretend play with augmented reality than non augmented reality [3].

Farhan Luthfi in 2015 conducted research on the use of leap motion on the human anatomy of three-dimensional. Users can study anatomy by moving their hands toward the model. Movement or gesture of the hand captured by leap motion is used as a control system. The hand gestures recognized by the system can change the user's point of view to anatomy, zoom-in / out, and display object names. Based on the results of questionnaires of 4.13 with Likert scale, users feel satisfied and agree this system is very interactive. [4].

In 2015, Gaoxia Zhu also conducted research on autism in the form of games using leap motion. In the game, the user must move the ball into a container that has the same color as the ball. Leap motion is used to read hand movements when moving the ball. Gaoxia measures the development of motor skills of autism sufferers from the accuracy of users moving the ball. Two children who become participant are noticed daily development when using this system. On the first day, the accuracy of moving the ball to the same color is 0%. However one child already has 100% accuracy in adjusting the ball to the same colored container as the ball on day 7. While the other child gets 100% accuracy on the 20th day [5].

Seiji Sugiyama in 2016 conducted research on the use of leap motion controller on augmented reality that is applied to display 3D textbooks. In that study, Seigi created an electronic book model on a web browser. The book can be controlled by hand without touching the surface of the screen or with the help of other control tools such as a mouse. Hand gestures are read by the system via leap motion controller [6].

## 4. Research methods

### *4.1. General architecture*

In this study consists of several steps, starting from the input process that captures the image and marker, augmented process and leap motion and its output. For more details can be seen in Figure 1.

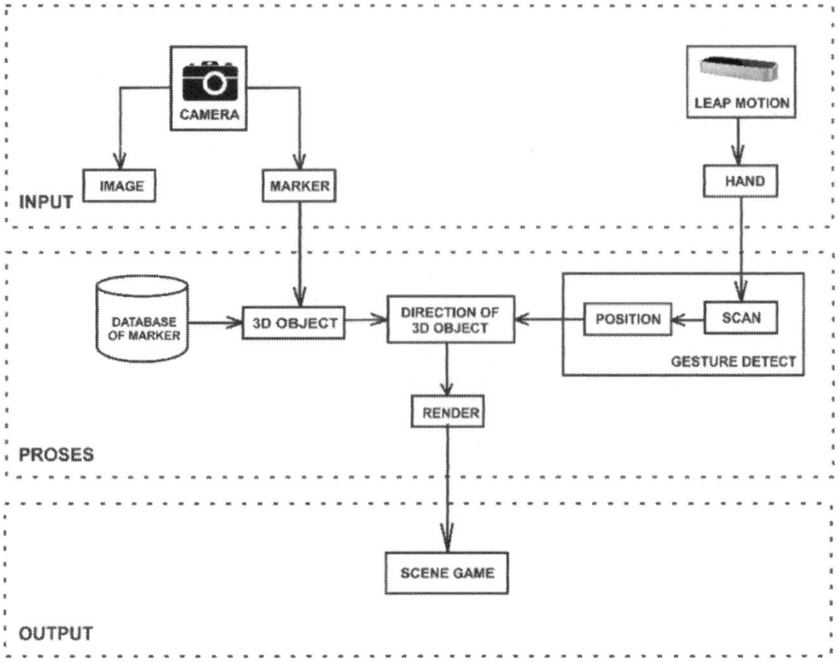


**Figure 1.** General architecture

*4.1.1. Input*

There are three inputs in this system namely image, marker and hand. The imagery and marker are captured by the camera. The imagery is used as a reality to be combined with a digital object and marker is used as a 3D object marker. While the hand is detected by leap motion controller.

*4.1.2. Process*

- 3D Object
  The system will display 3D Object from the captured marker of the camera and merged with the image captured by the camera as well. With this marker, 3D objects will be displayed in the form of augmented reality.
- Gesture Detect
  Gesture detect is done to track the movement of the user's hand in real-time. To track the movement of the user's hand done two processes namely scan and position. Scan is the process of detecting the user's hand. While the position is the process of finding the position of the user's hand.
- Direction of 3D Object
  3D objects displayed will show instructions / instructions in the form of audio to help children understand what to do on the system.
- Render
  Rendering is the final process of the whole process. At this point, the system will continue and display a new scene after the instruction in the previous 3D object has been completed by the child

*4.1.3. Output*

The system will display a game scene output for game users.

*4.2 Introduction of Hand Movement*

Leap motion recognizes the user's hand movement by reading the user's hand in the air just above the leap motion controller that has been connected to the computer via USB. Leap motion read movement

IOP Conf. Series: Journal of Physics: Conf. Series **978** (2018) 012109 doi:10.1088/1742-6596/978/1/012109

will be processed in unity using C # programming language. The movement of hands known by the system is:

- Hand movements such as sweep (swipe gesture) in the direction of x, y and z axis both using the right hand and left hand. However, the object moved / shifted with a gesture swipe in the direction of the z axis will not move according to the settings that the author made on unity.
- Hand gestures like holding objects to move objects more stable.

### *4.3 Marker Determination*
Marker is used as a marker for 3D objects to appearing the form of augmented reality. The marker used on this research is logo of University of Sumatera Utara like figure 2.

**Figure 2.** Markers used

## 5. Result and Discussion
Implementation of the system performed on the patient autism is a game scene that has four levels. The first level is a straight street, the second level is street-shaped turn, third level of branched and level street all four are labyrinths. In this game, the user must collect coins by moving the familiar elliptical object using hands that are moved in the air just above the leap motion as seen in figure 3.

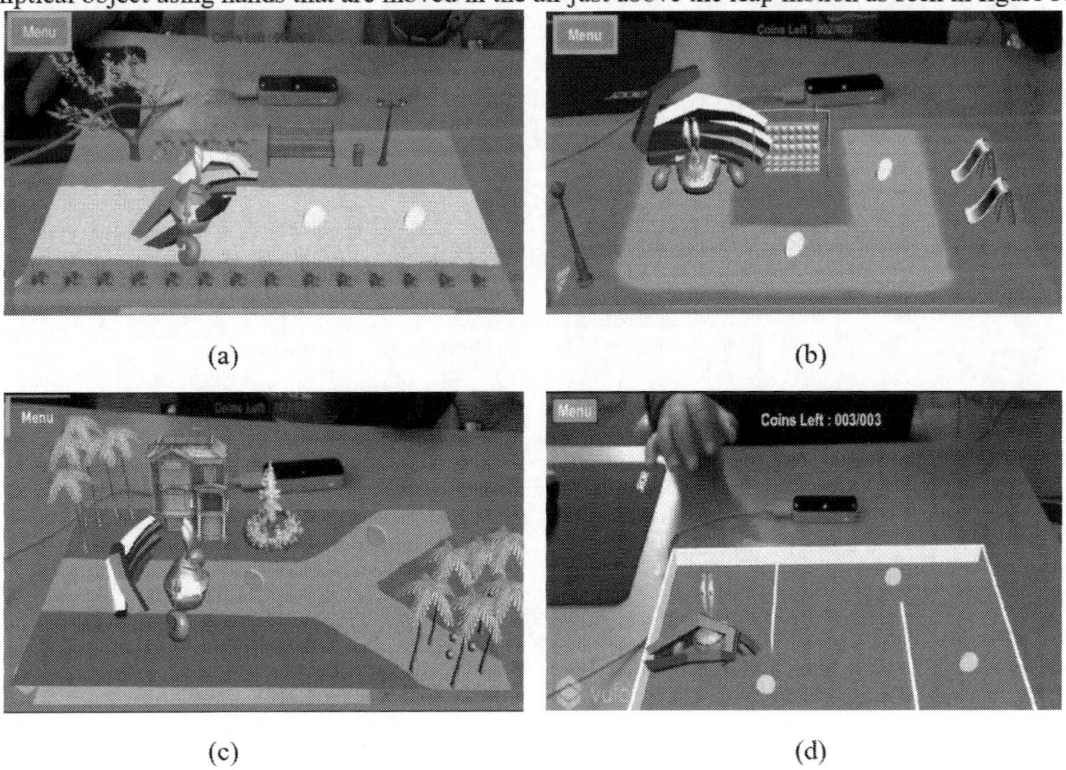


**Figure 3.** Scene Game(a),(b),(c)and(d)

### *5.1. Marker testing*

Marker testing is done to determine the farthest distance which the camera can read on the marker. In this research the authors make 3 test parameters for each test distance. While the test distance in this research as much as 6. Testing done 5 times each test distance. To earn a percentage in each description, the number of tests that are successfully divided by total testing (5) then multiplied by 100. Results system testing can be seen in table 1.

**Table 1.** Marker Test Results

| No | Test range | Number of Tests | Information | |
|---|---|---|---|---|
| | | | Detected | Stability |
| 1. | 20 cm | 5 | 100 % | 100 % |
| 2. | 40 cm | 5 | 100 % | 100 % |
| 3. | 50 cm | 5 | 100 % | 100 % |
| 4. | 60 cm | 5 | 100 % | 100 % |
| 5. | 65 cm | 5 | 60 % | 0 % |
| 6. | 70 cm | 5 | 0 % | - |

From table 1 information shows that the maximum distance for marker detected and stable is 60 cm. This happens because of the limited focus of the camera in recognizing that marker of 60 cm x 60 cm has reached the maximum limit.

*5.2. Testing the detection angle of detection*

Detection slope angle is done to determine the degree of maximum marker slope that can be read by camera. As marker distance testing, the detection slope angle testing is also performed 5 times with the number of slope angles tested is 5 corners. To obtain a percentage on accuracy, the number of successful tests divided by total testing (5) is then multiplied by 100. The results of system testing can be seen in table 2.

**Table 2**. Testing Tilt Angle Marker

| No | The angle of the marker | Number of tests | Accuracy |
|---|---|---|---|
| 1. | 0° | 5 | 100 % |
| 2. | 30° | 5 | 100 % |
| 3. | 45° | 5 | 100 % |
| 4. | 60° | 5 | 60 % |
| 5. | 90° | 5 | 0 % |

*5.3. Approval's level of system uage*

System testing performed on three autism patients who were accompanied by each expert child therapist. Testing of children can be seen in figure 4.

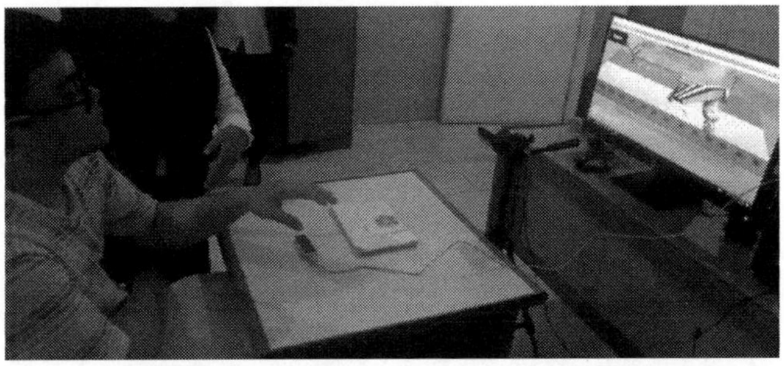

**Figure 4.** Testing of children with autism

The authors conducted a survey in the form of questionnaires conducted on respondents to see the level of system use approval. Questionnaires that the authors do in the form Likert scale with four scales of agreement as in table 3.

**Table 3.** Likert scale on the questionnaire

| Scale of agreement | Value |
|---|---|
| Strongly agree | 5 |
| Agree | 3.75 |
| Less agree | 2.5 |
| Disagree | 1.25 |

From the values on the scale can be obtained the average value of each statement by:

$$\text{Average value} = \frac{\Sigma\,(R \times NJ)}{NJM \times Rtotal} \times 100$$

Information:
R = Respondent
NJ = Answer Value
NJM = Maximum Answer Value
The system test was performed on three people with autism who were accompanied by each expert therapist (respondent). Based on the manner in which the calculations have been described previously, the average value of each of the statements at the level of system use agreement may be seen in figure 5.

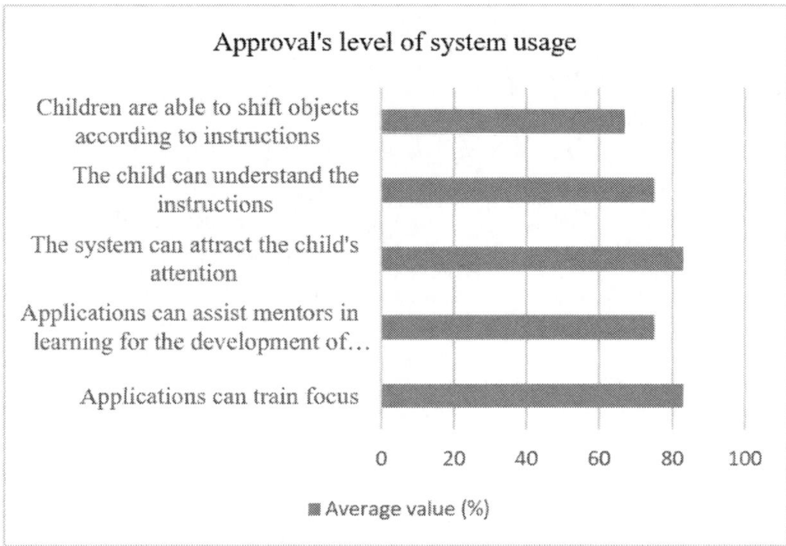


**Figure 5.** Level of System Use Approval

Based on table 5, the first statement that has an average percentage of 83% indicates that respondents feel strongly agree that the application can train the focus of the child.

## 6. Conclusion

After doing the implementation and testing phase of the system, then obtained some conclusions found in this study are :

- The system runs well when detecting markers and displaying 3D objects.
- Maximum distance marker detected by camera and 3D object shown with stable is 60 cm.
- The maximum tilt angle of the marker detected by the camera is 45°.
- With an average percentage of 83% on a Likert scale, respondents strongly agree that the app can train the child's focus.

## References

[1] Cahya L S 2013 Adakah ABK di Kelasku, Bagaimana Guru Mengenali ABK di Sekolah *Qoni Familia : Yogyakarta*

[2] Huzaemah 2010 *Kenali Autisme Sejak Dini* Pustaka Populer Obor : Jakarta.

[3] Bai Z, Blackwell AF, Coulouris G 2013 Thsrough the looking glass: Pretend play for children with autism *IEEE International Symposium on Mixed and Augmented Reality* : 49-58.

[4] Luthfi F 2015 *Anatomi manusia berdimensi tiga menggunakan leap motion controller* Skripsi Universitas Sumatera Utara.

[5] Zhu G 2015 A series of leap motion-based matching games for enhancing the fine motor skills of children with autism. *IEEE 15th Int. Conf. Adv. Learning Technol.* 430-431.

[6] Sugiyama S 2016 A study of displaying 3D electronic text using augmented reality via leap motion. *IEEE Int. Conf. Syst. Man, Cybernetics*: 3015-3020.

# Real-time monitoring system for elderly people in detecting falling movement using accelerometer and gyroscope

**B Siregar[1], U Andayani[1], R P Bahri[1], Seniman[1] and F Fahmi[2,*]**

[1]Faculty of Computer Science and Information Technology, Universitas Sumatera Utara, Medan Indonesia
[2]Department of Electrical Engineering, Faculty of Engineering, Universitas Sumatera Utara, Medan Indonesia

*Email: fahmimn@usu.ac.id

**Abstract .** Most of the elderly people is experiencing a decrease in physical quality, especially the weakness in the legs. This will cause elderly easy to fall and can have a serious impact on their health if not getting help very quickly. It is, therefore, necessary to take immediate action against the falling cases experienced by the elderly. One such action is by developing supervision and detecting of falling movements in real-time, which is then the connection to a member of the family. In this research, we used Arduino Uno as a microcontroller, sensor accelerometer, and gyroscope that serves to measure falling movement of the elderly person and supported by GPS technology Ublox Neo 6M to provide information about coordinates. The result was the high accuracy of delivering notification data to server and accuracy of data delivery to family notification equal to 93,75%. The system successfully detects the direction of falling: forward, backward, left or right and able to distinguish between unintentional falling and conscious falling like a bow or prostrate position.

## 1. Introduction

When a person has advanced age, the person is likely to experience a decrease in physical quality, one of the weaknesses (physical decline) experienced by the elderly is a weakness in his legs, and this will make them easier to fall. Falling is an event that causes a conscious subject to be on the ground unintentionally. Factors that cause falls are ill-informed like stroke [1,2], slippery or wet floors, holding places that are not strong or not easily held.

Supervision of the elderly is an important thing done because it allows the occurrence of accidents, such as falls. Falling experienced can have a very serious impact on the person's health if they do not get help quickly. Therefore, a solution is needed to monitor the movement of the elderly in real-time connected with interested parties, such as medical or family personnel.

Oversight of the elderly is something that is sometimes neglected by the family because they can not always be beside the person. Delays in handling fall often occur due to delays in information received, so unknown fall of elderly parents. It has shortcomings on the side of providing information to the family. With the application system it is expected to help to monitor the elderly and can detect the movement of the fall.

In the year 2012, there was research for Monitoring Against Physical Activity Users and Tracking Position with Real-time. This study aims to detect daily activities such as walking, running, and supine using the method of Decision Tree and naïve Bayesian. By using the accelerometer, data will be done

IOP Conf. Series: Journal of Physics: Conf. Series **978** (2018) 012110 doi:10.1088/1742-6596/978/1/012110

pattern recognition process to recognize user activity called Activity Recognition. Testing is done to test the functionality of the created system. Assaying accuracy on the Naïve Bayesian classification method resulted in a high accuracy rate (98.67%) compared to the Decision Tree method (92.33%) for all activities tested with several variations of sampling length [3].

Furthermore, another research in 2013 was done by putting the smartphone in the pocket user, and then system will detect the user's physical activity at the time of falling. After sensor data is obtained, the system will use windows sampling technique and overlapping method to process sensor data. Application of fall detection system has been implemented and able to detect the user's physical activity when dropped with the accuracy level of 91.75% and can send notification via e-mail [4]. In the year 2014 a study was conducted to calculate the acceleration quantity of the accelerometer sensor in measuring the total acceleration, the total difference of acceleration and the angle of the slope. Time of fall is detected based on the total value of acceleration and the amount of difference of total acceleration. This study made a comparison between techniques that used the total difference amount of acceleration and which did not use. In the test results showed both have the same sensitivity and specificity, which is 90% and 94% of 490 movements [5].

In 2014 a research was done by detecting falls using 3-axis acceleration sensors to measure and collect data. This research uses Zigbee-3G method and one-class SVM classification algorithm. This study analyzes the speed, angular displacement. An analysis is performed if the threshold range exceeds [6]. Another approach is based on image processing as implemented in medical device technology [7], another study shows the implementation of real-time application for water pollution has been referred[8].

In this study, we used accelerometer and gyroscope sensors to monitor and detect movement in the elderly. The input of this system is elderly parents. In the study we used accelerometer and gyroscope sensors that function to take every movement of the elderly and supported by GPS technology in getting the coordinates in order to be able to monitor the falling movement of elderly people.

## 2. Methodology

The general architecture of this study can be seen in Figure 1.

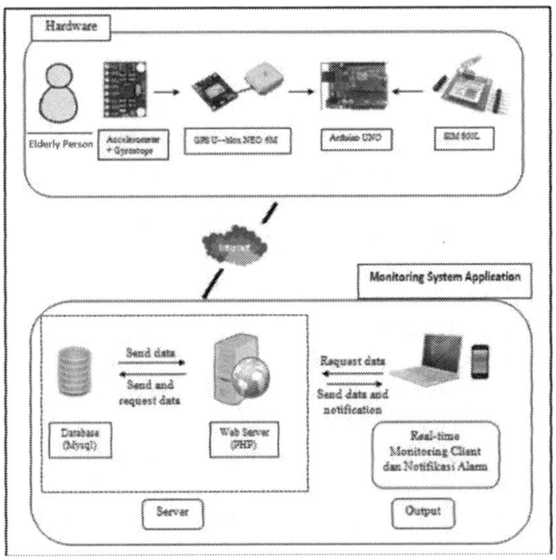


**Figure 1**. General Architecture

### *2.1. Hardware*

The first process of the general architecture is to place sensors on the right side of the waist, in the form of a box that has Accelerometer sensor and Gyroscope. Accelerometer and Gyroscope Sensor has 3 axis coordinates that are the axis X, Y, and Z. This device will detect the movement that occurs. Data experiments were detected as falling, falling forward, falling backward, falling right and falling left. When the user is on the move, the application will retrieve the Accelerometer and Gyroscope data. After that, the system receives input in the form of coordinates to be provided by GPS Module, and SIM 800L mounted on Arduino and sensor MPU6050. Elderly coordinates are obtained from GPS Modules that have been assembled on Arduino.

There are three modules that will be paired with an elderly person, Arduino, MPU6050, NEO-6M GPS Module, SIM 800L. Arduino as the main board and MPU6050 which functions as Accelerometer and Gyroscope data from the elderly activity, NEO-6M GPS module that serves to determine the coordinates of the elderly location, as well as SIM 800L to transmit data to the server. Arduino is still empty, will be uploaded a program to be able to run data obtained from Accelerometer, Gyroscope and to know the coordinate point of elderly location. After that, SIM 800L will send coordinate points to the server based on the server's IP address; the server will store the data aged activity and coordinate point into the database and send it to the user.

The user is a family or relative associated with an elderly person and who will monitor the location of the fall. Users can monitor elderly location through the elderly monitoring system. Arduino will convert the analog signal into a digital signal. The motion activity that has been obtained is sent to real-time monitoring application system using the SIM800L sensor.

*2.2 The Elderly Monitoring Application System*

Monitoring application system to be built is a web-based application system using PHP.

*2.2.1 Web Server*

The web server to be used is a self-built web server. The web server serves as a place of service and data processing between Arduino, database, and client. The web server will receive data from Arduino in the form of user movement activities such as walking, falling forward, falling backward, falling to the right, and falling to the left. This data will then be stored in the database and ready to be processed to show back to the client in the form of GPS coordinates.

*2.2.2 Client*

The client will access a web page on the web server for monitoring. This page will contain the GPS coordinates of each elderly activity, and these coordinates will always move and updated automatically every second without the client needing to refresh the web page. If the data received is beyond a predetermined threshold, the system will display a notification to alert the user of the system in the form of an SMS to the family.

*2.3 Fall Detection*

In determining the condition of the fall of the elderly can be seen from the angle of slope. The angle of the slope is determined from the X and Y axes (Table 1)

**Table 1.** Fall detection condition

| No. | angle (x, y) | Condition | Note |
|---|---|---|---|
| 1. | (0,25) | *StartTimer* | Forward |
| 2. | (0,50) | *End Timer* | |
| 3. | (0, -25) | *StartTimer* | Backward |
| 4. | (0,-50) | *End Timer* | |
| 5. | (25,0) | *StartTimer* | Right |
| 6. | (50, 0) | *End Timer* | |
| 7. | (-25, 0) | *StartTimer* | Left |
| 8. | (-50, 0) | *End Timer* | |

Start Timer – End Timer = 200ms

In this study determined the following conditions:

- If the time is greater than 200ms, then the condition is declared Normal.
- If the time is less than 200ms, then the condition is declared Fall.

## 3. Hardware Design

*3.1 Design of Arduino Uno and Sensor MPU6050*

The Serial Data Pin (SDA) of the MPU6050 is connected to the Analog 4 (A4) pin residing on the Arduino Uno. Pin Serial Clock (SCL) MPU6050 is connected to Analog 5 (A5) pin located on Arduino Uno. Ground Pin (GND) MPU6050 is connected to Ground resources. The VCC pin of MPU6050 is connected to the VCC resource.

*3.2 Design of Arduino Uno and SIM800L*

Pin Rx (Receiver) SIM800L is connected to Analog 2 (A2) pin located on Arduino Uno. Pin Tx (Transmitter) SIM80L is connected to Analog 3 (A3) pin located on Arduino Uno.Pin Ground SIM800L is connected to a Ground resource. Pin VCC SIM800L is connected to VCC resource.

*3.3 Arduino Uno Scheme to GPS NEO6M*

The NEO6M GPS Tx (Transmitter) pin is connected to the Digital 0 (PD0) pin that resides on the Arduino Uno. The NEO6M GPS pin Rx (Receiver) is connected to a Digital 1 (PD1) pin that resides on the Arduino Uno. NEO6M GPS Ground Pin is connected to Ground resources. VCC GPS pin NEO6M is connected to VCC resource.

*3.4 Power Scheme to LM2596*

The power supply is connected to the switch, and the positive voltage of the power supply (PS) is connected to the positive pin IN StepDown LM2596, while the negative PS voltage is connected to the negative pin IN. Then the positive LM2596 pinout is made as VCC power source and connected to positive LED foot; then the negative LED foot is connected to one of the legs of the resistor. The other leg of the resistor is connected to the Ground resource while the negative LM2596 pinout is used as a Ground resource. Resources are the source of StepDown LM2596.

The design of the hardware can be seen in Figure 2.

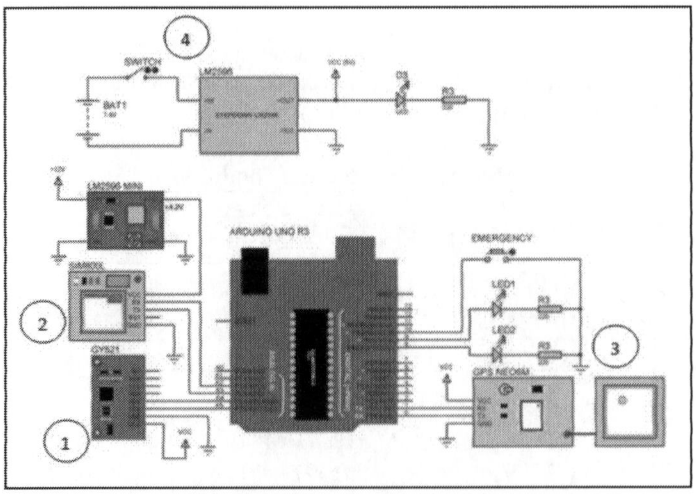


**Figure 2**. Hardware Design

## 4. Result and discussion

Prototype form that will be designed in the form of a box. The shape of the prototype that has been designed can be seen in Figure 3.

**Figure 3.** The Prototype Designed

Hardware that has been designed will be programmed through the program on the computer through a connecting cable. Initially, the program will run on SIM800L. The SIM card will search for the signal first. If the SIM card has got the network to transmit coordinate data obtained through the GPS module, it will display a notification in the form of the words "GPRS OK" on the computer monitor series. In the box, there are 3 pieces indicator lights. The light on the right will turn green indicating that the device is active. The light in the middle will turn red indicating that GSM has been active in the light in the middle of the blinking faster indicates that the internet is stable. The light on the left will turn red indicating that the GPS is active. On the right side, there is an emergency button that serves as an additional tool for notification to the family.

### *4.1 System Performance Testing*

At this stage discusses the problem of performance testing system or tool built to determine the performance of the system in doing the detection whether it is running well. The sensor is placed at the right waist in the elderly. To test the data delivery system from SIM800L to the server by using GSM / GPRS module. This study tested 4 different fall angles and monitored the testing of data transmission to the server and notification to the family in the form of SMS.

Testing by Angle by angle of front view (0,10) (0,15) (0,20) (0,25) and side view (10,0) (15,0) (20,0) (25,0). Test results at the angle (0.10) a box are placed on the right side; front view has front and rear sides. In this test the condition of elderly is still in a normal state, this is because the tool or sensor has not detected any reaction from the elderly. The view at the angle (0.10) can be seen in Figure 4a. Testing at the angle (0.25) a box is placed on the right side; front view has a front and back side. In this test, the elderly condition is still in the normal state, but at an angle (0.25) starts the initial process calculation to detect the fall but the system does not send a notification to the server or to the family. The view at the angle (0.25) can be seen in Figure 4b.

### *4.2 Testing Data Delivery*

Testing data delivery based on notification time on server and time of family notification. The application system testing can be seen in Table 2.

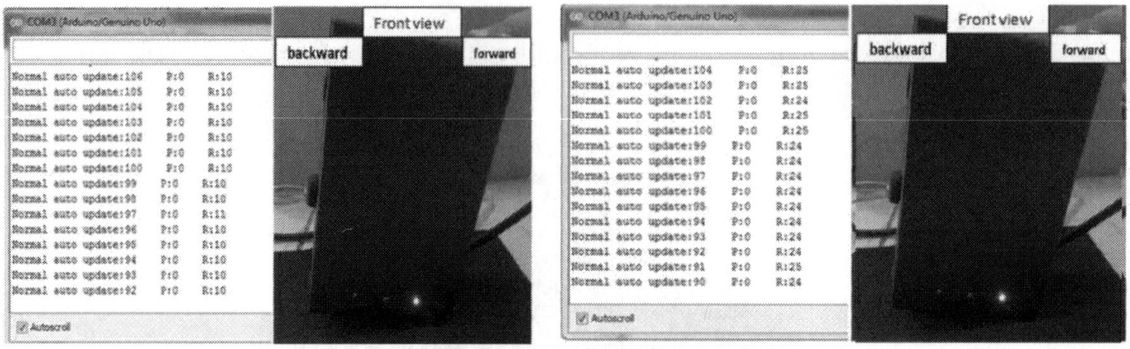


**Figure 4**. Testing at an angle (a) 0.10 and (b) 0.25

**Table 2**. Testing Data Delivery

| NO | Server HTTP | Notification |
|---|---|---|
| 1 | 18.55 | 18.55 |
| 2 | 19.00 | 19.00 |
| 3 | 19.07 | 19.07 |
| 4 | 19.12 | 19.12 |
| 5 | 19.17 | 19.17 |
| 6 | 19.22 | 19.22 |
| 7 | 19.29 | 19.29 |
| 8 | 19.34 | 19.34 |
| 9 | 19.41 | 19.41 |
| 10 | 19.46 | 19.46 |
| 11 | 19.51 | 19.51 |
| 12 | 19.58 | 19.58 |
| 13 | 20.03 | 20.03 |
| 14 | 20.10 | 20.10 |
| 15 | 20.17 | - |
| 16 | 20.23 | 20.23 |
| 17 | 20.30 | 20.30 |
| 18 | 20.35 | 20.35 |
| 19 | 20.41 | 20.41 |
| 20 | 20.46 | 20.46 |
| 21 | 20.52 | 20.52 |
| 22 | 20.59 | 20.59 |
| 23 | 21.04 | 21.04 |
| 24 | 21.11 | 21.11 |
| 25 | 21.16 | 21.16 |
| 26 | 21.23 | 21.23 |
| 27 | 21.28 | 21.28 |
| 28 | 21.33 | 21.33 |
| 29 | 21.40 | 21.40 |
| 30 | 21.46 | 21.46 |
| 31 | 21.51 | 21.51 |
| 32 | 22.00 | 22.00 |
| 33 | 22.07 | 22.07 |
| 34 | 22.08 | - |
| 35 | 22.13 | 22.13 |
| 36 | 22.20 | 22.20 |
| 37 | 22.26 | 22.26 |
| 38 | 22.32 | 22.32 |
| 39 | 22.37 | 22.37 |
| 40 | 22.43 | 22.43 |
| 41 | 22.50 | 22.50 |
| 42 | 22.56 | 22.56 |
| 43 | 23.02 | 23.02 |
| 44 | 23.07 | 23.07 |
| 45 | 23.15 | 23.15 |
| 46 | 23.21 | - |
| 47 | 23.27 | 23.27 |
| 48 | 23.32 | 23.32 |
| 49 | 23.40 | 23.40 |
| 50 | 23.46 | 23.46 |
| 51 | 23.52 | 23.52 |
| 52 | 23.58 | 23.58 |
| 53 | 00.04 | 00.04 |
| 54 | 00.10 | 00.10 |
| 55 | 00.15 | 00.15 |
| 56 | 00.21 | 00.21 |
| 57 | 00.28 | 00.28 |
| 58 | 00.29 | - |
| 59 | 00.35 | 00.35 |
| 60 | 00.40 | 00.40 |
| 61 | 00.46 | 00.46 |
| 62 | 00.52 | 00.52 |
| 63 | 00.58 | 00.58 |
| 64 | 01.04 | 01.04 |
| 65 | 01.10 | 01.10 |
| 66 | 01.16 | 01.16 |
| 67 | 01.21 | 01.21 |
| 68 | 01.27 | 01.27 |
| 69 | 01.32 | 01.32 |
| 70 | 01.38 | 01.38 |

| | | |
|---|---|---|
| 71 | 01.42 | 01.42 |
| 72 | 01.49 | 01.49 |
| 73 | 01.55 | 01.55 |
| 74 | 02.00 | 02.00 |
| 75 | 02.05 | 02.05 |
| 76 | 02.12 | 02.12 |
| 77 | 02.19 | - |
| 78 | 02.26 | 02.26 |
| 79 | 02.31 | 02.31 |
| 80 | 02.37 | 02.37 |

Table 2 shows the test results of sending data from the web server to the family by using SIM800L under normal conditions notifications will update automatically on the server every 2 minutes. In the elderly conditions in a state of falling so that the device sends a notification to the family to do the help and notification will be updated automatically on the server built, this is because of the network conditions on the tool and server in good condition.

At the time of the fall the family will receive notification in the form of SMS at the same time according to the old time declared in the fall state, but the notification does not go to the server because the internet connection on the server is not stable so the notification on the web cannot be updated in real-time.

### *4.3 Notification page*

Notification page is one page that aims to send a notification to the family if the user of the tool is dropped, the tool automatically sends a notification to the client (family), and notification will be sent to the client (family) in the form of SMS. The notification page view can be seen in Figure 5.

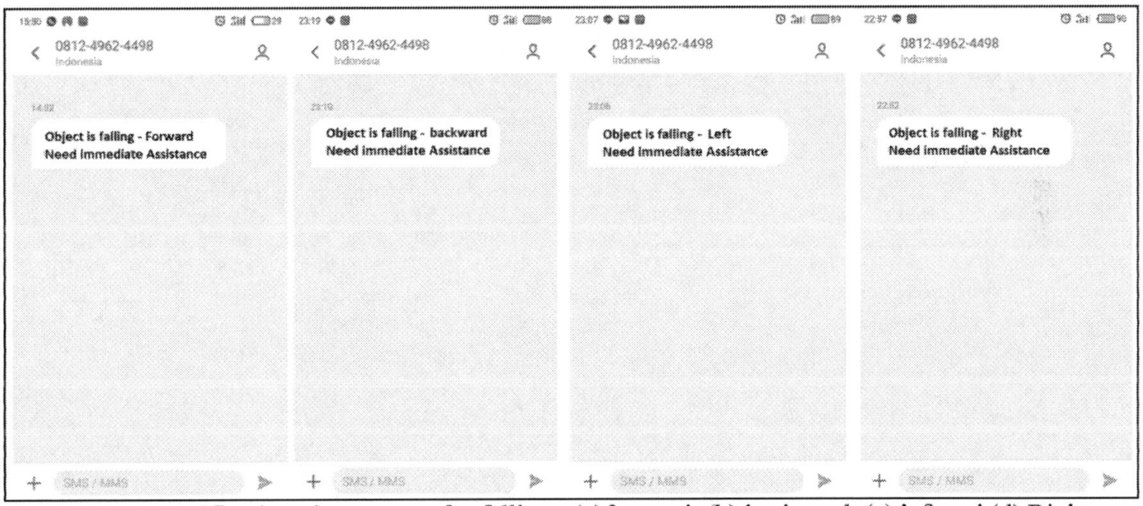


**Figure 5**. Notification alert system for falling : (a)forward, (b) backward, (c) left and (d) Right

## 5. Conclusions

This system uses Arduino as a microcontroller to monitor real-time movement in real-time and for data communication using GSM / GPRS network and provider card that serves to measure falling movement of the elderly person. The accuracy of sending notification data to the server is 100% and the accuracy of sending data to family notification in the form of SMS equal to 93.75%. The system successfully detects the direction of falling: forward, backward, left or right and able to distinguish between unintentional falling and conscious falling like a bow or prostrate position based on acceleration moment of falling. At the time of online testing, internet connection plays an important role in the smooth use of the system: some suggestions that can be considered for further research include more stable internet connection. Further development is recommended to determine the location of the more effective sensor installation.

**References**

[1] Fahmi F, Beenen L F M, Streekstra G J, Janssen N Y, de Jong H W, Riordan A and Marquering H A 2013 Head movement during CT brain perfusion acquisition of patients with suspected acute ischemic stroke *European journal of radiology*, ***82*(12),** 2334-2341

[2] Fahmi F, Riordan A, Beenen L F M, Streekstra G J, Jansse, N Y, de Jong H W and Marquering H A 2014 The effect of head movement on CT perfusion summary maps: simulations with CT hybrid phantom data *Medical & biological engineering & computing*, ***52*(2),** 141-147

[3] Hilman A and Wibisono W 2012 Sistem Monitoring dan Tracking Aktivitas Fisik User Bergerak Berbasis Sensor Accelerometer dan GPS pada Perangkat Mobile Berbasis Android Menggunakan Metode Klasifikasi Decision Tree dan Naive Bayesian, *Jurnal Teknik POMITS* Vol **1**(1) p1-6

[4] Arifin D N, Wibisono W and Pratomo BA 2013 Rancang Bangun Sistem Fall Detection untuk Pengguna Bergerak Berbasis Sensor Accelerometer dan Sensor Gyroscope pada Perangkat Mobile, *Jurnal Teknik POMITS* Vol **2**(1)

[5] Liandana M 2014 Deteksi Jatuh Untuk Lanjut Usia dengan Menggunakan Total Akselerasi dan Sudut Kemiringan Berbasis Smartphone Android, *Thesis Universitas Gadjah Mada*

[6] Ye Z, Li Y, Zhao Q and Liu X 2014 A Falling Detection System with wireless sensor for the Elderly People Based on Ergnomic, *Int.J. Smart Home* Vol **8**, p187-96

[7] Fahmi F , Marquering H , Streekstra G , Beenen L , Janssen N , Majoie C and vanBavel E 2014 Automatic detection of CT perfusion datasets unsuitable for analysis due to head movement of acute ischemic stroke patients *J. healthcare eng.* **5(1),** 67-78

[8] Rahmat R F, Athmanathan A, Syahputra M F and Lydia M S 2016 Real time monitoring system for water pollution in Lake Toba *Int. Conf. Informatics Comput. (ICIC)* pp. 383–388

IOP Conf. Series: Journal of Physics: Conf. Series **978** (2018) 012111 doi:10.1088/1742-6596/978/1/012111

# Clean water billing monitoring system using flow liquid meter sensor and SMS gateway

**F Fahmi[1,*], A Hizriadi[2], F Khairani[2], U Andayani[2], B Siregar[2]**

[1]Department of Electrical Engineering, Faculty of Engineering, Universitas Sumatera Utara, Medan, Indonesia
[2]Faculty of Computer Science and Information Technology, Universitas Sumatera Utara, Medan, Indonesia

fahmimn@usu.ac.id

**Abstract.** Public clean water company (PDAM) as a public service is designed and organized to meet the needs of the community. Currently, the number of PDAM subscribers is very big and will continue to grow, but the service and facilities to customers are still done conventionally by visiting the customer's home to record the last position of the meter. One of the problems of PDAM is the lack of disclosure of PDAM customers' invoice because it is only done monthly. This, of course, makes PDAM customers difficult to remember the date of payment of water account. Therefore it is difficult to maintain the efficiency. The purpose of this research is to facilitate customers of PDAM water users to know the details of water usage and the time of payment of water bills easily. It also facilitates customers in knowing information related to the form of water discharge data used, payment rates, and time grace payments using SMS Gateway. In this study, Flow Liquid Meter Sensor was used for data retrieval of water flowing in the piping system. Sensors used to require the help of Hall Effect sensor that serves to measure the speed of water discharge and placed on the pipe that has the same diameter size with the sensor diameter. The sensor will take the data from the rate of water discharge it passes; this data is the number of turns of the mill on the sensor. The results of the tests show that the built system works well in helping customers know in detail the amount of water usage in a month and the bill to be paid

## 1. Introduction

Public clean water company, PDAM (Perusahaan Daerah Air Minum) as a public service is designed and organized to meet the needs of the community. Water is a basic necessity in everyday life; water has an important role in supporting the prosperity and welfare of society. The availability of adequate water will encourage the development of the development sector in the community. Programs undertaken by PDAMs in both urban and rural areas have the objective of providing services to the community for obtaining clean and healthy water for both domestic and industrial purposes to support economic development and health status of the population. For now, the services provided by PDAM to the community include new installations, repairs, account payments, and others.

However, in the case of water bill payments, PDAM still occupied manual officers to visit each customer's home to check the water meter and provide a monthly bill of water consumption fee to the customer. The water bill payment system until now is still less open and less explanatory in the details of the bill to be paid by the customer. This is because customers can only know the details of the use of

IOP Conf. Series: Journal of Physics: Conf. Series **978** (2018) 012111 doi:10.1088/1742-6596/978/1/012111

water for one month only on the date of payment of the roll. In addition, customers also get into trouble because they always remember the date of payment of their monthly water account. For if it is too late in making payments, then the PDAM will provide a fine or even stop the distribution of water to customers.

The ease of accessing and knowing the water usage information specified is needed that we make an approach to know the details of water usage and the time of payment of water account easily via short message or SMS. The results of this study are expected to provide benefits especially for customers or users of PDAM water.

Some previous research, among others done in 2011, Dodik Juhanto through the design of information systems payments PDAM Ngawi water account using visual basic and mysql. The researcher designs the information system to facilitate more efficient management and account payments on PDAM Ngawi, collect data with observation, interview & library method then processed using Visual Basic and Mysql. [1].

In 2011, Yudi Wiharto designed an academic information system based on SMS gateway. In his research designed information systems for school academic processing, application programs run on a computer connected to a database, using a mobile phone that is connected via a USB port as a receiver (SMS) and will be responded automatically. [2],

In 2014, Zaratul Nisa Saputri designed a voice recognition application as a controller of UND's Arduino-based electrical equipment. Research conducted has a success rate of about 50% if receiving voice commands from different people. If the command is used in the form of a single syllable, the rate of occurrence is 98.30% [3],

In 2104 Suresh et al conducted a study of fluid monitoring and control flowing through industrial pipelines using a web server, with the help of Hall Effect Flow Sensors connected to the Arduino to measure fluid flow, while raspberry PI controlled the Solenoid Electro-Valve functioning for close and open the valve to control the fluid flow of the pipe. [4].

In 2016, Indera Surya Satria monitors and detects the location of piped water pipe leaks using flow liquid sensor on the network sensor. This research also used Arduino microcontroller based on Atmega328P chip which is used to detect leakage location on pipes by utilizing data of water discharge rate through it and stable working system to find the location of leakage location with maximum distance of 2 meter with water discharge rate 10 liter per minute to be able to determine the location of the leak with the closest detection results. [5]. Another study uses image-based monitoring to regularly check the water meter [6] and real-time monitoring for water pollution [7] also has been referred.

In this study, we used the Flow Liquid Meter Sensor to retrieve data from the flow of water that flows on the pipe. This sensor uses the help of hall effects sensor in it to measure the speed of water discharge and placed on the pipe that has the same diameter size with the sensor diameter. The sensor will retrieve data from the rate of water discharge through which it is the number of turns of the mill on the sensor. The goal is to provide information in the form of water debit data used, payment rates, and payment grace periods using SMS Gateway to the customers.

## 2. Method

The general architecture of this study can be seen in Figure 1.

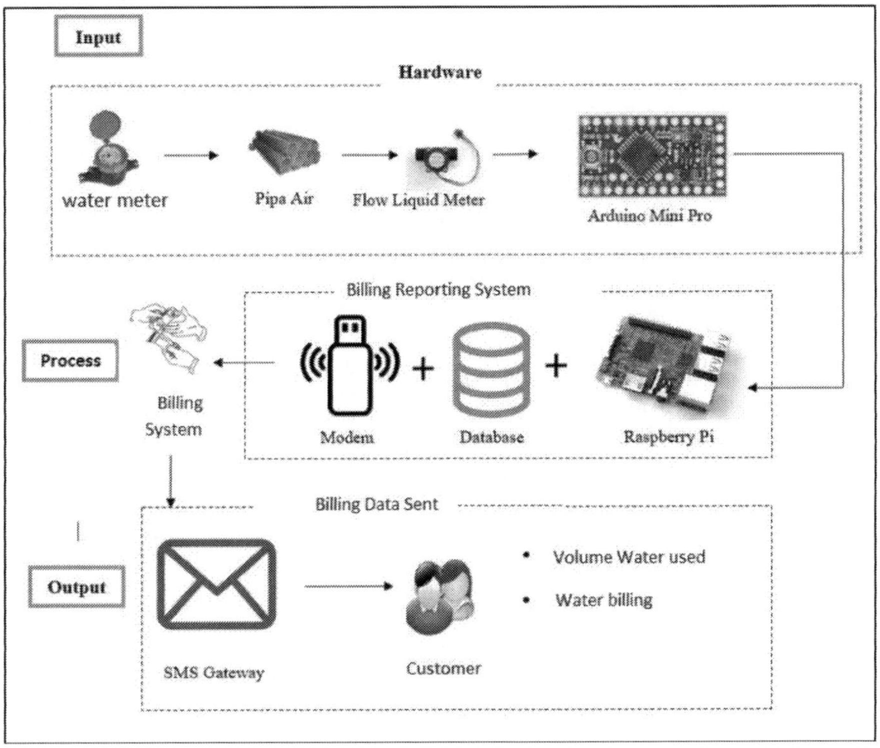


**Figure 1**. General Architecture

### *2.1. Hardware*

In hardware side, it is necessary to set the tools used and the process of taking data using sensors then sent to Arduino, then Arduino process on the data then send it to the customer billing information system using the help of Raspberry Pi.

The process starts from the water through the meter which then flows to the flow liquid sensor that has been installed in the pipe, the sensor will take data from the flow of water flowing in the pipe through the flow liquid sensor, the data will be obtained in the form of analog signal will be sent to Arduino for the process.

Arduino then receives data sent flow liquid sensor, the data received in the form of analog signal and then in the process into a digital signal, the water debit data that has been received being then sent via the internet to the customer billing information system to in though so that the tariff can be set to customer of PDAM. Data is sent using Raspberry Pi already connected to Arduino to help connect to the internet network.

### *2.2. Customer Billing Information System*

Process data sent via the internet will then be accepted customer billing information system, the data will be directly entered into the databases of data users of water, then the data is sent to the billing system to keep the tariffs on the wear based on the amount of water discharge used, then the data will be sent back to be summarized into the database as billing data and PDAM water usage. The billing data will be sent back to the billing system for sending at the end of each month through the process of sending customer billing data.

Submission of Billing Data Customer charges via SMS Gateway automatically to each customer at the end of the month in the form of PDAM water utility bill. SMS gateway delivery through BTS operator and directly into the mobile phone of each PDAM customer in the form of water debit data in use, payment rates and payroll time data.

### 2.3. Application System

The application system of customer billing statement to be built is a web-based application system using PHP. The web server used is a web server that will be built alone. This web server will serve as a place of service and data processing between Arduino, database, and client. The web server will receive the water debit data sent by the PC. This water debit data will then be stored in the database and ready to be processed to be represented back to the client in graphical form. These graphs will be displayed in a certain time interval and will always be updated automatically every second as long as the Arduino sends the water debit data to the system. The process of processing water debit data ready to be represented back to the client will always be done either when the client access the web server or when the client is not accessing the web server.

The client will access a web page on the web server to monitor, and only the specific client gets permission to access the page. This page will contain water debit information through the pipe and will always be updated automatically every second with no need for the client to refreshing the web page. In addition, this web page client can also add customer data or delete customer data and clients can also find out the total bill of water usage from the previous month.

The data used in this study is data obtained directly from the sensor. Arduino will receive data from the sensor which is the number of turns of the mill inside the sensor due to the flow of water through the sensor. The number of turns of this windmill is then processed using equation 1 so that obtained data of water debit passing through the sensor every second.

$$Q = \frac{n}{c} \qquad (1)$$

Where Q is the water discharge, n as the number of turns of the windmill and c is the calibration factor having a value of 7.5 depending on the type of sensor used.

After the water debit data through the sensor then the data will be monitored and conducted reporting the amount of water discharge. Any Arduino water discharge data will be sent to the server directly connected to Arduino via raspberry pi. This data is then sent to the web server and ready to be represented to the web client in the form of graphs and tables that are updated automatically every second and the client can see the amount of water discharge used and the total cost that needs to be paid.

Based on the results of monitoring obtained, every second of data discharge water that comes out will be stored in the database with the user in every month. Then on every 2nd weeks of each month, the flowing water flow data will be accumulated from the usage in the previous month. Then on a predetermined date, the subscriber will receive an SMS Gateway containing the total water usage for that month and a number of tuition that the customer must pay.

### 2.4. Design

The liquid meter flow sensor is placed between two pipes as can be seen in Figure 2.

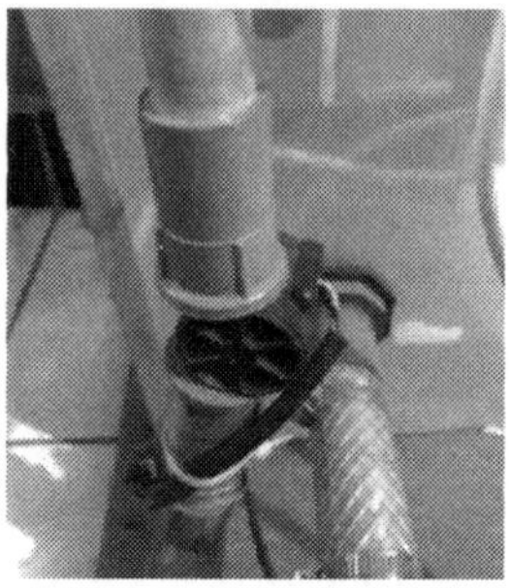

**Figure 2**. Flow liquid meter installed on water pipe

IOP Conf. Series: Journal of Physics: Conf. Series **978** (2018) 012111 doi:10.1088/1742-6596/978/1/012111

Arduino has several pins that serve as a place of data processing and power, on this system pins are used for processing data sent from the sensor. The liquid meter flow sensor will be connected to the digital pin 2 to receive data from the sensor, GND pin and 5V pin as power for the sensor, as in Figure 3.

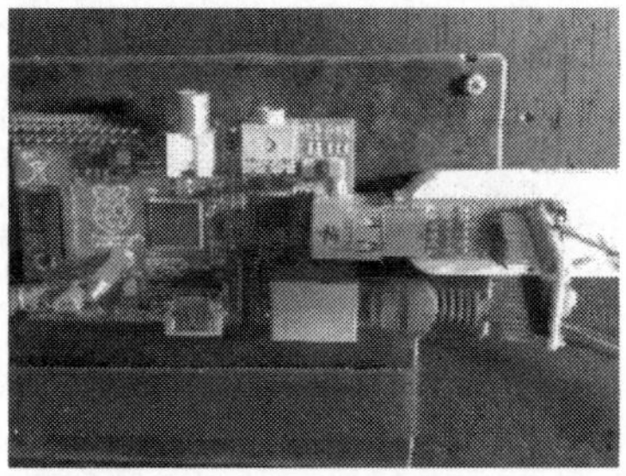

**Figure 3**. Arduino Meter Sensor and Flow Flow

Pi Raspberry is a microcomputer module that also has a digital output input port such as a microcontroller port. Raspberry Pi is used on Arduino to connect to PC using serial data. To obtain serial data from Arduino for raspberry pi using CP 2102 (USB serial) exchanger. This module will be stacked on the Arduino as shown in Figure 4 and Figure 5.

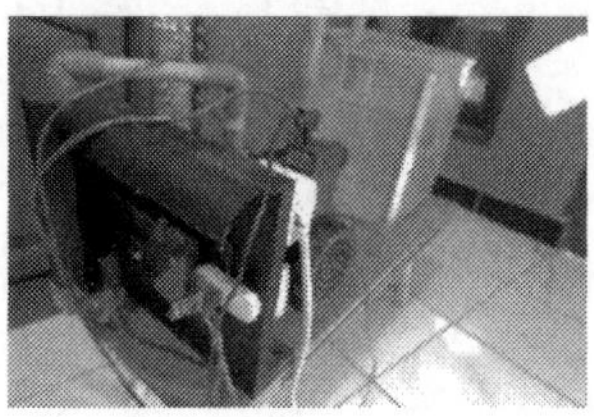

**Picture 4**. Raspberry Pi and Arduino Mini Pro

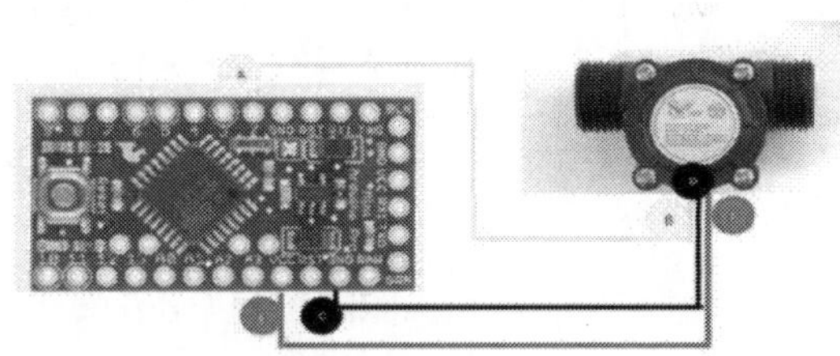

**Figure 5**. Prototype design

### 3. Results and discussion

At this stage we tested how much water flows during the specified time and the amount of bill that must be paid by registered customers. A display of the test results based on the length of the water flowing at the sensor fluid flow meter. Earlier, calibration was made between the numbers in the water meters and the numbers in the sensor to make it easier to find out if there was a running stream of water. The results of this test showed a system that works for 5 minutes, 10 minutes, 15 minutes, 20 minutes and 25 minutes. The 5-minute test can be seen in Figure 6. In figure 6 the value on the meter before running for 10 minutes is 1744 and the value on the meter after 10 minutes is 1952.6. So in 10 minutes, there are 208.6 liters of water flowing through the water meter. Furthermore, the value of the sensor before and after 10 minutes is shown in figure 7.

**Figure 6**. Value of Meters (a) Before and (b)After 5 Minutes

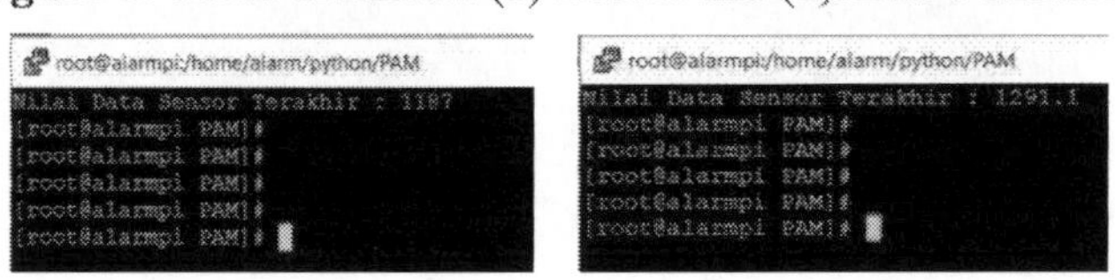

IOP Conf. Series: Journal of Physics: Conf. Series **978** (2018) 012111 doi:10.1088/1742-6596/978/1/012111

**Figure 7**. Sensor Value Before and After 5 Minutes

In figure 7, the value of the meter sensor before running for 5 minutes is 1187 in accordance with the calibration value that has been inserted so that the sensor and meter values are the same. While the value of the sensor after 5 minutes is 1291.1. So in 5 minutes, there are 104.1 liters of water that flows through sensor flow liquid meter. The results of the 5 minutes, 10 minutes, 15 minutes, 20 minutes and 25 minutes tests are shown in table 1.

**Table 1.** Sensor Flow Liquid Meter Based on Time

| No | Time (Minute) | Calibration | Meter (ltr) | Sensor Value (Ltr) | Gap (Ltr) |
|---|---|---|---|---|---|
| 1 | 5 | 1187 | 1291.4 | 1291.1 | 0.3 |
| 2 | 10 | 1294 | 1502.5 | 1502.3 | 0.2 |
| 3 | 15 | 1505 | 1817.9 | 1817.7 | 0.2 |
| 4 | 20 | 1820 | 2237.2 | 2237 | 0.2 |
| 5 | 25 | 2240 | 2752.5 | 2752.3 | 0.2 |

Based on the results of long time testing when used for 10 minutes, 15 minutes, 20 minutes and 25 minutes there is a difference of 0.2 and 0.3 liters of water. This is because the wheel rotates on the flow liquid meter is different from the rotation found in the water meter which is still read manually. And in this test, it shows if the tool will produce the same difference in one-time use.

On this test will display test results based on SMS Gateway can be sent according to the registered customer's phone number. Based on the tests that have been done, the amount of water used in SMS is obtained according to the data sent by the operator provider used. Example of SMS Gateway display successfully posted on the system seen in Figure 8.

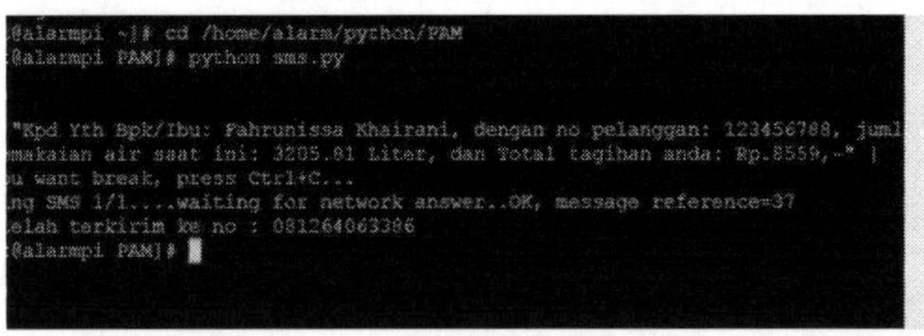


**Figure 8**. SMS Gateway Display Successfully Delivered Through The System To Customers

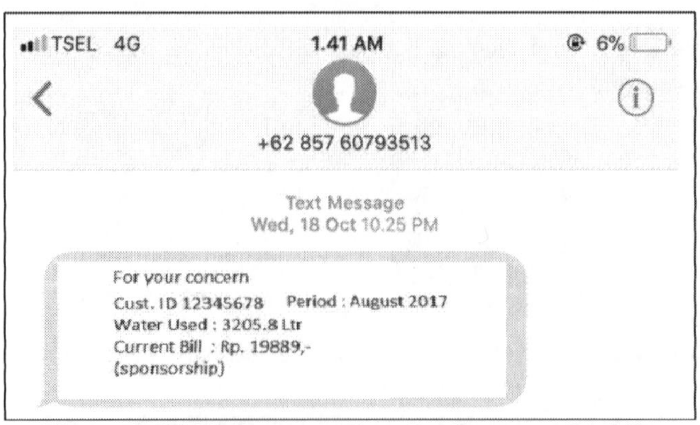


**Figure 9**. SMS Gateway View On Customer Mobile

In figure 9 there is a suitability of the message contained in the system with the contents of the message contained in the mobile phone for each registered customer.

## 4. Conclusion

Based on the testing and reporting system of customer bill based on water consumption using flow liquid meter, it is concluded that the built system can help the customer to know in detail the amount of water usage in the month and the bill to be paid. In the built system there is a difference of water liter that exits the sensor and the meter. And the contents of the SMS gateway received by the customer according to the contents of the message contained in the system.

## References

[1] Juhanto D 2011 Sistem Informasi Pembayaran Rekening Air PDAM Ngawi Menggunakan Visual Basic dan Mysql Skripsi Sekolah Tinggi Manajemen Informatika dan Komputer

[2] Wiharto Y 2011 Sistem Informasi Akademik Berbasis SMS Gateway *Jurnal Teknologi Informatika I* (TEKNOMATIKA) 3(1):1-28

[3] Saputri Z N 2014 Aplikasi Pengenalan Suara Sebagai Pengendali Peralatan Listrik Berbasis *ARDUINO UNO* Skripsi Universitas Brawijaya

[4] Suresh N , BalajiE , AntoK J and Jenith J 2014 Raspberry PI Based Liquid Flow Monitoring and Control *Int. J. Res. Eng. Technol. (IJRET)* **3 (7)**: 122-125

[5] Rahmat R F, Satria I S, Siregar B, Budiarto R 2016 Water Pipeline Monitoring and Leak Detection using Flow Liquid Meter Sensor In Electrical Engineering, Computer Science and Informatics (EECSI), *2016 Int. Conf. IAES*

[6] Triantoro T, Batubara F R and Fahmi F 2014 Image-based water gauge reading developed with ANN Kohonen *Electric. Eng. Comput. Scie. (ICEECS), 2014 Int. Conf.* (pp 74-78) IEEE

[7] Rahmat R F, Athmanathan A, Syahputra M F and Lydia M S 2016 Real time monitoring system for water pollution in Lake Toba *Int. Conf. Informatics Comput. (ICIC)* pp. 383–388 IEEE

# Korean letter handwritten recognition using deep convolutional neural network on android platform

**S Purnamawati[1*], D Rachmawati[2] , G Lumanauw[1], R F Rahmat[1], R Taqyuddin[1]**

[1]Department of Information Technology, Faculty of Computer Science and Information Technology, Universitas Sumatera Utara, Medan, Indonesia
[2]Department of Computer Science, Faculty of Computer Science and Information Technology, Universitas Sumatera Utara, Medan, Indonesia

sarah_purnamawati@usu.ac.id

**Abstract**. Currently, popularity of Korean culture attracts many people to learn everything about Korea, particularly its language. To acquire Korean Language, every single learner needs to be able to understand Korean non-Latin character. A digital approach needs to be carried out in order to make Korean learning process easier. This study is done by using Deep Convolutional Neural Network (DCNN). DCNN performs the recognition process on the image based on the model that has been trained such as Inception-v3 Model. Subsequently, re-training process using transfer learning technique with the trained and re-trained value of model is carried though in order to develop a new model with a better performance without any specific systemic errors. The testing accuracy of this research results in 86,9%.
**Keywords**: Word Recognition, Korean Alphabet Recognition, Handwritten Recognition, Deep Convolutional Neural Network, Transfer Learning

## 1. Introduction

Popularity of Korean culture or Korean wave (*hallyu*) has become a phenomenon that draws public attention nowadays, especially in Indonesia. Many young Indonesians are motivated to learn Korean language and to learn it, they try to understand Korean non-latin character (Hangul) that consists of vocal and consonant [1]. Bahasa Indonesia uses latin character that makes Indonesian learners have to deal with a set of difficulties in learning Hangul.

The used method in this research is Deep Convolutional Neural Network. This method showed an outstanding performance in image recognition, because DCNN has a good ability to extract high-level features. The paper was organized as follows: Sect. 2 reviewing several previous researches related to Hangul recognition and Deep Convolutional Neural Network. Section 3 describing the used method in recognizing Korean letter handwritten Section 4. explaining the result of this research. Section 5 discussing the summary and suggestions for future research.

## 2. Related work

There are a lot of researches about Korean character (Hangul) conducted with various determination and method. Kang used Hierarchical Stochastic Modelling for Hangul character recognition [2]. Sari proposed Template Matching and K-Nearest Neighbors method to recognize Hangul with accuracy rate 85% for Template Matching and 72% for K-Nearest Neighbors [3]. Our previous work also in character recognition is in [11]

There was Hangul handwritten recognition done by Nasution. She used Backpropagation Neural Network with accuracy rates for memorization and generalization were 98,2% and 85,3%. The output of this system was very simple because it was still using GNU GCC MinGW Compiler [4].

Prabowo conducted research on utilization API Gesture to practice writing Hangul in android smartphone. However, the result from the system was not processed for any specific purpose [5].

In [6], they used Deep Convolutional Neural Network for recognizing Hangul handwritten with accuracy rate 95,96% for SERI95a dataset and 92,92% for PE92 dataset. The output of this research was desktop application.

Other similar research was also done by Bui and Chang. They used Deep Convolutional Neural Network for recognizing non-MNIST dataset, which was digit handwritten dataset with more noise and hard to recognize. This research achieved best accuracy rate of 98% [7].

## 3. Methodology

The used method in this research consisted of several steps, namely pre-trained neural network with inception-v3 model as an outcome, retraining process using transfer learning technique, and word recognition process. Word recognition process was done with following steps, image acquisition, pre-processing and recognition using Deep Convolutional Neural Network.
Figure 1 describes the general architecture of the proposed method.

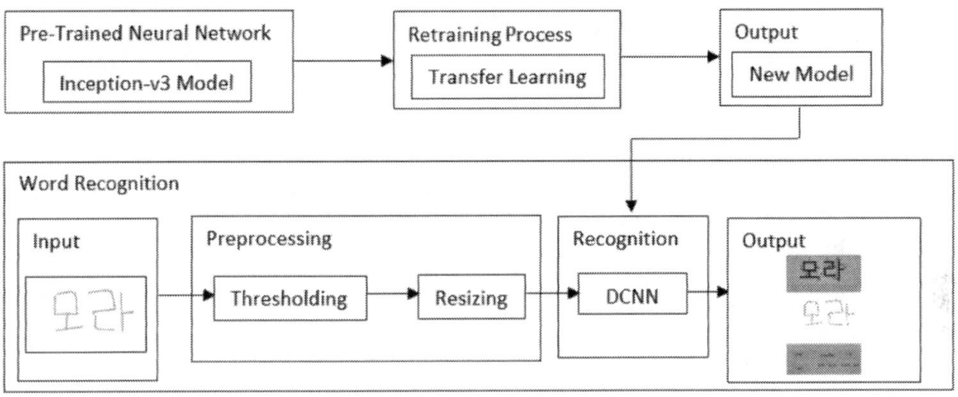


**Figure 1.** General Architecture

### *3.1 Pre-trained neural network*

Pre-Trained Neural Network phase was used to train ImageNet Large Scale Visual Recognition Challenge (ILSVRC) dataset which produced a deep learning model called Inception-v3 Model. This model had an awesome capability in recognizing image.

DCNN method was used in the training process since it had outstanding performance in image recognition. DCNN's capability to recognize image was supported by its feature to extract high-level features. In addition, convolutional layer and max pooling layer could recognize various object. DCNN could conduct every phase in image recognition, since DCNN received raw image as an input, thus it did not need feature extraction phase such as conventional classifier [6]. Figure 2. shows a DCNN's architecture.

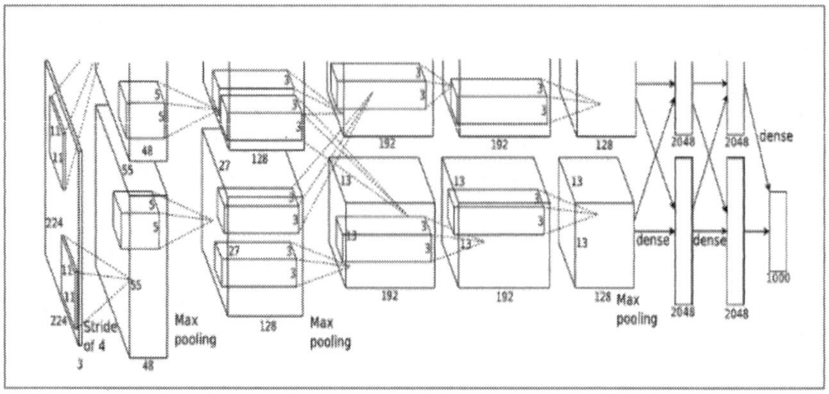

**Figure 2.** Deep Convolutional Neural Network's Architecture

DCNN consists of 5 convolutional layers and three fully-connected layers. The output of fully-connected layer is 1000-way-softmax and produces a distribution over 1000 class labels. Training process of each layer is shown in Table 1.

**Table 1.** Training Process of Each Layer

| Layer | Kernel | Size | Neuron |
|---|---|---|---|
| Input Image | - | 224x224x3x1 | 150.528 |
| 1st Convolutional Layer | 96 | 55x55x48x2 | 290.400 |
| 2nd Convolutional Layer | 256 | 27x27x128x2 | 186.624 |
| 3rd Convolutional Layer | 384 | 13x13x192x2 | 64.896 |
| 4th Convolutional Layer | 384 | 13x13x192x2 | 64.896 |
| 5th Convolutional Layer | 256 | 13x13x128x2 | 43.264 |
| Fully-Connected Layer | - | 2048x2 | 4096 |
| Fully-Connected Layer | - | 2048x2 | 4096 |
| Fully-Connected Layer (SoftMax Output) | - | 1000x1 | 1000 |

The input image was filtered by the first Convolutional Layer with 96 kernels of size 11x11x3 with a stride of 4 pixels. The output of the first Convolutional Layer was filtered by the second. Convolutional Layer with 256 kernels of size 5x5x48. The third, fourth and fifth Convolutional layer was connected to each other without intervensi pooling and normalization layers. The output of second Convolutional Layer connected to the third Convolutional Layer with 384 kernels of size 3x3x256. The fourth and fifth Convolutional Layer had 256 kernels of size 3x3x192.

*3.2 Retraining process (transfer learning)*

Inception-v3 Model could not recognize characters and words that written in Hangul, therefore a retraining process using transfer learning technique with a dataset of Hangul handwritten characters needed. In this phase, bottleneck value of each image was calculated. Bottleneck is an informal term aim to calculate values of each image that will be used for classification. Figure 3. shows an example bottleneck values of A-Ju word.

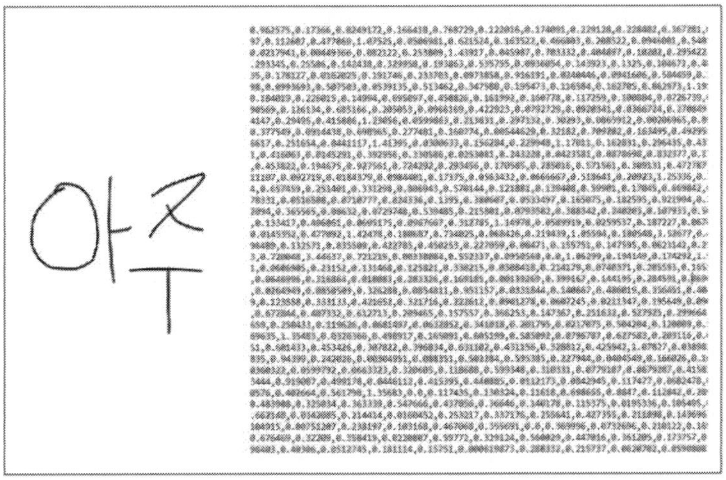


**Figure 3**. Example Bottleneck Values of A-Ju Word

### *3.3 Output (new model)*
The output of retraining process is a new model that able to recognize handwritten of Korean words with Hangul characters.

### *3.4 Word Recognition*
Word recognition has several steps as follows.

- Input: The input image is image that written on screen smartphone android.
- Preprocessing: There are 2 steps in preprocessing phase as follows.
  - Thresholding: Thresholding is a conversion of input image to binary image [9]. Thresholding formula can be described in formula 1.

$$g(x,y) = \begin{Bmatrix} 1 \; if \; f(x,y) \geq T \\ 0 \; if \; f(x,y) < T \end{Bmatrix} \qquad (1)$$

  - Resizing: The input image was resized to become 224 x 224.
- Recognition: In this phase, the result of image preprocessing was recognized by using DCNN method.
- Output: The output of this research was a word that successfully recognized with translation and pronunciation features.

## 4. Experimental results
The obtained results from the training and application testing were discussed in this section.

### *4.1. Training results*
Dataset in this research consisted of 11 Korean words with 1157 images, which 1100 of them were used as training data and the rest was used as testing data. The dataset of this research shows in Table 2.

**Table 2.** Dataset of This Research

| No | Words | Pronunciation | Translation |
|---|---|---|---|
| 1 | 아주 | *A-Ju* | Very |
| 2 | 아니 | *A-Ni* | No / Not |
| 3 | 안녕 | *An-Nyeong* | Hello / Hey |
| 4 | 보여 | *Bo-Yeo* | Can You See That |
| 5 | 어디 | *Eo-Di* | Where |
| 6 | 어서 | *Eo-Seo* | Soon |
| 7 | 고기 | *Go-Gi* | Meat |
| 8 | 입다 | *Ib-Da* | Using |
| 9 | 주다 | *Ju-Da* | Giving |
| 10 | 마자 | *Ma-Ja* | That's Right |
| 11 | 모라 | *Mo-la* | Don't Know |

The retrain process was done by using *Tensorflow*. Tensorflow is an open source library made by Google for machine learning that supports some of programming language [10]. Bottleneck value of each image was calculated first. The training was conducted with 4000 training steps. During training process, the 10 images were chosen randomly from the training set in each step, then bottleneck value was processed into final layer to get prediction value. There are 7 main parameters of this training, namely Training Steps, Learning Rate, Testing Percentage, Validation Percentage, Eval Step Interval, Train Batch Size dan Validation Batch Size. Table 3. shows main parameters in this training.

**Table 3.** Main Parameters

| PARAMETER | PARAMETER's VALUE |
|---|---|
| Training Steps | 4000 steps |
| Learning Rate | 0,01 |
| Testing Percentage | 10% |
| Validation Percentage | 10% |
| Eval Step Interval | 10 |
| Train Batch Size | 100 |
| Validation Batch Size | 100 |

Each training step indicated training accuracy, validation accuracy and cross entropy. Cross entropy is a loss function which shows how well the learning process is running. The lower the value of cross entropy, the better the learning process. If the cross entropy keeps going downwards, it means the learning process improved. If the training accuracy is high but the validation accuracy is low, that means the network is overfitting, which the network is memorizing some feature that are not useful for recognizing image. The final accuracy is displayed after the training process. Summary of the training process shows in Table 4.

**Table 4.** Summary of Training Process

| Training Steps | Training Accuracy (%) | Cross Entropy | Validation Accuracy (%) |
|---|---|---|---|
| 0 | 15.0 | 2.342418 | 11 |
| 250 | 100 | 0.336106 | 100 |
| 500 | 99 | 0.193195 | 92 |
| 750 | 99 | 0.112672 | 96 |
| 1000 | 100 | 0.115776 | 99 |
| 1250 | 100 | 0.082326 | 100 |
| 1500 | 100 | 0.067476 | 100 |
| 1750 | 100 | 0.074174 | 98 |
| 2000 | 100 | 0.056418 | 97 |
| 2250 | 100 | 0.044302 | 97 |
| 2500 | 100 | 0.056943 | 99 |
| 2750 | 100 | 0.044161 | 100 |
| 3000 | 100 | 0.030162 | 98 |
| 3250 | 100 | 0.036282 | 100 |
| 3500 | 100 | 0.032134 | 100 |
| 3750 | 100 | 0.022177 | 100 |
| 3999 | 100 | 0.035141 | 96 |
| **Final Accuracy: 100%** | | | |

Figure 4. shows training accuracy graph in this research that has value range accuracy [0,1] which value 0 means minimum of accuracy value and 1 means maximum of accuracy value in 4000 training steps.

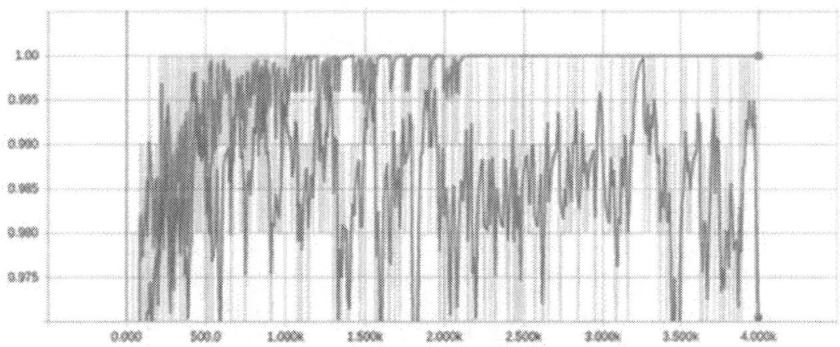


**Figure 4.** Training Accuracy Graph

Cross entropy graph of this 4000 training steps in this research is shown in Figure 5. In Figure 5, the value of cross entropy kept downwards, means the system had a better performance in recognizing Hangul.

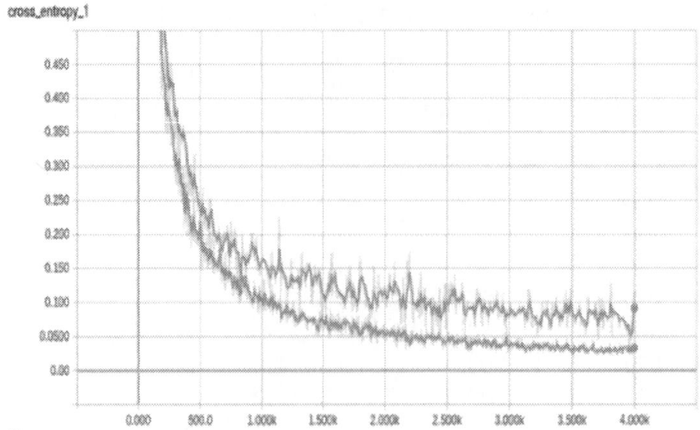


**Figure 5.** Cross Entropy Graph

### *4.2. Testing Results*

We tested Korean Letter Handwritten recognition with 57 images. In this application, the example of how to write the words was shown above the canvas. To start the recognition, firstly, the image was inputted by writing a word in a a screen of smartphone android and then the detect button pressed. The pronunciation and translation of the word would be displayed. If the play button pressed, the voice of pronunciation would be played. The recognition process is show in Figure 6.

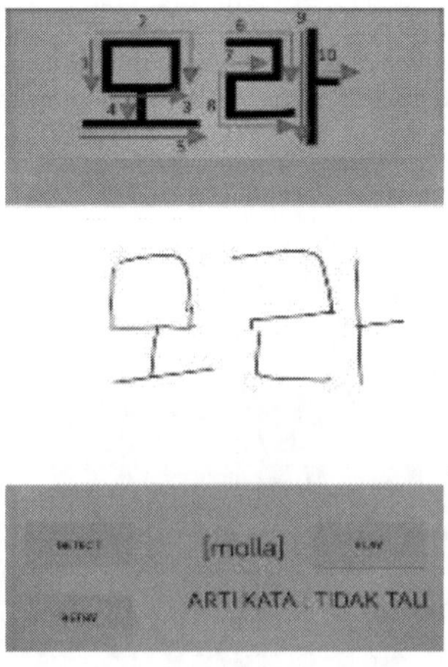


**Figure 6.** Recognition process

We conducted the application testing with 14 people. They wrote different words in the different numbers. The result of testing system is shown in Table 5.

**Table 5**. The Result of Testing System

| Image | Word | Accuracy (%) | Image | Word | Accuracy (%) |
|---|---|---|---|---|---|
| 1 | A-Ju | 97.8 | 29 | Eo-Seo | 94.3 |
| 2 | A-Ju | 67.5 | 30 | Eo-Seo | 55.4 |
| 3 | A-Ju | 69.3 | 31 | Eo-Seo | 99.3 |
| 4 | A-Ju | 90.7 | 32 | Eo-Seo | 94.9 |
| 5 | A-Ni | 66.5 | 33 | Go-Gi | 98.9 |
| 6 | A-Ni | 96.2 | 34 | Go-Gi | 63.6 |
| 7 | A-Ni | 81.1 | 35 | Go-Gi | 68.3 |
| 8 | A-Ni | 88.5 | 36 | Go-Gi | 95.4 |
| 9 | A-Ni | 90.4 | 37 | Go-Gi | 95.7 |
| 10 | A-Ni | 93.5 | 38 | Go-Gi | 94.5 |
| 11 | An-Nyeong | 61.0 | 39 | Ib-Da | 90.6 |
| 12 | An-Nyeong | 90.8 | 40 | Ib-Da | 98.5 |
| 13 | An-Nyeong | 96.4 | 41 | Ib-Da | 72.2 |
| 14 | An-Nyeong | 98.8 | 42 | Ib-Da | 92.9 |
| 15 | An-Nyeong | 99.5 | 43 | Ju-Da | 91.2 |
| 16 | An-Nyeong | 68.8 | 44 | Ju-Da | 72.1 |
| 17 | Bo-Yeo | 97.9 | 45 | Ju-Da | 86.1 |
| 18 | Bo-Yeo | 87.1 | 46 | Ju-Da | 95.9 |
| 19 | Bo-Yeo | 89.9 | 47 | Ju-Da | 87.3 |
| 20 | Bo-Yeo | 90.6 | 48 | Ma-Ja | 62.1 |
| 21 | Bo-Yeo | 70.2 | 49 | Ma-Ja | 98.4 |
| 22 | Eo-Di | 68.0 | 50 | Ma-Ja | 97.3 |
| 23 | Eo-Di | 96.1 | 51 | Ma-Ja | 96.6 |
| 24 | Eo-Di | 95.2 | 52 | Ma-Ja | 93.4 |
| 25 | Eo-Di | 90.3 | 53 | Ma-Ja | 97.5 |
| 26 | Eo-Seo | 71.4 | 54 | Mo-lla | 91.7 |
| 27 | Eo-Seo | 96.1 | 55 | Mo-lla | 93.1 |
| 28 | Eo-Seo | 95.8 | 56 | Mo-lla | 92.1 |
| | | | 57 | Mo-lla | 72.7 |
| | | | **Accuracy Average** | | **86.9%** |

Overall, the detection performed with accuracy average achieve of 86,9%. There were several problems affecting accuracy of each image. The first problem was the stroke shape of the words. The stroke shape would be conformed with the image training. The abrupt stroke of the words affected the accuracy value too.

## 5. Conclusion and Future Work

Based on testing result of Korean letter handwritten recognition using Deep Convolutional Neural Network on Android platform, the average accuracy of 86,9% was achieved. Accuracy value of input image was affected by the stroke shape of Hangul. 100 % of training accuracy was acquired in each word which had 100 images. The number of trained images in each word affected the training accuracy. In the future, the number of word classes are possible to be added. System can recognize more than two syllables and cursive letters

**References**

[1] Damarullah W, Hamzah A, and Lestari U 2013 *Aplikasi Pengenalan dan Pembelajaran Bahasa Korea (Hangul) Berbasis Android* Journal Script vol.1 No.1 pp **78** 88

[2] Kang K W and Kim J H 2003 *Utilization of Hierarchical, Stochastic Relationship Modelling for Hangul Character Recognition. IEEE Transaction On Pattern Analysis And Machine Intelligence* vol.26 no.9 pp **1185**-1196

[3] Sari K P, Tritoasmoro I I and Susatio E 2012 *Pengalih Citra Karakter Korea-Indonesia Menggunakan Klasifikasi K-Nearest Neighbour Dan Template Matching*. Skripsi. Telkom University

[4] Nasution R H 2014 *Pengenalan Kata Tulisan Tangan Huruf Korea Menggunakan Jaringan Saraf Tiruan Propagasi Balik*. Skripsi. Universitas Sumatera Utara

[5] Prabowo Y D, Supraptono E and Safi'I E 2016 *Efektivitas Aplikasi Berbasis Android Menggunakan API Gesture Untuk Mengenal Aksara Korea*. Didaktikum: Jurnal Penelitian Tindakan Kelas vol.17 no.2 pp **78** 82

[6] In-Jung K. and Xie X 2014 Handwritten Hangul Recognition using Deep Convolutional Neural Networks *Int. J. Document Analysis and Recognition (IJDAR)* vol.18 pp **1** 13

[7] Bui V and Chang L 2016 Deep Learning Architectures for Hard Character Classification *Int. Conf. Artificial Intelligence*: pp **108** 114.

[8] Khrizhevsky A, Sutskever I and Hinton G E 2012 *ImageNet Classification with Deep Convolutional Neural Networks. Advances in Neural Information Processing Systems (NIPS)* pp **1097** 1105

[9] Yadav S A, Sharma S and Kumar S R 2015 A Robust Approach For Offline English Character Recognition. *1st Int. Conf. on Futuristic Trend in Computational Analysis and Knowledge Management* pp **121** 126

[10] Devikar P 2016. Transfer Learning for Image Classification of Various Dog Breeds. *Int. J. of Advanced Res. Comput. Eng. Technol. (IJARCET)* vol 5 pp **2707** 2715

[11] Gunawan D, Arisandi D, Ginting F M, Rahmat R F and Amalia A 2017 Russian character recognition using Self-Organizing Map *J. Phys. Conf. Ser.* **801**, p. 12040

IOP Conf. Series: Journal of Physics: Conf. Series **978** (2018) 012113 doi:10.1088/1742-6596/978/1/012113

# Advertisement scheduling on commercial radio station using genetics algorithm

**S Purnamawati[1*], E B Nababan[1], B Tsani[1], R Taqyuddin[1], R F Rahmat[1]**

[1]Departement of Information Technology, Universitas Sumatera Utara, Jl. Alumni No. 3, Medan, North Sumatera 20155, Medan, Indonesia

Email: sarah_purnamawati@usu.ac.id

**Abstract**. On the commercial radio station, the advertising schedule is done manually, which resulted in ineffectiveness of ads schedule. Playback time consists of two types such as prime time and regular time. Radio Ads scheduling will be discussed in this research is based on ad playback schedule between 5am until 12am which rules every 15 minutes. It provides 3 slots ads with playback duration per ads maximum is 1 minute. If the radio broadcast time per day is 12 hours, then the maximum number of ads per day which can be aired is 76 ads. The other is the enactment of rules of prime time, namely the hours where the common people (listeners) have the greatest opportunity to listen to the radio, namely between the hours and hours of 4 am – 8 am, 6 pm – 10 pm. The number of screenings of the same ads on one day are limited to prime time ie 5 times, while for regular time is 8 times. Radio scheduling process is done using genetic algorithms which are composed of processes initialization chromosomes, selection, crossover and mutation. Study on chromosome 3 genes, each chromosome will be evaluated based on the value of fitness calculated based on the number of infractions that occurred on each individual chromosome. Where rule 1 is the number of screenings per ads must not be more than 5 times per day and rule 2 is there should never be two or more scheduling ads delivered on the same day and time. After fitness value of each chromosome is acquired, then the do the selection, crossover and mutation. From this research result, the optimal advertising schedule with schedule a whole day and ads data playback time ads with this level of accuracy: the average percentage was 83.79%.
**Keyword**: Radio ads Scheduling, Genetic Algorithms, random function.

## 1. Introduction

Nowadays technological advances are increasingly fast and sophisticated. The computer is used as a tool for management and control. On the electronic media, the commercial radio the operation of radio stations needs large costs and good management. In order to run with optimum, radio stations have to maximize the quality of broadcast to gain acceptance with advertising services provided by the company or individual products. Ads is a serving form and non-private paid promotion contains idea, product or service by identified sponsors [1]. Ads serving on electronic media require a good management which includes ordering broadcast schedule, broadcast hours and broadcast fees in accordance with the duration and frequency of ads. A bad management will result in a loss on the company because of the lack of interest of listener in ads serving. Research on the algorithm of scheduling the ads using Genetic algorithm has been used. Genetic algorithm is a computational approach to resolve the problem that is modeled in the process of biological evolution, including selection, crossover, and mutation [14]. Thus, this algorithm used to set scheduling a recurring activity including scheduling radio advertisement commercial. Based on the problem above, this research purpose is to build a system Advertising Scheduling on Commercial Radio Stations Using Genetic Algorithms.

## 2. Previous research

Genetics Algorithm before is not used to solve a problem, but a study of adaptation phenomenon in the nature and make attempts to apply it to computer system [2]. The genetics algorithm is different with conventional convergent. It using the nature selection mechanism and genetics study so the common names used in this algorithm same as genetics in biology. A solution that has been raised by the algorithm called chromosome. A collection of chromosome is called population. Every chromosome is evaluated and the success rate is called Fitness. [3]

Some previous research that is relevant to the subject matter of this discussion among others are: Entitled research about an automated scheduling system using genetic algorithms in the Faculty of science and Technology, concluded that genetic algorithms with the preparation can generate a schedule with the guarantee of a lecturer does not teach in a full day of classes semester package and not scheduled sequence in one day. Then, another research is Genetic Algorithm for solving scheduling the lecture and lab course. With genetic algorithm scheduling solution can search for lectures and practical work with the time more quickly. The test results with the input parameter values in the same or different, the process of scheduling yield generation and iteration. [4]

By using the genetic algorithm program can produce a schedule of lectures, based on the results of the test parameters, the method of selection is the best selection of tournaments, the best methods of cross-breeding is the cross-breeding of many points, without elitism generally more quickly resolve scheduling and better population figures for model of chromosome problems scheduling courses that have around 300 genes is around 100 individuals, but no sooner in time compared to the total population. [5]

## 3. General method

The method proposed in this research will be done by scheduling advertising on commercial radio stations using genetic algorithms. In the process of scheduling the ads, there are some important things to be done for the application of genetic algorithm such as encoding of chromosomes, evoking the initial population, the process of selection, crossover, mutation process until the process of looping regeneration is eligible or not [6][7]. The general architecture of this method can be seen at figure 1.

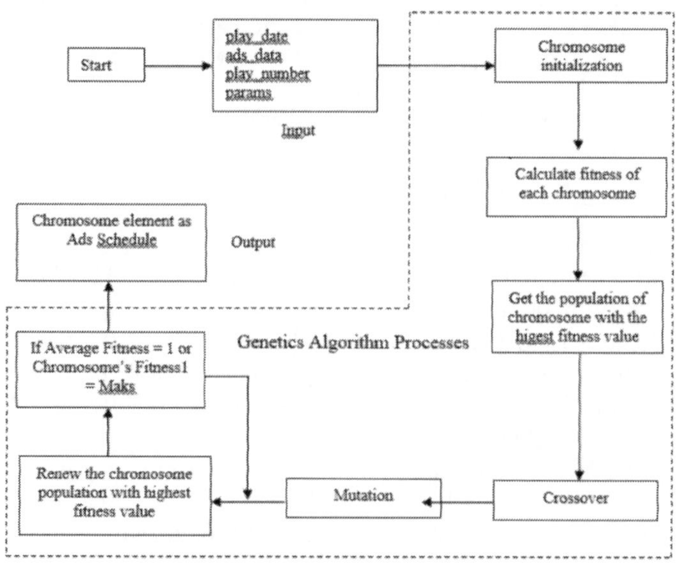


**Figure 1.** General Architecture

### *3.1 Chromosome representation*

Genetic algorithms in scheduling problems required chromosome representation. A chromosome is represented by the matrix m × n = [504] [3], where m or row in the matrix represents the number of ads, while the n or the columns of the matrix represents the chromosomes [8], namely IDIklan, IDWaktu and IDHari. Defines the number of chromosome representation of genes being used and can represent the solution of the problem. This parameter is named gene that banded together to form chromosomes. The chromosome representation can be seen on table 1.

**Table 1**. Chromosome Representation

| Kromosom | IDIklan | IDHari | IDWaktu |
|---|---|---|---|
| **1** | 0022 | 5 | 03 |
| **2** | 0006 | 3 | 12 |
| **3** | 0010 | 1 | 14 |
| . | . | . | . |
| **171** | 0171 | 7 | 04 |

For every 15 minutes is provided 3 slots in minutes to 15, 30 and 45 at any hour. On each ad unit performed playback 3 ads per ad with duration of maximum 1 minute. So, if the radio broadcast time per day there are 19, then the maximum number of ads per play date that can be aired is 19 hours x 3 x 3 slots ads into ad 171. So, the length of a chromosome itself from each individual is as much as 171 chromosomes from the number of ads will be broadcast. While the length of a chromosome is composed of 3 genes. the initial Population was later resurrected at random for the next evaluation process conducted every single chromosome.

The matrix of the Chromosome representation schedule broadcasting advertisements created at random in the following way:

- Determine the number of hours of playback according to the number of ad slots.
- Random values from 0001 to IDIklan gene 0171
- Random gene IDWaktu of value 01 to 99 (05:15-24:00)
- The number of Playback advertising on demand.

### *3.2 Initial population*

The process of initializing a population revoke a random population without concern to value of the fitness. This process is to encode information into the slot of chromosomes. Initial population made randomly from chromosome to slot some number of individuals in a single population [9][14]. As for the stages of the process from the initial population is as follows:

- Initialisation is created by entering parameter values of genetics that is the initial population.
- Initial population = n (number of individuals).
- Initial population made of chromosome representation as much as n.

### *3.3 Fitness Evaluation*

Fitness will be calculated based on the number of breaches that occur at each individual chromosome in a slot. The rules referred to in this research are:

- Rule 1 is not to be going on the existence of a number of screenings per ads more than 5 times in one day
- Rule 2 is the existence of two or more scheduling ads delivered on the same day and time.

As for the stages of the process of evaluation of the fitness is as follows:

- Each chromosome in one individual checks each value cause penalty or not by checking the way of chromosome 1 to 504 for the type of playback time Prime Time and 693 for the kind of Regular playing time Time.
- Count the number of chromosomes that exposed violations of rules divided by the total number of chromosomes or 504/693.

### *3.4 Selection*

Data selection method in this research using Wheel Roulette. Parents will be selected based on the value of fitness, the greater the value of the fitness will get a greater possibility for elected as parent. Probability selection process. Individual that has the greatest probability was chosen as the first master for the formation of new breeds to the process of chromosomal crossover [10][11]. Thus, individual with second great probability was chosen for the formation of new breeds on the election of individuals to the process of chromosomal crossover.

### *3.5 Crossover and mutation*

When the selection process has been implemented and has been elected to the new parent, then the next stages of the genetic algorithm is the crossover operator. Crossover is a way of combining the parent genes to produce new breeds. Crossover is done by way of exchanging value genes in the same gene position of both carriers [10]. The following are the stages of the process of chromosomal crossover stem with two individuals exchange value gen IDHari, IDWaktu:

Chromosome 1 Parent 1

| | | | |
|---|---|---|---|
| Chromosome 8 | 0006 | 3 | 22 |

Chromosome 1 Parent 2:

| | | | |
|---|---|---|---|
| Chromosome 9 | 0010 | 1 | 14 |

And here are the results of the generation process of stem from two individual crossovers:

Child 1 Parent 1:

| | | |
|---|---|---|
| 0006 | 1 | 14 |

Child 1 Parent 2:

| | | |
|---|---|---|
| 0010 | 3 | 22 |

The next stage after the crossover that is a mutation. The mutation is performed by selecting one of the genes randomly selected genes are then inserted into the other position. The mutation will be conducted randomly by selecting two positions of genes from chromosome and then its value is replaced with the value of the random Data display ads served to do the infusion information advertising into the database [12][13].

The value of IDWaktu and IDHari genes selected randomly as well. The mutation will be conducted randomly by selecting two positions of genes from chromosome and then its value is replaced with the value of the random too. The following are the stages of the process of chromosomal mutation in two individuals exchange genes with stem IDWaktu. The chromosomes are randomly selected as much as the value of the probability of mutations is as follows:

Random chromosome in IDWaktu:

| Chromosome | IDIklan | IDHari | IDWaktu |
|---|---|---|---|
| 2 | 0006 | 3 | 12 |

Random chromosome in IDIklan :

| Chromosome | IDIklan | IDHari | IDWaktu |
|---|---|---|---|
| 2 | 0006 | 3 | 51 |

Random chromosome in IDHari :

| Chromosome | IDIklan | IDHari | IDWaktu |
|---|---|---|---|
| 2 | 0006 | 3 | 12 |

## 4. Result and discussion

As for the result of the design of systems that have been constructed are as follows:

### *4.1. Scheduling Process*

Scheduling processes used to generate prime schedule for the ads. The process result can be seen on Figure 2.

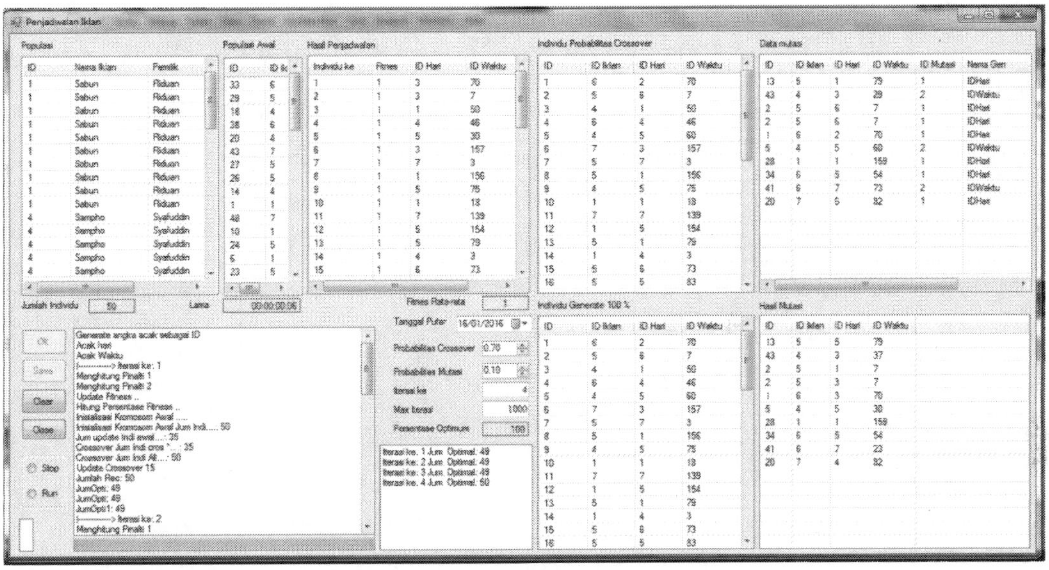

**Figure 2.** Scheduling Process Output

### *4.2. Display of Prime-Time Scheduling Report*

Prime Time Scheduling Schedule Display serves to display information on the results of radio ad scheduling process at Prime-Time time. The Prime-Time Scheduling Report view can be seen as in Figure 2.

JADWAL PEMUTARAN IKLAN RADIO PRIME TIME

Tanggal: 16-11-2015

| No. | Nama Iklan | Pemilik | Waktu |
|---|---|---|---|
| 1 | Sampho | Syafuddin | jam 6:15 |
| 2 | Rumah | Nandar Halim | jam 6:17 |
| 3 | Deterjen | Amirullah | jam 6:32 |
| 4 | Rumah | Nandar Halim | jam 6:46 |
| 5 | Rumah | Nandar Halim | jam 6:47 |
| 6 | Mobil | Gagan | jam 7:30 |
| 7 | Mobil | Gagan | jam 8:45 |
| 8 | Mobil | Gagan | jam 16:15 |
| 9 | Deterjen | Amirullah | jam 16:16 |
| 10 | Magnum Classic | Magnum | jam 16:32 |
| | | | |

**Figure 3.** Prime Time Scheduling Report

### *4.3. Display of Regular Time Scheduling Report*

Regular Time Scheduling Schedule Display serves to display information on the results of the radio ad scheduling process at Regular Time.

### *4.4. Scheduling Tests with Genetic Parameter Variations*

Genetic algorithms function to produce the best schedule. In the testing phase, to be able to produce the test is done by entering the input value of varied genetic parameters. The result of comparison of scheduling test with Regular Time time can be seen in Table 2.

**Table 2.** Genetics Parameter Test View at Team Regular Time

| No | Number of individuals | Average fitness | Cross-over Probability | Probability of Mutation | Generation | Max Generation | Optimum Percentage |
|---|---|---|---|---|---|---|---|
| **1** | 50 | 1 | 0,70 | 0,10 | 4 | 1000 | 100 |
| **2** | 100 | 0,97 | 0,70 | 0,10 | 285 | 1000 | 93 |
| **3** | 150 | 0,95 | 0,70 | 0,10 | 51 | 1000 | 90 |
| **4** | 200 | 0,94 | 0,70 | 0,10 | 317 | 1000 | 88,5 |
| **5** | 250 | 0,92 | 0,70 | 0,10 | 424 | 1000 | 83,6 |
| **6** | 300 | 0,92 | 0,70 | 0,10 | 107 | 1000 | 83,33 |
| **7** | 350 | 0,93 | 0,70 | 0,10 | 382 | 1000 | 79,43 |
| **8** | 400 | 0,88 | 0,70 | 0,10 | 1000 | 1000 | 76,75 |
| **9** | 450 | 0,86 | 0,70 | 0,10 | 1000 | 1000 | 73,33 |
| **10** | 500 | 0,85 | 0,70 | 0,10 | 1000 | 1000 | 71 |

### *4.5. Data Testing*

#### *4.5.1. Testing with 50 Ads for 12/01/2016*

There are 50 adult data processing or viewing on the time series that is regular time with genetic parameter of 70% crossover probability, 10% mutation probability, and maximum iteration of 1000 gained an average fitness of 1 and optimum percentage as much as 100% On the 4th iteration with a 6-second processing time.

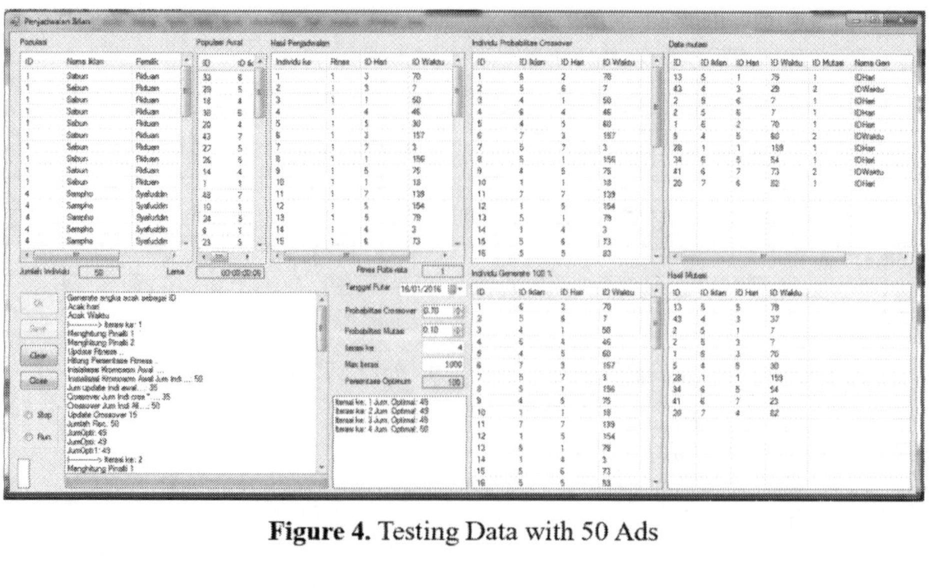

**Figure 4.** Testing Data with 50 Ads

#### *4.5.2. Testing with 300 Ads for 12/01/2015*

In Figure 5, there are 300 ad population processing or viewing on the showtime type ie regular time with genetic parameter 70% crossover probability, 10% mutation probability, and maximum iteration of 1000 gained an average fitness of 0.92 and the number of individuals Optimally as much as 83.33% at the 107th iteration with a processing time of 32 minutes 37 seconds.

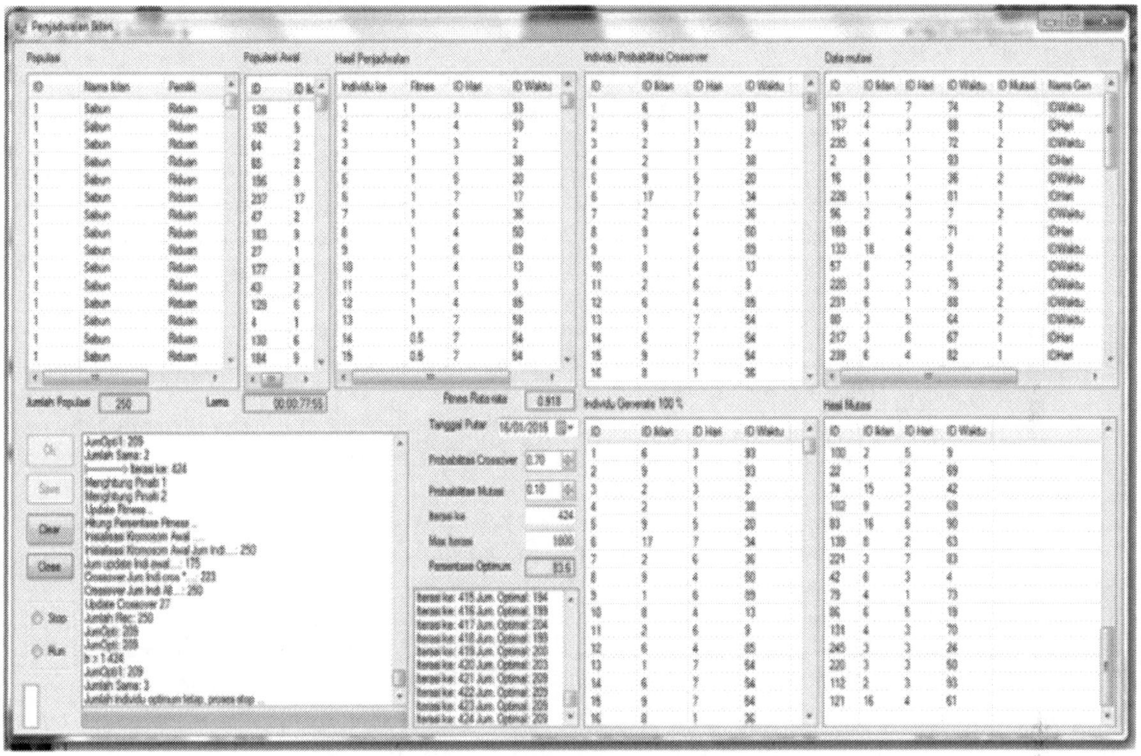

**Figure 5.** Testing Data with 300 Ads

## 5. Conclusions and recommendations

Based on the discussion and evaluation, this research concludes the genetic algorithm can be used for radio ad scheduling with prime-time type and regular time. Which are heavily influenced by random functions, so it is not always given the optimal result. The genetic algorithm can search for a solution of radio advertisement scheduling with faster time while the test results with the input values of the same or different genetic parameters, the scheduling process yields different optimum percentage values this is due to the random function. The larger the population or the scheduled ad data the longer scheduling time is due to more random values.

It's possibly for the next research and development is to compare the accuracy of genetic algorithms with other optimization algorithms such as Backpropagation neural network algorithm and Learning Vector Quantization.

## References

[1] P. Kotler and G. David 2004 Country as brand, product and beyond: a place marketing and brand management perspective *Destination branding: Creating the unique destination proposition 2,* pp. 40-45.

[2] Aryawarman Q S and Setiawan R 2007 Sistem Penjadwalan Iklan Pada Radio Megantara Fm Nganjuk Dengan Algoritma Dynamic Priority Schedulling *Prosiding Seminar Nasional Sistem & Teknologi Informasi (SNASTI)*

[2] Fariza A, Martiana E, Sucipto H 2006 *Aplikasi Algoritma Genetika Multi Obyektif pada Traveling*

IOP Conf. Series: Journal of Physics: Conf. Series **978** (2018) 012113 doi:10.1088/1742-6596/978/1/012113

*Salesman Problem*. Politeknik Elektronika Negeri Surabaya ITS

[3] Gen M, and Cheng R 1997 *Genetic Algorithms and Engineering Design*. Newyork: Jhon Wiley & Sons, Inc

[4] Goldberg DE, 1989 *Genetic Algorithms in Search, Optimization and Machine Learning*. Canada: Addison-Wesley Publishing

[5] Indira K and Kanmani, 2012 Performance Analysis of Genetic Algorithm for Mining Association Rules. *Int. J. Comput. Scie.* **9**.

[6] Karegowda A G, Manjunath A S, and Jayaram M A 2011 Application of genetic algorithm Optimized neural network Connection weights for medical Diagnosis of pima Indians diabetes, *Int. J. Soft Comput.* **2**

[7] Luksas S, Anwar T, Yuliani W 2005 Penerapan Algoritma Genetika untuk Traveling Salesman Problem dengan Menggunakan Metode Order Crossover dan Insertion Mutation. *Seminar Nasional Aplikasi Teknologi Informasi* : **1** 2.

[8] Nugraha I 2008 *Aplikasi algoritma genetik untuk optimasi penjadwalan kegiatan belajar mengajar dalam Strategi Algoritmik.* Institut Teknologi Bandung.

[9] Obitko M 1998 *Introduction to Genetic Algorithms*. University of Applied Sciences.

[10] Sivanandam S N and Deepa S N 2008 *Introduction to Genetic Algorithms*. New York: Springer-Verlag Berlin Heidelberg.

[11] Sofge D A 2002 Using Genetic Algorithm Based Variable Selection to Improve Neural Network Models for Real-World Systems. *Proc. 2002 Int. Conf. Machine Learning 7 Appl*. Washington, D.C., U.S.A.

[12] Soraya D T 2007 *Penjadwalan Perkuliahan Dengan Pendekatan Constraint Programming.* Skripsi, Fakultas Ilmu Komputer Universitas Indonesia.

[13] Tamba G.M.P 2004 *Sistem Penjadwalan Perkuliahan Menggunakan Algoritma Genetika (Studi Kasus Fakultas Matematika dan IPA IPB)*. Institut Pertanian Bogor.

[14] Syahputra M F, Velicia F, Rahmat R F and Budiarto R 2017 Scheduling diet for diabetes mellitus patients using genetic algorithm *Journal of Physics: Conference Series* **801**, p. 12033

IOP Conf. Series: Journal of Physics: Conf. Series **978** (2018) 012114 doi:10.1088/1742-6596/978/1/012114

# File text security using Hybrid Cryptosystem with Playfair Cipher Algorithm and Knapsack Naccache-Stern Algorithm

**Amalia[1], M A Budiman[2] and R Sitepu[3]**

[1,2,3] Department of Computer Science, Faculty of Computer Science and Information Technology, Universitas Sumatera Utara, Medan, 20155, Indonesia

Email: amalia@usu.ac.id, mandrib@usu.ac.id, ravi@students.usu.ac.id

**Abstract.** Cryptography is one of the best methods to keep the information safe from security attack by unauthorized people. At present, Many studies had been done by previous researchers to generate a more robust cryptographic algorithm to provide high security for data communication. To strengthen data security, one of the methods is hybrid cryptosystem method that combined symmetric and asymmetric algorithm. In this study, we observed a hybrid cryptosystem method contain Modification Playfair Cipher 16x16 algorithm as a symmetric algorithm and Knapsack Naccache-Stern as an asymmetric algorithm. We observe a running time of this hybrid algorithm with some of the various experiments. We tried different amount of characters to be tested which are 10, 100, 1000, 10000 and 100000 characters and we also examined the algorithm with various key's length which are 10, 20, 30, 40 of key length. The result of our study shows that the processing time for encryption and decryption process each algorithm is linearly proportional, it means the longer messages character then, the more significant times needed to encrypt and decrypt the messages. The encryption running time of Knapsack Naccache-Stern algorithm takes a longer time than its decryption, while the encryption running time of modification Playfair Cipher 16x16 algorithm takes less time than its decryption.

## 1. Introduction

Nowadays, the growth of data and information exchange on the internet makes us need to increase data protection from compromised security attacks by unauthorized people. One of the solutions is by implementing cryptography algorithm. Cryptography is a study based on mathematical techniques for information security such secrecy, data integrity, and entity authentication [1]. Cryptography can also be used for identification of individuals, authenticationnd and non-repudiation [2].

Cryptography algorithms can be classified into symmetric Key Algorithm and Asymmetric Key Algorithm. The Symmetric Algorithm use a private key [3] or a single key for encryption and decryption process, while the asymmetric algorithm uses public key [3] or use a different key for the encrypt and decrypt process. The advantage of using the symmetric algorithm is the faster time execution. Meanwhile, the most problem with this algorithm determines the efficient method to exchange the key safely. The asymmetric algorithm, based on a mathematic formula, is a robust algorithm because this algorithm is not easy to compromise, but takes longer processing time than the symmetric algorithm[3]. In this study, we combine asymmetric and symmetric algorithm to get advantages of these two various algorithms. We implemented Playfair Cipher algorithm which is a symmetric algorithm to encrypt and decrypt the ciphertext and Knapsack Naccache-Stern algorithm which is an asymmetric algorithm to encrypt and decrypt the key.

Playfair Cipher algorithm developed by Charles Wheatstone in 1854, but popularized by Lord Playfair [4]. This algorithm is substitution digraph which encrypts two letters at once that utilize 5x5 matrix for encryption and decryption process. Playfair Cipher with a 5 x 5 matrix has limitation characters to encrypt and encrypt process, not all ASCII characters are covered in this matrix. So, in this study, we developt Playfair Cipher matrix into a 16 x16 matrix so all of ASCII characters can be used efficiently for encryption and decryption process.

Knapsack Naccache-Stern algorithm is a deterministic public key algorithm based on knapsack problem. This algorithm has an advantage which it can produce a robust encryption result that cannot quickly to destruct. However, this algorithm has a deficiency which is has a long processing time for encryption [5]. Knapsack Naccache-Stern algorithm has three processes, which are key generation process, encryption process, and decryption process. We implemented Agrawal Kayal Saxena (AKS) algorithm, a deterministic algorithm in polynomial time, to generate and examine the prime number [6].

We combined these two cryptography algorithms to cover the weakness of each algorithm. For the experiment, we implement file text .doc, .docx and .thx as plaintext.

## 2. Method

To encrypt the plaintext with Playfair Cipher algorithm, we arrange the key into 16x16 matrix.
The encryption process according to the following rules[7]:

- If the pair of the letters appears on the same raw of the table, then the ciphertext of these letters are the letters which are in right position of them.
- If the pair of the letters appears in the same column of the table, then the ciphertext of the letters are the letter to their immediate below respectively.
- If the pair of the letters is not in the same row or column, then the ciphertext of the letters are the letter to their immediate right respectively but at the other pair of corners of the rectangle defined by the original pair. Notice that the order is essential, the first letter of the encrypted pair is the one that lies on the same row as the first letter of the plaintext pair.

The key of Playfair Cipher algorithm that used to encrypt the messages will be encrypted with Knapsack Naccache-Stern algorithm. The asymmetric algorithm has to generate a public and private key first, before encryption and decryption process. To generate the key for Knapsack Naccache-Stern algorithm according to the following rules [5]:

- Pick $p_i$ the prime number starts from $i = 0$ to $i=7$.
- Multiply all the prime numbers formed.
- Pick $p$, a prime number that greater than the result of the multiplying all of the prime number ($p_i$), such that:

$$p > \prod_{i=0}^{n} p_i$$

- Pick a secret integer $s<p-1$ such that GCD $(p-1, s) = 1$.
- Set $v_i$ with this rule $v_i^s \equiv p_i \bmod p$

The public key is $\{p, n, v_0, \ldots, v_n\}$and the private key is $s$.

In this research, for the key generation steps, we tried some of the prime numbers randomly. We implemented AKS to examine the validity of this prime number. AKS algorithm has the following rules [6]:

- Let $p$, a number that will be checked a prime or not.
- Pick a positive integer $z$ where $2 \leq z < p-2$ and GCD $(p,z) = 1$

then $p$ is a prime number if and only if:

$$(1+z)^p \equiv 1+z^p \pmod p$$

The next step is a process to encrypt the Playfair Cipher Algorithm's key with a preformed public key of Knapsack Naccahe-Stern. This public key is generated according to the following formula [5]:

$$c = \prod_{i=0}^{n} v_i^{m_i} \bmod p$$

where:
$c$ = ciphertext
$m_i$ = the $i$th bit of the message $m$
$n, p, v_i$ = public key

The decryption process of Knapsack Naccache-Stern algorithm is used to get the Playfair Cipher algorithm's key back. So the process is continued by using the preformed public key according to the following rules [5]:

$$m = \sum_{i=0}^{n} \frac{2^i}{p_i - 1} x \left(\gcd\left(p_{i,}, c^s \bmod p\right) - 1\right)$$

where:
$m$ = message
$c$ = ciphertext
$s$ = private key

A decrypted process can be done to get back the key of the Playfair Cipher algorithm. The rules for decryption process of Playfair Cipher algorithm has the same rules as its encryption process. We observed running time for all processes in each algorithm with various characters length and various length of the private key. We also noted size of ciphertext that is produced by these process. The aim of this process, we can see the relation between the length of plaintext with running time for encryption process of Playfair Cipher Algorithm.

## 3. Results and Discussion

### *3.1 Encryption Process for Playfair Cipher Algorithm*

To measure the running time for Playfair Cipher Algorithm encryption process, we do encryption experiments with various length of plaintext. In this study, we tried 10, 100, 1000, 100000 characters with the length of the symmetric key that are 10, 20, 30 and 40 characters. The result of this process can be described in Figure 1.

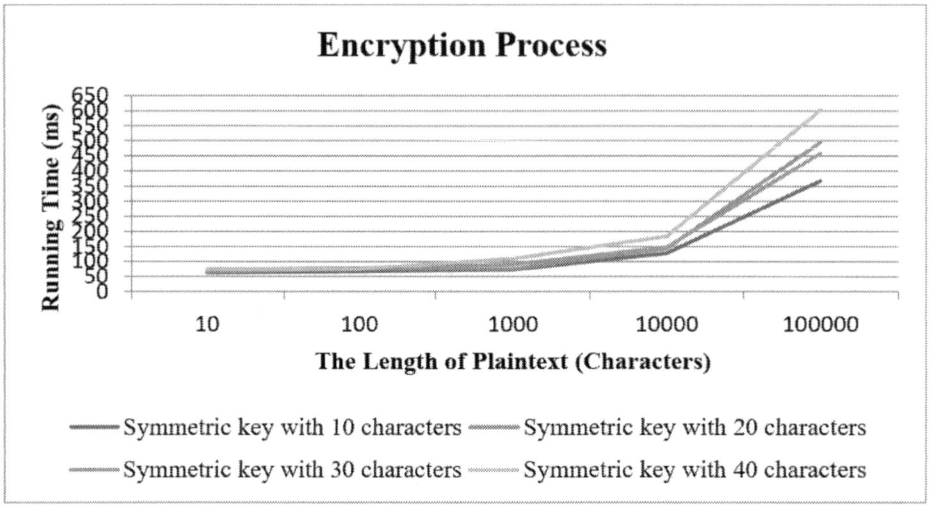


**Figure 1.** The Relation Between The Length of Plaintext with Running Time for Encryption Process of Playfair Cipher Algorithm

According to Figure 1, we can see the encryption processing time increase linearly with the increasing length of plaintext. Besides that, encryption running time process also affected by the length of the symmetric key. The longer symmetric keys used, then the longer times needed to encrypt.

### *3.2 Decryption Process for Playfair Cipher Algorithm*

For decryption process, we do the same experiment like encryption process. We tried various length of ciphertext with various length of the symmetric key like a. We decided 10, 100, 1000, 10000 and100000 characters of ciphertext and 10, 20, 30, and 40 characters of the symmetric key. The results are shown in figure 2.

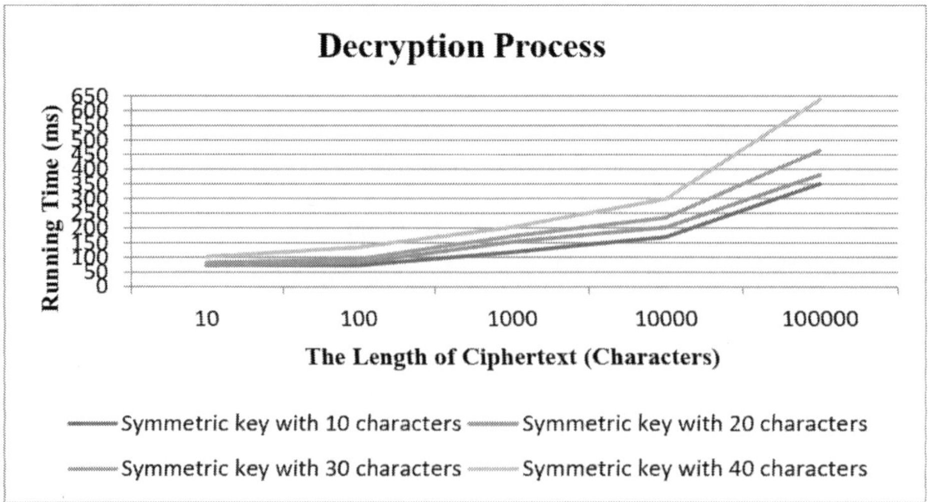


**Figure 2.** The Relation Between The Length of Ciphertext with Running Time for Decryption Process of Playfair Cipher Algorithm

Based on figure 2, we can see the result of decryption processing time has the same pattern like encryption processing time which is decryption processing time increase linearly with the increasing length of ciphertext and decryption running time process affected by the length of the symmetric key.

### *3.3 Encryption Process of Knapsack Naccache-Stern Algorithm*

In this study, we also calculated the running time for encryption process of Knapsack Naccache-Stern algorithm. We tried for some various length of the symmetric key which is 10, 20, 30, and 40 characters. The results are shown inthe figure 3.

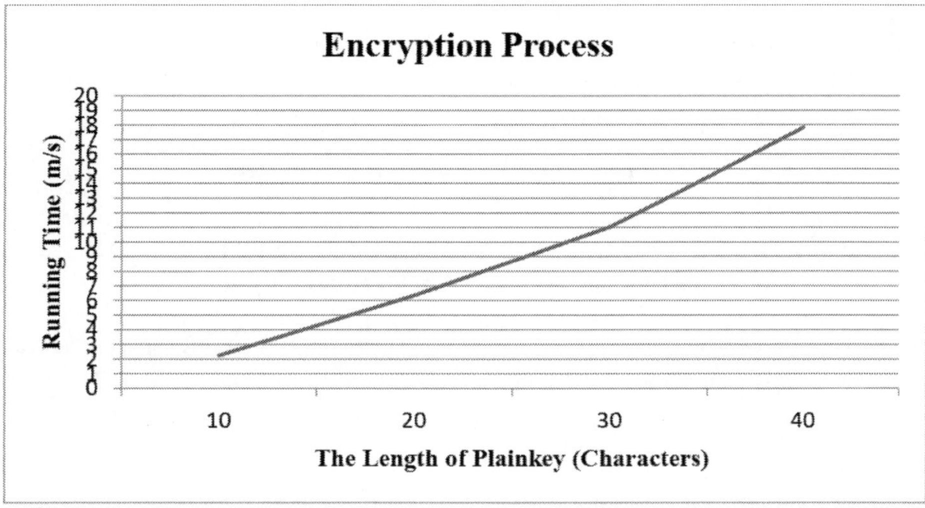


**Figure 3.** The Relation Between The Length of Plainkey with Running Time for Encryption Process of Knapsack Naccache-Stern Algorithm

Based on figure 3 the encryption processing time of plainkey using Knapsack Naccache-Stern algorithm is affected by its length. The longer characters of the plainkey, then the longer periods needed to encrypt.

### *3.4 Decryption Process for Knapsack Naccache-Stern Algorithm*

The running time for decryption process of Knapsack Naccache-Stern calculated with the cipher key that already formed on the previous stage. The results are shown onthe figure 4

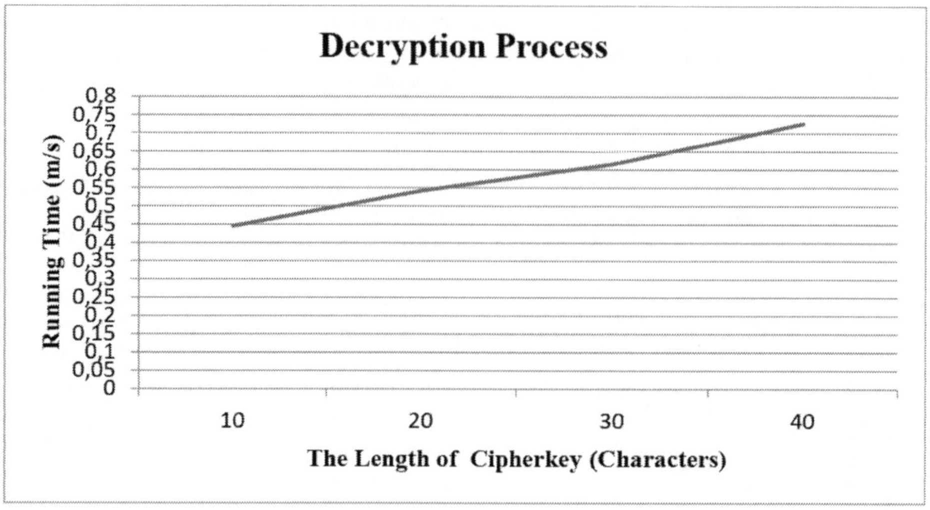


**Figure 4.** The Relation Between The Length of Cipherkey with Running Time for Decryption Process of Knapsack Naccache-Stern Algorithm

According to figure 4, the decryption processing time of cipherkey using Knapsack Naccache-Stern algorithm is also affected by its length. The longer characters of the cipherkey, then the longer periods needed to decrypt.

*3.5 Size of Ciphertext*

In this work, we also observe the relation between the size of plaintext with the size of ciphertext. We tested some of the various sizes of plaintext that is encrypted with 10 characters length of the symmetric key. The results are shown in figure 5.

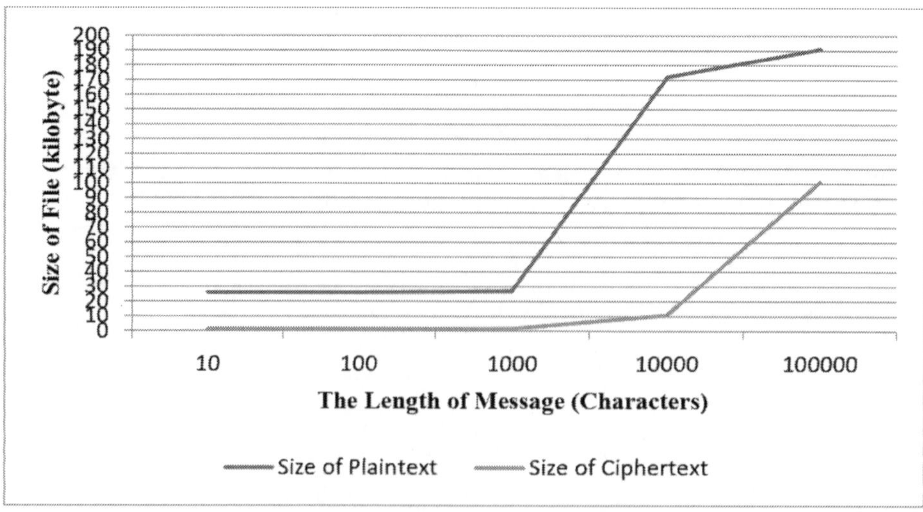


**Figure 5.** The Relation Between The Size of Plaintext with The Size of Ciphertext

Based on figure 5, we can see encrypted process produced smaller length size of the file. It means the size of ciphertext is smaller than the size of the plaintext.

IOP Conf. Series: Journal of Physics: Conf. Series **978** (2018) 012114 doi:10.1088/1742-6596/978/1/012114

## 4. Conclusion

Based on this research of cryptography using Playfair Cipher algorithm and Knapsack Naccache-Stern algorithm, it can be concluded:

- Implementation of file text security using hybrid cryptosystem with Playfair Cipher algorithm and Knapsack Naccache-Stern algorithm can be a choice to secure a plaintext.
- Modification of Playfair Cipher from 5x5 matrix into 16x16 matrix will increase the quality of cryptography because more characters of key combinations can be formed than using Playfair Cipher 5x5.
- The running time for encryption and decryption process each algorithm is linearly proportional, it means the longer messages character then, the more significant times needed to encrypt and decrypt the messages.
- The encryption running time of Knapsack Naccache-Stern algorithm takes a longer time than its decryption, while the encryption running time of modification Playfair Cipher 16x16 algorithm takes less time than its decryption.
- The size of ciphertext is smaller than the size of its plaintext.

## References

[1] W Mao and Wenbo *Modern cryptography : theory and practice*. Prentice Hall PTR, 2004.
[2] S Subasree, N K Sakthivel, and T Nadu 2010 Design of a New Security Protocol Using Hybrid Cryptography Algorithms *Comput.* vol. 2 pp. 95–103.
[3] E Fujisaki and T Okamoto 2013 Secure Integration of Asymmetric and Symmetric Encryption Schemes *J. Cryptol* **26**(1) pp. 80–101.
[4] Alam AA, Khalid BS, Salam CM 2013 A Modified Version of Playfair Cipher Using 7 x 4 Matrix *Int. J. Comput. Theory Eng.* **5**(4) pp. 626–628.
[5] D Naccache and J Stern 1997 A New Public-Key Cryptosystem Springer, Berlin, Heidelberg pp. 27–36.
[6] Agrawal M, Kayal N, Saxena N 2002 PRIMES is in P *Dep. Comput. Sci. Eng. Indian Inst. Technol. Kanpur Kanpur-208016.*

IOP Conf. Series: Journal of Physics: Conf. Series **978** (2018) 012115 doi:10.1088/1742-6596/978/1/012115

# PDF file encryption on mobile phone using super-encryption of Variably Modified Permutation Composition (VMPC) and two square cipher algorithm

**D Rachmawati[1] , M A Budiman[2] and F Atika[3]**
Departemen Ilmu Komputer, Fakultas Ilmu Komputer dan Teknologi Informasi, Universitas Sumatera Utara, Jl. Universitas No. 9-A, Kampus USU, Medan 20155, Indonesia

dian.rachmawati@usu.ac.id[1] , mandrib@usu.ac.id[2],
fadhilahatika@students.usu.ac.id[3]

**Abstract.** Data security is becoming one of the most significant challenges in the digital world. Retrieval of data by unauthorized parties will result in harm to the owner of the data. PDF data are also susceptible to data security disorder. These things affect the security of the information. To solve the security problem, it needs a method to maintain the protection of the data, such as cryptography. In cryptography, several algorithms can encode data, one of them is Two Square Cipher algorithm which is a symmetric algorithm. At this research, Two Square Cipher algorithm has already developed into a 16 x 16 key aims to enter the various plaintexts. However, for more enhancement security it will be combined with the VMPC algorithm which is a symmetric algorithm. The combination of the two algorithms is called with the super-encryption. At this point, the data already can be stored on a mobile phone allowing users to secure data flexibly and can be accessed anywhere. The application of PDF document security on this research built by Android-platform. At this study will also calculate the complexity of algorithms and process time. Based on the test results the complexity of the algorithm is $\theta$ (n) for Two Square Cipher and $\theta$ (n) for VMPC algorithm, so the complexity of the super-encryption is also $\theta$ (n). VMPC algorithm processing time results quicker than on Two Square Cipher. And the processing time is directly proportional to the length of the plaintext and passwords.

## 1. Introduction

Data security becomes one of the most significant challenges in the digital world. If someone has data that should be confidential but instead known by unauthorized parties, so it would be detrimental to those who have the data. There are so many things that can be done to ensure the security of data, one of them by encoding the data into data that can not be read or understood by others but can be read by the authorities. For these security problems, it needs a method to maintain data security, one of them with cryptography. Cryptography is a field of science that studies the process of encoding data using mathematical equations. Cryptography is an art and science that maintain data security [1].

In this study the writer will use PDF files to make it safe, because the PDF file is susceptible to damage to the security of the information contained within it, allowing one to know clearly and be able to modify the contents of the file.

In cryptography, there are several algorithms that can encode data, one of them is Two Square algorithm or another name Double Playfair is derived from Playfair Cipher algorithm.

For better security, Two Square Cipher algorithm with square 5x5 will be developed into a 16x16 square, so the length of each key in the square is 256 characters that can be loaded with a 256-character ASCII code. Development of keys to 16x16 is of course also useful in plaintext that is entered, so that it is not just an alphabet letter only. For better security the Two Square Cipher 16x16 algorithm will be combined with the Variably Modified Permutation Composition (VMPC) algorithm. VMPC is a modern symmetric algorithm and designed by Bartosz Zoltak, published in 2004 [2].

The combination of Two Square Cipher and Variably Modified Permutation Composition (VMPC) algorithm is intended for obtaining stronger ciphers than using only one cipher, so it is not easy to solve. This method is called super-encryption. Super-Encryption is algorithm that combines two of cipher [3]. Super encryption aims to increase the encryption complexity of a data so that unauthorized users can not decrypt the data, if some key lock is known [4].

In this study, the authors build applications based on Android. Because now the data can be stored on the mobile phone making it possible to secure data and more flexible in mobile.

Based on the background, the author tries to implement super encryption method with 16x16 Two Square Cipher algorithm which will be combined with Variably Modified Permutation Composition (VMPC) algorithm to create security application of PDF file based on Android.

## 2. Method

The first encryption process will be done with Two Square Cipher algorithm and then encrypted again using VMPC algorithm. And then the first decryption process will be done with the VMPC algorithm then Two Square Cipher algorithm.

Password will be randomized with KSA and PRGA in encryption and decryption process. The PRGA will be randomized along the ciphertext. Pseudo-Random Generation Algorithm (PRGA) is run after VMPC Key Scheduling Algorithm (KSA). The pseudocode of KSA and PRGA as shown on table 1.

**Table 1.** *Key Scheduling Algorithm* and *Pseudo-Random Generation Algorithm* [5].

| | |
|---|---|
| *1. x = 0* | *1. k = 0* |
| *2. for k from 0 to 255: S[k] = k* | *2. repeat step 3 until 6 along plaintext length:* |
| *3. for n from 0 to 767: repeat step 4 until 6:* | *3. x = S[(x + S[k]) modulo 256 ]* |
| *4. k = n modulo 256* | *4. Keystream= S[(S[S[x]] + 1) modulo 256 ]* |
| *5. x = S[(x + S[k] + Key[n mod keylength]) mod 256 ]* | *5. Temp = S[k]* |
| *6. Temp = S[k]* | *S[k] = S[x]* |
| *S[k] = S[x]* | *S[x] = Temp* |
| *S[x] = Temp* | *6. k = (k + 1) modulo 256* |

The Two Square Cipher algorithm is a derivative of the Playfair Cipher algorithm. This algorithm is a classical symmetric algorithm. In the Two Square Cipher algorithm there are two 5 x 5 squares placed side by side, where the first letter of each plaintext diagraph is placed on the left square, and the second letter in each plaintext diagram is placed on the right square. So that every character in the plaintext diagraph can not be in the same column, but they can be on the same line. [6].

In encrypting the message, the message to be encrypted is first set as follows:

1. If the number of characters in the message is odd, then inserted a space in the last character of the message to have a pair.
2. To create a square, the key is selected on each square, preferably using two different keywords. The keys are placed sequentially from the top left to the side, then down, without repetition. Next, all unused ASCII codes are entered sequentially into square.

Two Square Cipher Encryption Method

The encryption algorithm has rules include [7]:

Encryption algorithm has rules include (Widjayanti, 2010):

1. Separate plaintext into digraph by two characters.

2. One by one matched into the square. The first character digraph is inserted into the left square and the second character is positioned on the right.
3. If it is located on different rows and columns, look for the intersection of the character. the first characters that are listed from the encryption result are characters parallel to the first character plaintext. And if both characters are in the same row, slide to the right.

Two Square Cipher Decryption Method
For decryption is done the inverse of encryption, as described as follows [7]:
1. Separate the ciphertext into digraph by two characters.
2. The first character of the digraph is inserted into the right square and the second character is inserted into the left.
3. Search the intersection of the characters. If characters are in the same row, slide to left. If it are in a different line, the decrypted characters are characters parallel to the first character ciphertext.

## 3. Result and Discussion

The system was built using Android Studio 1.4.1. with the programming language is Java. This system is tested with Personal Computer with 2.7 GHz processor specification Intel Core i5, 8 GB Memory 1867 MHz DDR.

The result of the complexity of the algorithm Two Square Cipher is $\Theta(n)$ , and VMPC algorithm is. $\Theta(n)$. So the complexity value of super-encryption with Two Square Cipher algorithm and VMPC algorithm is $\Theta(n)$.

In this test, the parameters used are algorithm complexity and real running time. The system testing criteria as follows:
1. Plaintext in the form of PDF files with the total length 24, 50, 129, and 180 characters.
2. Passwords with overall length 10, 20, and 30 characters.
3. The calculation of real running time is done manually using Java programming language.

The result of processing time Two Square Cipher algorithm plaintext with 24, 50, 129, and 180 character lengths respectively is 16 ms, 25.6 ms, 26.6 ms, and 39.6 ms. For VMPC algorithm is 1 ms, 1.3 ms, 6.3 ms and 15 ms. And for super-encryption is 14.3 ms, 21 ms, 39 ms, and 64.6 ms. It can be seen that the process time of the VMPC algorithm is faster than the Two Square Cipher. The results can be illustrated in a graph of Figure 1

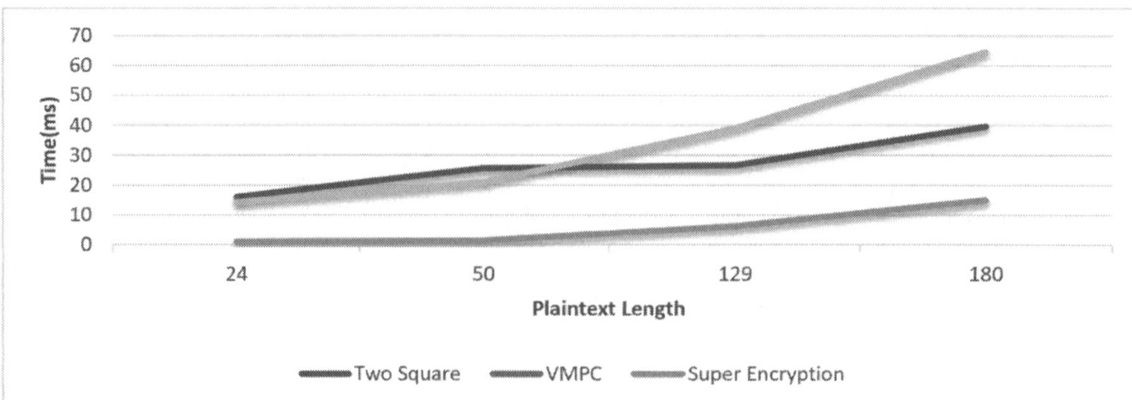


**Figure 1.** Graph of Plaintext Length Against the Process Time of Two Square Cipher, VMPC and Super-Encryption

Figure 1 shows a graph of plaintext length for the process time of the Two Square Cipher Algorithm, VMPC, and super encryption, where the length of the plaintext is directly proportional to the processing time that produces a linear state.

The result of processing time Two Square Cipher algorithm password with 10, 20, and 30 character lengths respectively is 8 ms, 8.3 ms, and 8.6 ms. For VMPC algorithm is 1 ms, 2 ms, and 2.3 ms. And for super-encryption is 13 ms, 22.67 ms, and 28.3 ms. It can be seen that the process time of the VMPC algorithm is faster than the Two Square Cipher. The results can be illustrated in a graph of Figure 2.

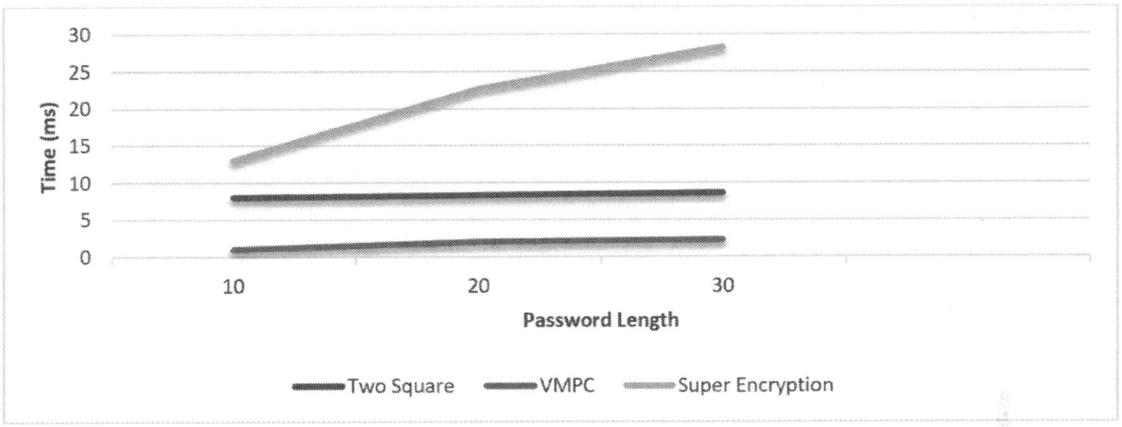


**Figure 2.** Graph of Password Length Against the Process Time of Two Square Cipher, VMPC, and Super-Encryption

Figure 2 shows a graph of plaintext length for the process time of the Two Square Cipher Algorithm, VMPC, and super encryption, where the length of the password is directly proportional to the processing time that produces a linear state.

## 4. Conclusion

- The results of encryption testing on the plaintext and passwords show a meaningful linear result If the length of characters plaintext and passwords is longer then the longer processing time required.
- The result of the complexity Two Square Cipher Algorithm is $\boldsymbol{\Theta(n)}$, VMPC algorithm is $\boldsymbol{\Theta(n)}$, So for the complexity of super-encryption is also $\boldsymbol{\Theta(n)}$.
- The results of processing time VMPC algorithm are faster than the Two Square Cipher algorithm although the results of complexity show the same thing.

## References

[1] Schneir, Bruce 1996, *Applied Cryptography 2nd ed Cryptography: Protocols, Algorithms, and Source Code in C* 15-20.

[2] Karahan, M. 2015. *New Attacks On RC4A and VMPC*. Bilkent University.

[3] Setyaningsih, E, et. al, December 2012, *Image Encryption on Mobile Phone using Super Encryption Algorithm*, TELKOMNIKA, Vol.10, No.4, pp. 837~845 ISSN: 1693-6930

[4] Gupta, H. & Sharma, V.K. 2011. *ROLE OF MULTIPLE ENCRYPTION IN SECURE ELECTRONIC TRANSACTION*. International Journal of Network Security & Its Applications (IJNSA), Vol.3, No.6. Amity Institute of Information Technology, Amity University Campus, Sector – 125, Noida (Uttar Pradesh), India.

[5] Zoltak, B. 2004. *VMPC One-Way Function and Stream Cipher*: Delhi, India.

[6] Churchhouse, R.F. 2004. *Codes and ciphers Julius Caesar, the Enigma and the internet*: Cambridge.

[7] Widjayanti, R. 2010. *Two SquareCipher*. Program Studi Teknik Informatika, Institut Teknologi Bandung.

IOP Conf. Series: Journal of Physics: Conf. Series **978** (2018) 012116 doi:10.1088/1742-6596/978/1/012116

# A comparative study of Message Digest 5(MD5) and SHA256 algorithm

**D Rachmawati[1*], J T Tarigan[1*] and A B C Ginting [1*]**

[1]Departemen Ilmu Komputer, Fakultas Ilmu Komputer dan Teknologi Informasi, Universitas Sumatera Utara, Jl. Universitas No. 9-A, Medan 20155, Indonesia

[*]Email: dian.rachmawati@usu.ac.id, jostarigan@usu.ac.id, realardi@gmail.com

**Abstract**. The document is a collection of written or printed data containing information. The more rapid advancement of technology, the integrity of a document should be kept. Because of the nature of an open document means the document contents can be read and modified by many parties so that the integrity of the information as a content of the document is not preserved. To maintain the integrity of the data, it needs to create a mechanism which is called a digital signature. A digital signature is a specific code which is generated from the function of producing a digital signature. One of the algorithms that used to create the digital signature is a hash function. There are many hash functions. Two of them are message digest 5 (MD5) and SHA256. Those both algorithms certainly have its advantages and disadvantages of each. The purpose of this research is to determine the algorithm which is better. The parameters which used to compare that two algorithms are the running time and complexity. The research results obtained from the complexity of the Algorithms MD5 and SHA256 is the same, i.e., Θ (N), but regarding the speed is obtained that MD5 is better compared to SHA256.

## 1. Introduction

Cryptography is the science and art which aims to maintain the security of the message [9]. The primary objectives of cryptography are authentication, integrity, and non-repudiation. The process of disguising the substance of a message called encryption. The result of encryption process is a cipher text. While the methods used to restore the cipher text to plain text is called decryption. There are three kinds of cryptographic function: hash function, private key functions, and public key functions[4].

One way hashing is the topic of cryptography [2]. One way hash function is an algorithm which takes a message of variable length as input and produces a fixed length string as output referred as hash code or merely hash of the input message [1]. A robust one-way hash function is usually expected to satisfy some requirements, namely collision resistance, preimage resistance, second preimage resistance[3]. There are a lot of hash function. Two of them are MD5 and SHA256.

The MD5 algorithm takes as input a message of arbitrary length and produces as output a 128-bit message digest of the input. The authentication algorithm computes a digest of the entire data of the secret message, used for authentication [6]. MD5 consists of 64 operations, grouped into four rounds of 16 operations [10]. The MD5 algorithm is designed to be quite fast on 32-bit machines. This algorithm found by Professor Ronald L. Rivest[2].

The SHA256 algorithm is a cryptography hash function and used in digital certificate as well as in data integrity. SHA256 is developed by N.I.S.T[5]. The SHA256 algorithm takes as input a message of arbitrary length that smaller than $2^{64}$ bits and produces as output a 256-bit message digest of the input[7].

## 2. Method

### *2.1. MD5 Algorithm*

*Step 1: Append padded bits*

The message is filled so that its length is congruent to 448, modulo 512. This padding is single 1 bit added to the end of the message, followed by as many zeros are required so that the length of bits equals 448 modulo 512.

*Step 2: Append length*

A 64-bit representation of the message's length is appended to the result. This stage to make the message length an exact multiple of 512 bits in length.

*Step 3: Divide the message*

MD5 processes the input string in 512-bit blocks, divided into 16 32-bit sub-blocks. The output of the algorithm is set of four 32-bit blocks, which concatenate to form single 128-bit hash value.

*Step 4 : Initialize MD Buffer*

Four 32-bit variable are initialized:
A = 0x01234567
B = 0x89ABCDEF
C = 0xFEBCDA98
D = 0x76543210
These are called chaining variables.

*Step 5: Process message*

The main loop of the algorithm begins and continues for as many 512-bit blocks as are in the message. The four copied into the different variable: *a* gets *A*, *b* gets *B*, *c* gets *C*, and *d* gets *D*. the main loop has four rounds, all very similar. Each series uses a different operation 16 times. Each operation performs a nonlinear function on three of *a*, *b*, *c*, and *d*. Then it adds that result to the right a variable number of bits and adds the result to one of *a*, *b*, *c*, and *d*. Finally, the result replaces one of *a*, *b*, *c*, and *d*.
There are four nonlinear functions:
$F(X,Y,Z) = (X \wedge Y) \vee ((\sim X) \wedge Z)$
$G(X,Y,Z) = (X \wedge Z) \vee (Y \wedge (\sim Z))$
$H(X,YZ) = X \oplus Y \oplus Z$
$I(X,Y,Z) = Y \oplus (X \vee (\sim Z))$
($\vee$ is OR, $\wedge$ is AND, $\oplus$ is XOR, $\sim$ is NOT)

*Step 6: Output*

The message digest produced as output is A, B, C, D. That is, output begins with low-order byte of A, and end with the high-order byte of D.

### *2.2. SHA256 Algorithm*

*Step 1: Append padded bits*

The message is filled so that its length is congruent to 448, modulo 512. This padding is single 1 bit added to the end of the message, followed by as many zeros are required so that the length of bits equals 448 modulo 512.

*Step 2: Append length*

A 64-bit representation of the message's length is appended to the result. This step to make the message length an exact multiple of 512 bits in length.

*Step 3:* *Parsing the message*
The padded message is parsed into N 512-bit message blocks, $M^{(1)}$, $M^{(2)}$,..., $M^{(N)}$, by appending 64-bit block.

*Step 4:* *Initialize Hash Value*
The initial hash value, $H^{(0)}$ is set, consist of eight 32-bit words, in a hexadecimal form.

*Step 5:* *Prepare the message schedule*
SHA256 uses a message schedule of sixty-four 32-bit words.The words of the message schedule are labeled $W_0, W_1, \ldots, W_{63}$.[8]

$$Wt = \begin{cases} M_t^{(t)} & 0 \leq t \leq 15 \\ \sigma_1^{(256)}(W_{i-2}) + W_{i-7} + \sigma_0^{(256)}(W_{i-15}) + W_{i-16} & 16 \leq t \leq 63 \end{cases}$$

Where:

$\sigma_1^{(256)}(W_{i-2}) = ((W_{i-2})\ ROTR\ 17) \oplus ((W_{i-2})\ ROTR\ 19) \oplus ((W_{i-2})\ SHR\ 10)$

$\sigma_0^{(256)}(W_{i-15}) = ((W_{i-15})\ ROTR\ 7) \oplus ((W_{i-15})\ ROTR\ 18) \oplus ((W_{i-15})\ SHR\ 3)$

*Step 6:* *Initialize the eight working variables, a, b, c, d, e, f, g, and h, with the(i-1)st hash value*
For $t$=0 to 63:
{

$T_1 = h + \Sigma_1^{(256)}(e) + Ch(e,f,g) + K_1^{(256)} + W_t$

$T_2 = \Sigma_0^{(256)}(a) + Maj(a,b,c)$

$H = G$

$G = F$

$F = E$

$E = d + T_1$

$D = C$

$C = B$

$B = A$

$A = T_1 + T_2$

}

Where:

$\Sigma_1^{(256)}(e) = (e\ ROTR\ 6) \oplus (e\ ROTR\ 11) \oplus (e\ ROTR\ 25)$

$\Sigma_0^{(256)}(a) = (e\ ROTR\ 2) \oplus (e\ ROTR\ 13) \oplus (e\ ROTR\ 22)$

$Ch(e,f,g) = (e \wedge f) \oplus (\sim e \wedge g)$

$Maj(a,b,c) = (a \wedge b) \oplus (a \wedge c) \oplus (b \wedge c)$

*Step 7 :* *Output*
After repeating steps one through four a total of N times, the resulting hash function is
$H_0^{(N)} \,||\, H_1^{(N)} \,||\, H_2^{(N)} \,||\, H_3^{(N)} \,||\, H_4^{(N)} \,||\, H_5^{(N)} \,||\, H_6^{(N)} \,||\, H_7^{(N)}$

IOP Conf. Series: Journal of Physics: Conf. Series **978** (2018) 012116 doi:10.1088/1742-6596/978/1/012116

## 3. Results and Discussions

The experiments were performed on Windows 10 Pro with Intel(R) Core(TM) i5-2430M CPU @ 2,40GHz (4CPUs),~2.4 GHz architecture, and 8.00 GB RAM. The development environment used for coding C# scripts is Visual Studio 2015. Each Algorithm was tested to three sample documents that are looped seven times. The results of the experiments of each set are presented in as follows.

### *3.1. MD5 Algorithm Trial*

#### *3.1.1. MD5 Algorithm First Trial*

File size = 11539 bytes
Hash Value = 98B62AB5CD7D3BBF13D529FE5BDAC629
Average Running Time = 3.3644 millisecond

#### *3.1.2. MD5 Algorithm Second Trial*

File size = 22528 bytes
Hash Value = 577CA8666C87EAE17E6A053FE3812E4C
Average Running Time = 3.674242857143 millisecond

#### *3.1.3. MD5 Algorithm Third Trial*

File size = 47104 bytes
Hash Value = 14206DAB6EA6E8092D536DC36C26E040
Average Running Time = 4.7833 millisecond

### *3.2. SHA256 Algorithm Trial*

#### *3.2.1. SHA256 Algorithm First Trial*

File size = 11539 bytes
Hash Value :
32E527EBBFD81F6B7739A87F88A4E4AB208DBD50D170DE0D0DF772A60C79529E
Average Running Time = 8.198342857143 millisecond

#### *3.2.2. SHA256 Algorithm Second Trial*

File size = 22528 bytes
Hash Value :
25B0F833564169D56741059DCCFA09ACCF891DCFC8042323090D83BD08D29FA4
Average Running Time = 10.25057142857 millisecond

#### *3.2.3. SHA256 Algorithm Third Trial*

File size = 47104 bytes
Hash Value :
42A0E1BCE1DFC2889BB0EC4A442BDFAF924C901971E187FFEFBD32734BF91CE7
Average Running Time = 16.8093142857143 millisecond

### *3.3. Comparison of Running Time between MD5 Algorithm and SHA256 Algorithm*

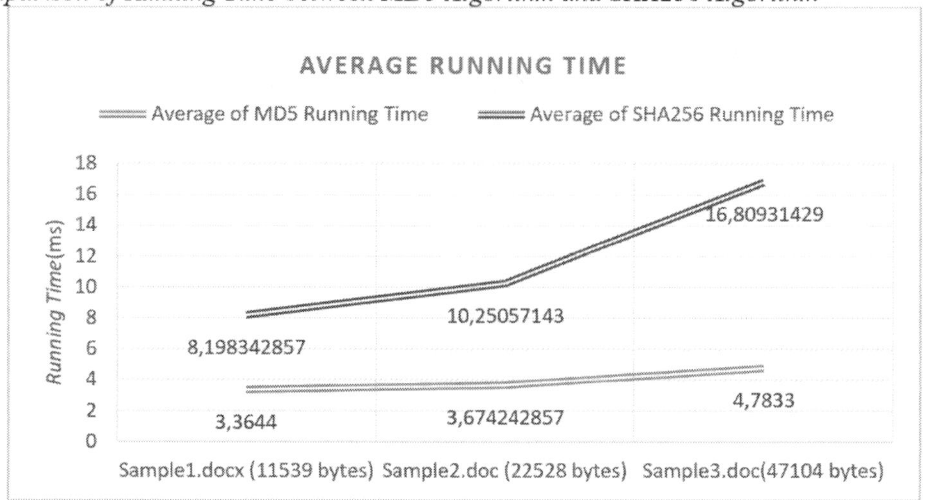


**Figure 1.** Graph about average running time MD5 and SHA256

It is clear that the running time of MD5 is faster than a SHA256 algorithm.

### *3.4. Complexity of Message Digest 5 (MD5) and SHA256*

**MD5 Complexity**

$\sum = T(n)$

$= (C1 + 73C2 + 64C3 + C4 + 3C5)N^0 + (456C2+64C3+C4+2C6+64C7)N$

$= \Theta\ (N)$

**SHA256 Complexity**

$\sum = T(n)$

$= (29C1 + 4C2)N^0 + (10140C1 + 2C3 + 64C4 + 2C5)N^1$

$= \Theta(N)$

Both MD5 and SHA256 have a same complexity that is $\Theta$ (N)

## 4. Conclusions

In conclusion, we know that complexity of the MD5 algorithm and SHA256 is equal and the value is $\Theta$(N), but the running time of MD5 is faster than SHA256.

## 5. Acknowledgments

The authors gratefully acknowledge that the present research is supported by Fund Dissemination IPTEKS Research Results for Lecturers / Researchers Universitas Sumatera Utara.

## References

[1] Gauravaram, Praveen 2007 *Cryptographic Hash Functions: Cryptanalysis Design and Application.* Ph.D. thesis, Information Security Institute, Faculty of Information Technology, Queensland University of Technology.

[2] Gupta, Piyush, and Kumar, Sandeep 2014 *A Comparative Analysis of SHA and MD5 Algorithm* International Journal of Computer Science and Information Technologies **5** (3) 4492-4495

[3] Handschub, H. and Gilbert, H. 2002 *Evaluation Report Security Level of Cryptography – SHA-256.* Technical Report, Issy-les-Moulineaux.

[4] Hossain, M.A, Islam, M. K, Das S. K., and Nashiry, M. A. 2012 *CryptanalyzingofMessage Digest Algorithms MD4 and MD5* International Journal on Cryptography and Information Security(IJCIS) **2**(1) 1-13

[5] Kasgar, A.K, Agrawal, J., and Sahu, S. 2012. *New Modified 256-bi MD5 Algorithm with SHA Compression Function.* International Journal of Computer Applications **42** (12) 47-51.

[6] Mishra, S, Mishra, S, and kumar, N. 2013. *Hashing Algorithm: MD5.* International Journal for Scientific Research and Development **1** (9) 1931-1933.

[7] NIST 2002 *Secure Hash Standard (SHS)*, FIPS PUB 180-2.

[8] Roshdy, R., Fouad, M., and Dahab, M. A. 2013 *Design and Implementation A New Security Hash Algorithm Based on MD5 And SHA-256.* International Journal of Engineering Sciences & Emerging Technologies 6 (1) 29-36.

[9] Schneir, Bruce 1996 *Applied Cryptography 2nd ed* Cryptography: Protocols, Algorithms, and Source Code in C"; 1.1 MD5 15-20.

[10] Thomas, C.G, and Jose, R.T. 2015 *A Comparative Study on Different Hashing Algorithms.* International Journal of Innovative Research in Computer and Communication Engineering **3**(7) 170-175.

# Determination of quality television programmes based on sentiment analysis on Twitter

**A Amalia[1], W Oktinas[2], I Aulia[2] and R F Rahmat[2]**

[1]Departement of Computer Science, Faculty of Computer Science and Information Technology, Universitas Sumatera Utara, Medan, Indonesia.
[2]Departement of Information Technology, Faculty of Computer Science and Information Technology, Universitas Sumatera Utara, Medan, Indonesia

*Email: amalia@usu.ac.id

**Abstract.** Public sentiment from social media like Twitter can be used as one of the indicators to determine the quality of TV Programmes. In this study, we implemented information extraction on *Twitter* by using sentiment analysis method to assess the quality of TV Programmes. The first stage of this study is *pre-processing* which consists of *cleansing, case folding, tokenizing, stop-word removal, stemming*, and *redundancy filtering*. The next stage is weighting process for every single word by using TF-IDF method. The last step of this study is the sentiment classification process which is divided into three sentiment category which is positive, negative and neutral. We classify the TV programmes into several categories such as news, children, or films/soap operas. We implemented *an improved k-nearest neighbor* method in classification 4000 twitter status, for four biggest TV stations in Indonesia, with ratio 70% data for training and 30% of data for the testing process. The result obtained from this research generated the highest accuracy with k=10 as big as 90%.

## 1. Introduction

Television Industries in Indonesia is growing so fast. It can be seen by many television stations established in Indonesia recently. The quantities of TV stations also affected with increasingly of TV programmes that can be classified into some categories like news, movie/soap opera, kids and else. The variety and quality of TV programmes often will be discussed by many people in social media like *Twitter*. *Twitter* as one of microblogging with more than 500 million account users and can generates 400 million tweets a day [1]. On *Twitter,* people can reveal his/her opinion mining including to assess the quality of TV programmes on TV stations. These set of opinions can be a consideration to TV stations to improve the quality of their TV programmes. Opinion mining or sentiment analysis is a computational study from people's opinion, sentiment and emotion that are expressed in the text [2] [3]. Sentiment analysis will classify polarity of text to mine whether the opinion is positive, negative or neutral opinion about something [4]. Recently, many researchers utilize *Twitter* to reveal society opinion about something because *Twitter* provides *Twitter* API to facilitate crawling data in large quantities [5]. Sentiment analysis is a classification process, but classification process on twitter data is more challenging than document classification because Twitter data contains unstructured text. The first stage of twitter classification is to classify whether the status is an opinion or not [2]. The next step is to organize the view into positives or negative sentiment. This study aims to determine the quality of TV programmes based on public opinion on Twitter. We implemented an Improved K-

Nearest Neighbor method for classification public opinion based on Twitter status in Bahasa Indonesia. *Improved k-Nearest Neighbor* is a modification of k-Nearest neighbor method. Improved k-Nearest Neighbor modify the k-values determination, for each category has various values. The diversity of k-values for each category is adjusted with training document in the category. The purpose of this modification is a to improve the k-Nearest Neighbor method that has issued in accuracy because all class has same k-values without considering the number of training document for each category. Meanwhile, distribution of training documents in training stage has not had same quantity.

## 2. Related Works

Many previous studies related to sentiment analysis and classification methods. Studies that utilize classification methods had been done by [6]. Earlier studies like marketing and social study utilize Twitter because Twitter can be used efficiently [7]. Study of public sentiment analysis about government policy had been done by [8], the study use comparison of some classification methods like k-Nearest Neighbor, Naïve Bayes and Support Vector Machine. The result of this study shows that Support Vector Machine raised a better accuracy which is 86% [9]. Another sentiment analysis study had been done by [10], this study implemented Naïve Bayes and unigram feature to observe sentiment analysis on President election in the United States. This study gets 59% accuracy [9]. Another sentiment analysis study had been done by [5], this study observed customer satisfaction based on emoticon on Twitter. This study compares three learning methods which are naïve Bayes, support vector machine and maximum entropy. This study use unigram feature and bigram, the research conducted the accuracy increase for Bigram feature for Naïve Bayes and maximum entropy. Sentiment analysis for public opinion in President election in India had been done by [11], this study use combination of *Naïve Bayes, Hidden Markov Model, and SentiWordNet*. This study gets 71.48% of accuracy.

## 3. Methodology

The Methodology in determining TV Programmes' quality based on opinion analysis can be described in Figure 1. This process contains four big stages. The stages are Crawling data, Pre-processing Stage, word weighting and classification stages.

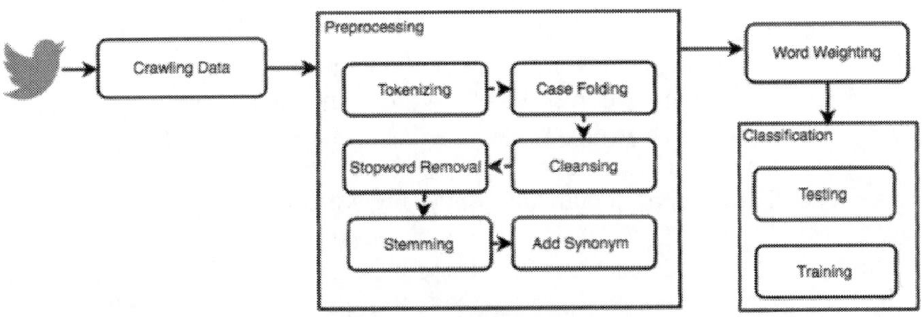


**Figure 1.** Methodology.

### *3.1. Crawling Data*

In this study, we observe public opinion in Bahasa Indonesia on twitter to four biggest TV station in Indonesia. We crawl all tweet-mention to the official account twitter of these five TV stations through search API. These accounts are @officialRCTI, @whatsonANTV, @GlobalTVseru and @official_MNCTV. We got 4000 data or 1000 twitter status for each account mention.

### *3.2. Preprocessing*

Preprocessing in this study is a process to prepare the data or tweets into 'a-bag-of-words' form. The purpose of this conversion is to reduce data dimension but still preserve the meaning of the tweets. We also eliminate a redundant data and fix inconsistent data from the dataset. This process contains tokenizing, case folding, cleansing, stopword removal, stemming and filter redundant tweets.

Tokenizing is a process to break the tweets into word by word. The next step is case folding process. This process is a process to convert these set of words into lower case. The aim of this action to convert all characters in the same way. The next process is Cleansing, which is a process to remove noise from tweets like hashtag(#), username(@username, URL address, emoticon and email address. The next stage is stopword removal. This stage is a process to eliminate un-important words like conjunctions, prepositions, and pronouns. We implemented Tala dictionary as a reference to un-important words. Tala dictionary is a list of words in Bahasa Indonesia that are not important (Tala 2013). However, we have not eliminated all words as Tala reference, because some words have essential meaning in opinion analysis, particularly for negative meaning. The next step is stemming, stemming is a process to convert a word into its basic words or remove the affix of the words. We implemented Nazief and adriani stemming algorithm. The last stage of preprocessing is adding synonym functions. This process is an action to check synonym of the words in 'a-bag-of-words' for each tweet. Some of these synonym words are slang or abbreviation words. Twitter is social media that only has 140 characters limitation for each tweet. This limitation makes some people used abbreviation words, for example, people use "ga" instead of "enggak" to represent "not". The additional of this synonym words will optimize frequency of occurrences calculation for words which have the same meaning.

### *3.3. Word Weighting*

Words weighting is a process to measure or evaluate how important a word is to a document. We implemented TF-IDF measurement. TF-IDF or Term Frequency-Inverse Document Frequency is TF value multiply to IDF value. TF is a measurement of word quantity occurrences in a document. Meanwhile, IDF is a process to calculate the word weighting based on documents quantities which contain a certain word. The greater TF-IDF value of a word weighting calculation the more important that word is. The equation of TF-IDF can be seen in equation 1 and 2.

$$IDF\ (w) = \log(\frac{N}{DF(w)}) \tag{1}$$

$$\text{TF -IDF (w,d)} = \text{TF (w,d) x IDF(W)} \tag{2}$$

Where :

| | |
|---|---|
| TF-IDF (w,d) | : weighting of a word in a whole document |
| Tf(w,d) | : Frequency occurrences of a word w in a document |
| IDF(w) | : inverse DF from a word |
| N | : Numbers of documents |
| DF (w) | : Numbers of documents containing word w |

### *3.4. Classification*

The next step is classification process. We implanted Improved K-Nearest Neighbor algorithm for classification. These steps contains training stage and testing stage. We divided dataset to training and testing process with ratio 7 : 3. The quantities of the divided dataset for each TV station can be seen in Table 1.

**Table 1.** Training Data Proportion

| TV Station | Quantity | |
|---|---|---|
| | Training Data | Testing Data |
| ANTV | 700 | 300 |
| Global TV | 700 | 300 |
| RCTI | 700 | 300 |
| MNCTV | 700 | 300 |

On training stage, 70% of the dataset is classified or labelled manually as positive, negative or neutral opinion. These classifications that will be used to create a model of sentiment analysis. 30% of the rest of data will be classified based on the model that got from training step. We implement an *improved k-nearest neighbor algorithm.* The first step, we calculate the similarity of two documents with cosine similarity method. Cosine similarity is a document vector (D) measurement with query vector (Q). The more similar D to Q, the more appropriate the document with the query. Cosine similarity equation can be seen in equation 3.

$$cosSim(X, dj) = \frac{\sum_i^m = 1\, xi.dji}{\sqrt{(\sum_i^m = 1\, xi)^2}\,.\sqrt{(\sum_i^m = 1\, dji)^2}} \quad (3)$$

Where X is testing document, dj is training document, xi and dji is word weighting value.
We classify the testing document with *Improved k-Nearest Neighbors algorithm.* We calculate similarity comparison of training data and testing data for each category and determine the category based on the best probability. Equation 4 shows the formula of this method.

$$p(x, c_m) = argmax_m \frac{\sum_{dj\in\, top\, n\, knn(cm)} sim(x,d_j)y(dj,c_m)}{\sum_{dj\in\, top\, n\, k\, NN(cm)} sim(x,d_j)} \quad (4)$$

where:
*p(x,cm)* : probability document X in category of cm
*sim(x,dj)* : similarity of document X with training document dj
*top n kNN* : top n neighbour
*y(dj.cm )* : attribute function from suitable category

we tried ten times trials, to find out the influence of training set proportion for each category and k value to classification effectiveness.

**Table 2.** Training Data Proportion

| Training Data Proportion | | | |
|---|---|---|---|
| Positive | Negative | Neutral | Quantity |
| 1385 | 700 | 724 | 2800 |

Table 2. Show differences proportion of training data quantity that is used in the classification process. We tried various of k values for 1200 test data. Average accuracy value for 10 times trial can be seen in Table 3. From 2800 of training data with proportion quantity 1385 positives, 700 negatives and 724 neutral. From table 3 show that the highest accuracy is 90% for k = 10 and the lowest accuracy is 85% for k = 60.

**Table 3. Testing System Based On K Values**

| k | n (new *k-values* ) | | | Accuracy |
|---|---|---|---|---|
| | Positive | Negative | Neutral | |
| 5 | 2 | 1 | 1 | 89% |
| 10 | 5 | 2 | 2 | 90% |
| 15 | 7 | 4 | 4 | 88% |
| 20 | 10 | 5 | 5 | 88% |
| 25 | 12 | 6 | 6 | 88% |
| 30 | 15 | 8 | 8 | 87% |
| 35 | 17 | 9 | 9 | 87% |
| 40 | 20 | 10 | 10 | 87% |
| 50 | 25 | 13 | 13 | 86% |
| 60 | 30 | 15 | 15 | 85% |

IOP Conf. Series: Journal of Physics: Conf. Series **978** (2018) 012117 doi:10.1088/1742-6596/978/1/012117

## 4. Result and Discussion

Based on the previous stage, we conclude that the best accuracy obtained for k =10. The excerpt of the result of the experiments for 1200 data testing with k = 10 can be seen in Table 4.

**Table. 4** System Test Result

| No | Training Data | Classification | |
|---|---|---|---|
| | | System | Manual |
| 1. | @whatsonANTV wajib di tonton Karena Alur ceritanya pasti bikin Baper dan selalu bikin penasaran, Jadi selalu tak in� https://t.co/Fql8XWcyYv | Positif | Positif |
| 2. | Arigatou buat yang udah nonton @merlynasun di @LensorANTV edisi pagi ini, wassalamu'alaikum dan BRAVO OLAHRAGA!!! https://t.co/wiQv6VHCSP | Positif | Positif |
| ... | ... | .. | .. |
| 1200 | Mantap jiwa dubbingnya keren. Walau tayangnya malem kalo nonton Jodha Akbar mata tetep melek @Official_MNCTV #JodhaAkbarMNCTV | Positif | Positif |

Based on this study, from 1200 data testing, the system can classify 1080 opinion mining categorization correctly. Accuracy classification of this study can be calculated :

$$Accuracy = \frac{1080}{1200} \text{ x } 100\%$$
$$= 90\%$$

Based on this study we can conclude accuracy of opinion classification using *Improved K-Nearest Neighbor* is 90%.

We also classify the favourite TV stations and TV programmes Category based on positives opinion. The quality of TV station and TV programmes can be determined by comparing the quantities of positives sentiment analysis is higher than negatives and neutral view. The result can be described in Figure 2.

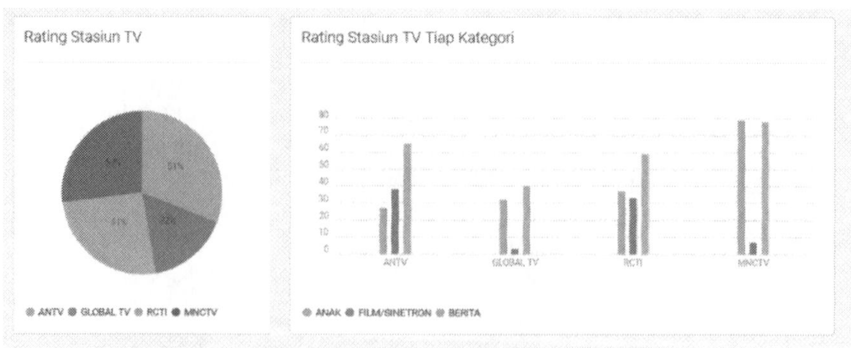


**Figure 2**. TV Stations Quality Based On Public Opinion

TV station that got the highest positive public opinion is ANTV with getting 61% of positives opinion. For TV programmes, MNCTV got two highest categories, for kid programmes category 80% and news category 79%.

## 5. Conclusion

Improved K-Nearest Neighbor (IKNN) method can be used to classify Indonesian sentiment into 3 categories: positive, negative and neutral sentiments. From the results of comparing positive opinion with negative and neural opinion, we can conclude the highest for news category is MNCTV, the highest result for the category of film / soap opera is ANTV, and the highest result for children category is MNCTV by 80%. The highest overall favourite for Station TV is ANTV. Sentiment analysis examination for Bahasa Indonesia with Improved K-Nearest Neighbor (IKNN) yield the highest accuracy for k =10 which is 90%.

## 6. Reference

[1] D. Farber, "Twitter Hits 400 Million Tweets per Day, Mostly Mobile. CNET," 2012.
[2] B. Liu, "Sentiment Analysis and Opinion Mining," *Synth. Lect. Hum. Lang. Technol.*, vol. 5, no. 1, pp. 1–167, May 2012.
[3] I. P. Cvijikj and F. Michahelles, "Understanding social media marketing," in *Proceedings of the 15th International Academic MindTrek Conference on Envisioning Future Media Environments - MindTrek '11*, 2011, p. 175.
[4] B. Pang, L. Lee, … S. V. the A.-02 conference on, and undefined 2002, "Thumbs up?: sentiment classification using machine learning techniques," *dl.acm.org*.
[5] A. Go, R. Bhayani, and L. Huang, "Twitter Sentiment Classification using Distant Supervision."
[6] Amalia, D. Gunawan, M. S. Lydia, and Charlie, "The Usage Evaluation of Official Computer Terms in Bahasa Indonesia in Indonesian Government Official Websites," *J. Phys. Conf. Ser.*, vol. 755, p. 11001, 2016.
[7] A. Pak and P. Paroubek, "Twitter as a Corpus for Sentiment Analysis and Opinion Mining."
[8] G. Stylios *et al.*, "Public Opinion Mining for Governmental Decisions," *Electron. J. e-Government*, vol. 8, no. 2, pp. 203–214, 2010.
[9] H. Wang, D. Can, A. Kazemzadeh, F. Bar, and S. Narayanan, "A system for real-time Twitter sentiment analysis of 2012 U.S. presidential election cycle," *Proceedings of the ACL 2012 System Demonstrations*. Association for Computational Linguistics, pp. 115–120, 2012.
[10] A. Wang, "Services-oriented URL filtering and verification Chao Wang *, Xi Li and Xuehai Zhou Nadia Nedjah," vol. 4, no. 4, 2013.
[11] R. Jose and V. S. Chooralil, "Prediction of election result by enhanced sentiment analysis on twitter data using classifier ensemble Approach," in *2016 International Conference on Data Mining and Advanced Computing (SAPIENCE)*, 2016, pp. 64–67.

# Data pre-processing in record linkage to find the same companies from different databases

**D Gunawan[1*], M S Lubis[1], D Arisandi[1] and B Azzahry[1]**

[1]Department of Information Technology, Universitas Sumatera Utara, Jl. dr. Mansur No. 9 Kampus USU Medan 20155

*Email: danigunawan@usu.ac.id

**Abstract**. As public agencies, the Badan Pelayanan Perizinan Terpadu (BPPT) and the Badan Lingkungan Hidup (BLH) of Medan city manage process to obtain a business license from the public. However, each agency might have a different corporate data because of a separate data input process, even though the data may refer to the same company's data. Therefore, it is required to identify and correlate data that refer to the same company which lie in different data sources. This research focuses on data pre-processing such as data cleaning, text pre-processing, indexing and record comparison. In addition, this research implements data matching using support vector machine algorithm. The result of this algorithm will be used to record linkage of data that can be used to identify and connect the company's data based on the degree of similarity of each data. Previous data will be standardized in accordance with the format and structure appropriate to the stage of preprocessing data. After analyzing data pre-processing, we found that both database structures are not designed to support data integration. We decide that the data matching can be done with blocking criteria such as company name and the name of the owner (or applicant). In addition to data pre-processing, the result of data classification with a high level of similarity as many as 90 pairs of records.

## 1. Introduction

Public services become more complex, especially in licensing, tax and retribution services. All those services cannot be separated and have to be integrated to embody a good governance. The problem arises in Medan city because the company data which have been registered in several public services databases are separated and there is no synchronization among those public services. This may lead to inefficient public services. Therefore, it is required to find the same company the existing records which lie in several databases. In this case, we observed two public services namely Badan Lingkungan Hidup (BLH) and the Badan Pelayanan Perizinan Terpadu (BPPT) of Medan city. This research will discuss the data pre-processing analysis before implementing data matching between two databases.

Previous experiment showed the ability of a data intruder to link records between distinct microdata sets by the co-presence of publicly available aggregate data [1]. This research uses Naïve Bayesian approach as well as used in other research, such as focused crawlers [2][3]. In this research, the Naïve Bayesian approach can avoid the cost for computational process with exact inference. According to this research, Naïve Bayesian algorithm is a useful tool to identify linkages that are most likely to be correct and might be useful for disclosure control.

Another research uses hidden Markov model (HMM) which is trained to standardize typical Australian name and address. These data are collected from health data collection. This research is intended to solve record linkage on set of simple names [4]. It is required to have comprehensive steps to process longer and complex string. For the address attribute, they found that hidden Markov model produced equal or even better accuracy than rule-based system.

Recent publication proposed secure record linkage for health data integration in developing country [5]. Their technique is claimed can anonymize identifiable private data of the patients. Their experimental results show that the proposed method has successfully linked records in absence of standard identification number in acceptable accuracy for noisy data.

The organization of this paper is as follow: first, the introduction about the background of the research and related works. Next, section two discusses the analysis of data pre-processing to support record linkage. Section three discusses the result of this research. The last but not least is the conclusion.

## 2. Data Pre-processing Analysis

Data pre-processing analysis was done by conducting several processes. The first process is analyzing the data sources. Both data sources contain many attributes that have different purposes. These attributes should be examined to obtain the similar purpose between them. The similar purpose attributes will be processed later as the parameters to obtain data matching. Second, the data cleaning process for the data in selected attributes from the previous process. Third, we conduct the indexing process or filtering by removing record pairs which impossible to be similar. Fourth, comparing the record pair by using edit distance algorithm. In addition, we conduct data matching by applying support vector machine algorithm. The details of the data pre-processing analysis are as follow:

### *2.1 Data Source*

The data are obtained from two public agencies, which are Badan Lingkungan Hidup (BLH) and Badan Pelayanan Perizinan Terpadu (BPPT) of Medan city. BLH has 500 records which are recorded since 1995 to 2014, meanwhile BPPT has 21,754 records since 2010 to 2014. Each agency has its own database with its own structure. As not all the attributes are related between two databases, we should examine their structure manually to obtain related attributes. Then, the unrelated attributes will be removed and the rest will be kept for the next process.

**Table 1.** The database structure of BLH

| Attribute | Description | Relevance |
|---|---|---|
| id_perusahaan | Unique number for company identification | No |
| nama_perusahaan | The company name | Yes |
| nama_pemilik | The owner or applicant who register the company to BLH | Yes |
| alamat_usaha | The company full address | Yes |
| nama_kecamatan | The company district | Yes |
| nama_kelurahan | The company sub-district | Yes |
| telp | The company official phone number | Yes |
| kodepos | The company postal code | Yes |
| no_rekomendasi | The recommendation number issued by related institution | No |
| no_ijin | The registration number issued by BLH of Medan City | No |

The database from BLH has ten attributes as shown in table 1. According to our examination, there are seven attributes that will be used for data matching process such as perusahaan (the company name), nama_pemilik (company owner or person who register the company to BLH), alamat_usaha (the company address), nama_kecamatan (the company district), nama_kelurahan (the company sub-district), telp (the company phone number), and kodepos (the company postal code). Attributes id_perusahaan (company unique number), no_rekomendasi (the recommendation number) and no_ijin (register number) are not used because of irregular format.

In the other hand, the structure of BPPT database has more attributes. As shown in table 2, the BPPT has 21 attributes. As well as the attributes in BLH database, not all of the attributes in BPPT database can be used. For example, id_perusahaan (company unique number), nofax (attribute to store the company's fax number), email (attribute to store the company's email address), jabatan (attribute to store the applicant position), tempatT (attribute to store the applicant residence), tanggal_proses (attribute to store the company registration date), tanggal_ambil (attribute to store the date of license pickup), nama_ambil (attribute to store the person who pick up the license), nomor_identitas (attribute to store the person's identification number), penomoran_ijin (attribute to store license number) are not used because there is no similar attribute (with the same purpose) in BLH database. Attributes which store the province and the city of the company (attribute prop and kota respectively) also removed because all the data only consist of the companies which are located in Medan. According to the difference between both database structures, both databases are not designed to support data integration each other.

**Table 2.** The database structure of BPPT

| Attribute | Description | Relevance |
|---|---|---|
| no | Process Id | No |
| id_perusahaan | Unique number for company identification | No |
| tanggal_proses | The registration date | No |
| tanggal_ambil | The date when the license is picked up | No |
| nama_ambil | The person who picked up the license | No |
| no_identitas | The applicant id number | No |
| penomoran_ijin | The license number | No |
| pemohon | The applicant name | Yes |
| jabatan | The applicant position in the company | No |
| tempatT | The applicant residence address | No |
| nama | The company name | Yes |
| badan_usaha | The type of company | Yes |
| alamat | The company full address | Yes |
| prop | The company province | No |
| kota | The company city | No |
| nama_kecamatan | The company district | Yes |
| nama_kelurahan | The company sub-district | Yes |
| kodepos | The company postal code | Yes |
| notelp | The company official phone number | Yes |
| nofax | The company official facsimile number | No |
| email | The company email address | No |

### *2.2 Data Cleaning*

We conduct three steps to clean the data before pre-processing step. Those are handling missing values, smoothing noisy values, and identifying and correcting inconsistent values [6].

#### *2.2.1 Handling missing values*
There are a few ways to handle missing values. One of them is removing the data which does not have important value in data matching. In this research we removed attribute which is empty or does not have any data. For example, there is no data for attribute kodepos (postal code) in BLH database, meanwhile the kodepos data are exist in BPPT. Therefore, we omitted the attribute kodepos to be included in data matching process. Another way to handle missing value is adding data to the attribute manually. However, the way can only be implemented in numerical data only. This step cannot be implemented in our case because the attribute of the data is string.

#### *2.2.2 Smoothing noisy value*
The existing data can contain the false data. For example, the rule for the date format in database is year, month and day. This attribute can consist of the date which has format month, day and year. After examining the data between BLH and BPPT databases manually, we did not found any noisy value that has to be fixed before data pre-processing.

#### *2.2.3 Identifying and correcting inconsistent values*
The example of inconsistent values is the sub-district name does not match with the corresponding district. This value should be corrected depending on the availability of supporting data. In several cases, we should prepare the external look-up table or formula/rule that can support data correction.

### *2.3 Text pre-processing*
Data pre-processing refers to process to convert raw data to formatted data which could enhance the data matching result to be more accurate and efficient [4]. As the data is in text format, then this process will include text pre-processing. Dataset is collection of data which depict attributes from real world entities [7]. As this research uses dataset from two databases which have different structure, we should convert them to maintain consistent format. To maintain consistent format in the dataset, we conduct a few steps such as removing unnecessary characters and symbols, normalization, attribute parsing, and verification. These steps are explained as follow:

#### *2.3.1 Removing unnecessary characters and symbols*
The first step of data pre-processing is related to the more detailed data cleaning. Values in the dataset might consist of characters, words, terms or abbreviations which do not have any information value. Those values can be removed from the attributes. We also remove unnecessary spaces, such as space in the beginning or the end of attribute values. Characters or symbols are required to be converted to standard form, including convert double spaces into single space. This is required to the normalization and tokenization process. The conversion requires look up table as shown in table 3.

**Table 3.** The example of look up table for characters, symbols and abbreviations

| Values | Conversion |
|---|---|
| ‘,’, ‘?’, ‘~’, ‘:’, ‘;’, ‘^’, ‘*’, ‘=’, ‘’’’, ‘-‘, ‘#’, ‘(‘, ‘{‘, ‘[‘, ‘<’, ‘)’, ‘}’, ‘]’, ‘>’, '#N/A', ‘N/A’, ‘NA’, ‘n/a’, ‘na’ | empty |
| ‘jl’, ‘jln’ | jalan |
| ‘kompleks’, ‘complex’, ‘comp’, ‘komp’, ‘perum’ | komplek |
| ‘no.’, ‘nmr’, ‘n’ | nomor |
| ‘block’, ‘bloc’, ‘blk’ | blok |
| ‘lr’, ‘lor’ | lorong |

| | |
|---|---|
| ‘gg’, ‘g’ | gang |
| ‘lt’, ‘l’ | lantai |
| ‘lingk’, ‘lk’ | lingkungan |

*2.3.2 Normalization and tokenization*
The next step of data preprocessing is normalization and tokenization process. In this step, we convert the value that contains spelling error or typo. For this purpose, we use look up table as shown in table 4. Thus, it is not necessary to consider multiword expressions [8] in normalization and tokenization process.

**Table 4.** The example of look up table to normalize token

| Spelling error or typo | Correct Term |
|---|---|
| 'a r hakim', 'ar hakim', 'a hakim' | arif rahman hakim |
| 'h m joni', 'hm joni', 'h m jhoni', 'hm jhoni' | moh joni |
| 'prof h m yamin', 'prof hm yamin', 'h m yamin', 'hm yamin' | moh yamin |
| 'sisingamaharaja', 'sisingamaraja', 'sm raja' | sisingamangaraja |
| 'letjend s parman', 'letjen s parman', 'jendral s parman', 'jend s parman', 'sparman' | s parman |

*2.3.3 Attribute parsing*
The next step is attribute parsing to form more detailed information. This step extracts an attribute that might contain more than one information. For example, the attribute that store company address information usually consists of more than one information such as street name, alley name, RT/RW (group of several houses within a small area), house/office number, and postal code.

The purpose of this process is obtaining the other information which are available within an attribute value. Parsed attribute value will be used in record pair comparison. Attribute parsing example is shown in table 5.

**Table 5.** The example of look up table to normalize token

| Spelling error or typo | Correct Term |
|---|---|
| 'a r hakim', 'ar hakim', 'a hakim' | arif rahman hakim |
| 'h m joni', 'hm joni', 'h m jhoni', 'hm jhoni' | moh joni |
| 'prof h m yamin', 'prof hm yamin', 'h m yamin', 'hm yamin' | moh yamin |
| 'sisingamaharaja', 'sisingamaraja', 'sm raja' | sisingamangaraja |
| 'letjend s parman', 'letjen s parman', 'jendral s parman', 'jend s parman', 'sparman' | s parman |

*2.4 Indexing*
The purpose of indexing step is to reduce the number of record pairs which will be compared. This step is done by removing the record pairs which are impossible to be matched. Common approach to index the records is by processing all the records and cluster them to several blocks based on certain criteria [9]. These criteria are called blocking key or sorting key. One of the important aspects to index the records does not lie on the method, but the definition of blocking key which cluster the records into similar block [9]. In this research, we used company name, company owner or applicant name, company address and phone number for blocking keys.

### *2.5 Record Pair Comparison*
The next step of data matching process is record pair comparison to calculate the similarity level of both records' attributes. To compare two attributes, we use Jaro-Wrinkler edit distance. Basically, edit distance is defined as smallest number of insertions, deletions and substitutions which are required to convert a string to another string. The number of editing steps between both strings will be the level of similarity.

### *2.6 Classification*
The last step of data matching process is classification from record pairs candidates as the result of indexing and record pair comparison. The classification is done based on similarity level which are saved in vector format. In this research we use support vector machine to classify records. Generally, the higher similarity level of both records, the higher possibility that both entities are the same.

## 3. Discussion
As mentioned earlier, the data are obtained from two institutions namely Badan Lingkungan Hidup (BLH) and Badan Pelayanan Perizinan Terpadu (BPPT) of Medan city. The implementation of this research includes data pre-processing, indexing and record comparison.

The data that will be used in data matching are different in structure and content. Thus, these raw data are required to be cleaned and standardized before commencing the next process. The pre-processing data includes removing unnecessary characters and symbols, standardization and tokenization, and attribute parsing.

Indexing process arranges BLH and BPPT data based on blocking key. We used perusahaan (company name) and pemilik (owner or applicant) as blocking key. Each records will be compared to the next step according to similar value in the attributes perusahaan (company) and pemilik (owner or applicant). The record pairs that are not included in indexing are considered as non-matched. After indexing process, all attributes of record pairs will be compared and yield similarity value. After all the data are ready, data classification can be commencing. In this research, we use support vector machine to classify the data to obtain data similarity. As the result, we obtain 90 record pairs generated from 500 BLH records and 21,754 BPPT records.

## 4. Conclusion
The Badan Lingkungan Hidup (BLH) and Badan Pelayanan Perizinan Terpadu (BPPT) of Medan city are known to share the same companies in both databases. To find the same company, the best way is linking the data with one standard unique key to determine an entity. However, in this case, the unique key to define the same company is not available. Therefore, this research is analyzing the data pre-processing before implementing data matching algorithm. We found that the data are not well prepared for data integration as we found so many differences among both databases. After analyzing data pre-processing we decide to use perusahaan (company name) and pemilik (owner of applicant) as blocking key to support data matching.

## References
[1] Smith D and Elliot M 2005 An Experiment in Naïve Bayesian Record Linkage *Proc. of the 55th Session of the Int. Statistical Institute*

[2] Amalia A, Gunawan D, Najwan A and Meirina F 2016 Focused crawler for the acquisition of health articles *2016 Int. Conf. on Data and Software Engineering (ICoDSE)* pp 1–6

[3] Gunawan D, Amalia A and Najwan A 2017 Improving Data Collection on Article Clustering by Using Distributed Focused Crawler *J. Comput. Appl. Informatics* **1** pp 39–50

[4] Churches T, Christen P, Lim K and Zhu J X 2002 Preparation of name and address data for record linkage using hidden Markov models *BMC Med. Inform. Decis. Mak.* **2** 1 p 9

[5] Khan S I and Hoque A S M L 2017 Health data integration with Secured Record Linkage: A practical solution for Bangladesh and other developing countries *2017 Int. Conf. on*

*Networking, Systems and Security (NSysS)* pp 156–161

[6] Christen P 2012 *Data Matching Concepts and Techniques for Record Linkage, Entity Resolution, and Duplicate Detection Springer* Springer-Verlag Berlin Heidelberg

[7] De Jonge E and Van Der Loo M 2013 *An introduction to data cleaning with R* Statistics Netherlands

[8] Gunawan D, Amalia A and Charisma I 2016 Automatic extraction of multiword expression candidates for Indonesian language *2016 6th IEEE Int. Conf. on Control System, Computing and Engineering (ICCSCE)* pp 304–309

[9] Christen P 2012 A Survey of Indexing Techniques for Scalable Record Linkage and Deduplication *IEEE Trans. Knowl. Data Eng.* **24** 9 pp 1537–1555

IOP Conf. Series: Journal of Physics: Conf. Series **978** (2018) 012119 doi:10.1088/1742-6596/978/1/012119

# Building automatic customer complaints filtering application based on Twitter in Bahasa Indonesia

**D Gunawan[1*], R P Siregar[1], R F Rahmat[1] and A Amalia[2]**

[1] Department of Information Technology, Universitas Sumatera Utara, Jl. dr. Mansur No. 9 Kampus USU Medan 20155

[2] Department of Computer Science, Universitas Sumatera Utara, Jl. dr. Mansur No. 9 Kampus USU Medan 20155

*Email: danigunawan@usu.ac.id

**Abstract**. Twitter has become a media to provide communication between a company with its customers. The average number of Twitter active users monthly is 330 million. A lot of companies realize the potential of Twitter to establish good relationship with their customers. Therefore, they usually have one official Twitter account to act as customer care division. In Indonesia, one of the company that utilizes the potential of Twitter to reach their customers is PT Telkom. PT Telkom has an official customer service account (called @TelkomCare) to receive customers' problem. However, because of this account is open for public, Twitter users might post all kind of messages (not only complaints) to Telkom Care account. This leads to a problem that the Telkom Care account contains not only the customer complaints but also compliment and ordinary message. Furthermore, the complaints should be distributed to relevant division such as "Indihome", "Telkomsel", "UseeTV", and "Telepon" based on the content of the message. This research built the application that automatically filter twitter post messages into several pre-defined categories (based on existing divisions) using Naïve Bayes algorithm. This research is done by collecting Twitter message, data cleaning, data pre-processing, training and testing data, and evaluate the classification result. This research yields 97% accuracy to classify Twitter message into the categories mentioned earlier.

## 1. Introduction

Customer care holds an important role in a company. Nowadays, along with the development of information technology, there are alternative ways to communicate between the company and customer. For example, Twitter has become the alternative communication to the customer care division as the number of its active users reach 330 million monthly (as per third quarter of 2017) [1]. Usually, a company has one official Twitter account to communicate between the customer and the company. In addition to the convenience offered by using Twitter, it has a drawback. For a well-known company, the twitter posts can reach thousands a day. The problem is the twitter posts to the customer care division are not only customer complaints, but also other twitter posts that are not related to customer care. This requires more manpower to manually select the related twitter posts manually. Therefore, the company requires an application that can automatically filter twitter posts to related division in customer care.

This research uses a twitter account called @TelkomCare which is owned by a well-known telecommunication company in Indonesia, PT Telkom. It has four divisions to handle the customer complaints such as "Indihome", "Telkomsel", "UseeTV", and "Telepon". This research will filter the twitter posts to account @telkomcare to related categories according to the message contents. To classify those messages, we use Naïve Bayes algorithm.

Naïve Bayes algorithm has been used in many applications or researches. For example, previous research [2] used Naïve Bayes to implement focused crawler. This algorithm was used to classify the web contents in order to obtain a specific topic. Later, the focused crawler was improved with distributed system to obtain more contents [3]. Both researches used bag of words method to classify the contents. Meanwhile, as in our research the number of the same keywords does not influence the meaning of the twitter post, then we do not calculate the number of keywords appearance to classify the contents. However, we do not consider the usage of multiword expression [4] to extract the keywords as used in content clustering research [5].

Previous study classified twitter users based on their posts into three classes, such as political, ethnical or business-oriented [6]. They used Gradient Boosted Decision Trees (GBDT) to classify the twitter posts. This research yield 76% accuracy.

Other study used Naïve Bayes classifier to filter twitter timeline (front page of each twitter account) to five categories such as news, opinions, deals, events and private messages [7]. This research used eight features selection to eliminate twitter posts which are not written in English, less than three words and only contain URL address without other information. The accuracy of the system with eight features selection reached 90%.

Recently, a study also used Naïve Bayes algorithm to classify emotion in twitter posts based on the contents [8]. There are six categories of emotion such as happiness, fear, sadness, anger, surprise, and disgust. The evaluation result of this research is 83.10%.

This paper is organized as follow: the first section is the introduction about the background of the conducted research. Then, section two discusses the research methodology to obtain the result. Section three discusses the result of this research and the last section is conclusion.

## 2. Research Methodology

This research was done by conducting several steps. The first step is collecting and analyzing the data sources. In this step we examined the pattern of twitter post by the users. After collecting and analyzing the data, the next step is extracting keywords to determine each customer care division. Next, we implement text pre-processing to twitter post in order to match the required format for data training. After implementing text pre-processing, we clean the unwanted data such as the twitter post which has only two or three words. Then the data will be divided into two types, data training and data testing. The last step, we built the application to show the demonstration of the twitter post classification. The details of the research methodology are as follow:

### 2.1 Data Source

We obtained the data from Telkom Care twitter that contain the posts from Telkom subscribers through Twitter. The data is divided into two parts, such as training data and testing data. Training data is used to train the system in order to classify new tweets which are not defined in certain class yet. Testing data is used to test the system in order to confirm the classification method. To define the classification label, we use several keywords that are determined by manual observation.

This research uses 120 tweets (user posts in Twitter) for training data. The training data are obtained from the official account of customer care PT Telkom Indonesia, @TelkomCare. The data are obtained by using Streaming Twitter API method. We were collecting tweet for 60 minutes.

After collecting tweets from the official account of PT Telkom, the next step is class definition. We obtained 250 tweets in 60 minutes. All this tweets are mixed between the potential data training or not. We selected 120 unique tweets manually which have potential content to represent the classification.

After manually selected the potential data training, we divided them into several categories according to the current condition of PT Telkom division. There are four classes such as Indihome, UseeTV, Telkomsel, Telepon. The number of training data for each class is shown in table 1. We did not use the data training for uncategorized tweets. The tweet that is not categorized in one of the four classes above will automatically defined as uncategorized. The example of data training is shown in table 2.

**Table 1.** Tweet pre-defined classification

| Class | Number of Tweets |
|---|---|
| Indi Home | 39 |
| UseeTV | 27 |
| Telkomsel | 26 |
| Telepon | 27 |

**Table 2.** Tweet pre-defined classification

| Tweet | Category |
|---|---|
| @TelkomCare cara berlangganan indihome gimana? | 39 |
| @TelkomCare harga untuk paket telkomsel flash reguler 1 gb berapa ya? | 27 |
| @TelkomCare min ini iflix gabisa dibuka yg di tv saya. Disuruh update. Gmana updatenya klo di tv? | 26 |
| @TelkomCare mohon dibantu saat ini saluran telpon di tempat saya tidak ada tone sehingga speedy di tempat saya mati. | 27 |

### 2.2 Extracting Keywords

Keywords is the parameter used to train the system in determining the suitable class for each tweet. This research had through several steps to determine the keywords. The first step was discussion with the twitter administrator about the words which would be the reference to determine subscriber's complaint. As the result, there are four main keywords to represent each product, such as Indihome, UseeTV, Telkom, and Telepon.

The first implementation of the classification using these keywords, the result is still messy. Generally, every tweet will include one of those keywords. However, there are a lot of cases, that the system cannot classify it because of lack of the keywords inside the tweet. And also, there are tweets classified as uncategorized, although if we observe the tweet semantically, it is included in one of the four classes. Therefore, other keywords are required to handle the issue.

The next step was conducting a new research by pulling the tweets as much as possible from the users who mention Telkom account. After that, the next step is extracting frequently used words which is potentially used to be supporting main keywords to determine tweet classification. From this process we obtain 202 keywords. By adding these keywords, we obtain much better result than the first implementation. However, for certain cases there are tweets which contain strong keywords. To solve the problem, the system requires a learning machine. In this research we implement Naïve Bayes classifier.

### 2.3 Text Pre-processing

Preprocessing step is required before we conduct training dataset and testing dataset. There are some steps to minimize noise by making the structure uniform and reducing widen dimension. After the dataset pass through some pre-process steps, then the dataset will be represented in matrix. The whole pre-processing steps are filtering, case folding, tokenizing, stemming, and normalization. Those are explained below:

IOP Conf. Series: Journal of Physics: Conf. Series **978** (2018) 012119 doi:10.1088/1742-6596/978/1/012119

A. Filtering

Twitter is an expressive social media. It means Twitter allows users to insert symbols such as emoticon and special characters in a post to express their feelings. These expressive symbols tend to cause the wrong interpretation in classification system. Thus, the special characters should be eliminated before processing the classification. For example, the characters such as # and @ are common characters in Twitter.

B. Case folding

Each twitter posting may contain uppercase or lowercase. This condition will influence text pre-processing because between "A" and "a" will be considered as different character. To overcome this condition, all the characters will be converted to lowercase.

C. Tokenizing

Tokenizing is the process to extract the tweet to separated words. This process is done as a preliminary process to obtain the keywords of each tweet. We use space as delimiter to split the tweet to words.

D. Stemming

Each token in tweet may contain the same root word in different format. For example, the root word may have suffix. These words may be recognized as different word because of different word form. To overcome this situation, we have to convert all words in to its root word. We use existing stemming algorithm for Bahasa Indonesia called Enhanced Confix Stripping Stemmer [9]. For example, the words "berlangganan" and "ditagihkan" become "langgan" and "tagih".

E. Spelling normalization

After normalization process is completed, the next step is the keyword search from the extracted tokens. Tokens will be checked with the previously defined keywords. If the token matches the keywords, then it will be served for the next process. The token that does not match the existing keywords will be eliminated. We found that there is a problem in matching the token with the available keywords. Some tokens are typo or abbreviated so that they are not perfectly match with the keywords. To overcome this problem, we did the spelling normalization by implementing levenshtein algorithm. This algorithm will test the similarity between the token with the keywords.

**Table 3.** Data modeling

| No. | Keywords | Label | Category (Probability) |
|---|---|---|---|
| 1 | saldo | x1 | Telkomsel |
| 2 | indihome | x2 | Indihome |
| 3 | maintenance | x3 | Indihome, Telepon, Telkomsel, UseeTV |
| 4 | telepon | x4 | Telepon |
| 5 | jaringan | x5 | Indihome, Telepon, Telkomsel, UseeTV |
| 6 | useetv | x6 | UseeTV |
| ... | ... | ... | ... |
| 201 | speed | x201 | Indihome, Telkomsel, UseeTV |
| 202 | grapari | x202 | Telkomsel |

IOP Conf. Series: Journal of Physics: Conf. Series **978** (2018) 012119 doi:10.1088/1742-6596/978/1/012119

2.4 Data Modeling

All the training and testing data are required to be modeled into matrix in order to form structured data. The structured data will avoid the mistake in calculation. The example of data modeling for training data is shown in table 3. The data modeling for testing data is similar with the example in table 3, unless the value in the class column is not determined yet.

2.5 Classifier

The existing problem is the customer complaint posts are divided manually by the human. This process consumes a lot of time because there are also many posts that are not related to the complaints. This will cause the complaints are not handled as soon as possible by the customer care staffs. There are many existing classifiers that can classify the posts. This research uses Naïve Bayes Classifier because of the its speed in classification process.

2.6 Evaluation

To evaluate the result of classification method, we use accuracy, precision, recall and F-score. As shown in formula (1), the accuracy is obtained by dividing the total number of correct classification data with the total number of data testing.

$$Accuracy = \frac{total\ number\ of\ correct\ classification}{total\ number\ of\ data\ testing} x\ 100\% \quad (1)$$

Precision and recall can be mapped in matrix as shown in table 4. According to the matrix, recall and precision are formulated as in (2) and (3).

**Table 4.** Precision and recall

| | Relevant | Irrelevant | Total |
|---|---|---|---|
| Found | a (hits) | b (noise) | a+b |
| Not found | c (misses) | d (rejected) | c+d |
| Total | a+c | b+d | a+b+c+d |

$$Recall = \frac{a}{a+c} x\ 100\% \quad (2)$$

$$Precision = \frac{a}{a+b} x\ 100\% \quad (3)$$

## 3. Result and Discussion

3.1 Classification Result

This research uses 660 twitter posts which mention @telkomcare as the data testing. These data testing are analyzed manually by human to obtain the actual classification. We found that Indihome category has 198 posts, UseeTV category has 122 posts, Telkomsel category has 110 posts, Telepon category has 110 post and there are 120 uncategorized posts. Next, we implemented text-preprocessing to all the data testing. Finally, we applied Naïve Bayes classifier to obtain the automatic classification of twitter posts. The result is shown in table 5.

IOP Conf. Series: Journal of Physics: Conf. Series **978** (2018) 012119 doi:10.1088/1742-6596/978/1/012119

**Table 5.** Classification result

| Category | Relevant | Irrelevant | Number of classified posts | Actual number of the data |
|---|---|---|---|---|
| Indihome | 178 | 2 | 180 | 198 |
| UseeTV | 118 | 2 | 120 | 122 |
| Telkomsel | 110 | 10 | 120 | 110 |
| Telepon | 112 | 8 | 120 | 110 |
| Uncategorized | 120 | 0 | 120 | 120 |

According to table 5, the system identifies 180 posts in Indihome category, 120 posts in UseeTV category, 120 posts in Telkomsel category, 120 posts in Telepon category and 120 posts in uncategorized. However, from the manual analysis, we found that 178 posts are related to Indihome category, 118 posts are related to UseeTV category, 110 posts are related to Telkomsel, 112 posts are related to Telepon category and 120 posts are uncategorized. The example of irrelevant posts classification is shown in table 6.

**Table 6.** Irrelevant posts classification

| Twitter post | Classification Result | Actual Classification | Status |
|---|---|---|---|
| @TelkomCare min saya sdh pakai indihome unlimited , tapi usetv nya kok masih ada yg terkunci ya , apa harus pasang usetv lg biar chanel ful | Indihome | UseeTV | False positive |
| @TelkomCare bgmn cara mendapatkan IMC? Sy bermigrasi dr speedy ke indihome, qualified tidak? Ato bisa tuker poin indihome tidak? | UseeTV | Indihome | False positive |
| @TelkomCare apakah paket internet saya bisa digunakan untuk mendaftar paket indihome telkomsel mania? | Telkomsel | Indihome | False positive |

After analyzing the classification result, we found 20 twitter posts are identified as "false positive". This result is influenced by the keywords and the number of data training. For example, the classification result in the table 6 shows that the twitter post in the third row is classified as Telkomsel category. However, the actual classification is Indihome. This is influenced by the number of keywords between the Telkomsel and Indihome, such as "internet", "paket", "indihome", and "telkomsel". Those keywords will produce almost the same probability between Telkomsel and Indihome category. The number of data training will influence Naïve Bayes algorithm to decide the classification result.

3.2 Evaluation

The system performance can be evaluated by using accuracy, precision, recall and F-score. According to the classification result, we calculate the accuracy according to the formula (1) as follow:

$$Accuracy = \frac{636}{660} x100\% = 96.3\%$$

As mentioned earlier, to determine the actual classification we use manual classification approach. This manual classification is obtained by discussion and interview with the person in charge. The result of evaluation result in table 7 shows that the average of precision value is 97%, recall is 97.6% and F-score is 97.2%. According to the theory by Pao [10], this system can be determined as an ideal system because of the ratio of recall and precision have nearly the same value.

**Table 7.** Evaluation result

| Category | Relevant (a) | Irrelevant (b) | Total (a+b) | Not found (c) | Total (a+c) | Recall [a/(a+c)] x100% | Precision [a/(a+b)] x100% | F-score |
|---|---|---|---|---|---|---|---|---|
| Indihome | 178 | 2 | 180 | 18 | 192 | 91% | 99% | 95% |
| UseeTV | 118 | 2 | 120 | 2 | 122 | 97% | 98.3% | 98% |
| Telkomsel | 110 | 10 | 120 | 0 | 110 | 100% | 92% | 96% |
| Telepon | 112 | 8 | 120 | 0 | 112 | 100% | 93.4% | 97% |
| Uncategorized | 120 | 0 | 120 | 0 | 120 | 100% | 100% | 100% |
| | | | | Average | | 97.6% | 97% | 97.2% |

## 4. Conclusion

According to the result, this application is able to classify twitter posts into four main classes, namely "Indihome","UseeTV", "Telkomsel" and "Telepon" with the accuracy of 97%. This research yields 97% and 97.6% of the precision and recall average respectively. According to the precision and recall, the F-score calculation yields 97.2%. This means this system can be determined as an ideal system because of the precision and recall almost have the same value.

## References

[1] Statista 2017 Number of monthly active Twitter users worldwide from 1st quarter 2010 to 3rd quarter 2017 (in millions) Available: https://www.statista.com/statistics/282087/number-of-monthly-active-twitter-users/

[2] Amalia A, Gunawan D, Najwan A, and Meirina F 2016 Focused crawler for the acquisition of health articles *2016 Int. Conf. on Data and Software Engineering (ICoDSE)* pp 1–6

[3] Gunawan D, Amalia A, and Najwan A 2017 Improving Data Collection on Article Clustering by Using Distributed Focused Crawler *J. Comput. Appl. Informatics* **1** pp 39–50

[4] Gunawan D, Amalia A, and Charisma I 2016 Automatic extraction of multiword expression candidates for Indonesian language *6th IEEE Int. Conf. on Control System, Computing and Engineering (ICCSCE)* pp 304–309

[5] Gunawan D, Amalia A, and Charisma I 2017 Clustering Articles in Bahasa Indonesia Using Self-Organizing Map *2017 Int. Conf. on Electrical Engineering and Informatics*

[6] Pennacchiotti M and Popescu A 2011 A Machine Learning Approach to Twitter User Classification *Proc. of the Fifth Int. AAAI Conf. on Weblogs and Social Media* pp 281–288

[7] Sriram B, Fuhry D, Demir E, Ferhatosmanoglu H, and Demirbas M 2010 Short Text Classification in Twitter to Improve Information Filtering *SIGIR '10 Proc. of the 33rd int. ACM SIGIR conf. on Research and development in information retrieval* pp 841–842

[8] Wikarsa L and Thahir S N 2015 A text mining application of emotion classifications of Twitter's users using Naïve Bayes method *2015 1st Int. Conf. on Wireless and Telematics (ICWT)* pp 1–6

[9] Adriani M, Asian J, Nazief B, and Tahaghoghi S M M 2007 Stemming Indonesian: A confix-stripping approach *ACM Trans. Asian Lang. Inf. Process.* **6** 4 pp 1–33

[10] Pao M L 1989 *Concepts of Information Retrieval* Libraries Unltd Incorporated

# The Implementation of Cosine Similarity to Calculate Text Relevance between Two Documents

**D Gunawan[1*], C A Sembiring[1], M A Budiman[2]**

[1] Department of Information Technology, Universitas Sumatera Utara, Jl. dr. Mansur No. 9 Kampus USU Medan 20155
[2] Department of Computer Science, Universitas Sumatera Utara, Jl. dr. Mansur No. 9 Kampus USU Medan 20155

*Email: danigunawan@usu.ac.id

**Abstract**. Rapidly increasing number of web pages or documents leads to topic specific filtering in order to find web pages or documents efficiently. This is a preliminary research that uses cosine similarity to implement text relevance in order to find topic specific document. This research is divided into three parts. The first part is text-preprocessing. In this part, the punctuation in a document will be removed, then convert the document to lower case, implement stop word removal and then extracting the root word by using Porter Stemming algorithm. The second part is keywords weighting. Keyword weighting will be used by the next part, the text relevance calculation. Text relevance calculation will result the value between 0 and 1. The closer value to 1, then both documents are more related, vice versa.

## 1. Introduction

The rapid stream of information emerges a new challenge to fetch specific information. As an abundance of web pages and documents are still increasing, focused crawler [1] is an alternative to gather the topic specific information. According to previously conducted research, filtering specific information can be done by using text relevance, classification or clustering algorithm. Besides focused crawling, text relevance can be used to find related web pages or documents that can be used to automatic text summarization [2], especially multi-documents summarization. Text relevance refers to the match between the goal related to the reader's goal [3].

Previous research implements web pages similarity in focused crawler [4][5]. This research utilizes a classification algorithm called Naïve Bayes to provide web pages' similarity. It does not calculate text relevance between web pages. This research classifies the web pages in the same category by providing bag of words.

In addition, a research about the efficient focused crawling strategy has been conducted [6]. This research proposes the combination of link structure and content similarity to crawl pages in certain topic. This research automatically extracted the topic feature vector and judging the topic similarity by using the combination of link structure and page content. Their experiment proves that their strategy has successfully increate the precision of fetching topic pages.

Another research utilizes clustering algorithm called self-organizing map to gather the similar web pages [7] which also can be used to recognize character [8]. This approach is a bit similar with [4] and [5] because it uses keywords to determine the web pages similarity. However, the keywords are not predetermined. The keywords are generated based on term frequency – inverse document frequency (TF-IDF). It also considers multiword expression (MWE) candidates [9] to determine the keywords.

This research is a preliminary research to obtain suitable text relevance algorithm to find related documents. It demonstrates the implementation of text relevance calculation between two documents by using cosine similarity. Cosine similarity is useful to measure the similarity between two documents based on terms of their subject matter [10]. The organization of this paper is as follow: the first section is the introduction about the preliminary research. Section two discusses the calculation of text relevance. Section three discusses the implementation of text relevance. The last section is conclusion of the research.

## 2. Methodology

This research is intended to demonstrate the implementation of text relevance calculation between a reference document with the second documents. Reference document is a document that is assigned as a reference to obtain the similarity between two documents. Second document is the document that will be compared to reference document to obtain its relevancy value with the reference document. We conduct several process to obtain text relevance between two documents. First process is text pre-processing. The result of this process is keywords of each document. The result will be used to calculate keywords weighting. The next process is text relevance calculation. In this process we calculate the relevance between two document by conducting cosine similarity.

Text pre-processing consists of several steps. The first step is punctuation removal. This punctuation will not influence the calculation as we do not include semantic. The next step is applying case folding that will turn the sentences into lower case. This action is required because there is no difference in meaning between “cell” and “CELL”. This process is continued with tokenizing. After tokenizing, every token will be compared to the words in stop word list. All matched words will be removed and the rest will be the keywords to the related document. By implementing the stop word removal, we might reduce the time to the next process. As we do not consider semantic, then the word which has prefix will be turned into its root word. This process is called stemming. In this research we use Porter Stemming algorithm [11].

The calculation of text relevance requires keyword weighting. To obtain keyword weighting, we count the number of the same words. However, as there is possibility that the same words come in many forms, we should transform the words to their root word. This issue has been covered in text pre-processing. Formula (1) shows the calculation of keyword weighting, where $w$ is weight of the keyword, $w_i$ is the weight of the keyword i, and $w_{max}$ is the weight of the keywords which has the maximum weight.

$$w = \frac{w_i}{w_{max}} \tag{1}$$

The text relevance is calculated by using cosine similarity. Formula (2) shows the calculation of text relevance between two documents, where $wkre$ is the weight of the reference document, which is obtained from the same keywords between reference document and second document, $wkse$ is the weight of the second document, which is obtained from the same keywords between reference document and the second document, $wkr$ is the weight of the keywords in reference document and $wks$ is the weight of the keywords in second document. As shown in formula (3), $wkre$ is obtained by dividing the sum of the weight of the same keywords between reference document and second document ($\sum kre$) with the maximum keyword weight in the reference document ($wkr_{max}$). Meanwhile, $wkse$ is obtained by dividing the sum of the weight of the same keywords between second document and reference document ($\sum kse$) with the maximum keyword weight in the second document ($wks_{max}$). This calculation is shown in formula (4). As shown in formula (5), $wkr$ is obtained by dividing the sum of the keywords' weight in reference document ($\sum kr$) with the maximum keyword weight in the reference document ($wkr_{max}$). Meanwhile, $wks$ is obtained by dividing the sum of the keywords' weight in second document ($\sum ks$) with the maximum keyword weight in the second document ($wks_{max}$). This calculation is shown in formula (6).

$$relevance = \frac{wkre\,.\,wkse}{\sqrt{wkr^2}\,.\,\sqrt{wks^2}} \tag{2}$$

$$wkre = \frac{\sum kre}{wkr_{max}} \tag{3}$$

$$wkse = \frac{\sum kse}{wks_{max}} \tag{4}$$

$$wkr = \frac{\sum kr}{wkr_{max}} \tag{5}$$

$$wks = \frac{\sum ks}{wks_{max}} \tag{6}$$

## 3. The Implementation

This section will discuss the implementation of text relevance approach to obtain related document. The implementation begins with text pre-processing. In the text pre-processing we will discuss every process involved. Next, we will discuss keywords extraction after the text has been pre-processed. Then we will discuss about normalization of the keywords. The last, we will discuss about text relevance calculation.

Text pre-processing is begun with punctuation removal. Then we apply case folding to set all the letters to lower case. After applying case folding, we tokenize the text into separated tokens. Some of these token will be removed if the term is listed in stop word list. After the tokens are clean without any meaningless term, then the next step is stemming. Stemming process is done to obtain root word that is used to calculate text relevance between two documents. In this research we use Porter Stemming algorithm to stem the tokens.

**Table 1.** Keywords extraction

| Document | Keywords | Category |
|---|---|---|
| Reference document | Adenocarcinoma:9; cell:9; scc:8; xenograft:8; region:7; distribut:6; cto:5; fdb:4; cuatsm:4; necrot:4; origin:3; intratumor:3; high:3; compar:3; overlap:3; accumul:3; live: 3; deriv:2; adduct:2; colon:2; cancer: 2; cucuatsm:2; characterist:2; pimonidazol:2; tumor:2; occur: 2; studi:2; observ:2; lung:2; dualtrac:1; | Biology and medical |
| Second document | Ffphpa:12; uptak:10; amino:8; radiotrac:5; ffet:4; imag:4; studi:4; prepar:3; cancer:3; pet:3; emt:3; compar:3; min:2; crosscoupl:2; asc:2; pdmediat:2; suvmin:2; maximum:2; transport:2; reach:2; cell:2; evalu:2; small:2; anim:2; reaction:2; high:2; radiochem:2; inhibit:2; cellular:2; | Biology and medical |

The result after stemming process is keywords extraction from a document. For example, the first record in table 1 is the topic document that is used as reference by the second document (second record in table 1) to show the relevancy between both documents. The keywords are written in root word and its number of appearance in a document (separated with colon). One keyword and another is separated with semicolon (";"). For example, "cell:9; intratumor:3; cancer:2;" means the term "cell", "intratumor", and "cancer" has been appeared in document as much as 9, 3 and 2 times respectively. These documents are grouped in to the same category namely "Biology and medical".

The next process is calculating weight of the keywords. For example, we will use the data in table 1 for weighting calculation. The weight of the term "cell" and "intratumor" from the reference document is 9 and 3 respectively. This weight is obtained from the number of term appearance in the text. Thus, the term "cell" is known as the maximum weight ($w_{max}$). According to formula (1), the keyword weight of the term "intratumor" ($w_i$) is:

$$w = \frac{3}{9}$$

$$w = 0.33$$

The weight of "intratumor" keyword is 0.33. By using the same formula, we can obtain the keyword weight of the reference document in table 1. The snippet result is shown in table 2.

**Table 2.** Keyword weighting

| **Keyword** | $w_i$ | $w_{max}$ | $w$ |
|---|---|---|---|
| adenocarcinoma | 9 | 9 | 1 |
| cell | 9 | 9 | 1 |
| scc | 8 | 9 | 0.89 |
| xenograft | 8 | 9 | 0.89 |
| region | 7 | 9 | 0.78 |
| distribut | 6 | 9 | 0.67 |
| intratumor | 3 | 9 | 0.33 |
| compar | 3 | 9 | 0.33 |
| pimonidazol | 2 | 9 | 0.22 |
| tumor | 2 | 9 | 0.22 |

Text relevance is calculated by utilizing cosine similarity function as shown in formula (2). The calculation involves the same keywords from the both documents (reference document and second document) to obtain the weight. For example, as shown in table 3, the same keywords from reference document and second document are: cell, high, compar, cancer and study.

**Table 3.** The same keywords between reference and second document

| **Keywords** | **Weight** | |
|---|---|---|
| | **Reference document** | **Second document** |
| Cell | 9 | 2 |
| High | 3 | 2 |
| Compar | 3 | 3 |
| Cancer | 2 | 3 |
| Study | 2 | 4 |

According to table 1, the maximum weight of the reference document is 9 and the maximum weight of the second document is 12. Then, according to formula (3), weight of the reference document ($wkre$) is calculated below:

$$wkre = \frac{9+3+3+2+2}{9} = 2.11$$

Then, the weight of the second document ($wkse$) is calculated according to formula (4). The calculation is shown below:

$$wkse = \frac{2+2+3+3+4}{12} = 1.67$$

Next, the keywords weight in reference document ($wkr$) is calculated according to formula (5). The calculation is shown below:

$$wkr = \frac{110}{9} = 12.22$$

After that, the keywords weight in reference document ($wks$) is calculated according to formula (6). The calculation is shown below:

$$wks = \frac{104}{12} = 8.67$$

After we obtain all the parameters ($wkre$, $wkse$, $wkr$, and $wks$), then we calculate the text relevance according to formula (2). The calculation is shown below:

$$relevance = \frac{2.11\,.\,1.67}{\sqrt{12.11^2}\,.\sqrt{8.67^2}}$$

$$relevance = \frac{2.46}{105.97}$$

$$relevance = \ 0.023$$

In the case of information retrieval, the result of cosine similarity will be between 0 and 1. If the value is closer to 1 (one), then it means the second document is very relevant with the reference document. In this result, the relevance value is 0.023. This means the value is closer to 0 (zero). Therefore, we can conclude that the second document is not relevant to the reference document.

## 4. Conclusion

Focused crawler can be used to filter web pages and documents into topic specific information. Topic specific information can be filtered by using classification, clustering or text relevance method. This research is a preliminary research to demonstrate the implementation of text relevance between two documents. In this research, text relevance is calculated by using cosine similarity, where the value which is closer to 1 is considered as more relevant, vice versa. In the future, research should conduct the text relevance implementation in Bahasa Indonesia.

**References**

[1] Chakrabarti S, van den Berg M, Dom B 1999 Focused crawling: a new approach to topic-specific Web resource discovery *Comput. Networks* **31** 11–16 pp 1623–1640

[2] Gunawan D, Pasaribu A, Rahmat R F, and Budiarto R 2017 Automatic Text Summarization for Indonesian Language Using TextTeaser *IOP Conf. Ser. Mater. Sci. Eng.* **190** 1 p 12048

[3] Hager P J *et al.* 2012 Text Relevance in *Encyclopedia of the Sciences of Learning* Springer US pp 3307–3310

[4] Amalia A, Gunawan D, Nazwan A, and Meirina F 2016 Focused crawler for the acquisition of health articles *2016 Int. Conf. on Data and Software Engineering (ICoDSE)* pp 1–6

[5] Gunawan D, Amalia, and Najwan A 2017 Improving Data Collection on Article Clustering by Using Distributed Focused Crawler *J. Comput. Appl. Informatics* **8** 1 pp 39–50

[6] Cheng Q, Beizhan W, and Pianpian W 2008 Efficient focused crawling strategy using combination of link structure and content similarity *Proc. 2008 IEEE Int. Symp. IT Med. Educ. ITME 2008*

[7] Gunawan D, Amalia A, and Charisma I 2017 Clustering Articles in Bahasa Indonesia Using Self-Organizing Map *2017 Int. Conf. on Electrical Engineering and Informatics*

[8] Gunawan D, Arisandi D, Ginting F M, Rahmat R F, and Amalia A 2017 Russian Character Recognition using Self-Organizing Map *J. Phys. Conf. Ser.* **801** 1 p 12040

[9] Gunawan D, Amalia A, and Charisma I 2016 Automatic extraction of multiword expression candidates for Indonesian language *2016 6th IEEE Int. Conf. on Control System, Computing and Engineering (ICCSCE)* pp 304–309

[10] Singhal A 2001 Modern Information Retrieval: A Brief Overview *Bulletin of the Technical Committee on Data Engineering* pp 35–43

[11] Porter M 2006 "The Porter Stemming Algorithm" Available: https://tartarus.org/martin/PorterStemmer/

IOP Conf. Series: Journal of Physics: Conf. Series **978** (2018) 012121 doi:10.1088/1742-6596/978/1/012121

# SSL/TLS Vulnerability Detection Using Black Box Approach

**D Gunawan*, E H Sitorus, R F Rahmat, A Hizriadi**

Department of Information Technology, Universitas Sumatera Utara, Jl. dr. Mansur No. 9 Kampus USU Medan 20155

*Email: danigunawan@usu.ac.id

**Abstract**. Socket Secure Layer (SSL) and Transport Layer Security (TLS) are cryptographic protocols that provide data encryption to secure the communication over a network. However, in some cases, there are vulnerability found in the implementation of SSL/TLS because of weak cipher key, certificate validation error or session handling error. One of the most vulnerable SSL/TLS bugs is heartbleed. As the security is essential in data communication, this research aims to build a scanner that detect the SSL/TLS vulnerability by using black box approach. This research will focus on heartbleed case. In addition, this research also gathers information about existing SSL in the server. The black box approach is used to test the output of a system without knowing the process inside the system itself. For testing purpose, this research scanned websites and found that some of the websites still have SSL/TLS vulnerability. Thus, the black box approach can be used to detect the vulnerability without considering the source code and the process inside the application.

## 1. Introduction

Security plays crucial role in data communication. Socket Secure Layer (SSL) and Transport Layer Security (TLS) are cryptographic protocols that are intended to secure data communication over a network by providing data encryption. However, security threat not only occurs over data communication. For example, heartbleed [1] as one of the most vulnerable SSL/TLS bugs in OpenSSL is caused by the absence of payload length checking by the server. By exploiting this vulnerability, an attacker can request up to 64,000 characters. This research focus on heartbleed case to demonstrate black box approach in detecting SSL/TLS vulnerability. In addition, this research also implements additional SSL verification to gather information about existing SSL in the server. Those are the type of hashing algorithm namely SHA1 and the self-signed status. According to Netcraft and Symantec, SHA1 is known as "insecure" hashing algorithm that is still being used by nearly a million of SSL certificates around the globe [2][3]. Moreover, the self-signed status of SSL certificate is known as a bad practice because it only provides encryption without authentication. There is no way to authenticate the server because the certificate is issued by the server itself, not by the trusted Certificate Authority (CA).

Recent research using black box approach [4] to detect weakness and to test system effectivity in detecting vulnerability such as cross-site scripting, SQL injection and so on. Another research utilized black box approach to identify web application vulnerability and group the attacking scenarios [5]. This method was proposed to reduce the number of false positives. By using black box method, this allow the application to find a new page in the web which may contain vulnerability.

Another research proposed clustering approach to find vulnerabilities in web applications [6]. This research is proved as the alternative utilization of clustering algorithm [7]. This research was not intended to search the website vulnerability. However, this research analyses the structure of a web application, group several application and then finding anomaly from the structure.

This paper is organized as follow: the first section is the introduction about the background of the conducted research. Then, section two discusses the process inside the black box approach. Section three discusses the implementation of the black box approach. The last section is conclusion.

## 2. Research Methodology

This research is intended to detect Socket Secure Layer (SSL) or Transport Layer Security (TLS) vulnerability in web application by applying black box approach. Black box approach will apply a series of procedures to verify and detect SSL/TLS vulnerability in web applications. This research will focus on heartbleed vulnerability and detecting SSL/TLS version. The black box testing includes two procedures in order to scan a web application, such as heartbleed scan and SSL verification.

The first procedure, scanning heartbleed requires three additional steps: handshake, attack scenario and vulnerability identification [8]. These steps are depicted in figure 1.

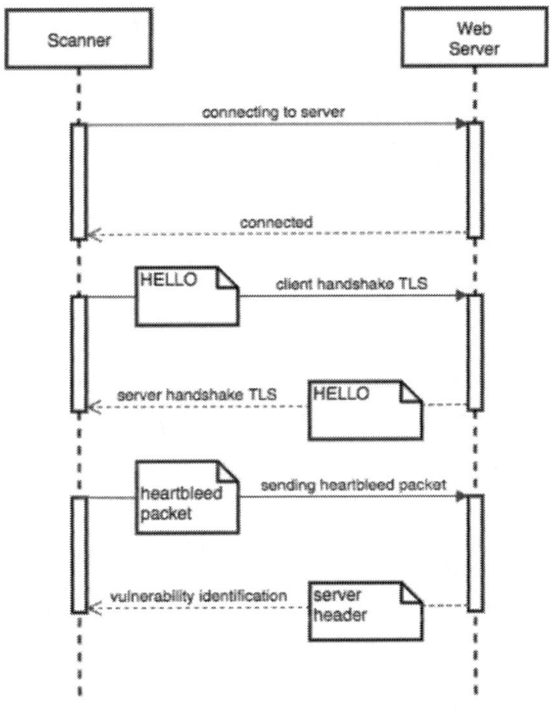


**Figure 1.** Heartbleed scanning process

A. Handshake

The first step is connecting to the server based on certain Uniform Resource Locator (URL). The handshake process will negotiate automatically and then dynamically determine parameters to build communication channel between normal entities before the communication through that channel is begun. This step is required to ensure the communication has been established between server and clients. In this step scanner will send a message to web server and ensure the channel or protocol that will be used between client and server when communicating. As shown in figure 1, the handshake process is started with a scanner sends "HELLO" message to the server. If the handshake process is successful, the server will response with "HELLO" message as well.

B. Attack Scenario

Figure 1 shows that attack scenario is commencing after handshake process and the server has responded it to the scanner. In this scenario, scanner will deliver heartbleed packet in hexadecimal format, which will be converted to byte before it is sent to the server. This heartbleed packet contains content type, TLS version, length, handshake type and payload length. The attack process is begun by deceiving heartbeat extension in the server. The scanner will send response in three bytes but require response in more than three bytes.

C. Vulnerability Identification

The next step is the scanner will read the web server response to the heartbleed packet that is sent previously. Web server will response header parameters which contain type, payload and length. Type and payload value will determine that the server is vulnerable to heartbleed or not. If the server is determined as vulnerable, the scanner (or in the real world is the attacker) may read the hexadecimal dump memory to view memory server contents.

The example of header data that will be sent by the server to the scanner is as follow:

- Type = 24 and payload > 3. This means that the server is vulnerable to heartbleed bug. Server will response with the memory server contents.
- Type = 24 and payload < 3. This means that the server is vulnerable because the server response the heartbleed request by the scanner.
- Type 21 or type = 0. This means server error or server good.

The second procedure is SSL verification. SSL is used to provide data encryption in data communication. This is an additional procedure which is used to gather the information about existing SSL in the server. By gathering this information, we will know the type of signature algorithm in the server, signature expiry date, and certificate chain. The type of signature algorithm determines the strength of data encryption. Signature expiry date will cause an error and warning that the service is not guaranteed to be secured. Certificate chain which is used by client and server also determines the strength the SSL security. Certificate can be issued by the Certificate Authority (CA) or by the server itself (also known as self-signed certificate). To ensure that a certificate is issued by the CA, the scanner should verify *subjectdn* and *issuerdn*.

Therefore, the scanner will try to apply private and public key verification to obtain the information about certificate chain status. SSL verification can also be used to obtain the type of cryptography algorithm which is applied in SSL certificate to provide secure communication channel. The information gathered by the second procedure can be used as analysis and supporting the first procedure. Although the result of the scanning shows that the website or web application is still using obsolete signature algorithm or self-signed certificate, this does not mean that the website or web application is very insecure. This means that the website or web application requires improvement in certificate algorithm or using the certificate issued by the CA.

## 3. The Implementation of SSL/TLS Vulnerability Detection

This section will discuss the implementation of the vulnerability detection by using the scanner. The vulnerability detection is tested to the secure websites and insecure websites. This experiment shows that the SSL/TLS vulnerability especially for heartbleed bug can be detected without knowing how the target's business process is running. Heartbleed occurs because of a bug in request handling by the heartbeat extension in OpenSSL. This extension is used to maintain client and server communication. Because of this bug, the server does not check the message and the length of the message which are requested by the client. Therefore, client can freely obtain the server memory contents by requesting certain payload length.

The result of the implementation of SSL/TLS vulnerability detection is shown in table 1. These data are taken in the late 2015. According to the data testing, there are several websites that are still suffering to heartbleed vulnerability. In addition, several websites are still using "insecure" SHA1 algorithm and using self-signed certificate.

**Table 1.** Scanner testing result

| Web Address | Heartbleed | SSL Validation | |
|---|---|---|---|
| | | SHA1 | Self-signed |
| https://www.koreabridge.net | ✓ | ✓ | ✓ |
| https://www.voobys.com | ✓ | ✓ | ✓ |
| https://worthytoshare.net | ✓ | ✓ | ✓ |
| https://www.uns.ac.id | - | ✓ | ✓ |
| https://www.ugm.ac.id | - | ✓ | - |
| https://www.ui.ac.id | - | ✓ | - |
| https://mail.usu.ac.id | - | ✓ | - |
| https://news.detik.com | - | ✓ | - |
| https://www.olx.co.id | - | - | - |
| https://www.its.ac.id | - | - | - |
| https://www.lazada.co.id | - | - | - |
| https://www.kaskus.co.id | - | - | - |
| https://www.bukalapak.com | - | - | - |

## 4. Conclusion

Socket Secure Layer (SSL) and Transport Layer Security (TLS) are used to secure the data communication. However, some websites or web applications might use obsolete version of SSL/TLS that may lead to security vulnerability. This research shows black box approach can be used to test or detect SSL/TLS vulnerability without knowing business process inside the websites or web applications. Black box approach can be used as complement to another testing scheme.

## References

[1] Durumeric Z *et al* 2014 The Matter of Heartbleed *Proc. of the 2014 Conf. on Internet Measurement Conference - IMC '14* pp 475–488

[2] Netcraft 2015 One million SSL certificates still using 'insecure' SHA-1 algorithm Available: https://news.netcraft.com/archives/2015/10/19/one-million-ssl-certificates-still-using-insecure-sha-1-algorithm.html

[3] Symantec 2014 SHA1 certificate shown as insecure or with mix content warning on Google Chrome 39 Available: https://www.symantec.com/connect/blogs/sha1-certificate-shown-insecure-or-mix-content-warning-google-chrome-39

[4] Bau J, Bursztein E, Gupta D, and Mitchell J 2010 State of the Art: Automated Black-Box Web Application Vulnerability Testing *2010 IEEE Symposium on Security and Privacy* pp 332–345

[5] Akrout R, Alata E, Kaaniche M, and Nicomette V 2014 An automated black box approach for web vulnerability identification and attack scenario generation *J. Brazilian Comput. Soc.* **20** 1 p 4

[6] Suhina V, Groš S, and Kalafatić Z 2008 Detecting vulnerabilities in Web applications by clustering Web pages *31st Int. Convention MIPRO 2008*

[7] Gunawan D, Amalia A, and Charisma I 2017 Clustering Articles in Bahasa Indonesia Using Self-Organizing Map *2017 Int. Conf. on Electrical Engineering and Informatics*

[8] Westpoint 2014 Understanding the Heartbleed Proof of Concept Available: http://www.westpoint.ltd.uk/blog/2014/04/14/understanding-the-heartbleed-proof-of-concept

IOP Conf. Series: Journal of Physics: Conf. Series **978** (2018) 012122 doi:10.1088/1742-6596/978/1/012122

# Standalone Mobile Application for Shipping Services Based on Geographic Information System and A-Star Algorithm

**D Gunawan[1*], I Marzuki[1], A Candra[2]**

[1] Department of Information Technology, Universitas Sumatera Utara, Jl. dr. Mansur No. 9 Kampus USU Medan 20155
[2] Department of Computer Science, Universitas Sumatera Utara, Jl. dr. Mansur No. 9 Kampus USU Medan 20155

*Email: danigunawan@usu.ac.id

**Abstract**. Geographic Information Systems (GIS) plays an essential role in shipping service related application. By utilizing GIS, the courier can find the route to deliver goods for its customer. This research proposes a standalone mobile application to provide the shortest route to the destinations by utilizing geographic information systems with A-Star algorithm. This application is intended to be used although the area has no Internet network available. The developed application can handle several drop off points then calculates the shortest route that passes through all the drop off points. According to the conducted testing, the number of drop off points that can be calculated is influenced by the specification of the smartphone. More destinations require more smartphone resources and time to process.

## 1. Introduction

Shipping service is closely related to Geographic Information System (GIS). One of main features of shipping service is finding the cheapest and fastest way to go to the destination. Because this will affect service charges to the customer. This condition is closely related to online retailers. Online retailers are suggested to strategically arrange a proper shipping charges to maximize profitability [1]. Current technology has supported these requirements. Maps are available and can be accessed online or offline. The algorithms to find the cheapest way, with various kind of parameters, are available and easy to access.

Previous research has conducted a research about food delivery system [2]. This research utilizes A-Star algorithm to determine the shortest path and location-based service (LBS) to expose its current location. Thus, a user can monitor the movement of the courier before and after he/she goes to the requested destination. Recent publication proposed the optimization of A-Star algorithm using Parallel Bidirectional Search (PBS) in hexagon-based environment [3]. They found that by implementing PBS A-Star, it can accelerate 68% faster compared to classic A-Star algorithm. A-Star algorithm is also used in network routing to obtain function effectiveness [4]. This research is intended to obtain the optimal routing path. According to conducted simulation, their proposed method shows improving result when the effectiveness of the evaluation function is also enhanced. Not only for routing function, A-Star algorithm can also be used as the solving tools to kinematic problems of robot arm [5].

Despite of utilizing A-Star algorithm which is especially designed to find the shortest path, another kind of algorithm, such as a clustering algorithm might be used to find shortest path. A research utilized traditional self-organizing map (SOM) to allow all agents exclusively search the real-time shortest path to go to their targets [6]. The SOM algorithm also can be used to cluster texts [7] or images [8].

This research is intended to build a standalone mobile application in Android platform for shipping services based on geographic information system and utilizing A-Star algorithm to obtain the shortest path. This mobile application is designed to calculate the shortest path in the device, not in the server. Therefore, the data communication can be eliminated to cover network issue if the application is used in rural area.

This paper is organized in four sections. The first section discusses the background, aim and related works of the conducted research. Section two focuses on the research methodology to cover the issue in the first section. The third section is the result of the implementation of standalone mobile application which utilizes GIS and A-Star algorithm. In the last section we draw a conclusion.

## 2. Research Methodology

This research is discussing the development of a standalone mobile application that will be used to assist shipping services to deliver goods by providing maps and the shortest path to the destination. The mobile application is built on Android platform. In this research, the shortest path calculation support more than one destination. Thus, the application will calculate the shortest path which pass through the whole destinations.

In client-server model, the business process or the calculation of the shortest path is done in server. In practice, the calculation requires high server specification and the ability to handle many requests from clients. Also, the data communication between client and server requires Internet network. However, lack of Internet network in rural area will cause this client-server model useless. Therefore, we proposed standalone mobile application. Standalone mobile application means this application is intended to be used in absence of client-server communication. Thus, the application is required to be operated without Internet network availability. However, due to the lack of offline maps available, in this experiment we are going to use an online maps service called Google Maps.

The application implements loosely coupled design. Therefore, the online map can be replaced with the offline map in the production phase. In addition, we are going to implement A-Star algorithm to provide the shortest path to the expected destination. However, we assume that there is no obstacle (that can cause cost) in the path. This research is divided into three parts. First of all is providing the coordinates. Second, implementing A-Star algorithm to provide the shortest path. Third, testing the application. These parts are depicted in figure 1.

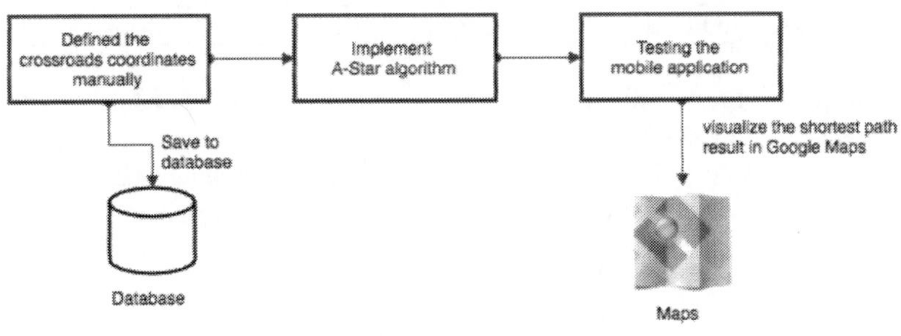


**Figure 1.** General architecture

Figure 1 shows that the coordinates are defined by manually capture coordinates from the Google Maps. This coordinates can also be captured on the spot by using Global Positioning System (GPS). To focus on the development of the standalone mobile application, we only capture the coordinates of every crossroads in Medan. This action is required because Google Maps does not provide coordinates of the crossroads. Thus, we have to define them by ourselves. This coordinates will be saved in the database.

Then, we also set the possible path from one node to others. This will be used to calculate the possible paths and compare them to obtain the shortest path.

Second, implementing A-Star algorithm to find the shortest path. This algorithm requires the cost of one point to another point by calculating the distance of two coordinates. The first coordinates will be defined by the GPS installed in the smartphone. Then the application will find the closest coordinates saved in the database. Next, the application calculates every possible path to obtain the shortest path to the destination. For example, we are going to calculate the shortest path from node A to G which is depicted in figure 2.

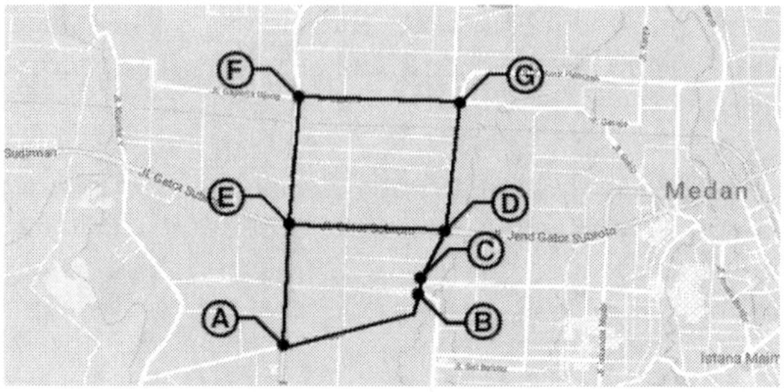


**Figure 2.** Node examples

The coordinates of the nodes in figure 2 are shown in table 1. The coordinates are taken manually from Google Maps.

**Table 1.** Example coordinates

| Node | Street Name | Coordinates |
|---|---|---|
| A | Jl. Ringroad/Jl. Sunggal | 3.5861425, 98.6325452 |
| B | Jl. Sunggal/Jl. Setiabudi | 3.5852696, 98.6368904 |
| C | Jl. Sunggal/Jl. Sei Batang Hari | 3.5844988, 98.6406202 |
| D | Jl. Gatot Subroto/Jl. Sunggal | 3.5894582, 98.6418203 |
| E | Jl. Ringroad/Jl. Gatot Subroto | 3.5902065, 98.6270023 |
| F | Jl. Asrama/Jl. Gaperta | 3.6042378, 98.6293834 |
| G | Jl. Kapten Muslim/Jl. Gaperta | 3.6026019, 98.6415906 |

To calculate the shortest path, we have to compare the calculation of the distance between two nodes (symbolized by $g(n)$) and estimation cost between two nodes by using heuristic function (symbolized by $h(n)$) as shown in formula (1). In this case, we use euclidean distance (formula (2)) to provide heuristic function. The distance between two nodes and the result of heuristic function calculation is shown in table 2.

$$f(n) = g(n) + h(n) \tag{1}$$

$$h(n) = \sqrt{(q1 - p1)^2 + (q2 - p2)^2} \tag{2}$$

**Table 2.** Shortest path calculation result

| Start Node | Destination Node | $g(n)$ (km) | $h(n)$ | $f(n)$ |
|---|---|---|---|---|
| A | B | 1.9 | 0.004432011 | 1,904432011 |
| A | E | 1.5 | 0.006873124 | 1,506873124 |
| B | C | 0.098 | 0.003808614 | 0,101808614 |
| C | D | 0.65 | 0.005102537 | 0,655102537 |
| D | E | 1.8 | 0.014836882 | 1,814836882 |
| D | G | 1.6 | 0.013145707 | 1,613145707 |
| E | F | 1.5 | 0.014231901 | 1,514231901 |
| E | D | 1.8 | 0.014836882 | 1,814836882 |
| F | G | 1.6 | 0.012316327 | 1,612316327 |

The illustration of finding the shortest path is shown in figure 2. Every possible paths will be calculated and compared each other. Then the results of every possible paths which toward to the same node will be compared and the lowest $f(n)$ will be chosen as the shortest possible path. According to the calculation of $f(n)$ in the table 2, the shortest path result is A-E-F-G. This result is depicted in figure 3.

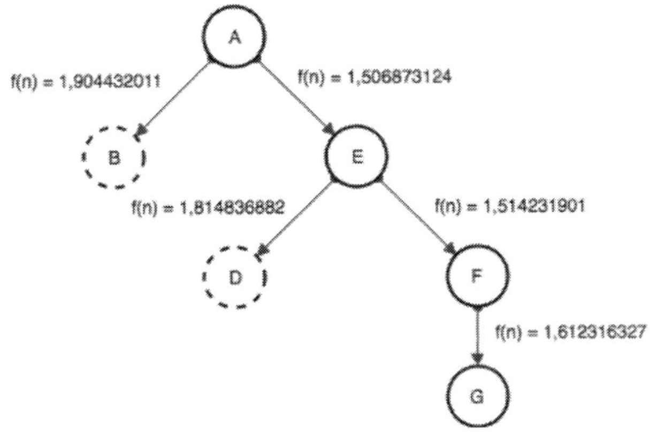


**Figure 3.** Shortest path result

Third, the result of shortest path calculation is visualized to the online map by using Google Maps. This maps can be replaced with the offline map in production phase. In addition, we test the application by using two kind of different smartphones. By conducting this test, we observe the influence of the number of destinations with the smartphone specification. In this research we experiment with five, seven and eight destinations. We observe the memory usage and required time to accomplish the task.

## 3. Result and Discussion

This section will discuss the implementation of geographic information system (GIS) and the A* algorithm to obtain the shortest path for shipping service application. In this research we did the experiment by using two smartphones. The specification of both smartphones are as shown in table 3.

Both smartphones' operating system are Android. Smartphone A and B uses Android version 2.3 and 4.0 respectively. Smartphone A uses less Central Processing Unit (CPU) compared to smartphone B. Smartphone A also lacks of Graphical Processing Unit (GPU) and chipset. Both smartphones use the same network. They use the same A-GPS for the GPS system. However, smartphone B added GLONASS in addition to A-GPS.

IOP Conf. Series: Journal of Physics: Conf. Series **978** (2018) 012122 doi:10.1088/1742-6596/978/1/012122

**Table 3.** Smartphone specification

| Feature | Smartphone A | Smartphone B |
|---|---|---|
| OS | Android v2.3 (Gingerbread) | Android v(4.0) Ice Cream Sandwich |
| CPU | 832 MHz | Quad-core 1.4 GHz Cortex-A9 |
| GPU | - | Mali-400MP4 |
| Chipset | - | Exinos 4412 Quad |
| Network | GSM/HSPA | GSM/HSPA |
| GPS | A-GPS | A-GPS, GLOASS |

This testing is done by selecting five, seven and eight destinations for each smartphone. The results of each phone are shown in figure 4. Figure 4 (a) and (b) shows the result of shortest path calculation for five and seven destinations respectively.

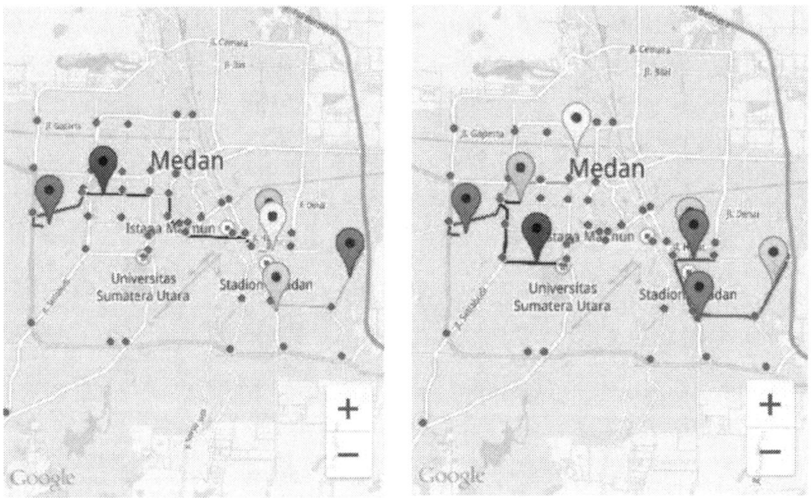


(a) (b)

**Figure 4.** The visualization of shortest path result

Both smartphones can successfully calculate five and seven destinations. As shown in table 4, both smartphones require more time and spend more memory to process more destination. However, smartphone A cannot calculate more than seven destinations. It stops when calculating eight destinations. This might be caused by many possible path in eight destinations which drain smartphone's resources.

**Table 4.** Testing result

| | Smartphone A | | | Smartphone B | | |
|---|---|---|---|---|---|---|
| | 5 DEST | 7 DEST | 8 DEST | 5 DEST | 7 DEST | 8 DEST |
| Memory (MB) | 19.54 | 21 | 51 | 23 | 24 | 35 |
| Execution Time (ms) | 1.32 | 1.32 | - | 1.03 | 1.05 | 2.05 |
| Distance (km) | 15 | 24 | - | 15 | 24 | 24 |

## 4. Conclusion

This research is building the application for shipping service by utilizing Geographic Information System (GIS) and calculate the shortest path by using A-Star algorithm. It proposes a standalone mobile application which do not require data communication between client and server. The absence of data communication should be considered if the application will be used in rural area. This research shows that more destinations requires more smartphone resources and time to process. It depends on the smartphone's specification.

## References

[1] Yao Y and Zhang J 2011 Pricing for Shipping Services of Online Retailers: Analytical and Empirical Approaches *2011 44th Hawaii Int. Conf. on System Sciences* pp 1–10

[2] Siregar B, Gunawan D, Andayani U, Lubis E S and Fahmi F 2017 Food Delivery System with the Utilization of Vehicle Using Geographical Information System (GIS) and A Star Algorithm *J. Phys. Conf. Ser.* **801** 1 p 12038

[3] Saian P O N, Suyoto and Pranowo 2016 Optimized A-Star algorithm in hexagon-based environment using parallel bidirectional search *2016 8th Int. Conf. on Information Technology and Electrical Engineering (ICITEE)* pp 1–5

[4] Septiana R, Soesanti I and Setiawan N A 2016 Evaluation function effectiveness in Wireless Sensor Network routing using A-star algorithm *2016 4th Int. Conf. on Cyber and IT Service Management* pp 1–5

[5] Chaichawananit J and Saiyod S 2016 Solving inverse kinematics problem of robot arm based on a-star algorithm *2016 13th Int. Joint Conf. on Computer Science and Software Engineering (JCSSE)* pp 1–6

[6] Wang C-H and Hung K-N 2012 Adaptive SOM-Based Fuzzy Neural Network controller design for multi-agent system dispatching and path planning *2012 IEEE Int. Conf. on Fuzzy Systems* pp 1–7

[7] Gunawan D, Amalia A and Charisma I 2017 Clustering Articles in Bahasa Indonesia Using Self-Organizing Map *2017 Int. Conf. on Electrical Engineering and Informatics*

[8] Gunawan D, Arisandi D, Ginting F M, Rahmat R F and Amalia A 2017 Russian Character Recognition using Self-Organizing Map *J. Phys. Conf. Ser.* **801** 1 p 12040

# Prime Numbers Comparison using Sieve of Eratosthenes and Sieve of Sundaram Algorithm

**D Abdullah[1], R Rahim[2], D Apdilah[3], S Efendi[4], T Tulus[5] and S Suwilo[5]**

[1]Department of Informatics, Universitas Malikussaleh, Aceh, Indonesia
[2]School of Computer and Communication Engineering, Universiti Malaysia Perlis, Malaysia
[3]Department of Informatics Engineering, AMIK Intel Com Global Inda, Indonesia
[4]Faculty of Computer Science and Information Technology, Universitas Sumatera Utara, Indonesia
[5]Department of Mathematics, Universitas Sumatera Utara, Indonesia

Email: dahlan@unimal.ac.id

**Abstract.** Prime numbers are numbers that have their appeal to researchers due to the complexity of these numbers, many algorithms that can be used to generate prime numbers ranging from simple to complex computations, Sieve of Eratosthenes and Sieve of Sundaram are two algorithm that can be used to generate Prime numbers of randomly generated or sequential numbered random numbers, testing in this study to find out which algorithm is better used for large primes in terms of time complexity, the test also assisted with applications designed using Java language with code optimization and Maximum memory usage so that the testing process can be simultaneously and the results obtained can be objective

## 1. Introduction

Prime numbers are the unique numbers because this number cannot be divided by other positive integers, except 1 and itself [1]. In fact, it is also mentioned that prime numbers are universal languages that can understand by all beings with a high level of intelligence [1] [2] [3]. Prime numbers have several benefits, one of which is concerning disguise messages or often called cryptograph [2]. Determine the prime numbers can be determined manually of course by counting one by one and this takes a very long time or generated by using certain algorithms, such as Sieve of Eratosthenes and Sieve of Sundaram algorithms.

Algorithm Sieve of Eratosthenes and Sieve of Sundaram is an algorithm used to eliminate numbers that do not prime from a collection of numbers that exist [4] [5], this process is quite good as an algorithm that could be applied to various processes of cryptographic algorithms [5], the research will be tested to compare the two algorithms In determining the prime of a large number and knowing which algorithm is safe for use on small and large numbers.

## 2. Theory

The basis of number theory is the number itself, and the core block of numbers is a prime number [6] [7], so prime numbers are often said to be central to number theory [6]. Prime numbers are numbers that have only two factors, namely the number one and the number itself. Numbers that have more than two factors are referred to as composite numbers. Numbers one has only one factor and is considered not prime and composite numbers [8].

When a composite number is written as the product of all the prime factors, it is said to be the main factorization of that number. Factorization of a number into its basic constituent element also called prime decomposition. Typically, factors are listed from the smallest to the largest [9] [10].

Sieve of Eratosthenes is a way to find all prime numbers between 1 and a number n. The algorithm discovered by Eratosthenes, an ancient Greek scientist. Sieve of Eratosthenes is the simplest way to find prime numbers [4].

Here is the step of the Sieve of Eratosthenes algorithm process [4] [5] [9]:

1. Determine the number to be tested it is primarily, from 2 to the largest desired number, call it this list as A list.
2. Mark the number two from list A and move it into another list that holds the primes to be searched.
3. Delete all numbers that are multiples of 2 in list A.
4. The next number found in list A is a prime number. Mark and move to list B.
5. Delete all numbers that are multiples of primes in list An earlier, numbers that have been removed from the list do not need to be in the process
6. Repeat process 4 and five until no numbers are left in list A.

Sieve of Sundaram algorithm is an algorithm generating prime numbers that begin with the value after 1.2.3 and conducted by n2 + 1, the function of determining prime numbers using the following formula:

$$i.j \in N, 1 \leq i \leq j$$

The above formula is used to process all numbers that are not primed in a set of values, to determine prime numbers using the following equation:

$$i. + j + 2ij \leq n$$

## 3. Result and Discussion

The method of testing prime number generator using the Sieve of Eratosthenes algorithm based on the steps previously described, here is the test to determine the primes of the numbers 1 – 100

1. First step show numbers 1 to 100

**Table 1**. Number 1 to 100

| | | | | | | | | | |
|---|---|---|---|---|---|---|---|---|---|
| 1 | 2 | 3 | 4 | 5 | 6 | 7 | 8 | 9 | 10 |
| 11 | 12 | 13 | 14 | 15 | 16 | 17 | 18 | 19 | 20 |
| 21 | 22 | 23 | 24 | 25 | 26 | 27 | 28 | 29 | 30 |
| 31 | 32 | 33 | 34 | 35 | 36 | 37 | 38 | 39 | 40 |
| 41 | 42 | 43 | 44 | 45 | 46 | 47 | 48 | 49 | 50 |
| 51 | 52 | 53 | 54 | 55 | 56 | 57 | 58 | 59 | 60 |
| 61 | 62 | 63 | 64 | 65 | 66 | 67 | 68 | 69 | 70 |
| 71 | 72 | 73 | 74 | 75 | 76 | 77 | 78 | 79 | 80 |
| 81 | 82 | 83 | 84 | 85 | 86 | 87 | 88 | 89 | 90 |
| 91 | 92 | 93 | 94 | 95 | 96 | 97 | 98 | 99 | 100 |

2. The second step, Removing the number 1 and multiples of 2 except the number 2 itself, the result show below

**Table 2**. Remove number multiple of 2

| | | | | | | | | | |
|---|---|---|---|---|---|---|---|---|---|
| 1 | 2 | 3 | 4 | 5 | 6 | 7 | 8 | 9 | 10 |
| 11 | 12 | 13 | 14 | 15 | 16 | 17 | 18 | 19 | 20 |
| 21 | 22 | 23 | 24 | 25 | 26 | 27 | 28 | 29 | 30 |
| 31 | 32 | 33 | 34 | 35 | 36 | 37 | 38 | 39 | 40 |
| 41 | 42 | 43 | 44 | 45 | 46 | 47 | 48 | 49 | 50 |
| 51 | 52 | 53 | 54 | 55 | 56 | 57 | 58 | 59 | 60 |
| 61 | 62 | 63 | 64 | 65 | 66 | 67 | 68 | 69 | 70 |
| 71 | 72 | 73 | 74 | 75 | 76 | 77 | 78 | 79 | 80 |
| 81 | 82 | 83 | 84 | 85 | 86 | 87 | 88 | 89 | 90 |
| 91 | 92 | 93 | 94 | 95 | 96 | 97 | 98 | 99 | 100 |

3. Third step, removing multiples of 3 except the number 3 itself

**Table 3**. Remove number multiple of 3

| | | | | | | | | | |
|---|---|---|---|---|---|---|---|---|---|
| 1 | 2 | 3 | 4 | 5 | 6 | 7 | 8 | 9 | 10 |
| 11 | 12 | 13 | 14 | 15 | 16 | 17 | 18 | 19 | 20 |
| 21 | 22 | 23 | 24 | 25 | 26 | 27 | 28 | 29 | 30 |
| 31 | 32 | 33 | 34 | 35 | 36 | 37 | 38 | 39 | 40 |
| 41 | 42 | 43 | 44 | 45 | 46 | 47 | 48 | 49 | 50 |
| 51 | 52 | 53 | 54 | 55 | 56 | 57 | 58 | 59 | 60 |
| 61 | 62 | 63 | 64 | 65 | 66 | 67 | 68 | 69 | 70 |
| 71 | 72 | 73 | 74 | 75 | 76 | 77 | 78 | 79 | 80 |
| 81 | 82 | 83 | 84 | 85 | 86 | 87 | 88 | 89 | 90 |
| 91 | 92 | 93 | 94 | 95 | 96 | 97 | 98 | 99 | 100 |

4. Fourth step, removing multiples of 5 except the number 5 itself

**Table 4**. Remove number multiple of 5

| | | | | | | | | | |
|---|---|---|---|---|---|---|---|---|---|
| 1 | 2 | 3 | 4 | 5 | 6 | 7 | 8 | 9 | 10 |
| 11 | 12 | 13 | 14 | 15 | 16 | 17 | 18 | 19 | 20 |
| 21 | 22 | 23 | 24 | 25 | 26 | 27 | 28 | 29 | 30 |
| 31 | 32 | 33 | 34 | 35 | 36 | 37 | 38 | 39 | 40 |
| 41 | 42 | 43 | 44 | 45 | 46 | 47 | 48 | 49 | 50 |
| 51 | 52 | 53 | 54 | 55 | 56 | 57 | 58 | 59 | 60 |
| 61 | 62 | 63 | 64 | 65 | 66 | 67 | 68 | 69 | 70 |
| 71 | 72 | 73 | 74 | 75 | 76 | 77 | 78 | 79 | 80 |
| 81 | 82 | 83 | 84 | 85 | 86 | 87 | 88 | 89 | 90 |
| 91 | 92 | 93 | 94 | 95 | 96 | 97 | 98 | 99 | 100 |

5. Fifth step, removing multiples of 7 except the number 7 itself

**Table 5**. Remove number multiple of 7

| | | | | | | | | | |
|---|---|---|---|---|---|---|---|---|---|
| 1 | 2 | 3 | 4 | 5 | 6 | 7 | 8 | 9 | 10 |
| 11 | 12 | 13 | 14 | 15 | 16 | 17 | 18 | 19 | 20 |
| 21 | 22 | 23 | 24 | 25 | 26 | 27 | 28 | 29 | 30 |
| 31 | 32 | 33 | 34 | 35 | 36 | 37 | 38 | 39 | 40 |
| 41 | 42 | 43 | 44 | 45 | 46 | 47 | 48 | 49 | 50 |
| 51 | 52 | 53 | 54 | 55 | 56 | 57 | 58 | 59 | 60 |
| 61 | 62 | 63 | 64 | 65 | 66 | 67 | 68 | 69 | 70 |
| 71 | 72 | 73 | 74 | 75 | 76 | 77 | 78 | 79 | 80 |
| 81 | 82 | 83 | 84 | 85 | 86 | 87 | 88 | 89 | 90 |
| 91 | 92 | 93 | 94 | 95 | 96 | 97 | 98 | 99 | 100 |

The end result of the Sieve of Eratosthenes algorithm process is as follows: 2, 3, 5, 7, 11, 13, 17, 19, 23, 29, 31, 37, 41, 43, 47, 53, 59, 61, 67, 71, 73, 79, 83, 89, and 97, for experiments using sieve of sundaram perform in a way that is similar to the sieve of Eratosthenes.

The above test is performed to show the process of determining the primes of each algorithm, to find out which algorithm is faster in determining the prime number from small numbers to large is by testing using application that develops by using Java, see the figure below

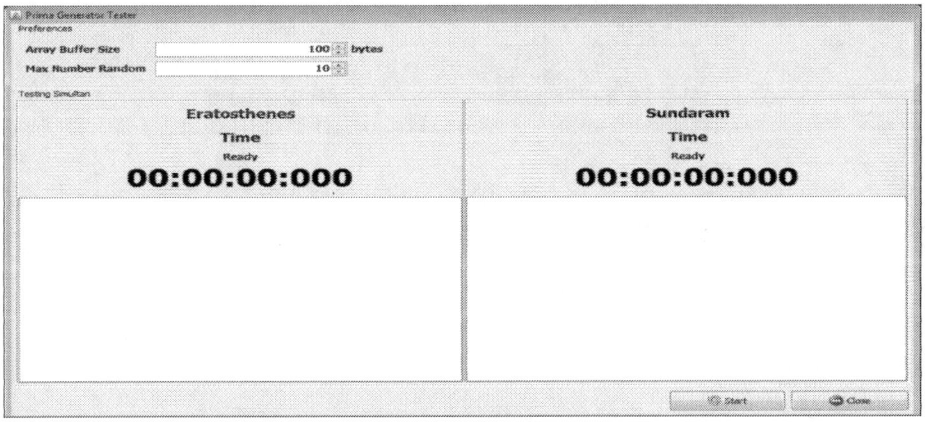


**Figure 1**. Prime Number Generator Java

The above application is made to perform simultaneous testing to generate random numbers and determining the prime numbers from a random number produced by the algorithm Sieve of Eratosthenes and Sieve of Sundaram, for example, as follows:

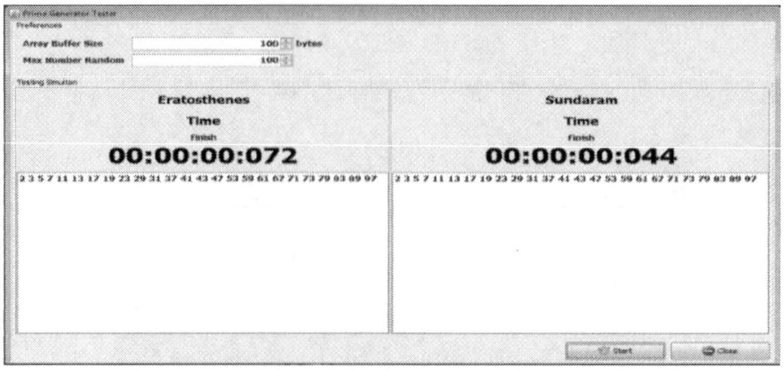


**Figure 2**. Prime Number Result with Application

The above test appears to display the primes of the numbers 1 - 100 with the time displayed simultaneously, for a detailed test can be seen from the table below:

**Table 6**. List Numbers

| No | Number | Sieve of Eratosthenes (Second) | Sieve of Sundaram (Second) |
|---|---|---|---|
| 1 | 10 | 0:0:0.168 | 0:0:0.083 |
| 2 | 100 | 0:0:0.056 | 0:0:0.004 |
| 3 | 1000 | 0:0:0.542 | 0:0:0.188 |
| 4 | 5000 | 0:0:3.986 | 0:0:2.927 |
| 5 | 50000 | 0:0:19.259 | 0:0:18.498 |
| 6 | 100000 | 0:0:25.983 | 0:0:27.349 |
| 7 | 300000 | 0:2:53.957 | 0:3:16.563 |
| 8 | 500000 | 0:5:28.661 | 0:8:23.901 |
| 9 | 1000000 | 0:11:43.813 | 0:16:45.890 |
| 10 | 5000000 | 0:23:42.651 | 0:31:29.401 |

From the table above obtained graph of testing for the number of primes produced by using Sieve of Eratosthenes and Sieve of Sundaram algorithm as follows:

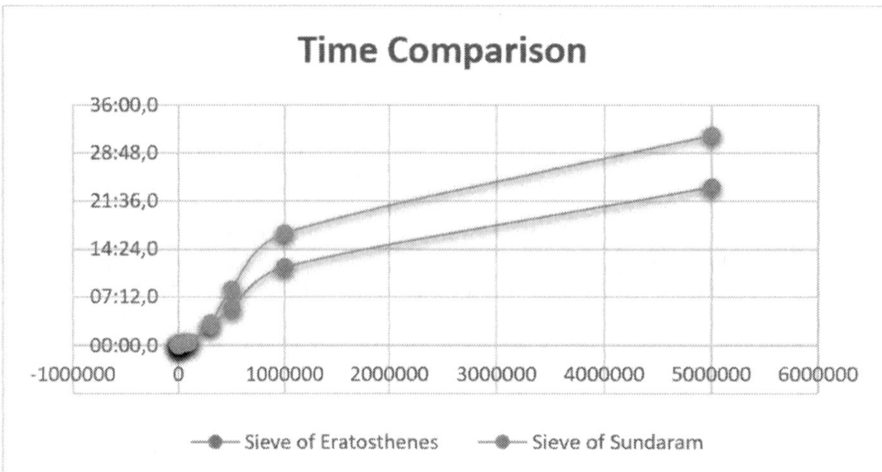


**Figure 3**. Time Graph Comparison

The graph (Figure 3) explains the greater the number of primes produced then the sieve of Eratosthenes is better than the Sieve of Sundaram algorithm.

## 4. Conclusion

The experiment to compare prime generator with Sieve of Eratosthenes and Sieve of Sundaram algorithms has found that for small prime numbers Sieve of Sundaram is better than Sieve of Eratosthenes; but to display a large prime number Sieve of Eratosthenes is better, experiment do with the help of applications built using Java with the optimization of code and use buffer memory to optimize process prime generator.

## References

[1] Kim YS, Shrestha S R and Jang JW 2011 Refined Algorithm for Prime Number Generation in Embedded Security Systems *IEEE Asia-Pacific Services Computing Conference (APSCC)* (South Korea: Jeju Island).

[2] Qing Z and Zhihua H 2011 The large prime numbers generation of RSA algorithm based on genetic algorithm *Int. Conf. on Intelligence Sc. and Information Eng.* (China: Wuhan).

[3] Igumnov V S 2004 Generation of the Large Random Prime Numbers *Proc. 5th Annual. 2004 Int. Siberian Workshop* (Russia: Erlagol, Altai).

[4] O'Neill M E 2009 The Genuine Sieve of Eratosthenes *J. of Functional Programming* **19** pp 95-106.

[5] Lambert M 2004 Calculating the Sieve of Eratosthenes *J. of Functional Programming* **14** pp 759-763.

[6] Sarukkai R R 1995 Prime Numbers and Output Codes *IEEE Int. Conf. on Neural Networks* (Australia: Perth, WA).

[7] Dey S, Abraham A and Sanyal S 2007 An LSB Data Hiding Technique Using Prime Numbers *Third Int. Symposium on Information Assurance and Security* (UK: Manchester).

[8] Liu D 2011 Iteration Distribution of Prime Numbers *J. of Mathematics Research* **3** pp 112-117.

[9] Jullien G A 1980 Implementation of Multiplication, Modulo a Prime Number, with Applications to Number Theoretic Transforms *IEEE Transactions on Computers* **29** pp 899-905.

[10] Fu Y and Xu S 2010 Generalized Fast Coding Steganographic Algorithm Based on Prime Number *IEEE Int. Conf. on Information Theory and Information Security (ICITIS)* (China: Beijing).

# AUTHOR INDEX


Abdolrazzagh-Nezhad, M. .......... 364
Abdullah, D .......... 708, 715, 732, 824
Abdurachman, E .......... 19, 542
Aditya, B R .......... 32
Adnan, A .......... 474
Adnan, S .......... 46
Adrian, H .......... 497
Aisyah, S .......... 121
Akbar, F .......... 678
Alam, C N .......... 310
Albra, W .......... 708, 715, 732
Algharem, M .......... 612
Alianto, H .......... 238
Alzahrani, M .......... 607, 660
Amalia, A .......... 536, 566, 771, 788, 801
Amalia, C .......... 708
Ambarita, H .......... 632, 638, 646, 652
Amri, M F .......... 323
Andayani, U .......... 232, 290, 429, 554, 594, 601, 708, 715, 722, 732, 739, 747
Ani, N .......... 303
Anindito .......... 38
Anthonius .......... 669
Anugrahwaty, R .......... 46
Apdilah, D .......... 824
Aries .......... 82
Arifin, Y .......... 542
Arisandi, D .......... 258, 271, 407, 685, 732, 794
Asnawi .......... 732
Astuti, T P .......... 212
Atika, F .......... 778
Aulia, I .......... 788
Azzahry, B .......... 794
Baba, S .......... 297, 508, 624
Badra, K .......... 420
Bahana, R .......... 13, 19
Bahri, R P .......... 739
Banimahendra, R D .......... 166
Basyuni, M .......... 297, 336, 508, 624
Buaton, R .......... 108
Buchari, M A .......... 524
Budi, I .......... 356
Budiarto, R .......... 46, 612
Budiman, M A .......... 345, 566, 578, 590, 678, 771, 778, 808
Candra, A .......... 818
Chairani, R .......... 715
Chayanie, N I .......... 566
Dewiyana, H .......... 351
Dhany, H W .......... 560
Dharmawan, T .......... 387
Dharshinni, N P .......... 187
Dinzi, R .......... 420
Dirgantoro, B P .......... 19
Dongoran, A .......... 219, 572
Dunia, A I M .......... 697
Dwiastuti, M .......... 531
Efendi, S .......... 414, 501, 824
Effendy, H .......... 115
Erliana, C I .......... 715
Erna, A .......... 63
Fahmi, F .......... 232, 429, 554, 594, 739, 747
Fahrani, O R .......... 54
Faisal, A .......... 492
Fajar, A N .......... 238, 265
Fauzi, R .......... 697
Fauziah .......... 129
Fhonna, R P .......... 95
Fikri, I .......... 387
Fuqara, N .......... 345
Gaol, F L .......... 19
Gellysa, U .......... 63
Ginting, A B C .......... 782
Gunawan, D .......... 794, 801, 808, 814, 818
Gunawan, W .......... 82, 238, 265
Hadhiatma, A .......... 180
Hadi, S .......... 1
Hadipurnawan, D A .......... 69
Harahap, M .......... 121
Hardi, S M .......... 436, 458, 464, 469
Hasan, S .......... 420, 492
Hasanah, S .......... 407
Hasibuan, Z A .......... 194
Hayati, R .......... 624
Hendradjaya, B .......... 524
Hendric, S W H L .......... 19
Herriyance .......... 212
Hizriadi, A .......... 669, 678, 747, 814
Husein, A M .......... 121
Hustinawati .......... 129
Hutagalung, G A .......... 207
Ikhsan, R B .......... 82
Iksal, M .......... 387
Inayatulloh .......... 238
Indrawan, I .......... 245
Indrianti, Y .......... 26, 156, 201
Iqbal, M M .......... 303, 387
Irfan, M .......... 115, 310
Irvanizam, I .......... 380
Iryanto .......... 207
Isa, S M .......... 542
Izhari, F .......... 560
Jannah, M .......... 150
Jaya, I .......... 436, 464, 469, 701
Jessica .......... 578
Karina, N A .......... 89
Kawai, H .......... 652
Kemit, L F .......... 271
Khairani, F .......... 747
Khairani, M .......... 207
Khodra, M L .......... 371

# AUTHOR INDEX


Kurniawan, M ..........594
Kusumandari, D E ..........323
Kusumasari, T F ..........251, 283
Lestantri, I D ..........447
Lestari, D P ..........400
Lubis, A R ..........251, 283
Lubis, M ..........251, 283, 794
Lubis, Y F A ..........207
Lumanauw, G ..........754
Lumbanbatu, A F ..........271
Lumbantobing, N P ..........722
Lumombo, L ..........7
Madenda, S ..........129
Mahfud, N ..........7
Manalu, S R ..........26, 69, 156, 201
Manurung, E B P ..........487
Marbun, J T ..........601
Mardiyanto, S ..........524
Maryani ..........38
Marzuki, I ..........818
Maulidevi, N U ..........514
Maulita, Y ..........108
Mawengkang, H ..........150, 187, 277, 290
Maylawati, D S ..........115
Meiyanti, R ..........345
Mercia ..........238
Moniaga, J V ..........69
Muchtar, M A ..........46, 290, 407, 669
Muhazir, A ..........121
Muliadi, V ..........102
Munandar, A. ..........584
Mustamiin, M ..........356
Nababan, E B ..........271, 317, 364, 763
Nababan, Eb ..........554
Nainggolan, F ..........232
Napitupulu, D ..........708, 715, 732
Napitupulu, F H ..........652
Naseer, M ..........548
Nasir, L A ..........156
Nasution, M K M ..........75, 95, 187, 225, 424, 469
Nasution, M Z F ..........394
Nasution, S ..........594
Nasution, T H ..........290, 536
Noah, S A ..........225
Novelan, M S ..........441
Nugraha, G A ..........13
Nugroho, E ..........19
Nurdin ..........708
Nurdin, A ..........514
Nurhudatiana, A ..........13
Oktinas, W ..........788
Oku, H ..........297, 508, 624
Omar, M H ..........612
Pardede, A M H ..........108
Parlindungan, M R ..........590
Pasha, M F ..........54
Paulus, E ..........1
Permadi, A ..........32
Permai, S D ..........26
Pranoto, H ..........594
Prasetio, A ..........143
Pratama, F ..........536
Priyadi, Y ..........143
Purba, W ..........121
Purnamawati, S ..........54, 258, 754, 763
Putra, S J ..........310
Putri, I L ..........258
Putrima ..........447
Rachman, G H ..........371
Rachmawati, D ..........578, 590, 754, 778, 782
Rahim, R ..........824
Rahmadani, R ..........277
Rahmadani, S ..........219, 572
Rahmah, A ..........194
Rahmat, R F ..........46, 54, 258, 290, 612, 669, 678, 685, 708, 715, 722, 754, 763, 788, 801, 814
Ramadhan, R ..........685
Rambe, A H ..........497, 697
Ramli, M ..........394
Riwinoto ..........7
Rizqyawan, M I ..........323
Ronowikarto, A D ..........632, 638
Rosmaini, E ..........251, 283
Royananda ..........46
Rumahorbo, Ja ..........554
Sabiq, A ..........447
Safrina, N ..........458
Sagami, H ..........297, 508, 624
Sahidi, F ..........69
Salambue, R ..........474
Santoso, H B ..........166, 173, 194, 356
Sari, P P ..........258, 732
Sarim, H M ..........364
Sasmoko ..........26, 156, 201
Satriawan, C H ..........400
Sawaluddin ..........531
Sefnides, I. Z. ..........531
Sembiring, C A ..........808
Seniman ..........407, 601, 715, 739
Sensuse, D I ..........173
Setiawan, M I ..........708, 715, 732
Setyawan, E Y ..........632, 638
Sfenrianto ..........329
Sheta, O ..........271, 685
Shiddiq, M ..........474
Siahaan, A P U ..........345
Sibagariang, Y P ..........652
Sihombing, H V ..........646
Sihombing, P ..........89, 212, 701
Simanjuntak, S E ..........290

# AUTHOR INDEX


Simbolon, A I ........ 323
Simbolon, Artha I. ........ 584
Siregar, A R ........ 351
Siregar, B ........ 232, 265, 429, 554, 594, 722, 739, 747
Siregar, I ........ 290, 536
Siregar, R E T ........ 632, 638
Siregar, R I ........ 245
Siregar, R P ........ 801
Siregar, Y M ........ 701
Sitepu, R ........ 771
Sitepu, S A ........ 414
Sitompul, D ........ 212
Sitompul, O S ........ 225, 290, 317, 394, 487
Sitorus, E H ........ 814
Situmorang, Z. ........ 414
Soelistijanto, B. ........ 102
Soewito, B ........ 19
Subandi, A ........ 345
Subiyakto, A ........ 310
Suhendra, M A ........ 323, 584
Suherlan, E ........ 447
Suherman ........ 420, 487, 492, 497, 697
Sulistiyono, N. ........ 508, 624
Sumadyo, M. ........ 173
Sumardi ........ 508, 624
Supangkat, S H. ........ 548
Suparto ........ 201
Supriadi, I. ........ 548
Suriati ........ 207, 531
Suryani, M. ........ 1
Sutarman ........ 277, 560
Suwilo, S ........ 824
Syahputra, M F ........ 258, 271, 290, 669, 678, 685, 708, 715, 722, 732
Syaliman, K U ........ 317
Syarif, M I. ........ 89
Syarip ........ 481
Tambunan, D R S. ........ 652
Tanjung, K. ........ 536
Tanuar, E. ........ 38
Taqiuddin, R ........ 46
Taqyuddin, R. ........ 754, 763
Tarigan, J T ........ 89, 436, 458, 464, 469, 701, 782
Tatiara, R. ........ 265
Togatorop, E. ........ 481
Tri, Y ........ 63
Triana, Y S ........ 387
Trianda, D ........ 429
Tsani, B ........ 763
Tulus ........ 95, 207, 441, 531
Tulus, T ........ 824
Turnip, A ........ 701
Utama, D N ........ 303, 387
Vista, C B ........ 400
Wahana, A ........ 115, 310
Wati, R ........ 297, 336, 508, 624
Wibowo, E P ........ 129
Widhoyoko, S A ........ 26, 156, 201
Widyantoro, D H ........ 371, 400
Wiguna, T ........ 19
Winanto, I D ........ 290
Wulandhari, L A ........ 542
Yassar ........ 481
Yel, M B ........ 329
Yohannes ........ 82
Yunus, U. ........ 38
Zaiwani, B E ........ 501
Zakarias ........ 219, 572
Zamzami, E M ........ 436, 441, 464
Zarlis, M. ........ 150, 219, 290, 501, 560, 572

**Institute of Physics**
Dirac House, Temple Back
Bristol BS1 6BE UK

ISSN: 1742-6588
ISBN 978-1-5108-5988-3
9 781510 859883

CURRAN ASSOCIATES INC.
proceedings
.com

# International Conference on Innovative Ideas in Science (IIS 2015)

IOP Conference Series: Materials Science and Engineering
Volume 144

Baia Mare, Romania
12 – 13 November 2015

IOP
Institute of Physics